W9-BWL-586

STARS AND STELLAR SYSTEMS

Compendium of Astronomy and Astrophysics

(IN NINE VOLUMES)

GERARD P. KUIPER, *General Editor*

BARBARA M. MIDDLEHURST, *Associate General Editor*

I
TELESCOPES

II
ASTRONOMICAL TECHNIQUES

III
BASIC ASTRONOMICAL DATA

IV
CLUSTERS AND BINARIES

V
GALACTIC STRUCTURE

VI
STELLAR ATMOSPHERES

VII
NEBULAE AND INTERSTELLAR MATTER

VIII
STELLAR STRUCTURE

IX
GALAXIES AND THE UNIVERSE

CONTRIBUTORS

G. O. Abell
Sidney van den Bergh
E. M. Burbidge
G. R. Burbidge
George B. Field
K. C. Freeman
Erik Holmberg
J. Kristian
David Layzer
R. Minkowski
Alan T. Moffet
Thornton Page
Manuel Peimbert
Morton S. Roberts
Allan Sandage
Peter A. G. Scheuer
Maarten Schmidt
C. D. Shane
Hyron Spinrad
G. de Vaucouleurs
A. E. Whitford

GALAXIES AND THE UNIVERSE

Edited by

ALLAN SANDAGE

MARY SANDAGE

and

JEROME KRISTIAN

Index Prepared by

GUSTAV A. TAMMANN

THE UNIVE[illegible]HICAGO PRESS
C[illegible]ONDON

This publication has been supported in part by the

NATIONAL SCIENCE FOUNDATION

The University of Chicago Press, Chicago 60637
The University of Chicago Press, Ltd., London

Published 1975
Printed in the United States of America

International Standard Book Number: 0-226-45961-6
Library of Congress Catalog Card Number: 74-7559

Preface to the Series

THE SERIES "Stars and Stellar Systems, Compendium of Astronomy," comprising nine volumes, was organized in consultation with senior astronomers in the United States and abroad early in 1955. It was intended as an extension of the four-volume "Solar System" series to cover astrophysics and stellar astronomy. In contrast to the "Solar System" series, separate editors have been appointed for each volume. The volume editors, together with the general editors, form the editorial board that is responsible for the over-all planning of the series.

The aim of the series is to present stellar astronomy and astrophysics as basically empirical sciences, co-ordinated and illuminated by the application of theory. To this end the series opens with a description of representative telescopes, both optical and radio (Vol. **1**), and of accessories, techniques, and methods of reduction (Vol. **2**). The chief classes of observational data are described in Volume **3**, with additional material being referred to in succeeding volumes, as the topics may require. The systematic treatment of astronomical problems starts with Volume **4**, as is apparent from the volume titles. Theoretical chapters are added where needed, on dynamical problems in Volumes **4**, **5**, and **9**, and on astrophysical problems in Volumes **6**, **7**, and **8**. In order that the chapters may retain a greater degree of permanence, the more speculative parts of astronomy have been de-emphasized. The level of the chapters will make them suitable for graduate students as well as for professional astronomers and also for the increasing number of scientists in other fields requiring astronomical information.

The undersigned wish to thank both the authors and the volume editors for their readiness to collaborate on this series, which it is hoped will stimulate the further growth of astronomy.

The editors wish to acknowledge the support by the National Science Foundation both in defraying part of the costs of the editorial offices and in providing a publication subsidy.

GERARD P. KUIPER
BARBARA M. MIDDLEHURST

Preface to Volume 9

ASTRONOMERS in this century have been privileged to witness and contribute to the beginning of a scientific understanding of the nature and the history of the universe on the largest scale—an understanding which men have sought for millennia.

This book appears at the fiftieth anniversary of Hubble's discovery of the Cepheids in M31. That event, which occurred within the professional lifetime of astronomers who are still active, can be said to mark the beginning of modern extragalactic astronomy. It definitively established the spiral nebulae as "island universes" similar to our own Milky Way Galaxy and led to further studies which soon showed (1) that the universe is uniformly populated by immense numbers of galaxies, and (2) that the entire assemblage is expanding, with the direct implication of a state of large density—a creation event—at a finite time in the past. At about the same time, the publication of Einstein's gravitational theory and the discovery by Friedmann of its expanding solutions provided a modern theoretical framework within which the observations could be understood.

The intervening years have been ones of preliminary reconnaissance in which the rough boundaries and time scales of the new world picture have been sketched out, and efforts made to understand the mechanisms by which galaxies and the universe reached the state in which we see them today.

The present volume is a product of the 1960s. It was first proposed in 1962, and the first manuscript was received in 1964. Several early manuscripts, however, were later updated or revised, and the final collection of articles was completed between 1965 and 1973, with a mean epoch of 1970. Although the latest represent up-to-date summaries, we are impressed by the enduring qualities of even the earliest manuscripts. Each was intended to be a comprehensive statement of the subject at the time it was written, and by and large none have been outmoded by more recent results. Rather, as we hope, these papers describe basic techniques and knowledge upon which the new generation of astronomers and students can build.

Certainly the most important new development of the past decade has been the flowering of radio astronomy and its impact on the study of galaxies as individuals (chapters 6, 7, 8, and 9) and on cosmology (chapter 18). But there have also been many productive efforts in applying more familiar optical techniques to unsolved problems such as the masses of galaxies (chapter 3) and their stellar content (chapter 2), and integrated properties such as color, surface brightness, and energy distribution (chapters 4 and 5).

Important advances were also made in problems of galaxy formation (chapter 10) and dynamic behavior (chapter 11). A combination of theoretical ideas and observations of the kinematics of individual stars in our own Galaxy have provided at least the beginnings of an understanding of the dynamical and chemical history of galaxies in terms of collapse from a pregalactic medium (chapter 11).

The discussion of galaxies as individuals (chapters 1–11) leads naturally to consideration of galaxies as members of the universe (chapters 12–17) and finally to the universe itself (chapters 18 and 19), where the rudiments of cosmology are sketched, although incompletely. The most serious flaw of the book, in our opinion, is the neglect of the many exciting developments in general relativity and theoretical cosmology in the past 10–15 years. These developments have not yet had a major impact on astronomy, but this is due primarily to our limited and rudimentary observational knowledge, and we have no doubt that their future importance will be great.

We feel that one of the major concerns in astronomy for the remainder of this century will be with extragalactic questions: the origin and evolution of galaxies, and the structure and past and future history of the universe. The construction of several new large telescopes and the development of greatly improved instrumentation offer the prospect of obtaining the basic observational data much more rapidly than seemed possible a decade ago. In addition, the recent extension of observations to longer and shorter wavelengths (radio and microwaves, X-rays and gamma rays) promise additional new insights, perhaps comparable in importance to the 3° primeval radiation.

One of the most exciting aspects of modern science in general, and of astronomy in particular, is the hope that it holds of being able to discern the structure and evolution of the universe since its creation. The first steps have been taken, but by far the larger part of the story remains to be discovered. As Schucking once remarked, "to describe the universe in terms of our present knowledge is to say almost nothing about everything."

The authors and editors know that much of the material presented here will prove to be wrong in detail, but we hope that it is not far off in broad outline, and that the book will be useful in the coming spectacular development of the subject.

Allan Sandage
Mary Sandage
Jerome Kristian

Table of Contents

1. CLASSIFICATION AND STELLAR CONTENT OF GALAXIES OBTAINED FROM DIRECT PHOTOGRAPHY 1

Allan Sandage

1. Introduction 1
2. Early Classification of Galaxies 2
3. Development of the Modern System 4
   3.1. Early Isolation of the Types 4
   3.2. The Early Hubble System 5
   3.3. Hubble's Major Modification between 1936 and 1950 6
   3.4. Finer Subdivision along the Sequence 8
4. Revision by de Vaucouleurs 8
   4.1. Extension of the Sequence beyond Sc 8
   4.2. Transitions Between Ordinary and Barred Spirals 8
   4.3. The r and s Varieties 9
   4.4. Graphical Representation of the Classification 9
   4.5. Additional Features of the de Vaucouleurs Revision 10
      4.5.1. Presence of Rings 10
      4.5.2. Increasingly Detailed Notation 11
5. Selected Illustrations of Galaxy Types 13
   5.1. Elliptical Galaxies 13
   5.2. Spirals with Thin Filamentary Arms 22
   5.3. Spirals with Massive Arms 22
   5.4. Transitions between SA and SB 23
   5.5. Pure Barred Galaxies 23
   5.6. Sdm, Sm, and Im 23
   5.7. Irr II Galaxies 24
   5.8. Interacting Galaxies 24
6. Van den Bergh's Classification 24
7. Morgan's Classification Based on the Luminosity Concentration of the Spheroidal Component 26
8. System of Vorontsov-Velyaminov 27
9. Comparison of the Classification Systems 28
10. Seyfert Galaxies, N Galaxies, and Quasars 28
11. Stellar Content Related to Type: Formation and Evolution 29

References 33

2. THE STELLAR AND GASEOUS CONTENT OF NORMAL GALAXIES AS DERIVED FROM THEIR INTEGRATED SPECTRA 37

Hyron Spinrad and Manuel Peimbert

1. The Stellar Content of Normal Galaxies 37
1.1. Introductory Remarks and a Historical Review 37
1.2. Slit Spectra of Galaxies—Selected Observations and Interpretations . 39
1.3. Photoelectric Narrow-Band Measures of Galaxies 44
1.4. The Spinrad-Taylor Galaxy Models 46
1.5. Synthesis of Other Types of Galaxies 61
1.6. Comments on the Stellar Content of Some Spiral Galaxies 62
1.7. Speculative Topics of Interest 62

2. Emission Lines from Normal Galaxies 63
2.1. Introduction 63
2.2. Normal H II Regions in Our Galaxy 64
2.3. Normal H II Regions in Other Galaxies 67
2.4. Nuclear H II Regions 71
2.5. Other H II Regions 76

References 77

3. THE MASSES OF GALAXIES 81
E. M. Burbidge and G. R. Burbidge

1. Introduction 81

2. Theories of Mass Determination 82
2.1. Rotations of Galaxies 82
2.1.1. A Mass Point 83
2.1.2. Homogeneous Spheroid 83
2.1.3. Two or More Homogeneous Spheroids; Mass Point or Sphere with Homogeneous Spheroids 84
2.1.4. Nonhomogeneous Spheroids 85
2.1.5. Disk: Models of Wyse and Mayall, Aller, Schwarzschild . . 87
2.1.6. Bottlinger's Model 88
2.1.7. Flat-Disk Approximation of Brandt and Belton 89
2.1.8. Toomre's Disk Model 90
2.1.9. Barred Spirals 91
2.1.10. Correction for Partial Support of Gas by Pressure in the Center 92
2.2. Masses of Spherical Galaxies from Velocity Dispersion of Stars . . . 93
2.3. Masses from Orbital Motion of Double Galaxies 94
2.4. Average Masses of Galaxies in Clusters Determined by Using the Virial Theorem 95

3. Methods of Observation 96
3.1. Rotation of Galaxies 96
3.1.1. Rotation Curves from Absorption Lines 96
3.1.2. Gradient of Rotation in the Centers of Galaxies, from Inclinations of Absorption Lines 98
3.1.3. Rotation Curves from Emission Lines 100
3.2. Velocity Dispersions and Potential Energies 102
3.3. Orbital Motions of Double Galaxies 104

4. Results 105
4.1. Distances 108
4.2. Luminosities 108

4.3. Uncertainties in Results. Differences between Radio and Optical Determinations . . . 109
4.4. M31 . . . 112
4.5. M33 . . . 112
4.6. Noncircular Motions . . . 113

5. Mass-to-Light Ratios and Masses of Galaxies in Clusters; Angular Momentum 115
5.1. Mass-to-Light Ratios . . . 116
5.2. Masses of Galaxies in Clusters . . . 117
5.3. Angular Momentum . . . 118

References . . . 119

4. Magnitudes, Colors, Surface Brightness, Intensity Distributions, Absolute Luminosities, and Diameters of Galaxies . . . 123

Erik Holmberg

1. Photometry of Galaxies . . . 123
1.1. Techniques and Practical Problems . . . 123
1.2. Results Obtained by Photographic Photometry . . . 124
1.3. Results Obtained by Photoelectric Photometry . . . 125

2. Absorption Effects in Magnitudes and Colors . . . 126
2.1. Internal Absorption in Galaxies . . . 126
2.2. Galactic Absorption . . . 131

3. Surface Magnitudes, Luminosity Densities, and Intrinsic Colors . . . 131

4. Internal Distributions of Luminosity and Color . . . 133
4.1. Distribution of Luminosity and Color in Elliptical Galaxies . . . 133
4.2. Distribution of Luminosity and Color in Spiral Systems . . . 137
4.3. Construction of Three-Dimensional Models . . . 144

5. Absolute Magnitudes . . . 145
5.1. Luminosities of Individual Galaxies . . . 145
5.2. The Luminosity Function . . . 147

6. Diameters . . . 151
6.1. Angular Dimensions . . . 151
6.2. Absolute Diameters . . . 153

References . . . 155

5. Integrated Energy Distribution of Galaxies . . . 159

A. E. Whitford

1. Introduction . . . 159

2. Color Indices . . . 160

3. Filter-Band Spectrophotometry . . . 160

4. Absolute Energy Curves from Scanners . . . 166

5. Inferences Regarding Stellar Content . . . 168

6. The K-Correction . . . 170

7. Evolutionary Effects . . . 172

References . . . 175

6. The Identification of Radio Sources 177
R. Minkowski

1. Discovery and Early Identifications 177
2. Surveys of Radio Sources 179
3. Identification of 3C and MSH Sources. 181
4. Precise Positions 184
5. The Identification of the 3CR Sources 185
6. The Identification of Parkes Sources 186
7. The Unidentified Sources 187
8. Radio Sources and Clusters of Galaxies 188
9. The Luminosity Function 190

References 195

Appendix to Chapter 6: Optical Identification of 3CR Sources (October 1972) 199
J. Kristian and R. Minkowski

Position References 210
Radio Positions and Maps 210
Optical Positions 210

7. Strong Nonthermal Radio Emission from Galaxies 211
Alan T. Moffet

1. Introduction 211
2. Synchrotron Radiation 213
   2.1. Emission from a Single Particle 213
   2.2. Emission from an Ensemble of Particles 218
   2.3. Modifications of the Power-Law Spectrum 221
      2.3.1. Radiation Transfer 221
      2.3.2. Thermal Absorption 224
      2.3.3. Synchrotron Self-Absorption 225
      2.3.4. The Tsytovitch Effect 229
      2.3.5. Modifications of the Electron Spectrum 232
   2.4. Luminosity and Energy Requirements for a Source of Synchrotron Radiation 233
   2.5. Compton Losses 237
   2.6. Expansion Losses 240
3. Observed Radio Characteristics 241
   3.1. Distribution 241
   3.2. Spectra 243
      3.2.1. Statistics of Spectra 243
      3.2.2. Correlations with Other Source Properties 243
      3.2.3. Classification of Spectra 245
   3.3. Angular Sizes and Brightness Distributions 247
      3.3.1. Techniques 247

3.3.2. Shapes and Diameters of Radio Galaxies 253
3.3.3. Wavelength Dependence 259
3.4. Polarization 259

4. Intrinsic Properties 263
4.1. Optical Identification and Redshifts 263
4.2. Luminosities 264
4.3. Linear Dimensions 269
4.4. Energy Requirements 270
4.5. Optical Properties of Radio Galaxies 271

5. Theories of Radio Galaxies 273
5.1. Energy Sources 273
5.2. Symmetrical Division 274
5.3. Source Containment 275
5.4. The End Game 276

References 277

8. QUASARS 283
Maarten Schmidt

1. Introduction 283

2. History 284

3. Identifications 285
3.1. Definition 285
3.2. Quasi-stellar Sources (Radio Quasars) 286
3.3. Quasi-stellar Objects (Quasars) 286
3.4. Sky Distribution 287

4. Emission Lines, Redshifts 288
4.1. Cosmological Redshifts 290
4.2. Doppler Redshifts 291
4.3. Gravitational Redshifts 292

5. Continuum Energy Distribution 293
5.1. Optical Continuum 293
5.2. Infrared Continuum 295
5.3. Radio Continuum 295

6. Nature of Quasars 296
6.1. Gravitational Hypothesis 296
6.2. Local-Doppler Hypothesis 297
6.3. Cosmological Hypothesis 298
6.4. Other Hypotheses 299
6.5. Conclusion 299

7. Absorption Lines 300
7.1. Observations 300
7.2. Interpretation 301

8. Cosmology, Evolution 302

9. Models, Lifetimes, and Energy Sources 305

References 306

9. Radio Observations of Neutral Hydrogen in Galaxies 309
Morton S. Roberts

1. Introduction 309
2. Determination of Kinematic Properties of Galaxies from 21-Centimeter Observations 310
3. Total Masses 318
4. Systemic Radial Velocity 320
5. The Hydrogen Content of Galaxies 323
6. The Distribution of Hydrogen Within a Galaxy 329
7. Integral Properties of Galaxies 335
   7.1. Total Mass and Absolute Photographic Luminosity 336
   7.2. Total Mass and Hydrogen Mass 342
   7.3. Hydrogen Mass and Absolute Photographic Luminosity 348
   7.4. Hydrogen Mass, Absolute Photographic Luminosity, and Color . . 349
   7.5. Ordinary and Barred Spirals 349
   7.6. Elliptical and Radio Galaxies 353
8. The Evolution of Galaxies 353

References 355

10. The Formation and Early Dynamical History of Galaxies 359
George B. Field

1. Introduction 359
2. Behavior of Density Perturbations 360
   2.1. Gravitational Instability with Radiation Absent 360
   2.2. Gravitational Instability with Radiation Present 369
   2.3. Other Types of Instability 375
   2.4. Galaxy Formation in Other Cosmological Models 378
   2.5. Critique of Gravitational Instability 380
3. Contraction of a Protogalaxy 381
   3.1. Initial Conditions 381
   3.2. Calculation of the Collapse 382
   3.3. Collapse of Protogalaxies 384
   3.4. Instability during Collapse 386
4. Early Evolution 388
   4.1. Fragmentation 388
   4.2. Formation of Elliptical Systems 391
   4.3. Formation of Disk Systems 392
   4.4. Differentiation between Ellipticals and Disks 398
5. Observational Data 398
   5.1. Intergalactic Matter 398
   5.2. Observation of Young Galaxies 399
   5.3. History of the Milky Way Galaxy 400
   5.4. Multiple Systems of Galaxies 402

References 405

11. Stellar Dynamics and the Structure of Galaxies 409
K. C. Freeman

1. Introduction 409

2. Foundations of Stellar Dynamics 409
 2.1. The Relaxation Time 409
 2.2. The Collisionless Boltzmann Equation 410
 2.3. The Hydrodynamical Equations 411
 2.4. Jeans's Theorem 412
 2.5. Some Simple Consequences of Jeans's Theorem 413
 2.6. Adiabatic Invariants 414
 2.7. The Virial Theorem 414

3. Some Basic Properties of the Galaxy 414
 3.1. The Rotation of the Galaxy 415
 3.2. Stellar Motions near the Sun 418
  3.2.1. The Asymmetric Drift 420
  3.2.2. The Total Mass Density 421
  3.2.3. The Density Gradient 421
  3.2.4. The Ratio $\sigma_{\varphi\varphi}/\sigma_{RR}$ 422
 3.3. Mass Models of the Galaxy 422
 3.4. Epicyclic Orbits 423
 3.5. Populations in the Galaxy 424

4. Some General Properties of Galaxies 425
 4.1. Morphological Classification of Galaxies 426
 4.2. Absolute Luminosities and Colors of Galaxies 428
 4.3. Luminosity and Color Distributions in Galaxies 429
  4.3.1. The Ellipticals 429
  4.3.2. The Spirals and Lenticulars 429
 4.4. Internal Motions and Masses 434
  4.4.1. The Normal Ellipticals 434
  4.4.2. The Spirals 436
  4.4.3. The Frequency Distribution of Masses 438
 4.5. The Content of Galaxies 438
 4.6. Some Problems 439
  4.6.1. Problems Associated with the Different Kinds of Galaxies . . 439
  4.6.2. Problems Associated with the Mass and Angular Momentum Distributions 439
  4.6.3. The Problem of Spiral Structure 440
  4.6.4. Some Other Problems 440

5. The Third Integral 440

6. The Self-Consistency Problem: Models for Globular Clusters and Elliptical Galaxies 446
 6.1. Two Simple Examples 447
  6.1.1. The Direct Problem 447
  6.1.2. The Inverse Problem 448
 6.2. Models of Globular Clusters and Elliptical Galaxies 448

7. The Collapse of the Galaxy 450

8. Collective Effects and Collisionless Relaxation in Stellar Systems 453
8.1. The Stability of Homogeneous Media 454
8.1.1. The Infinite Gravitating Fluid 454
8.1.2. Collisionless Stellar Systems 455
8.1.3. Landau Damping 455
8.1.4. Gas and Star Systems 455
8.2. Phase Damping 456
8.3. Violent Relaxation 458
8.4. Time-dependent Stellar Systems 460

9. The Disks of Spiral and Lenticular Galaxies 460
9.1. The Stability of the Disk 461
9.1.1. The Stabilizing Effect of Rotation 461
9.1.2. Toomre's Work on Disks of Stars 463
9.1.3. Hunter's Work on Cold Disks 464
9.1.4. Ostriker's and Peebles's Work 465
9.2. Mestel's Hypothesis 465
9.3. The Exponential Disk 466
9.4. The Bending of the Galactic Plane 471
9.5. Problems of Velocity Dispersion in the Disk 472
9.6. The Problem of Spiral Structure 474
9.6.1. Relevant Observational Data 474
9.6.2. Goldreich and Lynden-Bell's Theory 475
9.6.3. (i) The Density-Wave Theory: Basic Theory 476
9.6.3. (ii) Observational Studies of the Density-Wave Theory . . 480
9.6.3. (iii) Origin of the Spiral Structure 481
9.6.3. (iv) Shock Formation in Spiral Galaxies 482

10. Barred Spiral Galaxies 483
10.1. The Formation of the Bar 483
10.2. Some Properties of Bar-like Stellar Systems 483
10.3. Formation of Spiral Structure 484
10.4. Formation of Ring Structure 485
10.5. The Magellanic Barred Spirals 486

11. Formation and Evolution of Galaxies 491
11.1. The Metal-Enrichment Picture 492
11.2. Some Kinematic Problems 494
11.2.1. The Angular Momentum/Mass Ratio for the Halo 494
11.2.2. Disk Globular Clusters 494
11.2.3. The Origin of the Orbital Eccentricities 495
11.2.4. The Anisotropy of the Velocity Dispersion 496
11.3. Problems of the Large-Scale Mass Distribution 496
11.3.1. The Ellipticals 496
11.3.2. The Disk Galaxies 498
11.4. Morphological-Type Problems 500
11.4.1. Elliptical or Disk 500
11.4.2. The Types in the Disk Family 501
11.5. The Origin of Angular Momentum 504

References 504

12. THE EXTRAGALACTIC DISTANCE SCALE 509
Sidney van den Bergh
1. Introduction 509
2. Period-Luminosity Relation of Classical Cepheids 510
2.1. Assumption Underlying Use of the Period-Luminosity Relation. . . 510
2.2. The Period-Luminosity-Color Relation 511
2.3. Galactic Calibration of the Period-Luminosity-Color Relation . . . 514
2.4. Reddening Values and Apparent Distance Moduli for Members of the Local Group 515
2.4.1. The Large Magellanic Cloud 515
2.4.2. The Small Magellanic Cloud. 517
2.4.3. The Andromeda Nebula 517
2.4.4. M33 517
2.4.5. NGC 6822 517
2.4.6. IC 1613 518
3. Novae 518
3.1. Galactic Calibration 518
3.2. The Andromeda Nebula 518
3.3. The Magellanic Clouds 519
4. RR Lyrae Variables 521
4.1. Galactic Calibration 521
4.2. Distances of the Magellanic Clouds 521
5. W. Virginis Stars 522
5.1. The Small Magellanic Cloud 522
5.2. The Andromeda Nebula 522
6. Red Giants of Population II 523
7. Globular Clusters 523
8. Spectral Luminosity Determinations 524
9. Summary of Data on the Local Group 525
10. Distances beyond the Local Group 526
11. Third Brightest Cluster Galaxy 526
12. Surface Brightness of Galaxies 527
13. Luminosity Classification of Galaxies 528
14. Diameters of H II Regions 529
15. Mass-to-Light Ratios 530
16. Brightest Nonvariable Stars in Galaxies 530
17. Globular Clusters 531
18. Supernovae 532
19. Galaxy Diameters 533
20. Summary of Data on the Hubble Constant 534
21. Regional Variations of the Hubble Constant 534

22. Future Observations . . . 536
22.1. Desirable Observations within the Local Group . . . 536
22.2. Observations of Nearby Clusters . . . 536
22.3. Observations of Distant Galaxies . . . 537
References . . . 537

13. BINARY GALAXIES . . . 541
Thornton Page
1. Introduction . . . 541
2. Definition of Optical and Physical Pairs . . . 541
3. Observational Data . . . 543
4. Types and Magnitudes of Galaxies in Pairs . . . 544
5. Dimensions of Galaxies in Pairs . . . 547
6. Orientation of Axes in Pairs of Galaxies . . . 547
7. The Dynamics of Binary Galaxies . . . 549
8. Formation and Evolution of Binary Galaxies . . . 553
References . . . 555

14. NEARBY GROUPS OF GALAXIES . . . 557
G. de Vaucouleurs
1. Definition of a Group . . . 557
2. Census of Nearby Groups . . . 558
3. Distance Moduli . . . 558
3.1. Magnitudes . . . 558
3.2. Diameters . . . 559
3.3. Sampling Correction . . . 559
3.4. Absorption Correction . . . 559
4. Local Group . . . 560
5. The Nearer Groups within 10 Megaparsecs . . . 562
G1. Sculptor Group . . . 564
G2. M81 Group . . . 572
G3. Canes Venatici I Cloud . . . 572
G4. NGC 5128 Group . . . 573
G5. M101 Group . . . 573
G6. NGC 2841 Group . . . 573
G7. NGC 1023 Group . . . 574
G8. NGC 2997 Group . . . 574
G9. M66 Group . . . 574
G10. Canes Venatici II Cloud . . . 575
G11. M96 Group . . . 575
G12. NGC 3184 Group . . . 575
G13. Coma I Cloud . . . 575
G14. NGC 6300 Group . . . 576

6. Nearby Groups beyond 10 Megaparsecs 576
7. Statistical Properties of Nearby Groups 581
7.1. Completeness of Survey 581
7.2. Space Density of "Cluster" Centers 583
7.3. Frequency Function of Diameters 583
7.4. Luminosity Function 583
7.5. Statistical Masses 583
7.6. Population Types 585
8. Nearby Dwarf Galaxies 586
8.1. Definition 586
8.2. Dwarfs in the Local Group 586
8.3. Survey of Nearby Dwarfs 587
8.4. Completeness of Survey 589
9. Isolated Nearby Galaxies 589
10. Apparent and Space Distribution of Nearby Groups: Local Supercluster . . 592
References 596
Addendum (January 1973) 597

15. Clusters of Galaxies 601
G. O. Abell
1. Introduction 601
2. Numbers and Catalogs of Clusters 601
3. Observed Properties of Clusters 604
3.1. Types of Clusters 604
3.2. Galaxian Content of Clusters 607
3.3. The Luminosity Function and Colors of Cluster Galaxies 610
3.4. Populations of Clusters 619
3.5. Sizes and Structures of Clusters 621
3.6. Velocity Dispersions in Clusters 627
4. Dynamics of Clusters 629
4.1. General Considerations 629
4.2. Masses of Clusters 630
4.3. Formation and Evolution of Clusters 634
5. The Distribution of Clusters 636
5.1. The Evidence for the Local Supercluster 636
5.2. Other Evidence of Second-Order Clusters 637
5.3. The Large-Scale Distribution of Clusters and the Mean Density of Matter in the Universe 641
References 642

16. Distribution of Galaxies 647
C. D. Shane
1. Historical Note 647

2. Observations 648
3. Galactic Extinction 650
4. Clouds of Galaxies 654
5. Distribution in Depth 660
References 663

17. GALAXY CLUSTERING: ITS DESCRIPTION AND ITS INTERPRETATION 665
David Layzer

1. Introduction 665

2. The Analysis of Galaxy Counts 667
2.1. The Poisson Distribution 667
2.2. Statistically Homogeneous and Isotropic Distributions 668
2.3. The Counted Number 669
2.4. A Special Case 670
2.5. The True and the Projected Distributions; the Function $p(x)$ . . . 671
2.6. Harmonic Analysis; Smoothing 673
2.7. The Clustering Amplitude $\alpha_\lambda(\omega)$ 678
2.8. Estimating Space Averages 680
2.9. Abell's Analysis of Superclustering 682
2.10. The Lick Galaxy Counts 683
2.11. Other Analyses 686

3. Dynamics of Clustering 687
3.1. Field Equations and Equations of Motion for a Cosmic Distribution in the Newtonian Approximation 687
3.2. The Energy Equation and the Clustering Spectrum 689
3.3. The Virial Theorem for Cosmic Distributions 690

4. Origin of Clustering 693
4.1. The Phenomenon of Clustering 693
4.2. The Cosmological Principle 694
4.3. Initial Growth of Small-Amplitude Density Fluctuations 696
4.4. The Weak-Field Approximation 698
4.5. Initial Growth of Density Fluctuations in the Weak-Field Approximation 700
4.6. Thermal Instability 702
4.7. Early History of the Cold Universe 705
4.7.1. Thermal History of the Early Cold Universe 705
4.7.2. Helium Production in the Cold Universe 706
4.7.3. The Metallic Phase 706
4.8. Thermal Fluctuations Revisited 708
4.9. Gravitoturbulence 709
4.10. Generalized Pressure and Internal Energy 712
4.11. The Fluctuation Spectrum and the Binding-Energy Spectrum . . . 713
4.12. The Gravitoturbulent Regime 714
4.13. The Clustering Process 715
4.14. Effects of Evolution; Comparison with Observation 718

4.14.1. Simplification of the Hierarchy . . . 718
4.14.2. Stars and Galaxies . . . 718
4.14.3. Initial Diameters of Galaxies and Protostars . . . 719
4.14.4. Origin of Galactic Spin . . . 719
4.15. Star Formation and the Microwave Background . . . 720

References . . . 722

18. Radio Astronomy and Cosmology . . . 725
Peter A. G. Scheuer

1. Introduction . . . 725
1.1. The Mean Density of the Universe . . . 725
1.2. Residual Radiation . . . 726
1.3. The Radio Source Population . . . 726

2. Summary of Basic Formulae . . . 727

3. Attempts to Detect Intergalactic Atomic Hydrogen . . . 728
3.1. Absorption in the 21-Centimeter Line . . . 728
3.2. 21-Centimeter Line Emission from Intergalactic Atomic Hydrogen . 729
3.3. Absorption in the Lyman-α Line . . . 730

4. Attempts to Detect Intergalactic Ionized Hydrogen . . . 731
4.1. Thomson Scattering . . . 731
4.2. Free-free Emission and Absorption . . . 731
4.3. Dispersion . . . 734
4.4. Faraday Rotation . . . 734
4.5. Scattering by Irregularities . . . 734
4.6. Is There Any Intergalactic Gas? . . . 735

5. The Microwave Background . . . 737
5.1. The Observed Spectrum . . . 737
5.2. Interpretation . . . 738

6. Counts of Radio Sources . . . 739
6.1. Basic Ideas . . . 739
6.2. Observations . . . 740
6.2.1. Linearity of the Flux-Density Scale . . . 742
6.2.2. Effects of Confusion and Noise . . . 742
6.2.3. Resolution of Sources of Large Angular Size . . . 744
6.2.4. Statistical Uncertainties . . . 745
6.2.5. Isotropy . . . 745
6.3. Interpretation of Source Counts . . . 745
6.4. Are We in a Hole? . . . 751

7. The Evolution of the Luminosity Function of Quasi-stellar Objects . . . 752
7.1. The Luminosity Function . . . 752
7.2. The Luminosity-Volume Test . . . 754
7.3. More General Procedures . . . 756
7.4. Other Correlations . . . 757

References . . . 758

19. THE REDSHIFT . . . 761
Allan Sandage

1. Introduction . . . 761
2. Early Results on the Spectra and Redshifts of Galaxies . . . 761
   2.1. The First Observations . . . 761
   2.2. De Sitter's Static Solution . . . 763
   2.3. Hubble and Humason's Observational Extension of the Redshift Law . 765
3. Aids in the Practical Measurement of Redshifts by Photographic Methods . 768
   3.1. Lists and Charts of Comparison Spectra . . . 768
   3.2. Nightsky Contamination . . . 770
   3.3. Absorption and Emission Lines in Galaxies . . . 771
4. Redshifts of Bright Galaxies; Constancy of $\Delta\lambda/\lambda_0$; Systematic Errors . . 773
5. Corrections to Measured Redshifts . . . 774
6. Absolute Luminosity and Intrinsic Diameter as Functions of Observables in Specific Models . . . 775
   6.1. Two Approaches . . . 775
   6.2. The Redshift–Apparent-Magnitude Relation . . . 776
   6.3. The K-Correction to Measured Intensities . . . 779
   6.4. The Difference from Hubble's Procedure . . . 779
   6.5. Intrinsic Diameters from Angular Measurements . . . 780
7. The Observations and World Models . . . 781

References . . . 783

AUTHOR INDEX . . . 787

SUBJECT INDEX . . . 801

GALAXY INDEX . . . 812

CHAPTER 1

Classification and Stellar Content of Galaxies Obtained from Direct Photography

ALLAN SANDAGE
Hale Observatories, Carnegie Institution of Washington, California Institute of Technology

1. INTRODUCTION

THE FIRST step in the development of most sciences is a classification of the objects under study. Its purpose is to look for patterns from which hypotheses that connect things and events can be formulated by a method proposed and used by Bacon (1620). If the classification is useful, the hypotheses lead to predictions which, if verified, help to form the theoretical foundations of a subject.

Simple description, although not sufficient as a final system, is often an important first step. In the study of galaxies, Wolf's (1908) and Vorontsov-Velyaminov's (1962, 1963, 1964, 1968) descriptions are examples of a system of this first type. But as a classification develops, a next step is often to group the objects of a set into classes according to some continuously varying parameter. If the parameter proves to be physically important, then the classification itself becomes fundamental, and often leads quite directly to the theoretical concepts.

It is too early to judge if the classification of galaxies has reached this stage because no theory of their origin and evolution is yet certain. But the systems of Hubble (1926) (as extended by Holmberg 1958; de Vaucouleurs 1956, 1959*a*; van den Bergh 1960*a*, *b*; and others) and of Morgan (1958, 1959) are based on continuously varying parameters and therefore constitute classifications of the second kind.

The classification criteria for the Hubble system are (1) the size of the nuclear bulge relative to the flattened disk, (2) the character of the spiral arms, and (3) the degree of resolution into stars and H II regions of the arms and/or disk. Item (1) is likely to be related to the angular-momentum distribution of the original protogalaxy, and to the timing of earliest star formation relative to the collapse time. The other two criteria are

Received June 1973.

probably related to the present rate of conversion of gas into stars in a rotating galaxy.

Both the Hubble and the Morgan systems appear to be more than simple descriptions because many galaxian properties such as the integrated color, the composite spectral type, and the density of free H I gas vary systematically along the sequence of forms. These classifications may therefore be fundamental in the above sense because they provide connective relations, based on form alone, that may be related to the initial conditions of formation and subsequent temporal change.

2. EARLY CLASSIFICATION OF GALAXIES

The most extensive all-sky surveys before the introduction of photography were made by the Herschels from about 1780 to 1860 by visual methods. Star clusters, galactic nebulae, and galaxies were catalogued, and moderately extensive descriptions were given. The *General Catalogue of Nebulae*, incorporating the observations of 5079 objects, of which 4630 were discovered by the Herschels, was published by Sir John Herschel in 1864 in the *Philosophical Transactions.* This forms the largest single base for the *New General Catalogue* (NGC) of Dreyer which, together with the two *Index Catalogues*, incorporates most discoveries of nebulae to 1908. The descriptive symbols used by the Herschels are summarized by Dreyer in the Introduction to the NGC, and by Curtis (1933) in his view of galaxian research to 1934. Although the descriptions do not constitute a classification in the formal sense, they are still valuable as a supplement to the current systems, and have been used as recently as 1956 by de Vaucouleurs (1956) in his survey of southern galaxies.

The faintest structural features of galaxies could be detected only when photographic surveys came into general use about 1890. These features proved to be decisive in the classification problem because the presence or absence of spiral arms is what divides galaxies into the two major groups (E and S), and separates the spirals along a linear sequence by the character of the arm structure.

Wolf's (1908) system was a classification based on photographs taken at Heidelberg in which letters were used for various forms, and in which no distinction was made between galactic planetary nebulae and galaxies. Although the system is not now in general use, it was the first to use a linear sequence which proceeds from amorphous forms with no spiral patterns (type *d* to *k*), to fully developed spirals (types *r* to *w*). Extensive use of the Wolf classification was made as late as the 1940s. Because of this, and because it provides a more detailed description of the many variations of spiral patterns than does Hubble's, Wolf's sequence is shown in figure 1.

Although Hubble's system places galaxies in a physically significant linear sequence (or a series of such sequences as in de Vaucouleurs's extension) it is "simple" in that it gives no notational recognition to the great variety of spiral arms, as does Wolf's. For this reason, Danver (1942) preferred the Wolf types and commented: "As to the Hubble classes, these [appear to constitute] a division along a line of development. This is certainly a great advantage, but if a conception of the appearance of the object is desired, it is better to designate the types according to Wolf." It is quite possible that when the theory of spiral structure is more fully developed than at present, a new classification of the arm patterns *alone* may be needed, and the Wolf scheme might serve as a new point of departure. The modern work of Vorontsov-Velyaminov (§ 8) may be a step in this direction.

Among the more complete discussions using the Wolf symbols are Reinmuth's (1926) study of the Herschel galaxies, Lundmark's (1927) summary of galaxian research to 1926, Holmberg's (1937) work on double galaxies, Reiz's (1941) study of the surface distribution of galaxies, and Danver's (1942) work on the forms of spiral arms. Danver

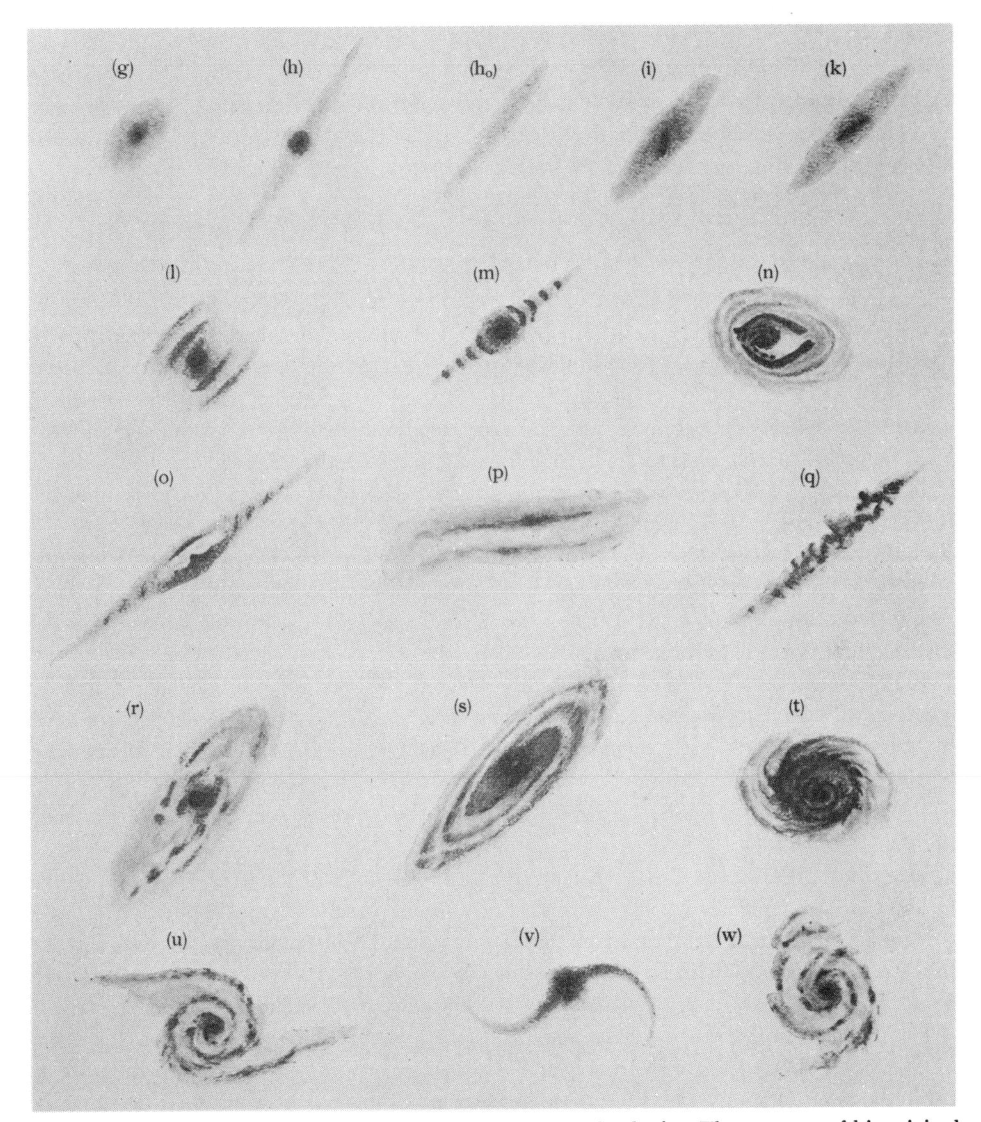

FIG. 1.—Wolf's 1908 descriptive classification system of galaxies. The top row of his original diagram, which showed galactic nebulae (mostly planetaries), is deleted here.

gives a comparison of the Hubble types with those of Wolf, extending a similar discussion by Shapley and Ames (1932, pp. 68–69).[1]

Lundmark (1926, 1927) proposed a classification based on a major division of galaxies into three groups (amorphous ellipticals, true spirals, and Magellanic Cloud–type irregulars), further dividing the groups according to the concentration of light toward the center. The division into E and S classes is similar to that made by Hubble (1926), but the parameter which divides galaxies within the classes (concentration by Lundmark and the nature of the spiral arms by Hubble) is different. Curtis (1933, appendix 5G), summarizes Lundmark's system, and compares it with the Wolf symbols.

Shapley (1928) also proposed a classification based on concentration, but included the nonintrinsic properties of apparent magnitude and apparent flattening in his notation. The system was used at Harvard in the early stages of the survey of galaxian distribution to $m_{pg} = 17$ (e.g., *Harvard Annals*, **88**), but is not now in general use.

3. DEVELOPMENT OF THE MODERN SYSTEM

3.1. Early Isolation of the Types

Reliable separation of galaxies into classes could successfully begin only after extensive photographic surveys were well under way, and the forms of galaxies discovered. This descriptive phase is not yet complete for the interacting galaxies (cf. Vorontsov-Velyaminov 1959; Zwicky 1959 and references therein), peculiar galaxies (Arp 1966), and perhaps even for the infinite detail of more regular systems (Vorontsov-Velyaminov, Krasnogorskaja, and Arkipova, in the four volumes of the *Morphological Catalog of Galaxies* 1962–1968). However, following the isolation of the S0 type by Hubble (cf. Spitzer and Baade 1951; Sandage 1961), of the dwarf ellipticals by Shapley (1938*a*, *b*) with the subsequent discovery of great numbers in the Local Group (cf. Harrington and Wilson 1950), and of dwarf spirals related to Magellanic Cloud–type Sm (cf. the Hubble Atlas, p. 40, for Dwf I and II in the NGC 1023 group), all the more common types of galaxies had been discovered by 1940.

If 1940 marks the end of the major survey for types, the beginning occurred in 1845 when spiral structure was first discovered visually in M51 by Lord Rosse with the 72-inch reflector at Burr Castle. According to Curtis (1933) spiral structure was subsequently identified by Rosse in perhaps 20 additional galaxies. The classical photographic work of Isaac Roberts from 1885 to 1904 confirmed the spiral types, and increased their number to about 40.

The photographic surveys of Keeler (1898–1900), Perrine (1901–1903), and Curtis (1909–1918; *Lick Pub.*, **13**, 1918) with the Lick Crossley reflector marks the beginning of research which led to the present classification. Mount Wilson surveys by Ritchey and by Pease (1917, 1920) with the 60-inch added to the material, as did the early systematic work of Knox-Shaw, Gregory, and Madwar with the Reynolds reflector at Helwan (Vols. **1** and **2** of *Helwan Observatory Bulletins*). Knox-Shaw (1915) and Reynolds (1920*a*) were among the first to call attention to amorphous galaxies with no trace of spiral arms (E systems). Curtis (1918) first isolated the barred spirals (called by him Φ type), and Lundmark (1926, 1927) emphasized the highly resolved Magellanic Cloud types, incorporating them as a separate group in his classification system.

[1] Hubble (1917) used Wolf's classification in his early work at Yerkes.

3.2. The Early Hubble System

From these data, supplemented by plates taken by Hubble, Humason, and Duncan with the Mount Wilson 60- and 100-inch reflectors, and building on the earlier proposals (§ 2), Hubble (1926) formulated the linear classification system which has evolved into the present standard systems. Galaxies are divided into three major classes (ellipticals, spirals, and irregulars). The spirals are separated into two families, the "normal" spirals and the barred; and are separated along each family into types a, b, and c according to the three criteria listed in § 1, which generally (but not always)vary together along the form sequence from early Sa (SBa) to late Sc (SBc). The system has been described in many places (e.g., Hubble 1926, 1936; de Vaucouleurs 1959*a;* Sandage 1961; Baade 1963, chap. 2; Hodge 1966) and need not be repeated here.

Soon after publication of Hubble's 1926 paper, Reynolds (1927*a*, *b*) criticized the system because of a supposed inadequate number of classification bins. Impressed by the enormous variety of galaxian structures, Reynolds wrote: "The problem I have always found in attempting a general classification of spiral nebulae is that one meets case after case where a special class is required for the individual object. Spectral classification of stars is a simple and straightforward matter compared with this."[2]

Wolf had earlier commented similarly in the first sentence of his 1908 paper "Es gibt kein zwei Nebelflecken am Himmel, die sich gleichen." But it is precisely because Hubble's system specifically ignores the multitude of superficial details of arm structure (for example), and concentrates on the gross characteristics of pattern according to broad criteria, that the system has such merit. Nearly all galaxies can be put into a specific classification bin (in the revised system) without forcing, because the bins are large. This is true even for most of the interacting and peculiar galaxies discussed by Vorontsov-Velyaminov (1959), Arp (1966), Burbidge, Burbidge, and Hoyle (1963), and others. The underlying Hubble type is usually visible, albeit with peculiar features denoted by pec or p after the basic type. Baade's comment (1963, chap. 2) that the Hubble system has great merit is shared by most classifiers, but not all would agree with his belief that the logical extensions by de Vaucouleurs are unnecessary—extensions which retain the simplicity of the scheme, but narrow the class sizes and give notational recognition to the transition cases between ordinary and barred spirals.

Genuinely peculiar objects do exist, such as M82, NGC 3077, NGC 520, NGC 2685, NGC 3718 (illustrated in the *Hubble Atlas* 1961). These fall outside the system, but constitute only a few percent of any random sample.

An example of the universality of Hubble's system is given by a trial classification of galaxies in the *Atlas of Peculiar Galaxies* (Arp 1966). Of the 338 illustrated galaxies, 45 are so abnormal that they could be given no underlying basic Hubble type (except peculiar), but this is only 13 percent of the total sample, which itself is already highly selected as regards peculiarity.

[2] Although Reynolds criticized Hubble's 1926 system as too "simple," he had himself, seven years earlier, proposed a classification (Reynolds 1920*b*) that was similar enough to Hubble's E, Sa, Sb, Sc, and Irr types (but called classes I–V), that, in the absence of his 1927 repudiation, Reynolds would now have been considered as an early originator of part of the modern classification.

3.3. Hubble's Major Modification between 1936 and 1950

The S0 class was not isolated observationally until after 1936, although Hubble had earlier come to believe such a class was necessary. In *The Realm of the Nebulae* he wrote, "The junction [between E and S types] may be represented by the more or less hypothetical class S0. Observations suggest a smooth transition between E7 and SBa [on the barred side of the tuning-fork diagram], but indicate a discontinuity between E7 and Sa [on the spiral side] in the sense that Sa spirals are always found with arms fully developed [whereas SBa in the 1936 system had no arms, by definition]." The difference, then, was primarily a matter of definition (cf. plate II of *Realm*), but it did represent an asymmetry between the ordinary and barred spiral families because no "armless ordinary spirals" had been found whereas armless barred spirals did exist (cf. NGC 2859, *Hubble Atlas*, p. 42; NGC 2950, p. 42; and NGC 4643, p. 42, all originally classed as SBa but now classed as SB0).

Because of this asymmetry, Hubble undertook a special search for nonbarred examples of "armless spirals" (the modern S0 class). They were subsequently found on long-exposure plates taken in the Mount Wilson survey of the Shapley-Ames galaxies north of $\delta = -15°$, carried out by Hubble between 1936 and 1950 with the 60- and 100-inch reflectors.

Elliptical and S0 galaxies are easily confused on small-scale plates, which explains why the S0 class remained unrecognized for so long. Neither class has arms, and both show smooth intensity distributions with no resolution into bright supergiant stars. The criterion that distinguishes the two classes is a difference in the radial intensity distribution $I(r)$. E galaxies have a steep intensity gradient (Hubble 1930, de Vaucouleurs 1948, 1953, 1959*a*), whereas S0 galaxies have an extensive exponential outer envelope, superposed on a central E-like distribution (cf. de Vaucouleurs 1956, 1959*a*; Liller 1960; Johnson 1961; Hodge and Webb 1964; Hodge and Merchant 1966). This envelope is similar to the underlying exponential disks of spirals (cf. de Vaucouleurs 1958, 1959*b*, 1962, 1963*a*, 1964; Freeman 1970).

Although it is clear that Hubble discovered the S0 class and studied some of its members, he published no discussion of the important phenomenon. The first literature references were made by Spitzer and Baade (1951), based on Baade's conversations with Hubble. Summaries were given by de Vaucouleurs (1956, 1959*a*) based on prepublication notes from the *Hubble Atlas*, and by Sandage (1961) where the class is discussed and illustrated.

A description of Hubble's revised system as it existed in 1950, written by Hubble between 1947 and 1950, is useful, and is reproduced from the *Hubble Atlas:*

> The sequence of classification, as originally presented, consisted of a series of elliptical nebulae ranging from globular (E0) to lenticular (E7) forms, and two parallel series of unwinding spirals, normal (S) and barred (SB). Each of the latter series was subdivided into three sections, termed early, intermediate, and late, and designated by the letters a, b, and c, respectively. Thus the early, normal spirals were represented by the symbol Sa, and the early barred spirals by SBa.
>
> The data available in 1936 seemed to indicate a smooth and continuous transition from elliptical nebulae to barred spirals, and, in fact, the first section of the latter

series, SBa, exhibited no spiral arms. The corresponding section of the normal series, Sa, contained so many nebulae with fully developed spiral arms, that, where arms could not be definitely recognized, their presence was assumed, and the failure to detect them was attributed to effects of orientation or other causes. The procedure was unsatisfactory because it introduced subdivisions in the parallel series of spirals that were clearly out of step. Moreover, the transition from E7 to Sa appeared so abrupt that, if real, it might be regarded as cataclysmic.

With accumulating data, and especially with the increasing number of good photographs with the 100-inch reflector, the situation has clarified. Numerous systems are now recognized which are later than E7 but which show neither bars nor spiral structure. These nebulae fill the supposed gap between E7 and Sa and remove the excuse for postulating a cataclysmic transition. [These transition galaxies are designated S0. They are actually found in nature and are no longer a hypothetical class, as was once believed; see *The Realm of the Nebulae*, pages 45–46, and the legend to figure 1, page 45, of the Yale University Press edition of 1936. A.S.]

A similar group of objects corresponds to the section [called SBa in the 1936 classification]. This situation emphasizes the desirability of redefining the sections of both series in a more comparable manner.

The revision might be made in various ways but only that actually adopted will be described. First, two new types, S0 and SB0, have been introduced to include objects later than E7 but with no trace of spiral structure. Second, the series of true spirals, as before, are subdivided into the three sections Sa, Sb, Sc, and SBa, SBb, SBc. In the case of the normal spirals, the change amounts to a subdivision of the former section Sa into the two sections S0 and Sa. In the case of the barred spirals the entire former section, SBa, is now termed SB0, and the former section SBb is subdivided into the two sections SBa and SBb. The revisions are summarized in the following table:

Old Class	New Class	Old Class	New Class
Sa	{S0, Sa}	SBa	SB0
		SBb	{SBa, SBb}
Sb	Sb		
Sc	Sc	SBc	SBc

The introduction of the new types leads to a revision of the original assignment of symbols. The original SBa nebulae are now described as SB0, and the original SBb nebulae are redistributed between SBa and SBb. Among the normal spirals, the Sa objects are redistributed between S0 and Sa. Otherwise the system remains unchanged.

The transition stages, S0 and SB0, are firmly established. In both sequences, the nebulae may be described as systems definitely later than E7 but showing no spiral structure. The next stages, Sa and SBa, are represented by nebulae which show incipient spiral structure. Fully developed spirals are distributed over the two later stages of each sequence according to the relative extent of the unresolved central region, and the degree to which the arms are resolved and unwound.

3.4. Finer Subdivision along the Sequence

Hubble's bin sizes among the spirals were large, and inspection of survey plates showed that they could usefully be narrowed. At various times, the Mount Wilson observers have used combination symbols S0, S0/a, a, ab, b, bc, c, c/Irr, and Irr to divide the linear sequence into nine groups instead of three along both the ordinary[3] and barred families.

Holmberg (1958) has used + and − symbols with the notation a, b−, b+, c−, c+, Irr I to divide the spirals into six groups.

De Vaucouleurs has used both notations in his revised classification by dividing the E's and S0's into early, intermediate, and late bins as E, E^+, $S0^-$, $S0^0$, $S0^+$, and by using the mixed notation along the spiral sequences as a, ab, b, bc, c.

4. REVISION BY DE VAUCOULEURS

4.1. Extension of the Sequence beyond Sc

Galaxies originally classed Sc on the Hubble system cover a large interval along the sequence, ranging from regular well-developed arms in the early Sc to nearly chaotic structures in the very late Sc. De Vaucouleurs has made a division and extension of the Sc and SBc families by introducing the cd, d, dm, m, and Im subdivisions. This important addition extends the Sd class of Shapley and Paraskevopoulos (1940) and Shapley (1950) toward even later types. The Sd type-example is NGC 7793 (photograph in Shapley 1961, p. 22), which would have been classed as a very late Sc on the Hubble system.

The further extension beyond Sd by inclusion of Magellanic Cloud types (Sm or SBm, or SAm and SBm in the revised notation) follows from the discovery of weak but definite spiral structure in the Large Magellanic Cloud (LMC) (de Vaucouleurs 1954, 1955*a*). Closely related galaxies, collectively denoted Irr by Hubble (as were the Sm types) but clearly following linearly beyond the Sm stages, are the dwarf Population I systems such as IC 1613, NGC 2366, Holmberg I and II, and IC 2574 (*Atlas*, pp. 39, 40). De Vaucouleurs's symbols I(m) and IB(m) replace Hubble's Irr class and show the connection with the Magellanic Cloud types. The revision brings the type back into the formal sequence as suggested by Lundmark (1927).

4.2. Transitions between Ordinary and Barred Spirals

Many galaxies combine features of pure ordinary spirals (such as NGC 628, *Atlas*, p. 29) and pure barred systems (e.g., NGC 1300, *Atlas*, p. 45). Examples are NGC 4579 (*Atlas*, p. 13), 5236 (*Atlas*, p. 28), and 3504 (*Atlas*, p. 46), which Hubble and Sandage denoted by the mixed symbols Sb/SBb, Sc/SBb, and SBb/Sb, respectively, where the leading symbol gives the dominant type. De Vaucouleurs uses a more convenient and symmetrical notation by (1) adding a type symbol A to the ordinary spirals, (2) retaining the symbol B for the barred spirals, and (3) adding the mixed symbol AB

[3] Nonbarred spirals were called "normal" by Hubble. De Vaucouleurs (1959*a*) has emphasized that barred and nonbarred spirals are equally normal because both appear with about the same frequency (de Vaucouleurs 1963*b*, table 5) if the transition types are neglected. He suggested use of the term "ordinary" rather than "normal," and we follow this notation in the remainder of this chapter.

for the transition cases. De Vaucouleurs's notation for the three examples quoted are SABb (4579), SABc (5236), and SABab (3504), where the dominant type is underlined.

4.3. The r and s Varieties

Barred spirals are of two dominant varieties: the r type in which the arms begin tangent to the external ring upon which the bar terminates (the purest example may be NGC 2523, *Atlas*, p. 48, and plate 6 of this chapter), and the s type in which the arms begin from the end of the bar (e.g., NGC 1300, *Atlas*, p. 45, and plate 6 here).

Recognition of the r and s varieties was made in the *Hubble Atlas* for all barred spirals by the notation SBb(r) for 2523, or SBb(s) for 1300. Mixed types such as 1073 (*Atlas*, p. 49, plate 6 here) were called sr or rs according to which variety dominates.

The same phenomenon is present in ordinary spirals, although it is somewhat more difficult to detect. Examples are 4274 Sa(r) (*Atlas*, p. 12), and 309 Sc(r) (*Atlas*, p. 32),

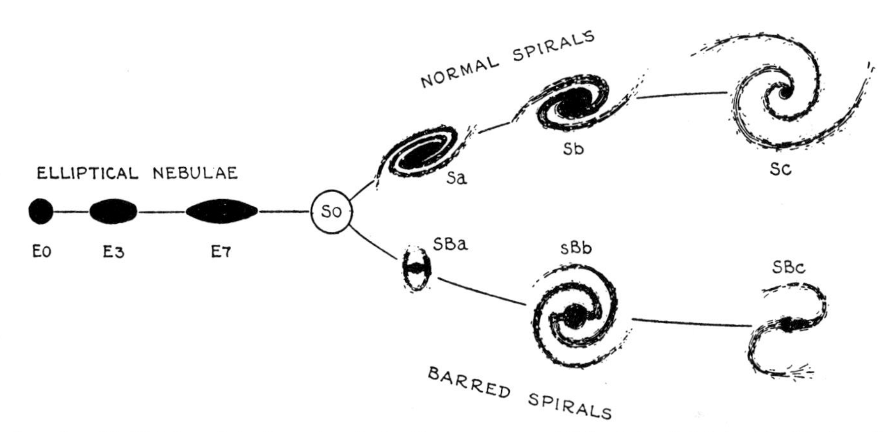

Fig. 2.—Hubble's original tuning-fork diagram as published in 1936 in his *Realm of the Nebulae*.

and 5457 (M101) Sc(s) (*Atlas*, pp. 27 and 31), and especially 4321 (M100) Sc(s) (Atlas, pp. 28 and 31). The r-type spirals had previously been noted by Shapley and Paraskevopoulos (1940), by Randers (1940), by Vorontsov-Velyaminov (1965), and undoubtedly by many others.

De Vaucouleurs adopted the r and s distinction for all non-E galaxies, underlining the r or s symbol in the mixed variety according to the dominant form. For example, the transition galaxy NGC 5236 is classified as SAB(s)c in the revised system, while NGC 4579 (*Atlas*, p. 13; plate 5 here) is SAB(rs)b [classed Sb/SBb(rs) in the *Hubble Atlas*].

4.4. Graphical Representation of the Classification

Visualization of the classification system has always been useful. The 1936 Hubble system is illustrated in the well-known tuning-fork diagram shown in figure 2, taken from *The Realm of the Nebulae*.[4]

[4] Some history can be read from fig. 2. The revision of criteria and the redistribution of the S0, Sa, and SBa types (§ 3.3) had not yet been made by Hubble when this diagram was prepared. This is shown by the nature of the galaxy labeled SBa, which has no spiral arms and would be

A more complicated diagram is needed to illustrate the separation of *each branch* of the tuning fork into the r and s varieties. A useful representation proved to be a three-dimensional figure such as figure 3, which is taken from Hodge's (1966) version of a less complete diagram given in the *Hubble Atlas* (p. 26). Mixed varieties rs and transition types SAB are not readily accommodated in this representation, but could form the other sides of the figure for the rs and sr varieties, and would fill the interior of the volume for the mixed SAB types if this representation were carried further.

The extension was made by de Vaucouleurs (1959*a*) in a remarkable generalization of the entire scheme by filling the interior of the "classification volume" with both the r and

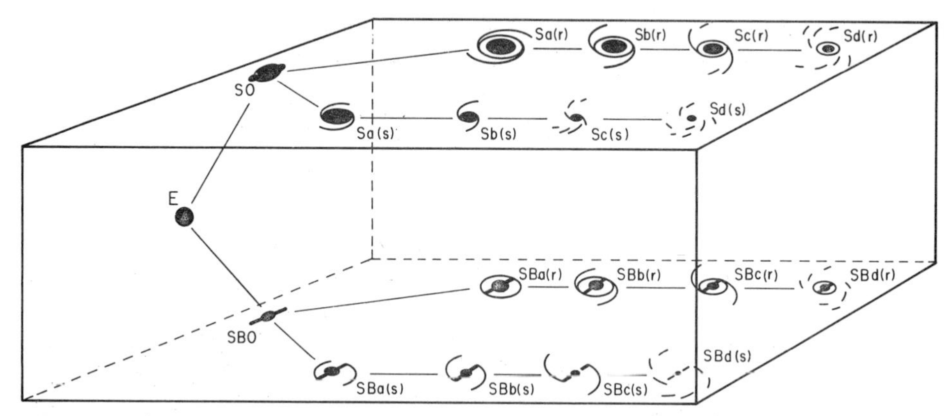

FIG. 3.—First stage of development of the concept of the *classification volume*. Here, the ordinary and barred families are separated onto opposite sides of a box. Within each family, a separation is made into the r and s strains, depending on whether the arms start from a ring or from the nucleus. Transition cases between the barred and ordinary families are not formally recognized in this visualization, but would fit in the interior of the box, a concept that leads into de Vaucouleurs's generalization shown in figs. 4 and 5. (Rendering of a diagram from Hodge 1966.)

s varieties and the A, B, and AB *families*. The basic linear sequence of *classes* is then separated *along* the axis of the volume. Figures 4 and 5 illustrate the revision, and are taken from de Vaucouleurs (1959*a*).

4.5. Additional Features of the de Vaucouleurs Revision

4.5.1. *Presence of rings*. Galaxies that possess external rings such as NGC 2859 [RSB(r)0+, *Atlas*, p. 42] are denoted by R preceding the designation of the basic type (SB in this case). The difficulty in designating rings in SB galaxies is that it is often very hard to distinguish a true ring from a form in which two spiral arms are tightly coiled and nearly touch after each has made a turn of 180°. Examples where it is certain that confusion would exist on small-scale plates between this form and true rings include NGC 3185 (*Atlas*, p. 43; note especially the description given), 2217 (*Atlas*, p. 43), and 3081 (*Atlas*, p. 11; note the description). More easily distinguished cases are 3504

---

a modified S0 type in modern notation. Consequently, Hubble isolated the S0 class only after 1936.

(*Atlas*, p. 46), and 4750 (*Atlas*, p. 21). None of these galaxies have true rings. They are classed as (R)[1] systems by de Vaucouleurs to denote "pseudo rings."

True rings do occur in such galaxies as NGC 2859 (*Atlas*, p. 42), but it is not certain that the rings are attached to the main body of the parent galaxy. Examples of definitely attached partial rings outside the regular spiral pattern include the faintest outer structure in NGC 2685 (*Atlas*, p. 7 insert), 4736 (*Atlas*, p. 16), 4457 (*Atlas*, p. 9), 3368 (*Atlas*, p. 12), 1068 (*Atlas*, p. 16), and 5101 (*Atlas*, p. 42). True rings appear to occur

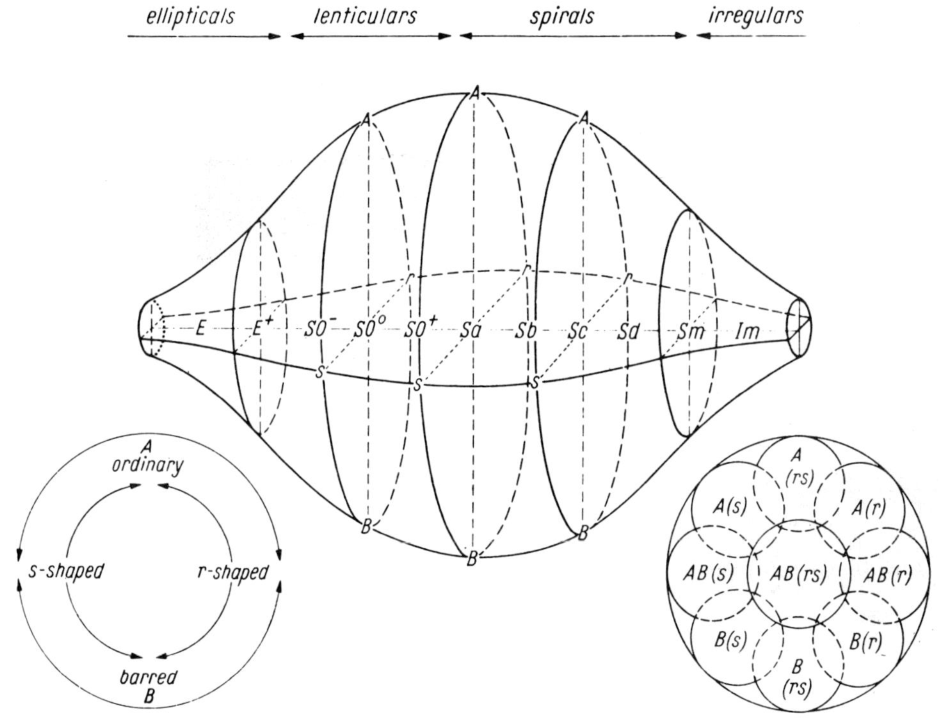

FIG. 4.—The *classification volume* of de Vaucouleurs. The division into gross types is made along the axis of the figure from left (E) to right (Sm), the division into the ordinary and the barred family is by position on the surface (from de Vaucouleurs 1959*a*).

predominantly in early-type systems, although cases such as NGC 4736 [(R)SA(r)ab, de Vaucouleurs 1963*b*; or Sb, Hubble-Sandage: *Atlas* 1961] exist. This case is particularly interesting because partial resolution of the ring into stars ($B \simeq 22$ or $M_B \simeq -8$) is seen on 200-inch plates.

4.5.2. *Increasingly detailed notation.* In his objections to Hubble's simplified system, Reynolds (1927*a*, *b*) pointed out that spiral arms differ in *character*. Some systems have "massive" arms, such as M33 (*Atlas*, p. 36), NGC 4567 (*Atlas*, p. 13), and M51 (*Atlas*, p. 26), while others have thin, delicate, filamentary arms, such as NGC 2841 (*Atlas*, p. 14), 488 (*Atlas*, p. 15), 628 (*Atlas*, p. 29), and 1232 (*Atlas*, p. 32).

This feature is undoubtedly important and, although not recognized in Hubble's notation, does form the basis of a division of galaxies into strains or groups from Sa to Sc, where the characteristics can be traced through the entire sequence. Such groupings are discussed in the *Hubble Atlas* (cf. the section on Sc galaxies where families with similarly shaped arms are isolated).

De Vaucouleurs has proposed to recognize these characteristics explicitly in the notation by adding symbols m (for massive) and f (for filamentary) after the type designation, with an additional symbol to indicate how many arms are present. For example, NGC 1232 (*Atlas*, p. 32), which has filamentary and highly branched segmented struc-

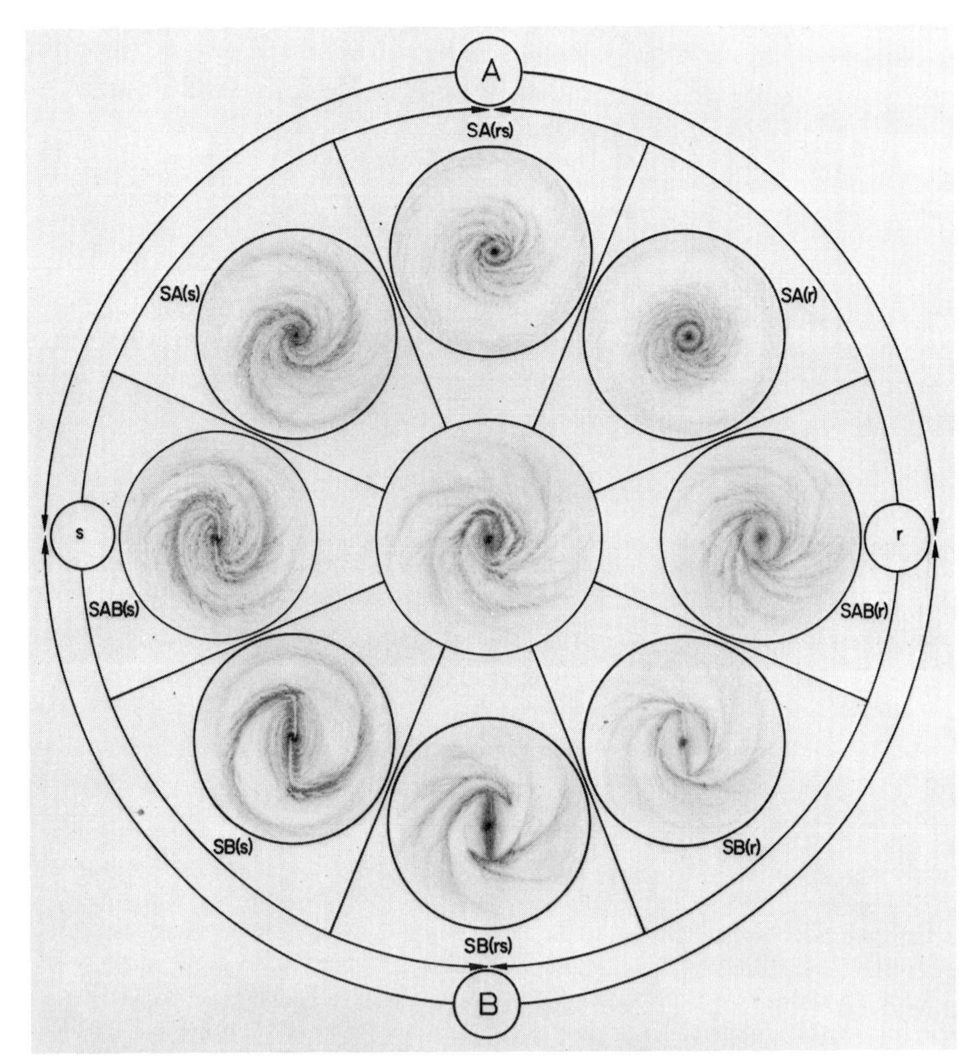

FIG. 5.—A cross-section of fig. 4 near the region of the Sb and SBb spirals, showing the manner in which the transition cases between ordinary (A) and barred (B) families, and the (r) and (s) strains can be accommodated (from the *Reference Catalogue of Bright Galaxies*, G. and A. de Vaucouleurs 1964).

tures starting from two main arms of the rs variety, would have a complete notation $SAB(rs)_{2^+}c_f$, where c denotes the arm character, and 2^+ denotes branching from two main arms.

By now the notation has become quite detailed, and is as complete as is likely to be useful. No classification system can describe the infinite variations among galaxies: this was Reynolds's objection. Most classifiers would agree with part of Baade's statement (1963, p. 19) that beyond this stage "if you want to study the variations on the theme Sc [for example], you simply have to take plates and examine them—only then do you get the full story. No code system can replace this. The code finally becomes so complicated that only direct inspection of plates helps."

But, as de Vaucouleurs points out, the virtue of the extended notation is that symbols can be progressively dropped until the notation becomes as simple as Hubble's original system. The classifier can exclude as much detail as he wishes and still remain within the standard classification at any desired level of complexity.

5. SELECTED ILLUSTRATIONS OF GALAXY TYPES

Classification on the Hubble–de Vaucouleurs system depends on subjective criteria; nevertheless, it works in practice, as evidenced by the fact that all classifiers are able to reproduce it well in the mean. Learning the system is best done by comparing photographs (such as those in de Vaucouleurs 1959*a*; the *Hubble Atlas* [Sandage 1961]; the *Cape Photographic Atlas* [Evans 1957]; Morgan 1958; Sersic 1968; and in other more scattered references) with several standard classification catalogs such as Humason, Mayall, and Sandage (1956 [HMS]), Morgan (1958, 1959), van den Bergh (1960c), de Vaucouleurs (1963*b*), the *Hubble Atlas*, and the *Reference Catalogue* (de Vaucouleurs and de Vaucouleurs 1964) decoded by its tables 1*b*, 2, 3, 4, and 5.

To reproduce all type-examples here would be an unwarranted duplication of material already in the literature, but a minimal selection is given in plates 1–8. The photographs are chosen to show progressive variation of the three classification criteria (and their conflict in some cases), and to illustrate the continuity of arm characteristics for the massive and the filamentary types—a continuity which can be traced throughout the sequence from Sa to Sc^+ for each strain separately.

5.1. Elliptical Galaxies

E types are not illustrated because nearly all are structureless and have similar appearance, showing smooth intensity distributions with relatively steep gradients. Except for the presence of globular clusters, detected in members of the Local Group (Hubble 1932) and in some of the nearby Virgo cluster ellipticals (Sandage 1961 in the *Hubble Atlas* descriptions; Vorontsov-Velyaminov 1966), no resolved components of the stellar content are usually present brighter than $M_V \simeq -3$ (Baade 1944). (NGC 205 is an exception where dust patches and a few early-type stars begin to resolve at about $M_V = -4$ [Baade 1951].)

E galaxies occur over a wide range of absolute luminosity extending from $M_V = -24$ for the brightest members of clusters of galaxies (based on a Hubble constant of 50 km s^{-1} Mpc^{-1}) to fainter than $M_V = -9$ for dwarf ellipticals (dE) of the Sculptor-Fornax

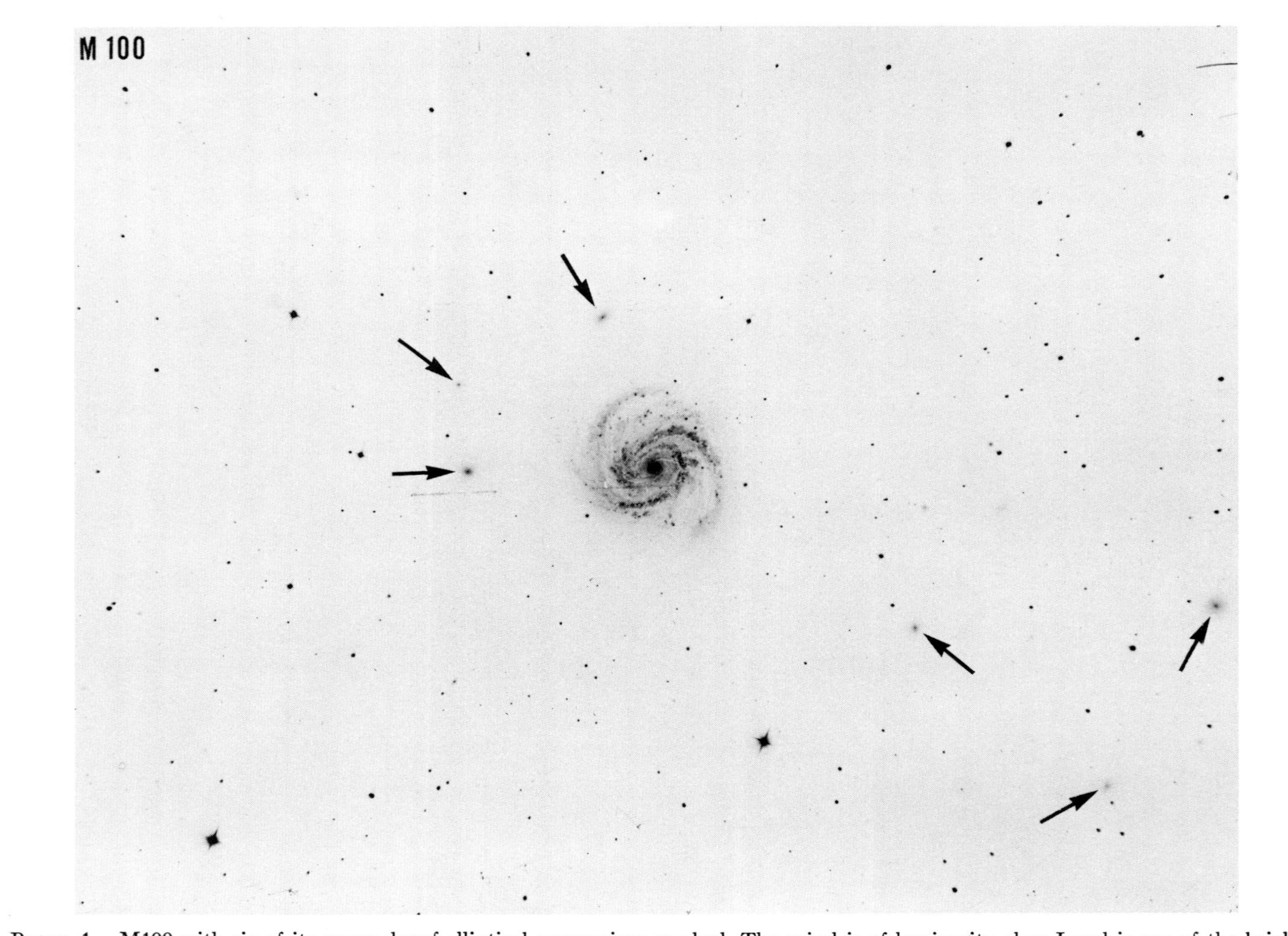

PLATE 1.—M100 with six of its many dwarf elliptical companions marked. The spiral is of luminosity class I and is one of the brightest galaxies in the Virgo cluster. Many dE galaxies are spread throughout the area (Reaves 1956). From a plate taken with the Mount Wilson 100-inch reflector.

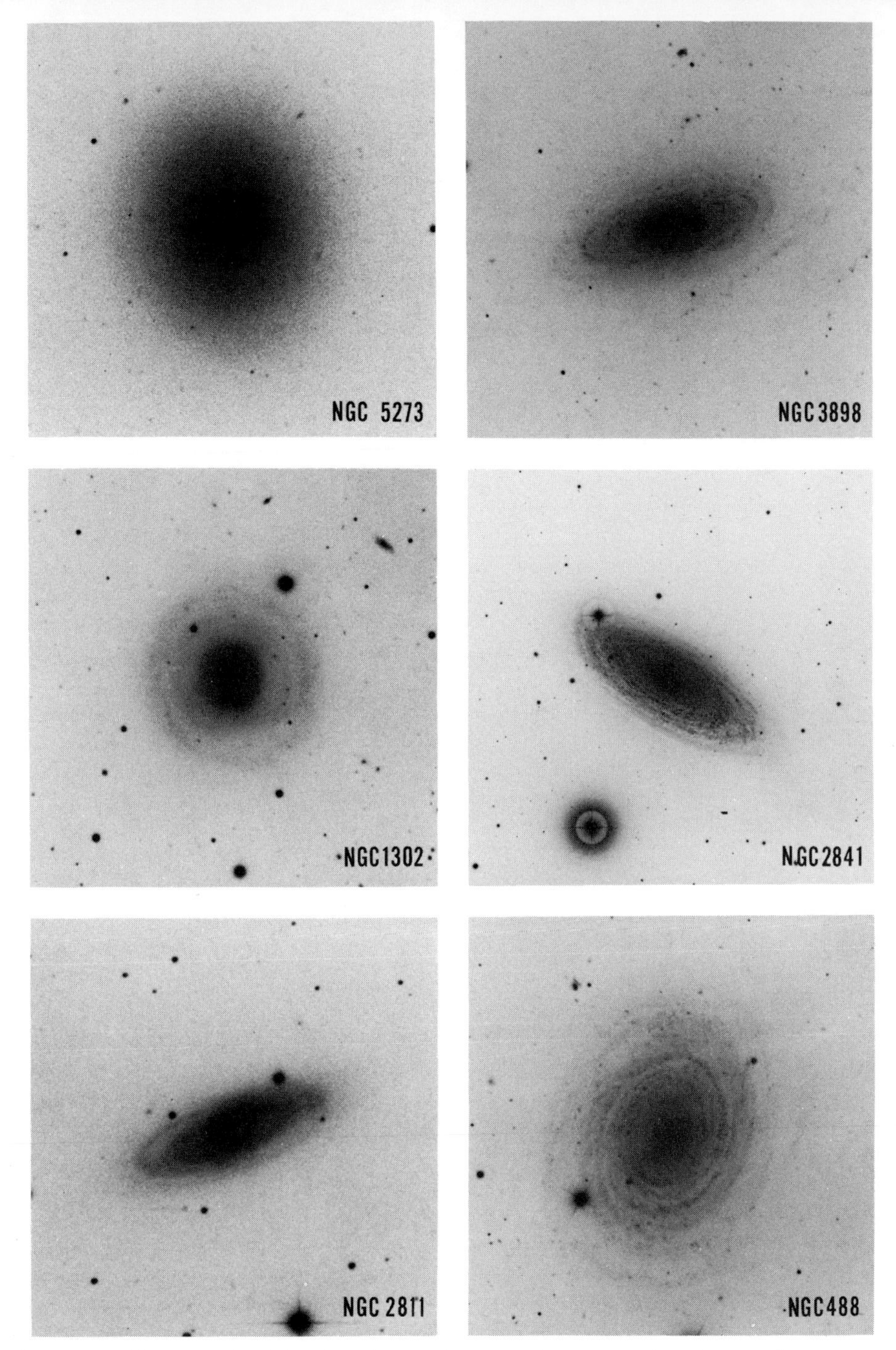

PLATE 2.—This and the following plate 3 illustrate the progressive change of appearance along the sequence of classification for thin, multiple-armed ordinary spirals from the earliest type $S0_2$/Sa(s) in the upper left (NGC 5273, *Atlas*, p. 8) to the latest Sm (or Sc/Irr Hubble) at the lower right of plate 3 for NGC 5204. The pictures are arranged in order of increasingly later type by following the left column from top to bottom, and then the right column similarly, in this and the following plate. The types of the galaxies can be found in table 2 of the text. Photographs taken with Mount Wilson or Palomar reflectors as described in the *Hubble Atlas* (Sandage 1961).

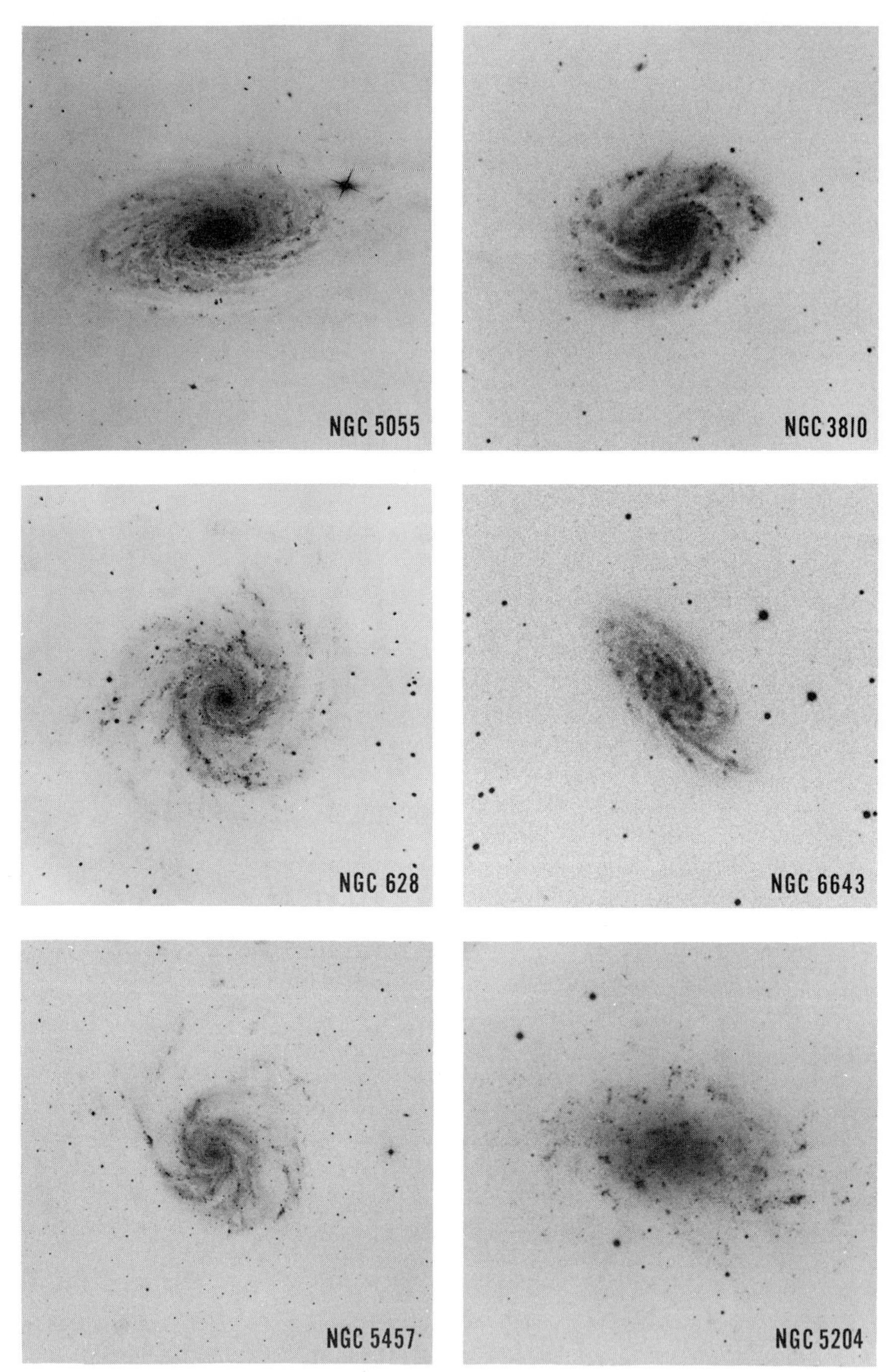

PLATE 3.—Continues the sequence of thin, multiple-armed ordinary spirals from plate 2 into the middle Sb to late Sd–Sm sections. The gradual and progressive changes of arm structure from Sa to late Sc form the continuous pattern visible in this sequence. The continuity is described in the *Hubble Atlas* in the text of the Sb section (*Atlas*, p. 16), and in the commentary to pictures on pp. 14 and 15 of the *Atlas* section (see in particular the description of NGC 5055). Mount Wilson-Palomar photographs.

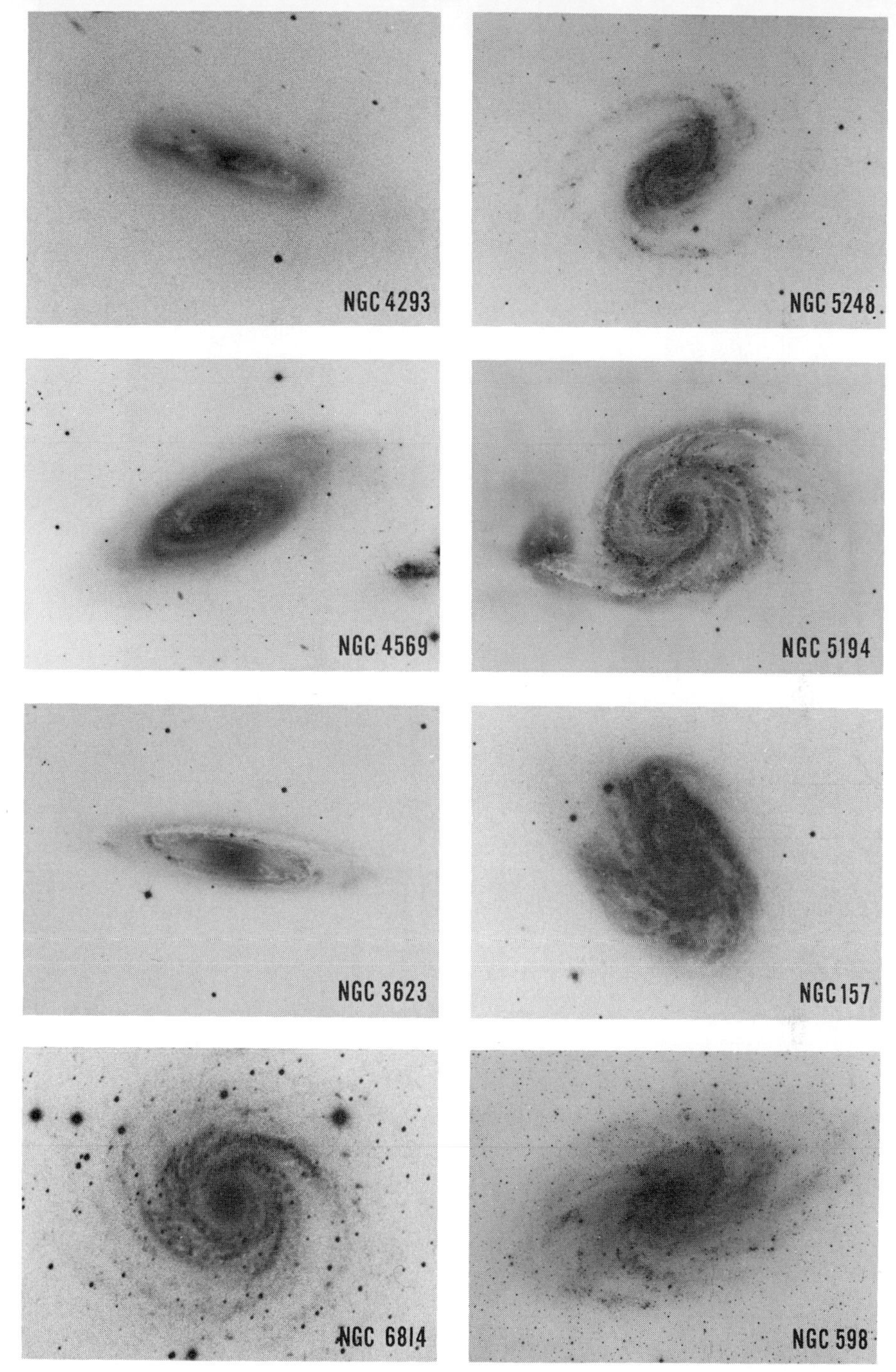

PLATE 4.—Progressive change of form of the massive-armed spirals from the earliest (NGC 4293, Sa) in the upper left, to the latest (M33, Scd) in the lower right. The nature of the arms changes continuously in the left column from top to bottom, and similarly in the right-hand column. Although all galaxies in this and the preceding two plates are of the same family (ordinary spirals), the difference in the arm structure pointed out by Reynolds (thin and filamentary in plates 2–3, and massive here) is evident. The division continues along the full length of the sequence from Sa to Sm. The plates were taken with the Mount Wilson or Palomar reflector.

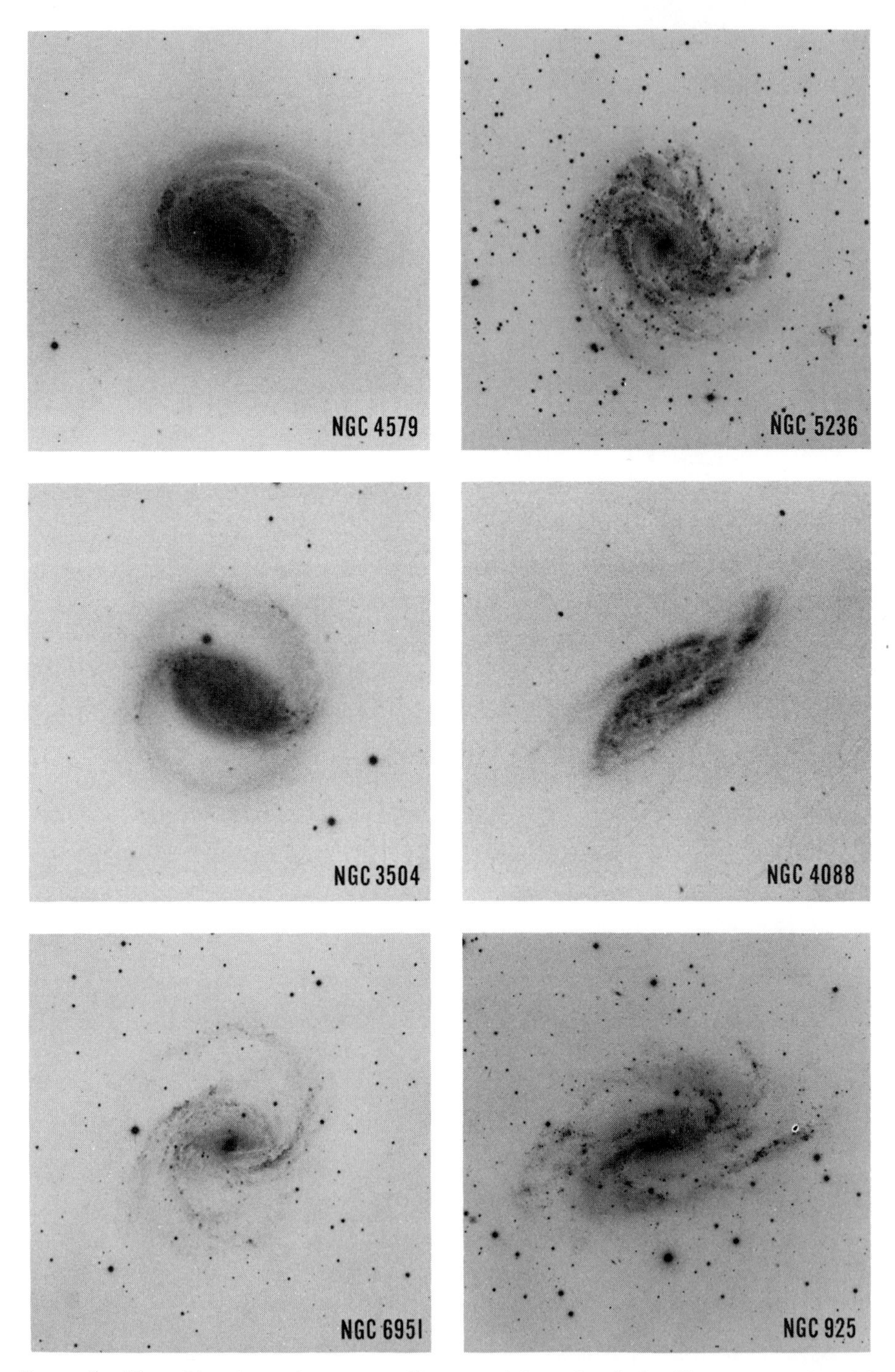

PLATE 5.—Transition types between ordinary and barred spirals. The types assigned by various classifiers are listed in table 2.

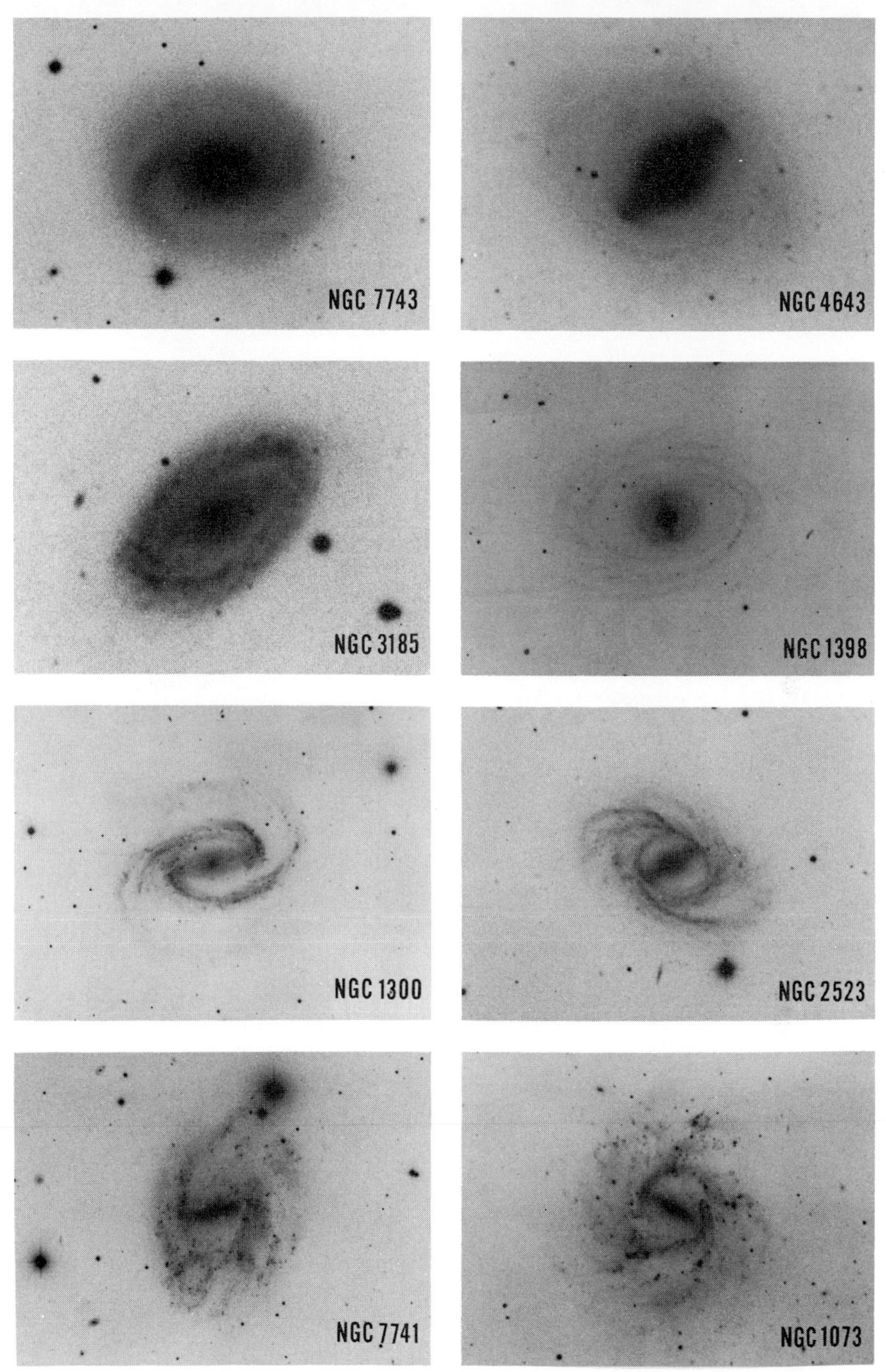

PLATE 6.—Barred spirals, showing the division into r and s types. Galaxies where the arms start from the end of the bar (s types) are shown on the left. Galaxies in which they begin tangent to an internal ring (r) are on the right. Within each column, the galaxies are arranged in the order from early to late. The one mixed type that is illustrated is NGC 1073 [SBc (rs)] at the lower right. NGC 3185 in the left column is of (s) variety, although it may appear to have a complete ring on small-scale plates. The arms do not touch to form a ring, but are very tightly wound after starting from the end of a bar.

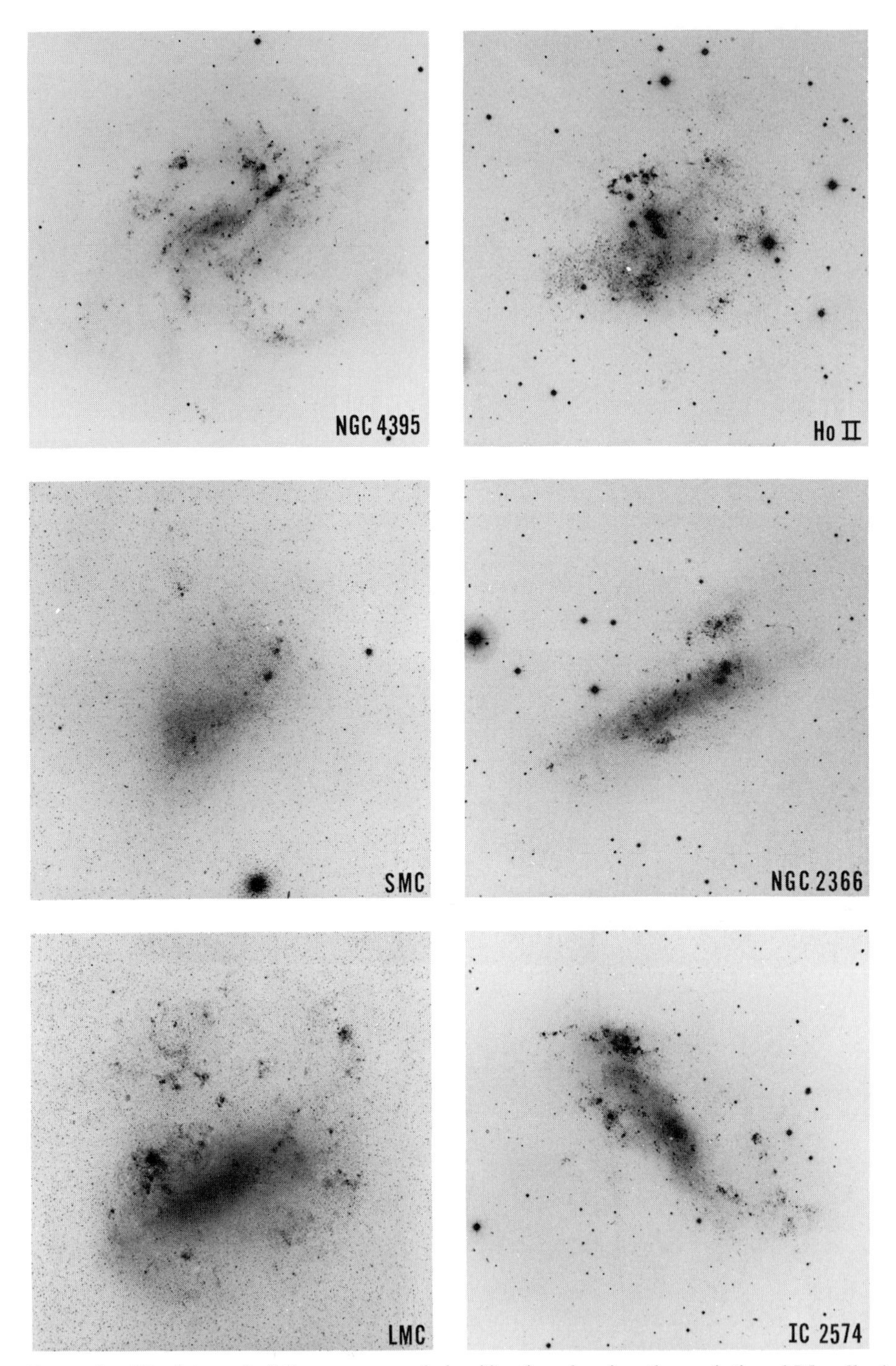

PLATE 7.—The late end of the sequence of classification showing the varieties of Magellanic Cloud–type systems, classified either as Sm or Im galaxies depending on the presence or absence of vague spiral arms. The plates of the SMC and LMC are by Henize using the Mount Wilson 10-inch refractor in South Africa. The other photographs are from Mount Wilson or Palomar reflector plates.

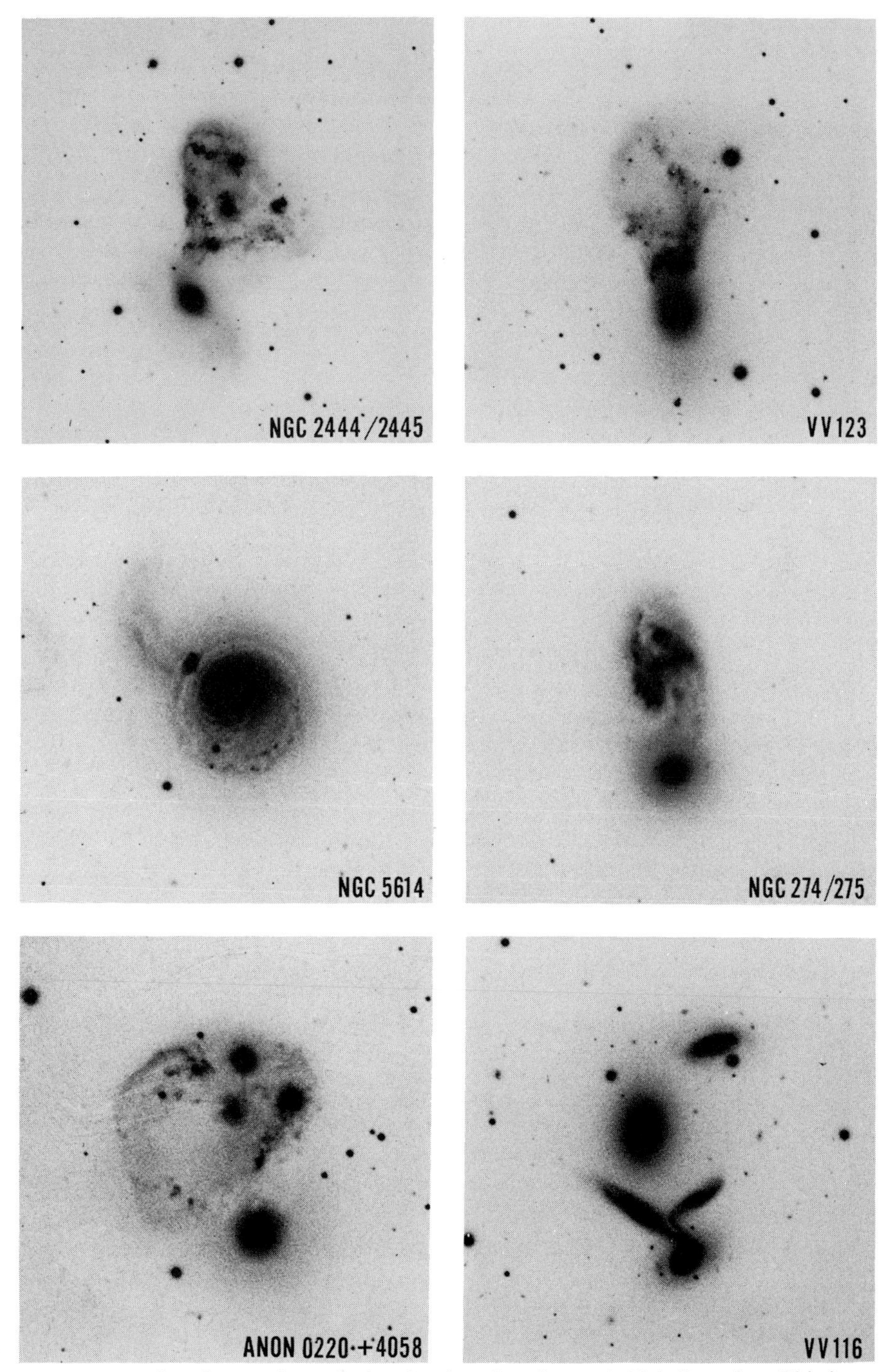

PLATE 8.—Various interacting galaxies showing the effects of tidal distortion. This is not a representative sample such as the compilations of Vorontsov-Velyaminov (1959) or of Arp (1966).

type (Shapley 1938*a*, *b*; Baade and Hubble 1939). In addition to the Sculptor ($M_V \simeq -13$) and Fornax ($M_V \simeq -12$) galaxies, other low-luminosity ellipticals exist in the Local Group, such as M32 ($M_V \simeq -16.3$), NGC 205 ($M_V \simeq -16.3$), NGC 185 ($M_V \simeq -15.1$), NGC 147 ($M_V \simeq -14.8$), Leo I ($M_V \simeq -11$), Leo II ($M_V \simeq -9.5$), Ursa Minor ($M_V \simeq -9$), and Draco ($M_V \simeq -8.5$). In addition, dwarf ellipticals are found as companions to many giant galaxies of all types. A particularly good example is NGC 4321 (M100) in the Virgo cluster which is surrounded by at least six dE galaxies all fainter than $M_V \simeq -15$. Plate 1 shows the subclustering about the giant Sc as reproduced from a 100-inch plate. Similar dE galaxies have been found in the Virgo cluster (Reaves 1956), in the M81 group (Sandage 1954), and in the Fornax cluster (Hodge 1959, 1960). The dE galaxies together with dwarf spirals and low-luminosity Im are undoubtedly the most frequent type of galaxy per unit volume of space (van den Bergh 1959; Holmberg, chap. 4 of this volume), although they form a totally insignificant fraction of catalogued galaxies to a given apparent magnitude because of their exceedingly low absolute luminosity. Indeed, it has only been since about 1940 (Baade 1944; Wilson 1955) that the dE galaxies have been recognized as a class.

5.2. Spirals with Thin Filamentary Arms

Such galaxies are shown in plates 2 and 3, arranged in order of increasingly later forms. The 12 illustrated galaxies form a continuously connected sequence. Along this sequence the size of the amorphous central region relative to the disk-length (see Freeman 1970; Sandage, Freeman, and Stokes 1970) generally decreases (there are exceptions) as the arms become more dominant and as the arms and disk become more highly resolved into knots and stars. For the first three galaxies illustrated (NGC 5273, 1302, and 2811) this spheroidal component, including the unresolved arms, covers a large fraction of the total area. However, by the time NGC 6643 [SA(rs)c] is reached along the sequence (plate 3) the spheroidal component has shrunk to an exceedingly small region, and by NGC 5204 [SA(s)m] has disappeared entirely. The arms are filamentary, and are generally segmented throughout the sequence. Galaxies in this characteristic subgroup are discussed more thoroughly in the *Hubble Atlas* (cf. section on Sb and the legend to NGC 5055 in *Atlas*, p. 15).

5.3. Spirals with Massive Arms

Massive-armed spirals form a second strain through the spiral sequence and are shown in plate 4. No dominant spheroidal component is present for the galaxies illustrated here, even among the earliest types such as NGC 4293 [Sa or RSB(s)0/a] and 3623 [Sa or SA$\underline{\text{B}}$(rs)a] in plate 4. The reproductions emphasize that Hubble's three criteria do not always agree. The most obvious examples are 4293 and 4569 [Sb or SAB(r$\underline{\text{s}}$)ab]. In such conflicting cases Hubble, Sandage, and de Vaucouleurs place most emphasis on the arm structure, and therefore classify such systems as Sa or early Sb despite the small spheroidal component. Besides the examples just mentioned, other such galaxies are NGC 4866 (*Atlas*, p. 11) and especially 4941 (*Atlas*, p. 10).

On the other hand, Morgan (cf. § 7) emphasizes the relative size of the spheroidal component as his primary criterion. Morgan's system has advantages for certain problems because Morgan and Mayall (1957) show that the degree of dominance of the spheroidal component correlates well with integrated spectral type.

5.4. Transitions between SA and SB

Transition galaxies are shown in plate 5. Those illustrated are among only the most obviously mixed cases previously called S/SB by Hubble and Sandage. De Vaucouleurs emphasizes the transitions more strongly than Hubble by using the SAB notation for galaxies where the bar is quite subtle. For example, five of the eight galaxies in plate 4 are called SAB by de Vaucouleurs, whereas none of these were called transition cases in the *Hubble Atlas*. De Vaucouleurs's procedure recognizes the fact that most galaxies possess an incipient bar, as discussed most explicitly by Lindblad and Langebartel (1953). The effect of the finer division in the revised system is that among the 1500 bright galaxies reclassified by de Vaucouleurs (1963*b*, table 5) there are about equal numbers of SA, SAB, and SB systems, whereas very few transition cases were isolated in the Hubble system (see, e.g., the types listed in HMS 1956).

5.5. Pure Barred Galaxies

Pure cases of the r and s varieties are shown in plate 6, s on the left, r on the right. As previously mentioned, NGC 3185 (*Atlas*, p. 43; and plate 6 here) is especially interesting because of the very tightly wound arms, which on casual inspection appear (incorrectly) to form a complete ring. The first galaxy in the upper left [NGC 7743; SBa(s) Hubble, or (R?)SB(s)0^+ de Vaucouleurs, which must be a misclassification because of the presence of distinct arms] has a similar pattern, but the arms are more clearly separated, perhaps due to a more favorable orientation of this galaxy to the line of sight. The separation of the end of the bar and the arm that begins at the opposite end of the bar increases progressively along the classification sequence, being clearly separated at the SB(s)b stage, such as NGC 1300 (*Atlas*, p. 45; and plate 6). This separation can be used as a secondary classification criterion for SBa and SBb types (cf. the discussion on p. 23 of the *Hubble Atlas* text).

As with the ordinary spirals, the resolution into stars increases, and the arms become more prominent relative to the amorphous central regions (the spheroidal component) as one progresses from the a to d subclasses. These monotonic changes are particularly evident in plate 6.

5.6. Sdm, Sm, and Im

These late-type galaxies form an important extension. Most such systems were called Irr by Hubble and Irr I by Holmberg, but they clearly follow naturally beyond the SAd and SBd forms. For example, Hodge and Hitchcock (1966) showed that galaxies of this type are highly flattened as a class (as in the earlier spirals), extending earlier work by de Vaucouleurs (1955*a*, *b*) for the Magellanic Clouds themselves. Furthermore, many studies show that the LMC is in a state of regular rotation (cf. Feast, Thackeray, and Wesselink 1961; Feast 1964, 1968 with extensive references therein), again similar to the earlier-type more regular spirals. And finally, ill-defined, subtle, but definite spiral structure has been detected (§ 4.1) by de Vaucouleurs (1954, 1955*a*).

Most galaxies of this type are of low surface brightness, and have much lower luminosity than earlier-type spirals (HMS 1956; van den Bergh 1959, 1960*a*; de Vaucouleurs 1963*b*). Van den Bergh (1960*b*) has discussed the Sm and Im types and has produced a catalog (1959), including the dE group, from the *Palomar Sky Survey* prints. Representatives are illustrated in plate 7 and in the *Hubble Atlas* (pp. 38–40).

5.7. Irr II Galaxies

Systems such as M82, NGC 520, and the companion to M51 (*Atlas*, p. 26; and plate 4 here) were often classified as peculiar by Hubble. This group represents a more restricted definition than that given by Arp (1966). As defined here, Irr II galaxies have an amorphous texture of the luminous form (M82: *Atlas*, p. 41; NGC 3077: *Atlas*, p. 41) and fall outside the standard system. Although peculiar, they should be distinguished from those galaxies whose peculiarities are caused by tidal interactions.

5.8. Interacting Galaxies

Galaxies that show the effects of mutual interaction need not be extensively discussed here. They clearly represent tidal perturbations (Toomre and Toomre 1972), and the classification system need not be changed to accommodate them. But the forms are so interesting in themselves that a few illustrations (plate 8) are useful.

An extensive compilation with photographs has been made by Vorontsov-Velyaminov (1959). Some such galaxies are illustrated in the series of photographs taken with large reflectors published in an atlas by Arp (1963). Studies of particular individual galaxies include papers by Zwicky (1956, 1959), Burbidge and Burbidge (1959), Sandage (1963), Sersic (1966), and others.

A general theory of interacting systems is discussed by Toomre and Toomre (1972), and for several particular systems by Limber (1965).

6. VAN DEN BERGH'S CLASSIFICATION

By inspecting *Palomar Sky Survey* prints, van den Bergh (1960*a*, *b*) made the important discovery that the appearance of the spiral arms was a steep function of the absolute luminosity of galaxies. He showed that galaxies of the highest luminosity have the longest and most highly developed arms, whereas fainter systems such as NGC 5204 (plate 3) have poorly developed arms.

The result is equivalent to stating that the appearance of galaxies varies systematically as one proceeds from left to right within the scatter of redshift–apparent-magnitude diagrams for field galaxies (see, for example, figs. 6, 7, and 9 of HMS).

By inspecting the character of the spiral arms alone, van den Bergh was able to divide Sb types into five luminosity classes (I, I–II, II, II–III, and III), which subsequent calibration showed were $\sim$0.5 mag apart in $\langle M_{pg} \rangle$. The Sc-Irr systems could be divided into eight half-classes (I to V with intermediates).

Van den Bergh's system is two-dimensional. It retains the Hubble types a, b, and c as a division along a linear sequence, and adds the luminosity class as the second parameter.

Additional symbols are used such as + and − (following Holmberg) to divide the Hubble classes more finely, n for nebulous arms such as NGC 3623 (plate 4) and 4569 (plate 4), an asterisk(*) for patchy arms such as NGC 157 (plate 4) and 4088 (plate 5), and t for indications of "tidal" distortion. Extreme characteristics are noted by double symbols; incipient features, by (n), (*), and (t).

Van den Bergh notes that most of the Hubble-type examples usually illustrated in textbooks, in the *Hubble Atlas*, and in plates 1–8 are supergiant galaxies, and that the early Hubble classification system defined by such bright galaxies cannot generally be applied to dwarf galaxies without modification. He emphasizes that all dwarf and subgiant galaxies classified in HMS (1956) are of type Sc, which shows that few if any dwarf Sa and Sb systems exist. This fact is recognized in van den Bergh's system for Sb systems where only class I, II, and III galaxies are present, whereas class V (dwarf) galaxies exist in the Sc to Irr classes.

The validity of the luminosity classes is shown by the clear separation of the classes in the redshift-magnitude relation (van den Bergh 1960*a*, fig. 4; 1960*b*, fig. 4). Van den Bergh's preliminary calibration of $\langle M_{pg} \rangle$ for each class was based on this separation, adopting $H = 100$ km s^{-1} Mpc^{-1}. The calibration is reproduced in table 1. The dispersions of $\langle M_{pg} \rangle$ were found to be $\sigma \simeq 0.3$–0.4 mag for the well-defined cases (van den Bergh 1960*a*, table 4), as obtained by comparing differences between the M_{pg} predicted from the luminosity classification and known values for particular galaxies in groups and clusters.

TABLE 1

VAN DEN BERGH'S CALIBRATION BASED ON $H_0 = 100$ km s^{-1} Mpc^{-1}

Type	M_{pg}	Type	M_{pg}
Sb I.........	−20.4	Sc I.................	−20.0
Sb I–II......	−19.9	Sc I–II..............	−19.7
Sb II........	−19.4	Sc and Irr II..........	−19.4
Sb II–III....	−18.6	Sc and Irr II–III......	−18.9
Sb III.......	−18.0	Sc and Irr III.........	−18.3
		Sc and Irr III–IV.....	−18.0:
		Sc and Irr IV.........	−17.3:
		Sc and Irr IV–V......	−16.1:

From *Pub. David Dunlap Obs.*, Vol. **2**, No. 6, 1960.

The importance of the luminosity classification, regardless of its calibration, is that relative distances can be obtained to large numbers of field spirals within an accuracy of $\delta r/r = 0.461\,\delta M$. This error is small when $\sigma(\delta M)$ is, say, 0.4 mag that arises from classification errors and from true cosmic spread. The distances can be changed to absolute values once H_0 is accurately known. (Table 1, based on an assumed value of H_0, is subject to modification. A calibration in 1972 by Sandage and Tammann gave $H_0 \simeq 55$, which would require the values in table 1 to be about 1 mag brighter). On a distance scale with $H_0 = 100$ km s^{-1} Mpc^{-1}, van den Bergh has given absolute moduli for many Shapley-Ames galaxies in his reclassification catalog (van den Bergh 1960).

The van den Bergh classification is defined by the many type-examples given in his catalog, as classified from the Palomar prints. Because of the burned-out nature of many galaxies on these prints, he could usually make no distinction between S0 and E systems, and many flattened systems were classed E with flattenings ranging to 8. Such large flattenings do not, of course, exist in the standard system because E galaxies are never flatter than E7, and very few of these exist (cf. Sandage, Freeman, and Stokes 1970). Many of the flattened E8 so classified in the catalog are actually edge-on spirals and S0 galaxies, which, for this reason, are not represented in the *DDO Catalogue*. Comparison of the DDO types with other classification lists is made by de Vaucouleurs (1963*b*, table 10 and fig. 5).

7. MORGAN'S CLASSIFICATION BASED ON THE LUMINOSITY CONCENTRATION OF THE SPHEROIDAL COMPONENT

A most important feature of galaxies is the great difference in stellar content between their spiral arms and the spheroidal component. The difference is evident from direct photographs: it was shown explicitly by Baade's resolution studies in 1944, and is apparent from the change of integrated colors along the classification sequence as the relative importance of the old (spheroidal) and young arm populations exchange dominance in going from S0 to Sm systems.

Because the difference is pronounced, it must be expected that the integrated spectral type should correlate with the light ratio of the spheroidal to young-disk populations, and therefore roughly with Hubble type, since this ratio is one of the classification parameters.

The earliest indication of such population differences was noted by Seares (1916*a*, *b*), who discovered that the central parts of galaxies are redder than the arm regions. After the start of the Mount Wilson and Lick systematic redshift programs, both Humason and Mayall noted gross spectroscopic differences among galaxies, and summarized their results in the discussion of their redshift catalog (HMS 1956). Starting in 1932, Humason regularly classified the dominant spectral class and showed that it correlated with Hubble type in the expected sense.

Morgan and Mayall (1957) reviewed these early results and advanced the work by showing that the composite spectral class correlates well with the degree of dominance of the spheroidal component alone, i.e., with the concentration of the luminosity toward the center.[5] Morgan (1958, 1959) developed this spectral-type concentration correlation into a classification system whose color-class notation explicitly isolates the spectral type that is expected on the basis of the correlations.

The Morgan system contains information on the *state of stellar evolution* in the central regions of galaxies. The classification complements information in the Hubble–de Vaucouleurs system which, in cases of conflicting criteria, emphasizes more strongly the strength of the spiral-arm population [see, e.g., NGC 4941, Sab: Hubble; SABab: de Vaucouleurs; *Atlas*, p. 10].

In the introduction to his 1958 paper, Morgan states: "The correspondence between form and spectral appearance is especially close for two categories of galaxies: (1) irregular systems of the Magellanic Cloud–type and spirals having an insignificant central concentration of luminosity [Sd, Sm], and (2) giant ellipticals such as those found in the Virgo cloud and spiral systems such as M31, in which the major share of the luminosity of the main body is due to a bright amorphous central region. In the first category, early-type stars and emission nebulosities have a profound effect on the spectrum in the blue and violet regions [producing strong hydrogen absorption lines and bright emission lines such as found in galactic H II regions]; in the second, the luminosity of the brighter parts is due principally to yellow giant stars.

[5] The difference in scatter between the Morgan-Mayall and the Humason spectral-type, galaxy-class correlation shows clearly that the size of the nuclear region does not vary uniquely and monotonically with Hubble type, although there is a general trend. The point has previously been made by Vorontsov-Velyaminov in his comments on the Hubble system (cf. Introduction, Vol. 2, *Morphological Catalog of Galaxies*, p. 4) and by Freeman (1970) in his plot of the spheroidal-to-disk length ratio as a function of Hubble type.

"Systems possessing an intermediate degree of central concentration of light (M51 [plate 4 here on NGC 5194]) give spectroscopic evidence of an intermediate kind of stellar population: the degree of compositeness is very high, and it appears likely that most of the luminosity is due to a mixture of F to G main-sequence stars and K giants."

Morgan's notation contains a *concentration* class ranging from a through af, f, fg, g, gk, and k, in the direction of increasing domination of the nuclear light (and consequently later composite spectral type according to the Morgan-Mayall correlation), and a *form* notation similar to Hubble's with E for ellipticals; S, spirals; B, barred spirals; and I for irregulars. Three new form classes are added: D for dustless systems dominated by the amorphous light; L for galaxies of low surface brightness (such as NGC 45 [*Atlas*, p. 37]); and N for systems having small, brilliant nuclei superposed on a considerably fainter background.

The addition of an inclination index ranging from 1 (circular) to 7 (spindle with $a/b \simeq 10$) and the symbol p denoting "peculiar" completes the notation. Examples of well-known galaxies classified on the Morgan system, many of which are shown in plates 2–7, are 5273 (gkD2 Morgan; S0/Sa Hubble), 488 (kS2 Morgan; Sb Hubble), 628 (fgS1 Morgan; Sc Hubble), 5204 (fI–fS4 Morgan; Sc/Ir Hubble; SAm de Vaucouleurs); M33 (fS3 Morgan; Sc Hubble) and 4449 (aI Morgan; Ir Hubble; IBm de Vaucouleurs).

There is a general correlation of the Morgan class with the Hubble–de Vaucouleurs types. A comparison has been made by de Vaucouleurs (1963*b*, table 13). The major difference is in the definition of the D systems. The closest Hubble type is S0, although there is a rather large extension into the E's on one side of the scatter and the early SA and SB types on the other. The difference is of some importance because Morgan classifies the brightest galaxies in rich clusters as cD (c for supergiant), whereas the Mount Wilson observers do not.

Following Matthews, Morgan, and Schmidt (1964), the cD notation has been generally adopted by radio astronomers for the radio ellipticals, but these galaxies are still classified as giant E galaxies on the Mount Wilson system. The Mount Wilson procedure follows naturally from the discovery of Hubble and Humason in the 1930s that rich clusters of galaxies are composed *almost entirely of E and S0 galaxies*. The brightest member galaxies are of the same general type as the fainter very numerous E systems (restricting to the top 4-mag range), differing only slightly if at all in the occasional presence of a extended envelope (it is not yet known if these outer regions follow a power intensity law as in true E galaxies, or if there is a small exponential component as in the disks of true S0's).

The introduction of a separate D class in an otherwise continuous sequence of E forms among the brighter cluster members is then the chief difference between the Mount Wilson and the Yerkes classification of galaxies in rich clusters.

A summary of the Morgan system and its significance for studies of stellar content is given by Morgan and Osterbrock (1969).

8. SYSTEM OF VORONTSOV-VELYAMINOV

The *Morphological Catalog of Galaxies* (MCG) in four volumes by Vorontsov-Velyaminov, Arhysova, and Krasnogorskaja contains positions, sizes, magnitudes, and descriptions of 29,000 galaxies on prints of the *Palomar Sky Survey* from the pole to $\delta = -30°$. The catalog is essentially complete to $m_{pg} = 15$ mag.

A new system of description was devised, using the symbols that are illustrated and discussed in the Introduction to Volume **1** of the MCG. In philosophy, the system is similar to Wolf's scheme, but more particularly to Reynolds's hope for a less simple system than Hubble's (§ 3.2), where the great diversity of pattern would be recognized in the notation. This hope is realized, because the combined symbols in the MCG are sufficient to describe a large variety of detailed structural features. But as yet the system does not constitute a classification of the second kind (§ 1), where continuously varying parameters are used to provide connective relations. This, however, was not the purpose of the Vorontsov-Velyaminov descriptions, and the system cannot be criticized on such grounds because the authors state in the Introduction to Volume **2**: "Our descriptions do not [constitute a] classification. They are merely a step forced by the diversity of galaxies revealed by the *Palomar Atlas* and by the peculiarity of this *Atlas*. [It] may be [that] our descriptions will help to elaborate a new [more] appropriate classification." Undoubtedly features such as the γ forms, the rings, pseudo-rings, and peculiar arm structures to which Vorontsov-Velyaminov has often called attention will prove to be important when the detailed dynamics of galaxies become better understood. At the moment, the other classification systems in general use emphasize the more gross aspects of galaxy systematics.

9. COMPARISON OF THE CLASSIFICATION SYSTEMS

To aid in learning the classifications, the types assigned by Hubble, Sandage, Holmberg, de Vaucouleurs, van den Bergh, and Morgan to galaxies illustrated in this chapter are listed in table 2. The illustrated examples are insufficient to show all complications, and reference should be made to various classification catalogs, supplemented by published photographs. Some of the more extensive classification lists are: (1) Pettit (1954) and the *Hubble Atlas* for the modified Hubble system, (2) Holmberg (1958), (3) de Vaucouleurs (1963*b*), (4) van den Bergh (1959, 1960*a*, *b*, *c*), and (5) Morgan (1958, 1959).[6]

10. SEYFERT GALAXIES, N GALAXIES, AND QUASARS

A group of galaxies that, to varying degrees, stand apart from the standard classification are those whose parts (Seyfert nuclei), or nearly the whole of their optical image (QSOs) are in a more highly condensed state than normal systems. As far as classification of their optical properties is concerned, they form a continuum which can be conveniently divided into three sections called Seyferts, N galaxies, and quasars. Although their radio properties were emphasized most strongly in the decade of the 1960s, radio emission appears to be superficial as regards the optical forms. Both radio-intense and radio-quiet examples exist in all three sections of the compact regime, and the radio-quiet (or radio-weak) objects form the majority of the sample per volume of space.

The degree of compactness varies along the sequence in the order Seyfert, N, QSO. Seyfert galaxies are almost normal in the Hubble sense, except for an intensely bright nucleus. Those Seyferts that are spirals (e.g., NGC 1068, NGC 4051, NGC 4151) have slightly abnormal outer arms that form nearly complete faint exterior "rings" beyond the

[6] Recall that Holmberg does not separate the ordinary and barred spirals and that van den Bergh does not distinguish E, S0, and edge-on spirals (spindles) because of the burned-out nature of the *Sky Survey* prints.

inner spiral pattern (cf. Hodge 1968), but otherwise (and aside from their nuclei) they are easily placed within the Hubble sequence. N galaxies are dominated to a larger extent by their compact, nonthermal, central region, whereas most quasi-stellar objects (QSOs or quasars) are completely stellar *(by definition)* on photographs with a scale of 10″ mm^{-1}.

The continuity between Seyferts, N galaxies, and QSOs in their classification has been discussed by many authors, and will not again be reviewed here (cf. Sandage 1971, 1973). Early classical papers on the Seyferts alone include the original discussion by Seyfert (1943), and the renaissance of the subject by Woltjer (1959). The N systems of Morgan (1958, 1959) were reemphasized by Matthews, Morgan, and Schmidt (1964) via the route of radio sources, and the connections between Seyferts and QSOs were summarized by Burbidge, Burbidge, and Sandage (1963).

Discussions of the classification system as it existed in mid-1971 were given by Morgan (1971) in the Pontifical Academy volume devoted to this subject (O'Connell 1971). A useful summary of the Seyfert problem, of the group characteristics of these compact systems taken as a whole, and of the effect of different spatial resolutions on the classification of a given object can be found in the *Proceedings of the Conference on Seyfert Galaxies and Related Objects* edited by Pacholczyk and Weymann (1968).

11. STELLAR CONTENT RELATED TO TYPE: FORMATION AND EVOLUTION

One of the most remarkable features of the standard classification system is that the stellar content of galaxies varies systematically along the linear sequence from E to Sm. In the order Sa to Sd, the progressive changes are (1) increasing absolute luminosity of the brightest stars in regions of the spiral arms, (2) increasing percentage of mass in the form of gas and dust, (3) increasing sizes and numbers of H II regions in the spiral arms, and (4) progressively bluer integrated $B - V$ and $U - B$ colors, indicating progressively earlier type stars that contribute most of the light.

The correlations represent a physical result because the classification is made principally not on the resolution of the galaxy into stars, but rather on the character of the nuclear bulge, and on the presence, the shape, and the regularity of the arm structure.

Some of the questions raised by the correlation of *content* and *form* have been discussed by Sandage, Freeman, and Stokes (1970) in a review of Hubble's (1926) problem to find the true flattening of elliptical galaxies from the observed distribution of apparent flattenings. In that study, flattenings were compared for ellipticals and spirals, and we concluded with Hubble, Holmberg (1946), and de Vaucouleurs (1959a) that all spiral and S0 galaxies are flatter than the flattest elliptical.

Because flattening is a dynamical property that, in an equilibrium configuration, cannot change in times less than the relaxation time ($\sim 10^{12}$–10^{14} years), the difference in intrinsic flattening between E and S galaxies shows that one type cannot evolve into the other. A basic difference must then have existed between E and S systems *at the time of their formation to cause* such different forms now. An identification and an understanding of this difference is necessary at the outset of even the most elementary discussion of galaxy evolution.

Flattening can occur on a *short* time scale only if (1) the relaxation time itself is short under conditions that prevailed before the system formed into stars, (2) the gravitational

TABLE 2

CLASSIFICATION OF THE GALAXIES ILLUSTRATED IN PLATES II-VII

Name NGC	Hubble	Holmberg	de Vaucouleurs	van den Bergh	Morgan
			Thin Multiple-Armed Spirals (Plate II)		
5273	S0$_2$/Sa(s)	—	SA(s)0^0	E 1(p?)	gKD2
1302	Sa	—	(R)SB(r)o/a	S(B)a	kBl
2811	Sa	—	SB(rs)a	Sb$^+$ II-III	—
3898	Sa	—	SA(s)ab	Sb$^-$ II	kD5-KS5
2841	Sb	Sb$^-$	SA(r:)b	Sb$^-$ I	kS5
488	Sb	Sb$^-$	SA(r)b	Sb$^-$ I	kS2
			Plate III		
5055	Sb	Sb$^+$	SA(rs)bc	Sb$^+$ II	gS4
628	Sc	Sc$^-$	SA(s)c	Sc I	fgSl
5457	Sc	Sc$^-$	SAB(rs)cd	Sc I	fSl
3810	Sc	Sc$^-$	SA(rs)c	Sc I	fS3
6643	Sc	Sc$^-$	SA(rs)c	Sc I-II	—
5204	Sc/Irr	Sc$^+$	SA(s)m	Ir$^+$ IV	fI-fS4
			Massive-Armed Spirals Plate IV		
4293	Sa	—	RSB(s)o/a	Pec	fg:S6
4569	Sb	Sb$^!$	SAB(rs)ab	Sb$^+$n	fS4p
3623	Sa	Sa	SAB(rs)a	Sbn II:	gS5
6814	Sb	—	SAB(rs)bc	Sb$^+$ I	—
5248	Sc	Sc$^-$	SAB(rs)bc	Sc I	fS4
5194	Sc	Sc$^-$	SA(s)bcp	Sc(t) I	fSl
157	Sc	Sc$^-$	SA(rs)bc	Sc(*) I	afS3
598	Sc	Sc$^+$	SA(s)cd	Sc II-III	fS3

Ordinary-Barred Transitions Plate V

4579	Sb/SBb	Sb$^-$	SAB(rs)b	Sbn	gkS3-gKB
3504	SBb(s)/Sb	—	SAB(s)ab	Sb(t?)	fg Bl?
6951	SBb(s)Sb	Sb$^+$	SAB(rs)bc	Sbp I-II	f:Slp-f:BpN
5236	Sc/SBb	—	SAB(s)c	—	fgSl
4088	SBc/Sc	Sc$^-$	SAB(s)bc	Sc* I-II	a:B:4
925	Sc/SBc	Sc$^+$	SAB(s)d	S(B)c II-III	afB

Barred Spirals r and s Types Plate VI

7743	SBa(s)	—	(R?)SB(s)0$^+$	Sa?	gkBl
3185	SBa(s)	Sa	(R)SB(r)a	S(B)b$^+$ III	fB
1300	SBb(s)	Sb$^+$	SB(rs)bc	SBb I	fB2
7741	SBc(s)	Sc$^+$	SB(s)cd	SBc II	afB
4643	SB0$_3$/SBa(r)	—	SB(rs)o/a	SBa	kB
1398	SBb(r)	—	(R')SB(r)ab	S(B)b$^-$ I	kB2
2523	SBb(r)	—	SB(r)bc	SBb$^-$ I	fgB
1073	SBc(sr)	Sc$^+$	SB(rs)c	S(B)c II	afBl

Magellanic Cloud and Irregular Types Plate VII

4395	Sc/Irr	Sc$^+$	Sa(s:)m	S$^+$ IV-V	aSl
SMC	Irr	—	IB(s)mp	III-IV	—
LMC	Irr	—	SB(s)m	Ir IV-V	—
Hol II	Irr	Ir I	Im	Ir IV-V	—
2366	Irr	Ir I	dIB(s)m	Ir IV-V	aL?
IC 2574	Irr	Ir I	SAB(s)m	Ir IV-V	—

potential energy of a stellar system is rapidly changing (cf. Lynden-Bell 1967), or (3) both. The difference in formation history of E and S galaxies must then be due to some difference, such as the angular-momentum distribution of the contracting protogalaxies, or any other agent that would *control the initial rate of star formation during the collapse time of the protogalaxy toward the fundamental plane*. Elliptical galaxies obviously *did not* collapse completely (they are not now highly flattened), and this shows that nearly *all star formation took place in times short compared to the free-fall time* ($\sim 10^9$ years), leaving no gas to interact by gas-gas collisions to damp into a fundamental plane that is characteristic of spirals. On the other hand, the existence of such a plane in all spirals seems to betray a *slower* initial conversion of gas to stars. Furthermore, much of the gas has remained in the plane over the lifetime of the system.

The required rapid star formation in E systems and in the spheroidal components of spirals and S0 galaxies finds considerable direct support from observations. Baade's resolution of parts of the Local Group galaxies M31, M32, NGC 205, NGC 185, and NGC 147 into stars at $M_V \simeq -3$, $B - V \simeq +1.5$ at the same intensity level as the globular clusters embedded in their galactic halos shows the antiquity of these resolved subsystems (Baade 1944). The argument is made stronger by the lack of bright young blue supergiants in the spheroidal regions of S0, Sa, and Sb galaxies, and in E systems. Constraints on the age are stringent (Sandage 1969, 1971) from these data.

Considering the flattening and the stellar content data, and arguing from the relaxation problem of Lynden-Bell (1967), Sandage, Freeman, and Stokes reached the following conclusions.

1. Stars in the spheroidal component of all galaxies were formed very rapidly on a time scale comparable to the collapse time of the protogalaxy ($\lesssim 10^9$ years). The argument is independent of that used by Eggen, Lynden-Bell, and Sandage (1962) which was based on the orbital eccentricities of galactic stars. It depends here only (*a*) on the absence of a fundamental plane both in E galaxies and in the spheroidal subsystems of spirals, and (*b*) on the observation that the stellar distribution itself appears to be relaxed, presumably by Lynden-Bell's (1967) mechanism.

2. The *halo* stars were formed from matter having low angular momentum per unit mass, i.e., the spheroidal component, during the collapse time. Other matter with higher angular momentum collapsed to a disk.

3. The galaxy type was determined by the amount of free gas left over in the disk after collapse. No appreciable evolution along the Hubble sequence has occurred since the galaxies were formed.

4. The dominance of the disk in spiral and S0 systems betrays their mean angular momentum per unit mass, higher than that which exists in less-flattened E galaxies. Does the higher angular momentum slow the rate of star formation, keeping uncondensed matter (gas and dust) in reserve for new generations to form continuously in the dusty disks?

5. All stars in the spheroidal component of galaxies should show an age distribution ΔT which is *small* compared to the total age T, such that $\Delta T/T \lesssim 0.1$. This expectation agrees with observational data for halo globular clusters in our own Galaxy, and with the uniformity of observed energy distributions $I(\lambda)$ for E galaxies and the centers of S0, Sa, and Sb systems.

Without collapse of matter from a wider configuration it is difficult to understand the presence of a dominant plane in spirals. But in ellipticals it is difficult to understand *its absence* without rapid and complete star formation on a time scale *short* compared with the free-fall time from the edges of a protogalaxy.

The comments in this last section may be more speculative than substantive, but they are discussed here because they follow naturally from certain characteristics of the Hubble classification. Because of this, it seems possible that the classification is fundamental in the sense discussed in § 1; i.e., the connective parameters of the system appear to contain part of the physics of galaxies. That the Hubble system has this property would follow if ideas of this section could be substantiated. And likewise, that the Morgan nuclear sequence might also be a classification of the second kind would follow if the stellar component of the spheroidal bulge is understood in terms of stellar evolution of the composite H-R diagram (chap. 2), as we now believe.

REFERENCES

Abell, G. O. 1955, *Pub. A.S.P.*, **67**, 258.
Arp, H. C. 1966, *Ap. J. Suppl.*, **14**, No. 123, 1.
Baade, W. 1944, *Ap. J.*, **100**, 137.
———. 1951, *Pub. Obs. Univ. Michigan*, **10**, 7.
———. 1963, *Evolution of Stars and Galaxies*, ed. C. Payne-Gaposchkin (Cambridge: Harvard University Press).
Baade, W., and Hubble, E. 1939, *Pub. A.S.P.*, **51**, 40.
Bacon, F. 1620, *Novum Organum*: numerous modern editions.
Bergh, van den. 1959, *Pub. David Dunlap Obs.*, Vol. **2**, No. 5.
———. 1960*a*, *Ap. J.*, **131**, 215.
———. 1960*b*, *ibid.*, p. 558.
———. 1960*c*, *Pub. David Dunlap Obs.*, Vol. **2**, No. 6.
Burbidge, E. M., and Burbidge, G. R. 1959, *Ap. J.*, **130**, 23.
Burbidge, E. M., Burbidge, G. R., and Hoyle, F. 1963, *Ap. J.*, **138**, 873.
Burbidge, G. R., Burbidge, E. M., and Sandage, A. 1963, *Rev. Mod. Phys.*, **35**, 947.
Curtis, H. D. 1918, *Lick Obs. Pub.*, Vol. **13**.
———. 1933, *Handbuch der Astrophysik*, Band V/2, chap. 6.
Danver, C. G. 1942, *Lund Obs. Ann.* Vol. **10**.
Eggen, O. J., Lynden-Bell, and Sandage, A. 1962, *Ap. J.*, **136**, 748.
Evans, D. S. 1957, *Cape Photographic Atlas of Southern Galaxies* (Cape Observatory).
Feast, M. W. 1964, *M.N.R.A.S.*, **127**, 195.
———. 1968, *ibid.*, **140**, 345.
Feast, M. W., Thackeray, A. D., and Wesselink, A. J. 1961, *M.N.R.A.S.*, **122**, 433.
Freeman, K. C. 1970, *Ap. J.*, **160**, 811.
Harrington, R. G., and Wilson, A. G. 1950, *Pub. A.S.P.*, **62**, 118.
Hodge, P. W. 1959, *Pub. A.S.P.*, **71**, 28.
———. 1960, *ibid.*, **72**, 188.
———. 1966, *The Physics and Astronomy of Galaxies and Cosmology* (New York: McGraw-Hill Book Co.).
———. 1968, *A.J.*, **73**, 846.
Hodge, P. W., and Hitchcock, J. L. 1966, *Pub. A.S.P.*, **78**, 79.
Hodge, P. W., and Merchant, A. 1966, *Ap. J.*, **144**, 875.
Hodge, P. W., and Webb, C. J. 1964, *Ap. J.*, **140**, 681.
Holmberg, E. 1937, *Lund Obs. Ann.*, No. **6**.
———. 1946, *Medd. Lund. Obs.*, Ser. 2, No. 117.
———. 1950, *ibid.*, No. 128.
———. 1958, *ibid.*, No. 136.
Hubble, E. 1917, *Yerkes Pub.* Vol. **4**, Part 2.
———. 1926, *Ap. J.*, **64**, 321.
———. 1930, *ibid.*, **71**, 231.
———. 1932, *ibid.*, **76**, 44.
———. 1936, *The Realm of the Nebulae* (New Haven: Yale University Press).
Humason, M. L., Mayall, N. U., and Sandage, A. R. 1956, *A.J.*, **61**, 97 (HMS).
Johnson, H. M. 1961, *Ap. J.*, **133**, 314.
Knox-Shaw, H. 1915, *Helwan Obs. Bull.* No. 15. p. 129.

Liller, M. H. 1960, *Ap. J.*, **132**, 306.
Limber, D. N. 1965, *Ap. J.*, **142**, 1346.
Lindblad, B., and Langebartel, R. G. 1953, *Ann. Stockholm Obs.*, **17**, No. 6.
Lundmark, K. 1926, *Ark. Math. Astr. Phys.*, Ser. B, Vol. **19**, No. 8.
———. 1927, *Medd Astr. Obs. Uppsala*, No. 30.
Lynden-Bell, D. 1967, *M.N.R.A.S.*, **136**, 101.
Matthews, T. A., Morgan, W. W., and Schmidt, M. 1964, *Ap. J.*, **140**, 35.
Morgan, W. W. 1958, *Pub. A.S.P.*, **70**, 364.
———. 1959, *ibid.*, **71**, 394.
———. 1971, in *The Nuclei of Galaxies*, ed. D. O'Connell (Vatican City: Pontifical Academy of Sciences).
Morgan, W. W., and Mayall, N. U. 1957, *Pub. A.S.P.*, **69**, 291.
Morgan, W. W., and Osterbrock, D. E. 1969, *A.J.*, **74**, 515.
O'Connell, D. J. U. 1971, *The Nuclei of Galaxies* (Vatican City: Pontifical Academy of Sciences).
Pacholczyk, A. G., and Weymann, R. 1968, *A.J.*, **73**, 836.
Pease, F. G. 1917, *Ap. J.*, **47**, 24.
———. 1920, *ibid.*, **51**, 276.
Pettit, E. 1954, *Ap. J.*, **120**, 413.
Randers, G. 1940, *Ap. J.*, **92**, 235.
Reaves, G. 1956, *A.J.*, **61**, 69.
Reinmuth, K. 1926, *Veroff. Sternw. Heidelberg*, Vol. **9**.
Reiz, A. 1941, *Lund Obs. Ann.*, Vol. **9**.
Reynolds, J. H. 1920*a*, *Observatory*, **43**, 377.
———. 1920*b*, *M.N.R.A.S.*, **80**, 746.
———. 1927*a*, *Observatory*. **50**, 185.
———. 1927*b*, *ibid.*, p. 308.
Sandage, A. 1954, *Carnegie Yrbk.*, No. **55**, p. 23.
———. 1961, *Hubble Atlas of Galaxies* (Washington: Carnegie Institution of Washington), Pub. 618.
———. 1963, *Ap. J.*, **138**, 863.
———. 1969, *Ap. J.*, **157**, 515.
———. 1971, in *The Nuclei of Galaxies*, ed. D. O'Connell (Vatican City: Pontifical Academy of Sciences), p. 601.
———. 1973, *Ap. J.*, **180**, 687.
Sandage, A., Freeman, K. C., and Stokes, N. R. 1970, *Ap. J.*, **160**, 831.
Sandage, A., and Tammann, G. A. 1974, *Ap. J.*, a series of six papers.
Seares, F. 1916*a*, *Proc. Nat. Acad. Sci.*, **2**, 553.
———. 1916*b*, *Pub. A.S.P.*, **28**, 123.
Sersic, J. L. 1966, *Zs. f. Ap.*, **64**, 202.
———. 1968, *Atlas de Galaxias Australes* (Córdoba, Argentina: Observatorio Astronómico Universidad Nacional de Córdoba).
Seyfert, C. K. 1943, *Ap. J.*, **97**, 28.
Shapley, H. 1928, *Harv. Bull.*, **849**.
———. 1938*a*, *Bull. Harv. Coll. Obs.*, No. 908.
———. 1938*b*, *Nature*, **142**, 715.
———. 1950, *Pub. Obs. Univ. Michigan*, **10**, 79.
———. 1961, *Galaxies* (Cambridge: Harvard University Press), p. 22.
Shapley, H., and Ames, A. 1932, *Harv. Ann.* Vol. **88**, No. 2.
Shapley, H., and Paraskevopoulos, J. S. 1940, *Proc. Nat. Acad. Sci. USA*, **26**, 31.
Spitzer, L., and Baade, W. 1951, *Ap. J.*, **113**, 413.
Toomre, A., and Toomre, J. 1972, *Ap. J.*, **178**, 623.
Vaucouleurs, G. de. 1948, *Ann. d'Ap.*, **11**, 247.
———. 1953, *M.N.R.A.S.*, **113**, 134.
———. 1954, *Observatory*, **74**, 23.
———. 1955*a*, *A.J.*, **60**, 126.
———. 1955*b*, *ibid.*, p. 219.
———. 1956, *Mem. Commonwealth Obs.* (Mount Stromlo), Vol. **3**, No. 13.
———. 1958, *Ap. J.*, **128**, 465.
———. 1959*a*, *Handbuch der Physik* (Berlin: Springer-Verlag), **53**, 275.
———. 1959*b*, *Ap. J.*, **130**, 728.
———. 1962, *ibid.*, **136**, 107.
———. 1963*a*, *ibid.*, **138**, 934.
———. 1963*b*, *Ap. J. Suppl.*, **8**, 31.
———. 1964, *Ap. J.*, **139**, 899.
Vaucouleurs, G. and A. de. 1964, *Reference Catalogue of Bright Galaxies* (Austin: University of Texas).
Vorontsov-Velyaminov, B. A. 1959, *Atlas and Catalogue of Interacting Galaxies* (Moscow: Sternberg Institute, Moscow State University).
———. 1965, *Soviet Astr.—A.J*, **8**, 649 [*Astr. Zh.*, **41**, 814, 1964].
———. 1966, *ibid.*, **10**, 184 [*Astr. Zh.*, **43**, 231].

Vorontsov-Velyaminov, B. A., Krasnogorskaja, A., and Arkipova, V. P. 1962, *Morphological Catalog of Galaxies* (Moscow), Vol. **1**.
———. 1963, *ibid.*, Vol. **2**.
———. 1964, *ibid.*, Vol. **3**.
———. 1968, *ibid.*, Vol. **4**.
Wilson, A. G. 1955, *Pub. A.S.P.*, **67**, 27.
Wolf, M. 1908, *Pub. Ap. Inst. König. Heidelberg*, Vol. **3**, No. 5.
Woltjer, L. 1959, *Ap. J.*, **130**, 38.
Zwicky, F. 1956, *Ergebn. exakt. Naturw.* **29**, 344.
———. 1959, *Handbuch der Physik* (Berlin: Springer-Verlag), **53**, 373.

CHAPTER 2

The Stellar and Gaseous Content of Normal Galaxies as Derived from Their Integrated Spectra

HYRON SPINRAD
Berkeley Astronomical Department, University of California
AND
MANUEL PEIMBERT
Instituto de Astronomía, Universidad Nacional de México and Observatorio de Tonantzintla

1. THE STELLAR CONTENT OF NORMAL GALAXIES

1.1. Introductory Remarks and a Historical Review

The study of the stellar content of galaxies is clearly an important problem; we wish to know what kinds of stars populate galaxies of different morphological types, the stellar luminosity function at the galaxy center, and if possible its variation throughout the observable regions of a galaxy. We also need quantitative information on how a distant stellar grouping will evolve with time. The search for data and unequivocal answers to these questions has been difficult for astronomers. One reason is that galaxies are very distant, and our observations of individual stars even in Local Group systems are very restricted. In the Magellanic Clouds we can study stars of $M_v \leq -5$ in some detail and measure colors with modest precision to $M_v = 0$. But even in the nearby Large Magellanic Cloud (LMC) and Small Magellanic Cloud (SMC) the vast underlying mass of main-sequence stars and most giants and subgiants lie beyond the reach of existing instruments. It is the study of the integrated light of these relatively fainter stars which we discuss in detail here, although individually they have never been detected at extragalactic distances.

In what follows we assume that the integrated light of the vast majority of *normal* galaxies and especially of their nuclei is completely dominated by starlight. We avoid the complications of possible nonthermal or at least nonstellar continua which may be important contributors to the optical radiation of Seyfert nuclei, N galaxies, and some

Received July 1970.

other strong radio sources (cf. Grewing, Demoulin, and Burbidge 1968), and possibly some compact systems.

Until recently the only *quantitative* measures of galaxy population parameters were broad-band photographic and photoelectric colors measured by Pettit, Stebbins and Whitford, Baum, Tifft, de Vaucouleurs, de Vaucouleurs and Solheim, Holmberg, Sandage, Johnson, Hodge, and others. While they are internally precise, these measures are only partly qualified to answer the questions of stellar mixtures in galaxies; clearly the colors which cover the longest-wavelength baseline would be most useful in examination of both possible cold and hot stellar constituents. The available broad-band colors have shown clearly that hot stars are important contributors to the blue light in the centers of Irr and Sc galaxies and the outer spiral-arm regions of many Sc, Sb, and possibly Sa systems. No early-type stars were needed to match the colors of the centers of most Sa, Sb, and E systems. We cannot expect more detail from these measures. The broad-band photometry of galaxies, per se, will be discussed and summarized elsewhere in this volume.

The first theoretical discussion of powerful *quantitative* techniques of galaxy population synthesis was by Whipple in 1935. With characteristic insight, Whipple outlined a quantitative approach in which he envisioned careful measures of galaxy colors and absorption-line equivalent widths, and similar measures for nearby galactic stars, to form the building blocks of the model. Whipple did not find many data available to actually attempt a population model in the 1930s, but more recent work by de Vaucouleurs, Wood, Moore, Spinrad and Taylor, and O'Connell all follow the principles he laid out three and one-half decades ago.

Many data of a qualitative nature are now available to "extragalactic observers," due to the early spectroscopic work of Humason (1936) and Humason, Mayall and Sandage (HMS) (1956). These observers quoted spectral types for many systems and correlated the appearance of low-dispersion spectra of galaxies with their morphological type; naturally their spectral classification was crude, as the spectra were of small scale and were generally exposed for redshift determination.

It was left to Morgan and Mayall (1957), Morgan (1959*a*), and Morgan and Osterbrock (1969) to recognize and discuss the argument that the equivalent blue spectral type of a galaxy center was related to the character of the stellar mix—in particular, Morgan's (1959*a*) young-star-*rich* and young-star-*deficient* groups are definitely a measure of the timing of the last burst of star formation and the main-sequence termination point of the stellar mix. Morgan and his collaborators have detailed and clarified the relation between spectral type and galaxy morphological class that was first brought out by HMS. Galaxy spectral types near 4000 Å range from B to K. Does this large range in type suggest an equal range in stellar main-sequence turnoff point? A visual inspection of the spectrum seems inadequate to answer this important question in detail. We need sharper criteria and a more quantitative approach. To satisfy this need the continuing curiosity of several astronomers has suggested a new approach.

A new type of quantitative approach to galaxy population studies has been recently introduced by Wood (1966), Spinrad (1966), McClure and van den Bergh (1968), and Spinrad and Taylor (1970). All of these authors set up different photoelectric narrow-band measurement systems of color and line indices, hopefully sensitive to stellar temperature, luminosity, and abundance differences, and employed them in constructing galaxy content models. The inherent photoelectric precision (better than a few percent in intensity measures) balances to some degree the lack of high spectral and geometric resolution (usually also lacking in low-dispersion spectral analysis).

The photoelectric wavelengths chosen (or rejected!) for continuum and line positions must be based upon information gathered from the inspection of slit spectra of many types of stars, galaxies, emission-line regions, and possibly other integrated sources such as globular clusters. The success or failure of the technique is very dependent on this point, so that the number and positioning of bandpasses in future galaxy synthesis should be based upon the experience and intuition of previous workers.

Another approach, conceptually similar and providing many data of somewhat lower internal precision, would be an analysis based upon actually measured equivalent widths from medium-dispersion spectra of stars and galaxies. One problem here will be the difficulty in obtaining good spectrograms of all stars and galaxies over the entire wavelength region accessible to photography (3300–11,000 Å); the infrared end is especially difficult, although Whitford (1966) has pioneered this effort to 8800 Å (the I-N plate limit) on some bright galaxies. The large line widths in massive galaxies will preclude precise W_λ measurement, since all lines will be broad and shallow, and blending will be serious at $\lambda < 6000$ Å. Comparisons of sharp-lined stars and broad-lined galaxies are a particular annoyance and a probable source of systematic error in this technique. The complete task would be a worthwhile and enormous venture; crude beginnings have been made by the de Vaucouleurs on the LMC (1959*a*, *b*) (a rather narrow-lined system) and by Moore (1968) on several E galaxies (M32, NGC 3115, 3379 and 4472). However, no lines beyond 4900 Å were measured by either de Vaucouleurs or Moore; we consider this wavelength truncation an unfortunate feature of their analysis—and one that makes it very difficult to learn much about late-type stars in the galaxies studied only with blue-sensitive emulsions.

To conclude this general historical review, we shall mention that important extensions of galaxy spectra to both long and short wavelengths seem imminent as these pages are being written. The Wisconsin group (Code 1969) now has OAO satellite photometry of several galaxies to below 2000 Å in the far-ultraviolet, while medium-resolution infrared spectra of the brightest systems are planned by the Berkeley group and also at Caltech, initially around the 2-micron region.

We will now look back at previous results of this type of research on stellar populations and critically review the work; as in many review papers, the clarity of hindsight enables the reviewers to be righteously critical of their colleagues. It is likely that this contribution will also suffer with time if the field we discuss advances at a healthy pace in the future.

1.2. Slit Spectra of Galaxies—Selected Observations and Interpretations

Humason (1936) and Humason, Mayall, and Sandage (HMS) (1956) indicated, in a rough way, the dependence of blue spectral type on galaxy morphological type. Although Humason's type estimates were compromises (based upon the strength of Hγ and Hδ, the G-band, and λ4227 of Ca I) (see fig. 1, showing M31 and M32 at relatively high dispersion), he recognized the problem of composite spectra. Both these investigations noted that the E, Sa, SBa, and most Sb galaxies were clustered near spectral type G5 whereas Sc systems covered the range from G5 to A5. There was no substantial discussion of the spectral types in terms of galaxy stellar content; however, it became clear to research workers following HMS (partly *due* to their early work) that the stellar content of the centers of E, Sa, and Sb systems was mostly stars cooler than the Sun. Conversely, the early spectral types of Sc and many Irr galaxies definitely indicated the presence of hot stars. We note, parenthetically, that since most of the HMS spectra were obtained for redshift measurement, the exposures are often not optimized for spectral classification, so individual galaxy spectral types should be taken with some reserve.

The de Vaucouleurs' (1959) study of the LMC bar was the first real attempt to synthesize the content of a galaxy through color measures supplemented by the equivalent widths of absorption lines. Indeed, until 1969 this pioneering effort was the only

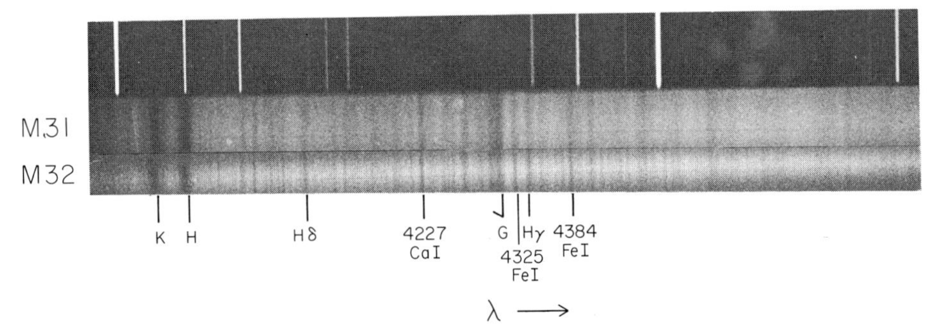

FIG. 1.—A comparison of relative high-dispersion spectra of M31 and M32 in the blue region. The overall spectral types are about late G. Note the width of the lines in the M31 spectrum. The outstanding spectral features are marked; note that the Balmer lines are slightly stronger in M32 than in M31. These two spectra are by Minkowski at 85 Å mm^{-1} original dispersion.

detailed study of the H-R diagram in an Irr galaxy. (O'Connell [1970] has recently observed and synthesized M82.) The de Vaucouleurs measured the Balmer lines, H and K of Ca II, the G-band, λ4227 of Ca I, selected Fe I lines, and, importantly, He I λ3820. The LMC spectrum is quite composite (see fig. 2); He I indicates the presence of B5 or earlier stars, Hγ and Hδ are contributed by B, A, and F dwarfs, while the G-band and λ4384 of Fe I must come from stars of type G and K. The model proposed has a van Rhijn luminosity function, the main sequence of a young open cluster like the Pleiades (or younger perhaps), and evolved giants and bright giants. There seems to be a large Hertzsprung gap. If LMC spectra of 150 Å mm^{-1} or better could be obtained in the red and near-infrared, it would be useful to measure the b triplet of Mg I (λλ5175–5184), Na I D, and the red TiO bands at λλ6180 and 7054. These cool-star spectral features would give information on the temperature and luminosities of the K and M stars which begin to enhance the low-excitation Fe I lines in the blue and green. The existence of a "young-star-rich" population in the center of the LMC is no surprise: with resolved Cepheids, hot and cold supergiants, and plenty of H I we might be surprised if the mass of fainter stars had the H-R diagram aspect of an older group of stars. The great luminosity and substantial numbers of absolutely bright K stars make inferences of the number of K and M dwarfs difficult; it will be hard to determine a stellar $\mathfrak{M}/L$ ratio from the integrated-light spectral details in the LMC. However, even crude spectroscopic limits would be useful desiderata to compare with the dynamical data now available. A general problem associated with Morgan A-type systems like NGC 4449 is that the hot blue stars totally swamp out light from all fainter stars at $\lambda < 6000$ Å.

A most important qualitative discussion on galaxy spectra and galaxy content was made by Morgan and Mayall (1957). These authors examined the spectra of many systems; the nuclear regions of E, Sa, and Sb galaxies were found to show noticeable CN bands (the blue band at λ4125 and the violet one at λ3883) at even low dispersion. The CN criterion has been classically thought to indicate uniquely the presence of gK stars

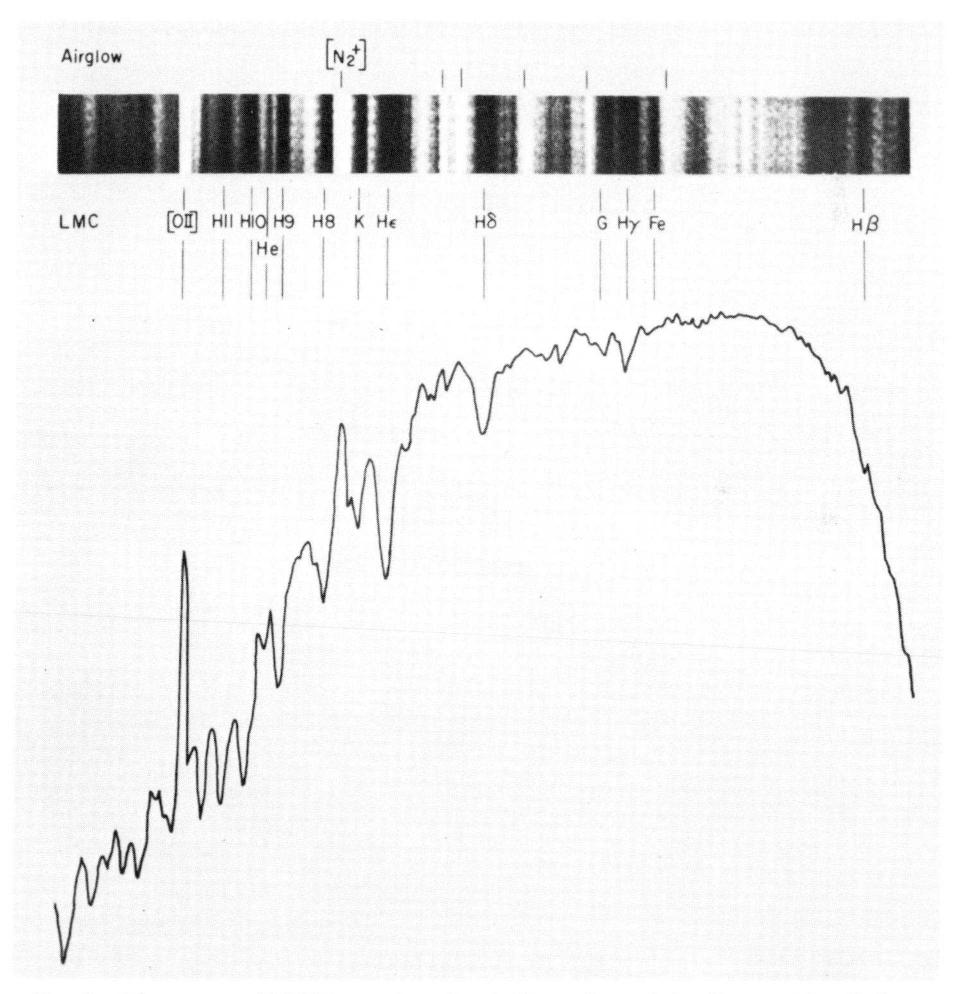

FIG. 2.—The spectra of LMC bar region after de Vaucouleurs. A few lines are identified; note the dominance of an early-type spectrum with strong H lines, He absorption, and the usual emission line of [O II] λ 3727.

(CN giants, G8–K3 III); and thus has grown the idea of K-giant "domination" of the blue spectra of E, Sa, and Sb galaxy centers. As we shall later see, the question of the origin of the CN bands is probably more complex, but Morgan and Mayall did convincingly show that galaxy spectra had strong metal lines—a situation quite different from the metal-poor halo globular clusters whose spectra were described earlier by Morgan (1956). This step was an important advance at that time.

The H-R diagram suggested by Morgan and Mayall for M31 and the larger E galaxies is the standard old-open-cluster picture: a main-sequence termination point near F8 V, with red giants and subgiants. As emphasized by Morgan and Mayall, the available

TABLE 1

GALAXY CATEGORIES AND SPECTRA

Category	Description	Spectroscopic Identification	Typical Galaxy
Orion..........	H II regions, blue stars; often irregular shaped galaxy	Strong emission lines like the Orion Nebula, a hot star continuum, He I abs. line, other indicators of types B–F in the *blue*. [λ3820 of He is a good indicator of early B stars.]	NGC 4214, NGC 4449, LMC bar, M82 core
Intermediate...	Nuclear regions of Sc galaxies, main bodies of giant spirals. Yields types f and fg.	Blue sp-type near F8, very composite spectrum. λ3727 [O II] emission common.	NGC 5194, NGC 4321
Amorphous....	Centers of big Sb, Sa systems. Main bodies of giant E galaxies.	The type K0 in most cases, type closer to M0 in the deep red. Emission lines weak.	M31, M81, NGC 4472
Weak-lined.....	Metal-poor population of dwarf E system	Globular-cluster-like; H lines intermediate, metals weak, no emission lines	Dwarf E's like NGC 205, possibly NGC 5195?

blue-region spectra of galaxies yield little information on the reddest stellar constituents; naturally red-region spectra are necessary to rectify this inadequacy.

Morgan and his associates also have gone further than HMS in describing "young-star-rich" galaxies with very early spectral types. Morgan and Osterbrock (1969) have recently summarized and amplified the low-dispersion approach to the stellar content determination; following them we have listed in table 1 the spectroscopic categories suggested by Morgan and Osterbrock and further interpreted by us.

One important new point suggested by the integrated spectra discussed by Morgan and Osterbrock is the *possible* metal-poor, weak-lined nature of the Ep galaxy NGC 5195, the companion of M51. This peculiar system is too large to be regarded as a dwarf E of the type shown to be metal-poor by their C-M diagrams; and to add to the mystery, NGC 5195 is an extremely red galaxy (unpublished photometry by O'Connell and by Whitford and Sears). This exciting new suggestion will require independent confirmation; in particular, the Balmer lines in the 5195 spectrum are quite strong and may suggest a sizable contribution by A stars. Hγ and Hδ are never this strong in metal-deficient globular clusters in our Galaxy.

One controversial point in the Morgan and Osterbrock paper is the spectral-type gradient suggested for M31 (center versus edge of the amorphous central lens) and our own Galaxy (transparent region in the Sagittarius star cloud versus an obscured location). Minkowski has pointed out to us the influence of the airglow night-sky spectrum on low-surface-brightness regions of galaxies; in particular, from the airglow spectra of Chamberlain and Oliver (1953) and the Kitt Peak group (Broadfoot and Kendall 1968) we see that at low dispersion Hδ absorption will be greatly accentuated in an airglow-contaminated G-type spectrum, while Fe I λλ4384–4406 will tend to be submerged. The net effect is to make the resultant spectrum appear earlier—perhaps like an F star. Such

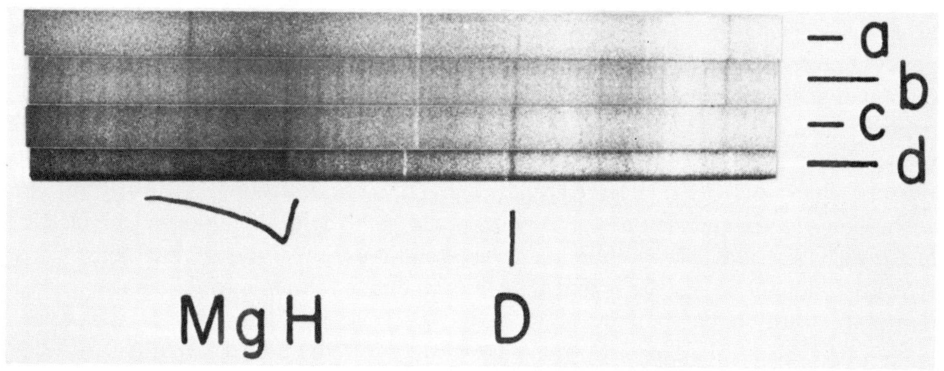

FIG. 3.—Low-dispersion F plate spectra (250 Å mm^{-1}) of the following composite sources: (*a*) globular cluster M15; (*b*) dwarf E galaxy M32; (*c*) nucleus of Sb galaxy M31; and (*d*) nucleus of Sa galaxy NGC 4594 (after Spinrad and Wood 1965). Note the strengths of Mg and MgH and the Na D lines as we go from the top to bottom in the montage. The interpretation of increasing D and MgH with increasing galaxy luminosity is one of added metal abundance and increasing contribution of lower-main-sequence dwarf stars from (*a*) to (*d*) in this illustration.

an occurrence would seem to us unavoidable in a conventional photographic blue-region spectrum of a low-surface-brightness object. For example, M31 some 20′ from the nucleus on the major axis has $B = 22.0$ mag (arc sec)$^{-2}$, about equal to the average night-sky level (de Vaucouleurs 1958) at dark U.S. sites. The situation for the Sagittarius *obscured* regions is probably even worse, so it is our opinion that the earlier spectral types suggested by Morgan and Osterbrock (1969) for these faint regions are probably results of airglow contamination, and careful subtraction of the sky spectrum (probably photoelectrically) will be required to reach a firm decision on the run of stellar mix beyond $r = 15'$ or $20'$ in M31. A small telescope in a very dark sky site is ample for this type of investigation.

Spinrad (1962) and Deutsch (1964) and very recently Sandage (1969) have obtained galaxy spectrograms in the yellow and red regions, on D and F emulsions and with a panchromatic-response image tube. All authors have noticed strong atomic Na D and Mg b lines in massive E and Sa and Sb galaxies with TiO bands usually visible near λ5847 and λ6180 (see fig. 3). In fact, the scans by Whitford (1971) and Oke and Sandage

(1968) show MgH and TiO very nicely in E galaxies—see the illustrations in chapter 5 of this volume. These features all indicate the presence of K and M stars; from the absolute strengths of D and the magnesium triplet, both Spinrad and Deutsch suggested that the luminosities of these cold stars were low. Normal K giants just do not have strong enough lines. Since the D lines may be partially interstellar in some systems, we feel that their detection alone is insufficient evidence for K and M dwarfs, although the correlation between D and Mg b or that between D and MgH (Spinrad and Wood 1965) is suggestive of a stellar dominance even at the sodium resonance lines. We note, parenthetically, that TiO λ5175 will contribute almost nothing to the Mg I plus MgH absorption in this region; the wavelength is too short for M stars to matter.

Deutsch (1964) examined the dependence of D on galaxy M_v; Spinrad (1962) had previously suggested that the D lines were strongest in the highest-luminosity E and Sa, Sb systems. Deutsch showed that Spinrad's estimates of sodium-line intensities were partly confused by night-sky D emission; correction for the airglow fill-in of the galaxian sodium doublet tends to make the D lines intrinsically stronger in the galaxies. Deutsch demonstrated that the sodium blend was definitely stronger in NGC 4649 in the Virgo cluster and in the M31 nucleus than in M32. In the least luminous systems like the dwarf E galaxy NGC 205, a large airglow correction for D is required. Clearly the correction for airglow emission in the D lines should be made by simultaneous monitoring of the night-sky line. The D lines are a very good criterion of metal abundance. Their strength in luminous galaxies is partly physically due to sodium overabundance. It may be possible with Sandage's new and extensive spectroscopic material to use the galaxy luminosity dependence shown by the magnesium and sodium lines for future distance-scale determination. Certainly the correlation of central-region CN strength (McClure and van den Bergh 1968; Spinrad unpublished) with M_v for Sa, Sb, E, and S0 galaxies appears to have promising future application.

The situation on stellar content for "young-star-deficient" E and Sb galaxies seemed quite confused and contradictory in 1964: Morgan and Mayall had shown that CN-marked K giants "dominated" the blue part of the spectrum, while Spinrad suggested that K dwarfs had a strong influence in the visual and red. Shortly afterwards Whitford (1966) failed to locate the infrared Na I doublet (λλ8183.3, 8194.8) on 200 Å mm^{-1} spectrograms of several bright galaxies; the lines should have been detectable if K5–M8 dwarfs contributed more than about one-third of the λ8190 radiation (see fig. 4 with Wing's infrared spectra). Recent work has shown that Whitford's 1 Å upper limit to these Na I subordinate lines in galaxy spectra is consistent with their scanner detection in M31 by Spinrad and Taylor (1970). These authors suggest an equivalent width of about 0.6 Å for the λ8190 doublet. The infrared sodium lines would be easier to observe if they were well removed from nearby strong telluric H_2O lines; special care has proven necessary in scanner photometry near λ8190.

Since 1964 astronomers have attempted new photoelectric measurements of galaxy spectra over longer-wavelength baselines, partly to resolve the giant-versus-dwarf controversy.

1.3. Photoelectric Narrow-Band Measures of Galaxies

Photoelectric photometry of continua and spectral lines in galaxies is a relatively new approach to our problem. The advantages over slit spectroscopy are (1) a long-wavelength region is available with photomultipliers covering the wavelength region 3300–11,000 Å and PbS photometers extending the range to at least 3.4 μ (Sandage, Becklin, and Neugebauer 1969); (2) intensity measures may be obtained with high precision—1 percent accuracy is possible, although often much observing is required; (3) galaxy spectral lines broadened by modest velocity dispersions are measured as well as sharp stellar features, especially if $\sigma \leq 300$ km s^{-1}, so photoelectric resolution in the spectrum is adequate; (4) emission lines arising in the galaxy (see § 2 of this chapter), or in the Earth's atmosphere, may be analytically removed from the galaxy spectrum, if so desired.

One major disadvantage, alluded to before, is that only the "preprogrammed" features will be observed, and surprises will often be missed without the general re-

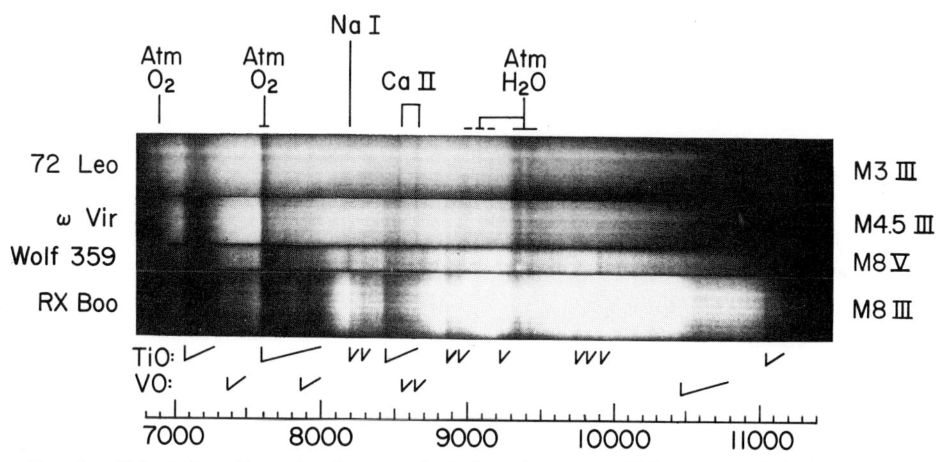

FIG. 4.—Wing's low-dispersion image-tube infrared spectra of M giants and the late M dwarf Wolf 359. Note the strong TiO and VO in all the cold (>M4) stars and the appearance of the Na I subordinate doublet blend at λ8190 in Wolf 359. These lines are not apparent in giant spectra at low resolution; Spinrad and Taylor have detected them in galaxies with difficulty.

connaissance feature of seeing the spectrum in detail. Thus the photoelectric observations may be best utilized when combined with a more conventional spectroscopic attack on galaxy content. Some other broad-band (50 Å) scans have been made to show overall galaxy energy distributions. They are discussed by Whitford in chapter 5 of this volume.

We arbitrarily concentrate our discussion here to narrow- and intermediate-band photometry. The broad-band multicolor approaches (Johnson 1966; Tifft 1963; Sandage, Becklin, and Neugebauer 1969) will be discussed elsewhere in this volume; in general, their solutions are less specific in interpretation than those narrow-band techniques which include galaxy line measures.

Wood (1966) has published narrow-band fixed-filter measures of galaxy lines and continua from 3400 to 7300 Å. The work was restricted to low-radial-velocity systems (with typical filter $\Delta\lambda \simeq 100$ Å or less). Many kinds of galactic stars were observed on Wood's 12-color system; they were used for calibration and model construction. Wood's synthetic model galaxies, constructed to match mainly large E galaxies, followed the model-making rules: (*a*) match the colors; (*b*) match line indices, especially Mg plus MgH; (*c*) use a "reasonable" and continuous color-magnitude; (*d*) be consistent with Morgan's nuclear spectral types. They had the following general properties: (1) They resembled old stellar groups like M67—no young stars were allowed. (2) Many lower-main-sequence K and M stars were needed; at *V* wavelength about 40 percent of the light came from the unevolved main sequence (this is the number appropriate for Wood's model of NGC 4406). (3) Some blue stars, assumed to be old horizontal-branch types, were required. The number of such hot stars could easily depend critically on the amount of interstellar reddening assumed in the direction of the galaxy; Wood used a cosecant law to find his color excesses.

A most interesting problem suggested by Wood's photometry and model fitting was a factor of *three* discrepancy between typical galaxy mass-to-light ratios ($\mathfrak{M}/L$) and the derived model stellar ratio. The model ratios were much lower. Where was the "missing mass"? If confirmed, such a discrepancy would open a large Pandora's box of nonstellar constituents for galaxy nuclei, such as rocks, H_2, invisible collapsed massive objects (cf. Wolfe and Burbidge 1970), and the like.

In any case, Wood's try at a *quantitative* solution partially confirmed Spinrad's (1962) assertion that main-sequence stars, G0 V–M V, contributed a substantial fraction of the galaxy light at $\lambda > 5000$ Å. Wood's stellar luminosity function was dwarf-enriched compared with the solar-vicinity van Rhijn function.

Another general attack, using the same weapons—augmenting Wood's photometry with the David Dunlap Observatory five-color system—was made by McClure and van den Bergh (1968). Of the several color indices measured in stars, globular clusters, and galaxies, McClure and van den Bergh found their blue CN index, $C(41 - 42)$, to be rather weaker in dwarf E systems than in giant ellipticals. The CN found near the nuclei of large E, Sa, and Sb systems was so strong that the authors suggested abundance anomalies—SMR (super-metal-rich) giants were needed, as had been previously hinted by Spinrad (1966). We will say more on abundance gradients in galaxies later.

McClure and van den Bergh suggested a rather different picture for the small dwarf companion to Andromeda, M32 (NGC 221). They found a good match with NGC 6356, a nuclear globular cluster with a small metal deficiency, plus main-sequence G and K stars. In the case of the giant E system NGC 4406, a turnoff point near F5–F8 V, typical of a dwarf-rich, metal-rich old cluster, was found to fit best. These models naturally predicted relatively low $\mathfrak{M}/L$ ratios for E galaxies as they employed Wood's photometry for their long-wavelength colors and magnesium indices. One other valuable aspect of this contribution is the large number of data on colors and CN strengths presented for 56 galaxies in and outside of large clusters.

1.4. The Spinrad-Taylor Galaxy Models

In 1965 Spinrad and Taylor began a long program of galaxy synthesis observations at Lick Observatory. Thirty-six wavelengths, 3300–10,700 Å, were observed with 16 and 32 Å resolution, using the Wampler prime-focus scanner at the Crossley and the 120-inch reflector. Many galactic stars of different types were scanned along with some globular clusters to provide potential building blocks for synthetic galaxies. The stars were collected into about 30 different groups used in the computer synthesis. The hope was to blend the stars into an appropriate mixture to satisfy the 36 $I(\lambda)$ observations for the galaxies; these scanner wavelengths included continua positions and metallic absorption lines of Ca I, Mg I, Na I, and Ca II, and the hydrogen Balmer lines Hα, Hγ, and Hδ. Molecular bands of CN, CH, MgH, CaH, and TiO were also included. If necessary, the galaxy scans were corrected for interstellar reddening in our own galaxy; the M31 data were unreddened by $E(B - V) = 0.11$ mag. A Whitford (1958) reddening law was assumed.

The galaxies observed satisfactorily to date include the nuclei of M31, M81, and M32; partial observations—sufficiently precise and complete enough to use for some model building constraints—are available for NGC 4594, NGC 205, NGC 5194, NGC 3379, NGC 4472, and NGC 3115. The procedures adopted for the observations and reductions and for model building are quite sophisticated, but are so demanding of prime dark observing time and high-quality photometric nights that the time scale for completion of the project has become very long. Some early results were mentioned by Spinrad (1966, 1967), and models for M31, M81, and M32 have been reported by Spinrad and Taylor (1971). The "rules of the game" are also discussed by these authors. We next summarize their conclusions concerning models for the nucleus of M31, a case with good observational data and few ambiguities.

Spinrad and Taylor were able to obtain satisfactory M31 nuclear models only with a heavily dwarf-enriched lower main sequence using SMR stars at all possible H-R diagram locations (see fig. 5, schematic H-R diagrams). With the SMR giants and dwarfs included, the blue and ultraviolet CN bands can be matched with those in giant galaxies. Even so, the M31 center has TiO bands too strong to duplicate with any combination of nearby galactic stars. The nearby stars are not strong-lined enough! In principle, this one remaining substantial deficit (see residual plot of fig. 6) may be erased by the use of TiO-rich K giants in the old SMR open clusters M67, NGC 188, and NGC 6791 (Spinrad and Taylor 1969, 1971). These K3–K5 III stars have rather strong TiO (measured at λλ6180, 7100)—much more than is normal for their red colors. However, most of these stars are much too faint for observations and inclusion in the synthesis star groups—especially at infrared wavelengths, where the detector sensitivity is fairly low.

The strong blue CN bands would now originate in a few SMR giants and many SMR dwarfs, not just in the normal CN giants. This is a substantical change from the earlier ideas.

In the best M31 and M81 models (see table 2 for a complete model printout) the main-sequence turnoff point lies near G0 V; thus, the stellar mix is old—i.e., no very young stars are present—but star formation apparently continued to about 2–3 billion years ago. This shorter than classical time scale for a G0 V H-R position turnoff is due to the high internal Z assumed for SMR stars (Aizenman, Demarque, and Miller 1969; Torres-Peimbert 1971). Thus the equivalent "ages" of the nuclei of these two large Sb galaxies are much smaller than the age of the oldest stars in our own system and probably in M31 and M81. However, star formation has apparently ceased in M31 now: the detection of a few OB main-sequence stars is quite easy with the models discussed here. The sensitivity of the near-ultraviolet continua from 3300 to 3880 Å is such that 50 O9 V stars would produce an intolerably large intensity residual at the shortest wavelengths scanned (see fig. 6) in models which have 5×10^5 red giants, 10^6 G0 V stars, and 10^9 M5 V stars.

One satisfactory product of the Spinrad-Taylor (abbreviated ST hereafter) models for M31 and M81, and likely other large galaxies as well, is their high stellar $\mathfrak{M}/L$ ratio. Enough M dwarfs are present to make the model $\mathfrak{M}/L = 40$ (uncertainty 40% or so), while the dynamic ratio *for the nucleus* is of this order (King and Minkowski 1972) and the dynamic ratio for the entire body of a large E system may be at the $30 \leq \mathfrak{M}/L \leq 100$ level (Page 1962; Fish 1964). Models for at least the M31 nucleus with $\mathfrak{M}/L \geq 100$ are too red (see fig. 7). Also, models with $\mathfrak{M}/L = 7$ are too blue at long wavelengths.

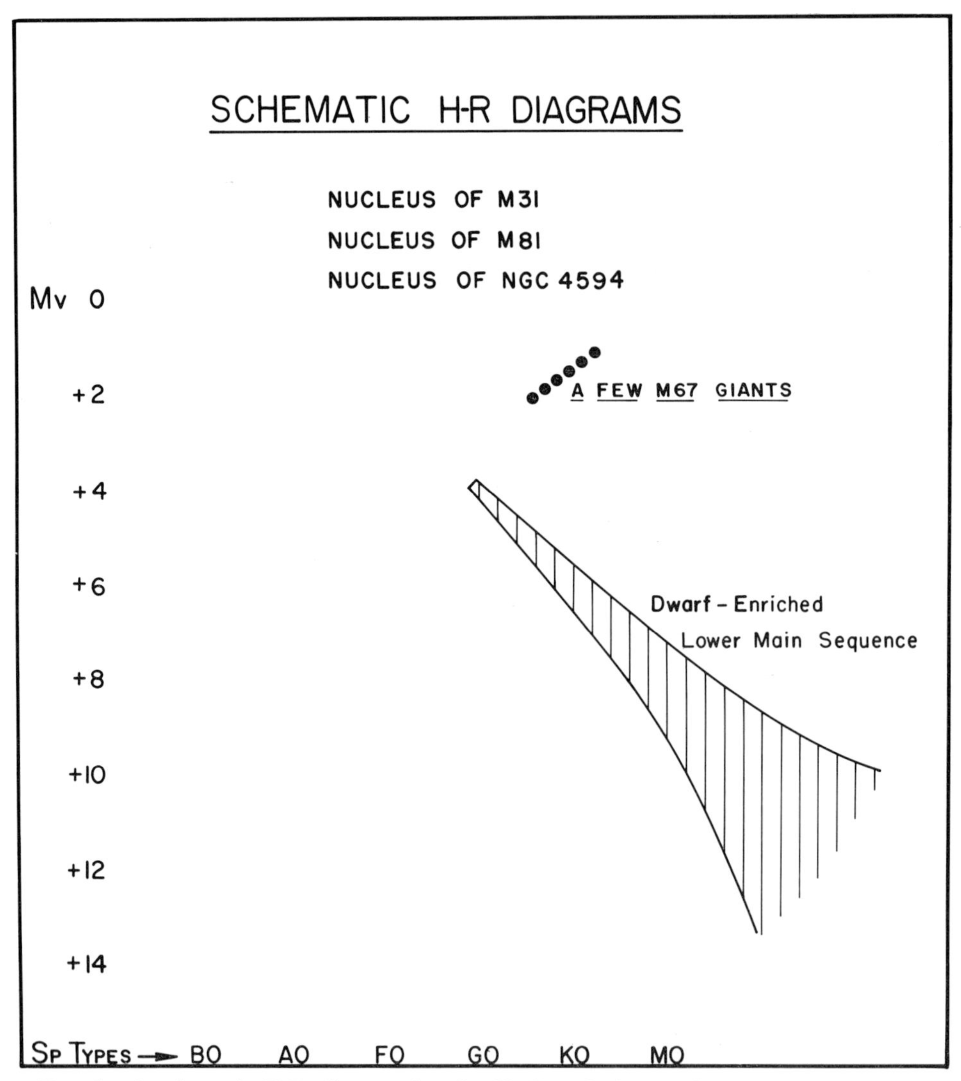

FIG. 5.—A schematic H-R diagram for the Sb–Sa galaxies synthesized; no hot stars are allowed in the successful models for these systems. After Spinrad (1962).

While common stars may not be the *only* residents of galaxy centers, we do not now require appreciable amounts of extra mass there. In hindsight, with respect to Wood's (1966) efforts, we would suggest that since stellar model $\mathfrak{M}/L$ ratios depend entirely on *late* M *dwarfs*, not K5–M3 V stars, one cannot get decisive information on $\mathfrak{M}/L$ without observations beyond 8000 or perhaps even 10,000 Å. This caution has been also noted by Tinsley (1968). Spinrad and Taylor (1971) suggest galaxy models with no dynamic mass discrepancy at all for M31 and perhaps none for large E galaxies; their models have many

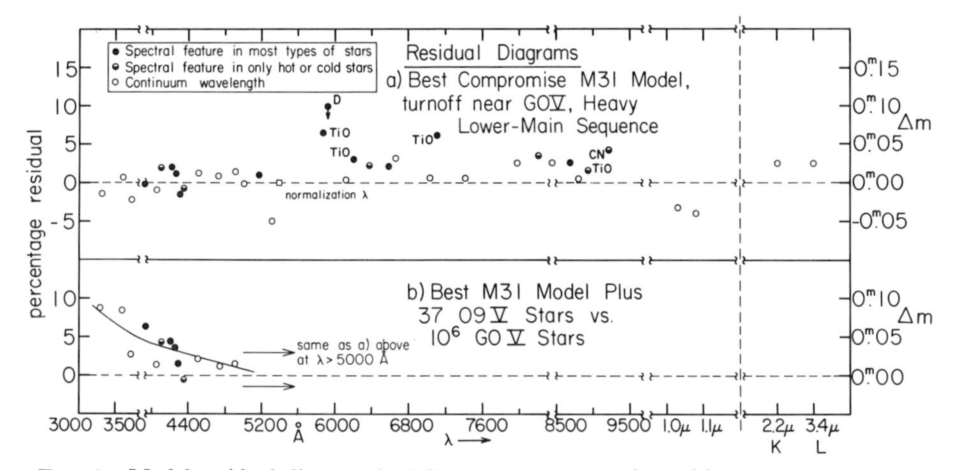

FIG. 6.—Model residual diagram for M31 nucleus observations: (*a*) The best model of ST (1970). Plotted is the percentage residual, model-observation/observation intensity at each wavelength. Note the general continuum agreement (*open circles*) over the wavelength interval 3300 Å–3.4 μ. Even in the best model the strong absorption feature of the Na I D lines and the red TiO bands are too weak in the model to match the real galaxy (or above dotted line in residual diagram—*intensity in feature* is too great in the model). (*b*) Same model plus 37 O9 V hot stars which lead to a peculiar and undesirable slope in the near ultraviolet region. This modest addition of young stars can be early ruled out; thus, the numerical constraint against O–B stars may be made very stringent in K-nuclei galaxies. Arrow at D indicates the amount of correction that might be applied to compensate for interstellar sodium absorption in our own Galaxy (~600 mÅ).

little M7 V and M8 V stars. Inclusion of numerous white dwarfs will not appreciably raise the $\mathfrak{M}/L$ ratio for the ST models; however, their inclusion in an alternative giant-dominated Morgan and Mayall type model does raise $\mathfrak{M}/L$ from about 1 to approximately 4. On the other hand, the ST (1971) model $\mathfrak{M}/L$ ratio for M32, a slightly metal-poor dwarf E, is significantly lower than the derived M31 ratio of 40—$\mathfrak{M}/L = 8 \pm 5$ for M32. This result agrees reasonably well with the dynamic ratios computed by Burbidge, Burbidge, and Fish (1961) and by King and Minkowski (1972). No simple power law adequately fits the derived mass spectrum, although the analytic approach of Tinsley (1968) and of Reddish (1968) could be employed as a rough representation of the lower main sequence.

The documentation of these statements may be partially illustrated by presenting alternative galaxy models. First of all, to show the necessity of the inclusion of many K

M31 NUCLEUS (GALACTIC REDDENING = 0.11 MAGNITUDES)

BEST COMPROMISE MODEL (1969)

COMPUTED MASS TO VISUAL LIGHT RATIO = 43.5

Fractional Light Contributions at λ =

Kind of Star	Percent Light	Percent Mass	3300	3600	3700	3880	4040	4100	4200	4227	4300	4340
O-B Dwarfs	0.	0.	0.	0.	0.	0.	0.	0.	0.	0.	0.	0.
Late B Dwarfs	0.	0.	0.	0.	0.	0.	0.	0.	0.	0.	0.	0.
Early A Dwarfs	0.	0.	0.	0.	0.	0.	0.	0.	0.	0.	0.	0.
Late A Dwarfs	0.	0.	0.	0.	0.	0.	0.	0.	0.	0.	0.	0.
Early F Dwarfs	0.	0.	0.	0.	0.	0.	0.	0.	0.	0.	0.	0.
Middle F Dwarfs	0.	0.	0.	0.	0.	0.	0.	0.	0.	0.	0.	0.
Late F Dwarfs	.99	.02	2.79	2.67	2.28	3.25	2.17	1.76	2.12	2.07	2.04	1.37
Early G Dwarfs	6.44	.15	14.57	14.02	12.64	13.96	11.43	10.66	11.37	11.32	10.09	8.51
G5V - K0V	4.26	.22	6.92	6.57	6.34	5.85	5.66	5.83	5.75	5.62	5.21	5.18
K1V - K2V	4.26	.25	5.30	5.02	4.99	4.36	4.40	5.02	4.74	4.35	4.41	4.79
K3V - K4V	3.17	.28	2.51	2.39	2.55	2.18	2.44	2.89	2.79	2.10	2.79	3.18
K5V - K7V	1.49	.47	.57	.63	.70	.75	.79	.93	.99	.55	1.04	1.14
M0V - M2V	1.19	1.09	.32	.43	.48	.60	.55	.60	.65	.29	.71	.69
M3V - M4V	1.19	3.90	.23	.36	.39	.59	.55	.55	.57	.20	.59	.55
M5V - M6V	1.19	12.70	.23	.35	.28	.41	.45	.41	.40	.15	.45	.43
M7V	.99	13.65	0.	0.	0.	.46	.37	.42	.36	.15	.40	.43
M8V	.50	66.92	0.	0.	0.	.15	.06	.16	.04	.03	.06	.19
Early G Subgiants	11.78	.12	27.11	26.11	23.29	27.71	22.19	19.32	21.51	21.58	20.19	15.45
Late G Subgiants	9.60	.10	15.36	15.66	15.75	14.70	14.10	13.78	13.99	14.29	12.52	11.84
SMR K0 - K1IV	13.86	.09	10.30	10.84	12.03	9.74	12.67	13.35	12.32	13.51	12.71	14.44
SMR K2III - IV	18.12	.03	8.25	9.00	10.53	8.63	12.26	13.17	12.21	13.02	13.71	16.15
SMR K3III	16.04	.02	4.71	5.03	6.55	5.51	8.28	9.17	8.30	0.09	10.57	12.71
SMR K4 - 5III	4.95	.00	.81	.91	1.21	1.16	1.64	1.97	1.88	1.68	2.50	2.97
Hyades G8III	0.	0.	0.	0.	0.	0.	0.	0.	0.	0.	0.	0.
Normal G8 - K0III	0.	0.	0.	0.	0.	0.	0.	0.	0.	0.	0.	0.
Normal K0 - K1IV	0.	0.	0.	0.	0.	0.	0.	0.	0.	0.	0.	0.
K1III - K3III	0.	0.	0.	0.	0.	0.	0.	0.	0.	0.	0.	0.
K4III - K5III	0.	0.	0.	0.	0.	0.	0.	0.	0.	0.	0.	0.
M0III - M2III	0.	0.	0.	0.	0.	0.	0.	0.	0.	0.	0.	0.
M3III - M4III	0.	0.	0.	0.	0.	0.	0.	0.	0.	0.	0.	0.
M5III - M6III	0.	0.	0.	0.	0.	0.	0.	0.	0.	0.	0.	0.
M8III - M9III	0.	0.	0.	0.	0.	0.	0.	0.	0.	0.	0.	0.
Carbon Stars	0.	0.	0.	0.	0.	0.	0.	0.	0.	0.	0.	0.
WD A	0.	0.	0.	0.	0.	0.	0.	0.	0.	0.	0.	0.
Globulars - M15 + M92	0.	0.	0.	0.	0.	0.	0.	0.	0.	0.	0.	0.
Globulars - M3 + M5	0.	0.	0.	0.	0.	0.	0.	0.	0.	0.	0.	0.
Globular - NGC 6356	0.	0.	0.	0.	0.	0.	0.	0.	0.	0.	0.	0.
Observed Integrated Spectrum			182.31	309.53	455.96	393.01	1015.53	1013.01	1062.06	1065.54	922.76	1314.04
Computed Integrated Spectrum			179.17	311.85	445.90	392.42	1003.39	1034.61	1082.97	1076.26	906.50	1296.54
Percentage Residuals			-1.72	.75	-2.21	-.15	-1.20	2.13	1.97	1.01	-1.76	-1.33

Fractional Light Contributions at λ =

Kind of Star	4500	4715	4900	5000	5175	5300	5360	5864	5892	6110
O-B Dwarfs	0.	0.	0.	0.	0.	0.	0.	0.	0.	0.
Late B Dwarfs	0.	0.	0.	0.	0.	0.	0.	0.	0.	0.
Early A Dwarfs	0.	0.	0.	0.	0.	0.	0.	0.	0.	0.
Late A Dwarfs	0.	0.	0.	0.	0.	0.	0.	0.	0.	0.
Early F Dwarfs	0.	0.	0.	0.	0.	0.	0.	0.	0.	0.
Middle F Dwarfs	0.	0.	0.	0.	0.	0.	0.	0.	0.	0.
Late F Dwarfs	1.40	1.29	1.17	1.23	1.38	1.06	.99	.89	.97	.83
Early G Dwarfs	8.45	7.89	7.38	7.60	8.32	6.75	6.44	5.99	6.45	5.67
G5V - K0V	5.08	4.79	4.66	4.68	4.64	4.42	4.26	4.10	4.25	3.92
K1V - K2V	4.67	4.58	4.44	4.46	3.92	4.30	4.26	4.31	4.31	4.15
K3V - K4V	3.24	3.31	3.13	2.98	2.32	3.11	3.17	3.39	3.12	3.33
K5V - K7V	1.28	1.41	1.31	1.22	1.02	1.41	1.49	1.72	1.45	1.81
M0V - M2V	.83	.97	.95	.89	.83	1.12	1.19	1.31	1.00	1.55
M3V - M4V	.71	.81	.84	.80	.73	1.16	1.19	1.02	.73	1.39
M5V - M6V	.62	.76	.82	.75	.66	1.16	1.19	.88	.58	1.35
M7V	.53	.62	.69	.61	.53	.97	.99	.65	.48	1.05
M8V	.17	.19	.29	.21	.20	.49	.50	.28	.22	.44
Early G Subgiants	15.35	14.28	13.39	13.96	15.59	12.41	11.78	10.80	11.67	10.22
Late G Subgiants	11.79	10.95	10.58	10.77	11.49	9.95	9.60	9.10	9.71	8.68
SMR K0 - K1IV	14.04	14.09	14.25	14.11	14.18	13.92	13.86	13.99	14.37	13.67
SMR K2III - IV	15.98	16.58	17.33	17.24	16.38	17.84	18.12	19.01	19.01	18.78
SMR K3III	12.67	13.78	14.73	14.47	14.04	15.32	16.04	17.12	16.63	17.32
SMR K4 - 5III	3.20	3.69	4.03	4.03	3.77	4.60	4.95	5.44	5.06	5.84
Hyades G8III	0.	0.	0.	0.	0.	0.	0.	0.	0.	0.
Normal G8 - K0III	0.	0.	0.	0.	0.	0.	0.	0.	0.	0.
Normal K0 - K1IV	0.	0.	0.	0.	0.	0.	0.	0.	0.	0.
K1III - K3III	0.	0.	0.	0.	0.	0.	0.	0.	0.	0.
K4III - K5III	0.	0.	0.	0.	0.	0.	0.	0.	0.	0.
M0III - M2III	0.	0.	0.	0.	0.	0.	0.	0.	0.	0.
M3III - M4III	0.	0.	0.	0.	0.	0.	0.	0.	0.	0.
M5III - M6III	0.	0.	0.	0.	0.	0.	0.	0.	0.	0.
M8III - M9III	0.	0.	0.	0.	0.	0.	0.	0.	0.	0.
Carbon Stars	0.	0.	0.	0.	0.	0.	0.	0.	0.	0.
WD A	0.	0.	0.	0.	0.	0.	0.	0.	0.	0.
Globulars-M15 + M92	0.	0.	0.	0.	0.	0.	0.	0.	0.	0.
Globulars-M3 + M5	0.	0.	0.	0.	0.	0.	0.	0.	0.	0.
Globular-NGC 6356	0.	0.	0.	0.	0.	0.	0.	0.	0.	0.
Observed Integrated Spectrum	1465.69	1277.28	991.05	1029.26	887.56	1080.59	1000.00	881.16	765.88	970.09
Computed Integrated Spectrum	1484.61	1287.16	1007.40	1029.89	899.38	1025.98	1000.00	941.99	843.73	977.92
Percentage Residuals	1.29	.77	1.65	.06	1.33	-5.05	-.00	6.90	10.16	.81

Fractional Light Contributions at λ =

Kind of Star	6180	6386	6564	6620	7000	7100	7400	7980	8190	8400
O-B Dwarfs	0.	0.	0.	0.	0.	0.	0.	0.	0.	0.
Late B Dwarfs	0.	0.	0.	0.	0.	0.	0.	0.	0.	0.
Early A Dwarfs	0.	0.	0.	0.	0.	0.	0.	0.	0.	0.
Late A Dwarfs	0.	0.	0.	0.	0.	0.	0.	0.	0.	0.
Early F Dwarfs	0.	0.	0.	0.	0.	0.	0.	0.	0.	0.
Middle F Dwarfs	0.	0.	0.	0.	0.	0.	0.	0.	0.	0.
Late F Dwarfs	.85	.80	.69	.75	.70	.72	.61	.57	.54	.51
Early G Dwarfs	5.77	5.51	4.91	5.19	4.90	5.08	4.34	4.11	3.98	3.75
G5V - K0V	4.03	3.85	3.61	3.69	3.51	3.60	3.14	3.00	2.89	2.78
K1V - K2V	4.28	4.15	3.98	4.03	3.88	4.01	3.57	3.48	3.37	3.24
K3V - K4V	3.43	3.37	3.36	3.36	3.25	3.40	3.03	3.02	2.90	2.83
K5V - K7V	1.82	1.83	2.01	1.98	1.97	2.04	1.99	2.08	1.98	2.00
M0V - M2V	1.38	1.54	1.95	1.87	1.93	1.92	2.24	2.43	2.36	2.44
M3V - M4V	.94	1.46	2.10	1.85	2.32	1.93	3.24	3.59	3.72	3.90
M5V - M6V	.82	1.44	2.29	1.97	2.49	2.05	3.92	4.43	4.60	4.97
M7V	.63	1.24	1.94	1.50	2.35	1.64	3.72	4.11	4.42	4.98
M8V	.25	.67	1.47	.89	1.91	.99	4.08	4.38	6.39	7.31
Early G Subgiants	10.33	9.88	8.76	9.31	8.73	8.99	7.71	7.26	6.98	6.60
Late G Subgiants	8.85	8.47	7.83	8.16	7.71	7.99	6.92	6.64	6.39	6.09
SMR K0 - K1IV	14.05	13.53	13.20	13.30	12.80	13.23	11.70	11.39	11.04	10.66
SMR K2III - IV	19.31	18.89	18.62	18.64	18.18	18.77	17.17	16.86	16.39	16.10
SMR K3III	17.63	17.38	17.20	17.40	17.01	17.40	16.22	16.05	15.56	15.41
SMR K4 - 5III	5.62	5.99	6.08	6.14	6.37	6.22	6.42	6.58	6.49	6.45
Hyades G8III	0.	0.	0.	0.	0.	0.	0.	0.	0.	0.
Normal G8 - K0III	0.	0.	0.	0.	0.	0.	0.	0.	0.	0.
Normal K0 - K1IV	0.	0.	0.	0.	0.	0.	0.	0.	0.	0.
K1III - K3III	0.	0.	0.	0.	0.	0.	0.	0.	0.	0.
K4III - K5III	0.	0.	0.	0.	0.	0.	0.	0.	0.	0.
M0III - M2III	0.	0.	0.	0.	0.	0.	0.	0.	0.	0.
M3III - M4III	0.	0.	0.	0.	0.	0.	0.	0.	0.	0.
M5III - M6III	0.	0.	0.	0.	0.	0.	0.	0.	0.	0.
M8III - M9III	0.	0.	0.	0.	0.	0.	0.	0.	0.	0.
Carbon Stars	0.	0.	0.	0.	0.	0.	0.	0.	0.	0.
WD A	0.	0.	0.	0.	0.	0.	0.	0.	0.	0.
Globulars - M15 + M92	0.	0.	0.	0.	0.	0.	0.	0.	0.	0.
Globulars - M3 + M5	0.	0.	0.	0.	0.	0.	0.	0.	0.	0.
Globular - NGC 6356	0.	0.	0.	0.	0.	0.	0.	0.	0.	0.
Observed Integrated Spectrum	892.52	808.18	657.75	637.46	337.21	246.70	109.99	87.59	80.97	75.42
Computed Integrated Spectrum	919.60	827.27	671.48	658.20	338.86	261.41	110.89	90.03	83.87	77.62
Percentage Residuals	3.03	2.36	2.09	3.25	.49	5.96	.82	2.79	3.58	2.91

Kind of Star	Percent Light	Percent Mass	Fractional Light Contributions at λ = 8662	8800	8900	9200	10300	10700	V-K	V-L	Individual Stellar Parameter Mass	Light	Model Luminosity Function Number of Stars
O-B Dwarfs	0.	0.	0.	0.	0.	0.	0.	0.	0.	0.	1.00E+01	2.50E+03	0.
Late B Dwarfs	0.	0.	0.	0.	0.	0.	0.	0.	0.	0.	3.00E+00	1.10E+02	0.
Early A Dwarfs	0.	0.	0.	0.	0.	0.	0.	0.	0.	0.	2.00E+00	2.80E+01	0.
Late A Dwarfs	0.	0.	0.	0.	0.	0.	0.	0.	0.	0.	1.70E+00	1.10E+01	0.
Early F Dwarfs	0.	0.	0.	0.	0.	0.	0.	0.	0.	0.	1.40E+00	4.00E+00	0.
Middle F Dwarfs	0.	0.	0.	0.	0.	0.	0.	0.	0.	0.	1.30E+00	2.50E+00	0.
Late F Dwarfs	.99	.02	.49	.47	.46	.45	.37	.34	.19	.17	1.20E+00	1.70E+00	9.05E+04
Early G Dwarfs	6.44	.15	3.57	3.43	3.42	3.25	2.84	2.66	1.42	1.26	1.00E+00	1.00E+00	1.00E+06
G5V - K0V	4.26	.22	2.72	2.59	2.59	2.49	2.17	2.05	1.12	1.04	9.00E-01	4.00E-01	1.65E+06
K1V - K2V	4.26	.25	3.15	3.01	3.06	2.93	2.61	2.56	1.58	1.29	7.00E-01	2.70E-01	2.45E+06
K3V - K4V	3.17	.28	2.78	2.66	2.73	2.65	2.41	2.36	1.86	1.54	6.50E-01	1.70E-01	2.90E+06
K5V - K7V	1.49	.47	2.00	1.92	1.96	1.97	1.83	1.82	1.39	1.15	6.00E-01	4.40E-02	5.24E+06
M0V - M2V	1.19	1.09	2.54	2.43	2.44	2.56	2.46	2.48	2.53	2.52	4.00E-01	1.00E-02	1.85E+07
M3V - M4V	1.19	3.90	4.11	4.08	3.97	4.46	4.34	4.50	5.14	4.76	3.00E-01	2.10E-03	8.79E+07
M5V - M6V	1.19	12.70	5.32	5.38	5.30	6.03	6.01	6.39	10.11	9.37	2.00E-01	4.30E-04	4.29E+08
M7V	.99	13.65	5.32	5.50	5.12	6.46	6.65	6.73	10.34	11.01	1.50E-01	2.50E-04	6.15E+08
M8V	.50	66.92	7.74	8.96	8.10	11.97	13.80	14.73	18.33	26.33	1.00E-01	1.70E-05	4.52E+09
Early G Subgiants	11.78	.12	6.27	6.03	5.99	5.81	4.89	4.49	2.61	2.30	1.20E+00	2.70E+00	6.78E+05
Late G Subgiants	9.60	.10	5.78	5.58	5.59	5.31	4.61	4.43	3.90	3.23	1.20E+00	2.70E+00	5.53E+05
SMR K0 - K1IV	13.86	.09	10.50	10.21	10.40	9.39	9.08	8.96	7.42	6.27	1.20E+00	4.30E+00	5.01E+05
SMR K2III - IV	18.12	.03	16.00	15.73	16.21	14.41	14.55	14.24	11.59	9.96	1.20E+00	1.70E+01	1.66E+05
SMR K3III	16.04	.02	15.31	15.49	15.88	13.78	14.72	14.56	13.83	12.01	1.20E+00	2.80E+01	8.90E+04
SMR K4 - 5III	4.95	.00	6.40	6.53	6.78	6.08	6.67	6.71	6.65	5.77	1.20E+00	4.30E+01	1.79E+04
Hyades G8III	0.	0.	0.	0.	0.	0.	0.	0.	0.	0.	2.00E+00	2.80E+01	0.
Normal G8 - K0III	0.	0.	0.	0.	0.	0.	0.	0.	0.	0.	2.00E+00	2.80E+01	0.
Normal K0 - K1IV	0.	0.	0.	0.	0.	0.	0.	0.	0.	0.	1.20E+00	4.30E+00	0.
K1III - K3III	0.	0.	0.	0.	0.	0.	0.	0.	0.	0.	2.00E+00	4.30E+01	0.
K4III - K5III	0.	0.	0.	0.	0.	0.	0.	0.	0.	0.	2.00E+00	7.00E+01	0.
M0III - M2III	0.	0.	0.	0.	0.	0.	0.	0.	0.	0.	1.50E+00	7.00E+01	0.
M3III - M4III	0.	0.	0.	0.	0.	0.	0.	0.	0.	0.	1.50E+00	7.00E+01	0.
M5III - M6III	0.	0.	0.	0.	0.	0.	0.	0.	0.	0.	1.50E+00	4.30E+01	0.
M8III - M9III	0.	0.	0.	0.	0.	0.	0.	0.	0.	0.	1.50E+00	2.80E+01	0.
Carbon Stars	0.	0.	0.	0.	0.	0.	0.	0.	0.	0.	1.50E+00	7.00E+01	0.
WD A	0.	0.	0.	0.	0.	0.	0.	0.	0.	0.	9.00E-01	3.00E-03	0.
Globulars - M15 + M92	0.	0.	0.	0.	0.	0.	0.	0.	0.	0.	1.00E+05	1.00E+05	0.
Globulars - M3 + M5	0.	0.	0.	0.	0.	0.	0.	0.	0.	0.	1.00E+05	1.00E+05	0.
Globular - NGC 6356	0.	0.	0.	0.	0.	0.	0.	0.	0.	0.	1.00E+05	1.00E+05	0.
Observed Integrated Spectrum			66.68	73.55	72.50	60.47	38.24	22.76	3.05	3.35	←Sandage et al. (1969)		
Computed Integrated Spectrum			68.62	73.91	73.69	63.11	37.01	21.87	3.08	3.38			
Percentage Residuals			2.91	.49	1.64	4.36	-3.22	-3.92	Colors in Magnitudes				

Line Index	M31 Observed		Computed Model	
	R	W	R	W
1. Ultraviolet CN (4040) / (3880)	2.58	.37	2.56	.37
2. H Delta (4040) / (4100)	1.00	.00	.97	-.03
3. Blue CN ((4040) + (4500)) / (2*(4200))	1.17	.20	1.15	.18
4. Calcium I ((4040) + (4500)) / (2*(4227))	1.16	.19	1.16	.19
5. CH ((4040) + (4500)) / (2*(4300))	1.34	.29	1.37	.31
6. H Gamma ((4040) + (4500)) / (2*(4340))	.94	-.01	.96	.01
7. MG I + MG H ((5000) + (5300)) / (2*(5175))	1.19	.28	1.14	.25
8. D Lines (5864) / (5892)	1.15	.13	1.12	.10
9. Titanium Oxide (6110) / (6180)	1.09	.05	1.06	.03
10. Calcium Hydride ((6110) + (6620)) / (2*(6386))	.99	.02	.99	.02
11. H Alpha ((6110) + (6620)) / (2*(6564))	1.22	.18	1.22	.18
12. Titanium Oxide ((7000) + (7400)) / (2*(7100))	.91	.11	.86	.06
13. Sodium I ((7980) + (8400)) / (2*(8190))	1.01	.01	1.00	-.00
14. Calcium II ((8400) + (8800)) / (2*(8662))	1.12	.10	1.10	.09
15. Titanium Oxide (8800) / (8900)	1.01	.03	1.00	.02
16. Infrared CN ((8800) + (10300)) / (2*(9200))	.92	.05	.88	-.00

and M dwarfs to the M31 nuclear model, we compare the intensity residuals of the best model (fig. 6*a*) and a giant-dominated model with a normal luminosity function at M_v's fainter than the Sun, whose overall colors match the M31 data reasonably well. Figure 8 *c* and *d* shows the comparison; clearly the luminosity-sensitive magnesium and sodium lines are much more poorly fitted in the giant-rich model; comparisons between residual diagrams 6*a* and 8 show that the $\mathfrak{M}/L$ of the best fit is "desirably" large, while

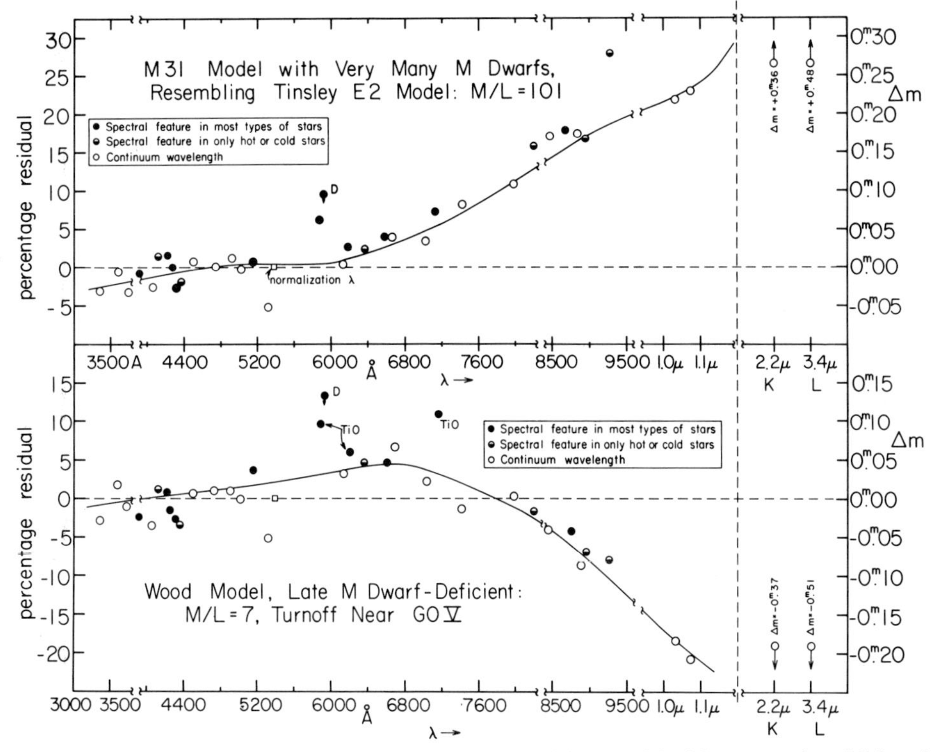

FIG. 7.—Another residual diagram for poorer models. (*a*) A model with so many late M dwarfs that $\mathfrak{M}/L = 101$. This is ruled out—it is far too red at $\lambda > 7000$ Å. (*b*) A model with too few late M dwarfs ($\mathfrak{M}/L = 7$); it is far too blue at $\lambda > 8000$ Å.

the TiO deficit discussed earlier is shown in each model. Another way to show the need for K and M dwarfs is the interplay of the red colors ($\lambda > 8000$ Å) and the strength of the infrared Na I doublet at λ8190, as discussed by Spinrad (1966). See his illustration of Na I λ8190 index in cool stars (fig. 4 of Spinrad 1966 and fig. 4 here). Table 3 lists some representative values of *w*(Na I 8190),[1] I(λλ8000, 10700), and $V - K$ color (in magnitudes). M giants produce a weak (or ***apparently*** negative) *w*(Na I) blocking fraction with good color match to M31, K giants are too blue and are "neutral" in producing *w*(Na) changes, while K and M dwarfs neatly do the trick. The M dwarfs must run as late as M8 V. We also list in tabular form (table 4) the difference between normal-abundance giant model line strengths and SMR giants—again compared with line *w*'s at the center

[1] With the notation of Spinrad and Taylor (1969).

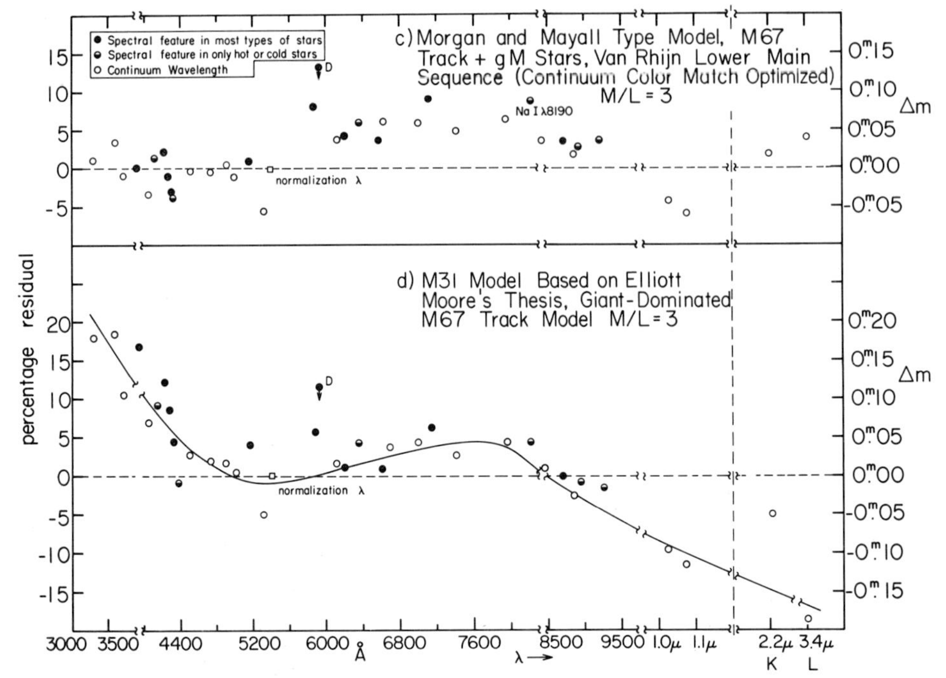

FIG. 8.—M31 model residual diagrams for the following: (*c*) A giant-light dominated model, patterned after Morgan and Mayall (1957). This has a reasonably good continuum fit, but has rather weaker lines than demanded. Note the poor residual at the infrared sodium doublet, λ8190. (*d*) Another one patterned after Moore's (1968) thesis on E galaxy synthesis—again with a large residual at D and a rather poor continuum match.

TABLE 3

w(Na) AND INFRARED COLORS FOR COLD STARS AND FOR THE CENTER OF M31

Star	Type	*w*(Na I)	*I*(8000)	*I*(10700)	*V−K* (mag)
HR 6806	K2 V	−0.00	51.2	13.1	2.1:
61 Cyg B	K7 V	+0.03	104.4	29.9	+3.30
61 Cyg A	K5 V	+0.01	77.1	21.1	2.83
Yale 4009	M3 V	+0.04	168.6	51.7	4.5:
μ Leo	SMR, K2 III	+0.01	65.6	17.6	2.66
ν Peg	K4 III	+0.00	91.2	27.6	3.2:
β And	M0 III	+0.00	127.2	39.5	3.88
83 UMa	M2 III	−0.01	158.8	50.1	4.25
ρ Per	M4 III	−0.14	307.4	105.0	5.32
M31	Corrected for galactic reddening, $E=0.11$ mag	+0.01	73.6	22.8	3.05

of M31. The M31 lines are even stronger than the same temperature models with SMR stars included.

Another different sort of stellar mix is indicated by presently rather crude ST models for the center of the large Sc galaxy M51 (NGC 5194). Here we have a main sequence extending up to about B5 V, or B8 V, as indicated by the ultraviolet continuum and the emission-corrected Hδ and Hγ lines (see fig. 9). The lower main sequence is probably present, but its luminosity function cannot be well determined until accurate infrared scans are obtained. However, cool stars are clearly needed, as the λ7054 (head) TiO band is rather strong in M51. The nucleus of the system may have an H-R diagram like

TABLE 4

M31 MODELS AND ABUNDANCES

Feature	$w(F)$, M31	$w(F)$, Normal Giant Model	$w(F)$, SMR	Further Abundance Factor Needed to Match M31
UV CN	+0.37	+0.36	+0.37	....
Bl CN	0.20	0.17	0.18	1.2 N
Ca I λ4227	0.19	0.17	0.19	....
Mg and MgH	0.27	0.23	0.25	1.1×Mg
Na I D	0.12*	0.08	0.10	1.4×Na
TiO λ6180	0.05	0.03	0.03	~2×Ti
TiO λ7054	0.11	0.06	0.06	~3×Ti
IR CN	+0.05	−0.01	0.00	~3×N

* Corrected for 0.3 Å interstellar absorption in our Galaxy.

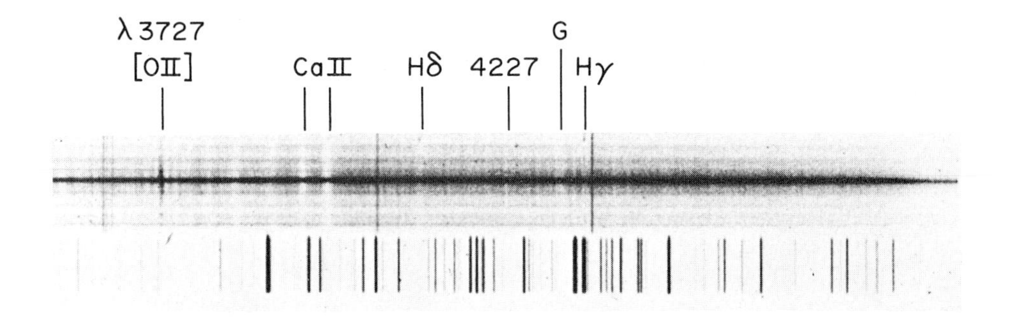

FIG. 9.—A slit spectrogram of the center of NGC 5194 (main M51 spiral) taken by Dr. E. M. Burbidge at the Lick 120-inch prime focus. Note the composite nature of the spectrum; in particular, modestly strong λ4227 of Ca I, the G-band of CH, plus strong Balmer lines of H (Hγ partly filled by an emission component).

the Pleiades or M11; star formation has continued to almost the present epoch. In § 2 of this review we discuss the probable nitrogen-enrichment of the nuclear interstellar gas in M51.

Saslaw and Spinrad have obtained partially complete scans in the blue for a number of relatively nearby Sc galaxies; as expected, their nuclei are quite blue and $H\delta$ is usually a strong absorption feature. The qualitative interpretation is a stellar mix dominated by young stars (perhaps like the Pleiades). The old stellar constituents (which could be either an old disk population, like the solar neighborhood, or even a mildly metal-poor group) have not been detected in these scans; the requisite data must be obtained in the infrared.

It will be an important future task to obtain scans or slit spectra of Sc nuclei at $\lambda >$ 7000 Å.

There are several "second order" parameters that may be derived from the ST population model for M31. Three of them are: (1) the fractional mass of gas generated by star deaths; (2) galaxy luminosity evolutionary corrections for cosmology; and (3) the rate of stellar collisions in the 1–3 pc core of the galaxy. Here we wish to compute the change in magnitude, at V or in $M_{\rm bol}$, of a galaxy as a function of time, as we look back to the past by studying systems at increasing redshift. The systems observed by Sandage (1968) and Baum (1962) in the great distant clusters are all giant E and S0 galaxies; nevertheless, our M31 synthesis is very relevant because the *central* parts of large Sa and Sb spirals are both dynamically and spectroscopically a very close match to giant E galaxies.

The mass loss into the interstellar medium through star deaths is not discussed physically, but we assume that each giant star goes to a 0.6 $\mathfrak{M}_\odot$ white dwarf with the mass difference staying as interstellar gas. The ratio of $\mathfrak{M}_{\rm gas}/\mathfrak{M}_{\rm stars}$, effectively $\mathfrak{M}_{\rm gas}/\mathfrak{M}_{\rm total}$, for our E and Sa–Sb systems depends upon the slope of the stellar luminosity function (see fig. 10). Many luminous stars, or even a Hyades cluster or van Rhijn luminosity function, will lead to a large ratio of $\mathfrak{M}_{\rm gas}/\mathfrak{M}_{\rm stars}$ (where "large" means of the order 10^{-1}), whereas the Salpeter original luminosity function yields a ratio of 0.4. The result for the best ST nuclear M31 model is $\mathfrak{M}_{\rm gas}/\mathfrak{M}_{\rm total} = 10^{-3}$ (uncertain by a factor 2 or 3); thus we predict rather little observable gas in the center of M31 and, in analogy, little H I in large E galaxies. The observed ratio of $\mathfrak{M}(\text{H I})/\mathfrak{M}_{\rm total}$ in NGC 4472 (Robinson and Koehler 1965) is about 2×10^{-4} (see Roberts's paper, chapter 9 of this volume).

The second calculation, the dL/dt determination, really should be made with data from a model computed for the *integrated total light* of a large E or S0 galaxy. This would be appropriate for the bright systems which populate rich, remote clusters of galaxies to which the magnitude-redshift relation observations actually refer (cf. Sandage 1968). The ST (1971) nuclear models are approximately correct here, but a more exact solution involves a study of the stellar content variation with radial distance in a giant E galaxy. Such a program, rather crude in its present state, has been begun by Spinrad, Gunn, Taylor, McClure, and Young (1971). These authors, following earlier work by McClure (1969), studied variations in radial line strengths in Sb and E galaxies. In all galaxies observed, the CN band strengths (usually giant-dominated and very abundance-sensitive in gK stars) rapidly decrease outward from the nucleus, while Mg I and Na I (dwarf-dominated features, somewhat abundance-sensitive) decrease slowly outward in radial distance. In both M31 and the luminous Virgo E galaxy NGC 4472, the stellar content 200–500 pc from the nucleus may be described as a normal (solar) abundance, very dwarf-enriched, old open cluster type. The $\mathfrak{M}/L$ ratio may be slightly greater here than at the nucleus. At $r \sim 1000$ pc the metal abundance is lower, perhaps one-half to one-fifth solar, and the K dwarfs remain important. Unfortunately, the present data do not extend to large enough radial distances to tell us much about the luminosity function in the halo of M31 or NGC 4472. Is there much of a V light contribution by extreme

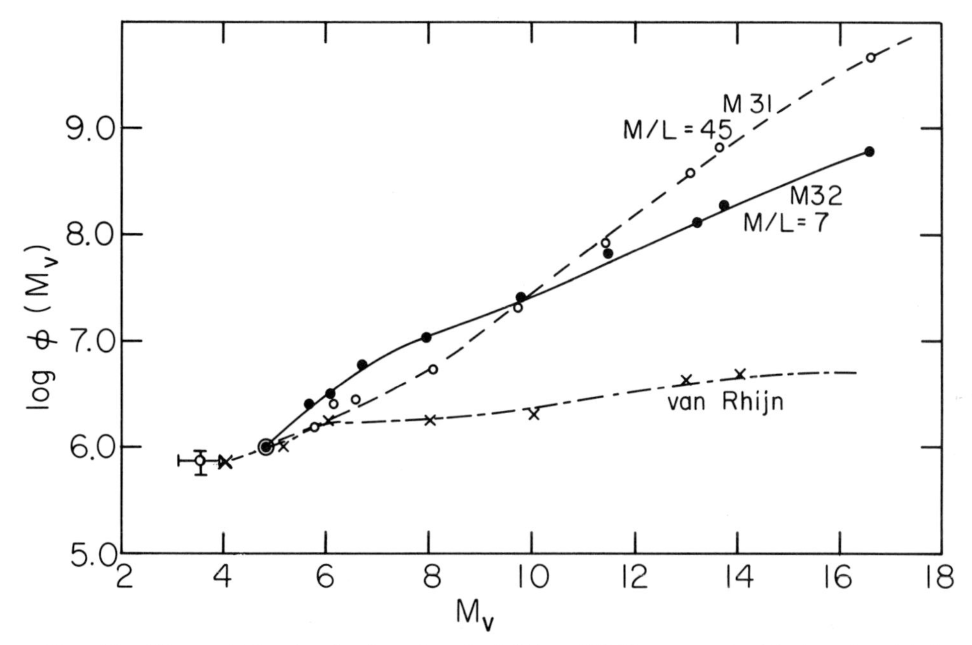

FIG. 10.—The model luminosity functions for M31 and M32 compared with the solar vicinity van Rhijn luminosity function at the faint end (M_v below 4). The M31 symbol with double error bars is an extrapolation of the giant branch in M31 back to the star's original main-sequence position.

Population II stars here? In the M31 halo Baum (1959) and van den Bergh (1969) suggest an *old Population I* model, not an M3 halo prototype mix, from the count/brightness ratios some 5–10 kpc from the center.

In any case, we are basing the present evolutionary change computation upon the M31 disk model—favored by Spinrad *et al.* (1971)—slightly differing from the nuclear model of ST. Tables 5 and 6 indicate the type of computation needed, based upon stellar evolutionary tracks following M67 and NGC 188 in the C-M diagram. An interpolated evolutionary track with a time interval of 3 billion years was used to predict the past luminosities of the old stellar content of the galaxy model. We suggest a change of *less than* some $+0.10$ mag in V over 3×10^9 years, the galaxy actually being very slightly *fainter* in the past. (The last column of table 6 shows it to be 94% as bright then, so ΔM_v is $+0.06$ mag.) The nuclear model correction (ST 1971) was -0.08 mag, the galaxy

TABLE 5

COMPARISON OF THE M31 NUCLEAR POPULATION WITHIN THE M31 DISK, 200 PARSECS SOUTH

STELLAR GROUP	PERCENTAGE *V* LIGHT CONTRIBUTED AT	
	M31 Nucleus	M31, 200 pc S
F8 V	1.0	0
G0–G4 V	6.4	9.8
G5–K0 V	4.3	9.8
K1–K2 V	4.3	9.8
K3–K4 V	3.2	7.4
K5–K7 V	1.5	2.5
M0–M2 V	1.2	2.0
M3–M4 V	1.2	2.0
M5–M6 V	1.2	1.3
M7 V	1.0	0.8
M8 V	0.:5	0.5
G0 IV	11.8	9.8
G5 IV	9.6	9.8
K0 IV	13.9 (SMR)	16.4 (Normal)
K1–K3 III–IV	34.1 (SMR)	14.8 (Normal)
K4–K5 III	5.0 (SMR)	3.3 (Normal)

TABLE 6

AN EVOLUTIONARY CORRECTION MODEL FOR THE M31 DISK POPULATION GROUP AS EXPECTED 3×10^9 YEARS AGO

Group Name	Present % *V*	Linear Brightness Level Ratio -3×10^9 yrs	Relative Number, Ratio*	% *V*,† -3×10^9 Years
F8 V	0.1	0.7×present	200.0	14.0
G0 V–G4 V	9.8	1.0	1.0	9.8
G5–K0 V	9.8	1.0	1.0	9.8
K1–K2 V	9.8	1.0	1.0	9.8
K3–K4 V	7.4	1.0	1.0	7.4
K5–K7 V	2.5	1.0	1.0	2.5
M0–M2 V	2.0	1.0	1.0	2.0
M3–M4 V	2.0	1.0	1.0	2.0
M5–M6 V	1.3	1.0	1.0	1.3
M7 V	0.8	1.0	1.0	0.8
M8 V	0.5	1.0	1.0	0.5
G0 IV	9.8	1.58×present	0.5	7.9
G5 IV	9.8	1.45×present	0.45	6.5
K0 IV	16.4	1.58×present	0.4	10.4
K1–K3 III–IV	14.8	1.32×present	0.4	7.8
K4–K5 III	3.3	1.32×present	0.3	1.3

* In sense, Number 3×10^9 years ago/present number, using disk model luminosity function and extrapolation.

† % *V* = present % *V* × Brightness ratio × number ratio.

nucleus being brighter 3 billion years ago. In summary, a compromise change of zero would be believable and probably within the estimated error-bar in each determination. So, let us suggest $dL/dt = \Delta M_v = 0.0$ in 3×10^9 years, but clearly the last word in this sort of calculation has not yet been uttered.

The need for an accurate estimate of the evolutionary correction has been outlined by Sandage (1961, 1968), Minkowski (1962), by Whitford in chapter 5 of this volume, and also by Chambers and Roeder (1969) and Tinsley (1970). We hope to use the Hubble diagram (M_{bol} versus z) to decide between different cosmological models. If evolutionary effects are really as negligible as our models with steep luminosity functions predict, the observed linearity in the relationship between galaxy bolometric magnitude and the logarithm of the redshift would indicate an evolving model universe with a deceleration constant, $q_0 \simeq +1$. With a Hubble constant $H_0 = 75$ km s^{-1} Mpc^{-1}, this implies a mean cosmological density excess of some 70 times, and an "age" of the present expansion of 8×10^9 years compared to an uncertain 12–15 $\times 10^9$ years for the oldest objects in our Galaxy. To push q_0 down toward zero and eliminate the apparent density discordance, an evolutionary decrease in M_{bol} of at least 0.4 mag would be needed; and this would seem unlikely. Of course, other parameter changes, such as a smaller H_0 or possible variations in the K-correction (see chap. 5 by Whitford), would also modify the final q_0 and ρ_0 values derived.

The third computation, the rate of star collisions, is discussed in ST (1971); suffice it to mention here that collisions do occur in the inner 1–3 pc of M31 on this model, and their overall rate is about one per 10^2 years. However, an *individual* M dwarf can survive without a direct encounter for more than 10^{11} years, so that the concept of an M-dwarf-heavy luminosity function at present can be maintained.

1.5. Synthesis of Other Types of Galaxies

Not all galaxies studied recently have been interpreted as consisting of old, normal-abundance stellar groups. The peculiar irregular galaxy M82 is being studied by R. O'Connell (1970). After a large correction for reddening, he has found the stars in the M82 *nuclear* region to resemble somewhat the H-R diagram of the young cluster h and χ Persei. There are B dwarfs and supergiants and K and M supergiants across the Hertzsprung gap. A van Rhijn lower main sequence is adequate here, although not well determined at $M_v > +10$. Away from the M82 nucleus O'Connell finds high-luminosity stars to be very scarce.

The dwarf E galaxy NGC 205 presents another sort of stellar group: it appears to be metal-poor overall, rather like the halo globular clusters M3 and M5. It is distinctly weaker in metallic lines than is M32 (McClure and van den Bergh 1968). Such a result for NGC 205 is not unexpected, as the count/brightness ratio data of Baum and Schwarzschild (1954) and Baum (1959) had pointed in the Population II direction. However, the Lick scans show rather stronger Hγ and Hδ than do galactic halo globular clusters, so some young stars may be present throughout NGC 205; the detection of a few B supergiants in NGC 205 had been previously noted by Baade (1951). Unfortunately, deep red and infrared scanner data which might give us information on K and M stars in NGC 205 (which might or might not have TiO, depending on their abundances) are still unavailable. They would seem to be an important desideratum.

1.6. Comments on the Stellar Content of Some Spiral Galaxies

Although the detailed stellar synthesis described in §§ 1.4 and 1.5 has not been attempted for many galaxies, one may get some inference on galaxy content from modern spectroscopic observations originally gathered with a different intent. One example is the extensive series of papers on (mostly) Sc galaxies by the Burbidges (and see chap. 3 in this volume). We have organized the published $\mathfrak{M}/L$ ratios for their galaxies with reasonably good rotation-curve masses into two groups: one consists of galaxies with Na I D lines strong enough to measure at low dispersion, and the other consists of galaxies without measurable D lines. In most instances, strong D will mean a large K dwarf contribution (and perhaps a high Na/H ratio), so a relatively large stellar $\mathfrak{M}/L$ might be expected unless some of the W_λ of the D lines is interstellar. Such an alternative seems unlikely in most dust-free galaxies, but of course it *might* be important in dusty systems like M82 and NGC 3227 (Rubin and Ford 1968; and the recent work by DeMoulin 1969). The average mass of the (11) *strong* D-line galaxies is $\langle\mathfrak{M}\rangle = 12 \times 10^{10}\, \mathfrak{M}_\odot$ with $\langle\mathfrak{M}/L\rangle = 6.0$; the (18) *weak* D-line galaxies have $\langle\mathfrak{M}\rangle = 3.2 \times 10^{10}\, \mathfrak{M}_\odot$ and $\langle\mathfrak{M}/L\rangle = 2.0$. These averaged $\mathfrak{M}/L$ ratios are crude but probably meaningfully different. They were derived from rotation-curve masses and thus should not be compared with early-type galaxy nuclei or models. We feel that most massive Sc galaxies are likely to be dwarf-enriched—much as Deutsch and Spinrad suggested for E, Sa, and Sb galaxies. The total mass differentiation of the "D" (strong D lines like dwarfs) and "G" (weak D and Mg lines, like K giants) galaxies, suggested by Spinrad (1962), is probably still partly valid, but the criteria should be made more quantitative.

Although we do not discuss them here, we refer and recommend to the reader a pair of theoretical analytic discussions of the stellar content of spiral and E galaxies by Reddish (1968) and Tinsley (1968).

1.7. Speculative Topics of Interest

Old E galaxies are now a matter-of-fact conclusion from our data and the data of many previous workers. But are there any young E systems? Certainly we know of none with an A or even F-type spectrum. But NGC 3379 and NGC 5866 are definitely bluer than average for relatively large systems. Spinrad and Taylor's preliminary scans of NGC 3379 also show stronger-than-normal Hδ and Hγ absorption, so some stars hotter than G0 V are likely to be present in the 3379 nucleus. They could be F5 main-sequence stars with a young age, or alternatively they might be horizontal-branch F giants. Thus, in the first interpretation the centers could not be primeval, since their ages are one-third to one-tenth the age of the universe. If future evidence leans toward the presence of many horizontal-branch stars in NGC 3379, our interpretation would be different. Such evolved stars would be old, and could arise from either an M3 or M67 type array.

The *a priori* evidence here is that an E0 galaxy which looks very well organized *might* not have only old stars. The *possible* existence of moderately young E galaxies probably would not really be a critical blow either for steady-state or against Friedmann models. However, it will be necessary to pin down the degree to which metal-poor halo stars contribute before the more conservative alternative of horizontal-branch stars may be discarded. This will be a difficult task because the outer regions of NGC 3379 have a very low surface brightness.

Are there any large metal-poor galaxies—overgrown halo globular clusters with masses of $\sim 10^{10}$ $\mathfrak{M}_\odot$? If so, why have the local enrichment processes failed? The low-mass Sculptor-type dwarf spherical systems in the Local Group (cf. van den Bergh 1968) have globular-cluster–like C-M arrays and are apparently very metal deficient (Baade and Swope 1961), NGC 205 is metal-poor by a very uncertain factor of 5–10, and M32 has a slight deficiency. This ordering of metal poorness is also an approximate mass ordering, so one might extrapolate to normal abundance for all galaxies with $\mathfrak{M} \geq 10^{10}$ $\mathfrak{M}_\odot$ (a very simplified approach, but tolerable in our present ignorance). A parallel trend of abundance versus galaxian mass was also noted by Peimbert and Spinrad (1970*c*) for S and Irr galaxies. A possible violation of the trend is NGC 5195, the moderately large and very red companion to NGC 5194. It might be weak-lined according to Morgan and Osterbrock (1969) as mentioned earlier, but the system could also be a heavily reddened Ep with hotter normal stars. Higher-dispersion slit spectra for velocity-dispersion measurement, accurate spectral scans, and some independent measure of the color excess seem requisite for a decision—none of these desiderata are beyond our present capabilities. This case takes on some added importance in light of Arp's (1969) recent suggestion that NGC 5195 may be an ejectum from NGC 5194 (but Toomre and Toomre 1972 offer a more conventional explanation for the M51 pair).

2. EMISSION LINES FROM NORMAL GALAXIES

2.1. Introduction

We now turn from the stellar content to the gaseous content of the medium between the stars. Although the techniques of analysis are rather different, the two subjects are physically related since mass ejected by stars enriches the interstellar gas, while some of that gas probably takes part in star formation again. The chemical abundances of both constituents might therefore be related, and for some elements it is easier to derive the abundances from the emission lines originating in H II regions than from the integrated light of the stellar content; furthermore, the degree and the extent of the ionization of the interstellar material are related to the stellar ultraviolet energy distribution, as well as to the number of stars responsible for the ionization. That is, the absolute emission-line flux provides an excellent clue as to the number of O and B stars present in a given H II region.

The gaseous interstellar content of galaxies has been found to be appreciable. Modern neutral hydrogen 21-cm observations have been summarized by Roberts (1969), and the presence of optical emission lines has been established in a large fraction of galaxies (see, for example, the catalog on radial velocities by HMS 1956). There are two excellent review papers on emission lines in galaxies by Morgan and Osterbrock (1969) and by Osterbrock (1969), and there is another on the nuclei of galaxies by Burbidge (1970).

In § 2.2 we will discuss the main properties of normal galactic H II regions that can be compared with extragalactic H II regions; by "normal" we mean those H II regions ionized by massive young O and B stars. In § 2.3 we will discuss normal H II regions in other galaxies; in § 2.4 we will study H II regions associated with the nuclei of galaxies where in many cases the present rate of star formation is negligible and the ionization might be due either to collisions, to a weak nonthermal radiation field, or to stars of lower luminosity than in the case of normal H II regions; in § 2.5 we will mention some extragalactic H II regions that do not exactly fit into the scheme of normal spiral-arm or nuclear H II regions.

2.2. Normal H II Regions in our Galaxy

Many of the physical processes in H II regions and planetary nebulae are similar and have been described by Seaton (1960), and Aller and Liller (1966). Also there have been several review articles on H II regions in our Galaxy—for example, those by Pottasch (1965), Osterbrock (1967), and H. M. Johnson (1968)—where many references to previous work may be found.

Emission-line intensities are affected by interstellar extinction; therefore, the observed and theoretical Balmer and Paschen decrements are combined with the normal extinction law to derive a reddening correction. It is well known that there are deviations from the normal law (see, for example, H. L. Johnson 1968). The largest deviations occur in the red and infrared regions, and the use of the normal (Whitford) law would produce an underestimation of the total absorption; however, the line intensity ratios in the visual region are in general not affected by deviations from the normal law.

The mass of an H II region can be obtained from the electron density $N_e(3727)$, derived by means of the [O II] $\lambda\lambda 3729/3726$ intensity ratio, and is given by

$$\mathfrak{M}(3727) = m_{\rm H}\left[1 + \frac{4N({\rm He})}{N({\rm H})}\right]\int f N_e(3727)\,dV\,, \tag{1}$$

where $m_{\rm H}$ is the mass of the hydrogen atom, f is the proton-to-electron number ratio, and V is the volume. In equation (1) only the contributions of hydrogen and helium have been considered.

From radio observations or from absolute fluxes of hydrogen emission lines and by assuming a given geometrical shape, it is possible to derive the root mean square density, $N_e(\text{rms})$, and the mass can be obtained from

$$\mathfrak{M}(\text{rms}) = m_{\rm H}\left[1 + \frac{4N({\rm He})}{N({\rm H})}\right]\int f N_e(\text{rms})\,dV\,. \tag{2}$$

If there are no spatial density fluctuations, equations (1) and (2) should yield the same mass. From direct photographs of H II regions, it is clear that they show extreme density fluctuations; consequently equation (1) should, and indeed does, yield higher masses than equation (2). Under the assumption that only a fraction α of the volume is filled with a density $N_e(3727)$ and the rest of it is empty, we have

$$\mathfrak{M}(\alpha) = m_{\rm H}\left[1 + \frac{4N({\rm He})}{N({\rm H})}\right]\int \alpha f N_e(3727)\,dV\,; \tag{3}$$

and since

$$\alpha N_e^2(3727) = N_e^2(\text{rms})\,,$$

then

$$\mathfrak{M}(\alpha) = m_{\rm H}\left[1 + \frac{4N({\rm He})}{N({\rm H})}\right]\int \alpha^{1/2} f N_e(\text{rms})\,dV\,. \tag{3'}$$

$\mathfrak{M}(\alpha)$ is a lower limit to the real mass since a smaller value of $N_e(3727)$ would be in contradiction with the $\lambda\lambda 3726/3729$ observations; alternatively $\mathfrak{M}(\text{rms})$ is an upper limit to the real mass since a higher value for $N_e(\text{rms})$ would violate the absolute flux observations; therefore the real mass should lie between $\mathfrak{M}(\alpha)$ and $\mathfrak{M}(\text{rms})$ since density fluctuations are not as extreme as assumed in equation (3). Most mass determinations are based on expressions of the type of equation (2), and a few are based on expressions similar to equation (1) and are therefore upper limits to the real mass.

Several methods are used to determine electron temperatures. In the best observed H II regions, different methods yield different values of the temperature. This fact has been interpreted as being due to spatial temperature fluctuations (Peimbert 1967*a*). These temperature fluctuations are produced by both density and ionization fluctuations. The temperatures derived by the different methods are weighted means over the observed volume. In particular the temperatures determined from the ratio of two forbidden lines that originate at different levels weigh preferentially the regions of higher electron temperature since the forbidden-line emission increases with temperature, and consequently the resulting temperature is higher than average. On the other hand, the temperatures determined from the ratio of the Balmer continuum to a Balmer line favor the low-temperature regions since the Balmer emission increases when the temperature decreases, and consequently this method yields temperatures lower than average.

Once the temperature is known, it is possible to determine chemical abundances from emission-line ratios. From the helium and hydrogen recombination coefficients computed by Robbins (1968) and Pengelly (1964), respectively, and assuming a constant temperature over the observed volumes, we have

$$\frac{N(\mathrm{He}^+)}{N(\mathrm{H}^+)} = 9.8 \times 10^{-2} T_e^{0.22} \frac{I(5876)}{I(\mathrm{H}\beta)} = 6.5 \times 10^{-1} T_e^{0.13} \frac{I(5876)}{I(\mathrm{H}\alpha)}\,; \qquad (4)$$

this ratio does not depend sensitively on the temperature or on spatial temperature fluctuations; an error of 1000° K in the electron temperature yields an error in the abundance ratio of 2 percent if Hβ is used or of 1 percent if Hα is used.

Seaton (1968) has proposed not only that the permitted lines of heavy elements observed in gaseous nebulae are produced by recombination but also that the excitation can be produced by absorption of stellar radiation in spectral lines (case C of Aller, Baker, and Menzel 1939). Due to the latter possibility and to the weakness of heavy-element permitted lines, it is very difficult to determine chemical abundances from these lines.

Nevertheless, chemical abundances can also be determined from nebular forbidden lines. The relative abundances of the number of atoms in the pth state of ionization, $N(X^{+p})$, to the number of ionized hydrogen atoms are given by

$$\frac{N(X^{+p})}{N(\mathrm{H}^+)} = C T_e^{-0.327} \exp\,(\Delta E/KT_e)\,\frac{I(X^{+p}, \lambda mn)}{I(\mathrm{H}\beta)}\,, \qquad (5)$$

where it has been assumed that collisional de-excitation of the upper level is negligible and that there are no spatial temperature fluctuations; C is a constant that depends on atomic parameters only, and ΔE is the energy difference between the ground and excited levels. Type (5) equations are very sensitive to the electron temperature; thus a small error in temperature produces a large error in the chemical abundances. Moreover, if spatial temperature fluctuations are not taken into account, the chemical abundances derived from forbidden lines might be underestimated by factors as large as 3.

The intensity ratio given by equation (5) depends on the abundance of the element relative to hydrogen and also on the electron temperature. Unfortunately these two parameters are related to each other in the sense that any underabundance decreases the cooling efficiency and increases the electron temperature; thus the intensity ratios do not vary significantly. In fact, gaseous nebulae of very different chemical composition have similar ratios of nebular to recombination-line intensities. Therefore to derive the chemical abundance it is necessary to know the electron temperature accurately.

To obtain the total abundance of a given element, it is necessary to correct for the amount present in ionization stages that are not observed. The observed ionic abundances of each element are thus multiplied by ionization correction factors larger than unity which take into account the unobserved stages. For example, it is well known that there is practically no helium twice ionized in normal H II regions since very low upper

TABLE 7

CHEMICAL ABUNDANCES*

Object	H	He	C	N	O	Ne	S	Source
Orion	12.00	11.02	8.71	7.63	8.79	7.86	7.50	1
M8	12.00	11.01	....	7.50	8.69	....	7.14	1
M17	12.00	11.00	....	7.43	8.84	....	7.76	1
Sun	12.00	....	8.72	7.89	8.96	....	7.30	2
Sun	12.00	....	8.55	7.93	8.77	....	7.21	3,4

* Given in Log N.

SOURCE.—(1) Peimbert and Costero 1969; (2) Goldberg, Müller, and Aller 1960; (3) Lambert 1968; (4) Lambert and Warner 1968.

limits have been established with radio and optical methods (Palmer, Zuckerman, Penfield, Lilley, and Mezger 1969; Peimbert 1970 unpublished); however, it has been found that the $N(He^+)/N(H^+)$ ratio changes from point to point in the Orion Nebula, thus implying the presence of considerable amounts of *neutral helium* within H II regions (Peimbert and Costero 1969). Therefore, the total helium-to-hydrogen abundance ratio is given by

$$\frac{N(\mathrm{He})}{N(\mathrm{H})} = i(\mathrm{He}) \frac{N(\mathrm{He}^+)}{N(\mathrm{H}^+)} , \tag{6}$$

where i(He) is the ionization correction factor and can be obtained from the abundance ratios of different ions of a given element, like sulfur or oxygen. From these ratios the degree of ionization of the nebula is also determined. Chemical abundance determinations have been carried out by Aller and Liller (1959), Méndez (1968), Rubin (1969), and Peimbert and Costero (1969).

In table 7 we compare the chemical abundances of Orion, M8, and M17 with the solar abundances derived from the solar atmosphere by Goldberg, Müller, and Aller (1960) and by Lambert (1968). It can be seen that the abundances are similar. The solar helium abundance has been obtained by two different methods; from theoretical evolutionary models (Bahcall, Bahcall, and Shaviv 1968; Torres-Peimbert, Simpson, and Ulrich

1969) and by combining cosmic-ray data (Fichtel and McDonald 1967) with photospheric observations (Lambert 1967); the helium-to-hydrogen abundance ratio by number determined by these two methods ranges from 0.06 to 0.08. The values are smaller than 0.10–0.11 which are the values derived from galactic H II regions.

2.3. Normal H II Regions in Other Galaxies

There have been many investigations using direct photography to detect H II regions in other galaxies, based on the comparison of plates taken with different filters or with the use of an objective prism (Haro 1950; Shajn and Haze 1951, 1952*a*, 1952*b*); this technique was applied by Henize (1956) to the Magellanic Clouds, to M31 by Baade and Arp (1964), to M33 by Courtès and Cruvellier (1965), and more recently by Hodge (1966, 1967, 1969*a*, *b*, *c*) to a large group of galaxies. From these studies it was found that in spiral galaxies the H II regions are in general concentrated along the optical spiral arms. From the large-scale distribution of ionized gas in spiral galaxies it has been found that in general there is a peak in the H II region space density at approximately one-fourth of the distance from the center to the outermost H II region (Hodge 1969*b*); furthermore in Sc galaxies it has been found that the ionized gas is more centrally located than the neutral gas and that the density peak of the neutral gas is exterior to the main optical spiral structure (Roberts 1968; Hodge 1969*b*).

From the observed fluxes, and assuming a geometrical shape, Dickel (1965) has derived the rms masses for several H II region complexes in the Magellanic Clouds; Peimbert and Spinrad (1970*a*, *c*) made crude estimates on the $\mathfrak{M}$(rms) of the H II region complexes in NGC 5461 and NGC 5471, as well as of a large H II region in NGC 4449 and an H II region in NGC 6822: typical values of the masses derived from these studies are in the $10^5\,\mathfrak{M}_\odot$ range. Feast (1964) determined the filling factor, α, for 19 objects in the Large Magellanic Cloud and found an average value of $\alpha = 0.14$; this value is larger than those derived by Osterbrock and Flather (1959) and Peimbert (1966) for the Orion Nebula by factors of 4 and 7, respectively. Aller, Czyzak, and Walker (1968) found $\mathfrak{M}(\text{rms}) \sim 2 \times 10^6\,\mathfrak{M}_\odot$ for NGC 604, while Johnson (1959) and Faulkner (1963) estimated the mass of 30 Doradus to be in the neighborhood of $10^6\,\mathfrak{M}_\odot$. The extragalactic H II regions for which some data have been gathered are, in general, the largest in nearby galaxies and considerably bigger than the H II regions in the solar neighborhood; they might only be comparable to W49 in our Galaxy (Mezger, Schraml, and Terzian 1967; Mezger and Henderson 1967).

Burbidge and Burbidge (1962), for their study of internal motions and rotations of galaxies, obtained a large number of spectrograms in the Hα region and, as a by-product, the Hα/[N II] line intensity ratios in a large number of objects; they found that H II regions in spiral arms and in irregular galaxies show Hα/[N II] ratios similar to those found in normal galactic H II regions.

Considerable work has been devoted to derive line intensities from large normal H II regions in nearby galaxies: we note Aller (1942), Page (1952), Johnson (1959), Feast (1961), Mathis (1962, 1965*a*, *b*), Aller and Faulkner (1962), Faulkner and Aller (1965), Aller, Czyzak, and Walker (1968), Wares and Aller (1968), and Peimbert and Spinrad (1970*a*, *c*). We will compare emission-line intensity ratios of galactic H II regions, given in table 8 (from Peimbert and Costero 1969), with some of the best photoelectric ob-

TABLE 8

LINE INTENSITIES OF GALACTIC H II REGIONS*

Wavelength (Å)	Identification	Ori I	Ori II	Ori III	Ori IV	M8 I	M8 II	M17 I	M17 II	M17 III	M17 IV
3727	[O II]	+0.02	+0.05	+0.33	+0.21	+0.29	+0.35	+0.05	+0.08	+0.09	+0.43
4102	Hδ	−0.58	−0.59	−0.59	−0.62	−0.57	−0.59	−0.59	−0.62	...	−0.58
4340	Hγ	−0.33	−0.34	−0.31	−0.34	−0.32	−0.29	−0.33	−0.34	−0.33	−0.32
4363	[O III]	−1.92	−1.89	−2.28	...	−2.43	...	−1.87	...	...	...
4861	Hβ	0.00	0.00	0.00	0.00	0.00	0.00	0.00	0.00	0.00	0.00
5007	[O III]	+0.58	+0.51	+0.11	−0.80	+0.20	+0.28	+0.53	+0.54	+0.64	+0.40
5876	He I	−0.87	−0.89	−0.99	<−1.88†	−0.96	−0.90	−0.88	−0.90	−0.76	−0.87
6300	[O I]	−2.03	−2.07	−1.93	...	...	...	...	...	...	...
6563	Hα	+0.43	+0.42	+0.42	+0.43	+0.46	+0.47	+0.45	+0.44	+0.45	+0.46
6584	[N II]	−0.41	−0.31	−0.09	+0.05	−0.16	−0.07	−0.60	−0.59	−0.66	−0.14
6716, 6731	[S II]	−1.21	−1.25	−0.90	...	−0.86	...	−1.10	...	...	−0.52

* Given in log $I(\lambda)/I(H\beta)$.

† Upper limit since the line was not detected.

servations of extragalactic H II regions which are given in table 9. The extragalactic measurements have been corrected for reddening, using the Balmer lines, case B, and the normal extinction law; the reddening derived is larger than expected from the local galactic reddening alone. This implies that part of the reddening is produced within the observed external galaxy. As we mentioned above, the observations refer to extragalactic H II regions more luminous than those of the solar neighborhood; however, since they are farther away, their observed fluxes are weaker; furthermore, the background stellar continuum superposed on the emission lines makes the measurement of weak emission lines more difficult than for H II regions in our Galaxy. Therefore, the comparison is restricted to forbidden nebular lines, permitted lines of hydrogen and helium, and the auroral line λ4363 of [O III]; the auroral lines and recombination lines of other elements are very weak and have been measured in only a few cases.

TABLE 9

EXTRAGALACTIC LINE INTENSITIES

Wave-length (Å)	NGC 604	NGC 604	NGC 604	NGC 2070	NGC 4449	NGC 5461	NGC 5471	NGC 6822
3727	+0.32	+0.44	+0.42	+0.15	+0.47	+0.29	+0.09	+0.31
4102	−0.55	−0.56	...	−0.55	...	−0.56	−0.56	−0.55
4340	−0.32	−0.29	−0.31	−0.32	−0.27	−0.32	−0.34	−0.30
4363	...	...	−1.97	−1.38	...	...	...	−1.36
4861	0.00	0.00	0.00	0.00	0.00	0.00	0.00	0.00
5007	+0.28	+0.28	+0.32	+0.68	+0.57	+0.51	+0.74	+0.73
5876	...	−0.88	...	...	−0.96	−0.94	−0.89	−0.90
6563	...	+0.46	...	...	+0.47	+0.47	+0.43	+0.50
6584	...	−0.40	...	...	−0.79	−0.42	−1.09	−1.00
6716, 6731	...	...	...	...	...	...	...	−0.73
$C(H\beta)$	0.30	0.22	...	0.41	0.20	0.20	0.20	0.50
Source	1	2	3	4	2	2	2	5

SOURCE.—(1) Aller, Czyzak, and Walker 1968; (2) Peimbert and Spinrad 1970*a*; (3) Peimbert 1970; (4) Faulkner and Aller 1965; (5) Peimbert and Spinrad 1970*c*.

Peimbert (1970) and Peimbert and Spinrad (1970*c*) have compared NGC 604 with Ori III and NGC 6822 with Ori I and M17 I, respectively, and have found that the oxygen abundances in NGC 604 and in Orion are similar, while NGC 6822 V is underabundant in nitrogen and oxygen by factors of 6 and 1.7 with respect to Orion and M17.

From the observations of Faulkner and Aller (1965) and the computations of Eissner, Martins, Nussbaumer, Saraph, and Seaton (1969) an electron temperature T_e of 11,400° K is found from the [O III] lines for NGC 2070; and from an analysis similar to that of Peimbert and Spinrad (1970*c*) it is found that NGC 2070 is underabundant in the oxygen-to-hydrogen ratio with respect to Orion and M17 by a factor of 2. Additional evidence in favor of a higher electron temperature for NGC 2070 than those of galactic H II regions is obtained from radio observations by comparing LTE electron temperatures derived from the H 109α line (Gardner, Milne, Mezger, and Wilson 1970).

The abundance ratio of ionized helium to ionized hydrogen is known with higher accuracy than the $N(O)/N(H)$ ratio since the former ratio is not very sensitive to the temperature. Table 10 contains the $N(He^+)/N(H^+)$ ratios obtained by means of equation (4), using $T_e = 8000°$ K for all entries except NGC 6822 V for which $T_e = 11{,}100°$ K was used. To estimate the ionization correction factor and thus the total $N(He)/N(H)$ ratios given in table 10, Peimbert and Spinrad (1970*b*, *c*) made use of the λλ5007/3727 line intensity ratio. The $N(He)/N(H)$ abundance ratios lie very close to 0.10 and seem to imply that most of the helium is of primeval origin. These values are in the neighborhood of those predicted by Gamow's big-bang theory for the case of no neutrino or antineutrino degeneracy (Peebles 1966; Wagoner, Fowler, and Hoyle 1967). From the values of table 10, NGC 604 and the Small Magellanic Cloud are respectively slightly overabundant and slightly underabundant in helium. The differences are small, however, and it is difficult to ascertain whether they are real or are due to observational errors; if

TABLE 10

EXTRAGALACTIC HELIUM ABUNDANCES

Object	Hubble Type	i_{cf}	$N(He^+)/N(H^+)$	$N(He^0)/N(H^+)$	$N(He)/N(H)$	Observer
NGC 346......	Irr	1.08	0.077	0.006	0.083	1
NGC 604......	Sc	1.33	0.076	0.025	0.101	2
NGC 604......	Sc	1.30	0.132	0.040	0.172	3
NGC 604......	Sc	1.37	0.094	0.035	0.129	4
NGC 2070.....	Irr	1.08	0.081	0.007	0.088	5
NGC 2070.....	Irr	1.08	0.105	0.008	0.113	6
NGC 2070.....	Irr	....	0.17±0.09	....	....	7
NGC 4449.....	Irr	1.25	0.078	0.018	0.096	4
NGC 5461.....	Sc	1.20	0.082	0.016	0.098	4
NGC 5471.....	Sc	1.06	0.092	0.006	0.098	4
NGC 6822.....	Irr	1.08	0.087	0.007	0.094	8
NGC 7679.....	S0	1.32	0.078	0.025	0.103	4
M51..........	Sc	1.25	<0.345*	<0.086*	<0.431*	4
M81..........	Sb	1.38	<0.208*	<0.079*	<0.287*	4
M82..........	?	2.00	0.065	0.065	0.130	9

* Upper limits since the helium lines were not detected.

OBSERVER.—(1) Aller and Faulkner 1962; (2) Mathis 1962; (3) Aller, Czyzak, and Walker 1968; (4) Peimbert and Spinrad 1970*a;* (5) Faulkner and Aller 1965; (6) Mathis 1965*b;* (7) Gardner *et al.* 1970; (8) Peimbert and Spinrad 1970*c;* (9) Peimbert and Spinrad 1970*b*.

these differences are real, they would indicate a different rate of stellar helium enrichment.

The $N(He^+)/N(H^+)$ ratio derived by Gardner *et al.* (1970) is the first ratio determined from radio recombination-line observations in an extragalactic H II region. It will be very important to reduce the observational error to find out if there is a real difference compared to optical values.

With respect to the question of whether or not there is a helium-abundance radial gradient within a given galaxy, we have available quantitative information only for two galaxies, and the answer is negative. In our Galaxy, Peimbert and Costero (1969) found about the same helium abundance in the Orion Nebula, which belongs to the Orion arm, as in M8 and M17 which belong to the Sagittarius arm. Shipman and Strom (1970) have determined the helium abundance of 94 B stars of Population I; they obtain a mean value for $N(He)/N(H) = 0.115$. Their data suggest that this ratio varies by as much as a factor of 1.2 from one OB association to another, but there is no apparent abundance variation associated with distance to the center of the Galaxy. From table 10, both NGC 5461 and NGC 5471, whose projected distances to the center of M101 are 5.3 and 13.4 kpc, respectively (adopting a distance of 4 Mpc between our Galaxy and M101), show the same helium abundance.

Aller (1942, 1956) noticed that in M33 the higher-ionization nebulae tend to be found at greater distances from the galactic center than the lower-ionization nebulae; the same result is reported by Ford and Rubin (1969) for M31; Searle (1971) has made an extensive study of H II regions in several galaxies and has found that there is a gradient in three intensity ratios [N II] λ6584/Hα, Hβ/[O III] λ5007, and [N II] λ6584/[O II] λ3727; these ratios decrease from the center outward. The correlation is very good and extends to the whole galaxy, and it holds for H II regions very close to the nucleus as well as for H II regions many kiloparsecs away from it. The changes in the line intensity ratios might be due to the variation of four physical parameters in the H II regions: electron temperature, chemical composition, degree of ionization, and electron density; Peimbert and Spinrad (1970*b*) also noted that NGC 5471 shows a higher degree of

TABLE 11

MASS, DENSITY, AND DENSITY FLUCTUATIONS OF IONIZED HYDROGEN

Object	N_e(3727) (cm^{-3})	N_e(rms) (cm^{-3})	α	$\mathfrak{M}$(3727) ($\mathfrak{M}_\odot$)	$\mathfrak{M}$(rms) ($\mathfrak{M}_\odot$)	$\mathfrak{M}(\alpha)$ ($\mathfrak{M}_\odot$)	r (pc)
M51...	500	2.14	1.83×10^{-5}	1.85×10^{8}	7.9×10^{5}	3.39×10^{3}	153
M81...	1000	16.9	2.86×10^{-4}	5.8×10^{6}	9.8×10^{4}	1.66×10^{3}	38
NGC 4278.	....	8.7	....	....	5.6×10^{5}	....	85

ionization than NGC 5461. The degree of ionization of many H II regions in our Galaxy is known, and there is no ionization gradient present, but the observations are confined to a very small range in distances; furthermore, the sizes of the observed H II regions are in general considerably smaller than those of extragalactic H II regions. We feel that this subject is an area to be profitably explored in the near future.

2.4. NUCLEAR H II REGIONS

From the first spectrograms of galaxies, it was found that λ3727 appeared in emission in a substantial fraction of them. Mayall (1958) investigated the occurrence of λ3727 in the different spectral types and summarized the statistics up to 1956. Osterbrock (1960) and later Miller (1969) showed that with spectrograms taken for the specific purpose of detecting λ3727, the percentage of elliptical galaxies with λ3727 in emission becomes considerably higher than the value derived by Mayall.

Minkowski and Osterbrock (1959) studied in detail the λ3727 line profile in NGC 1052 and found that it corresponds to $N_e(3727) < 300$ cm^{-3}. Münch (1959) derived $N_e(3727) \simeq 1000$ for M81, and Boesgaard (1967) obtained $N_e(3727) \simeq 500$ for M51.

In table 11 we have the masses estimated by Peimbert (1968) and Osterbrock (1960) for M51, M81, and NGC 4278, based on the observed λλ3726/3729 line intensity ratios and the Hα intensities. The masses of the H II regions are considerably smaller than the stellar masses comprising the nucleus of each object. The gas of the nuclear H II regions might be the result of mass loss from the stars or the remnant of past stellar formation, or a combination of both.

Burbidge and Burbidge (1962, 1965) found that in the nuclei of elliptical galaxies, as well as in the nuclei of some Sa, Sb, and even Sc galaxies, the [N II]/Hα ratio is different from that in normal H II regions. Their observations are summarized in table 12. Two galaxies that show the [N II]/Hα ratio larger than unity in their nuclei, NGC 4939 and NGC 5194 (M51), are presented in figures 11 and 12.

In table 13 we show the emission-line intensities derived for M51, M81, and NGC 4278 (M82 and NGC 7679 will be discussed later). The singly and doubly ionized oxygen line intensities are similar to those of galactic H II regions—for example, M51 is similar to M17 IV: and M81 to Ori III—however, the nitrogen line intensities are stronger in these galaxies by factors of 10 and 5, respectively. Although the line-intensity determinations for NGC 4278 are not as accurate, it appears that the nitrogen line is about 2 to 3 times

TABLE 12

PERCENTAGES OF VARIOUS TYPES OF GALAXIES HAVING [N II] λ6584 STRONGER THAN Hα*

Hα/[N II] λ6584	E	S0	Spirals	Irregulars
≤ 1.........	100	81	55	0
< 1.........	91	69	25	0

* From Burbidge and Burbidge 1965.

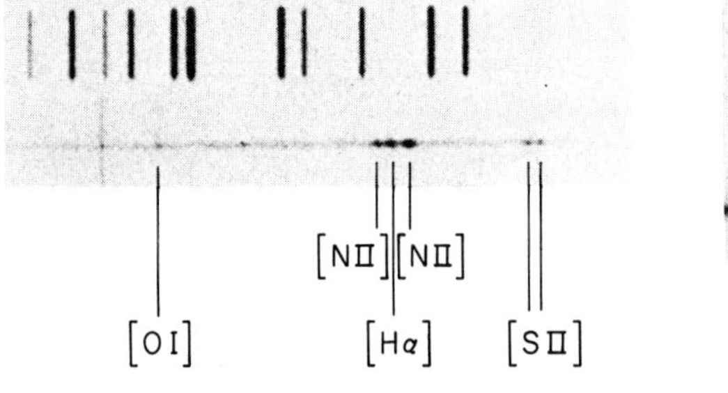

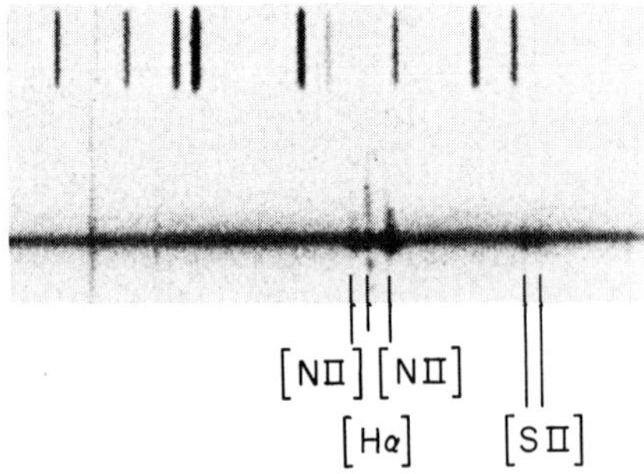

FIGS. 11 and 12.—Spectra of the nuclear regions of NGC 4939 and NGC 5194 (M51) respectively, taken with the prime-focus spectrograph on the 120-inch telescope at Lick Observatory. Note the strong [N II] emission lines flanking Hα, and the weaker forbidden lines of neutral oxygen and ionized sulfur.

stronger than in the Orion regions of comparable ionization. Generally, these nuclear regions have been considered to be dust-free; this seems to be an ad hoc assumption, and internal $E(B - V)$ values should be decided for each object individually. In the observations of M51 and M81, only our local galactic reddening was taken into account. Support for assuming this small reddening correction comes from the observed Balmer decrement in M81.

The following argument has been given in the literature to prove that planetary nebulae are not responsible for most of the line emission: when the ionization in planetary nebulae and nuclear H II regions is similar, the densities are at least a factor of 10 higher in planetary nebulae; alternatively, when the densities are similar in both types of objects, the degree of ionization is considerably higher in planetary nebulae. Even if most of the emission does not originate in planetary nebulae, they might be one of the most important *sources* of interstellar gas.

TABLE 13

INTERSTELLAR EMISSION LINES IN NUCLEAR H II REGIONS

λ \ Object	M51*	M81	M82	NGC 4278	NGC 7679
3727	+0.45	+0.30	+0.16	+0.46	+0.13
4340	...	−0.29	−0.29	...	...
4861	...	+0.04	0.00	0.00	0.00
5007	+0.50	+0.13	−0.42	−0.12	+0.04
5876	<−0.32†	<−0.54†	−1.06	...	−0.96
6300	−0.31	−0.17	−1.33	...	...
6563	+0.46	+0.46	+0.52	+0.11	+0.46
6584	+0.86	+0.61	+0.25	+0.04	+0.13
6716, 6731	+0.35	+0.12	−0.27	...	...
$C(H\beta)$	0.08	0.08	1.69	...	0.27
Observer	1	1	2	3	4

* Centered in the galaxy with a diaphragm 3″.5 in radius.

† Upper limit since the line was not detected.

OBSERVER.—(1) Peimbert 1968; (2) Peimbert and Spinrad 1970*b*; (3) Osterbrock 1960; (4) Peimbert and Spinrad 1970*a*.

Since the ionization in most nuclear H II regions is not due to massive O and B stars or planetary nebulae, two other ionization mechanisms have been proposed: collisional ionization and radiative ionization, the latter produced by relatively faint stars possibly belonging to the horizontal branch. Based on the high turbulent velocities in the nuclei of NGC 1052 and NGC 4278, which amount to several hundred kilometers per second, Minkowski and Osterbrock (1959) and Osterbrock (1960) have suggested collisional ionization as the dominant mechanism. However, since the turbulent velocity decreases from the center outward in NGC 4278, Osterbrock concluded that the ionization had to be radiative, at least in the outer edges. Alternatively, by scaling the globular cluster M3 to the size of the nucleus of NGC 1052, Minkowski and Osterbrock obtain 10 times more horizontal-branch stars than the number needed to explain the line emission. The emission-line intensities vary considerably from one elliptical galaxy to another; under the assumption of radiative ionization, this can be due to either variations in the amount of gas or variations in the amount of ionizing radiation. Since the mass of the nuclear H II regions is very small, Minkowski and Osterbrock suggested that neutral gas might be present in considerable amounts and that the change in the line intensities could be due to the radiation field and not to the amount of gas; as support for their hypothesis, they mention that the number of horizontal-branch stars changes considerably from globular cluster to globular cluster.

Rubin and Ford (1970) from their spectroscopic work on M31 report that the two [N II] lines are seen along the major axis from SW 233 pc to NE 1800 pc and along the minor axis from SE 4.1 kpc to NW 2.3 kpc; these lines originate in an extended tenuous H II region centered in the nucleus that has not been identified photographically (there are no photographically identified emission regions within 16′, or 3.2 kpc, of the nucleus of M31). Spinrad *et al.* (1971) has recently detected [O III] λ5007 emission at 1′ south of the M31 nucleus, but it is very weak there and can be seen only on continuum-corrected scans. Rubin and Ford further report that the [N II] line is present in all their plates and that it can be traced even farther away from the center than the range of their nuclear observations. [O II] λ3727 is present in all plates; He I λ5876 and Hα are also present in some of the plates. The relative intensity of the [N II] lines decreases with increasing distance from the nucleus ([N II]/Hα > 1 for $r < 5$ kpc and [N II]/Hα < 1 for $r > 5$ kpc), and the [N II] line is narrower than their instrumental profile which corresponds to 5 Å or 250 km s^{-1}.

To explain the different [N II]/Hα line intensity ratio in nuclear H II regions, Burbidge and Burbidge (1962) and Burbidge, Gould, and Pottasch (1963) suggested collisional ionization as well as high electron temperatures ranging from 10,000° to 40,000° K. Burbidge and Burbidge (1965) stressed the possibility that the chemical abundances in nuclear and normal H II regions might be different. Morgan and Osterbrock (1969) also suggested that the effect might be due to a difference in ionization, nitrogen being mostly in the N^{++} and N^{+++} stages in normal H II regions, and mostly N^{+} in nuclear regions. Peimbert (1968), from the [O I], [O II], and [O III] line intensities and assuming that they are formed in the same region, found that collisions by thermal electrons are not responsible for the ionization in M51 and M81. This result, coupled with the Balmer decrement observed in M81, led him to the adoption of radiation as the main mechanism of ionization. However, under the hypothesis of radiative ionization, it is difficult to explain the relatively high [O I]/Hα line intensity ratios (see below). Peimbert also found that solar abundances cannot explain the relative line intensities in the nuclei of M51 and M81: if the nitrogen abundance is assumed to be solar, then oxygen is found to be underabundant by factors from 4 to 20; if oxygen is assumed to be solar, then nitrogen is overabundant by factors from 2 to 6. The latter possibility was accepted as the most plausible, based on the following arguments: If the nitrogen abundance is solar, then the temperature determined from the [N II]/Hα line intensity ratio would be considerably higher than the temperature obtained from heating and cooling processes. Moreover, the interstellar gas might be contaminated with material already processed in stellar interiors. From the computations by Caughlan and Fowler (1962) it can be seen that the CNO equilibrium values are very different from the solar values, nitrogen being enriched at the expense of carbon and oxygen. Both the predictions from theoretical models for red giants (Iben 1964, 1967; Torres-Peimbert 1969) and the observations of some red giants and planetary nebulae (Spinrad and Vardya 1966; Aller 1964; Peimbert 1967*b*) indicate a moderate nitrogen enrichment but an almost constant abundance of oxygen. There is also direct evidence that the stellar CN features are very strong in the core of M81 and other large galaxies (McClure and van den Bergh 1968; Spinrad *et al.* 1971), suggesting higher than solar nitrogen abundances.

The main-sequence turnoff point in M81 is about F8 to G0 (Spinrad and Taylor 1971), which implies that star formation (at least on a large scale) ceased a long time ago. If the ionization is due to blue stars, these could be members of the horizontal branch or of the blue descending branch. Recent OAO-2 satellite observations of the nucleus of M31 by Code (1969) indicate that the radiant energy received from the nucleus *may* increase with decreasing wavelength below 2000 Å. This observation might be connected with the radiative source needed to ionize the nuclear H II region in M31, although the source need not necessarily be hot stars.

It is possible that a significant fraction of the radio emission that originates in the nuclei of galaxies is related to the optical emission lines, although it has not been established. In the cases of NGC 1052, NGC 4278, and M81, radio components smaller than 2″ containing a significant fraction of the total flux have been detected (Heeschen 1968, Wade 1968); on the other hand, the size of the regions where the optical emission lines originate is at least 5″ in diameter.

Heeschen (1968) and Wills (1968) have shown that the radio emission in NGC 1052 and NGC 4278 is not thermal; moreover, Wills has suggested that the low-frequency cutoff in NGC 1052 might be due to synchrotron self-absorption and consequently that the size of the source would have to be about 7×10^{-4} arc sec. In the case of NGC 4278 Wills suggests that the cutoff might be due to thermal absorption and thus $N_e(\mathrm{rms}) \simeq 55$–$80\ \mathrm{cm}^{-3}$; however, from the H$\beta$ intensity measured by Osterbrock (1960) $N_e(\mathrm{rms}) = 8.7\ \mathrm{cm}^{-3}$ for the observed volume (see table 11), which is significantly smaller; consequently we suggest that the turnoff might also be due to synchrotron self-absorption. This possibility also implies a very small diameter for the radio source.

Wade (1968) from observations at a wavelength of 11.1 cm has found that M81 shows two components: 60 percent of the emission takes place in a region larger than 2′ in diameter, while the remaining emission originates in a region smaller than 2″ in diameter. From the Hα emission measured by Peimbert (1968) for the nucleus of M81 it is found that the thermal contribution at 11.1 cm is two orders of magnitude smaller than the observed flux; furthermore, the thermal component is optically thin at this wavelength. It is desirable to obtain observations at other radio wavelengths of the compact component of M81.

In several spiral galaxies the [N II]/Hα ratio decreases from the center outward (an example being M51, see fig. 12), which implies a variation with position in the gaseous nitrogen abundance. This variation can be due to two causes: first, that the metallicity of the stars that are producing the enrichment through mass loss decreases from the center outward, as has been observed by Spinrad *et al.* (1971) and by McClure and van den Bergh (1968) in M31; or second, that the chemical composition of the nitrogen-rich ejected material is virtually the same over the region in which the [N II]/Hα ratio changes but that it is added to interstellar material with normal abundances, possibly a remnant of star formation, which is not as concentrated to the center as the ejected material. This might be the case in M51, or a combination of both effects could be responsible for the observed ratio variation. Peimbert and Spinrad (1970*c*) have suggested that the decrease from the center outward of the metal abundance extends to H II regions located at several kiloparsecs from the nuclei of spiral galaxies.

We will next turn our attention to the strengths of the [O I] and [S II] lines in nuclear regions. In most galactic H II regions and planetary nebulae the [O I] lines are at least two orders of magnitude fainter than Hα and one order of magnitude fainter than He I λ5876; similarly the [S II] lines are about one order of magnitude fainter than Hα (see table 8), while in nuclear H II regions these lines are considerably stronger. In the nucleus of NGC 4258, an Sb galaxy, Burbidge and Burbidge (1962) reported the presence of emission lines of [O III], [O II], [S II], [N II], and Hα in about this order of intensity. Rubin (1968) found that [S II]/Hα is larger than unity only near the nucleus of M31. For NGC 1052 (E3), NGC 3998 (S0), and NGC 4438 (S0p), Burbidge and Burbidge (1965) report that the [S II] lines are very strong and furthermore that [O I] λ6300 is present. From table 13 it is clear that in the nuclei of M51 and M81 the [S II] and [O I] lines are about an order of magnitude stronger than in galactic spiral-arm H II regions of similar ionization. The ionization potential of neutral oxygen is almost the same as of hydrogen, and consequently the [O I] lines should originate in the boundaries of H II regions or in the H I regions themselves; the [S II] lines might originate in H I regions or in the lowest ionization points of H II regions, because almost all sulfur must be singly ionized in H I regions, since its ionization potential is only 10.36 eV. If the assumptions are made that the excitation of the [O I] and [S II] lines is collisional, that both lines originate in the same region and that collisional de-excitation is negligible, then the intensity ratio can be expressed as

$$\frac{I(S^{+})}{I(O^{0})} \propto \frac{N(S^{+})}{N(O^{0})} \exp\left(\frac{1.40 \times 10^{3}}{T}\right). \tag{7}$$

For typical H II region temperatures equation (7) is essentially temperature-independent. Therefore, since the observed [S II]/[O I] line intensity ratios are similar in both nuclear and normal H II regions, it implies that the abundance ratios are also similar; and thus from the fact that the [S II] and [O I] lines are relatively stronger than the [O II], [O III], and Hα lines, it follows either that the lowest ionization regions are relatively more important in nuclear than in normal H II regions or that turbulence is effective in maintaining a temperature of several thousand degrees in H I regions and consequently that the [O I] and [S II] lines originate in relatively hot H I regions. More observations of line intensity ratios coupled with emission-line widths are needed to decide between these two possibilities.

2.5. Other H II Regions

Carranza, Courtès, Georgelin, Monnet, Pourcelot, and Astier (1968) from an interferometric study of ionized hydrogen in M33 found that the Hα/[N II] ratio of the H II regions located in the spiral arms is similar to that of normal H II regions. They also found a general field of gas emission in the interarm region, where the Hα/[N II] ratio is about 0.5; this disk H II region extends at least 15′ from the center. The velocity dispersions of the disk and of the H II regions in the arms are 5 and 9 km s^{-1}, respectively. The Hα emission in the disk is about two to three times fainter than in the arms; the resolution of their instrument does not allow us to decide whether the disk radiation is uniform or whether it is due to several clouds of size 15 pc or smaller. The root mean square electron density of the disk is estimated at 4 to 6 cm^{-3} and the mass $\mathfrak{M}$(rms) at $3 \times 10^5 \mathfrak{M}_\odot$. This mass is negligible compared with the mass contained within $7'.5$ of the center which, from dynamical considerations, is found to be $1 \times 10^9 \mathfrak{M}_\odot$. The variation in the nitrogen line intensity might be due to different electron temperatures, to a variation in the nitrogen abundance, or to a different degree of ionization in the interarm region. Observations in other emission lines are needed to decide between these alternatives. In particular, the strength of λ3727 might help to decide whether or not the electron temperature in the interarm region is considerably higher than in normal H II regions. In any case, it seems unlikely that cloud collisions play an important role in the ionization, since the velocity dispersion of the disk is very small.

The line intensities of M82 and NGC 7679 are given in table 13. Apparently, these H II regions are different from the other H II regions we have been discussing. Their total fluxes at Hα are considerably stronger than in normal spiral-arm H II regions, the ionization is smaller than that of the normal giant H II regions, and the [O I], [S II], and [N II] line intensities are slightly stronger than those of normal small H II regions of similar degree of ionization, but considerably weaker than in the nuclear regions of M51 and M81. NGC 7679 is number 216 in the *Atlas of Peculiar Galaxies* (Arp 1966) and has been classified as a galaxy with an adjacent loop; Minkowski (1969) notes that it shows an A5 spectrum and that NGC 7679 probably is an Sc galaxy, contrary to the S0 classification given by Humason, Mayall, and Sandage (1956).

It is our pleasure to acknowledge the comments, unpublished data, and figures from many astronomers; in particular we want to mention Drs. S. van den Bergh, E. M. Burbidge, G. R. Burbidge, A. D. Code, P. W. Hodge, I. R. King, R. D. McClure, P. G. Mezger, J. E. Miller, R. Minkowski, R. O'Connell, D. E. Osterbrock, V. C. Rubin, A. R. Sandage, L. Searle, B. J. Taylor, S. Torres-Peimbert, G. de Vaucouleurs, and A. E. Whitford. A portion of Spinrad's research was supported by the ONR and the NSF.

REFERENCES

Aizenman, M. L., Demarque, P., and Miller, R. H. 1969, *Ap. J.*, **155**, 973.
Aller, L. H. 1942, *Ap. J.*, **95**, 52.
———. 1956, *Gaseous Nebulae* (London: Chapman & Hall), p. 99.
———. 1964, *Pub. A.S.P.*, **76**, 279.
Aller, L. H., Baker, J. G., and Menzel, D. H. 1939, *Ap. J.*, **89**, 587.
Aller, L. H., Czyzak, S. J., and Walker, M. F. 1968, *Ap. J.*, **151**, 491.
Aller, L. H., and Faulkner, D. J. 1962, *Pub. A.S.P.*, **74**, 219.
Aller, L. H., and Liller, W. 1959, *Ap. J.*, **130**, 45.
———. 1966, in *Nebulae and Interstellar Matter*, ed. B. M. Middlehurst and L. H. Aller (Chicago: University of Chicago Press), p. 483.
Arp, H. C. 1966, *Ap. J. Suppl.*, **14**, 1.
———. 1969, *Astr. and Ap.*, **3**, 418.
Baade, W. 1951, *Pub. Univ. Mich. Obs.*, **10**, 7.
Baade, W., and Arp, H. C. 1964, *Ap. J.*, **139**, 1027.
Baade, W., and Swope, H. 1961, *A.J.*, **66**, 300.
Bahcall, J. N., Bahcall, N. A., and Shaviv, G. 1968, *Phys. Rev. Letters*, **20**, 1209.
Baum, W. 1959, *Pub. A.S.P.*, **71**, 106.
———. 1962, in *Problems of Extra-Galactic Research*, ed. G. C. McVittie (New York: Macmillan), p. 390.
Baum, W., and Schwarzschild, M. 1955, *A.J.*, **60**, 247.
Bergh, S. van den. 1968, *J.R.A.S. Canada*, **62**, 1.
———. 1969, *Ap. J. Suppl.*, No. 171, **19**, 145
Boesgaard, A. M. 1967, unpublished.
Broadfoot, A. L., and Kendall, K. R. 1968, *J. Geophys. Res.*, **73**, 426.

Burbidge, E. M., and Burbidge, G. R. 1962, *Ap. J.*, **135**, 694.
———. 1965, *ibid.*, **142**, 634.
Burbidge, E. M., Burbidge, G. R., and Fish, R. 1961, *Ap. J.*, **133**, 1092.
Burbidge, G. R. 1970, *Ann. Rev. Astr. and Ap.*, **8**, 369.
Burbidge, G. R., Gould, R. J., and Pottasch, S. R. 1963, *Ap. J.*, **138**, 945.
Carranza, G., Courtès, G., Georgelin, Y., Monnet, G., Pourcelot, A., and Astier, N. 1968, *Ann. d'Ap.*, **31**, 68.
Caughlan, G. R., and Fowler, W. A. 1962, *Ap. J.*, **136**, 453.
Chamberlain, J. W., and Oliver, N. J. 1953, *Ap. J.*, **118**, 197.
Chambers, R. H., and Roeder, R. C. 1969, *Ap. and Space Sci.*, **3**, 530.
Code, A. 1969, *Pub. A.S.P.*, **81**, 475.
Courtès, G., and Cruvellier, P. 1965, *Ann. d'Ap.*, **28**, 683.
Demoulin, M.-H. 1969, *Ap. J.*, **157**, 81.
Deutsch, A. J. 1964, *Ap. J.*, **139**, 532.
Dickel, H. R. 1965, *Ap. J.*, **141**, 1306.
Eissner, W., Martins, P., Nussbaumer, H., Saraph, H. E., and Seaton, M. J. 1969, *M.N.R.A.S.*, **146**, 63.
Faulkner, D. J. 1963, Ph.D. thesis, Australian National University.
Faulkner, D. J., and Aller, L. H. 1965, *M.N.R.A.S.*, **130**, 393.
Feast, M. W. 1961, *M.N.R.A.S.*, **122**, 1.
———. 1964, *ibid.*, **128**, 327.
Fichtel, C. E., and McDonald, F. B. 1967, *Ann. Rev. Astr. and Ap.*, **5**, 525.
Fish, R. 1964, *Ap. J.*, **139**, 284.
Ford, W. K., and Rubin, V. C. 1969, *Bull. A.A.S.*, **1**, 188.
Gardner, F. F., Milne, D. K., Mezger, P. G., and Wilson, T. L. 1970, *Ap. Letters*, **6**, 87.
Goldberg, L., Müller, E. A., and Aller, L. H. 1960, *Ap. J. Suppl.*, **5**, 1.
Grewing, M., Demoulin, M.-H., and Burbidge, G. R. 1968, *Ap. J.*, **154**, 447.
Haro, G. 1950, *A.J.*, **55**, 66.
Heeschen, D. S. 1968, *Ap. J. (Letters)*, **151**, L135.
Henize, K. G. 1956, *Ap. J. Suppl.*, **2**, 315.
Hodge, P. W. 1966, *An Atlas and Catalog of* H II *Regions in Galaxies* (Seattle: Astronomy Department, University of Washington).
———. 1967, *A.J.*, **72**, 129.
———. 1969*a*, *Ap. J. Suppl.*, **18**, 73.
———. 1969*b*, *Ap. J.*, **155**, 417.
———. 1969*c*, *ibid.*, **156**, 847.
Humason, M. L. 1936, *Ap. J.*, **83**, 18.
Humason, M. L., Mayall, N. U., and Sandage, A. R. 1956, *A.J.*, **61**, 97 (HMS).
Iben, I., Jr. 1964, *Ap. J.*, **140**, 1631.
———. 1967, *ibid.*, **147**, 624.
Johnson, H. L. 1966, *Ap. J.*, **143**, 187.
———. 1968, in *Nebulae and Interstellar Matter*, ed. B. M. Middlehurst and L. H. Aller (Chicago: University of Chicago Press), p. 167.
Johnson, H. M. 1959, *Pub. A.S.P.*, **71**, 425.
———. 1968, in *Nebulae and Interstellar Matter*, ed. B. M. Middlehurst and L. H. Aller (Chicago: University of Chicago Press), p. 65.
King, I. R., and Minkowski, R. 1972, in *External Galaxies and Quasi-Stellar Objects*, ed. D. S. Evans (Dordrecht: Reidel Publ. Co.), p. 87.
Lambert, D. L. 1967, *Observatory*, **87**, 199.
———. 1968, *M.N.R.A.S.*, **138**, 143.
Lambert, D. L., and Warner, B. 1968, *M.N.R.A.S.*, **138**, 181.
McClure, R. D. 1969, *A.J.*, **74**, 50.
McClure, R. D., and Bergh, S. van den. 1968, *A.J.*, **73**, 313.
Mathis, J. S. 1962, *Ap. J.*, **136**, 374.
———. 1965*a*, *Pub. A.S.P.*, **77**, 90.
———. 1965*b*, *ibid.*, **77**, 189.
Mayall, N. U. 1958, in *Proceedings I.A.U. Symposium No. 5* (New York: Cambridge University Press), p. 23.
Meinel, A. B. 1953, *Ap. J.*, **118**, 200.
Méndez, M. 1968, *Bol. Obs. Tonantzintla y Tacubaya*, **4**, 240.
Mezger, P. G., and Henderson, A. P. 1967, *Ap. J.*, **147**, 471.
Mezger, P. G., Schraml, J., and Terzian, Y. 1967, *Ap. J.*, **150**, 807.
Miller, J. E. 1969, private communication.
Minkowski, R. 1962, *Problems of Extra-Galactic Research* (I.A.U. Symposium No. 15), ed. G. C. McVittie (New York: Macmillan Co.), p. 379.
———. 1969, private communication.
Minkowski, R., and Osterbrock, D. E. 1959, *Ap. J.*, **129**, 583.
Moore, E. 1968, Ph.D. thesis, University of Arizona.

Morgan, W. W. 1956, *Pub. A.S.P.*, **68**, 509.
Morgan, W. W. 1959*a*, *Pub. A.S.P.*, **71**, 92.
———. 1959*b*, *ibid.*, p. 394.
Morgan, W. W., and Mayall, N. U. 1957, *Pub. A.S.P.*, **69**, 291.
Morgan, W. W., and Osterbrock, D. E. 1969, *A.J.*, **74**, 515.
Münch, G. 1959, *Pub. A.S.P.*, **71**, 101.
O'Connell, R. 1970, thesis, California Institute of Technology.
Oke, J. B., and Sandage, A. R. 1968, *Ap. J.*, **154**, 21.
Osterbrock, D. E. 1960, *Ap. J.*, **132**, 325.
———. 1967, *Pub. A.S.P.*, **79**, 523.
———. 1969, *Colloque International d'Astrophysique, Liège*, **17**, 391.
Osterbrock, D. E., and Flather, E. 1959, *Ap. J.*, **129**, 26.
Page, T. L. 1952, *Ap. J.*, **116**, 63.
———. 1962, *ibid.*, **136**, 685.
Palmer, P., Zuckerman, B., Penfield, H., Lilley, A. E., and Mezger, P. G. 1969, *Ap. J.*, **156**, 887.
Peebles, P. J. E. 1966, *Ap. J.*, **146**, 542.
Peimbert, M. 1966, *Ap. J.*, **145**, 75.
———. 1967*a*, *ibid.*, **150**, 825.
———. 1967*b*, unpublished Ph.D. thesis, University of California, Berkeley.
———. 1968, *Ap. J.*, **154**, 33.
———. 1970, *Pub. A.S.P.*, **82**, 636.
Peimbert, M., and Costero, R. 1969, *Bol. Obs. Tonantzintla y Tacubaya*, **5**, 3.
Peimbert, M., and Spinrad, H. 1970*a*, *Ap. J.*, **159**, 809.
———. 1970*b*, *ibid.*, **160**, 429.
———. 1970*c*, *Astr. and Ap.*, in press.
Pengelly, R. M. 1964, *M.N.R.A.S.*, **127**, 145.
Pottasch, S. R. 1965, in *Vistas in Astronomy*, Vol. **6**, ed. A. Beer (New York: Pergamon Press), p. 149.
Reddish, V. 1968, *Quart. J.R.A.S.*, **9**, 409.
Robbins, R. R. 1968, *Ap. J.*, **151**, 497.
Roberts, M. S. 1968, in *Interstellar Ionized Hydrogen*, ed. Y. Terzian (New York: W. A. Benjamin), p. 641.
———. 1969, *A.J.*, **74**, 859.
Robinson, B. J., and Koehler, S. A. 1965, *Nature*, **208**, 993.
Rubin, R. H. 1969, *Ap. J.*, **155**, 841.
Rubin, V. C. 1968, in *Interstellar Ionized Hydrogen*, ed. Y. Terzian (New York: W. A. Benjamin), p. 641.
Rubin, V. C., and Ford, W. K. 1968, *Ap. J.*, **154**, 431.
———. 1970, *ibid.*, **159**, 379.
Sandage, A. R. 1961, *Ap. J.*, **134**, 916.
———. 1968, *Observatory*, **88**, 91.
———. 1969, private communication.
Sandage, A. R., Becklin, E., and Neugebauer, G. 1969, *Ap. J.*, **157**, 55.
Searle, L. 1971, *Ap. J.*, **168**, 327.
Seaton, M. J. 1960, *Rept. Progr. Phys.*, **23**, 313.
———. 1968, *M.N.R.A.S.*, **139**, 129.
Shajn, G. A., and Hase, V. T. 1951, *Pub. Crimean Ap. Obs.*, **7**, 87.
———. 1952*a*, *ibid.*, **8**, 3.
———. 1952*b*, *Trans. IAU*, **8**, 693.
Shipman, H. L., and Strom, S. E. 1970, *Ap. J.*, **159**, 183.
Spinrad, H. 1962, *Ap. J.*, **135**, 715.
———. 1966, *Pub. A.S.P.*, **78**, 367.
———. 1967, oral presentation to Commission 28, at the Prague IAU meeting.
Spinrad, H., Gunn, J., Taylor, B. J., McClure, R., and Young, J. 1971, *Ap. J.*, **164**, 11.
Spinrad, H., and Taylor, B. J. 1969, *Ap. J.*, **157**, 1279.
———. 1971, *Ap. J. Suppl.*, **22**, 445.
Spinrad, H., and Vardya, M. S. 1966, *Ap. J.*, **146**, 399.
Spinrad, H., and Wood, D. B. 1965, *Ap. J.*, **141**, 109.
Tifft, W. G. 1963, *A.J.*, **68**, 302.
Tinsley, B. M. 1968, *Ap. J.*, **151**, 547.
———. 1970, *Ap. and Space Sci.*, **6**, 344.
Toomre, A., and Toomre, J. 1972, *Ap. J.*, **178**, 623.
Torres-Peimbert, S. 1969, private communication.
———. 1971, *Bol. Obs. Tonantzintla y Tacubaya*, **6**, 3.
Torres-Peimbert, S., Simpson, E., and Ulrich, R. K. 1969, *Ap. J.*, **155**, 957.
Vaucouleurs, G. de. 1958, *Ap. J.*, **128**, 465.
Vaucouleurs, G. de, and Vaucouleurs, A. de. 1959*a*, *Lowell Obs. Bull.*, No. 92.
———. 1959*b*, *Pub. A.S.P.*, **71**, 83.

Wade, C. M. 1968, *A.J.*, **73**, 876.
Wagoner, R. V., Fowler, W. A., and Hoyle, F. 1967, *Ap. J.*, **148**, 3.
Wares, G. W., and Aller, L. H. 1968, *Pub. A.S.P.*, **80**, 568.
Whipple, F. L. 1935, *Harvard Obs. Circ.*, No. 404.
Whitford, A. E. 1958, *A.J.*, **63**, 201.
———. 1966, in *Spectral Classification and Multicolor Photometry* (I.A.U. Symposium No. 24), ed. K. Loden, L. O. Loden, and V. Sinnerstead (London and New York: Academic Press), p. 19.
———. 1971, *Ap. J.*, **169**, 215.
———. 1974, chap. 5 of this volume.
Wills, D. 1968, *Ap. Letters*, **2**, 187.
Wolfe, A., and Burbidge, G. R. 1970, *Ap. J.*, **161**, 419.
Wood, D. B. 1966, *Ap. J.*, **145**, 36.

CHAPTER 3

The Masses of Galaxies

E. M. BURBIDGE AND G. R. BURBIDGE
University of California, San Diego

1. INTRODUCTION

THE ONLY direct method of determining total masses or mass distributions within galaxies is to consider the acceleration of stars, gas, or whole galaxies under the action of the gravitational field exerted by the object to be studied. Many galaxies occur as binary systems; for these, a sample of average total masses can be determined. Galaxies also tend to occur in clusters ranging in size from small groups with only a few members, to large clusters like the Coma cluster; and if a cluster is assumed to be in a stationary state, the virial theorem can be used to estimate average masses of the galaxies in it.

The method that has had the widest application for external galaxies is that in which the rotational velocity of one constituent subsystem of the galaxy is measured (e.g., the stars or the neutral or ionized gas component), as a function of distance from the center of the galaxy. This has the great advantage of giving mass distributions, which, upon being integrated, yield total masses, so one obtains more information than by those methods that yield only total masses.

Certain basic assumptions are made in the study of mass distributions from rotations of galaxies. Accelerations can never be measured directly in a galaxy; they must be deduced from observations of the velocity field at one instant of time. Velocities must therefore be assumed to be independent of time, or at least, only slowly changing with time; i.e., the galaxy is assumed to be in a steady state. Further, in all galaxies but our own, three-dimensional velocities cannot be determined for any of the constituents, because only the line-of-sight velocity component can be measured. Therefore it is necessary to assume axial symmetry of the galaxy. Fortunately, in regular normal spirals and S0 systems this appears to be a valid assumption. It is clearly not valid for barred spirals and irregulars, which present a difficult problem. Barred spirals can be assumed to have twofold symmetry about an axis perpendicular to their equatorial planes, while all spirals have complete rotational symmetry with respect to their axes.

Finally, it is assumed that Newtonian gravitation is the dominant force, so that, specifically, even in the case of the ionized gas component, magnetic forces are neglected.

For galaxies which do not appear to have much angular momentum, mass determina-

Received February 1969.

tions can be made by measuring the velocity dispersion of the stars in the central region and estimating the mass from the virial theorem. Most of the mass determinations for elliptical galaxies have been made by this method.

With a number of direct determinations of mass in hand, some workers have correlated properties depending on the mass (e.g., densities, mass-to-light ratios) with other observable properties (e.g., *U*, *B*, *V* colors), and have then used these correlations to estimate masses for a larger sample of galaxies. Such indirect estimations may be used in various investigations; it must always be remembered that they are only indirect and should be used with caution.

In § 2 the means of determining masses by the various methods will be described. Section 3 will deal with the observations needed for application of the various methods. Section 4 will give the results of the various methods, together with a discussion of the mass-to-light ratio in various galaxies and a few comments on the angular momentum. Finally, complications will be discussed, such as deviations of the velocity field from circular rotational motions, and lack of agreement between mass determinations for cluster galaxies by means of the virial theorem and mass determinations by the more direct methods.

2. THEORIES OF MASS DETERMINATION

2.1. Rotations of Galaxies

The stars and the gas move in a gravitational field produced by the total content of the galaxy—stars, gas, and dust together. In an axisymmetric galaxy let (ϖ, θ, z) be a cylindrical coordinate system with origin at the center of the galaxy, where ϖ is the radial coordinate in the equatorial plane and z is the coordinate perpendicular to this plane. Let the components of the attraction, or attractive force per unit mass, be Q_ϖ, Q_z in the ϖ and z directions, respectively. If Φ is the gravitational potential, then

$$Q_\varpi = \partial\Phi/\partial\varpi$$

and

$$Q_z = \partial\Phi/\partial z \;. \tag{1}$$

If the rotational velocity is u_θ, then

$$Q_\varpi \varpi = -u_\theta^2(\varpi) \;. \tag{2}$$

For axisymmetric galaxies in general the line-of-sight velocities, measured along the major axis and corrected for the inclination of the equatorial plane to the line of sight, can be used to derive u_θ as a function of ϖ. For all galaxies but our own, however, there is in general no observational evidence on velocities in the z-direction. Most of the observational work on rotations has been on galaxies that contain an appreciable component of ionized gas, so that $u_\theta(\varpi)$ can be obtained from emission lines in H II regions. Some nearby galaxies can be studied by means of the 21-cm line from neutral hydrogen, and $u_\theta(\varpi)$ can be obtained from this. In either case, analogy with our Galaxy suggests that it is reasonable to assume that the gas—the "test body" on which the gravitational field acts—lies in a thin disk in the equatorial plane.

Rotation curves giving u_θ as a function of ϖ can then be computed for different models of the mass distribution, of varying degrees of complexity. These can then be fitted to the observed rotation curve of a galaxy by an appropriate choice of parameters, and hence the mass distribution and the total mass can be obtained. Perek (1962) has given a detailed account of published models for the distribution of mass in oblate stellar systems and a unified mathematical treatment of them. Here we shall discuss only those models applicable to extragalactic systems, with appropriate formulae and without the full mathematical analysis. Perek's review should be consulted for further details, and for the models involving stellar dynamics in our Galaxy. We start with the simplest model and proceed in order of complexity.

2.1.1. *A mass point.* The simplest possible model of a galaxy is a mass point. This model is of course applicable to the derivation of masses of binary, multiple, and cluster galaxies. If the only observational data consist of a rotational velocity measured at a point near the outside of a galaxy, then a crude Keplerian estimate of the mass interior to this point can be made from the equation

$$M = \varpi u_\theta^2/G \,, \tag{3}$$

where ϖ is the distance from the center of the galaxy of the point where the velocity u_θ is measured. Masses estimated in this way tend to come out larger than those estimated from detailed rotation curves; assumption of a mass point at the center for the gravitational potential gives the highest possible degree of central concentration and will clearly overestimate the mass.

2.1.2. *Homogeneous spheroid.* A model possessing rotational symmetry which comes next to a mass point in simplicity is a single homogeneous spheroid. Let the major and minor axes be of length $2a$ and $2c$, let the uniform density be ρ, and use the cylindrical coordinate system defined above. If the eccentricity is $e = (1 - c^2/a^2)^{1/2}$, and if we introduce an angle β such that inside the spheroid

$$\sin \beta = e \,, \tag{4}$$

while for a point outside the spheroid

$$\varpi^2 \sin^2 \beta + z^2 \tan^2 \beta = a^2 e^2 \,, \tag{5}$$

then the components of the attractive force per unit mass in the ϖ and z directions, Q_ϖ and Q_z, are given by

$$Q_\varpi = 2\pi G\rho\varpi e^{-3}(1 - e^2)^{1/2}(\beta - \sin \beta \cos \beta) \,, \tag{6}$$

$$Q_z = 4\pi G\rho z e^{-3}(1 - e^2)^{1/2}(\tan \beta - \beta) \,, \tag{7}$$

the Q's now being defined as positive in the direction of decreasing ϖ and z.

Suppose the rotational velocity is observed at the outermost point in the equatorial plane of a galaxy whose axial ratio, c/a, can be observed. Then $\varpi = a$, and equations (4) and (6) give

$$Q = \frac{3}{2} \frac{GM}{a^2 - c^2} \left[\frac{a}{(a^2 - c^2)^{1/2}} \operatorname{arc} \cos^{-1} \left(\frac{c}{a} \right) - \frac{c}{a} \right] \tag{8}$$

since $\rho = 3M/4\pi a^2 c$, where M is the mass of the spheroid. From equation (2) we have therefore

$$M = a u_\theta^2 / G\alpha \,, \tag{9}$$

where

$$\alpha = \frac{3}{2}\frac{a^2}{a^2 - c^2}\left[\frac{a}{(a^2 - c^2)^{1/2}}\,\text{arc cos}^{-1}\left(\frac{c}{a}\right) - \frac{c}{a}\right]. \tag{10}$$

In the interior of a homogeneous spheroid, the mass contained within any similar spheroid, i.e., one that has the same axial ratio c/a as the outer boundary, is

$$M(\varpi) = \frac{4\pi\rho}{3}\frac{c}{a}\varpi^3 ,$$

where ϖ is the distance from the center, in the equatorial plane. Thus from equation (5) we have

$$u_\theta = \left(\frac{4\pi G\rho\alpha}{3}\frac{c}{a}\right)^{1/2}\varpi ; \tag{11}$$

i.e., the rotation curve u_θ plotted against ϖ is linear and its slope is proportional to $\rho^{1/2}$. Thus if the mass distribution is represented by a homogeneous spheroid, the body will rotate as a solid body and the rotation curve will be a straight line. Rotation curves for the inner parts of a galaxy are often approximately linear; and if this is the only region that can be observed, this approximation is all that the observations warrant.

Equations (9) and (10) always yield a smaller mass than that which would be derived if the mass were concentrated in a mass point at the center. As $c/a \to 1$, $\alpha \to 1$, so that for a spherical mass, equation (9) reduces to equation (3).

Burbidge, Burbidge, and Prendergast (1959*a*, 1960*b*) used the homogeneous spheroidal model to analyze NGC 1068 and NGC 3556. Several galaxies listed in § 4 with incomplete mass determinations have been analyzed by this method. In all of these cases, the observed rotation curves were approximately linear, so that no more refined analysis was warranted. Several galaxies for which isolated measures of u_θ in their outer parts are available have had masses determined by method (1).

2.1.3. *Two or more homogeneous spheroids*; *mass point or sphere with homogeneous spheroids.* A composite of two or more superposed concentric homogeneous spheroids, or a mass point with one or more homogeneous spheroids, or a sphere with homogeneous spheroids, can be used to form a model in which density varies with distance from the center. A mass point with a homogeneous spheroid was used by Oort (1927) as one example of possible mass models which would yield the observed Oort constants A and B for our Galaxy. Oort also gave a solution consisting of two homogeneous spheroids; he showed that the observations could be approximately satisfied either by a single spheroid with a mass point at the center, giving a high degree of central concentration of mass for the Galaxy, or by two concentric spheroids, giving a lesser central concentration.

The attractions from a set of spheroids simply add, so that if the rotational velocity at a point ϖ in the equatorial plane produced by each spheroid is u_i, the total rotational velocity is given by

$$u^2 = \Sigma_i\, u_i^2 .$$

For our own Galaxy, various authors have constructed models consisting of combinations of homogeneous spheroids (see Perek 1962). We shall not discuss these here, since we are concerned only with external galaxies, and these models are too detailed to be applicable to the more restricted observational data available for galaxies other than our own.

purpose of this review to discuss in detail models of the Galaxy; we mention this model because it will be interesting to apply it to external galaxies.

2.1.5. *Disk: Models of Wyse and Mayall, Aller, Schwarzschild.* For highly flattened galaxies a model consisting of a disk of variable density can be a good approximation. A plane disk can be regarded as a limiting case of an extremely flat spheroid, where the surface density σ is the projection of ρ onto the equatorial plane. The potential and attraction can be derived at internal and external points in the plane of a homogeneous disk, and the galaxy can be treated as made up of one or more nonhomogeneous disks, each of which is considered to be composed of infinitesimal homogeneous disks of different radii and surface densities, $d\sigma$.

The mathematics is more complicated than that for spheroids. For a homogeneous plane disk the attraction for points internal and external to the disk, respectively, in the plane of the disk, is given by (Wyse and Mayall 1942)

$$Q_{\varpi} = 4G\sigma a/\varpi(K - E) \qquad \text{(interior)} , \tag{21}$$

$$Q_{\varpi} = 4G\sigma(K - E) \qquad \text{(exterior)} , \tag{22}$$

where $K(k)$, $E(k)$ are Legendre's complete elliptic integrals of the first and second kinds, respectively, and the modulus k is ϖ/a in equation (21) and a/ϖ in equation (22).

Wyse and Mayall (1942) analyzed the rotation curves of M31 and M33 by treating them as nonhomogeneous disks. Each infinitesimal homogeneous disk of density $d\sigma$ provided a contribution dQ of the form given by equations (21) and (22). The density was expressed as a series

$$\sigma = A + B\varpi/a + c(\varpi/a)^2 + \ldots . \tag{23}$$

The analysis and numerical computations were lengthy and laborious. A fit to Babcock's rotation curve for M31 required two nonhomogeneous disks: in one the density was expressed by equation (23) and the series terminated at the fifth power in ϖ/a, while the other was a small disk with density expressed by the first two terms only of (23), i.e., linearly decreasing with increasing ϖ/a. The inner disk was needed to account for the maximum in Babcock's rotation curve 3′ from the center and the subsequent steep decrease in rotational velocity to a minimum at 10′.

For M33, either a disk with density proportional to the thickness of a homogeneous spheroid, or a disk of nearly constant density, gives a rotation curve consistent with the observed curve of Mayall and Aller, which had large error bars on the observed points.

Aller (1942) also used nonhomogeneous disk models to analyze M33, but he obtained the potential and attraction by considering the disk as made up of elementary rings, so that the analysis involved spherical harmonics. He made two models, one with a constant density out to about 8′ from the center and a density proportional to the surface brightness beyond this, and one with a constant density out to 15′.5 and zero beyond this.

In the models of Wyse and Mayall the surface densities for both M31 and M33, when compared with the light distributions in the galaxies, yield mass-to-light ratios that increase steeply outward from the center. Schwarzschild (1954) noted this, and considered that it was unlikely to accord with physical reality. He therefore proposed a simple disk model in which the mass-to-light ratio was taken to be constant throughout the galaxy. The light distribution was represented by

$$I(\varpi) = \sum_n A_n(1 - \varpi/a_n) , \tag{24}$$

where the coefficients A_n, a_n can be obtained by fitting the observed data. If f is the constant mass-to-light ratio, the surface density is then

$$\sigma(\varpi) = f\Sigma A_n(1 - \varpi/a_n) . \tag{25}$$

The attraction and thence the rotation curve can then be derived from equation (25) by making use of the functions calculated and tabulated by Wyse and Mayall; f is the only adjustable parameter.

A new rotation curve for M31 had been published by Mayall (1950). This was derived from H II regions only and did not cover the central regions; consequently there was no sharp maximum and minimum in the rotational velocity to be taken account of, as there had been in Babcock's curve. Schwarzschild fitted this curve to his computed curve and thus obtained f. For both M31 and M33, the summation in equation (24) was taken over five terms. Integration of equation (25) then gave the total mass.

2.1.6. *Bottlinger's model.* In our Galaxy, the fact that three-dimensional velocities and the distribution function for the various stellar populations can be measured leads to the possibility of constructing models for the mass distribution by using the velocity and spatial distributions. The basic principle used is the equation of continuity, and the detailed equations and various solutions can be found in books on stellar dynamics. Since this attack is not applicable to external galaxies, we do not discuss it here. However, there is one model of interest which arose as a modification of the model of Oort for rotation of our Galaxy. This is Bottlinger's force law, in which an expression of the form

$$Q_\varpi = -\frac{a\varpi}{1 + b\varpi^3} \tag{26}$$

is assumed for the attraction in the radial direction in the equatorial plane (Bottlinger 1933). This expression is the simplest form of a general expression

$$Q_\varpi = -\frac{a_1\varpi + a_2\varpi^2 + \ldots}{1 + b_1\varpi^3 + b_2\varpi^4 + \ldots} , \tag{27}$$

which has the advantage that for large distances ϖ from the center it gives $Q_\varpi \propto \varpi^{-2}$ in accordance with physical reality. In fact, for large ϖ, equation (27) becomes

$$Q \approx -a/b\varpi^2 ;$$

and, since the attraction at large distances tends toward the value for a mass point M, i.e., $Q \approx -GM/\varpi^2$, we have a formula for the total mass

$$M = a/bG . \tag{28}$$

The rotation velocity u_θ is given by $u_\theta^2/\varpi = -Q$; so if a rotation curve has been obtained from observations of a galaxy, the constants a and b can be found by making a least-squares fit of the form

$$u_\theta = \varpi\left(\frac{a}{1 + b\varpi^3}\right)^{1/2} \tag{29}$$

to the observed rotation curve. Equation (28) then yields M. For small ϖ, the rotation curve given by equation (29) is linear, and this is often a good approximation to the observations, particularly if the central parts of a galaxy were observed with low spatial resolution. It may not, however, be necessarily a good approximation to the true form of the rotation curve.

Alternatively, differentiation of equation (29) with respect to ϖ yields equations for the maximum velocity on the rotation curve, $u_{\theta\,\max}$, and the value of ϖ where it occurs, $\varpi_{\max}$, in terms of a and b:

$$u_{\theta\,\max} = (a/3)^{1/2}\varpi_{\max}\,, \qquad \varpi_{\max} = (2/b)^{1/3}$$

so that equation (28) gives

$$M = \frac{3}{2}\frac{u_{\theta\,\max}{}^2\varpi_{\max}}{G} \tag{30}$$

from the observable quantities $u_{\theta\,\max}$ and $\varpi_{\max}$.

Lohmann (1954) applied this method to obtain the masses of M31 and M33 from the rotation curves of Mayall (1950) and Mayall and Aller (1942), respectively. Epstein (1964) used the 21-cm line profiles of a number of spiral and irregular galaxies to determine their masses by means of equation (30). He obtained $u_{\theta\,\max}$ by taking one-half the total velocity extent of the 21-cm line profile, and increased this by cosec ξ, where ξ is the inclination of the plane of the galaxy to the plane of the sky, to take account of the projection of the rotation on the plane of the sky. For $\varpi_{\max}$, the distance of the point of maximum rotational velocity from the center, Epstein took 0.3 times the total radius of the galaxy as given by Holmberg (1958).

Seielstad and Whiteoak (1965) and Rogstad, Rougoor, and Whiteoak (1967) also used this method for analysis of 21-cm rotation curves of galaxies. Instead of obtaining $\varpi_{\max}$ from the optical data, they used the slope of the rotation curve at the center of the galaxy to obtain the parameter a from equation (29); the quantity $u_{\theta\,\max}$ was obtained from one-half the width of the line profile, as was done by Epstein. The masses they obtained were increased by 10 percent to allow for finite thickness of spiral galaxies.

2.1.7. *Flat-disk approximation of Brandt and Belton.* If rotation curves fitted to the measured velocities are used directly to determine the run of density and the mass by integration of ρ, the results will apply only to that part of the galaxy lying interior to the farthest measured velocity. Clearly the mass will not be a good approximation to the total mass unless the rotation curve has reached a maximum and begun to decrease again, in the manner of a Keplerian velocity curve where the total mass lies interior to the point considered and can be treated as a point mass. Quite often, solutions to equation (15) made by the method of Burbidge, Burbidge, and Prendergast for galaxies whose observed rotation curves extend past the maximum, yield $\rho(a)$ which has dropped to a very small value by the last measured point. Because of the characteristics of the solution—involving subtraction of large coefficients, and the tendency for the computed velocity curve to follow every kink in the observations in regions where the measured points are sparse—the last points sometimes yield a negative density, which is clearly unreal, so that the points outside this should be discarded in computing the total mass.

At other times, the rotation curve may terminate at a point where the density has not become vanishingly small, and some attempt must be made to take into account the mass lying outside this point. Burbidge, Burbidge, Prendergast, and colleagues did this for some galaxies in an empirical way described in § 4. The Bottlinger-Lohmann equation (30) allows for mass lying outside the last measured points by assuming a functional form for the rotation curve that allows it to be extrapolated. Brandt (1960) and Brandt and Belton (1962) were the first to draw attention to the possibility of the mass calculated from equation (15) being underestimated, and derived a functional form for the rotation curve of the same general mathematical form as the Bottlinger-Lohmann model, but derived essentially from equation (15). The galaxy is taken to be the limiting case of a very flattened object ($c/a \to 0$), and then equation (15) can be rearranged in terms of a surface density, $\sigma(a)$, instead of $\rho(a)$. This method therefore treats the galaxy as a flat disk, but the mathematics derives from that applicable to a nonhomogeneous spheroid. In terms of $\sigma(a)$, and with $c/a = 0$, equation (15) becomes

$$u_\theta^2(\varpi) = G\int_0^\varpi \sigma(a) \frac{a da}{(\varpi^2 - a^2)^{1/2}}; \tag{31}$$

and this, as was first noted by Kuzmin (1952), is of the form of Abel's integral equation which can be inverted explicitly to give σ as an integral of u_θ^2. The mass interior to the point ϖ is then

$$M(\varpi) = \frac{2}{G\pi}\int_0^\varpi \frac{u_\theta^2(a) a da}{(\varpi^2 - a^2)^{1/2}}. \tag{32}$$

Now Brandt assumed a functional form for $u_\theta(\varpi)$, which represents the observations satisfactorily and which gives an attraction Q that is a generalization of Bottlinger's force law (§ 2.1.6) as follows:

$$u_\theta(\varpi) = \frac{A\varpi}{(1 + B^n\varpi^n)^{3/2n}}. \tag{33}$$

For very large ϖ, the galaxy must appear as a point mass, i.e., $u_\theta^2 = GM_T/\varpi$, where M_T is the total mass. Substitution of this in equation (33) gives, for the limit when ϖ is very large,

$$M_T = A^2/GB^3; \tag{34}$$

or, analogously to the case with Bottlinger's force law, M_T can be expressed in terms of the maximum rotational velocity and the value of ϖ where it occurs as

$$M_T = \left(\frac{3}{2}\right)^{3/n} \frac{u_{\theta\,\max}^2 \varpi_{\max}}{G}. \tag{35}$$

For both M31 and NGC 5055, Brandt and Belton used $n = 3/2$ in equations (33) and (35). In the Bottlinger-Lohmann expression, $n = 3$. Investigation of the effect of neglecting the finite thickness of a galaxy, which will be particularly appreciable in the central regions, led Brandt (1960) to estimate that this effect could be allowed for, for average spirals, by increasing the derived mass by 10 percent.

The total mass of M31 calculated in this way was in agreement with various other determinations; for NGC 5055 it was some 50 percent larger than the value of Burbidge *et al.* The validity of the method depends entirely upon how good a fit equation (33) is to the observed rotation curve, and how well the unseen velocity curve outside the last point would follow the extrapolation.

2.1.8. *Toomre's disk model.* Toomre (1963) suggested a disk model similar to that of

Brandt and Brandt and Belton, for application to highly flattened galaxies. The potential Φ is represented by an expression involving the Bessel function $J_0(k\varpi)$,

$$d\Phi(\varpi, z) = J_0(k\varpi) \exp(-k|z|)dk , \tag{36}$$

with surface density $\sigma(\varpi)$ given by

$$d\sigma(\varpi) = \frac{k}{2\pi G} J_0(k\varpi)dk ; \tag{37}$$

this form for the potential satisfies Poisson's equation. It can then be shown that σ can be expressed in terms of the rotation velocity u_θ by

$$\sigma(\varpi) = \frac{1}{2\pi G} \int_{k=0}^{\infty} J_0(\varpi k)k \int_{s=0}^{\infty} u_\theta^2(s)J_1(ks)ds dk . \tag{38}$$

Setting $u_\theta(0) = u_\theta(\infty) = 0$, and integrating the inner integral by parts, the order of integration may be reversed and we obtain

$$\sigma(\varpi) = \frac{1}{2\pi G} \int_{s=0}^{\infty} \frac{du_\theta^2}{ds} H(s, \varpi)ds , \tag{39}$$

where H is a weighting function of known mathematical form. The density equation can actually be integrated explicitly for a class of analytic functions for $u_\theta^2(\varpi)$.

2.1.9. *Barred spirals.* Since the barred spirals are not axisymmetric, none of the methods that have been discussed for obtaining mass distributions are strictly applicable. It is very difficult in most cases to find the spatial orientation of the galaxy. However, some barred spirals have light distributions that are much more axisymmetric than is the case in the "classic" barred spirals, and the mass determinations may not be too much in error for them.

Besides the absence of axial symmetry, it may not be valid to assume that the system is in a quasi-steady state or that the gravitational attraction of the whole galaxy is the dominant force acting on the gas—for example, magnetic fields might be important. Prendergast (1962) was the first to consider the flow pattern of the gas in a barred spiral, solely under the action of gravitation. If the galaxy is in a quasi-steady state, it can be so only in a coordinate system rotating with the bar, and the bar must rotate with constant angular velocity. The simplest assumption to make for the "classical" barred spirals is that most of the mass lies in the bar, and that this can be approximated as a homogeneous prolate spheroid rotating end over end. The forces acting on the gas and governing its velocity field are the Coriolis, gravitational, and centrifugal forces.

Let ω be the angular velocity of the bar, u the velocity of the gas, and Ω a potential that replaces the Φ used for other galaxies—this Ω represents the combined effect of gravitational and centrifugal forces. Let R be a local radius of curvature of curves of constant Ω. Then an approximate solution for u which can be numerically evaluated was derived by Prendergast as follows:

$$u = -\omega R + \left[R^2\omega^2 - R\left(\frac{\partial\Omega}{\partial s}\right)\right]^{1/2} , \tag{40}$$

where $\partial\Omega/\partial s$ is the force perpendicular to the streamline of the gas. The gas density can be computed in this model.

Many cases were evaluated to map the velocities and positions reached by gas that would flow along the bar that is itself rotating, and would circulate in the bar and leave it by its endpoints, to stream out and become the (trailing) spiral arms. Observations which could be compared with these were made for NGC 7479 (Burbidge, Burbidge, and Prendergast 1960*d*). Here indeed a linear rotation curve was observed for the bar, consistent with the model of a uniform prolate spheroid rotating end over end, and the velocity dropped sharply at a point just beyond the end of the bar, as predicted. It was therefore possible to compute the mass of the bar (see table 1 in § 4). Observations were made for several other barred spirals, but it proved possible only to estimate the mass of the nuclear region in these cases (see table 4 in § 4). If a barred spiral approximates fairly well to an ordinary spiral in its luminosity distribution, the methods applicable to ordinary spirals can be used to determine masses.

The method outlined above for tackling the problem of the gas flow was followed up by Freeman (1965, 1966) in a series of detailed mathematical papers. Observations were made in NGC 4027 (de Vaucouleurs, de Vaucouleurs, and Freeman 1968); in this galaxy the bar lies at right angles to the line of nodes, and this simplified the problem of constructing the velocity field from line-of-sight measures only. There should be no measured velocity along the bar in this case if there is no gas flow along the bar; the situation is similar to that along the minor axis of an ordinary spiral. But velocities were indeed observed, amounting to about 100 km s^{-1} away from the center of the bar, and this suggests that gas is streaming along the bar, as had been originally proposed.

If the bars of at least some barred spiral galaxies can indeed be approximately represented by uniform prolate spheroids, they may be subject to instability. Chandrasekhar (1963) discussed in detail the equilibrium and stability of the Roche ellipsoids; a uniform prolate spheroid is the limiting form of the Jacobi ellipsoid, and in an object such as NGC 7479 the tidal action of an external mass which is 1 percent of the mass of the bar will induce an instability which will become manifest in 10^9 years.

2.1.10. *Correction for partial support of gas by pressure in the center.* The stars in the central region of a galaxy have velocities of the order of hundreds of kilometers a second; and if there is a considerable amount of gas in the central regions, as is often the case in Sc and sometimes in Sb galaxies, the stars may interact sufficiently with the gas so that the gas will possess appreciable random motions. This will produce a pressure-like support for the gas, which will act in the same sense as the centrifugal force in opposing gravitational attraction. The rotational velocity for the gas in the central regions will be reduced below what would have been needed to support the gas against gravitation if there were no such pressure. The velocity curve may thus develop a concavity in the center; and if this is analyzed in terms of gravitational forces alone, the resulting density distribution will be found to decrease in the central regions.

This situation was found in NGC 2146 (Burbidge, Burbidge, and Prendergast 1959*b*), and it was corrected by putting in a pressure term which acts in the central region. If ρ is the density of the gas, we have

$$-\frac{u_\theta^2}{\varpi} = \frac{\partial \Phi}{\partial \varpi} - \frac{1}{\rho}\frac{\partial p}{\partial \varpi}\,. \tag{41}$$

If $\langle u^2 \rangle$ is the mean square random or "turbulent" velocity of the gas in the central region, the pressure caused by it can be expressed as $p = \frac{1}{3}\rho\langle u^2 \rangle$. Assume u_θ to be approximately linear in ϖ near the center, with slope α, and take $\langle u^2 \rangle$ to be constant over the region of interest. Then equation (41) can be integrated to give ρ in terms of the central value ρ_0:

$$\rho = \rho_0 \exp \left[\frac{3(\Phi - \Phi_0)}{\langle u^2 \rangle} + \frac{3\alpha^2 \varpi^2}{2\langle u^2 \rangle} \right] . \tag{42}$$

Near the center the potential is approximately $-4\pi G \rho^* \varpi^2/3$, where ρ^* is some average total density (stars plus gas) over the region considered. Then the scale height λ of the material in the ϖ direction is given by

$$\frac{1}{\lambda^2} = \frac{8\pi G \rho^*}{\langle u^2 \rangle} - \frac{3\alpha^2}{\langle u^2 \rangle} ,$$

or

$$\rho^* = \frac{1}{8\pi G} \left[3\alpha^2 + \frac{\langle u^2 \rangle}{\lambda^2} \right] . \tag{43}$$

In NGC 2146 the emission lines were appreciably broad in the center; the velocity necessary to give this was roughly estimated, and λ was taken to be half the distance in the center over which the lines were appreciably broadened. It was then found that ρ^* in the center was some 200 times the value obtained from analyzing the rotation curve alone. While the correction in the central density is therefore very large, the total mass of the galaxy was increased only by about 5 percent.

2.2. Masses of Spherical Galaxies from Velocity Dispersion of Stars

Poveda (1958) and others have shown that the mass of a nonrotating elliptical galaxy can be obtained by means of the virial theorem, if the velocity dispersion of the stars in the center of a galaxy is known. For a galaxy in equilibrium, and composed of stars having the same mass, the virial theorem states that $2T + \Omega = 0$ or

$$M\langle V^2 \rangle + \Omega = 0 , \tag{44}$$

where the potential energy Ω is given by

$$\Omega = -G \int_0^R \frac{M(r)dM}{r} , \tag{45}$$

M, R are the total mass and radius of the galaxy, $M(r)$ is the mass contained within a sphere concentric with the galaxy and of radius r, and $\langle V^2 \rangle$ is the average of the square of the space velocities relative to the center of mass of the galaxy. It is customary, in deriving $\langle V^2 \rangle$ from observations, to use the assumption made by Poveda that the orbits of individu al stars are so highly eccentric that the stars can be thought of as performing anharmonic oscillations through the center of the galaxy. The dispersion σ^2 can then be substituted for $\langle V^2 \rangle$. In fact, only the dispersion in one dimension, that in the line-of-sight component of velocity, σ_r, is accessible to observation; but if spherical symmetry is assumed, then

$$\sigma^2 = 3\sigma_r^2 . \tag{46}$$

It is assumed that the distribution function for the line-of-sight velocity is Gaussian. Then the fraction of stars with line-of-sight velocities between v_r and $v_r + dv_r$ is

$$\phi(v_r)dv_r = \frac{1}{(2\pi)^{1/2}\sigma_r} \exp\left(-\frac{v_r^2}{2\sigma_r^2} \right) dv_r . \tag{47}$$

The virial theorem can be applied in this way only to a nonrotating stellar system. Any elliptical other than E0 can be presumed to be rotating, and King (1961) pointed out that the kinetic energy, T, in the virial theorem (eq. [44]) must include both random and rotational motions. For a galaxy of small ellipticity, such as M32 which is of type E2, King estimated the rotational energy by means of the following formula:

$$\frac{T_{\rm rot}}{T} = \frac{8\epsilon}{5}, \tag{48}$$

where T and $T_{\rm rot}$ are, respectively, the total internal kinetic energy and the rotational kinetic energy, and $\epsilon = (a - c)/a$ is the ellipticity (0.2 in the case of M32). If the kinetic energy of random motion is $T_{\rm rand}$, then

$$\frac{T_{\rm rot}}{T_{\rm rand}} = \left(\frac{5}{8\epsilon} - 1\right)^{-1}. \tag{49}$$

Shortly afterward, rotation in the nucleus of M32 was directly measured by Walker (1962).

The observations necessary to derive the potential energy Ω and the line-of-sight velocity dispersion v_r are described in § 3.2. Results of mass determinations of elliptical galaxies by this method are given in § 4, table 2.

2.3. Masses from Orbital Motion of Double Galaxies

Galaxies often occur in pairs; and if these are bound double systems, masses can be determined from the orbital motion, as in binary stars. Unlike binary stars, however, the only quantities that can ever be measured are the instantaneous line-of-sight velocity component of each member of the pair and their separation projected on the plane of the sky. There is no way of determining the inclination of the orbital plane or the line of centers of the galaxies to the line of sight. Therefore, individual masses cannot be determined. However, by assuming that the orbital planes and the lines joining the centers are oriented at random, with respect to the line of sight, and by taking a large sample of double galaxies distributed over the sky, it is possible to determine average masses. If the galaxies can be grouped into morphological types, then average masses can be determined separately for spirals, irregulars, ellipticals, and S0 types.

This method has been developed and extensively used by Page (1952, 1960, 1961, 1962, 1965, and a full account is given in his chapter 13 of this volume). The method is particularly valuable for E and S0 galaxies for which very few individual mass determinations have been made, and for irregulars, where individual masses determined from rotation curves are often subject to errors due to uncertainty in the inclination of the equatorial plane. It is also valuable for spiral galaxies, because it will give total masses including the contribution from very faint outer parts, whereas, as noted in § 2.1.7, rotation curves can seldom be measured out to the effective boundary of a galaxy and will give only the mass interior to the last measured point, so that some method of taking account of mass outside this has to be added to the analysis and may lead to errors. Thus a comparison between masses of spirals determined individually, for which there are many data available now, and average masses determined by Page's method, is particularly valuable.

Two galaxies of a binary system are assumed to move in circular orbits about their center of mass. The measured difference in line-of-sight velocities, Δv, is the sum of the orbital velocities of the two galaxies reduced by the projection factor due to the inclination of the orbital plane to the line of sight and the inclination of the line joining the centers to the line of nodes. The angular separation measured in the plane of the sky is s. This can be converted to a projected linear separation by using the mean velocity of recession of the pair, v, and the Hubble relation, to give the distance of the system.

If ϕ is the projection angle of the true distance between the centers of the galaxies in the plane of their orbit, having an unknown value between 0 and $\pi/2$, and ψ is another angle involved in the projection of the velocity, having some value between 0 and 2π, then Page's expression for the sum of the masses M_1 and M_2 of the two galaxies is

$$M_1 + M_2 = \frac{\text{const.}}{G} sv(\Delta v)^2 \sec^3 \phi \sec^2 \psi \,. \tag{50}$$

The numerical constant in equation (50) involves only the conversion of units and the Hubble constant which determines the distance from v and hence the scale of s. As in § 4, we use the value $H = 75$ km s^{-1} Mpc^{-1}. Sandage (1968) has obtained $H = 75.3$ in these units, and this is essentially the same as the value of Sandage (1958) which was used in many of the mass determinations listed in § 4. Holmberg (1964) derived $H =$ 77 km s^{-1} Mpc^{-1} in these units. If we allow for an uncertainty in H by setting $H = 75h$ km sec^{-1} Mpc^{-1}, and if v, Δv are measured in kilometers per second and s in minutes of arc, Page's formula for the gravitational equation becomes, for M_1, M_2 in solar masses,

$$M_1 + M_2 = \frac{901}{h} sv(\Delta v)^2 \sec^3 \phi \sec^2 \psi \,. \tag{51}$$

Page (1965) used the distribution function of r determined by Holmberg (1954), and assumed that ϕ, ψ, M, and r are independent of each other and that the errors in Δv have a normal distribution. Careful statistical analysis of 52 double galaxies, divided into various groups according to morphological type, yielded the results shown in table 5 in § 4.

2.4. Average Masses of Galaxies in Clusters Determined by Using the Virial Theorem

Early in this section the virial-theorem method applied to individual elliptical galaxies was described. For a cluster of galaxies in a stationary state the same method can be applied to obtain the total mass of the cluster, and hence average masses for its members. This method was used more than 30 years ago to obtain masses by treating the galaxies as components of the larger system and by measuring their velocity dispersion about the mean for the cluster and their spatial distribution in the cluster (Zwicky 1933; Tuberg 1942). A related method, used by Smith (1936), assumed that the galaxies with the largest velocities relative to the mean are moving at the velocity of escape from the cluster.

If the cluster is in a stationary state, equation (44) is applicable. In this case the redshifts of individual galaxies in the cluster can be measured and the dispersion in the line-of-sight velocity $\langle v_r^2 \rangle$ can be determined directly. Here $v_r = V_r - \langle V_r \rangle$, where V_r is the individual recession velocity and $\langle V_r \rangle$ is the mean recession velocity. If galaxies of different luminosities and/or types are to be considered, the three-dimensional velocity dispersion can be taken as $3\langle v_r^2\rangle$ and the first term in the virial equation is

$$2E = 3\,\Sigma_i\, M_i v_r^2\,, \tag{52}$$

where the sum is taken over all cluster members. The potential energy Ω is given by

$$\Omega = -\frac{2}{\pi}\sum_{\text{pairs}}\frac{GM_iM_j}{r_{ij}}\,, \tag{53}$$

where r_{ij} is the projected distance between every pair of galaxies in the cluster, M_i and M_j are the masses, and the factor $2/\pi$ accounts in an average way for projection of the true separations between galaxies onto the plane of the sky.

In practice, it has not been found possible to measure the line-of-sight velocity of every galaxy in a rich cluster, so a suitable sample is measured and scaled up. The difference in mass between galaxies can be taken account of if magnitude estimates and rough morphological classifications can be made, because it can be assumed, for example, that all ellipticals and S0 galaxies have some unknown but constant mass-to-light ratio, and similarly spirals have a mass-to-light ratio that is a factor 10 less than that of the ellipticals (such an assumption is justified by the results of individual mass determinations and the statistical results from binaries, summarized in § 4). Small groups of galaxies can be studied in the same way, although the fewer the number in the group, the greater the chance of error in using average projection factors for the velocities and separations.

Masses of galaxies in clusters or groups, derived in this way, have almost always come out much greater than determinations for individual galaxies or binaries. The reason for this is thought to be either that much mass is hidden in clusters, or else that the systems have positive total energy and are disrupting. For these reasons we shall not give masses of galaxies obtained by this method. However, more discussion of this problem will be given in § 5.

3. METHODS OF OBSERVATION

Three general methods of obtaining masses of galaxies have been described in the preceding section: rotation curves of single galaxies, the virial theorem applied to the stars in a single galaxy, and orbital motion of pairs of galaxies. We shall now consider the observations needed for each method, and the various instrumental procedures, the advantages and disadvantages of each, and the accuracy attainable in the results.

3.1. Rotation of Galaxies

Rotations can be measured by means of observations of the spectral Doppler shifts in three constituents of galaxies: the stellar component, from measurements of the absorption lines (mainly Ca II H and K or the Balmer lines), the ionized gas component, from measurements of emission lines (mainly Hα and [O II] λ3727), and the neutral atomic hydrogen component, from 21-cm measures. The 21-cm method is described elsewhere in this volume, so it will be mentioned only incidentally here, and only results of mass determinations using it will be given.

3.1.1. *Rotation curves from absorption lines.* Most E and S0 and many Sa galaxies do not have measurable emission lines in their integrated spectra, so that, if rotations are to be measured, absorption lines must be used. Spectrographic observations can be made with a long slit set in the direction of the major axis, in which case the spectrograms must be measured in two coordinates, so that the velocity can be calculated at various points at right angles to the direction of the dispersion, corresponding to given angular distances from the center of the galaxy. Alternatively, spectrograms can be taken with the slit set parallel to the minor axis, intersecting the major axis at a number of points whose positions are accurately known by offsetting from the center of the galaxy. If the former method is used, it will be necessary to take many spectrograms, with increasing exposures; and since the surface brightness decreases as the square of the distance from the center, long exposures with low dispersion will be needed for the outer parts, on which the inner part of the galaxy will be very heavily overexposed. It is therefore best to set the slit parallel to the minor axis for the outer parts of the galaxy, and to combine these measures with spectrograms of the central region taken along the major axis with higher dispersion and greater scale perpendicular to the dispersion. Because of the steep intensity gradient, it will always be desirable to make observations in conditions of good seeing.

There are various difficulties inherent in the use of absorption lines which explain why few results have so far been obtained:

i) Absorption lines are much less easy than emission lines to measure accurately.

ii) An exposure must be sufficiently long to produce a reasonably strong continuum spectrum, against which the absorption lines have to be measured; if emission lines are present, these will often show up outside the nucleus of the galaxy, on exposures that give only a barely detectable continuum or none at all.

iii) The stars in a galaxy, even in the central portions, are sufficiently far apart that one star does not shadow another along a line of sight passing through the galaxy. Therefore, when we observe the spectrum in a small element of surface projected on the sky, we are integrating the light coming from all the stars in a cylinder cutting through the galaxy. Different points along the axis of the cylinder, at different distances from the center of the galaxy, will have different rotational velocities, and the projection factors converting these to line-of-sight velocities will also be a function of distance along the axis of the cylinder.

Clearly, an iterative procedure is necessary to derive the true circular velocity as a function of distance from the center. The problem is somewhat similar to that encountered by the radio astronomers, working with 21-cm observations either in our Galaxy, where similar integration along the line of sight occurs, or in external galaxies where the beam size is not small compared with the angular size of the galaxy, leading to considerable integration in each data point.

Since density is a decreasing function of distance from the center, the greatest contribution to the light received from a cylinder cutting through the galaxy about the line of sight will come from the point nearest the center. As a first approximation, therefore, a velocity curve can be constructed on the assumption that the line-of-sight velocity measured corresponds to that of the point nearest the center. A density distribution can then be calculated from the velocity curve, by one of the methods given in § 2. The contribution to each data point from points farther from the center can then be calculated, and corrections can be computed to the line profile from whose central point the original velocity curve and resulting density distribution were derived. An improved approximation can thus be obtained.

If the galaxy being studied by absorption-line measures is of type Sa or Sb, or if it is one of the peculiar S0 galaxies with dust lanes, additional complications can arise due to absorption by the dust. If the dust lies close to the equatorial plane, starlight from the far side will suffer more extinction than will the light emitted from the near side. If the dust has a more extended or an irregular distribution, then at various points in the galaxy one may be looking to different effective depths into the galaxy. Corrections will be hard to estimate; in general, all one can do is note that such mass determinations in systems where much dust is present are of lower weight than determinations in dust-free systems.

Since dust is much more prevalent in Sc and Sb galaxies with plenty of uncondensed gas, the problem just mentioned is actually more severe in studies of rotations obtained from emission-line measures. We shall return to this point later. In E and S0 systems with no dust, it is impossible to determine the orientation of the equatorial plane to the line of sight, and the resulting correction which must be applied to the line-of-sight velocities to obtain true rotational velocities. Therefore, if mass distributions, masses, and mass-to-light ratios are desired, only the most elongated objects should be studied. In these it is reasonable to assume that the line of sight lies in or very close to the equatorial plane. In spirals that are studied by absorption-line measures, the orientation can usually be derived from the form of the spiral arms; it is that angle for which deprojection converts the arms into the best approximation to equiangular spirals.

Detailed studies of rotations, mass distributions, and total masses by means of absorption-line measures have been made for M31 (Babcock 1939; Mayall 1950; Lallemande, Duchesne, and Walker 1960), and unpublished rotation curves for NGC 4111 and NGC 3115 by Minkowski and Humason have been derived.

The original rotation curve for M31 by Babcock obtained from absorption lines showed a steep rise from the center followed by a drop to zero (see fig. 1). The curve in the very center of M31, derived by Lallemand, Duchesne, and Walker also from absorption lines, showed a steep rise in the innermost few parsecs followed by a drop to zero. These drops to zero have always provided severe constraints on theoretical models, and checking of them is very desirable. We also need to know the rotation curve in these innermost regions as it is demonstrated by the ionized gas, from emission lines, so that the rotation of the stellar component can be compared with that of the gas component.

3.1.2. *Gradient of rotation in the centers of galaxies, from inclinations of absorption lines.* We have described the difficulties involved in obtaining complete rotation curves and mass distributions from absorption-line velocities. Useful information can, however, be obtained from the inclinations of the absorption lines in the central regions, where reasonably high spectral and spatial resolution can be used because there is plenty of light. The rapid rotation in the innermost few parsecs of M31, referred to above, found by Lallemand, Duchesne, and Walker (1960), was detected on electronic-camera spectrograms taken at the coudé focus of the Lick 120-inch telescope, as was the rotation in the center of M32 (Walker 1962). Apart from this large-scale, moderately high dispersion work, inclinations of the absorption lines in many spiral galaxies of various types were measured by Mayall (1960) following an earlier study (Mayall 1948). Rotation in the centers of some of the spiral galaxies studied by Burbidge, Burbidge, and Prendergast were measured by means of the Na I D absorption lines as well as the Ca II H and K

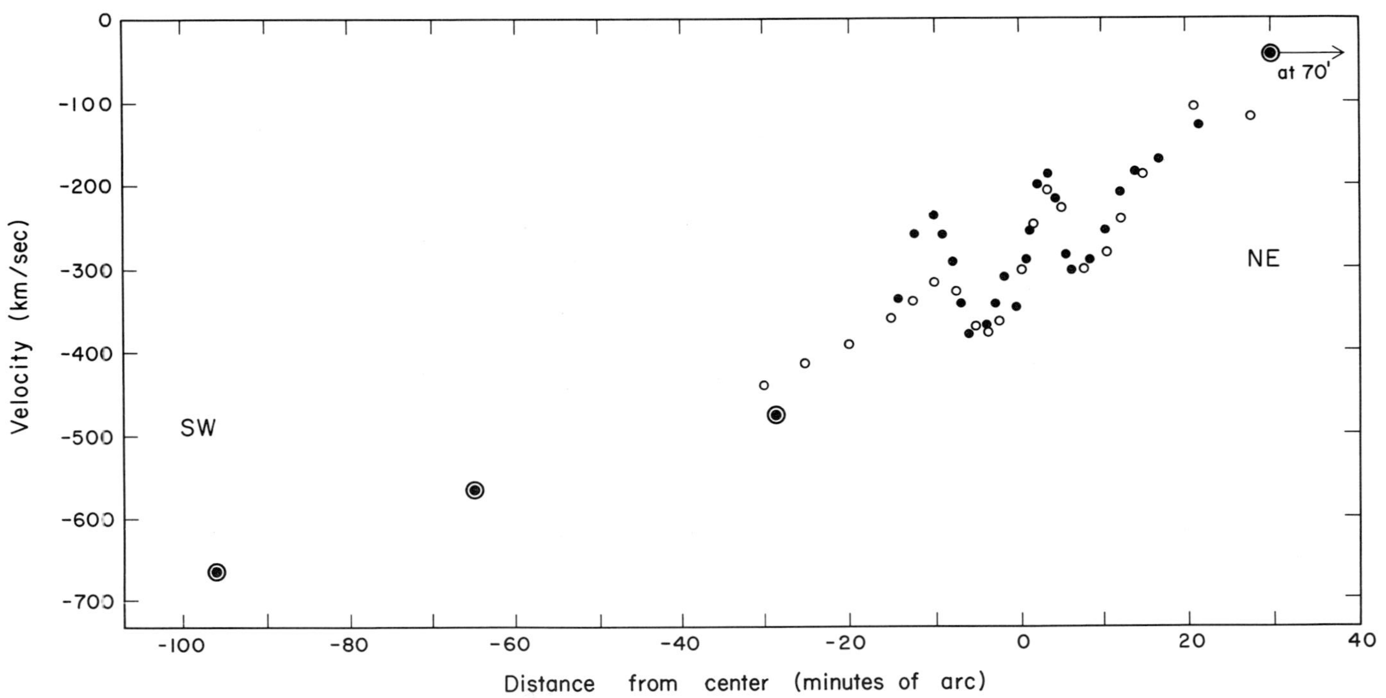

FIG. 1.—Original rotation curve of M31 by Babcock (1939), showing steep rise and drop in inner region derived from absorption lines.

lines. King and Minkowski (1966) measured rotation in the inner regions of two giant ellipticals in the Virgo cluster, NGC 4621 (E3) and NGC 4697 (E5). Apart from their value in demonstrating the existence of rotation in these systems, such measures can be used to give information on the average central density, and can be combined with measures of the line-of-sight velocity dispersion if one wishes to estimate total kinetic energies including random and rotational velocities.

3.1.3. *Rotation curves from emission lines.* By far the greatest part of our knowledge of masses of individual galaxies comes from studies of rotation curves derived by measurement of emission lines. The lines that can be used are the low-ionization lines arising in the general diffuse H II regions in and sometimes between the spiral arms. In decreasing order of strength these are Hα, [O II] $\lambda\lambda$3729, 3726, [N II] $\lambda\lambda$6583, 6548, and [S II] $\lambda\lambda$6731, 6717. Such lines can be very well studied by the long-slit technique in galaxies whose angular sizes are not too large. This technique is very much more successful for emission-line objects than for galaxies with only absorption lines, because the ionized gas in a galaxy is usually distributed in a thin disk about the equatorial plane, and the strengths of the emission lines are often rather uniform throughout most of the galaxy.

Galaxies should be selected whose equatorial planes are inclined to the tangent plane of the sky at angles not less than about 45°. If a galaxy is viewed too nearly face-on, the factor by which the observed velocity has to be corrected for projection becomes very large and errors are magnified. However, in contrast to the situation in absorption-line objects with no obscuring dust, one should not select edge-on objects for emission-line measurements, unless one is interested only in a Keplerian approximation to the total mass. The dust in spiral galaxies is usually close to the equatorial plane, and if in addition the Hα emission line is used, where an optical depth of unity may be reached for only a small linear penetration into the galaxy, then only the outermost parts of the galaxy will be observed at all points along the major axis—like the rim of a wheel. In this case it is obvious that a plot of line-of-sight velocity against distance from the center will give a straight line and will give no information on the rotational velocity as a function of distance from the center, and consequently no information on the density distribution. However, the total mass can be estimated, from the greatest velocity which should be found at the farthest point from the center. This state of affairs was found in the edge-on Sc-irregular galaxy NGC 3556 (Burbidge, Burbidge, and Prendergast 1960*b*).

In the long-slit technique, the slit is set along the major axis of the galaxy, and velocities as a function of distance from the center or from some other point of reference are measured by means of a two-coordinate measuring machine. If the slit of the spectrograph is linear, its image on the spectrogram will be curved, of parabolic form, and corrections for this must be applied to the measures. Minkowski (1942) has derived the form of the correction curve; it may be determined empirically for any spectrograph with given grating-camera combination once and for all by photographing the comparison spectrum in the normal position of the object, and measuring the lines across the whole window. Most long-slit spectrograms of galaxies show airglow or other night-sky lines. These can be measured along with the galaxy lines, and the curvature corrections can be applied and the reductions made in the same way as for the galaxy lines; the absence of systematic errors will be demonstrated if the velocities are zero along the whole length of these check lines.

The long-slit technique was originally developed at Lick Observatory, where the prime-focus spectrograph on the Crossley telescope has a slit 6′.2 long, curved to obviate the need for curvature corrections, and where the absorption-line studies of the rotation curve of M31 and the tilts of inclined absorption lines in many galaxies, already referred to, were made by Babcock and Mayall. This technique has been used most extensively for emission-line rotation curves, at McDonald with the 82-inch reflector, at Lick with the 120-inch, and also at Haute Provence with the 76-inch and at Asiago with the 50-inch. Results and references are given in § 4. The errors in individual measured points usually lie in the 10–40 km s^{-1} range; several closely spaced measures on the extended emission lines can produce mean velocity points in which the errors are reduced.

Apart from the choice of a linear or curved slit, already mentioned, the main difference in the various spectrographs is the way in which a curved focal plane is to be reproduced on the photographic emulsion. In the *B*-spectrograph at McDonald, photographic film is used, curved in both coordinates over a metal button on the camera. Additional complications in the reductions arise because of this two-dimensional bending, but they can easily be handled in the computations. Measures with the Lick prime-focus spectrograph are made using either flat glass plates, pressed in contact with a flat glass button, or film or very thin glass plates curved in the direction of the dispersion only. In the former case it is preferable to use a slit no longer than 3 arc minutes and a small spectral region; in the latter case a longer spectral region can be used, but the slit must be even shorter because this arrangement is used with the camera of longer focal length. A field flattener can also be used, but in practice has not been for rotation-curve work. At Palomar, Münch has used a plateholder with a field flattener; this requires care in its use because small shifts could introduce errors.

For nearby galaxies which have angular sizes too large for the long-slit method to be conveniently used, individual H II regions can be picked out from survey work on direct photographs with filter-plate combinations selecting the wavelengths around Hα, and the spectra of these individual regions can be obtained. This has been done for M31 (Babcock 1939; Mayall 1950; Mayall and Aller 1942); for M81 (Münch 1959); for the Large Magellanic Cloud (de Vaucouleurs 1960); for NGC 55 (de Vaucouleurs 1961); and for M33 (Brandt 1965).

In general such H II regions will not lie on the major axis, and the rotational velocity at the given point must be obtained from the measured line-of-sight velocity by a consideration of the geometry of the galaxy with respect to the plane of the sky. Let (ϖ, θ) be a polar coordinate system in the plane of the galaxy, and (s, α) be a polar coordinate system in the plane of the sky, with common origin at the nucleus and with the angles measured from the line of nodes. The latter can be determined from inspection of direct photographs, as can also ξ, the angle between the axis of symmetry of the galaxy and the line of sight. Let the rotational velocity be $u_\theta(\varpi)$, and the line of sight velocity be $v(s, \alpha)$. For any point at which we measure v, s, and α, we have

$$\tan\theta = \sec\xi \tan\alpha\,, \tag{54}$$

$$u_\theta(\varpi) = v(s, \alpha) \operatorname{cosec}\xi \sec\theta \tag{55}$$

$$\varpi^2 = s^2[\cos^2\alpha + \sec^2\xi \sin^2\alpha]\,. \tag{56}$$

In individual prominent H II regions it may often be possible to measure more emission lines than the low-ionization lines arising in the diffuse H II regions. For example, [O III] λλ5007, 4959 and Hβ may often be emitted by these denser complexes.

A new and promising instrumental method has been developed and applied to a few galaxies of large angular diameter and rich in H II regions by Courtès and his collaborators at the Haute Provence Observatory. The instrument used is a Fabry-Perot interferometer. The advantages of this method over the use of a long slit are twofold; first, a two-dimensional field is obtained on one plate whereas with a single slit, the velocity field across the face of a galaxy has to be built up by means of single cuts across the galaxy. Second, the Hα line, which is the best line to use for this work, is intrinsically quite narrow in most galaxies in H II regions outside the nucleus itself, and often in the nucleus also. Thus quite high dispersions—an order of magnitude higher than traditional nebular dispersions—can be used without loss of luminosity, and such a dispersion is provided by a Fabry-Perot etalon.

The most detailed instrumental description has been given by Courtès in 1964. The interference rings are projected on to the area of sky being studied, and their images are formed in Courtès's instrument by using a reducing camera giving an F/0.95 focal ratio from the original F/5 of the Haute Provence 193-cm telescope. The field that he obtains is 15′ in diameter, and the interferometer works essentially as a multislit spectrograph with 15–20 circular slits, each 2″.5 in width. Differences in velocity are measured by the changes in radius of the interference fringes. Hydrogen tube calibrations are impressed on the plate before and after exposure on the object to give the line-of-sight velocities over the field.

Fringes of higher order are removed either by using an Hα interference filter with 4–10 Å passband or, if one wishes to work with both Hα and the [N II] λ6583 line, by means of ordinary red filters. If an interference filter is used, the continuum radiation is much reduced.

With this instrument, velocities can in principle be measured to ± 2 km s^{-1} accuracy. The main source of error is misadjustment of the etalon during the course of the exposure, and modern developments in mounting the etalon and making the spacers have reduced the risk of this. Such errors are systematic and do not affect the internal consistency of a plate; systematic effects can be detected and removed by comparing emission regions that are common to two or more plates. The most detailed study by this method has been made on M33 by Carranza, Courtès, Georgelin, Monnet, and Pourcelot (1968).

3.2. Velocity Dispersions and Potential Energies

To determine the mass of a spherical elliptical galaxy by means of the virial theorem, two quantities must be obtained from the observations (in addition, of course, to the distance of the galaxy): the line-of-sight velocity dispersion of the stars in the center, and the distribution of light projected on the plane of the sky, from which the potential energy must be derived. We consider the velocity dispersion first.

Suppose any one of the stars giving the bulk of the luminosity in the center of a galaxy has a spectrum whose distribution of intensity, $I(\lambda)$, is known in the neighborhood of some strong absorption line or other feature such as molecular bands or blends of absorption lines. Then, if the velocity distribution given by equation (47) in § 2.2 is imposed on this, the resulting spectral distribution $I(\lambda_0)$ is, with $v_r = c(\lambda - \lambda_0)/\lambda_0$,

$$I(\lambda_0) = \frac{c}{(2\pi)^{1/2}\lambda_0\sigma_r} \int_{-\infty}^{+\infty} I(\lambda) \exp\left[-\frac{c^2(\lambda - \lambda_0)^2}{2\lambda_0^2\sigma_r^2}\right] d\lambda \,. \tag{57}$$

$I(\lambda)$ can be taken to be a strip of spectrum from a standard star, of spectral type matching as well as possible the integrated spectral type of the galaxy, and photographed with the same spectrograph as that used for the galaxy. If λ represents a spectrum already affected by a velocity dispersion σ_1, and a further dispersion σ_2 is applied, then $I(\lambda_0)$ will have the dispersion

$$\sigma = (\sigma_1^2 + \sigma_2^2)^{1/2} \,. \tag{58}$$

The first determination of a velocity dispersion was for M32 (NGC 221), by Minkowski (1954). He used the coudé spectrograph of the Palomar 200-inch telescope as an analog computer; the regular slit was replaced by glass plates with a nonuniform deposit of aluminum made in such a way that the transmission in the direction of dispersion varies as a Gaussian function. Standard stars were photographed through sets of slits corresponding to a range of velocity dispersions, and the best match to the spectrum of M32 was determined by inspection.

Burbidge, Burbidge, and Fish (1961*a*, *b*) reevaluated σ_r for M32 by measuring $I(\lambda)$ for a standard star of type K0 III, δ Tauri, and computing $I(\lambda_0)$ from equation (57) for a range of values of σ_r and determining that which gave the best fit to $I(\lambda)$ measured for the galaxy. The dispersion they derived in the first paper was in error by a factor $1/\sqrt{2}$; this was pointed out and corrected in the second. In a third paper, Burbidge, Burbidge, and Fish (1961*c*) determined the velocity dispersion in NGC 3379, using the spectrum of M32 as the standard; equation (58) then gave the appropriate dispersion for NGC 3379.

Minkowski (1961) determined dispersions in 12 additional galaxies, mostly from unwidened spectra which could not be microphotometered because of the variation of intensity at right angles to the dispersion. He compared the spectra, therefore, with synthetic spectra reconstituted photographically from the computed intensity distributions for various σ_r.

We now turn to the determination of the potential energy Ω. A full discussion was given by Poveda (1958). If the mass-to-light ratio is assumed to be constant throughout a galaxy, then the light distribution as measured in the plane of the sky can be assumed to give the projected mass distribution, from which the three-dimensional mass distribution can be recovered. As Hubble originally showed, many elliptical galaxies have light distributions that suggest the basic model is similar, so that the light distribution in one galaxy can be converted to that in another by suitable scaling quantities. This can be expressed as a relation between the surface brightness, $s(r)$, at a distance r from center of the form

$$\log s(r) = -Ar^{1/4} + B \,, \tag{59}$$

where A and B are arbitrary constants (de Vaucouleurs 1948). For spherical galaxies, on the assumption that the spatial distribution of light derivable from integration of equation (59) can be converted to the spatial distribution of mass by a factor that is constant throughout the galaxy, it can be shown that the potential energy is given in cgs units by (de Vaucouleurs 1953; Poveda 1958)

$$\Omega = -0.33GM^2/R' , \tag{60}$$

where R' is an "effective" radius—the radius of a circle enclosing half the light of the galaxy. The quantities G, M, R' are all in cgs units. If a nonspherical elliptical is to be studied, and if R' is determined from photometry along the minor axis, the Ω should be multiplied by the factor $(a/b)^{2/3}$, where a/b is the apparent major-to-minor axial ratio; further, to compensate for the fact that the a/b of spherical galaxies seen in projection averages 16 percent less than the axial ratios, the coefficient 0.33 in equation (60) should be increased to 0.35 (Fish 1964). According to King (1961), allowance should also be made for the rotational kinetic energy in nonspherical systems, but this has been done only for M32.

The best way to determine R' from photometry of elliptical galaxies is to plot log $s(r)$ against $r^{1/4}$ (Fish 1964). Integration of equation (59) shows that the radius R' is defined by log $[s(r)/s(0)] = -3.33$. From such a plot, the value of $r^{1/4}$ can be read off at which $s(r)$ has fallen 3.33 logarithmic units below the extrapolated surface brightness of the center, $S(0)$.

A compilation of R' for 26 ellipticals was given by Fish (1964). Most of these have not had their central velocity dispersions measured; when this is done, there will be a large increase in our knowledge of masses of ellipticals. The few that have had $\langle\sigma^2\rangle$ measured are given in table 2 of § 4; the other masses tabulated by Fish were not determined by this method, but they were obtained from the luminosities by assuming a value for the mass-to-light ratio.

3.3. Orbital Motions of Double Galaxies

The spectroscopic observations needed for the determination of mass from orbital motions are relatively simple, and the success of the method depends initially on being able to select pairs of galaxies that really are physically bound binary systems. The first study of this problem and resulting catalog of double galaxies was by Holmberg (1937). More discussion was given by Holmberg (1954), and a detailed analysis of the problem has been made by Page, Dahn, and Morrison (1961). The effect of including optical pairs of galaxies that are not physical pairs will be to increase the derived average masses over their true values. The reason for this is that two galaxies at different distances from the observer and seen close together by chance juxtaposition will have different Hubble recessional velocities, in addition to their random field velocities, and this velocity difference will be wrongly interpreted as due to orbital motion.

Practically all the spectroscopic observations yielding the velocity differences Δv have been obtained by Page (see chapter 13 of this volume); the spectrograph he used at McDonald Observatory is the same as that used for many of the rotation curves discussed in § 3.1, and so the errors are of the same order of magnitude. Recent measurements by Page in the southern hemisphere have been made with a similar spectrograph using an image tube.

The numerical constant in equation (51), determining the masses from the measured separations s and velocity differences, Δv, differs from the value of 675 given by Page

(1965), simply because of the change in the standard Hubble constant from the value 100 km s^{-1} Mpc^{-1} used by him, to Sandage's 1968 value of 75 km s^{-1} Mpc^{-1}. At present it seems unlikely that H is as large as 100.

4. RESULTS

Masses of spiral galaxies, ellipticals, and irregular galaxies determined by the various methods described in §§ 2 and 3 are listed in tables 1, 2, and 3. In table 1 the spirals have been grouped into Sc, Sb, and Sa; otherwise the galaxies are arranged in order of right ascension. Incomplete masses, referring to the central parts of galaxies for which complete rotation curves could not, for various reasons, be determined are tabulated in table 4.

The columns of tables 1, 2, and 3 give the following information: column (2) gives the Hubble morphological classification (Sandage 1961). Column (3) gives the measured photographic magnitude, uncorrected for absorption either in our Galaxy or internally in the galaxy in question. Wherever possible, this magnitude has been taken from Holmberg (1958), otherwise from other sources compiled by de Vaucouleurs and de Vaucouleurs (1964) or from Humason, Mayall, and Sandage (1956). Column (4) gives the distance of the galaxy, and this is discussed under a separate heading below. Column (5) gives the mass in units of 10^{10} $M_{\odot}$, and column (6) gives the mass-to-light ratio (photographic) in solar units; the luminosities used here are discussed in a separate heading below. Column (7) gives a letter designating the method used in calculating the mass according to the following scheme:

a = mass point
b = one homogeneous spheroid
c = several homogeneous spheroids
d = inhomogeneous spheroids
e = disk (Wyse and Mayall)
f = disk (Brandt and Belton)
g = Bottlinger-Lohmann formula
h = virial theorem
i = 21-cm using half line width
j = rotation of stellar component (S0 and E).

Finally, column (8) gives the references.

Table 4, giving the incomplete mass determinations, follows the same scheme except that mass-to-light ratios in column (6) are replaced by the radius out to which the partial mass is determined.

For some galaxies, more than one mass determination has been given because results from independent determinations give an idea of the errors that may be involved. The galaxies M31 and M33, which have been studied by many workers, are discussed individually below, and the general question of comparison between different methods and their uncertainties is also discussed below.

A complete listing of galaxies for which masses and mass-to-light ratios have been obtained from 21-cm measures is not given here, to avoid overlap with the chapter by M. S. Roberts in this volume. In particular, many of the irregulars have been omitted, for two reasons. First, in several cases the orientation and hence the projection factor necessary to obtain the mass is not known. Second, noncircular motions are suspected to be present in some irregular galaxies, and can lead to a spurious mass. The general question of comparison between masses obtained from optical and from 21-cm measures is discussed below.

TABLE 1

MASSES AND MASS-TO-LIGHT RATIOS OF SPIRAL GALAXIES

Galaxy (1)	Type (2)	m_{pg} (3)	D (Mpc) (4)	Mass (units of $10^{10} M_{\odot}$) (5)	Mass-to-Light (pg) (6)	Method (7)	Reference (8)
NGC 157.	Sc	11.2	24	6.0	1.5	d	Burbidge, Burbidge, and Prendergast 1961*d*
NGC 247.	Sc	9.5	2.4	1.8	4	i	Epstein 1964
NGC 253.	Sc	6.9	4.0	20	5	a	Burbidge, Burbidge, and Prendergast 1962*b*
NGC 300.	Sc	8.7	2.4	1.7	6	g	Epstein 1964
	"	"	"	4.5	16	g	Rogstad, Rougoor, and Whiteoak 1967
M33.	Sc	6.2	0.63	1.8	4	a	Volders 1959
	"	"	"	1.3	3	e	Dieter 1962
	"	"	"	3.4	8	f	Brandt 1965
NGC 628.	Sc	9.7	9.2	4.6	2	g	Rogstad, Rougoor, and Whiteoak 1967
	"	"	"	7.5	3	i	Epstein 1964
NGC 925.	Sc	10.5	6.3	4.5	5	g	Rogstad, Rougoor, and Whiteoak 1967
NGC 1084	Sc	11.1	19	1.6	0.8	d	Burbidge, Burbidge, and Prendergast 1963*a*
NGC 1792	Sc	10.7	14	1.8	1	d	Rubin, Burbidge, and Burbidge 1964
NGC 2146	Sbc pec	11.3	12	1.8	3	d	Burbidge, Burbidge, and Prendergast 1959*b*
NGC 2403	Sc	8.8	3.2	12	12	i	Epstein 1964
	"	"	"	5.6	6	g	Rogstad, Rougoor, and Whiteoak 1967
NGC 3432	Sc	11.6	8.6	0.63	5	b	Bertola 1966*a*
NGC 3556	Sc–Ir	10.6	10	1.7	1.6	b	Burbidge, Burbidge, and Prendergast 1960*b*
NGC 3646	Sc pec	11.8	55	27	3.4	a	Burbidge, Burbidge, and Prendergast 1961*c*
NGC 4027	SBc	11.7	12.5	1.3	1.8	d	de Vaucouleurs, de Vaucouleurs, and Freeman 1968
NGC 4236	Sc	10.1	2.5	2.3	2	g	Rogstad, Rougoor, and Whiteoak 1967
NGC 4244	Sc	10.5	3.3	6	20	i & g	Epstein 1964; Rogstad, Rougoor, and Whiteoak 1969
NGC 4490	Sc	10.1	8.6	0.20	0.4	c	Bertola 1966*a*
NGC 5194	Sc	8.9	4.0	5	6	a	Burbidge and Burbidge 1964
	"	"	"	2.8	3	g	Rogstad, Rougoor, and Whiteoak 1967
NGC 5236	Sc	7.5	3.6	12	4	i	Epstein 1964
NGC 5248	Sc	10.4	15.4	5	3	a	Burbidge, Burbidge, and Prendergast 1962*a*
NGC 5457	Sc	8.2	4.6	16	11	g	Rogstad, Rougoor, and Whiteoak 1967; see also Volders 1959; Dieter 1962; Epstein 1964
NGC 6503	Sc	10.8	5	0.13	0.7	d	Burbidge, Burbidge, Crampin, Rubin, and Prendergast 1964*a*
NGC 6946*. .	Sc	9.7	4.2	1.9	2	i	Epstein 1964
The Galaxy.	Sb	. . . . . .	. . . . . .	13	. . . . . .	c & d	Innanen 1966*a*
M31.	Sb	4.3	0.69	34	8.4	c & d	Schmidt 1957
	"	"	"	31	8.0	f	Roberts 1966
NGC 972.	Sb	12.1	22	1.2	1.2	d	Burbidge, Burbidge, and Prendergast 1965*c*
	"	"	"	3	3	b	Demoulin 1966
NGC 1808	Sb	11.1	10	2.7	2	d	Burbidge and Burbidge 1968
NGC 2903	Sbc	9.5	7.9	4.9	4	d	Burbidge, Burbidge, and Prendergast 1960*c*
VV 116D†	Sb	. . . . . .	86	3.3		a	Burbidge and Burbidge 1961*b*

* Mass by Rogstad *et al.* is about one order of magnitude greater than that by Epstein.

† Quintet containing this galaxy appears to be bound (satisfies virial equation).

TABLE 1—*Continued*

Galaxy (1)	Type (2)	m_{pg} (3)	D (Mpc) (4)	Mass (units of $10^{10} M_{\odot}$) (5)	Mass-to-Light (pg) (6)	Method (7)	Reference (8)
NGC 3031‡..	Sb	7.8	3.2	15	7	‡	Münch 1959
NGC 3227§..	Sb	11.3	15	3.5			Rubin and Ford 1968
NGC 3521	Sb	10.1	8.5	8	4	d	Burbidge, Burbidge, Crampin, Rubin, and Prendergast 1964*b*
NGC 4258	Sb	8.9	7.8	10	6.5	d	Burbidge, Burbidge, and Prendergast 1963*c*
	"	"	"	8.5	5.5		Duflot 1965
NGC 5005	Sb	8.1	14	10	6.3	d	Burbidge, Burbidge, and Prendergast 1961*a*
NGC 5055	Sb	9.3	7.4	5.5	2.8	d	Burbidge, Burbidge, and Prendergast 1960*a*
	"	"	"	11	5.6	f	Brandt and Belton 1962
NGC 7331	Sb	10.3	14.4	14	3	d	Rubin, Burbidge, Burbidge, Crampin, and Prendergast 1965
NGC 7479	SBb	11.6	35	2.2		b	Burbidge, Burbidge, and Prendergast 1960*d*
NGC 7640	Sb	11.3	6.4	8	7	g	Rogstad, Rougoor, and Whiteoak 1967
NGC 681.	Sa	12.8	23	1.9	3.6	d	Burbidge, Burbidge, and Prendergast 1965*a*
NGC 3623	Sa	10.2	9.3	20	7	a	Burbidge, Burbidge, and Prendergast 1961*b*

‡ Stars in center have larger velocity dispersion than gas.
§ Pronounced noncircular motions within 5″ of center.

TABLE 2

MASSES AND MASS-TO-LIGHT RATIOS OF ELLIPTICAL GALAXIES

Galaxy (1)	Type (2)	m_{pg} (3)	D (Mpc) (4)	Mass (units of $10^{10} M_{\odot}$) (5)	Mass-to-Light (pg) (6)	Method (7)	Reference (8)
M32............	E2	9.1	0.69	0.36	27	h & j	King 1961; see also Burbidge, Burbidge, and Fish 1961*a*
NGC 3115.......	E7/S0	10.1	6	9.6	19	j	Poveda 1961
NGC 3379.......	E0	10.5	9.7	1.3	12	h	Fish 1964; Burbidge, Burbidge, and Fish 1961*c*
NGC 4111.......	S0	11.6	11	5.3	11	j	Poveda 1961
NGC 4406.......	E3	10.1	14	130	29	h	Fish 1964
NGC 4472.......	E1	9.3	14	150	14	h	Fish 1964
NGC 4486.......	E0 pec	9.6	14	350	44	h	Fish 1964; Poveda 1961
NGC 4486 B.....	E0 pec	14.2	14	5	80	h	Rood 1965
NGC 5128.......	E0 pec	8.0	5	20	10	a	E. M. Burbidge and G. R. Burbidge 1959

4.1. Distances

The mass of a galaxy, whether it is determined from rotation, the virial theorem, or orbital motion, varies directly in proportion to the distance taken for the galaxy. The mass-to-light ratio, because of the inverse square law for luminosity, depends inversely on the distance. Whenever the distance given in column (4) of the tables differs from the distance used by the authors of the papers quoted, their values for the mass and mass-to-light ratios have been scaled to the values corresponding to the distances adopted here. Our distances have in general been those adopted by Sandage. The distance taken for the Virgo cluster is 14.8 Mpc (Sandage 1968). The distance for the

TABLE 3

Masses and Mass-to-Light Ratios of Irregular Galaxies

Galaxy (1)	Type (2)	m_{pg} (3)	D (Mpc) (4)	Mass (units of $10^{10} M_{\odot}$) (5)	Mass-to-Light (pg) (6)	Method (7)	Reference (8)
NGC 55.....	Irr–Sc	7.9	2.4	2–4	3–6	d	de Vaucouleurs 1961
	"	"	"	2–3	2	g	Epstein 1964; Rogstad, Rougoor, and Whiteoak 1967
SMC........	Irr	2.7	0.06	0.15	3		Hindman 1967
LMC........	Irr–S	0.5	0.055	1.2	4	a	de Vaucouleurs 1960
	"	"	"	0.5–1	2–4	g	Feast, Thackeray, and Wesselink 1961; Feast 1964
	"	"	"	0.6	3		Hindman 1967
NGC 3034...	Irr–Sc	9.2	3.2	1.0		a	Mayall 1960; Burbidge, Burbidge, and Rubin 1964
	"	"	"	2.5	5	a & i	Volders and Högböm 1961; Epstein 1964
	"	"	"	1.5	6	b	Duflot 1964, 1965
NGC 4605...	Irr	11.0	3.2	0.07	2	c	Bertola 1966*a*
NGC 4631...	Irr–Sc	9.8	4.4	2	2.3	d	de Vaucouleurs and de Vaucouleurs 1963
NGC 6822...	Irr	9.2	0.5	0.15	11	a, g, & i	Volders and Högböm 1961; Epstein 1964; Rogstad, Rougoor, and Whiteoak 1967
VV 254......	Irr–S	......	61	13		a	Burbidge and Burbidge 1963

M81 group (which includes M81, M82, NGC 2403, and NGC 4236) is 3.25 ± 0.20 Mpc (Tammann and Sandage 1968). The distance for the Ursa Major I cloud (which includes NGC 4111 and NGC 4258) is 3.3 ± 0.2 Mpc (Sersic 1960; Hodge 1966; Hitchcock and Hodge 1968).

Finally, the value of the Hubble constant adopted for all larger distances is 75 km s^{-1} Mpc^{-1} (Sandage 1968).

4.2. Luminosities

Some of the differences in mass-to-light ratios obtained by various workers are due to the different luminosities used. Holmberg (1964) lists absolute magnitudes corrected for absorption in our Galaxy by the formula $\Delta m_{pg} = 0.25$ cosec b^{II} mag, and corrected for absorption in the galaxy in question, which is a function of the type and the inclination. Holmberg also lists the distance moduli that he used, so his luminosities can be adjusted to any other distance. However, not all the galaxies with mass determinations are included in his list, and not all workers use his corrections for absorption within the galaxy in question. Ideally, for a comparison of mass-to-light ratios with those obtained empirically from stellar populations matching the spectral features, one should use absorption-free luminosities. At the present time, however, the mass-to-light ratios are not on a very consistent basis.

4.3. Uncertainties in Results. Differences between Radio and Optical Determinations

That there are considerable uncertainties in the determinations of masses and mass-to-light ratios is admitted by all workers in the field. We will consider in turn some of the main causes of error.

a) If the distance taken for the galaxy is wrong, the masses and mass-to-light ratios will be wrong in direct and inverse proportion, respectively. Many galaxies studied lie at distances not greater than that of the Virgo cluster; and for these, direct calibration

TABLE 4

Incomplete Mass Determinations

Galaxy (1)	Type (2)	m_{pg} (3)	D (Mpc) (4)	Mass (units of $10^{10} M_{\odot}$) (5)	Radius (kpc) (6)	Method (7)	Reference (8)
NGC 613.	SBc	11.0	20	0.9	0.7	b	Burbidge, Burbidge, Rubin, and Prendergast 1964
NGC 1068	Sb	9.6	16	2.7	2	b	Burbidge, Burbidge, and Prendergast 1959*a*; Duflot 1963*b*; Bertola 1965
NGC 1097	SBb	10.4	16	1.0	0.8	b	Burbidge and Burbidge 1960
NGC 1365	SBc	10.6	20	3	1.5	b	Burbidge and Burbidge 1960
NGC 2782*..	Sa pec	12.5	34	15	1.7	b	Duflot 1962
NGC 3310	Sb	11.2	14	1.8	1.7	d	Walker and Chincarini 1967
	"	"	"	0.2	0.6	b	Bertola 1966*a*
NGC 3504†..	SBb	11.6	20	0.9	4.8	d	Burbidge, Burbidge, and Prendergast 1960*e*
	"	"	"	0.2	0.6	b	Burbidge, Burbidge, and Prendergast 1960*e*
NGC 4676N†	S0 pec		92	3.2	4	b	Burbidge and Burbidge 1961*a*
NGC 4676S†.	S0 pec	"	"	4.5	6	b	Burbidge and Burbidge 1961*a*
NGC 4736	Sb	8.9	10	1.1–2.1	2.5	d	Chincarini and Walker 1967*a*; see also Duflot 1963*a*
NGC 4826	Sb	9.3	8.5	0.9	2	d	Rubin, Burbidge, Burbidge, and Prendergast 1965
NGC 5383	SBb	12.4	32	0.6	1.6	b	Burbidge, Burbidge, and Prendergast 1962*c*
NGC 6181	Sc	12.3	33	4.8	3.5	a	Burbidge, Burbidge, and Prendergast 1965*b*
NGC 7469	Sa	14.3	68	0.4	0.9	b	Burbidge, Burbidge, and Prendergast 1963*b*
	"	"	"	1.1	3.6	d	

* Velocity has big gradient; object rather face-on, so a big correction for inclination leads to a large mass. If noncircular motions are present, this could be spurious.

† Orientation, hence factor $\operatorname{cosec}^2 \xi$ in mass, uncertain.

of distance indicators such as Cepheids, H II regions, and globular clusters are gradually converging on what appear to be reliable distances. Beyond the Virgo cluster, the Hubble relation has to be used, and published masses have usually been obtained with either $H = 75$ or 100 km s^{-1} Mpc^{-1}. As the work of Sandage and his collaborators gradually improves the accuracy of H, the relatively few mass determinations for which the recession velocity exceeds 4000 km s^{-1} become more secure. Between the Virgo cluster and a distance corresponding to 4000 km s^{-1}, departures from an isotropic expansion velocity field may introduce errors in using a mean Hubble relation. It remains to determine whether this is a serious cause of error in distances or not.

b) For the mass-to-light ratios, uncertainty in the absorption-free luminosity leads to errors, as discussed in § 4.2 above.

c) Mass derived from rotation models involving similar spheroids and flat disks could be in error because there may be a substantial fraction of the mass contained in spheroids of lesser ellipticity—for example, in the nuclear bulge and the halo population, where the stars will be traversing elliptical orbits. While this component may not be very pronounced for Sc galaxies, which predominate in the mass determinations from rotation, undoubtedly it should not be neglected in Sb and Sa galaxies. The more complex models, such as those for our Galaxy and for M31 because of its nearness, can take this into consideration, but for most of the galaxies the observational material is not accurate enough or detailed enough to warrant the construction of models containing large numbers of parameters.

d) Mass estimates and mass-to-light ratios derived from optical measurements of rotation may be systematically low because considerable mass may lie farther out than the last observed velocity on the rotation curve. For example, the luminous galaxy may be embedded in a cocoon of neutral hydrogen which is dynamically part of the galaxy but optically unobservable. Attempts to overcome this difficulty by using an extrapolation formula to fit the rotation curve, e.g., the Bottlinger-Lohmann or the Brandt-Belton formula, yield valid results only if the unseen rotation curve does actually obey these formulae. One does not know how much the tail wags the dog.

e) Mass estimates derived from 21-cm measures may be systematically too high for several possible reasons. First, the extrapolation formulae used to represent the rotation curve may not be valid in the outer parts. Second, noncircular motions in a galaxy can give a spread in velocity which produces too large an apparent rotation. In the analysis of Rogstad, Rougoor, and Whiteoak (1967), terms are included to represent such noncircular motions, but nevertheless these effects may still be present. It is for this reason that masses determined by the 21-cm method for NGC 4736 and the barred spiral NGC 3359 have been omitted from the tables; noncircular motions are known to occur in these galaxies. Third, neutral hydrogen around the outermost parts of a galaxy may not be in equilibrium with the galaxy as a whole. For example, the high-velocity clouds of hydrogen seen in high latitudes in our own Galaxy might be due to incoming clouds from intergalactic regions in the Local Group, and possibly there are clouds of hydrogen around galaxies, which are being either accreted or ejected.

It is interesting to compare the optical and 21-cm rotation measurements and mass determinations for NGC 5055. The optical rotation curve derived by Burbidge, Burbidge, and Prendergast (1960*a*) is shown in figure 2. It has a very well-defined maximum velocity or turnover point 1ʹ.8 from the center, and the density curve derived from the rotation curve by the method of inhomogeneous spheroids has dropped to a very low value by the outermost measured point. Extrapolation beyond this last derived point would add relatively little mass. But Rogstad, Rougoor, and Whiteoak (1967) have measured a 21-cm rotation profile, and, fitting this by a Bottlinger-Lohmann velocity curve, they have derived a radius for a turnover of the velocity curve 5ʹ.0 from the

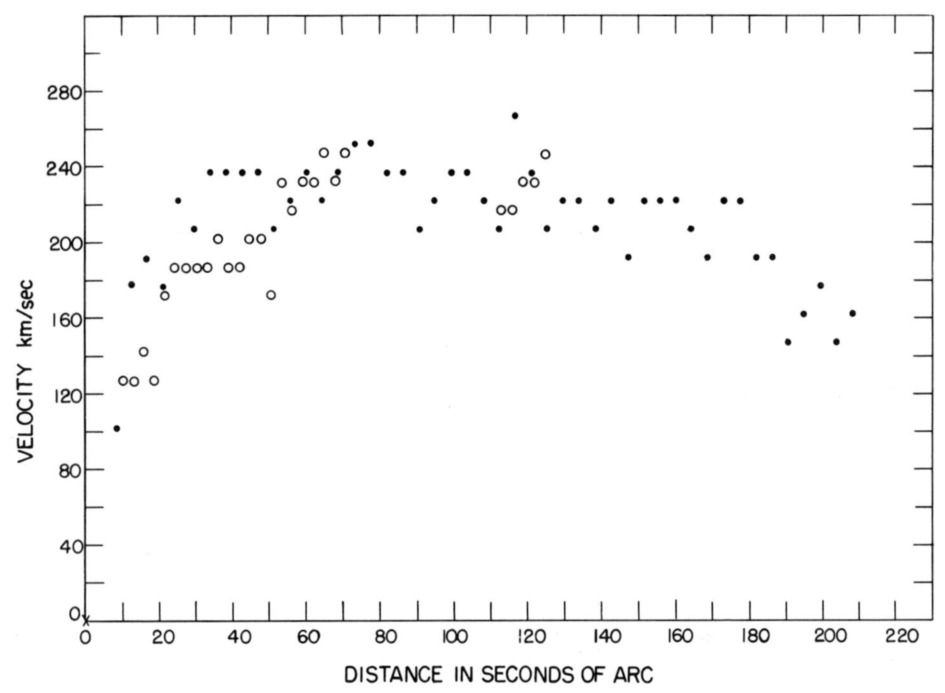

FIG. 2.—Rotational velocities in NGC 5055, measured from Hα and [N II] λ6583. *Abscissa*, distance from center along major axis in seconds of arc. *Ordinate*, velocity relative to center, uncorrected for orientation of galaxy to line of sight. *Dots*, measures on east side of center; *open circles*, measures on west side of center, reflected about center; *cross*, central velocity determined mainly from absorption lines (Burbidge, Burbidge, and Prendergast 1960*a*).

center, a factor of nearly 3 farther out than the optical turnover point. The reason for this discrepancy is not known. Seielstad and Whiteoak (1965) studied a number of irregular and Sc galaxies by the 21-cm method, and found that in all of them the 21-cm rotation curves reached maximum past the optical limits of the galaxy. Optical rotation curves are not available for these particular galaxies (NGC 300, 2403, 3109, 4214, and 4244), so direct comparison here cannot be made. On the other hand, Rogstad, Rougoor, and Whiteoak (1967) found that in most of the galaxies they studied, the turnover radii, where the velocity reached maximum, lay within the optical limits of the galaxy.

In some of the optical mass determinations, when the last measured velocity point lay still well within the visible optical limits of the galaxy and the derived density distribution had not dropped to a vanishingly small value at that point, an estimate of the total mass has been made by supposing that the density remains constant at the last derived value, out to the last visible traces of the galaxy.

Perhaps the best answer will come with the accumulation of more double-galaxy measures, because the masses derived from them are total masses and must always include the matter that is not optically visible. The present situation is that the double-galaxy results yield average mass-to-light ratios in better agreement with the averages from optical measures than with those from the radio measures.

4.4. M31

Because it is so close, more attention has been paid to the rotation of M31 than to any other galaxy with the possible exception of M33. The mass given in table 1 is that obtained from the 21-cm measurements, which are reliable in this case because the galaxy has an angular size large compared with the beamwidth of the telescope. However, from the 21-cm method it is not possible to obtain a rotation curve for the central region. The rotation of the nuclear region has been studied at high dispersion by Lallemand, Duchesne, and Walker (1960), who found that the rotation curve rises steeply in the innermost few parsecs. For the central part outside the nuclear region the only published measurements are still those of Babcock (1939). These are shown in figure 1. Most workers appear to have left aside the absorption-line measures of Babcock, probably because it has been felt that ultimately it would be possible to measure the rotation curve in this region by using emission lines from the ionized gas.

For the nuclear region Kinman has considered models, using the observed light distribution, the rotation curve of Lallemand *et al.*, and the velocity dispersion measures of Minkowski (1961). First he constructed a model consisting of 33 homogeneous spheroids that would reproduce the light distribution within 1.6 kpc of the center. Then he showed that a simple mass model based on this could not reproduce the observed rotation curve. The true curve is probably even steeper than the observed curve because the observations were not corrected for the light contribution from parts of the galaxy along the line of sight and not in the equatorial plane. Further, the random velocity dispersion of 225 km s^{-1} observed by Minkowski has to be taken into account in a valid model. Kinman found that a model in which the stellar population of the nucleus consists primarily of stars with moderately eccentric orbits can be constructed; this combines, by the stellar dynamical ellipsoidal hypothesis, an isotropic velocity dispersion with rotation, and can result in a mass-to-light ratio that is in reasonable agreement with that deduced by Spinrad (1966) for the central region (about 16). It should be mentioned that both mass-to-light ratios quoted for the whole galaxy in table 1 do include the correction for internal absorption within M31, according to Holmberg. Kinman pointed out that in this model for the central region, gas produced by evolving stars could not stay in a dynamically steady state in the outer parts of the nucleus, and this might account for the outflow observed by Münch and discussed under noncircular motions, below.

4.5. M33

Brandt (1965) collected all previous mass determinations, scaled to a distance of 630 kpc, to compare with the value which he had obtained. The comparison showed clearly the effect of taking different radii; Brandt's value is an extrapolation beyond the last observed point, and it is probably safer to use the values of Volders or Dieter. The mass-to-light ratios in table 1 are obtained by using Holmberg's absorption-free luminosity; the values given in Brandt's paper are not corrected for internal absorption.

Walker looked for rapid rotation like that found in M31 and M32, but found none. Carranza *et al.* (1968), using the Fabry-Perot interferometer designed by Courtès, were able to measure emission-line velocities with high accuracy in M33. They looked for a component of line-of-sight velocity along the minor axis, like that found by Münch in M31 which was interpreted as a radial outflow of gas from the central region. To a high degree of accuracy, however, they found no such motions along the minor axis. However, they did find a circular velocity given by the gas in H II regions in the spiral arms of M33 which differs from that given by the more diffuse gas belonging to the general disk, between the arms. The velocity discontinuity between the disk and the arms amounted to about 15 km s^{-1}.

4.6. Noncircular Motions

Observations of line-of-sight velocities made for the construction of rotation curves can reveal noncircular motions in the following ways: (*a*) A component of velocity relative to the central velocity may be found along the minor axis, where there should be no contribution from rotation. (*b*) Bumps or discontinuities may be found in the velocity curve measured along or near the major axis. (*c*) Velocities measured by the long-slit method in various position angles may not give consistent results when projected onto the major axis. (*d*) Very broad lines or other manifestations of a peculiar velocity field may be found in the nucleus, e.g., in Seyfert galaxies.

Galaxies in which noncircular motions have been found in the gaseous component are: M31 (Münch 1960, 1961); NGC 253 (Burbidge, Burbidge, and Prendergast 1962*b*); NGC 925 (Höglund and Roberts 1965); NGC 1068 (Walker 1968); NGC 1275 (Minkowski 1957; Burbidge and Burbidge 1965); NGC 1365 (Burbidge, Burbidge, and Prendergast 1962*d*); NGC 3034 (M82) (Lynds and Sandage 1963; Burbidge, Burbidge, and Rubin 1964); NGC 3227 (Rubin and Ford 1968); NGC 4027 (de Vaucouleurs, de Vaucouleurs, and Freeman 1968); NGC 4258 (Burbidge, Burbidge, and Prendergast 1963*c*; Chincarini and Walker 1967*b*); NGC 4486 (Walker and Hayes 1967); NGC 4631 (de Vaucouleurs and de Vaucouleurs 1963); NGC 4736 (Chincarini and Walker 1967*a*); NGC 5194 (M51) (Burbidge and Burbidge 1964); NGC 5383 (Burbidge, Burbidge, and Prendergast 1962*c*); NGC 6181 (Burbidge, Burbidge, and Prendergast 1965*b*).

While the Seyfert galaxies NGC 1068, 1275, 3227, and 4151 rank as the first in which such motions were found, our Galaxy and M31 show evidence for smaller noncircular motions. Münch (1960, 1961) interpreted the velocities he measured along the minor axis of M31 as radial outflow of gas from the center. The observations of the central region of our Galaxy, made by 21-cm measurements, are more detailed and reveal several kinds of motion: outward expansion, fast inner rotation, nonaxisymmetric outflow, and possibly outflow in directions inclined to the equatorial plane (Rougoor 1964; Oort 1968).

The barred spirals can always be expected to show some streaming motions, and these have been found wherever looked for. Good cases are NGC 4027 (de Vaucouleurs, de Vaucouleurs, and Freeman 1968) and NGC 5383 (Burbidge, Burbidge, and Prendergast 1962*c*); the velocities in the latter are shown in figure 3. The nuclear region of NGC

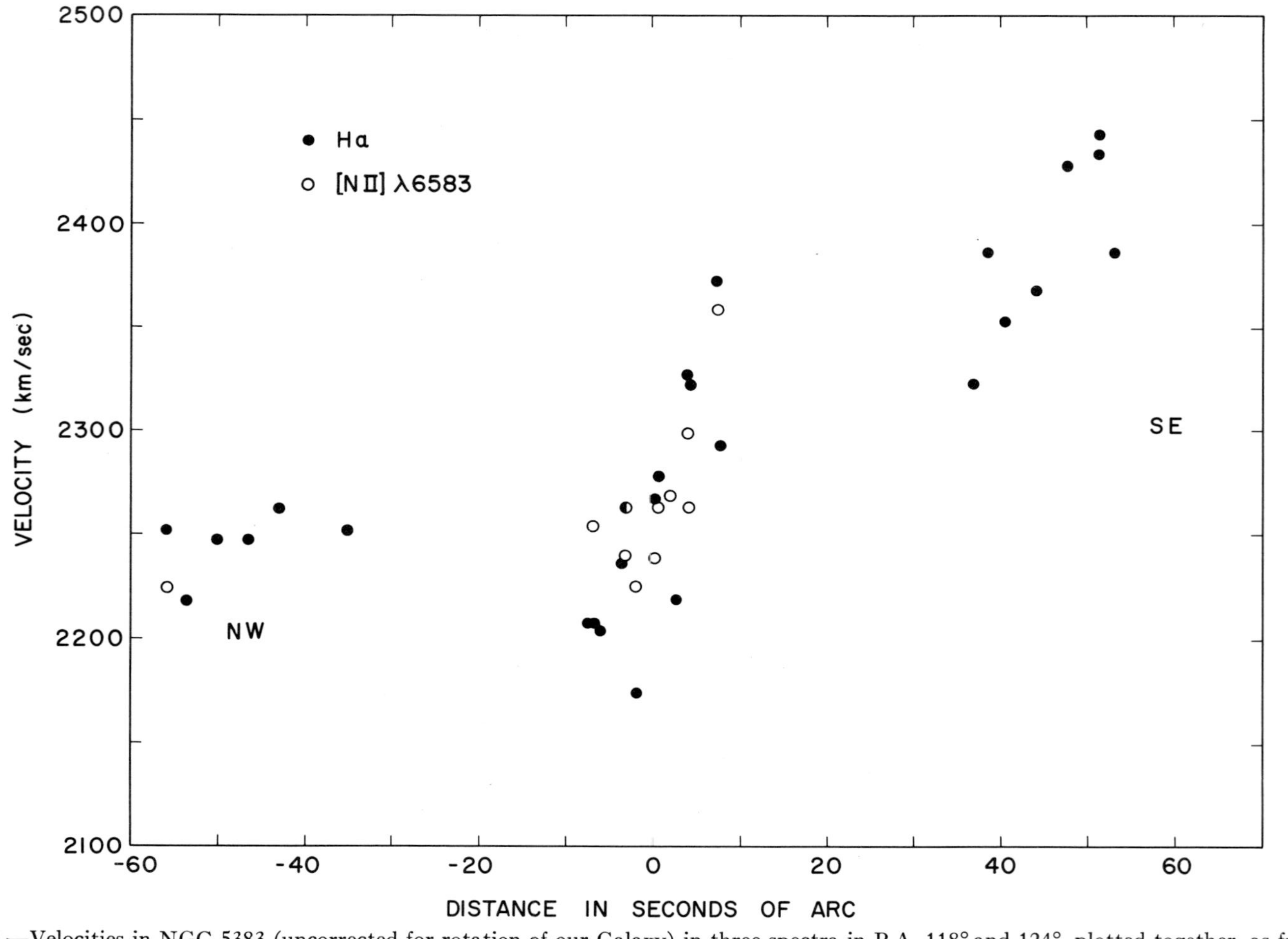

FIG. 3.—Velocities in NGC 5383 (uncorrected for rotation of our Galaxy) in three spectra in P.A. 118° and 124°, plotted together, as function of distance from center (Burbidge, Burbidge, and Prendergast 1962*d*).

1365 also has noncircular motions. This is shown by the fact that long-slit measures in various position angles could not be projected into a consistent rotation curve (Burbidge, Burbidge, and Prendergast 1962*d*).

The radio galaxy M82 provided the first clear-cut evidence for an explosive event occurring in the nuclear region of a galaxy; this was demonstrated by the outward flow of gas perpendicular to the equatorial plane found by Lynds and Sandage (1963) and studied also by Burbidge, Burbidge, and Rubin (1964). The very strange velocity field in NGC 1275, found by Minkowski (1954), when mapped in many position angles was interpreted in the same way by Burbidge and Burbidge (1965). The complex motions found by Walker (1968) in NGC 1068, and by Rubin and Ford (1968) in NGC 3227, are also manifestations of violent events in their nuclei, and this applies also to NGC 4486 and NGC 4151.

The noncircular motions found in other galaxies are smaller, but they commonly occur; in NGC 925 they were found from the 21-cm measurements by Höglund and Roberts (1965), and in NGC 4258 and NGC 4736 they are demonstrated by bumps on the rotation curve. Such effects merge into the small velocity discontinuities across spiral arms found in our Galaxy (see, e.g., Shane and Bieger-Smith 1966) and in M33 (Carranza *et al.* 1968) which can probably be accounted for by the detailed dynamics of spiral-arm formation.

It is not known how widespread is the occurrence of noncircular motions in the nuclear regions of galaxies. They were found in M51 (NGC 5194) and NGC 253 because those galaxies are large, bright, and fairly near; perhaps they occur rather generally in spiral galaxies. On the other hand, there is a small-diameter, rather weak, radio source in the center of NGC 253, so the motions there may be connected with this.

The fact that noncircular motions can occur, and are particularly likely to occur in radio galaxies or galaxies otherwise showing violent events, means that a single measurement of an inclined emission line cannot always be taken as a measurement of pure rotation. For distant galaxies one is tempted to make this assumption, but it may not be valid.

5. MASS-TO-LIGHT RATIOS AND MASSES OF GALAXIES IN CLUSTERS; ANGULAR MOMENTUM

In the previous section we have described the uncertainties and selection effects which affect the mass determinations given in tables 1–4. Despite the very extensive work which has been done, largely in the last 15 years, there are many gaps in our knowledge. While we have a large enough sample of late-type spiral galaxies to be able to derive average masses, the sample of ellipticals, S0, and Sa types is so small that averages obtained from individual masses are not well determined. For these types the best average masses are obtained by using the double galaxies. Comparison of table 5 with tables 1–4 shows that while the average masses derived for spirals and irregulars by Page are in reasonably good agreement with the results for individual galaxies, the results for the ellipticals give an average which appears to be large compared with the few individual determinations which have been made. Page (1965) has discussed the selection effects which may have influenced the selection of pairs of galaxies which have been studied. Selection of higher-luminosity pairs is to be expected because they might be thought to be easier to study spectroscopically. If the more luminous galaxies tend to be the more massive objects, this would lead to high average masses. A similar selection effect might be present in the individual galaxies chosen by optical astronomers for rotation studies, though other selection effects, particularly the presence of ionized gas and a suitable orientation, are more important. Also, there is no reason to believe that the more luminous and massive spiral galaxies have been chosen for studies by the 21-cm technique.

Consequently, the results given in tables 1–5 are likely to be reasonably secure as far as the spirals are concerned.

Similar selection effects—the choice of the most massive pairs—in the study of the ellipticals may also be present. In this case we have such a small sample of single ellipticals—which have also been chosen because they are the easiest to study, either because they are intrinsically very bright or because they are close by—that we cannot make a meaningful comparison between masses determined by the two methods. One suspects that the average elliptical masses derived by Page may tend to be high compared with a proper sample, but one does not know.

TABLE 5

MASSES AND MASS-TO-LIGHT RATIOS OF DOUBLE GALAXIES
(From Page; $H = 75$ km s^{-1} Mpc^{-1})

Type	Mean Mass (units of $10^{10}\ M_{\odot}$	Mass-to-Light (pg)
S and Irr:		
(16 systems)	5.3	2.4
(13 high-weight systems)	2.0	1.0
E and S0:		
(18 systems)	88	74
(13 high-weight systems)	79	68

5.1. MASS-TO-LIGHT RATIOS

The results show that the average mass-to-light ratios for spirals and ellipticals are very different. The average value for spirals given by Page is very small. The inhomogeneous data given in tables 1–3 are generally in agreement with this result, though the mass-to-light ratio in some spirals—e.g., M31—have values 10–20. For the ellipticals, the average given by Page is about 70. For comparison, the mass-to-light ratio for the solar neighborhood is about unity, for globular clusters it is about 0.7, while for old clusters of metal-rich stars such as M67 values as low as 0.1 are indicated. Since the mass determinations and the estimates of total luminosity given in this article are not likely to be very far wrong, it is clear that the constituents in elliptical galaxies, and in some spirals, must differ very greatly from the types of stellar population found in the solar neighborhood and in star clusters in our own Galaxy. They must differ in the sense that there is a large population of objects which contribute a great deal to the mass but which contribute very little light. Since the integrated spectra of the nuclei of M31 and some bright ellipticals show that the light comes largely from a population of stars of M67 type (Morgan and Mayall 1957), it is to a population of this type that a large mass fraction contributing little light must be added. Since there is not thought to be a large amount of diffuse matter in galaxies with high mass-to-light ratios, this mass is thought to be in the form of very small mass stars (red dwarfs) or in the form of highly evolved stars (white dwarfs or neutron stars), or collapsed mass. Spinrad (1966) and others have investigated possible models which will do this. They have favored the view that there is a very large population of red dwarfs. This is not a unique answer, and a case can also be made for the possibility that much mass in the central region of an elliptical is mass close to or inside its Schwarzschild radius so that it cannot be seen. But detailed discussion of this possibility lies outside the scope of this article.

5.2. Masses of Galaxies in Clusters

In § 2 the method of mass determination using the virial theorem applied to the galaxies in groups or clusters was described. However, no masses derived in this way are given in this article. The reason is that in nearly all cases the masses so derived are very much greater than the masses obtained by the other methods. We have no reason to suppose that the masses of galaxies in groups and clusters really are systematically much greater than those derived by other methods; in some cases, galaxies which have been studied individually or in pairs lie in the Virgo cluster which can be studied using the virial theorem, and the average masses so derived are far higher than the masses measured directly.

There are therefore two possible explanations for this effect: (1) There is enough unseen matter lying between the galaxies in a cluster or group to bind it. (2) The systems are not stable; they have positive total energy and are expanding as was originally proposed by Ambartsumian (1956, 1961).

A conference centered about this problem was held in 1961 (Neyman, Page, and Scott 1961), and the question has not been resolved. For example, the average mass of the elliptical galaxies in the Virgo E cluster would need to be about $3 \times 10^{12} M_{\odot}$ if the cluster were stable and contained only these galaxies, and their average mass-to-light ratio would need to be about 500. For the Coma cluster the average mass would need to be about $2 \times 10^{12} M_{\odot}$, and the mass-to-light ratio would be about 700. These values are to be compared with those given in tables 1–5. It is important to realize that if such clusters are bound, a very large amount of unseen matter between the galaxies must be present *over and above* the mass contained in the galaxies which is emitting little light, and which is also not accounted for.

For some clusters which contain spirals, e.g., the Hercules cluster (G. R. Burbidge and E. M. Burbidge 1959), the average mass required for stability is about $10^{12} M_{\odot}$. In some very small groups similar situations prevail. For centrally condensed clusters like the Coma cluster one suspects that the system is stable, and that therefore some 75 percent of the mass is present in unseen form. However, for widely extended clusters with little central condensation, the possibility that the cluster is expanding must be taken seriously, despite the time-scale problems which are involved. In small groups it is often difficult to see how a large amount of uncondensed matter could be hidden; and if it is present, it is probably collapsed. On the other hand, a number of small groups are known (e.g., Stephan's Quintet) in which one member has a very different redshift from the others. This is not likely to be because it is a foreground or background galaxy, and it must be concluded that the object is being explosively ejected from the group. A full discussion of these problems falls outside the subject matter of this article. Most of the data are still those given at the conference organized by Neyman, Page, and Scott (1961), and the reader is referred to this for a full account of the different possibilities.

5.3. Angular Momentum

A quantity which can in principle be derived from the rotation curve and the mass distribution of a galaxy is the angular momentum. The total angular momentum, the angular momentum per unit mass, and the distribution of angular momentum within a galaxy are all relevant to consideration of the evolutionary history of a galaxy.

Observationally there is an inherent uncertainty in determinations of the total angular momentum because the rotation curve is usually least well determined in the outermost parts of a galaxy where the light intensity is lowest, and therefore the mass distribution here is subject to the greatest uncertainty, yet the outermost parts contribute substantially to the total angular momentum. The fitting of analytical expressions of the Bottlinger-Lohmann or Brandt-Belton type to the rotation curves gives the appearance of circumventing this difficulty, but will do so only insofar as these extrapolatory functions really do represent well the outermost rotation curve.

Brosche (1963) obtained the angular momenta of the Galaxy, the Large Magellanic Cloud, and a number of spiral galaxies including M31 and M33; he used the various published rotations and masses and then found that the total angular momentum p, when plotted against the mass M on a log-log plot, showed a good correlation of the form

$$p \propto M^2 \,. \tag{61}$$

In order to test whether galaxies could have collapsed from uniform clouds without large-scale turbulent mixing, Crampin and Hoyle (1964) obtained the distribution of angular momentum in some spiral galaxies from the published rotation curves and mass distributions, and compared it with that of a homogeneous, spheroidal, uniformly rotating gas cloud. These galaxies had been analyzed by the method of Burbidge, Burbidge, and Prendergast. If the two distributions are the same, then the equation

$$2\pi\sigma(\varpi)\left(\frac{u_\theta}{\varpi}+\frac{du_\theta}{d\varpi}\right)^{-1} = C_1\left(1-\frac{\varpi u_\theta}{C_2}\right)^{1/2} \tag{62}$$

should hold for all ϖ. Here $C_1 = 2\pi b\rho_0/\omega = 1.5M_c/\omega a^2$; a and b are the semimajor and semiminor axes of the initial spheroidal cloud; ρ is its initial (uniform) density; M_c is its mass; and ω is its angular velocity of rotation. In the final galaxy $u_\theta(\varpi)$ is the rotation curve; $\sigma(\varpi)$ is the mass per unit column normal to the galactic plane, as a function of ϖ, the distance from the center; and $C_2 = Ru_\theta(R)$, where R is the outer radius of the collapsed gas cloud, i.e., the radius of the galaxy. Crampin and Hoyle found that equation (62) did indeed hold except in the central regions. On the other hand, Innanen (1966*b*) found less good agreement for our Galaxy upon making this calculation for various mass models of the Galaxy.

To look into the relation between angular momentum and mass, Ozernoy (1967) took the model of a homogeneous spheroid rotating with constant angular velocity, since

Crampin and Hoyle had found reasonable agreement between this and actual models of galaxies. For the angular momentum per unit mass, K, he derived a theoretical relation

$$K = \frac{2}{5}\left(\frac{4\pi}{3}\right)^{-2/3} G^{1/2} A^{1/2}(e) e^{-2/3} \rho^{-1/6} M^{2/3}, \qquad (63)$$

where A is a geometrical expression depending only on the eccentricity e through the equation

$$A(e) = 2\pi e(1 - e^2)^{-3/2}[\arc\sin (1 - e^2)^{1/2} - e(1 - e^2)^{1/2}]. \qquad (64)$$

On the hypothesis that the average density within the subclasses Sa, Sb, and Sc is approximately constant, one would then expect that for each class

$$K \propto M^{2/3}. \qquad (65)$$

For the total angular momentum, one would have $p \propto M^{5/3}$, instead of Brosche's empirical relation (61). Ozernoy simplified equations (63) and (64) by taking $e = 1$; the angular momentum per unit mass is then

$$K = \frac{2}{5}(GMR)^{1/2}, \qquad (66)$$

where R is the observed semimajor axis. Taking a large number of galaxies from the compilation of Holmberg (1964), in which masses have been estimated from a mean relation between intrinsic color index and mass-to-luminosity ratio derived from the fairly small number of actual masses available in 1964, Ozernoy found that the relation (65) was reasonably well satisfied.

Research in extragalactic astronomy is supported at UCSD in part by the National Science Foundation and by the National Aeronautics and Space Administration.

REFERENCES

Aller, L. H. 1942, *Ap. J.*, **95**, 48.
Ambartsumian, V. A. 1956, *Izv. Acad. Nauk Armenian SSR*, **9**, 23.
———. 1961, *A.J.*, **66**, 536.
Babcock, H. W. 1939, *Bull. Lick Obs.*, **19**, 41.
Bertola, F. 1965, *Ann. d'Ap.*, **28**, 574.
———. 1966*a*, *Contr. Obs. Ap. Asiago*, No. 172.
———. 1966*b*, *ibid.*, No. 186.
Bottlinger, K. F. 1933, *Veröff. Sternw. Babelsberg*, Vol. **10**, No. 2.
Brandt, J. C. 1960, *Ap. J.*, **131**, 293.
———. 1965, *M.N.R.A.S.*, **129**, 309.
Brandt, J. C., and Belton, M. J. 1962, *Ap. J.*, **136**, 352.
Brandt, J., and Scheer, L. S. 1965, *A.J.*, **70**, 471.
Brosche, P. 1963, *Zs. f. Ap.*, **57**, 143.
Burbidge, E. M., and Burbidge, G. R. 1959, *Ap. J.*, **129**, 271.
———. 1960, *ibid.*, **132**, 30.
———. 1961*a*, *ibid.*, **133**, 726.
———. 1961*b*, *ibid.*, **134**, 248.
———. 1963, *ibid.*, **138**, 1306.
———. 1964, *ibid.*, **140**, 144.
———. 1965, *ibid.*, **142**, 1351.
———. 1968, *ibid.*, **151**, 99.
Burbidge, G. R., and Burbidge, E. M. 1959, *Ap. J.*, **130**, 629.
Burbidge, E. M., Burbidge, G. R., Crampin, D. J., Rubin, V. C., and Prendergast, K. H. 1964*a*, *Ap. J.*, **139**, 539.
———. 1964*b*, *ibid.*, p. 1058.

Burbidge, E. M., Burbidge, G. R., and Fish, R. A. 1961*a*, *Ap. J.*, **133**, 393.
———. 1961*b*, *ibid.*, p. 1092.
———. 1961*c*, *ibid.*, **134**, 251.
Burbidge, E. M., Burbidge, G. R., and Prendergast, K. H. 1959*a*, *Ap. J.*, **130**, 26.
———. 1959*b*, *ibid.*, p. 739.
———. 1960*a*, *ibid.*, **131**, 282.
———. 1960*b*, *ibid.*, p. 549.
———. 1960*c*, *ibid.*, **132**, 640.
———. 1960*d*, *ibid.*, p. 654.
———. 1960*e*, *ibid.*, p. 661.
———. 1961*a*, *ibid.*, **133**, 814.
———. 1961*b*, *ibid.*, **134**, 232.
———. 1961*c*, *ibid.*, p. 237.
———. 1961*d*, *ibid.*, p. 874.
———. 1962*a*, *ibid.*, **136**, 128.
———. 1962*b*, *ibid.*, p. 339.
———. 1962*c*, *ibid.*, p. 704.
———. 1962*d*, *ibid.*, p. 119.
———. 1963*a*, *ibid.*, **137**, 376.
———. 1963*b*, *ibid.*, p. 1022.
———. 1963*c*, *ibid.*, **138**, 375.
———. 1965*a*, *ibid.*, **142**, 154.
———. 1965*b*, *ibid.*, p. 641.
———. 1965*c*, *ibid.*, p. 649.
Burbidge, E. M., Burbidge, G. R., and Rubin, V. C. 1964, *Ap. J.*, **140**, 942.
Burbidge, E. M., Burbidge, G. R., Rubin, V. C., and Prendergast, K. H. 1964, *Ap. J.*, **140**, 85.
Carranza, G., Courtès, G., Georgelin, Y., Monnet, G., and Pourcelot, A. 1968, *Ann. d'Ap.*, **31**, 63.
Chandrasekhar, S. 1963, *Ap. J.*, **138**, 1182.
Chincarini, G., and Walker, M. F. 1967*a*, *Ap. J.*, **147**, 407.
———. 1967*b*, *Ap. J.*, **149**, 487.
Courtès, G. 1964, *A.J.*, **69**, 325.
Crampin, D. J., and Hoyle, F. 1964, *Ap. J.*, **140**, 99.
Demoulin, M.-H. 1966, *Pub. Haute Provence Obs.*, Vol. **8**, No. 1.
Dieter, N. H. 1962, *A.J.*, **67**, 317.
Duflot, R. 1962, *Pub. Haute Provence Obs.*, Vol. **5**, No. 14.
———. 1963*a*, *ibid.*, Vol. **6**, No. 21.
———. 1963*b*, *ibid.*, No. 23.
———. 1964, *ibid.*, Vol. **7**, No. 7.
———. 1965, *ibid.*, Vol. **8**, No. 16.
Epstein, E. E. 1964, *A.J.*, **69**, 490.
Feast, M. W. 1964, *M.N.R.A.S.*, **127**, 195.
Feast, M. W., Thackeray, A. D., and Wesselink, A. J. 1961, *M.N.R.A.S.*, **122**, 433.
Fish, R. A. 1961, *Ap. J.*, **134**, 880.
———. 1964, *ibid.*, **139**, 284.
Freeman, K. 1965, *M.N.R.A.S.*, **130**, 183.
———. 1966, *ibid.*, **134**, 1.
Hindman, J. V. 1967, *Australian J. Phys.*, **20**, 147.
Hitchcock, J. L., and Hodge, P. W. 1968, *Ap. J.*, **152**, 1067.
Hodge, P. W. 1966, *Atlas and Catalogue of* H II *Regions in Galaxies* (Seattle: University of Washington).
Höglund, B., and Roberts, M. S. 1965, *Ap. J.*, **142**, 1366.
Holmberg, E. 1937, *Lund Ann.*, No. 6.
———. 1954, *Medd. Lunds. Obs.*, Ser. 1, No. 186.
———. 1958, *ibid.*, Ser. 2, No. 136.
———. 1964, *Ark. f. Ap.*, **3**, 387.
Hulst, H. van de, Raimond, E., and Woerden, H. van. 1957, *B.A.N.*, **14**, 1.
Humason, M. L., Mayall, N. U., and Sandage, A. R. 1956, *A.J.*, **61**, 97.
Innanen, K. A. 1966*a*, *Zs. f. Ap.*, **64**, 158.
———. 1966*b*, *Ap. J.*, **143**, 150.
Kerr, F. J., and Vaucouleurs, G. de. 1956, *Australian J. Phys.*, **9**, 90.
King, I. 1961, *Ap. J.*, **134**, 272.
King, I., and Minkowski, R. 1966, *Ap. J.*, **143**, 1002.
Kuzmin, G. 1952, *Pub. Astr. Obs. Tartu*, **32**, 211.
Lallemand, A., Duchesne, M., and Walker, M. F. 1960, *Pub. A.S.P.*, **72**, 76.
Lohmann, W. 1954, *Zs. f. Ap.*, **35**, 159.
Lynds, C. R., and Sandage, A. R. 1963, *Ap. J.*, **137**, 1005.
Mayall, N. U. 1948, *Sky and Tel.*, Vol. **8**, No. 3.
———. 1950, *Pub. Obs. Univ. Michigan*, **10**, 19.

———. 1960, *Ann. d'Ap.*, **23**, 344.
Mayall, N. U., and Aller, L. H. 1942, *Ap. J.*, **95**, 5.
Minkowski, R. 1942, *Ap. J.*, **96**, 306.
———. 1954, *Carnegie Inst. Yearbook*, **53**, 26.
———. 1957, *IAU Symposium No. 4*, ed. R. N. Bracewell (Palo Alto: Stanford University Press), p. 107.
———. 1961, *IAU Symposium No. 15*, ed. G. C. McVittie (New York: Macmillan), p. 112.
Morgan, W. W., and Mayall, N. U. 1957, *Pub. A.S.P.*, **69**, 291.
Münch, G. 1959, *Pub. A.S.P.*, **71**, 101.
———. 1960, *Ap. J.*, **131**, 250.
———. 1961, *IAU Symposium No. 15*, ed. G. C. McVittie (New York: Macmillan), p. 119.
Neyman, J., Page, T., and Scott, E. L. 1961, *A.J.*, **66**, 533.
Oort, J. H. 1927, *B.A.N.*, **4**, 79, 91.
———. 1968, in *Galaxies and the Universe*, ed. L. Woltjer (New York: Columbia University Press), p. 1.
Ozernoy, L. M. 1967, *Astr. Tsirk.*, No. 407, 1.
Page, T. 1952, *Ap. J.*, **116**, 63.
———. 1960, *ibid.*, **132**, 910.
———. 1961, *Proc. 4th Berkeley Symposium Math. Statistics*, **3**, 277.
———. 1962, *Ap. J.*, **136**, 685.
———. 1965, *Smithsonian Ap. Obs. Spec. Rept.*, No. 195.
Page, T., Dahn, C. C., and Morrison, F. F. 1961, *A.J.*, **66**, 614.
Perek, L. 1958, *Bull. Astr. Inst. Czechoslovakia*, **9**, 208, 212.
———. 1962, *Adv. Astr. and Ap.*, **1**, 165.
Poveda, A. 1958, *Bol. Obs. Tonantzintla y Tacubaya*, No. 17.
———. 1961, *Ap. J.*, **134**, 910.
Prendergast, K. H. 1962, *Interstellar Matter in Galaxies*, ed. L. Woltjer (New York: Columbia University Press), p. 217.
Roberts, M. S. 1966, *Ap. J.*, **144**, 639.
Rogstad, D. H., Rougoor, G. H., and Whiteoak, J. B. 1967, *Ap. J.*, **150**, 9.
Rood, H. J. 1965, *A.J.*, **70**, 689.
Rougoor, G. W. 1964, *B.A.N.*, **17**, 381.
Rubin, V. C., Burbidge, E. M., and Burbidge, G. R. 1964, *Ap. J.*, **140**, 80.
Rubin, V. C., Burbidge, E. M., Burbidge, G. R., Crampin, D. J., and Prendergast, K. H. 1965, *Ap. J.*, **141**, 759.
Rubin, V. C., Burbidge, E. M., Burbidge, G. R., and Prendergast, K. H. 1965, *Ap. J.*, **141**, 885.
Rubin, V. C., and Ford, W. K. 1968, *Ap. J.*, **154**, 431.
Sandage, A. R. 1958, *Ap. J.*, **127**, 513.
———. 1961, *Hubble Atlas of Galaxies* (Washington: Carnegie Institution of Washington).
———. 1968, *Ap. J. (Letters)*, **152**, L149.
Schmidt, M. 1956, *B.A.N.*, **13**, 15.
———. 1957, *ibid.*, **14**, 17.
———. 1965, in *Galactic Structure*, ed. A. Blaauw and M. Schmidt (Chicago: University of Chicago Press), p. 513.
Schwarzschild, M. 1954, *A.J.*, **59**, 273.
Seielstad, G. A., and Whiteoak, J. B. 1965, *Ap. J.*, **142**, 616.
Sersic, J. L. 1960, *Zs. f. Ap.*, **50**, 168.
Shane, W. W., and Bieger-Smith, G. P. 1966, *B.A.N.*, **18**, 263.
Smith, S. 1936, *Ap. J.*, **83**, 23.
Spinrad, H. 1966, *Pub. A.S.P.*, **78**, 367.
Tammann, G. A., and Sandage, A. R. 1968, *Ap. J.*, **151**, 825.
Toomre, A. 1963, *Ap. J.*, **138**, 385.
Tuberg, M. 1942, *Ap. J.*, **98**, 501.
Vaucouleurs, G. de. 1948, *Ann. d'Ap.*, **11**, 247.
———. 1953, *M.N.R.A.S.*, **113**, 134.
———. 1960, *Ap. J.*, **131**, 265.
———. 1961, *ibid.*, **133**, 405.
Vaucouleurs, G. de, and Vaucouleurs, A. de. 1963, *Ap. J.*, **137**, 363.
———. 1964, *Reference Catalogue of Bright Galaxies* (Austin: University of Texas).
Vaucouleurs, G. de, Vaucouleurs, A. de, and Freeman, K. 1968, *M.N.R.A.S.*, **139**, 425.
Volders, L. 1959, *B.A.N.*, **14**, 323.
Volders, L. and Högböm, E. 1961, *B.A.N.*, **15**, 307.
Walker, M. F. 1962, *Ap. J.*, **136**, 695.
———. 1968, *ibid.*, **151**, 71.
Walker, M. F., and Chincarini, G. 1967, *Ap.* J., **147**, 416.
Walker, M. F., and Hayes, S. 1967, *Ap. J.*, **149**, 481.
Wyse, A. B., and Mayall, N. U. 1942, *Ap. J.*, **95**, 24.
Zwicky, F. 1933, *Helv. Phys. Acta*, **6**, 110.

CHAPTER 4

Magnitudes, Colors, Surface Brightness, Intensity Distributions, Absolute Luminosities, and Diameters of Galaxies

ERIK HOLMBERG
Uppsala University Observatory

1. PHOTOMETRY OF GALAXIES

1.1. Techniques and Practical Problems

Since the basic principles involved in the photometry of extended objects have been discussed elsewhere in this compendium (Vol. 2, chap. 17), this section is confined to a short summary of the special problems encountered in the photometry of galaxies.

The total magnitudes of most galaxies are difficult to measure by any technique on account of the faint surface luminosity and low luminosity gradients of the outer parts. It is, in fact, possible to determine the integrated magnitude only by a more or less approximate extrapolation procedure. The difficulty may be overcome by stipulating that the magnitude refer to the light contained within a certain isophote. If the limiting surface brightness is sufficiently low, the magnitude thus defined will approach the total magnitude of the galaxy.

In *photographic photometry*, the calibration of the plate may, as regards the comparison stars, be based on (*a*) in-focus images, (*b*) out-of-focus images, (*c*) *schraffierkasette* images, or (*d*) Fabry images. The first method (*a*) is used only for estimates (by visual inspection of the plate) of magnitudes, especially in survey work involving a very large number of galaxies. Because of the considerable difference in plate density between stars and galaxies, the estimated magnitudes may be affected by large systematic errors. The second method (*b*) demands extrafocal stellar images that are well defined and have a uniform density—a requirement that is not met by all telescopes. With good images, the method gives results of high accuracy. It also permits a transfer of the magnitude scale from one part of the sky to another, the definition of the extrafocal images being practically independent of changes in the atmospheric seeing. With a *schraffierkasette*

Received January 1966.

(*c*), the focal (or slightly out-of-focus) images are moved in a zigzag pattern over small squares on the plate during the exposure. Since each part of the square is exposed intermittently, the intermittency effect being different for stars and galaxies, the procedure is not entirely satisfactory. In the Fabry method (*d*), the lens (mirror) of the telescope is imaged on the photographic plate by means of a small auxiliary lens mounted near the focus. The optical arrangement is similar to that generally used in photoelectric observations. The method shares the drawbacks of the photoelectric technique when applied to galaxies of large extensions.

With methods (*c*) and (*d*) the galaxies and the comparison stars are treated in the same way. The galaxies cannot be separated from the superposed stars. With method (*b*), the extrafocal exposure of the standard stars is combined with an in-focus exposure of the galaxy, preferably on the two halves of the same plate. The superposed stars are easily eliminated in the analysis of the microphotometer tracings. The method seems to be the only one that permits photometry of galaxies of very low surface brightness, down to about 1 percent of the night-sky brightness.

In *photoelectric photometry*, the calibration presents a minimum of problems. The integrated magnitudes measured for galaxies refer to the focal diaphragm used, and cannot be reduced to a standard isophote. For technical reasons, the diaphragms are usually of a rather small size. The disturbance caused by the superposed star field is of minor importance in the case of concentrated objects of high or medium surface brightness. On the other hand, it is very difficult, if not impossible, to measure galaxies of low surface luminosity by the usual photoelectric technique.

Attempts have been made (cf. van Houten, Oort, and Hiltner 1954) to overcome the difficulties of superposed stars in the photoelectric work by tracing the galaxy at the telescope. The laboratory measuring work (photographic method [*b*]) is thus shifted to the telescope. It does not seem possible to apply this procedure to any large number of objects, on account of the long observation times involved. The tracings would also, especially in the case of galaxies of low surface luminosity, be disturbed by the accidental variations in the sky brightness.

1.2. Results Obtained by Photographic Photometry

The results presented in this section refer mainly to the integrated magnitudes of galaxies. The surface-luminosity distributions are discussed in § 4.

Since it includes the entire sky, the Shapley-Ames (1932) catalog of galaxies brighter than the 13th magnitude is still of importance in many respects as a survey catalog. The magnitude estimates, made on small-scale plates, are based on comparisons between in-focus images of galaxies and SA stars. The Shapley-Ames magnitude may be reduced to a correct photographic (pg) system by a correction that depends mainly on the surface magnitude of the object, and on the galactic latitude (de Vaucouleurs 1956; Holmberg 1958). The correction sometimes exceeds 1 mag; the mean error of the corrected magnitudes is about 0.3 mag. Down to the 12th magnitude, the statistical distribution of the corrected magnitudes corresponds to a constant space density of galaxies.

In the pioneer work by Hubble (1930), referring to 15 elliptical galaxies, the luminosity distributions on a relative scale were determined for various cross-sections from microphotometer tracings of the photographic images. By a similar method, Redman

(1936) and Redman-Shirley (1938) derived the integrated photographic magnitudes for a number of galaxies of the same type. The integrations were based on the luminosity distributions along the major and minor axes of the objects. The plates were calibrated by a tube sensitometer, the zero-point being determined from extrafocal exposures of stars.

By means of the Mount Wilson 60-inch and 100-inch telescopes, Holmberg (1950*a*, 1958) has derived the photographic and photovisual (pv) magnitudes of 300 galaxies, mostly nearby spiral systems. Each galaxy was photographed in focus on one half of the plate, the other half carrying an extrafocal exposure of north polar sequence (NPS) stars or stars of SA 57. Both the scale and the zero-point were determined from the extrafocal stars; the two telescopes give out-of-focus images of excellent quality. The magnitude was obtained by a numerical integration, the number of measured cross-sections ranging from 10 to over 20, depending on the size of the galaxy. Since the limiting isophote corresponds to 26.5 mag (pg) per square second, the magnitudes may be assumed to approach the total luminosities. The results are in very good agreement with modern photoelectric measurements (cf. § 1.3).

The *schraffierkasette* procedure has been applied by Hubble (1936*a*, *b*) to a number of bright galaxies. The technique has been thoroughly tested by Sandage (1956), in connection with the determination of magnitudes for members of distant clusters of galaxies (200-inch telescope). The most extensive use of the method is found in the *Catalogue of Galaxies and of Clusters of Galaxies* by Zwicky *et al.* (1961, 1963, 1965, 1966, 1968), which includes all galaxies down to an apparent photographic magnitude of 15.5. The telescope (Palomar 18-inch Schmidt) was guided in a regular pattern, so as to make the out-of-focus images of galaxies and stars describe squares approximately 1 minute of arc across. These squares were photometrically evaluated by using a step scale consisting of similar squares, the scale being calibrated by means of SA stars. The imperfections inherent in the method have been partly overcome by the application of systematic corrections to the magnitudes. According to Abell (1964), the Zwicky magnitude scale may need some additional corrections.

The Fabry method has been used by Bigay (1951) to determine integrated photographic magnitudes for 175 galaxies. A comparison with other magnitude lists indicates that the results are free from any appreciable systematic errors. On account of the disturbing effects of the superposed star fields, the accidental errors are, however, comparatively large.

1.3. Results Obtained by Photoelectric Photometry

Photoelectric photometry of galaxies was started at Mount Wilson in 1936, when Whitford published the magnitudes measured for 11 nearby objects. In a subsequent investigation, by Stebbins and Whitford (1937), photometric data were derived for a considerably larger material. A new list by the same observers (1952) gives photographic magnitudes for 176 galaxies, and pg − pv colors for 97. Since larger diaphragms were used, and the zero-point of the magnitude scale was improved (NPS stars), the later results are the most accurate ones. The focal diaphragms measured up to 8′.6 (60-inch telescope) or 5′.1 (100-inch telescope).

The Mount Wilson photoelectric work was continued by Pettit (1954), whose aim was to measure all galaxies included in the redshift programs. The results have been analyzed by Sandage (1956), who has tried to reduce the magnitudes to a homogeneous system by applying corrections for the aperture effect. The corrections were determined by means of the Palomar 48-inch survey plates, the Pettit magnitude being reduced to a standard isophote corresponding to 2.5 times the radius of the object found by visual inspection of the plate. The final list of corrected photographic magnitudes comprises 576 galaxies.

A comparison of the three largest magnitude lists—those of Holmberg (1958), Stebbins-Whitford (1952), and Pettit-Sandage—indicates a very satisfactory systematic agreement, both as regards zero-point and scale; the Stebbins-Whitford magnitudes should, however, be given an aperture correction of about −0.1 mag. The dispersion in the differences between the magnitudes of the first two lists is 0.11 mag, the greater part of which is probably caused by the superposed stars included in the photoelectric data. A comparison with the Pettit-Sandage magnitudes yields a somewhat larger dispersion. The Holmberg colors are in excellent agreement with the Stebbins-Whitford colors, the dispersion in the differences being only 0.05 mag.

In three lists, Bigay *et al.* (1953) and Bigay and Dumont (1954, 1955) have published pg magnitudes and pg − pv colors for about 70 galaxies, as measured by means of the 48-inch telescope at Haute-Provence. Except for a small systematic magnitude difference, the magnitudes and colors are in good agreement with the above results. Later on, a new series of measurements of *U, B, V* magnitudes of galaxies was started (Bigay 1964).

A comprehensive list of integrated magnitudes and colors of bright galaxies in the *U, B, V* system has been published by de Vaucouleurs (1959, 1961). Most of the results are based on observations with the Lowell Observatory 21-inch reflector.

Among more recent contributions, attention is called to the magnitude and color lists published by Tifft (1961, 1963), and by Hodge (1963*a*).

In their *Reference Catalogue of Bright Galaxies*, G. and A. de Vaucouleurs (1964) have attempted to reduce all available measurements of magnitudes and colors to a uniform system.

2. ABSORPTION EFFECTS IN MAGNITUDES AND COLORS

The observed total magnitudes and integrated colors of galaxies suffer from two kinds of extinction effects: the absorption produced by internal obscurations in the galaxies, and the galactic absorption. Both effects can be determined, at least in a statistical sense, by an analysis of the observational data. According to all the evidence available, the intergalactic absorption, if any, is negligible.

2.1. Internal Absorption in Galaxies

External stellar systems, with the exception of E and S0 objects, undoubtedly have a composition more or less similar to that of the Galactic system: the stars are mixed with clouds of dust and gas. The dust particles are recognized by the extinction effects, which are clearly noticeable, especially in spiral galaxies with an edgewise orientation. As in the Galaxy, the dark matter in a spiral system seems to be concentrated in a thin disk located in the principal plane.

It should be noted here that the extinction effects may be the combined result of absorption and of light-scattering. Since the scattering of the light does not, on the average, change the observed total magnitudes and integrated colors of the galaxies, the results to be discussed below will be interpreted as referring to absorption effects. In any case, the results will permit the reduction of the observed magnitudes and colors to a homogeneous system.

For obvious reasons, the methods used to investigate the absorption in the Galaxy cannot be applied to external systems. Except for the nearest galaxies, the distributions of surface magnitude and color over the projected image are the only photometric data at our disposal. Because of the limited observational material, the conclusions have, in fact, to be based mainly on the observed total magnitudes and integrated color indices. In individual cases, these quantities give a very incomplete description of the internal absorption. However, certain mean results may be derived if the galaxies of a given morphological type are treated as a uniform group, that is, if the galaxies are assumed to be more or less identical, as regards type of stellar content and internal absorption. It is thus possible to determine the two most important absorption effects: the loss in total luminosity, and the change in integrated color corresponding to different inclinations to the line of sight. The inclination of the principal plane is derived from the apparent diameter ratio.

Since the absolute luminosities are not accurately known, an investigation of the total absorption as a function of inclination has to be based on the integrated surface magnitude. This quantity is, in fact, a very stable parameter, the intrinsic dispersion being less than the dispersion in the absolute magnitude. The surface magnitude S may conveniently be defined as

$$S = m + 5 \log a \,, \tag{1}$$

where m is the total apparent magnitude, and a the apparent major diameter (in minutes of arc). The magnitude m should be corrected for galactic absorption; the changes in the diameter due to absorption and to an inclination effect (cf. § 6.1) may be neglected, these changes being small compared with the unavoidable selection effects in the material (in the following analysis the inclination effect in the diameter is probably counterbalanced by the selection effects). The quantity S is, apart from a constant, equal to the observed integrated surface magnitude, if the galaxy has a face-on orientation. It would be independent of the inclination if the objects investigated contained no obscuring matter. With absorption, the variation in S with inclination will be a reflection of the absorption effect in the total magnitude.

Since the absorption in the Galactic system can be statistically described by a cosecant law, at least down to latitudes of about 15°, a similar law may be expected in external spiral systems: the latitude is replaced by the inclination to the line of sight. If the light emitted from each element of volume in the galaxy suffers a relative loss proportional to the cosecant of the inclination angle, then a cosecant law should be approximately applicable to the absorption effect in the total magnitude. The variation in surface magnitude with inclination i would thus be expressed by the equation

$$S = S_* + \alpha \operatorname{cosec} i \,, \tag{2}$$

where α represents the absorption in total magnitude. Both α and the constant term S_* may be assumed to vary from one type of galaxy to another.

Figure 1 gives the results that are obtained if the above procedure is applied to the photometric data given by Holmberg (1958). The analysis is based on the pg total magnitudes and the photometric diameters, as determined for 53 spirals of type Sa-Sb and 66 spirals of type Sc. The surface magnitude ($\Delta S = S - S_*$) has been plotted against the diameter ratio b/a. Although the dispersion in the individual values is comparatively large, the means (*open circles*) corresponding to the different intervals

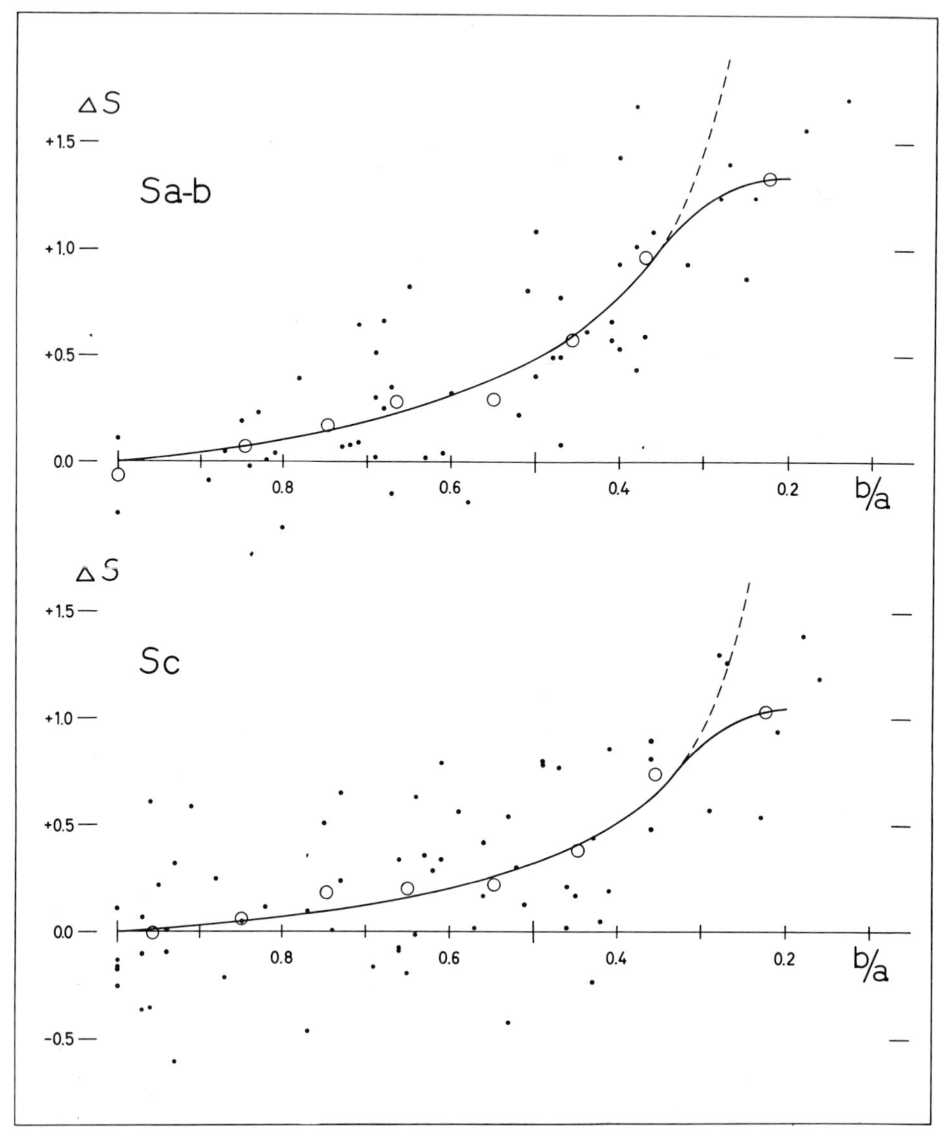

FIG. 1.—Variation in the pg surface magnitude with inclination, as found for two groups of spiral galaxies. The inclination is represented by the apparent diameter ratio. The class means are denoted by open circles.

of b/a agree very closely with the adopted curves. In the interval $b/a > 0.35$, these curves have been derived by least-squares solutions based on the above relation (the inclination is obtained from eq. [14]).

For both classes, Sa-Sb and Sc, the absorption seems to obey a cosecant law for $b/a > 0.35$ ($i > 17°$). The parameter α, or the total pg absorption corresponding to inclination 90°, is equal to 0.43 and 0.28 mag, respectively. For small diameter ratios, the two curves approach definite maximum values, about 1.33 mag (Sa-Sb) and 1.03 mag (Sc); for inclination 0° the absorption would thus be 1.76 and 1.31 mag, respectively. The total pg absorption corresponding to different inclinations is listed in table 1.

Attention will now be turned to the color excess produced by the internal absorption. The inclination effect is found by plotting the observed integrated color indices (freed

TABLE 1

TOTAL pg ABSORPTION, AND pg − pv COLOR EXCESS (in parentheses), DUE TO INTERNAL ABSORPTION, AS DERIVED FOR SPIRAL GALAXIES WITH DIFFERENT INCLINATIONS TO THE LINE OF SIGHT

DIAMETER RATIO		INCLINATION (degrees)	ABSORPTION (mag) (color excess)	
Photometric	Shapley-Ames		Type Sa-Sb	Type Sc
1.00	1.00	90	0.43 (0.11)	0.28 (0.07)
0.90	0.87	64	0.48 (0.13)	0.31 (0.08)
0.80	0.74	52	0.54 (0.14)	0.35 (0.09)
0.70	0.62	43	0.63 (0.16)	0.41 (0.10)
0.60	0.50	35	0.74 (0.18)	0.48 (0.11)
0.50	0.38	28	0.92 (0.20)	0.60 (0.12)
0.40	0.27	21	1.22 (0.22)	0.79 (0.13)
0.30	0.17	13	1.65 (0.24)	1.16 (0.14)
0.20	0.09	0	1.76 (0.26)	1.31 (0.15)

from galactic absorption) against the diameter ratio. In this case, it seemed advisable not to bind the analysis to a cosecant law; the regression curve has, in fact, a somewhat different shape.

Figure 2 gives the results obtained from the pg − pv color indices of Holmberg (1958). As far as can be judged from the figure, the observed color is a linear function of the diameter ratio: $C = \text{const.} + \beta(1 - b/a)$. Most of the means (*open circles*) deviate from the linear relation by less than the corresponding probable errors. Least-squares solutions give the results, $\beta = 0.19$ mag (Sa-Sb) and 0.10 mag (Sc).

The figure also includes 32 galaxies of types E and S0, and 19 galaxies of type Ir I. For types E-S0 there is no variation in color with diameter ratio, which seems to permit the conclusion that the internal absorption is negligible; the same result is obtained if the S0 systems are analysed separately. Also the Ir I galaxies exhibit no variation in color. Since these galaxies undoubtedly contain absorbing matter, the interpretation is in this case different: the three-dimensional bodies of Ir I systems have a more or less irregular shape, and the apparent diameter ratio cannot be used as an indicator of the orientation in space.

Since the above determinations of total absorption and of color excess are independent of each other, a comparison offers an interesting check on the results. The 62 Sa-Sb-Sc spirals with b/a larger than 0.60 have a mean ΔS of 0.12 mag and a mean ΔC of 0.030 mag; the ratio of differential total to selective absorption is thus 4.0. For $b/a \leq 0.60$ (57 objects) the corresponding means are 0.67 and 0.079 mag, respectively; the ratio is 8.5. In the pg − pv system the ratio to be expected is presumably close to 4.0, that is, if the absorbing matter is located as a screen in front of the luminous matter. In a galaxy, where absorbing and luminous matter are mixed in space, the ratio is always *larger* than the "screen ratio," a fact that can be easily demonstrated; however, in this case also the ratio approaches the "screen ratio" as the total absorption decreases. The absorption ratios derived for the spiral galaxies are, in fact, in good agreement with the ratios to be expected, an indication that the results are free from any appreciable systematic errors.

The analysis of the colors did not permit a determination of the selective absorption for spiral systems oriented in the celestial plane. The problem can be solved by accepting

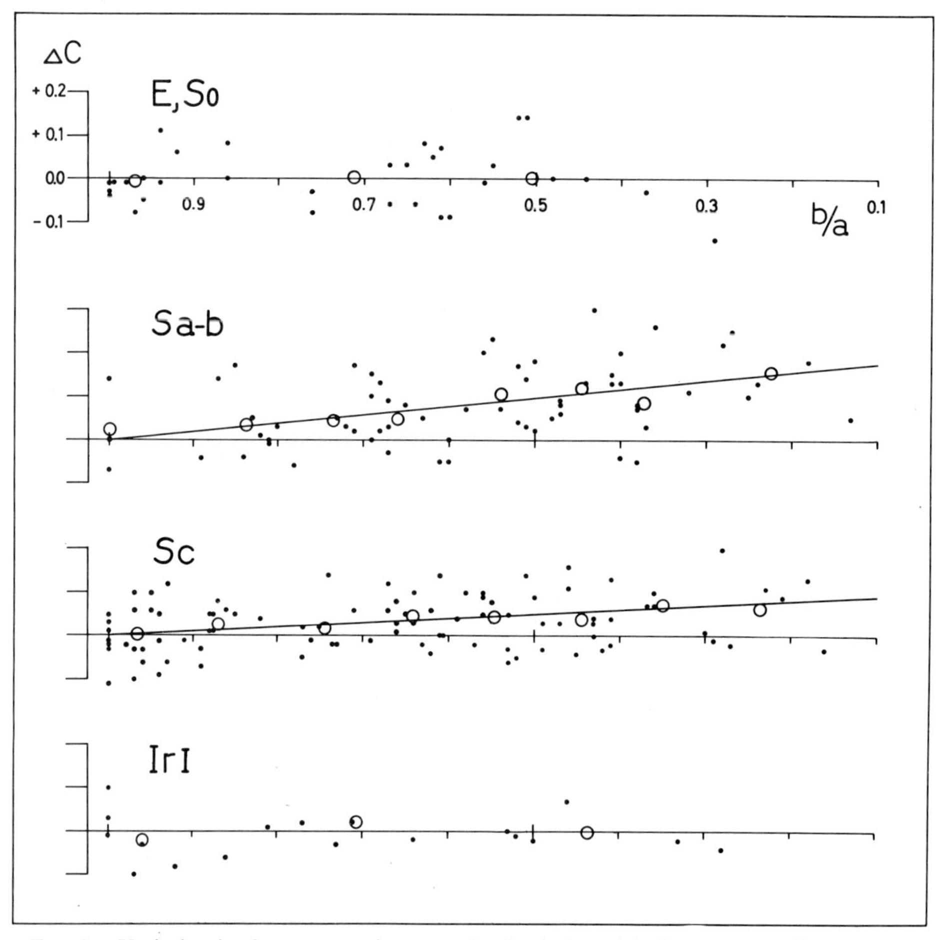

FIG. 2.—Variation in the pg − pv integrated color index with the apparent diameter ratio, as found for four different type groups. The class means are denoted by open circles.

the above absorption ratio. Accordingly, the color excess corresponding to inclination 90° would for Sa-Sb and Sc spirals be equal to one-quarter of 0.43 and 0.28 mag, respectively. The total color excess corresponding to different inclinations is tabulated in table 1.

2.2. Galactic Absorption

Since the problem of galactic absorption is treated elsewhere in this compendium, this section is confined to a summary of the results to be obtained from the material discussed above.

A determination of the galactic pg absorption may conveniently be based on the surface magnitudes of the Sa-Sb-Sc spirals. A plot of these magnitudes, corrected for internal absorption, against the cosecant of galactic latitude indicates a linear relation. A least-squares solution, based on 113 spirals with latitudes numerically larger than 20°, gives the result:

$$\text{galactic pg absorption} = 0.26 \text{ cosec } |B| \text{ mag.}$$

The absorption constant is in good agreement with the classical result by Hubble (1934).

The determination of the selective galactic absorption can be based on galaxies of all types. The observed colors of Sa-Sb-Sc spirals are corrected for internal absorption; no corrections are applied for E, S0, and Ir I objects. A least-squares solution, based mainly on the same material as that included in figure 2, gives the result

$$\text{galactic (pg} - \text{pv) absorption} = 0.062 \text{ cosec } |B| \text{ mag.}$$

The relation is practically linear. The mean error in the absorption constant amounts to only 0.007 mag. It may be noted that the selective absorption is close to one-quarter of the total absorption.

3. SURFACE MAGNITUDES, LUMINOSITY DENSITIES, AND INTRINSIC COLORS

The analyses of the preceding paragraphs give, for each type of galaxy, two important by-products: the mean integrated surface magnitude and the mean integrated color index, both freed from all absorption effects. The absorption-free quantities are denoted by S_* and C_*, respectively. The surface magnitude, defined by equation (1), may be used to determine the mean luminosity density in galaxies.

In the following discussion, the morphological types will be referred to a slightly modified Hubble system. The type classes are: E, S0, Sa, Sb−, Sb+, Sc−, Sc+, Ir I, and Ir II. The Sb and Sc spirals have thus been divided into "early" and "late" subtypes, denoted by − and +, respectively. Typical specimens are NGC 224, 3031 (Sb−), NGC 3953, 3992 (Sb+), NGC 5194, 5457 (Sc−), and NGC 598, 2403 (Sc+). The irregular galaxies are separated into two groups, Ir I (blue objects of the Magellanic Cloud type) and Ir II (red objects like NGC 3034). In most cases, the application of the revised system does not meet any serious difficulties.

Column (2) of table 2 gives the means of S_* for six classes of spirals. It may be recollected that the results refer to the magnitude and diameter system of Holmberg (1958). If the listed quantities are corrected by −0.26 mag, they will be expressed in magnitudes per square minute (galaxy oriented in the celestial plane). The surface luminosity reaches a maximum for Sb− spirals, and minimum values for S0 and Sc+ objects. The low mean luminosity found for type S0 may seem surprising; it is explained by the weak surface luminosity of the extended outer regions. As column (3) shows, the dispersion in S_* is for all classes comparatively small (about 0.3 mag); it is considerably less than the dispersion in total absolute magnitude. It should be noted that the results refer to a selected material (high-luminosity galaxies). As will be demonstrated later (§ 6.2), the surface magnitude becomes fainter with decreasing total luminosity.

The mean density of luminosity in the galaxies may be derived from S_*, the latter quantity representing the projected spatial density. If the spiral systems are assumed

TABLE 2

MEAN INTEGRATED pg SURFACE MAGNITUDE (and dispersion), MEAN LOG pg LUMINOSITY DENSITY (solar units per pc³), AND MEAN INTEGRATED pg − pv COLOR INDEX (and dispersion), AS DERIVED FOR DIFFERENT TYPES OF GALAXIES

Type (1)	$\bar{S}_*$ (mag) (2)	Dispersion (mag) (3)	Log lum. density (4)	$\bar{C}_*$ (mag) (5)	Dispersion (mag) (6)
E	....	....	....	+0.77	0.057
S0	+15.10	0.36	−2.3	0.77	0.076
Sa	14.42	0.26	−2.0	0.53	0.090
Sb−	14.19	0.29	−2.0	0.48	0.071
Sb+	14.57	0.35	−2.1	0.38	0.069
Sc−	14.76	0.34	−2.1	0.24	0.059
Sc+	15.03	0.32	−2.1	0.17	0.056
Ir I	....	....	....	0.12	0.057

NOTE.—The data are corrected for all absorption effects.

to be oblate spheroids, with a ratio of smallest to largest axis of 0.20 (cf. § 6.1), it is easily found that

$$\text{log luminosity density} = -0.4S_* - \log A + 8.20\,, \qquad (3)$$

where the absolute major diameter (pc) is denoted by A. The density is expressed in solar units per pc³ (absolute pg magnitude of the Sun = +5.37). From the catalog by Holmberg (1964) the following mean values of log A are derived: 4.42 (S0), 4.43 (Sa), 4.56 (Sb−), 4.51 (Sb+), 4.41 (Sc−), and 4.30 (Sc+). The means refer to the same material as that analyzed above; the distance scale is based on a Hubble parameter $H = 80$ km s^{-1} per Mpc. The resulting logarithmic luminosity densities are listed in column (4) of table 2. It is quite surprising to find that the pg luminosity density is approximately the same for all types of high-luminosity spiral systems (about 0.01 solar unit per pc³). The luminosity densities may be used as a starting point to determine the mass densities in galaxies.

The right-hand part of table 2 gives information on the mean integrated pg − pv color indices. Since corrections have been applied for galactic and internal absorption, the indices presumably represent the intrinsic colors of the stellar contents. In the case of Ir I galaxies, the mean color excess due to internal absorption has been estimated at 0.08 mag, corresponding to a total pg absorption of 0.3 mag. It should be added that a correction for redshift effect, for all the types averaging −0.02 mag, has also been applied. Galaxies of type Ir II are not included in the table. Because of their limited number, a statistical evaluation of the internal absorption (possibly very high) has not been attempted; six objects in the Holmberg (1958) catalog have a mean color, corrected for galactic absorption and redshift effect, of 0.76 mag.

The intrinsic color indices range from +0.12 to +0.77 mag. These two limiting values presumably represent the integrated colors of stellar populations of type I and type II, as defined by Baade (1944). For comparison, it may be noted that the integrated pg − pv color index of the population in the solar neighborhood appears to be about +0.3 mag. The dispersion in the color indices varies for the different type classes from 0.06 to 0.09 mag.

A comprehensive study of the integrated colors in the U,B,V system has been published by de Vaucouleurs (1961). Photoelectrically measured $B - V$ and $U - B$ colors are listed for altogether 148 bright galaxies. The colors have been reduced to an inclination of 90° to the line of sight. The maximum corrections for Sa-Sb spirals amount to $\Delta(U - B) = 0.19$, and $\Delta(B - V) = 0.13$ mag; the latter result is in good agreement with the data of table 1. Analyses of the $U - B$, $B - V$ relations give diagrams with well-established sequences, even when all the morphological types are taken together. A detailed investigation has been made of the aperture effect or the relation between observed integrated color and size of the diaphragm. Galaxies of types E to Sc are redder in the central parts, whereas systems of the Magellanic Cloud type are bluer. A study of the Virgo cluster shows that the integrated color found for E and S0 systems depend on the absolute luminosity. Dwarf systems of these types appear to be systematically bluer than giant systems; a close correlation is indicated between color and absolute magnitude for $M > -17.5$.

4. INTERNAL DISTRIBUTIONS OF LUMINOSITY AND COLOR

The preceding sections have been confined to a discussion of integrated photometric data. We now turn to the distributions within the galaxies of luminosity and color (and absorbing matter). The projected distributions, referring to the two-dimensional image, are determined by direct observations (§§ 4.1 and 4.2). In certain cases, a theoretical analysis may lead to the three-dimensional distribution (§ 4.3).

4.1. Distribution of Luminosity and Color in Elliptical Galaxies

In the classical study by Hubble (1930) intensity curves were determined for 15 elliptical galaxies; since the zero point was not known, only relative intensities could be obtained. Hubble came to the conclusion that the luminosity distribution, along various axes of the different galaxies, could be represented by the simple relation

$$I = \frac{I_0}{(r/a + 1)^2}, \tag{4}$$

where I is the light intensity (per unit area), I_0 the central intensity, r the distance from the center, and a a scale factor. The curve reproduces the Hubble data remarkably well out to $r/a = 14$; beyond this distance the accuracy of the observations was not deemed to be sufficiently high.

From photoelectric observations of about a dozen E0 galaxies, Baum (1955) arrived at the following, slightly modified relation:

$$I = \frac{I'}{r/a(r/a + 1)} . \quad (5)$$

Baum remarks that, since it is unlikely that all the E0 galaxies investigated are flattened to the same degree (along the line of sight), the similarity of their profiles suggests that the relationship between I and r is not sensitive to the intrinsic oblateness.

A somewhat different analytical expression has been proposed by de Vaucouleurs (1948, 1953):

$$\log I = \log I_a - A[(r/a)^{1/4} - 1] . \quad (6)$$

Here a is the radius within which half the total light of the galaxy is emitted, and I_a the intensity at a; the coefficient A, with a numerical value of about 3.3, is supposedly more or less the same from one system to another. The formula has been successfully applied to a number of galaxies, among them three measured by de Vaucouleurs.

From an investigation of about a dozen E and S0 galaxies, Abell (1962) concluded that the projected light distribution in most cases can be represented satisfactorily by the interpolation formulae

$$I = \frac{I_0}{(r/a + 1)^2} \quad \text{for } r/a \leq 21.4 ;$$
$$I = \frac{22.4 I_0}{(r/a + 1)^3} \quad \text{for } r/a > 21.4 . \quad (7)$$

This form for the distribution function was considered suitable for a determination of the total magnitude by integration. Further comments on the distribution function are given in a later paper by Abell and Mihalas (1966).

In addition to the results already mentioned, attention is called to the luminosity distribution data published by Oort (1940), Evans (1951, 1952), Dennison (1954), van Houten, Oort, and Hiltner (1954), Liller (1960), van Houten (1961), Miller and Prendergast (1962), and Hodge (1963*c*).

Since the luminosity distribution is almost the only observational evidence about the internal structure and the dynamics of elliptical galaxies, the formulation of a representative luminosity law is of considerable interest. As regards the Hubble formula, it is quite clear that it cannot be applied to the outer parts of the galaxies. In NGC 3379 (Dennison 1954), to take one example, the formula fits the observations well out to $r/a = 20$; beyond this distance the observed luminosity curve declines more steeply than the extrapolated Hubble curve. Similar objections have been raised against the de Vaucouleurs formula. According to the above investigation by van Houten, the two theoretical laws give comparable results when tested on 19 galaxies.

As an illustration, relations (4) and (6) are in figures 3*a* and 3*b* applied to the photometric results obtained by Liller (1960) for 14 elliptical galaxies (one peculiar object omitted). The observed quantity $\log I$ has been plotted against $\log (r + a)$ and $r^{1/4}$, respectively; the theoretical relations are in both cases straight lines (the parameter a in Hubble's formula has been determined by analysis of the dependence of $I^{-1/2}$ on r).

The galaxies are, from top to bottom, arranged in order of increasing ellipticity, from NGC 4374, E1.0, to NGC 4570, E6.5 (classification by Liller). For the last four objects, E5.0 to E6.5, that are missing in figure *3a*, the parameter a could not be evaluated, since the Hubble formula breaks down completely. A study of the two figures shows that in most cases the theoretical laws fail to reproduce the luminosity distributions in the outer parts of the galaxies. The failure is most accentuated for the most elongated objects (bottom of fig. *3b*). In fact, the goodness of fit seems to be a function of the ellipticity. For E1.0–1.5 the Hubble formula fits well out to $r/a = 24$, for E2.0–2.5 out to $r/a = 13$, and for E3.0–4.0 out to $r/a = 7$, whereas the formula breaks down for a type index of 5.0 and larger.

The above results show that a two-parameter law is not flexible enough to permit a complete description of the luminosity distribution in elliptical galaxies. A successful formula should be able to differentiate between galaxies of different intrinsic oblateness (and different apparent ellipticity). A formula with at least three independent parameters

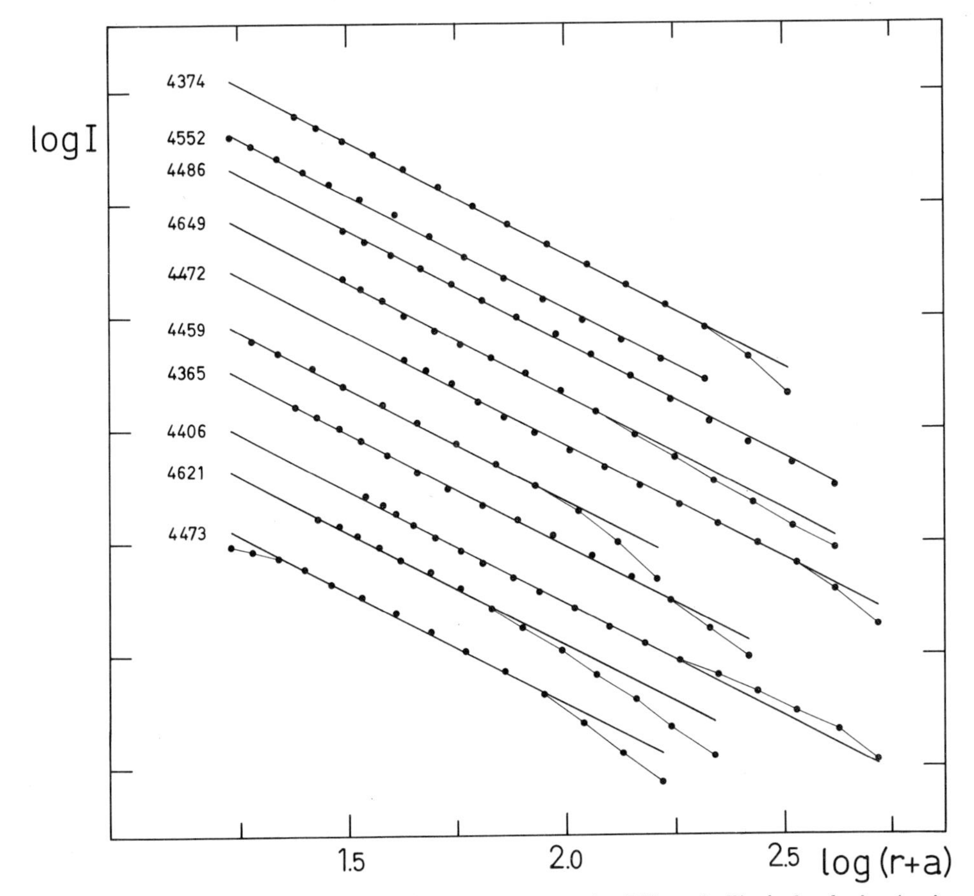

FIG. *3a*.—Hubble's law applied to the measurements by Liller of elliptical galaxies (major axis). On the vertical scale one division corresponds to $\Delta \log I = 1.0$. The numbers refer to the NGC.

seems to be indicated. Attention is called to the semiempirical three-parameter law that has been successfully used by King (1962) to describe the density distribution in globular clusters:

$$\text{Density} = k\left\{\frac{1}{[1+(r/r_c)^2]^{1/2}} - \frac{1}{[1+(r_t/r_c)^2]^{1/2}}\right\}^2 . \tag{8}$$

Here r_t is the total radius, that is, the value of r at which the density reaches zero; the parameter r_c is a scale factor that may be called the core radius. An application of the

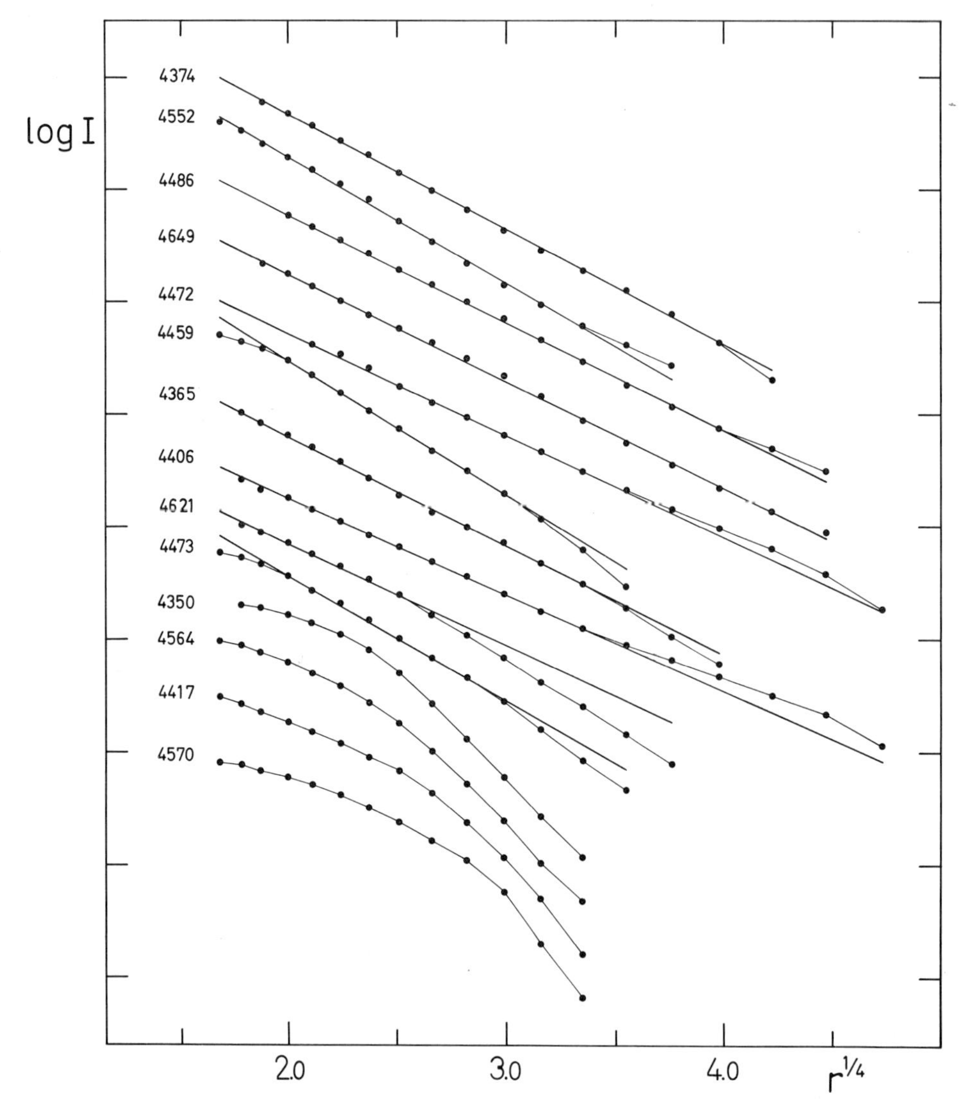

FIG. 3*b*.—Formula by de Vaucouleurs applied to Liller's measurements. The vertical scale is the same as in fig. 3*a*. The numbers refer to the NGC.

formula to the elliptical galaxies of figure 3 shows that the luminosity distributions are very satisfactorily represented by a three-parameter law of this type.

In a number of papers, Hodge has analyzed the distributions of stars in the dwarf elliptical systems of the Local Group; the star counts refer to the Fornax (1961*a*), Sculptor (1961*b*), Leo I (1962), Leo II (1963*b*), Ursa Minor (1964*a*), and Draco (1964*b*) systems. The density distribution curves do not agree with Hubble's law, nor with de Vaucouleurs's law. The very steep density gradients in the outer parts of the systems may, however, in this case be explained as a result of gravitational disturbance by the Galaxy. The tidal action puts an upper limit to the radius, depending on the total mass of the system; the disturbance is the same as that affecting the globular clusters (cf. von Hoerner 1957). The above three-parameter law by King is able to reproduce the observed density curves very well.

As regards the color distribution in elliptical galaxies, available observations indicate that the color index is nearly constant in the projected image. According to de Vaucouleurs (1961), who has analyzed photoelectrically measured $B - V$ (and $U - B$) indices, there is a slight reddening of the color, by a few hundredths of a magnitude, in the central parts of the objects. A similar result is obtained from the Holmberg (1958) material.

4.2. Distribution of Luminosity and Color in Spiral Systems

The luminosity distribution in galaxies of type S0 has sometimes been found to be rather similar to that characteristic of type E, and may thus be approximately represented by the distribution laws discussed in the preceding section; however, the correctness of the classification should perhaps be reexamined in these cases. As a rule, there are noticeable differences between the two classes. According to Johnson (1961), the distribution in the three S0 systems measured by him is not a smoothly falling curve but is rather segmented, so that the logarithmic surface brightness is approximately a linear function of the radial distance, with step changes in the coefficients. Similar conclusions can be drawn from a study of the photometric data of Liller (1960) and van Houten (1961).

It seems in most cases possible, within the limits of the accidental errors of observation, to divide the distribution along the major axis into only two parts, one referring to the nuclear region and the other to the outer region, each being represented by a linear relation between $\log I$ and r. As is demonstrated below, a composite distribution of this kind is found also for spirals of later types. The S0 galaxy can thus be treated as a lenticular disk, in which a more or less spherical nucleus is embedded. Some typical distribution curves, based on the photometric investigations mentioned above, are given in figure 4.

For comparison, the figure also includes the luminosity distribution (Liller 1960) for the E5.5 galaxy NGC 4417. This distribution, which could not be reproduced by the laws given by Hubble and by de Vaucouleurs (cf. fig. 3*b*), is very well represented by the S0-type curve. Some of the lenticular objects classified as type E apparently are transition cases between E and S0.

The color distribution in S0 galaxies seems to be very similar to that found for type E (cf. the references given above).

Although narrow absorption lanes are sometimes observed, most galaxies of type S0 may be assumed to be essentially free from obscuring matter (cf. § 2.1; see also *The Hubble Atlas of Galaxies*, Sandage 1961).

We shall now deal with spiral galaxies of types Sa, Sb, and Sc. The internal structure in these systems is dominated by spiral arms and absorption lanes, and by a more or less pronounced nucleus. The type groups are not as homogeneous as those discussed above. Even within the same group there are great variations in structure from one

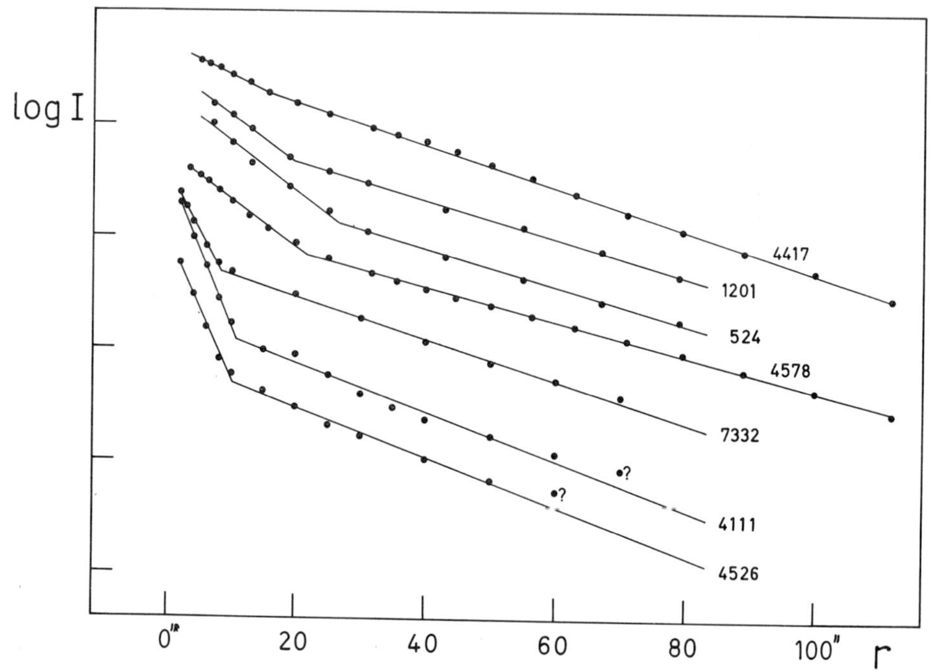

FIG. 4.—Luminosity distributions (major axis) in some typical S0 galaxies. On the vertical scale one division corresponds to $\Delta \log I = 1.0$. The observations refer to van Houten (NGC 4111, 4526, 7332), Johnson (NGC 524, 1201), and Liller (NGC 4578). For comparison, the distribution (Liller) in the E5.5 galaxy NGC 4417 has been included.

object to another; the luminosity distribution determined for an individual object is not necessarily representative of the "average" system. Generally, the integrated surface brightness and the integrated color index decreases along the sequence Sa-Sb-Sc (cf. table 2); a decrease is also found in the relative size and the surface brightness of the nuclear region. The colors of the nuclei will be discussed at the end of this section.

Investigations of the luminosity and color distributions in Sa and Sb spirals have been published by Holmberg (1945), de Vaucouleurs (1958), Lyngå (1959), Fish (1961), van Houten (1961), Simkin (1967), and others. The observational material is still rather limited. An analysis aiming at the determination of mean distribution parameters will have to be postponed until more data are available.

As regards type Sc, an attempt has been made by the writer (not published elsewhere) to construct a representative model. The results are based on eight typical Sc− spirals (cf. § 3) of large angular dimensions and with a face-on orientation, the luminosity and color distributions being derived from the Holmberg (1958) plate material.

In figure 5 the full curves give the photographic surface-brightness distributions (means of two plates) in central cross-sections parallel either to the right ascension or the declination circle; the results are corrected for galactic absorption (0.25 cosec $|B|$ mag) and redshift effect, but not for internal absorption. On account of the limited space, the

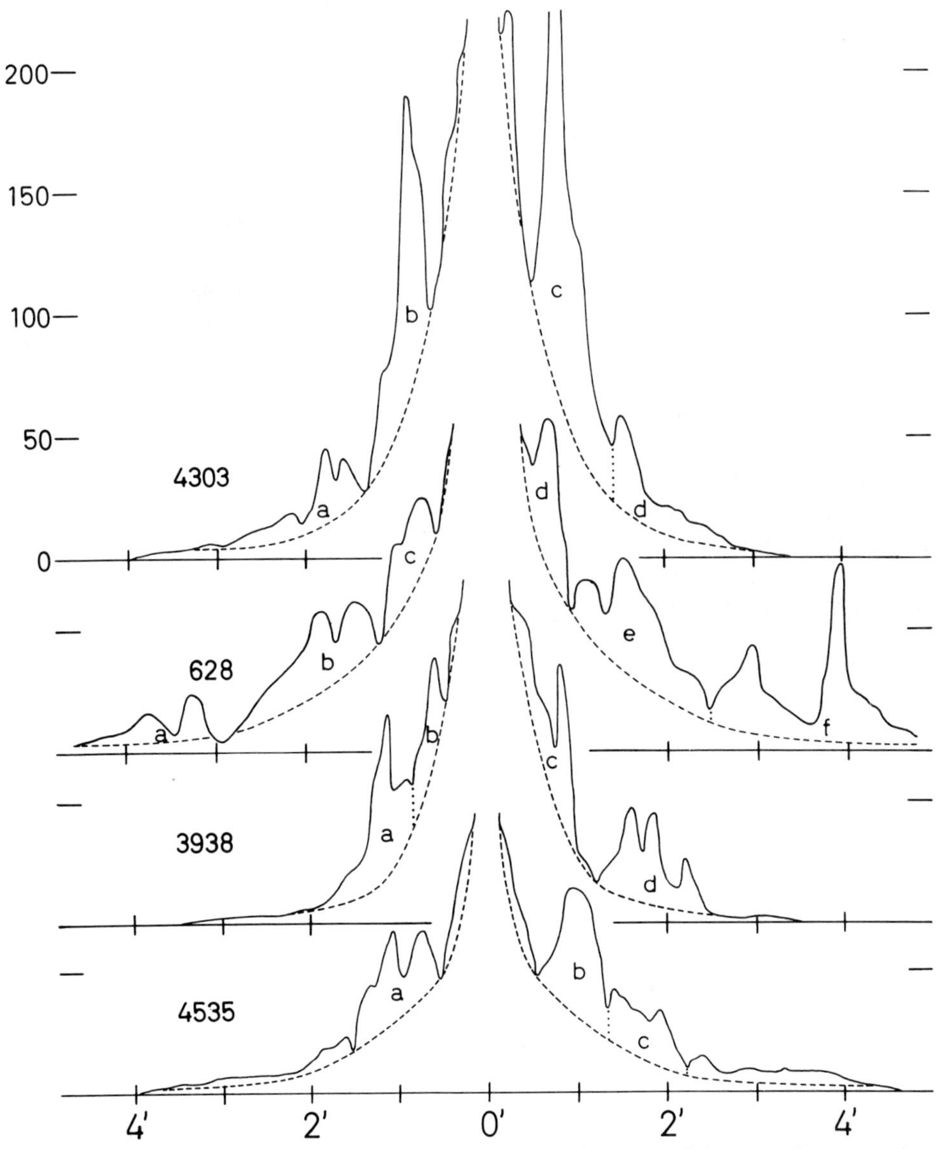

FIG. 5.—Photographic surface-brightness distributions (*full curves*) in central cross-sections of eight spiral systems of type Sc— with a face-on orientation. The vertical scale, the same for all the objects, gives the surface brightness in units of one star of magnitude 26.5 per square second. The numbers refer to the NGC. (*Fig. 5 is continued on p. 140.*)

corresponding curves referring to the photovisual region (means of two plates) are not reproduced here. The broken curves, which are drawn through the minima between the spiral arms and which for each object have the same shapes on the left-hand and the right-hand sides, represent an attempt to determine the distributions referring to the disk population; similar curves have been constructed for the photovisual distributions. It is very interesting to find that the resulting integrated color of the disk population is practically independent of the central distance; the broken curves representing the disks have, in fact, been slightly adjusted in such a way as to give a distance-independent color index. The pg − pv index ranges from +0.35 mag (NGC 4303) to +0.68 mag (NGC 1042), the mean being $C = +0.50$ mag.

Figure 6 presents the mean results, corresponding to the 16 half-sections investigated.

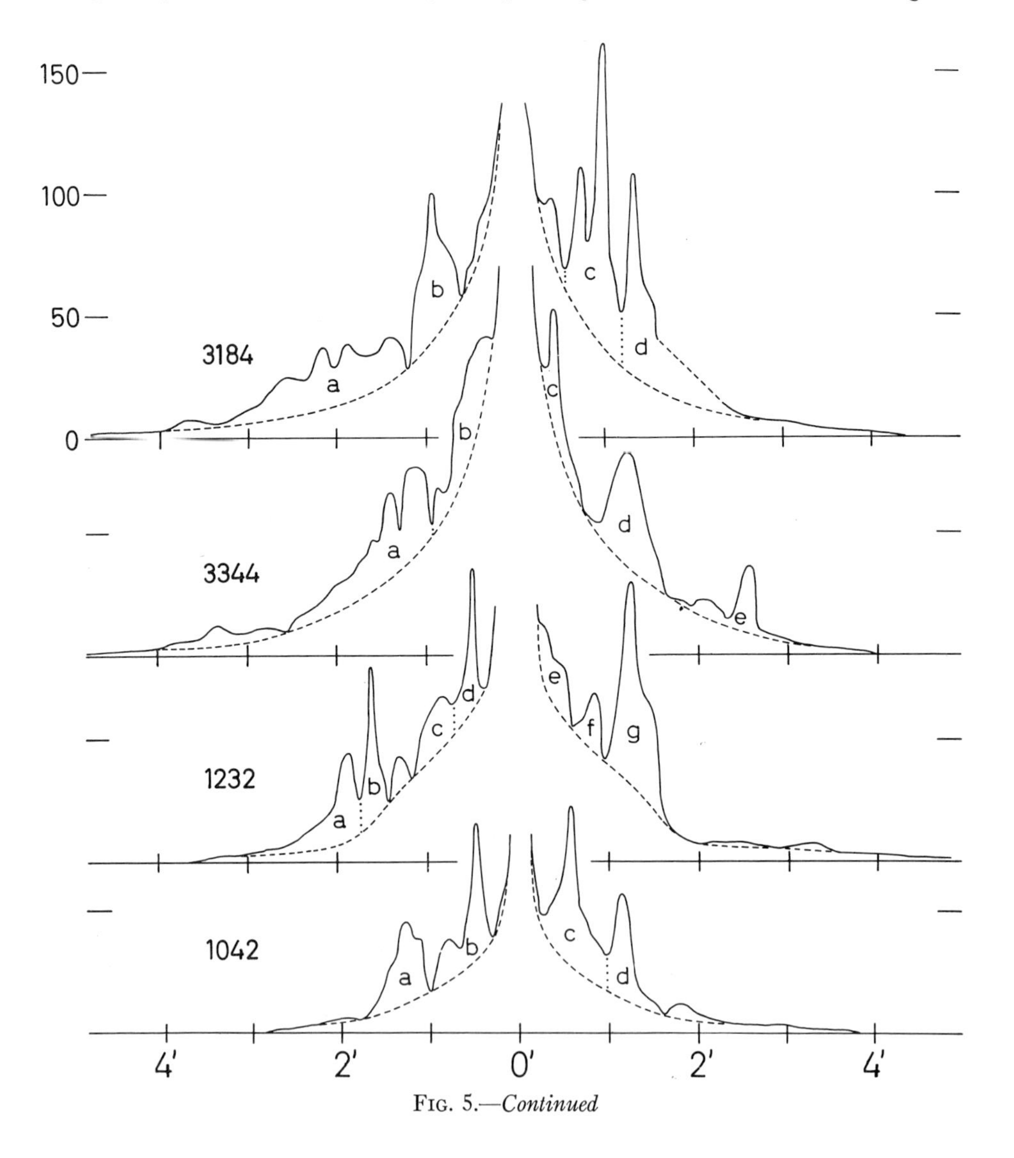

FIG. 5.—*Continued*

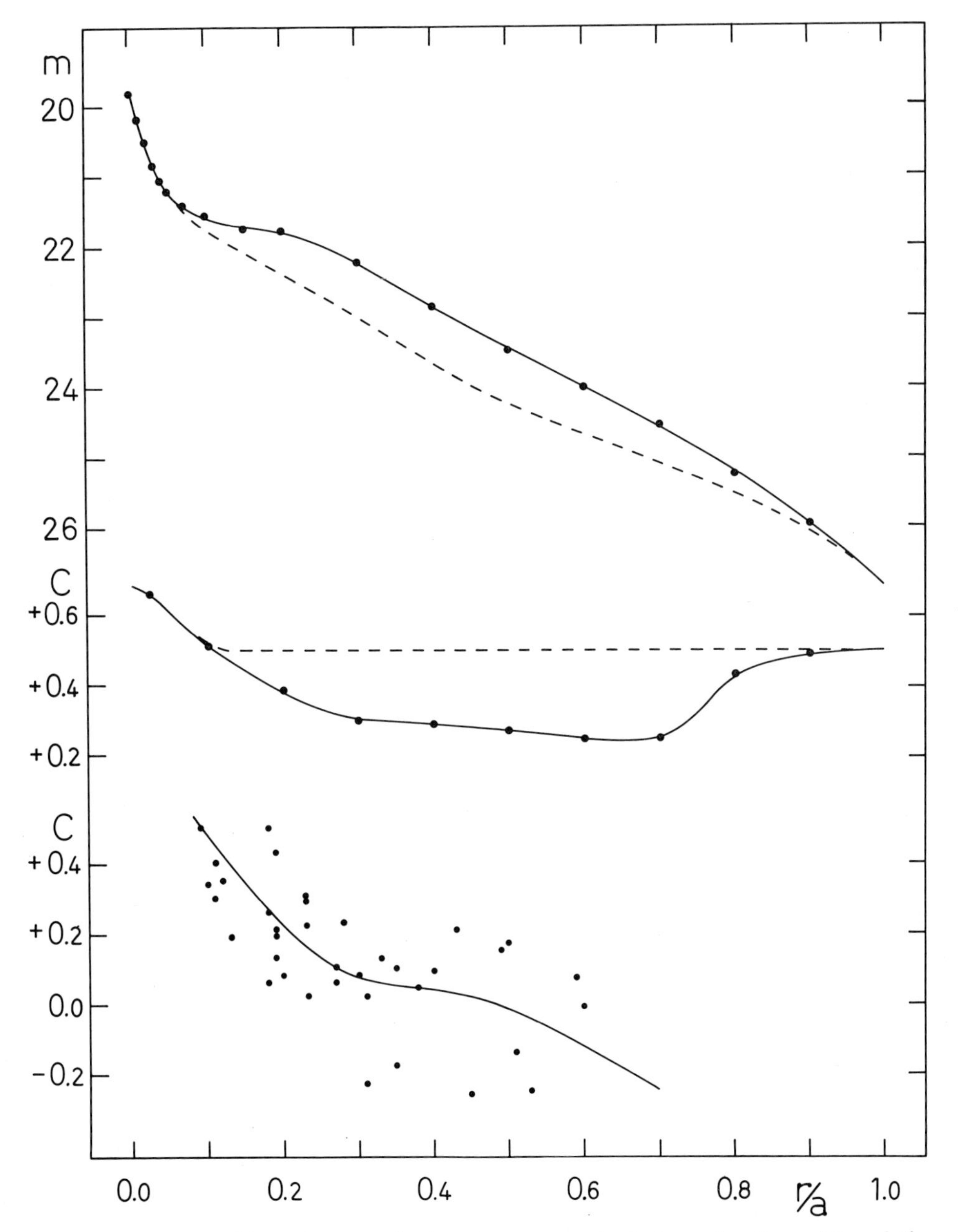

Fig. 6.—Photographic surface magnitude and pg − pv color index as functions of the relative distance from the center (mean results for 8 Sc− systems); the broken curves refer to the disk. At the bottom, the color indices of individual spiral arms are plotted against the distance.

The distance from the center is expressed in units of the photometrically measured half-diameter a. The upper part gives the mean photographic luminosity curve (m = mag per square second), and the middle part the mean pg − pv color index curve; as before, the broken lines refer to the disk. The lower part of the figure gives the color indices of the individual spiral arms.

The mean photographic luminosity curve referring to the disk can be approximately represented by the linear relation

$$m = 21.3 + 5.5r/a \,. \tag{9a}$$

The maximum deviations amount to $\Delta m = \pm 0.2$. If the disk (as represented by this formula) is subtracted, the distribution referring to the nucleus is very closely described by another linear relation,

$$m = 20.15 + 50r/a \,. \tag{9b}$$

TABLE 3

INTEGRATED pg − pv COLOR INDICES OF INDIVIDUAL SPIRAL ARMS*

Spiral Arm	C (mag)	Spiral Arm	C (mag)	Spiral Arm	C (mag)
628a.......	−0.01	1232d.......	+0.30	3938a.......	+0.13
b.......	+0.02	e.......	+0.50	b.......	+0.43
c.......	+0.19	f.......	+0.50	c.......	+0.08
d.......	+0.40	g.......	+0.23	d.......	−0.14
e.......	+0.06				
f.......	+0.07	3184a.......	+0.21	4303a.......	+0.17
		b.......	+0.21	b.......	+0.22
1042a.......	+0.04	c.......	+0.19	c.......	+0.02
b.......	+0.06	d.......	+0.08	d.......	+0.15
c.......	+0.26				
d.......	+0.10	3344a.......	−0.23	4535a.......	+0.31
		b.......	+0.35	b.......	+0.29
1232a.......	−0.26	c.......	+0.34	c.......	+0.09
b.......	−0.18	d.......	+0.10		
c.......	+0.13	e.......	−0.25		

* See fig. 5.

The nucleus extends to about $r/a = 0.10$. The combined distribution referring to disk + nucleus is thus of the same type as that found for S0 galaxies (cf. fig. 4). It may be noted that the disk is responsible for about 57 percent of the observed total luminosity, and the nucleus (disk subtracted) for about 2 percent; the remaining 41 percent refers to the spiral arms. As stated above, the color index of the disk ($C = +0.50$ mag) is independent of the central distance. The color of the nucleus (disk subtracted) varies slightly from the center to the outer parts; the mean color, $C = +0.77$ mag, is the same as that found for elliptical galaxies (cf. table 2).

In figure 5 the most prominent spiral arms have been denoted by the letters *a–g*. The integrated color indices of these arms (disk subtracted), as listed in table 3, range from −0.26 to +0.50 mag, the arithmetical mean being +0.14 mag. The color index is a function of the distance of the arm from the center; as is found from figure 6, the mean color changes from about +0.5 mag for $r/a = 0.1$ to about −0.1 mag for $r/a = 0.6$. The outermost arms thus contain younger populations of stars than the arms near the nucleus, an indication that the gas density in the arms varies with the distance.

As regards the distribution of absorbing matter in the spiral systems Sa-Sb-Sc, the information available is rather limited. On account of the large individual variations in internal structure, and the complications introduced by possible light-scattering, the procedure used in § 2 to determine the total absorption and color excess cannot be successfully applied to separate parts of the galaxies. It is true that the obscuration lanes observed in systems with small inclinations to the line of sight, and the asymmetry effects in systems of medium inclinations, give certain clues, as regards the general spatial arrangement of luminous and dark matter. Theoretical studies of these effects have been undertaken by Holmberg (1947), Elvius (1956), and Fish (1961). Since new stars are formed mainly in the spiral arms, it is very likely that the gas, and presumably also the dust, are concentrated to the arms. Observational evidence for the existence of heavy absorption in spiral arms has been presented by Holmberg (1950*b*). The intrinsic luminosity of the arms may thus be considerably higher than that observed.

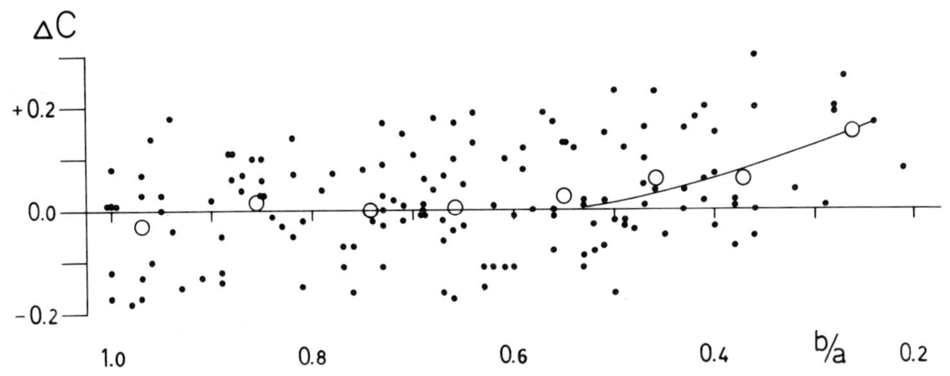

FIG. 7.—Variation in the pg − pv color indices of the nuclei of spiral galaxies with inclination, as represented by the ratio of the apparent (total) diameters.

For one special part of the spiral systems, the nuclear region, it seems possible to get a significant result, as regards the dependence of the color on inclination. The colors of the central parts of the nuclei have been measured by Holmberg (1969) for 172 spiral galaxies, mainly the same material as that analyzed in § 2. The mean pg − pv color index varies with the morphological type: $C = +0.77$ mag (S0), 0.76 mag (Sa, Sb−), 0.74 mag (Sb+), 0.62 mag (Sc−), and 0.38 mag (Sc+); the colors are corrected for galactic absorption and redshift effect. In figure 7 the color excess ΔC (=observed color *minus* mean color corresponding to the type) has been plotted against the diameter ratio b/a. The plot may be compared with those given in figure 2. In the interval $b/a =$ 0.5–1.0 the observed color is independent of the inclination of the spiral system to the line of sight, which seems to indicate that no absorption exists in the nuclear region; it should, however, be remarked that absorbing matter may be present if it is assumed to have a spherical distribution within the nucleus. The color excess found for small inclinations, $b/a < 0.5$, was to be expected, since in these cases the nuclei are partly screened off by dust clouds in the outer parts of the systems.

It may be noted that there is a rather pronounced correlation between the nuclear color index and the spectral class, as listed in the redshift catalog by Humason, Mayall, and Sandage (1956). In those cases in which the spectrograph slit has been put across the nucleus, the spectral classification apparently refers to the nuclear part of the galaxy. The linear regression line gives a mean color index of 0.44 mag for F0, 0.57 mag for F5, 0.70 mag for G0, and 0.82 mag for G5, the dispersion in the residuals being 0.087 mag.

In conclusion, it may be of interest to examine whether there is any dependence of the nuclear color on the physical state of the nucleus, especially in view of the current discussion on the possibility of explosive events in the nuclear parts of galaxies. The above material includes only two Seyfert galaxies, NGC 4051 and 4258; they both have large negative color excesses, −0.11 and −0.09 mag, respectively. From an inspection of the Hubble plate collection, Sérsic and Pastoriza (1967) have selected 20 galaxies that appear to have abnormal nuclei; eight of these objects are included in the above material, their mean nuclear color excess amounting to −0.06 mag. The presence of the emission line O II λ3727 in the spectrum also seems to affect the nuclear color. When this line is present in measurable strength, as indicated in the above redshift catalog, the mean color excess is −0.03 mag; when the line is not present, the mean excess is +0.03 mag.

4.3. Construction of Three-dimensional Models

In certain cases, the three-dimensional luminosity distribution in a galaxy can be reconstructed from the observed projected distribution. The most important conditions to be satisfied are (*a*) that the internal absorption is negligible, and (*b*) that the distribution sought can be represented by a simple model.

The least complicated case is presented by E-type galaxies of a spherical shape. With an isotropic model, in which the luminosity density is a smoothly decreasing function of the radial distance, the solution is easily found either by a numerical integration or by analytical methods. The problem is analogous to that of determining the space distribution of stars in a globular cluster from star counts (cf. Vol. **4** of this compendium). In the above-mentioned photometric study of elliptical galaxies, Hubble (1930) has derived the spatial distribution in a spherical system as corresponding to the projected distribution defined by equation (4). A three-dimensional distribution, supposedly referring to a spherical system, has been constructed also by Liller (1960). A detailed analysis of the two models, which are in good agreement, leads to the rather surprising result that the spatial distribution of light in a globular E-type galaxy can be very well represented by the simple formula

$$L = \frac{L_0}{(\rho/\alpha + 1)^3}, \tag{10}$$

where L is the luminosity per unit volume at the distance ρ from the center, and α is a scale factor. Except for the central portion of the nuclear region and the outermost parts (where the observations are less accurate), each of the two models deviates from this formula by less than 0.04 in log L.

In another special case, a lenticular system with an edgewise orientation, the three-dimensional distribution can be derived by similar methods. Liller (1960) has given the solution for a highly flattened system, the observational data referring to three E5 galaxies which presumably have inclinations to the line of sight close to zero. It is very interesting to find that the solution differs considerably from the result obtained for a globular system. In the central part the decrease in L is slower than that corresponding to equation (10), whereas in the outer regions L is inversely proportional to the fifth power of the central distance. Apart from a change in the scale factor, the distribution is apparently the same along the various axes of the ellipsoid.

In the general case, the three-dimensional distribution in an E galaxy cannot be completely recovered since the inclination (or the intrinsic oblateness) of the system is not known. Provided that the model has the same luminosity distribution, apart from a variable scale factor, along the various axes—that is, that the model is made up of concentric, oblate, spheroidal shells of constant luminosity density—a theoretical analysis (cf. Fish 1961) shows that it is possible to derive the *relative* intensities associated with the successive shells. The solution is independent of the inclination of the system.

The internal structure of systems of type S0 does not appear to be quite as homogeneous as that of E galaxies. The presence of rudimentary spiral arms, and sometimes weak absorption patches, introduces a complication in the determination of the three-dimensional distribution. A comparison between the models derived by Johnson (1961) for NGC 524 and by van Houten (1961) for NGC 4111, 4762, and 7332 shows that the solution may vary considerably from one object to another. An attempt to construct a model representative of the average S0 system leads to the result that the luminosity density in the principal plane (outside the nuclear region) is, in a first approximation, inversely proportional to the second power of the central distance.

In the case of spiral systems of types Sa-Sb-Sc the heavy internal absorption makes a recovery of the three-dimensional luminosity distribution very difficult, if not impossible. It was found in the preceding section that the surface-magnitude distribution in the disk of an Sc system is similar to that found for type S0. If the disk, as defined by the minima between the spiral arms, is assumed to be unaffected by absorption, then the spatial distribution of light in the disk will be approximately the same as the distribution derived for S0 systems.

5. ABSOLUTE MAGNITUDES

5.1. Luminosities of Individual Galaxies

The distance moduli needed for the determination of absolute magnitudes may be derived by different methods: in the first place, by analyses of the stellar contents of galaxies. The distance determinations will not be discussed here, a detailed summary of all problems connected with the distance scale being given in chapter 12.

Without any knowledge of the distance, it is possible to get a fairly good estimate of the absolute luminosity of a galaxy by means of the observed distance-independent photometric parameters—in the first place, the integrated surface magnitude. As is shown in § 6.2, the surface magnitude of a galaxy is related to the absolute total magnitude (or the absolute size). An analysis of the Holmberg (1958) material leads to an empirical relation between the photographic absolute magnitude M_*, the surface magnitude S_*, and the integrated color index C_*, all quantities being corrected for internal absorption (cf. § 2.1). For spiral systems of types Sa-Sb-Sc a least-squares solution gives the following result (Holmberg 1964):

$$M_* = 3S_* + 1.30C_* - 64.42, \tag{11}$$

the estimated mean error in M_* being about 0.6 mag. It should be noted that the analysis is based on selected material (high-luminosity galaxies) and that the Hubble parameter needed for the calibration is assumed to be $H = 80$ km s^{-1} per Mpc. A list of "photometric" distance moduli derived by this method is given in Holmberg (1964).

Another method for a direct estimate of the absolute magnitude has been developed by van den Bergh (1960 *a, b*). His system of luminosity classification, as applied to Sb-Sc spirals and irregular objects, is based on the correlation between the absolute magnitude and the degree of development of the spiral arms. The most strongly developed spiral structure occurs in spirals of the highest absolute luminosity. The roman numerals I–V, as used by van den Bergh in addition to the symbol for the Hubble type, refer to supergiant, bright giant, giant, subgiant, and dwarf galaxies, respectively. The luminosity classes have been calibrated by means of redshift distances. Distance moduli, based on this classification, have been derived for a considerable number of the northern Shapley-Ames galaxies (van den Bergh 1960*c*). Except for a few large deviations, these moduli agree reasonably well with those determined by Holmberg.

The absolute luminosities of galaxies vary within very wide limits. The faintest known dwarf galaxies, the Draco and Ursa Minor systems in the Local Group, have absolute photographic magnitudes of about $M = -8$ (van den Bergh 1968). The determination of the total magnitudes of dwarf systems of this type is very difficult, on account of the exceedingly low surface brightness. The limit $M = -8$ is no doubt set by observational selection effects; it seems safe to assume that the faintest dwarf galaxies in the Local Group still remain undiscovered. It does not appear possible at present to try to establish a definite lower limit for the luminosity of stellar systems.

The determination of an upper luminosity limit presents a more easily solved problem. We are in this case independent of the Local Group, and have access to a much larger sample. According to Sandage (1956), the brightest members (rank 1) of 18 selected galaxy clusters have a mean absolute photographic magnitude of -21.5 (redshift distances; $H = 80$). In individual cases, the luminosity may be up to 0.5 mag brighter, which indicates an upper limit of $M = -22$. From the Holmberg (1964) catalog a similar result is obtained. The most luminous system among nearly 200 nearby field galaxies has an absolute photographic magnitude of -22. For comparison, it may be noted that the brightest member of the Virgo cluster has $M = -21.4$.

The mean absolute magnitudes of galaxies of different morphological types, corresponding both to a given volume of space and to a given class of apparent magnitude, will be discussed in the next section.

There is a very pronounced relation between the absolute magnitudes and the logarithmic absolute major diameters of galaxies. This correlation is examined in § 6.2.

As regards correlations between the absolute magnitudes and other observable quantities, attention is directed to the magnitude-color diagrams that have been constructed, for instance, by Chester and Roberts (1964). In a diagram of this kind, the different morphological classes are arranged in a special way. The diagram would no doubt permit more interesting conclusions if the luminosities were reduced to unit mass, that is, if the luminosities were divided by the corresponding masses. Unfortunately, reliable mass determinations are available only for a limited number of galaxies.

5.2. The Luminosity Function

The classical study by Hubble (1936*a*, *b*) led to the conclusion that the luminosity function, or the statistical distribution of the absolute photographic magnitudes of all galaxies in a given volume of space, could be reproduced by a normal error-curve having a dispersion of 0.85 mag. The curve was derived from studies of the brightest resolved stars in Sb-Sc-Ir galaxies, and of the residuals in the magnitude-redshift relation.

From an analysis based on the Local Group of galaxies, and the groups around M81 and M101, Holmberg (1950*a*) arrived at a luminosity function with a considerably larger dispersion (almost 2 mag). The curve was no longer symmetrical; since the material included a comparatively large number of dwarf systems, the maximum of the curve was displaced in the direction of fainter magnitudes.

A radically different luminosity function was later suggested by Zwicky (1957). Investigations of clusters of galaxies led to the conclusion that there is no observable maximum in the luminosity distribution. All the way down to the limiting magnitude the class frequency increases exponentially as the luminosity grows fainter. The logarithmic distribution function is a straight line, with an inclination coefficient of approximately 0.2:

$$\phi(M) = \text{const. } 10^{M/5} . \tag{12}$$

The great majority of galaxies would thus be very small dwarf systems.

Similar results, also based on clusters of galaxies, have been obtained by Abell (1962, 1964). In this case, however, the inclination of the logarithmic distribution curve was somewhat larger; except for the brightest tail end of the curve, the coefficient was found to range from about 0.2 to about 0.3.

The gradual change that has taken place in our conception of the galaxian luminosity function is illustrated by figure 8 (reproduced from Zwicky 1957). The three curves refer to the investigations mentioned above. The right-hand part of the Holmberg curve agrees fairly well with the curve suggested by Zwicky; the abrupt decline of the former curve at the fainter end is probably explained by incompleteness in the material. It should, however, be emphasized that the luminosity function derived by Zwicky (and the function found by Abell) refers to dense clusters of galaxies, and that a function of this type is not necessarily representative of a galaxian population in the field outside the clusters.

A reliable determination of the luminosity function for noncluster galaxies would be possible if the analysis could be based on a large number of galaxy groups, like the Local Group, and if completeness of the material could be attained down to a reasonably faint limiting magnitude. An attempt to secure such material has been made by Holmberg (1969). Examinations (on the *Palomar Sky Atlas*) of 160 groups, centered on more or less nearby spiral systems, have resulted in a list of altogether 274 physical companions. Because of the large disturbances from the background field it was found necessary to limit the survey to circular areas with a radius of 50 kpc around the central galaxies All companions with absolute major diameters larger than 0.6 kpc were picked up, and the diameters were transformed to an absolute scale by means of the distances estimated for the central spiral systems. In addition to the diameters, the morphological types could be estimated for the medium-sized and large satellites.

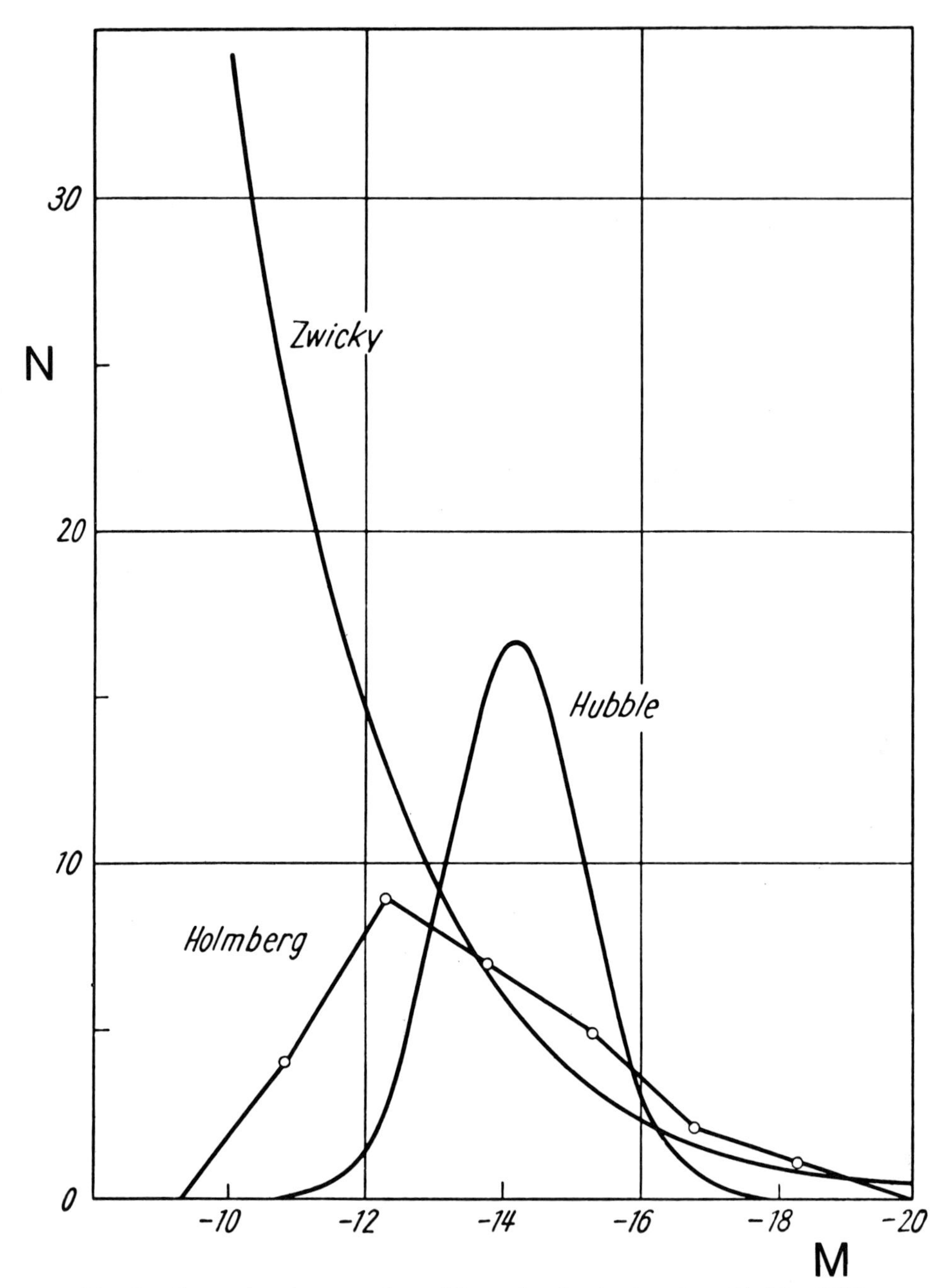

FIG. 8.—Comparison between three different luminosity functions (Zwicky 1957).

The diameters of the 274 group members having been transformed into absolute magnitudes by means of equation (15), Holmberg arrived at the luminosity functions reproduced in figure 9. The left-hand part of the distribution is not included in the figure; the curve rises exponentially all the way down to the estimated limit $M = -10.6$. The full curve, based on the class frequencies defined by the large open circles, refers to the entire material, whereas the broken curve (*small open circles*) refers to the E-S0-Ir

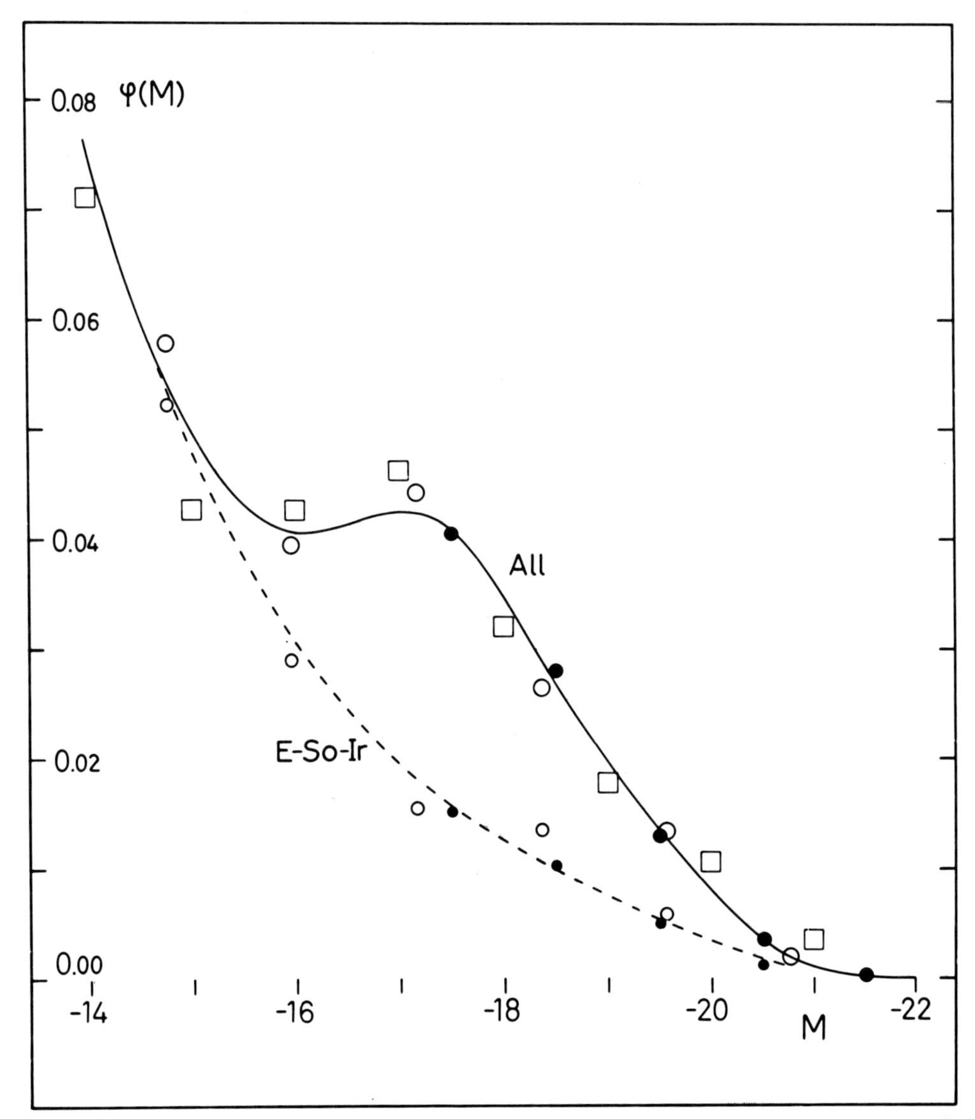

FIG. 9.—Luminosity functions for the members of 160 physical groups (Holmberg 1969). The ordinate gives an absolute calibration, number of galaxies per magnitude class per Mpc^3; it corresponds to a space density of 0.17 per Mpc^3 ($M < -15.0$), derived from analyses of magnitudes, diameters, and redshifts.

galaxies alone; the area between the two curves thus represents Sa-Sb-Sc spirals. With a good degree of approximation, the distribution referring to the latter objects can be described by a normal error-curve with a mean magnitude $M = -17.7$, and a dispersion of 1.2 mag. There seems to be a lower limit to the luminosity of spirals of about $M = -15$ (absolute diameter = 5 kpc). The distribution referring to the E-S0-Ir galaxies agrees with the curve suggested by Zwicky; the logarithmic distribution is represented down to $M = -10.6$ by a straight line having an inclination of 0.195.

In figure 9 two interesting comparisons are made with results available from other sources. The open squares refer to the Local Group and the M81 group, the up-to-date lists of members probably being complete down to $M = -13.5$. The class frequencies are, except in the last class, overlapping means. In spite of the small total number, there is good agreement with the full curve. The large filled circles represent all the galaxies brighter than the 12th apparent magnitude in the available redshift lists, whereas the small filled circles refer to the E-S0-Ir galaxies alone. It is very satisfactory to find that the luminosity curves from the redshift material are in perfect agreement with the

TABLE 4

RELATIVE FREQUENCIES OF GALAXIES OF DIFFERENT TYPES IN A GIVEN VOLUME OF SPACE ($M < -15.0$)

Type	Relative number (%)	$\langle M_m \rangle$
E-S0	30	−19.8
Ir I	17	−17.5
Ir II	7	(−19.0)
Sa-Sb	12	−20.0
Sc	34	−19.2

curves derived from the physical groups, especially in view of the fact that in the brighter magnitude classes the former curves are based on a much larger sample. It should be explained here that, for $m > 12$, or $M > -17$ ($H = 80$), it does not seem possible to get any reliable information on the luminosity functions from the available redshifts. The results would, as in the case of Kiang (1961), be falsified by the very serious selection effects in the material.

From the above analysis, the total luminosity function (pg) referring to all galaxies in 1 Mpc^3 (outside the big clusters) can be described by the expression

$$\phi(M) = 40 \times 10^{0.195M} + 0.025\, e^{-0.35(M+17.7)^2}\,, \tag{13}$$

where the first part refers to types E-S0-Ir and the second part to types Sa-Sb-Sc. The formula represents the distribution from $M = -10.6$ to $M = -19$ or -20; beyond this limit the curve rapidly approaches zero. It should be noted that the luminosity function is based on a Hubble parameter $H = 80$ km s^{-1} per Mpc.

In conclusion, some results will be given that refer to different classes of morphological type. In the interval $M < -14.2$ there are 34 E-S0 objects, 19 Ir I objects, and eight Ir II objects. In the interval $M < -15.0$, where Sa-Sb-Sc spirals are more fully represented, the relative frequencies range from 34 percent (Sc) to 7 percent (Ir II), as appears from table 4. The mean absolute magnitude corresponding to a given volume of space cannot be determined for E-S0-Ir galaxies; for the Sa-Sb-Sc spirals the mean magnitude is $M = -17.7$, as stated above. It is, however, possible to compute, by means of the luminosity functions, the mean absolute magnitude that would correspond to a given class of apparent magnitude. The results are listed in the last column of table 4. The mean magnitudes are to some extent based on a combination of the above data and the Holmberg (1964) catalog, which is essentially complete, as regards galaxies of bright apparent magnitudes.

6. DIAMETERS

6.1. Angular Dimensions

Because of the low density gradients in the outermost parts of galaxies, the definition of the major and minor diameters encounters certain difficulties. The problem may be simplified by stipulating either that the diameters should be referred to a given isophote or that the boundary defined by the diameters should enclose a certain portion of the total luminosity. The first definition supposedly applies to all diameters that are derived from direct measurements of photographic images, provided that the plate material is homogeneous (constant limiting magnitude). The second definition demands a determination of the surface luminosity distribution in each object.

Apparent diameters that are measured directly on photographs may be divided into two groups: those that are determined by visual inspection of the plate (here called *estimated* diameters), and those that are derived from photometer tracings (*photometric* diameters). Among the catalogs that give estimated diameters, the list by Reinmuth (1926) and the survey by Shapley-Ames (1932) should be mentioned. Photometric diameters, referring to a given isophote (26.5 mag pg per square second), have been determined by Holmberg (1958).

The second of the above definitions has been utilized by de Vaucouleurs (1948). The *effective* major and minor diameters define an ellipse enclosing half the total luminosity of the galaxy. The geometrical mean of the two semidiameters is called the effective radius. A general application of this definition would involve a considerable amount of photometric work. The method also meets with practical difficulties in the case of galaxies of small angular dimensions.

Since for the great majority of galaxies only estimated diameters are available, attention is called to the systematic errors that exist in these data. Laboratory experiments by Holmberg (1946), based on artificial galaxy images, have clearly demonstrated that the human eye is an imperfect instrument in recording diameters. The ratio of estimated to true (photometric) diameter appears to be a function not only of the slope of the luminosity distribution curve but also of the elongation of the image. The eye overestimates the length of an elongated image and underrates its width. The effect is well established by a comparison between the Shapley-Ames diameters and the photometric measurements by Holmberg. The result is shown in table 5, where the photometric and the estimated diameters are denoted by a, b and α, β, respectively. For round objects, the ratio of estimated to photometric diameter is about 0.46. As the elongation becomes more pronounced, α/a increases to 0.66 whereas β/b decreases to 0.38. Although the ellipticity of the isophotes may, in most galaxies, vary to some extent with the distance from the center, the major part of this systematic effect must be ascribed to errors in the estimated diameters. The smoothed average values of the Shapley-Ames ratio β/α, corresponding to different ratios b/a, have already been listed in table 1.

It should be noted that, although the photometric diameters referring to a given isophote are presumably free from systematic errors, they nevertheless include certain systematic effects. The diameters are to some extent dependent on the galactic absorption, and on the inclination of the galaxy to the line of sight. The absorption reduces the surface brightness, and the diameters are thus referred to a more central isophote. With the slope of the luminosity distribution curve that has been found for the outermost parts of Sc− galaxies (fig. 6), a differential galactic absorption of 0.25 mag would change the diameters by 2.5 percent. In spite of the internal absorption, the surface brightness of a spiral galaxy undoubtedly increases as the inclination to the line of sight is changed from 90° to 0°. The resulting change in the measured major diameter would be about 7 percent, if the surface brightness is assumed to increase by a factor of 2.

Spiral galaxies may, with a good degree of approximation, be treated as oblate spheroids. If the relative thickness of the three-dimensional disk is known, the measured diameters can be used to determine two important parameters: the inclination of the principal plane and the volume occupied by the stellar content.

A determination of the average flattening of the spiral disk must be based on galaxies for which an edgewise orientation is assured—for instance, by absorption lanes being projected against the central part of the nucleus. The six galaxies listed in table 6 have

TABLE 5

COMPARISON BETWEEN THE ESTIMATED DIAMETERS (α, β) OF THE SHAPLEY-AMES CATALOG AND PHOTOMETRIC DIAMETERS (a, b)

b/a	$\langle\alpha/a\rangle$	$\langle\beta/b\rangle$	$\langle\beta/b\rangle:\langle\alpha/a\rangle$
0.81–1.00.......	0.47	0.45	0.96
0.61–0.80.......	0.50	0.44	0.88
0.41–0.60.......	0.59	0.42	0.71
≤0.40.......	0.66	0.38	0.58

TABLE 6

FLATTENING OF THE DISK, AS DETERMINED FOR SPIRAL GALAXIES WITH AN EDGEWISE ORIENTATION

OBJECT	TYPE	PHOTOMETRIC		ESTIMATED	
		Diameters	Ratio	Diameters	Ratio
NGC 891.........	Sb+	15:0×3:8	0.25	11:5×0:8	0.07
NGC 4244........	Sc+	18. ×2.9	0.16	12.8×1.2	0.09
NGC 4565........	Sb+	20. ×3.6	0.18	14.5×1.0	0.07
NGC 4631........	Sc+	19. ×4.4	0.23	13.4×1.5	0.11
NGC 5746........	Sb−	9.0×2.4	0.27	6.8×0.6	0.09
NGC 5907........	Sb+	15.7×2.0	0.13	11.2×0.8	0.07
Mean..........	....	....	0.20	....	0.083

been selected by Wyse and Mayall (1942) as typical cases. The photometric diameters (Holmberg 1958) give a mean ratio (smallest to largest axis of the spheroid) of 0.20. A study of more comprehensive material leads to the same result. On the other hand, the estimated diameters give a mean ratio of only 0.08. Since the latter result seems to have been rather widely adopted as a measure of the flattening of the average spiral system, attention is called again to the unreliability of estimated diameters.

Among the remaining types of galaxies, only elliptical systems possess rotational symmetry. According to the well-known study by Hubble (1926), which is based on the assumption that the principal planes have a random orientation with respect to the line of sight, the flattening of the elliptical disk varies from about 0.3 to 1.0. According to Hodge and Hitchcock (1966), most irregular galaxies have axial ratios in the range 0.2–0.4, the mean ratio being about 0.3.

If a galaxy can be treated as an oblate spheroid, the inclination i of the principal plane to the line of sight is given by the relation

$$\sin^2 i = \frac{(b/a)^2 - f^2}{1 - f^2}, \tag{14}$$

where b/a is the apparent diameter ratio, and f the flattening of the disk or the ratio of the smallest to the largest axis (Hubble 1926).

With a random orientation with respect to the line of sight, the statistical distribution of the inclination angle i can be determined: the relative distribution function will be equal to $\cos i$. The assumption of a random orientation is confirmed by the observation data available for spiral galaxies (Holmberg 1946). It may be noted that a number of earlier studies have led to the conclusion that there is an excess of spiral systems with an edgewise orientation. This erroneous result is a reflection of the systematic errors in the estimated diameters.

6.2. Absolute Diameters

The absolute dimensions of galaxies have a very wide range. The diameters of the smallest dwarfs known are in the interval 0.1–1 kpc, whereas the largest giant systems may have diameters exceeding 50 kpc.

As regards the lower limit, information is in the first place obtained from a study of the Local Group of galaxies; the smallest dwarf systems discovered so far are naturally nearby objects. With the distance moduli listed by van den Bergh (1968) and the angular diameters derived by Hodge (1966; see also the references at the end of § 4.1) from star counts, the smallest members of the group are the Draco system (diameter = 1.0 kpc), the Leo II system (diameter = 1.3 kpc), and the Leo I system (diameter = 1.8 kpc). It may be noted that the Hodge diameter is defined by the parameter r_t of equation (8), and that it thus represents the extrapolated total extension of the object. With the Holmberg (1958) scale, the diameters of Leo II and Leo I are 0.7 and 0.8 kpc, respectively (Draco system not included). A still smaller galaxy has recently been found by Hodge (1967). It is an extremely blue, irregular dwarf system, the color-magnitude diagram of the resolved stars being like that of a very young, not very populous, open cluster. With a provisional distance of 320 kpc, the maximum detectable angular extension corresponds to an absolute diameter of about 0.2 kpc. Attention is called also to the compact galaxies studied by Zwicky (1964*a*, 1966); some of these peculiar objects seem to have diameters less than 1 kpc.

Information on the upper diameter limit may be obtained from the Holmberg (1964) catalog, which gives absolute major diameters for 245 more or less nearby galaxies. The distances, partly derived from redshifts, correspond to a Hubble parameter $H = 80$ km s^{-1} per Mpc. Although the catalog does not represent a given volume of space, it is essentially complete, as regards large-diameter systems, out to a distance modulus of about 30.5. Within this volume, the five largest galaxies in each type group have the following mean logarithmic diameters (pc): 4.59 (E-S0), 4.65 (Sa, Sb), 4.57 (Sc), and 4.15 (Ir I; one exceptional object omitted). Spirals of Sa-Sb types thus have the largest

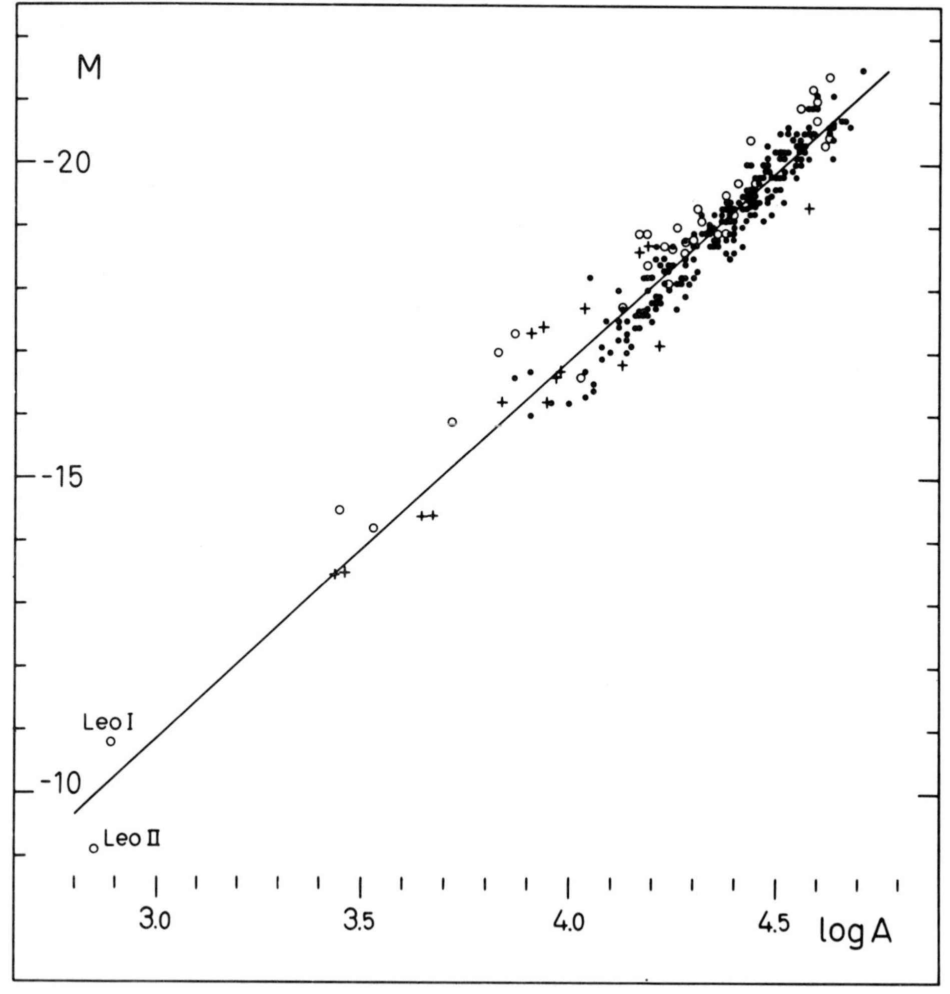

FIG. 10.—Correlation between absolute photographic magnitude and log absolute major diameter (pc), as derived for galaxies of type E-S0 (*open circles*), types Sa-Sb-Sc (*filled circles*), and type Ir (*crosses*).

maximum diameter (20 percent larger than that of Sc spirals and about 3 times that of irregular objects).

It is of considerable interest to find that there is a very well-defined relation between the logarithmic absolute major diameters and the absolute magnitudes of galaxies. The correlation diagram reproduced in figure 10 is based on the data listed in the above-mentioned catalog, and it thus refers to the Holmberg magnitude and diameter system. The diagram includes 189 galaxies of the types Sa-Sb-Sc, 36 of the types E-S0, 16 of the type Ir I, and 5 of the type Ir II, in all 246 objects. The absolute photographic magnitudes have not been corrected for internal absorption, with the exception of the Sa-Sb-Sc group, for which *differential* absorption corrections have been applied. If the absorption-free magnitudes, as listed for the latter objects, are corrected by $+1.0$ mag (Sa-Sb) or $+0.7$ mag (Sc), they are reduced to a random orientation of the spirals with respect to the line of sight. The absolute magnitudes range from -9.1 (Leo II system) to -21.5, and the absolute diameters from 0.7 to 50 kpc. The relation is very pronounced, and apparently linear, the coefficient of correlation being -0.962 ± 0.005 (m.e.). The regression line drawn in the figure is represented by the equation

$$M = -6.00 \log A + 7.14 . \tag{15}$$

The relation indicates that the surface magnitude gets fainter with decreasing size, the total variation being about 2 mag from the largest galaxies to the smallest dwarfs. The dispersion in M around the regression line amounts to only 0.40 mag. It is of special interest to note that the different type groups do not show any significant systematic deviations from the regression line; the only exception is the E-S0 group, with a deviation of about -0.3 mag.

Because of the pronounced correlation, and the linear shape of the regression line, the statistical distribution of the logarithmic absolute diameters should be nearly the same as the distribution of the absolute magnitudes. This fact has been made use of in § 5.2 for the determination of the luminosity function from the diameters. It will be recalled that the distribution of the logarithmic diameters, as derived for the members of 160 physical groups, can be separated into two parts: an exponential curve for E-S0-Ir galaxies, and a normal error-curve for Sa-Sb-Sc spirals. As regards the former curve, no maximum is reached, although the material includes diameters down to less than 1 kpc. On the other hand, a minimum diameter of about 5 kpc is indicated for the Sa-Sb-Sc spiral systems. In a given volume of space, the great majority of galaxies would thus be exceedingly small dwarf galaxies.

REFERENCES

Abell, G. O. 1962, *Problems of Extra-Galactic Research, IAU Symposium* **15**, ed. G. C. McVittie, (New York: Macmillan Co.), p. 213.
———. 1964, *Ap. J.*, **140**, 1624.
Abell, G. O., and Mihalas, D. M. 1966, *A.J.*, **71**, 635.
Baade, W. 1944, *Ap. J.*, **100**, 137.
Baum, W. A. 1955, *Pub. A.S.P.*, **67**, 328.
Bergh, S. van den. 1960*a*, *Ap. J.*, **131**, 215.
———. 1960*b*, *ibid.*, **131**, 558.
———. 1960*c*, *Pub. David Dunlap Obs.*, Vol. **2**, No. 6.
———. 1968, *J.R.A.S. Canada*, **62**, 145, 219 = *Comm. David Dunlap Obs.*, No. 195.
Bigay, J. H. 1951, *Ann. d'Ap.*, **14**, 319.
———. 1964, *ibid.*, **27**, 170.

Bigay, J. H., Dumont, R., Lenouvel, F., and Lunel, M. 1953, *Ann. d'Ap.*, **16**, 133.
Bigay, J. H., and Dumont, R. 1954, *Ann. d'Ap.*, **17**, 78.
———. 1955, *ibid.*, **18**, 141.
Chester, C., and Roberts, M. S. 1964, *A.J.*, **69**, 635.
Dennison, E. 1954, thesis, University of Michigan (unpublished).
Elvius, A. 1956, *Stockholm Obs. Ann.*, Vol. **18**, No. 9.
Evans, D. S. 1951, *M.N.R.A.S.*, **111**, 526.
———. 1952, *ibid.*, **112**, 606.
Fish, R. A. 1961, *Ap. J.*, **134**, 880.
Hodge, P. W. 1961*a*, *A.J.*, **66**, 249.
———. 1961*b*, *ibid.*, **66**, 384.
———. 1962, *ibid.*, **67**, 125.
———. 1963*a*, *ibid.*, **68**, 237.
———. 1963*b*, *ibid.*, **68**, 470.
———. 1963*c*, *ibid.*, **68**, 691.
———. 1964*a*, *ibid.*, **69**, 438.
———. 1964*b*, *ibid.*, **69**, 853.
———. 1966, *Ap. J.*, **144**, 869.
———. 1967, *ibid.*, **148**, 719.
Hodge, P. W., and Hitchcock, J. L. 1966, *Pub. A.S.P.*, **78**, 79.
Hoerner, S. von. 1957, *Ap. J.*, **125**, 451.
Holmberg, E. 1945, *Medd. Lunds Astr. Obs.*, Ser. 2, No. 114.
———. 1946, *ibid.*, Ser. 2, No. 117.
———. 1947, *ibid.*, Ser. 2, No. 120.
———. 1950*a*, *ibid.*, Ser. 2, No. 128.
———. 1950*b*, *ibid.*, Ser. 1, No. 170.
———. 1958, *ibid.*, Ser. 2, No. 136.
———. 1964, *Ark. f. Astr.*, **3**, 387 = *Uppsala Astr. Obs. Medd.*, No. 148.
———. 1969, *ibid.*, **5**, 305 = *Uppsala Astr. Obs. Medd.*, No. 166.
Houten, C. J. van. 1961, *Bull. Astr. Inst. Netherlands*, **16**, 1.
Houten, C. J. van, Oort, J. H., and Hiltner, W. A. 1954, *Ap. J.*, **120**, 439.
Hubble, E. 1926, *Ap. J.*, **64**, 321 = *Contr. Mt. Wilson Obs.*, No. 324.
———. 1930, *ibid.*, **71**, 231 = *Contr. Mt. Wilson Obs.*, No. 398.
———. 1934, *ibid.*, **79**, 8 = *Contr. Mt. Wilson Obs.*, No. 485.
———. 1936*a*, *ibid.*, **84**, 158 = *Contr. Mt. Wilson Obs.*, No. 548.
———. 1936*b*, *ibid.*, p. 270 = *Contr. Mt. Wilson Obs.*, No. 549.
Humason, M. L., Mayall, N. U., and Sandage, A. R. 1956, *A.J.*, **61**, 97.
Johnson, H. M. 1961, *Ap. J.*, **133**, 314.
Kiang, T. 1961, *M.N.R.A.S.*, **122**, 263.
King, I. 1962, *A.J.*, **67**, 471.
Liller, M. H. 1960, *Ap. J.*, **132**, 306.
Lyngå, G. 1959, *Medd. Lunds Astr. Obs.*, Ser. 2, No. 137.
Miller, R. H., and Prendergast, K. H. 1962, *Ap. J.*, **136**, 713.
Oort, J. H. 1940, *Ap. J.*, **91**, 273.
Pettit, E. 1954, *Ap. J.*, **120**, 413.
Redman, R. O. 1936, *M.N.R.A.S.*, **96**, 588.
Redman, R. O., and Shirley, E. G. 1938, *M.N.R.A.S.*, **98**, 613.
Reinmuth, K. 1926, *Veröff. Sternw. Heidelberg*, Bd. 9.
Sandage, A. R. 1956, see Humason, Mayall, and Sandage.
———. 1961, *The Hubble Atlas of Galaxies* (Carnegie Institution of Washington, Pub. No. 618).
———. 1962, *Problems of Extra-Galactic Research, IAU Symposium 15*, ed. G. C. McVittie (New York: Macmillan Co.), p. 359.
Sérsic, J. L., and Pastoriza, M. 1967, *Pub. A.S.P.*, **79**, 152.
Shapley, H., and Ames, A. 1932, *Ann. Harv. Coll. Obs.*, **88**, 43.
Simkin, S. M. 1967, *A.J.*, **72**, 1032.
Stebbins, J., and Whitford, A. E. 1937, *Ap. J.*, **86**, 247.
———. 1952, *ibid.*, **115**, 284.
Tifft, W. G. 1961, *A.J.*, **66**, 390.
———. 1963, *ibid.*, **68**, 302.
Vaucouleurs, G. de. 1948, *Ann. d'Ap.*, **11**, 247.
———. 1953, *M.N.R.A.S.*, **113**, 134.
———. 1956, *A.J.*, **61**, 430.
———. 1958, *Ap. J.*, **128**, 465.
———. 1959, *Lowell Obs. Bull.*, **4**, 105.
———. 1961, *Ap. J. Suppl.*, **5**, 233.
Vaucouleurs, G. de, and Vaucouleurs, A. de. 1964, *Reference Catalogue of Bright Galaxies* (Austin: University of Texas Press) = Univ. Texas Monogr. Astr., No. 1.

Whitford, A. E. 1936, *Ap. J.*, **83**, 424.
Wyse, A. B., and Mayall, N. U. 1942, *Ap. J.*, **95**, 24.
Zwicky, F. 1957, *Morphological Astronomy* (Berlin: Springer-Verlag).
———. 1964*a*, *Ap. J.*, **140**, 1467.
———. 1964*b*, *ibid.*, **140**, 1626.
———. 1966, *ibid.*, **143**, 192.
Zwicky, F., Herzog, E., and Wild, P. 1961, *Catalogue of Galaxies and of Clusters of Galaxies*, Vol. **1** (Pasadena: California Institute of Technology).
Zwicky, F., and Herzog, E. 1963, *ibid.*, Vol. **2**.
Zwicky, F., Karpowicz, M., and Kowal, C. T. 1965, *ibid.*, Vol. **5**.
Zwicky, F., and Herzog, E. 1966, *ibid.*, Vol. **3**.
Zwicky, F., and Herzog, E. 1968, *ibid.*, Vol. **4**.
Zwicky, F., and Kowal, C. T. 1968, *ibid.*, Vol. **6**.

CHAPTER 5

Integrated Energy Distribution of Galaxies

A. E. WHITFORD
Lick Observatory, University of California, Santa Cruz

1. INTRODUCTION

THE light from galaxies is quite generally the integrated effect of a population of unresolved stars. Even in spirals close enough for the prominent arms to be partially resolved, Holmberg (1950) showed that most of the light comes from an underlying "main body" or smooth continuum, now called the disk population. For Sb spirals he found that the arms contribute only about 10 percent of the total light; for extreme Sc spirals the light of arms is between a quarter and a third of the whole. The red giants of Population II first resolved by Baade (1944) in galaxies of the Andromeda group add only a thin veneer to the general unresolved background of the disk. From the count-brightness ratios determined by Baum and Schwarzschild (1955; further discussion by Baum 1959) it may be concluded that the resolved stars which stood out above the counting threshold $M_v = -2.0$ ($m_v \simeq 22.2$) contributed only a few percent to the total surface brightness of the disk of M31.

The Magellanic Clouds and a few nearby dwarf ellipticals of pure Population II offer the only opportunities for attempts at relating integrated properties to a stellar census. Studies of the latter class of galaxies include that of the Draco system by Baade and Swope (1961) and those of six dwarf ellipticals in the Local Group by Hodge (1966). Though such studies are of considerable intrinsic interest, the types of galaxies represented are statistically unimportant in the general field.

For giant elliptical systems the observer is completely dependent on the analysis of the integrated radiation for information on stellar content; not a single example of this type of galaxy is near enough for any individual stars to have been resolved. These systems have assumed special importance because as the brightest members of large clusters of galaxies they are used as high-luminosity standard objects in observational tests of the linearity of the law of redshifts and attempts to determine the cosmological deceleration parameter q_0 (Hubble 1936; Humason, Mayall, and Sandage 1956; Sandage 1968).

This chapter treats the observational measurements of the spectral energy distribution of the integrated light from normal galaxies. Galaxies which show an extraordinary

Received December 1969; revised February 1974.

radiation pattern because of recent violent events are discussed in chapter 7. The inferences regarding the makeup of an unresolved stellar population that can be deduced from the distribution curve are also considered in this chapter. The K-correction to adjust the observed magnitude of a redshifted galaxy to what would be seen by an observer in the velocity frame of the distant object is derived; this quantity is essential in the determination of photometric distances. Finally, the possible effect of evolutionary changes in stellar populations on energy distribution curves is considered, since the long light-travel times involved mean that distant and nearby galaxies are being observed at significantly different fractions of their life up to the present epoch.

2. COLOR INDICES

If galaxies were composed of stars of a single type, then the concept of a color temperature would give a good approximation to the spectral energy distribution. Though such a hypothesis seems unlikely from analogy with the composite population within the Milky Way, some early information of value was obtained from color indices. For lack of any more exact measurements, Hubble (1936) adopted the working assumption that the light of the average galaxy could be approximated by a 6000° blackbody. The justification was the resemblance of Humason's low-dispersion spectra of galaxies to the spectra of dwarf G stars of solar type.

Stebbins and Whitford (1937) took advantage of the ability of a photoelectric cathode to integrate all the light falling on it, irrespective of the intensity pattern in the focal-plane aperture, to obtain color indices of a considerable number of representative galaxies. The first system, called C_2, was based on bandpasses of about 250 Å equivalent width and effective wavelengths of 4340 and 4670 Å. The colors of E galaxies and the central regions of Sb galaxies were found to be like those of dK1 stars, whose color temperatures are considerably less than 6000° K. These results were extended later (Stebbins and Whitford 1952) on the C_p color system, which later became the $P - .V$ system. Comparisons with the $B - V$ system are given by Johnson (1963) and de Vaucouleurs (1961). The results of color-index measurements of galaxies by various techniques are treated in detail in chapter 4 of this volume.

The most important results of these early studies were not the exact numerical color indices obtained, but rather the demonstration that galaxies of various types of the Hubble sequence have rather small dispersion in color and that there is a regular progression of color along the sequence. The small scatter in the color of 32 E galaxies in the 1952 study ($C_p = +0.86 \pm 0.06$ m.e.) strengthened the assumption that such systems could be used as standard objects. Observations of 148 galaxies of all types on the UBV system were used by de Vaucouleurs (1961) in an extensive study of color vs. type and of uniformity in each class. The $U - B$ versus $B - V$ plots clearly showed the effects of compositeness. Additional discussion of homogeneity among ellipticals is reserved for the next section.

3. FILTER-BAND SPECTROPHOTOMETRY

Although the spectral distribution of the energy radiated by galaxies has been investigated almost entirely by photoelectric methods, the first spectrophotometry over an extended range of wavelengths was by photographic methods. Greenstein (1938) took advantage of the favorable geometry of the McDonald Observatory hillside spectrograph to compare the central region of M31 with stars which had been standardized on the Greenwich gradient system (Greaves, Davidson, and Martin 1934). About 3′ near the nucleus could be accepted without appreciable loss of spectral purity. The galaxy was found to fit the gradient of a blackbody of 4200° ± 200° K over the range 3900–6500 Å. Correction for absorption at galactic latitude $b = -20°$ raised the color temperature to 4400° K. The deviations from a blackbody that were noted were in the same sense as those noted in the sun: an excess at 4500 Å and a deficiency in the 3850–4000 Å region ascribed to the H and K lines and to the CN band at 3883 Å. Greenstein found no clear evidence of compositeness over the observed range.

The much larger range of wavelengths covered by the six-color photoelectric spectrophotometry of Stebbins and Whitford (1948) clearly showed the composite nature of the radiation from eight representative galaxies of types ranging from E to Sc; there were no examples of type S0 or Sa. The six filter bands had effective wavelengths for an equal energy spectrum ranging from 3530 Å in the ultraviolet to 10,300 Å in the infrared. A cesium oxide cathode, of the type more recently designated S1, gave adequate sensitivity over the entire set of filter bands. The bands were rather broad: in general $0.1 < \Delta\lambda/\lambda < 0.2$, where $\Delta\lambda/\lambda$ is defined as (equivalent width)/(effective wavelength). The resultant smoothing of the energy curves was later found to be a disadvantage.

Whitford and Sears (1970) measured additional galaxies on the six-color system, including a number of giant ellipticals in the Virgo cluster. Table 1 shows the results for galaxies from both series for which the aperture A was a considerable fraction of the isophotal diameter D_0 as given by de Vaucouleurs (1959, 1961). The six-color indices are referred to a mean dG6 star (Stebbins and Whitford 1945), and normalized so that $[B] + [G] + [R] = 0$.

These galaxy measurements may be reduced to an absolute standard through the 25 six-color stars calibrated by Stebbins and Kron (1964) in terms of a 5500° K blackbody. Intercomparison with the measurements of these same stars on the normal six-color system (Stebbins and Kron 1956) shows that the transformation from colors referred to a mean dG6 star to the absolute 5500° K system is not quite universal; there are appreciable differences with spectral type arising from the use of broad-band filters on objects having a wide range of temperatures and blanketing characteristics.

The steps in the computation are shown in table 2. For the spiral galaxy, it is satisfactory to adopt the relation between the two systems established from the mean of five stars of average type G0 used by Stebbins and Kron (1964) for placing the sun on the 5500° K system. The second line of the first section of the table lists the color differences thus determined; when combined with the magnitude differences read at the proper reciprocal wavelengths from a 5500° K blackbody curve, the galaxy observations are transformed to an absolute standard. The reciprocal wavelengths listed are those found by Stebbins and Kron to be appropriate for a 4500° K blackbody.

For the giant ellipticals in the second section of table 2, the slightly different relation given by the three giant stars of average type K0 common to the two lists (η Her, ϵ Vir, α Boo) appears preferable and is used on the second line of this section.

Baum (1959, 1962) used an eight-color narrow-band system, reduced to absolute energy by reference to standard stars, for an attempt at galaxy synthesis and for

TABLE 1

SIX-COLOR OBSERVATIONS OF GALAXIES
(Stebbins and Whitford 1948; Whitford and Sears 1970)

NGC	Messier	Type	D_0	A/D_0	$1/\lambda$ (μ^{-1}) Telescope (inches)	2.83 [U]	2.37 [V]	2.05 [B]	1.75 [G]	1.39 [R]	0.97 [I]	Remarks
221	32	E2	3′.2	0.75	100	+0.54	+0.45	+0.27	+0.03	−0.29	−0.91	
224	31	Sb	103	0.06	60	+0.87	+0.52	+0.28	+0.04	−0.31	−0.83	1
2841	....	Sb	5.2	0.34	36	+0.68	+0.40	+0.23	+0.02	−0.26	−0.76	
3031	81	Sb	19.0	0.12	36	+0.81	+0.50	+0.29	+0.02	−0.32	−0.86	1
3115	....	E7	2.8	0.63	36	+0.77	+0.43	+0.23	+0.03	−0.26	−0.79	
3368	96	Sa	4.7	0.38	36	+0.61	+0.36	+0.23	+0.02	−0.26	−0.78	
3379	105	E0	2.3	0.77	36	+0.73	+0.42	+0.24	+0.02	−0.26	−0.78	
4374	84	E1p	2.4	0.74	36	+0.71	+0.42	+0.23	+0.03	−0.26	−0.77	2
4406	86	E3	2.9	0.61	36	+0.74	+0.39	+0.23	+0.01	−0.25	−0.75	2
4472	49	E1	4.5	0.25	36	+0.80	+0.48	+0.24	+0.01	−0.26	−0.80	2
4486	87	E0p	3.7	0.48	36	+0.75	+0.42	+0.25	+0.01	−0.26	−0.83	2
4594	104	Sb	4.6	0.38	36	+0.92	+0.52	+0.27	+0.03	−0.31	−0.88	3
4649	60	E2	3.2	0.55	36	+0.83	+0.54	+0.23	+0.04	−0.28	−0.80	2
4736	94	Sb	6.8	0.50	100	0.00	+0.13	+0.14	+0.06	−0.19	−0.58	
4826	64	Sb	5.9	0.58	100	+0.17	+0.21	+0.17	+0.03	−0.20	−0.60	
5194	51	Sc	8.9	0.38	100	−0.30	−0.05	+0.06	+0.06	−0.11	−0.58	
5866	....	S0	2.3	0.97	36	+0.55	+0.33	+0.22	+0.03	−0.24	−0.70	

NOTES TO TABLE 1.—Type is Mount Wilson classification from Humason, Mayall, and Sandage 1956. D_0 is reduced face-on diameter to common limiting isophote (de Vaucouleurs and de Vaucouleurs 1964). A/D_0 is ratio of aperture to D_0. Telescopes: 100-inch and 60-inch at Mount Wilson Observatory; 36-inch Crossley reflector at Lick Observatory.

REMARKS.—(1) Nuclear region only of giant SB systems, for comparison with E systems. (2) Bright member of Virgo cluster. (3) Highly inclined system with prominent dust lane; internal reddening influences colors.

TABLE 2

REDUCTION OF SIX-COLOR OBSERVATIONS TO ABSOLUTE ENERGY

Color $(1/\lambda)_{eff}$ (μ^{-1})	[U] 2.79	[V] 2.36	[B] 1.99	[G] 1.70	[R] 1.35	[I] 1.00	Remarks
				Spiral Galaxy			
(1) M51(Sc)–dG6	−0.30	−0.05	+0.06	+0.06	−0.11	−0.58	From table 1
(2) dG6–5500° K BB	+0.16	−0.02	−0.07	0.00	+0.08	+0.05	Mean 5 solar-type stars, SK
(3) 5500° K BB, Δ mag	+0.46	+0.13	+0.01	+0.03	+0.27	+0.84	Rel. to max. at 5360 Å
(4) M51, abs. Δ mag	+0.32	+0.06	0.00	+0.09	+0.24	+0.31	Sum of (1), (2), and (3)
(5) Normalized, [G]=0	+0.23	−0.03	−0.09	0.00	+0.15	+0.22	Final abs. energy, Δ mag
				Giant Elliptical Galaxy			
(1) Mean E–dG6	+0.76	+0.44	+0.24	+0.02	−0.26	−0.79	7 E's from table 1, M32 excluded
(2) dG6–5500° K BB	+0.20	+0.06	−0.10	0.00	+0.09	+0.11	Mean 3 giant stars, SK
(3) 5500° K BB, Δ mag	+0.46	+0.13	+0.01	+0.03	+0.27	+0.84	Rel. to max. at 5360 Å
(4) Mean E abs. Δ mag	+1.42	+0.63	+0.15	+0.05	+0.10	+0.16	Sum of (1), (2), and (3)
(5) Normalized, [G]=0	+1.37	+0.58	+0.10	+0.00	+0.05	+0.11	Final abs. energy, Δ mag

NOTE.—BB = Blackbody; SK = Stebbins and Kron (1964, 1956).

measurement for redshifts by translation of plotted energy curves. Tifft (1961, 1963) has given detailed measurements of galaxies of various types of an eight-color intermediate-band system, also reduced to an absolute standard.

H. L. Johnson (1966*a*) extended the spectrophotometry of galaxies out to 3.4 μ in the infrared on the *UBVRIJKL* broad-band system used for extensive measures of stars. The mean colors of elliptical galaxies derived from measurements of five bright systems are given in table 3, along with the reciprocal effective wavelengths of the bands. The average aperture-diameter ratio A/D_0 for the group was 0.31. The various indices $V - R$, $V - I$, etc., are all zero for an A0 star as in the original *UBV* system (Johnson and Morgan 1953). The absolute calibration of magnitude 0.0 in each color band has been derived by Johnson (1965). These calibrations, converted to magnitudes relative to the V-magnitude, are used in table 3 to deduce the absolute energy curve of a giant elliptical galaxy over a very wide range of wavelengths.

Multicolor filter observations of galaxies from the OAO-2 satellite were used to extend the energy curves of galaxies into the deep ultraviolet (Code, Welch, and Page 1972). In some cases the observations reached as far as 1330 Å. Down to 2460 Å the results were consistent with projections from ground-based data. On a color-color plot of $m(2460 - 4250)$ versus $B - V$ the band across the diagram occupied by galaxies of all types is more widely separated from the line representing single stars than it is on the $U - B$ versus $B - V$ plot (de Vaucouleurs 1961) because the hottest stars in the composite population become a more dominant fraction of total light at a shorter ultraviolet wavelength. For the Irr I and Sc galaxies the absolute energy curve continues nearly level or rises slightly down to $\lambda = 2000$ Å because of the contribution of the OB supergiants. The energy curves of E and S0 galaxies show a color in the 1980–2460 Å region like that of early F stars.

Below 2000 Å an unexpectedly sharp upturn was observed in the energy curves of galaxies of all types. For the E galaxies the measurements at these wavelengths were marginal; but the central bulge of M31, which resembles an E galaxy in many ways, showed the steep rise clearly. The gradient is steeper than that expected from a black-body at infinite temperature. A reasonable explanation of the upturn was found in the scattering of the light of hot stars by interstellar or circumstellar particles having the properties required to explain ultraviolet extinction and scattering in the Milky Way: an absorption maximum at $\lambda = 2200$ Å and a steeply rising albedo for $\lambda < 2000$ Å.

Lasker (1970) investigated the homogeneity of the spectral energy curves of 100 E and S0 galaxies in order to test the validity of a universal K-correction (§ 6) for giant systems of these types. Eight intermediate bandwidth filters with central wavelengths from 3474 to 7953 Å avoided strong spectral features. The two types were found to be indistinguishable, and for $M_v < -21$ the dispersion in the color indices was small. The total range of color containing half of the galaxies depended on the wavelength interval of the index selected: $\Delta C(3474 - 5488$ Å$) = 0.11$; $\Delta C(4698 - 5488$ Å$) = 0.04$; $\Delta C(5488 - 7953$ Å$) = 0.08$. At least part of the scatter was shown to be intrinsic, not observational error. These small departures from complete homogeneity of the energy curves were shown to have little influence on the value of the deceleration parameter q_0 determined from the redshift relation.

The restriction to giant systems was necessary because of the color-luminosity effect

TABLE 3

REDUCTION OF *UBVRIJKL* COLORS OF MEAN E GALAXY TO ABSOLUTE ENERGY

Color $1/\lambda$ (μ^{-1})	*U* 2.78	*B* 2.27	*V* 1.82	*R* 1.43	*I* 1.11	*J* 0.80	*K* 0.46	*L* 0.29	Remarks
(1) Mean E–A0 star........	+1.56	+1.02	0.00	−0.92	−1.67	−2.23	−3.04	−3.48	Mean 5 giant E's
(2) $F(\lambda)$, A0 star, m=0.00...	43.5	72.0	39.2	17.6	8.3	3.4	0.39	0.081	Unit=10^{-13} W m^{-2} s^{-1} μ^{-1}
(3) Δ mag. rel. to *V*.........	−0.11	−0.66	0.00	+0.87	+1.69	+2.66	+5.00	+6.71	From line (2)
(4) Mean E, abs. Δ mag...	+1.44	+0.36	0.00	−0.05	+0.02	+0.43	+1.96	+3.23	(1)+(3)

in elliptical galaxies found by Baum (1959) and de Vaucouleurs (1961). With decreasing luminosity the colors become bluer, and for the faintest systems they approximate those of halo-type globular clusters. Lasker's measures, which included a few systems of lower luminosity, showed the trend, and agreed with de Vaucouleurs's finding that the color effect is most pronounced in the $U - B$ index.

Sandage (1972) studied the colors of 44 E and S0 galaxies in the Virgo and Coma clusters on the *UBV* system. In the Virgo cluster the newly measured faint galaxies extended the de Vaucouleurs results by 3 mag to give a total range of 8.5 mag. After an appropriate allowance for the relative distances of the two clusters, a single plot of all $U - B$ colors showed a strong correlation between $U - B$ and V, the apparent visual magnitude. A linear relation fitted to the data gave a slope $\Delta(U - B)/\Delta M_v \approx -0.085$, with a dispersion $\sigma(V) = 0.72$ mag. Data from other clusters and groups are required to establish the effect as being quite general, and to determine the reliability of a possible universal color-luminosity relation that could be used in finding distances.

Faber (1972, 1973) investigated color differences among 31 elliptical galaxies using measurements on a 10-color intermediate-band system. The luminosities extended 6 mag downward from the giant ellipticals of the Virgo and Perseus I clusters. The filter system was a hybrid of the systems developed by McClure and van den Bergh (1968) and by Wood (1966) for the study of stellar populations in galaxies. Band centers ranged from 3450 to 7415 Å and included both continuum points and prominent spectral features. Final colors were reduced to an absolute energy standard.

Both the colors and the line-strength indices were found to be closely correlated with absolute magnitude. The continuum color indices for all wavelength intervals, including those wholly in the red, showed a similar pattern: a smooth decline to bluer colors over a 4-mag interval from the bright end (roughly $-18 > M_v > -22$), and a sharp decrease between $M_v = -18$ and -16 to the three galaxies with colors like those of globular clusters; this is similar to the break found by Baum (1959) in his $B - V$ colors. The line-strength indices increased monotonically with increasing luminosity. A composite (CN + Mg) index was found to be a reddening-free indicator of absolute magnitude over the whole range, with a standard deviation $\sigma(M_v) = 0.56$ about a nearly linear least-squares fit to the data. A variation in mean metal abundance was shown to be an acceptable explanation of the observed trends of continuum colors and line indices as a function of luminosity.

4. ABSOLUTE ENERGY CURVES FROM SCANNERS

Photoelectric scanning spectrometers (scanners) have brought considerably higher resolution to the delineation of the energy curves of galaxies. The linearity of the photoelectric process makes the subtraction of the sky background straightforward. As the number of spectral resolution elements increases, the time for sequential one-channel scanning becomes quite long; but multichannel instruments (Oke 1969) can reduce the time considerably. Scanners with large collimators and with gratings producing high angular dispersion can tolerate a large entrance aperture without loss of purity, and thus accept a larger fraction of the total light of the galaxy. For a scanner with a given optical configuration, the smaller the telescope the larger the angular aperture on the sky for the same purity, but the photon yield is reduced. The resolution of scans for the determinations of energy curves of galaxies is typically of the order of 50 Å; but for study of spectral features in the nuclei of bright galaxies, this can be reduced to 16 Å (Spinrad 1966).

The first scan of an E galaxy (Code 1959) showed how serious had been the smoothing of the early energy curves derived from broad-band filter photometry. The curve for M32 was found to be a close match to the G8 III star η Piscium. The sharp decline in the violet and ultraviolet caused by the well-known heavy blanketing by metallic lines in giant stars resulted in a curve which was far from any blackbody curve.

Oke (1962) undertook a systematic program of galaxy scans that included the brightest ellipticals in the Virgo cluster and the nuclear regions of spirals. The results (Oke and Sandage 1968) showed that the energy curves of giant ellipticals are quite homogeneous over the range of wavelengths covered (3400–11,000 Å). The energy curve of the nucleus of M31 matched that of the Virgo ellipticals. M32 was found to be definitely bluer over the whole range.

Whitford (1971) made a parallel series of scanner measurements on five giant elliptical galaxies. For $\lambda < 5100$ Å the resolution was 50 Å, the same as in the Oke-Sandage scans. In general shape and finer details Whitford's mean absolute energy curve closely resembled that of Oke and Sandage (1968), but showed a definite tilt in the direction of a higher temperature or bluer colors. The difference was explained by two effects:

1. A color-aperture effect was quite apparent in comparisons of scans at average aperture-diameter ratios A/D_0 of 0.2 and 0.6; it was of the order of magnitude expected from de Vaucouleurs's (1961) results on the *UBV* colors of E galaxies. Oke and Sandage (1968) had speculated that their measures at an average $A/D_0 \simeq 0.07$ might require corrections for this effect. Sandage, Becklin, and Neugebauer (1969) studied color effects on a much finer scale in the central region of M31; equivalent spatial resolutions in the Virgo ellipticals would not be possible. Color measures with *UBVRK* filters on areas 5 to 7″ in diameter over a radius of 60″ from the nucleus ($A/D_0 \simeq 0.012$) showed a variation only in $U - B$ with the nucleus 0.19 mag redder. The absence of a central maximum in $V - K$ was taken as evidence against nonthermal radiation from the nucleus.

Since *UBV* color effects for $A/D_0 > 0.6$ are quite small, Whitford adopted the mean curve for this aperture as representing a whole galaxy, and therefore comparable with what would be observed in measuring the light of a distant galaxy of small angular extent.

2. Whitford adopted as a reference standard the absolute energy calibration of Vega and other early-type stars by Hayes (1970). The higher effective temperature for Vega found by Hayes relative to the earlier calibration by Oke (1964) carried over into the energy curve for galaxies.

Figure 1 shows the Oke-Sandage and the Whitford energy curves for giant elliptical galaxies.

Oke and Schild (1970) recalibrated Vega by means of a scanner comparison with three terrestrial standards. The results showed excellent agreement with Hayes (1970) between 4000 and 8000 Å, but in the ultraviolet below the Balmer jump and in the infrared beyond the Paschen discontinuity the Oke-Schild result showed Vega about 0.05 mag brighter than found by Hayes. The new calibration was used by Schild and

Oke (1971) in deriving an absolute energy curve for giant elliptical galaxies that referred to the total light. The scanner was attached to a 4-inch telescope, and the entrance aperture was 6′.2, corresponding to $A/D_0 > 1$ for all Virgo ellipticals. The resolution was 100 Å, and the wavelengths covered were 3400–8000 Å. The resulting energy curve, based principally on NGC 4486, showed quite close agreement with Whitford's curve over the range 4000–8000 Å, but averaged slightly higher in the ultraviolet.

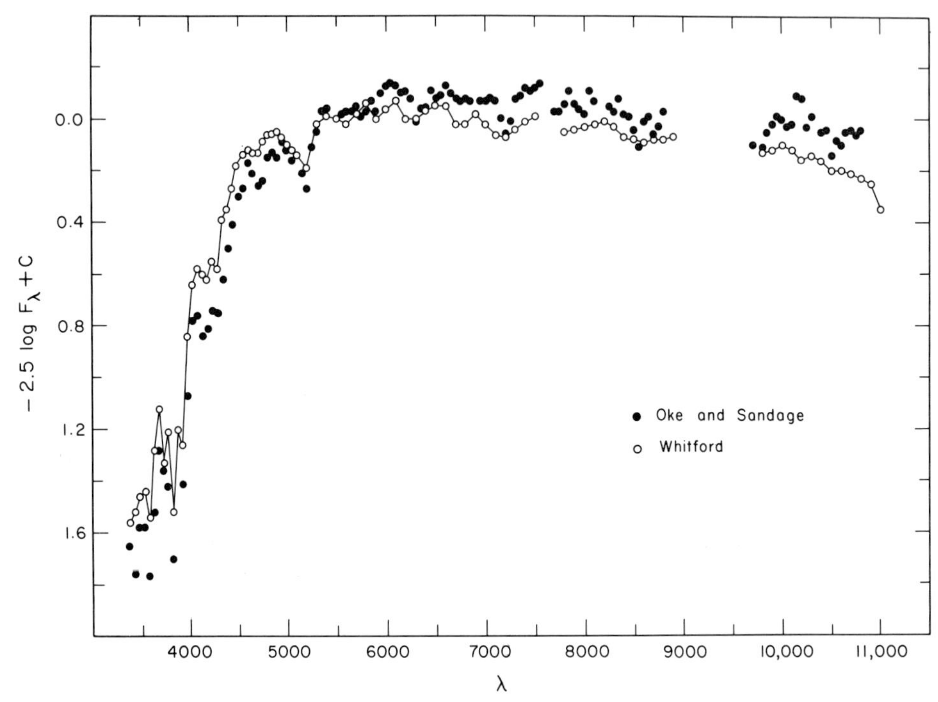

FIG. 1.—Mean absolute energy curves of giant elliptical galaxies, derived from scanner observations.

5. INFERENCES REGARDING STELLAR CONTENT

The results on the absolute energy curves of galaxies derived from broad-band filter observations (tables 2 and 3) are plotted in figure 2. Also shown are the scanner results of Whitford, taken from figure 1. The six-color points, the Johnson eight-color wide-range photometry, and the scanner results represent three sets of data obtained in different ways and reduced to an absolute standard by independent methods. The agreement is nevertheless rather satisfactory.

Certain conclusions regarding the stellar content of an elliptical galaxy can be read from such an energy curve:

1. The steep drop in intensity at 4000 Å is the signature of strong metallic line blanketing in late-type stars of normal metal content. The marked CN dip shortward of 3883 Å (visible on the scanner curve) is the feature on low dispersion spectrograms (Morgan and Mayall 1957; Morgan and Osterbrock 1969) that led to the conclusion that this part of the spectrum is dominated by K0 III giants.

2. The ultraviolet part of the spectrum from 3400 to 3800 Å shows more energy than a K0 III giant. The smoothed energy curve of a K0 III giant is shown in figure 2; the data were taken from Johnson's (1966*b*) tables, and reduced to absolute energy by the procedure illustrated in table 3. The scanner curve agrees with this conclusion from broad-band spectrophotometry. Morgan and Mayall ascribed the departure of this part of the spectrum from K0 III characteristics to the light of F8–G5 stars. Johnson (1966*a*) included a few A and B stars in his population synthesis to match the observed broad-band ultraviolet intensity. The scanner curve cannot distinguish between these suggested composite populations which combine hotter stars with K0 III giants from the energy curve of G8 III stars alone (Whitford 1970).

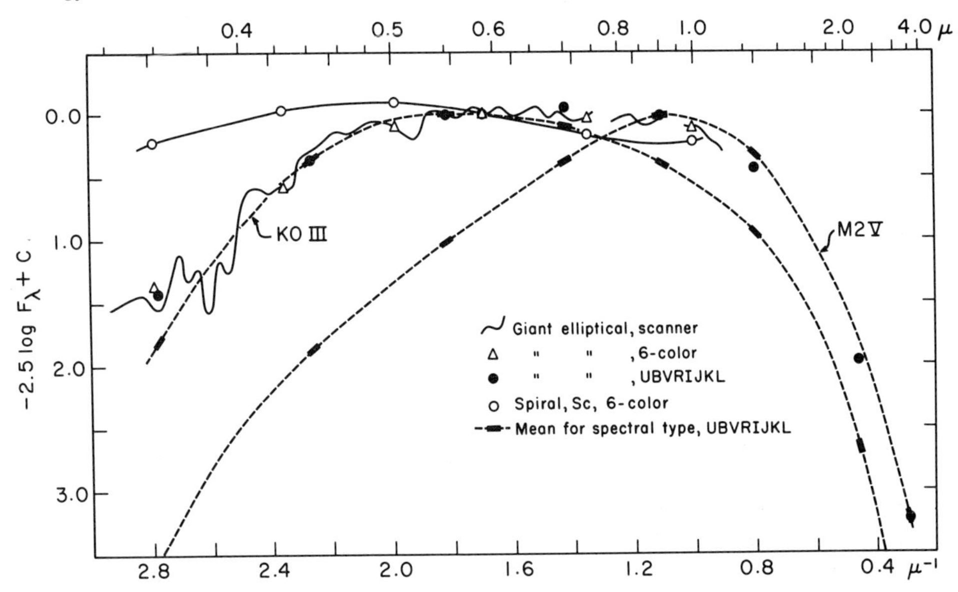

FIG. 2.—Absolute energy curve for giant elliptical galaxies compared with stars of selected spectral types, and a spiral galaxy. The scanner curve is from Whitford (1971), the other galaxy data from tables 2 and 3, and the star data from Johnson (1966*b*). Except for the scanner, plotted points are from broad-band filter observations, and curves are necessarily smoothed.

3. The presence of radiation sources cooler than a K0 III star is shown by the high infrared intensity of the scanner curve, and by the Johnson *IJKL* photometry. The latter is closely matched by an early M star. Johnson (1966*a*) chose an M0 III star for his synthesis of an elliptical galaxy. An M2 V dwarf has nearly the same infrared indices and is shown in figure 2, normalized at the *I* point.

Although the shape of the energy curve does give some clues and does place some constraints on the population model describing the stellar content of an elliptical galaxy, the final solution must also use information from spectrum features that are sensitive to temperature and stellar luminosity. Evidence on count-brightness ratios (Baum 1959) and on the strong-line giant-like spectra of E galaxies (Morgan and Mayall 1957) showed that the population of such systems must resemble the old Population I of the Milky Way, typified by the old clusters M67 and NGC 188. This model explains the absence of supergiant and early main-sequence stars. The strong hydrogen lines characteristic of the latter are not seen in the spectra.

A major issue that cannot be settled from the energy curves is the proportion of dwarfs in the population. The mass-luminosity ratio of giant elliptical systems from dynamical determinations (Roberts 1963; also chapter 3 of this volume) is in the range $15 < \mathfrak{M}/L_{pg} < 50$, in solar units. The $\mathfrak{M}/L_{pg}$ ratio of the stellar population in the solar neighborhood, characterized by the van Rhijn luminosity function, is approximately 2 (Oort 1965). Although the deep violet part of the energy curve is dominated by giants, the low-temperature components needed to match the infrared part can be simulated by a nonunique mixture of stars of various temperatures. The lack of any strong, broad absorption features in the red and infrared part of the spectrum makes the use of broad-band spectrophotometry adequate for delineating this part of the energy curve of galaxies; there is nothing like the heavy blanketing in giant stars that is so prominent in the deep violet. An example of this nonuniqueness is given by Johnson's (1966*a*) conclusion that his very wide range energy curve can be formally represented by a model in which there are no lower main-sequence stars. Spinrad (1966) pointed out, however, that for this model $\mathfrak{M}/L < 0.1$, and emphasized that the discrimination needed to determine the balance between giants and dwarfs must come from narrower luminosity-sensitive spectrum features and from the requirement of a reasonable $\mathfrak{M}/L$ ratio for the whole population. Results from such analyses are reviewed in chapter 2 of this volume.

Much less attention has been given to studying the energy curves and total stellar content of spiral and irregular galaxies. G. and A. de Vaucouleurs (1959) used photographic techniques to relate the stellar population of the Large Magellanic Cloud as deduced from a low-resolution spectrum to the *UBV* colors. They noted the lack of red and infrared photometry which might have indicated the presence of late-type main-sequence stars.

The energy curve of the Sc spiral M51 shown in figure 2 reflects the contribution of the bright blue spiral-arm stars to the [*U*], [*V*], and [*B*] points of the six-color photometry. The upturn of the curve in the red and infrared shows the presence of the cooler stars of the underlying disk population.

6. THE *K*-CORRECTION

In his pioneer investigations of the distribution of galaxies in space and of the velocity-distance relation, Hubble (1936) took account of the effect of a redshift on the apparent magnitude of a galaxy. He introduced corrections for a "number effect," for an "energy effect," and for a third term K which allowed for the fact that the shift of the energy curve of the distant galaxy to the red brought a different band at shorter wavelengths (measured in the rest frame of the receding object) into the spectral acceptance band of the observer's photographic plate.

The first two of these corrections have since been assimilated (Sandage 1961*a*) into Robertson's (1938) very general equation for the apparent bolometric luminosity of a redshifted galaxy. There remains the K-correction as a purely technical effect that would be eliminated if during an observation of a galaxy with a redshift z the response at each wavelength in a filter acceptance band could be shifted from its normal position λ to a longer wavelength $\lambda(1 + z)$. Both the effective wavelength and the equivalent bandwidth would be increased by the same factor $(1 + z)$, and the measurement would be made in the rest-frame wavelengths of the distant object.

In his multicolor photometry of faint galaxies Baum (1962) adopted a first approximation to the ideal procedure by constructing an $F(\lambda)$ curve (necessarily somewhat smoothed) from his narrow-band filter observations, and then reading off the magnitude at the proper rest-frame wavelength for the distant system. Only bandwidth corrections remained. In principle the correct hypothetical procedure can be very closely approached with a spectrum scanner. Since, however, nearly all the intensity measurements of faint galaxies have been made on rather broad-band heterochromatic magnitude systems fixed in the observer's velocity frame, appropriate K-corrections are essential in deriving photometric distances and in evaluating cosmological parameters.

The expression for the K-correction has two terms (Humason, Mayall, and Sandage 1956; Oke and Sandage 1968): (1) The acceptance band is narrower in the rest frame of the distant object by a factor $(1 + z)$, independent of wavelength or of the shape of the energy curve; the apparent magnitude of the object is thereby increased by an amount $2.5 \log (1 + z)$. (2) At each wavelength λ in the acceptance band the intensity received is not the $F(\lambda)$ read from the standard energy curve measured in the observer's velocity frame, but $F[\lambda/(1 + z)]$. The apparent magnitude will be increased if the energy curve falls toward shorter wavelengths, or decreased if it rises. For the ith filter band with a response function $S_i(\lambda)$ the K-correction is

$$K_i = 2.5 \log (1 + z) + 2.5 \log \frac{\int_0^\infty F(\lambda) S_i(\lambda) d\lambda}{\int_0^\infty F[\lambda/(1 + z)] S_i(\lambda) d\lambda},$$

where the flux $F(\lambda)$ is in energy per unit wavelength interval, and all wavelengths are those of the observer's velocity frame.

The first systematic calculation of K-corrections by Humason, Mayall, and Sandage (1956), based on the six-color energy curve of M32 derived by Stebbins and Whitford (1948), gave results which were much too small, mainly because of the smoothing effect of the broad filter bands (discussed in the next section). Also, multicolor and scanner observations (e.g., Oke and Sandage 1968) later showed that M32 is bluer than the giant elliptical galaxies that are the standard objects in cosmological investigations; it would therefore have a small K-correction. Later Sandage (1967) used Code's (1959) and Oke's (1962) preliminary scans of NGC 4374 for a much improved evaluation.

The values of the K-corrections for giant elliptical galaxies derived from the Oke-Sandage (1968) energy curves proved to be slightly too large when revised energy curves became available. These later curves measured the total light of galaxies and were based on new calibrations of the reference stars (§ 5). Sandage (1973) adopted K-corrections that combined the generally accordant values derived by Schild and Oke (1971) and by Whitford (1971). These K-corrections for the B, V, and R bands and for the $(B - V)$ and $(V - R)$ colors are shown in the last five columns of table 4. In the fourth column the values of K_B do not go beyond $z = 0.28$ because the calculation requires knowledge of the energy curve for $\lambda < 3400$ Å not provided by observations of nearby giant ellipticals. Oke (1971) obtained information on the energy at shorter wavelengths from scanner observations of galaxies with large redshifts, including 3C295, for which $z = 0.46$. The rest wavelengths $2350 < \lambda_0 < 3400$ Å were brought into the observable range. The third column lists the K_B values derived by Oke from this extended energy curve. The divergent values of K_B for $0.20 < z < 0.28$ in the third and fourth columns reflect the difference in the ultraviolet parts of the Schild-Oke and the Whitford energy curve.

Sandage (1973) adopted the K_{R_s} values in table 4, calculated for the R_s system of Sandage and Smith (1963) as being applicable to the R system of Johnson, Mitchell, Iriarte, and Wisniewski (1966). K_R is insensitive to the exact form of the R response curve because the K_R correction is dominated by the nonselective bandwidth term, shown in the second column of table 4. The contribution of the energy-curve–dependent

TABLE 4

K-CORRECTIONS FOR *B*, *V*, AND *R* MAGNITUDES OF GIANT ELLIPTICAL GALAXIES

z (1)	2.5 log $(1+z)$ (2)	K_B Oke (3)	K_B (4)	K_V (5)	K_{R_s} (6)	$K_B - K_V$ (7)	$K_V - K_R$ (8)
0.00	0.00	...	0.00	0.00	0.00	0.00	0.00
0.04	0.04	...	0.19	0.06	0.03	0.13	0.03
0.08	0.08	...	0.40	0.13	0.08	0.27	0.05
0.12	0.12	...	0.62	0.20	0.12	0.42	0.08
0.16	0.16	...	0.82	0.27	0.16	0.55	0.11
0.20	0.20	0.96	1.03	0.39	0.22	0.64	0.17
0.24	0.23	1.11	1.22	0.53	0.27	0.69	0.26
0.28	0.27	1.26	1.40	0.70	0.31	0.71	0.39
0.32	0.30	1.41	...	0.88	0.35	...	0.53
0.36	0.33	1.55	...	1.08	0.40	...	0.68
0.40	0.37	1.69	...	1.25	0.48	...	0.77
0.44	0.40	1.83	...	1.41	0.57	...	0.84
0.48	0.43	1.96	...	1.54	0.66	...	0.88

second term rises to 30 percent of the total only at $z = 0.48$. This insensitivity, which stems from the flatness of the energy curve over the relevant wavelengths, gives the red magnitude system an obvious advantage in cosmological studies.

7. EVOLUTIONARY EFFECTS

In an early analysis of the interpretation of the redshift, Hubble and Tolman (1935) noted that the radiation from a galaxy may not be constant over the light-travel time involved in cosmological comparisons of the most distant observable objects with nearby examples. They recognized that the light-travel time can be a moderate fraction of the expansion age for the Universe, a conclusion that remains unchanged following the revision of the distant scale that has since increased the time intervals in question manyfold. Knowledge of the evolutionary tracks of various types of stars now gives a basis for estimating the rate of change of the integrated radiation if a satisfactory model for the stellar population of a galaxy can be established.

The first measurements of the colors of distant galaxies by Stebbins and Whitford (1948; also Whitford 1952) showed them to be much redder than was predicted from the information then available for calculating *K*-corrections. An elliptical galaxy with a redshift $z = 0.13$ was 0.52 mag redder on the C_p system than nearby ellipticals, as opposed to only 0.22 mag calculated from an absolute energy curve for M32 derived by standardizing the six-color system against the sun via dG2 stars. The unexpectedly large discrepancy, which became known as the Stebbins-Whitford effect, was ascribed to evolutionary changes during the light-travel time. Rapid evolution of galaxies in little more than a billion years would have been required.

G. de Vaucouleurs (1948) suggested that the broad-band filters of the six-color and C_p systems had inadequate spectral resolution to take account of the sharp decline in the energy curve of the sun and the galaxies near 4000 Å. As the redshift moves such a sharp dropoff across the blue filter band, the resulting K_B correction is much larger than that calculated from the smoothed energy curve based on broad-band measurements. Further filter measurements by Whitford (1957) calibrated against a B star showed a negligible effect. The first energy curves of galaxies by the scanner method (Code 1959) showed that the de Vaucouleurs explanation was the correct one (Whitford 1962).

Later measurements have demonstrated that the colors of the brightest members of distant clusters of galaxies do in fact agree with those expected from calculations of K-corrections calculated from scanner energy curves. Oke and Sandage (1968) and Schild and Oke (1971) made the comparison for $B - V$ colors for redshifts $z \leq 0.20$. Sandage (1973) discussed $B - V$ and $V - R$ colors from about 40 clusters and groups with $z \leq 0.46$. For $z < 0.20$ the $B - V$ colors were in satisfactory agreement with the values in table 4. At redshifts $z > 0.20$, $B - V$ passes through a maximum because the energy dropoff near 4000 Å transfers its influence from the B to the V band. This effect on V explains the steep rise of $V - R$ in table 4 for $0.20 < z < 0.40$; a rounding off in $V - R$ similar to that in $B - V$ at $z = 0.20$ then sets in. Sandage's data included only three objects in the range $0.20 < z < 0.46$, but the fit to predictions was satisfactory. These observational results indicate that evolutionary changes in the energy curves of elliptical galaxies over the light-travel time corresponding to $z = 0.46$ are too small to be detected from studies of broad-band colors.

An independent test was provided by Oke's (1971) multichannel scanner curves of three galaxies with redshifts $z = 0.20$, 0.38, and 0.46. A direct comparison with the shape of the energy curve of nearby giant ellipticals could be made, without resort to K-corrections. Over the range of rest wavelengths $3400 < \lambda_0 < 7600$ Å no discernible difference was found. The extension of calculated K-corrections to large redshifts (§ 6) did require the assumption that there had likewise been no evolutionary change in the energy curve for $2350 < \lambda_0 < 3400$ Å.

The relation between the light-travel time τ and the redshift z depends on the deceleration parameter q_0. Equations for world models with the cosmological constant $\Lambda = 0$ have been given by Sandage (1961*b*); tables give τ in terms of the Hubble time H_0^{-1} and of the time t_0 since the beginning of the expansion. For $z = 0.46$, and the two cases $q_0 = 0$ and $q_0 = +1$, $\tau/H_0^{-1} = 0.316$ and 0.270, respectively. For the same parameters $\tau/t_0 = 0.316$ and 0.473. If the Hubble constant $H_0 = 75$ km s^{-1} Mpc^{-1} the corresponding times are 4.0 and 3.4×10^9 years.

In order to estimate the effect of evolution over such time intervals on the magnitude and color of an elliptical galaxy, Sandage (1961*b*) took the old open cluster NGC 188 as a population model. A metal-rich population was assumed to have been formed at a single epoch, with the present main-sequence break point at $M_V \simeq +4$. The empirical fact that the giant branches of the two old clusters NGC 188 and M67 are very nearly parallel on the $[M_V, (B - V)_0]$-plane leads to the conclusion that the total luminosity of the giant branch at a given stage of evolution is proportional to the luminosity of the main-sequence stars then at the break point. Sandage argued that since lifetime on the main sequence $t =$ const. $\mathfrak{M}/L$, and the mass-luminosity relation for stars of about 1 $\mathfrak{M}_\odot$ is $L \simeq \mathfrak{M}^4$, $t =$ const. $L^{-3/4}$. The age t_1 of a distant object when the light was emitted was $t_1 = t_0 - \tau$, where t_0 is the present age. For $z = 0.46$ and $q_0 = +1$, the light-travel time $\tau = 0.473 t_0$ (from Sandage 1961*b*) and $t_1/t_0 = 0.527$. Then

$$L_1/L_0 = (t_1/t_0)^{-4/3} = 0.527^{-4/3} = 2.35\ ,$$

or at the beginning of the light-travel time the luminosity of stars at the break point in the distant galaxy was 0.93 mag brighter than similar stars in a nearby system.

If stars were evolving off the main sequence at an equal rate at the two epochs, and if the giant branch were responsible for substantially all the light at the V wavelength, then the evolutionary correction would be $\Delta M_V = -0.93$ mag. If the unevolved dwarfs account for half of the visual light at t_0, however, the correction would be reduced accordingly. But since galaxy synthesis models have in general a rather steeply rising luminosity function to match observed $\mathfrak{M}/L$ ratios (§ 5 of this chapter), there would have been fewer stars per magnitude interval at epoch t_1. Whether this reduction in number at the break point at earlier times is sufficient to compensate or even over-compensate for the greater luminosity of a single star ascending to giant branch depends on the slope of the luminosity function in the range $+3 < M_V < +5$. Sandage (1968, 1970) concluded that the evolutionary correction is probably small, and could possibly be in the sense that luminosity of elliptical galaxies increases with time.

Tinsley (1968, 1971) and Tinsley and Spinrad (1971) have developed models for galaxy evolution in which the initial store of gas was used up gradually, and star birth continued at a decreasing rate over most of the life of the galaxy. The calculated change in intrinsic $B - V$ color for ellipticals was 0.1–0.2 mag for the look-back time corresponding to $z = 0.46$, where Oke (1971) found no observable change.

A later discussion by Tinsley (1972) adopts single-epoch star formation and analyzes the change of luminosity with age, in order to ascertain first-order effects on the derivation of q_0 from the observed magnitude-redshift relation. A homology argument (see also Wielen 1964) is used find the rate at which stars evolve through the break point. For a mass function $N(\mathfrak{M}) \propto \mathfrak{M}^{-x}$, the luminosity function is $N(L_V) = N_0 L_V{}^{-\beta+1}$, where $\beta \approx 1 + 0.2x$, corresponding to $L_V \approx \mathfrak{M}^5$ in solar units. The inverse relation of main-sequence lifetime and luminosity may be written $t^{-\gamma} = L_V$, where $\gamma \approx 1$, and t is measured in units of the solar main-sequence lifetime. For the assumptions of the Sandage NGC 188 model $\gamma = 1.33$, and for the same assumptions in the Tinsley calculation $\gamma = 1.25$.

Then the rate at which stars evolve through the break point as a function of L_V at that point is

$$\frac{dN}{dt} = \frac{dN}{dL_V}\frac{dL_V}{dt} = \text{const. } \gamma(\beta - 1)t^{\gamma(\beta-1)-1}\,.$$

From a comparison of calculated and observed colors of galaxies at an age of the order of 10^{10} years Tinsley finds that models with x outside the range $1 < x < 3$ are excluded. If these limits are accepted, the exponent of t in the rate equation $\gamma(\beta - 1) - 1 = 0.2\gamma x - 1 < 0$, and the total luminosity of the giant branch is a decreasing function of time. The integrated luminosity of the dwarf branch can only decrease. Spinrad and Taylor's (1971) model for the nucleus of M31 has a luminosity function rising much more steeply than the van Rhijn function and the value of $x \approx 3.5$. The resulting $\mathfrak{M}/L_V = 43.5$, and the giant branch gives about 75 percent of the visual light. Here the exponent of t, while formally still negative, is not sufficiently certain to exclude a constant rate of evolution of stars onto the giant branch. But even in this case the model does not provide the increasing number of giants considered by Sandage to offset approximately the decrease in total luminosity of the giant branch with time as calculated from separation between parallel evolution tracks. Tinsley considers differences between giant evolution tracks to be negligible. The parallelism argument does provide a qualitative explanation of the very slow change in the color of ellipticals over the last 4 billion years.

Tinsley's result that in general the number of giants decreases with time means that as the break point evolves down the main sequence, the increase in the number of stars in each succeeding step ΔL_V of the luminosity function is more than compensated by the increasingly longer time required to pass through each step.

The amount by which the luminosity of elliptical galaxies decreases with time can be calculated for various choices of the parameter x, the value of which is still rather uncertain. Values of $x = 4$ to 5 would lead to luminosity increasing with time. High values of x enrich the faint end of the main sequence, with the result that late M dwarfs contribute an important fraction of the total radiation in the near-infrared and infrared (§ 5, this chapter). Atomic lines and molecular bands in this part of the spectrum that are sensitive to stellar luminosity are therefore important in giving direct observational clues regarding the mass function and the rate of luminosity evolution in elliptical galaxies.

REFERENCES

Baade, W. 1944, *Ap. J.*, **100**, 137.
Baade, W., and Swope, H. 1961, *A.J.*, **66**, 300.
Baum, W. A. 1959, *Pub. A.S.P.*, **71**, 106.
———. 1962, in *Problems of Extragalactic Research* (*IAU Symposium No. 15*), ed. G. C. McVittie (New York: Macmillan Co.), p. 390.
Baum, W. A., and Schwarzschild, M. 1955, *A.J.*, **60**, 247.
Code, A. D. 1959, *Pub. A.S.P.*, **71**, 118.
Code, A. D., Welch, G. A., and Page, T. L. 1972, in *Scientific Results from the Orbiting Astrophysical Observatory* (*OAO-2*), NASA SP-310.
Faber, S. M. 1972, *Astr. and Ap.*, **20**, 361.
———. 1973, *Ap. J.*, **179**, 731.
Greaves, W. M. H., Davidson, C., and Martin, E. 1934, *M.N.R.A.S.*, **94**, 488.
Greenstein, J. L. 1938, *Ap. J.*, **88**, 605.
Hayes, D. 1970, *Ap. J.*, **159**, 165.
Hodge, P. W. 1966, *Ap. J.*, **144**, 869.
Holmberg, E. 1950, *Medd. Lunds*, Ser. II, No. 128.
Hubble, E. 1936, *Ap. J.*, **84**, 517.
Hubble, E., and Tolman, R. C. 1935, *Ap. J.*, **82**, 302.
Humason, M., Mayall, N. U., and Sandage, A. 1956, *A.J.*, **61**, 97.
Johnson, H. L. 1963, in *Stars and Stellar Systems* Vol. **3**, ed. K. Aa. Strand (Chicago: University of Chicago Press), p. 216.

Johnson, H. L. 1965, *Comm. Lunar Planetary Lab.*, Vol. **4**, No. 53.
———. 1966*a*, *Ap. J.*, **143**, 187.
———. 1966*b*, *Ann. Rev. Astr. and Ap.*, **3**, 193.
Johnson, H. L., Mitchell, R. I., Iriarte, B., and Wisniewski, W. Z. 1966, *Comm. Lunar and Planetary Lab.*, Vol. **4**, No. 63.
Johnson, H. L., and Morgan, W. W. 1953, *Ap. J.*, **117**, 313.
Lasker, B. M. 1970, *A.J.*, **75**, 21.
McClure, R. D., and Bergh, S. van den. 1968, *A.J.*, **73**, 313.
Morgan, W. W., and Mayall, N. U. 1957, *Pub. A.S.P.*, **69**, 291.
Morgan, W. W., and Osterbrock, D. E. 1969, *A.J.*, **74**, 515.
Oke, J. B. 1962, *Problems of Extragalactic Research* (*IAU Symposium No. 15*), ed. G. C. McVittie (New York: Macmillan Co.), p. 34.
———. 1964, *Ap. J.*, **140**, 689.
———. 1969, *Pub. A.S.P.*, **81**, 11.
———. 1971, *Ap. J.*, **170**, 193.
Oke, J. B., and Sandage, A. 1968, *Ap. J.*, **154**, 21.
Oke, J. B., and Schild, R. 1970, *Ap. J.*, **161**, 1015.
Oort, J. H. 1965, in *Stars and Stellar Systems* Vol. **5**, ed. A. Blaauw and Maarten Schmidt (Chicago: University of Chicago Press), p. 473.
Roberts, M. S. 1963, *Ann. Rev. Astr. and Ap.*, **1**, 149.
Robertson, H. P. 1938, *Zs. f. Ap.*, **15**, 69.
Sandage, A. 1961*a*, *Ap. J.*, **133**, 355.
———. 1961*b*, *ibid.*, **134**, 916.
———. 1967, *IV Centenario d. Nascita Galileo* II, (Firenze: G. Barbera), **3**, p. 104.
———. 1968, *Observatory*, **88**, 91.
———. 1970, *Physics Today*, **23**, No. 2, 34.
———. 1972, *Ap. J.*, **176**, 21.
———. 1973, *ibid.*, **183**, 711.
Sandage, A., Becklin, E. E., and Neugebauer, G. 1969, *Ap. J.*, **157**, 55.
Sandage, A., and Smith, L. L. 1963, *Ap. J.*, **137**, 1057.
Schild, R., and Oke, J. B. 1971, *Ap. J.*, **169**, 209.
Spinrad, H. 1966, *Pub. A.S.P.*, **78**, 367.
Spinrad, H., and Taylor, B. J. 1971, *Ap. J. Suppl.*, **22**, 445.
Stebbins, J., and Kron, G. E. 1956, *Ap. J.*, **123**, 440.
———. 1964, *ibid.*, **139**, 424.
Stebbins, J., and Whitford, A. E. 1937, *Ap. J.*, **86**, 247.
———. 1945, *ibid.*, **102**, 318.
———. 1948, *ibid.*, **108**, 413.
———. 1952, *ibid.*, **115**, 284.
Tifft, W. G. 1961, *A.J.*, **66**, 390.
———. 1963, *ibid.*, **68**, 302.
Tinsley, B. M. 1968, *Ap. J.*, **151**, 547.
———. 1971, *Ap. and Space Sci.*, **12**, 394.
———. 1972, *Ap. J.* (*Letters*), **173**, L93.
Tinsley, B. M., and Spinrad, H. 1971, *Ap. and Space Sci.*, **12**, 118.
Vaucouleurs, G. de 1948, *C.R.*, **227**, 466.
———. 1959, *A.J.*, **64**, 397.
———. 1961, *Ap. J. Suppl.*, **5**, 233.
Vaucouleurs, G. de, and Vaucouleurs, A. de. 1959, *Pub. A.S.P.*, **71**, 83.
———. 1964, *Reference Catalogue of Bright Galaxies* (Austin: University of Texas Press).
Whitford, A. E. 1952, *Ap. J.*, **120**, 599.
———. 1957, *Sky and Tel.*, **16**, 222.
———. 1962, in *Problems of Extragalactic Research* (*IAU Symposium No. 15*), ed. G. C. McVittie (New York: Macmillan Co.), p. 27.
———. 1970, unpublished.
———. 1971, *Ap. J.*, **169**, 215.
Whitford, A. E., and Sears, R. L. 1970, unpublished.
Wielen, R. 1964, *Zs. f. Ap.*, **59**, 129.
Wood, D. 1966, *Ap. J.*, **145**, 36.

CHAPTER 6

The Identification of Radio Sources

R. MINKOWSKI
Radio Astronomy Laboratory, University of California, Berkeley

IN THE two decades since the discovery of radio sources, the search for identification with optical objects has progressed from the initial phase when each new identification was an achievement to the present stage where hundreds of sources have been identified. Individual identifications as such are no longer of great importance, and no attempt will be made here to give a complete record of all identifications that have been suggested. The early history of the discovery of radio sources and their identification has been depicted in some detail by Shklovskii (1960). Sandage (1966*a*) has given a summary of the development of this field.

1. DISCOVERY AND EARLY IDENTIFICATIONS

The discovery of discrete radio sources dates back to the pioneering survey of Reber (1944), who found two areas of enhanced intensity in Cygnus and Cassiopeia in addition to the region of high intensity which peaks at the galactic center and is clearly connected with the Galaxy as a whole. Reber considered the areas of enhanced intensity to be details of galactic structure. Reber's result was confirmed by Hey, Phillips, and Parsons (1946), who discovered a little later (Hey, Parsons, and Phillips 1946) rapid intensity fluctuations in the Cygnus area. The analysis of the observation led to the explanation of the fluctuations as a result of ionospheric scintillations and to the conclusion that the enhanced intensity in the Cygnus region arises in a discrete source with a diameter of less than 2°.

Two years later, Bolton and Stanley (1948) were able to establish that the source is smaller than 8 arc min and to obtain a radio position of moderately high precision with the seacliff interferometer. A little later, Bolton (1948) announced the discovery of six new sources. Ryle and Smith (1948) confirmed the observations of the Cygnus source and detected several new variable sources—among them Cassiopeia A, which at low frequencies is the strongest source in the sky.

One year later, Bolton and Stanley (1949) were able to suggest the identification of the source Taurus A with the Crab Nebula, the remnant of the supernova of A.D. 1054, and Bolton, Stanley, and Slee (1949) could identify the source Virgo A with the galaxy

Received August 1968.

NGC 4486 (M87) and the source Centaurus A with the galaxy NGC 5128. It was thus clear from the earliest stages of the attempts to identify radio sources that a variety of objects can be strong radio sources, including supernova remnants and peculiar galaxies. To these classes were added somewhat later galactic emission nebulae when Haddock, Mayer, and Sloanaker (1954) discovered radio emission from the Orion Nebula and other H II regions. The peculiarity of NGC 4486, the now well-known jet, had long been recognized (Curtis 1918), but was relatively inconspicuous. The peculiarity of NGC 5128 is conspicuous. The debate on the nature of this object (see Evans 1949) was settled only when spectroscopic observations after its identification with the radio source showed a type of spectrum and a radial velocity that confirmed the extragalactic nature (Baade and Minkowski 1954*b*).

The two strongest sources, Cassiopeia A and Cygnus A, could be identified only after Smith (1951) had determined precise radio positions with uncertainties of $\pm 1^s$ in right ascension and $\pm 40''$ in declination (Baade and Minkowski 1954*a*). Cassiopeia A turned out to be an emission nebulosity whose properties are so unusual that the initial incomplete observations led Baade and Minkowski to reject the interpretation as a supernova remnant which was eventually established by Baade's observation of the motions and Minkowski's observations of the radial velocities (Minkowski 1957, 1958*a*). Cygnus A was found to be a faint galaxy whose nucleus shows two distorted components with unusually strong emission lines.

It was now clear that extragalactic sources were connected with galaxies of various kinds and degrees of peculiarity. Nearby bright galaxies without obvious peculiarities were not represented among the strong sources. This led Baade and Minkowski (1954*b*) to designate galaxies without optical peculiarities and with weak radio emission as "normal" galaxies. The term has become ambiguous because it is now certain that galaxies with strong radio emission do not necessarily show pronounced and easily recognizable optical peculiarities (Matthews, Morgan, and Schmidt 1964). There seems to be a fairly sharp division in the types of galaxies that are strong radio emitters with total emitted power above about 10^{40} ergs s^{-1} usually designated as radio galaxies, and those that are weak emitters with power below that limit; spheroidal galaxies prevail among radio galaxies, spirals and irregulars are mainly weak emitters, and the bulk of spheroidal galaxies are too weak to be observable with present means. But both optically normal and peculiar galaxies can be radio galaxies.

The first indication that normal bright galaxies are observable sources came in the First Cambridge (1C) survey (Ryle, Smith, and Elsmore 1950). The 1C catalog listed 50 sources with positions having probable errors of several minutes in right ascension and about 1° or more in declination. Identifications of four sources with bright galaxies were suggested. None of these identifications were fully convincing, and two of them were indeed spurious. The source 1C 07.01, suggested to be NGC 598 (M33), is the quasi-stellar source 3C48, and 1C 14.01, suggested to be NGC 5457 (M101), is 3C 295, at this time the most distant radio galaxy known. The suggested identification of 1C 01.01 with NGC 224 (M31) was soon confirmed by Brown and Hazard (1951). It was now established that bright normal galaxies are observable sources. Most of these are too weak to be included in the surveys which are the basis of identification work. Since they are bright objects, their identification presents no particular difficulty provided radio data are precise enough to exclude the possibility that a source found close to a bright galaxy is not actually an intrinsically strong but distant source. Information on the radio emission of bright galaxies with weak emission is obtained mainly from a search for sources coinciding with bright galaxies, not from the search for optical objects coinciding with a radio survey source which characterizes identification work. Examples of investigations of bright galaxies are such studies as that of 20 spirals by Brown and Hazard (1961), and of 37 galaxies mainly in the southern sky by Matthewson and Rome (1963). The most extensive investigation is that of 515 galaxies in the Shapley-Ames catalog by Heeschen and Wade (1964).

The number of known sources increased steadily during the decade following the initial discoveries. Important surveys during that period are by Brown and Hazard (1953), listing 20 sources observed at 159 MHz with a parabolic reflector of 218 feet diameter; by Mills (1952), listing 77 sources observed at 101 MHz with a Michelson-type interferometer; and by Bolton, Stanley and Slee (1954), listing 104 sources observed at 100 MHz. The positional accuracy reached in these surveys was low, roughly of the order of a few minutes in right ascension and of a degree in declination. Very few identifications were made on the basis of these positions. The discussion of the distribution of the sources led Mills to the important conclusion that the sources could be divided in two classes, Class I consisting of strong sources with high concentration toward the galactic plane and thus mainly galactic, and Class II with random distribution over the sky, and presumably extragalactic.

2. SURVEYS OF RADIO SOURCES

An extensive survey with a four-element interferometer at 91 MHz, completed at Cambridge in 1955 (Ryle and Hewish 1955) marks the end of the initial phase. The 2C catalog (Shakeshaft, Ryle, Baldwin, Elsmore, and Thomson 1955) lists 1936 sources between declination $-3°$ and $+83°$ with positional accuracies of about $\pm 2'$ in right ascension and $\pm 12'$ in declination to a limiting flux density of 10 flux units (f.u.) at 91 MHz [1 f.u. is 10^{-26} W m^{-2} Hz^{-1}]. Shortly after the 2C catalog results of a preliminary survey with the Mills cross at 85.5 MHz were made available (Mills and Slee 1957). A total of 383 sources were listed in an area between right ascension 0^h and 8^h and declination $+10°$ and $-20°$ with position accuracies of about $\pm 5'$ in right ascension and $\pm 6'$ in declination. Comparison of the two surveys showed very poor general correlation. Analysis of the results showed that both surveys suffered from confusion, the 2C catalog for sources fainter than 40 f.u., the Mills-Slee list for sources fainter than 20 f.u.

The 2C survey was followed by a third survey at Cambridge with a four element interferometer at 159 MHz (Edge, Shakeshaft, McAdam, Baldwin, and Archer 1959). With almost twice the frequency used for 2C, the 3C survey has almost four times the resolution and is much less subject to the effects of confusion which limited the usefulness of the 2C survey. The 3C catalog lists 471 sources between declination $-22°$ and $+71°$ to a flux density of 8 f.u., with accuracies of about $\pm 1'$ in right ascension and $\pm 6'$ in declination. The reliability of the catalog has been discussed by Edge, Scheuer, and Shakeshaft (1958) and later by Bennett and Smith (1961). In these discussions "reliability" is discussed mainly from the point of view of source counts, which require that all sources be recorded with correct flux densities. Positional errors are of more concern in the search for identifications. The substantial position errors that arise from placing the source in the wrong lobe of the interference pattern are less frequent in the 3C catalog than they were in 2C, but about 20 percent of the 3C sources were found to suffer from this defect, which in general prohibits identification.

Combination of the 3C survey with new surveys at 178 MHz by Leslie (1961) led to a revision of the 3C catalog. The revised 3C (3CR) catalog (Bennett 1962*a*, *b*) lists 328 sources between declinations $-5°$ and $+90°$ to a flux density of 9 f.u. with positional accuracies from $\pm 1^{s}.0$ in right ascension and 15″ in declination for positions determined from interferometric observations to $\pm 6'$ in right ascension and $\pm 45'$ in declination for positions from total-power records. Subsequent investigations by several observers, in particular the survey by Pauliny-Toth, Wade, and Heeschen (1966), show that the 3CR catalog is close to 100 percent complete for small sources down to the limit of the flux density. Missing are a few large sources near the galactic plane, presumably supernova remnants, such as the sources HB 9 and HB 21 (Brown and Hazard 1953). This is irrelevant for the identification of extragalactic sources. The 3CR catalog is now the main finding list for the northern hemisphere.

The fourth Cambridge survey with an aperture-synthesis telescope at 178 MHz (Scott, Ryle, and Hewish 1961) has now been completed. The 4C catalog (Pilkington and Scott 1965; Gower, Scott, and Wills 1967) lists 4843 sources in the area between declination $-7°$ and 80° to a flux density of 2 f.u. with positional accuracies of about $\pm 0'.5$ in right ascension and $\pm 3'$ in declination. A survey of the north polar region north of declination 86° at 178 MHz by Ryle and Neville (1962) explores an extension of the principle of aperture synthesis preparatory to the operation of the "one-mile" telescope at Cambridge (Ryle, Elsmore, and Neville 1965). The north polar survey lists 87 sources to a flux density of 0.25 f.u. with positional accuracy of $\pm 15''$ in both coordinates. The same technique has been used by Branson (1967) for a survey of the sky north of declination 70° at 81.5 MHz. The catalog lists 558 sources to a flux density of 1 f.u. with positional accuracies ranging from $\pm 1'$ to $\pm 3'$. A fifth Cambridge survey with the one-mile telescope simultaneously at 408 MHz and 1407 MHz has been started. The 5C1 catalog (Kenderdine, Ryle, and Pooley 1966) lists 110 sources with a positional accuracy of about 5″ in each coordinate in an area approximately 4° in diameter, centered at $\alpha = 9^{h}40^{m}$, $\delta = +50°$ (1950.0), to an equivalent flux density at 178 MHz of 0.05×10^{-26} f.u. in the center of the area. The 5C2 catalog (Pooley and Kenderdine) lists similarly 207 sources in an area 4° in diameter centered at $\alpha = 11^{h}00^{m}$, $\delta = +49°40'$.

Other surveys are in progress at the Ohio State University to flux densities of 2 f.u. at 600 MHz and 0.3 f.u. at 1415 MHz (Kraus, 1964; 1966), at the Vermilion River Observatory (MacLeod, Swenson, Yang, and Dickel 1965) to a flux density of 0.8 f.u. at 610 MHz, and at Bologna (Braccesi *et al.* 1965) to a flux density of 1 f.u. at 408 MHz.

The results of the deepest surveys show that the number of observable sources exceeds 10^{5} per steradian. This number is reached at the limit of 0.01 f.u. of the 5C2 survey (Pooley and Ryle 1968).

The first extensive coverage of the southern sky was given by the survey by Mills, Slee, and Hill (MSH) (1958, 1960, 1961) with the Mills cross. The MSH catalog lists 2270 sources between declination $+10°$ and $-80°$ to a flux density of 7 f.u. at 86 MHz.

The positional accuracy is about 5′ in right ascension, 6′ in declination. Investigations and discussions of the reliability of the MSH catalog by Bolton (1960), Kellermann and Harris (1960), Bennett and Smith (1961), and Hill and Mills (1962) show that the catalog begins to become unreliable below 20 f.u. Nearly half of the MSH sources with flux densities above 40 f.u. are extended. The absence of most of them from the 3C catalog were the main cause of the discordance between source counts from 3C and MSH. An investigation with the 14′ beam width of the 210-foot radio telescope at Parkes at 1400 MHz by Milne and Scheuer has shown that these extended sources at moderate galactic latitudes tend to be diffuse patches which are probably features of the galactic background. At high galactic latitudes, many of the extended sources turned out to be blends of small sources.

The survey of the sky south of declination +20° with the 210-foot telescope at Parkes (Bolton, Gardner, and Mackey 1964; Price and Milne 1965; Day, Shimmins, Ekers, and Cole 1965; Shimmins, Day, Ekers, and Cole 1966) is now the main source of information for the southern sky. The Parkes catalog, which still excludes the area close to the galactic plane, lists 1736 sources observed to a flux density of 4 f.u. at 408 MHz, with added observations at 1410 and 2650 MHz with a positional accuracy of about 1′.0 in both coordinates.

3. IDENTIFICATION OF 3C AND MSH SOURCES

Imperfect as the 3C and the MSH catalog are, they provided nevertheless a basis for some progress in the identification of sources. Summaries and discussions of the attempts at identification were given by Minkowski (1958*a* and *b*, 1960, 1961, 1962, 1963), Dewhirst (1959), Mills (1959, 1960) and Bolton (1960). For only a small fraction of the sources could identifications be suggested. Many of these could be regarded as not more than tentative, and quite a few turned out later to be spurious. During that period it became evident that the majority of galaxies identified as sources show no conspicuous peculiarities and that highly peculiar galaxies are usually not strong radio sources. Prevalent are E and S0 galaxies of high luminosity, among them galaxies with very extended envelopes designated D galaxies by Morgan. Some sources were found to be galaxies with brilliant starlike nuclei with faint envelopes, showing strong and narrow emission lines, designated N galaxies (Matthews, Morgan, and Schmidt 1964). Close double elliptical galaxies in a common envelope were thought to be frequently radio sources. This was later found to be the result of a selection effect in the search for identifications (Minkowski 1962, 1963). A correlation between clusters of galaxies in Abell's catalog (1958) and sources in the MHS catalog was first discovered byMills (1960) and confirmed for 3C sources by van den Bergh (1961). The correlation was also found to exist for identified sources (Minkowski 1962, 1963).

The reasons for the slow progress of the identification work were clearly recognized during that period. The identification of Cygnus A, the strongest extragalactic source, with a relatively faint galaxy implies the existence of more distant objects which are observable sources at radio wavelengths though optically beyond the reach of even the 200-inch telescope. The flux density of Cygnus A at 178 MHz is 8100 f.u. The galaxy has the redshift $\Delta\lambda/\lambda_0 = 0.0570$, its apparent visual magnitude $V = 14.36$, and its apparent blue magnitude $B = 15.51$, after correction for galactic absorption and reduction to an isophote of 25 mag per square second of arc (Sandage 1966*a*). At a distance corresponding to the redshift 1.5, the flux density would be 9.7 f.u., well above the limit of the 3C catalog, but the visual magnitude would be fainter than 24.6 mag, beyond the limit of the 200-inch telescope. These data are based on a cosmological model with $q_0 = +1$, the radio K-correction (Roeder and McVittie 1963) for a spectral index -0.80, and an optical K_v correction which for such a large redshift cannot be estimated reliably, but must be substantially larger than 3 mag to judge from the values of K_v found by Sandage (1966*a*) from the photometric data by Oke (1962).

The number of sources thus inaccessible to optical observation, however, would be small, probably much smaller than the crudely estimated upper limit of 10 to 20 percent of all sources stronger than 4 f.u. at 158 MHz (Minkowski 1958*a*). But the search for identifications has been carried out almost exclusively with the 48-inch Schmidt telescope on Palomar, not with the 200-inch telescope. An object like Cygnus A with a redshift of 0.7 would still have the relatively high flux density of 50 f.u. at 178 MHz, but would have a visual magnitude 21.3 which is below the limit of the 48-inch Schmidt telescope. Thus a substantial number of sources may be inaccessible for identification with that telescope.

The main difficulty for identification work based on the 3C and MSH catalogs was the low positional accuracy which did not permit identification with galaxies fainter than 17 mag (Minkowski 1958*a*). If coincidence of a radio source and a galaxy of magnitude m within twice the probable errors $\pm p_\alpha$ and $\pm p_\delta$ in right ascension and declination, respectively, of the radio position is to be considered as significant, the statistically expected number of galaxies as bright as or brighter than m in the error area $16 p_\alpha p_\delta$ is a measure of the chance that the coincidence is spurious. If this chance is to be less than 5×10^{-2}, and if $N(m)$ is the number of galaxies per square degree which are as bright or brighter than m, then the probable errors in p in seconds of arc must obey the condition $p_\alpha p_\delta \leq 4 \times 10^4 N(m)^{-1}$. The galaxy counts by Shane and Wirtanen (1967) show at high galactic latitudes an average of about 50 galaxies per square degree to photographic magnitude 19. To permit the identification of galaxies of this magnitude, the product $p_\alpha p_\delta$ must be smaller than 10^3 square seconds of arc. Thus, the limiting magnitude for identification of MSH sources is about 15 mag; that for 3C sources, about 16.5 mag. The difficulty of finding identifications for these surveys thus is readily understandable. To identify galaxies of 21 mag, near the limit of the 48-inch Schmidt telescope, probable errors not larger than 10″ are required. Once this accuracy was achieved, the number of identifications indeed increased rapidly, and quasi-stellar sources were added to the variety of objects that are strong radio sources.

Coincidence of positions is a necessary condition for an identification. Difficulties can arise from the difference between the structures of the source and of the galaxy with which the source is to be identified. This difference was first shown by interferometric observations of Cygnus A by Jennison and Das Gupta (1953, 1956) and by Jennison (1957), who discovered that the source is double with components separated by 88″ and located outside of the galaxy. Eventually it was established that a large fraction of all extragalactic sources is double, some with components of unequal intensity (Maltby and Moffet 1962). Identifications which seem to be beyond doubt establish that the position of the centroid of the radio source agrees with the position of the galaxy. The description of a source as a double on the basis of interferometric observations implies not more than that the visibility function can be adequately represented with the aid of that description. Improved techniques now permit the direct observation of the brightness distribution (Ryle, Elsmore, and Neville 1965; Macdonald, Neville, and Ryle 1966; Kenderdine, Ryle, and Pooley 1966; Macdonald, Kenderdine, and Neville 1968). The description of a source as double usually turns out to be a fair approximation. There are sources, however, whose structure is more complicated. Particularly for sources of this kind in a cluster of galaxies, doubts may seem justified as to whether a suggested identification is valid at all or whether more than one galaxy in the cluster is a radio source. An example is the source 3C 66. There is no reason to believe, however, that many identifications are unreliable for such reasons.

A difficulty that has indeed led to erroneous identifications is presented by quasi-stellar sources (QSSs). These, of course, were not known in the early phases of the identification work. Apparently quite convincing identifications with galaxies have been invalidated later by the discovery of a QSS in the proper position. An example is the 3CR source 3C 275.1 which seemed convincingly identified with NGC 4651, a 12th mag peculiar spiral galaxy with a jet and counterjet (Longair 1965), but found to be a QSS located in the outskirts of the galaxy (Sandage, Véron, and Wyndham 1965). Again, there is no reason to believe that such coincidences cause more than occasional spurious identifications.

Since apparently satisfactory agreement of positions is not always sufficient proof of the validity of an identification, confirmation of identifications by additional observations is desirable, but no generally useful and reliable criterion seems to exist for galaxies. Strong peculiarity, so conspicuous in early identifications, is not typical for radio galaxies, as their classification by Matthews, Morgan, and Schmidt (1964) shows. Careful investigation of radio galaxies which appear normal at first sight reveals peculiarities in 75 percent of the objects (Matthews 1966). Many of these peculiarities, however, are not restricted to radio galaxies (Vorontsov-Velyaminov 1965, 1966). An investigation of a random sample of galaxies which are not radio sources is needed to decide which, if any, peculiarities are typical for radio galaxies. In any case these peculiarities cannot serve as support for identifications with faint galaxies that cannot be observed with adequate resolution.

The presence of strong emission lines in the spectra of many radio galaxies is to some extent supporting evidence. Spheroidal galaxies which are not strong radio emitters do not tend to show strong emission lines. Weak emission of [O II] is not uncommon: it is shown by 18 percent of the E galaxies and by 48 percent of the S0 galaxies (Humason, Mayall, and Sandage 1956), but rich emission spectra showing lines of H and [O III] are rare. Of 46 radio galaxies for which data are available (Maltby, Matthews, and Moffet 1962; Schmidt 1965), only nine show no emission lines, 13 show [O II] lines (in many cases with unusually high intensity), and 24 have spectra rich in emission lines. Even the presence of a rich emission spectrum, however, is not necessarily reliable supporting evidence. For instance, the suggested identification of 3C 348 (Her A) with a faint galaxy having an emission-line spectrum containing high-excitation emission lines (Minkowski 1957) had to be replaced later on the basis of improved radio positions by the identification with an even fainter galaxy (Maltby, Matthews, and Moffet 1962) whose spectrum shows not more than strong [O II] λ3727 (Greenstein 1962).

Identifications with galaxies thus are based mainly on unsupported coincidences of positions. In contrast to the identifications with galaxies, identifications as QSSs can be confirmed reliably by observations of colors and spectra. Observations of $(U - B)$ and $(B - V)$ colors are not adequate to differentiate between QSSs and white dwarfs, but observations in the near-infrared may be useful to decide on the nature of a suspected object (Braccesi 1967).

4. PRECISE POSITIONS

With the exception of the north polar survey and the 5C survey, the catalog positions do not have sufficient accuracy for an effective search for identifications. When the need for high precision became obvious, investigations turned to the determination of precise positions for individual sources from previous surveys.

The first step in this direction was the determination of positions for 64 3C sources at 178 MHz with a transit interferometer with one fixed and one movable aerial by Elsmore, Ryle, and Leslie (1959). A positional accuracy of about $\pm 20''$ in right ascension and $\pm 90''$ in declination was achieved. Interferometers with movable aerials have been used extensively for the determination of positions for the sources in the 3C and 3CR catalogs. Use of higher frequencies has led to increased accuracy. Observations at the Owens Valley Observatory at 960 MHz by Read (1963), at 1400 MHz by Fomalont, Matthews, Morris, and Wyndham (1964), Wyndham and Read (1965), and Fomalont, Wyndham, and Bartlett (1967) give positions with an accuracy of about $\pm 10''$ in both coordinates for almost all 3CR sources. These observations have been extended by Olsen (1967) to 490 sources in the declination zone $+20°$ to $+40°$ of the 4C catalog (Pilkington and Scott 1965). Observations at the National Radio Astronomy Observatory at 2695 MHz by Wade, Clark, and Hogg (1965) reach accuracies of $\pm 1''$ to $\pm 5''$. Positions for 150 sources at 610 MHz by Adgie and Gent (1966) have accuracies of about $\pm 2''$, with lower accuracies down to $\pm 5''$ in right ascension and $\pm 24''$ in declination for a small fraction of the sources. Wills and Parker (1966) give right ascensions for 74 sources observed at 178 MHz with an accuracy of $\pm 8''$, but declinations with lower and according to present standards not useful accuracy of $\pm 45''$. Parker, Elsmore, and Shakeshaft (1966) give positions with accuracies of about $\pm 2''$ for 22 3CR sources observed at 1407 MHz with the Cambridge one-mile telescope.

Use of large parabolic reflectors at high frequencies can lead to positions which rival interferometric positions in accuracy, but they do not quite reach the highest precision. All 3C and 3CR sources have been observed with the 300-foot transit telescope at Green Bank at 750 and 1400 MHz with accuracies of about $\pm 30''$ in both coordinates by Pauliny-Toth, Wade, and Heeschen (1966). Positions with an accuracy of $\pm 10''$ have been determined by Shimmins, Clarke, and Ekers (1966) with the 210-foot telescope at Parkes for 644 sources of the Parkes catalog. These positions are at this time basic for identifications in the southern sky. A supplementary list of 160 positions with an accuracy of $\pm 15''$ has been given by Shimmins (1968).

Observations of lunar occultations provide the most accurate method of determining positions of radio sources with accuracies of the order of $\pm 1''$. The method was used first by Hazard (1961, 1962) for a determination of the position of 3C 212. Simultaneously with the position, detailed information on the size and structure is obtained (Hazard 1962; Scheuer 1962). An important result of the method is the determination of the position and structure of 3C 273 (Hazard 1962, 1965; Hazard, Mackey, and Shimmins 1963) which led to its identification with a stellar object by Schmidt (1963) and the recognition of the QSSs as a new class of objects which includes the previously identified source 3C 48. The spectrum of this latter source defied interpretation until the observations of 3C 273 demonstrated the occurrence of large redshifts that are typical for the QSSs (Greenstein and Matthews 1963). Few results from observations of occultations have become available; recent examples are observations by De Jong (1966, 1967), Taylor (1966), and Hazard, Gulkis and Bray (1966). The last investigation shows that, with the aid of the 1000-foot telescope at Arecibo, occultations of sources as weak as 0.5 f.u. at 400 MHz can be observed.

After positions of adequate precision became available, the search for identifications made rapid progress. The most spectacular achievement was the discovery of the QSSs (Matthews, Bolton, Greenstein, Münch, and Sandage 1961), made at a very early stage of the investigations of the 3CR sources. The results of these investigations, which offer the closest approach to a statistically complete sample now available, will be discussed in detail in § 5 below.

About 400 identifications have been suggested for sources in the Parkes catalog. The most complete sample that has been investigated contains 383 sources between declination $+20°$ and $-30°$ for which accurate positions are available (Bolton 1966). These results will be discussed in § 6 below.

Attempts to identify sources in the 4C catalog have been made by Pilkington (1964), Scheuer and Wills (1966), Wills (1966*a*, *b*), Aĭzu (1966), and Caswell and Wills (1967). Since the positions in the 4C catalog have at most moderate accuracy from the point of view of finding identifications, the proposed identifications are not very reliable and many may be spurious. The accurate positions that are now becoming available (Olsen 1967; Wills 1967) will eventually enable astronomers to obtain a valid sample of identified 4C sources. Many 4C sources south of declination $+27°$ are included in the Parkes survey.

5. THE IDENTIFICATION OF THE 3CR SOURCES

The 3CR catalog is at this time the best example of a catalog that is complete for a large part of the sky to a defined flux density and for which the search for identifications has reached a high degree of completeness so that the results have some statistical significance.

Many investigations have concerned themselves with the search for optical identifications of the 3CR sources. The principal contributors to the search are Matthews (whose unpublished results are included partly in the list by Matthews, Morgan, and Schmidt 1964), Matthews and Sandage (1963), Longair (1965), Sandage and Wyndham (1965), Sandage, Véron, and Wyndham (1965), Wyndham (1965), Shakeshaft and Longair (1965), and Wills and Parker (1966). Eventually the entire 3CR catalog was inspected by Véron (1966) and independently by Wyndham (1966). The optical basis of the search were the plates and prints of the *National Geographic Society–Palomar Observatory Sky*

Survey with the 48-inch Schmidt telescope. The limit to the brightness of the objects selected as identifications was 20 mag in the blue and 19 mag in the red, 1 mag above the plate limit of the *Sky Survey* (Minkowski and Abell 1963).

The results are shown in figure 1 for different ranges of flux density. The identification is nearly complete only for the 18 sources with flux densities above 40 f.u. The fraction of unidentified and not definitely identified sources increases, as should be expected,

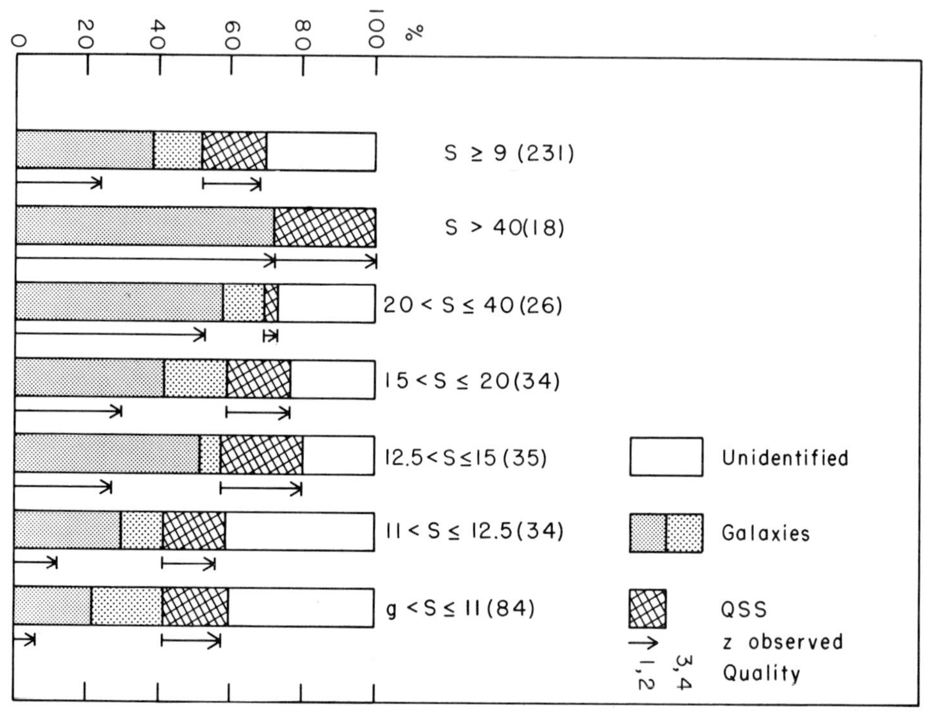

FIG. 1.—The identification of the 3CR sources with $|b^{II}| \geq 15°$ (1967 July)

with decreasing flux density. For flux densities below 11 f.u., 35 percent are unidentified. Only 22 percent have certain or highly probable identifications; for a similar fraction probable or possible identifications are listed. Twenty percent are QSSs.

[A more recent review of 3CR identifications is given in the Appendix to this chapter.]

6. THE IDENTIFICATION OF PARKES SOURCES

Bolton (1966) has discussed the results of the search for identifications of 383 sources in the Parkes catalog with flux densities above 1.6 f.u. at 1410 MHz for which precise positions are available in the area between declination +20° and −30°, excluding the neighborhood of the galactic plane. For sources with average spectral index, the limit of flux density corresponds roughly to that of 9 f.u. at 178 MHz of the 3CR catalog. All sources are believed to be extragalactic. The area searched is that which is common to the 48-inch *Sky Survey* and the Parkes catalog. It should be noted that in the declination zone $-27°$ to $-35°$ the limiting magnitude of the Sky Survey is about 0.5 mag fainter than at high declinations (Minkowski and Abell 1963). This might have some effect on the completeness of identifications at the lowest declinations.

Of the 243 identifications made, 60 either were previously well known or have been identified independently and are in the list of identified 3CR sources by Véron and Wyndham (see § 5). Altogether 143, or 37 percent of the sources, are identified with galaxies. For 100, or 26 percent, identification as QSSs is suggested; of these, 34 are fully confirmed by spectroscopic observation, 26 more by observations of colors. Of the 383 sources, 140, or 37 percent, remain unidentified.

The result is very similar to the results of the search for identifications of the 3CR sources in figure 1. For the unidentified sources and the identifications of galaxies, the difference is probably not larger than that which may arise from differences in judgment whether a source should be counted as unidentified or as probably or possibly identified with a galaxy. There are relatively more QSSs in the Parkes list. Bolton ascribes this to the effect of the difference of the survey frequencies used in the 3CR and the Parkes catalog and the difference in radio spectra for QSSs and galaxies. Seventy-four identifications have been suggested by Bolton, Shimmins, and Merkelin (1968). Half of these sources are identified with galaxies; the other half, as possible QSSs.

7. THE UNIDENTIFIED SOURCES

The question of the nature of the unidentified sources has been discussed by Véron (1966) and by Bolton (1966) with divergent results.

Véron concludes that the unidentified sources are mainly QSSs. The main argument is based on a plot of the number N_g of radio galaxies brighter than visual magnitude m_v. The ($\log N_g$, m_v) curve for certain and highly probable identifications flattens out at faint magnitudes in a way which indeed suggests strongly that it is unlikely that many sources could be connected with galaxies fainter than $m_v = 21$ mag. If, however, the probable and possible galaxies are added to the count, the ($\log N$, m_v) curve flattens out much less, as Véron's plot shows, and admits without difficulty that all or most of the unidentified sources are connected with the galaxies. Véron's conclusion thus seems to be valid only if the majority of the probable and possible identifications are invalid and chance coincidences.

Véron also considers the counts of the number N_S of sources stronger than flux density S as a function of S. The ($\log N$, $\log S$) curves are straight lines, with slopes of -1.85 for the extragalactic 3CR sources in agreement with Ryle and Neville (1963), -1.55 ± 0.5 for radio galaxies, -2.2 for the definitely and tentatively identified QSSs, and -2.2 for the sum of all QSSs and all unidentified sources. Véron considers the lower slope for radio galaxies and the equality of the steeper slopes for the QSSs and for the sum of QSSs and unidentified sources as an additional argument in favor of the interpretation that most unidentified sources are QSSs. The source counts are obviously consistent with that interpretation, but they do not prove it. If the unidentified sources were galaxies, their addition would increase the slope of the source count for galaxies. It is not unreasonable to think that among the unidentified sources are objects with larger redshifts than those of the identified radio galaxies whose redshifts are with few exceptions not larger than 0.2. It corresponds to reasonable expectation that the slope

−1.55 for radio galaxies is close to the value of −1.5 expected for nearby objects for which cosmological and evolutionary effects are negligible. What happens to the slope when the counts are extended to larger redshifts is an important and interesting question. The steepening of the slope that would arise from the addition of the unidentified sources to the galaxies may be not more than the unknown effect of extending the count to larger redshifts. It cannot be regarded as evidence proving that the unidentified sources are not galaxies.

Bolton's argument is based on the fact that the spectral index between 1410 and 2650 MHz is different for radio galaxies and QSSs. The spectra of the QSSs are flatter, and the dispersion of spectral indices is larger for the QSSs than for galaxies. With regard to both spectra and the dispersion of the spectral indices the unidentified sources are identical to the radio galaxies and different from the QSSs. No other interpretation seems possible than that the great majority of the unidentified sources are galaxies.

The source counts at 1410 and 2650 MHz show a picture similar to that at 178 MHz found by Véron. The slope of the ($\log N$, $\log S$) relation is −1.8 for the QSSs at 1410 MHz, and somewhat flatter at 2650 MHz owing to the high fraction of sources with flat spectra among the QSSs and the selection criteria for the source sample. The radio galaxies do not show a constant slope, but the ($\log N$, $\log S$) relation flattens toward the faint end. This may be an effect of increasing incompleteness of identifications with decreasing flux density. The unidentified sources show a slope of −2.4 at both frequencies. If the unidentified sources are added to the radio galaxies, the combined ($\log N$, $\log S$) relation has a slope of −1.85. This is quite similar to the slope for the QSSs. Bolton's interpretation of this result is that the addition of the unidentified sources extends the counts of radio galaxies to redshifts as large as those found in the QSSs, and that consequently the cosmological and evolutionary effects play a similar role. This interpretation would fail if the redshifts of the QSSs are not cosmological. The conclusion that the unidentified sources are mainly galaxies is in no way dependent on the interpretation of the source counts, but rather is a direct consequence of the observed properties of the radio spectra between 1410 and 2650 MHz of QSSs, radio galaxies, and unidentified sources.

8. RADIO SOURCES AND CLUSTERS OF GALAXIES

Three of the earliest identifications are with galaxies in clusters of galaxies: Cygnus A (3C 405) with the brightest galaxy in an anonymous cluster; Virgo A (3C 274) with NGC 4486, the third brightest galaxy in the Virgo cluster; and Perseus A (3C 84) with NGC 1275, the brightest galaxy in the Perseus cluster. Later Hercules A (3C 348) and 3C 295 were identified with the brightest galaxies in anonymous clusters. Thus, five of the eight strongest extragalactic sources in the 3CR catalog are in clusters of galaxies.

The association of clusters of galaxies with radio sources was discovered by Mills (1960), who found that 55 sources of the MSH catalog between declination +10° and −20° coincide with clusters in the catalog of clusters of galaxies by Abell (1958). The expected number of chance coincidences is 16. A similar result for sources of the 3C catalog was obtained by van den Bergh (1961), who found 27 coincidences for 282 sources with $b^{\mathrm{I}} \geq 25°$. The expected number of chance coincidences is only 9. From a discussion of the distribution of the coincidences over clusters of different richness he concludes that the probability of the occurrence of a radio source in a cluster is independent of the cluster population, but that the number of coincidences is too small to exclude the possibility that the probability is proportional to the richness. Similar studies were made by Tovmassian and Schachbazian (1961).

Attempts to observe radio emission from clusters of galaxies presents a different approach to the problem. Sholomitskii and Kokin (1965) searched for sources near 15 clusters at 920 MHz and found radio emission with flux densities between 0.4 and 2.8 f.u. in all but one of them. In three clusters, more than one source was suspected. The positional accuracy seems to have been relatively low: some of the sources may not be in the clusters to which they were ascribed. A more extensive investigation, which followed a preliminary survey by Rogstad, Rougoor, and Whiteoak (1965), was undertaken by Fomalont and Rogstad (1966) who surveyed all clusters within 230 Mpc, from Abell's catalog, at 1420 MHz. Of these clusters, 43 percent have radio emission greater than 0.2 flux units. A possible correlation of detection probability with cluster richness is noted. From a comparison of the radio luminosities with the luminosity function by Longair (see § 9) the conclusion is reached that a large fraction of the strong radio galaxies are in condensed clusters of the type catalogued by Abell. Tovmassian and Moiseev (1967*a*) report observations of 137 clusters of distance group 5 in Abell's catalog. A total of 25 radio sources was detected within $5'$ of the centers of these clusters, to be compared with an expected two or three chance coincidences. Most of the radio sources have diameters of less than $1'$ while the clusters have diameters of about $25'$. This means that, as a rule, a single galaxy is responsible for the radio emission. No correlation between the presence of the radio emission and the richness of the cluster was found.

For the identified sources the association of radio galaxies and clusters was demonstrated by Minkowski (1962, 1963) who found that 42 percent of the identified MSH sources and 33 percent of the identified 3C sources are in Abell's clusters which contain only 10 percent of all galaxies. Tovmassian and Moiseev (1967*b*) have determined more precise positions of 33 radio sources previously found to be in clusters catalogued by Abell. Of 30 sources in the MSH catalog, 19 sources were found to be within $5'$ of the cluster centers, and three sources outside the central area. Five sources are outside the clusters, and three sources were not detected.

For a sample of 41 identified radio galaxies, Matthews, Morgan, and Schmidt (1964) find that about 20 percent are in single galaxies or in very small groups of galaxies, about 50 percent are in clusters of richness ≥ 0, and about 35 percent in clusters of richness ≥ 1. The most commonly encountered optical form of the radio galaxies are galaxies which Morgan designates as type D. Galaxies of this type (Morgan and Lesh 1965) are centrally located in clusters, of which they are outstandingly the brightest members; they are never highly flattened, and have a characteristic appearance with bright central parts similar to elliptical galaxies on photographs of moderate scale and extended amorphous envelopes. The connection between radio galaxies and D galaxies contains the answer to the problem of the association of clusters of galaxies and radio sources. A comparison by Sandage (1966*b*) of the redshift-magnitude relations for the brightest member of clusters and for 39 identified radio sources shows that they are identical and that the radio galaxies do not differ significantly in luminosity from the brightest cluster galaxies. This shows that a galaxy must be a supergiant in luminosity, and probably in mass, to be a strong radio emitter. The association of radio sources with clusters arises probably for no deeper reason than that the supergiant D galaxies are typically the brightest galaxies of clusters.

9. THE LUMINOSITY FUNCTION

The space density of radio sources and the distribution of their radio luminosities are basic information for an understanding of the physical properties of the sources and their evolution. Attempts to derive the space density have not intentionally differentiated between radio galaxies and the QSSs. There are, however, only a few known radio galaxies with the high radio luminosities typical for the QSSs if they are assumed to be at the great distances following from the cosmological interpretation of their redshifts. Discussions of the space density thus were automatically limited to a range of radio luminosity which contains only radio galaxies. The discussion of the QSSs is a separate subject.

The luminosity function $\rho(M)$ specifies conventionally the number of sources per unit volume per unit interval of absolute magnitude M. [Apparent radio magnitudes are defined by $M_R(\nu) = 53.45 - 2.5 \log S(\nu)$.] In the past, confusion has arisen from the use of the term "luminosity function" for the number $n(M_R, S)$ or $n(M_R, m_R)$ of sources per unit interval of absolute radio magnitude M_R at a given flux density S or apparent radio magnitude m_R. The designation "luminosity distribution" for $n(M_R, S)$ is now coming into use. The luminosity distribution has shown itself to be a suitable basis for the interpretation of the source counts, which is outside the scope of this chapter. Only the "local" luminosity function will be discussed here.

The use of radio magnitudes has been found convenient because they provide a simple way to correlate radio data with the catalogued magnitudes of galaxies. More recently the use of radio luminosity $P(\nu)$ instead of absolute magnitude $M_R(\nu)$ is beginning to find acceptance. Different definitions of $P(\nu)$ are in use. If d is the distance, P may be given in W Hz^{-1} sterad^{-1} ($P = d^2 s$) or in W Hz^{-1} ($P = 4\pi d^2 S$).

The derivation of the radio luminosity function of galaxies requires a sample of sources in a defined area of q steradians where the identifications are complete for all sources connected with radio galaxies to a limiting apparent radio magnitude m_1. Ideally, all redshifts z should be known. The magnitude m_1 should be high enough so that the number of sources stronger than m_1 is sufficiently large for statistical analysis. If z is small enough for all sources to render cosmological effects negligible, the distance is given by $d = 10^6\, czH_0^{-1}$ pc where H_0 is the Hubble constant in kilometers per second per Mpc. The absolute radio magnitudes $M_R = m_R - 5 \log d + 5$ can then be obtained. It is now possible by simple counting to find the number $n(M_R, m_1)$ of sources per unit interval of M_R that have absolute radio magnitude M_R. These sources have distances for which $\log d_1 \leq 1 + 0.2(m_1 - M_R)$. They are thus contained in the volume $V(M_R, m_1) = (q/3)d_1^3$. The luminosity function is then given by $\rho(M_R) = (M_R, m_1)/V(M_R, m_1)$. If the sample contains sources for which the redshift has not yet been determined, then distance and thus M_R may be obtained, with restricted accuracy, by using the now well established fact that radio galaxies are a group of galaxies with high and well-defined mean optical luminosity. This was first noticed by Bolton (1960) and confirmed by Maltby, Matthews, and Moffet (1962). From a sample of 55 radio galaxies for which photometry and redshifts are available, Sandage (1966*b*) finds for the mean optical luminosity $\langle M_v \rangle = -21.60$, $\langle M_B \rangle = -20.77$, with a dispersion of $\delta_v = 0.44$, for a Hubble constant of 75 km s^{-1} per Mpc. With these values, the distance modulus $m - M$ can be obtained from the optical magnitudes of the radio galaxies.

If the redshift is not small enough throughout the sample of sources to make cosmological effects negligible, the optical and the radio magnitudes must be corrected by application of the K-correction for the effect of redshifting the energy distribution through the fixed passband of the observations. The optical K-correction has been discussed by Sandage (Appendix B of Humason, Mayall, and Sandage 1956; Sandage 1966*a*). The radio K-correction has been given by Roeder and McVittie (1963). A cosmological model must then be selected and be used to replace the distance d by the luminosity distance D and the volume V by the value appropriate for the cosmological model. The relevant relations have been given by Roeder and McVittie for homogeneous isotropic model universes having zero cosmological constant. In the derivation of the luminosity function it is assumed that it does not vary with distance. This assumption restricts the validity of the luminosity function to a local region in which evolutionary changes of the sources are negligible. Only as far as distances are based on optical magnitudes and a fixed average luminosity of radio galaxies is it also implied that there are no evolutionary changes of the optical luminosities of galaxies.

The first attempt to determine the luminosity function was made by Mills (1960). Lacking adequate data on the distances of galaxies in a list of probable and possible identifications, he used the difference $m_R - m_p$ as a measure of the radio luminosity. If, as Mills suggested and as is now established, the mean of M_p for radio galaxies has a well-defined value $\langle M_p \rangle$, one has $M_R = m_R - m_p + \langle M_p \rangle$. Mills determined the probability of a galaxy, chosen at random, having a specified value of $m_R - m_p$. Mills considered the available data to be inadequate for the final step of deriving the luminosity function by reducing the data to a unit volume for all sources.

From a sample of 23 identified sources of the MSH survey in an area of about 3 steradians with $S(85) \geq 20$ f.u., corresponding to $M_R(85) \leq 9.2$, and a sample of 27 identified sources of the 3C survey in an area of about 3 steradians with $m_R(159) \leq 8.5$, the luminosity function was derived by Minkowski (1961). In that discussion, the effects of statistical incompleteness were considered. Since the redshifts for all sources, with the exception of a few sources, were not large enough to cause appreciable cosmological effects, these were disregarded. The same sample of sources was investigated by Roeder and McVittie (1963), who took all cosmological effects including the K-corrections into account. Several approximations for K_p were used; the case $K_p = 4z$ comes sufficiently close to $K = 5z$ which fits the values of K_p derived by Sandage (1966*a*) from the photometry by Oke (1962). Cosmological models with zero cosmological constant and the values $q_0 = 1$ and $q_0 = 3.8$ for the deceleration parameter q_0 were used. To obtain the radio correction K_R, all sources were assumed to have identical spectra with the spectral index -1.05. This leads to a negligibly small value for K which for $z < 0.5$ never exceeds a few tenths of a magnitude. The results are insensitive to the K_p correction, and sensitive to the value of q_0 only at the highest luminosities. The luminosity function of Roeder and McVittie differs little from that of Minkowski; the difference is mainly due to improvements of the values for z for some galaxies and to differences of the values of

$\langle M_p \rangle$. It has now become evident that the sample of MSH sources contains a relatively large fraction of spurious identifications. The result derived from those sources should be disregarded. The sample of 3C sources has three spurious identifications, 3C 47 and 3C 196, both QSSs which are not to be identified with the galaxies thought to be the identifications, and 3C 280 for which no valid identification exists. Removing these sources from the sample has little effect on the result. The values for the luminosity function by Roeder and McVittie are plotted in figure 2 for comparison. Oort (1962) has assumed the luminosity function to be Gaussian and has calibrated it with a few strong sources. The result agrees not unsatisfactorily with that of Roeder and McVittie (see their fig. 2).

Recently, the luminosity function has been determined by Caswell and Wills (1967). The luminosity function was derived with $H_0 = 100$ km s^{-1} per Mpc from a sample

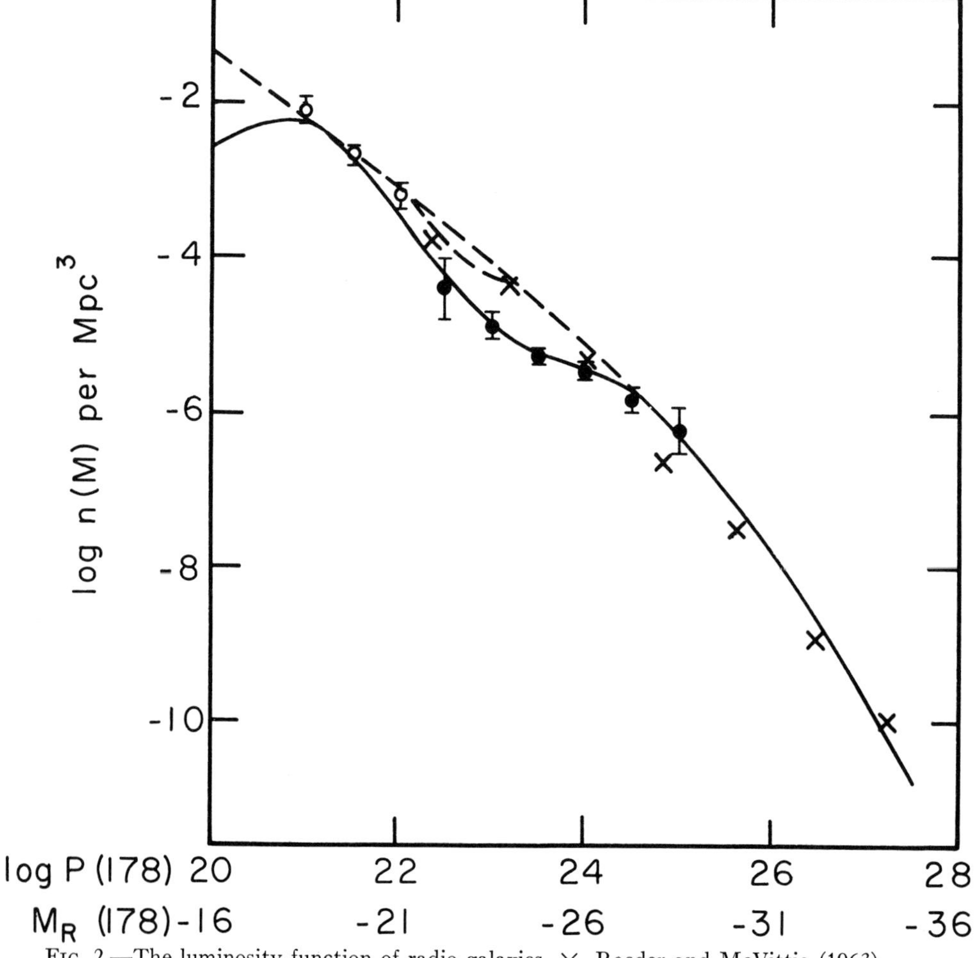

FIG. 2.—The luminosity function of radio galaxies. ×, Roeder and McVittie (1963). —— and ●, Caswell and Wills (1967), with ○ data from Heeschen and Wade (1964). – – –, model of Longair (1966). $H_0 = 100$ km s^{-1} per Mpc. $P(178)$ in watts Hz^{-1} sterad^{-1}.

consisting of bright galaxies observed by Heeschen and Wade (1964) and of 36 identified sources with flux densities above 2 f.u. at 178 MHz from an aperture synthesis survey by Crowther and Clarke (1966). No cosmological corrections were applied. The result is shown in figure 2, where the values have been adjusted to a unit volume of 1 Mpc^3. Also shown is a model luminosity function derived by Longair (1966). The values by Roeder and McVittie, which have been adjusted to $H_0 = 100$ km s^{-1} Mpc^{-1} and to 178 MHz agree satisfactorily except below $M_R = -26$, where they are unreliable owing to the small number of fainter sources in the sample. The luminosity function by Caswell and Wills shows an inflection near $M_R = -23$ which indicates that normal galaxies and radio galaxies are two distinct classes of sources. The occurrence of a minimum in the frequency of occurrence of radio sources near that radio luminosity has been suggested earlier by Matthews, Morgan, and Schmidt (1964).

The optical and the radio luminosity function of the QSSs in the 3CR catalog has been derived by Schmidt (1968) from a sample of 33 QSSs on the assumption that the red-shifts are cosmological. Since the redshifts are large, the flux densities observed at a fixed frequency are emitted at individually different higher frequencies over a very wide range. The first step of the analysis is the reduction of the observed flux densities to flux

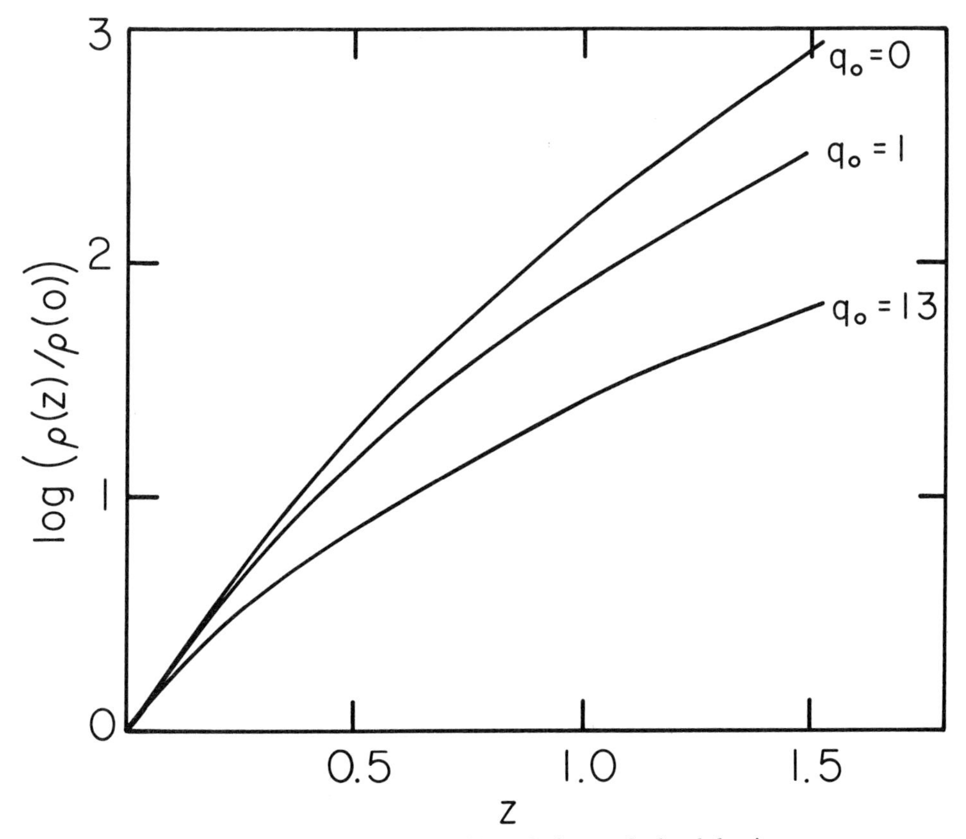

FIG. 3.—Space density of QSSs, relative to the local density

densities emitted at some arbitrarily selected fixed frequency. The adoption of a cosmological model is required. Two relativistic cosmological models with zero pressure, zero cosmological constant, a Hubble constant of 100 km s^{-1} Mpc^{-1}, and deceleration parameters q_0 were used, with $q_0 = 0$ or $q_0 = 1$. Optical flux densities were reduced to an emitted frequency of 1.2×10^{15} Hz, corresponding to 2500 Å, and radio flux densities to an emitted frequency of 500 MHz; from these, the absolute optical and radio luminosities were derived.

Since it cannot be assumed that the space density remains constant over the large range of redshifts and thus of luminosity distances, the second step is the investigation of the spatial distribution. Since the sample of sources is limited in flux density, it is possible to compute the maximum volume V_m over which a given source could be observed and the volume $V(z)$ which is included within the luminosity distance of the source. Uniform space distribution requires that all values of $V(z)/V_m$ between 0 and 1 are equally probable or that the mean value $\langle V(z)/V_m \rangle = \frac{1}{2}$. This turns out to be far from true: only six of the 33 sources used have $V(z)/V_m < 0.50$. A strong increase of space density with increasing z is indicated. A function $V'(z)$ can be derived which determines apparent volumes for which V'/V_m' shows uniform distribution. The ratio of density $\rho(z)$ at the redshift z to the local density $\rho(0)$ can now be determined; values

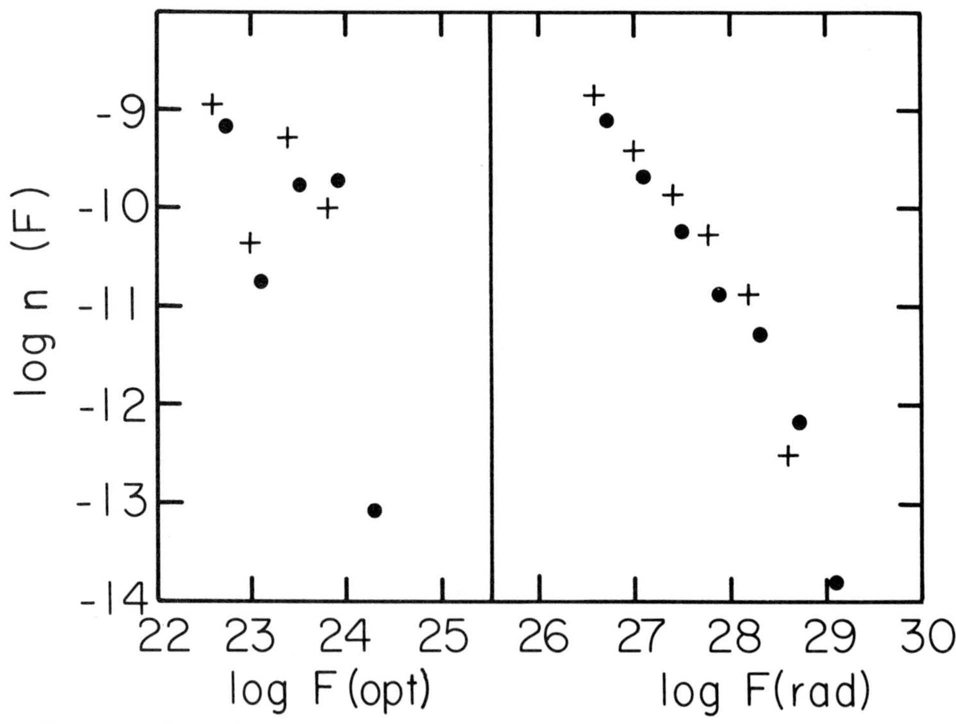

Fig. 4.—Optical and radio luminosity functions of QSSs. ●, $q_0 = 0$; +, $q_0 = 1$; $n(F)$, number per Mpc^3 of sources with luminosity in the range log $F \pm 0.2$ log F; F(opt), luminosity in watts Hz^{-1} at emitted frequency 1.2×10^{15} Hz (2500 Å). F(rad), luminosity in watts Hz^{-1} at emitted frequency 500 MHz.

are shown in figure 3. The result is insensitive to the value of q_0. There are not fully statistically significant indications that the increase of density with redshift might be slightly larger for weak than for strong sources.

Finally, the radio and the optical luminosity functions, defined as the local density of sources of luminosity F per interval $\Delta \log F = 0.4$ (= 1 mag), are obtained by collecting the sources in groups of 1 mag or 0.4 in log F and summing in each group the values of $(V_m')^{-1}$ which give the contribution of the individual sources to the space density. The optical and radio luminosity functions are shown in figure 4. The optical luminosity function is not well determined. The radio luminosity function shows a smooth increase towards lower luminosities by a factor of about 3–4 per mag.

The total local space density of QSSs in the luminosity range covered by the data in figure 4 is 1.0×10^{-9} Mpc^{-3} for $q_0 = 0$ and 1.7×10^{-9} Mpc^{-3} for $q_0 = 1$.

A lower limit of about 70 QSSs is derived for the total number in the 3CR catalog. This limit agrees with the results of the identification work. If the ratio of three galaxies to one QSS for the identified sources with $|b^{II}| \geq 15°$ is valid for the whole catalog, 74 QSSs would be expected.

Extension of the radio luminosity function by 1.5 mag toward lower luminosities is sufficient to account for the sky density of radio-quiet quasi-stellar objects estimated by Sandage and Luyten (1967).

REFERENCES

Abell, G. O. 1958, *Ap. J. Suppl.*, **3**, 211.
Adgie, R. L., and Gent, H. 1966, *Nature*, **209**, 549.
Aïzu, K. 1966, *Pub. Astr. Soc. Japan*, **8**, 219.
Baade, W., and Minkowski, R. 1954*a*, *Ap. J.*, **119**, 206.
———. 1954*b*, *ibid.*, **119**, 215.
Bennett, A. S. 1962*a*, *M.N.R.A.S.*, **125**, 75.
———. 1962*b*, *Mem. R.A.S.*, **68**, 163.
Bennett, A. S., and Smith, F. G. 1961, *M.N.R.A.S.*, **122**, 71.
Bergh, S. van den. 1961, *A.J.*, **66**, 562.
Bolton, J. G. 1948, *Nature*, **162**, 141.
———. 1960, *Obs. Calif. Inst. Tech. Radio Obs.*, No. 5.
———. 1966, *Nature*, **211**, 917.
Bolton, J. G., Gardner, F. F., and Mackey, M. B. 1964, *Australian J. Phys.*, **17**, 340.
Bolton, J. G., Shimmins, A. J., and Merkelin, J. 1968, *Australian J. Phys.*, **21**, 81.
Bolton, J. G., and Stanley, G. J. 1948, *Nature*, **161**, 312.
———. 1949, *Australian J. Sci. Res.*, **2A**, 139.
Bolton, J. G., Stanley, G. J., and Slee, O. B. 1949, *Nature*, **164**, 101.
———. 1954, *Australian J. Phys.*, **7**, 110.
Braccesi, A. 1967, *Nuovo Cimento*, **49**, 148.
Braccesi, A., Ceccarelli, M., Fanti, R., Gelato, G., Giovanni, C., Harris, D., Rosatelli, C., Sinigaglia, G., and Volders, L. 1965, *Nuovo Cimento*, **40B**, 267.
Branson, N. J. B. A. 1967, *M.N.R.A.S.*, **135**, 149.
Brown, R. Hanbury, and Hazard, C. 1951, *M.N.R.A.S.*, **111**, 357.
———. 1953, *ibid.*, **113**, 123.
———. 1961, *ibid.*, **122**, 479; **123**, 279.
Caswell, J. L., and Wills, D. 1967, *M.N.R.A.S.*, **135**, 231.
Crowther, J. H., and Clarke, R. W. 1966, *M.N.R.A.S.*, **131**, 425.
Curtis, H. D. 1918, *Pub. Lick Obs.*, **13**, 9.
Day, G. A., Shimmins, A. J., Ekers, R. D., and Cole, D. J. 1966, *Australian J. Phys.*, **19**, 35.
De Jong, M. L. 1966, *A.J.*, **71**, 373.
———. 1967, *Ap. J.*, **148**, 651.
Dewhirst, D. W. 1959, *Paris Symposium on Radio Astronomy*, ed. R. N. Bracewell (Stanford: Stanford University Press), 507.
Dieter, N. H. 1967, *Ap. J.*, **150**, 435.
Edge, D. O., Scheuer, P. A. G., and Shakeshaft, J. R. 1958, *M.N.R.A.S.*, **118**, 183.

Edge, D. O., Shakeshaft, J. R., McAdam, W. B., Baldwin, J. E., and Archer, S. 1959, *Mem. R.A.S.*, **68**, 37.
Elsmore, B., Ryle, M., and Leslie, P. R. R. 1959, *Mem. R.A.S.*, **68**, 61.
Evans, D. S. 1949, *M.N.R.A.S.*, **109**, 94.
Fomalont, E. B. 1967, *Obs. Owens Valley Radio Obs.*, No. 7.
Fomalont, E. B., Matthews, T. A., Morris, D., and Wyndham, J. D. 1964, *Ap. J.*, **69**, 772.
Fomalont, E. B., and Rogstad, D. H. 1966, *Ap. J.*, **146**, 528.
Fomalont, E. B., Wyndham, J. D., and Bartlett, J. F. 1967, *A.J.*, **72**, 445.
Gower, J. F. R. 1966, *M.N.R.A.S.*, **133**, 151.
Gower, J. F. R., Scott, P. F., and Wills, D. 1967, *Mem. R.A.S.*, **71**, 49.
Greenstein, J. L. 1962, *Ap. J.*, **135**, 679.
Greenstein, J. L., and Matthews, T. A. 1963, *Nature*, **197**, 1041.
Haddock, F. T., Mayer, C. H., and Sloanaker, R. M. 1954, *Ap. J.*, **119**, 456.
Hazard, C. 1961, *Nature*, **191**, 58.
———. 1962, *M.N.R.A.S.*, **124**, 343.
———. 1965, *Quasi-stellar Sources and Gravitational Collapse*, ed. I. Robinson, A. Schild, and E. L. Schucking (Chicago: University of Chicago Press), p. 135.
Hazard, C., Gulkis, S., and Bray, A. D. 1966, *A.J.*, **71**, 857.
Hazard, C., Mackey, M. B., and Shimmins, A. J. 1963, *Nature*, **197**, 1037; **198**, 19.
Heeschen, D. S., and Wade, C. M. 1964, *A.J.*, **69**, 277.
Hey, J. S., Parsons, S. J., and Phillips, J. W. 1946, *Nature*, **158**, 234.
Hey, J. S., Phillips, J. W., and Parsons, S. J. 1946, *Nature*, **157**, 296.
Hill, E. R., and Mills, B. Y. 1962, *Australian J. Phys.*, **15**, 437.
Humason, M. L., Mayall, N. U., and Sandage, A. R. 1956, *A.J.*, **61**, 97.
Jennison, R. C. 1959, *Paris Symposium on Radio Astronomy*, ed. R. N. Bracewell (Stanford: Stanford University Press), p. 309.
Jennison, R. C., and Das Gupta, M. K. 1953, *Nature*, **172**, 996.
———. 1956, *Phil. Mag.*, **1**, 55.
Kellermann, K. I., and Harris, D. E. 1960, *Observations Calif. Inst. Tech. Radio Obs.*, No. 7.
Kenderdine, S., Ryle, Sir Martin, and Pooley, G. G. 1966, *M.N.R.A.S.*, **133**, 189.
Kraus, J. D. 1964, *Nature*, **202**, 269.
———. 1966, *A.J.*, **71**, 861.
Leslie, P. R. R. 1961, *M.N.R.A.S.*, **122**, 51.
Longair, M. S. 1965, *M.N.R.A.S.*, **129**, 419.
———. 1966, *ibid.*, **133**, 421.
Macdonald, G. H., Kenderdine, S., and Neville, A. C. 1968, *M.N.R.A.S.*, **138**, 259.
Macdonald, G. H., Neville, A. C., and Ryle, Sir Martin. 1966, *Nature*, **211**, 1241.
MacLeod, J. M., Swenson, G. W., Jr., Yang, K. S., and Dickel, J. R. 1965, *A.J.*, **70**, 756.
Maltby, P., Matthews, T. A., and Moffet, A. T. 1963, *Ap. J.*, **137**, 153.
Maltby, P., and Moffet, A. T. 1962, *Ap. J. Suppl.*, **7**, 141.
Matthews, T. A. 1966, *A.J.*, **71**, 169.
Matthews, T. A., Bolton, J. G., Greenstein, J. L., Münch, G., and Sandage, A. R. 1961, *Sky and Tel.*, **21**, 148.
Matthews, T. A., Morgan, W. W., and Schmidt, M. 1964, *Ap. J.*, **140**, 35.
Matthews, T. A., and Sandage, A. R. 1963, *Ap. J.*, **138**, 30.
Matthewson, D. S., and Rome, J. M. 1963, *Australian J. Phys.*, **16**, 360.
Mezger, P. G., and Höglund, B. 1967, *Ap. J.*, **147**, 490.
Mills, B. Y. 1952, *Australian J. Sci. Res.*, **5A**, 266.
———. 1959, in *Handbuch der Physik*, ed. S. Flügge (Berlin: Springer-Verlag), **53**, 565.
———. 1960, *Australian J. Phys.*, **13**, 550.
Mills, B. Y., and Slee, O. B. 1957, *Australian J. Phys.*, **10**, 162.
Mills, B. Y., Slee, O. B., and Hill, E. R. 1958, *Australian J. Phys.*, **11**, 360.
———. 1960, *ibid.*, **13**, 676.
———. 1961, *ibid.*, **14**, 497.
Milne, D. K., and Scheuer, P. A. G. 1964, *Australian J. Phys.*, **17**, 106.
Minkowski, R. 1957, in *Radio Astronomy*, ed. H. C. van de Hulst (Cambridge: Cambridge University Press), p. 107.
———. 1958*a*, *Paris Symposium on Radio Astronomy*, ed. R. N. Bracewell (Stanford: Stanford University Press), p. 315.
———. 1958*b*, *Pub. A.S.P.*, **70**, 143.
———. 1960, *Proc. Nat. Acad. Sci.*, **46**, 13.
———. 1961, *Proc. Fourth Berkeley Symposium on Mathematical Statistics and Probability*, ed. J. Neyman, (Berkeley: University of California Press), **3**, 245.
———. 1962, *Problems of Extragalactic Research*, ed. G. C. McVittie (New York: Macmillan), p. 201.
———. 1963, *Proc. Nat. Acad. Sci.*, **49**, 779.
Minkowski, R., and Abell, G. O. 1963, in *Basic Astronomical Data*, ed. K. A. Strand (Chicago: University of Chicago Press), p. 481.

Morgan, W. W., and Lesh, J. R. 1965, *Ap. J.*, **142**, 1364.
Oke, J. B., 1962, *Problems of Extragalactic Research*, ed. G. C. McVittie (New York: Macmillan), p. 34.
Olsen, E. T. 1967, *Obs. Owens Valley Radio Obs.*, No. 6.
Oort, J. H. 1962, *B.C.A.P.*, Memo 13.
Parker, E. A., Elsmore, B., and Shakeshaft, J. R. 1966, *Nature*, **210**, 22.
Pauliny-Toth, I. I. K., Wade, C. M., and Heeschen, D. S. 1966, *Ap. J. Suppl.*, **13**, 65.
Pilkington, J. D. H. 1964, *M.N.R.A.S.*, **128**, 103.
Pilkington, J. D. H., and Scott, P. F. 1965, *Mem. R.A.S.*, **69**, 183.
Pooley, G. G., and Kenderdine, S. 1968, *M.N.R.A.S.*, **139**, 529.
Pooley, G. G., and Ryle, M. 1968, *M.N.R.A.S.*, **139**, 515.
Price, R. M., and Milne, D. K. 1965, *Australian J. Phys.*, **18**, 329.
Read, R. B. 1963, *Ap. J.*, **138**, 1.
Reber, G. 1944, *Ap. J.*, **100**, 279.
Roeder, R. C., and McVittie, G. C. 1963, *Ap. J.*, **138**, 899.
Rogstad, D. H., Rougoor, G. W., and Whiteoak, J. B. 1965, *Ap. J.*, **142**, 1965.
Ryle, M., Elsmore, B., and Neville, A. C. 1965, *Nature*, **207**, 1024.
Ryle, M., and Hewish, A. 1955, *Mem. R.A.S.*, **67**, 97.
Ryle, M., and Neville, A. C. 1963, *M.N.R.A.S.*, **125**, 39.
Ryle, M., and Smith, F. G. 1948, *Nature*, **162**, 462.
Ryle, M., Smith, F. G., and Elsmore, B. 1950, *M.N.R.A.S.*, **110**, 508.
Sandage, A. R. 1965, *Ap. J.*, **141**, 1560.
———. 1966*a*, *Rend. Scuola Internat. de Fisica "E. Fermi"* (New York: Academic Press), Course 35.
———. 1966*b*, *Atti Convegno Sulla Cosmologia* (Firenze: G. Barbèra), p. 104.
———. 1966*c*, *Ap. J.*, **145**, 1.
Sandage, A. R., and Luyten, W. J. 1967, *Ap. J.*, **148**, 767.
Sandage, A. R., Véron, P., and Wyndham, J. D. 1965, *Ap. J.*, **142**, 1307.
Sandage, A. R., and Wyndham, J. D. 1965, *Ap. J.*, **141**, 328.
Scheuer, P. A. G. 1962, *Australian J. Phys.*, **15**, 333.
Scheuer, P. A. G., and Wills, D. 1966, *Ap. J.*, **143**, 274.
Schmidt, M. 1963, *Nature*, **197**, 1040.
———. 1965, *Ap. J.*, **141**, 1.
———. 1968, *ibid.*, **151**, 393.
Scott, P. F., Ryle, M., and Hewish, M. 1961, *M.N.R.A.S.*, **139**, 781.
Shakeshaft, J. R., and Longair, M. S. 1965, *Observatory*, **85**, 30.
Shakeshaft, J. R., Ryle, M., Baldwin, J. E., Elsmore, B., and Thomson, J. H. 1955, *Mem. R.A.S.*, **67**, 106.
Shane, C. D., and Wirtanen, C. A. 1967, *Pub. Lick Obs.*, **22**, 1.
Shimmins, A. J. 1968, *Australian J. Phys.*, **21**, 65.
Shimmins, A. J., Clarke, M. E., and Ekers, R. D. 1966, *Australian J. Phys.*, **19**, 649.
Shimmins, A. J., Day, A. G., Ekers, R. D., and Cole, D. J. 1966, *Australian J. Phys.*, **19**, 837.
Shklovskii, S. I. 1960, *Cosmic Radio Waves* (Cambridge: Harvard University Press).
Sholomitskii, G. B., and Kokin, Y. F. 1965, *Astr. Zh. (USSR)*, **42**, 674 (English transl. in *Soviet Astr.—A.J.*, **9**, 517).
Smith, F. G. 1951, *Nature*, **168**, 555.
Taylor, J. H. 1966, *Ap. J.*, **146**, 646.
Tovmassian, G. M., and Moiseev, I. V. 1967*a*, *Australian J. Phys.*, **20**, 715.
———. 1967*b*, *ibid.*, **20**, 725.
Tovmassian, G. M., and Shachbazian, R. K. 1961, *Izv. Akad. Nauk Arm. SSR*, **14**, 121.
Vaucouleurs, G. de, and Vaucouleurs, A. de, 1964, *Reference Catalogue of Bright Galaxies* (Austin: University of Texas Press).
Véron, P. 1966, *Ann. d'Ap.*, **29**, 231.
Vorontsov-Velyaminov, B. 1965, *Observatory*, **85**, 212.
———. 1966, *Astr. Zh. (USSR)*, **43**, 1318 (English transl. in *Soviet Astr.—A.J.*, **10**, 1057).
Wade, C. M., Clark, B. G., and Hogg, D. E. 1965, *Ap. J.*, **142**, 406.
Wills, D. 1966*a*, *Observatory*, **86**, 140.
———. 1966*b*, *ibid.*, p. 246.
———. 1967, *M.N.R.A.S.*, **135**, 339.
Wills, D., and Parker, E. A. 1966, *M.N.R.A.S.*, **131**, 503.
Wyndham, J. D. 1965, *A.J.*, **70**, 120.
———. 1966, *Ap. J.*, **144**, 459.
Wyndham, J. D., and Read, R. B. 1965, *A.J.*, **70**, 120.

APPENDIX TO CHAPTER 6

Optical Identification of 3CR Sources (October 1972)

J. KRISTIAN
Hale Observatories, Carnegie Institution of Washington, California Institute of Technology
AND
R. MINKOWSKI
Radio Astronomy Laboratory, University of California, Berkeley

THE ORIGINAL manuscript of this chapter by Minkowski contained a table of 3CR identifications as of 1968 August. Because of the substantial improvements in the years immediately following in source positions and maps, as well as accumulated new data on individual sources, it seemed useful to update the table. This was done by Kristian in the autumn of 1972. Since the conclusions drawn by Minkowski in the body of the text are not qualitatively changed by the revision, and in keeping with the general philosophy of presenting each chapter as the state of thinking in a subject at the time of its receipt, the chapter itself was not changed, and the revised table is given separately here.

Table 1 lists the sources in order of decreasing 3CR flux density, as given in column 2. Column 1 is a running serial number. (For convenience in locating particular sources, table 2 lists the serial numbers in order of 3CR number.) Column 3 gives 3CR numbers, with galactic sources in italics. Columns 4 and 5 list the galactic coordinates.

Column 6 describes the identifications. The designations used are:

SN, supernova remnant

H II, emission nebula

Gxy, galaxy (also N gxy, gxy cluster, etc.)

QSS, quasi-stellar source as established by redshift measurement

Nature uncertain, objects which are the identifications based on good position agreement, but which are too faint to distinguish as distant galaxies or faint stellar objects, and for which photometry or spectroscopy are unavailable

EF, empty field—high-latitude source with good radio position at which nothing is visible to the plate limit

No ID, no good identification available at present (obscured fields, poor radio or optical positions, complex or extended radio sources, etc.).

TABLE 1

THE 3CR CATALOG

Serial Number	Intensity (f.u.)	3CR Number	l^{II}	b^{II}	Identification	Quality	m_v	z	Remarks
1.....	11000	*461*	112	− 2	SN				Cas A
2.....	8100	405	76	+ 6	Gxy	2	15.14	0.057	Cyg A
3.....	1420	*144*	185	− 6	SN				Crab Nebula
4.....	970	274	284	+75	Gxy	1	8.74	0.004	M87 = Vir A. In Virgo cluster.
5.....	540	*400*	50	0	H II	3			W51. Double source. OH emission. Faint obscured optical emission.
6.....	410	*410.1*	78	+ 2					Part of Cyg X complex
7.....	400	*392*	35	0	SN?				W44. Extended, nonthermal. Optically obscured.
8.....	325	348	96	+44	Gxy	3	16.90	0.154	Her A
9.....	230	*409.1*	82	+ 5					Part of Cyg X complex
10.....	210	*157*	189	+ 3	SN				IC 443
11.....	203	353	21	+20	Gxy	3	15.36	0.031	
12.....	175	123	171	−12	cDgxy	2	~19.5		Faint cluster
13.....	160	*416.2*	81	+ 1	H II				Part of Cyg X complex
14.....	134	*10*	120	+ 1	SN				Tycho SN (1572)
15.....	110	*390.2*	30	0	H II				W43. H and OH emission. Optically obscured.
16.....	90	*363.1*	31	+16	No ID				
17.....	74	*419.1*	82	0	H II				Part of Cyg X complex
18.....	73	295	61	+73	Gxy	1	20.11	0.461	Brightest in faint cluster
19.....	....	409	63	− 6	No ID				EG. Obscured
20.....	67	273	290	+64	QSS	2	12.80	0.158	
21.....	66	134	168	− 2	No ID				EG. Obscured
22.....	60	*416.1*	85	+ 4	H II				Part of Cyg X complex
23.....	59	196	226	+17	QSS	1	17.60	0.871	
24.....	58	84	151	−13	Gxy	2	11.87	0.018	NGC 1275. Perseus cluster (A426)
25.....	....	147	162	+10	QSS	1	16.90	0.545	
26.....	57	111	162	− 9	No ID				EG. Obscured. Widely separated triple source.
27.....	57	380	77	+24	QSS	1	16.81	0.691	
28.....	....	*398*	41	0	No ID				W49. Thermal+nonthermal sources. Highly obscured.
29.....	52	433	67	−24	Gxy	3	16.24	0.102	Dumbbell. m_v of brightest component.
30.....	51	310	39	+60	Gxy group	3	15.24	0.054	m_v of brightest member
31.....	....	390.1	38	+ 4	No ID				
32.....	....	*403.2*	68	+ 2	No ID				W56. Thermal.
33.....	49	33	124	+10	Gxy	2	15.19	0.060	
34.....	....	452	98	−17	Gxy	3	16.00	0.081	Probably in cluster.
35.....	47	48	134	−29	QSS	1	16.20	0.367	
36.....	44	219	174	+45	Gxy	2	17.26	0.174	Brightest member of rich cluster
37.....	....	270	282	+67	Gxy	1	10.4	0.007	NGC 4261. In Virgo cluster.
38.....	....	298	352	+61	QSS	1	16.79	1.439	
39.....	....	390.3	112	+27	N gxy	3	14.80	0.057	
40.....	43	317	9	+50	Gxy	1	13.44	0.035	Cluster A2052.
41.....	41	20	122	−11	Gxy	2	~19		

TABLE 1—*Continued*

Serial Number	Intensity (f.u.)	3CR Number	l^{II}	b^{II}	Identification	Quality	m_v	z	Remarks
42.....	....	98	180	−31	Gxy	2	(14.45)	0.031	
43.....	....	*372.1*	26	+ 6	No ID				Extended, probably galactic
44.....	....	338	63	+44	Gxy	1	12.63	0.030	NGC 6166. Brightest gxy in A2199.
45.....	40	327	12	+38	Gxy	3	15.88	0.104	
46.....	37	438	89	−13	No ID				EG
47.....	35	465	103	−33	Gxy cluster	3	13.29	0.030	NGC 7720 in A2634
48.....	34	468.1	117	+ 3	No ID				EG. Obscured
49.....	35	66	140	−17	Gxy cluster		12.90	0.022	Complex extended source in A347
50.....	32	410	69	− 4	No ID				EG. Obscured
51.....	....	*396.1*	32	− 5	No ID				SN?
52.....	29	61.1	123	+24	Gxy cluster?				Multiple source extends over 3′. Possible 19-mag cluster near center.
53.....	29	234	200	+53	N gxy	3	17.27	0.185	In group of compact galaxies
54.....	....	430	100	+ 8	Gxy	2	~15	0.055	
55.....	28	83.1	150	−13	Gxy	3	~12.5	0.018	NGC 1265. In Perseus cluster (A426).
56.....	....	227	229	+42	N gxy	3	16.33	0.086	
57.....	....	313	9	+52	No ID				
58.....	....	397	41	0	No ID				Nonthermal. Obscured.
59.....	27	65	141	−20	EF	1	≳21.5		Fair 200-inch red plate empty
60.....	26	403	42	−12	Gxy	2	15.42	0.059	
61.....	25	103	157	− 7	No ID				EG.
62.....	....	225	220	+44	Two gxys?	2			EG. Double radio with 6′ separation, each with gxy near.
63.....	24	27	123	+ 6	No ID				EG. Obscured.
64.....	....	40	142	−63	Gxy cluster		12.28	0.018	Cluster A194. No good ID with single gxy. m_v is for brightest gxy = NGC 545.
65.....	....	*58*	131	+ 3					Thermal. Probable SN remnant.
66.....	....	79	164	−34	N gxy	2	18.56	0.256	In cluster?
67.....	....	237	232	+47	Nature uncertain	1	~20		
68.....	....	391	32	0	No ID				Obscured
69.....	....	264	236	+73	Gxy	2	12.74	0.021	NGC 3862
70.....	....	327.1	12	+37	No ID				
71.....	....	330	99	+41	Gxy	2	~21		Brightest gxy in faint cluster
72.....	....	386	47	+11	Gxy	3	~13	0.018	Foreground star superposed on nucleus.
73.....	....	*415.1*	82	+ 2					Part of Cyg X complex
74.....	23	18	119	−53	Gxy	2	~18.5		
75.....	....	69	136	− 1	No ID				EG. Obscured.
76.....	....	75	170	−45	Gxy in cluster	3	13.62	0.024	Dumbbell in A400
77.....	....	171	162	+22	N gxy	1	18.89	0.239	
78.....	....	427.1	111	+19	No ID				EG.
79.....	....	*434.1*	94	+ 1					Not a source, but a background ridge
80.....	....	445	62	−47	N gxy	3	15.77	0.057	
81.....	22	388	75	+20	Gxy	2	15.32	0.092	Dumbbell. m_v of brightest component. Cluster.
82.....	....	*396*	39	0	No ID				SN? Highly obscured.

TABLE 1—*Continued*

Serial Number	Intensity (f.u.)	3CR Number	l^{II}	b^{II}	Identification	Quality	m_v	z	Remarks
83		*400.2*	54	−2	No ID				SN? Highly obscured.
84		459	83	−51	N gxy	1	17.55	0.220	
85	21	17	115	−65	Gxy	3	18.02	0.220	
86		55	140	−32	No ID				EG
87		86	144	−1	No ID				EG. 3 components.
88		129	160	0	No ID				Very extended, low-latitude source
89		230	238	+39	Nature uncertain	3	15.51		Object has colors of a galactic star. Not good position agreement.
90		286	57	+81	QSS	1	17.30	0.846	
91		389	29	0	No ID				Obscured
92		431	92	0	No ID				EG. Obscured
93	20	47	137	−41	QSS	2	18.10	0.425	
94		*139.1*	173	−1	IC 410				W8. Thermal.
95		249	256	+51	EF	1	$\gtrsim$22		200-inch *V*-plate empty. Possible obscuration.
96		280	120	+70	Gxy pair	2	$\approx$21.5		
97		382	61	+17	Gxy	2	14.66	0.058	
98		442	75	−34	A, gxy in group; B, no ID	2	13.66	0.026	2 sources; A = NGC 7236/7237
99	19.5	109	182	−28	N gxy	2	17.76	0.306	
100		133	178	−10	No ID				EG. Obscured
101		*153.1*	189	+1	H II				NGC 2174/2175
102		192	198	+26	Gxy cluster		15.46	0.060	Triple source. m_v is for brightest gxy in cluster
103		268.1	128	+44	No ID				EG. Complex geometry.
104	19.0	154	186	+4	No ID				EG.
105		244.1	151	+51	No ID				
106		254	173	+66	QSS	1	17.98	0.734	
107		390	41	+6	No ID				EG. Obscured.
108		399.2	46	−1	No ID				Obscured
109		401	93	+18	Gxy	2	~18		Cluster?
110		428	91	+1	No ID				EG. Obscured.
111	18.5	89	186	−43	Gxy	3	~15.5		In small group
112		138	187	−11	QSS	1	17.90	0.759	
113		216	178	+43	QSS?	1	18.48		No z, but object is stellar, blue
114	18.0	142.1	198	−15	No ID				EG
115		265	192	+75	Gxy group	3	~20		
116		272.1	278	+74	Gxy	1	9.36	0.003	M84 = NGC 4374. In Virgo cluster.
117	17.5	153	165	+13	Gxy	1	~18	0.277	Cluster?
118		263.1	227	+74	Gxy	1	~20		Requires confirmation.
119		315	39	+58	Gxy	3	16.80	0.108	Complex source. m_v of brightest component of dumbbell. Cluster?
120	17.0	68.1	146	−24	No ID				
121		309.1	111	+41	QSS	1	16.78	0.903	
122		319	88	+51	Gxy	1	~18.5		Brightest in group
123		337	69	+44	Gxy group	2	~21.5		Local absorption patch ?

TABLE 1—*Continued*

Serial Number	Intensity (f.u.)	3CR Number	l^{II}	b^{II}	Identification	Quality	m_v	z	Remarks
124.....	16.5	198	218	+23	Gxy	3	~17	0.081	May be in small group
125.....	....	255	263	+53	EF	1	≳22		200-inch *V*-plate empty
126.....	....	400.1	69	+ 9	No ID				
127.....	16.0	6.1	121	+17	No ID				EG
128.....	....	88	181	−42	Gxy	2	13.95	0.030	
129.....	....	135	200	−21	A, gxy; B, no ID	2	17.00	0.127	2 sources. A in small group? Radio trail.
130.....	....	175	205	+10	QSS	2	16.60	0.768	
131.....	....	208	214	+33	QSS	3	17.42	1.110	
132.....	....	228	221	+46	EF	1	≳20.5		48-inch IIIaJ empty. Earlier ID probably a plate flaw
133.....	....	275.1	293	+79	QSS	2	19.00	0.557	
134.....	....	300	18	+68	Gxy	3	~18		
135.....	15.5	31	127	−30	Gxy	2	12.14	0.017	NGC 383. Brightest in small group.
136.....	....	212	214	+35	N gxy	1	~19		
137.....	....	247	171	+62	No ID				EG
138.....	15.0	2	99	−61	QSS	1	19.35	1.037	
139.....	....	9	112	−47	QSS	1	18.21	2.012	
140.....	....	29	127	−64	Gxy	3	14.07	0.045	
141.....	....	63	167	−57	Gxy		~18.5		
142.....	....	78	175	−45	Gxy	1	12.84	0.029	NGC 1218
143.....	....	105	188	−34	EF	1	≳21		200-inch yellow image-tube plate empty
144.....	....	119	161	− 4	QSS?	1	≳20		
145.....	....	274.1	270	+83	Gxy	3	~20		
146.....	....	436	80	−19	Gxy	3	18.18	0.215	
147.....	....	437.1	70	−30	No ID				4 unresolved sources
148.....	....	454.3	86	−39	QSS	1	16.10	0.860	
149.....	14.5	15	115	−64	Gxy	2	15.34	0.073	
150.....	....	22	123	−12	Nature uncertain	2	~19		
151.....	....	*147.1*	206	−16	H II				IC 434.
152.....	....	196.1	226	+17	Gxy	1	~17.5		
153.....	....	220.3	129	+31	EF	3	≳20.5		*Sky Survey* empty
154.....	....	223	188	+49	Gxy group	3	17.06	0.137	m_v is for brightest gxy
155.....	....	288	86	+75	Gxy pair	1	~16.5		Cluster?
156.....	....	356	78	+34	No ID				EG. Two galactic stars near. Complex geometry.
157.....	14.0	14.1	121	− 3	No ID				EG. Obscured.
158.....	....	28	124	−37	Gxy	1	17.54	0.196	In cluster A115
159.....	....	158	197	0	No ID				EG. Obscured.
160.....	....	172	191	+13	Nature uncertain	3			Possible faint filaments. Low latitude but galaxies seen.
161.....	....	180	219	+ 7	Gxy	2	~19		
162.....	....	287	23	+81	QSS	1	17.67	1.054	
163.....	....	368	38	+15	No ID				EG
164.....	....	411	83	−15	No ID				EG
165.....	13.5	71	172	−52	Gxy	1	8.91	0.004	M77=NGC 1068. Brightest Seyfert gxy.

TABLE 1—*Continued*

Serial Number	Intensity (f.u.)	3CR Number	l^{II}	b^{II}	Identification	Quality	m_v	z	Remarks
166		131	172	−8	No ID				EG. Obscured.
167		186	182	+26	QSS	1	17.60	1.063	
168		226	225	+43	Gxy	3	~19.5		
169		267	255	+70	No ID				EG
170		305	103	+49	Gxy	1	13.74	0.041	
171		336	41	+42	QSS	2	17.47	0.927	
172		449	95	−16	Gxy	3	13.20	0.017	Brightest in small group
173	13.0	136.1	180	−8	Gxy	3	~17		
174		141	175	−1	No ID				EG? Obscured.
175		166	193	+8	Gxy	3	~19.5		
176		181	204	+15	QSS	1	18.92	1.382	
177		200	194	+33	Gxy	1	~20		
178		231	141	+40	Gxy	1	8.39	0.001	M82 = NGC 3034
179		238	234	+47	EF	1	≳21.5		48-inch III aJ empty
180		263	227	+74	QSS	2	16.32	0.652	
181		402	83	+13	No ID				Complex source
182		441	85	−21	Nature uncertain	3	~19		
183		455	85	−41	QSS	1	~19.5	0.543	Previous misidentification with galaxy
184		456	86	−46	Gxy	1	18.54	0.233	
185	12.5	35	126	−13	Gxy	3	~14.5	0.068	Complex source
186		91	148	−4	No ID				EG
187		130	156	+5	No ID				
188		132	179	−13	Gxy	1	~18.5		Also stellar object 5″ away
189		205	160	+37	QSS	2	17.62	1.534	
190		208.1	213	+34	Nature unknown	1	~20		EG
191		277	123	+67	No ID				EG
192		296	354	+62	Gxy	1	12.21	0.024	NGC 5532
193		303	91	+58	N gxy	3	17.1	0.141	Two fainter candidates. N gxy is brightest in group.
194		306.1	352	+47	No ID				
195		321	37	+54	Nature uncertain.	2			EG near 200″ red limit. Star-like, but may be red or variable.
196		381	76	+23	Gxy	2	17.24	0.161	Brightest in group
197		399.1	63	+9	No ID				EG?
198		403.1	39	−14	Gxy	3	~16	0.056	In group
199		418	89	+6	Nature uncertain. QSS?	1	~21.7		Star-like in good seeing. Reddish, but low latitude.
200		424	54	−22	Gxy	1	~18		In group
201	12.0	33.1	124	+10	No ID				Extended multiple source
202		68.2	147	−26	EF?	2			15-mag galactic star is only possible ID on 48-inch plates
203		173.1	140	+27	No ID				
204		184.1	134	+29	Gxy	2	~17	0.119	In group
205		217	185	+43	EF	1	≳20.5		48-inch plates empty
206		220.1	132	+33	No ID				

TABLE 1—*Continued*

Serial Number	Intensity (f.u.)	3CR Number	l^{II}	b^{II}	Identification	Quality	m_v	z	Remarks
207	....	239	171	+53	No ID				EG
208	....	252	185	+67	No ID				EG
209	....	270.1	167	+81	QSS	2	18.61	1.519	
210	....	277.1	123	+60	QSS	1	17.93	0.320	
211	....	293	54	+76	Gxy	1	14.32	0.045	
212	....	349	73	+38	Ambiguous	2			Gxy and stellar object both near radio position
213	....	394	45	+ 4	No ID				EG? Obscured.
214	....	435.1	118	+24	No ID				3 sources
215	11.5	11.1	121	+ 1	No ID				EG? Obscured.
216	....	16	118	−50	No ID				
217	....	42	133	−33	Nature uncertain	2	~20		EG
218	....	43	134	−39	QSS?	3	~20		No z; but variable
219	....	49	141	−47	Nature uncertain	1	~21.3		
220	....	175.1	202	+12	EF?	2			Only object on 48-inch plate is galactic star
221	....	249.1	256	+51	QSS	2	15.71	0.311	
222	....	277.3	72	+89	Gxy	1	15.94	0.086	Coma A
223	....	287.1	326	+63	N gxy	3	18.27	0.216	
224	....	303.1	115	+38	Gxy	2	~18		In group
225	....	305.1	115	+39	Gxy?	1	~21		Near limit of 200″ red plate
226	....	320	57	+55	Gxy		~18		In group
227	....	324	35	+49	Gxy	1	~21		In faint cluster
228	....	432	68	−23	QSS	1	17.96	1.804	
229	....	469.1	120	+18	No ID				
230	11.0	34	128	−31	No ID				Several objects near
231	....	44	141	−35	No ID				EG.
232	....	107	193	−35	EF?	2	$\gtrsim$21.3		Poor 200-inch V plate empty except for faint smudge
233	....	139.2	178	− 4	No ID				EG. Obscured.
234	....	152	190	− 1	EF	1	$\gtrsim$22		48-inch IIIaJ, 200-inch V image-tube plates empty. EG.
235	....	165	191	+ 9	No ID				
236	....	184	145	+30	EF	1	$\gtrsim$20.5		48-inch plates empty
237	....	187	218	+13	Gxy	3	$\approx$19.5		
238	....	190	208	+22	QSS?	1	~21		$m_B=21.25$
239	....	213.1	196	+40	Gxy	1	~19		In group
240	....	250	213	+67	EF?				EG. Earlier suggested ID's are too far away.
241	....	280.1	115	+77	QSS	2	19.44	1.659	
242	....	299	77	+67	Gxy	1	~20	0.367	Brightest member of faint cluster
243	....	318.1	11	+49	No ID				EG
244	....	340	41	+41	Nature uncertain	3	~19.5		
245	....	346	35	+36	Gxy	1	~16		
246	....	351	90	+36	QSS	2	15.28	0.371	
247	....	352	72	+36	Nature uncertain	1	$\gtrsim$22		Very faint, diffuse.
248	10.5	13	121	+ 1	No ID				EG

TABLE 1—*Continued*

Serial Number	Intensity (f.u.)	3CR Number	l^{II}	b^{II}	Identification	Quality	m_v	z	Remarks
249	...	129.1	161	0	No ID				
250	...	191	212	+21	QSS	1	18.4	1.946	
251	...	210	198	+39	EF	1	≳21		Poor 200-inch *V* plate empty. Earlier ID is *Sky Survey* flaw.
252	...	284	38	+86	Gxy	3	~18.5	0.239	In group
253	...	285	103	+73	Gxy	2	16.00	0.080	Brightest in group
254	...	300.1	346	+53	No ID				Earlier misidentified with gxy.
255	...	325	96	+44	No ID				
256	...	332	53	+45	Gxy	2	~16	0.152	In group
257	...	341	47	+42	Gxy	2	~19.5		
258	...	434	67	−24	Gxy	3	20.84		
259	...	437	71	−28	EF	1	≳21.5		48-inch IIIaJ plate empty
260	...	454.2	111	+ 5	No ID				EG. Obscured.
261	10.0	41	131	−29	No ID				EG
262	...	52	131	− 8	Gxy Cluster	3	~18.5		
263	...	54	135	−18	No ID				Possible extended object at plate limit.
264	...	67	147	−31	Gxy	1	~18		
265	...	93.1	160	−16	Gxy	1	~19		In group
266	...	99	190	−37	Gxy	1	19.1		Possible cluster
267	...	114	177	−22	No ID	3			Confused radio and optical geometry. Previous ID has wrong position.
268	...	173	179	+18	Nature uncertain	1	~18		
269	...	207	214	+30	QSS	1	18.15	0.683	
270	...	215	212	+37	QSS	1	18.27	0.411	
271	...	241	213	+56	EF	1	≳20		48-inch plates empty
272	...	318	30	+55	Gxy	1	≈19.5		Faint group
273	...	326	33	+48	No ID				
274	...	334	33	+41	QSS	3	16.41	0.555	
275	...	345	63	+41	QSS	1 •	15.96	0.594	
276	...	415.2	90	+ 8	No ID				EG
277	...	458	83	−50	Gxy	3	~20		Some doubt of ID because of position.
278	9.5	14	118	−44	Nature uncertain	1	~20		
279	...	36	128	−17	Nature uncertain. QSS?	1	~21.7		Compact, relatively blue
280	...	46	132	−24	No ID				Extended double +structure. Several galaxies near (cluster?).
281	...	76.1	163	−36	Ambiguous				15-mag gxy and 14-mag star
282	...	93	182	−38	QSS?	1	18.09		No z, but has UV excess
283	...	124	195	−28	EF	1	≳21.3		Poor 200-inch red and *V* plates empty
284	...	125	164	− 4	No ID				Nonthermal. Obscured.
285	...	169.1	171	+19	No ID				EG
286	...	177	203	+14	No ID				Two sources (4C 16.21, 4C 15.19)
287	...	197.1	173	+35	Gxy	1	~16.5		In group
288	...	204	150	+36	QSS	2	18.21	1.112	

TABLE 1—*Continued*

Serial Number	Intensity (f.u.)	3CR Number	l^{II}	b^{II}	Identification	Quality	m_v	z	Remarks
289	....	220.2	188	+47	EF?	1			Galactic star at position
290	....	223.1	183	+49	Gxy	3	(16.4)	0.108	In group?
291	....	245	233	+56	QSS	1	17.25	1.029	
292	....	256	218	+69	Gxy	1	~21		In group
293	....	268.3	131	+52	Gxy	1	~21		In cluster
294	....	272	141	+74	Nature uncertain. QSS?	1	~21.5		May be blue or variable.
295	....	275	299	+58	Gxy	1	~21		In cluster
296	....	277.2	306	+78	No ID				EG
297	....	288.1	112	+56	QSS	1	18.12	0.961	
298	....	289	102	+65	EF	1	≳22		200-inch red image-tube plate empty. Earlier ID has wrong position.
299	....	293.1	0	+72	Gxy	1	~19		
300	....	297	340	+52	EF	1	≳20.5		EG
301	....	322	89	+49	No ID				
302	....	323	94	+46	Nature uncertain	2			
303	....	343	94	+39	QSS?	1	20.61		No z, but QSS colors
304	....	371	100	+29	N gxy	1	14.81	0.051	In group
305	....	379.1	105	+28	Gxy	2	~18		
306	....	435	61	−30	Nature uncertain	2	~19.5		
307	....	454.1	114	+11	No ID				Nonthermal
308	....	460	98	−35	Gxy	2	~19		Near edge of cluster. Early searches confused by *Sky Survey* flaw.
309	9.0	19	120	−30	Gxy	1	~20.5		May be in cluster
310	....	21.1	122	+ 5	No ID				Nonthermal. Obscured.
311	....	33.2	125	+ 7	No ID				Nonthermal. Obscured.
312	....	137	189	+ 3	No ID				EG
313	....	194	178	+32	Probable gxy	2	~19.5		Red, fuzzy but compact.
314	....	222	230	+38	EF	1	≳21.5		48-inch IIIaJ plate empty
315	....	236	190	+54	Gxy	1	15.97	0.099	
316	....	257	255	+60	No ID				
317	....	258	230	+69	Gxy	2	~19.5		In cluster
318	....	266	148	+64	No ID				
319	....	268.2	188	+78	Gxy	3	~19		
320	....	268.4	147	+71	QSS	1	18.42	1.400	Probably in cluster
321	....	294	62	+72	No ID				Image of bright galactic star overlaps position
322	....	314.1	109	+42	Gxy Cluster	3	~17		
323	....	323.1	34	+49	QSS	3	16.69	0.264	
324	....	326.1	34	+47	EF		≳22		EG. 200-inch red image-tube plate empty
325	....	343.1	93	+39	Gxy	1	20.75		No z, but has gxy colors
326	....	357	56	+31	Gxy	2	~15.5	0.167	
327	....	454	87	−36	QSS	1	18.40	1.756	
328	....	470	113	−18	Gxy	2	~19.5		

TABLE 2

SERIAL NUMBERS OF TABLE 1 ARRANGED IN ORDER OF 3CR NUMBER

3CR Number	Serial Number	3CR Number	Serial Number	3CR Number	Serial Number	3CR Number	Serial Number
2	138	98	42	200	177	275	295
6.1	127	99	266	204	288	275.1	133
9	139	103	61	205	189	277	91
10	14	105	143	207	269	277.1	210
11.1	215	107	232	208	131	277.2	296
13	248	109	99	208.1	190	277.3	222
14	278	111	26	210	251	280	96
14.1	157	114	267	212	136	280.1	241
15	149	119	144	213.1	239	284	252
16	216	123	12	215	270	285	253
17	85	124	283	216	113	286	90
18	74	125	284	217	205	287	162
19	309	129	88	219	36	287.1	223
20	41	129.1	249	220.1	206	288	155
21.1	310	130	187	220.2	289	288.1	297
22	150	131	166	220.3	153	289	298
27	63	132	188	222	314	293	211
28	158	133	100	223	154	293.1	299
29	140	134	21	223.1	290	294	321
31	135	135	129	225	62	295	18
33	33	136.1	173	226	168	296	192
33.1	201	137	312	227	56	297	300
33.2	311	138	112	228	132	298	38
34	230	139.1	94	230	89	299	242
35	185	139.2	233	231	178	300	134
36	279	141	174	234	53	300.1	254
40	64	142.1	114	236	315	303	193
41	261	144	3	237	67	303.1	224
42	217	147	25	238	179	305	170
43	218	147.1	151	239	207	305.1	225
44	231	152	234	241	271	306.1	194
46	280	153	117	244.1	105	309.1	121
47	93	153.1	101	245	291	310	30
48	35	154	104	247	137	313	57
49	219	157	10	249	95	314.1	322
52	262	158	159	249.1	221	315	119
54	263	165	235	250	240	317	40
55	86	166	175	252	208	318	272
58	65	169.1	285	254	106	318.1	243
61.1	52	171	77	255	125	319	122
63	141	172	160	256	292	320	226
65	59	173	268	257	315	321	195
66	49	173.1	203	258	317	322	301
67	264	175	130	263	180	323	302
68.1	120	175.1	220	263.1	118	323.1	323
68.2	202	177	286	264	69	324	227
69	75	180	161	265	115	325	255
71	165	181	176	266	318	326	273
75	76	184	236	267	169	326.1	324
76.1	281	184.1	204	268.1	103	327	45
78	142	186	167	268.2	319	327.1	70
79	66	187	237	268.3	293	330	71
83.1	55	190	238	268.4	320	332	256
84	24	191	250	270	37	334	274
86	87	192	102	270.1	209	336	171
88	128	194	313	272	294	337	123
89	111	196	23	272.1	116	338	44
91	186	196.1	152	273	20	340	244
93	282	197.1	287	274	4	341	257
93.1	265	198	124	274.1	145	343	303

TABLE 2—*Continued*

3CR Number	Serial Number	3CR Number	Serial Number	3CR Number	Serial Number	3CR Number	Serial Number
343.1......	325	390.1......	31	409.1......	9	437........	259
345........	275	390.2......	15	410........	50	437.1......	147
346........	245	390.3......	39	410.1......	6	438........	46
348........	8	391........	68	411........	164	441........	182
349........	212	392........	7	415.1......	73	442........	98
351........	246	394........	213	415.2......	276	445........	80
352........	247	396........	82	416.1......	22	449........	172
353........	11	396.1......	51	416.2......	13	452........	34
356........	156	397........	58	418........	199	454........	327
357........	326	398........	28	419.1......	17	454.1......	307
363.1......	16	399.1......	197	424........	200	454.2......	260
368........	163	399.2......	108	427.1......	78	454.3......	148
371........	304	400........	5	428........	110	455........	183
372.1......	43	400.1......	126	430........	54	456........	184
379.1......	305	400.2......	83	431........	92	458........	277
380........	27	401........	109	432........	228	459........	84
381........	196	402........	181	433........	29	460........	308
382........	97	403........	60	434........	258	461........	1
386........	72	403.1......	198	434.1......	79	465........	47
388........	81	403.2......	32	435........	306	468.1......	48
389........	91	405........	2	435.1......	214	469.1......	229
390........	107	409........	19	436........	146	470........	328

For extragalactic sources, column 7 lists the quality of the identification given in column 6 based on agreement of the optical and radio positions (for extended and double radio sources, the center is taken as the radio position for this comparison). These are: (1) The combined optical and radio position uncertainties are less than 4″ and the positions agree to this accuracy. (2) Combined uncertainties and position agreement better than 10″. (3) Combined uncertainties and agreement about 10″–20″, or agreement to better than 10″ but marginally larger than combined uncertainty, or complex radio source, or crowded optical field.

In the absence of independent criteria, particularly for galaxies, agreement of radio and optical positions is the best measure of the reality of an identification. Virtually all of the identifications of quality 1 and 2 are expected to be valid, and a small but significant fraction of those of quality 3 are expected to be coincidental.

Column 8 lists the visual magnitudes of the identified sources where measured. Where no measurements are available, estimates from plates are supplied, with the symbol ~. Most of these are by Wyndham (1966) from *Sky Survey* plates, and may be uncertain by a magnitude or more. For empty fields, approximate lower limits from the estimated limiting magnitudes of the best available plates are given. Column 9 lists redshifts where available. Most of the magnitudes and redshifts given are from the extensive recent compilations by Sandage (*Ap. J.*, **178**, p. 1 and p. 20, 1972).

Column 10 gives remarks on individual sources. EG in this column is used for unidentified sources which are judged to be extragalactic on the basis of size and nonthermal spectrum.

A summary of the nature of the identifications is given in table 3, for all sources, and separately for sources at low ($|b| < 15°$) and high galactic latitudes. The most striking feature of table 3 is the high proportion of all 3CR sources which have no secure identifications (35% of all sources, 29% of high-latitude sources). For $|b| \geq 15°$, there are few galactic sources and few obscured fields. Identifications and optical data, however, are nearly complete only for the 24 high latitude sources brighter than 28 f.u.

TABLE 3

NATURE OF 3CR IDENTIFICATIONS IN TABLE 1

Object	$\|b^{II}\|<15°$	$\|b^{II}\|>15°$	All b^{II}	% of Total
All sources	95	233	328	
Galactic sources	23	2	25	7.6
Galaxy, z measured	6	58	64	19.5
Galaxy, z not measured	11	46	57	17.4
Quasar, z measured	3	37	40	12.2
Quasar suspect, no z	1	6	7	2.1
Nature uncertain	3	17	20	6.1
Empty fields	2	20	22	6.7
No ID	46	47	93	28.4

POSITION REFERENCES

The references are not intended to be exhaustive, but are mainly limited to the most recent and most extensive lists of accurate measurements. Additional references can be found in those given and in the references in the body of the chapter.

RADIO POSITIONS AND MAPS

Adgie, R. L., and Gent, H. 1966, *Nature*, **209**, 549.
Adgie, R. L., Crowther, J. H., and Gent, H. 1972, *M.N.R.A.S.*, **159**, 233.
Elsmore, B., and Mackay, C. D. 1969, *M.N.R.A.S.*, **146**, 361.
Fomalont, E. B. 1971, *A.J.*, **76**, 513.
Fomalont, E. B., and Moffet, A. T. 1971, *A.J.*, **76**, 5.
MacDonald, G. H., Kenderdine, S., and Neville, A. C. 1968, *M.N.R.A.S.*, **138**, 259.
Mackay, C. D. 1969, *M.N.R.A.S.*, **145**, 31.
Parker, E. A., Elsmore, B., and Shakeshaft, J. R. 1966, *Nature*, **210**, 22.
Wade, C. M. 1970, *Ap. J.*, **162**, 381.
Wade, C. M. and Miley, G. K. 1971, *A.J.*, **76**, 101.

OPTICAL POSITIONS

Barbieri, C., Capaccioli, M., Granz, R., and Pinto, G. 1972, *A.J.*, **77**, 444.
Griffin, R. 1963, *A.J.*, **68**, 421.
Hunstead, R. W. 1971, *M.N.R.A.S.*, **152**, 277.
Kristian, J. and Sandage, A. R. 1970, *Ap. J.*, **162**, 391.
Kristian, J., Sandage, A. R., and Katem, B. N. 1974, *Ap. J.*, **191**, 43.
Murray, C. A., Tucker, R. H., and Clements, E. D. 1969, *Nature*, **221**, 1229.
Véron, P. 1966, *Ap. J.*, **144**, 861.
Wyndham, J. D. 1966, *Ap. J.*, **144**, 459.

CHAPTER 7

Strong Nonthermal Radio Emission from Galaxies

ALAN T. MOFFET
Owens Valley Radio Observatory, California Institute of Technology

1. INTRODUCTION

The first 20 years of radio astronomy produced many startling results, of which the most unexpected and exciting was the discovery of radio galaxies. They were found in the years immediately following World War II, first by the tens and soon by the hundreds. As often happens, the true nature and the great importance of these objects were not immediately apparent. They were dubbed "radio stars," although it was very soon clear that no radio source agrees in position with any of the brightest stars and that the radio emission from the Sun is too weak to be detected over interstellar distances.

The early work on "radio stars" has been described by Minkowski in the preceding chapter. We recall that of the first half-dozen discrete sources to be detected, three were immediately identified (Bolton, Stanley, and Slee 1949) with well-known optical objects: Taurus A with a galactic supernova remnant, the Crab Nebula; and Centaurus A and Virgo A with the peculiar galaxies NGC 5128 and M87, respectively. There, for several years, the process stopped, although the number of known "radio stars" continued to increase. It was not at all clear just what these objects might be. Furthermore, in all cases, even in the few discrete radio sources with known optical counterparts, the mechanism for generating the radio emission remained a mystery.

Two events which occurred in 1952 and 1953 brought to an end the radio-star era. These were (1) the identification by Baade and Minkowski (1954) of the very intense source Cygnus A with a galaxy having strong emission lines red shifted by $z = \Delta\lambda/\lambda_0 = 0.06$, and (2) the demonstration by the Russian school of theoreticians that the nonthermal radio emission from discrete sources and from the Galaxy is almost certainly synchrotron radiation (Ginzburg 1951; Shklovskii 1952, 1953).

From the properties of Cygnus A the reason for the great difficulty in identifying other radio sources with their optical counterparts was easily understood. If Cygnus A were at all typical, the faintest sources then known—about 400 times less intense than

Received December 1972.

Cygnus A—would be 20 times more distant; i.e., they would have redshifts of the order of 1.0. The most distant object then known had a redshift of 0.28. The search for optical counterparts of radio galaxies promised to be difficult but exciting.

Cygnus A has proved to be an ideal prototype radio galaxy. In one of the first interferometric studies capable of determining the structure of radio sources it was found to consist of a pair of bright regions separated by several times their own diameter and symmetrically disposed on either side of the associated galaxy (Jennison and Das Gupta 1953). We now know that this type of structure is found in the majority of radio galaxies. Plate 1 shows the contours of radio brightness for Cygnus A measured with

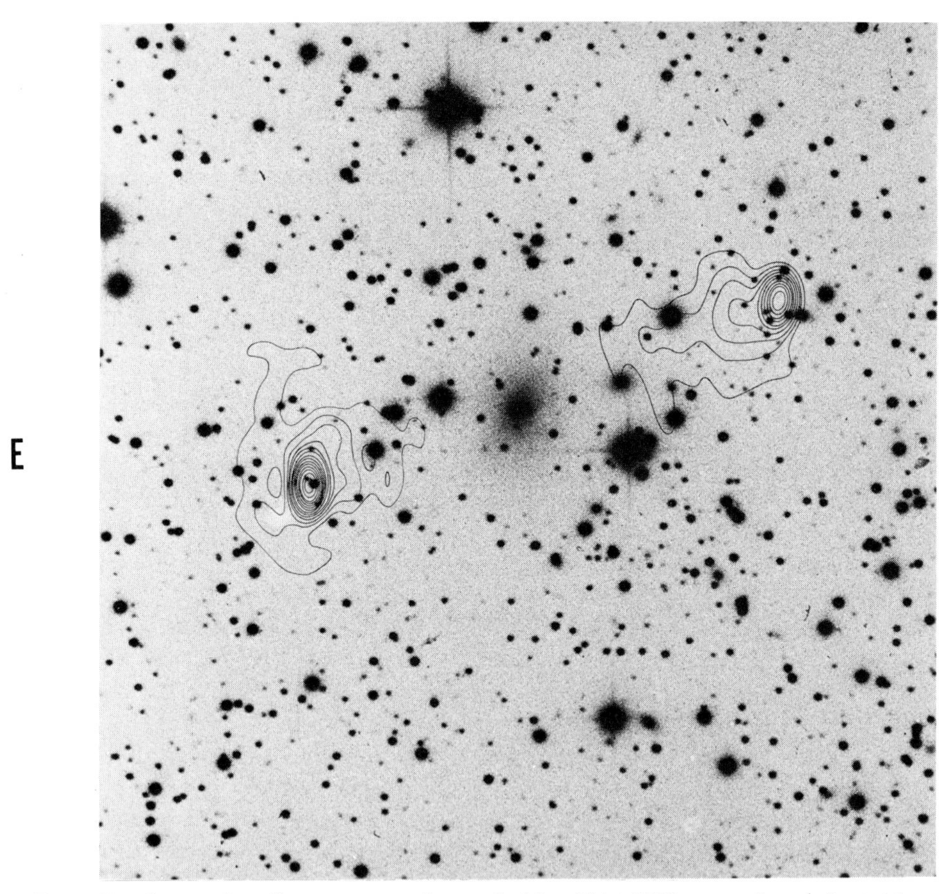

PLATE I.—Cygnus A, radio contours as observed with a 6″ × 10″ beam at 6 cm (adapted from Mitton and Ryle 1969) superposed on a photograph taken by Baade with the Hale reflector (Baade and Minkowski 1954). The field is 4′ × 6′.

the Cambridge three-element interferometer (Mitton and Ryle 1969) superposed on the optical field as photographed with the Hale 200-inch reflector. The optical counterpart of the radio source is clearly seen near the centroid of the radio emission.

The identification of radio galaxies with their optical counterparts has been a prerequisite to the understanding of these objects. This is because there is no way to obtain the distance to a radio galaxy from radio measurements alone. Without a measure of distance it is impossible to convert the apparent properties of the radio source, such as flux density and angular extent, into its intrinsic properties—luminosity and physical size. However, once these intrinsic properties and the mechanism of emission are known, it is possible to estimate the physical conditions within the source—magnetic field strength, particle density and energy distribution, and total energy requirements.

The process of making optical identifications has been discussed by Minkowski in the previous chapter. This chapter will describe the observed properties of radio galaxies (§ 3) and the inferred intrinsic properties and physical conditions (§ 4). Current notions about the evolution of radio galaxies will then be summarized (§ 5). It is our current belief that all the observed radio emission from these objects is produced by the synchrotron mechanism. Thus an understanding of the theory of synchrotron radiation is of fundamental importance, and the next portion of this chapter (§ 2) will be devoted to that subject.

2. SYNCHROTRON RADIATION

Synchrotron radiation is emitted whenever an electron having relativistic velocity encounters a magnetic field with a component normal to its path. The phenomenon was first described by Schott (1912), and modern treatments have been given by Westfold (1959) and Le Roux (1961). In the Soviet literature it is often called magneto-bremsstrahlung or acceleration radiation. The name commonly used in the western literature comes from the fact that electrons accelerated and confined in the strong magnetic field of a synchrotron radiate strongly via this mechanism.

A series of reviews of synchrotron radiation has been given by Ginzburg and Syrovatskii (1964*a*, 1965, 1969) and Ginzburg, Sazonov, and Syrovatskii (1968), and the reader may wish to refer to them or to the text by Pacholczyk (1970) for a more complete treatment of the subject.

2.1. Emission from a Single Particle

An electron having a velocity $\boldsymbol{v}$ and moving in a magnetic field of induction $\boldsymbol{B}$ will experience an acceleration

$$\dot{\boldsymbol{v}} = \frac{e}{mc}\sqrt{(1 - v^2/c^2)}\,\boldsymbol{v} \times \boldsymbol{B} = \frac{ec}{E}\boldsymbol{v} \times \boldsymbol{B}\,, \tag{2.1}$$

where E is the total energy of the particle and m its rest mass. Since $\dot{\boldsymbol{v}}$ is normal to $\boldsymbol{v}$, the magnitude of the velocity does not change. The path of the electron is a helix with its axis parallel to $\boldsymbol{B}$. If the electron were nonrelativistic ($v \ll c$), it would radiate like a dipole rotating at the electron cyclotron frequency

$$\nu_m = \frac{eB\sin\theta}{2\pi mc} = \frac{eB_\perp}{2\pi mc}\,, \tag{2.2}$$

where $B_\perp = B \sin \theta$ is the component of the magnetic field perpendicular to the electron's path. The angle θ is the angle between $\boldsymbol{v}$ and $\boldsymbol{B}$ (see fig. 1). It is useful to remember that $\nu_m \approx 2.80$ MHz per gauss.

The gyration frequency for a relativistic electron will be reduced because of the increased effective mass of the particle:

$$\nu_g = \frac{eB_\perp}{2\pi mc}\sqrt{(1 - v^2/c^2)} = \nu_m\sqrt{(1 - v^2/c^2)} \,. \qquad (2.3)$$

If the electron is highly relativistic, the transformation from its rest system to the system of the observer will compress the forward lobe of the dipole radiation pattern into a small cone with its axis along $\boldsymbol{v}$ and with a half-angle ξ of the order

$$\xi = \sqrt{(1 - v^2/c^2)} = mc^2/E \,. \qquad (2.4)$$

The velocity vector $\boldsymbol{v}$ sweeps out a cone of half-angle θ around $\boldsymbol{B}$. An observer situated within an angle ξ of that cone will detect a pulse of synchrotron radiation as the vector $\boldsymbol{v}$ sweeps past him. The energy radiated at angles much greater than ξ away from the cone swept out by $\boldsymbol{v}$ is quite negligible. Thus we will express the direction toward the observer as a small angle ψ by which that direction differs from θ, as is shown in figure 1.

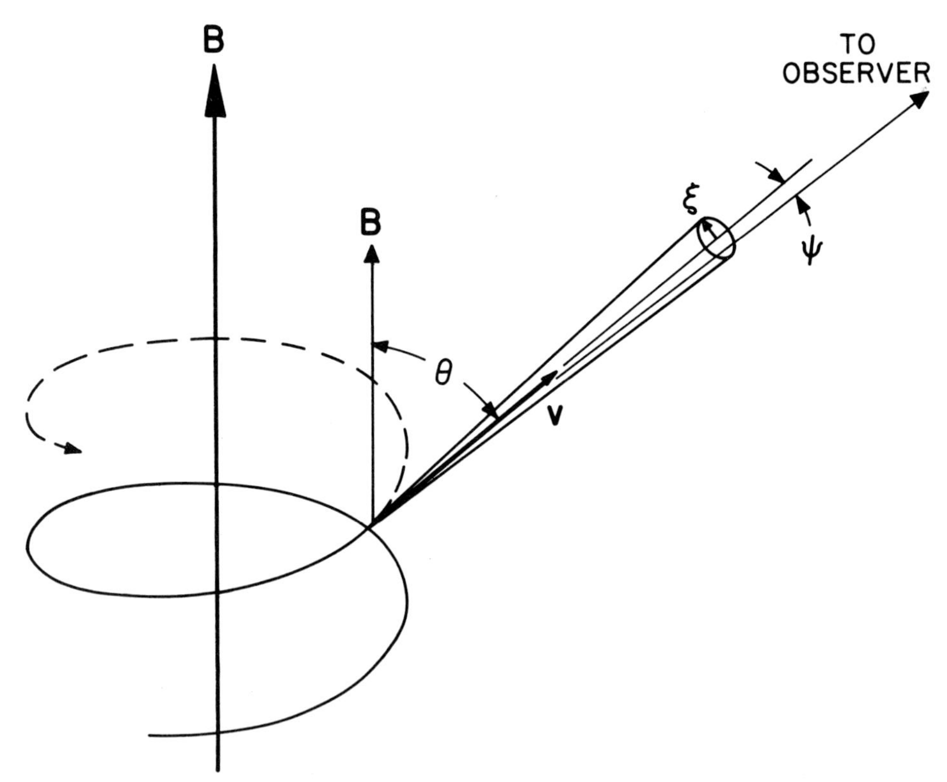

FIG. 1.—Pencil of synchrotron radiation from an electron with velocity $\boldsymbol{v}$ making an angle θ with the magnetic field $\boldsymbol{B}$. The observer's direction is given by ψ. Almost all the radiation is emitted in a cone about $\boldsymbol{v}$ with half-angle ξ.

For a very distant observer the motion of the electron in a direction parallel to $\boldsymbol{B}$ will not affect the angle between $\boldsymbol{B}$ and the line of sight toward the electron. The observer remains on the cone swept out by $\boldsymbol{v}$, and he detects a succession of synchrotron pulses repeating at the Doppler-shifted gyration frequency

$$\nu_g' = \nu_g/(1 - v \cos^2 \theta/c) \approx \nu_g/\sin^2 \theta \,. \tag{2.5}$$

Here $v \cos \theta$ is the electron's velocity of translation parallel to $\boldsymbol{B}$ and $v \cos^2 \theta$ is the component of that velocity in the direction of the observer. (It is assumed that ψ, the difference between θ and the actual angle the line of sight makes with $\boldsymbol{B}$, is negligibly small). Since the electron is highly relativistic, we can take $v/c \approx 1$ and make the approximation that $(1 - v \cos^2 \theta/c) \approx \sin^2 \theta$.

It is interesting to note that this Doppler shift of the gyration frequency and the consequences which ensue from it were neglected in all treatments of synchrotron radiation previous to 1967 (Epstein and Feldman 1967; Ginzburg, Sazonov, and Syrovatskii 1968). Serious errors were not introduced by this omission, since the effect of the Doppler shift cancels out in the computation of the emission from an ensemble of electrons confined within a stationary source region (i.e., a source which is not expanding relativistically).

The width of each synchrotron pulse seen by the observer is approximately equal to the time required for the electron to move through an angle 2ξ about $\boldsymbol{B}$, as seen by the observer at the moment when the electron is moving toward him. To the observer the electron's angular velocity about $\boldsymbol{B}$ does not appear to be constant. On the average it has the value $\dot{v}/v = 2\pi\nu_g$, but as the electron approaches the observer this rate is "blueshifted" by a factor $1/(1 - v/c) \approx (1 + v/c)/(1 - v^2/c^2) \approx 2/(1 - v^2/c^2)$. Thus the pulse width appears to be

$$\Delta t \approx \frac{2\xi}{(\dot{v}/v)} \tfrac{1}{2}(1 - v^2/c^2) = \frac{\xi(1 - v^2/c^2)}{2\pi\nu_g} = \frac{(1 - v^2/c^2)^{3/2}}{2\pi\nu_g} \,. \tag{2.6}$$

Because the pulses recur with frequency ν_g', the frequency spectrum of the radiation is a series of spikes at all harmonics of ν_g'. Most of the energy will be radiated in those harmonics which yield frequencies of the order of $(2\pi\Delta t)^{-1}$, i.e., at frequencies

$$\sim \frac{1}{2\pi\Delta t} \approx \frac{\nu_g}{(1 - v^2/c^2)^{3/2}} = \nu_g' \sin^2 \theta \left(\frac{E}{mc^2}\right)^3 \approx \frac{\nu_m}{(1 - v^2/c^2)} = \nu_m \left(\frac{E}{mc^2}\right)^2 . \tag{2.7}$$

The order of these harmonics is proportional to E^3, but since $\nu_g' \propto E^{-1}$, the actual value of the frequency of maximum emission is proportional to E^2. For $E \gg mc^2$ the harmonics are so closely spaced that the spectrum is essentially a continuum. The frequency near which the emission is a maximum is called the *critical frequency* for synchrotron emission and is conventionally defined as 3/2 of the above expressions, or

$$\nu_c = \frac{3e}{4\pi mc} B_\perp \left(\frac{E}{mc^2}\right)^2 = C_1 B_\perp E^2 \,. \tag{2.8}$$

The constant C_1 equals 6.266×10^{18} in cgs (Gaussian) units[1] or 16.08 for "practical" units in which ν_c is expressed in MHz, $B_\perp$ in microgauss, and E in GeV. Thus a 3-GeV

[1] Unless otherwise indicated, Gaussian units are used. In rationalized mks units the value of C_1 would be 6.266×10^{36}.

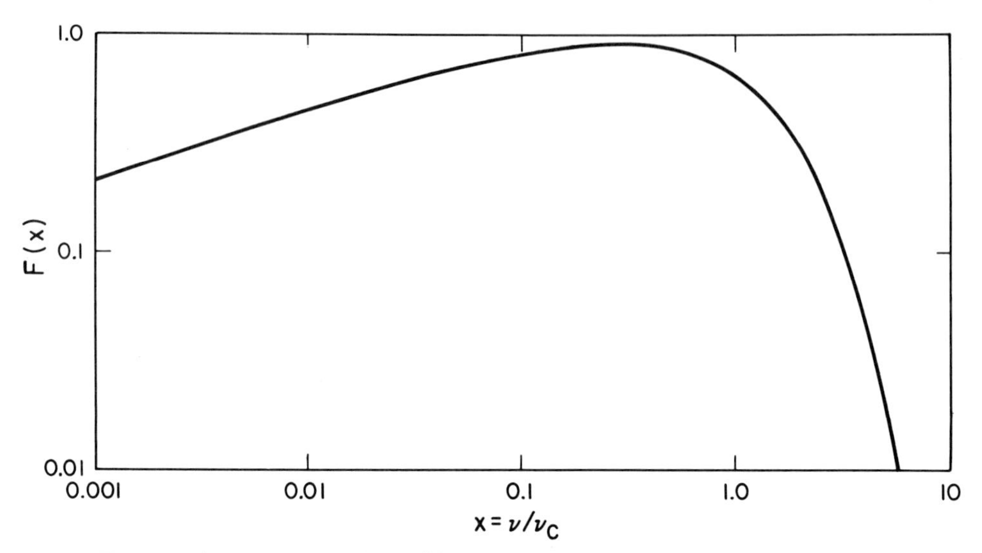

FIG. 2.—The spectral function $F(x)$ which characterizes synchrotron radiation.

electron in a field of 10 microgauss would radiate most of its energy at frequencies near 1500 MHz.

The power spectrum of the emission from a single electron is calculated by taking the Fourier transform of the electric field in the synchrotron radiation spikes. The field is computed from the Lienard-Wiechert potentials produced by the moving electron. For the details of this calculation the reader is referred to the articles mentioned at the beginning of this section, especially the reviews by Ginzburg and his colleagues. The result for $p(E, \nu)d\nu$, the total power emitted by an electron with energy E, in all directions and polarizations, between frequencies ν and $\nu + d\nu$ is

$$p(E, \nu)d\nu = \frac{3^{1/2}e^3}{mc^2} B_\perp \nu/\nu_c \int_{\nu/\nu_c}^{\infty} K_{5/3}(\eta)d\eta d\nu = 4\pi C_2 B_\perp F(\nu/\nu_c)d\nu \,, \tag{2.9}$$

where

$$F(x) = x\int_x^{\infty} K_{5/3}(\eta)d\eta \,. \tag{2.10}$$

The frequency dependence of the spectrum is contained in the function $F(\nu/\nu_c)$, which is an integral of the modified Bessel function $K_{5/3}(\eta)$. The constant C_2 has the value 1.865×10^{-23} in cgs units.[2] Values of the spectral function $F(x)$ have been tabulated by Westfold (1959), and its behavior is shown in figure 2. Series expansions for large and small values of x are given by Westfold (1959) and by Ginzburg and Syrovatskii (1965); for $x < 0.01$, $F(x) \approx 2.13x^{1/3}$, and for $x > 20$, $F(x) \approx (\pi x/2)^{1/2}e^{-x}$ with errors of 5 percent or less.

The rate at which the electron loses energy is the integral of equation (2.9) over all frequencies,

$$-\frac{dE}{dt} = \int_0^{\infty} p(E, \nu)d\nu = 4\pi C_2 B_\perp \int_0^{\infty} F(\nu/\nu_c)d\nu = 4\pi C_2 B_\perp \nu_c \int_0^{\infty} F(x)dx \,. \tag{2.11}$$

[2] In mks units C_2 would be $3^{1/2}\ \mu_0\ ce^3/(16\pi^2 m) = 1.865 \times 10^{-26}$; μ_0 is the permeability of free space.

The definite integral at the right in equation (2.11) is equal to $8\pi/(9\sqrt{3})$, so

$$-\frac{dE}{dt} = 4\pi C_2 B_\perp \nu_c \frac{8\pi}{9\sqrt{3}} = \frac{32\pi^2}{9\sqrt{3}} C_1 C_2 B_\perp{}^2 E^2 = C_3 B_\perp{}^2 E^2 \,, \tag{2.12}$$

where the constant C_3 equals 2.368×10^{-3} in cgs units. The solution of the differential equation (2.12) for the electron energy as a function of time is

$$E(t) = [C_3 B_\perp{}^2 (t + t_{1/2})]^{-1} = E_0(1 + t/t_{1/2})^{-1} \,, \tag{2.13}$$

where $t_{1/2}$, the time required for the electron to lose half of its initial energy E_0, is

$$t_{1/2} = (C_3 B_\perp{}^2 E_0)^{-1} \,. \tag{2.14}$$

Note that the decay is not exponential, so $t_{1/2}$ is not a "half-life" in the usual sense. The constant $C_3{}^{-1}$ has a value in practical units of 8.352×10^9 years (microgauss)^2 GeV. Thus an electron with an initial energy of 3 GeV, trapped in an effective field of 10 microgauss, will radiate half its energy in a bit less than 3×10^7 years.

The radiation from a single electron is, in general, elliptically polarized. The electric vector of the radiation field has its maximum value in the same direction as the projection of the acceleration $\dot{\boldsymbol{v}}$ on the plane normal to the line of sight. From equation (2.1), $\dot{\boldsymbol{v}} \propto \boldsymbol{v} \times \boldsymbol{B}$, so the direction of the maximum in that electric vector is also perpendicular to the projection of $\boldsymbol{B}$ on that plane.

The ellipticity of the polarization varies in a complex manner with frequency and with the position of the observer relative to the particle orbit. The position of the observer is described by the angle ψ (see fig. 1). The polarization becomes linear for $\psi = 0$, i.e., for an observer located exactly on the cone swept out by $\boldsymbol{v}$. Inside and outside this cone the rotation of the electric vector around the polarization ellipse occurs in opposite directions.

In astrophysical situations it is assumed that the electrons have a distribution of pitch angles θ, and an average is taken over all values of ψ. If the number of electrons with different pitch angles, $n(\theta)$, changes only slowly with θ, then there are about equal numbers with $\psi < 0$ and with $\psi > 0$, and radiation in the two senses of elliptic polarization cancels. The only possibility of observing an elliptically polarized component would be to look nearly along the field lines ($\theta \approx 0$) or in barely relativistic situations (E only a few times mc^2) where ξ is not exceedingly small and $n(\theta)$ may change appreciably as θ changes by the order of ξ (Legg and Westfold 1968).

The result of this cancellation is an emission which has partial linear polarization, with the maximum in the electric vector still perpendicular to the projection of $\boldsymbol{B}$ on the plane normal to the line of sight. If the maximum and minimum in intensity are $p_{\max}$ and $p_{\min}$, respectively, the degree of polarization Π is given by

$$\Pi = \frac{p_{\max} - p_{\min}}{p_{\max} + p_{\min}} \,. \tag{2.15}$$

The degree of polarization is large for $\nu \gg \nu_c$, where it goes as $1 - 2\nu_c/3\nu$; it is about 0.7 at $\nu = \nu_c$ and decreases to a constant value of $\frac{1}{2}$ for $\nu \ll \nu_c$. The value of Π is often expressed as a percent polarization.

2.2. Emission from an Ensemble of Particles

The synchrotron radiation which an observer detects from a particular element of volume in a radio source comes from all the electrons with the same pitch angle.[3] Thus the spectrum of the emission from that volume element is the integral of equation (2.9) over the energy distribution of electrons with the appropriate pitch angle. If the *volume emissivity* $\epsilon_\nu(\theta)$ is the emission at frequency ν and in direction θ (per unit of volume, frequency, and solid angle) and if $n(E, \theta)$ is the number density of electrons with energy E and pitch angle θ (per unit of energy and solid angle), we have

$$\epsilon_\nu(\theta) = \int_0^\infty n(E, \theta) p(E, \nu) dE . \quad (2.16)$$

It is assumed that the angular spread ξ of the emission from a single electron is negligibly small. The total number density of electrons with energy E is the integral of $n(E, \theta)$ over all solid angles, $n(E) = \int n(E, \theta)\, d\Omega$. If the electrons have an isotropic distribution of pitch angles, we may replace $n(E, \theta)$ in equation (2.16) by $n(E)/4\pi$. A further assumption is that the source is stationary so that $n(E)$ does not change rapidly with time; this is equivalent to requiring that the source not be expanding with relativistic velocities and not have turbulent motion with relativistic velocities.

We will see that the emission from a radio galaxy typically varies with frequency according to a power law. This is theoretically convenient, since it is easy to show that a power-law synchrotron emission spectrum is produced if the electron energy distribution is also given by a power law. Let the number of electrons per unit volume of the source and having energies between E and $E + dE$ be given by

$$n(E) dE = n_0 E^{-\gamma} dE . \quad (2.17)$$

It is clear that such a distribution can only prevail over a limited energy range, say $E_1 < E < E_2$, if we are to avoid divergences at one or the other extreme in the total number density of electrons,

$$n_e = \int_{E_1}^{E_2} n(E) dE \quad (2.18)$$

and in the total electron energy density,

$$u_e = \int_{E_1}^{E_2} n(E) E dE . \quad (2.19)$$

If we combine equations (2.9), (2.16), and (2.17) and assume an isotropic pitch-angle distribution, the emission at frequency ν and in direction θ per unit volume, frequency, and solid angle is

$$\epsilon_\nu(\theta) = \int_{E_1}^{E_2} p(E, \nu) \frac{n(E)}{4\pi} dE = \frac{3^{1/2} e^3}{4\pi m c^2} B_\perp n_0 \int_{E_1}^{E_2} E^{-\gamma} F(\nu/\nu_c) dE . \quad (2.20)$$

[3] More precisely, it comes from all those electrons whose pitch angles differ by less than ξ from the angle between the magnetic field and the direction of the observer. This small difference can usually be neglected, and we will use θ interchangeably for either the latter angle or for the angle between $\boldsymbol{v}$ and $\boldsymbol{B}$.

In this integral, ν_c and E must be related through equation (2.8). If a change of variables is made from E to ν/ν_c, equation (2.20) becomes

$$\epsilon_\nu(\theta) = \frac{1}{2}\frac{3^{1/2}e^3}{4\pi mc^2}\left(\frac{3e}{4\pi m^3c^5}\right)^{(\gamma-1)/2} n_0 B_\perp^{(\gamma+1)/2}\nu^{-(\gamma-1)/2}\int_{\nu/\nu_2}^{\nu/\nu_1} x^{(\gamma-3)/2}F(x)dx \quad (2.21)$$

or

$$\epsilon_\nu(\theta) = \tfrac{1}{2}C_2 n_0 B_\perp^{(\gamma+1)/2}(C_1/\nu)^{(\gamma-1)/2}G(\nu/\nu_1, \nu/\nu_2, \gamma)\,, \quad (2.22)$$

where $G(x_1, x_2, \gamma)$ is the definite integral on the right in equation (2.21) and is a function of the electron spectral index and of the distance of the observed frequency ν from the cutoff frequencies ν_1 and ν_2 (the critical frequencies corresponding to energies E_1 and E_2). We see that the power-law electron distribution with exponent $-\gamma$ has given rise to a power-law emission spectrum with exponent $-(\gamma - 1)/2$,

$$\epsilon_\nu \propto \nu^{-(\gamma-1)/2} = \nu^\alpha\,, \quad (2.23)$$

where α is the emission spectral index.[4]

The function $G(x_1, x_2, \gamma)$ is a measure of the contributions of the electrons with various critical frequencies to the emission at the particular frequency ν. Thus G reflects the behavior of $F(\nu/\nu_c)$. If, for instance, $F(\nu/\nu_c)$ were a δ-function at $\nu = \nu_c$, G would be identically equal to one. Because $F(\nu/\nu_c)$ drops off rapidly for $\nu > \nu_c$ (see fig. 2), G is not very sensitive to its first argument, ν/ν_1, provided $\nu > 3\nu_1$. On the other hand, $F(\nu/\nu_c)$ decreases quite slowly for $\nu < \nu_c$, and for this reason G is very sensitive to its second argument if ν approaches the upper cutoff ν_2. The difference between $G(\nu/\nu_1, \nu/\nu_2, \gamma)$ and the limiting case $G(\infty, 0, \gamma)$ is an indication of how much the emission is reduced because of the "missing" electrons having energies below or above the cutoff energies, E_1 and E_2.

The behavior of G has been discussed by Westfold (1959). In general it cannot be evaluated except by numerical integration. However, the limiting case $G(\infty, 0, \gamma)$, which we will define as $g(\gamma)$, can be expressed as a product of gamma functions:

$$G(\infty, 0, \gamma) = g(\gamma) = \int_0^\infty x^{(\gamma-3)/2}F(x)dx$$
$$= 2^{(\gamma-3)/2}\frac{\gamma + 7/3}{\gamma + 1}\,\Gamma\left(\frac{3\gamma - 1}{12}\right)\Gamma\left(\frac{3\gamma + 7}{12}\right). \quad (2.24)$$

The above expression is valid for $\gamma > \frac{1}{3}$; for $\gamma \leq \frac{1}{3}$ the integral diverges.

Values of $g(\gamma)$ are given in table 1 together with quantities x_1 and x_2 which indicate the range of frequency over which g may safely be used to replace G, i.e., the range over which the emission spectrum shows negligible departure from a power law. The value

[4] There is good agreement in the literature that the electron energy distribution is to be written as $n(E) \propto E^{-\gamma}$. The negative sign is introduced in the exponent to make $\gamma > 0$ for a distribution in which n decreases as E increases. It is uncommon for an inverted distribution, calling for $\gamma < 0$, to arise in an astrophysical situation; thus the choice for the sign of γ causes no ambiguities and allows us the convenience of working with positive values.

The situation is quite different for the radio spectral index α, since it may have either positive or negative values. Thus I have chosen the convention that the spectral index equals the slope of the log flux-log frequency relationship, i.e., $S_\nu \propto \nu^\alpha$. The typical nonthermal source will have a spectral index $\alpha \approx -0.7$, but a blackbody will radiate with $\alpha = +2$.

of G will be 10 percent less than that of g if $\nu/\nu_1 = x_1$ or if $\nu/\nu_2 = x_2$. Thus if $x_1\nu_1 < \nu < x_2\nu_2$, the error in replacing $G(\nu/\nu_1, \nu/\nu_2, \gamma)$ by $g(\gamma)$ will always be less than 20 percent. The values of x_1 and x_2 are from Ginzburg and Syrovatskii (1964*a*). Also included in table 1 are several other quantities related to $g(\gamma)$ which will be explained in due course.

TABLE 1

SPECTRUM-DEPENDENT QUANTITIES FOR SYNCHROTRON RADIATION

FUNCTION	γ									
	0.5	1.0	1.5	2.0	2.5	3.0	3.5	4.0	4.5	5.0
$g(\gamma)$..	23.9	5.24	2.88	2.08	1.74	1.61	1.60	1.69	1.86	2.13
$g'(\gamma)$..	19.8	4.12	2.16	1.50	1.21	1.07	1.03	1.05	1.13	1.26
$f'(\gamma)$..	0.274	1.09	1.95	2.87	3.85	4.90	6.00	7.15	......	......
$x_1(\gamma)$..	....	0.80	1.3	1.8	2.2	2.7	......	3.4	......	4.0
$x_2(\gamma)$..	....	0.00045	0.011	0.032	0.10	0.18	......	0.38	......	0.65

The emission described by equations (2.20)–(2.22) is highly polarized, with degree of polarization

$$\Pi = (\gamma + 1)/(\gamma + 7/3) \tag{2.25}$$

independent of ν, provided $\nu \ll \nu_2$. Furthermore the emission is anisotropic, varying as $(\sin\theta)^{(\gamma+1)/2}$ by virtue of the factor $\sin\theta$ in the effective magnetic field $B_\perp$. It is, of course, impossible to measure the anisotropy of the emission from a radio galaxy, but some estimate of the degree of homogeneity of the magnetic field can be gained from the net polarization of the radiation. Since this is usually found to be much less than the values given by equation (2.25), the conclusion reached is that in most radio galaxies the field is fairly well tangled. Thus in the absence of more specific information, the emissivity given in equation (2.22) should probably be averaged over all values of the inclination of the magnetic field. The effect is to multiply the function G of equation (2.22) by the geometrical factor

$$\frac{1}{2}\int_0^\pi (\sin\theta)^{(\gamma+1)/2}\sin\theta d\theta = \tfrac{1}{2}\pi^{1/2}\frac{\Gamma[(\gamma+5)/4]}{\Gamma[(\gamma+7)/4]}\,. \tag{2.26}$$

Included in table 1 are values of $g'(\gamma)$, which are equal to $g(\gamma)$ times the above factor. It can be seen that the effect of this averaging on the total intensity is small; however, the averaging causes the polarization to disappear completely. The final expression for the mean emissivity is

$$\epsilon_\nu = \tfrac{1}{2}C_2 n_0 B^{(\gamma+1)/2}(C_1/\nu)^{(\gamma-1)/2}G'(\nu/\nu_1, \nu/\nu_2, \gamma)\,, \tag{2.27}$$

where G' equals the previously defined integral G multiplied by the factor given in equation (2.26).[5]

[5] Ginzburg and Syrovatskii (1964*a*, 1965) tabulate a function $a(\gamma)$ which differs from the function $g'(\gamma)$, as defined above, by a factor $3^{1/2}/8\pi$. The definitions used here of G and of the functions derived from it given in table 1 conform to the earlier derivation by Westfold (1959), and there are minor differences between these and the analogous functions tabulated by Ginzburg and Syrovatskii.

2.3. Modifications of the Power-Law Spectrum

A radio source consisting of a randomly oriented magnetic field which contains a power-law distribution of relativistic electrons, $n(E) = n_0 E^{-\gamma}$, has a spectrum which is proportional to the emissivity ϵ_ν of equation (2.27), provided the source region is optically thin at all frequencies of interest. Furthermore, if the cutoff frequencies are sufficiently far removed from the frequencies of observation, the cutoff function G' will be independent of frequency, and the spectrum will follow a strict power law, $\epsilon_\nu \propto \nu^\alpha$, with $\alpha = -(\gamma - 1)/2$.

Although the spectra of radio galaxies do typically follow a power law, they frequently show departures from this relationship at one end or the other of the observable frequency range (see § 3.2). We must now consider ways in which the power-law emission spectrum may be modified. These may involve processes which modify the electron energy distribution as well as conditions which affect the radio spectrum directly.

One way to produce departures from the power-law spectrum is already expressed in equation (2.27), namely, in the cutoff function $G'(\nu/\nu_1, \nu/\nu_2, \gamma)$. For instance, if there should be a sharp low-energy cutoff in the electron spectrum at an energy corresponding to a critical frequency ν_1, then if the observed frequency were well below ν_1 the intensity of the radio emission would be proportional to $\nu^{1/3}$, independent of the value of γ. It is not clear that this situation has ever been observed, however, probably because a sharp low-energy cutoff in the electron spectrum seldom, if ever, occurs. At the other end of the electron energy spectrum, if there were an abrupt high-energy cutoff, the radio emission would drop exponentially for frequencies above the critical frequency corresponding to the cutoff energy, reflecting the behavior of $F(\nu/\nu_c)$, which defines the shape of the single-particle emission spectrum. The effect on the radio emission spectrum of such hypothetical upper and lower energy cutoffs in the electron distribution is shown in figure 3 for several values of the electron spectral index γ.

It is obvious that changes of slope can be introduced in the emission spectrum by suitable changes of slope at corresponding energies in the electron distribution. A change in the electron spectral index of $\Delta\gamma$ will introduce a change $\Delta\alpha = -\frac{1}{2}\Delta\gamma$ in the spectral index of the emission. Several ways to modify the electron spectrum are discussed in § 2.3.5.

The frequency dependence of $\nu^{1/3}$ (i.e., $\alpha = +\frac{1}{3}$), seen at the left in figure 3, is the sharpest low-frequency cutoff that can be obtained in a "pure" synchrotron spectrum, unmodified by absorption or other processes. Since a number of radio sources show low-frequency spectra which rise more steeply than this, say with $\alpha \approx +2$, some combination of these other processes must alter the observed intensity distribution. Three such processes have been suggested (see Ginzburg and Syrovatskii 1964*a*, 1965; Scheuer 1965; Scheuer and Williams 1968), namely: (1) thermal absorption, (2) synchrotron self-absorption, and (3) the "Tsytovitch effect."

2.3.1. *Radiation transfer.*—The two absorption processes can best be understood with the aid of some concepts from the theory of radiation transfer. If I_ν is the specific intensity of a pencil of radiation (with units of ergs s^{-1} cm^{-2} Hz^{-1} sterad^{-1}), the equation of transfer specifies the change in I_ν as the pencil traverses a distance ds within the source,

$$dI_\nu = (\epsilon_\nu - \kappa_\nu I_\nu) ds \,, \tag{2.28}$$

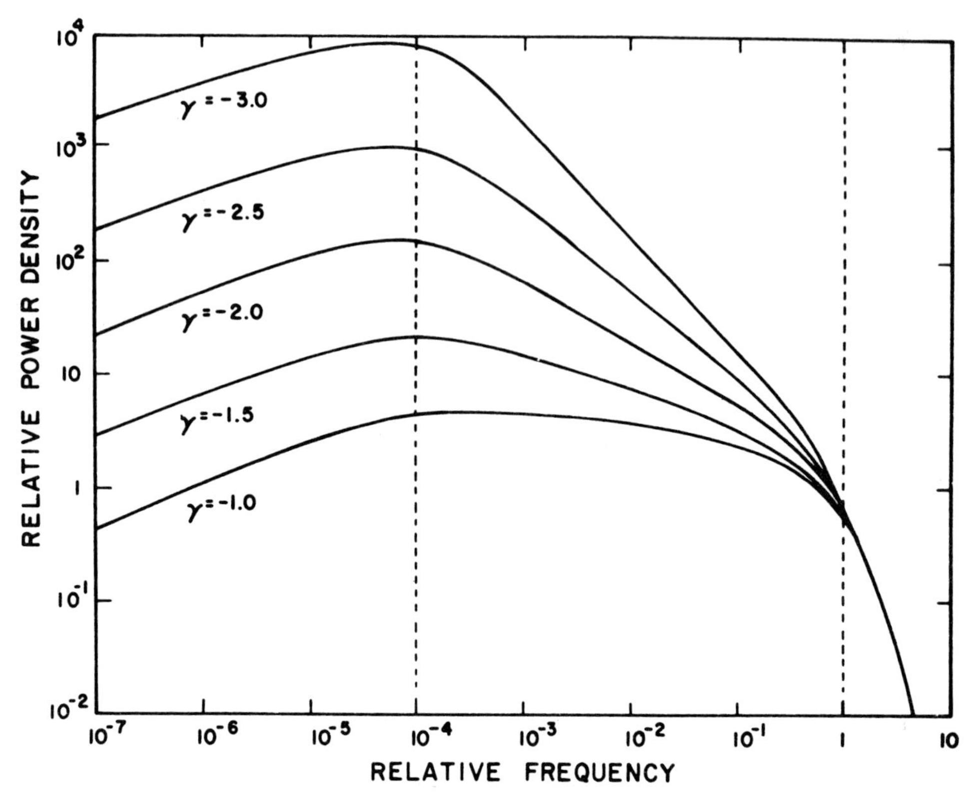

FIG. 3.—Theoretical synchrotron emission spectra from power-law electron energy distributions with sharp high- and low-energy cutoffs (from Kellermann 1964).

where ϵ_ν is the *emissivity* per unit volume, frequency, and solid angle, and κ_ν is the *absorption coefficient*, or the absorption per unit length. In general, the intensity, emissivity, and absorption coefficient may each include an angular dependence as well as a dependence on position within the source region. The solution of the transfer equation is

$$I_\nu(s) = I_\nu(0)e^{-\tau(s,0)} + \int_0^s \epsilon_\nu e^{-\tau(s,s')}ds' \tag{2.29}$$

where τ, the optical depth, is defined by the integral of the absorption,

$$\tau(s, s') = \int_{s'}^s \kappa_\nu ds'' . \tag{2.30}$$

As an example, consider the case shown in figure 4 of a homogeneous slab with thickness l, in which ϵ and κ are constant and isotropic. The optical depth is then

$$\tau(s, s') = \kappa_\nu \times (s - s') , \tag{2.31}$$

and for propagation directly through the slab the solution to the transfer equation is

$$I_\nu(l) = I_\nu(0) \exp(-\kappa_\nu l) + \frac{\epsilon_\nu}{\kappa_\nu} [1 - \exp(-\kappa_\nu l)] . \tag{2.32}$$

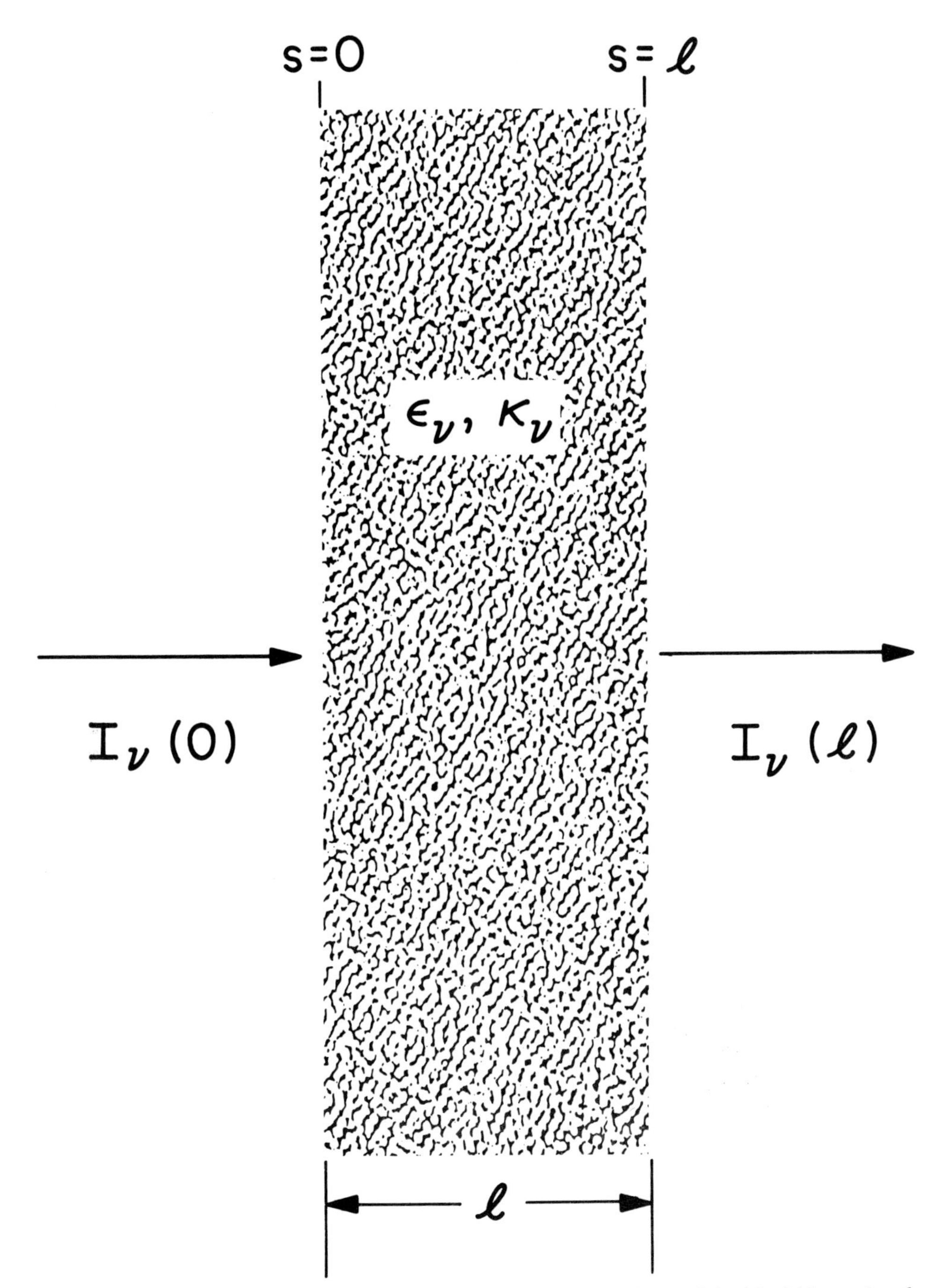

FIG. 4.—Transfer of radiation through a homogeneous slab of material with thickness l and with emissivity ϵ_ν and absorption coefficient κ_ν.

Suppose that $I_\nu(0) = 0$, i.e., that no radiation enters from the far side of the source region. Then if the source is optically thin ($\tau = \kappa_\nu l \ll 1$),

$$I_\nu(l) \approx \epsilon_\nu l \, . \tag{2.33}$$

If the radiation passes through a purely absorbing layer, in which the emission is negligibly small, its intensity is reduced by the factor exp $(-\tau)$, which equals exp $(-\kappa_\nu l)$ for the example considered above.

On the other hand, if the source is optically thick ($\tau = \kappa_\nu l \gg 1$), then

$$I_\nu(l) = \epsilon_\nu / \kappa_\nu \, . \tag{2.34}$$

If we apply these results to radio galaxies, we see that in the absence of any absorption the intensity of the emitted radiation I_ν would be equal to the integral of the emissivity ϵ_ν through the source, and the source would have the pure synchrotron spectrum discussed in § 2.2. However, the presence of absorption, either within the source region or between the source and the observer, would modify the spectrum.

2.3.2. *Thermal absorption.*—One mechanism giving rise to absorption is the scattering of thermal electrons by the ions in a plasma. In such a scattering event, a photon may be emitted or absorbed as the electron is accelerated by the Coulomb field of the ion. This mechanism produces the thermal radio emission from H II regions such as the Orion Nebula in our own Galaxy. In radio galaxies thermal emission is negligible compared with synchrotron emission; however, thermal absorption may significantly modify the radiation received from synchrotron sources within or behind the main body of the associated galaxy or from sources whose radiation must pass through ionized regions in our own Galaxy.

An expression for the thermal absorption coefficient at radio frequencies has been derived by Scheuer (1960) and Oster (1961) and is

$$\kappa_\nu = \frac{16\pi Z^2 e^6 n_t n_i}{3(2\pi m k)^{3/2} c} T^{-3/2} \nu^{-2} \ln \left[\left(\frac{2kT}{\gamma^* m} \right)^{3/2} \frac{m}{\pi \gamma^* Z e^2 \nu} \right] , \tag{2.35}$$

where n_t is the density of thermal electrons with kinetic temperature T, n_i is the density of ions having mean charge Ze, k is the Boltzmann constant, and γ^* is exp $(\gamma) =$ 1.781 . . . , where $\gamma = 0.5772$ . . . is the Euler constant. This expression is valid for frequencies well above the plasma frequency $\nu_p = (n_t e^2/\pi m)^{1/2} \approx 9.0 \times 10^3 n_t^{1/2}$. If we are dealing with an electrically neutral hydrogen plasma, we may set $Z = 1$ and $n_t = n_i = n$. Then the absorption coefficient may be written

$$\kappa_\nu = C_4 n^2 T^{-3/2} \nu^{-2} \ln (C_5 T^{3/2}/\nu) \, , \tag{2.36}$$

where

$$C_4 = \frac{16\pi e^6}{3(2\pi m k)^{3/2} c} = 9.786 \times 10^{-3} (\text{cgs}) \, ,$$

$$C_5 = \left(\frac{2k}{\gamma^* m} \right)^{3/2} \frac{m}{\pi \gamma^* e^2} = 4.954 \times 10^7 \, , \tag{2.37}$$

and

$$\ln (C_5) = 17.72 \, .$$

The frequency dependence of the thermal absorption coefficient is seen to be $\kappa_\nu \propto \nu^{-2} \times [\text{const.} - \ln(\nu)]$. If the plasma temperature $T \approx 10^4$ ° K, the logarithmic term, known as the Gaunt factor, takes on the values shown in table 2.

If $T \sim 10^4$ ° K, the effect of the logarithmic term at radio frequencies may be approximated by altering the exponent of ν from -2 to -2.1, giving

$$\kappa_\nu \approx 100 C_4 n^2 T^{-3/2} \nu^{-2.1} \approx n^2 T^{-3/2} \nu^{-2.1} . \tag{2.38}$$

It is convenient to define a characteristic cutoff frequency ν_t for thermal absorption as the frequency at which the optical depth $\kappa_\nu l$ of the absorbing layer equals 1. Below this frequency the intensity of the source would fall exponentially, as $e^{-\tau} \approx \exp[-(\nu/\nu_t)^{-2.1}]$. Thus thermal absorption can produce a sharp low-frequency cutoff in the spectrum of a radio source. From equations (2.31) and (2.38)

$$\nu_t = [n^2 T^{-3/2} l]^{1/2.1} . \tag{2.39}$$

For a rough estimate of ν_t we may ignore the difference between $\nu^{2.1}$ and ν^2 and take a mean value of the Gaunt factor. Then in practical units

$$\nu_t\ (\text{MHz}) \approx \tfrac{1}{2} n\ (\text{cm}^{-3}) l^{1/2}\ (\text{pc}) . \tag{2.40}$$

As an example, suppose the signal from a radio galaxy must pass for 10 kpc through the plane of our own Galaxy where the rms electron density is 0.1 cm^{-3}. Then the thermal

TABLE 2

RADIO FREQUENCY GAUNT FACTORS

ν (Hz)	$\ln(10^6\ C_5/\nu)$
10^6	17.72
10^7	15.42
10^8	13.11
10^9	10.81
10^{10}	8.51
10^{11}	6.21

absorption would be important for frequencies of the order of 5 or 10 MHz. This effect is observed in the spectra of radio sources lying at very low galactic latitudes.

If a large number of thermal electrons are mixed with the relativistic plasma in a synchrotron source, the low-frequency behavior is given by equation (2.34) using the synchrotron emissivity of equation (2.27) and the thermal absorption coefficient of equation (2.38),

$$I_\nu \approx \epsilon_\nu / \kappa_\nu \propto \nu^{-(\gamma-1)/2} / \nu^{-2.1} . \tag{2.41}$$

If we let $\alpha_0 = -(\gamma - 1)/2$ be the spectral index at frequencies well above ν_t, where the source is optically thin, then the low-frequency spectral index will be $\alpha = 2.1 + \alpha_0$. It is assumed that the density of thermal electrons is not so large that the Tsytovitch effect inhibits the synchrotron emission (see § 2.3.4.).

2.3.3. *Synchrotron self-absorption.*—If the intensity of synchrotron radiation within a source becomes sufficiently high, then reabsorption of the radiation through the synchrotron mechanism may become important. If it does take place, synchrotron self-absorption will drastically modify the spectrum of the source at low frequencies. The possibility of reabsorption of synchrotron radiation was first suggested by Twiss (1954), and the absorption coefficient has been calculated by Le Roux (1961); Wild, Smerd, and Weiss (1963); and Ginzburg and Syrovatskii (1965).

For any process the absorption at frequency ν is the difference between the upward transitions from states with energies $E - h\nu$ to states with energy E and the stimulated downward transitions between the same states. At given energy E the upward rate is $I_\nu(\theta)B_{12}n(E - h\nu, \theta)$ and the downward rate is $I_\nu(\theta)B_{21}n(E, \theta)$ transitions per unit of volume, frequency, and solid angle. Note that in general the Einstein coefficients B_{12} and B_{21} depend on the direction of the incident radiation, its polarization, and the direction of motion of the electrons. In the treatment here, the polarization dependence will be neglected. The radiation is coupled to only those electrons moving in a direction differing by less than the angle $\xi \approx mc^2/E$ from the direction of the radiation, so the angular dependence of B_{12} and B_{21} will be assumed to be a delta function, coupling radiation propagating at angle θ with respect to the magnetic field with electrons having pitch angles θ. More complete treatments of synchrotron self-absorption including the polarization dependence have been given by Pacholczyk and Swihart (1967), Ginzburg, Sazonov, and Syrovatskii (1968), and Pacholczyk (1970).

Each transition adds or subtracts one photon of energy $h\nu$ from the pencil of radiation $I_\nu(\theta)$. The absorption coefficient κ_ν is the integral of the transition rates over all energies, divided by the intensity, or

$$\kappa_\nu(\theta) = -\frac{1}{I_\nu(\theta)}\frac{dI_\nu(\theta)}{ds} = h\nu\int_0^\infty [B_{12}n(E - h\nu, \theta) - B_{21}n(E, \theta)]dE\,. \quad (2.42)$$

The Einstein coefficients are related in the following way:

$$w(E)B_{21} = w(E - h\nu)B_{12}\,, \qquad A_{21}/B_{21} = 2h\nu^3/c^2\,,$$

where the $w(E)$ are the statistical weights and A_{21} is the coefficient for spontaneous emission. The latter may be related to the radiation rate $p(E, \nu)$ for a single electron from equation (2.9) by noting that the volume emissivity is just

$$\epsilon_\nu(\theta) = h\nu\int_0^\infty A_{21}n(E, \theta)dE\,, \quad (2.43)$$

and identifying this with the previously defined emissivity from equation (2.16),

$$\epsilon_\nu(\theta) = \int_0^\infty p(E, \nu)n(E, \theta)dE\,. \quad (2.16)$$

Thus $A_{21} = p(E, \nu)/h\nu$. Since $p(E, \nu)$ depends on the pitch angle θ through the factor $B_\perp$, the Einstein coefficients will also depend on pitch angle.

The statistical weight of an electron with energy between E and $E + dE$ is equal to the volume in momentum space which it can occupy. For an electron with momentum P this volume equals $4\pi P^2 dP$; for a relativistic electron $P \approx E/c$, so the statistical weight is given by

$$w(E)dE = 4\pi E^2 dE/c^3\,. \quad (2.44)$$

From equation (2.8) and figure 2 we can see that emission takes place only when $h\nu \ll E$, so we may approximate $w(E - h\nu) = w(E)\,(1 - 2h\nu/E)$. The difference in the electron densities may be approximated in the same way, $n(E - h\nu, \theta) = n(E, \theta) - h\nu(dn/dE) = n(E, \theta)\,[1 - (h\nu/n)\,(dn/dE)]$. Then, to first order in $h\nu/E$, the expression for the absorption coefficient becomes

$$\kappa_\nu(\theta) = \frac{c^2}{2h\nu^3}\int_0^\infty n(E,\theta)\left[\frac{2h\nu}{E} - \frac{h\nu}{n(E,\theta)}\frac{dn(E,\theta)}{dE}\right]p(E,\nu)dE$$

$$= -\frac{c^2}{2\nu^2}\int_0^\infty E^2\frac{d}{dE}\left[\frac{n(E,\theta)}{E^2}\right]p(E,\nu)dE\,. \qquad (2.45)$$

As was the case in calculating the synchrotron emission from an ensemble of particles, it is convenient now to assume that the electrons have an isotropic velocity distribution and a power-law spectrum extending over a limited range of energies $E_1 < E < E_2$. If so, we may replace $n(E, \theta)$ by $n(E)/4\pi$ and reduce the limits of the integral to E_1 and E_2. Thus we have $n(E) = n_0E^{-\gamma}$, $d[n(E)/E^2]/dE = -(\gamma + 2)n_0E^{-(\gamma+3)}$, and the absorption coefficient becomes

$$\kappa_\nu(\theta) = \frac{c^2}{8\pi\nu^2}(\gamma + 2)n_0\int_{E_1}^{E_2}E^{-(\gamma+1)}p(E,\nu)dE\,. \qquad (2.46)$$

This integral is very similar to that for the synchrotron emission from a similar ensemble of electrons, equation (2.20), except that the exponent of E in the integrand is reduced by one, and there is an added factor of $(\gamma + 2)c^2/2\nu^2$. Making the same change of variables as was used to go from equation (2.20) to (2.21), we obtain the absorption coefficient in terms of the cutoff frequencies ν_1 and ν_2 as

$$\kappa_\nu(\theta) = \frac{c^2}{2}\frac{3^{1/2}e^3}{4\pi mc^2}\left(\frac{3e}{4\pi m^3c^5}\right)^{\gamma/2}n_0B_\perp^{(\gamma+2)/2}\nu^{-(\gamma+4)/2}(\gamma + 2)\int_{\nu/\nu_1}^{\nu/\nu_2}x^{(\gamma-2)/2}F(x)dx\,. \qquad (2.47)$$

It is evident that the integral in equation (2.47) is identical to that in equation (2.21) evaluated with $\gamma + 1$ in place of γ. Thus it is equal to the function $G(\nu/\nu_1, \nu/\nu_2, \gamma + 1)$ which is defined by equations (2.20) and (2.21). Making use of this function, we can write the absorption coefficient as

$$\kappa_\nu(\theta) = \tfrac{1}{2}c^2C_2C_1^{\gamma/2}n_0B_\perp^{(\gamma+2)/2}\nu^{-(\gamma+4)/2}(\gamma + 2)G(\nu/\nu_1, \nu/\nu_2, \gamma + 1)\,. \qquad (2.48)$$

If the critical frequencies corresponding to the cutoff energies are sufficiently far removed from the frequency of observation, we may again substitute $g(\gamma + 1) = G(\infty, 0, \gamma + 1)$ for $G(\nu/\nu_1, \nu/\nu_2, \gamma + 1)$. With this in mind, table 1 has been carried out to values of γ as large as 5, although the observed spectra of radio sources show little evidence for values of γ greater than 3.5.

The absorption coefficient of equation (2.48) refers to radiation traveling in direction θ with respect to a homogeneous magnetic field $\boldsymbol{B}$. As before, if the magnetic field is tangled in a random way, the mean absorption coefficient is obtained by multiplying by a geometrical factor. The correct factor is given by equation (2.26) with $\gamma + 1$ in place of γ. Thus we may use the function $G'(\nu/\nu_1, \nu/\nu_2, \gamma + 1)$ or, where appropriate, $g'(\gamma + 1)$ in the expression for the mean absorption coefficient,

$$\kappa_\nu = \tfrac{1}{2}c^2C_2C_1^{\gamma/2}n_0B^{(\gamma+2)/2}\nu^{-(\gamma+4)/2}(\gamma + 2)G'(\nu/\nu_1, \nu/\nu_2, \gamma + 1)\,. \qquad (2.49)$$

The absorption coefficient has a frequency dependence of $\nu^{-(\gamma+4)/2}$ while that of the emissivity is $\nu^{-(\gamma-1)/2}$. A source of diameter l will become optically thick at a frequency such that $\kappa_\nu l \approx 1$, and at frequencies below that cutoff the emission of the source will go as $\epsilon_\nu/\kappa_\nu \propto \nu^{5/2}$. Below the cutoff for synchrotron self-absorption the spectral index is

$\alpha = +2.5$, independent of γ. This differs from the blackbody spectral index of $+2$ because we have specified a power-law distribution instead of a Maxwell-Boltzmann distribution for the electron energies.

Slish (1963) and Williams (1963) have pointed out that the conditions under which synchrotron self-absorption occurs may be stated directly in terms of the observable properties of a radio source, i.e., its angular dimensions and flux density at the Earth. From equation (2.33) we see that at frequencies well above cutoff, the specific intensity I_ν of a source in the form of a homogeneous slab with thickness l is $\epsilon_\nu l$. The flux density of the radiation received at the Earth is the product of I_ν times the solid angle subtended by the source,

$$S_\nu = I_\nu \Delta\Omega = \epsilon_\nu l \Delta\Omega \,. \tag{2.50}$$

For frequencies well above cutoff, the flux obeys the power law $S_\nu = S_0(\nu/\nu_0)^\alpha$, where $\alpha = -(\gamma - 1)/2$, and ν_0 is some frequency in the region above cutoff.

We will define the cutoff frequency for synchrotron self-absorption ν_s as the frequency for which $\kappa_\nu l = 1$. In the absence of absorption, flux at this frequency would be $S_s^* = S_0(\nu_s/\nu_0)^\alpha$. Then at the cutoff frequency, equation (2.50) may be rewritten as

$$\epsilon_\nu/\kappa_\nu = S_s^*/\Delta\Omega \qquad (\text{if } \nu = \nu_s) \,. \tag{2.51}$$

The emissivity and opacity are given in equations (2.27) and (2.49). Taking their ratio we have

$$\frac{1}{c^2} C_1^{-1/2} B^{-1/2} \nu_s^{5/2} \left[\frac{G'(\nu_s/\nu_1, \nu_s/\nu_2, \gamma)}{(\gamma + 2) G'(\nu_s/\nu_1, \nu_s/\nu_2, \gamma + 1)} \right] = \frac{S_s^*}{\Delta\Omega} \,, \tag{2.52}$$

or

$$\nu_s^{5/2} = c^2 C_1^{1/2} S_s^* B^{1/2} F'/\Delta\Omega \,, \tag{2.53}$$

where $1/F'$ is the function in brackets in equation (2.52). If the energy cutoff frequencies ν_1 and ν_2 are sufficiently remote, the functions G' may be replaced by their limits g'; F' will then depend on γ alone. The function $f'(\gamma) = (\gamma + 2)\, [g'(\gamma + 1)/g'(\gamma)]$ is included in table 1; its value is approximately equal to $2\gamma - 1$.

Slish and Williams noted that the intensities of many small-diameter radio sources decrease at frequencies below a few hundred megahertz. They attributed this to synchrotron self-absorption and used the theory outlined above to predict angular diameters for these sources. If $\Delta\Omega = \theta^2$, where θ is measured in seconds of arc, ν_s is in megahertz, B in gauss, and S in flux units of 10^{-26} W m^{-2} Hz^{-1}, we may rewrite equation (2.53) as

$$\theta^2 \approx 900 S_s^* B^{1/2} F' \nu_s^{-2.5} \,. \tag{2.54}$$

The observable quantities ν_s and S_s^* determine the value of $\theta/B^{1/4}$; the dependence of θ on the assumed value of B is weak. Figure 5 is a plot of flux density versus cutoff frequency for several different angular diameters, assuming that $B = 10^{-4}$ gauss and that $F' = 3$. Also shown are curves representing the spectra of three homogeneous model sources with $\gamma = 1$, 2, and 3, but having the same cutoff frequency, 100 MHz, and equal flux densities S_s^* of 10 flux units. These sources would have angular dimensions of about $0\overset{''}{.}05$.

As has been pointed out by Williams (1963), the cutoff frequency for synchrotron self-absorption is that at which the surface brightness temperature of the source becomes comparable to the kinetic temperature of the electrons which emit the radiation. For

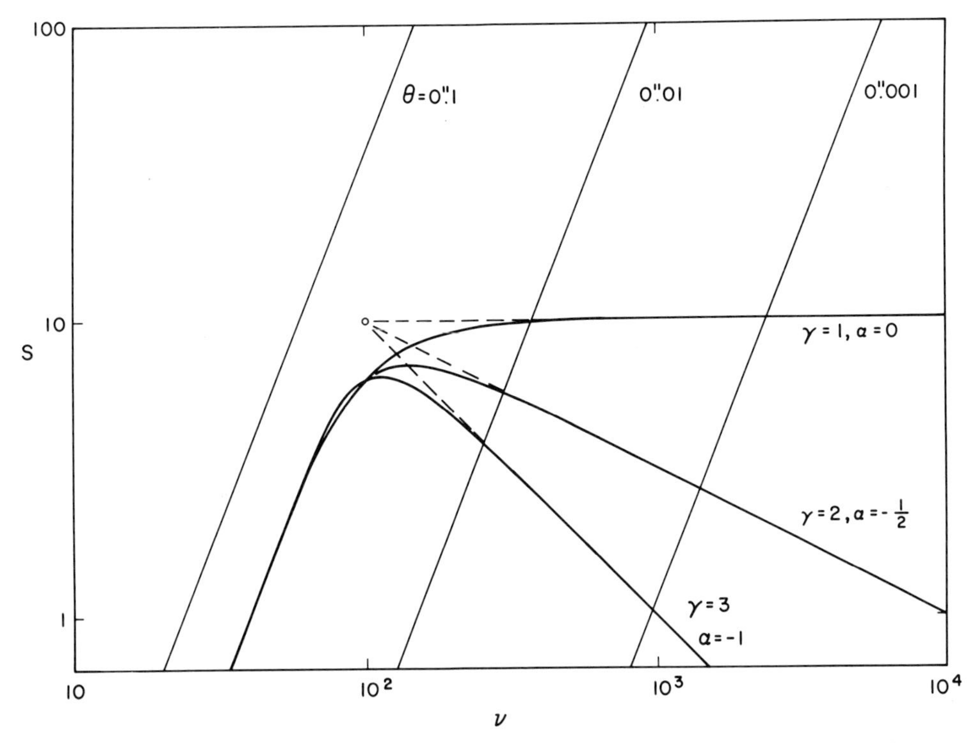

FIG. 5.—Theoretical synchrotron spectra from optically thick sources having different electron energy distributions but with a common cutoff frequency of 10^2 MHz and a common cutoff flux density of 10 f.u. Values of the angular diameter for other fluxes and frequencies are shown by the inclined lines. A magnetic field of 10^{-4} gauss is assumed.

non-Maxwellian electron energy distributions we define this kinetic temperature as $kT \sim E$. From the Rayleigh-Jeans law, the brightness temperature T_b is given by

$$2kT_b\nu^2/c^2 = S_0(\nu/\nu_0)^\alpha/\Delta\Omega\,, \tag{2.55}$$

or

$$kT_b = c^2 S_0(\nu/\nu_0)^\alpha/(2\nu^2\Delta\Omega)\,. \tag{2.56}$$

The relevant electrons are those with critical frequencies equal to the cutoff frequency, so from equation (2.8) we may relate E and ν_s,

$$kT \sim E \sim (\nu_s/C_1 B)^{1/2}\,. \tag{2.57}$$

Setting $kT = kT_b$, we have

$$\left(\frac{\nu_s}{C_1 B}\right)^{1/2} \sim \frac{c^2 S_0(\nu_s/\nu_0)^\alpha}{2\nu_s{}^2\Delta\Omega} = \frac{c^2 S_s{}^*}{2\nu_s{}^2\Delta\Omega}\,, \tag{2.58}$$

which agrees with equation (2.53) in functional dependence and in order of magnitude.

2.3.4. *The Tsytovitch effect.*—The characteristic properties of synchrotron radiation are mainly due to the fact that high-energy electrons move at nearly the speed of light. A relativistic electron almost keeps up with, and thus strongly reinforces, electromagnetic waves moving parallel to its path. This is true even though the curvature of the electron's orbit means that it can move nearly parallel to a given direction only for a very short time. However, if the electron is captured in a region where the phase velocity of the waves differs from the velocity of light, the nature of the radiation emitted will be drastically changed.

The phase velocity of electromagnetic waves differs from c, its value in vacuum, whenever the index of refraction η of the medium in which the waves are propagating differs from unity. If $\eta > 1$, so that the phase velocity is less than c, a relativistic particle will emit Cerenkov radiation and will lose energy very rapidly. On the other hand, in plasmas the index of refraction is generally less than unity, and the phase velocity is greater than c. If η becomes appreciably less than unity, a relativistic electron can no longer keep in phase with the waves it generates, and the intensity of synchrotron emission is very much reduced. This fact was first pointed out by Tsytovitch (1951); a complete theoretical analysis including gyrotropic effects has been given by Eidman (1958). A discussion specifically directed toward radio astronomy was given by Razin (1960), and the effect is often called the *Razin effect* (see also Tsytovitch 1973).

The index of refraction of a plasma, for frequencies well above the plasma frequency ν_p, is given by

$$\eta^2 = 1 - \frac{n_t e^2}{\pi m \nu^2} = 1 - \frac{\nu_p{}^2}{\nu^2}\,, \tag{2.59}$$

where n_t is the number density of thermal electrons. The magnetic field, which must be present to cause synchrotron radiation, splits the index of refraction into slightly different values for the ordinary and extraordinary modes of propagation. For the weak fields presumed to exist in radio galaxies this splitting is small and may be neglected.

The critical frequency for synchrotron radiation in vacuum is, from equations (2.7) and (2.8),

$$\nu_c = \tfrac{3}{2}\nu_g[1 - (v/c)^2]^{-3/2}\,. \tag{2.60}$$

If the refractive index is not equal to unity, we replace c by c/η and obtain

$$\nu_c{}' = \tfrac{3}{2}\nu_g[1 - (\eta v/c)^2]^{-3/2} \tag{2.61}$$

or

$$\nu_c{}' = \tfrac{3}{2}\nu_g\left[1 - \left(1 - \frac{\nu_p{}^2}{\nu^2}\right)\frac{v^2}{c^2}\right]^{-3/2}, \tag{2.62}$$

$$\nu_c{}' \approx \nu_c\left[1 + \frac{\nu_p{}^2/\nu^2}{1 - (v^2/c^2)}\right]^{-3/2} = \nu_c\left[1 + \left(\frac{E}{mc^2}\right)^2\left(\frac{\nu_p}{\nu}\right)^2\right]^{-3/2} \tag{2.63}$$

Thus the critical frequency is itself a *function* of frequency and may be greatly reduced at low frequencies. But we can see from figure 2 that radiation is emitted only when $\nu \lesssim \nu_c$. For certain values of energy, field, and plasma density, this condition may hold over only a limited range of frequencies, as illustrated in figure 6. The intensity of synchrotron radiation will be appreciably reduced whenever the bracketed term in equation (2.63) becomes much greater than unity. We may use this fact to derive a characteristic lower cutoff frequency for the Tsytovitch effect $\nu_{\rm Ts}$, given by the condition

$$(E/mc^2)^2\nu_p{}^2/\nu_{\rm Ts}{}^2 = 1\,. \tag{2.64}$$

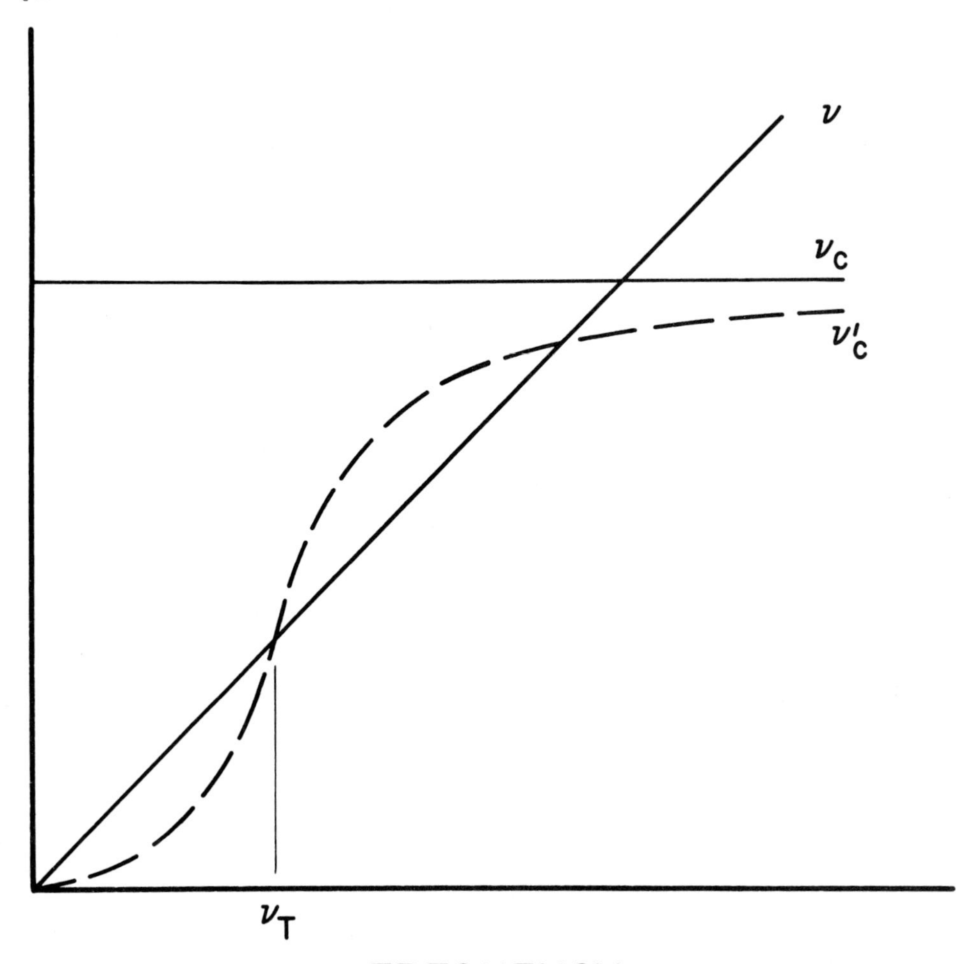

FIG. 6.—Illustrating the Tsytovitch effect. Synchrotron radiation is possible only when the effective critical frequency ν_c' exceeds the observed frequency ν. In the absence of a plasma ν_c is a constant.

Equation (2.8) may be used (with $\nu_c = \nu_{\mathrm{Ts}}$) to express the energy in terms of frequency. The plasma frequency is $\nu_p = [n_t e^2/(\pi m)]^{1/2}$. Then equation (2.64) becomes

$$\frac{4\pi mc\nu_{\mathrm{Ts}}}{3eB_\perp}\,\frac{n_t e^2}{\pi m\nu_{\mathrm{Ts}}{}^2} = 1\;, \tag{2.65}$$

giving the Tsytovitch cutoff frequency as

$$\nu_{\mathrm{Ts}} = \frac{4cn_t e}{3B_\perp} \approx 20\,\frac{n_t}{B_\perp}\,. \tag{2.66}$$

It is of interest to compare this frequency with the thermal-absorption cutoff frequency ν_t. From equation (2.40) $\nu_t \approx 5 \times 10^5 n_t l^{1/2}$, where l is the dimension of the absorbing region measured in parsecs. Thus $\nu_{Ts}/\nu_t \approx 4 \times 10^{-5}/(B_\perp l^{1/2})$, and the Tsytovitch cutoff will lie below the thermal cutoff whenever

$$B_\perp \gtrsim 4 \times 10^{-5} l^{-1/2} \text{ (pc)} . \tag{2.67}$$

Thus the Tsytovitch effect should be important only in sources with very small size and weak magnetic fields.

2.3.5. *Modifications of the electron spectrum.*—The previous three sections have discussed ways in which the synchrotron emission spectrum may be modified at low frequencies so that it differs from a strict power-law dependence on frequency. As was mentioned in § 2.3, the spectrum of a radio source will also be modified by departures of the electron distribution from a power-law dependence on energy. A change of $\Delta\gamma$ in the electron spectral index will be reflected as a change $\Delta\alpha = -\Delta\gamma/2$ in the radio spectral index, and of course more complicated departures from a power law may also occur.

Even if the electron distribution in a radio source follows a strict power law at some initial time, the distribution will sooner or later depart from the power law, since electrons with different energies lose energy at different rates. The principal loss mechanisms are (1) losses by synchrotron emission and by inverse Compton effect, both proportional to E^2, (2) losses by radiation during collisions with heavy particles in the plasma (bremsstrahlung losses) and losses due to expansion of the source, both of which are proportional to E, and (3) losses by ionizing atoms or molecules of the plasma, which are almost independent of energy.

The rate of change of $n(E, t)$, the density of electrons having energy E at time t, is obtained by counting up the gains and losses in an energy interval dE during a time interval dt. The result is

$$\frac{\partial n(E, t)}{\partial t} = -\frac{\partial}{\partial E}\left[\frac{dE}{dt} n(E, t)\right] + q(E, t) , \tag{2.68}$$

where $q(E, t)$ is the rate at which electrons of energy E are supplied to the volume element under consideration. Kardashev (1962) has considered the behavior of the solutions of this equation in great detail, including several additional loss mechanisms as well as the possibility of energy gains through statistical acceleration. Detailed numerical calculations of several of Kardashev's cases have been carried out by de la Beaujardiere (1966). We will consider only a few illustrative examples.

If we confine ourselves to the loss mechanisms mentioned above, the rate of energy loss per electron can be written

$$-dE/dt = a + bE + cE^2 . \tag{2.69}$$

If electrons are continuously being injected with a power-law spectrum $q(E) = KE^{-\gamma_0}$, an equilibrium energy distribution is eventually reached with

$$n(E, t) = \frac{K}{\gamma_0 - 1} \frac{E^{-(\gamma_0 - 1)}}{a + bE + cE^2} . \tag{2.70}$$

Over restricted ranges of energy this function may approximate a simple power law with an index of $-(\gamma_0 - 1)$, $-\gamma_0$, or $-(\gamma_0 + 1)$, depending on which type of energy loss mechanism predominates. If a, b, and c had appropriate values, the radio spectrum emitted by such a distribution might show two changes of slope with a spectral index of $\alpha_1 = -(\gamma_0 - 2)/2$ at low frequencies, $\alpha_2 = -(\gamma_0 - 1)/2$ in some intermediate range of frequencies and $\alpha_2 = -\gamma_0/2$ at high frequencies, each time changing by $-\frac{1}{2}$.

Another simple case of considerable interest is one in which the particle injection is not continuous, but in which the source is initially supplied with electrons having a power-law energy distribution of index $-\gamma_0$ and an isotropic velocity distribution. Then if synchrotron radiation losses predominate, the emission spectrum will develop a discontinuity, changing from $\alpha = \alpha_0 = -(\gamma_0 - 1)/2$ at low frequencies to $\alpha = (\frac{4}{3}\alpha_0 - 1) = -(2\gamma_0 + 1)/3$ at high frequencies. Kardashev (1962) showed that the frequency at which the discontinuity appears will decrease with time t in years as

$$\nu \approx 3.4 \times 10^8 B^{-3} t^{-2} , \tag{2.71}$$

where B is the magnetic field in the source region, measured in gauss.

A source formed by the injection of monoenergetic electrons into a magnetic field will develop a power-law electron spectrum with index -2 and with a lower energy cutoff which decreases with time according to equation (2.13). There is, of course, a fixed upper energy cutoff at the injection energy.

2.4. Luminosity and Energy Requirements for a Source of Synchrotron Radiation

It is of great interest to determine, if possible, the energy requirements for a radio source, and this can be done for synchrotron sources provided the luminosity, the spectral distribution of the emission and the intensity of the magnetic field are all known. The spectral distribution comes directly from radio measurements, and the luminosity follows once the distance to the source is determined. The latter usually comes from the redshift of the associated optical object. The magnetic field may not be measured directly, but its value may sometimes be inferred.

As in § 2.2, we assume that the source consists of a volume V containing a tangled magnetic field with average strength B, in which there are electrons with density $n(E) = n_0 E^{-\gamma}$ between energies E_1 and E_2. The electron energy density is given by equation (2.19),

$$u_e = \int_{E_1}^{E_2} n(E) E dE ,$$

and the total electron energy $U_e = u_e V$. To simplify matters let $N_0 = n_0 V$. Then

$$U_e = N_0 (E_2^{2-\gamma} - E_1^{2-\gamma})/(2 - \gamma) \quad (\text{if } \gamma \neq 2) . \tag{2.72}$$

It remains to evaluate N_0 in terms of the luminosity and the magnetic field.

The rate of emission, summed over all frequencies, for a single electron is given by equation (2.12),

$$-\frac{dE}{dt} = C_3 B^2 E^2 , \tag{2.12}$$

where we neglect the geometric factors and assume an average effective magnetic field equal to B. Integrating over the electron spectrum, we obtain the total luminosity of the source,

$$L = \int_{E_1}^{E_2} C_3 B^2 E^2 N_0 E^{-\gamma} dE . \tag{2.73}$$

Performing the integral and solving for N_0, we find the total energy in electrons to be

$$U_e = \frac{L}{C_3 B^2} \frac{3-\gamma}{2-\gamma} \frac{E_2^{2-\gamma} - E_1^{2-\gamma}}{E_2^{3-\gamma} - E_1^{3-\gamma}} \quad (\text{if } \gamma \neq 2 \text{ or } 3) . \tag{2.74}$$

Thus the total electron energy is proportional to L/B^2 times a shape factor which depends on γ and the energy span.

If evidence of a cutoff is really observed at one end or the other of the radio spectrum, it is better to express the total electron energy in terms of cutoff frequencies ν_1 and ν_2 and the spectral index α. If we allow the approximation that each electron radiates only at its critical frequency, we may use equations (2.8) and (2.74) to obtain

$$U_e = \frac{LC_1^{1/2}}{C_3 B^{3/2}} \frac{2\alpha+2}{2\alpha+1} \frac{\nu_2^{\alpha+1/2} - \nu_1^{\alpha+1/2}}{\nu_2^{\alpha+1} - \nu_1^{\alpha+1}} \quad (\text{if } \alpha \neq -\tfrac{1}{2} \text{ or } -1) . \tag{2.75}$$

The electron energy now depends on $LB^{-3/2}$ and on a shape factor in frequency. It is clear from equation (2.72) that for rather flat spectra ($\alpha \geq -\frac{1}{2}$) an upper cutoff is required to avoid a divergence. For steep spectra ($\alpha \leq -\frac{1}{2}$) a lower cutoff is required. The lower cutoff *energy* may always be taken to equal the electron rest-mass energy, since an electron with that energy is certainly not relativistic and will not produce synchrotron radiation. Many radio sources show departures from a power-law spectrum in the range of ten to a few hundred MHz, and a lower cutoff frequency of 10^7 Hz is often assumed. It is hardly coincidental that this is about the lowest frequency at which observations of discrete sources have been made.

Electrons may not be the only energetic particles within the source. Appreciable amounts of energy may be stored in protons and other heavy particles which emit negligible amounts of radiation because they are accelerated much less by the Lorentz force. To obtain the total particle energy for a radio source we must estimate the relative amounts of energy in relativistic electrons and in energetic baryons. The only relevant observation at our disposal is the fact that electrons account for about 2 percent of the energy in the primary cosmic-ray spectrum observed at high altitudes above the Earth. Thus if we set the total particle energy $U_p = aU_e$, the value of a near the Earth might be 50. Various authors (Burbidge 1956, 1959; Ramaty and Lingenfelter 1966) have estimated a from models for the production and subsequent decay of energetic electrons in radio sources and in our Galaxy. These estimates suggest a value of about 100 for a.

Energy is stored in the magnetic field as well as in the particles. The field energy density is equal to $B^2/8\pi$, so the total magnetic energy $U_m = VB^2/8\pi$. The total energy of the source is

$$U_T = U_p + U_m = aALB^{-3/2} + VB^2/8\pi \tag{2.76}$$

if we take U_e from equation (2.75) and represent the constants and the shape factor by A. The factor $C_1^{1/2}/C_3$ which appears in A has the value 1.057×10^{12} in cgs units.

The energies U_T, U_p, and U_m are shown as functions of B in figure 7. We see that the magnetic energy varies as B^{+2} so that it dominates when the magnetic field is large. The particle energy varies as $B^{-3/2}$ (it would be B^{-2} if we chose to regard the cutoff energies as fixed instead of the frequencies); thus it dominates for small field strengths. The total energy has a minimum near the value of B for which U_p and U_m are equal.

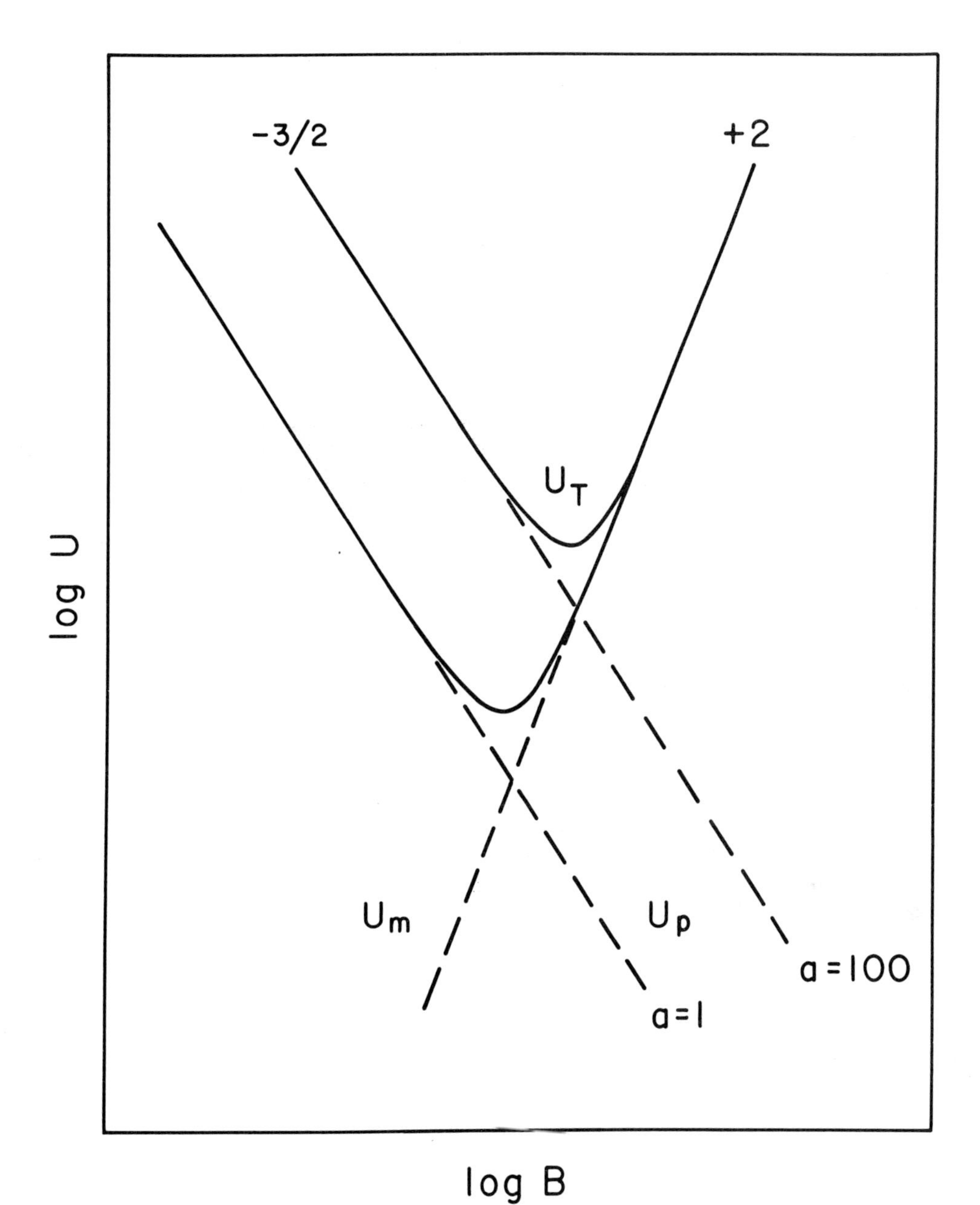

FIG. 7.—Energy relations in a source of synchrotron radiation, showing the dependence of energy on magnetic field and on the ratio a of energy in heavy particles to energy in electrons.

Since even the minimum energies for radio galaxies are embarrassingly large, it is customary to make the best of a difficult situation and to assume that the magnetic field is actually equal to the "equipartition" value, which gives minimum energy.

The minimum energy condition is then

$$U_{\rm min} = 0.50(aAL)^{4/7}V^{3/7}\,,$$
$$B(U_{\rm min}) = (6\pi aAL/V)^{2/7} = 2.3(aAL/V)^{2/7}\,. \tag{2.77}$$

Thus estimates of $U_{\rm min}$ will be in error by a factor of only about 4 even if our guess about the value of a is in error by an order of magnitude. The estimate is also not too sensitive to errors in the measured source dimensions.

As an example of the use of the synchrotron theory in determining the properties of a radio source we will go through a calculation using the observed data for Fornax A (Maltby, Matthews, and Moffet 1963). The source is discussed in more detail in §§ 3 and 4; it is illustrated in Plate II. There are two radio source regions, each subtending 18′, separated by 29′, center-to-center. The redshift is $z = \Delta\lambda/\lambda_0 = 0.0058$, giving a distance of $d = 17$ Mpc if a Hubble parameter of 100 km s^{-1} Mpc^{-1} is assumed. The

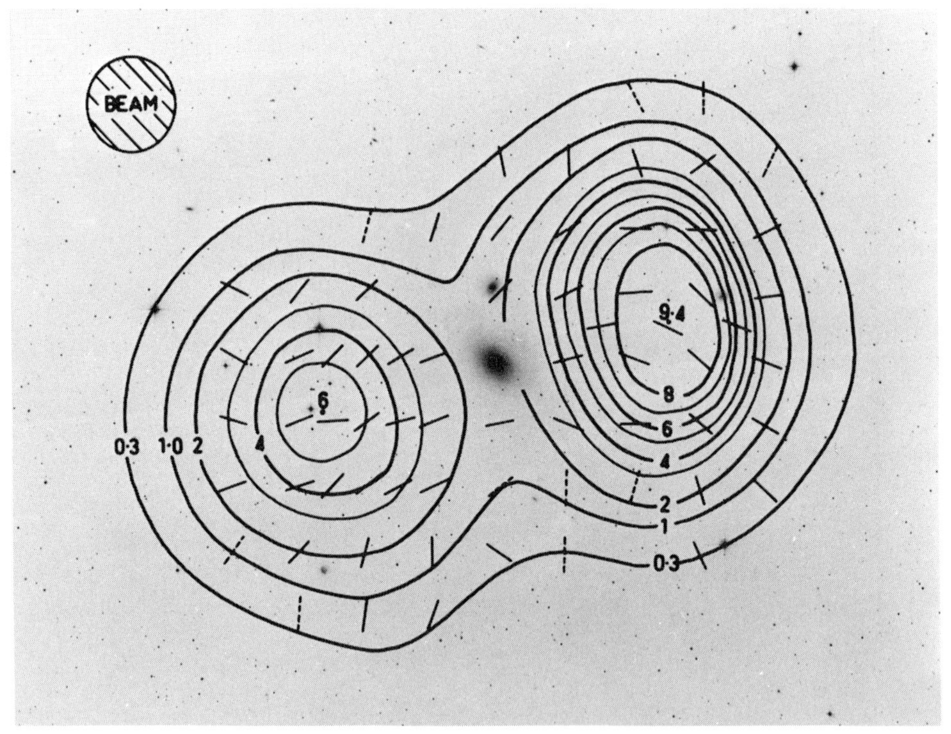

PLATE. II.—Fornax A. The contours show the source brightness as observed at 11.3 cm by Gardner and Price (from Gardner and Whiteoak 1966); the vectors show the direction of polarization (maximum electric field) corrected for Faraday rotation. The angular distance between the centers of the two components is 29′. The photograph of NGC 1316 is from a 48-inch Schmidt plate by Zwicky (from Moffet 1966).

flux density at 1400 MHz is 108 flux units, and the spectral index is $\alpha = -0.75$. We will assume lower and upper limits to the radio spectrum of $\nu_1 = 10^7$ and $\nu_2 = 10^{10}$ Hz. Within these limits the flux $S = \int S_\nu d\nu = 8.1 \times 10^{-12}$ ergs s^{-1} cm^{-2}. Assuming isotropic radiation, the total radio luminosity is $L = 4\pi d^2 S = 2.8 \times 10^{41}$ ergs s^{-1}.

The radio source regions have diameters of 89 kpc and are separated by 140 kpc. Assuming spherical source regions, the total volume of the radio source is $V = 7.4 \times 10^5$ kpc^3 $= 2.2 \times 10^{70}$ cm^3. With the spectral index and cutoff frequencies given above, the factor A in equations (2.76) and (2.77) has the value

$$\begin{aligned} A &= \frac{C_1^{1/2}}{C_3}\frac{2\alpha + 2}{2\alpha + 1}\frac{\nu_2^{\alpha+1/2} - \nu_1^{\alpha+1/2}}{\nu_2^{\alpha+1} - \nu_1^{\alpha+1}} \\ &= (1.057 \times 10^{12})(-1)(-5.6 \times 10^{-5}) \\ &= 5.9 \times 10^7 . \end{aligned} \tag{2.78}$$

We will assume that the ratio of total particle to total electron energy is $a = 100$. Then from equations (2.77) the minimum energy and the equipartition field are $U_{\min} = 1.3 \times 10^{59}$ and $B(U_{\min}) = 8 \times 10^{-6}$ gauss. The electron energies E_1 and E_2 which would give critical frequencies equal to ν_1 and ν_2 in this magnetic field are $E_1 = 280$ MeV and $E_2 = 9$ GeV. The total number of relativistic electrons is $10^{60.3}$, giving a density of relativistic electrons of 10^{-10} cm^{-3}. The lifetime of 9-GeV electrons in an 8 microgauss field is, from equation (2.14), $t_{1/2} = 1.4 \times 10^7$ years.

2.5. Compton Losses

In a synchrotron source of extremely high luminosity and small size it is possible for a significant number of collisions to occur between relativistic electrons and radio-frequency photons. The interaction is identical to that of the familiar Compton effect, except that in the laboratory we observe collisions between energetic photons and electrons at rest. Thus the process in radio sources is often called the *inverse* Compton effect, even though this is something of a misnomer. The process has been discussed by Ginzburg and Syrovatskii (1964*b*) and Felten and Morrison (1966). It may be of importance either because it contributes significantly to the energy losses of the relativistic electrons (as discussed in § 2.3.5.) or because the scattered photons may be detected in another part of the electromagnetic spectrum.

The details of the Compton collision depend on the relative angles with which the electron and photon come together. However, in the case of an isotropic flux of both photons and electrons the average result is that a photon with frequency ν which collides with an electron of energy E is increased in energy so that its frequency becomes

$$\nu' = \tfrac{4}{3}\nu(E/mc^2)^2 , \tag{2.79}$$

which is valid provided that $(h\nu E/mc^2) < mc^2 \ll E$. These conditions hold for radio-frequency photons interacting with the relativistic electrons in a synchrotron source. The factor $(E/mc^2)^2$ by which the photon energy is increased may be thought of as coming from two successive Lorentz transformations, the first from the observer's frame of reference to the center-of-momentum frame of the electron-photon pair and the second from that frame back to the observer's frame. Since the collision reverses the direction of the photon in the center-of-momentum system, the photon gains energy by a factor of (E/mc^2), the electron Lorentz factor, in each of the two successive transformations.

The spectrum of the scattered radiation depends strongly on the energy spectrum of the scattering electrons and only weakly on the initial photon spectrum. For a power-law electron spectrum with isotropic velocity distribution as given by equation (2.17), $n(E) = n_0 E^{-\gamma}$, the volume emissivity of the scattered radiation is

$$\epsilon_{\nu'} = \frac{n_0 n_{\rm ph} \sigma_{\rm T} hc}{8\pi} (mc^2)^{-(\gamma-1)} \left(\frac{4\bar{\nu}}{3\nu'}\right)^{(\gamma-1)/2} k(\gamma), \tag{2.80}$$

where $n_{\rm ph}$ is the total density of incident photons, also assumed isotropic, σ_T is the Thomson scattering cross-section of the electron $\sigma_T \equiv (8\pi/3)(e^2/mc^2)^2$, and $k(\gamma)$ is a slow function of the electron spectral index γ, having a value of order unity (Ginzburg and Syrovatskii 1964*b;* Baylis, Schmid, and Lüscher 1967). The mean incident frequency $\bar{\nu}$ is defined by

$$\bar{\nu} = \int \nu I_\nu d\nu \Big/ \int I_\nu d\nu\,, \tag{2.81}$$

where the range of the integral includes all the frequencies which may make an important contribution to the scattering. From equation (2.80) it may be seen that the scattered radiation has a power-law spectrum with a spectral index identical to that of the synchrotron emission from the electrons.

The rate at which an electron collides with the photons is $\sigma_T c n_{\rm ph}$ collisions per second. The photon number density $n_{\rm ph}$ may be expressed in terms of the photon energy density $u_{\rm ph}$ and the mean frequency $\bar{\nu}$ as $n_{\rm ph} = u_{\rm ph}/h\bar{\nu}$. From equation (2.79) the average energy of the scattered photons is $\frac{4}{3}\, h\bar{\nu}\, (E/mc^2)^2$. Thus the rate at which the electron energy decreases due to scattering of photons is

$$-\frac{dE}{dt} = \tfrac{4}{3} h\bar{\nu}(E/mc^2)^2 \sigma_T c u_{\rm ph}/h\bar{\nu} = \tfrac{4}{3}\sigma_T c u_{\rm ph}(E/mc^2)^2\,. \tag{2.82}$$

This expression for the rate of energy loss due to Compton scattering is exactly analogous to that given in equation (2.12) for the rate of energy loss due to synchrotron radiation. In fact, it is possible to consider synchrotron radiation as Compton scattering of the virtual photons present in a static magnetic field. The combined expression for both rates of energy loss is

$$-\frac{dE}{dt} = (8\pi C_3)\tfrac{2}{3}(u_m + u_{\rm ph})E^2\,, \tag{2.83}$$

where C_3 is the constant of equation (2.12) and u_m is the magnetic field energy density, $B^2/8\pi$. The factor $\frac{2}{3}$ multiplying both energy densities is geometrical and comes from the assumption of isotropic electron and photon fluxes and a uniform magnetic field. In the latter case it is just the mean value of $B_\perp^2$: $\langle B_\perp^2\rangle = B^2 \int \sin^2\theta d\Omega/4\pi = \frac{2}{3}B^2$.

The importance of equation (2.83), aside from its very pleasing symmetry between photon and magnetic energy densities, is that in an optically thin source the ratio of the luminosity of Compton-scattered photons to the luminosity of synchrotron photons is just

$$L_C/L_S = u_{\rm ph}/u_m\,. \tag{2.84}$$

If the photon energy density is assumed to be mainly due to the synchrotron photons themselves, then the photon energy density may be estimated from the synchrotron luminosity and surface area of the source as $u_{\rm ph} \approx L_S/(c \times \text{Area})$. Thus if the radius of the source is r and its distance is d, giving an angular size $\theta \approx r/d$, we have

$$\frac{L_C}{L_S} = \frac{L_S/4\pi r^2 c}{B^2/8\pi} = \frac{S_{\bar{\nu}}\Delta\nu 4\pi d^2}{B^2 4\pi r^2 c/8\pi} = \frac{S_{\bar{\nu}}\Delta\nu}{B^2\theta^2 c/8\pi}, \tag{2.85}$$

where $\Delta\nu$ and $S_{\bar{\nu}}$ represent an effective bandwidth and flux density chosen such that the flux $S \equiv \int S_{\bar{\nu}} d\nu = S_{\bar{\nu}}\Delta\nu$.

From equation (2.85) it can be seen that Compton scattering will be most important in strong sources of small angular size. But this was just the condition which led to a large synchrotron opacity, as discussed in § 2.3.3. Thus it is the sources with spectra showing the effects of self-absorption, and particularly the variable intensity sources, which might be significantly affected by Compton scattering. For such sources we might well adopt $S_{\bar{\nu}} = S_s^*$ and $\Delta\nu = \nu_s$ using the notation of § 2.3.3. Then, making use of equations (2.56)–(2.58) one may transform equation (2.85) to

$$\frac{L_C}{L_S} \approx \frac{2(kT_b)^5 C_1^2 \nu_s}{c^3} \approx (T_b/10^{10.3})^5 \nu_s, \tag{2.86}$$

where T_b is now the maximum surface brightness temperature attained by the optically thick source, which will be the value near the intensity peak at $\nu \approx \nu_s$.

In the optically thick sources seen so far $\nu_s \sim 10^9$ to 10^{10}, so if we substitute a mean value of $\nu_s = 10^{9.5}$ into equation (2.86) we obtain $L_C/L_S \sim (T_b/10^{12.2})^5$. This expression was derived by Kellermann and Pauliny-Toth (1969) who used it to explain an apparent upper limit of $\sim 10^{12}$ °K to the observed surface brightness temperatures of a number of sources with optically thick components. They argued that this limit will be imposed, since in any source in which the ratio in equation (2.86) approaches unity there will be a rapid energy loss as Compton-scattered photons approach the energy density of the original radio photons. In this latter case higher-order scatterings become important, and there is a very heavy energy drain on the relativistic electrons.

If Compton losses become comparable with synchrotron losses in a source, then the scattered photons might be detectable in the infrared, visible, or ultraviolet spectrum. Suppose the magnetic field in a source with a synchrotron cutoff at $\nu_s = 10^{10}$ Hz were $B = 10^{-2}$ gauss. Then from equation (2.8) we would be dealing with electrons with energies $E \sim 250$ MeV, or with Lorentz factors of about 500. Thus we would find the scattered radiation emerging with $\nu' \sim \nu_s(E/mc^2)^2 \sim 2.5 \times 10^{15}$ Hz, in the ultraviolet. Higher magnetic fields could bring the scattered radiation into the visible or the infrared.

If such scattered radio photons did constitute a significant part of the optical luminosity of a compact radio source such as a Seyfert galaxy or a quasi-stellar source, the optical radiation should share the polarization properties of the radio emission (Baylis, Schmid, and Lüscher 1967). In addition the variations in light intensity should be correlated with the radio variations. No such connection has yet been demonstrated.

It is important to note that the ratio in equation (2.86) was derived on the assumption of isotropic fluxes of electrons and photons. The amount of Compton-scattered radiation may be markedly reduced if the electrons stream radially outward from some central source. This type of electron flux has been proposed in some models of quasi-stellar sources and may also be applicable to the radio sources in Seyfert galaxies (Woltjer 1966; see also Burbidge and Burbidge 1967, chap. 11).

2.6. Expansion Losses

The observational material discussed in §§ 3 and 4 shows that the radio sources associated with radio galaxies have a very large range of linear sizes. It is commonly assumed that these sources originate in the nuclei of galaxies and move outward and expand with time until they attain dimensions much larger than those of the parent galaxy. It is easy to show that this expansion, which may involve an increase in source component size of 10^3 times or more, cannot be adiabatic, since adiabatic expansion would quickly extinguish the emission from the source. The problem was first considered by Shklovskii (1960) in connection with the expansion of supernova remnants.

Let a cloud of well-tangled magnetic field, of strength B and containing relativistic electrons, expand isotropically and adiabatically from radius r to radius $r' = fr$. Then the product Br^2 will be conserved since total magnetic flux is conserved. If the expansion is slow compared with the gyration time of the relativistic particles, the magnetic flux enclosed by each particle orbit will also be conserved, requiring a change in electron energy $E \propto B^{1/2}$. (The orbit radius for a relativistic particle is proportional to E/B, so the enclosed flux is proportional to E^2/B = const.) If the energy spectrum of the electrons is

$$n(E)dEdV = n_0 E^{-\gamma} dEdV \,, \tag{2.87}$$

then the effect of the expansion on E, dE, and dV will cause n_0 to vary as $r^{-(\gamma+2)}$. Thus the following changes occur:

$$\begin{aligned} r \rightarrow r' &= fr \,, \quad & B \rightarrow B' &= f^{-2}B \,, \\ E \rightarrow E' &= f^{-1}E \,, \quad & n_0 \rightarrow n_0' &= f^{-(\gamma+2)}n_0 \,. \end{aligned} \tag{2.88}$$

If the source is optically thin, the luminosity at constant frequency will change as

$$L_\nu = \tfrac{4}{3}\pi r^3 \epsilon_\nu \propto r^3 n_0 B^{(\gamma+1)/2} \propto f^3 f^{-(\gamma+2)} f^{-(\gamma+1)} \,, \tag{2.89}$$

thus

$$L_\nu \rightarrow L_\nu' = f^{-2\gamma} L_\nu = f^{4\alpha-2} L_\nu \,. \tag{2.90}$$

The decrease of luminosity with size is very great; since α is typically -0.75, we have $L_\nu \propto f^{-5}$. However, we see sources with dimensions $\sim$100 kpc and sources with dimensions $\ll 1$ kpc all having about the same maximum luminosities. Thus if the sources of various sizes represent an evolutionary sequence, the expansion cannot take place under adiabatic conditions. This will be discussed briefly in § 5.

There is reason to believe that adiabatic expansion may take place in the very small, optically thick sources which are observed to vary in intensity with time scales ranging upward from a few weeks. In the region where such a source is optically thick it has

$$L_\nu = 4\pi r^2 B_\nu = 4\pi r^2 \epsilon_\nu / \kappa_\nu \propto r^2 B^{-1/2} \propto f^{+3} \,, \tag{2.91}$$

where the ratio ϵ_ν/κ_ν comes from equation (2.52), and the changes in r and B due to the expansion are given by equations (2.88). Thus at frequencies well below the cutoff due to synchrotron self-absorption the luminosity at a given frequency rises in proportion to

f^{+3}, while at frequencies where the source is optically thin we have seen that $L_\nu \propto f^{-2\gamma}$. The change in cutoff frequency may also be calculated by substituting the relations of equations (2.88) into equation (2.49) and requiring that $k_\nu r \approx 1$ at $\nu = \nu_s$. The result is

$$\nu_s \propto f^{-(4\gamma+6)/(\gamma+4)} . \tag{2.92}$$

The flux density near the cutoff frequency, $S_s{}^*$, may be found by a similar substitution in equation (2.53) to vary as

$$S_s{}^* \propto f^{-(7\gamma+3)/(\gamma+4)} . \tag{2.93}$$

The shape of the spectrum remains unchanged; it is merely translated to lower intensities and lower frequencies.

The observed changes in the spectra of variable sources, both radio galaxies and quasi-stellar sources, have in many instances been found to fit these relations (Kellermann and Pauliny-Toth 1968).

3. OBSERVED RADIO CHARACTERISTICS

This section contains a summary of the observed *radio* properties of radio galaxies. In some cases the related properties of quasi-stellar radio sources will also be mentioned, although a discussion of the quasi-stellar sources in general is the subject of chapter 8 of this volume. Radio telescopes and the techniques of radiometry are discussed in chapters by Bolton and Drake in Volume **1** of this series. Certain techniques have been developed more extensively since those articles were written, and where necessary these will be described briefly. More thorough reviews of the instrumental aspects of radio astronomy are given by Bracewell (1962), Kraus (1966), and Christiansen and Högbom (1969).

3.1. Distribution

The existing surveys of radio sources are described in chapters 6 and 18 of this volume, and in the latter chapter the statistics of the source counts and their cosmological implications are covered in detail. Table 3 gives a summary of the most important of these surveys, which have provided information on $\sim 10^4$ sources. The surveys which are now in progress will increase this number by nearly an order of magnitude. Except for a small number of sources located within our own Galaxy and another small number associated with nearby normal galaxies, these radio sources are all thought to be either radio galaxies or quasi-stellar radio sources. From studies of the optical counterparts of bright radio sources (Wyndham 1966; Véron 1966) and from studies of the spectra and diameters of fainter, unidentified sources (Bolton 1966) it is concluded that about two-thirds of the 1000 brightest extragalactic sources are radio galaxies while one-third are quasi-stellar sources.

The number of sources $N(S_\nu)$ with flux density greater than S_ν at frequency ν increases rapidly with decreasing values of the limiting flux density S_ν. For flux densities greater than 4 flux units (f.u.) at 408 MHz it is found that $N(S_\nu) \propto S_\nu^{-1.85}$. The slope of this relation becomes less steep for fainter sources. The interpretation of this relationship in terms of cosmological models is the subject of chapter 18 of this volume (see also Ryle 1968).

As far as is now known, radio galaxies are isotropically distributed in the sky, provided the surveys are corrected for the effects of galactic sources seen at low galactic latitudes. Although the large-scale distribution of radio galaxies is apparently isotropic, there is some controversy over the question of clustering of sources. Ceccarelli and Grueff (1967) claimed to find evidence of clustering on a scale of a degree or less, and Wagoner (1967) found statistical effects which could also be interpreted in this way. However, Holden (1966) and Payne (1967) find no evidence for clustering on any scale greater than a few minutes of arc.

Some small amount of clustering is to be expected, since we know of three nearby clusters of galaxies each containing several radio sources. The Virgo cluster has three

TABLE 3

MAJOR SURVEYS OF RADIO SOURCES

Survey*	No. of Sources	(MHz)	Limit of Completeness (f.u.)	Region
MSH	2270	85	20	$\delta<+10°$
3CR	328	178	9	$-5°<\delta$
PKS	2133	410 (1410,2650)	2 to 5	$\delta<+27°$
PKS 2700	800	2700	0.35	$-4°<\delta<+4°$
	618		0.32	$-75°<\delta<-33°$; $03^h, 11^h, 19^h, 23^h$
4C	4842	178	2	$-7°<\delta<+80°$
5C	715	408 (1407)	0.012 to 0.025	3°.8 diameter patches centered on 4 fields
DA	615	1420	3	$-5°<\delta<+70°$
BDFL	424	1400	2	$-5°<\delta<+70°$; $\|b\|>5°$
VRO	1520	611	1	$+15°<\delta<+31°.5$; $+40°<\delta<+44°$
Ohio	11300	1415	0.3	$-36°<\delta<+63°$; some gaps
B1	654	408	1	$-30°<\delta<-20°$
B2	3235	408	0.2	$+29°18'<\delta<+34°02'$

* REFERENCES. MSH, Mills, Slee, and Hill 1958, 1960, 1961. 3CR, Bennett 1962. PKS, CSIRO Radiophysics Staff 1969; Shimmins and Day 1968. PKS 2700, Wall, Shimmins, and Merkelijn 1971; Shimmins 1971. 4C, Pilkington and Scott 1965; Gower, Scott, and Wills 1967. 5C, Kenderdine, Ryle, and Pooley 1966; Pooley and Kenderdine 1968; Pooley 1969; Willson 1971. DA, Galt and Kennedy 1968. BDFL, Bridle, Davis, Fomalont, and Lequeux 1972. VRO, MacLeod, Swenson, Yang, and Dickel 1965; Dickel, Yang, McVittie, and Swenson 1967; Wendker, Dickel, Yang, and staff 1970; Dickel, Webber, Yang, and staff 1971. Ohio, Dixon and Kraus 1968; Fitch, Dixon, and Kraus 1969; Ehman, Dixon, and Kraus 1970; Brundage, Dixon, Ehman, and Kraus 1971. B1, Braccesi *et al.* 1965. B2, Colla *et al.* 1970.

prominent members, NGC 4261, 4374, and 4486 (3C 270, 272.1, and 274), while the Coma cluster has two prominent sources NGC 4869 and 4874 (5C4.81 and 85) and several fainter galaxies detected as radio sources (Willson 1971). The Perseus cluster has NGC 1265 and 1275 which have been known for some time as 3C 83.1 and 3C 84; more recently Ryle and Windram (1968) detected radio emission from IC 310, and Miley, Perola, van der Kruit, and van der Laan (1972) detected three fainter cluster members. The relationship among these Perseus cluster sources was examined by Ryle and Windram (1968). From the fact that NGC 1265 and IC 310 have "tails" extending roughly away from the much stronger source NGC 1275, they concluded that the fainter sources have been excited by energetic plasma ejected from NGC 1275. Hill and Longair (1971) elaborated on this idea, finding additional cases in which a faint radio galaxy may have been excited by a stream of energetic plasma from a nearby strong source. Miley *et al.* (1972) find one faint source in the Perseus cluster with a tail pointing *toward* NGC 1275. They prefer the explanation that the tails are caused by the streaming out of excited plasma behind a radio galaxy moving through a fairly dense intracluster gas.

An apparent anisotropy in the sense of a spatial association between strong radio sources (often showing large redshifts in the spectra of their optical counterparts) and relatively nearby peculiar or distorted galaxies has been claimed by Arp (1966, 1967, 1968*a*, *b*, 1971). The reality of such an association is difficult to test in a statistically meaningful way, since the number of cases involved is still quite small. An evaluation by van der Laan and Bash (1968) suggests that the association claimed by Arp is not statistically significant; however, work on this problem is continuing. It is of great importance since an association between objects of very different redshifts would cast doubt on the validity of Hubble's law and on our entire picture of the Universe. Gunn (1971) has found two instances in which a quasi-stellar object is associated with a cluster of galaxies and has the same redshift as the brightest cluster galaxies. This indicates that at least some quasi-stellar objects have normal cosmological redshifts.

3.2. Spectra

The earliest studies of radio galaxies showed that the radio spectra of these objects could not be explained by thermal emission but were fairly well approximated by power laws with indices of -0.5 to -1.0. Synchrotron radiation provides a ready explanation of such nonthermal spectra; as we have seen in § 2.3, they may arise from power-law electron energy distributions with indices between -2 and -3.

3.2.1. *Statistics of spectra.*—The spectra of radio galaxies are remarkably similar. Not only do they nearly all follow a power law, but the observed range in spectral index is very small. The remarkable uniformity of radio source spectra was first demonstrated by Harris and Roberts (1959), and it has since been confirmed in more extensive studies by Conway, Kellermann, and Long (1963), Kellermann (1964, 1966), Bolton, Gardner, and Mackey (1964), Pauliny-Toth, Wade, and Heeschen (1966), Long, Smith, Stewart, and Williams (1966), Williams and Stewart (1967), Olsen (1967), Williams, Collins, Caswell, and Holden (1968), and Kellermann, Pauliny-Toth, and Williams (1969).

Figure 8, taken from Kellermann's (1966) paper, shows the narrow range observed for the spectral index. The mean value of α for the 184 sources in the histogram is -0.77. The distribution is well fitted by a Gaussian with a dispersion of 0.14. The mean value of α would correspond to an electron spectral index $\gamma = 2.54$. The same mean spectral index is found for all levels of flux density at frequencies between about 100 and 1000 MHz. At frequencies above 1000 MHz the dispersion of the spectral index distribution increases markedly, indicating that the spectra of radio sources are not so uniform at decimeter wavelengths (Bolton, Gardner, and Mackey 1964; Bolton 1966; Williams *et al.* 1968).

3.2.2. *Correlations with other source properties.*—Kellermann (1964) and Lequeux (1965) examined a large number of radio sources in search of a correlation between spectral index and any of the other properties of these objects. The only such correlation they found to have possible statistical significance was a relation between the linear extent of the source and its spectral index, in the sense that the smaller sources tend to have flatter spectra. Kellermann (1966) proposed a model to explain this behavior.

He suggested that sources are formed with a small diameter by injection, into a magnetic field, of relativistic electrons having an energy spectral index $\gamma = 1.5$, producing a flat radio spectrum with $\alpha = -0.25$. According to Kellermann's model the typical radio galaxy is formed when such a source expands to a size of 10 kpc or greater. Energy losses due to the expansion are made up by recurrent injection of more electrons with $\gamma = 1.5$. The situation is similar to one of Kardashev's models discussed in § 2.3.5.; the recurrent injections form a quasi-continuous electron supply, and the radiation losses cause the equilibrium electron spectrum to have an index $\gamma = \gamma_0 + 1 = 2.5$. The radio spectrum then has $\alpha = -0.75$, as observed for the vast majority of sources. In a later stage, or at very high frequencies where the injection cannot be considered quasi-continuous, radiation losses force a steeper spectrum with $\alpha = -1.33$, as predicted by Kardashev's one-shot injection model.

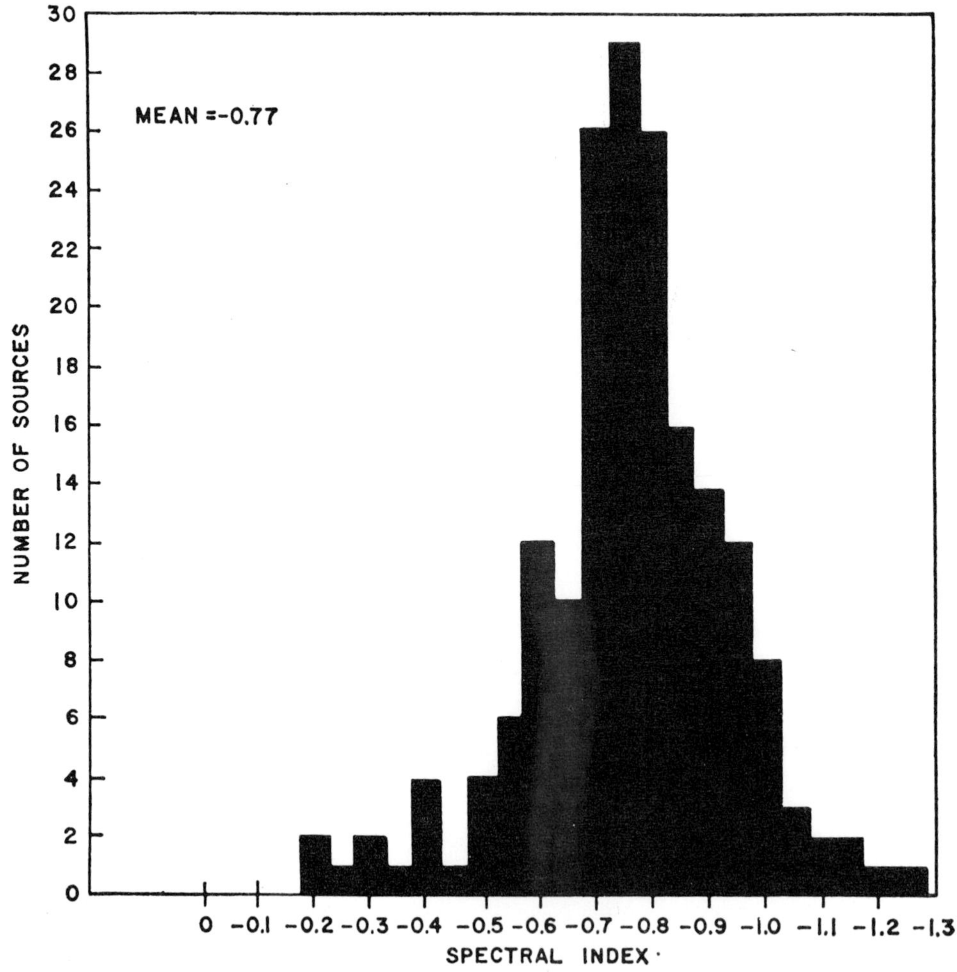

Fig. 8.—Histogram of mean spectral indices over the range 38–1400 MHz excluding sources near the galactic equator, $|b| < 10°$ (from Kellermann 1966).

This model seems a little artificial in its assumption of a universal electron injection spectrum with $\gamma = 1.5$. There are very few sources with spectra as flat as $\alpha = -0.25$, and almost all of these are flattened by the effects of synchrotron self-absorption. The Crab Nebula is the only such flat-spectrum source in which the spectrum is clearly unaffected by absorption. An examination of the data on which Kellermann bases his model shows that all the flat-spectrum sources have linear extents less than 10 kpc, which suggests that absorption effects may well be important in shaping the spectra of these objects. If there is a "universal" electron spectrum, it most probably has $\gamma \approx 2.5$.

Van der Laan and Perola (1969) have criticized Kellermann's quasi-continuous injection model on the grounds that radiation losses, if they are an important factor in determining the energy spectrum of electrons in a source, should produce many sources with steep spectra at high frequencies. However, few if any such sources are observed.

A correlation between spectral index and luminosity has been suggested by several authors (Heeschen 1960; Conway, Kellermann, and Long 1963). Kellermann, Pauliny-Toth, and Williams (1969) find that, for radio galaxies only, there is a change in mean spectral index at high frequencies from -0.5 for the faintest to -1.0 for the intrinsically brightest.

Any satisfactory theory of radio galaxies must explain the presence of a limited range of spectral index in sources with a tremendous range of luminosity and extent. If we include galactic sources such as supernova remnants, then the range of luminosity over which the "universal" nonthermal spectrum prevails is 11 orders of magnitude.

3.2.3. *Classification of spectra.*—The spectra of radio galaxies show marked differences, despite their general similarity. Kellermann, Pauliny-Toth, and Williams find that about 40 percent show no departure from a simple power law over the range 40–5000 MHz and designate these straight-line spectra as Class S. Examples of these are shown in figure 9*a*. Many sources have convex spectra, with more negative spectral index at high frequencies; these are called Class C− and are shown in figure 9*b*. A subgroup of these have a sharp decrease in intensity at low frequencies, as is shown in figure 9*c*. Finally a few sources have concave spectra with a positive change in the spectral index at high frequencies; examples of this Class C+ are shown in figure 9*d*.

The change in spectral index at high frequencies in the C− sources may be due to a gradual steepening of the relativistic electron spectrum at high energies. The sharp cutoffs at low frequencies, as shown in figure 9*c*, are almost certainly due to synchrotron self-absorption within the sources. These objects are all compact, with high surface brightness, a fact which was first noted by Kellermann, Long, Allen, and Moran (1962). Kellermann, Pauliny-Toth, and Williams (1969) note that at least a portion of the radiation from many sources originates in compact components (Little and Hewish 1968; Cohen 1969) and that these could become optically thick at frequencies below 1000 MHz, causing the overall spectrum of the source to flatten at low frequencies.

As an example of the reabsorption process, the spectrum of the quasi-stellar source 3C 147 drops in intensity at frequencies below about 100 MHz. Extrapolation of the linear part of its spectrum down to that frequency gives a flux density of about 100 flux units. From figure 5 or equation (2.54) it can be seen that the angular diameter of the source must be about 0\".2 if synchrotron self-absorption is responsible for the

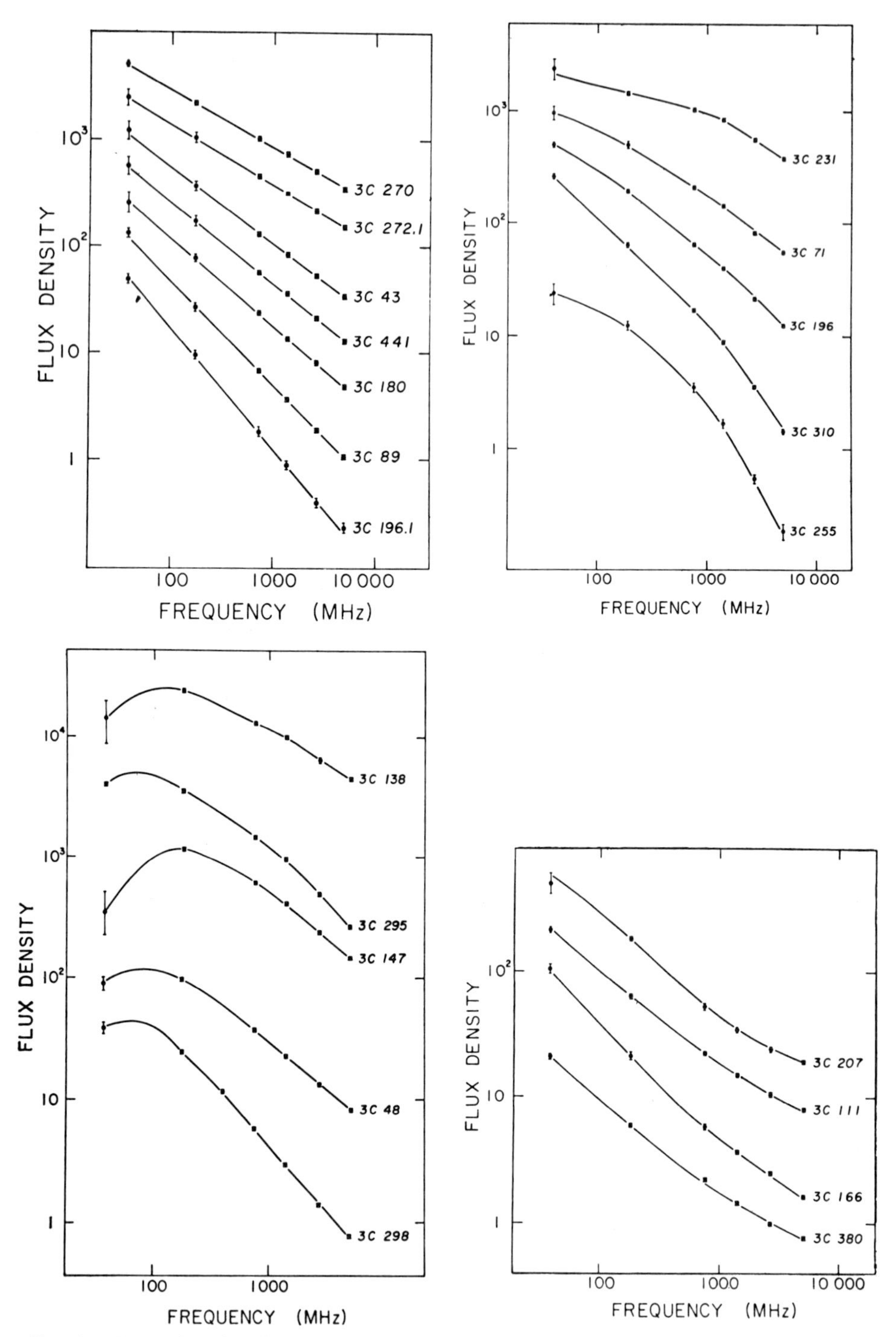

FIG. 9.—Examples of radio source spectra: (*a*) class S, simple power-law spectra; (*b*) class C−, sources with convex spectra, curving downward at high frequencies; (*c*) sources showing abrupt cutoffs at low frequencies; (*d*) class C+, sources with concave spectra (from Kellermann, Pauliny-Toth, and Williams 1969).

intensity decrease. The measured angular diameter of 3C 147 is in agreement with this prediction (Adgie *et al.* 1965; Anderson and Donaldson 1967), although it must be noted that the diameter measurements were made at somewhat higher frequencies.

The sources with C+, or inverted, spectra are almost all quasi-stellar sources, but a small minority are radio galaxies, of which the best known are NGC 1275 = 3C 84, 3C 120, and 3C 371. The anomalous spectrum of NGC 1275 was first discovered by Dent and Haddock (1965), who found that while the meter-wavelength emission from this source is normal and follows a rather steep power law, the flux density at wavelengths shorter than about 10 cm rises, being about 3 times greater at 3 cm than at 10 cm. Soon after this discovery Dent (1965, 1966) found that the centimeter-wavelength flux from NGC 1275 and from several quasi-stellar radio sources was time variable (see figure 10), indicating a size for the emitting region of the order of a parsec at most.

NGC 1275 has a distance of about 50 Mpc, so a source 1 pc in diameter would subtend about $0\overset{''}{.}004$. As may be seen in figure 5, a source of that diameter having a flux density of 20 flux units would become optically thick at about 1000 MHz.

3C 120 is a Seyfert galaxy with an extremely variable radio source at centimeter wavelengths (see fig. 10). Pauliny-Toth and Kellermann (1968) report multiple outbursts, one of which increased the 2-cm flux from 3C 120 from 6 to 19 flux units within a period of 10 weeks. The small sizes of the variable-intensity components in 3C 84 and 3C 120 have been confirmed by very long baseline (VLB) interferometry (Cohen 1969); changes in the structure of these small components have also been seen (Cohen 1969; Shaffer, Cohen, Jauncey, and Kellermann 1972).

A further type of radio galaxy with an anomalous spectrum has been found by Heeschen (1968, 1970), who discovered that many elliptical galaxies contain low-luminosity radio sources with either very small angular size or with size comparable to the extent of the optical galaxy. The latter have normal spectra, but those with small angular size usually have inverted spectra, characteristic of self-absorbed synchrotron sources. At least two of these sources, in NGC 1052 and NGC 4486, have been shown by VLB interferometry to have angular sizes $\leq 0\overset{''}{.}001$ (Cohen *et al.* 1969, 1971).

The general subject of radio source spectra has been reviewed in detail by Scheuer and Williams (1968) and van der Laan and Perola (1969).

3.3. Angular Sizes and Brightness Distributions

The earliest explicit observation of a radio galaxy was made by Bolton and Stanley (1948). Using a sea interferometer, they determined that Cygnus A had an angular extent of less than 8′. One of the major efforts in radio astronomy since that time has been to measure the angular extent of as many radio sources as possible, and further, to determine the distribution of brightness within the sources.

3.3.1. *Techniques.*—The most straightforward way to investigate the size and structure of a radio source is to scan it with an antenna having a primary beam pattern much smaller in angular extent than the source itself. A scan of the source region provides information with which one may construct a contour map or even build up a television-like image of the source. Unfortunately even the best existing antennas have beamwidths of at least 1′ to 10′, and only a very few of the nearest radio galaxies can be resolved

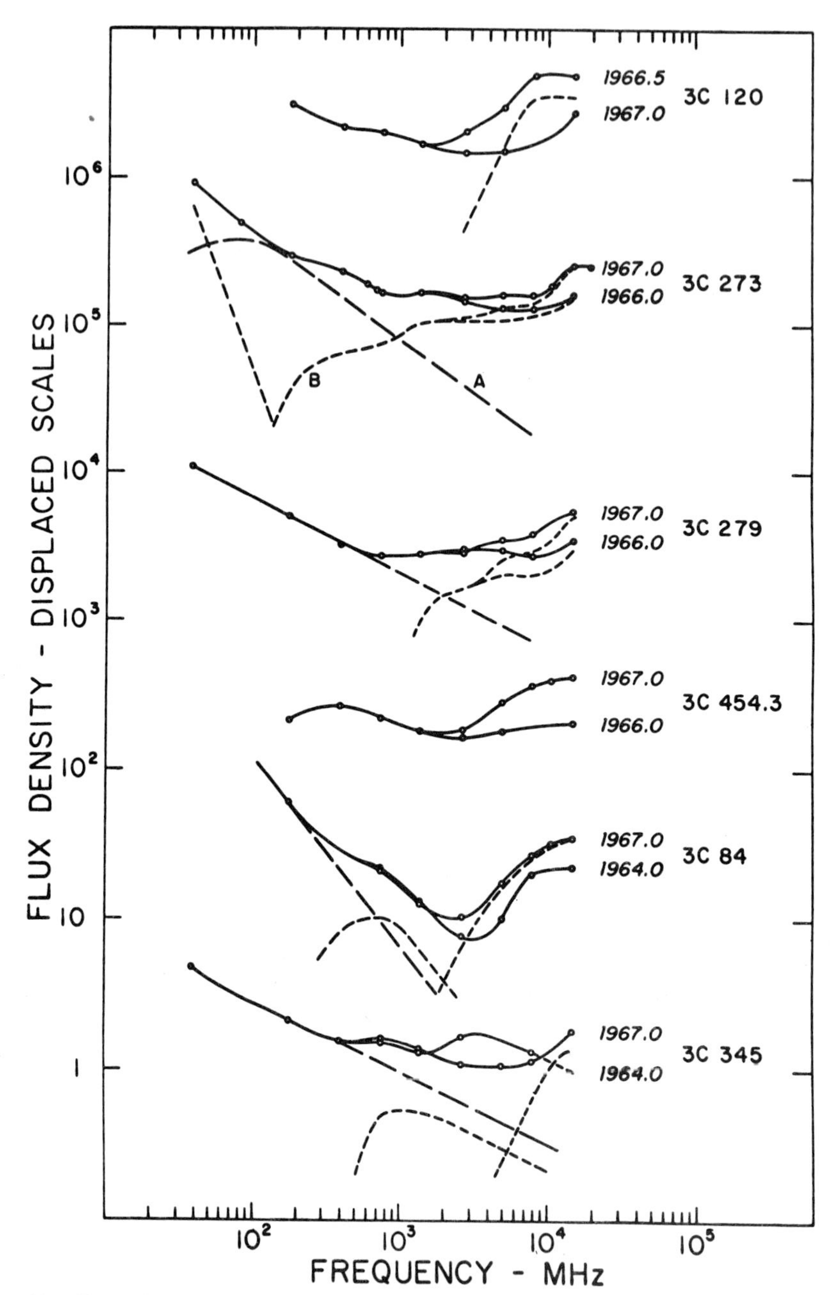

FIG. 10.—Examples of intensity variations at high frequencies in several sources including the radio galaxies 3C 120 and 3C 84. The full lines show the spectra at various epodes, while the dashed lines show possible separation into individual components with different cutoff frequencies for synchrotron self-absorption (from Kellermann and Pauliny-Toth 1968).

with beams of this size.[6] These nearby radio galaxies are extremely important, because our knowledge of their radio and optical properties is more nearly complete than for any others.

Plates II and III show radio and optical pictures of two radio galaxies which are well resolved with the largest available pencil-beam antennas. Centaurus A has an angular

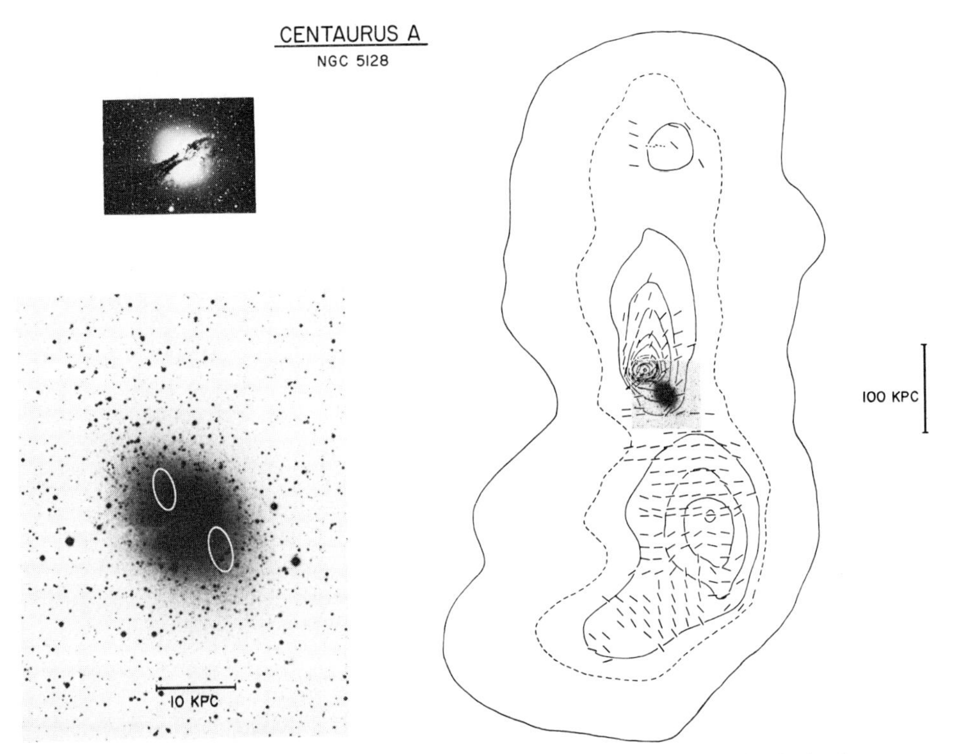

PLATE III.—Centaurus A. At right are shown the intensity contours and polarization vectors (orthogonal to the direction of maximum electric field) for the extended radio source, with an inset photograph of NGC 5128. At lower left is an enlarged photograph of the galaxy showing the position of the central double source. At upper left is a positive photograph showing the absorbing lane. The composite illustration was prepared by T. A. Matthews (from Moffet 1966; radio data from Cooper, Price, and Cole 1965).

extent of 10°; Fornax A, nearly 1°. In each case the radio source has two major components located on opposite sides of a large elliptical galaxy. This double structure, first discovered in Cygnus A, is now known to be characteristic of almost all radio galaxies. Centaurus A is a compound double source, since two additional components lie within the central galaxy. These are shown in the insert in Plate III. Additional properties of Centaurus A and Fornax A will be described in §§ 3.4 and 4.

Various clever techniques have been developed to obtain relatively fine pencil beams

[6] There are a few antennas capable of higher resolution, but only at very short wavelengths; they have insufficient collecting area for study of any but a few of the brightest radio galaxies.

at long wavelengths, where the steerable paraboloids commonly used at centimeter and decimeter wavelengths have very poor resolution. The Mills Cross, tee, and Christiansen grating antennas are well described by Bolton in Volume 1 of this series. A novel antenna, the *radio heliograph*, has since been developed by Wild (1967). Using 96 small paraboloids on the circumference of a 3-km-diameter circle, it achieves a resolution of 3′ over a 2° field at 3.75 m wavelength. While designed for solar radio observations, this instrument has been used to study some of the brightest radio galaxies (Morimoto and Lockhart 1968, Lockhart and Sheridan 1970).

Other methods must be used to obtain diameter and brightness distribution information for the smaller and more distant radio galaxies. Fortunately the techniques of *interferometry* permit us to obtain high resolution with antennas of moderate size. Interferometric observations have given us the bulk of our present information on the sizes and shapes of radio galaxies.

An interferometer in its simplest form consists of two separate antennas whose outputs are brought to a common point and multiplied together. If both antennas receive energy from the same radio source, the two signals will be at least partially coherent, and this coherence will be detected in the multiplication process. Radio interferometers are discussed in basic papers by Ryle (1952) and Blum (1959).

The response of a two-element interferometer to a point source of strength S is

$$S \cos(2\pi w \sin\theta) \approx S \cos(2\pi w\theta) ,$$

where w is the baseline length, or the separation between the elements, measured in wavelengths. The angle θ is the angle between the direction of the point source and the plane which bisects the baseline and is perpendicular to it; for simplicity we assume that θ is small, in which case the approximation $\sin\theta \approx \theta$ is valid. The response is sinusoidal with a *spatial frequency* of w oscillations per radian. If the source distribution has a finite angular diameter, the response of the interferometer at any particular spacing is determined by the corresponding spatial frequency component in the brightness distribution of the source. Such a separation into components of different spatial frequency is equivalent to a Fourier analysis of the source distribution; thus the response of the two-element interferometer at any spacing is a measure of the Fourier transform of the source brightness distribution evaluated at the corresponding spatial frequency.

In general the Fourier transform is a complex function, the amplitude being proportional to the amplitude of the interference fringes and the phase being equal to the shift of the fringes away from the position of the centroid of the source distribution. When normalized to its value at small antenna spacings, the transform is called the *visibility function* of the source. If the visibility function is known in sufficient detail, it may be inverted to obtain the original source distribution. If the maximum spacing attained is w, the resolution achieved will be $\sim 1/w$.

The number of observations which must be made with different lengths and orientations of the interferometer baseline is determined by the size of the source. Let the sources have maximum angular diameters of X and Y radians along the axes of a Cartesian coordinate system centered on the source. Then if u and v are the projections of the baseline on these axes, such that $w^2 = u^2 + v^2$, the visibility function of the source is completely determined by its values on a regular lattice of points at intervals

$\leq 1/X$ in u and $\leq 1/Y$ in v, according to the two-dimensional sampling theorem (Bracewell 1958). Sampling at intervals greater than these *critical spacings* cannot give an unambiguous description of the source structure; however, sampling at even very much larger intervals can give useful information about the angular diameters of various components within the source.

Interferometric studies of several hundred discrete sources have been done with a fairly complete sampling of the visibility function out to spacings of several thousand wavelengths along one or two lines in the spacing plane, or (u, v)-plane as it is often termed. Sampling along a single line in this plane is equivalent to observation of the source with a fan-beam antenna; i.e., the source is resolved in only one direction. This type of observation is sometimes called a *fan-beam synthesis*.

Sampling at more widely spaced intervals in order to determine rough source diameters has been carried out to spacings of 2×10^6 wavelengths with interferometers in which the signal from one antenna is brought to the correlator via microwave relay (Palmer *et al.* 1967). Using the technique of independent interferometry, or *very long baseline interferometry* (VLBI) (Bare *et al.* 1967; Broten *et al.* 1967), in which the signals from the two antennas are tape recorded and correlated later after the tapes have been brought to a common point, baselines of up to 10^9 wavelengths have been achieved. This technique has shown that several radio sources have features $\lesssim 0\overset{''}{.}0001$ in angular diameter.

The procedures of variable-spacing interferometry are presented in detail by Moffet (1962). The interpretation of measured visibility functions in terms of source structure is discussed by Maltby and Moffet (1962), Rowson (1963), Arsac (1966), Bash (1968*a*, *b*), and Fomalont (1968, 1969, 1971).

If measurements are carried out with a variable-spacing interferometer at a sufficiently dense lattice of points out to a maximum spacing of D in all directions, the resulting information about the brightness distribution in the source region will be equivalent to that which would be obtained with a filled aperture of diameter D. This was first noted by Ryle, who called the process *aperture synthesis*. Ryle and his co-workers have constructed a number of very successful radio telescopes making use of the aperture synthesis principle (Ryle and Hewish 1960, Ryle and Neville 1962, Ryle 1962, 1972).

Figure 11 shows the 6-cm intensity contours of the radio galaxy 3C 295 obtained with the Cambridge 5-km synthesis radio telescope (Ryle 1972), which gives a resolution of 2″ in right ascension by 2″ cosec δ in declination. In this picture the ordinate is contracted by a factor of cosec δ to give a circular beam shape; the small L in the corner indicates 2″ in each coordinate. This instrument makes use of the Earth's rotation to fill in an ellipse in the (u, v) plane using antennas physically spaced along a one-dimensional baseline. This technique is called *Earth rotation synthesis* (Ryle 1962). Other instruments designed as Earth rotation synthesis interferometers are in use at Stanford, California; Fleurs, Australia; and Westerbork, Netherlands. A very large array (VLA) using this principle is under construction in New Mexico by the National Radio Astronomy Observatory, as are several other smaller instruments.

In recent years important information has been gained about the shapes and positions of several radio sources through observations of *lunar occultations* of these sources. If there were no diffraction at the limb of the Moon, a point source would seem to disappear instantaneously as the Moon passed in front of it. Since the Moon's movements are known very accurately, the time of each such disappearance or reappearance may be used to find an arc (representing the position of the limb at that time) on which the source must lie. The intersection of several such arcs will provide an accurate position for the source.

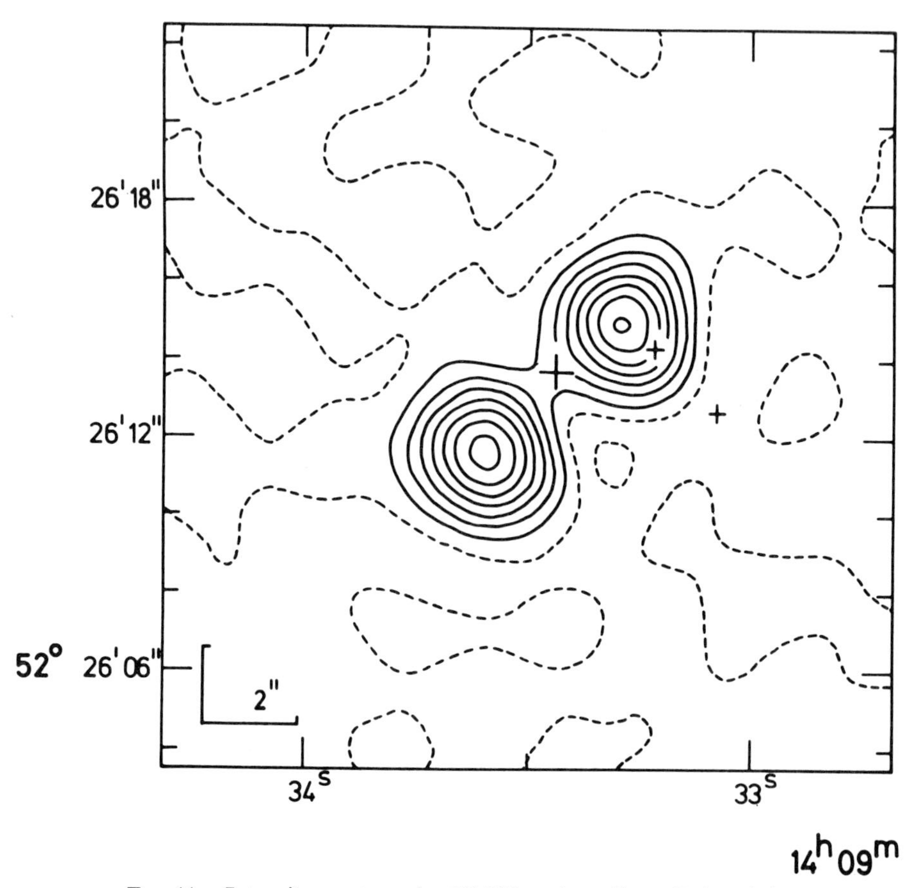

Fig. 11.—Intensity contours for 3C 295 at 6 cm (from Ryle 1972).

In the absence of diffraction a source of finite angular diameter would appear to disappear gradually. Differentiation of the curve of intensity versus time during disappearance or reappearance would yield a one-dimensional profile of the source brightness distribution. However, the finite wavelength of radio waves assures that diffraction will take place at the Moon's limb. The apparent intensity of a point source will begin to oscillate as the limb approaches, will drop to a quarter of its normal intensity at the moment of geometrical occultation, and will then decrease exponentially as the limb advances beyond that point. The angular width of the first Fresnel diffraction fringe is proportional to the square root of the wavelength and is about 6".6 for $\lambda = 21$ cm.

Fortunately the Fresnel diffraction is mathematically well behaved, and its effects may be removed by a method discovered by Scheuer (1962). The procedure is to convolve the observed occultation curve with a *restoring function* obtained from the second derivative of the diffraction pattern which would be observed from a point source. This second derivative oscillates at one end of its range, with increasing amplitude and frequency, but the divergence may be subdued by convolution with a smoothing function, usually a Gaussian. The width of this Gaussian equals the angular resolution obtained in the restored profile of the occulted source.

Using Scheuer's method, angular resolutions as small as 0″.3 have been obtained (Scheuer 1965), and source positions have been determined to better than 0″.5 (von Hoerner 1966; Hazard, Gulkis, and Bray 1967). The positional accuracy is limited by our incomplete knowledge of the profile of the Moon's limb. The limiting factor in resolution is the signal-to-noise ratio of the original observation. The average rate of motion of the Moon's limb is 0″.5 per second of time. Thus the signal which defines the intensity of any 1″ strip of the source is received in about 2^s. High-sensitivity is usually obtained in radio astronomy by using long integration times for each observation; in occultations this is not possible. Thus successful observation of occultations of faint radio sources demands large antennas which can yield good signal-to-noise ratios even with integration times of only a fraction of a second. The limiting factors for occultation observations have been discussed by von Hoerner (1964).

There is one additional method which has been used to give information about radio source diameters. At relatively low frequencies the signals from small-diameter radio sources are observed to scintillate, particularly during the part of the year when the path from the source to the Earth comes nearest to the Sun (Hewish, Scott, and Wills 1964). This *interplanetary scintillation* occurs when radio waves from the source are scattered by irregularities in the interplanetary plasma flowing out from the Sun, often called the solar wind. The angle of scattering is of the order of 1″. Sources smaller than about 1″ in diameter are caused to scintillate, while signals from large-diameter sources are not noticeably affected. Any radio source which shows marked interplanetary scintillation must have at least one major component with a diameter of 0″.5 or less. The theory of interplanetary scintillation has been investigated in detail by Little and Hewish (1966), Salpeter (1967), and Budden and Uscinski (1970, 1971). Major studies of source diameters using this technique have been carried out by Cohen, Gundermann, and Harris (1967) and by Little and Hewish (1968).

The techniques and results of high-resolution observations, including interferometry, occultations, and interplanetary scintillation, have been reviewed in detail by Cohen (1969).

3.3.2. *Shapes and diameters of radio galaxies.*—We have remarked that the spectra of radio galaxies are surprisingly similar. The same may be said of the shapes of these objects. The radio sources shown in Plates I–IV each have two prominent components situated on opposite sides of a galaxy. This same type of double structure has been found in the majority of mature radio galaxies—taking as "mature" all those in which the radio source is of galactic dimensions, say greater than 20 kpc.

Our information about the structure of radio sources comes mainly from interferometric investigations (and from the three-element aperture-synthesis telescope at Cambridge, which is a special type of interferometer). Fan-beam syntheses with resolution in one or more directions have been carried out on about 500 northern sources. These include studies at 31 cm by Maltby and Moffet (1962), at 21 cm by Lequeux (1962) and by Fomalont (1968, 1969, 1971), and at 11 cm by Bash (1968*a*, *b*). Studies of 50 southern sources at 75 and 21 cm have been made by Ekers (1969). Most of these observations were limited to spacings of 2000 wavelengths or less, giving resolutions of the order of 1′. However, Lequeux examined a small number of sources at spacings out to 10,000 wavelengths, and Bash observed 234 sources at intervals of 2000 wavelengths between 10,000 and 20,000 wavelengths, permitting resolution of details as small as 5″ in some sources.

Observations of 82 sources which have been resolved with the Cambridge interferometer are described by Macdonald, Kenderdine, and Neville (1968). A further 60 sources are described by Mackay (1969). Figure 12, taken from the former paper, shows

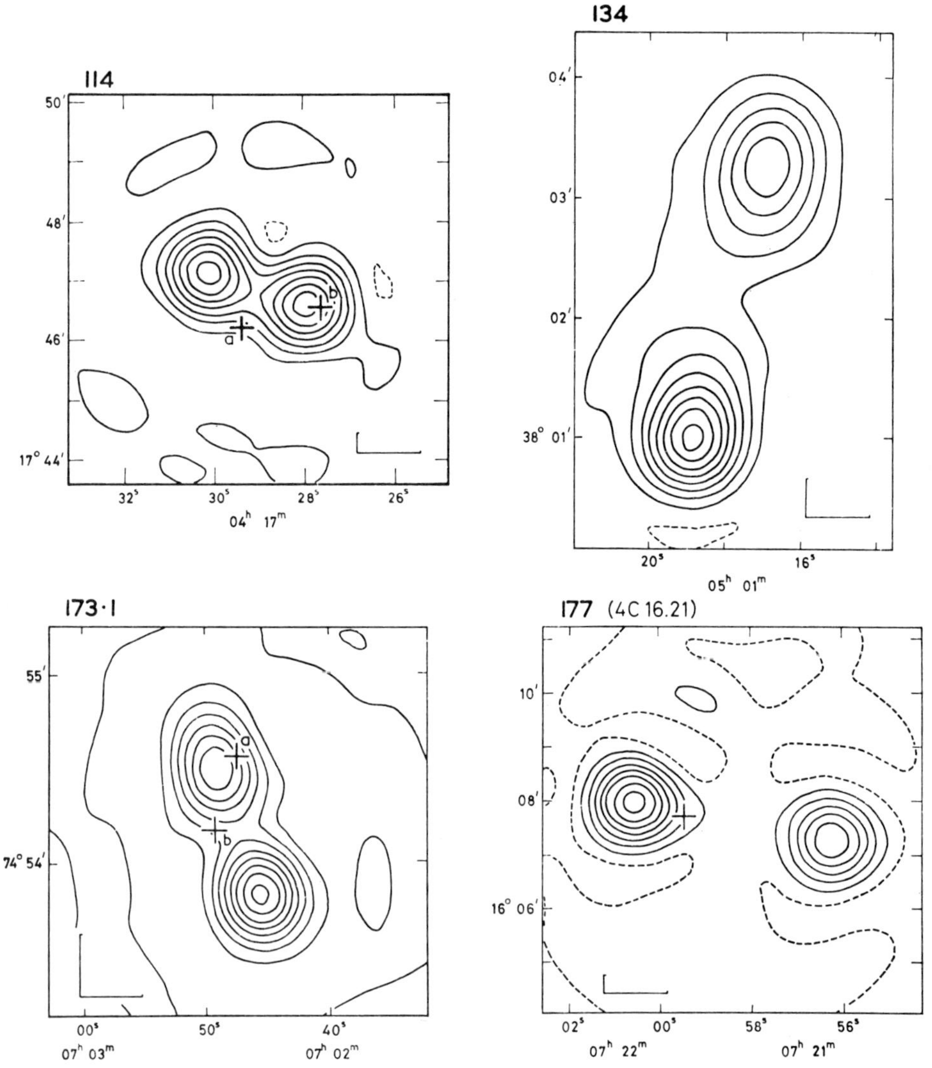

FIG. 12.—Intensity contours for four 3C sources at 21 cm showing the characteristic double structure of radio galaxies (from Macdonald, Kenderdine, and Neville 1968).

a typical group of double sources.[7] Not all sources present such a simple picture, though. Figure 13 shows 21-cm contours for 3C 465, one of the most complex sources resolved with the Cambridge instrument. A number of centers of emission are clearly present, spread out over an arc nearly 10′ in length. (Beware of the distortion of the vertical scales in figs. 11–14.)

Slightly more complicated than the ordinary double sources are a number which show variations on this theme. Some may be triple, with a prominent central component; others may have a bridge of emission connecting the components or sometimes more than one pair of symmetrical components. Figure 14 shows 3C 452, which combines both of these features. When more than one pair of components is present, the outermost pair always seems to be the most intense (Macdonald, Kenderdine, and Neville 1968). There are also some triple sources, in which a central component is situated between an outer pair.

Typically the components of the double sources have diameters of a third to a fifth of their separation. However, there are notable exceptions with much larger ratios of separation to diameter. One of the most interesting is 3C 33, shown in Plate IV. The schematic representation of the radio source regions in Plate IV was based on interferometry with a resolution of about 10″ (Moffet 1964). However, more recent work with higher resolution indicates that each component probably has a more complex structure consisting of a bright core with diameter 2″ to 5″ and an extended region with length ~15″ (Bash 1968*b;* Mitton 1970). The Cambridge 21-cm synthesis map (Macdonald, Kenderdine, and Neville 1968) shows a faint third source region extending up from the southern component of the double to about the position of the central galaxy. This feature is absent at 5 cm (Mitton 1970). The separation of the double source is 244″, giving a ratio of separation to diameter of about 25 if the mean diameter of the components is taken as 10″.

Not all radio galaxies show the basic double structure, even allowing for variations such as bridges or central components. A substantial minority are "core and halo" objects, in which a small component of high surface brightness is superposed on an extended, fainter source. The prototype core-halo source is M87 = NGC 4486 = 3C 274, which has an elliptical halo about 6′ × 10′ in diameter and a core which has a complex brightness distribution within an angular region of 30″ × 70″ (Hogg, Macdonald, Conway, and Wade 1969). The core source coincides in position with, and is aligned parallel to, the optical jet which extends out of the nucleus of M87 for a distance of 21″ in position angle 290° (Baade 1956; de Vaucouleurs, Angione, and Fraser 1968). Hogg *et al.* (1969) showed that considerable structure was present in the radio core of M87, with two components less than 2″ in diameter, one coinciding with the nucleus of the galaxy and one with the brightest part of the optical jet. The nuclear source has been shown by Cohen *et al.* (1969) to have a component $\lesssim 0''.002$ in angular size, corresponding to a linear diameter of 6 light-months.

The most nearly complete sample of radio source brightness distributions is in the

[7] The information about these double sources which is conveyed by such contour maps is very little more than that obtained in the simpler fan-beam synthesis studies, namely, the diameter of the components and their orientation and separation. The many contour lines which are shown give a somewhat misleading impression of detail which is not, in fact, present in the observations.

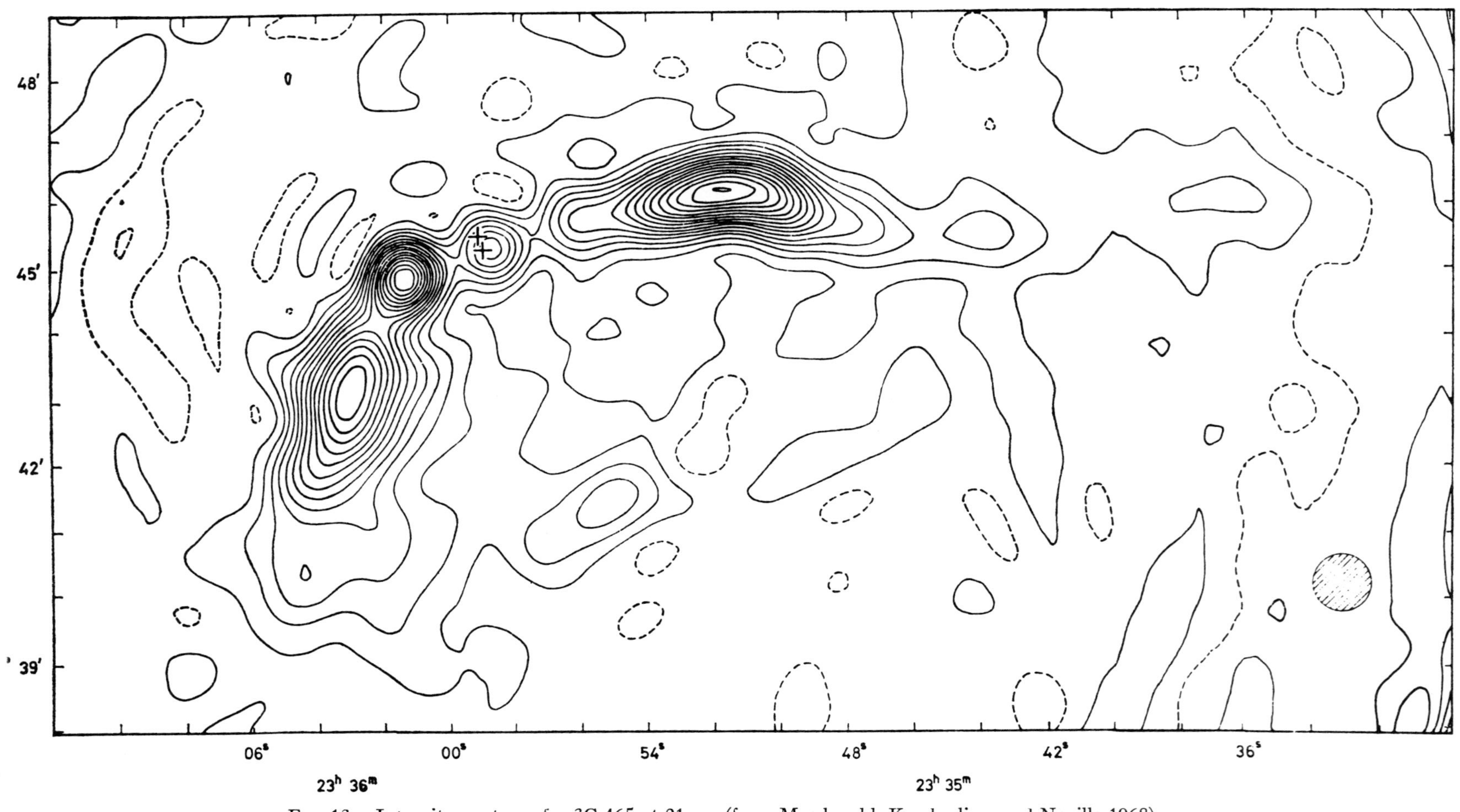

FIG. 13.—Intensity contours for 3C 465 at 21 cm (from Macdonald, Kenderdine, and Neville 1968).

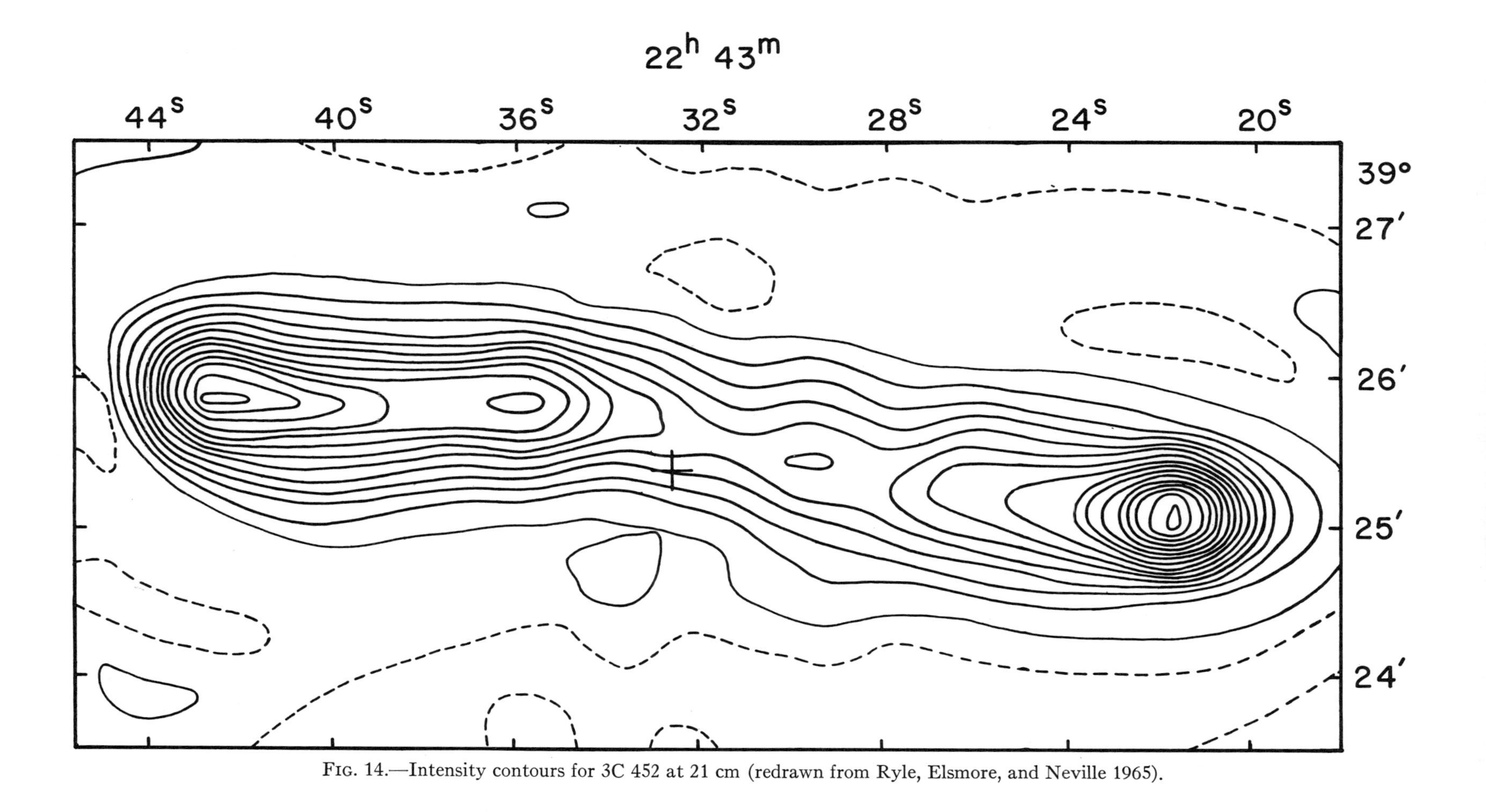

Fig. 14.—Intensity contours for 3C 452 at 21 cm (redrawn from Ryle, Elsmore, and Neville 1965).

study of Fomalont (1968, 1969), who made an east-west fan-beam synthesis of all sources known in 1965 to have flux density greater than 2 flux units at 21 cm and $\delta >$ $-48°$. Of the 532 sources he examined about half were completely unresolved, with angular size less than 15″. Eighty sources were large enough that details of their structure could be established with confidence. Table 4 shows the distribution of structural types for these sources. The symmetrical doubles, triples, and core-halos have already been described. The asymmetrical doubles might be called displaced core-halos, since they contain two components of markedly different size but separated by more than the sum of their diameters. Eleven percent of the sources had brightness distributions too complex to be determined by the observations. Remarkably, only 6 percent appeared to have a simple, symmetrical brightness distribution, and these few could easily be accounted for as chance alignments of doubles. If the first two categories are lumped

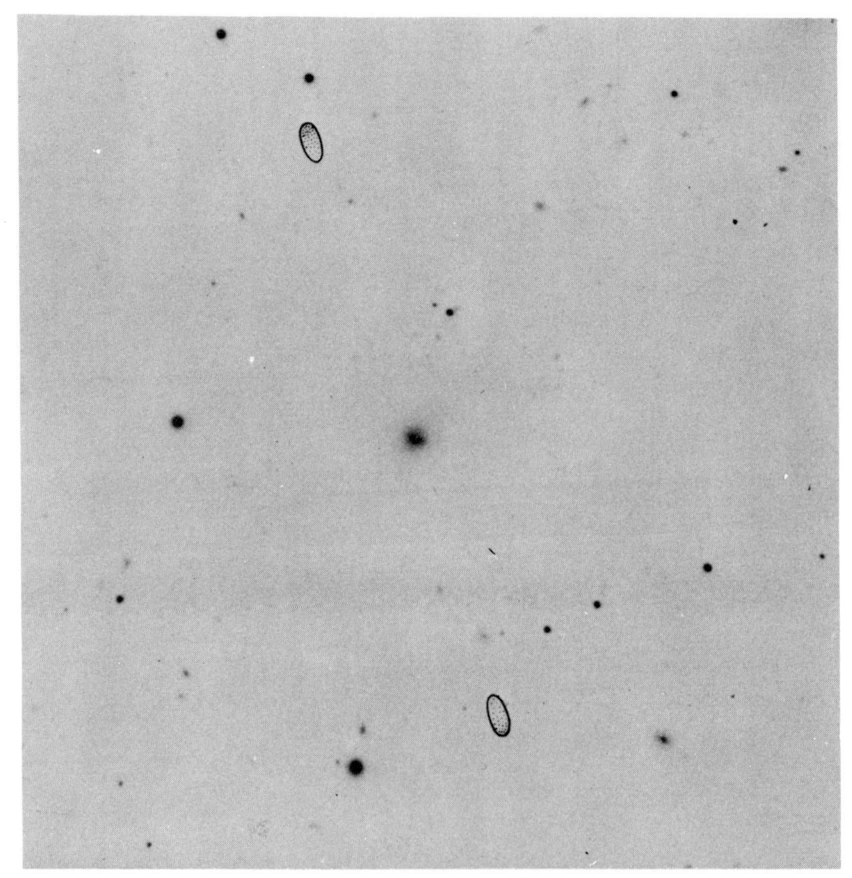

PLATE IV.—3C 33. Schematic representation of the radio source showing the large separation of the radio components relative to their size and the size of the parent galaxy. The southern component has an intensity about 2.5 times greater than the northern one. The photograph is from a plate taken by Minkowski with the Hale reflector. The field is 5′.8 square (from Moffet 1966).

together, it appears that a symmetric ejection of the radio source components occurs in at least half of all cases.

Although the large-scale features of many sources are rather similar, consisting of a relatively symmetrical pair of components separated by several times their own diameter, it seems likely that many sources will have a more complicated fine structure when viewed with higher resolution. A hint of this may be seen in parts of Centaurus A (Plate III). Another indication comes from studies of Cygnus A and Hercules A with resolution of a few seconds of arc (Wade 1966; Hogg *et al.* 1969) in which much of the 11-cm emission is found to come from small "hot spots." Allen, Brown, and Palmer (1962) and Bash (1968*a*) have shown that sources with diameters in the range 2″ to 20″ are at least as complex as those of larger angular size.

TABLE 4

STRUCTURAL DISTRIBUTION OF RESOLVED SOURCES*

Type of Structure	Number	%
Symmetrical double	27	34
Triple	13	16
Asymmetrical double	10	13
Core-halo	16	20
Complex	9	11
Simple	5	6

* After Fomalont 1969.

3.3.3. *Wavelength dependence.*—In general the appearance of a radio galaxy seems to vary only slowly with wavelength. For instance, the eastward component of Cygnus A is 20 percent stronger than its companion over a range of at least 10:1 and probably 60:1 in wavelength (Zakharenkov, Kaidanovskii, Pariiskii, and Prozorov 1963). Morimoto and Lockhart (1968) find very slight changes in the brightness distribution of Fornax A between 80 MHz and 2650 MHz.

However, some real changes do take place. Macdonald, Kenderdine, and Neville (1968) find significant differences in structure between 408 and 1421 MHz for 15 double sources. The bridging emission, if present, always has a steeper spectrum than does the emission from the main components; thus it is more prominent at lower frequencies. This may be part of a general tendency for the more extended components of a source to have steeper spectra (Mackay 1971). In the core-halo sources the halo spectrum is often steeper than that of the core (Moffet and Palmer 1965).

3.4. POLARIZATION

The radio emission from radio galaxies typically shows some degree of linear polarization at centimeter and short-decimeter wavelengths. The fact that linear polarization is commonly observed in radio galaxies strengthens the belief that radio emission from these objects is produced by the synchrotron mechanism. The subject of polarization in radio astronomy has been reviewed by Gardner and Whiteoak (1966).

Polarization of radio sources is usually detected by rotating a linearly polarized feed at the focus of a parabolic antenna. The intensity of a randomly polarized source will not appear to change when the feed is rotated, but any linearly polarized component will produce a variation as it is alternately accepted and rejected by the feed. Intensity maxima will recur each time the feed is rotated through 180°. Another method is to use two antennas as an interference polarimeter. This latter method eliminates certain instrumental problems caused by pickup of polarized radiation from objects in the vicinity of the radio telescope. If the spacing between the two antennas can be varied, the interference polarimeter may be used to determine the brightness distribution of polarization within a source. The method and some early measurements are discussed by Morris, Radhakrishnan, and Seielstad (1964*a*, *b*).

Subsequent to the review by Gardner and Whiteoak (1966), catalogs of polarization data have been given by Gardner and Davies (1966*a*, *b*), Maltby and Seielstad (1966), Sastry *et al.* (1967), Berge and Seielstad (1967), Gardner and Whiteoak (1969, 1971), Gardner, Morris, and Whiteoak (1969*a*, *b*), Gardner, Whiteoak, and Morris (1969), Davies and Gardner (1970), and Whiteoak and Gardner (1971). Two-dimensional polarization maps have been made of Fornax A by Gardner and Price (in Gardner and Whiteoak 1966) and of Centaurus A by Cooper, Price, and Cole (1965); these are shown in Plates II and III. One-dimensional polarization brightness distributions for 68 radio galaxies have been given by Seielstad and Weiler (1969).

The polarization angle, or the position angle in which the electric field vector in the radiation from a source has its maximum, is generally found to vary with the wavelength of observation. In almost all cases the polarization angle changes linearly with the square of the wavelength, which is the dependence to be expected for Faraday rotation in an ionized plasma.

If a linearly polarized signal with wavelength λ passes through a region containing an ionized gas and a magnetic field, its plane of polarization will be rotated by an angle

$$\theta = \frac{e^3}{2\pi m^2 c^4} \lambda^2 \int n_t(s) B_{\parallel}(s) ds = 2.631 \times 10^{-17} \lambda^2 \int n_t(s) B_{\parallel}(s) ds , \tag{3.1}$$

where the integral follows the path of the signal from the source to the receiver, $n_t(s)$ is the density of thermal electrons at each point along the path, and $B_{\parallel}(s)$ is the component of magnetic field parallel to the path. The integral in equation (3.1) is called the *rotation measure* and is usually expressed in units of radians per square meter (Gardner and Whiteoak 1963). Thus

$$\theta = \mathrm{RM} \times \lambda^2 , \tag{3.2}$$

with λ expressed in meters, and where

$$\mathrm{RM} = 0.8118 \int n_t B_{\parallel} ds . \tag{3.3}$$

The coefficient in equation (3.3) is correct if n_t is in units of cm^{-3}, $B_{\parallel}$ in microgauss and ds in parsecs. A positive value of the rotation measure implies a magnetic field directed toward the observer, causing the position angle of the electric field vector to increase with increasing wavelength.

Observed magnitudes of the rotation measure range from zero to about 1000, with a typical value of 20 radians m^{-2}. The magnitude of the rotation measure tends to be small for sources observed at high galactic latitudes, suggesting that most of the Faraday rotation takes place in passing through the plane of the Galaxy. If typical values in the plane of the Galaxy were $n_t = 0.02$ electrons cm^{-3} and $B_{\parallel} = 1$ microgauss, then a 1-kpc path length would produce a rotation measure of 16. Studies of the variation in rotation measure as a function of galactic latitude and longitude have given considerable information about the geometry of the galactic magnetic field (Morris and Berge 1964*a*; Gardner and Davies 1966*a*; Berge and Seielstad 1967; Gardner, Morris, and Whiteoak 1969*b*).

The degree of polarization seen in the net radiation from a radio source (i.e., summed over the whole of the source) is typically a few percent at wavelengths in the vicinity of 10 cm. In the typical radio galaxy the degree of polarization decreases with increasing wavelength, with the polarization virtually disappearing for wavelengths greater than 30–50 cm. There are exceptional objects in which the degree of polarization is independent of wavelength or even increases with increasing wavelength. A few sources have appreciably higher polarization; the present champion is 3C 76.1, which has a net polarization of 18 percent at 10.6-cm wavelength (Maltby and Seielstad 1966).

As was explained in § 2, synchrotron radiation is strongly polarized at emission, but there are many ways in which the degree of polarization may be reduced. If the magnetic field within the source is sufficiently tangled, no net polarization will be observed at any wavelength; this condition was assumed at several points in § 2. Alternatively, the radiation from various parts of the source may suffer from different amounts (and senses) of Faraday rotation, which will tend to reduce the net degree of polarization. At very short wavelengths the rotations are all small, and the net polarization is hardly affected; at long wavelengths the rotation may be several radians, and if there are differences of the order of radians in the Faraday rotation of radiation from different parts of the source, the net polarization will be very much reduced. This *Faraday depolarization* may occur either within the source volume itself or at any point along the path from the source to the observer.

Bologna, McClain, Rose, and Sloanaker (1965) found that the degree of polarization at 21-cm wavelength was inversely correlated with the absolute value of the galactic latitude. From this they concluded that a portion of the Faraday depolarization occurs in the Galaxy. However, Maltby (1965) examined the rate of depolarization between 10 and 21 cm and concluded that the major part of the depolarization must occur in the sources themselves, except for sources with very low galactic latitudes ($|b^{\rm II}| < 6°$).

Internal Faraday depolarization will occur even in a homogeneous source, since radiation from different depths within the source will be rotated by different amounts. In a homogeneous source the degree of polarization would vary as $\sin\theta/\theta$, where θ is the *Faraday depth*, or the amount of rotation suffered by radiation passing entirely through the source. Thus radiation from a source will be markedly depolarized when the Faraday depth of the source is greater than about 1 radian.

When sufficient sensitivity and resolution are attained, study of the polarization in different parts of a radio source will tell us a great deal about the structure of the magnetic field within the source. As yet, we have detailed information about the distribution of polarization in only a relatively small number of sources. Cooper, Price, and Cole (1965) have completed a detailed study of Centaurus A, from which the radio contours in Plate III were taken. By observing at several wavelengths they were able to correct for Faraday rotation and determine the *intrinsic polarization* direction in various parts of the source, i.e., the direction of the maximum electric vector when extrapolated to zero wavelength according to a λ^2 law. The short bars in Plate II are drawn perpendicular to the local direction of intrinsic polarization and therefore indicate the direction of the magnetic field. There are large areas in which the magnetic field seems to have a uniform direction, nearly perpendicular to the major axis of the source. The degree of polarization varies from place to place, but is typically 10–20 percent at 11 cm. The rotation measure varies from -36 to -82 radians m^{-2}, but over large areas of the source it lies between -55 and -65. Cooper, Price, and Cole concluded from the rate of depolarization that the internal Faraday rotation is small, amounting to less than one-sixth of the total.

Fornax A, shown in Plate II, has been studied by Gardner and Whiteoak (1966). The short bars indicate the direction of the maximum electric vector at 11 cm. The rotation measure is very small for this source, so the direction of the magnetic field will be roughly perpendicular to these lines. Once again the field is fairly uniform and approximately perpendicular to the major axis of the source. The degree of polarization is about 10 percent at 11 cm.

In some cases the brightness distribution of the polarized component of the radiation from a source differs markedly from that of the unpolarized component. Thus in the fainter component of Fornax A, the polarized radiation comes from a smaller region than does the unpolarized radiation. Even more marked differences occur in the central double source of Centaurus A (Cooper, Price, and Cole 1965) and in several other radio galaxies (Golnev and Soboleva 1965). Seielstad and Weiler (1969) in their studies of one-dimensional polarization distributions in 68 radio galaxies at 21-cm wavelength find that the distribution of linearly polarized emission usually is quite different from that of the unpolarized emission except in the case of the halo component of the core-halo sources. In these objects the halo is often fairly uniformly polarized; however, the cores are much less polarized.

A number of attempts have been made to correlate the net polarization with other properties of radio sources, generally with little success. The one striking relation is between the intrinsic position angle of the net polarization of double radio sources and the position angle of the major axis of these sources. The difference between these two angles tends to be zero for sources of low surface brightness ($T_b < 3000°$ at 21 cm) and 90° for sources of high surface brightness (Morris and Berge 1964*b;* Mitton 1972). This implies that the ordered part of the magnetic field must tend to be parallel to the major axis of high-brightness sources and perpendicular to the axis of low-brightness sources. A model which might account for this behavior is discussed by Gardner and Whiteoak (1966).

In general the relatively low luminosity extended features of a radio galaxy show higher polarization than do the compact, high-brightness regions. This indicates either that the field is more tangled in the compact regions or that the Faraday depolarization is greater there. Miley *et al.* (1972) report polarization as great as 60 percent in the faintest parts of the tail of 3C 129/3C 129.1. With the high sensitivity of the Westerbork synthesis telescope other faint extended features have been found to have high polarization (van der Laan 1972).

Circular polarization at a level of less than 0.5 percent has been detected in a few compact quasi-stellar sources (Biraud and Véron 1968; Seaquist 1971; Conway, Gilbert, Raimond, and Weiler 1971; Roberts, Ribes, Murray, and Cooke 1972). The only radio galaxies in which circular polarization has been detected are 3C 84 and 3C 120, and these are similar to quasi-stellar sources in many respects (Berge and Seielstad 1972). As was explained in § 2.1, the circularly polarized component of synchrotron radiation usually cancels out. Only in cases where the magnetic field is very strong, so that the radio emission comes from electrons with $E \sim mc^2$, would one expect to see a circularly polarized component (Sciama and Rees 1967; Legg and Westfold 1968; Melrose 1971*a*, *b*).

4. INTRINSIC PROPERTIES

Radio measurements alone do not give a very complete picture of radio galaxies, for they cannot be used to determine the precise distances of these objects. It is true that radio sources which appear small and faint are likely to be farther away than those which are intense and of large angular diameter, but we know that the dispersion in both luminosity and size is very great. Thus, we have accurate information about the intrinsic properties of only those radio sources identified with objects with distances known from optical observations. Progress in measuring the distances to radio galaxies has been materially hastened by the fortunate circumstance that the spectra of most radio galaxies show strong emission lines; thus their redshifts may be measured even when they are exceedingly distant.

4.1. Optical Identification and Redshifts

The identification of radio sources with their optical counterparts is the subject of chapter 6 of this volume. The number of reliable identifications has grown very rapidly since about 1960; up until that time the measurements of radio source positions were not accurate enough to be certain of identifications in any except the most obvious cases.

The work is most nearly complete in the northern sky, where the positions of all objects in the Revised 3C catalog have been examined (Wyndham 1966; Véron 1966). This catalog is known to be essentially complete to a lower intensity level of ten flux units at 178 MHz, so a homogeneous sample is available for statistical studies. Wyndham finds that 69 percent of the 236 3CR sources having $|b^{II}| \geq 15°$ may be confidently identified with objects at least 1 mag above the plate limit of the Palomar–National Geographic Society *Sky Survey*, that is, with objects brighter than about 20 mag in the blue and 19 mag in the red. Of these, about one-quarter are quasi-stellar.

Another large group of identifications has come from the work of Bolton and his several collaborators using positions from the Parkes radio telescope. Although the Parkes survey covers the whole sky south of $\delta = +27°$, the Parkes identification work has been concentrated in the part of the sky which can be photographed with the Palomar Schmidt camera, i.e., for $\delta > -44°$. Within this range several hundred identifications have been made. For references to these see the Parkes catalogs (CSIRO Radiophysics Staff 1969; Wall, Shimmins, and Merkelijn 1971; Shimmins 1971). These extend to fainter radio sources than are present in the 3CR catalog, and there is some indication that the proportion of galaxies among the fainter identifications is higher than the two-thirds found in the 3CR. In addition, of course, the proportion of blank fields, where there is no candidate visible above the limit of the photographic plate, is also higher.

The first Parkes surveys were done at 408 MHz. Following these, limited areas were surveyed with the Parkes telescope at 2700 MHz (Wall, Shimmins, and Merkelijn 1971; Shimmins 1971). A somewhat higher fraction of the sources (about 40%) found in this survey were identified with quasi-stellar objects. In a survey at 5 GHz (Pauliny-Toth *et al.* 1972) a similar result was obtained. The increased fraction of quasi-stellar objects found in the high-frequency surveys is a consequence of the fact that these objects typically have flatter spectra than do the radio galaxies.

As of the end of 1972, optical distance measurements were available for 173 radio galaxies.[8] The redshifts range from virtually zero to $z = \Delta\lambda/\lambda_0 = 0.46$. Table 5 summarizes the redshifts and the inferred luminosities of the radio galaxies with known distances. The table includes only objects with total radio luminosities, integrated over the range 10^7 Hz $< \nu < 10^{10}$ Hz, of at least 10^{40} ergs s^{-1}. The threshold of 10^{40} ergs s^{-1} has been chosen to exclude normal late-type spiral and irregular galaxies, which often have radio luminosities of 10^{38} or 10^{39} ergs s^{-1}.

Included in table 5 are the radio and optical names of each object, the redshift, the galaxy type, the observed visual magnitude (or the V photoelectric magnitude), and the observed radio flux density at 1400 MHz. From these the absolute optical magnitude M and the total radio luminosity L have been calculated; for the latter quantity a power-law spectrum has been assumed with the spectral index equal to the measured value, if available, for each source at decimeter wavelengths. A Hubble constant of 50 km s^{-1} Mpc^{-1} has been assumed.

A few sources of lower absolute radio luminosity are of interest in discussions of radio galaxies, and some of these are listed in table 6, including NGC 253, M82, and several of the E and S0 galaxies studied by Heeschen (1970), which have been mentioned in § 3.2.3.

4.2. Luminosities

It can be seen from the data in table 5 that the radio luminosities of radio galaxies range upward from the rather arbitrary threshold of 10^{40} ergs s^{-1} to more than 10^{45} ergs s^{-1}. The range of radio luminosities for all galaxies is much larger. As was mentioned above, late-type spirals and irregulars typically have luminosities in the range 10^{37}–10^{39} ergs s^{-1} while normal ellipticals and early-type spirals have luminosities of less than 10^{36} ergs s^{-1}. An intermediate range of luminosities, 10^{39}–10^{40} ergs s^{-1}, is also found in the peculiar E and S0 galaxies included in table 6. Since the bulk of the strong radio galaxies are also ellipticals, we find a range of 10 orders of magnitude in the radio emission from this type of galaxy.[9]

[8] *Note added in proof.* One omission from table 5 has been called to my attention, namely 4C 35.6, a multiple-nucleus galaxy observed by Peterson (1970), for which $z = 0.0473$, $m_V = 14.70$, $S_{1400} = 0.82$, $M_V = -23.1$, and $\log L = 42.0$.

[9] Ekers (at the IAU General Assembly, 1973) has demonstrated the continuity of the radio luminosities of elliptical galaxies over this wide range. With the great sensitivity of the Westerbork array it is possible to detect many sources of intermediate (10^{37}–10^{40} ergs s^{-1}) luminosity. The distinction between "normal elliptical" and "radio galaxy" is thus very indistinct, emphasizing the arbitrariness of a cutoff at 10^{40} ergs s^{-1}.

TABLE 5

REDSHIFTS AND LUMINOSITIES FOR 172 RADIO GALAXIES

	Source Names		Type	z	m_V	S_{1400}	M_V	log L	Ref.
(1)	(2)	(3)	(4)	(5)	(6)	(7)	(8)	(9)	(10)
	PKS0023-33		E	0.0497	16.7	1.3	-20.8	42.4	27
WK10				0.0728	18.	(0.1)	-20.6	41.4	5
3C15	PKS0034-01		E1	0.0733	15.34	4.1	-23.0	42.9	2
3C17	PKS0035-02		E	0.2201	18.02	6.0	-23.0	43.9	19
4C03.01	PKS0036+03	NGC193	E2	0.0145	14.3	1.5	-20.4	41.1	28
5C3.100				0.0718	16.	(0.02)	-22.6	40.7	3
	PKS0043-42		E	0.053	18.	7.2	-19.6	42.9	27
5C3.175			PEC	0.1343	16.	(0.01)	-24.1	40.9	5
3C26	PKS0051-03		E	0.2106	17.94	2.0	-23.0	43.5	19
3C28	PKS0053+26		E	0.1959	17.54	1.8	-23.3	43.6	19
3C29	PKS0055-01		E0	0.0450	14.07	4.6	-23.2	42.5	18
3C31		NGC382/3	E/E	0.0169	12.14	5.0	-23.1	41.7	8
3C33	PKS0106+13		D	0.060	15.19	13.2	-22.7	43.2	19
	PKS0108-14.2		E	0.0518	15.8	1.2	-21.8	42.4	27
3C35			D3	0.0677	15.	2.1	-23.8	42.6	5
4C31.04	OC328		S0/E	0.0590	14.5	3.0	-23.5	42.6	5
3C40	PKS0123-01	NGC545/7	DB	0.0177	12.28	4.4	-22.9	42.1	14
4C18.06	PKS0124+18		E	0.0437	15.5	1.4	-21.7	41.9	22
	PKS0131-36		S0	0.030	14.2	5.6	-22.1	42.2	25
4C05.10	PKS0153+05	NGC741	E	0.0188	13.2	0.9	-22.1	41.0	8
4C05.09	PKS0203+05		DB	0.1300	17.	0.5	-22.7	42.4	5
4C35.03			D	0.0374	14.5	2.2	-22.6	42.0	22
3C61.1	RN8		CL	0.18	19.	6.3	-21.8	44.0	10
3C66			ED2	0.0215	12.90	9.7	-23.2	42.2	14
4C08.11	PKS0238+08	NGC1044	DB	0.0216	14.8	1.2	-20.8	41.4	22
3C71	PKS0240-00	NGC1068	SY	0.00344	8.91	4.9	-22.7	40.3	8
3C75	PKS0255+05		DB	0.0241	13.62	5.8	-22.3	42.2	14
4C34.09		NGC1167		0.0203	14.	2.0	-21.8	41.4	28
3C76.1	PKS0300+16		DE3	0.0328	14.86	3.0	-21.8	42.0	18
3C78	PKS0305+03	NGC1218	D	0.0289	12.84	7.1	-23.5	42.3	19
3C79	PKS0307+16		N	0.2561	19.	4.9	-22.7	44.0	19
4C39.11		NGC1233		0.0163	14.	2.1	-21.5	41.2	28
3C83.1		NGC1265	ED3	0.018	12.5	8.4	-23.4	41.9	8
3C84	PER A	NGC1275	ED2	0.0176	11.87	13.5	-24.0	42.1	8
FOR A	PKS0320-37	NGC1316	S0	0.0058	8.90	108.	-23.8	42.0	8
3C88	PKS0325+02		D	0.0302	13.95	4.4	-22.5	42.1	19
4C39.12				0.0209	15.	0.5	-21.2	41.0	28
	PKS0336-35	NGC1399	E	0.0043	10.9	(2.)	-21.2	40.1	8
	PKS0349-27		E	0.066	16.8	5.2	-21.3	43.0	21
4C21.13	PKS0349+21		CD	0.1325	16.	0.8	-24.0	42.7	5
3C98	PKS0356+10		E	0.0306	14.45	10.0	-22.1	42.5	19
3C109	PKS0410+11		N	0.3057	17.	4.1	-25.4	44.1	9
	PKS0427-53		DB	0.0392	13.2	4.4	-23.8	42.3	24
3C120	PKS0430+05	BW TAU	SY	0.0333	15.	4.1	-21.8	42.3	2
	PKS0431-13.5		E	0.0360	16.3	(0.7)	-20.6	41.6	27
	PKS0449-17		E	0.0313	13.7	0.8	-22.9	41.5	27
	PKS0453-20		E	0.0354	14.0	3.7	-22.9	42.2	4
3C135	PKS0511+00		N	0.1270	17.00	2.0	-23.0	43.0	18
PIC A	PKS0518-45		D	0.0342	15.7	52.1	-21.1	43.3	19
	PKS0521-36		N	0.061	16.8	14.7	-21.3	43.2	26

TABLE 5 (continued)

(1)	Source Names (2)	(3)	Type (4)	z (5)	m_V (6)	S_{1400} (7)	M_V (8)	log L (9)	Ref. (10)
3C153				0.2771	18.5	4.1	-24.0	44.1	5
	PKS0618-37		DB	0.0326	16.6	2.4	-20.3	42.1	4
	PKS0620-52		D	0.0502	15.5	2.7	-22.3	42.4	22
	PKS0634-20		E	0.056	16.8	7.0	-21.7	43.0	21
3C171			N	0.2387	19.	3.8	-22.6	43.8	19
4C23.18	PKS0658+23		D	0.0905	15.7	1.2	-23.9	42.5	5
	PKS0718-34		E	0.0297	16.5	1.7	-20.9	41.7	4
3C178	PKS0722-09		SC	0.0073	13.	1.4	-23.6	40.5	2
3C184.1				0.1187	17.	3.3	-22.7	43.2	5
4C56.16				0.0356	16.	3.8	-20.9	42.3	28
4C37.21		NGC2484	SO	0.0433	14.9	2.2	-22.5	42.3	3
3C192	PKS0802+24		E	0.0596	15.46	4.8	-22.7	42.8	16
3C197.1			DE	0.1302	16.5	0.3	-23.3	42.9	3
3C198	PKS0819+06		E	0.0809	16.78	2.1	-22.1	42.7	19
4C24.18			I	0.0423	17.	0.8	-20.3	41.8	3
	PKS0843-33	NGC2663	E3	0.0073	12.5	1.6	-22.4	40.5	4
4C54.17		NGC2656	E	0.0449	15.	(2.)	-22.3	42.1	22
3C218	PKS0915-11		D	0.065	14.8	36.3	-23.5	43.7	13
3C219			D	0.1745	17.22	7.2	-23.3	43.9	19
3C223			E2	0.1367	17.06	3.1	-22.8	43.3	16
3C223.1			E5	0.1075	16.36	1.9	-22.9	43.8	16
3C227	PKS0945+07		N	0.0855	17.0	7.3	-21.8	43.3	18
3C234			N	0.1846	17.5	5.3	-23.1	43.9	19
3C236			E3	0.0988	15.97	4.3	-23.1	43.1	18
4C14.37		NGC3367	SBC	0.0092	11.6	0.4	-22.1	40.2	8
4C29.41			ED	0.0481	14.	2.0	-23.4	42.2	5
3C255	PKS1116-02		SP	0.0238	15.5	1.6	-20.8	41.9	2
	PKS1123-35		E	0.0336	16.0	2.1	-20.9	42.0	4
4C25.35	PKS1125+26	NGC3689	SC	0.0088	12.9	1.0	-20.7	40.4	28
4C-03.43	PKS1130-037		E	0.0482	15.5	(1.)	-21.9	42.0	5
4C21.32	PKS1131+21		E	0.066	16.	0.7	-22.1	42.4	21
4C17.52	PKS1137+12	NGC3801	E3	0.0105	13.3	1.5	-20.7	40.8	28
3C264	PKS1142+19		E	0.0206	12.74	5.1	-22.8	41.9	19
4C03.41	PKS1215+03		E/D	0.076	17.	2.0	-21.4	42.9	2
3C270	PKS1216+06	NGC4261	E	0.0070	10.40	15.3	-22.7	41.4	8
3C272.1	PKS1222+13	M84	E	0.00293	9.36	6.1	-21.9	40.2	8
3C274	PKS1228+12	M87	E	0.0041	8.74	197.	-23.2	42.0	8
	PKS1245-41	NGC4696	E	0.0086	12.0	3.2	-21.9	40.9	23
	PKS1250-10		E2	0.0138	12.	0.9	-22.6	41.1	27
3C277.3	COM A		D2	0.0857	15.94	3.1	-22.8	42.9	19
3C278	PKS1252-12	NGC4782/3	DB	0.0140	11.46	6.6	-23.2	41.7	7
5C4.51		NGC4839	E	0.0249	14.2	(0.08)	-21.7	40.4	23
5C4.81		NGC4869	E3	0.0224	13.9	0.37	-21.8	41.2	8
5C4.85		NGC4874	SO	0.0239	12.4	0.18	-23.4	40.5	8
3C284				0.2334	18.	2.0	-23.3	43.7	5
3C285			D?	0.0797	15.99	2.3	-22.5	42.7	18
4C36.24		NGC5141	D	0.0174	14.	1.0	-21.1	41.0	22
CEN A	PKS1322-42	NGC5128	DE3	0.00090	6.98	912.	-22.1	41.7	8
4C31.42				0.0241	15.5	0.7	-20.4	41.1	28
3C287.1	PKS1330+02		N	0.2156	19.0	2.7	-22.0	43.6	16

TABLE 5 (continued)

Source Names (1)	(2)	(3)	Type (4)	z (5)	m_V (6)	S_{1400} (7)	M_V (8)	log L (9)	Ref. (10)
	PKS1333-33	IC4296	E1	0.0123	11.9	(3.)	-22.7	41.3	6
4C05.57	PKS1340+05		N	0.1333	18.	1.8	-21.7	43.1	2
	PKS1345+12		S0	0.1218	17.	4.9	-22.5	43.4	3
3C293			D5?	0.0453	14.32	4.5	-22.9	42.5	16
	PKS1358-11		E	0.025	15.	2.1	-21.0	41.7	21
	PKS1400-33		E	0.0143	12.8	(2.)	-22.1	41.3	4
OQ208			SY	0.0768	14.	(1.)	-24.4	42.4	5
4C-01.32				0.0249	14.8	(0.4)	-21.1	41.0	28
3C295			D	0.4614	20.11	23.1	-23.6	45.2	11
3C296	PKS1414+11	IC5532	E4	0.0237	12.21	4.4	-23.6	41.9	16
	PKS1417-19		N	0.1192	17.5	1.7	-22.1	42.9	2
3C305			SA?	0.0416	13.74	3.2	-23.3	42.2	16
3C306	PKS1452+16		E3	0.0458	14.9	0.7	-22.4	42.0	22
3C310	PKS1502+26		DB	0.0543	15.24	7.8	-22.4	43.1	19
3C315	PKS1511+26		DB	0.1086	16.80	3.9	-22.5	43.2	19
4C00.56	PKS1514+00		E	0.053	16.5	2.5	-21.1	42.4	21
3C317	PKS1514+07		E	0.0351	13.50	5.2	-23.2	42.6	19
4C-06.41			N	0.1280	18.	(0.5)	-21.7	42.6	5
3C327	PKS1559+02		D	0.1041	15.88	9.8	-23.4	43.7	19
4C17.66		NGC6047	E0	0.0320	15.1	0.5	-21.4	41.4	8
	PKS1610-60.8		E3	0.0176	12.8	(45.)	-23.9	42.8	27
3C332			DE3	0.1520	16.	2.6	-24.1	43.4	5
3C338		NGC6166	D4	0.0303	12.63	3.6	-23.8	42.7	12
	PKS1637-77		D	0.0438	16.	5.3	-21.6	42.5	4
3C348	PKS1648+05		D	0.1533	16.90	43.	-23.4	44.6	7
4C02.44	PKS1650+024		PEC	0.0250	14.	(0.6)	-22.2	41.2	27
	PKS1655-77		E0	0.0663	17.	2.0	-21.5	42.5	27
3C353	PKS1717-00		D	0.0307	15.36	49.	-21.4	43.2	19
3C357			ED4	0.1670	15.5	2.5	-25.0	43.5	5
4C55.33.1		NGC6454	D	0.0312	13.	0.5	-23.6	41.3	22
3C371			N1	0.0508	14.2	2.7	-23.5	42.5	17
3C381			ND?	0.1614	17.24	3.9	-23.3	43.5	18
3C382			D3	0.0586	14.66	5.8	-23.7	42.8	19
	PKS1834+19		E	0.0166	14.	1.7	-21.8	41.2	22
3C388			DB	0.0917	15.32	5.7	-23.9	43.0	19
3C390.3			N	0.0569	14.	12.3	-24.0	43.1	16
	PKS1928-34		E	0.0981	17.	0.9	-22.4	42.6	27
	PKS1934-63		E	0.182	16.	13.0	-24.8	43.9	15
3C402			DB	0.0247	14.	3.1	-22.6	41.8	5
3C403	PKS1949+02		S0	0.059	15.42	5.8	-23.2	43.0	9
3C403.1	PKS1949-01		ED2	0.0559	16.	1.5	-22.4	42.3	5
	PKS1954-55		E	0.0598	16.3	5.5	-21.8	42.8	27
3C405	CYG A		D3	0.0570	15.14	1255.	-24.5	45.2	1
	PKS2006-56		S0	0.0585	16.	(3.)	-22.0	42.7	22
	PKS2014-55		E1	0.0605	15.5	1.4	-22.6	42.3	27
RN73			N	0.047	17.5	(0.2)	-20.1	41.3	20
	PKS2040-26		E	0.0406	13.5	1.8	-23.6	42.0	27
	PKS2048-57		S0	0.0110	13.	2.1	-21.2	41.0	27
	PKS2058-28		E	0.0394	14.8	1.7	-22.2	42.0	27
	PKS2058-13		E3	0.0296	15.2	1.3	-21.3	41.9	27

TABLE 5 (concluded)

(1)	Source Names (2)	(3)	Type (4)	z (5)	m_V (6)	S_{1400} (7)	M_V (8)	log L (9)	Ref. (10)
3C430			ED4	0.0549	15.	7.6	-24.0	43.0	5
3C433	PKS2121+24		DB	0.1025	16.24	11.9	-23.3	43.7	19
	PKS2130-53		DB	0.0763	14.	1.3	-24.5	42.5	27
3C436				0.2154	18.18	3.3	-23.2	43.7	19
	PKS2152-69		D	0.0266	13.8	25.9	-22.3	42.8	24
3C442	PKS2212+13	NGC7236/7	DB	0.0270	13.66	3.4	-22.6	42.0	7
3C445	PKS2221-02		N	0.0568	17.	5.1	-20.8	42.7	19
3C449			E2	0.0181	13.20	3.7	-22.6	41.6	18
3C452			ED1	0.0820	16.00	10.7	-23.1	43.4	19
	PKS2247+11	NGC7385	E	0.0268	14.4	2.5	-21.8	41.8	8
3C455	PKS2252+12	NGC7413	E4	0.0331	14.00	3.0	-22.6	42.1	18
	PKS2300-18		N	0.129	18.3	1.7	-21.4	43.0	21
4C07.61	PKS2308+07	NGC7503	E1	0.0448	13.9	1.6	-23.3	42.4	8
3C456	PKS2309+09		E	0.2337	18.54	2.4	-22.7	43.7	19
3C459	PKS2313+03		N	0.2205	18.	4.3	-23.1	40.5	19
	PKS2318+07	NGC7626	E1	0.0119	12.8	0.7	-21.5	40.5	8
M23-1/12	PKS2322-12		E5	0.0825	15.46	1.9	-23.2	42.7	19
3C465	PKS2335+26	NGC7720	D4	0.0301	13.29	7.7	-23.2	42.4	19
4C-00.61	PKS2349-01		N	0.174	17.5	1.6	-22.9	43.2	21
	PKS2354-35		D	0.0487	14.4	1.0	-23.0	42.3	27
	PKS2356-61		D	0.0959	16.	19.2	-23.0	43.9	27
	PKS2357+00		DB	0.0844	16.	0.5	-22.7	42.1	22

NOTES TO TABLE 5

Cols. (1)–(2).—Radio source names, generally from the catalogs listed in table 3. Sources from other catalogs are as follows: WK 10, Windram and Kenderdine 1969; RN 73, Ryle and Neville 1962; 4C 55.33.1, Caswell and Wills 1967.

Col. (3).—Name from NGC or IC.

Col. (4).—Galaxy type as estimated by various observers.

Col. (5).—Redshift z is corrected for solar motion of 300 km s^{-1} toward $l^{II} = 270°$.

Col. (6).—Apparent magnitude m is usually m_V although some estimates are m_{pg}. Values given to two places after the decimal are from Sandage 1972*b*.

Col. (7).—Apparent flux density S is at 1400 MHz. Values in parentheses are estimated from other frequencies using a spectral index of -0.8.

Col. (8).—Absolute magnitude M includes correction for galactic absorption in the manner described by Sandage 1972*a;* K-corrections from Whitford (1971) are also included; $H = 50$ km s^{-1} Mpc^{-1}.

Col. (9).—Log_{10} of absolute radio luminosity; no K-corrections are included; $H = 50$ km s^{-1} Mpc^{-1}.

Col. 10.—References for redshift: (1) Baade and Minkowski 1954; (2) E. M. Burbidge 1967; (3) Burbidge 1970; (4) Burbidge and Burbidge 1972; (5) Burbidge and Strittmatter 1972; (6) Evans 1963; (7) Greenstein 1962; (8) Humason, Mayall, and Sandage 1956; (9) Lynds (quoted by Sandage 1972*b*); (10) Miller (quoted by Schmidt 1972); (11) Minkowski 1960; (12) Minkowski 1961*a;* (13) Minkowski 1961*b;* (14) Minkowski (quoted by Maltby, Matthews, and Moffet 1963); (15) Peterson and Bolton 1972; (16) Sandage 1966; (17) Sandage 1967*a;* (18) Sandage 1967*b;* (19) Schmidt 1965; (20) Schmidt (quoted by Penston 1971); (21) Searle and Bolton 1968; (22) Tritton 1972; (23) de Vaucouleurs and de Vaucouleurs 1964; (24) Westerlund (quoted by Burbidge and Burbidge 1972); (25) Westerlund and Smith 1966; (26) Westerlund and Stokes 1966; (27) Whiteoak 1972; (28) Wills 1967.

Many objects have been observed by more than one person, but only one (generally the first) reference is given. References to identifications, finding charts, etc., are generally given in papers listed for the redshifts.

TABLE 6

REDSHIFTS AND LUMINOSITIES FOR 10 GALAXIES OF INTERMEDIATE RADIO POWER

Source Names			Type	z	m_V	S_{1400}	M_V	log L	Ref.
(1)	(2)	(3)	(4)	(5)	(6)	(7)	(8)	(9)	(10)
	PKS0045-25	NGC253	SC	0.00086	7.0	5.0	-21.6	39.1	4
		NGC1052	E4	0.0048	11.2	0.5	-21.1	39.6	8
4C78.06		NGC2146	SA	0.00322	10.8	(0.6)	-20.9	39.4	8
		NGC2911	SO	0.0099	12.7	0.27	-21.2	39.9	8
3C231		M82	I	0.00107	8.39	8.7	-20.7	39.5	8
4C55.19		NGC3079	SC	0.0041	10.8	(0.6)	-21.2	39.7	8
5C2.203		NGC3583	S	0.0072	12.	(0.05)	-21.2	39.1	22
	PKS1302-49	NGC4945	SC	0.00184	9.	5.8	-21.9	39.8	4
		NGC5077	E3	0.0084	11.6	0.22	-21.9	39.8	8
4C47.36.1		NGC5194	SC	0.00182	8.4	(1.0)	-21.8	39.2	8

For explanation see text and notes to Table 5.

The optical luminosities of radio galaxies are remarkably similar. The mean value of M in table 5 is -22.5 with a standard deviation of 1.0. Sandage (1972*b*), using accurately measured photoelectric magnitudes for a smaller number of objects, finds a mean of -22.98 with a standard deviation of only 0.49. Because of this narrow dispersion in absolute magnitude, the optical apparent magnitude of a radio galaxy may be used as a fairly reliable distance indicator in the absence of a measured redshift. The mean absolute magnitude found for radio galaxies is almost the same as that found for the brightest members of rich clusters of galaxies, -22.26 (Sandage 1972*a*). Thus the objects which become radio sources are almost all extremely luminous; by inference they are also extremely massive.

4.3. Linear Dimensions

From the distance and angular brightness distribution of a radio galaxy its linear dimensions may be derived. This has been done for about 150 sources resolved by variable-spacing interferometry or aperture synthesis (Maltby, Matthews, and Moffet 1963; Macdonald, Kenderdine, and Neville 1968; Mackay 1969), and for a lesser number for which diameter estimates are available from scintillation studies or from VLB interferometry (Cohen 1969). The results show that double sources associated with galaxies have component separations from less than 10 kpc to nearly 1 Mpc, with the typical component size being about one-third to one-fifth of the separation. Since the diameter of a giant elliptical galaxy may be 50 kpc, it is evident that the largest radio sources have sizes more than 10 times as great as the associated galaxies. Indeed, a distance of 1 Mpc is typical of the intergalactic spacing in dense clusters of galaxies, where this type of radio source is often found.

Smaller linear sizes are found for components within extended sources, such as the hot spots found in Cygnus A and Hercules A (Hogg *et al.* 1969; Miley and Wade 1971) and the cores of several core-halo objects such as M87 (Fomalont 1969). These have typical sizes $\sim$1 kpc. It seems very likely that detailed structure on this scale will be found in many sources when observations with sufficiently high resolution become possible.

There are also smaller-diameter radio sources associated with galaxies. In particular the Seyfert galaxies NGC 1275 = 3C 84 and 3C 120 have components with linear dimensions of less than 0.4 pc as was described in § 3.2.3. Another Seyfert galaxy NGC 1068 = 3C 71, has a component ≤ 20 pc in size (Cohen, Gundermann, and Harris 1967), although most of the radio emission comes from a region about 600 pc in diameter (Bash 1968*b*). Both 3C 84 and 3C 120 have large-diameter components also, but in these two sources the halos are ~ 10 kpc in diameter (Fomalont 1968; Ryle and Windram 1968). Thus the core-halo type of brightness distribution seems to be common to those Seyfert galaxies which are radio sources.

4.4. Energy Requirements

Energy requirements for a synchrotron source of known volume, spectrum, and luminosity are estimated in the manner outlined in § 2.4. The estimate depends on the magnetic field strength in the source, and it is customary to assume equipartition between energy in particles and in magnetic fields, which gives the minimum energy required to account for the radiation from the source. These minimum energies are very large, ranging from $\sim 10^{56}$ ergs for the compact components in core-halo objects to 10^{60} ergs for the extended sources in objects such as Centaurus A and Cygnus A.

These energies are embarrassingly large compared with the sources from which they might plausibly be derived. The total mass-energy of a giant galaxy with a mass of $\sim 10^{12}\ M_{\odot}$ would be $\sim 10^{65}$ ergs. The gravitational binding energy of such a galaxy, assuming an effective radius of 1 kpc, would be $\sim 10^{62}$ ergs. With an efficiency of $\sim 10^{-2}$ for conversion of mass to energy through thermonuclear reactions, the total thermonuclear energy available throughout the life of the stars in such a galaxy would be $\sim 10^{63}$ ergs. Put in a different way, a Type II supernova explosion is estimated to yield $\sim 10^{51}$ ergs, so the energy required for a large radio galaxy is that of $\sim 10^{8}$ to 10^{9} supernovae, even assuming unit efficiency in conversion of the energy of explosion into the very special form in which we now see it, namely, energetic particles and magnetic fields.

It should be noted that the largest estimated energies are associated with radio sources occupying large volumes. It is possible to reduce the estimated energies for these sources by assuming that the effective radiating volume is much smaller than that given by the overall dimensions of the source regions. However, the reduction factor, given in equation (2.77), is only $V^{3/7}$, so the effective volume must be very greatly reduced if the energy problem is to be helped very much. G. R. Burbidge (1967) has made a radical suggestion along these lines, proposing that the radiation comes from tiny regions with a total volume 10^{-6} to 10^{-10} of the overall volume of the source This would reduce the total energy requirements by factors 10^{3} to 10^{4} but would not be consistent with the observed spectra of the sources, since such tiny subsources would become optically thick through synchrotron self-absorption at frequencies as high as 200 MHz.

The magnetic fields which are estimated from the equipartition condition range from 10^{-3} to 10^{-5} gauss, with the smaller fields associated with the sources of larger volume, as indicated by equation (2.77).

Minimum energies and equipartition fields for a number of sources have been given by Maltby, Matthews, and Moffet (1963), Macdonald, Kenderdine, and Neville (1968), and Mackay (1969). Several examples are given in table 7. The first five sources are all double, Virgo A is the prototype core-halo, and M82 is included to represent a small source of low luminosity.

4.5. Optical Properties of Radio Galaxies

The optical properties of radio galaxies have been investigated most extensively by Matthews, Morgan, and Schmidt (1964). They found that almost all strong radio sources are associated with some type of elliptical galaxy. There is a considerable range in the degree of compactness of these objects, but all are extremely luminous and massive.

TABLE 7

Equipartition Energies and Fields

Source	Size Diameter × Separation (kpc)	log L (ergs s^{-1})	log V (cm^3)	log U (ergs)	log B (gauss)
Cen A:					
Halo........	150×300	41.6	71.0	59.9	−5.3
Core........	4×10	41.1	66.9	57.4	−4.2
Cyg A.........	26×200	45.2	68.7	60.6	−3.4
3C 33.........	18×440	43.2	68.2	59.5	−3.6
3C 219........	100×700	43.9	70.5	60.5	−4.5
3C 295........	28×60	45.2	68.8	60.4	−3.6
Vir A:					
Halo........	50	41.7	69.6	58.6	−4.5
Core........	2.5	41.7	65.7	57.1	−3.7
Nucleus.....	0.0002	39.7	53.3	50.7	−2.6
M82..........	0.9	39.5	64.3	55.2	−4.1

The most common type of radio galaxy is the *D system*, which is an elliptical galaxy with a very extensive envelope of stars, sometimes having a diameter (as seen on the prints of the Palomar *Sky Survey*) as great as 60 kpc. Those supergiant D systems, which are found as the central members of rich clusters of galaxies, are given the notation cD and have been discussed separately by Morgan and Lesh (1965). The prototype is NGC 6166, associated with 3C 338.

Giant ellipticals or *E galaxies* are somewhat smaller than D systems, lacking the extensive envelope. Their diameters might range from 10 to 30 kpc. A typical example would be M87.

N galaxies are ellipticals with brilliant, starlike nuclei containing most of the luminosity of the system. The nuclear diameter is $\lesssim$ 1 kpc. Because most of the light comes from such a small volume they may be included in the *compact galaxy* classification of Zwicky (1964*b*). An example is the object associated with 3C 234.

A small minority of radio galaxies are dumbbell or *db galaxies*. These might be called D systems with double nuclei, in which two elliptical nuclei share a common extended envelope. A good example is 3C 278 = NGC 4782–4783. The details of the interactions within such multiple-nuclei systems are not at all clear at the present time.

A frequent characteristic of radio galaxies is that their optical spectra show emission lines. Sometimes these lines are strong and numerous, as in Cygnus A and 3C 33; in other cases only one or two lines are present, and some radio galaxies show no emission lines at all. In general the emission lines are strongest for the most intense radio emitters. The widths of the emission lines are usually small, indicating velocity dispersions in the emitting regions less than $\sim$600 km s^{-1}. A discussion of the optical spectra of radio galaxies is given by Schmidt (1965).

The optical features and spectra of the *Seyfert galaxies* are rather different, in that they are typically spiral galaxies having within their nuclei strong sources of optical continuum and line emission. The line widths are very broad, indicating velocity dispersions of 3000–8000 km s^{-1}. Much of the light from a Seyfert galaxy comes from the nucleus, and at a great distance the outer structure of the galaxy might not be recognized. In such a situation the object might be classified as an N galaxy with broad emission lines (Morgan 1968; Burbidge 1968).

Other peculiarities are observed in those few radio galaxies near enough to be examined in detail. Thus NGC 5128 = Centaurus A shows a broad dust lane across its face, as may be seen in the upper insert of Plate II. This dust lane may be traced out into a helical feature extending toward the outer radio source regions. The central double radio source lies obliquely across this lane, as is shown in the lower insert. The axis of rotation of the galaxy is roughly perpendicular to the dust lane.

Absorption features may also be seen in NGC 1316 = Fornax A, although they are less prominent and do not show up in Plate III. In this object they seem to trace out a spiral pattern, which might also be the case in NGC 5128 if we were able to view it from another angle. Faint extensions from the galaxy toward the radio source regions are also seen in NGC 1316 (Arp 1964). An absorption feature may well be present in Cygnus A, giving rise to the impression of a double nucleus, as can be seen in Plate I.

Some radio galaxies show rather direct evidence of the ejection of material from their nuclei. Most prominent of these is M87, which has a luminous jet extending from its nucleus about 21″, corresponding to 1100 pc if we take the distance to the Virgo cluster to be 11 Mpc. The spectrum of the jet shows a featureless continuum which is partially linearly polarized (Baade 1956; Hiltner 1959), and this light apparently comes from a number (6 to 10) of unresolved "knots" along the jet, having angular diameters less than 0″.3 or linear diameters less than 20 pc (de Vaucouleurs, Angione, and Fraser 1968). This continuum radiation is generally presumed to be synchrotron radiation and is the subject of an analysis by Felten (1968). A principal difficulty is posed by the short lifetimes of electrons which would emit at light wavelengths in magnetic fields as strong or stronger than those given for the M87 core radio source by the equipartition argument. In table 7 we have $B \approx 10^{-3.7}$ gauss for the radio source. The field in the optical jet could hardly be less than this value. Equation (2.8) gives an energy $\approx$ 1 erg $\approx 10^{12}$ eV for an electron with a critical frequency $\nu_c = 10^{15}$ Hz in a field of 10^{-4} gauss. Then from equation (2.14) the lifetime of such an electron is $\sim$1000 years. For a fixed critical frequency the lifetime of an electron would scale with magnetic field as $t_{1/2} \propto B^{-3/2}$.

The conclusion reached by Felten is that there must be a distributed source of energetic electrons within the jet, since the lifetime of the "optical" electrons is so short as to preclude their transport over more than a fraction of the jet's length. Alternatively it may be possible to interpret the light as Compton-scattered radio photons. A choice between these two mechanisms may be possible when we have a radio map[10] of M87 with a resolution of $\sim 1''$.

Another galaxy with evidence of ejection of material is M82 = NGC 3034 = 3C 231. This is an irregular galaxy with angular dimensions of about $2' \times 8'$. Filter photography in the Hα line shows filaments extending for several minutes of arc along the minor axis of the galaxy and on both sides of the galactic nucleus. Lynds and Sandage (1963) conclude that this material is being ejected from the galaxy with velocities $\sim$1000 km s^{-1}. Polarized light is observed near the inner parts of these filaments, over a region $\lesssim 1'$ on each side of the nucleus (Zwicky 1964*a*). Elvius (1962) finds a degree of polarization as high as 16 percent in one region on the south edge of the galaxy. The direction of polarization is more or less along the major axis, indicating a magnetic field parallel to the minor axis of the galaxy.

5. THEORIES OF RADIO GALAXIES

Because of the remarkably similar appearance of radio galaxies having a very wide range of intrinsic sizes, it is generally assumed that the observed radio galaxies represent something like an evolutionary sequence. Large amounts of energy are released in one or more explosive events in the nuclei of massive galaxies. Soon after the initial event the energy is channeled in opposite directions out of the nucleus, forming two roughly similar plasma clouds. These clouds then travel out and away from the parent galaxy until finally they fade away, having in some cases attained metagalactic size before they become undetectably faint.

While this evolutionary concept of radio galaxies is qualitatively satisfactory, efforts to make quantitative evolutionary models have only had limited success. The theoretical problems may roughly be divided into four stages of radio source development: (1) the source energy, (2) the early phases of development, in particular the reason for the double nature of the sources, (3) the containment and dynamics of the sources as they grow to sizes $\sim$100 kpc, and (4) the late phases of development and dissolution of the sources.

5.1. Energy Sources

There is no generally accepted explanation for the way in which energy is supplied to strong radio sources. Historically, the first suggestion was that of Baade and Minkowski (1954), who attributed all such objects to collisions of galaxies within clusters. However some radio galaxies are not members of clusters; and many, even in clusters, show no

[10] High resolution interferometric studies of the core of M87 have been made by P. N. Wilkinson (1975, *Nature*, **253**) and B. D. Turland (1975, *M.N.R.A.S.*, **170**). These show a very close correspondence between compact radio and optical components in the jet, although the radio components have typical diameters of $0\overset{''}{.}5$ to $2''$. Most of the flux from the core comes from a symmetrical double source of lower surface brightness extending about $30''$ on each side of the nucleus. This source was first mapped by MacDonald, Kenderdine, and Neville (1968) at 1.4 GHz and by Hogg *et al.* (1969) at 2.7 GHz, but it is shown much more clearly in the new 5 GHz map by Turland. Turland favors a model in which relativistic electrons are continually produced in each knot and are not contained but stream out at nearly the velocity of light. These electrons then maintain the radiation from the more extended components of the radio source.

evidence that a collision is taking place or has recently occurred. Thus the collision hypothesis seems unsatisfactory.

Other more recent suggestions have assumed that a quasi-stellar radio source is the initial phase of a radio galaxy, thus consolidating the energy supply problem for all the strong extragalactic radio sources. The variable sources which are associated with quasi-stellar objects and with some radio galaxies (e.g., 3C 84 = NGC 1275 and 3C 120) apparently experience an energy release $\gtrsim 10^{51}$ ergs every year or so into a volume much less than a light year in size (Kellermann and Pauliny-Toth 1968). The nature of this energy source is still fairly mysterious, but a source is clearly present which could account for the subsequent development of a radio galaxy.

One mechanism which has somewhat fallen out of fashion, but which still is in the running, is a multiple supernova outburst in the nucleus of a large galaxy (Cameron 1962; Field 1964), perhaps triggered in an awesome chain reaction (Burbidge 1961) or by collisions in a very dense nuclear star cloud (Gold, Axford, and Ray 1965; Colgate 1967). The energy release in a single supernova of the types we are familiar with is $\sim 10^{51}$ ergs, which is of the right order to account for individual outbursts in variable sources. However very many of these would have to occur in a limited interval (perhaps 10^4 years) to provide the energy for a large radio galaxy.

A second mechanism which has provoked much speculation is the release of energy when an object of large mass undergoes gravitational collapse. This was first suggested in the context of the radio galaxies by Hoyle and Fowler (1963). Much purely theoretical work has since been done on such objects (e.g., Bardeen and Wagoner 1969), but it is still not clear how they could form without breaking up into numerous less massive pieces which would then be gravitationally stable (and which might then resemble a galaxy or a globular cluster). Bardeen (1970) offers hope that direct conversion of the energy of collapse into electromagnetic forms might be possible.

Hydromagnetic mechanisms have also been popular. Generally these have sought to convert the rotational energy of a galaxy into energetic particles and magnetic fields by twisting up a magnetic field. This might be either the field of the galaxy itself, twisted through differential rotation (Hoyle 1964) or an intergalactic field frozen into the rotating matter of the galaxy (Piddington 1964, 1967; Sturrock 1965, 1966).

The discovery of pulsars and the realization that a rotating magnet can efficiently accelerate fast particles (Gunn and Ostriker 1969; Goldreich and Julian 1969) has led to a number of pulsar-like theories of radio galaxies. The energy source might be a group of ordinary pulsars in the nucleus of a galaxy (Ostriker, Arons, Kulsrud, and Gunn 1971) or a single, very massive rotating magnetized object dubbed a "spinar" (Ozernoy 1966; Morrison 1969; Cavaliere, Pacini, and Setti 1969; Piddington 1970; Rees 1971).

5.2. Symmetrical Division

There have been few quantitative efforts to account for the early division of a radio source into two rather symmetrical components. Qualitative suggestions have invoked an ambient dipole magnetic field of the galaxy with the clouds of energetic particles moving most easily out along the poles of the field. Since the energy density in the plasma clouds as they leave the nucleus of a radio galaxy must be many orders of magnitude greater than the energy density in a typical galactic (or intergalactic) magnetic field, this seems a little like guiding a cannon shell with a rainbow. Alternatively there are some hand-waving explanations involving the two ends of the axis of rotation of a galaxy. The spinar models have a closer connection with a well-defined axis of rotation. Unfortunately no ordinary supernova remnant is a symmetrical double. Thus the pulsar phenomenon cannot simply be scaled up by 10 orders of magnitude to make a radio galaxy.

The problem of symmetrical division is obviously very fundamental to an understanding of the development of radio sources, yet all we know of it so far could have been developed in a single coffee-break discussion. Hopefully the observations now in progress, using VLB interferometry, of the changes in shape of variable sources will shed some new light on the earliest phases of radio source development.

5.3. Source Containment

If a double radio source were to expand freely from its point of origin in the nucleus of a galaxy, we would expect the rate of increase of the component radius to be of the same order as the velocity of ejection of each component. However, we see well-defined doubles such as 3C 33, in which the component separation is almost 100 times the component diameter. This necessitates a mechanism for containment of the components as they move out over distances of $\sim$100 kpc from their point of origin and over times of perhaps $\sim 10^6$ years. Four mechanisms have been suggested: (*a*) external magnetic fields, (*b*) inertial confinement, (*c*) dynamic, or ram-pressure, confinement, and (*d*) gravitational confinement.

Shaping of a double source by the ambient intergalactic magnetic field was suggested by van der Laan (1963) and by Piddington (1964). The difficulty that this mechanism must overcome is that the strength of the containing field must be at least as strong as the field in the source region, which is generally estimated to be at least 10^{-5} to 10^{-4} gauss. No one seriously contemplates extragalactic fields of this order because of the tremendous energy that would be stored in such a field extending over all of space or even over the volume of a large cluster of galaxies.

The expansion of a source region can be slowed by loading it with a great deal of cold gas which is frozen into the magnetic field of the source. The inertia of this gas will slow the source expansion caused by the pressure of the relativistic gas (Scheuer 1967). It seems difficult to hold together very compact source components in this way.

Dynamic containment seems the most promising of the several mechanisms which have been proposed. A cloud of relativistic gas which is ejected from a galaxy at velocity greater than the sound speed in the intergalactic medium will produce a shock wave which will tend to confine the moving cloud in the same way that the Earth's bow shock confines the Earth's magnetosphere. This effect was first treated quantitatively by De Young and Axford (1967) and has been elaborated on by Mills and Sturrock (1970) and De Young (1971). It seems possible to account for the general shape of many radio sources with this mode of containment, provided that ejection velocities are $0.1c$ to $0.3c$ and the intergalactic (or at least the intracluster) medium has a density $\gtrsim 10^{-29}$ g cm^{-3}. Blake (1972) has shown that Rayleigh-Taylor and Kelvin-Helmholtz instabilities will be important in limiting the lifetime and minimum size of components confined in this way. At the moment it seems difficult to account for the most compact of the components which have been observed.

Rees and Setti (1968) have noted that the density of the intergalactic medium in an expanding universe should be a function of redshift and have suggested that the sizes and lifetimes of extended sources should depend on z if ram pressure is the mechanism for source containment. It is not clear how the density of the intracluster medium changes in an expanding universe, however. The observation of radio tails associated with spiral radio galaxies which are members of dense clusters (Miley *et al.* 1972) gives strong evidence for an intracluster medium with density $\sim 10^{-28}$ g cm^{-3}. It is argued that in these objects (3C 129 and several sources in the Perseus cluster) the ejection velocity was not sufficient to contain the source very well by ram pressure. Studies of low-luminosity features, which at the moment can only be done with the Westerbork synthesis telescope, should be of great help in refining our ideas of source dynamics.

G. R. Burbidge (1967) has taken a very different approach, suggesting that the gravitational field of a very massive object confines the relativistic gas. One might elaborate on his idea and suppose that such an object is a spinar which continually supplies additional fast particles to the source region. Some ram-pressure confinement would presumably shape the outer parts of the source. The strongest argument against this model is that the massive object should have a substantial optical luminosity, and no optical counterpart has ever been detected in the position of the compact components of a radio galaxy such as 3C 33 or Cygnus A. The ejection of a symmetrical pair of massive objects from a galactic nucleus must also present some problems.

Any of the models for containment of a source must provide for a supply of energy to combat adiabatic losses while the components grow to sizes $\sim$10–100 kpc. The necessity of this can be seen by computing the resultant luminosity if the very large sources which we see were adiabatically compressed to 5 kpc component size: Cygnus A would have a luminosity of 10^{48} ergs s^{-1}, whereas Hercules A would have 10^{51} ergs s^{-1}—yet the greatest known radio luminosity is $\sim 10^{45}$ ergs s^{-1}.

5.4. The End Game

This brings us to the last phase of the development of radio galaxies, the final expansion and dissolution of the source. Ryle and Longair (1967) have shown that for sources with overall radius $\gtrsim 100$ kpc the luminosity tends to decrease with size in a way which is consistent with adiabatic expansion. Such a source should simply fade away on a time scale of 10^6 or 10^7 years. Semiquantitative discussions of this process have been given by Shklovskii (1960), van der Laan (1963), and Pacholczyk (1965). Van der Laan and Perola (1969) have shown that the relative absence of radio sources with steep spectra at high frequencies means that the old sources must die out by diffusion of their energetic particles into regions of low magnetic field before radiating away a major portion of their energy. In fact, they show that if a source is to be kept alive for longer than $\sim 10^7$ years, multiple injections of energetic particles must take place. In comparison with the earlier phases, this aspect of source evolution is fairly well understood

I am grateful for the advice of many colleagues, especially J. Gunn, M. Longair, and M. Schmidt. The difficult text of § 2 was carefully typed by D. Rapchak. A preliminary version of table 5 was compiled by J. Ekers. Portions of the chapter were written while I was a visitor at the University of Washington. Although written originally for this chapter, a portion of the section on synchrotron radiation, § 2, has appeared elsewhere (Moffet 1969) and is used here with the permission of Gordon and Breach, Science Publishers. The program of research in radio astronomy at Cal Tech is supported by the U.S. Office of Naval Research and the National Science Foundation.

REFERENCES

Adgie, R. L., Gent, H., Slee, O. B., Frost, A. D., Palmer, H. P., and Rowson, B. 1965, *Nature*, **208**, 275.
Allen, L. R., Brown, R. H., and Palmer, H. P. 1962, *M.N.R.A.S.*, **125**, 57.
Anderson, B., and Donaldson, W. 1967, *M.N.R.A.S.*, **137**, 81.
Arp, H. C. 1964, *Ap. J.*, **139**, 1378.
———. 1966, *Science*, **151**, 1214.
———. 1967, *Ap. J.*, **148**, 321.
———. 1968*a*, *ibid.*, **152**, 633.
———. 1968*b*, *Ap. J. (Letters)*, **153**, L33.
———. 1971, *Ap. Letters*, **7**, 221.
Arsac, J. 1966, *Fourier Transforms and the Theory of Distributions* (Englewood Cliffs, N.J.: Prentice-Hall), §§ 6.8 and 12.8.
Baade, W. 1956, *Ap. J.*, **123**, 550.
Baade, W., and Minkowski, R. 1954, *Ap. J.*, **119**, 206.
Bardeen, J. M. 1970, *Nature*, **226**, 64.
Bardeen, J. M., and Wagoner, R. V. 1969, *Ap. J. (Letters)*, **158**, L65.
Bare, C., Clark, B. G., Kellermann, K. I., Cohen, M. H., and Jauncey, D. L. 1967, *Science*, **157**, 189.
Bash, F. N. 1968*a*, *Ap. J.*, **152**, 375.
———. 1968*b*, *Ap. J. Suppl.*, **16**, 373.
Baylis, W. E., Schmid, W. M., and Lüscher, E. 1967, *Zs. f. Ap.*, **66**, 271.
Beaujardiere, O. de la. 1966, *Ann. d'Ap.*, **29**, 345.
Bennett, A. S. 1962, *Mem. R.A.S.*, **68**, 163.
Berge, G. L., and Seielstad, G. A. 1967, *Ap. J.*, **148**, 367.
———. 1972, *A.J.*, **77**, 810.
Biraud, F., and Véron, P. 1968, *Nature*, **219**, 254.
Blake, G. M. 1972, *M.N.R.A.S.*, **156**, 67.
Blum, E. J. 1959, *Ann. d'Ap.*, **22**, 140.
Bologna, J. M., McClain, E. F., Rose, W. K., and Sloanaker, R. M. 1965, *Ap. J.*, **142**, 106.
Bolton, J. G. 1966, *Nature*, **211**, 917.
Bolton, J. G., Gardner, F. F., and Mackey, M. B. 1964, *Australian J. Phys.*, **17**, 340.
Bolton, J. G., and Stanley, G. J. 1948, *Nature*, **161**, 312.
Bolton, J. G., Stanley, G. J., and Slee, O. B. 1949, *Nature*, **164**, 101.
Braccesi, A., Ceccarelli, M., Fanti, R., Gelato, G., Grovanni, C., Harris, D., Rosatelli, C., Sinigaglia, G., and Volders, L. 1965, *Nuovo Cimento*, **40**, 267.
Bracewell, R. N. 1958, *Proc. I.E.E.E.*, **46**, 97.
———. 1962, *Handbuch der Physik*, **54**, 42.
Bridle, A. H., Davis, M. M., Fomalont, E. B., and Lequeux, J. 1972, *A.J.*, **77**, 405.
Broten, N. W., Legg, T. H., Locke, J. L., McLeish, C. W., Richards, R. S., Chisholm, R. M., Gush, H. P., Yen, J. L., and Galt, J. A. 1967, *Science*, **156**, 1592.
Brundage, R. K., Dixon, R. S., Ehman, J. R., and Kraus, J. D. 1971, *A.J.*, **76**, 777.
Budden, K. G., and Uscinski, B. J. 1970, *Proc. Roy. Soc.*, **A316**, 315.
———. 1971, *ibid.*, **A321**, 15.
Burbidge, E. M. 1967, *Ap. J. (Letters)*, **149**, L51.
———. 1970, *ibid.*, **160**, L33.
Burbidge, E. M., and Burbidge, G. R. 1972, *Ap. J.*, **172**, 37.
Burbidge, E. M., Burbidge, G. R., Solomon, P. M., and Strittmatter, P. A. 1971, *Ap. J.*, **170**, 233.
Burbidge, E. M., and Strittmatter, P. A. 1972, *Ap. J. (Letters)*, **172**, L37.
Burbidge, G. R. 1956, *Ap. J.*, **124**, 416.
———. 1959, *ibid.*, **129**, 849.
———. 1961, *Nature*, **190**, 1053.
———. 1967, *ibid.*, **216**, 1287.
———. 1968, *Ann. Rev. Astr. and Ap.*, **8**, 369.
Burbidge, G. R., and Burbidge, E. M. 1967, *Quasi-stellar Objects* (San Francisco: W. H. Freeman).
Cameron, A. G. W. 1962, *Nature*, **194**, 963.
Caswell, J. L., and Wills, D. 1967, *M.N.R.A.S.*, **135**, 231.
Cavaliere, A., Morrison, P., and Pacini, F. 1970, *Ap. J. (Letters)*, **162**, L133.

Cavaliere, A., Pacini, F., and Setti, G. 1969, *Ap. Letters*, **4**, 103.
Ceccarelli, M., and Grueff, G. 1967, *Nuovo Cimento*, **48B**, 425.
Christiansen, W. N., and Högbom, J. A. 1969, *Radiotelescopes* (Cambridge: Cambridge University Press).
Cohen, M. H. 1969, *Ann. Rev. Astr. and Ap.*, **7**, 619.
Cohen, M. H., Cannon, W., Purcell, G. H., and Shaffer, D. B. 1971, *Ap. J.*, **170**, 207.
Cohen, M. H., Gundermann, E. J., and Harris, D. E. 1967, *Ap. J.*, **150**, 767.
Cohen, M. H., Moffet, A. T., Shaffer, D., Clark, B. G., Kellermann, K. I., Jauncey, D. L., and Gulkis, S. 1969, *Ap. J.* (*Letters*), **158**, L83.
Colgate, S. A. 1967, *Ap. J.*, **150**, 163.
Colla, G., Fanti, C., Fanti, R., Ficarra, A., Formiggini, L., Gandolfi, E., Grueff, G., Lari, C., Padrielli, L., Roffi, G., Tomasi, P., and Vigotti, M. 1970, *Astr. and Ap., Suppl.* **1**, 281.
Conway, R. G., Gilbert, J. A., Raimond, E., and Weiler, K. A. 1971, *M.N.R.A.S.*, **152**, 1P.
Conway, R. G., Kellermann, K. I., and Long, R. J. 1963, *M.N.R.A.S.*, **125**, 261.
Cooper, B. F. C., Price, R. M., and Cole, D. J. 1965, *Australian J. Phys.*, **18**, 589.
CSIRO Radiophysics Div. Staff. 1969, ed. J. A. Ekers, *Australian J. Phys., Ap. Suppl.*, No. 7.
Davies, R. D., and Gardner, F. F. 1970, *Australian J. Phys.*, **23**, 59.
Dent, W. A. 1965, *Science*, **148**, 1458.
———. 1966, *Ap. J.*, **144**, 843.
Dent, W. A., and Haddock, F. T. 1965, *Nature*, **205**, 487.
De Young, D. S. 1971, *Ap. J.*, **167**, 541.
De Young, D. S., and Axford, I. 1967, *Nature*, **216**, 129.
Dickel, J. R., Webber, J. C., Yang, K. S., and Staff. 1971, *A.J.*, **76**, 294.
Dickel, J. R., Yang, K. S., McVittie, G. C., and Swenson, G. W., Jr. 1967, *A.J.*, **72**, 757.
Dixon, R. S., and Kraus, J. D. 1968, *A.J.*, **73**, 381.
Ehman, J. R., Dixon, R. S., and Kraus, J. D. 1970, *A.J.*, **75**, 351.
Eidman, V. Ia. 1958, *Zh. Eksp. Teor. Fiz.*, **34**, 131 (English transl. in *Soviet Phys.—JETP*, **7**, 91).
Ekers, R. D. 1969, *Australian J. Phys., Ap. Suppl.*, No. 6.
Elvius, A. 1962, *Lowell Obs. Bull.*, **5**, 281.
Epstein, R. I., and Feldman, P. A. 1967, *Ap. J.* (*Letters*), **150**, L109.
Evans, D. S. 1963, *M.N.R.A.S., So. Africa*, **22**, 140.
Felten, J. E. 1968, *Ap. J.*, **151**, 861.
Felten, J. E., and Morrison, P. 1966, *Ap. J.*, **146**, 686.
Field, G. B. 1964, *Ap. J.*, **140**, 1434.
Fitch, L. T., Dixon, R. S., and Kraus, J. D. 1969, *A.J.*, **74**, 612.
Fomalont, E. B. 1968, *Ap. J. Suppl.*, **15**, 203.
———. 1969, *Ap. J.*, **157**, 1027.
———. 1971, *A.J.*, **76**, 513.
Galt, J. A., and Kennedy, J. E. D. 1968, *A.J.*, **73**, 135.
Gardner, F. F., and Davies, R. D. 1966*a*, *Australian J. Phys.*, **19**, 129.
———. 1966*b*, *ibid.*, 441.
Gardner, F. F., Morris, D., and Whiteoak, J. B. 1969*a*, *Australian J. Phys.*, **22**, 79.
———. 1969*b*, *ibid.*, **22**, 813.
Gardner, F. F., and Whiteoak, J. B. 1963, *Nature*, **197**, 1963.
———. 1966, *Ann. Rev. Astr. and Ap.*, **4**, 245.
———. 1969, *Australian J. Phys.*, **22**, 107.
———. 1971, *ibid.*, **24**, 899.
Gardner, F. F., Whiteoak, J. B., and Morris, D. 1969, *Australian J. Phys.*, **22**, 821.
Ginzburg, V. L. 1951, *Doklady Akad. Nauk SSSR*, **76**, 377.
Ginzburg, V. L., Sazonov, V. N., and Syrovatskii, S. I. 1968, *Usp. Fiz. Nauk*, **94**, 63 (English transl. in *Soviet Phys.—Usp.*, **11**, 34).
Ginzburg, V. L., and Syrovatskii, S. I. 1964*a*, *The Origin of Cosmic Rays* (Oxford: Pergamon Press).
———. 1964*b*, *Zh. Eksp. Teor. Fiz.*, **46**, 1865 (English transl. in *Soviet Phys.—JETP*, **19**, 1255).
———. 1965, *Ann. Rev. Astr. and Ap.*, **3**, 297.
———. 1969, *ibid.*, **7**, 375.
Gold, T. 1965, *Proc. 9th Int. Conf. Cosmic Rays*, London (Inst. of Phys. and Phys. Soc.), p. 132.
Gold, T., Axford, I., and Ray, E. C. 1965, in *Quasi-stellar Sources and Gravitational Collapse*, ed. I. Robinson *et al.* (Chicago: University of Chicago Press), p. 93.
Goldreich, P., and Julian, W. H. 1969, *Ap. J.*, **157**, 869.
Golnev, V. Ya., and Soboleva, N. S. 1965, *Astr. Zh.*, **42**, 694 (English transl. in *Soviet Astr.—AJ*, **9**, 537).
Gower, J. F. R., Scott, P. F., and Wills, D. 1967, *Mem. R.A.S.*, **71**, 49.
Greenstein, J. L. 1962, *Ap. J.*, **135**, 679.
Gunn, J. E. 1971, *Ap. J.* (*Letters*), **164**, L113.
Gunn, J. E., and Ostriker, J. P. 1969, *Phys. Rev. Letters*, **22**, 728.
Harris, D. E., and Roberts, J. A. 1959, *Pub. A.S.P.*, **72**, 237.
Hazard, C., Gulkis, S., and Bray, A. D. 1967, *Ap. J.*, **148**, 669.
Heeschen, D. S. 1960, *Pub. A.S.P.*, **72**, 368.
———. 1968, *Ap. J.* (*Letters*), **151**, L135.

———. 1970, *Ap. Letters*, **6**, 49.
Hewish, A., Scott, P. F., and Wills, D. 1964, *Nature*, **203**, 1214.
Hill, J. M., and Longair, M. S. 1971, *M.N.R.A.S.*, **154**, 125.
Hiltner, W. A. 1959, *Ap. J.*, **130**, 340.
Hoerner, S. von. 1964, *Ap. J.*, **140**, 65.
———. 1966, *ibid.*, **144**, 483.
Hogg, D. E., Macdonald, G. H., Conway, R. G., and Wade, C. M. 1969, *A.J.*, **74**, 1206.
Holden, D. J. 1966, *M.N.R.A.S.*, **133**, 225.
Hoyle, F. 1964, *Nature*, **201**, 804.
Hoyle, F., and Fowler, W. A. 1963, *M.N.R.A.S.*, **125**, 169.
Humason, M. L., Mayall, N. U., and Sandage, A. R. 1956, *A.J.*, **61**, 97.
Jennison, R. C., and Das Gupta, M. K. 1953, *Nature*, **172**, 996.
Kardashev, N. S. 1962, *Astr. Zh.*, **39**, 393 (English transl. in *Soviet Astr.—AJ*, **6**, 317).
Kellermann, K. I. 1964, *Ap. J.*, **140**, 969.
———. 1966, *ibid.*, **146**, 621.
Kellermann, K. I., Long, R. J., Allen, L. R., and Moran, M. 1962, *Nature*, **195**, 692.
Kellermann, K. I., and Pauliny-Toth, I. I. K. 1968, *Ann. Rev. Astr. and Ap.*, **6**, 417.
———. 1969, *Ap. J.* (*Letters*), **155**, L71.
Kellermann, K. I., Pauliny-Toth, I. I. K., and Williams, P. J. S. 1969, *Ap. J.*, **157**, 1.
Kenderdine, S., Ryle, M., and Pooley, G. G. 1966, *M.N.R.A.S.*, **134**, 189.
Kraus, J. D. 1966, *Radio Astronomy* (New York: McGraw-Hill).
Laan, H. van der. 1963, *M.N.R.A.S.*, **126**, 535.
———. 1972, report to 17th Gen. Assembly URSI, Warsaw.
Laan, H. van der, and Bash, F. N. 1968, *Ap. J.*, **152**, 621.
Laan, H. van der, and Perola, G. C. 1969, *Astr. and Ap.*, **3**, 468.
Legg, M. P. C., and Westfold, K. C. 1968, *Ap. J.*, **154**, 499.
Lequeux, J. 1962, *Ann. d'Ap.*, **25**, 221.
———. 1965, *ibid.*, **28**, 360.
———. 1966, *ibid.*, **29**, 533.
Le Roux, E. 1961, *Ann. d'Ap.*, **24**, 71.
Little, L. T., and Hewish, A. 1966, *M.N.R.A.S.*, **134**, 221.
———. 1968, *ibid.*, **138**, 393.
Lockhart, I. A., and Sheridan, K. V. 1970, *Proc. Astr. Soc. Australia*, **1**, 344.
Long, R. J., Smith, M. A., Stewart, P., and Williams, P. J. S. 1966, *M.N.R.A.S.*, **134**, 371.
Lynds, C. R., and Sandage, A. R. 1963, *Ap. J.*, **137**, 1005.
Macdonald, G. H., Kenderdine, S., and Neville, A. C. 1968, *M.N.R.A.S.*, **138**, 259.
Mackay, C. D. 1969, *M.N.R.A.S.*, **145**, 31.
———. 1971, *ibid.*, **154**, 209.
MacLeod, J. M., Swenson, G. W., Jr., Yang, K. S., and Dickel, J. R. 1965, *A.J.*, **70**, 756.
Maltby, P. 1965, *Ap. J.*, **144**, 219.
Maltby, P., Matthews, T. A., and Moffet, A. T. 1963, *Ap. J.*, **137**, 153.
Maltby, P., and Moffet, A. T. 1962, *Ap. J. Suppl.*, **7**, 141.
Maltby, P., and Seielstad, G. A. 1966, *Ap. J.*, **144**, 216.
Melrose, D. B. 1971*a*, *Ap. Letters*, **8**, 35.
———. 1971*b*, *Ap. and Space Sci.*, **12**, 172.
Miley, G. K., Perola, G. C., Kruit, P. C. van der, and Laan, H. van der. 1972, *Nature*, **237**, 269.
Miley, G. K., and Wade, C. M. 1971, *Ap. Letters*, **8**, 11.
Mills, B. Y., Slee, O. B., and Hill, E. R. 1958, *Australian J. Phys.*, **11**, 360.
———. 1960, *ibid.*, **13**, 676.
———. 1961, *ibid.*, **14**, 497.
Mills, D. M., and Sturrock, P. A. 1970, *Ap. Letters*, **5**, 105.
Minkowski, R. 1960, *Ap. J.*, **132**, 908.
———. 1961*a*, *A.J.*, **66**, 558.
———. 1961*b*, *Proceedings Fourth Berkeley Symp. on Math. Statistics and Probability*, ed. J. Neyman (Berkeley: University of California Press), **4**, 245.
Mitton, S. 1970, *Ap. Letters*, **5**, 207.
———. 1972, *M.N.R.A.S.*, **155**, 373.
Mitton, S., and Ryle, M. 1969, *M.N.R.A.S.*, **146**, 221.
Moffet, A. T. 1962, *Ap. J. Suppl.*, **7**, 93.
———. 1964, *Science*, **146**, 764.
———. 1966, *Ann. Rev. Astr. and Ap.*, **4**, 145.
———. 1969, in *Astrophysics and General Relativity*, ed. M. Chrétien, S. Deser, and J. Goldstein (New York: Gordon & Breach), **1**, 217.
Moffet, A. T., and Palmer, H. P. 1965, *Observatory*, **85**, 45.
Morgan, W. W. 1968, *Ap. J.*, **153**, 27.
Morgan, W. W., and Lesh, J. R. 1965, *Ap. J.*, **142**, 1364.
Morimoto, M., and Lockhart, I. A. 1968, *Proc. Astr. Soc. Australia*, **1**, 99.

Morris, D., and Berge, G. L. 1964*a*, *Ap. J.*, **139**, 1388.
———. 1964*b*, *A.J.*, **69**, 641.
Morris, D., Radhakrishnan, V., and Seielstad, G. A. 1964*a*, *Ap. J.*, **139**, 551.
———. 1964*b*, *ibid.*, **139**, 560.
Morrison, P. 1969, *Ap. J.* (*Letters*), **157**, L73.
Olsen, E. T. 1967, *A.J.*, **72**, 738.
Oster, L. 1961, *Rev. Mod. Phys.*, **33**, 525.
Ostriker, J., Arons, J., Kulsrud, R., and Gunn, J. 1971, *Bull. A.A.S.*, **3**, 237.
Ozernoy, L. M. 1966, *Astr. Zh.*, **43**, 300 (English transl. in *Soviet Astr.—AJ*, **10**, 241).
Pacholczyk, A. G. 1965, *Ap. J.*, **142**, 1141.
———. 1970, *Radio Astrophysics* (San Francisco: W. H. Freeman), chap. 3.
Pacholczyk, A. G., and Swihart, T. L. 1967, *Ap. J.*, **150**, 647.
Palmer, H. P., Rowson, B., Anderson, B., Donaldson, W., Miley, G. K., Gent, H., Adgie, R. L., Slee, O. B., and Crowther, J. H. 1967, *Nature*, **213**, 789.
Pauliny-Toth, I. I. K., and Kellermann, K. I. 1968, *Ap. J.* (*Letters*), **152**, L169.
Pauliny-Toth, I. I. K., Kellermann, K. I., Davis, M. M., Fomalont, E. B., and Shaffer, D. B. 1972, *A.J.*, **77**, 265.
Pauliny-Toth, I. I. K., Wade, C. M., and Heeschen, D. S. 1966, *Ap. J. Suppl.*, **13**, 65.
Payne, A. D. 1967, *Australian J. Phys.*, **20**, 291.
Penston, M. V. 1971, *Ap. J.*, **170**, 395.
Peterson, B. A. 1970, *A.J.*, **75**, 695.
Peterson, B. A., and Bolton, J. G. 1972, *Ap. J.* (*Letters*), **173**, L19.
Piddington, J. H. 1964, *M.N.R.A.S.*, **128**, 345.
———. 1967, *Planet. and Space Sci.*, **15**, 1625.
———. 1970, *M.N.R.A.S.*, **148**, 131.
Pilkington, J. D. H., and Scott, P. F. 1965, *Mem. R.A.S.*, **69**, 183.
Pooley, G. G. 1969, *M.N.R.A.S.*, **144**, 101.
Pooley, G. G., and Kenderdine, S. 1968, *M.N.R.A.S.*, **139**, 529.
Ramaty, R., and Lingenfelter, R. E. 1966, *J. Geophys. Res.*, **71**, 3687.
Razin, V. 1960, *Radiofizika*, **3**, 584.
Rees, M. J. 1971, *Nature*, **229**, 312.
Rees, M. J., and Setti, G. 1968, *Nature*, **219**, 127.
Roberts, J. A., Ribes, J. C., Murray, J. D., and Cooke, D. J. 1972, *Nature Phys. Sci.*, **236**, 3.
Rowson, B. 1963, *M.N.R.A.S.*, **125**, 177.
Ryle, M. 1952, *Proc. Roy. Soc.*, **A211**, 351.
———. 1962, *Nature*, **194**, 517.
———. 1968, *Ann. Rev. Astr. and Ap.*, **6**, 249.
———. 1972, *Nature*, **239**, 435.
Ryle, M., Elsmore, B., and Neville, A. C. 1965, *Nature*, **207**, 1024.
Ryle, M., and Hewish, A. 1960, *M.N.R.A.S.*, **120**, 220.
Ryle, M., and Longair, M. S. 1967, *M.N.R.A.S.*, **136**, 123.
Ryle, M., and Neville, A. C. 1962, *M.N.R.A.S.*, **125**, 39.
Ryle, M., and Windram, M. D. 1968, *M.N.R.A.S.*, **138**, 1.
Salpeter, E. E. 1967, *Ap. J.*, **147**, 433.
Sandage, A. R. 1966, *Ap. J.*, **145**, 1.
———. 1967*a*, *Ap. J.* (*Letters*), **150**, L9.
———. 1967*b*, *ibid.*, L177.
———. 1972*a*, *Ap. J.*, **178**, 1.
———. 1972*b*, *ibid.*, 25.
Sastry, Ch. V., Pauliny-Toth, I. I. K., and Kellermann, K. I. 1967, *A.J.*, **72**, 230.
Scheuer, P. A. G. 1960, *M.N.R.A.S.*, **120**, 231.
———. 1962, *Australian J. Phys.*, **15**, 333.
———. 1965, in *Quasi-stellar Sources and Gravitational Collapse*, ed. I. Robinson *et al.* (Chicago: University of Chicago Press), p. 373.
———. 1967, in *Plasma Astrophysics, Proc. E. Fermi Internat. School of Physics, Course 39*, ed. P. A. Sturrock (New York: Academic Press), p. 262.
Scheuer, P. A. G., and Williams, P. J. S. 1968, *Ann. Rev. Astr. and Ap.*, **6**, 321.
Schmidt, M. 1965, *Ap. J.*, **141**, 1.
———. 1972, *Nature*, **240**, 399.
Schott, G. A. 1912, *Electromagnetic Radiation* (Cambridge: Cambridge University Press).
Sciama, D. W., and Rees, M. J. 1967, *Nature*, **216**, 147.
Seaquist, E. R. 1971, *Nature Phys. Sci.*, **231**, 93.
Searle, L., and Bolton, J. G. 1968, *Ap. J.* (*Letters*), **154**, L101.
Seielstad, G. A., and Weiler, K. W. 1969, *Ap. J. Suppl.*, **18**, 85.
Shaffer, D. B., Cohen, M. H., Jauncey, D. L., and Kellermann, K. I. 1972, *Ap. J.* (*Letters*), **173**, L147.
Shimmins, A. J. 1971, *Australian J. Phys.*, *Ap. Suppl.*, No. 21.
Shimmins, A. J., and Day, G. A. 1968, *Australian J. Phys.*, **21**, 377.

Shklovskii, I. S. 1952, *Astr. Zh.*, **29**, 418.
———. 1953, *Doklady Akad. Nauk SSSR*, **90**, 983.
———. 1960, *Astr. Zh.*, **37**, 945 (English transl. in *Soviet Astr.—AJ*, **4**, 885).
Slish, V. I. 1963, *Nature*, **199**, 682.
Sturrock, P. 1965, *Nature*, **205**, 861.
———. 1966, *ibid.*, **211**, 697.
Tritton, K. P. 1972, *M.N.R.A.S.*, **158**, 277.
Tsytovitch, V. N. 1951, *Vestnik. Mosk. Univ., Ser. Phys.*, **4**, 27.
———. 1973, *Ann. Rev. Astr. and Ap.*, **11**, 363.
Twiss, R. Q. 1954, *Australian J. Phys.*, **11**, 564.
Vaucouleurs, G. de, Angione, R., and Fraser, C. W. 1968, *Ap. Letters*, **2**, 141.
Vaucouleurs, G. de, and Vaucouleurs, A. de. 1964, *Reference Catalogue of Bright Galaxies* (Austin: University of Texas Press).
Véron, P. 1966, *Ann. d'Ap.*, **29**, 231.
Wade, C. M. 1966, *Phys. Rev. Letters*, **17**, 1061.
Wagoner, R. V. 1967, *Nature*, **214**, 766.
Wall, J. V., Shimmins, A. J., and Merkelijn, J. K. 1971, *Australian J. Phys., Ap. Suppl.*, No. 19.
Wendker, H. J., Dickel, J. R., Yang, K. S., and Staff. 1970, *A.J.*, **75**, 148.
Westerlund, B. E., and Smith, L. F. 1966, *Australian J. Phys.*, **19**, 181.
Westerlund, B. E., and Stokes, N. R. 1966, *Ap. J.*, **145**, 354.
Westfold, K. C. 1959, *Ap. J.*, **130**, 241.
Whiteoak, J. B. 1972, *Australian J. Phys.*, **25**, 233.
Whiteoak, J. B., and Gardner, F. F. 1971, *Australian J. Phys.*, **24**, 913.
Whitford, A. E. 1971, *Ap. J.*, **169**, 215.
Wild, J. P. 1967, *Proc. I.R.E.E. Australia*, **28**, 279.
Wild, J. P., Smerd, S. F., and Weiss, A. A. 1963, *Ann. Rev. Astr. and Ap.*, **1**, 291.
Williams, P. J. S. 1963, *Nature*, **200**, 56.
Williams, P. J. S., Collins, R. A., Caswell, J. L., and Holden, D. J. 1968, *M.N.R.A.S.*, **139**, 289.
Williams, P. J. S., and Stewart, P. 1967, *M.N.R.A.S.*, **135**, 319.
Wills, D. 1967, *Ap. J. (Letters)*, **148**, L57.
Willson, M. A. G. 1971, *M.N.R.A.S.*, **151**, 1.
Windram, M. D., and Kenderdine, S. 1969, *M.N.R.A.S.*, **146**, 265.
Woltjer, L. 1966, *Ap. J.*, **146**, 597.
Wyndham, J. D. 1966, *Ap. J.*, **144**, 459.
Zakharenkov, V. F., Kaidanovskii, N. L., Pariiskii, Yu. N., and Prozorov, V. A. 1963, *Astr. Zh.*, **40**, 216 (English transl. in *Soviet Astr.—AJ*, **7**, 167).
Zwicky, F. 1964*a*, *Ap. J.*, **139**, 1394.
———. 1964*b*, *ibid.*, **140**, 1467.

CHAPTER 8

Quasars

MAARTEN SCHMIDT

Hale Observatories, California Institute of Technology, Carnegie Institution of Washington

1. INTRODUCTION

QUASARS or quasi-stellar objects are objects of starlike appearance whose spectra show large redshifts. Many quasars show variations in light, some with characteristic times that are as short as a few days. Radio quasars or quasi-stellar radio sources are quasi-stellar objects that have been detected at radio frequencies. Radio quasars often contain radio components of very small angular diameter and sometimes show rapid variations at high radio frequencies.

The nature of the quasars[1] is still quite mysterious. No significant proper motion has been detected in any quasar. Many astronomers accept the cosmological nature of the large observed redshifts. However, there is no direct confirmation of the corresponding very large distances. The recognition of many different redshifts in the rich absorption-line spectra of some quasars has only served to deepen the mystery.

The quasars have contributed to renewed interest in cosmology, in the study of collapsing objects in general relativity, and in the study of compact, N-type, and Seyfert galaxies.

The early developments in the field are covered in the proceedings of the 1963 Dallas conference (Robinson, Schild, and Schucking 1965). Proceedings of Texas conferences in 1964 and 1967 have been published (Schild and Schucking 1969; Maran and Cameron 1969). Several books covering quasars have appeared, among which that of Burbidge and Burbidge (1967*a*) is to be mentioned. The literature on quasars has developed in explosive fashion: more than 500 references are covered in two review articles (Burbidge 1967; Schmidt 1969) to which the reader is referred for detailed literature information.[2]

Received June 1969, revised January 1974. Developments since 1969 have caused some sections of the text to be out-of-date. The author has attempted to cure the main inadequacies by the addition of footnotes referring to recent developments and literature.

[1] The name "quasar" proposed by H. Chiu in 1964, when only quasi-stellar radio sources were known, has found wide acceptance. We shall use the term quasar for *all* starlike objects exhibiting large redshifts (see § 3.1). Radio quasars are those that are detected as radio sources, also called quasi-stellar radio sources, quasi-stellar sources, or QSSs. Quasars that are not detected or not selected as radio sources are also known as radio-quiet quasi-stellar sources, quasi-stellar objects (QSOs), quasi-stellar galaxies (QSGs), or—if unconfirmed—blue stellar objects, BSOs, or interlopers.

[2] See also *I.A.U. Symp.*, No. 44, 1972 (ed. D. S. Evans).

2. HISTORY

The first identification of a radio source with a starlike optical object was made by Matthews and Sandage (1963) in 1960. They were attempting to identify sources of high radio surface brightness with distant galaxies but, instead, found that the position of 3C 48 coincided with that of a 16th magnitude stellar object with some associated faint nebulosity. Photometry showed the stellar object to possess an ultraviolet excess compared with ordinary main-sequence stars; it also turned out to vary by 0.4 mag. in a year. The spectrum of the stellar object showed broad emission lines that could not be identified.

Subsequently, Matthews and Sandage (1963) identified the radio sources 3C 196 and 3C 286 with fainter starlike objects, whose colors were similar to those of 3C 48. The spectrum of the starlike object 3C 286 showed one broad emission line (Schmidt 1962) at a wavelength different from those of the lines in 3C 48. Spectra by Schmidt of a starlike object identified with 3C 147 by Matthews showed several emission lines, again at wavelengths different from those seen in the other objects.

It was the identification of the strong radio source 3C 273 that led to the solution of the mystery of the emission lines. Moon occultations of this source were observed in 1962 by Hazard, Mackey, and Shimmins (1963) from which they determined accurate positions of the two components, A and B, of this source. Component B which had a flat radio spectrum coincided with a 13th magnitude star. Component A was placed at the end of a jetlike optical feature at 20″ from the star. Spectra of the starlike object

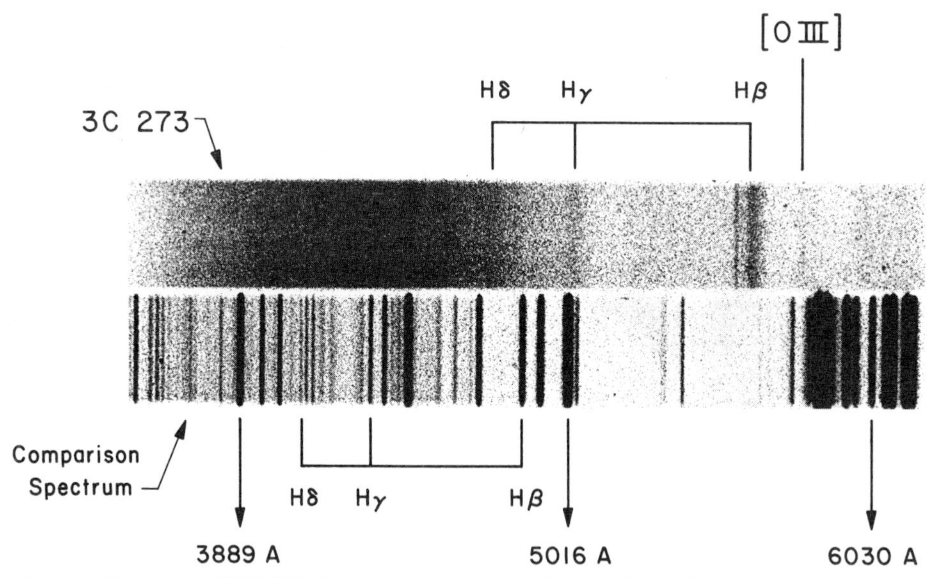

FIG. 1.—Spectrum of 3C 273 showing the hydrogen Balmer lines with a redshift $z = 0.158$. Part of the emission indicated as [O III] is due to Fe II according to Wampler and Oke (1967). The sharp line to the left of Hβ is night-sky emission of [O I] λ5577. At the bottom is a comparison spectrum of hydrogen, helium, and neon, with some wavelengths indicated.

taken in December 1962 showed six broad emission lines. Spectrophotometry by Oke (1963) in the infrared yielded another strong emission line at 7590 Å. It took until February 1963 before it was realized that most of the observed lines could be explained as hydrogen Balmer lines (fig. 1) with a redshift $z = \Delta\lambda/\lambda_0$ of 0.158 (Schmidt 1963). The remaining lines could be satisfactorily interpreted as [O III] λ5007 and Mg II λ2798. An immediate reevaluation of the spectrum of 3C 48 allowed its interpretation with a redshift $z = 0.367$ (Greenstein and Matthews 1963). Somewhat larger redshifts were found a year later for 3C 47 and 3C 147 (Schmidt and Matthews 1964).

Spectra of many starlike objects subsequently identified with 3C radio sources remained unexplained until 1965 when it became possible to identify far-ultraviolet lines with very large redshifts (Schmidt 1965). The process was one of iteration where, first, a new ultraviolet line was detected in a spectrum of large redshift; this line was used subsequently in the establishment of a larger redshift in a different object, leading to the discovery of a deeper ultraviolet line, etc. In this way, it was possible to successively establish the occurrence of emission from semiforbidden C III] λ1909, from the C IV doublet at λ1549 and from Lyman-α (Lα) at λ1216, leading to a redshift $z = 2.01$ for 3C 9 (Schmidt 1965).

The identification of starlike or quasi-stellar objects with radio sources, whose positions were sometimes not sufficiently accurately determined, was facilitated by the ultraviolet excess. Ryle and Sandage (1964) used a technique in which a single photographic plate of a field is doubly exposed, once to blue light and once to ultraviolet light. In the course of further identification work, Sandage (1965) noted that many objects with ultraviolet excess were present that are not associated with radio sources. Further study of these objects and spectra for some of them led to the establishment of the radio-quiet or radio-weak quasi-stellar objects (Sandage 1965).

3. IDENTIFICATIONS

3.1. Definition

We tentatively define a quasar as a starlike object with large redshift. "Starlike" is meant to indicate that a major part of the optical image is within a circle of 1″ diameter. This allows weak extended nebulosity such as that observed near 3C 48 and 3C 196. The requirement of a "large" redshift separates quasars from ordinary galactic stars that show a maximum redshift of 0.002. The smallest quasar redshift to date is 0.06.

Our requirement of a "starlike" image does not discriminate clearly against distant N-type galaxies. An N-type galaxy such as 3C 371 has a variable nucleus (Oke 1967*b*; Sandage 1967*a*); at maximum light it may get close to fulfilling our "starlike" criterion. Colors do not entirely segregate quasars from N-type and compact galaxies (see fig. 2). An example of possible confusion between distant N-type or Seyfert galaxies and quasars is provided by Ton 730. Sandage (1965) finds that this object, which has a redshift of 0.088, looks stellar to the eye at the 200-inch telescope, yet the image looks like an extremely compact galaxy on long-exposure plates.[3]

[3] For a discussion of the nature of the objects Ton 256 and B 264, see Arp (*Ap. J.*, **162**, 811, 1970). Nebulosity around or near a quasar has been searched for, and in several cases detected, by Kristian (*Ap. J.* [*Letters*], **179**, L61, 1973).

3.2. Quasi-stellar Sources (Radio Quasars)

Several hundreds of radio sources have been identified with starlike objects. In a few cases these identifications have been made from positions with an accuracy of about 1″ determined from occultations by the Moon. Most identifications are made on the basis of positions obtained with interferometers that are good to about 10″ or 20″. Since all confirmed radio quasars appear to be blue from a comparison of the blue and the red prints of the *Sky Atlas*, some weight has been attached to this property in the identification of further sources. The relatively few suggested nonblue quasar candidates have, on spectroscopic grounds, all turned out to be misidentifications, till now.

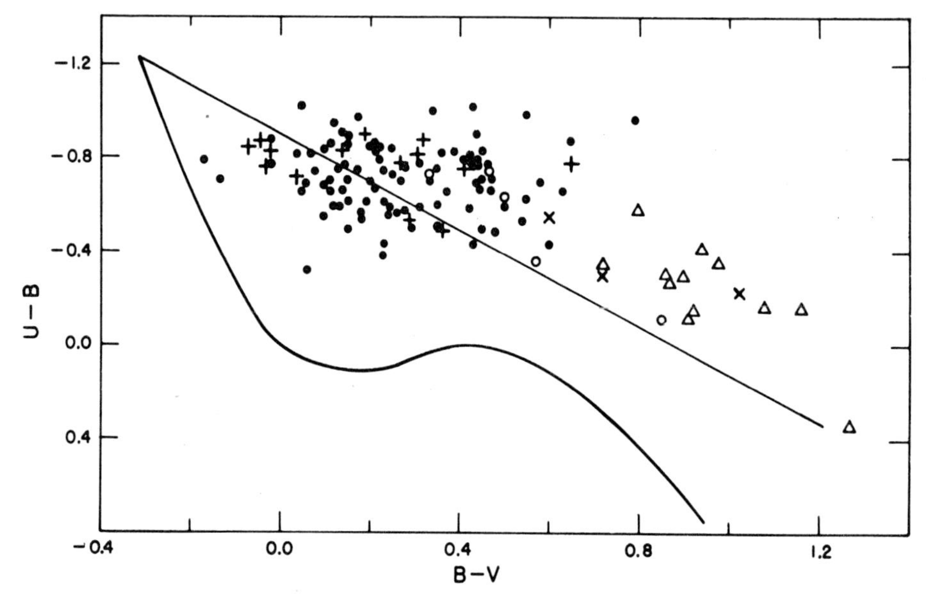

Fig. 2.—Two-color diagram for radio quasars with photometry known to September 1967 (*dots*), radio-quiet quasars with known redshifts (*crosses*), Seyfert nuclei (*open circles*), nonradio blue compact galaxies (×), and radio N galaxies (*triangles*). From Sandage (1967*b*).

Systematic identifications have been summarized for the 3C revised catalog by Wyndham (1966) and Véron (1966). Identifications in the 4C and Parkes catalogs are in progress; for references see Burbidge (1967) and Schmidt (1969). The total number of confirmed identifications in the 3C revised catalog is 44 (Schmidt 1968); this sample is essentially complete to apparent visual magnitude 18.4.

Spectroscopic work usually confirms between 50 and 80 percent of the quasar candidates. The remaining objects show zero-redshift absorption-line spectra typical of galactic stars; some examples have been discussed by Wlerick and Véron (1967) and Mackay (1967).

3.3. Quasi-stellar Objects (Quasars)

Quasars that are not observed as radio sources have been exclusively found on the basis of their colors. Photoelectric photometry of blue stellar objects, found in surveys of faint blue stars, allows selection of those near the blackbody line (fig. 2) in the $[U - B,$

$B - V$]-diagram (Sandage and Luyten 1969). Spectroscopic work shows that a large fraction of these show the quasi-stellar characteristics, most of the other stars being white dwarfs. The success rate of the photoelectric selection of quasars is even larger if photometry in the near-infrared is included (Braccesi, Lynds, and Sandage 1968). The radio flux density of optically discovered quasars is usually undetectably small.

Preliminary results of a survey of several fields at high galactic latitudes have been summarized by Sandage and Luyten (1969). They find that the number $N(B)$ of quasars brighter than magnitude B per square degree is given by

$$\log N(B) = 0.75\,B - 14.0\,. \tag{1}$$

This relation applies for B brighter than $19\frac{1}{2}$ and possibly down to $21\frac{1}{2}$. In the latter case there would exist of the order of 10^7 quasars brighter than $B = 22$.

The coefficient 0.75 in relation (1) corresponds to a slope in the ($\log N$, $\log S$)-relation of 1.87, almost identical to that found for radio sources. The significance of this agreement will be further discussed in § 8.

The number of quasars brighter than $B = 18$ per square degree is around 0.3. For 3C revised radio quasars this number is only 0.001 per square degree. Hence, only one in 300 quasars is a radio source sufficiently strong to be included in the 3C revised catalog.

3.4. Sky Distribution

The distribution of extragalactic radio sources around the sky is quite uniform (Ryle 1968). Since some 20 to 30 percent of these sources are quasars the distribution of radio quasars must also be quite uniform.

A claim by Strittmatter, Faulkner, and Walmsley (1966) that high-redshift quasars are concentrated in two regions in the sky (near the north galactic pole, and near R.A. $= 0^h$, decl. $= 0°$) is based on quite heterogeneous material; the effect is not shown by systematic identification work on 3C revised quasars.[4]

No quasars have been found to be associated with single galaxies of equal redshift.[5] Arp (1967) has claimed a statistical correlation between peculiar galaxies and radio sources. The suggested associations involve very large redshift differences between the peculiar galaxy and the radio source (radio galaxy or quasar). For references to statistical evaluations of the effect, see Schmidt (1969). It is not clear, though, that the statistical significance of an unpredicted effect can be meaningfully derived.[6]

No cluster of galaxies has been found to be associated with 3C 273 and 3C 48 (Sandage and Miller 1966). No clusters have been noticed around small-redshift quasars to date.[7]

[4] Wills finds on the basis of a complete sample of 4C quasars no evidence for deviations from a uniform distribution around the sky (*Nature Phys. Sci.*, **234**, 168, 1971).

[5] Stockton (*Nature Phys. Sci.*, **246**, 25, 1973), has found a galaxy with redshift 0.3736 very close to the quasi-stellar radio source 4C 37.43 (redshift 0.3708).

[6] This also applies to recent claims of associations between bright galaxies and quasars. For discussion and references, see Burbidge, O'Dell, and Strittmatter (*Ap. J.*, **175**, 601, 1972); Bahcall, McKee, and Bahcall (*Ap. Letters*, **10**, 147, 1972); and Hazard and Sanitt (*Ap. Letters*, **11**, 77, 1972).

[7] Much work on the association between clusters and quasars has been carried out since 1969. See Oemler, Gunn, and Oke, *Ap. J.* (*Letters*), **176**, L48, 1972; Bahcall, Bahcall, and Schmidt, *Ap. J.*, **183**, 777, 1973; and Hazard, Jauncey, Sargent, Baldwin, and Wampler, *Nature*, **246**, 205, 1973; and other references given in these articles.

No clustering of quasars has been observed. There are, however, two cases of a pair of similar radio sources or quasars. The radio sources 3C 343 and 3C 343.1 have almost identical spectra, are both small, but are separated by 29′ (Moffet 1965). A similar case is that of 3C 345 and NRAO 512, where the sources have similar radio spectra and are both variable (Kellermann and Pauliny-Toth 1968*a*).[8]

4. EMISSION LINES, REDSHIFTS

The optical spectra of the quasars are characterized by broad emission lines. In the far-ultraviolet the emission lines of Lα and C IV λ1549 are always strong (fig. 3). The intercombination line C III] λ1909 is weaker and sometimes absent. Emission from Mg II λ2798 is usually medium to strong. In the visible part of the spectrum the usual emission lines of the hydrogen Balmer series and forbidden lines of [O II], [O III], [Ne III], and [Ne V] are seen, as in the spectra of planetary nebulae. In some quasars the forbidden lines are weak or absent.

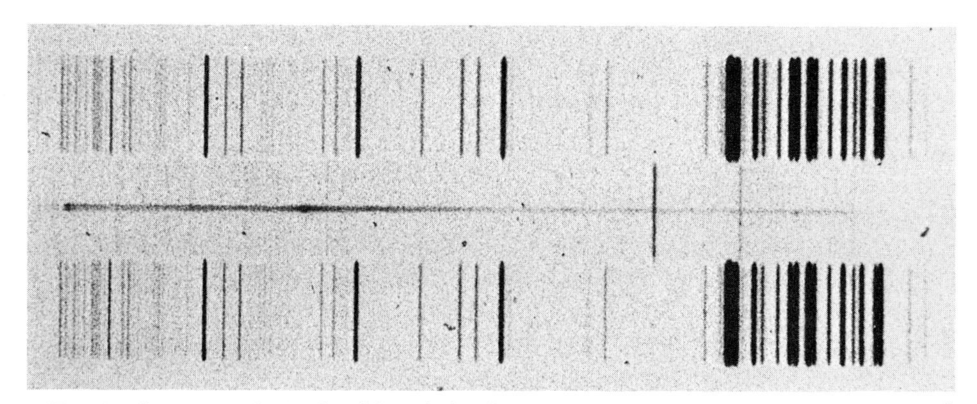

FIG. 3.—Spectrum of 3C 454 with emission lines of Lα at extreme left, and C IV at 4271 Å corresponding to a redshift $z = 1.757$. At top and bottom is a comparison spectrum of hydrogen, helium, and neon; for wavelengths see fig. 1.

The widths of the lines are typically observed to be 40 Å. In some cases the forbidden lines are narrow (width < 10 Å); in a very few cases all the lines observed are narrow.

The redshift of a quasar can usually be determined from the observed wavelengths of two well-established emission lines, following a procedure used by Schmidt (1965). A single redshift explains practically all the emission lines in a quasar. Time variations in the emission-line spectra have been suspected but have not yet been demonstrated beyond doubt. Published redshifts are summarized in table 2 of Burbidge (1967) and table 1 of Schmidt (1969). These tables contain 28 optically selected quasars and 123 radio quasars. The observed redshifts range from 0.06 to 2.2 for the quasars and from 0.16 to 2.4 for the radio quasars.[9]

[8] Very close pairs of quasars, in each case with dissimilar redshifts, have been isolated by Stockton (*Nature Phys. Sci.*, **238**, 37, 1972) and by Wampler, Baldwin, Burke, Robinson, and Hazard (*Nature*, **246**, 203, 1973).

[9] Recent catalogs are given by De Veny, Osborn, and Janes (*Pub. Astr. Soc. Pac.*, **83**, 611, 1971) and Setti and Woltjer (*Proc. Sixth Texas Symp. on Relativistic Astrophysics*, New York, 1972 December). The largest observed redshifts to date (1974 January) are 3.40 for the quasi-

Many quasar spectra show at first only one certain emission line, and considerable effort may be required to establish the existence of at least one further line with certainty. A redshift larger than 1.8 can be immediately determined from a single spectrum because Lα appears above 3400 Å and C IV is easily accessible. For redshifts less than 1.6, Lα is unobservable; beyond C IV the main accessible line is C III] λ1909, but it is often weak and sometimes absent. The next emission line, Mg II λ2798, becomes available below 6800 Å at a redshift of 1.4, but the C IV emission becomes unobservable below a redshift of 1.1. At lower redshifts, more of the "visual" part of the spectrum with forbidden and Balmer lines is observable.

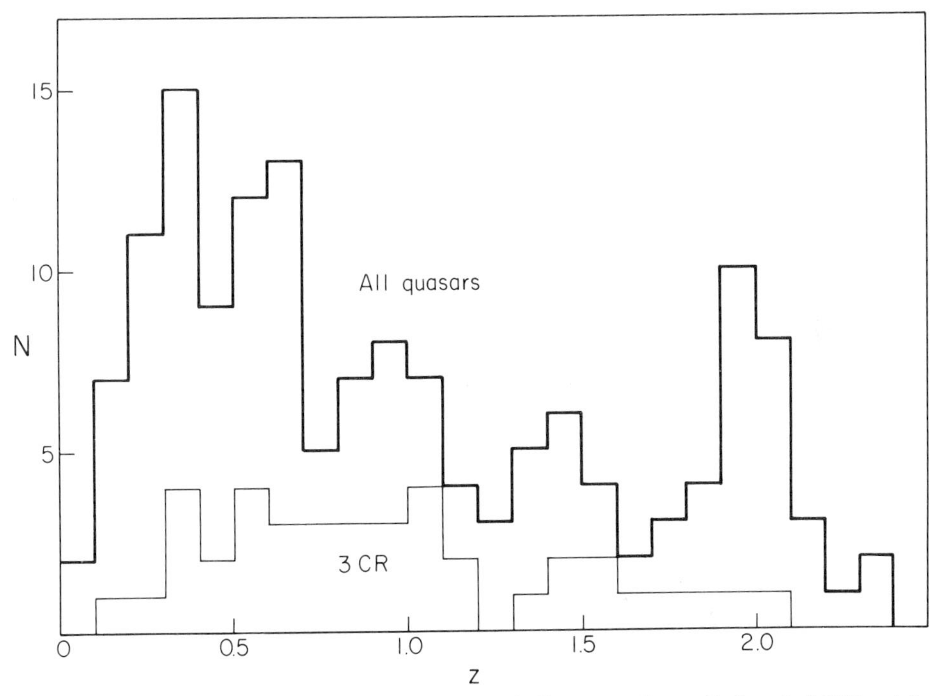

FIG. 4.—Distribution of emission-line redshifts of all quasars (*heavy line*) and of 3CR catalog quasars (*thin line*).

Hence, the probability of a rapid determination of a redshift is generally decreasing with increasing redshift, except for redshifts larger than 1.8 for which this probability is large. It is quite conceivable that the strong concentration of redshifts around 2.0 seen in figure 4 for all quasars is caused by the above effect. This possibility is confirmed by the absence of a concentration of redshifts around 2.0 among 3CR quasars: in this case the combined effort of the spectroscopic observers in the search for faint emission lines has led to an essentially complete sample, for which the above-mentioned selection effects do not apply.

Spectra of quasars show emission lines of H, He, C, N, O, Ne, Na, Mg, Si, and Fe.

---

stellar radio source OH 471 (Carswell and Strittmatter, *Nature*, **242**, 394, 1973) and 3.53 for OQ 172 (Wampler, Robinson, Baldwin, and Burbidge, *Nature*, **243**, 336, 1973).

Analyses of the emission-line spectra by Greenstein and Schmidt (1964), Shklovskii (1964), Osterbrock and Parker (1966), Wampler and Oke (1967), and Bahcall and Kozlovsky (1969) show no large deviations from a "normal" solar composition, except for a suspected underabundance of helium by a factor of perhaps 10.[10]

The forbidden lines observed in the spectra of most low-redshift quasars suggest the presence of a low-density plasma as in gaseous nebulae and planetary nebulae. We assume that the source of ionization of the plasma is the ultraviolet flux from a centrally located source of continuum radiation. The mass and radius of the gaseous line-emitting envelope may then be estimated as follows.

The observed flux $F(\mathrm{H}\beta)$ emission is

$$F(\mathrm{H}\beta) = \frac{P(\mathrm{H}\beta)}{4\pi r^2}, \tag{2}$$

where $P(\mathrm{H}\beta)$ is the total Hβ energy emitted by the quasar and r its distance. The emissivity $E(\mathrm{H}\beta)$ per unit volume follows from the recombination rate of protons and electrons; from Pengelly (1964) we find

$$E(\mathrm{H}\beta) \simeq 10^{-20.9} N_e^2 T_e^{-1}, \tag{3}$$

in cgs units, where we have approximated the tabulated dependence on T_e by a simple inverse power. Hence, if V is the total volume of the plasma, we have

$$F(\mathrm{H}\beta) \simeq 10^{-22.0} N_e^2 T_e^{-1} V r^{-2}. \tag{4}$$

The observed Hβ fluxes are 2×10^{-12} ergs cm^{-2} s^{-1} for 3C 273 (Oke 1965; Wampler and Oke 1967) and 1×10^{-13} ergs cm^{-2} s^{-1} for 3C 48 (Oke 1966). Considering the absence of [O II] and the weakness of the [O III] lines relative to Hβ in 3C 273 (Wampler and Oke 1967), the electron density N_e is probably 10^7 cm^{-3}, at least. This is confirmed by the absence of [Fe II] lines combined with the presence of permitted Fe II lines (Wampler and Oke 1967). In 3C 48 strongly forbidden [O II] emission is present and N_e is probably not more than 10^4 cm^{-3}. We shall use these values for illustrative purposes, together with an electron temperature T_e of 20,000° K.

The distances of the quasars have been the subject of much discussion. We discuss the properties of the gaseous envelope for various hypotheses concerning the nature of the redshifts. (Further related discussion of the nature of quasars and the redshifts is given in § 6.)

4.1. Cosmological Redshifts

Distances are large and given by the redshift in a given cosmological model. Table 1 gives the results for 3C 48 and 3C 273 assuming distances to be simply $cH^{-1}z$ with H = 100 km s^{-1} Mpc^{-1}. The value of R has been computed on the assumption of a spherical volume V. The optical depth to the center for electron scattering τ_e is larger than unity for 3C 273. Since the gas cloud has a radius of a light-year, this would allow observation of light fluctuations of the central continuum source with characteristic times of 10 or more years only. As variations with a characteristic time of less than a year have been seen, the value of τ_e has to be reduced to well below unity. This may be achieved by

[10] For further analyses, see Williams (*Ap. J.* [*Letters*], **167,** L27, 1971), Davidson (*Ap. J.*, **171,** 213, 1972; and **181,** 1, 1973), and MacAlpine (*Ap. J.*, **174,** 11, 1972).

distributing the gas, with the assumed density of 10^7 cm^{-3}, in filaments scattered over a larger volume (Schmidt 1964) or by confining the gas to a shell of thickness ΔR at radius R (Shklovskii 1964). In the former case the total volume will have a radius R_{tot} of $1.2\tau_e^{-1/2}$ pc.

If the low intensity of [O II] and [O III] in 3C 273 is due to depopulation of the upper levels by photoexcitation, then the electron density is not necessarily high (Burgess 1967). If N_e were 10^4 cm^{-3}, the radius of the uniform gas sphere would be 1.3×10^{20} cm, or 40 pc, similar to that of 3C 48 (see table 1).

TABLE 1

PROPERTIES OF THE LINE-EMITTING GAS ENVELOPES OF 3C 273 AND 3C 48 ON THE COSMOLOGICAL INTERPRETATION OF THE REDSHIFTS

Parameter	3C 273	3C 48
z	0.158	0.367
r (Mpc)	470	1100
N_e (cm^{-3})	10^7	10^4
V (cm^3)	8×10^{54}	2×10^{60}
M ($M_\odot$)	7×10^4	2×10^7
R (cm)	1.3×10^{18}	8×10^{19}
τ_e	9	0.5
R_{tot} (cm)	$3.6\times10^{18}\tau_e^{-1/2}$	$5.6\times10^{19}\tau_e^{-1/2}$

4.2. DOPPLER REDSHIFTS

If the redshifts are due to Doppler effects from "local" high-velocity objects (see § 6), then the volume and mass of the gas cloud are reduced proportional to r^2. If the objects were ejected from our Galaxy 5×10^6 years ago, then typical distances are 2000 times smaller than on the cosmological hypothesis. For 3C 48 the radius R of the gas cloud would be 0.5×10^{18} cm. The optical depth for electron scattering is then only 0.003; hence electron scattering cannot be responsible for the large widths of the emission lines in this case.

We can derive information on the total mass of the quasar on the reasonable assumption that the line widths are due to mass motions (Setti and Woltjer 1966; Schmidt 1967). We use 3C 48 as a typical example. If the gas envelope of the quasar does not escape, then the total mass M must be

$$M \approx G^{-1}(\Delta v)^2 R\,.$$

For $\Delta v = 10^8$ cm^{-1}, $R = 0.5 \times 10^{18}$ cm, we find $M = 4 \times 10^7\, M_\odot$.

If, on the other hand, the gas envelope is not bound to the quasar, then there is loss of mass amounting to

$$\text{Mass loss} \approx \rho(\Delta v)4\pi R^2 T\,,$$

where $\rho = 2 \times 10^{-24}\, N_e$ g cm^{-3} and T the age of the explosion. For $N_e = 10^4$ and $T = 5 \times 10^6$ years the mass loss is $2 \times 10^5\, M_\odot$.

The mass loss is reduced (Setti and Woltjer 1966) if only a fraction η of the emitting region is filled with gas of density N_e; in fact, $R \sim \eta^{-1/3}$ and mass loss $\sim\eta^{1/3}$. There is, however, a limit to the value of η imposed by the angular diameter. At the distance for 3C 48 of 500 kpc corresponding to $T = 5 \times 10^6$ years, the uniformly filled gas cloud of diameter 10^{18} cm appears at an angular diameter of 0.″13. Since the angular extent of quasars in their emission lines is unobservably small, i.e., less than 1 or 2 seconds of arc, we can increase R by at most a factor of 10, so $\eta > 10^{-3}$. The minimum mass loss is then $10^4\ M_\odot$. It is of interest that this estimate is independent of T, since $R \sim T$ from the angular-diameter argument.

4.3. Gravitational Redshifts

If the redshifts are gravitational, then

$$z = \Phi/c^2\,, \tag{5}$$

for small redshifts, where Φ is the gravitational potential of the line-emitting gas.

If the massive body responsible for the gravitational field is interior to the gas cloud, then at distance R, the potential $\Phi = GMR^{-1}$. The emitting gas exhibits a limited range of redshift corresponding to the observed line width w. Hence $\Delta R/R = w/\lambda$ and the volume V of the emitting gas is $4\pi R^2 \Delta R$.

TABLE 2

Maximum Distances of 3C 48 for Various Assumed Values of Its Mass, Based on the Gravitational Interpretation of the Redshift

M ($M_\odot$)	r	M ($M_\odot$)	r
1	10 km	10^{11}	10 kpc
10^5	2 a.u.	10^{12}	300 kpc
10^{10}	0.3 kpc		

Alternatively, consider the other extreme case that the gas cloud is inside the massive body. For a spherical distribution of mass in a shell with interior radius R, the potential in the inner region is constant at $\Phi = GMR^{-1}$. The gas cloud may fill the entire volume of $4/3\pi R^3$.

We may conveniently characterize both cases by adopting

$$z = Gc^{-2}MR^{-1} = 7.4 \times 10^{-29} MR^{-1}\,, \tag{6}$$

and

$$V = \tfrac{4}{3}\pi f R^3\,, \tag{7}$$

where f is the fraction of the volume of radius R filled by the gas cloud. Combination of equations (4), (6), and (7) yields

$$r = 10^{-52.9} N_e T_e^{-1/2} f^{1/2} M^{3/2} z^{-3/2} F(\mathrm{H}\beta)^{-1/2}\,. \tag{8}$$

This formula allows the determination of r for 3C 48 for different assumed masses M. Table 2 shows the results for $f = 1$; hence the distances given are maximum values. Clearly, the gravitational acceleration for masses less than $10^{11}\ M_\odot$ is forbiddingly large. Even at $10^{11}\ M_\odot$ 3C 48 would at 10 kpc rival the gravitational field of our Galaxy.

For a mass of $10^{12}\ M_\odot$ 3C 48 would be at 300 kpc; the 10^6 or so existing quasars would then reach out to 30 Mpc. This suggests (Hoyle and Fowler 1967) as a further argument that the average mass density contributed by the quasars should not exceed the critical cosmological density $\rho_{\rm cosm}$ of around 2×10^{-29} g cm^{-3}. We may approximate this condition by

$$M = \tfrac{4}{3}\pi r_n{}^3 \alpha \rho_{\rm cosm} , \tag{9}$$

if r_n is the distance of the nearest quasar and $\alpha < 1$. Introduction of this mass in equation (8) yields

$$r_n{}^7 = 10^{103.9} N_e{}^{-2} T_e f^{-1} \alpha^{-3} \rho_{\rm cosm}{}^{-3} z^3 F({\rm H}\beta) . \tag{10}$$

We adopt for the average quasar hypothetically $N_e = 10^4$, $T_e = 2 \times 10^4$, $z = 1$, and for the nearest quasar $F({\rm H}\beta) = 10^{-12}$, all in cgs units. For $f = 1$ and $\alpha = 1$, we find $r_n = 10^{24.9}$ cm or 4 Mpc. The corresponding mass is 2×10^{13} $M_\odot$. The nearest 10^6 quasars would have distances up to 400 Mpc.

We see from equation (10) that r_n is particularly insensitive to the adopted values of T_e and $F({\rm H}\beta)$. Since at least half of the quasars in which [O II] λ3727 is accessible do show this strongly forbidden transition, the electron density N_e in these cases is at most 10^4, so that r_n is a minimum value. Also, we have used maximum values of unity for f and α, again leading to an underestimate of r_n.

Spectroscopic arguments thus show that the gravitational hypothesis of the redshifts requires the existence of condensed objects of supergalactic masses at extragalactic distances. Other arguments pertaining to the gravitational hypothesis are discussed in § 6.

5. CONTINUUM ENERGY DISTRIBUTION

5.1. Optical Continuum

Broad-band *UBV* photometry of quasars shows that all have $U - B < -0.4$ (except for a few cases near the galactic plane that are obviously reddened). The $B - V$ colors are usually confined in the range 0.0–0.6, but there are a few notable deviations either way.

Both colors $U - B$ and $B - V$ show some correlation with z, first found by McCrea (1966). Figure 5 shows data for 64 quasars. The systematic variation of color with redshift can be understood only if different quasars exhibit a rather similar energy distribution. The color-redshift variation is then a difference curve of the energy distribution obtained by colors at varying emitted wavelengths. The energy distribution that corresponds to the color variations has been derived by Sandage (1966).

The possibility that most of the color variations are caused by the emission lines moving through the color filter ranges for varying redshift was suggested by Lynds (1966) and Strittmatter and Burbidge (1967). The continuum may to a first approximation be represented by $F(\nu) \sim \nu^{-1}$, but the line strengths required to explain the color variations are not too well established (compare Strittmatter and Burbidge 1967; Schmidt 1968).

This is perhaps not surprising, since the basic assumption that all quasars have a similar continuum is not strongly supported by spectrophotometry at higher resolution. Figure 6, kindly prepared by Dr. J. B. Oke, shows the results of spectrum scanner observations (Oke, Neugebauer, and Becklin 1970) of 15 quasars. Individual sources

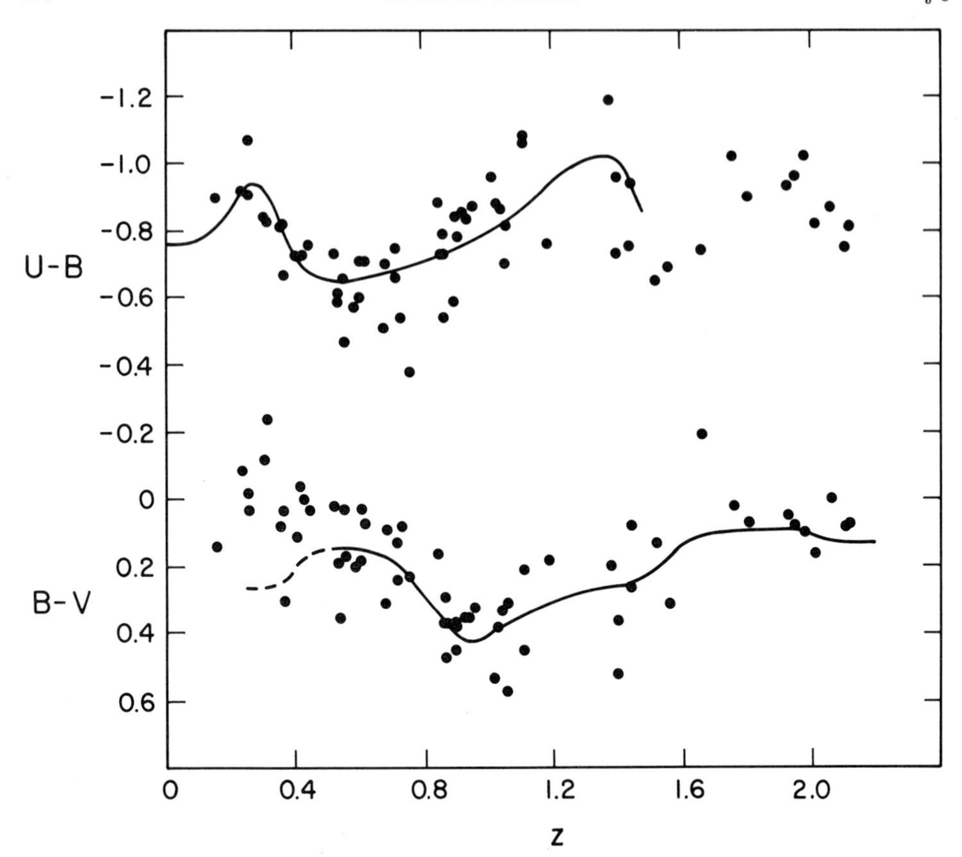

FIG. 5.—$U - B$ and $B - V$ colors for 64 quasars plotted against redshift z. The lines drawn in correspond to particular emission-line strengths; see Schmidt (1968).

clearly show large deviations from the ν^{-1} law. This figure illustrates that the concept of a standard continuum spectrum is ill defined and should be used with caution.

The equivalent width of Hβ in quasar spectra is at most 100 Å (Searle and Sargent 1968), much smaller than the value expected for a recombination spectrum (Oke and Sargent 1968). Most of the optical continuum observed must then be of a nonthermal nature. Linear polarization up to 10 percent has been measured by Appenzeller and Hiltner (1967) and Visvanathan (1968).

Many quasars show brightness variations of around 0.5–1 mag over time intervals of months or years; typical cases are 3C 48 (Matthews and Sandage 1963) and 3C 273 (Smith 1965). The quasars 3C 279, 3C 345, and 3C 446 have been particularly active in recent years. Changes of 0.25 mag within 24 hours are reported in 3C 279 by Oke (1967*a*). Variations of more than 3 mag are exhibited by 3C 446 (Sandage, Westphal, and Strittmatter 1966). No variations in the absolute strength of the emission lines in 3C 446 have been detected by Oke (1967*a*) and Wampler (1967).

Large and variable polarization has been observed in the optically active quasars

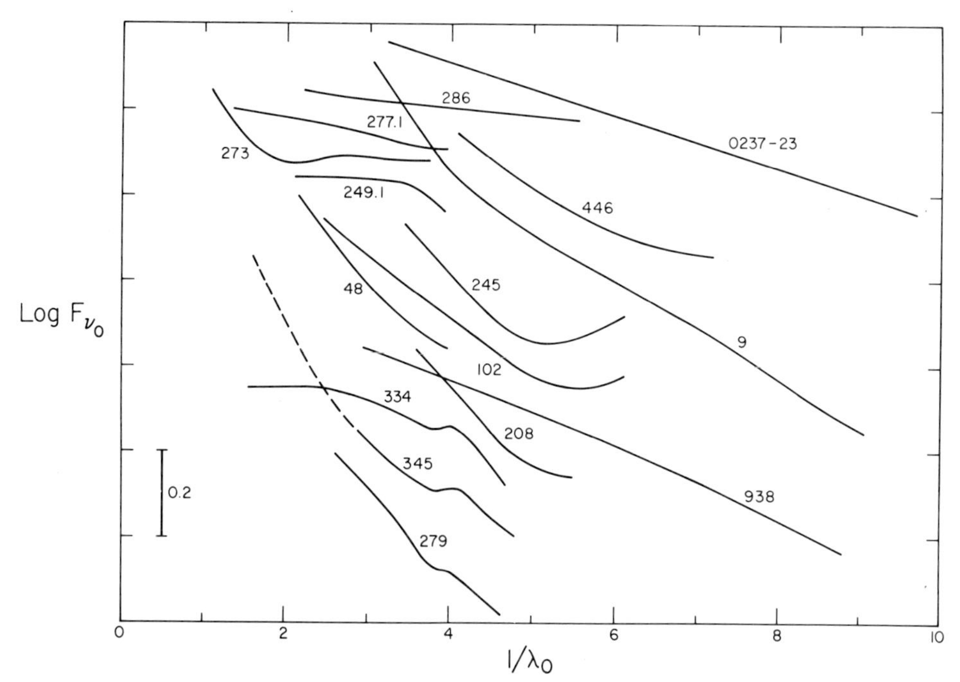

FIG. 6.—Continuum energy distributions $F(\nu_0)$ of 15 quasars, where ν_0 is the rest frequency. Individual spectra have been shifted vertically by arbitrary amounts. Prepared by Dr. J. B. Oke.

3C 446 (Kinman, Lamla, and Wirtanen 1966), 3C 279, and 3C 345 (Kinman 1967; Kinman *et al.* 1968).

5.2. Infrared Continuum

Observations in the infrared, from 1 to 22 microns, have shown an unexpectedly large flux from 3C 273 (Low and Johnson 1965). In fact, the infrared flux is at least 10 times the flux in the visual part of the spectrum. A similar dominance of infrared flux over visual flux is observed in Seyfert galaxies (Low and Kleinmann 1968).

Infrared photometry at 2.2 μ of 40 quasars has been obtained by Oke, Neugebauer, and Becklin (1970). In most cases the infrared brightness is close to that expected from a smooth extrapolation of the visual continuum. Observations at longer wavelengths will be required to prove or disprove the existence of a large infrared excess such as shown by 3C 273. Variations at 2.2 μ similar to the variations in the visual continuum were observed in 3C 279 and 3C 345.

5.3. Radio Continuum

The radio spectrum of a quasar is typically a power law, $F(\nu) \sim \nu^{\alpha}$, with $\alpha \simeq -0.75$ over the well-observed range 38–5000 MHz (Scheuer and Williams 1968). Concave spectra, such as that of 3C 273, have been explained as due to a superposition of two or more sources with different spectral index α.

Variability in flux density has been observed at high frequencies in a number of quasars (for details see the review by Kellermann and Pauliny-Toth 1968*a*). Variable polarization has been detected in several quasars (Aller and Haddock 1967; Hobbs, Hollinger, and Marandino 1968). All variable radio sources show at high frequencies a flattening of the spectrum and in some cases one or two maxima. The variations may be understood if they are caused by synchrotron radiation from an expanding cloud of relativistic electrons which is initially optically thick at centimeter wavelengths. As the cloud expands, it becomes optically thin at progressively longer wavelengths. Kellermann and Pauliny-Toth (1968*a*) have described recent outbursts in four quasars and two radio galaxies. Spectra of sources containing opaque components are discussed by Kellermann and Pauliny-Toth (1969). The opaque components show a maximum brightness temperature of around 10^{12} ° K, probably set by inverse Compton cooling.

The observed variations suggest that quasars contain very small components at high frequencies. This is confirmed by intercontinental interferometry at 6 cm wavelength, showing that some quasars have strong components with characteristic sizes of less than 0".001 (Kellermann *et al.* 1968). Many quasars exhibit scintillation caused by the interplanetary medium at meter wavelengths, indicating diameters of 1" or less. A few quasars, such as 3C 47, show a double structure with a separation of about a minute of arc, quite similar to that of a typical radio galaxy. Cohen (1969) has reviewed the available data on angular diameters of radio sources.[11]

The reader is referred to chapter 7 in this volume by Moffet for a discussion of radio radiation mechanisms and energy requirements for radio galaxies and quasars.

6. NATURE OF QUASARS

We discuss three hypotheses concerning the origin of the large observed redshifts, as well as the possibility of an as yet unknown origin.

6.1. Gravitational Hypothesis

The gravitational hypothesis of the redshifts is an attractive one at first sight. The rapid light variations suggest components in the object as small as 10^{16}–10^{17}cm. At such a distance from a central mass of around 10^{11} $M_{\odot}$ one would indeed observe a large gravitational redshift. Hence, a galactic mass condensed to a size corresponding to the light fluctuations would exhibit a gravitational redshift similar to the redshifts observed in quasars.

However, spectroscopic arguments presented in § 4, based on the frequent observation of strongly forbidden [O II] λ3727 emission, show that such an object could not be more distant than 10 kpc, and its gravitational acceleration would rival that of our Galaxy. In fact, the discussion in § 4 showed that individual masses of quasars have to be at least 2×10^{13} $M_{\odot}$, and that the 10^6 quasars known to exist would have to be distributed out to distances of 400 Mpc, at least.

The original arguments concerning gravitational redshifts based on the occurrence of forbidden emission lines were given by Greenstein and Schmidt (1964) for the case of a

[11] Miley (*M.N.R.A.S.*, **152**, 477, 1971) has collected information on angular structures of quasars and discussed the correlation between angular size and redshift. Rapid variability in radio structure has been detected by Whitney et al., *Science*, **173**, 225, 1971; see also Shaffer et al., *Ap. J.* (*Letters*), **173**, L147, 1972.

condensed central body. In this case the emission lines can originate from a thin layer only (such that the variation of gravitational redshift in the layer does not exceed that corresponding to the observed line widths) leading to a small value of the volume fraction f in equation (10), hence a large value of r_n, the distance of the nearest quasar. Hoyle and Fowler (1967) then proposed a model for which f is close to unity. It consists of a large number of compact objects distributed in a cluster such that gas assembled at the central parts of the cluster is directly visible from the outside. Since an appreciable part of the cluster is at almost constant potential, the gas may be spread over a larger volume. However, equations (8) and (10) show that changing f from the Greenstein-Schmidt value of around 10^{-2} to approximately unity for the Hoyle-Fowler model makes relatively little difference. Also, it should be emphasized that numerical values given above and in § 4 are for the case $f = 1$; for $f < 1$ the distance r_n of the nearest quasar becomes larger.

Another reason for the introduction of the Hoyle-Fowler cluster model was that the gravitational redshift of the center may have any value, while the maximum gravitational redshift of a photon from the surface of a stable, gravitationally bound fluid is around 0.62 (Bondi 1964). There is, at present, no proof that a Hoyle-Fowler cluster with a central redshift of around 2 will be stable for a reasonable length of time. Zapolsky (1968) has derived a collisional relaxation time of the cluster of only 1000 years, after which it is expected to become thermalized. Zapolsky notes that it seems artificial to require such a specialized stage of their evolution if the quasars are major contributors to the cosmological density.[12]

6.2. Local-Doppler Hypothesis

In this hypothesis the redshifts correspond to the Doppler effect but, contrary to the cosmological hypothesis, distances are not very large. This hypothesis was proposed by Terrell (1964) to avoid some of the problems that were believed to exist if quasars were at very large distances.

If quasars were ejected by many different galaxies throughout the universe, the observed spectral shifts would be mostly blueshifts (Noerdlinger, Jokipii, and Woltjer 1966; Faulkner, Gunn, and Peterson 1966).

Since no blueshifts are observed or suspected for any quasars, all velocities are away from us, indicating an explosion. Terrell suggested as the nearest possible explosion center the nucleus of our Galaxy. He estimated the minimum distance of 3C 273 at 200 kpc from the absence of a detectable proper motion, and the minimum age of the explosion at 5×10^6 years (Terrell 1964), but he has recently proposed somewhat smaller values (Terrell 1968).

The very large speeds (up to 0.84 c) make the kinetic energy of the exploding cloud of quasars very large: 10^6 quasars will carry a total kinetic energy of around 10^{60} M ergs, where M is the average quasar mass in solar-mass units. The mass M may be estimated on the basis of the large widths of the emission lines, as discussed in § 4. There it was

[12] The minimum required quasar mass of 2×10^{13} $M_\odot$ has a Schwarzschild radius of 6 light-years. It is unlikely that substantial variations of such an object could be observed over intervals of weeks or months. Conversely, this argument suggests that, for any interpretation of the redshift, the masses of rapidly variable quasars are not much more than 10^{10} $M_\odot$.

found that the minimum mass of the average quasar must be $10^4\ M_\odot$, independent of the exact age T of the local explosion.

Hence the total kinetic energy is at least 10^{64} ergs, or the rest-mass energy of $10^{10}\ M_\odot$, or 10 percent of the mass of the Galaxy. This excessive requirement essentially eliminates the local-Doppler hypothesis.

6.3. Cosmological Hypothesis

In the cosmological hypothesis the redshifts are due to the expansion of the Universe. At small redshifts the corresponding distance is given by $r = cH^{-1}z$, where H is the Hubble constant; at large redshifts the concept of "distance" needs redefinition (McVittie 1965), and its relation to redshift will be dependent on the cosmological model.

At the very large distances corresponding to the observed redshifts the absolute luminosities of the quasars become very large, at wavelengths around 4000 Å as much as 100 times larger than that of the brightest galaxies known. At first sight it is astonishing that such a large luminosity can be exhibited by a body small enough to show light variations with characteristic times as short as a few days. More precisely, the body should contain components as small as a few light-days that can contribute substantially to the total luminosity for a period of at least a few days.

Hoyle, Burbidge, and Sargent (1966) assumed that the optical radiation is of synchrotron origin and showed that inverse Compton losses by the synchrotron electrons on the synchrotron-generated photons will dominate over the synchrotron losses, unless the magnetic field strength is very large. The lifetimes of the synchrotron electrons become shorter than their travel times to the boundary of the source. This consideration constituted a serious objection to the cosmological hypothesis until optical variability in nuclei of N-type and Seyfert galaxies was discovered by Oke (1967*a*) and Sandage (1967*a*). The variations have characteristic times similar to those of quasars. It may be shown that in these nuclei of galaxies Compton losses dominate over synchrotron losses, too, although quantitatively less severely than in the quasars. In the case of the galaxies, however, there can be no doubt about the cosmological nature of their redshift. Possibly, the problem can be avoided by assuming a radial magnetic-field configuration as suggested by Woltjer (1966) for the quasars, or by considering synchrotron radiation from relativistic protons rather than electrons. In any case, our knowledge of the radiation mechanism in quasars is clearly not sufficiently thorough to allow the establishment of a reliable upper limit to their distances.

The cosmological hypothesis is indirectly supported by similarity arguments.[13] The similarity between quasars and Seyfert galaxies is remarkable, indeed; both kinds of objects show similar optical variations and radio variations, similar optical spectra and energy distributions. The infrared luminosity of the Seyfert radio galaxy 3C 120 is 10^{46} ergs s^{-1}, equal to the luminosity of many quasars on the cosmological hypothesis (Low

[13] Direct support is provided by the observed association of quasars and clusters of galaxies at the same redshift (Oemler, Gunn, and Oke, *Ap. J.* [*Letters*], **176**, L48, 1972) but the statistical significance of the associations is being debated (Burbidge and O'Dell, *Ap. J.* [*Letters*], **182**, L47, 1973). Also, the nebulosity detected by Kristian (*Ap. J.* [*Letters*], **179**, L61, 1973) around or near a number of quasars is compatible with the appearance of galaxies assuming the quasar redshift is cosmological.

and Kleinmann 1968). Some quasars show a large-scale double radio structure indistinguishable from that of radio galaxies;[14] also, maximum radio luminosities of 10^{45} ergs s^{-1} for radio galaxies and quasars on the cosmological hypothesis are identical (see Heeschen 1966 for more detailed arguments). Small-sized radio components are seen in a number of quasars and radio galaxies; similar components of lower radio luminosity have been observed in some elliptical galaxies (Heeschen 1968).

6.4. Other Hypotheses

Barnothy and Barnothy (1968) have proposed that a quasar is a very distant Seyfert nucleus seen through a massive foreground galaxy acting as a gravitational lens. The number of favorably aligned cases is entirely insufficient to explain the at least 10^6 existing quasars.

Arp (1967) has claimed a correlation between peculiar galaxies and radio sources including quasars. Burbidge and Burbidge (1969) believe that there are periodicities in the redshifts of emission-line objects. Burbidge and Burbidge (1967) earlier pointed out the high frequency of redshifts near 1.95 and suggested that this represents an "intrinsic" redshift of quasars. These and other statistical findings and their interpretation are subject to controversy. Solid confirmation of any of these findings does not exist at present; for further references see Burbidge (1967) and Schmidt (1969).[15]

Burbidge and Hoyle (1967) have suggested that the large redshifts of quasars are mostly gravitational and the small redshifts mostly cosmological on the arbitrary basis that quasars and radio galaxies in a radio catalog should occupy the same volume of space.

A redshift-magnitude diagram containing spectroscopically similar objects such as nuclei of Seyfert galaxies, compact galaxies, and quasars published by Arp (1968) according to Hoyle (1969) shows evidence for large gravitational redshifts in the quasars. This assertion is based on the steep slope of the line connecting the Seyfert nuclei, the compact galaxies, and the quasars. It has apparently been assumed by Hoyle that the three kinds of objects have equal absolute luminosities. There is no proof for this assumption whatsoever.

6.5. Conclusion

The local-Doppler hypothesis requires a forbiddingly large amount of kinetic energy, as we have seen in § 6.2. The gravitational-redshift hypothesis requires the existence of extremely massive clusters of condensed objects, whose stability and lifetimes may be insufficient for them to exist in the requisite numbers. Such massive condensed clusters would be quite unlike any astronomical body known today.

The latter argument was originally used against the cosmological hypothesis, but the observation of fast variation in the nuclei of some galaxies and of large infrared fluxes in some galaxies invalidates this argument to a considerable degree. It is only in the cosmological hypothesis that the similarity in radio properties between radio galaxies and quasars is plausible.

[14] See Miley (*M.N.R.A.S.*, **152**, 477, 1971) for the correlation between angular separation and redshift.

[15] Further arguments against the cosmological hypothesis have been reviewed by Arp (*Science*, **174**, 1189, 1971) and Burbidge (*Nature Phys. Sci.*, **246**, 17, 1973).

The author would rate the cosmological hypothesis as probable, the gravitational hypothesis as improbable, and the local-Doppler hypothesis as impossible. It has to be admitted, however, that, except for the similarity arguments, most of the evidence for the cosmological case consists of the arguments against the two other cases. No such arguments can, of course, be made against an unknown cause of redshift, since it is specifically assumed in this case that this cause lies outside present-day physics. In particular, Hoyle (1969) has suggested that general relativity is wrong, so as to allow large gravitational redshifts that are essentially constant over kiloparsecs. Burbidge and Burbidge (1967*b*) suggested an "intrinsic" redshift of 1.95 and periodicities in the distribution of redshifts of emission-line objects (Burbidge and Burbidge 1969). Clearly, the metaphysical redshift hypothesis should be added as "possible" in the author's subjective list of ratings given above.

If the cosmological hypothesis is correct, it is remarkable indeed that no single independent confirmation of the large distances exists as yet. Studies of absorption phenomena have up to now led to inconclusive results, as described in the next section.

7. ABSORPTION LINES

On the cosmological hypothesis it was expected that intergalactic neutral hydrogen would essentially absorb all quasar radiation at wavelengths below the Lα emission (Scheuer 1965). The continuum below Lα was observed in 3C 9, however, leading Gunn and Peterson (1965) to derive a very low neutral hydrogen density, around 6×10^{-11} atoms per cm^3 at a redshift of 2. This is actually an upper limit, as subsequent observations of various large-redshift sources have shown no evidence for any substantial depression of the continuum below Lα.

7.1. Observations

Discrete absorption lines are observed in the spectrum of many quasars. The typical "simple" quasar absorption-line spectrum is that of 3C 191 shown in figure 7 (Burbidge, Lynds, and Burbidge 1966; Stockton and Lynds 1966; Bahcall, Sargent, and Schmidt 1967). The absorption spectrum, at a redshift (1.946) slightly less than the emission

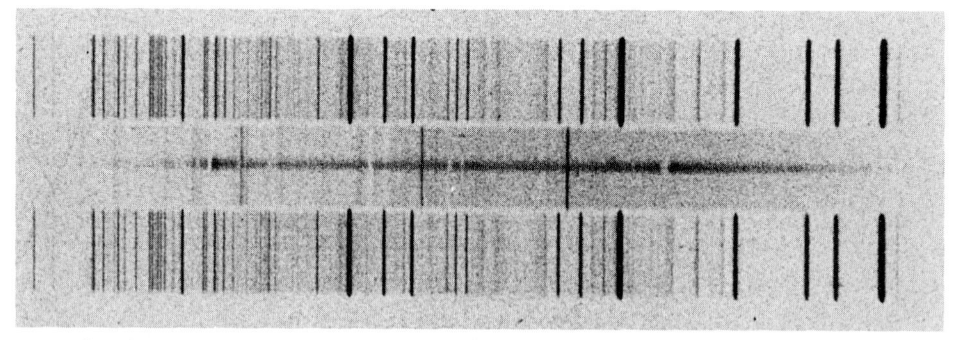

FIG. 7.—Spectrum of 3C 191 showing emission lines with a redshift $z_{em} = 1.957$ and absorption lines with $z_{abs} = 1.946$. The prominent emission lines are due to Lα and C IV, both of which show absorption in their blue wings. At top and bottom is a comparison spectrum of hydrogen, helium, and neon; for wavelengths see fig. 1.

redshift (1.957), shows lines of Lα, Si II, Si III, Si IV, C II, C IV, and N V. The fine-structure constant at redshift 1.95, as determined from the wavelengths of lines originating on fine-structure levels, agrees with the laboratory value (Bahcall, Sargent, and Schmidt 1967).

Some quasars such as PKS 0237−23 show a multitude of absorption lines. It has been shown that in these cases several redshift systems are present in the absorption spectrum. In PKS 0237−23 the work of Burbidge, Lynds, and Stockton (1968), Bahcall (1968), and Bahcall, Greenstein, and Sargent (1968) has isolated five redshifts and two more possible redshifts (for discussion see Schmidt 1969). The absorption redshifts range from 1.364 to 2.202, the emission redshift being 2.223.

The main basis on which the multiple-redshift interpretation is considered to be valid is the result of a statistical test applied by Bahcall (1968). He finds that the same identification procedure applied to a simulated spectrum having the same number of lines and the same distribution of line strengths as those observed in PKS 0237−23 allows identification of 0.7 redshifts on the average. This number is to be compared to the five redshifts found in the real spectrum of PKS 0237−23.

Several other quasars show multiple absorption redshifts, such as Ton 1530 with three or four redshifts, ranging from 1.89 to 1.98 (Bahcall, Osmer, and Schmidt 1969) and PHL 938 with absorption redshifts of 0.61 and 1.91 (Burbidge, Lynds, and Stockton 1968), and others.

In a few cases an absorption redshift is found that is a little larger than the emission redshift. An example is 4C 25.5 ($z_{\rm em} = 2.358$) in which an absorption redshift $z_{\rm abs} = 2.3683$ is present (Schmidt and Olsen 1968).

In all multiple-redshift systems the line widths are close to the resolving power of the spectrographs, around 1–3 Å. Single-redshift spectra have often wider lines, such as 7 Å for 3C 191 and 20 Å for 3C 270.1. Of great interest is the extreme case of PHL 5200 in which the absorption to the violet of the emission lines is around 100 Å wide, with possible substructure (Lynds 1967).

7.2. Interpretation

The origin of the absorption lines is still rather mysterious. For the case of 3C 191 it is practically certain that the absorbing material is very close to the quasar, as Bahcall has shown (Bahcall, Sargent, and Schmidt 1967) that the observed fine-structure transitions require an electron density of at least 10^3 cm^{-3} or a maximum distance of 10^2 pc from the quasar. By analogy it seems reasonable to assume that this also applies to other absorption redshifts that are slightly smaller than the emission redshifts. Apparently, we see in these cases absorption by material ejected from the quasar.

In the few cases where the absorption redshift is a little larger than the emission redshift, it is conceivable that the ejected material did not escape and is falling back into the quasar.

A fascinating problem is presented by the narrow-line multiple-redshift absorption systems. If these, too, are caused by material ejected by the quasars, then the enormous ejection velocities (up to $0.6c$) and the small internal velocity dispersion (100 km s^{-1}) of the absorbers is hard to reconcile. Alternatively, these absorption lines may be caused by material at smaller cosmological redshifts, concentrated in clusters of galaxies

(Bahcall and Salpeter 1966), or in galaxies (Wagoner 1967; Peebles 1968; Bahcall and Spitzer 1969). The latter hypothesis allows a statistical prediction of the distribution of absorption redshifts, both in a single quasar and among quasars in general (Bahcall and Peebles 1969); the required spectroscopic material of uniform quality has not been collected yet.

8. COSMOLOGY, EVOLUTION

The redshift-magnitude diagrams of quasars show a large scatter, indicating on the basis of the cosmological hypothesis a large range in absolute magnitudes. This is particularly the case for radio magnitudes, as seen in the radio redshift-magnitude diagram of quasars in the 3C revised (3CR) catalog (fig. 8). The large scatter of radio absolute magnitudes is emphasized, of course, by the existence of the radio-quiet quasars, which would fall far to the right in figure 8.

The scatter is less in figure 9, where the optical redshift-magnitude diagram of 3CR quasars is given. Optically fainter quasars do have, on the average, larger redshifts as expected on the cosmological hypothesis. The use of diagrams such as figure 9 for testing cosmological models is limited by the large range of absolute magnitudes and is further complicated by the effects of radio selection and by evolution effects.

The distribution of quasar redshifts in figure 4 shows an apparent ceiling in the observed redshifts around 2.4. Since the blue nature of a quasar as judged from the *Sky Atlas* prints would probably disappear only at a redshift near 3 (see Schmidt 1969), there is a growing suspicion that this ceiling is due to a real relative scarcity of quasars of larger redshifts.[16]

At very large distances only the intrinsically brightest (optical or radio) sources can be observed. At smaller distances intrinsically less bright sources become observable. This unavoidable selection effect leads to an apparent increase of absolute brightness with distance, which has often been erroneously interpreted as an evolutionary effect (see Schmidt 1969).

The steep increase of the number of quasars with magnitude, $d \log N/dm = 0.75$ as shown by equation (1) in § 3.3, exceeds that expected in uniformly filled Euclidean space. This slope corresponds to $-d \log N/d \log S = 1.87$ in the terminology of radio astronomy, remarkably similar to the slope of 1.8 shown by extragalactic radio sources in the 3C revised catalog. The latter has been interpreted in terms of an increase in number or luminosity of sources at earlier epochs (Longair 1966).

Evolutionary effects of radio quasars (in contrast to those of extragalactic sources in general, or quasars in general) cannot be studied through $N(S)$ counts. The reason is that only those radio sources in a catalog are recognized as radio quasars that are identified and spectroscopically confirmed; this requires that their optical brightness exceed some minimum value. In fact, the $N(S)$ slope of identified radio quasars is easily seen to be smaller than that of all radio quasars in a radio catalog.

A method that avoids this problem was used for 3CR identified quasars by Schmidt (1968). Since identifications are believed to be complete to $m_V \simeq 18.4$, one can compute for each individual quasar the maximum redshift z_m at which it would still be in the 3CR catalog and brighter than the optical limit. If we define V as the volume in the

[16] At least five quasars with redshifts exceeding 2.4 are now known (1974 January). See also n. 18.

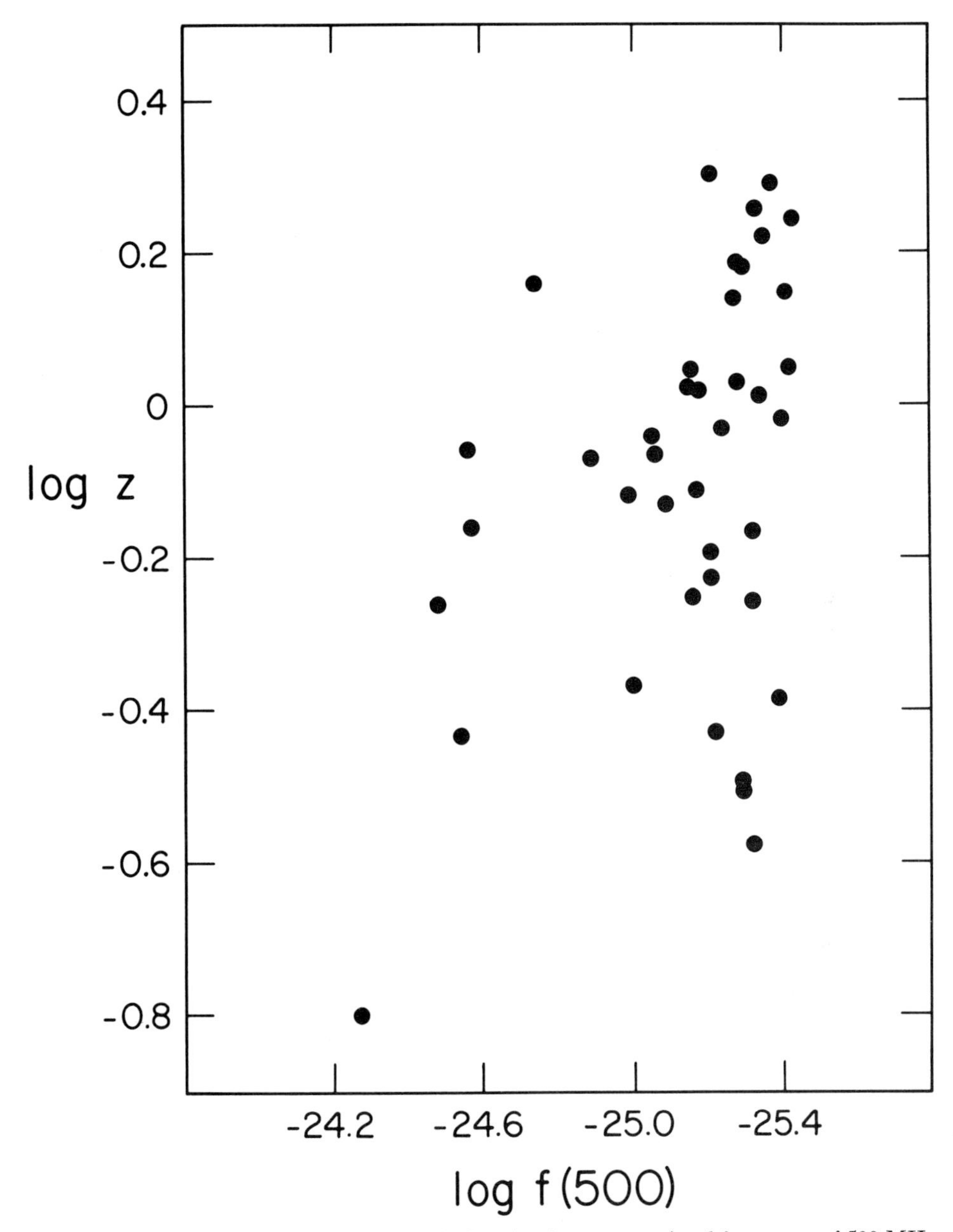

FIG. 8.—Diagram of redshift z versus radio flux density at an emitted frequency of 500 MHz for 3CR catalog quasars (Schmidt 1968).

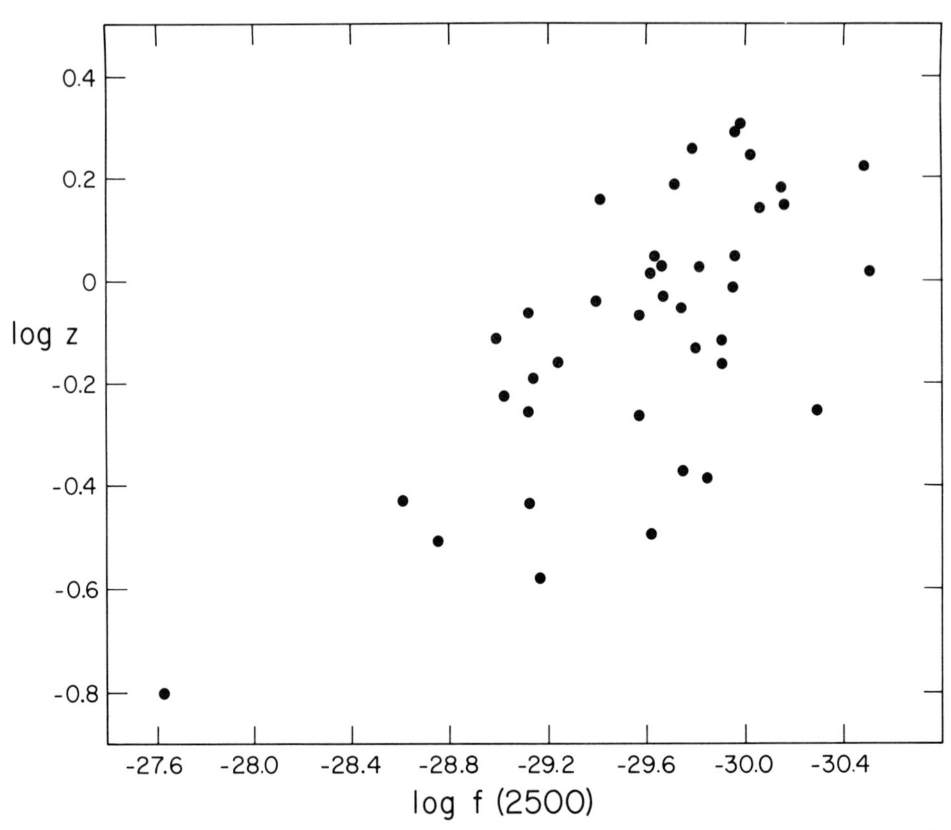

FIG. 9.—Diagram of redshift z versus optical flux density at an emitted wavelength of 2500 Å for 3CR catalog quasars (Schmidt 1968).

universe out to redshift z, and V_m the volume out to z_m, then for a uniform distribution in space the average value of V/V_m for the complete sample should be 0.50. The actual value found for the sample of 3CR sources was 0.70, only weakly dependent on the cosmological model. The corresponding increase of radio quasar space density is by a factor of around 100 at $z = 1$ for q_0 between 0 and 1. The $N(S)$ slope corresponding to this density increase is around 2.0.[17]

If the scarcity of quasars at redshifts above 2.4 is real, then apparently the number of quasars reached a maximum at an epoch corresponding to a redshift around 2, after which a steep decline in their number took place.[18]

[17] Optically selected quasars show a density increase that is similar or equal to that of the radio quasars (Schmidt, *Ap. J.*, **162**, 371, 1970). The precise dependence of density on redshift remains poorly known: a power law of $(1 + z)$ fits the available observations as well as an exponential law of cosmic time.

[18] There may be no need for such a maximum, since the observed scarcity of large-redshift quasars is consistent with a co-moving space density of quasars decreasing exponentially with cosmic time, with a characteristic time of about one billion years (Schmidt, *Ap. J.*, **176**, 273, 1972).

The local ($z = 0$) density of quasars is around 10^{-8} Mpc^{-3} (Schmidt 1968).[19] This estimate does not include quasars as intrinsically weak as B234 (Braccesi, Lynds, and Sandage 1968) which has an absolute visual magnitude of only −19. These weak quasars are probably much more numerous than the strong quasars.

9. MODELS, LIFETIMES, AND ENERGY SOURCES

On the cosmological hypothesis at least four components of a quasar can be distinguished: (*a*) a presumably central source of continuum radiation, containing components not exceeding 10^{16}–10^{17} cm; (*b*) an atmosphere consisting of filaments or bits of ionized gas, distributed over a volume at least 10^{19}–10^{20} cm in diameter; (*c*) the high-frequency, often variable, part of the radio source containing components of not more than 10^{19}–10^{20} cm; (*d*) the steady low-frequency part of the radio source of a typical size of 10^{21}–10^{24} cm.

It has been assumed that the entire structure is essentially concentric, the central source being also the seat of the unknown origin of the energy. The sizes given for the various components are approximate, of course, and vary much from source to source. It seems likely that components (*b*) and (*c*) occupy the same volume of space, although not necessarily in detail.

The lifetime of quasars can only be estimated indirectly. The largest observed dimension of a quasar is that of the separation of the double radio source in 3C 47, corresponding to about 300 kpc on the cosmological hypothesis. This suggests a minimum age of 10^6 years, but the source may not have been a quasar all the time. If it is assumed that quasars are a stage in the life of a radio galaxy, then the ratio of space densities should roughly equal the ratio of lifetimes. Assuming a lifetime of 10^9 years for radio galaxies (Schmidt 1966), we obtain a lifetime of 10^6 years for quasars in general, and around 10^4 years for strong radio quasars such as in the 3CR catalog.

The maximum radio luminosity of quasars is about equal to that found in radio galaxies. The optical and infrared luminosity of quasars is usually some 10 or 100 times that of radio galaxies. With a quasar lifetime that is some 1000 times shorter than that of radio galaxies, it appears that the total energy radiated by quasars is probably less than that by radio galaxies. Hence the "energy problem" for quasars seems superficially less severe than that for radio galaxies.

The high rate of energy production from a small volume has given rise to much speculation as to its origin. It is outside the scope of this chapter to discuss these theories in detail, but we will conclude with a brief summary.[20]

1. Massive superstars in gravitational collapse have been considered by Hoyle and Fowler (1963*a*, *b*). It is not clear in what form the gravitational energy is released. The time scale of the release will be very short, much shorter than the lifetime of quasars.

[19] The local density depends very much on the shape of the density law. For a power law of $(1 + z)$, the local density is about 2×10^{-7} Mpc^{-3}; and for an exponential law of cosmic time, about 10^{-8} Mpc^{-3} (Schmidt, *Ap. J.*, **176**, 303, 1972).

[20] Models containing a large pulsar-like object or a large number of ordinary pulsars have been discussed by Morrison (*Ap. J.* [*Letters*], **157**, L73, 1969); Cavaliere, Pacini, and Setti (*Ap. Letters*, **4**, 103, 1969); Gunn and Ostriker (*Ap. J.*, **160**, 979, 1970); and Rees (*Nature*, **229**, 312, 1971).

Fowler (1966) has considered thermonuclear energy release from a star of mass around $10^8 M_\odot$. Hydrogen burning in such an object will produce a luminosity of 2×10^{46} ergs s^{-1}. He finds that such a star may be stabilized against gravitational collapse for as long as 10^6 years by rotation, turbulence, and magnetic fields.

2. Stellar collisions have been considered by Gold, Axford, and Ray (1965), Woltjer (1964), Ulam and Walden (1964), and Spitzer and Saslaw (1966). Inelastic collisions of stars in a galactic nucleus of sufficient star density will lead to accelerated shrinkage of the nucleus, until stars are completely disrupted. The violent phase of the energy release probably has a much shorter time-scale than the life time of quasars.

3. Supernova explosions in quasars have been proposed by Colgate and Cameron (1963). They consider that massive 50 $M_\odot$ stars will be formed by inelastic collisions in a nucleus of high star density. These stars will become supernovae after around 10^6 years. Each supernova would yield some 10^{52} ergs in kinetic energy according to a model of Colgate and White (1966) in which the gravitational collapse and subsequent explosion is followed in detail. In a dense gaseous atmosphere most of the 10^{52} ergs s^{-1} would heat the gas and give rise to the high luminosity of quasars and account for their variability.

4. Massive lagging cores due to inhomogeneities in the original cosmological expansion of the universe have been studied by Novikov (1964), Ne'eman (1965), and Ne'eman and Tauber (1967). Pacini (1966) has suggested that quarks may be a major constituent of quasars in this case.

5. The role of magnetic fields in converting released gravitational energy in collapsing objects into relativistic particle energy has been discussed by many authors (see Sturrock and Feldman 1968; Ozernoy and Chertoprud 1967; and Piddington 1966).

6. Matter-antimatter annihilation as an energy source of quasars has been proposed by Alfvén (1965) and by Teller. The usual difficulty associated with the requirement that matter and antimatter be kept separated for long times has been discussed by Alfvén on the basis of a cosmological model in which there is no high-density phase.

Most of the above speculative theories will eventually turn out to have no relevance to the problem of the quasars. We have no assurance that any of the approaches mentioned is essentially correct. There is a strong suspicion that understanding rapidly variable, high-redshift sources will be of fundamental importance. The attainment of this understanding continues to be an exciting challenge to the eye and to the mind.

REFERENCES

Alfvén, H. 1965, *Rev. Mod. Phys.*, **37**, 652.
Aller, H. D., and Haddock, F. T. 1967, *Ap. J.*, **147**, 833.
Appenzeller, I., and Hiltner, W. A. 1967, *Ap. J.* (*Letters*), **149**, L17.
Arp, H. C. 1967, *Ap. J.*, **148**, 321.
———. 1968, *Ap. J.* (*Letters*), **153**, L33.
Bahcall, J. N. 1968, *Ap. J.*, **153**, 679.
Bahcall, J. N., Greenstein, J. L., and Sargent, W. L. W. 1968, *Ap. J.*, **153**, 689.
Bahcall, J. N., and Kozlovsky, B. 1969, *Ap. J.*, **155**, 1077.
Bahcall, J. N., Osmer, P. S., and Schmidt, M. 1969, *Ap. J.* (*Letters*), **156**, L1.
Bahcall, J. N., and Peebles, P. J. E. 1969, *Ap. J.* (*Letters*), **156**, L7.
Bahcall, J. N., and Salpeter, E. E. 1966, *Ap. J.*, **144**, 847.
Bahcall, J. N., Sargent, W. L. W., and Schmidt, M. 1967, *Ap. J.* (*Letters*), **149**, L11.
Bahcall, J. N., and Spitzer, L. 1969, *Ap. J.* (*Letters*), **156**, L63.
Barnothy, J., and Barnothy, M. F. 1968, *Science*, **162**, 348.
Bondi, H. 1964, *Proc. Roy. Soc. London*, *A*, **282**, 303.
Braccesi, A., Lynds, R., and Sandage, A. 1968, *Ap. J.* (*Letters*), **152**, L105.

Burbidge, E. M. 1967, *Ann. Rev. Astr. and Ap.*, **5**, 399.
Burbidge, E. M., Lynds, C. R., and Burbidge, G. R. 1966, *Ap. J.*, **144**, 447.
Burbidge, E. M., Lynds, C. R., and Stockton, A. N. 1968, *Ap. J.*, **152**, 1077.
Burbidge, G. R., and Burbidge, E. M. 1967*a*, *Quasi-stellar Objects* (San Francisco and London: W. H. Freeman and Company).
———. 1967*b*, *Ap. J.* (*Letters*), **148**, L107.
———. 1969, *Nature*, **222**, 735.
Burbidge, G. R., and Hoyle, F. 1967, *Nature*, **216**, 351.
Burgess, D. D. 1967, *Nature*, **216**, 1092.
Cohen, M. H. 1969, *Ann. Rev. Astr. and Ap.*, **7**, 619.
Colgate, S. A., and Cameron, A. G. W. 1963, *Nature*, **200**, 870.
Colgate, S. A., and White, R. H. 1966, *Ap. J.*, **143**, 626.
Faulkner, J., Gunn, J. E., and Peterson, B. 1966, *Nature*, **211**, 502.
Fowler, W. A. 1966, *Ap. J.*, **144**, 180.
Gold, T., Axford, I., and Ray, E. C. 1965, in *Quasi-stellar Sources and Gravitational Collapse*, ed. I. Robinson, A. Schild, and E. L. Schucking (Chicago: University of Chicago Press), p. 93.
Greenstein, J. L., and Matthews, T. A. 1963, *Nature*, **197**, 1041.
Greenstein, J. L., and Schmidt, M. 1964, *Ap. J.*, **140**, 1.
Gunn, J. E., and Peterson, B. A. 1965, *Ap. J.*, **142**, 1633.
Hazard, C., Mackey, M. B., and Shimmins, A. J. 1963, *Nature*, **197**, 1037.
Heeschen, D. S. 1966, *Ap. J.*, **146**, 517.
———. 1968, *Ap. J.* (*Letters*), **151**, L135.
Hobbs, R. W., Hollinger, J. P., and Marandino, G. E. 1968, *Ap. J.* (*Letters*), **154**, L49.
Hoyle, F. 1969, *Quart. J.R.A.S.*, **10**, 10.
Hoyle, F., Burbidge, G. R., and Sargent, W. L. W. 1966, *Nature*, **209**, 751.
Hoyle, F., and Fowler, W. A. 1963*a*, *M.N.R.A.S.*, **125**, 169.
———. 1963*b*, *Nature*, **197**, 533.
———. 1967, *ibid.*, **213**, 373.
Kellermann, K. I., Clark, B. G., Bare, C. C., Rydbeck, O., Ellder, J., Hansson, B., Kollberg, E., Hoglund, B., Cohen, M. H., and Jauncey, D. L. 1968, *Ap. J.* (*Letters*), **153**, L209.
Kellermann, K. I., and Pauliny-Toth, I. I. K. 1968*a*, *Ann. Rev. Astr. and Ap.*, **6**, 417.
———. 1968*b*, *A.J.*, **73**, 874.
———. 1969, *Ap. J.* (*Letters*), **155**, L71.
Kinman, T. D. 1967, *Ap. J.* (*Letters*), **148**, L53.
Kinman, T. D., Lamla, E., Ciurla, T., Harlan, E., and Wirtanen, C. A. 1968, *Ap. J.*, **152**, 357.
Kinman, T. D., Lamla, E., and Wirtanen, C. A. 1966, *Ap. J.*, **146**, 964.
Longair, M. S. 1966, *M.N.R.A.S.*, **133**, 421.
Low, F. J., and Johnson, H. L. 1965, *Ap. J.*, **141**, 336.
Low, F. J., and Kleinmann, D. E. 1968, *A.J.*, **73**, 868.
Lynds, C. R. 1966, communicated at IAU Symp. No. 29 at Byurakan Obs.
———. 1967, *Ap. J.*, **147**, 396.
Mackay, C. D. 1967, *Nature*, **216**, 1091.
McCrea, W. H. 1966, *Ap. J.*, **144**, 516.
McVittie, G. C. 1965, *General Relativity and Cosmology* (Urbana, Ill.: University of Illinois Press).
Maran, S. P., and Cameron, A. G. W. (eds.) 1969, *Topics in Relativistic Astrophysics*, proceedings of the 1967 Texas Symposium in New York (New York: Gordon & Breach).
Matthews, T. A., and Sandage, A. R. 1963, *Ap. J.*, **138**, 30.
Moffet, A. T. 1965, *Ap. J.*, **141**, 1580.
Ne'eman, Y. 1965, *Ap. J.*, **141**, 1303.
Ne'eman, Y., and Tauber, G. 1967, *Ap. J.*, **150**, 755.
Noerdlinger, P., Jokipii, J., and Woltjer, L. 1966, *Ap. J.*, **146**, 523.
Novikov, I. D. 1964, *Astr. Zh. USSR*, **41**, 1075.
Oke, J. B. 1963, *Nature*, **197**, 1040.
———. 1965, *Ap. J.*, **141**, 6.
———. 1966, *ibid.*, **145**, 668.
———. 1967*a*, *ibid.*, **147**, 901.
———. 1967*b*, *Ap. J.* (*Letters*), **150**, L5.
Oke, J. B., Neugebauer, G., and Becklin, E. E. 1970, *Ap. J.*, **159**, 341.
Oke, J. B., and Sargent, W. L. W. 1968, *Ap. J.*, **151**, 807.
Osterbrock, D. E., and Parker, R. A. R. 1966, *Ap. J.*, **143**, 268.
Ozernoy, L. M., and Chertoprud, V. E. 1967, *Astr. Zh. USSR*, **44**, 537.
Pacini, F. 1966, *Nature*, **209**, 389.
Peebles, P. J. E. 1968, *Ap. J.* (*Letters*), **154**, L121.
Pengelly, R. M. 1964, *M.N.R.A.S.*, **127**, 145.
Piddington, J. H. 1966, *M.N.R.A.S.*, **133**, 163.
Robinson, I., Schild, A., and Schucking, E. L. (eds.) 1965, *Quasi-stellar Sources and Gravitational Collapse*, proceedings of the 1963 Texas Symposium in Dallas (Chicago: University of Chicago Press).
Ryle, M. 1968, *Ann. Rev. Astr. and Ap.*, **6**, 249.

Ryle, M., and Sandage, A. 1964, *Ap. J.*, **139**, 419.
Sandage, A. R. 1965, *Ap. J.*, **141**, 1560.
———. 1966, *Ap. J.*, **146**, 13.
———. 1967*a*, *Ap. J.* (*Letters*), **150**, L9.
———. 1967*b*, *ibid.*, L177.
Sandage, A. R., and Luyten, W. J. 1969, *Ap. J.*, **155**, 913.
Sandage, A. R., and Miller, W. C. 1966, *Ap. J.*, **144**, 1240.
Sandage, A. R., Westphal, J. A., and Strittmatter, P. A. 1966, *Ap. J.*, **146**, 322.
Scheuer, P. A. G. 1965, *Nature*, **207**, 963.
Scheuer, P. A. G., and Williams, P. J. S. 1968, *Ann. Rev. Astr. and Ap.*, **6**, 321.
Schild, A., and Schucking, E. L. (eds.) 1969, *Quasars and High-Energy Astrophysics*, proceedings of the 1964 Texas Symposium in Austin (New York: Gordon & Breach).
Schmidt, M. 1962, *Ap. J.*, **136**, 684.
———. 1963, *Nature*, **197**, 1040.
———. 1964, paper read at Second Texas Conference on Relativistic Astrophysics, Austin (see Schild and Schucking 1969).
———. 1965, *Ap. J.*, **141**, 1295.
———. 1966, *ibid.*, **146**, 7.
———. 1967, *Relativistic Theory and Astrophysics* in *Lectures in Applied Mathematics*, ed. J. Ehlers (Providence, R.I.: American Mathematical Society), **8**, 203.
———. 1968, *Ap. J.*, **151**, 393.
———. 1969, *Ann. Rev. Astr. and Ap.*, **7**, 527.
Schmidt, M., and Matthews, T. A. 1964, *Ap. J.*, **139**, 781.
Schmidt, M., and Olsen, E. T. 1968, *A.J.*, **73**, S117.
Searle, L., and Sargent, W. L. W. 1968, *Ap. J.*, **153**, 1003.
Setti, G., and Woltjer, L. 1966, *Ap. J.*, **144**, 838.
Shklovskii, I. S. 1964, *Astr. Zh. USSR*, **41**, 801.
Smith, H. J. 1965, in *Quasi-stellar Sources and Gravitational Collapse*, ed. I. Robinson, A. Schild, and E. L. Schucking (Chicago: University of Chicago Press), p. 221.
Spitzer, L., and Saslaw, W. C. 1966, *Ap. J.*, **143**, 400.
Stockton, A. N., and Lynds, C. R. 1966, *Ap. J.*, **144**, 451.
Strittmatter, P. A., and Burbidge, G. R. 1967, *Ap. J.*, **147**, 13.
Strittmatter, P. A., Faulkner, J., and Walmsley, M. 1966, *Nature*, **212**, 1441.
Sturrock, P. A., and Feldman, P. A. 1968, *Ap. J.* (*Letters*), **152**, L39.
Terrell, J. 1964, *Science*, **145**, 918.
———. 1968, *Phys. Rev. Letters*, **21**, 637.
Ulam, S. M., and Walden, W. E. 1964, *Nature*, **201**, 1202.
Véron, P. 1966, *Ann. d'Ap.*, **29**, 231.
Visvanathan, V. 1968, *Ap. J.* (*Letters*), **153**, L19.
Wagoner, R. V. 1967, *Ap. J.*, **149**, 465.
Wampler, E. J. 1967, *Ap. J.* (*Letters*), **148**, L101.
Wampler, E. J., and Oke, J. B. 1967, *Ap. J.*, **148**, 695.
Wlerick, G., and Véron, P. 1967, *Ann. d'Ap.*, **30**, 341.
Woltjer, L. 1964, *Nature*, **201**, 807.
———. 1966, *Ap. J.*, **146**, 597.
Wyndham, J. D. 1966, *Ap. J.*, **144**, 459.
Zapolsky, H. S. 1968, *Ap. J.* (*Letters*), **153**, L163.

CHAPTER 9

Radio Observations of Neutral Hydrogen in Galaxies

MORTON S. ROBERTS

National Radio Astronomy Observatory,[1] *Green Bank, West Virginia*

1. INTRODUCTION

THE first detection of extragalactic hydrogen[2] was made in 1953 (Kerr and Hindman 1953; Kerr, Hindman, and Robinson 1954), two years after the discovery of the 21-cm line from Galactic hydrogen. This first detection was from the Magellanic Clouds which at a distance of ~0.05 megaparsecs are the two galaxies nearest the Milky Way. A 36-foot paraboloid was used in this initial study. Further observational work on extragalactic hydrogen was carried on at Leiden with a 25-meter antenna and at Harvard with a 60-foot telescope. During the following five years information on a dozen galaxies was obtained. The power incident on an antenna in these extragalactic H I studies is very small, typically 10^{-17} watts; the fact that successful measurements were made so early in the history of this spectral line is a tribute to radio engineering and the perseverance of radio astronomers.

Unlike observations of continuum radiation, where a wide frequency band may be used to improve the signal-to-noise ratio, line observations should be restricted to at most the bandwidth (i.e., the velocity range) of the emitted radiation. For a detailed radial-velocity picture, even narrower bandwidths must be employed. The basic equation which describes the noise fluctuations of a comparison radiometer is

$$(\Delta T)_{\mathrm{rms}} = CT_{\mathrm{sys}}(bt)^{-1/2} , \qquad (1)$$

where $(\Delta T)_{\mathrm{rms}}$ is a measure of the rms noise fluctuations, which we wish to minimize, T_{sys} is the overall system noise temperature (receiver, antenna, etc.), b is the bandwidth in Hertz, and t the integration time in seconds. The constant C will generally have a

Received January 12, 1966; revised November 1971.

[1] The National Radio Astronomy Observatory at Green Bank, West Virginia, is operated by Associated Universities, Inc., under contract with the National Science Foundation.

[2] Throughout the following we shall use "hydrogen" (and H I) to mean neutral atomic hydrogen unless specifically stated otherwise. Similarly, "21-cm" radiation will always refer to hydrogen-line radiation, as contrasted to continuum radiation at this wavelength.

value between 1 and 2, depending on the type of radiometer used. The initial phase of extragalactic hydrogen studies employed radiometers whose system noise temperatures were greater than 1000° K. Starting about 1960, major improvements were made in both receivers and antennas, resulting in significant additions to our knowledge of galaxies. Solid-state, low-noise preamplifiers (masers and parametric amplifiers) lowered the system noise temperature to 100° K for maser systems and to 200° K for parametric amplifiers. These values are continually being lowered by improved techniques, and a system noise $\lesssim$100° K is now quite common. At about the same time large paraboloids, up to 300 feet in diameter, and large tiltable plane-standing parabolas came into operation. The increased collecting area of these large antennas together with low-noise receivers allowed more distant and fainter galaxies to be studied. At present, close to 200 galaxies have been detected in hydrogen, and reliable hydrogen masses are available (as of 1971 June) for 130 systems.

The use of interferometric techniques to increase the resolving power was the next developmental step. This technique offers great promise and is the basis of much activity at present. Such observations are planned or in progress at Cambridge, Green Bank, Jodrell Bank, Owens Valley, and Westerbork.

The following information may be derived from 21-cm hydrogen-line studies of galaxies: (1) the distribution of hydrogen; (2) the total hydrogen content; (3) the radial-velocity field, i.e., a map of loci of constant radial velocity as seen on the projected image of the galaxy; (4) the systemic radial velocity; (5) an estimate of the total mass from (*a*) the velocity profile or (*b*) the rotation curve. Assuming that 21-cm line radiation is detectable, we find the determination of items (2), (4), and (5*a*) to be essentially independent of the antenna aperture or beam size, β. The success of determining the other parameters is directly related to the ratio δ/β, where δ is the angular size of the galaxy; we would like $\delta/\beta \gg 1$. Further, the precision in determining the systemic velocity is also dependent on δ/β since it is best determined from the velocity field and rotation curve. The necessary assumptions as well as the uncertainties and errors in deriving the above information are discussed in §§ 2–6. The properties of galaxies derived from 21-cm studies are presented in § 7. The final section deals with the evolution of galaxies.

2. DETERMINATION OF KINEMATIC PROPERTIES OF GALAXIES FROM 21-CENTIMETER OBSERVATIONS

The accurate measurement of frequency is a well-developed technique in electronics. Standard components are available which will produce or measure frequencies with an accuracy greater than 10^{-7} of the operating frequency, 1420 MHz. Thus, given a sufficiently strong, narrow signal, its frequency and corresponding Doppler velocity can be measured to an accuracy better than 0.1 km s^{-1} (at this frequency 1 km s^{-1} $\approx$ 5 kHz). For extragalactic hydrogen observations, filters with halfwidths corresponding to 20 km s^{-1} have frequently been used; the center frequencies of these filters can be determined with high precision, but velocity details finer than the filter width are obviously lost. For the nearer galaxies—the Magellanic Clouds and M31—filters narrower than 10 km s^{-1} have been employed. For comparison, the typical resolution of a nebular spectrograph ($\sim$300 Å mm^{-1}) used in optical observations introduces an uncertainty of

~20 km s^{-1} for measurements having an accuracy of 1 micron. Additional systematic errors of this order may also occur because of line curvature and line blends, the latter error resulting from an uncertainty in the rest value of the effective wavelength.

Positional as well as velocity information is needed for a complete study of the motions within a system, and here the optical data are far superior to radio measurements. Optical images on photographic plates are typically 1″ to 2″ in diameter, and simple measurements yield relative positions having a corresponding accuracy. For comparison the full width at half-power of the antenna beam at 21-cm for the Parkes 210-foot telescope is 13.′5; for the Green Bank 300-foot telescope, 10′; and for the Nancay antenna it is $4' \times \geq 24'$ (the latter value increasing with declination). The Cambridge interferometer has synthesized a $1.'5 \times 3.'0$ beam (Baldwin 1970), recent Owens Valley observations are with synthesized 2′ and 4′ beams (Rogstad and Shostak 1971), and 25″ resolution will be obtained with the Westerbork array. The position of a radiating source can, of course, be measured to a higher precision than the beam size. Thus, the location of the peak of a drift curve may be determined with an accuracy of better than 1′ from observations with a 10′ beam and a signal-to-noise ratio of $\gtrsim 10$. The problem is rather one of beam smearing; i.e., the observed data refer to the convolution of the actual source distribution by the antenna-beam pattern.

The radiating region sensed by the antenna-radiometer system is determined by both the antenna pattern and the filter width. The latter is involved since it restricts the radial velocity range and hence the spatial distribution of the radiating H I. This follows from the assumption that the gross motions within a galaxy are ordered. Under such conditions, the hydrogen-line radiation in the radial-velocity interval $V_r \pm \Delta V_r$ must arise from a very specific region of the galaxy, a region whose shape is determined by the rotation law.

As an illustration, consider the case of the symmetrical rotation curve in the upper half of figure 1. The radial velocity is related to the circular velocity $V_c(R)$, by

$$V_r = S + V_c(R) \sin i \cos \theta \,, \tag{2}$$

where S is the systemic radial velocity and θ the polar coordinate in the plane of the galaxy of a point at R; $\theta = 0$ on the major axis. For a specified inclination i, equation (2) together with the rotation curve of figure 1 will define loci of constant radial velocity. These loci for $S = 0$ and an inclination of 30° are shown in the lower half of figure 1. The position angle of the major axis is a free parameter, and this figure need only be rotated to correspond to any desired value.

For an infinitely narrow filter and for zero velocity dispersion, a radiating source at the indicated radial velocities has the shape of the loci shown in figure 1. The loci will appear broader if observed with a filter of finite width or if random motions exist. The hatched region in figure 1 is drawn to indicate a filter width of 20 km s^{-1} or random motions of ± 10 km s^{-1} about the indicated radial-velocity line. A combination of both of these will act in the same manner to increase the width. Thus a radio telescope, regardless of its antenna beam size, will sense radiation from a restricted region of a galaxy when the filter width is smaller than the total range of radial velocities (comprising both ordered and random motions). Information about the shape of the radiating source, the widened loci of figure 1, is lost, however, if its angular dimensions are com-

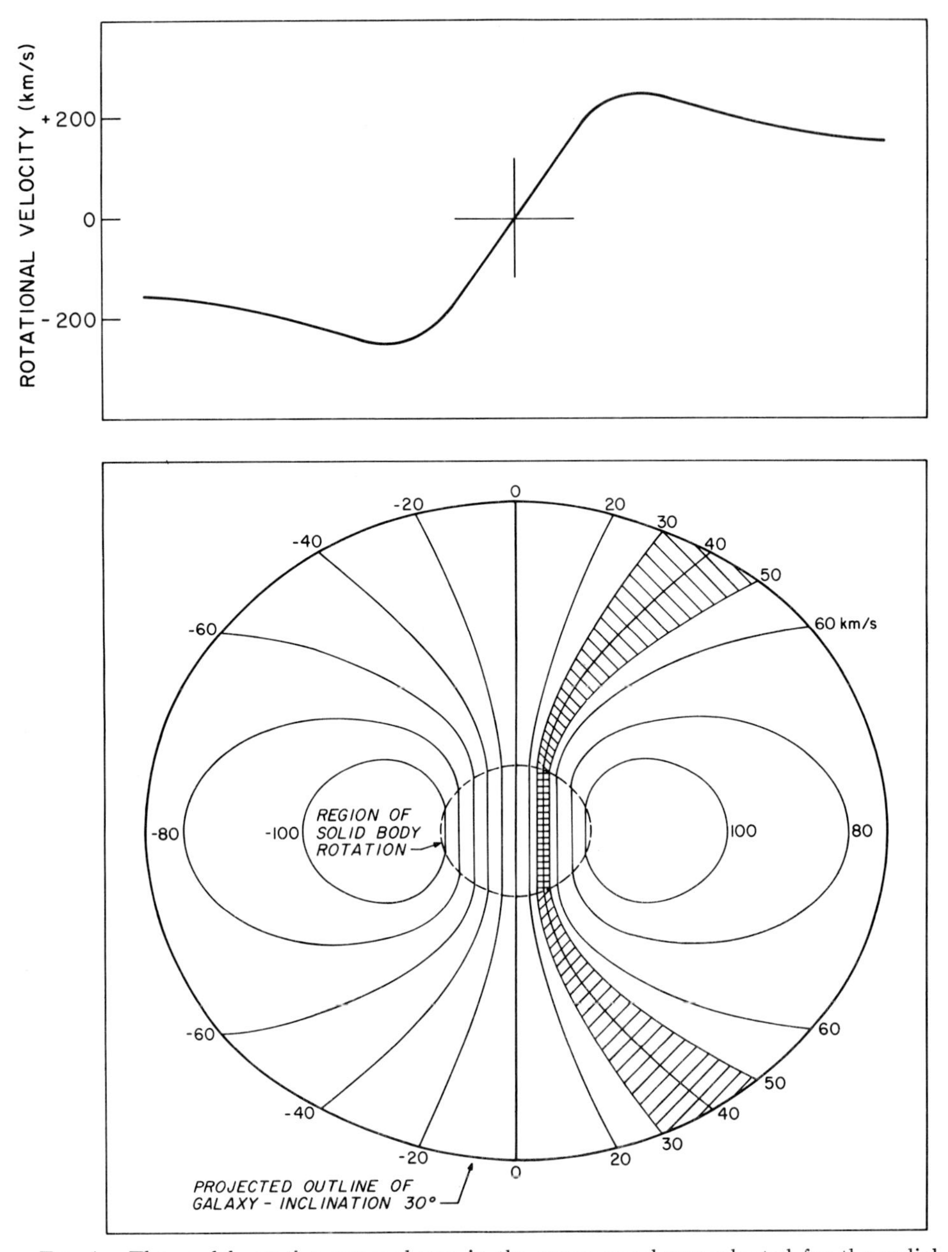

FIG. 1.—The model rotation curve shown in the upper panel was adopted for the radial-velocity map in the lower panel. The horizontal axis is distance from the center. The radial-velocity map shows lines of constant radial velocity as seen on the projected image of the model galaxy taken here to have an inclination of 30°. The locus marked 0 km s^{-1} corresponds to the systemic velocity and the minor axis. The hatched region represents that area whose radial velocity is $+40 \pm 10$ km s^{-1}.

parable to or smaller than the beam. It is this shape which yields information on the rotation curve and random motions of the H I. The optical "long-slit" method of measuring radial velocities involves placing the spectrograph slit on the image of the galaxy, i.e., on the locus pattern of figure 1. Measurements in the dispersion direction yield radial velocities, and since the image is not trailed, measurements along the spectral line yield positional information.

A more general case of equation (2) and figure 1 would also include an expansion (or contraction) term as well as z-motion (motion perpendicular to the principal plane of the galaxy). The expression for the radial component then becomes

$$V_r = S + V_c(R) \sin i \cos \theta + E(R) \sin i \sin \theta + Z(R, \theta) \cos i \,. \tag{3}$$

Here, the expansion term is assumed to be only a function of R, while the z-motions, though ordered, may be dependent on both polar coordinates. An expansion term causes the loci in figure 1 to rotate with a resultant crossing of the minor axis. A diagonal symmetry will also appear: loci in one pair of diagonally opposite quadrants will open while crowding more closely together in the other pair of quadrants. Note that measurements on the major axis, $\theta = 0$, will have no component of radial motion while data referring to the minor axis, $\theta = 90°$, will have no circular-motion component.

The actual source shape as seen by a given filter reflects both the kinematics within the system and the H I distribution. This is shown schematically in figure 2 where both the hydrogen distribution and radial-velocity loci are displayed. A 60° inclination is adopted here and two filters are considered, one centered at $V_r = 0$ km s^{-1}, the other at -80 km s^{-1}. The shaded, elliptically shaped region represents an assumed ring distribution of H I as seen at this inclination. The heavily shaded areas represent the intersection of the filter width (and velocity dispersion) with the hydrogen distribution. The resultant beam-smeared response is shown on the right side of figure 2 as contour diagrams, one for each filter. The separation in the contour diagram along the minor axis ($V_r = 0$) into two concentrations assumes adequate relative resolution. A poorer resolution would yield only one apparent concentration whose peak would actually lie in a region of low H I surface density. It is clear from this figure that the shapes of drift curves and velocity profiles are determined by both the actual hydrogen distribution *and* the radial-velocity field. Thus, a uniform distribution of hydrogen will also yield the double-peaked contours shown on the right because of the flaring out of the radial-velocity loci about the minor axis. The H I distribution and the rotation curve cannot, in general, be derived from the observational data unless one or the other is known or assumed and their interaction taken into account or unless both are derived simultaneously through model fitting.

A more specific consideration of the general problem is the derivation of rotation curves. Unless the ratio of source to beam size, δ/β, is very large, the effects of beam smearing will make positional information derived from drift curves or velocity profiles ambiguous and misleading. An example of this is shown in figure 3, where the results of computer-generated data are displayed. A model patterned after NGC 4631 (Roberts 1968*b*) having a hydrogen distribution made up of a ring of diameter 11′.2 superposed on a uniform disk of H I of diameter 22′ was smeared with a 10′ beam. The model rotation curve serving as input is shown in the figure. Ideally, this rotation curve should be recoverable from the data. Drift curves (the output response as a function of position for a given filter as the source passes through the beam) and velocity profiles (the output response for all filters for a given position of the beam on the source) were computed and three different types of position-velocity measurements were made. To avoid confusion with actual rotation curves, the derived data when plotted in the format of a rotation curve will be called "position-velocity" (*p-v*) curves. Figure 3 shows *p-v* curves based on (1) the peaks of drift curves, (2) the peaks of velocity profiles, and (3) the median values of velocity profiles. They do not agree among themselves or with the input rotation curve. Although this example is for a particular set of model-galaxy parameters it is typical of the problem encountered in analyzing data where δ/β is small. It is difficult to quantitatively define "small" as used in this context. For any given galaxy, the actual discrepancy between a *p-v* curve and the true rotation curve will depend on a variety of parameters—e.g., filter width, velocity dispersion within the galaxy, H I distribution—but *p-v* curves can be expected to vary significantly from the true rotation curves when $\delta/\beta < 10$. A number of *p-v* curves, often labeled rotation curves, have appeared in the literature; they should be used with caution. Their shapes cannot be corrected to the true rotation curve without knowledge of the hydrogen distribution and kinematics

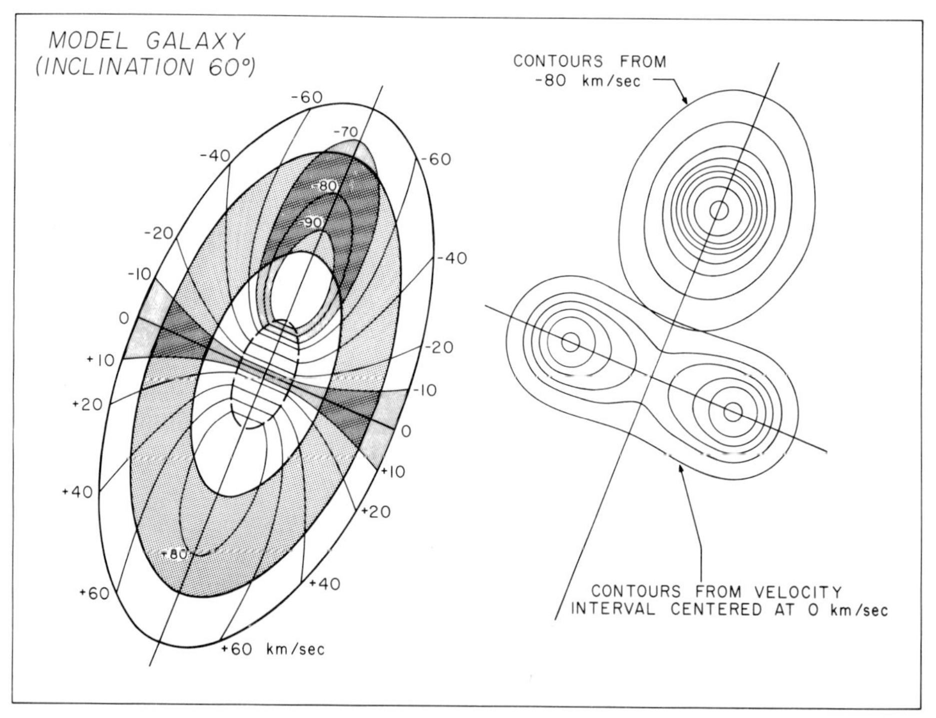

FIG. 2.—A radial-velocity map for a galaxy of 60° inclination is shown on the left. The lightly shaded region is a schematic representation of a ring distribution of H I seen in projection. Two sample velocity regions ± 10 km s^{-1} wide are shown, one centered at -80 km s^{-1} and the other at 0 km s^{-1}. The heavily shaded regions are the areas bounded by the sample velocity range and the hydrogen distribution; they represent the actual source shapes at each velocity. These sources, convolved with a telescope beam, are shown on the right as contour diagrams.

within the system. To varying degrees of accuracy they will, however, yield total mass estimates of the system. An approach in which the observed data are approximated through a model of both the H I distribution and the kinematics within the galaxy will give a more realistic representation of the rotation curve. This method is discussed later in this section.

Rogstad, Rougoor, and Whiteoak (1967) use the term "position profile" to describe the position-velocity curves they have derived from interferometric 21-cm observations of galaxies. Their data, as well as those of Seielstad and Whiteoak (1965), were obtained with only one interferometer spacing (two 90-foot antennas separated east-west by 100 feet) and suffer from the same effects described above. They obtained total mass estimates through model-fitting to their observed *p-v* curves.

The various *p-v* curves shown in figure 3 point up another factor in the discussion of 21-cm derived rotation curves and radial-velocity maps, namely, which parameter of the observational data gives the best representation of the kinematics within a galaxy? It is obvious that for this problem we require a parameter that is free of any weighting effects introduced by the amount of hydrogen within the system. Such a situation exists

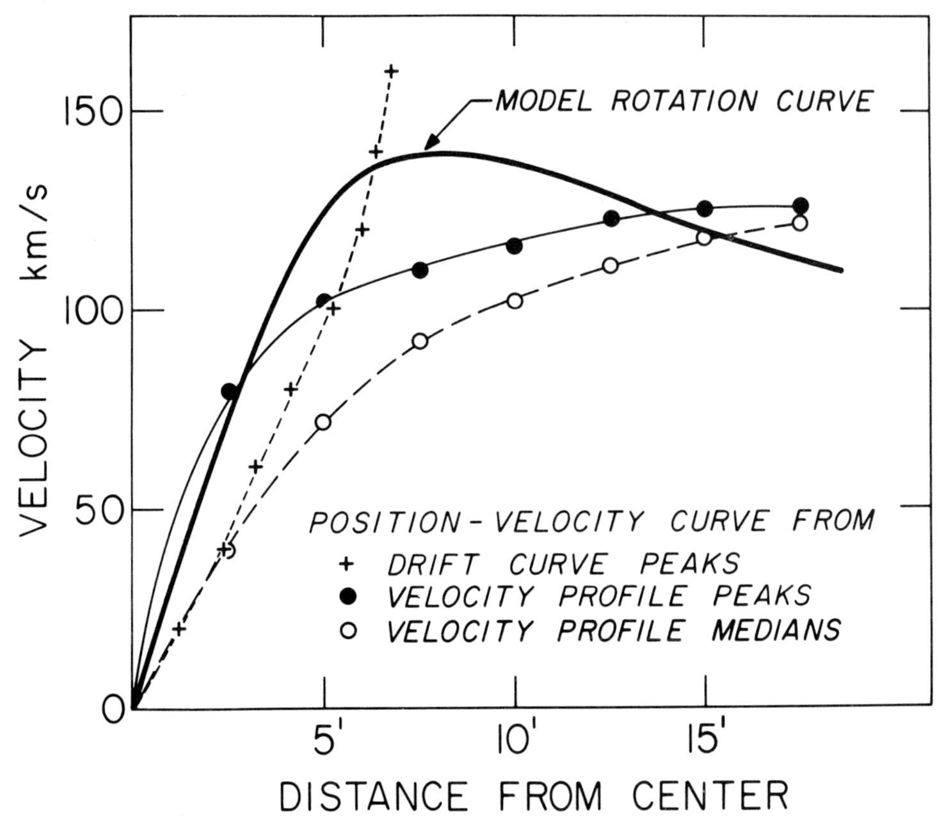

FIG. 3.—Comparison of a model rotation curve and the position-velocity curves derived from measurements of drift curves and velocity profiles. The heavy solid line served as input to a computer program which generated the drift curves and velocity profiles used to construct the position-velocity curves.

in optical studies where stars or H II regions are used as test probes of the velocity field. The brightness of the star or H II region does not enter into the kinematic study (the signal-to-noise ratio affecting the errors of observation is incidental to the present consideration). As noted previously, 21-cm observations represent the beam-averaged combination of both the hydrogen distribution and the velocity field. Thus the peak, mean, or median of a velocity profile gives a velocity measure which is weighted by the brightness temperature at various radial velocities included in the beam. An example of how the *p-v* curves differ for these three different types of measurements is given for the Small Magellanic Cloud by Hindman (1967, fig. 18). His data were obtained with the Parkes 210-foot telescope. The three *p-v* curves he derives differ significantly even though $\delta/\beta \simeq 20$.

Another example is found in the different analyses of M31. Figures 4 and 5 display

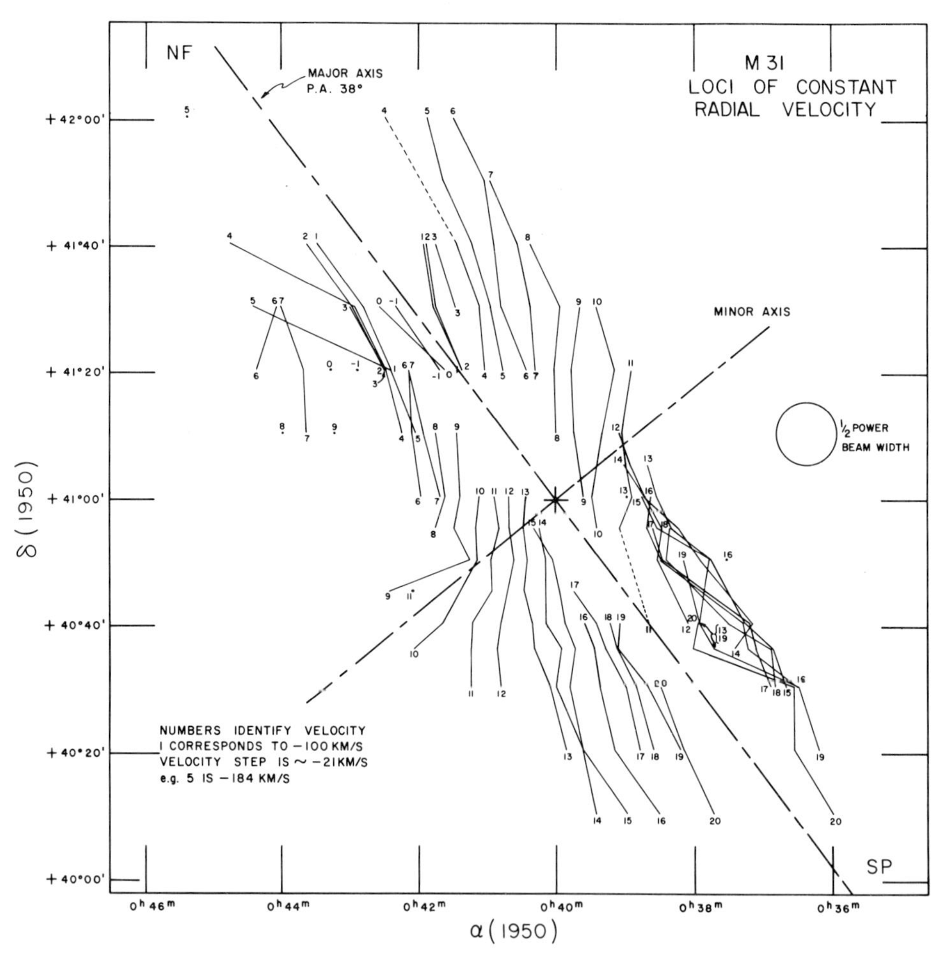

FIG. 4.—The radial-velocity field of M31 derived from the peaks of drift curves. Compare with fig. 5.

radial-velocity maps of M31 derived from 300-foot observations. The data used are essentially the same except for the somewhat greater north-south coverage in figure 5. The constant radial-velocity loci in figure 4 are defined by the peaks of drift curves (the contour step is 21 km s^{-1}) while those in figure 5 are defined by the midpoints of velocity profiles (the contour step is 50 km s^{-1}). This latter figure is similar to the radial-velocity map derived by Argyle (1965, fig. 6) who used mean radial velocities, and the map obtained by Gottesman and Davies (1970, fig. 5) who used median velocities. An ordered variation of the velocity loci in fig. 4 is seen in only two, diagonally opposite, regions: the northwest and the southeast quadrants. The other two quadrants are disordered in that many of the loci are crowded together in figure 4. This crowding indicates a high velocity dispersion in the complex regions. Such information is lost when only the midpoint (or mean or median) of a velocity profile is used. The use of drift curves emphasizes

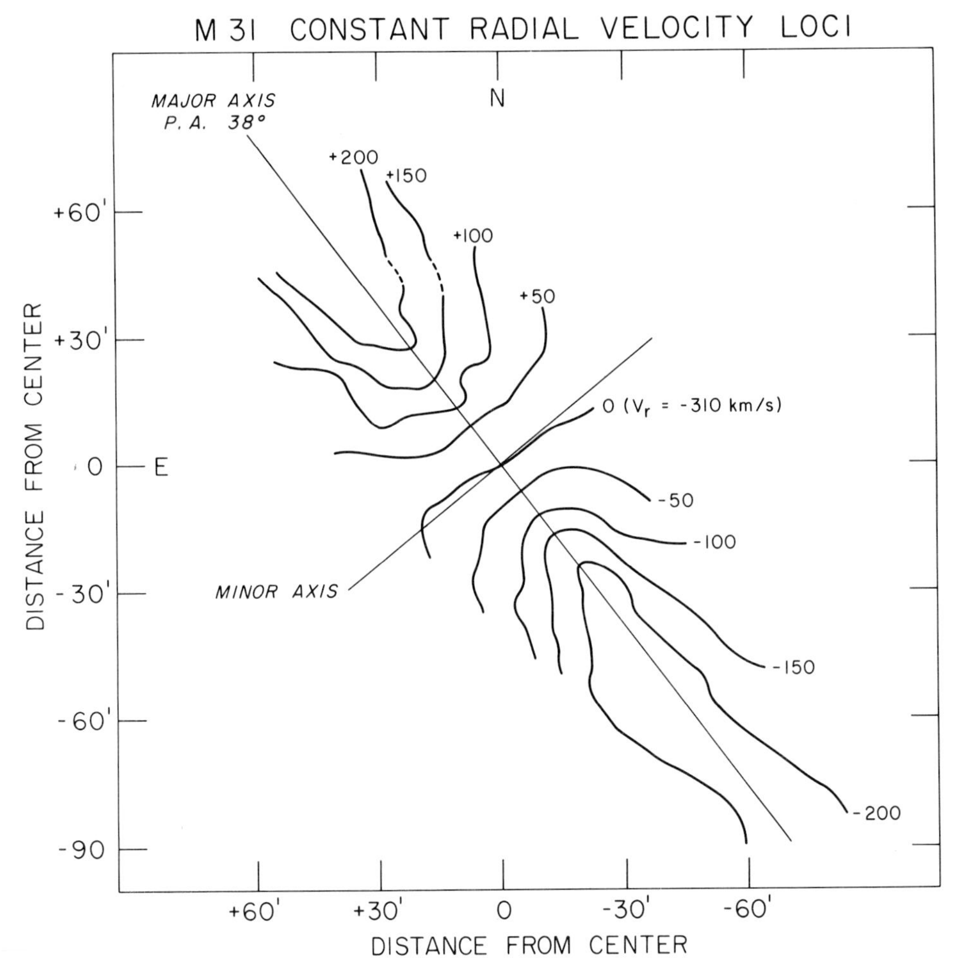

FIG. 5.—The radial-velocity field of M31 derived from the midpoints of velocity profiles. Compare with fig. 4.

such velocity pecularities but may mask the general motion in the region. To varying amounts both methods suffer from a weighting of the velocity data by the H I distribution. The differing information content of these various approaches is the source of some confusion in the literature (e.g., Gottesman and Davies 1970). It is interesting to note that both figures 4 and 5 as well as Argyle's data show peculiar features along the minor axis which can only be interpreted as noncircular motion.

The model-fitting approach to rotation curve and hydrogen distribution studies attempts to account for all of the factors responsible for the shape of a drift curve or velocity profile. It is thus free of the objections raised above but is correspondingly more complicated and is only tractable with the aid of large computers. Work along this line by the Owens Valley radio group was noted previously. A more detailed approach is being made at Nancay and at Green Bank. A computer-generated model of a simplified but reasonable representation of a galaxy is convolved with the antenna beam and a set of velocity filters. The resultant velocity profile (or drift curve) is then compared with the observed velocity profiles. A best, but not necessarily unique, set of galaxy parameters is then determined in terms of the deviations, observed minus computed, evaluated at fixed intervals along the velocity profile. The parameters which describe the model and which must be determined or fixed are (1) rotation curve, (2) distribution of neutral hydrogen, (3) random motions, which may include large-scale ordered deviations from pure rotational motion such as expansion or z-motion, (4) inclination, (5) position angle of the major axis, (6) systemic radial velocity, and (7) location of the center of the galaxy. Examples of the application of this procedure are given by Roberts (1968*b*); Gordon, Remage and Roberts (1968); Guélin and Weliachew (1969); Burns and Roberts (1971); and Gordon (1971). This approach appears to be the only satisfactory method for studying the velocity field and H I distribution in galaxies from data obtained with presently available filled-aperture antennas. A major improvement in the relative resolution of 21-cm measurements is becoming available through interferometric observations. However, model-fitting may still be required for the analysis of many galaxies if we recall the high relative resolution available for the Small Magellanic Cloud and the three different "rotation curves" (actually, position-velocity curves) obtained by Hindman (1967). Hindman's observations are with velocity resolutions of 7.8 and 2.1 km s^{-1} and a linear resolution of 0.2 kpc. For comparison, $2'$ (a typical synthesized beam in current studies) corresponds to 0.4 kpc at the distance of M31 and M33, and 1.9 and 2.7 kpc at the distance of NGC 2403 and 5457, respectively. Because of the inclination of a galaxy, projection effects will alter a circular beam to an elliptical outline in the plane of the galaxy and correspondingly increase the area sensed by the beam. Nevertheless, such observations represent a fivefold increase in resolution over the 300-foot telescope, and a corresponding improvement in modeling analyses can be expected.

3. TOTAL MASSES

Various methods have been employed to fit a rotation curve to the observed data and to derive the resultant mass and mass distribution; see, e.g., Kerr and de Vaucouleurs (1956); Burbidge, Burbidge, and Prendergast (1960*b*); Brandt and Scheer (1965) and references therein; McGee and Milton (1966*a*); Hindman (1967); Roberts (1966*a*, *b*); and Gottesman and Davies (1970).

The majority of the galaxies which have been detected in H I are of small angular size, and no meaningful rotation curves can be derived from the 21-cm data. However, total mass estimates can be made from the width of the velocity profile (Volders and Högbom 1961; Roberts 1962*a*, *b*; Epstein 1964*a*; Bottinelli, Gougenheim, Heidmann, and Heidmann 1968). This procedure involves several assumptions: (1) the width of the velocity profile is due primarily to rotation. The half-width at zero intensity then corresponds to the radial component of the maximum of the rotation curve (a correction for filter width and random motions must be made). (2) This velocity maximum occurs at a radial distance R_m such that the ratio of R_m to the linear dimension of the galaxy is constant. Roberts (1962*a*, *b*) and Epstein (1964*a*) adopt $R_m = \delta'/6$, where δ' is the angular diameter of the galaxy on Holmberg's (1958) system of measurements. (3) The rotation curve may be described by the Bottlinger-Lohmann formula. With these assumptions the total mass is given by

$$M_T/M_\odot = 1.7 \times 10^{-2} D\delta' V^2/\sin^2 i \,, \tag{4}$$

where D is the distance in parsecs, δ' is the diameter in arc minutes, and V (km s^{-1}) is one-half the full width of the velocity profile at zero intensity.

For irregular galaxies, an approach first suggested by Volders and Högbom (1961) has been used. The virial theorem is applied to a system in which a Gaussian density distribution is assumed:

$$\rho = \rho_0 \exp(-r^2/2a^2) \,. \tag{5}$$

For a stationary ensemble, the total mass is then given by

$$M_T/M_\odot = 2.7 \times 10^{-2} D\theta'(\Delta V)^2 \,, \tag{6}$$

where D is the distance in parsecs, θ' is the average optical diameter of the galaxy in arc minutes, and ΔV (km s^{-1}) is the full width of the velocity profile at half-intensity. In equation (6) the relative density, ρ/ρ_0, has been fixed at 0.05 at a point corresponding to the average optical radius.

Equation (6) is for a model in which the motions are primarily random with a Gaussian velocity distribution. However, many irregular-type galaxies have a radial-velocity field that shows that these systems are rotating, in which case equation (4) would be more appropriate than equation (6). If the velocity profile is approximately Gaussian in shape, both equations will give similar mass estimates to within the factor $\operatorname{cosec}^2 i$.

Total masses for galaxies which have been detected in H I are given in table 1 (p. 337). These masses have been derived by a variety of methods, but the majority are based on the application of equation (4) or (6). In a few cases optically derived masses are listed. These are identified by an asterisk. A comparison among those galaxies whose mass has been determined by different methods (optical, 21-cm model fitting, and 21-cm velocity profile) has been made by Roberts (1969). He finds good agreement, in general, with 21-cm determined masses higher in the mean, by about a factor of 2. Because of the different assumptions made in deriving total masses from optical and from 21-cm measurements, it is not clear which, if either, approach yields the true total mass.

4. SYSTEMIC RADIAL VELOCITY

The systemic radial velocity, S, of a galaxy can be determined with relatively high precision from 21-cm observations; the uncertainty is often less than 30 km s^{-1}. Optical determinations of S have uncertainties of from 50 to 100 km s^{-1} because of the low dispersion generally used in nebular spectroscopy. Only high-dispersion spectra of the nuclear region of a galaxy or a rotation-curve study of the entire system can significantly reduce this error. The sizes of these errors are insignificant in terms of the overall red-shift relation which involves velocities up to several orders of magnitude larger than the measurement error.

Of basic importance is the fundamental system of radial velocities used in extragalactic studies. The spectral lines from the Sun have been used to establish such a fundamental system for stellar radial velocities. The planets are used as an intermediate calibrator since their apparent magnitudes are comparable to stellar magnitudes and their radial velocities are accurately known. This is then extended to other spectral types by using members of open clusters which contain, as a tie-in, a solar-type star. In this manner the uncertainties in laboratory wavelengths and line blends may be allowed for. Such a procedure has not been possible for optical measurements of extragalactic radial velocities; in fact, the standard wavelength adopted at Lick and at Mount Wilson for lines in the blue differ by amounts of up to almost an angstrom, $\sim$70 km s^{-1} (de Vaucouleurs and de Vaucouleurs 1963). Fortunately, velocities determined from 21-cm measurements do define a fundamental system since (1) the laboratory wavelength is well determined and (2) there are no blends with other spectral lines. Possible contributions from undiscovered lines due to other elements or molecules, or recombination lines from H II regions, are insignificant.

The uncertainties in 21-cm radial velocities are primarily due to the signal-to-noise ratio and to approximations made in determining the velocity of the center of mass of the entire system. Thus, the assumption that the radial velocity of the centroid of the hydrogen distribution is identical to the systemic velocity is incorrect in general, and a small rotational component will be present in the deduced values of S. The few percent of the total mass of a galaxy that is in the form of hydrogen need not have its centroid coincident with that of the total mass distribution. The determinations of systemic velocities from the median value of a velocity profile have employed this assumption. The generally small errors previously quoted for these values (Roberts 1962*b*; Epstein 1964*a*) do not take this difference into account and should be increased by a factor of about 3.

As in the study of rotation curves, we require a measure of S that is unweighted by the brightness-temperature distribution. The midpoint of a velocity profile is the best such measure. An example in which weighting has an extreme effect on the derived velocity may be found in the velocity profile of M82 (Volders and Högbom 1961). This profile is highly asymmetrical; it has a strong Gaussian-like component plus a long, low-intensity extension at higher velocities. The 21-cm systemic velocity initially derived for M82 is 184 km s^{-1} with respect to the Sun and is based on "the main part of the observed [velocity] profile" (Volders and Högbom 1961). The optically derived systemic velocity of M82 is based on rotation curves and is of correspondingly high weight. Two values are available: 275 km s^{-1} (Mayall 1960) and 294 km s^{-1} (Burbidge, Burbidge, and Rubin 1964). The apparent discrepancy of $\sim$100 km s^{-1} is quite large, and several explanations have been proposed (e.g., Volders and Högbom 1961; Elvius 1964). However, if the midpoint of the Leiden profile is taken as the systemic velocity, the 21-cm value is increased to $\sim$240 km s^{-1}. The great asymmetry in the global velocity profile of M82 is thus far unique. It is used here to emphasize the problem of finding the "best" estimate of S.

Systemic radial velocities determined from the midpoint of velocity profiles generally have errors of 20 to 30 km s^{-1} due primarily to poor signal-to-noise ratios at the ends of the profiles. The availability of observations at various positions on the projected image of the galaxy will reduce the uncertainty. This may be done by locating the locus of constant radial velocity which corresponds to the minor axis or through model-fitting to the data. Fixing S as that value which makes the two halves of the rotation (or p-v) curve most symmetrical involves an assumption we know to be incorrect. Both optical and radio studies have shown that rotation curves are not in general symmetrical.

Observations of wavelength shifts of the same objects made at two widely separated spectral regions, $\sim 4 \times 10^{-5}$ cm and 21 cm, allow a test of the form of the Doppler expression over a wavelength range of $\sim 5 \times 10^{5}$. If optical and radio values agree, we expect the regression line for these two sets of values to have a zero intercept and slope of unity. A comparison of 21-cm and optical determinations of systemic velocities for 135 galaxies is shown in the central part of figure 6. These data are based on velocities available in the literature through 1970 December as well as unpublished Green Bank values. All velocities are with respect to the Sun, the usual convention for extragalactic studies.

The correlation between optical and 21-cm velocities over the range ~ -400 to $\sim +5000$ km s^{-1} is quite good, except between $\sim$1200 and $\sim +2200$ km s^{-1} where optical values determined from blue-sensitive spectra appear systematically too large by $\sim +100$ km s^{-1} (see insert graph in fig. 6 which contains 40 additional points in the interval 1200–2200 km s^{-1}.) This discrepancy is not present when values derived from *red-sensitive* spectra are used. This separation on wavelength region of the optical data makes it very unlikely that the difference is attributable to the 21-cm measurements. A comparison between red and blue optical values for all galaxies for which such data are available shows this same discrepancy. The red data are primarily from the Burbidges, Page, Rubin and Ford, and de Vaucouleurs; the measurements are usually based on Hα and [N II] emission lines. The blue data are primarily from Humason, Mayall, and Sandage (1956) and Mayall and de Vaucouleurs (1962) and are usually based on measurements of the λ3727 emission line and the H and K absorption lines.

The cause for the discrepant velocities is most likely due to contamination of the H and K galaxian lines by night-sky H and K absorption lines.

Radial velocities determined from 21-cm observations are derived from a frequency difference whereas optically derived radial velocities are based on a wavelength difference. Since $\Delta\lambda/\lambda_0$ does not equal $\Delta\nu/\nu_0$, the 21-cm radial velocities have been converted to the conventional optical expression $c\Delta\lambda/\lambda_0$:

$$c\left(\frac{\Delta\lambda}{\lambda_0}\right)_{21} = c\left(\frac{V_{21}}{c - V_{21}}\right), \tag{7}$$

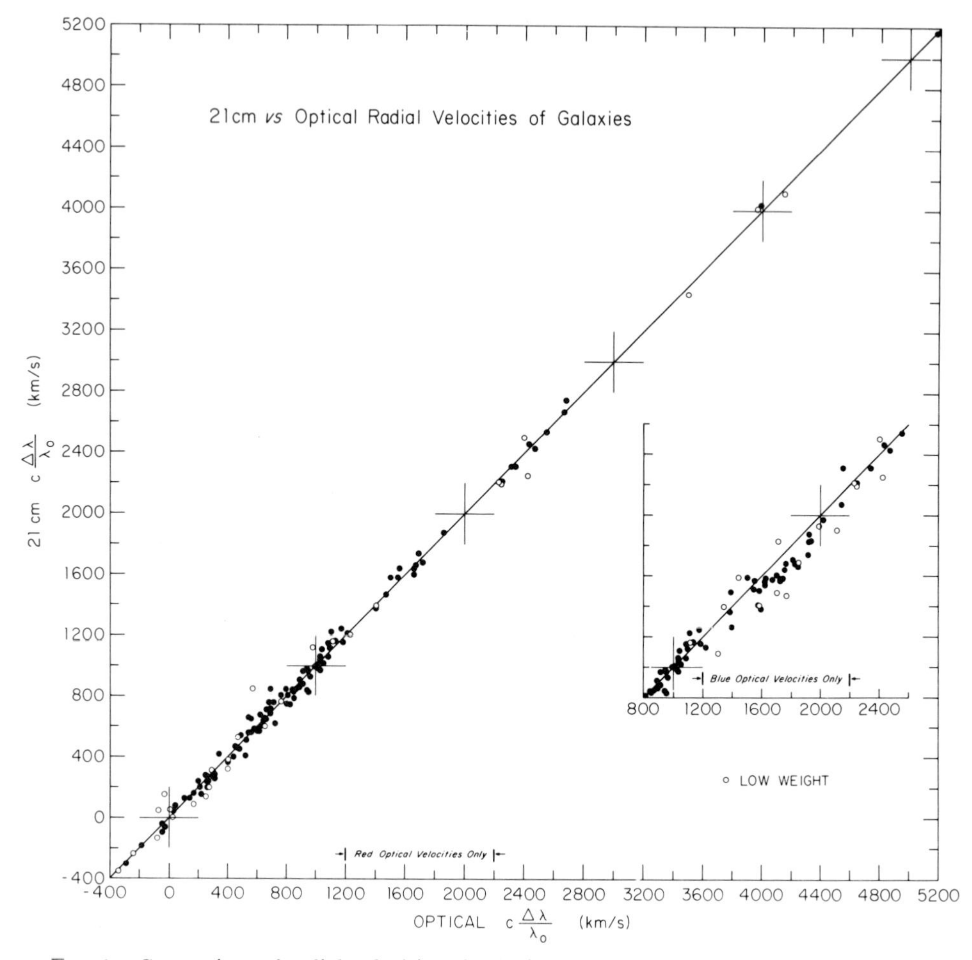

FIG. 6.—Comparison of radial velocities of galaxies measured at 21 cm and at optical wavelengths. The straight line is the mean of the two regressions obtained from a least-squares fit. Open circles mark low-weight points. To aid in reading the graph, the intersections of the coordinates at 1000 km s^{-1} intervals are shown. The insert graph is drawn at the same y-coordinate position as the main graph. Velocities in the range $1200 \leq V_r \leq 2200$ km s^{-1} whose optical spectra were obtained with blue-sensitive plates are shown in the insert. They are not plotted in the main graph. The points outside this velocity interval are common to both graphs. The majority of these points imply too high a radial velocity by ~100 km s^{-1}, an effect not shown by spectra exposed in the red. The main graph in this velocity interval has only "red" spectra values. The explanation of this discrepancy on the blue spectra is most likely due to blends with absorption night-sky H and K lines. The velocity interval over which this effect takes place is noted for both graphs.

where V_{21} is the radial velocity as usually given for 21-cm observations (i.e., $c\Delta\nu/\nu_0$). For the velocity range considered here the conversion correction is small, reaching 85 km s^{-1} at 5000 km s^{-1}. Its importance lies in its being a systematic effect in a test of the form of the Doppler expression.

The two regression lines from least squares solutions for 135 points in the main body of figure 6 (which is based on red optical velocities only for the range 1200–2200 km s^{-1}, and any optical measurement over the rest of the range) are:

$$c(\Delta\lambda/\lambda_0)_{\text{opt}} = -0.3(\pm 7.2) + 0.997(\pm 0.006)(c\Delta\lambda/\lambda_0)_{21}\,,$$

$$c(\Delta\lambda/\lambda_0)_{21} = 5.9(\pm 7.2) + 0.999(\pm 0.005)(c\Delta\lambda/\lambda_0)_{\text{opt}}\,. \tag{8}$$

Standard errors are quoted. A solution in which the low-weight values are given half-weight gives essentially the same results. A mean line of the above two regressions is plotted in this figure. To within the accuracy of the solution, the optical and 21-cm determinations of the systemic velocities for the 135 galaxies in this sample agree. We conclude that the form of the Doppler expression is applicable over a wavelength range of 5×10^5 for the velocity interval ~ -400 to $\sim +5000$ km s^{-1}.

The 21-cm determination of the redshift of Cygnus A (Lilley and McClain 1956) is not included here since this result was later withdrawn (Lilley 1963) after more extensive observations.

5. THE HYDROGEN CONTENT OF GALAXIES

The number, N, of neutral hydrogen atoms in an optically thin column of 1 cm^2 cross-section centered at sky coordinates (θ, ϕ) is given by

$$N(\theta, \phi) = 1.823 \times 10^{13} \int_{-\infty}^{+\infty} T_B(\theta, \phi, V_r)\,dV_r \quad \text{(cgs)}\,, \tag{9a}$$

where the integration is over all radial velocities, V_r. In frequency units (Hz) we have

$$N(\theta, \phi) = 3.848 \times 10^{14} \int_0^{\infty} T_B(\theta, \phi, \nu)\,d\nu\,, \tag{9b}$$

where T_B, the brightness temperature, is proportional to the specific intensity, I_ν, and is defined by

$$I_\nu = 2kT_B(\nu)/\lambda^2\,. \tag{10}$$

In equation (10), k is the Boltzmann constant and λ the wavelength of the hydrogen line. The numerical constants in equations (9) come from an evaluation of the absorption coefficient for atomic hydrogen and include both true absorption and stimulated emission terms (see e.g., Wild 1952; van de Hulst, Muller, and Oort 1954). To evaluate these constants we adopt for the rest frequency of the hydrogen hyperfine transition $\nu_0 = 1420.40575$ MHz, derived from frequency comparisons between a hydrogen maser and a cesium beam tube frequency standard. (The most recent determination of this frequency by Vessot *et al.* [1966] is 1,420,405,751.7864 ± 0.0017 hertz.) The corresponding wavelength, λ_0, is 21.10612 cm; and the nonrelativistic Doppler shift for 1 km s^{-1}, $\Delta\nu = \nu_0/c$, is 4.737963 kHz. Slightly different values for the coefficients in equations (9) will be found in some of the literature due to the use of earlier, less accurate values of ν_0 and other physical constants.

The total hydrogen content in an optically thin source is given by

$$\iint_{\text{source}} N(\theta, \phi) \cos \phi d\theta d\phi = 3.848 \times 10^{14} \iiint_{\text{source},\nu} T_B(\theta, \phi, \nu) \cos \phi d\theta d\phi d\nu . \quad (11)$$

For a small region of the celestial sphere, we may choose the coordinate system such that $\cos \phi \simeq 1$, $\theta = x/D$, and $\phi = y/D$, where D is the distance to the source in centimeters. Then

$$\iint_{\text{source}} N(x, y) dx dy = 3.848 \times 10^{14} D^2 \iiint_{\text{source},\nu} T_B(\xi, \eta, \nu) d\xi d\eta d\nu . \quad (12)$$

To determine the total mass of hydrogen in a source we relate T_B in equation (12) to the parameter obtained in radio measurements, T_A, the antenna temperature. To do this we must first introduce the following terms.

Let $A(\theta, \phi)$ be the effective area of the antenna where A is its value in the direction of the principal axis of the antenna beam ($\theta = 0$, $\phi = 0$) and $f(\theta, \phi)$ is the power pattern normalized with respect to its peak value. The beam efficiency, η_B, another parameter of the telescope, enters in the relation between T_A and T_B and is defined by

$$\eta_B = \eta_R \Omega_M / \Omega , \quad (13)$$

where η_R is the radiation efficiency and allows for losses in the antenna. In the frequency range of interest here, $\eta_R \approx 0.99$. The quantities Ω_M and Ω are the main-lobe solid angle and antenna solid angle, respectively, defined by

$$\Omega_M = \iint_{\text{main beam}} f(\theta, \phi) d\Omega , \quad (14a)$$

and

$$\Omega = \iint_{4\pi} f(\theta, \phi) d\Omega . \quad (14b)$$

Finally, we have from antenna theory

$$\frac{A}{\lambda^2} = \frac{\eta_R}{\Omega} . \quad (15)$$

The antenna temperature is the convolution of the true brightness distribution and the effective area of the antenna. This may be expressed as

$$T_A(\xi, \eta, \nu) = \frac{A}{\lambda^2} \iint_{\text{source}} f(\xi - \xi_0, \eta - \eta_0) T_B(\xi, \eta) d\xi_0 d\eta_0 . \quad (16)$$

The integrated antenna temperature from an extended source is

$$\iiint_{\nu,\text{conv.}} T_A(\xi, \eta, \nu) d\xi d\eta d\nu = \frac{\eta_R}{\Omega} \iiint_{\nu,\text{source}} T_B(\xi, \eta, \nu) d\xi d\eta d\nu \iint_{\text{main beam}} f(\xi, \eta) d\xi d\eta . \quad (17)$$

The integration on the left is over all frequencies, ν, as well as the convolution of the main beam and the source denoted "conv.". This integral may be evaluated numerically from the observed contours. The first integral on the right is over the source, and the second is over the main beam and yields the main-lobe solid angle, Ω_M. From equations (13, 14a, and 17) we obtain

$$\iiint_{\nu,\text{conv.}} T_A(\xi, \eta, \nu) d\xi d\eta d\nu = \eta_B \iiint_{\text{source}} T_B(\xi, \eta, \nu) d\xi d\eta d\nu . \quad (18)$$

An assumption, not explicitly stated above, has been made in deriving equation (18), namely, that sidelobes of the antenna do not fall on the source. This is a valid assumption for extragalactic hydrogen measurements, since any sidelobes that might fall on the source would in general see a different velocity region of the galaxy (see § 2) and therefore make no contribution to the particular frequency being measured by the main beam. A relation of the form of equation (18) can be derived without making the above assumption; in such a case η_B may be redefined to include some of the forward sidelobes.

Substituting equation (18) into equation (12), we obtain

$$\iiint_{\text{source}} N(x, y)dxdy = \frac{3.848 \times 10^{14} D^2}{\eta_B} \iiint_{\nu,\text{conv.}} T_A(\xi, \eta, \nu)d\xi d\eta d\nu \quad (\text{cgs}) . \tag{19}$$

Rewriting equation (19) in units of solar masses and parsecs, we obtain

$$\frac{M_{\text{H I}}}{M_\odot} = \frac{3.082 \times 10^{-6}}{\eta_B} D^2 \iiint_{\nu,\text{conv.}} T_A(\xi, \eta, \nu)d\xi d\eta d\nu . \tag{20}$$

The right side of equation (20) is determined from the observed contour diagram and is often in the form $T_B(\xi, \eta) = \eta_B{}^{-1} \int T_A(\xi, \eta, \nu)\, d\nu$. The observational data are frequently in units of km s^{-1} and minutes of arc rather than hertz and radians. In this case, equation (20) takes the form

$$\frac{M_{\text{H I}}}{M_\odot} = \frac{1.236 \times 10^{-9}}{\eta_B} D^2 \iiint_{V_r,\text{conv}} T_A(\xi', \eta', V_r)d\xi' d\eta' dV_r . \tag{21}$$

Equation (21) has been derived with the assumption of small optical depth, $\tau \ll 1$, throughout the observed source. Evidence from our own Galaxy as well as the Magellanic Clouds shows that this is not the case for certain localized regions, and equation (21) therefore gives only a lower limit to the hydrogen content of a galaxy. There are no quantitative data presently available which would indicate how close the actual hydrogen content lies to this lower limit. It is generally assumed that most of the hydrogen is optically thin.

Special cases of overall high optical depth can occur when the radial velocity gradient over a region is very small and/or when the line of sight through the 21-cm emitting region is particularly long. The latter situation may occur in galaxies seen at high inclination. Allowance for a nonnegligible optical depth is made in the more general expression

$$\frac{M_{\text{H I}}}{M_\odot} = 1.236 \times 10^{-9} D^2 F \iiint_{V_r,\text{conv}} T_B(\xi', \eta', V_r)d\xi' d\eta' dV_r , \tag{22}$$

where F is given by

$$F = \langle\tau\rangle/(1 - e^{-\langle\tau\rangle}) . \tag{23}$$

In general, τ, and therefore F, will vary over the projected image of the galaxy and should be evaluated at all ξ', η'. In such cases the changes in equations (22) and (23) are obvious. At present, only approximate methods for evaluating τ are available. These are discussed later in this section.

For a source that is small with respect to the main beam, we have from equations (10) and (12)

$$\iint_{\text{source}} N(x, y)dxdy = \frac{3.848 \times 10^{14}\lambda^2 D^2}{2k} \iiint_{\nu,\text{source}} I(\xi, \eta, \nu)d\xi d\eta d\nu \,. \tag{24}$$

Using the definition of flux density,

$$S_\nu = \iint_{\text{source}} I(\xi, \eta, \nu)d\xi d\eta \,, \tag{25}$$

and changing units as before, we obtain from equation (24)

$$\frac{M_{\text{H I}}}{M_\odot} = 2.356 \times 10^{19} D^2 \int_{-\infty}^{+\infty} S_\nu dV_r \,. \tag{26}$$

In equation (26) D is in parsecs, S_ν is the flux density in mks units (watts m^{-2} Hz^{-1}), and V_r is in km s^{-1}. The flux density may be derived without knowledge of the telescope efficiency by calibrating recorder deflection against known source fluxes. The antenna temperature may be inserted in equation (26) through the relation

$$P = kT_A = \tfrac{1}{2}S_\nu A = \tfrac{1}{2}S_\nu \eta_A A_g \,, \tag{27}$$

where P is the power per unit bandwidth. The ratio of the effective aperture of the antenna, A, to its geometric aperture, $A_g = \pi d^2/4$, is the aperture efficiency, η_A. The beam and aperture efficiencies are related by

$$\eta_A = \frac{4\lambda^2}{\pi \Omega_M d^2} \eta_B \,. \tag{28}$$

Epstein (1964*b*) has considered the optical depth for a highly simplified model. He finds that τ will become large for his model galaxy when the inclination is close to 90° (edge-on). A more realistic, but still approximate, estimate of the average optical depth may be made from the galaxy parameters derived through model fitting. The optical depth τ is given by

$$\tau = -\ln(1 - T_B/T_S) \,, \tag{29}$$

where T_B is the brightness temperature and T_S the spin temperature. The galaxy model allows T_B to be evaluated; some average value for the spin temperature must be assumed. (The uncertainty in T_S as well as its possible variation over the galaxy is a serious limitation to the accurate determination of τ.) The relation between brightness temperature and the observed quantity, antenna temperature, is given by equation (16). A more convenient form is

$$T_B = \frac{1}{\eta_B} \frac{\Omega_M}{\Omega_S} T_A \,, \tag{30}$$

where η_B and Ω_M have been defined previously and Ω_S is the effective source solid angle (Baars, Mezger, and Wendker 1965). For a velocity interval $\pm\Delta V_r$ about V_r, the effective source solid angle is

$$\Omega_S(V_r) = \iint B(\xi, \eta, V_r) f(\xi, \eta) d\xi d\eta \,, \tag{31}$$

for a brightness distribution given by $T_B(\xi, \eta, V_r) = T_B B(\xi, \eta, V_r)$ and a similarly normalized antenna power pattern $f(\xi, \eta)$.

To evaluate equation (31) we require the H I brightness distribution as a function of radial velocity. This is directly obtainable from the model galaxy parameters. A set of such parameters will give a velocity-hydrogen distribution picture similar to that shown in figure 2. The combination of velocity loci and hydrogen distribution defines the source size and shape, $B(\xi, \eta, V_r)$, for various radial-velocity intervals. The numerical integration of equation (31) may be simplified when $B(\xi, \eta, V_r)$ is small with respect to the beam by assuming that $f(\xi, \eta)$ is constant over the source. This procedure has been applied (Roberts 1968*b*) to NGC 4631, an edge-on galaxy, where the average optical depth is found to be $\langle\tau\rangle = 0.65$. The correction factor in equation (22) is then $F = 1.35$; i.e., the total hydrogen mass calculated for small optical depth must be increased by 1.35.

Such a detailed approach is not possible for most hydrogen-mass measurements. Instead a statistical approach may be taken by using suitably normalized hydrogen masses determined for galaxies of different inclination. That such a correction is required as well as the amount of correction is shown in figure 7. Plotted in the upper half of this figure is the ratio of the observed hydrogen mass, $M_H/M_\odot$, to the absolute photographic luminosity, $(L_0/L_\odot)_{pg}$ as a function of inclination. The luminosity has been corrected for foreground galactic extinction and extinction due to the inclination of the galaxy. The details of these corrections are described by Holmberg (1958) and Roberts (1969). That no inclination effect remains in the corrected luminosities is shown by the lower plot of figure 7, which displays surface brightness as a function of inclination. No trend is evident while a systematic decrease in the hydrogen-to-luminosity ratio at high inclinations is present. The hydrogen content of high-inclination systems is underestimated, and the relation in figure 7*a* is used to derive a statistical correction for this underestimation. The line shows the form of the adopted correction. The data (97 values) have been averaged over inclination intervals of 10°, and the mean and its standard error are shown. The data were first analyzed as a function of the different structural types, and all showed a decrease in the ratio starting at $i \sim 60°$, except type I0 (too few) and types Sa and Sab for which no systems with $i \gtrsim 60°$ were available in the present sample. The amount of the decrease and hence the size of the optical-depth correction appeared to differ among the various types, but the data are too few to derive such a correction as a function of type. (This point will be discussed again in § 7.) Instead a zero-point correction based on the average value of M_H/L_0 for each type and for $i < 60°$ was applied such that all the data were on the scale of the Sd–Sm systems. The zero-point corrections are noted in the figure. Such a procedure for deriving optical depth corrections has been described previously by Roberts (1969). The present values are somewhat different because of a larger set of data and because of improvements to the observed hydrogen masses of Bottinelli, Gougenheim, Heidmann, and Heidmann (1968) given by Gouguenheim (1969).

The hydrogen masses discussed later include a correction for optical depth for those systems with $i > 60°$ except for types S0, S0a, I0, and Ir.

It should be noted that Galactic hydrogen measurements indicate that cool clouds with a high optical depth ($\tau > 1$) are relatively common. Further, the velocity dispersions of absorption and most emission profiles differ significantly. It has been suggested (Clark 1965) "that the hydrogen seen in emission is not entirely the same hydrogen as

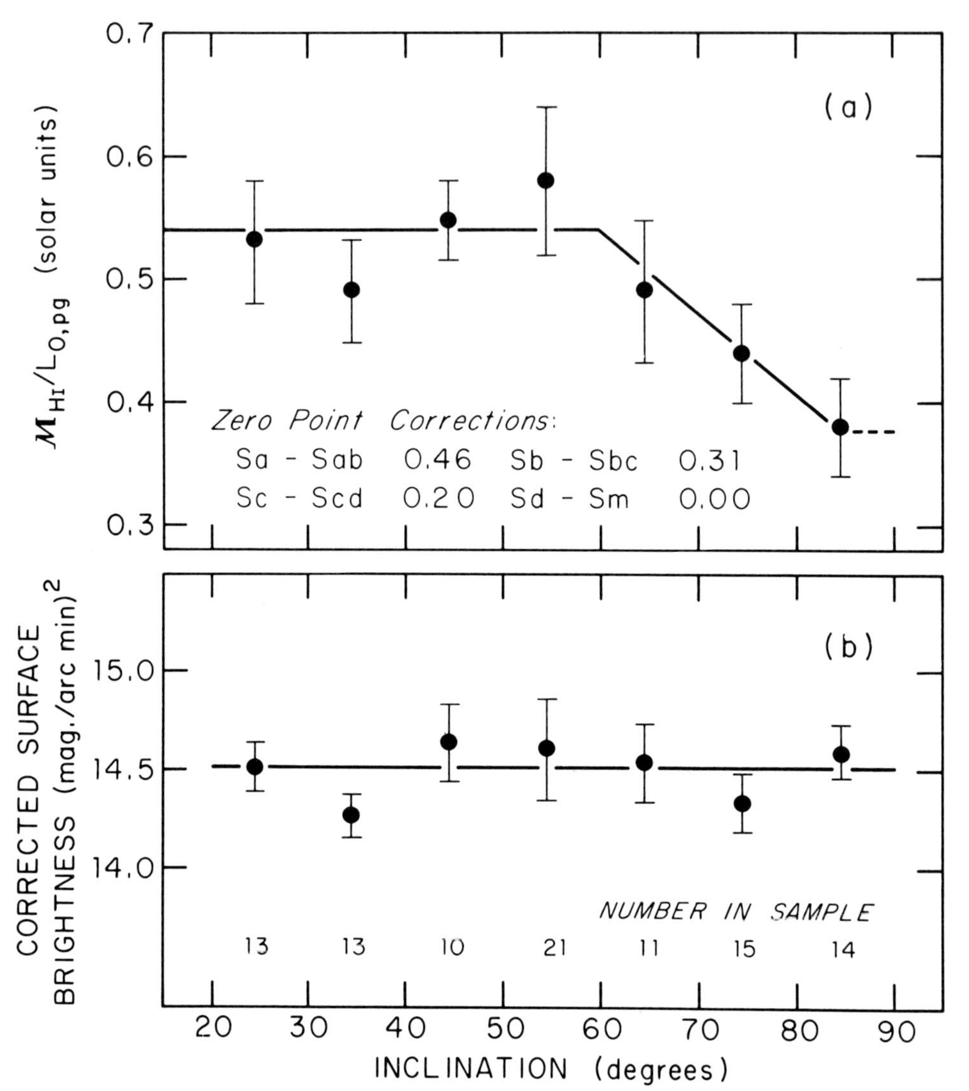

FIG. 7.—(*a*) Values of the ratio: observed hydrogen mass to corrected photographic luminosity, shown as a function of inclination. This ratio is dependent on structural type. To improve the statistics, values for all types of spirals are combined with the zero-point corrections shown. The error bars represent the standard deviation of the mean. The line drawn in this figure represents the form of the inclination correction applied to the observed hydrogen masses. There is a suggestion that the correction may be dependent on structural type. (*b*) Corrected surface brightness versus inclination. The luminosities used in forming the ratio have been corrected for inclination. This panel shows that the decrease in the ratio in panel (*a*) is not due to a residual effect in the luminosity correction.

that seen in absorption." Clark proposes a model in which hot (>1000° K), low-opacity hydrogen is distributed as a continuous medium surrounding cooler clouds whose temperature is less than 100° K. The lack of realistic models for the detailed hydrogen distribution, including the above one, prevents an evaluation of the uncertainty in present hydrogen-mass determinations; a factor as large as 2 may not be unreasonable.

Another assumption made in deriving the above expressions for the hydrogen content of a system is that the effects of interaction between line and continuum radiation are negligible. Epstein (1964*a*) has evaluated this assumption. He considers three cases: (1) an H I region in a surrounding continuum source; (2) a continuum source in front of the H I source; and (3) a continuum source behind the H I source. The conclusions are strongly dependent on the detailed geometry (solid angles of the antenna beam, H I source, and continuum source) and the optical depth of the continuum source. For "normal" galaxies, i.e., intrinsically weak radio emitters, the effects of continuum radiation will generally be slight. The derived H I mass will also be essentially correct for those weak "radio galaxies" where the continuum source is small with respect to the optical dimensions of the system, although exceptional cases may occur.

6. THE DISTRIBUTION OF HYDROGEN WITHIN A GALAXY

For the larger galaxies, filled-aperture observations supply sufficient relative resolution to outline the gross features of the H I distribution. These features may be categorized as (1) main-body distribution; (2) extent; (3) exterior concentration or companions; (4) a warp or bending of the plane; and (5) bridges.

That there is much structural detail in the H I distribution in a galaxy is evident from the extensive studies of our own Galaxy and the Magellanic Clouds. The latter systems yield the highest linear resolution of any extragalactic hydrogen observations made thus far. At the distance of the Magellanic clouds the half-power beamwidth of the Parkes 210-foot antenna corresponds to 0.2 kpc. Radio maps of the H I brightness distribution (McGee and Milton 1966*a*, *b*; Hindman 1967; Hindman and Balnaves 1967) allow meaningful comparisons with optical and radio continuum emission, H II regions, and stellar concentrations. Detailed velocity comparisons are also possible. Hindman (1964) finds that optically measured radial velocities of stars in the Small Magellanic Cloud (SMC) are in good agreement with the peaks of H I velocity profiles in the directions of these stars. McGee (1964) obtains a similar result for stars in the LMC, although a small percentage of the comparisons show relatively large differences. McGee suggests that these latter stars may be moving in a direction perpendicular to the plane of the LMC while the majority of the stars in his sample have motions in the plane of the galaxy. McGee also compares optical velocities of H II regions with 21-cm velocities. The agreement is excellent, with an average difference of 1.3 km s^{-1} and a standard deviation of 6.4 km s^{-1}.

Double-peaked profiles are found in one region of the LMC and in several areas of the SMC. Hindman and Sinclair (1965) have interpreted the latter (SMC) as due to three expanding shells of H I with diameters ranging from 1 to 2 kpc. Each shell contains an H I mass of between 0.5 and 1.0 $\times 10^7 M_\odot$, and each is expanding with a velocity of 20–25 km s^{-1}. McGee (1965) finds that the hydrogen distribution of the LMC consists mainly of 52 rather large concentrations grouped into four principal bodies. He feels that the various data relating to the LMC can be interpreted in terms of "two quite large spiral arms in one plane and two smaller, shorter arms in another plane."

The linear resolution available for the Andromeda Nebula (M31) with a filled aperture is an order of magnitude poorer than for the Magellanic Clouds. The half-power beam-width of the 300-foot antenna corresponds to 2 kpc at the distance of this system, and only gross features in the H I distribution may be observed. (Note that foreshortening along the minor axis direction of a galaxy results in a significant line-of-sight integration along the plane of the galaxy.) A striking feature in the H I distribution in Andromeda

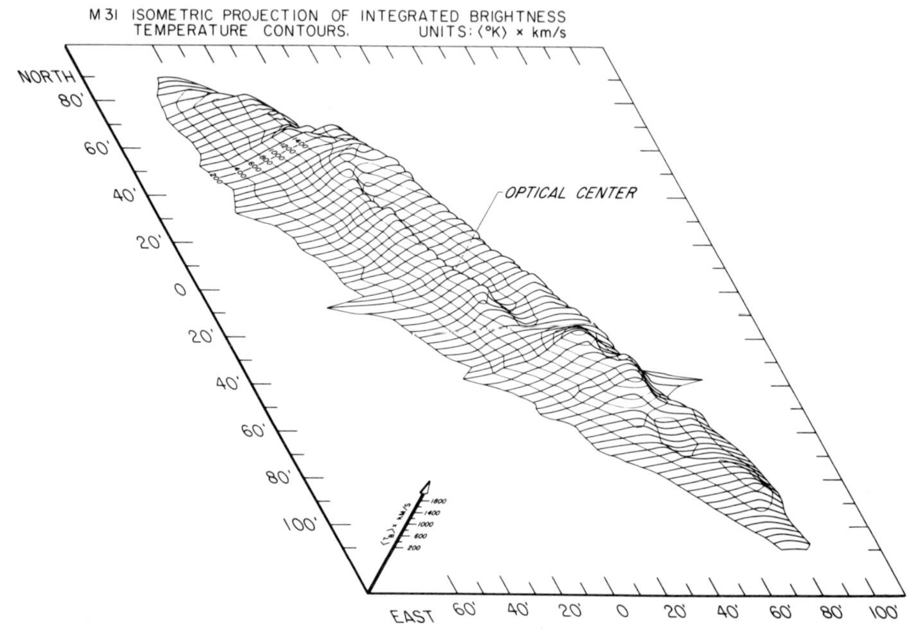

FIG. 8.—An isometric projection of integrated brightness temperature contours for M31. The "height" above the plane is in units °K km s^{-1} and for small optical depths is proportional to the surface density of hydrogen. The ridge line defining the hydrogen ring in M31 is clearly evident. The observing beam width is 10′.

is the pronounced minimum in the central region of this system (Burke, Turner, and Tuve 1964; Roberts 1966*a*, 1967; Brundage and Kraus 1966; Gottesman, Davies, and Reddish 1966). This is clearly visible in figure 8, which is an isometric presentation of the integrated brightness temperature contours as mapped by a 10′ beam.

To simplify terminology, a distribution with a central minimum as in M31 will be referred to as a "ring." The prominent optical spiral-arm features in M31 are embedded in this H I ring. This can be seen in figure 9, where the locations of optical spiral arms crossing the major axis (Baade 1958) are plotted on a major-axis profile of the integrated H I distribution. There is surely much detail in the H I distribution that is smeared out by the antenna beam. Similarly, the depth of the central minimum is underestimated. Presently available resolution can give only the general picture for M31 of a thick ringlike distribution of hydrogen enveloping the prominent optical arms.

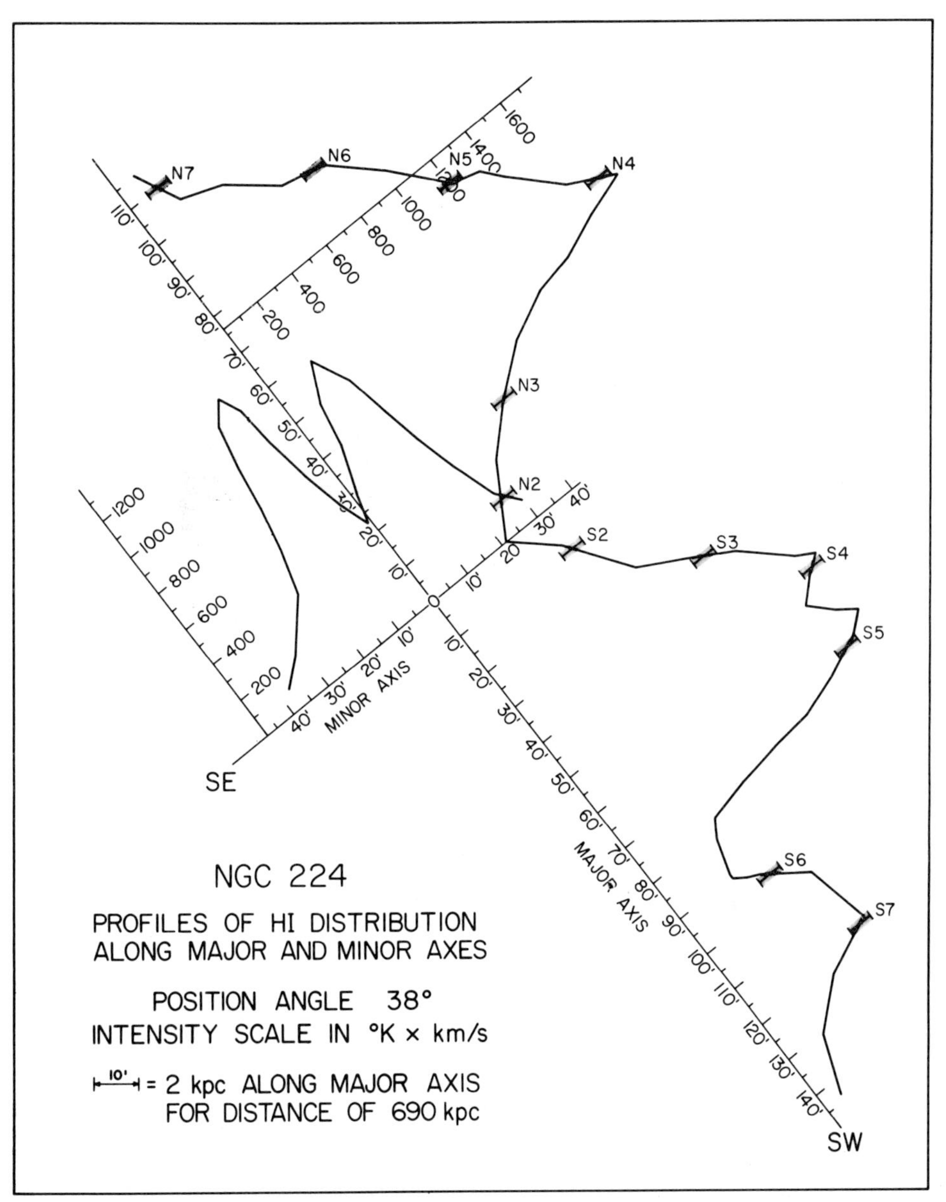

FIG. 9.—Profiles of the integrated brightness temperature along the major and minor axes of M31. The vertical bars on the major axis profile mark the locations of spiral arms crossing this axis (Baade 1958).

A similar central minimum in the H I surface density distribution exists in our own Galaxy (von Woerden as quoted by Oort 1965). This ring, which is quite thick in a radial direction, shows a maximum surface density at about 10 kpc from the center. The full width at half-intensity is also about 10 kpc. It should be noted that this Galactic surface density distribution includes the projection of nonplanar hydrogen. The location of the prominent optical spiral arms of the Galaxy is less certain. There are optical spiral features located in the general vicinity of the Sun. However, the giant regions of ionized hydrogen, as defined by radio recombination-line observations, appear concentrated at $\sim$4 to $\sim$6 kpc from the center (Reifenstein, Wilson, Burke, Mezger, and Altenhoff 1970). The distribution of thermal continuum radiation also implies a similar central location. Presumably, both of these data refer to a prominent spiral arm of the Galaxy. If this is so, then the "optical" arm lies significantly interior to the peak of the H I surface density distribution—a situation quite different from M31.

A search in other galaxies for this gross feature of the H I distribution is of obvious importance. In M31 the ring radius is $52' \simeq 10$ kpc in the plane of the galaxy; as seen on the sky the radii are $52' \times 11'$. Using a Gaussian curve to approximate the profile of the ring gives $\sigma = 26'$ as a measure of its thickness.

Beam smearing, on the scale of 10′, will make the detection of similar ring structures more difficult in galaxies of angular size significantly smaller than M31. In M33, a system whose angular dimensions are about half those of M31, a ring is visible in the integrated H I contours mapped with a 10′ beam. However the ringlike structure is not nearly so obvious in these data, and its presence is indicated by only a slight depression in the contour levels of the central region (Gordon 1971). Burke (1964), Davies (1964), and Meng and Kraus (1966) have also called attention to the ring distribution in M33. However, Baldwin (1970), using a $1'.5 \times 3'.0$ synthesized beam, finds that the H I distribution in the central region of M33 lacks the central dip seen in broader-beam observations. He suggests that a more nearly uniform distribution of hydrogen holds over this region.

Other galaxies showing a central minimum in their hydrogen distribution include NGC 2403 (Burns and Roberts 1971, Shostak 1971) and M101 (Roberts 1968*a*; Guélin and Welichew 1970; Rogstad and Shostak 1971). These two systems are similar to the Galaxy in that the peak of the hydrogen distribution is more distantly placed than the *prominent*, optical spiral arms. However, these arms do lie within part of the ring since it has such great width. If followed to faint brightness levels, spiral arms can often be traced to large distances from the center, and sections may lie near the peak of the H I ring. The prominent arms referred to here are the *high surface brightness* arms. These are the ones that lie interior to the peak of the H I ring in M101 and presumably in our Galaxy. In contrast to these examples of central minima the latest-type spirals, the irregular galaxies, appear to have a centrally concentrated hydrogen distribution that is essentially coincident with the optical image.

Although the sample is woefully small and contains only one intermediate-type galaxy, M31, there is a suggestion of a correlation of (1) hydrogen distribution with type, and/or (2) hydrogen distribution relative to optical arms with type. However, an equally good case can be made for a correlation with total mass rather than type since, in this sample, M31 is the most massive, the irregular galaxies the least massive, and the others intermediate in mass.

Several clarifying points regarding the ring distribution are required. These involve terminology and possible observational effects. First, a smooth circular distribution in the H I ring is not implied. Rather, large density fluctuations and a noncircular distribution may be expected; the H I ring may well be H I spiral arms seen with poor relative resolution. Further, a complete lack of H I is not implied in those instances where concentrations of ionized hydrogen (i.e., the prominent optical spiral arms) lie interior to the H I peak. Rather, we may expect local concentrations of neutral hydrogen near and about H II regions, as shown by Orion in our Galaxy (Menon 1958; Gordon 1970) and the H II regions in the Large Magellanic Cloud (McGee 1964; McGee and Milton 1966*a*).

Although there are some H II regions in the ring, an obvious question is: Why haven't the prominent spiral-arm features formed in this higher-density region? Lin (1970) has suggested that the density wave and its related shock front that he proposes for spiral-arm structure does not reach out this far (except in M31). Another possibility would have the volume density of H I smaller in the ring (i.e., the ring is thicker in the z-direction) than in the interior portions.

The appearance of a central minimum in the H I distribution of spirals could be due to (1) an observational effect related to the small-scale distribution of hydrogen, (2) the hydrogen being in some form other than neutral and atomic, or (3) a true deficiency. A brightness contour map is based on the observational quantity of integrated (over velocity) beam-averaged brightness temperature. The minimum in the central region could arise from an observational effect in which much of the H I there is concentrated into small regions of high optical depth. Beam dilution would then give the appearance of a minimum in the derived contours. Possibility (2) would have the hydrogen primarily in a molecular or an ionized form. Regardless of the correct explanation, we are forced to the conclusion that the distribution of amount and/or form of hydrogen differs with galaxy type or total mass.

The neutral-hydrogen distribution of a galaxy extends beyond its usually measured optical dimensions. Comparably normalized dimensions are significantly larger in neutral hydrogen than those measured at optical wavelengths. This was first clearly described by Dieter (1962*a*, *b*) for M33 and M101. Similar conclusions have been derived for those other galaxies where the relative resolution is adequate. A striking example is the irregular galaxy NGC 6822 (Roberts 1971). Here the hydrogen extent is more than 4 times greater than the optical size measured to a brightness level of 26.5 mag per square arc second. Bottinelli (1971) has compared the east-west hydrogen and optical extent of 56 galaxies. She finds that the ratio of these two dimensions varies systematically with structural type in the sense that a relatively larger H I size is found in later-type galaxies. This correlation balances the systematic trend of average H I surface density derived from optical dimensions (Epstein 1964*a*; Roberts 1969; Gougenheim 1969). Instead, an essentially constant H I surface density is suggested by Bottinelli: one independent of type over the range Sb–Ir.

The first extragalactic hydrogen measurements, those of the Large and Small Magellanic Clouds, showed a considerable displacement of the centroid of H I from both the centroid of bright stars and the core center (Kerr, Hindman, and Robinson 1954). This effect has been noticed in many other galaxies and summaries are given by Guélin and Weliachew (1970) and Bottinelli (1971). Beale and Davies (1969) suggest that this asymmetry is due to tidal interaction with a neighboring galaxy. However, both Guélin and Weliachew (1970) and Bottinelli (1971) are unable to find possible perturbers for at least some of the H I asymmetric galaxies. Bottinelli's criterion for a potential neighbor is one having at least 10 percent of the luminosity and nearer than 7 times the optical diameter of the galaxy considered.

The hydrogen asymmetry in M101 contains significant structural detail. A comparison of the 4′ map obtained by Rogstad and Shostak (1971) with a 10′ map (Roberts 1971) emphasizes the effects of angular resolution in our understanding of these asymmetries. Even better relative resolution is achieved for the 10′-M31 combination (figs. 8, 9). An asymmetry in the southwest direction, along the major axis, is clearly evident. Here the asymmetry is due to a concentration of $\sim 10^8 M_\odot$ of hydrogen: a feature first delineated by Burke, Turner, and Tuve (1963). For a poorer relative resolution, as in the case of M101, such a feature would appear only as an asymmetry in the outer brightness temperature contours. Similar asymmetries or distortions are observed in NGC 300 (Shobbrook and Robinson 1967), NGC 2403 (Burns and Roberts 1971), NGC 5236 (Lewis 1968) and M33 (Gordon 1971). In all these cases the shift of the H I major axis with respect to the optical major axis is in the sense of rotation of the galaxy.

It would appear that the H I asymmetries and the concentrations may well be the same phenomenon—the distinction being only observational. There are several possible explanations for such features other than tidal interaction. These are: (*a*) a warp or bending of the plane similar to that observed in the Galaxy, (*b*) a nonaxisymmetric expansion of the hydrogen in the plane with a stable region at the outer edge where the hydrogen collects, (*c*) hydrogen companions. Gordon's (1971) study of M33 with the 300-foot telescope gives the most detailed description of such hydrogen features. In this case the available radial-velocity information places severe constraints on the first two possible explanations above. Gordon finds that an expansion velocity of ~ 70 km s^{-1} at a distance of ~ 12 kpc is required, or alternatively a warp of the plane with a semiamplitude of 7 kpc at a similar distance from the center. For the latter case a significant expansion term is still necessary to account for the observed velocity field. A pure z-motion term would have to be of the order of 80 km s^{-1}. These numbers all seem too high, the corresponding time scale seems too short, and the driving mechanism remains a serious unknown.

These various objections are removed if the asymmetries are not physically connected to the plane, i.e., a model in which the concentrations are self-contained masses of neutral hydrogen in arbitrary orbit about the galaxy. The asymmetry in the brightness distribution would then be due to beam smearing and the distortion in the radial-velocity map attributable to the radial component of the orbital velocity. Alternatively, such companions could be coplanar with the parent galaxy and have a high expansion term. A similar situation exists for the bright optical companion of M51, NGC 5195, where the radial-velocity difference of these two systems can be explained only by an expansion term or by an orbit out of the plane of M51 (Roberts and Warren 1970) if the two are in a closed orbit at all.

Examples have been found of a "bridge" or "link" of hydrogen existing between two, obviously physically close, galaxies. The first such bridge discovered is that between the Large and Small Magellanic Clouds (Hindman, Kerr, and McGee 1963). Significant detailed structure has been found within this bridge (Turner 1970). There are also suggestions that a link exists between our Galaxy and the Magellanic Clouds. Other examples include NGC 4631 and 4656 (Roberts 1968*b*; Weliachew 1969) and the triple system M81–M82–NGC 3077 (Roberts 1971) where hydrogen appears to envelop all three galaxies.

Two edge-on ($i \simeq 90°$) galaxies have been searched for hydrogen at high-z distances, i.e., for halo hydrogen (Roberts 1966*b*). The systems are NGC 4244 and NGC 7640. Only upper limits were found. For $z \geq 6$ kpc, the number density is $< 10^{-3}$ cm^{-3}. This value was derived within the framework of the following two assumptions: (1) the radial-velocity range in the halo is similar to that in the plane of the galaxy since the integration over velocity was for a range only a bit larger than the observed range in the plane; and (2) the spin temperature is larger than the background temperature.

To summarize:

1. The neutral hydrogen in spirals is not centrally concentrated. Rather, a minimum in the projected H I density is found at the center. The maximum occurs beyond the location of the prominent (high surface brightness) optical spiral features in Sc-type systems. In M31, an Sb, the arms are approximately coincident with the H I maximum. In irregular-type galaxies the projected H I density is centrally concentrated.
2. The hydrogen extent of a late-type galaxy is significantly larger than its comparable optical dimensions.
3. Many (possibly all) spirals show distortions in the outer parts of their hydrogen distribution and radial-velocity field. There are a variety of possible explanations including hydrogen companions and/or a warp in the plane.
4. Hydrogen bridges or links have been found between galaxies whose separation is of the order of a few tens of kiloparsecs.

7. INTEGRAL PROPERTIES OF GALAXIES

There are four integral properties of galaxies to which we can assign quantitative values. These are:

1. luminosity $\propto$ (distance)^2,
2. color, independent of distance (for low-redshift systems),
3. total mass $\propto$ (distance),
4. total hydrogen mass $\propto$ (distance)^2.

Their dependence on distance is noted above, and it is this dependence which is often the basis for the greatest uncertainty in individual values. Additional uncertainty is introduced via observational errors, the assumptions inherent in deriving these values, and the statistical nature of the corrections made to the data to place them onto a consistent system, i.e., independent of inclination, galactic latitude, etc. For these reasons less weight should be placed on individual values than on the average value for a given structural type.

A summary of presently available data on spiral- and irregular-type galaxies is given in table 1. These are from a variety of sources as noted in the references (col. [12]). Both optical and 21-cm determinations of the total mass are included. In a number of cases the former include a correction (Roberts 1969) to yield a total mass from the derived observed mass (which generally refers only to the mass interior to the farthest measured point). Corrections have also been applied to the Nancay and Jodrell Bank hydrogen masses to place them onto a system consistent with other determinations. Luminosities and colors are on the international photometric system and refer to a face-on orientation (zero inclination) and have been corrected for Galactic extinction and reddening (Holmberg 1958; Roberts 1969). Details as to distances, structural types, etc., will be found in Roberts (1969).

The various interrelations among these integral properties are discussed below. They refer only to spiral and irregular-type galaxies. As a matter of convenience the word "galaxies" will refer only to these types. Only when necessary will a type designation be introduced.

The data in table 1 and the relations derived from them suffer from serious selection effects. Many of these galaxies have been chosen for study on the basis of apparent magnitude and/or structural type. The sample is therefore biased toward the bright end of the galaxian luminosity function and toward later-type systems (which have a higher percentage of their total mass in the form of neutral hydrogen and are therefore easier to detect). The sample is not representative in any parameter of a unit volume of space. However, by its very nature it is representative of the "typical" normal galaxy. This distinction must be borne in mind in making use of these data.

7.1. Total Mass and Absolute Photographic Luminosity

The total mass-luminosity relation for galaxies is shown in figure 10. Several features stand out. (1) There is a well-defined relation in the sense that more massive galaxies are more luminous. This is discussed further in § 7.3. (2) The scatter in both coordinates is quite large, often a factor of 10. Much of this is due to the uncertainties in determining total masses and luminosities as well as in the distances adopted. (A change in distance will move a point along a line of slope $\frac{1}{2}$.) It would be surprising if all of the scatter is due to such uncertainties, and there may well be an intrinsic width to the M_T/L_0 diagram. (3) With the exception of the irregular and possibly the Sm systems, there is no clear-cut separation according to structural type. The irregulars are clearly low-mass, low-luminosity systems; they do, however, intermix at the upper end with spirals.

The slope of these data is the mass-to-luminosity ratio. Values are given in table 2 and displayed in figure 11. There is no striking variation in this ratio with structural type, although the earlier-type galaxies may have lower values in the average. This is contrary to previous views (Roberts 1963), based on a far smaller sample, in which it was thought that the values of M_T/L_0 increased for earlier-type galaxies. (Some elliptical galaxies, the more luminous ones, do have a high M_T/L_0; see e.g., Poveda 1961.)

Total masses as a function of type are shown in figure 12. This figure shows more clearly the absence of any trend of mass with type with the exception of the irregulars and probably the Sm galaxies (which on the Hubble classification system would usually be called irregulars also). The frequently expressed view that the early-type spirals are the most massive is not supported by these data. The lack of low-mass spirals (several $\times 10^9 M_\odot$) in these data most likely reflects the selection effects noted above.

TABLE 1

CATALOG OF INTEGRAL PROPERTIES

NGC	Type	Distance (Mpc)	Inclination (°)	Color C_0	$L_{0,pg}$ ($10^9 L_\odot$)	M_H ($10^9 M_\odot$)	M_T ($10^{10} M_\odot$)	M_H/L_0	M_T/L_0	M_H/M_T	Ref.
45	Sdm	2.4	50	...	0.54#	...	2.1	...	39.0	...	1
55	SBm	2.4	84	...	20. #	5.3	2.9	0.26	1.4	0.18	2
157	Sbc	15.1	59	0.55	18.	...	5.0*	...	2.8	...	3
224	Sb	0.69	77	0.62	74.	7.6	21.	0.10	2.8	0.036	4, 5
247	Sd	2.4	75	0.43#	3.5	0.83	1.8	0.24	5.1	0.046	6
253	Sc	2.4	78	...	19. #	2.4	12. *	0.13	6.3	0.020	7, 8
300	Sd	2.4	42	...	1.7#	3.4	3.2	2.0	19.0	0.11	9
428	Sm	7.9	38	0.21	2.5	1.2	2.1	0.48	8.4	0.057	10
598	Scd	0.72	55	0.29	4.2	1.6	4.9	0.38	12.0	0.033	11
613	SBbc	15.0	47	0.52#	26. #	...	13. *	...	5.0	...	12
628	Sc	7.8	35	0.29	16.	11.	3.9	0.69	2.4	0.28	1, 13
681	Sab	17.3	81	0.59#	16. #	...	1.8*	...	1.1	...	14
772	Sb	16.6	55	0.48	27.	8.5	23.	0.31	8.5	0.037	10
925	Sd	6.8	53	0.21	8.5	4.4	4.9	0.52	5.8	0.090	15
972	I0	16.5	68	0.55#	15. #	2.8	1.3*	0.19	0.87	0.22	16
1023	SB0	6.3	77	0.76	7.3	0.34	7.4	0.047	10.0	0.0046	17
1055	Sb	11.0	77	0.60	22.	4.2	9.7	0.19	4.4	0.043	17
1068	Sb	11.0	39	0.59	36.	1.3	21.	0.036	5.8	0.0062	18
1084	Sc	14.2	65	0.28#	20. #	...	1.5*	...	0.75	...	19
1097	SBb	12.1	50	0.52#	31. #	10.	81.	0.32	26.0	0.012	17
1140	Ir	15.0	62	0.15#	5.5#	3.8	...	0.69	...	...	10
1156	Ir	6.3	52	0.23	1.9	1.1	0.63	0.58	3.3	0.17	10
1291	SB0a	8.0	49	...	19. #	6.6	...	0.35	...	...	57
1300	SBbc	13.8	55	0.37	17.	1.6	...	0.094	...	...	20
1326	SB0	16.5	47	...	15. #	2.3	...	0.15	...	...	20
1332	S0	14.0	76	0.88#	13. #	1.3	...	0.10	...	...	20
1365	SBb	15.1	66	0.32#	84. #	15.	31.	0.18	3.7	0.048	17
1507	SBm:	8.0	85	...	2.9#	0.70	1.3	0.24	4.5	0.054	17
1518	SBdm	8.0	61	0.30#	2.3#	1.5	...	0.65	...	...	20
1532	SBab:	10.5	86	...	26. #	3.0	14.	0.12	5.4	0.021	17
1569	Ir	2.5	65	0.23	0.83	0.17	...	0.20	...	...	10
1637	Sc	5.6	35	0.35	2.3	0.72	1.3	0.31	5.7	0.055	10
1744	SBd	10.0	49	0.17#	5.0#	5.1	9.2	1.0	18.0	0.055	17
1792	Sbc	10.4	64	...	15. #	1.7	1.8*	0.11	1.2	0.094	20, 21
1808	S0a	7.4	67	...	11. #	...	2.4*	...	2.2	...	22

TABLE 1—*Continued*

NGC	Type	Distance (Mpc)	Inclination (°)	Color C_0	$L_{0,pg}$ ($10^9 L_\odot$)	M_H ($10^9 M_\odot$)	M_T ($10^{10} M_\odot$)	M_H/L_0	M_T/L_0	M_H/M_T	Ref.
1964.....	Sb	15.0	71	0.43	27.	8.2	....	0.30	....	....	20
2139.....	SBcd	15.0	33	0.02#	9.7#	2.7	....	0.28	....	....	20
2188.....	SBm	5.0	79	....	1.8#	0.63	0.30	0.35	1.7	0.21	17
2217.....	SB0	15.0	36	0.55#	21. #	0.66	....	0.031	....	....	20
2280.....	Scd	15.0	62	....	....	8.7	....	....	....	....	20
2366.....	Ir	3.3	64	0.25	0.94	1.0	....	1.1	....	....	17
2403.....	Scd	3.3	60	0.24	7.0	4.6	10.	0.66	14.0	0.049	23
2500.....	SBd	6.0	25	0.33	1.1	0.39	....	0.35	....	....	17
2541.....	Scd	6.0	73	0.23	3.5	1.3	2.7	0.37	7.7	0.048	17
2613.....	Sb	7.0	70	0.08#	23. #	0.40	....	0.017	....	....	20
2683.....	Sb	5.8	85	0.61	14.	1.3	2.8	0.093	2.0	0.046	10
2776.....	Sc	26.2	20	0.25#	8.9#	4.0	12.	0.45	13.0	0.033	24
2835.....	SBc	7.6	43	....	7.1#	1.4	4.1	0.20	5.8	0.034	17
2841.....	Sb	6.0	67	0.55	12.	1.0	4.2	0.083	3.5	0.024	10
2903.....	Sbc	7.0	70	0.29	25.	4.3	6.1*	0.17	2.4	0.070	10, 25
2997.....	Sc	7.6	32	....	14. #	4.3	....	0.31	....	....	20
3027.....	SBd:	12.0	58	....	4.4	3.9	....	0.89	....	....	20
3031.....	Sab	3.3	55	0.69	19.	2.3	19. *	0.12	10.0	0.012	26, 27
3034.....	I0	3.3	82	0.71	10.	....	1.0*	....	1.0	....	28
3079.....	SBm	12.0	83	0.37	26.	6.4	21.	0.25	8.1	0.030	17
3109.....	Ir	2.2	90	....	2.6#	0.94	0.60	0.36	2.3	0.16	29
3115.....	S0	7.0	82	0.80#	12. #	0.30	5.0	0.025	4.2	0.0060	17
3169.....	Sap	9.0	34	0.55	5.4	0.62	....	0.11	....	....	20
3198.....	SBc	9.6	73	0.21	15.	8.5	6.0	0.57	4.0	0.14	10
3227.....	Sap	16.5	51	0.64#	18. #	0.50	3.6*	0.028	2.0	0.014	18, 30
3310.....	Sbcp	13.2	36	0.10#	16. #	3.4	....	0.21	....	....	20
3319.....	SBcd	9.2	58	0.18	4.2	2.0	1.5	0.48	3.6	0.13	10
3344.....	Sbc	7.9	25	0.22	8.3	2.0	3.4	0.24	4.1	0.059	10
3359.....	SBc	10.0	51	0.23	9.7	4.8	10.	0.49	10.0	0.048	1, 10
3432.....	SBm	9.6	89	0.19	10.	3.9	3.2	0.39	3.2	0.12	10, 31
3504.....	SBa	16.5	60	0.47#	19. #	....	1.0*	....	0.53	....	32
3521.....	Sbc	6.9	66	0.66	15.	4.5	8.6*	0.30	5.7	0.052	17, 33
3556.....	SBcd	8.2	84	0.32	19.	5.3	4.3	0.28	2.3	0.12	10
3621.....	Sd	5.0	49	....	8.3#	5.4	6.1	0.65	7.3	0.089	17
3623.....	SBa	7.0	76	0.55	26.	....	10.	....	3.8	....	34

TABLE 1—*Continued*

NGC	Type	Distance (Mpc)	Inclination (°)	Color C_0	$L_{0,pg}$ ($10^9 L_\odot$)	M_H ($10^9 M_\odot$)	M_T ($10^{10} M_\odot$)	M_H/L_0	M_T/L_0	M_H/M_T	Ref.
3627.....	SBb	7.6	57	0.49	20.	0.84	19.	0.042	9.5	0.0044	17
3628.....	Sbp	7.6	89	0.52	29.	10.	4.3	0.34	1.5	0.23	10
3631.....	Sc	14.5	32	0.35	18.	3.7	3.4	0.21	1.9	0.11	10
3646.....	Sbcp	20.9	60	0.33	19.	....	19. *	....	10.0	....	35
3718.....	SBap	14.5	57	0.41	17.	3.0	8.4	0.18	4.9	0.036	10
3726.....	Sc	11.5	46	0.24	12.	6.3	....	0.52	....	....	10
3938.....	Sc	8.9	28	0.27	7.2	2.2	1.3	0.31	1.8	0.17	10
3992.....	SBb	14.5	51	0.57	27.	3.7	....	0.14	....	....	10
4051.....	Sbc	8.0	46	0.49	6.1	0.83	9.6	0.14	16.0	0.0086	18
4144.....	Scd:	9.1	86	....	6.6#	1.1	....	0.17	....	....	17
4151.....	Sab:	11.5	52	0.44#	11. #	0.93	3.0	0.084	2.7	0.031	18
4214.....	Ir	3.8	45	0.19	2.6	1.3	0.38	0.50	1.5	0.34	1, 6
4236.....	SBdm	3.3	75	0.11	4.2	....	3.0	....	7.1	....	1
4244.....	Scd	3.8	86	0.17	4.3	4.0	7.1	0.93	16.0	0.056	1, 36
4258.....	Sbc	4.0	64	0.37	13.	....	7.5*	....	5.8	....	37
4303.....	Sbc	14.6	48	0.23	29.	11.	15.	0.38	5.2	0.073	24
4321.....	Sbc	14.6	25	0.52	23.	2.0	22.	0.087	9.6	0.0090	24
4449.....	Ir	3.8	51	0.13	3.3	3.2	2.4	0.97	7.3	0.13	6
4490.....	SBdp	8.0	47	0.19	12.	5.9	6.2	0.49	5.2	0.095	17
4559.....	SBcd	9.6	67	0.18	19.	12.	10.	0.63	5.3	0.12	17
4605.....	SBcp	3.3	67	0.13#	1.5#	....	0.10*	....	0.67	....	31
4631.....	SBd	4.0	85	0.31	9.6	4.0	6.8	0.42	7.1	0.059	38
4656.....	SBmp	4.0	85	0.13	3.7	1.9	1.5	0.51	4.1	0.13	38
4736.....	Sab	4.0	35	0.52	8.6	....	31.	....	36.0	....	1
5005.....	Sbc	8.7	66	0.47	15.	....	5.9*	....	3.9	....	39
5033.....	Sc	9.5	59	0.29	12.	6.8	6.5	0.57	5.4	0.10	10
5055.....	Sbc	4.6	59	0.47	11.	....	4.3*	....	3.9	....	40
5102.....	S0	4.0	69	....	3.9#	0.51	....	0.13	....	....	20
5194.....	Sc	4.6	35	0.43	11.	1.2	6.4*	0.11	5.8	0.019	41, 42
5204.....	Sm	4.6	53	0.16	1.0	0.64	0.40	0.64	4.0	0.16	10
5236.....	Sc	4.0	46	0.43#	27. #	7.3	28.	0.27	10.0	0.026	43
5248.....	Sbc	10.5	55	0.29	17.	....	4.2*	....	2.5	....	44
5301.....	Sb	18.0	77	....	23. #	5.8	....	0.25	....	....	20
5371.....	Sbc	26.7	40	0.37#	25. #	3.2	47.	0.13	19.0	0.0068	24
5383.....	SBbp	24.0	40	0.38#	22.	....	4.6*	....	2.1	....	45

TABLE 1—*Continued*

NGC	Type	Distance (Mpc)	Inclination (°)	Color C_0	$L_{0,pg}$ ($10^9 L_\odot$)	M_H ($10^9 M_\odot$)	M_T ($10^{10} M_\odot$)	M_H/L_0	M_T/L_0	M_H/M_T	Ref.
5457	Sc	4.6	27	0.21	21.	9.3	16.	0.44	7.6	0.058	1, 46
5474	Scdp	4.6	20	0.23	1.3	0.46	5.2	0.35	40.0	0.0088	17
5523	Scd:	10.0	79	...	5.7#	4.8	...	0.84	...	...	17
5585	Sd	4.6	50	0.25	1.4	0.82	1.7	0.59	12.0	0.048	10
5668	Sd	15.0	20	0.42#	7.0#	4.3	5.6	0.61	8.0	0.077	47
5713	SBbc	13.2	22	0.34#	7.5#	3.2	...	0.43	...	...	10
5774	Sd	16.0	22	0.34	4.0	7.5	...	1.9	...	...	17
5899	SBc	26.8	82	0.42#	27. #	12.	57.	0.44	21.0	0.021	24
5907	Sc:	4.6	87	0.43	3.9	1.6	16.	0.41	41.0	0.010	17
5921	SBbc	15.0	34	0.53#	14. #	3.5	...	0.25	...	...	17
5962	Sc	18.0	36	0.34#	10. #	1.6	...	0.16	...	...	20
6015	Scd	9.8	67	0.27	6.0	1.6	1.4	0.27	2.3	0.11	10
6181	Sc	24.0	60	0.25#	18. #	...	6.0*	...	3.3	...	48
6207	Sc	15.0	54	0.19#	9.4#	2.2	...	0.23	...	...	20
6217	SBbc	16.5	40	0.33#	12. #	5.8	2.4	0.48	2.0	0.24	10
6340	S0a	16.5	58	0.61#	11. #	1.0	...	0.091	...	...	20
6503	Scd	4.6	74	0.38	4.5	1.5	1.0	0.33	2.2	0.15	10
6643	Sc	16.5	61	0.30	19.	2.4	...	0.13	...	...	10
6814	SBbc	20.0	25	0.53#	20. #	2.3	...	0.12	...	...	20
6822	Ir	0.50	43	...	0.17	0.15	0.24	0.88	14.0	0.062	1, 27
6835	SBa:	17.0	75	...	12. #	2.6	...	0.22	...	...	20
6946	Scd	4.2	22	0.40	13.	3.9	23.	0.30	18.0	0.017	49
7013	S0a	11.0	67	...	4.4#	0.94	...	0.21	...	...	20
7137	Sc	18.0	25	0.35#	5.0#	2.8	...	0.56	...	...	20
7217	Sab	11.0	36	0.55	15.	0.39	8.6	0.026	5.7	0.0045	17
7218	SBcd	15.0	57	...	5.2#	2.2	...	0.42	...	...	20
7314	Sbc	16.0	56	0.38#	16. #	2.7	...	0.17	...	...	20
7331	Sbc	7.9	69	0.44	26.	5.0	6.1*	0.19	2.3	0.082	10, 50
7361	Sc:	13.0	79	...	8.1#	3.6	...	0.44	...	...	17
7448	Sbc	20.0	55	0.17#	20. #	5.3	...	0.26	...	...	20
7640	SBc	4.4	89	0.00	4.3	3.0	5.7	0.70	13.0	0.053	10
7741	SBcd	10.1	45	0.27	5.4	7.6	4.2	1.4	7.8	0.18	10
7793	Sd	2.4	47	0.49#	1.1#	0.46	...	0.42	...	...	17
IC 10	Ir	1.25	...	...	...	0.31	0.12	...	...	0.26	51
IC 342	Sd	3.3	35	...	...	9.1	5.2	...	...	0.18	52

TABLE 1—*Continued*

NGC	Type	Distance (Mpc)	Inclination (°)	Color C_0	$L_{0,pg}$ ($10^9 L_\odot$)	M_H ($10^9 M_\odot$)	M_T ($10^{10} M_\odot$)	M_H/L_0	M_T/L_0	M_H/M_T	Ref.
IC 1613	Ir	0.66	32	0.36	0.072	0.051	0.025	0.71	3.5	0.20	6
IC 2574	SBm	3.3	69	0.17	1.5	1.6	1.7	1.1	11.0	0.094	1, 6
Ho II	Ir	3.3	40	0.21	0.89	0.91	0.96	1.0	11.0	0.095	6
LMC	SBm	0.052	27	0.30#	3.1#	0.54	1.0*	0.17	3.2	0.054	53, 54
Sex A	Ir	1.0	36	0.12	0.052	0.070	0.14	1.3	27.0	0.050	6
SMC	Ir	0.06	60	0.24#	0.67#	0.48	0.15	0.72	2.2	0.32	55
Peg	Ir	0.6	65	0.35	0.010	0.025	...	2.5	...	...	17
IZW 114	...	8.6	...	...	0.11#	0.13	...	1.2	...	...	56
IIZW 40	...	7.3	...	−0.15#	0.67#	0.30	...	0.45	...	...	56

NOTE.—The distances are from a variety of sources with first priority generally given to values derived from group membership. A Hubble constant of 100 km s^{-1} Mpc^{-1} is used in those instances where the only distance indicator is the radial velocity. A # following the color (C_0) or luminosity ($L_{0,pg}$) entries indicates a conversion from colors and magnitudes given in the *Reference Catalog* to Holmberg's (1958) photometric system. See Roberts (1969) for the details. An asterisk in the total mass (M_T) column denotes an optically derived mass. Such masses have been converted to a "total" mass where necessary (Roberts 1969).

REFERENCES TO TABLE 1.—(1) Rogstad, Rougoor, and Whiteoak 1967. (2) Robinson and van Damme 1966. (3) Burbidge, Burbidge, and Prendergast 1961*d*. (4) Argyle 1965. (5) Rubin and Ford 1970. (6) Epstein 1964*a*. (7) Roberts 1964. (8) Burbidge, Burbidge, and Prendergast 1962*b*. (9) Shobbrook and Robinson 1967. (10) Roberts 1968*c*. (11) Gordon 1971. (12) Burbidge, Burbidge, Rubin, and Prendergast 1964. (13) Roberts 1966*c*. (14) Burbidge, Burbidge, and Prendergast 1965*a*. (15) Höglund and Roberts 1965. (16) Burbidge, Burbidge, and Prendergast 1965*c*. (17) Gouguenheim 1969. (18) Allen, Darchy, and Lauqué 1971. (19) Burbidge, Burbidge, and Prendergast 1963*a*. (20) Bottinelli, Chamaraux, Gouguenheim, and Lauqué 1970. (21) Rubin, Burbidge, Burbidge, and Prendergast 1964. (22) Burbidge and Burbidge 1968. (23) Burns and Roberts 1971. (24) McCutcheon and Davies 1970. (25) Burbidge, Burbidge, and Prendergast 1960*b*. (26) Münch 1959. (27) Volders and Högbom 1961. (28) Burbidge, Burbidge, and Rubin 1964. (29) van Damme 1966. (30) Rubin and Ford 1968. (31) Bertola 1966. (32) Burbidge, Burbidge and Prendergast 1960*c*. (33) Burbidge, Burbidge, Crampin, Rubin, and Prendergast 1964. (34) Burbidge, Burbidge, and Prendergast 1961*b*. (35) Burbidge, Burbidge, and Prendergast 1961*c*. (36) Roberts 1962*b*. (37) Burbidge, Burbidge, and Prendergast 1963*b*. (38) Roberts 1968*b*. (39) Burbidge, Burbidge, and Prendergast 1961*a*. (40) Burbidge, Burbidge, and Prendergast 1960*a*. (41) Burbidge and Burbidge 1964. (42) Roberts and Warren 1970. (43) Lewis 1968. (44) Burbidge, Burbidge, and Prendergast 1962*a*. (45) Burbidge, Burbidge, and Prendergast 1962*c*. (46) Volders 1959. (47) Roberts 1965. (48) Burbidge, Burbidge, and Prendergast 1965*b*. (49) Gordon, Remage, and Roberts 1968. (50) Rubin, Burbidge, Burbidge, Crampin, and Prendergast 1965. (51) Roberts 1962*a*. (52) Dieter 1962*b*. (53) Feast 1964. (54) McGee and Milton 1966*a*. (55) Hindman 1967. (56) Chamaraux, Heidmann, and Lauqué 1970. (57) Lewis 1970.

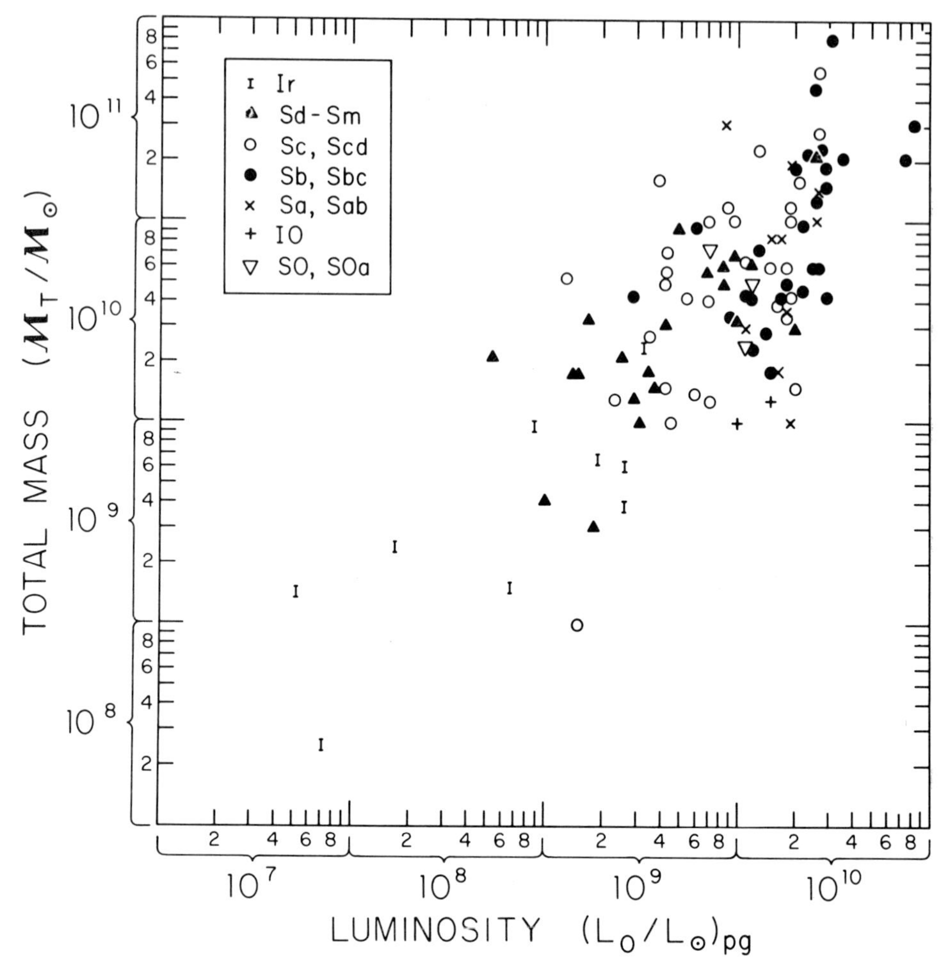

FIG. 10.—The relation between total mass and photographic luminosity for spiral galaxies.

7.2. Total Mass and Hydrogen Mass

Only three quantitatively measurable parameters are known to correlate with structural type. These are (i) the fractional hydrogen content of a galaxy, $M_{\text{H I}}/M_T$; (ii) the hydrogen-to-luminosity ratio $M_{\text{H I}}/L_0$; and (iii) the color, C_0. The correlation of (i) is shown in figure 13 and given in table 2. There is a systematic increase in $M_{\text{H I}}/M_T$ with later type going from ~1 percent for the earliest spirals to ~18 percent for the latest spirals, the irregulars. If we include in the gaseous component of a galaxy the helium content whose number ratio $N(\text{He})/N(\text{H I}) \simeq 0.1$ (Peimbert and Spinrad 1970; Roberts 1971), we find that, in the mean, at least one-fourth of the total mass of irregular systems is in the form of interstellar gas.

The fractional hydrogen content of our Galaxy is about 5 percent (Kerr and Westerhout 1965). The earlier value of 1–2 percent (van de Hulst, Raimond, and van Woerden 1957) has been increased by a more detailed analysis.

TABLE 2

AVERAGE VALUES OF INTEGRAL PROPERTIES

Type	$\left\langle \frac{M_{\text{H I}}/M_\odot}{(L_0/L_\odot)_{\text{pg}}} \right\rangle$	σ(mean)	*N*	$\left\langle \frac{M_T/M_\odot}{(L_0/L_\odot)_{\text{pg}}} \right\rangle$	σ(mean)	*N*	$\left\langle \frac{M_{\text{H I}}}{M_T} \right\rangle$	σ(mean)	*N*
	1. Ordinary and Barred Combined								
S0+SB0	0.081	±0.022	6	7.1	± 2.9	2	0.0053	±0.0007	2
S0a+SB0a	0.217	±0.075	3	2.2	...	1	...	...	...
Sa+SBa	0.134	±0.042	4	2.8	± 0.8	4	0.025	±0.011	2
Sab+SBab	0.088	±0.022	4	10.2	± 5.3	6	0.017	±0.0056	4
Sb+SBb	0.172	±0.031	14	6.4	± 2.1	11	0.048	±0.021	10
Sbc+SBbc	0.220	±0.029	17	6.0	± 1.2	17	0.069	±0.021	10
Sc+SBc	0.370	±0.037	23	8.4	± 2.2	19	0.074	±0.018	16
Scd+SBcd	0.500	±0.080	16	11.0	± 3.1	12	0.085	±0.016	12
Sd+SBd	0.770	±0.160	13	9.7	± 1.8	9	0.085	±0.013	10
Sdm+SBdm	0.650	...	1	23.0	±16.0	2	...	...	...
Sm+SBm	0.440	±0.087	10	4.9	± 1.0	10	0.109	±0.019	10
Ir	0.880	±0.158	13	8.0	± 2.8	9	0.179	±0.032	10
I0	0.190	...	1	0.93	± 0.07	2	0.220	...	1
	2. Ordinary and Barred Combined and Averaged over Several Type Intervals								
S0–Sab	0.119	±0.021	17	3.0	± 2.8	13	0.016	±0.0042	8
Sb–Sbc	0.201	±0.021	31	6.1	± 1.1	28	0.059	±0.012	20
Sc–Scd	0.430	±0.040	39	9.4	± 1.8	31	0.078	±0.012	28
Sd–Sm	0.630	±0.096	24	8.7	± 1.8	21	0.097	±0.011	20
Ir	0.880	±0.158	13	8.0	± 2.8	9	0.179	±0.032	10
I0	0.190	...	1	0.93	± 0.07	2	0.220	...	1
	3. Ordinary versus Barred; Averaged over Several Type Intervals								
S0–Sab	0.092	±0.017	10	8.0	± 4.1	8	0.014	±0.0048	5
SB0–SBab	0.157	±0.041	7	4.9	± 1.5	5	0.020	±0.0090	3
Sb–Sbc	0.187	±0.021	22	5.6	± 0.97	22	0.055	±0.0013	16
SBb–SBbc	0.228	±0.055	9	8.0	± 3.8	6	0.076	±0.056	4
Sc–Scd	0.386	±0.040	28	10.3	± 2.4	21	0.071	±0.016	19
SBc–SBcd	0.535	±0.098	11	7.4	± 1.9	10	0.094	±0.018	9
Sd–Sm	0.800	±0.195	10	12.0	± 3.7	9	0.095	±0.016	9
SBd–SBm	0.505	±0.079	14	6.2	± 1.3	12	0.098	±0.017	11

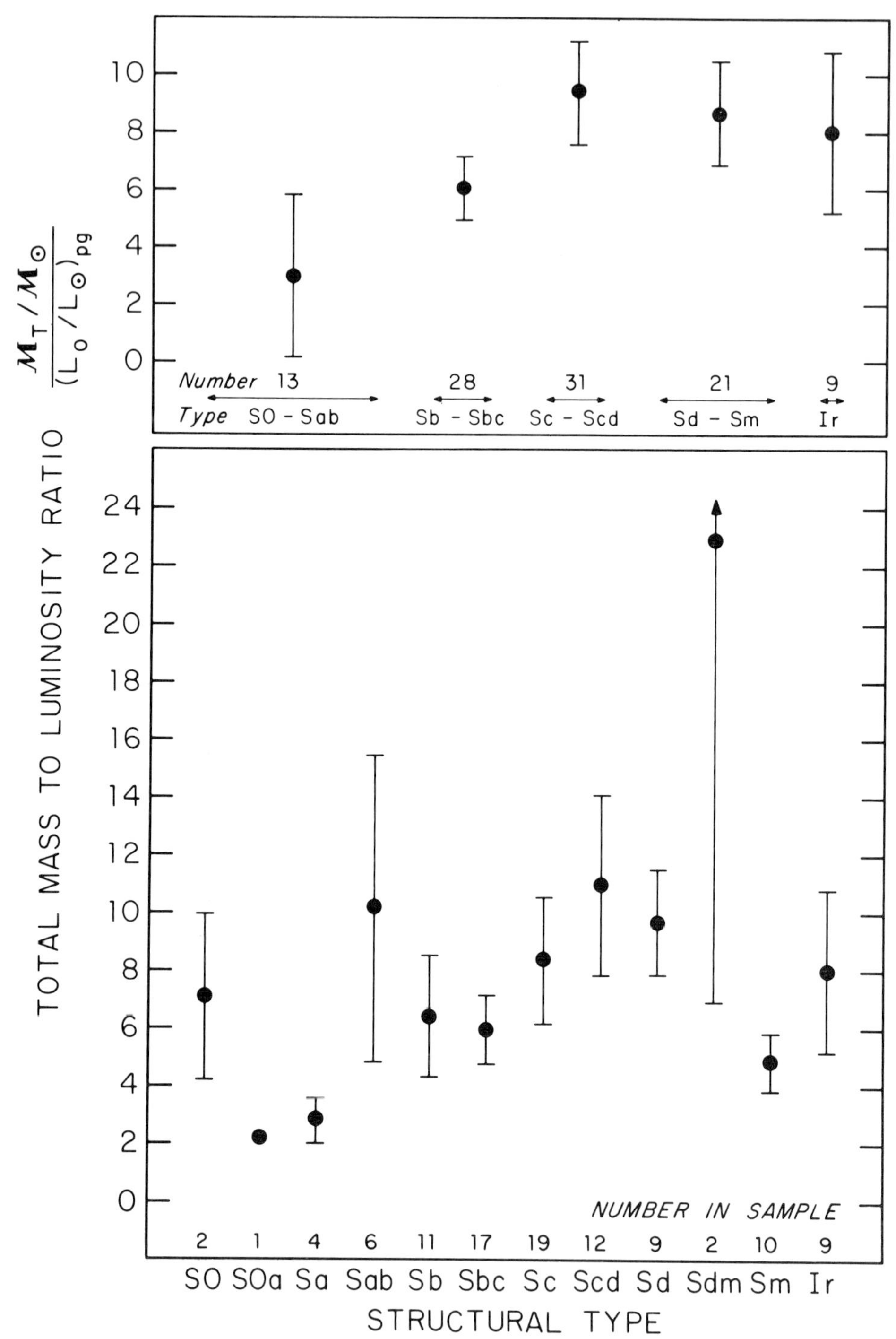

FIG. 11.—Average values of the total mass-to-luminosity ratio for spiral galaxies of different structural types (regular and barred systems are grouped together). The lower panel is for the individual types. These are grouped into broader type intervals in the upper panel. The error bars represent the standard deviation of the sample mean.

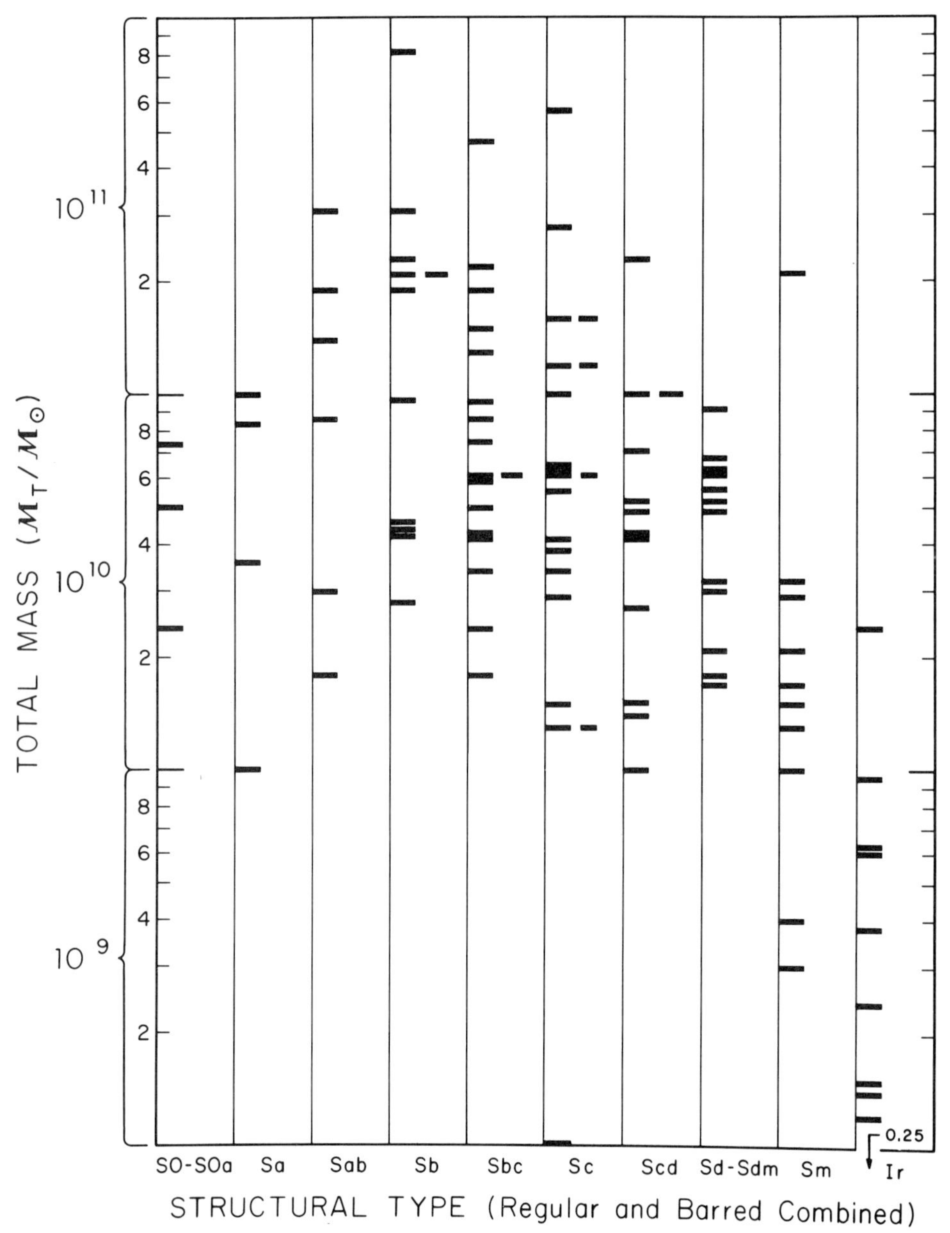

FIG. 12.—A graphical display of all available total mass measurements of spiral galaxies as a function of structural type. The absence of low-mass systems in the early and mid-type range may be a selection effect. Both 21-cm and optical determinations are included.

The detailed trend of $M_{H\ I}$ with M_T for different galaxy types is shown in figure 14. To within a factor of 2, there appears to be an upper limit of $10^{10}\ M_{\odot}$ for the hydrogen content of a galaxy regardless of type. Figure 15, a plot of M_H versus distance, also shows this rather abrupt upper limit. This figure also illustrates the bias in the hydrogen data against low hydrogen masses, i.e., less than $10^8\ M_{\odot}$. Only three systems, all within

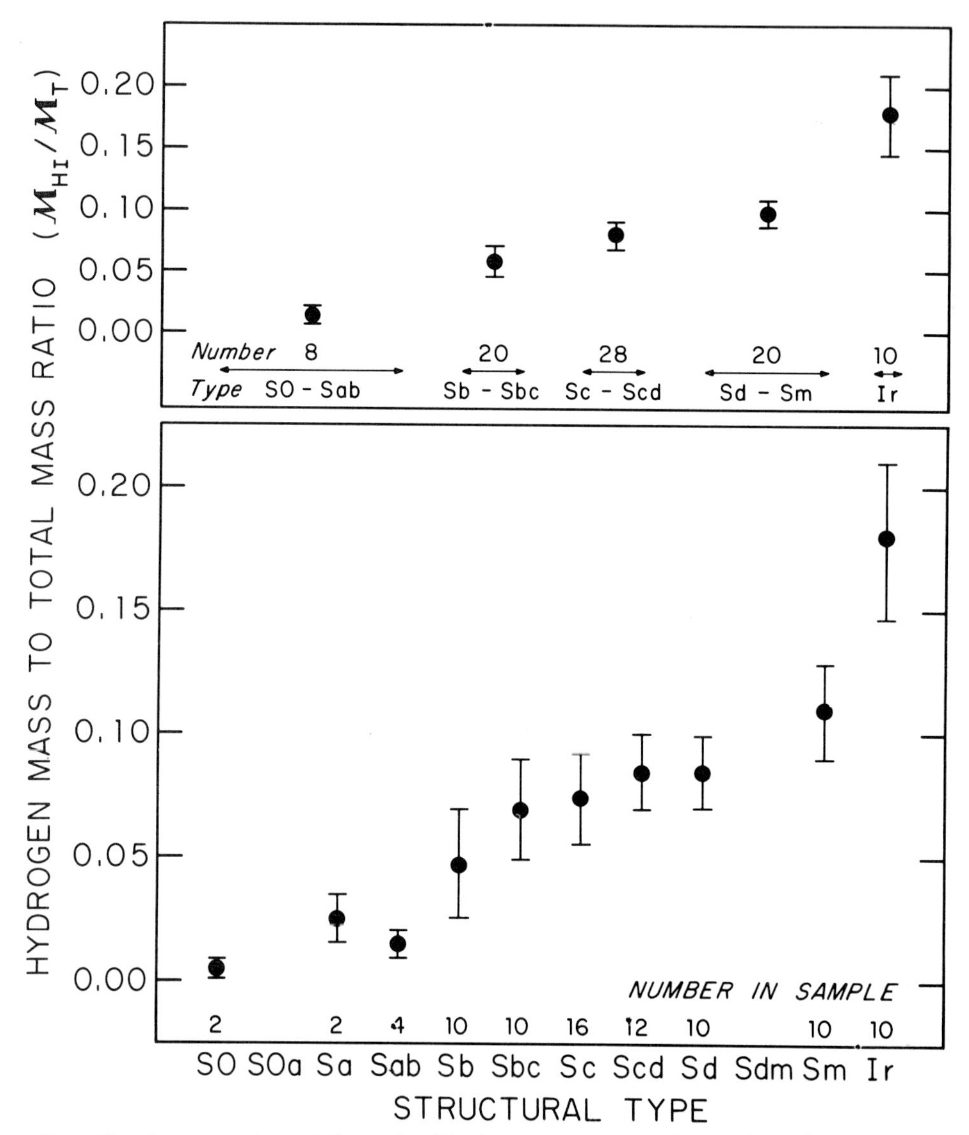

FIG. 13.—Average values of the ratio of hydrogen mass to total mass for spiral galaxies of different structural types (regular and barred systems are grouped together). The lower panel is for individual types. These are grouped into broader type intervals in the upper panel. The error bars represent the standard deviation of the sample mean.

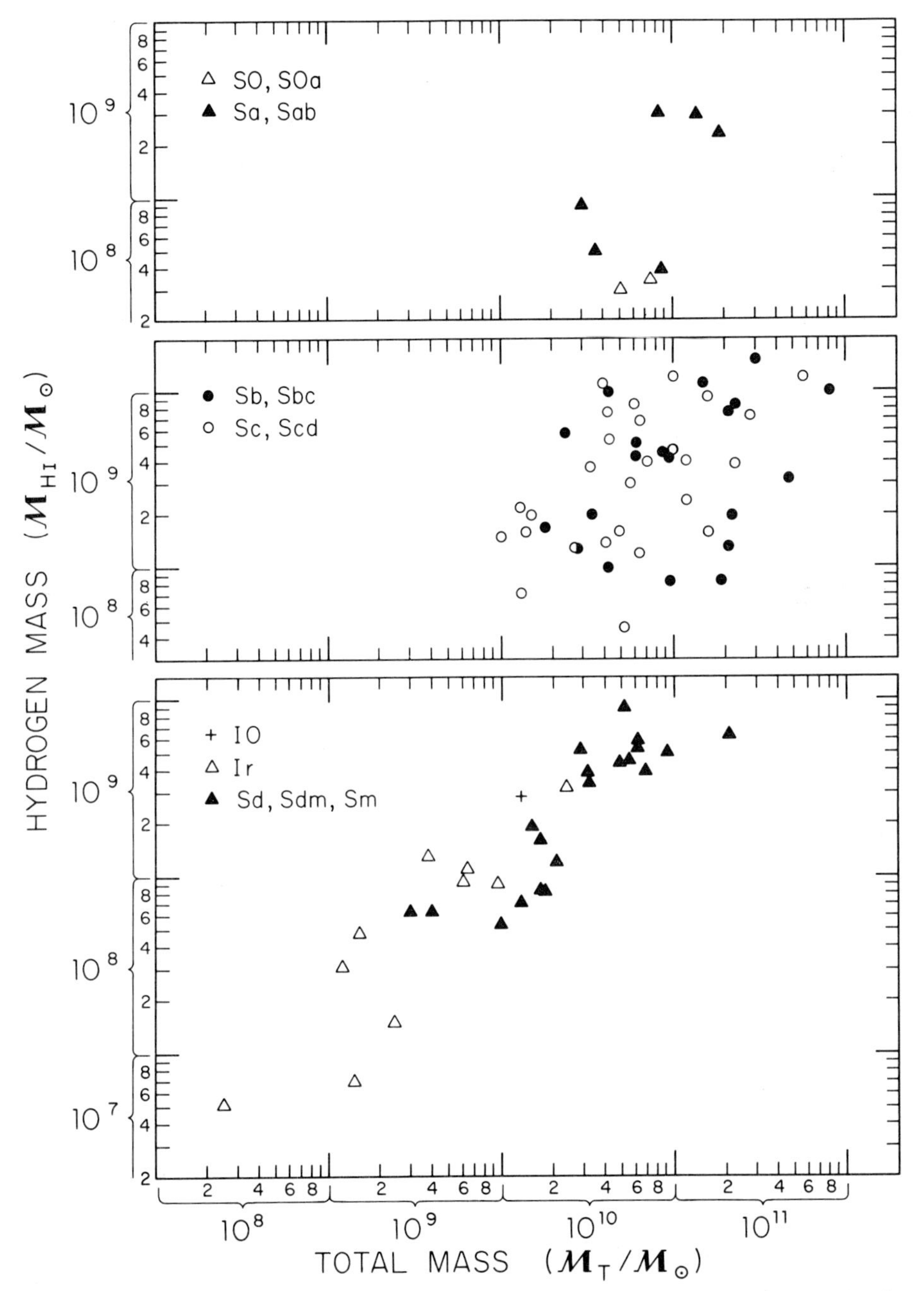

FIG. 14.—The relation between hydrogen mass and total mass for spiral galaxies. The slope s dependent on type (see fig. 13), and three different groups are shown: early, mid, and late type.

1 Mpc, have been measured with such low hydrogen content. As the distance increases, the lower limit for detection increases. Presently available instruments are capable of measuring temperatures corresponding to the lower right part of this diagram, and we may expect this region to be filled in the near future.

7.3. Hydrogen Mass and Absolute Photographic Luminosity

Both the hydrogen mass and the optical luminosity of a galaxy are derived by measuring a flux (or in radio terminology a flux density) at the appropriate wavelength. (Since

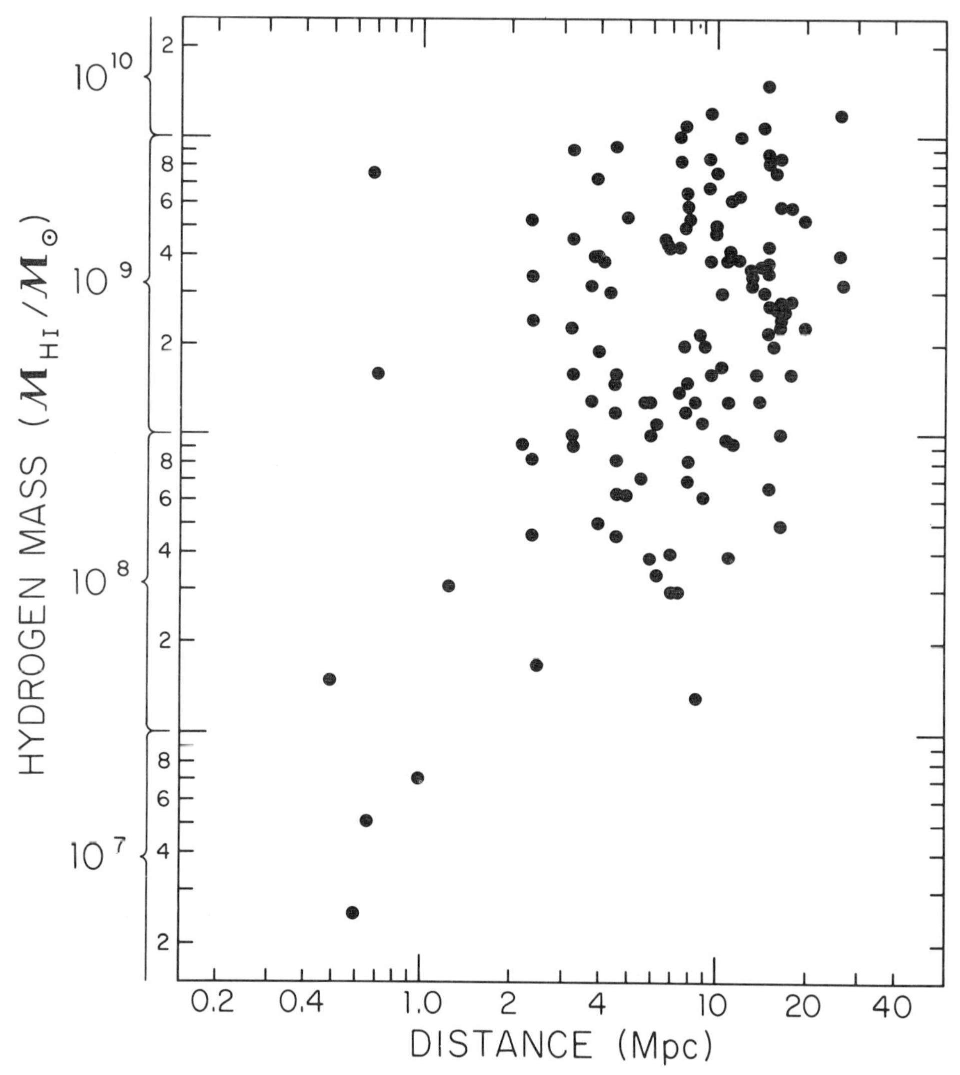

Fig. 15.—The hydrogen content of spiral galaxies as a function of their distance. This figure displays (1) the selection effect against low values of 21-cm flux, the lower right boundary and (2) the rather abrupt upper limit to the hydrogen content of a spiral.

the source is extended, a more precise statement would refer to the measurement of a specific intensity or its radio equivalent, a brightness temperature. To obtain the total radiation, these values are integrated over the solid angle of the source, e.g., $\int T_B d\Omega$, which yields the flux density.) Thus the ratio $M_{\text{H I}}/L_0$ is independent of distance. Values of this ratio are given in table 2 and displayed in figure 16. There is a pronounced trend of this ratio with type; the later systems have a larger $M_{\text{H I}}/L_0$. Average values of $M_{\text{H I}}/L_0$ for the different types show a range of about a factor of 10. Values of $M_{\text{H I}}$ versus L_0 for different structural classes are shown in figure 17.

The sample of Sm galaxies appears to go counter to the general trend of increasing $M_{\text{H I}}/L_0$ with type. It is worth noting that of the 10 galaxies in this sample, seven have inclinations greater than 60°, of which five have $i \geq 81°$. Too low an optical-depth correction for inclination may have been applied. The possible variation of the optical-depth–inclination correction with type noted earlier (§ 5) may be reflected in the apparently low average value of $M_{\text{H I}}/L_0$ for this galaxy type. Without additional data it would be premature to invert the problem by requiring a steadily increasing value of $M_{\text{H I}}/L_0$ and thus obtain a better hold on the optical-depth correction as a function of type.

Most of the mass of a galaxy is contributed by late-type dwarfs, while the principal source of the photographic luminosity (in spirals and irregulars) is from early-type stars. It might therefore appear surprising that there is a well-defined mass-luminosity relation for galaxies. The explanation lies in the relation between hydrogen content and total mass (the more massive systems contain intrinsically more H I) and the additional relation between luminosity and hydrogen content (the more luminous systems also contain more H I). Without implying any causative relation, one may say that the total mass fixes the scale of the hydrogen content and therefore the total photographic luminosity.

7.4. Hydrogen Mass, Absolute Photographic Luminosity, and Color

The colors of galaxies vary systematically with structural type (Holmberg 1958; de Vaucouleurs 1961) indicating a definite variation in the spectral-luminosity function among these systems. There is evidence that this variation also depends on the absolute magnitude of the system (Chester and Roberts 1964).

The color-type relationship for galaxies allows a more quantitative comparison between hydrogen content and type than the subjectively derived type classification. Such a comparison is shown in figure 18, where the ratio of $M_{\text{H I}}/L_0$ is plotted against the corrected color. This diagram is similar in concept to the color-color diagrams derived in stellar photometry but with one of the wavelength baselines representing two widely separated "magnitudes." Within a given structural class the bluer systems tend to have a higher value of $M_{\text{H I}}/L_0$.

7.5. Ordinary and Barred Spirals

The summary of integral properties given in table 2 is in three sections: (1) for each type, but with ordinary and barred spirals combined; (2) ordinary and barred combined and averaged over several type intervals to improve the statistics; and (3) ordinary and barred separately, but averaged over several type intervals. Parts 1 and 2 are plotted

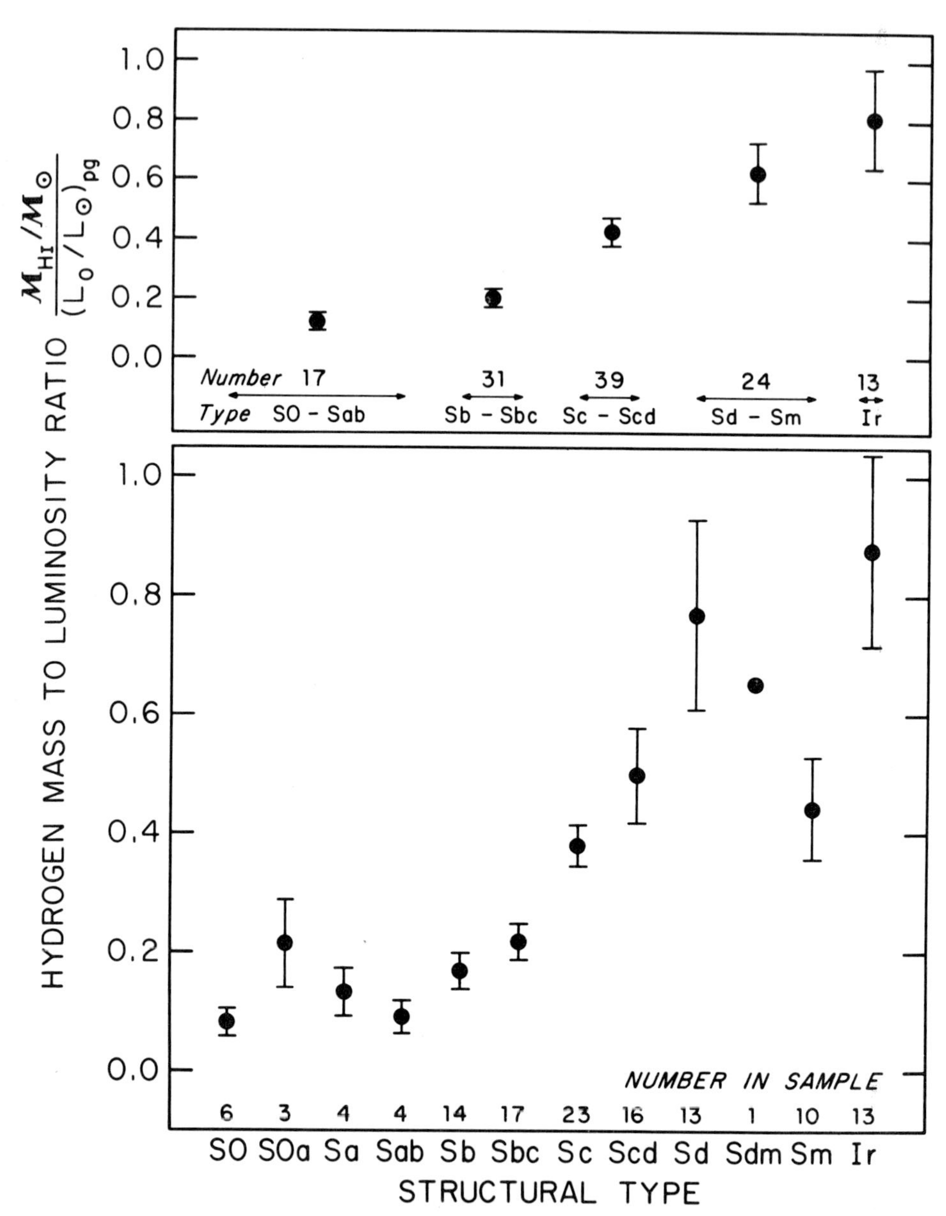

FIG. 16.—Average values of the ratio of hydrogen mass to photographic luminosity for spirals of different structural types (regular and barred systems are grouped together). This ratio is independent of distance. The lower panel is for individual types. These are grouped into broader type intervals in the upper panel. The error bars represent the standard deviation of the sample mean.

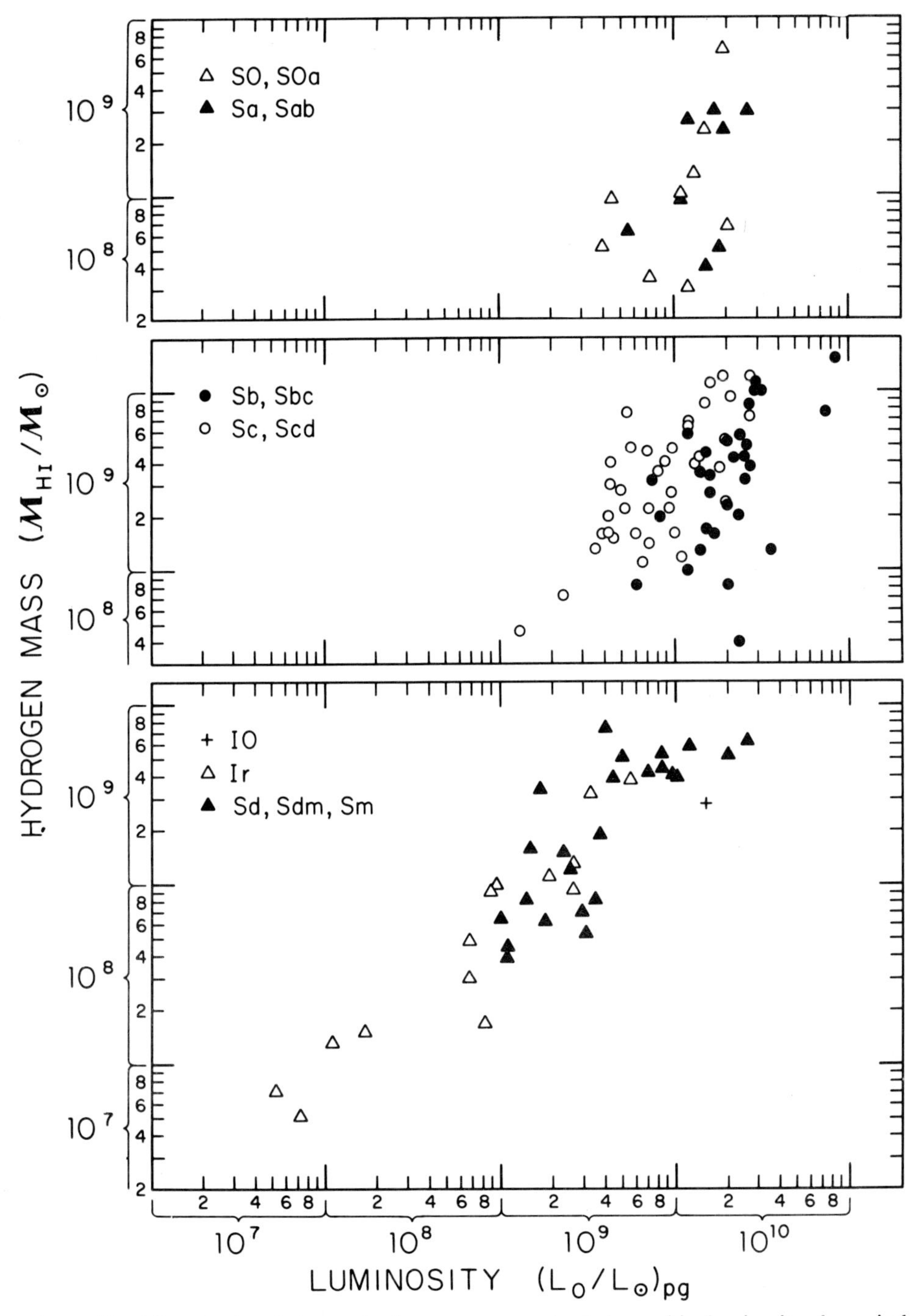

FIG. 17.—The relation between hydrogen mass and photographic luminosity for spiral galaxies. The slope is dependent on type (see fig. 16), and three different groups are shown: early, mid, and late type.

in the appropriate figures. The third section shows the ratio $\langle M_{\text{H I}}/M_T \rangle$ to be higher for barred than for ordinary spirals of the same type. However, the sample sizes for both groups are small, and the standard deviation of the mean indicates an overlap in $\langle M_{H\ \text{I}}/M_T \rangle$ for all types. The small differences between the mean values should thus be given low weight.

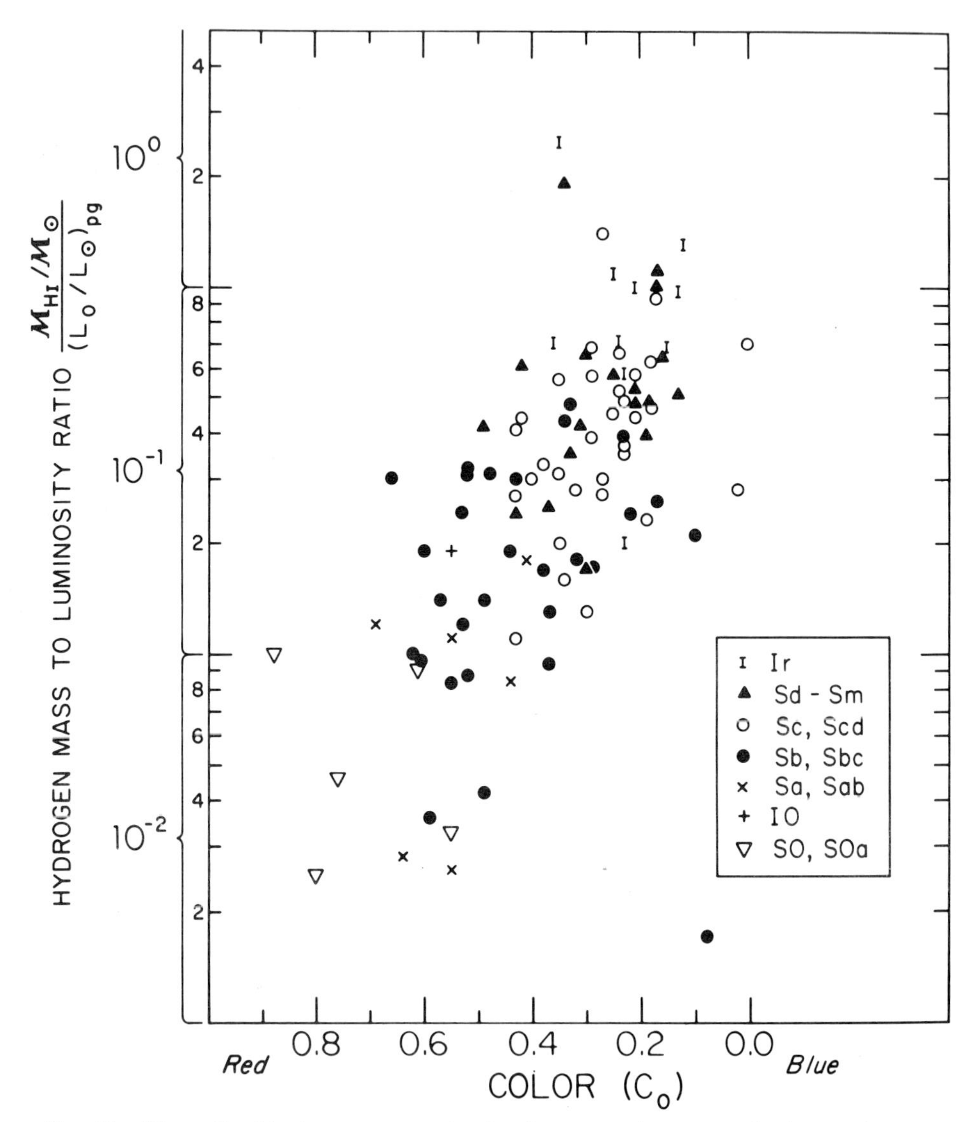

FIG. 18.—The ratio of hydrogen mass to luminosity versus the corrected color. Both parameters are independent of distance. This diagram is similar to color-color presentations in optical photometry. Note that bluer systems of the same type tend to have a higher ratio of hydrogen mass to luminosity.

The present data allow us to conclude only that there is no striking and systematic difference between ordinary and barred spirals in any of the integral property ratios involving hydrogen mass, total mass, and luminosity.

7.6. Elliptical and Radio Galaxies

Only one detection of 21-cm line radiation from an elliptical galaxy, NGC 4472, has been reported. Robinson and Koehler (1965), using the Parkes 210-foot telescope, measured, in two sets of observations, antenna temperatures of $T_A = 0.016 \pm 0.010°$ K and $0.015 \pm 0.008°$ K. The errors are standard deviations derived from repeated ON-OFF measurements. The corresponding hydrogen mass is $3 \times 10^8 M_\odot$ for a distance of 12 Mpc; $M_{\text{H I}}/L_0 = 0.007$, and the fractional mass content is 0.02 percent for an assumed total mass of $15 \times 10^{11} M_\odot$.

We shall distinguish between normal and radio galaxies by adopting the arbitrary value of $\geq 10^{40}$ ergs s^{-1} for the integrated continuum flux (Matthews, Morgan, and Schmidt 1964) for radio galaxies. With this characterizing value we find that only two radio galaxies have been detected in neutral hydrogen: NGC 5128 (Roberts 1971) in absorption and NGC 1068 (Allen, Darchy, and Lauqué 1971) in emission. Data for the latter are given in table 1. Little can be said about the total hydrogen content of NGC 5128 because the geometry and hence the optical depth of the absorbing features are unknown.

8. THE EVOLUTION OF GALAXIES

Most of the theories and suggestions relating to the evolution of galaxies may be classed into three principal categories. (1) All galaxies formed at (essentially) the same time, with the presently observed structural type reflecting a continuum of initial conditions under which they formed. In this view, the galaxies are all of the same *chronological* age. (2) All galaxies formed at the same time, and the present structural-type sequence represent an *evolutionary* sequence; i.e., one type of galaxy evolves into another type. In this scheme, the early-type galaxies are usually considered to be the most evolved. (3) The galaxies formed at different times. The various structural types then reflect different chronological and evolutionary ages. Again, the usual view here is that one type of galaxy will evolve into another type. This class of theories could also have initial conditions fix the structural-type sequence.

The properties of galaxies summarized in the preceding section do not appear to favor any particular category; they do place constraints on the theories. Thus, it is difficult to see how the presently observed sample of irregular-type galaxies can evolve into spirals *as we know them to be now* because of their great difference in mass. The rather well-bounded regions in figures 10 (M_T versus L_0), 14 ($M_{\text{H I}}$ versus M_T), and 17 ($M_{\text{H I}}$ versus L_0) would argue against evolutionary tracks in directions other than those occupied by the presently observed galaxies unless this evolution is very rapid and examples therefore rare. In this context it is important to keep in mind the selection effects of these data discussed in § 7. Any successful theory must account for the close interrelations among total mass, hydrogen content, and luminosity that are found to hold for normal galaxies. To a rough approximation, they are all linearly related.

One aspect of galactic evolution particularly related to hydrogen measurements is the variation of the relative hydrogen content $M_{\text{H I}}/M_T$ with structural type. The increase of this ratio with later-type systems has prompted the suggestion (Reddish 1961) that an evolutionary time scale could be associated with structural type. The systematic change in the relative number of bright stars and H II regions with structural type would appear to favor this view (Sandage 1962). Such a scheme would have the irregular galaxies the youngest, with a steady increase in evolutionary age as we go from Sc to Sb, etc. The basic reasoning is that the "surplus" of hydrogen now evident in the later-type systems will be used up as the galaxy ages. Obviously, star formation depletes the interstellar hydrogen in an isolated system, but the critical factor is the *rate* of this depletion. For a steady rate of star formation, the ratio $M_{\text{H I}}/L_0$ (whose dimension is time) is a measure of the time scale for hydrogen depletion.

An analysis (Roberts 1963) of the rate at which hydrogen (and helium) is consumed in forming stars has shown that the Ir, Sc, and Sb systems can maintain their present rate of star formation for periods $\gtrsim 10^{10}$ years. Further, this rate of star formation could also have continued for the preceding 10^{10} years if the presently observed hydrogen content was approximately doubled. These calculations do not argue against a temporal scale for the various structural types, but they do point up the fact that high proportions of gas and early-type stars in Sc and Ir galaxies need *not* imply that such systems are young.

The data on the hydrogen content of galaxies supply another aspect to the problem of the helium-to-hydrogen abundance ratio. This ratio appears to be surprisingly constant not only in different H II regions in our Galaxy but also in other galaxies ranging from type S0 through Ir. However, we find galaxies of the same structural type differing by a factor of ~ 10 in their hydrogen content and therefore, presumably, in their helium content. We also find a similar range, going from one morphological class to another, in the percentage of the total mass of a galaxy that is in the form of neutral hydrogen and therefore in helium. We assume here that the helium is well mixed with the hydrogen throughout the total extent of the hydrogen. This assumption is necessary since the helium-to-hydrogen ratio is determined only from H II regions which are not uniformly distributed throughout the region of neutral hydrogen.

Of the two usual proposals invoked to account for the helium abundance—(i) primordial, or (ii) nucleosynthesis in stars—the former would appear the more attractive in terms of the different amount of helium required in different galaxies. If we wish to invoke the nucleosynthesis explanation, we have the constraint of relating several parameters to the uncycled primordial plus cycled hydrogen (i.e., the present interstellar H I). These parameters are the primordial mass function, the rate (and type) of later star formation, and the amount and composition of the gas lost by evolving stars. In essence we require that the hydrogen and stellar content of a galaxy be related.

We have seen that the hydrogen content of a galaxy is well correlated with such galaxian properties as luminosity, total mass, color, and structural type. These are, in various ways, related to the stellar content of a galaxy, and so the general requirement of a hydrogen-star relationship for the nucleosynthesis explanation is satisfied. However, the detailed relationships among the quantities listed above are not available without an adequate theory of star and galaxy formation and evolution. The approach may possibly be inverted and the relationships derived, or at least indicated, from the hydrogen and helium data.

REFERENCES

Allen, R. J., Darchy, B. F., and Lauqué, R. 1971, *Astr. and Ap.*, **10**, 198.
Argyle, E. 1965, *Ap. J.*, **141**, 750.
Baade, W. 1958, *Stellar Populations*, ed. D. J. K. O'Connell (New York: Interscience), p. 3.
Baars, J. M. W., Mezger, P. G., and Wendker, H. 1965, *Ap. J.*, **142**, 122.
Baldwin, J. 1970, paper presented at the 14th General Assembly IAU.
Beale, J. S., and Davies, R. D. 1969, *Nature*, **221**, 531.
Bertola, F. 1966, *Mem. Soc. Astr. Italiana*, **37**, 433.
Bottinelli, L. 1971, *Astr. and Ap.*, **10**, 437.
Bottinelli, L., Chamaraux, P., Gougenheim, L., and Lauqué, R. 1970, *Astr. and Ap.*, **6**, 453.
Bottinelli, L., Gouguenheim, L., Heidmann, J., and Heidmann, N. 1968, *Ann. d'Ap.*, **31**, 205.
Brandt, J. C., and Scheer, L. S. 1965, *A.J.*, **70**, 471.
Brundage, W. D., and Kraus, J. D. 1966, *Science*, **153**, 411.
Burbidge, E. M., and Burbidge, G. R. 1964, *Ap. J.*, **140**, 1445.
———. 1968, *ibid.*, **151**, 99.
Burbidge, E. M., Burbidge, G. R., Crampin, D. J., and Prendergast, K. H. 1965, *Ap. J.*, **141**, 759.
Burbidge, E. M., Burbidge, G. R., Crampin, D. J., Rubin, V. C., and Prendergast, K. H. 1964, *Ap. J.*, **139**, 1058.
Burbidge, E. M., Burbidge, G. R., and Prendergast, K. H. 1960*a*, *Ap. J.*, **131**, 282.
———. 1960*b*, *ibid.*, **132**, 640.
———. 1960*c*, *ibid.*, p. 661.
———. 1961*a*, *ibid.*, **133**, 814.
———. 1961*b*, *ibid.*, **134**, 232.
———. 1961*c*, *ibid.*, p. 237.
———. 1961*d*, *ibid.*, p. 874.
———. 1962*a*, *ibid.*, **136**, 128.
———. 1962*b*, *ibid.*, p. 339.
———. 1962*c*, *ibid.*, p. 704.
———. 1963*a*, *ibid.*, **137**, 376.
———. 1963*b*, *ibid.*, **138**, 375.
———. 1964, *ibid.*, **140**, 80.
———. 1965*a*, *ibid.*, **142**, 154.
———. 1965*b*, *ibid.*, p. 641.
———. 1965*c*, *ibid.*, p. 649.
Burbidge, E. M., Burbidge, G. R., and Rubin, V. C. 1964, *Ap. J.*, **140**, 942.
Burbidge, E. M., Burbidge, G. R., Rubin, V. C., and Prendergast, K. H. 1964, *Ap. J.*, **140**, 85.
Burke, B. F. 1964, in *IAU Symposium No. 20*, ed. F. J. Kerr and A. W. Rodgers (Canberra: Australian Academy of Science), p. 102.
Burke, B. F., Turner, K. C., and Tuve, M. A. 1963, *Annual Report of the Director, 1962–1963, Dept. of Terrestrial Magnetism* (Washington: Carnegie Institution of Washington).
———. 1964, *Annual Report of the Director, 1963–1964, Dept. of Terrestrial Magnetism* (Washington: Carnegie Institution of Washington).
Burns, W. R., and Roberts, M. S. 1971, *Ap. J.*, **166**, 265.
Chamaraux, P., Heidmann, J., and Lauqué, R. 1970, *Astr. and Ap.*, **8**, 424.
Chester, C., and Roberts, M. S. 1964, *A.J.*, **69**, 635.
Clark, B. G. 1965, *Ap. J.*, **142**, 1398.
Davies, R. D. 1964, in *IAU Symposium No. 20*, ed. F. J. Kerr and A. W. Rodgers (Canberra: Australian Academy of Science), p. 102.
Dieter, N. H. 1962*a*, *A.J.*, **67**, 217.
———. 1962*b*, *ibid.*, p. 313.
Elvius, A. 1964, *Nature*, **201**, 171.
Epstein, E. E. 1964*a*, *A.J.*, **69**, 490.
———. 1964*b*, *ibid.*, p. 521.
Feast, M. W. 1964, *M.N.R.A.S.*, **127**, 195.
Gordon, C. P. 1970, *A.J.*, **75**, 914.
Gordon, K. J. 1971, *Ap. J.*, **169**, 235.
Gordon, K. J., Remage, N. H., and Roberts, M. S. 1968, *Ap. J.*, **154**, 845.
Gottesman, S. T., and Davies, R. D. 1970, *M.N.R.A.S.*, **149**, 263.
Gottesman, S. T., Davies, R. D., and Reddish, V. C. 1966, *M.N.R.A.S.*, **133**, 359.
Gouguenheim, L. 1969, *Astr. and Ap.*, **3**, 281.
Guélin, M., and Weliachew, L. 1969, *Astr. and Ap.*, **1**, 2.
———. 1970, *ibid.*, **7**, 141.
Hindman, J. V. 1964, *Nature*, **202**, 377.
———. 1967, *Australian J. Phys.*, **20**, 147.
Hindman, J. V., and Balnaves, K. M. 1967, *Australian J. Phys.*, *Ap. Suppl.*, No. 4.
Hindman, J. V., Kerr, F. J., and McGee, R. X. 1963, *Australian J. Phys.*, **16**, 570.

Hindman, J. V., and Sinclair, M. W. 1965, *Symposium on the Magellanic Clouds* (Mount Stromlo Observatory), p. 76.
Höglund, B., and Roberts, M. S. 1965, *Ap. J.*, **142**, 1366.
Holmberg, E. 1958, *Medd. Lund*, Series II, No. 136.
Hulst, H. C. van de, Muller, C. A., and Oort, J. H. 1954, *B.A.N.*, **12**, 117.
Hulst, H. C. van de, Raimond, E., and van Woerden, H. 1957, *B.A.N.*, **14**, 1.
Humason, M. L., Mayall, N. U., and Sandage, A. R. 1956, *A.J.*, **61**, 97.
Kerr, F. J., and Hindman, J. V. 1953, *A.J.*, **58**, 218.
Kerr, F. J., Hindman, J. V., and Robinson, B. J. 1954, *Australian J. Phys.*, **7**, 297.
Kerr, F. J., and Vaucouleurs, G. de. 1956, *Australian J. Phys.*, **9**, 90.
Kerr, F. J., and Westerhout, G. 1965, in *Galactic Structure*, ed. A. Blaauw and M. Schmidt (Chicago: University of Chicago Press), p. 167.
Lewis, B. M. 1968, *Proc. Astr. Soc. Australia*, **1**, 104.
———. 1970, *Observatory*, **90**, 264.
Lilley, A. E. 1963, private communication.
Lilley, A. E., and McClain, E. F. 1956, *Ap. J.*, **123**, 172.
Lin, C. C. 1970, invited discourse, 14th General Assembly, IAU.
McCutcheon, W. H., and Davies, R. D. 1970, *M.N.R.A.S.*, **150**, 337.
McGee, R. X. 1964, *Australian J. Phys.*, **17**, 515.
———. 1965, *Symposium on the Magellanic Clouds* (Mount Stromlo Observatory), p. 34.
McGee, R. X., and Milton, J. A. 1966*a*, *Australian J. Phys.*, **19**, 343.
———. 1966*b*, *Australian J. Phys.*, *Ap. Suppl.*, No. 2.
Matthews, T. A., Morgan, W. W., and Schmidt, M. 1964, *Ap. J.*, **140**, 35.
Mayall, N. U. 1960, *Ann. d'Ap.*, **23**, 344.
Mayall, N. U., and Vaucouleurs, A. de. 1962, *A.J.*, **67**, 363.
Meng, Y. S., and Kraus, J. D. 1966, *A.J.*, **71**, 70.
Menon, T. K. 1958, *Ap. J.*, **127**, 28.
Münch, G. 1959, *Pub. A.S.P.*, **71**, 101.
Oort, J. H. 1965, *Transactions IAU*, **12A** (New York: Academic Press) p. 789.
Peimbert, M., and Spinrad, H. 1970, *Ap. J.*, **159**, 809.
Poveda, A. 1961, *Ap. J.*, **134**, 910.
Reddish, V. C. 1961, *Observatory*, **81**, 19.
Reifenstein, E. C., Wilson, T. L., Burke, B. F., Mezger, P. G., and Altenhoff, W. J. 1970, *Astr. and Ap.*, **4**, 357.
Roberts, M. S. 1962*a*, *A.J.*, **67**, 431.
———. 1962*b*, *ibid.*, p. 437.
———. 1963, *Ann. Rev. Astr. and Ap.*, **1**, 149.
———. 1964, unpublished 60-foot observations.
———. 1965, *Ap. J.*, **142**, 148.
———. 1966*a*, *Ap. J.*, **144**, 639.
———. 1966*b*, *Phys. Rev. Letters*, **17**, 1203.
———. 1966*c*, unpublished 300-foot observations.
———. 1967, *IAU Symposium No. 31*, ed. H. van Woerden (New York: Academic Press), p. 189.
———. 1968*a*, in *Interstellar Ionized Hydrogen*, ed. Y. Terzian (New York: W. A. Benjamin), p. 617.
———. 1968*b*, *Ap. J.*, **151**, 117.
———. 1968*c*, *A.J.*, **73**, 945.
———. 1969, *ibid.*, **74**, 859.
———. 1970, *Ap. J.* (*Letters*), **161**, L9.
———. 1971, in *External Galaxies and Quasi-stellar Objects*, ed. D. S. Evans (Dordrecht: Reidel), p. 12.
Roberts, M. S., and Warren, J. L. 1970, *Astr. and Ap.*, **6**, 165.
Robinson, B. J., and Koehler, J. A. 1965, *Nature*, **208**, 993.
Robinson, B. J., and van Damme, K. J. 1966, *Australian J. Phys.*, **19**, 111.
Rogstad, D. H., Rougoor, G. W., and Whiteoak, J. B. 1967, *Ap. J.*, **150**, 9.
Rogstad, D. H., and Shostak, G. S. 1971, *Astr. and Ap.*, **13**, 99.
Rubin, V. C., Burbidge, E. M., Burbidge, G. R., Crampin, D. J., and Prendergast, K. H. 1965, *Ap. J.*, **141**, 759.
Rubin, V. C., Burbidge, E. M., Burbidge, G. R., and Prendergast, K. H. 1964, *Ap. J.*, **140**, 80.
Rubin, V. C., and Ford, W. K., Jr. 1968, *Ap. J.*, **154**, 431.
———. 1970, *Ap. J.*, **159**, 379.
Sandage, A. R. 1962, in *Symposium on Stellar Evolution*, ed. J. Sahade (Argentina: La Plata Observatory), p. 1.
Seielstad, G. A., and Whiteoak, J. B. 1965, *Ap. J.*, **142**, 616.
Shobbrook, R. R., and Robinson, B. J. 1967, *Australian J. Phys.*, **20**, 131.
Shostak, G. S. 1971, private communication.
Turner, K. C. 1970, *Annual Report of the Director, 1969–1970, Department of Terrestrial Magnetism* (Washington: Carnegie Institute of Washington).
van Damme, K. J. 1966, *Australian J. Phys.*, **19**, 687.

Vaucouleurs, G. de. 1961, *Ap. J. Suppl.*, **5**, 233 (No. 48).
Vaucouleurs, G. de, and Vaucouleurs, A. de. 1963, *A.J.*, **68**, 96.
Vessot, R., Peters, H., Vanier, J., Beehler, D., Harrach, R., Allan, D., Glaze, D., Snider, C., Baines, J., Cutler, L., and Bodily, L. 1966, *IEEE Trans. Instrumentation and Measurement*, IM-15, p. 165.
Volders, L. 1959, *B.A.N.*, **14**, 323.
Volders, L., and Högbom, J. A. 1961, *B.A.N.*, **15**, 307.
Weliachew, L. 1969, *Astr. and Ap.*, **3**, 402.
Wild, J. P. 1952, *Ap. J.*, **115**, 206.

CHAPTER 10

The Formation and Early Dynamical History of Galaxies

GEORGE B. FIELD
Center for Astrophysics, Harvard College Observatory and Smithsonian Astrophysical Observatory

1. INTRODUCTION

JEANS (1928) conjectured that galaxies condense from a gaseous background, which he assumed is uniform and at rest. Density perturbations of wavelength greater than the Jeans length,

$$\lambda_J = c_s \left(\frac{\pi}{G\rho}\right)^{1/2} \tag{1}$$

where c_s is the sound speed and ρ the density, are unstable toward gravitational collapse. Angular momentum leads to formation of a disk. Efficient radiation of the energy of collapse permits subunits to become unstable, and the system fragments into stars. The Jeans hypothesis still dominates discussion of galaxy formation (Oort 1964). Jeans's error of assuming a static, uniform background, which does not satisfy his basic equations, has been corrected by considering the gas to be expanding according to a relativistic cosmological model. Gravitational instability still exists in modified form.

Jeans's hypothesis is also applied to star formation (Spitzer 1967), where the background is taken to be a nonuniform distribution of gas in motion, as indicated by observations of interstellar gas. Because observational data on intergalactic gas are more uncertain than those on interstellar gas, the discussion of galaxy formation is more uncertain than that of star formation. In particular, since the analysis depends on conditions early in the evolution of the Universe, it is necessary to make cosmological assumptions at the outset. It is hoped that ultimately the study of galaxy formation will yield cosmological information. The present chapter is primarily a review of recent work based upon the Jeans hypothesis, which is far more extensive than work based on other hypotheses. The latter, which include the steady-state theory, origin in dense configurations, and gravitational clustering (Layzer 1964), are discussed where appropriate.

Received May 1968, at which time the author was affiliated with the Department of Astronomy, University of California, Berkeley, California.

Section 2 deals with gravitational instability, § 3 with contraction and effects of angular momentum, § 4 with fragmentation in elliptical and disk systems, and § 5 with observational data. Formation of multiple systems of galaxies, although it poses basic problems, has received relatively little attention from researchers, and is dealt with in § 5.4.

2. BEHAVIOR OF DENSITY PERTURBATIONS

2.1. Gravitational Instability with Radiation Absent

The linear theory of density perturbations in an expanding Universe was treated by Lifshitz (1946) in the framework of general relativity. His work will be considered in § 2.2, but it is helpful first to consider Bonnor's (1957) Newtonian treatment, which agrees with Lifshitz in the important case of no radiation. Bonnor treats spherical perturbations of a uniform expanding gas, including the effect of gas pressure. The density ρ, radial velocity u, pressure p, and radial gravitational acceleration g depend on the time t and the radial coordinate r through the equations

$$\frac{d\rho}{dt} + \frac{\rho}{r^2}\frac{\partial}{\partial r}(r^2 u) = 0 , \tag{2}$$

$$\frac{du}{dt} + \frac{1}{\rho}\frac{\partial p}{\partial r} - g = 0 , \tag{3}$$

$$\frac{d}{dt}\ln\left(\frac{p}{\rho^\gamma}\right) = 0 , \tag{4}$$

and

$$\frac{1}{r^2}\frac{\partial}{\partial r}(r^2 g) + 4\pi G\rho = 0 , \tag{5}$$

where $d/dt = \partial/\partial t + u\partial/\partial r$. Adiabatic conditions are assumed. The validity of Newtonian equations in cosmology has been discussed by McCrea (1955) and by Callan, Dicke, and Peebles (1965); they are valid for a uniform gas if $r \ll$ radius of Universe, $u \ll c$, and $p \ll \rho c^2$. Irvine (1965) has shown that they are also valid for spherical perturbations, because gravitational forces due to distant matter cancel by symmetry. A zero-order solution corresponding to uniform expansion is obtained by letting

$$u_0(r, t) = \frac{r}{R}\frac{dR}{dt} , \tag{6}$$

where R is a function of t alone. Equation (6) implies that $r(t)$ = constant $\times R(t)$, where the constant depends on the shell of particles under consideration. From equations (2) and (6) one finds that

$$\frac{d\rho_0}{dt} + \frac{3\dot{R}\rho_0}{R} = 0 ; \quad \rho_0(t) = \rho_0(t_1)\left[\frac{R(t_1)}{R(t)}\right]^3 . \tag{7}$$

Hence if $\rho_0(t_1)$ is independent of r, so is $\rho_0(t)$, and the expansion preserves uniform density. It follows from equation (4) that

$$p_0(t) = p_0(t_1)\left[\frac{R(t_1)}{R(t)}\right]^{3\gamma} \tag{8}$$

remains uniform (no pressure gradients) if $p_0(t_1)$ is independent of r. Equations (3), (6), (7), and (8) yield

$$\ddot{R}/R = g_0/r \,, \tag{9}$$

while from equation (5)

$$g_0/r = -4\pi G\rho_0/3 = -4\pi G\rho_0(t_1)R_1^3/3R^3 \,. \tag{10}$$

From equations (9) and (10)

$$\ddot{R}/R + 4\pi G\rho_0(t_1)R_1^3/3R^3 = 0 \,, \tag{11}$$

which has the integral

$$\tfrac{1}{2}\dot{R}^2 - 4\pi G\rho_0(t_1)R_1^3/3R = -\tfrac{1}{2}kc^2 \,, \tag{12}$$

where the constant has been written to agree with relativistic cosmology, where k/R^2 ($k = \pm 1$ or 0) is three-dimensional curvature (Adler, Bazin, and Schiffer 1965). Since $4\pi\rho_0(t_1)R_1^3/3$ is the mass $\mathfrak{M}$ within R_1 at time t_1, which is constant according to equation (7), equation (12) becomes

$$\tfrac{1}{2}\dot{R}^2 - G\mathfrak{M}/R = -kG\mathfrak{M}/R_M \,, \tag{13}$$

where $R_M \equiv 2G\mathfrak{M}/c^2$. Equation (13) expresses the conservation of energy for a particle at radius R. If $k = +1$, the energy is negative, and the expansion reverses at $R = R_M$, its maximum value. If $k = -1$, R_M marks the transition to unaccelerated motion, while if $k = 0$, R_M has no physical significance.

Following Layzer (1964), we nondimensionalize by letting

$$R' = \frac{R}{R_M} \,, \quad t' = \left(\frac{2G\mathfrak{M}}{R_M^3}\right)^{1/2} t \,, \tag{14}$$

and obtain

$$\left(\frac{dR'}{dt'}\right)^2 - \frac{1}{R'} = -k \,. \tag{15}$$

The resulting equation,

$$dt' = \left(\frac{R'}{1 - kR'}\right)^{1/2} dR' \,, \tag{16}$$

is solved by the substitutions

$$\begin{aligned} R' &= \sin^2\alpha \,, & k &= +1 \,, \\ &= \alpha^2 \,, & k &= 0 \,, \\ &= \sinh^2\alpha \,, & k &= -1 \,, \end{aligned} \tag{17}$$

to obtain

$$\begin{aligned} t' &= \alpha - \sin\alpha\cos\alpha \,, & k &= +1 \,, \\ &= \tfrac{2}{3}\alpha^3 \,, & k &= 0 \,, \\ &= \sinh\alpha\cosh\alpha - \alpha \,, & k &= -1 \,. \end{aligned} \tag{18}$$

For all values of k, R' and t' vanish at $\alpha = 0$. For $k = +1$, $R' = 1$ when $\alpha = \pi/2$ and $t' = \pi/2$. This corresponds to

$$t = t_M = \left(\frac{3\pi}{32G\rho_M}\right)^{1/2} \,, \tag{19}$$

where ρ_M is the density when $R = R_M$. R is symmetric about $t = t_M$, approaching zero again at $\alpha = \pi$, $t = 2t_M$. For $k = 0$,

$$R = \left(\frac{9G\mathfrak{M}t^2}{2}\right)^{1/3} \tag{20}$$

For $k = -1$, R behaves like equation (20) for $R \ll R_M$, but becomes proportional to t for $R \gg R_M$ (fig. 1).

To study perturbations, we let

$$\rho = \rho_0(1 + s)\,, \quad u = u_0 + u_1\,, \quad p = p_0 + p_1\,, \quad \text{and} \quad g = g_0 + g_1 \tag{21}$$

and drop higher-order terms in the perturbations (subscript 1) to obtain

$$\frac{ds}{dt} + \frac{1}{r^2}\frac{\partial}{\partial r}(r^2 u_1) = 0\,, \tag{22}$$

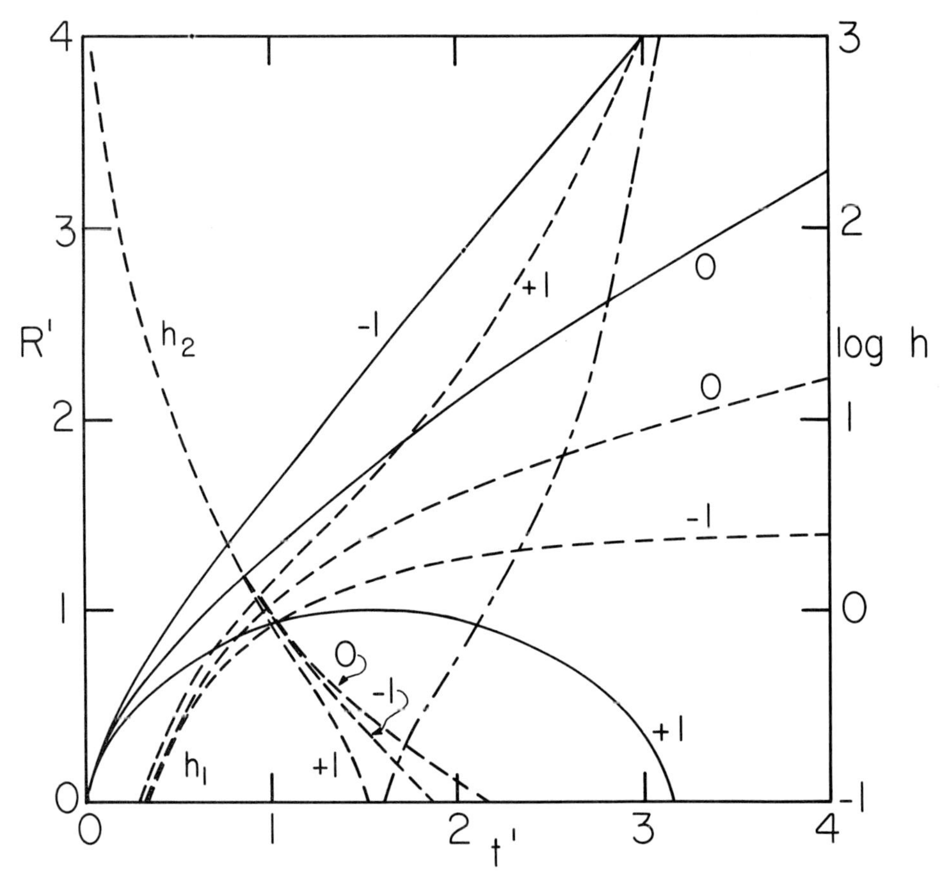

FIG. 1.—Variation of the scale factor and the amplitude of various perturbations in a dust-filled Friedmann model. Scale and time are normalized (eq. [14]). The curves are labeled by the value of the curvature index k. Growing perturbations (h_1) and decaying ones (h_2) are shown. When $k = +1$, $h_2 < 0$ for $t' > \pi/2$ (indicated by alternating short and long dashes).

$$\frac{du_1}{dt} + u_1 \frac{\partial u_0}{\partial r} + \frac{1}{\rho_0} \frac{\partial p_1}{\partial r} - g_1 = 0 , \tag{23}$$

$$\frac{d}{dt}\left(\frac{p_1}{p_0} - \gamma s\right) = 0 , \tag{24}$$

and

$$\frac{1}{r^2} \frac{\partial}{\partial r} (r^2 g_1) + 4\pi G \rho_0 s = 0 , \tag{25}$$

where now $d/dt = \partial/\partial t + u_0 \partial/\partial r$. Equation (24) indicates that $(p_1/p_0) - \gamma s$ is constant for each particle; the constant vanishes for isentropic perturbations. In terms of the dimensionless Lagrangian coordinate

$$x = r/R , \tag{26}$$

which is constant on unperturbed particle trajectories according to equation (6),

$$\frac{d}{dt} = \left(\frac{\partial}{\partial t}\right)_r + u_0 \left(\frac{\partial}{\partial r}\right)_t = \left(\frac{\partial}{\partial t}\right)_x ; \quad \left(\frac{\partial}{\partial r}\right)_t = \frac{1}{R}\left(\frac{\partial}{\partial x}\right)_t . \tag{27}$$

Equations (22), (23), and (25) become

$$R \frac{\partial s}{\partial t} + \frac{1}{x^2} \frac{\partial}{\partial x} (x^2 u_1) = 0 , \tag{28}$$

$$\left(\frac{\partial}{\partial t} + \frac{\dot{R}}{R}\right) u_1 + \frac{\gamma p_0}{\rho_0 R} \frac{\partial s}{\partial x} - g_1 = 0 , \tag{29}$$

and

$$\frac{1}{x^2} \frac{\partial}{\partial x} (x^2 g_1) + 4\pi G \rho_0 R s = 0 . \tag{30}$$

Application of $(\partial/\partial t + \dot{R}/R)$ to equation (28) and use of equations (29) and (30) gives

$$\begin{aligned} R \frac{\partial^2 s}{\partial t^2} &+ 2\dot{R} \frac{\partial s}{\partial t} + \frac{1}{x^2} \frac{\partial}{\partial x} \left[x^2 \left(\frac{\partial u_1}{\partial t} + \frac{\dot{R}}{R} u_1\right)\right] \\ &= R\left(\frac{\partial^2 s}{\partial t^2} + 2 \frac{\dot{R}}{R} \frac{\partial s}{\partial t}\right) + \frac{1}{x^2} \frac{\partial}{\partial x} \left[x^2 \left(g_1 - \frac{\gamma p_0}{\rho_0 R} \frac{\partial s}{\partial x}\right)\right] \\ &= R\left[\frac{\partial^2 s}{\partial t^2} + 2 \frac{\dot{R}}{R} \frac{\partial s}{\partial t} - 4\pi G \rho_0 s - \frac{\gamma p_0}{\rho_0 R^2 x^2} \frac{\partial}{\partial x} \left(x^2 \frac{\partial s}{\partial x}\right)\right] = 0 , \end{aligned} \tag{31}$$

which has the separable solution

$$s(x, t) = \frac{\sin \kappa x}{\kappa x} h(t) , \tag{32}$$

where $h(t)$ satisfies

$$\frac{d^2 h}{dt^2} + 2 \frac{\dot{R}}{R} \frac{dh}{dt} - 4\pi G \rho_0 h + \frac{\gamma p_0 \kappa^2}{\rho_0 R^2} h = 0 . \tag{33}$$

The wavelength $\lambda = 2\pi R/\kappa$ expands with the Universe. The mass of the perturbation $\mathfrak{M}_1$ and the zero-order mass $\mathfrak{M}_0$ contained within the first zero of $\sin \kappa x$ are related by

$$\mathfrak{M}_0 = \frac{\pi \rho_0 \lambda^3}{6} ; \quad \mathfrak{M}_1 = \frac{3\mathfrak{M}_0 h(t)}{\pi^2} . \tag{34}$$

In effect, Jeans took $R(t)$ constant in equation (33), a procedure which is inconsistent with equation (11). The second term vanishes; and there are solutions of the form e^{nt}, where

$$n^2 = 4\pi G\rho_0 - \frac{\gamma p_0 \kappa^2}{\rho_0 R^2} = 4\pi G\rho_0 - \frac{4\pi^2 c_s^2}{\lambda^2}, \tag{35}$$

since $\gamma p_0/\rho_0 = c_s^2$. One root of equation (35) is therefore real and positive (instability) if $\lambda > \lambda_J$, given by equation (1). The expansion modifies this result by introducing the second term of equation (33), and by making ρ_0, p_0, and R depend on t. The nondimensional form of equation (33) is

$$\frac{d^2h}{dt'^2} + \frac{2}{R'}\frac{dR'}{dt'}\frac{dh}{dt'} - \frac{3}{2R'^3}\left[1 - \frac{\kappa^2 c_M{}^2 (R')^{4-3\gamma}}{4\pi G\rho_M R_M{}^2}\right]h = 0\,, \tag{36}$$

where

$$c_M \equiv (\gamma p_M/\rho_M)^{1/2} \tag{37}$$

is the sound speed when $R = R_M$. Using equations (17) and (18), we can write equation (36) as

$$\frac{d^2h}{d\alpha^2} + 2\cot\alpha\frac{dh}{d\alpha} - 6\operatorname{cosec}^2\alpha[1 - \sigma(\sin^2\alpha)^{4-3\gamma}]h = 0\,, \quad (k = +1)\,, \tag{38}$$

$$\frac{d^2h}{d\alpha^2} + 2\alpha^{-1}\frac{dh}{d\alpha} - 6\alpha^{-2}[1 - \sigma(\alpha^2)^{4-3\gamma}]h = 0\,, \quad (k = 0)\,, \tag{39}$$

and

$$\frac{d^2h}{d\alpha^2} + 2\coth\alpha\frac{dh}{d\alpha} - 6\operatorname{cosech}^2\alpha[1 - \sigma(\sinh^2\alpha)^{4-3\gamma}]h = 0\,, \quad (k = -1)\,, \tag{40}$$

where

$$\sigma \equiv \frac{\kappa^2 c_M{}^2}{4\pi G\rho_M R_M{}^2} = \frac{\pi c_M{}^2}{G\rho_M \lambda_M{}^2}\,. \tag{41}$$

Here $\lambda_M = 2\pi R_M/\kappa$ is the wavelength when $R = R_M$. Solutions are known for $\sigma = 0$ (zero pressure or infinite wavelength). Those which are regular at $\alpha = 0$ are

$$\begin{aligned} h_1 &= \tfrac{5}{2}(3\operatorname{cosec}^2\alpha - 3\alpha\operatorname{cosec}^2\alpha\cot\alpha - 1)\,, && k = +1, \\ &= \alpha^2\,, && k = 0\,, \\ &= \tfrac{5}{2}(3\operatorname{cosech}^2\alpha - 3\alpha\operatorname{cosech}^2\alpha\coth\alpha + 1)\,, && k = -1\,, \end{aligned} \tag{42}$$

which all behave like α^2 as $\alpha \to 0$. Since R' also behaves this way,

$$h_1 = R' + O(R'^2)\,. \tag{43}$$

Hence gravitational instability exists for sufficiently long wavelengths in an expanding universe. However, the growth is like a power of the time rather than exponential, and it should be noted that even though the density contrast $s = \rho_1/\rho_0$ increases, ρ_1 itself actually decreases. The required second solutions to equations (38)–(40) are not regular at $\alpha = 0$:

$$\begin{aligned} h_2 &= \operatorname{cosec}^2\alpha\cot\alpha\,, && k = +1\,, \\ &= \alpha^{-3}\,, && k = 0\,, \\ &= \operatorname{cosech}^2\alpha\coth\alpha\,, && k = -1\,, \end{aligned} \tag{44}$$

which all behave like α^{-3} as $\alpha \to 0$. Both h_1 and h_2 are plotted in figure 1, following Harrison (1967*a*), who reviews the material of this section.

While linear combinations of these solutions have been discussed by Lifshitz (1946), Bonnor (1957), Adams, Mjolsness, and Wheeler (1958), Hunter (1962), and Savedoff and Vila (1962), solutions of the general equations (38)–(40) have been given for arbitrary values of σ and γ only in certain asymptotic limits. There is a simple solution for arbitrary σ if $k = 0$ and $\gamma = 4/3$:

$$h = \alpha^m , \quad m = -\frac{1}{2} \pm \frac{5}{2}\left(1 - \frac{24\sigma}{25}\right)^{1/2} . \tag{45}$$

The minus sign yields a stable mode, which corresponds to h_2 when $\sigma = 0$. The plus sign yields an unstable mode for $\sigma < 1$, which corresponds to h_1 when $\sigma = 0$. The instability criterion for this case is therefore $\sigma < 1$ or

$$\lambda_M > \lambda_{M,J} = c_M \left(\frac{\pi}{G\rho_M}\right)^{1/2} , \tag{46}$$

exactly as in the classical Jeans criterion, but with the parameters evaluated when $R = R_M$. The ratio of the pressure to the gravitational term in equation (33) is, in general, $\pi c_s^2/\lambda^2 G\rho_0$, which is independent of α if $\gamma = 4/3$ according to equation (39), and equal to σ. The criterion $\sigma < 1$ is therefore equivalent to $\pi c_s^2/\lambda^2 G\rho_0 < 1$, or to equation (1), evaluated at any time. If $\gamma \neq 4/3$, equation (1) is still the criterion that the pressure term is smaller than the gravitational; and as Bonnor shows, it is therefore the criterion of instability. Now, however, both terms of the inequality vary with α, so the criterion is not simply $\sigma < 1$, independent of α. Thus, if $\gamma > 4/3$, pressure forces decrease faster than gravitational ones, so the coefficients of σ in equations (38)–(40) decrease with α. Hence, a value of κ which is initially stable can become unstable when the pressure forces have dropped sufficiently. The transition occurs at $\alpha = \alpha_*$, defined as the moment when $\lambda = \lambda_J$. From equations (38)–(40) one can express α_* as

$$\begin{aligned} \alpha_* &= \sin^{-1}\sigma_* , & k &= +1 , \\ &= \sigma_* , & k &= 0 , \\ &= \sinh^{-1}\sigma_* , & k &= -1 , \end{aligned} \tag{47}$$

where

$$\sigma_* \equiv \sigma^{(3\gamma-4)/2} . \tag{48}$$

The conditions for instability are given in table 1. If $k = 0$ or -1, every mode undergoes a transition if $\gamma > 4/3$; while if $k = +1$, a transition occurs only if $\sigma < 1$.

TABLE 1

CONDITIONS FOR GRAVITATIONAL INSTABILITY

γ	$k = +1$	$k = 0$	$k = -1$
$<4/3$......	$\sigma<1$, all α $\sigma>1$, $\alpha<\alpha_*$	$\alpha<\alpha_*$	$\alpha<\alpha_*$
$=4/3$......	$\sigma<1$	$\sigma<1$	$\sigma<1$
$>4/3$......	$\sigma<1$, $\alpha>\alpha_*$	$\alpha>\alpha_*$	$\alpha>\alpha_*$

In order to apply equation (1) or its equivalent, table 1, to the Universe, its temperature-density history is needed. Although this information is lacking in detail, we shall see in § 2.2 that the present analysis begins to be valid at $R \simeq 10^{-3} R_1$, where R_1 refers to the present epoch. At earlier times radiation effects, which we have so far neglected, become important. At this epoch $T_0 = 3000°$ K and $\rho_0 = 2 \times 10^{-20}$ g cm^{-3}, so that $c_s =$ 6.5 km s^{-1}, $\lambda_J = 3.2 \times 10^{19}$ cm, and the Jeans mass is

$$\mathfrak{M}_J = \tfrac{1}{6}\pi \rho_0 \lambda_J{}^3 = 1.6 \times 10^5 \,\mathfrak{M}_\odot \,. \tag{49}$$

Perturbations of galactic order, $10^{11}\,\mathfrak{M}_\odot$, would have been unstable and would therefore begin to contract like the pressure-free solutions h_1 (Peebles 1965).

The linear theory gives information only about the early stages of the process, when the density differs very little from the background. Since the galaxies are now much (10^6 times) more dense than the background, it gives little information about the later stages. However, Peebles (1967*a*) has treated the nonlinear regime by the following argument. Label by G a spherical cloud which is ultimately to become a galaxy; it has mass $\mathfrak{M}_G$, radius R_G, and density (assumed uniform) $\rho_G = 3\mathfrak{M}_G/4\pi R_G{}^3$. In the early stages it can be described by the linear analysis and hence can be built up out of modes with various wavelengths. The wavelengths involved will mostly be of the order of R_G, which as we have seen is much greater than λ_J. The important modes are therefore essentially pressure-free; as the sphere contracts relative to the background, it can be followed by the pressure-free equations (11), (13), (14), (17), and (18) even into the nonlinear range. Consider another sphere having the same mass, but different radius and density, which represents the background. This typical cloud is described by the same equations, and hence differs from the first cloud only in the value of k/R_M (energy) and in the value of the arbitrary constant (phase) obtained in integrating equation (16) to find equation (18). Obviously $k = +1$ for the galactic cloud, but k for the typical cloud depends on the cosmological model. If the clouds differ little in energy and in phase, they differ little in density, and so the density difference ρ_1 should obey equations (42) and (44). To show this, consider $k = +1$ for the Universe as an example. Take

$$R = R_M \sin^2 \alpha \tag{50}$$

for the typical cloud, and define the phase so that

$$t = \left(\frac{R_M{}^3}{2G\mathfrak{M}_G}\right)^{1/2} (\alpha - \sin \alpha \cos \alpha) \,. \tag{51}$$

For the galactic cloud,

$$R_G = R_{M,G} \sin^2 \alpha_G \tag{52}$$

and

$$t = \left(\frac{R_M{}^3{}_{,G}}{2G\mathfrak{M}_G}\right)^{1/2} (\alpha_G - \sin \alpha_G \cos \alpha_G + \delta) \,, \tag{53}$$

where δ is an arbitrary phase shift. Hence

$$\alpha_G - \sin \alpha_G \cos \alpha_G + \delta = \left(\frac{\rho_{M,G}}{\rho_M}\right)^{1/2} (\alpha - \sin \alpha \cos \alpha) \tag{54}$$

and

$$h \equiv \frac{\rho_G - \rho}{\rho} = \frac{\rho_{M,G} \sin^6 \alpha}{\rho_M \sin^6 \alpha_G} - 1 \,. \tag{55}$$

Using the fact that the differences of energy and phase are small, one has

$$\delta \ll 1\,, \qquad \epsilon \equiv \frac{\rho_{M,G} - \rho_M}{\rho_M} \ll 1\,, \qquad \Delta \equiv \alpha_G - \alpha \ll 1\,, \tag{56}$$

so that equations (54) and (55) expand to give

$$\Delta = -\tfrac{1}{2}\delta \operatorname{cosec}^2 \alpha - \tfrac{1}{4}\epsilon(\cot \alpha - \alpha \operatorname{cosec}^2 \alpha) \tag{57}$$

and

$$h = \epsilon - 6\Delta \cot \alpha\,. \tag{58}$$

When equations (57) and (58) are combined, we obtain

$$h = \tfrac{1}{5}\epsilon h_1 + 3\delta h_2\,, \tag{59}$$

where h_1 and h_2 are given by equations (42) and (44). It is therefore apparent that the solution h_1 to the hydrodynamic equations represents the effect of a perturbation in energy. The galactic cloud has a smaller R_M and a larger ρ_M, and lags increasingly far behind the typical cloud and collapses before it does, so that $h_1 \to \infty$ as $\alpha \to \pi$. The other solution, h_2, represents a phase perturbation in which the galactic cloud is denser in the expansion phase and less dense in the contraction phase.

We have seen that neglect of pressure is an adequate approximation when $R = 10^{-3} R_1$. Furthermore, it improves as time progresses if $\gamma > 4/3$ (eqs. [38]–[40]) up to the moment $t_{M,G}$. Thereafter the galactic cloud contracts with resulting increase in pressure. Even if this pressure rise is maximum ($\gamma = 5/3$), the approximation is adequate almost up to $2t_{M,G}$, when the cloud reaches its original density again. In § 3.4 we show that radiative cooling during collapse probably improves the approximation. The energy of a point at the edge of the cloud is $-G\mathfrak{M}_G/R_{M,G}$ ergs g^{-1}. Averaged over the whole cloud it is $-3G\mathfrak{M}_G/5R_{M,G}$, which in view of equation (19) can be written

$$\langle E \rangle = -\frac{3}{10}\left(\frac{\pi G\mathfrak{M}_G}{t_{M,G}}\right)^{2/3}. \tag{60}$$

As the cloud can only lose energy in forming a galaxy, the observed energy of galaxies should, if anything, be lower. Fish (1964) found that for elliptical galaxies there is the empirical relation $\langle E \rangle = -2.4 \times 10^{-8}\, \mathfrak{M}^{1/2}$ ergs g^{-1}. Combining these expressions, we find that

$$t_{M,G} \geq 3.5 \times 10^7 (\mathfrak{M}_{11})^{1/4} \text{ years}\,, \tag{61}$$

where $\mathfrak{M}_{11} = \mathfrak{M}/10^{11}\, \mathfrak{M}_\odot$. If galaxies formed before this time, they would be more tightly bound than is observed (Weymann 1966*a*). Note that the dependence on $\mathfrak{M}$ is weak, so that if in fact all galaxies formed at $2t_{M,G} = 7 \times 10^7$ years without subsequent energy loss, the relation deduced by Fish would be approximately explained.

We may leave open the question of the actual value of $t_{M,G}$ and ask: What relative density perturbation h at the time when $R = 10^{-3} R_1$ is required to form a galaxy at time $2t_{M,G}$? For a galactic cloud,

$$\rho_G = \frac{\rho_{M,G}}{\sin^6 \alpha_G} \simeq \frac{\rho_{M,G}}{\alpha_G{}^6}(1 + \alpha_G{}^2)\,, \tag{62}$$

where we have assumed that $\alpha_G(10^{-3} R_1) \ll 1$ and have expanded to the first nontrivial term. If ρ_G and $\rho \to \infty$ as $t \to 0$ ($\delta = 0$), then

$$t' = \alpha_G - \sin \alpha_G \cos \alpha_G \simeq \tfrac{2}{3}\alpha_G{}^3(1 - \tfrac{1}{5}\alpha_G{}^2) \tag{63}$$

to the same accuracy. From the definitions of t' and of t_M, $t' = \pi t/2t_{M,G}$, so equation (63) can be solved to give

$$\alpha_G = \left(\frac{3\pi t}{4t_{M,G}}\right)^{1/3}\left[1 + \frac{1}{15}\left(\frac{3\pi t}{4t_{M,G}}\right)^{2/3}\right], \tag{64}$$

so that

$$\rho_G = \frac{1}{6\pi G t^2}\left[1 + \frac{3}{5}\left(\frac{3\pi t}{4t_{M,G}}\right)^{2/3}\right], \tag{65}$$

where we have used equation (19). For a cosmological model k with $\alpha = 0$ at $t = 0$,

$$\rho = \frac{1}{6\pi G t^2}\left[1 + \frac{3}{5}k\left(\frac{3\pi t}{4t_M}\right)^{2/3}\right] \tag{66}$$

to the same accuracy. Hence

$$h = \frac{\rho_G - \rho}{\rho} \simeq \frac{3}{5}\left(\frac{3\pi t}{4t_{M,G}}\right)^{2/3}\left[1 - k\left(\frac{t_{M,G}}{t_M}\right)^{2/3}\right], \tag{67}$$

where a correction term of order $(t/t_M)^{2/3}$ has been ignored. If $H \equiv \dot{R}/R$ and $q \equiv -R\ddot{R}/\dot{R}^2$ represent the present values of the Hubble constant and deceleration parameter, the relation

$$H^2 = 4\pi G \rho_1/3q\,, \tag{68}$$

involving the present density ρ_1, follows from equation (11). One can then show that

$$(H^2 t^2)^{1/3} = \left(\frac{2}{9q}\right)^{1/3}\frac{R}{R_1}; \qquad (H^2 t_M{}^2)^{1/3} = \left(\frac{\pi^2}{8q}\right)^{1/3}\frac{R_M}{R_1}. \tag{69}$$

Equations (11) and (13) give

$$\frac{R_1}{R_M} = k^{-1}\left(1 - \frac{1}{2q}\right) \tag{70}$$

for $k = \pm 1$. Using equations (69) and (70), t and t_M can be eliminated from equation (67) to give

$$h = \frac{3R}{5R_1}\left[\left(\frac{2^{3/2} H q^{1/2} t_{M,G}}{\pi}\right)^{-2/3} - 1 + \frac{1}{2q}\right]. \tag{71}$$

The case $k = 0$ is handled separately, and gives the first term in brackets. Since $q = \frac{1}{2}$ for this case, equation (71) is correct for all k. Taking $R = 10^{-3} R_1$, $H^{-1} = 10^{10}$ years, and three possible values of q, we give in table 2 the values of h required to form galaxies at various times. Oort's (1958) determination of the density of luminous matter yields a lower limit $q = 0.02$, corresponding to $k = -1$. The value $q = \frac{1}{2}$ corresponds to $k = 0$, while $q = 2$ is an upper limit from the observed magnitude-redshift relation ($k = +1$). The values of $t_{M,G}$ are consistent with equation (61); a mass of $10^{11}\,\mathfrak{M}_\odot$ is assumed. The required perturbations are roughly 1 percent, with larger values for earlier formation and in open models. We turn to the question of whether such perturbations would arise naturally before $R = 10^{-3}R_1$ in § 2.2.

2.2. Gravitational Instability with Radiation Present

We have considered the growth of density disturbances when radiation is absent. This approximation is adequate at the present epoch, since the energy density of radiation ($\sim 10^{-12}$ ergs cm^{-3}) does not exert significant forces on the matter. However, Gamow (1948) pointed out that if the Universe had been very dense in the past, it might also have been very hot, as required to produce the chemical elements in his view. Radiation pressure would therefore have been high, and the previous analysis inapplicable. Gamow predicted (1956) that blackbody radiation with a temperature of 6° K should be present in the Universe now. Dicke, Peebles, Roll, and Wilkinson (1965) suggested a search for it, and radiation having the required isotropy and lack of polarization was detected by Penzias and Wilson (1965) at a wavelength of 7.4 cm. Evidence obtained at seven other wavelengths (see, e.g., Shakeshaft and Webster 1968) confirms the presence of such radiation and indicates that its temperature is about 2.7° K.

TABLE 2

Fractional Density Perturbations Required When $R = 10^{-3} R_1$ to Form Clouds Which Contract at Various Times $t_{M,G}$

$t_{M,G}$ (years)	$-\langle E\rangle$ (ergs g^{-1})	Required Density Perturbation, h		
		$q = 0.02$	$q = 0.5$	$q = 2$
$10^{7.5}$	3.5×10^{14}	0.12	0.038	0.024
10^{8}	1.7×10^{14}	0.065	0.017	0.011
$10^{8.5}$	7.8×10^{13}	0.038	0.0081	0.0056
10^{9}	3.5×10^{13}	0.025	0.0037	0.0028

Several attempts (Layzer 1966, 1968; Kaufman 1965; Hoyle and Wickramasinghe 1967; Narlikar and Wickramasinghe 1967) have been made to explain the observations other than by a primordial radiation field. In this chapter, however, we shall assume that Gamow's interpretation is correct, and take $T = 3°$ K.

A radiation field stabilizes density perturbations if it is strong and well coupled to the gas. Peebles (1965) and Weymann (1966*b*) show that coupling is strong before $R = 10^{-3}R_1$, when recombination occurs ($T = 3000°$ K, $\rho = 2 \times 10^{20}$ g cm^{-3}). After decoupling, radiation and matter behave independently, so only gas pressure opposes gravitational instability (§ 2.1). Before decoupling, the wavelengths of interest for galaxies are optically thick, so that matter and radiation are well coupled and motions are nearly adiabatic. Field and Shepley (1968) generalized the work by Lifshitz (1946) on adiabatic perturbations to include both matter and radiation and arbitrary values of k. A mixture of gas and radiation has the stress-energy tensor

$$T_{ij} = \left(\rho + \frac{p}{c^2}\right) u_i u_j + \frac{p}{c^2} g_{ij} , \tag{72}$$

where

$$\rho = \rho_R + \rho_D , \quad p = p_R = \tfrac{1}{3}\rho_R c^2 . \tag{73}$$

Here u_i is the velocity and g_{ij} the metric tensor. The radiation has mass density ρ_R and pressure p_R, and the matter has mass density ρ_D and pressure $p_D = 0$ (hence D for dust). This is valid because $p_D/p_R < 10^{-7}$. Conventionally, one treats either radiation-filled ($\rho_D = 0$) or dust-filled ($\rho_R = 0$) models, but Chernin (1965) and Nariai, Tomita, and Kato (1967) showed how to include both in unperturbed and perturbed models, respectively. The unperturbed model is described by the line element

$$ds^2 = -R^2 d\eta^2 + R^2[d\chi^2 + F^2(\chi)(d\theta^2 + \sin^2\theta d\varphi^2)] , \tag{74}$$

where

$$R(\eta)d\eta/c = dt \tag{75}$$

is the proper-time interval. The variation of R describes the expansion just as in § 2.1, with the identification $\alpha \leftrightarrow \frac{1}{2}\eta$. The geometry of the space t = constant is described by $F(\chi)$, which is $\sin\chi$, χ, and $\sinh\chi$ for $k = +1$, 0, and -1, respectively. Einstein's equations are

$$R_{ij} - \tfrac{1}{2}Rg_{ij} = \kappa T_{ij} , \tag{76}$$

where here R is the Ricci scalar and $\kappa = 8\pi G/c^2$. From them are derived the equations governing $R(\eta)$, analogous to equations (11) and (12):

$$\tfrac{1}{3}\kappa R^2\rho = \alpha^2 + k ; \quad \kappa R^2 p/c^2 = -2\alpha^2 - k - \alpha' , \tag{77}$$

where

$$\alpha \equiv \frac{R'}{R} = \frac{1}{c}\frac{dR}{dt} , \tag{78}$$

and the prime denotes $d/d\eta$. If

$$\beta \equiv k + \alpha^2 - \alpha' = \tfrac{1}{2}\kappa R^2(\rho + p/c^2) , \tag{79}$$

where the last step follows from equation (77), and

$$\sigma \equiv -\frac{1}{\alpha}\left(\frac{\beta'}{\beta} + \alpha\right) = -\frac{(R\beta)'}{\alpha(R\beta)} , \tag{80}$$

then one can show that

$$\sigma = \frac{3}{c^2}\frac{dp}{d\rho} = 3\frac{c_s{}^2}{c^2} , \tag{81}$$

where $p(\rho)$ is assumed given and c_s is the sound speed in the absence of gravitation. One can show from equation (77) that

$$d(\rho c^2 R^3) + p d(R^3) = 0 , \tag{82}$$

verifying that $p(\rho)$ is the adiabatic relation. If the perturbations are adiabatic, we can put $\delta p = (dp/d\rho)\delta\rho$ and calculate it from equations (80) and (81). Neglecting matter pressure requires that we also neglect internal energy. Nucleon conservation then requires that $\rho_D R^3$ = constant = $3R_0/\kappa$, say. Equation (82) then implies that $\rho_R R^4$ = constant = $3r^2R_0{}^2/\kappa$, where r characterizes the ratio of radiation to matter ($\rho_R/\rho_D = r^2R_0/R$). These relations are used with equation (77) to obtain the solutions

$$\begin{aligned} R/R_0 &= \sin^2(\tfrac{1}{2}\eta) + r\sin\eta , & k &= +1 , \\ &= (\tfrac{1}{2}\eta)^2 + r\eta , & k &= 0 , \\ &= \sinh^2(\tfrac{1}{2}\eta) + r\sinh\eta , & k &= -1 , \end{aligned} \tag{83}$$

which reduce to equation (17) when $r = 0$. One calculates that

$$\sigma = \left(1 + \frac{3R}{4r^2R_0}\right)^{-1}, \tag{84}$$

so that when $R \ll r^2R_0$, $\rho_R \gg \rho_D$, $\sigma \to 1$, and $c_s \to 3^{-1/2}c$. When $R \gg r^2R_0$, $\sigma \to 0$.

From equation (77) and the fact that the Hubble constant $H = c\alpha/R$, k is determined by the sign of $\rho - \rho_c$, where the cosmological density $\rho_c = 3H^2/\kappa c^2 = 1.8 \times 10^{-29}$ g cm^{-3}. The density of luminous matter is much less than ρ_c, while that of nonluminous matter is unknown, so that k cannot be determined. Alternative methods based on red-shift observations (chap. 19) are not conclusive, so that we shall use $k = 0$ as an example. For small η, when radiation is important, equation (83) indicates that differences among models are small anyway. The solution for $k = 0$ can be scaled so that $r = 1$ by letting $\eta^* = s\eta$, $R_0{}^* = R_0/s^2$, and $r^* = sr$. One then finds that

$$\frac{R}{R_0} = \eta(1 + \tfrac{1}{4}\eta)\,; \quad \alpha = \frac{1 + \frac{1}{2}\eta}{\eta(1 + \frac{1}{4}\eta)}\,; \quad \beta = \frac{2 + (3/2)\,\eta + \frac{3}{8}\eta^2}{\eta^2(1 + \frac{1}{4}\eta)^2}\,;$$

$$\sigma^{-1} = 1 + \tfrac{3}{4}\eta + \tfrac{3}{16}\eta^2\,; \quad H = \frac{c}{R_0}\frac{1 + \frac{1}{2}\eta}{\eta^2(1 + \frac{1}{4}\eta)^2}\,;$$

$$\rho_R = \frac{3H^2}{\kappa c^2}(1 + \tfrac{1}{2}\eta)^{-2}\,; \quad \rho_D = \frac{3H^2}{\kappa c^2}\left(\frac{1 + 4\eta^{-1}}{1 + 4\eta^{-1} + 4\eta^{-2}}\right). \tag{85}$$

Using the present value $\rho_R = 6.8 \times 10^{-34}$ g cm^{-3}, we find that now $\eta = 480$, $R/R_0 = 5.7 \times 10^4$, $R_0 = 6.5 \times 10^{20}$ cm, and $R = 3.7 \times 10^{25}$ cm. The epoch $\eta = 1$ marks the transition from a radiation-filled to a dust-filled model; then $R = 5R_0/4$, $\rho_R = 4\rho_D/5$, and $\sigma = 16/31$. Decoupling occurs at $\eta = 13$, when $R/R_0 = 57$, $T_R = T_D = 3000°$ K, and $\rho_D = 1.8 \times 10^{-20}$ g cm^{-3}. R_0 is about equal to the distance to the horizon (cH^{-1}) at the time $\eta = 1$.

Lifshitz showed that density perturbations are accompanied by metric perturbations whose time components may be taken to vanish and whose space components $h^\nu{}_\xi$ are of the form

$$h_{(n)}{}^\nu{}_\xi = \tfrac{1}{3}\mu_{(n)}(\eta)Q_{(n)}(\chi, \theta, \varphi)\delta^\nu{}_\xi + \lambda_{(n)}(\eta)P_{(n)}{}^\nu{}_\xi(\chi, \theta, \varphi)\,, \tag{86}$$

where $Q_{(n)}$ satisfies the wave equation with eigenvalues $E_{(n)} = (n^2 - k)/3$ and where $P_{(n)}{}^\nu{}_\xi$ is a tensor formed from $Q_{(n)}$. Lifshitz introduced the auxiliary scalar functions of η, λ, and μ. Equation (86) is analogous to the separation of variables in equation (32). Lifshitz gives the form of the functions $Q_{(n)}$; for $k = 0$, $Q_{(n)} = e^{in\xi}$, where ξ is a Cartesian coordinate.

The linear analysis is performed by perturbing equation (76) and expressing the results in terms of $h^\nu{}_\xi$, $\delta\rho$, δu_μ, and δp by using equation (72). Then δp is eliminated by using the adiabatic relation. If one defines the new auxiliary functions $\Psi(\eta)$ and $\Phi(\eta)$ by

$$\Psi(\eta) = \frac{R}{\alpha\beta}[\alpha D\mu + (E - k)(\lambda + \mu)] \tag{87}$$

and

$$\Phi(\eta) = \frac{R}{E\beta}[ED\mu + (E - k)D\lambda] \tag{88}$$

with $D = d/d\eta$, $h^\nu{}_\xi$ can be written in terms of these functions, using equation (86), with the result that the remaining relations among $h^\nu{}_\xi$, $\delta\rho$, and δu_μ can be written

$$\frac{\delta\rho}{\rho} \equiv Qh = Q\frac{\alpha\beta\Psi}{3R(\alpha^2 + k)}\,; \qquad \delta u_\mu = \tfrac{1}{2}EP_\mu\Phi\,, \tag{89}$$

where $P_\mu = \frac{1}{3}E^{-1}Q_{;\mu}$ and where (n) is dropped. One readily generalizes relations found by Lifshitz to arbitrary k:

$$D^2\lambda + 2\alpha D\lambda - E(\lambda + \mu) = 0\,, \tag{90}$$

and

$$D^2\mu + (2 + \sigma)\alpha D\mu + (1 + \sigma)(E - k)(\lambda + \mu) = 0\,. \tag{91}$$

Field and Shepley show that μ can be eliminated from equations (87) and (88) to give

$$\Psi = \frac{1}{\beta E}\left(D^2 + \frac{E - k}{\alpha}D - \beta\right)RD\lambda \tag{92}$$

and

$$\Phi = \frac{1}{\beta E}(D^2 - \beta)RD\lambda = \Psi - \frac{E - k}{E}\frac{DRD\lambda}{\alpha\beta}\,. \tag{93}$$

It can be shown that equations (90) and (91) imply

$$\left(D + 2\frac{\alpha'}{\alpha} + \alpha\sigma\right)\beta D\Psi = -(E - k)\frac{\sigma}{\alpha}DRD\lambda\,. \tag{94}$$

If $\sigma = 0$ (large η), Ψ therefore satisfies

$$(D + 2\alpha'/\alpha)\beta D\Psi = 0\,, \tag{95}$$

the solutions to which are equations (42) and (44), when the relation between Ψ and h is used. This shows that the Newtonian analysis gives correct results when $p = 0$. Field and Shepley show, more generally, that equation (94) reduces to equation (33) when the pressure is small but finite, provided that $n \gg 1$ (short wavelength), demonstrating that § 2.1 is a valid special case.

In the present discussion, radiation is important, so the right-hand side of equation (94) must be eliminated by using previous relations. The result is

$$[(D - \sigma'/\sigma + \alpha)(D + 2\alpha'/\alpha + \alpha\sigma) + E\sigma]\beta D\Psi = 0\,, \tag{96}$$

a third-order equation. The solution Ψ = constant is not covariant, and can be eliminated by a coordinate transformation which reduces $h^\nu{}_\xi$ to zero. There are only two covariant solutions. If one defines $H(\eta)$ by

$$\Psi = \int \frac{H}{\alpha}\left(\frac{\sigma}{\beta}\right)^{1/2} d\eta\,, \tag{97}$$

equation (96) takes the simple form

$$(D^2 + \Theta)H = 0\,, \tag{98}$$

where

$$\Theta \equiv \sigma E - \frac{Z''}{Z}\,; \qquad Z = \frac{R}{\alpha}\left(\frac{\beta}{\sigma}\right)^{1/2}. \tag{99}$$

The stability criterion is

$$\Theta > 0\,, \tag{100}$$

since in that case H oscillates and so do Ψ (eq. [97]) and h (eq. [89]). One finds that for radiation-filled models with $k = 0$, $\Theta = E$, implying stability; the solutions for this case were given by Sachs and Wolfe (1967). A small amount of matter changes this situation according to Field and Shepley (1968). For $k = 0$ one finds that

$$\Theta = \frac{E}{1 + \frac{3}{4}\eta + (3/16)\eta^2} - \frac{1 + (3/2)\,\eta + \frac{3}{8}\eta^2}{\eta(1 + \frac{1}{4}\eta)(1 + \frac{1}{2}\eta)^2}\,, \tag{101}$$

so that Θ tends to go from negative values (instability) to positive values (stability) as η increases. This change occurs later the smaller E is; if $E < 9/8$, the change does not occur at all. This behavior is explained as follows. If $E \simeq 1$, one can show that the corresponding wavelength is about equal to the distance to the horizon at $\eta \simeq 1$, the transition epoch. If $E \ll 1$, a condensation at the origin grows under self-gravitation up to $\eta \simeq 1$. Sound waves traveling at about $3^{-1/2}c$ have not had time to propagate inward pressure gradients from the horizon (only attainable with speed c). After $\eta \simeq 1$, the pressure drops, c_s drops below $3^{-1/2}c$, and sound waves from distant points never reach the origin, so a pressure oscillation is not possible. If $E \gg 1$, on the other hand, the condensation starts to undergo pressure oscillations when the wavelength is somewhat less than the distance to the horizon. These oscillations continue up to $\eta \simeq 1$, at which time the pressure drops and the behavior switches to that of a dust-filled model. Equations (89) and (96) have exact solutions in the limiting case of maximum instability, $E \ll 1$. One of the solutions corresponds to a damped mode and so is of little interest for galaxy formation. The other is

$$h = \eta^2\left[\frac{(1 + \frac{3}{4}\eta + (3/16)\eta^2)(1 + (3/10)\eta + (1/40)\eta^2)}{(1 + \frac{1}{4}\eta)^2(1 + \frac{1}{2}\eta)^2}\right]. \tag{102}$$

The factor in brackets approaches 1 as $\eta \to 0$ and 3/10 as $\eta \to \infty$, so that h is essentially η^2 for all η. This corresponds to growth like R^2 for $\eta \ll 1$ and like R for $\eta \gg 1$, in agreement with equation (43). The scale $E = 9/8$ corresponds to 60 Mpc or $10^{16}\,\mathfrak{M}_\odot$ at the present time, and may be related to the existence of superclusters (Abell 1961). We conclude that the analysis of gravitational instability in § 2.1 is applicable at very early phases only for very large scales. Smaller scales, corresponding to individual galaxies, undergo oscillations during the radiation-dominated era before becoming unstable in the matter-dominated era. Recently it has been shown that nonadiabatic effects not considered above actually exert a profound effect during the early phases, damping the density oscillations and thereby not permitting them to survive until instability sets in. Ordinary sound waves are damped by the molecular diffusion of momentum into heat. Here too, the photons which are driving the wave diffuse relative to the matter and cause damping. Before decoupling, the coupling of radiation and matter is due to Compton scattering. The critical perturbation wavelength whose optical depth is unity is therefore $\lambda_c = 1/\kappa\rho_D = 2.5/\rho_D$ for pure hydrogen. Just before decoupling this corresponds to the critical mass

$$\mathfrak{M} = \tfrac{1}{6}\pi\rho_D\lambda_c{}^3 = 1.2 \times 10^7\,\mathfrak{M}_\odot\,. \tag{103}$$

As the Jeans mass is $1.6 \times 10^5\,\mathfrak{M}_\odot$, most scales which are unstable after decoupling are optically thick up to the moment of decoupling. The scales between 10^5 and $10^7\,\mathfrak{M}_\odot$ become optically thin somewhat before decoupling. The appropriate theory for treating the radiation diffusion effect is therefore that valid for large optical depths (Eddington approximation). Silk (1967) based a study of this problem on the relativistic hydrodynamic equations of Misner and Sharp (1965), which include radiative diffusion. Silk's results recapitulate much of what has been said previously about gravitational instability, but we can simplify them to bring out the new points. Ignoring the expansion, one obtains the dispersion relation for waves of the form exp $[i(\omega t - \boldsymbol{k}\cdot\boldsymbol{r})]$:

$$\omega^2 - \tfrac{1}{3}ik^2c^2\sigma\left(\frac{\omega}{\omega_d}\right) - \tfrac{1}{3}k^2c^2\sigma + 4\pi G\rho_D\left(\frac{1}{1-\sigma} - \tfrac{1}{3}i\frac{k^2c^2\sigma}{\omega\omega_d}\right) = 0\,, \tag{104}$$

where

$$\omega_d = \tfrac{4}{3}\kappa c\rho_R \tag{105}$$

is the coefficient in the expression for the radiative drag force, $-\omega_d \boldsymbol{v}$ (Peebles 1967*b*). The last term is gravitational, and is responsible for the instability of large wavelengths as explained above. Since $G\rho_D/(1-\sigma) \simeq t^{-2}$ in a cosmological model, this term is much smaller than $k^2c^2\sigma$ if λ is much less than the distance to the horizon, ct. Hence under these conditions one obtains acoustic waves with velocity $3^{-1/2}c$ from equation (104), in agreement with our previous discussion. However, now the waves are damped, and if $\omega_d \gg \omega$, the damping is weak, and is characterized by the time constant

$$\tau = \frac{1}{\mathrm{Im}\,(\omega)} = \frac{8\kappa(\rho_R + \frac{3}{4}\rho_D)}{k^2c}\,, \tag{106}$$

which agrees with expressions given by Ozernoy (1964) and Peebles (1967*b*) in the limit $\rho_D \gg \rho_R$ where their work is valid. When $\rho_D \simeq \rho_R$, the number of oscillations before damping sets in is roughly equal to the optical depth in one wavelength. If $\rho_R \gg \rho_D$, the time is increased by the factor ρ_R/ρ_D. Silk shows that the intuitive criterion that significant damping occurs before cosmic time t, $\tau < t$, is correct. If $\tau = t$, one obtains a critical mass of dust from the equation $\mathfrak{M}_c = \pi\rho_D\lambda^3/6$, where $\lambda = 2\pi/k$ is found from equation (106) with $\tau = t$. Applied to the time t_* when $\rho_R = \rho_D$ ($\eta \simeq 1$), this gives

$$\mathfrak{M}_{c*} = 2 \times 10^5 q^{-5}\,\mathfrak{M}_\odot\,. \tag{107}$$

Applied to the time of decoupling (4200° K according to Silk), equation (106) with $\rho_D \gg \rho_R$ gives

$$\mathfrak{M}_{cd} = 10^{12} q^{-5/4}\,\mathfrak{M}_\odot\,. \tag{108}$$

The strong dependence on q in equation (107) is due to a proportionality to $\rho_R^{-5/4}$ at time t_*, which is proportional to T_*^{-5} and hence to ρ_D^{-5}. Table 3 gives the critical masses

TABLE 3

SMALLEST MASSES WHICH SURVIVE DAMPING

q	$\mathfrak{M}_{c*}/\mathfrak{M}_\odot$	$\mathfrak{M}_{cd}/\mathfrak{M}_\odot$
0.02	5×10^{13}	2×10^{14}
0.5	5×10^{6}	3×10^{12}
2	5×10^{3}	5×10^{11}

at the times t_* and t_d for the values of q considered in table 2. As τ is decreasing as t increases, most of the damping occurs at the end of the time interval. Smaller masses than the critical ones begin to damp earlier, while larger ones never damp significantly. From the table it appears that most damping occurs between t_* and t_d. Most models would predict that masses $<10^{11}\,\mathfrak{M}_\odot$ are strongly damped. This is of importance for galaxy formation, since perturbations of smaller scales, even supposing they were once excited, damp out by this mechanism.

Up to now we have been considering the effect of damping on the modes which were discussed by Lifshitz. These modes are (almost) adiabatic, being characterized by a positive correlation between matter density and radiation pressure. Equation (104) points to the existence of a third mode which does not exist in the adiabatic analysis of Lifshitz. For large ω_d and $\lambda \ll \lambda_J$, it has

$$i\omega = 4\pi G\rho_D/\omega_d\ , \tag{109}$$

which implies slow growth. Physically, this is a mode in which a density condensation causes an inward acceleration, which is balanced by the viscosity between the inward-moving matter and the outward-streaming radiation. It is referred to as isothermal by Zel'dovich (1966) because the background radiation field, which determines the temperature, remains uniform. Such disturbances are not significantly amplified during the radiation-dominated era because

$$i\omega t = \frac{\rho_D}{\rho_R}\frac{4\pi G\rho_R t^2}{\omega_d t} \simeq \frac{\rho_D}{\rho_R\omega_d t}\ ; \tag{110}$$

the last step follows because $G\rho_R t^2 \simeq 1$ in that phase. Since $\rho_D < \rho_R$, and $\omega_d t \gg 1$ according to Peebles (1967*b*), $i\omega t \ll 1$. On the other hand, such disturbances, once initiated, are not damped like the adiabatic disturbances.

In summary, the analysis of perturbations including the effects of radiation leads to these conclusions: (i) adiabatic perturbations of all scales increase until the horizon extends to include them; (ii) thereafter they execute damped oscillations up to decoupling, with the result that only those with masses exceeding $10^{11}\,\mathfrak{M}_\odot$ survive; (iii) very large masses ($>10^{16}\,\mathfrak{M}_\odot$) never oscillate, but continue to grow from the earliest times up to the present; (iv) at decoupling, the spectrum of adiabatic perturbations consists of masses greater than $10^{11}\,\mathfrak{M}_\odot$, those with $\mathfrak{M} > 10^{16}\,\mathfrak{M}_\odot$ having been amplified particularly. In addition, isothermal perturbations of all scales may be present. In view of these results, and the fact that any density variations of about 1 percent should lead to the formation of galaxies at about 10^8 years, it is possible that galaxies were formed by gravitational instability acting on perturbations which formed much earlier. Zel'dovich (1967) has stressed the role of isothermal perturbations, as they are not subject to the same damping that the adiabatic ones are.

2.3. Other Types of Instability

The weakness of gravitational forces in amplifying density disturbances has encouraged investigation of other forces. Gamow (1949) pointed out that the "mock gravitational force" which results from mutual shadowing effects in a strong radiation field (Spitzer 1941) might be important in a radiation-filled universe. As a result of recent work it is possible to evaluate this possibility quantitatively. According to Weymann (1966*b*), the tendency for matter to cool on a $\gamma = 5/3$ adiabat tends to keep it slightly cooler than the radiation ($\gamma = 4/3$). He finds that if $T_R = T_D(1 + \delta)$, then

$$\delta = \frac{4\pi G m_e t}{\sigma_c c} = \frac{t}{8.5 \times 10^{11} \text{ years}}, \tag{111}$$

where σ_c is the Compton cross-section. As a result there is a net flow of energy from the radiation into the matter, the effective cross-section for this process being

$$\sigma_a = (4kT_R/m_e c^2)\delta\sigma_c . \tag{112}$$

The absorption actually occurs through the recoil experienced by the electron in a Compton scattering. Consider two electron-ion pairs at a distance r. Electron A absorbs energy from the radiation field, resulting in a reduction in the momentum flux which would ordinarily reach electron B, of magnitude $aT_R^4\sigma_a/4\pi r^2$. This causes a force on electron B toward electron A equal to $\sigma_c^2 kaT_R^5\delta/\pi m_e c^2 r^2$, whose inverse-square behavior warrants the name "mock gravity." Since the corresponding gravitational force between the ion pairs is Gm_p^2/r^2, one may define the "constant of mock gravitation,"

$$G' = 2\sigma_c kaT_R^5 Gt/m_p^2 c^3 , \tag{113}$$

where we have substituted δ from equation (111). Since in a radiation-dominated universe

$$T_R = \left(\frac{3c^2}{32\pi Gat^2}\right)^{1/4}, \tag{114}$$

one finds that

$$G'/G = (t/t')^{-3/2} , \tag{115}$$

where

$$t' = \frac{(45)^{1/6} e^{8/3} h^{1/2}}{4\pi (m_p m_e)^{4/3} c^{5/2} G^{5/6}} = 80 \text{ years} . \tag{116}$$

The Gamow effect may therefore be of importance for $t < 80$ years. For example, $T_R = 10^9$ ° K when $t = 400$ seconds. At this temperature, the electrons are only mildly relativistic and can be treated by these equations. Hence from equations (115) and (116) we obtain $G' = 3 \times 10^{10}\, G$. Spitzer showed that in the case of interstellar dust, this effect can lead to instability with a growth rate of $(G'\rho)^{1/2}$. Here, however, the motion of ion pairs is greatly reduced by radiation drag, owing to the high albedo of the electrons. No conclusions can be drawn until such effects are evaluated.

Another instability which might be important in certain circumstances is the thermal instability first described by Parker (1953) and applied to cosmology by Field (1965). A relativistic treatment has been given by Kato, Nariai, and Tomita (1967). In this mechanism, hot intergalactic gas is unstable toward constant-pressure perturbations, because regions of higher density cool more rapidly and become subject to compression by surrounding low-density regions which are still hot. A uniformly expanding gas cooling by free-free emission is unstable according to

$$h = (1 + b/t)^{-9/5} , \tag{117}$$

where $b = \mu\rho_1 t_1^3 C/3A$ and where C is the constant in the free-free emission formula $L = C\rho^2 T^{1/2}$ ergs cm^{-3} s^{-1}. Here $A = \lim_{t\to\infty} (T^{1/2}t)$, a meaningful limit because the expansion becomes adiabatic as $t \to \infty$. If now $\rho = 1.8 \times 10^{-29}$ g cm^{-3} and $T = 4.6 \times 10^4$ ° K, $b = \mu b_1 t_1^3 C/3RA$. At that time T was 7×10^7 ° K, and at earlier times it was decreasing more rapidly than t^{-2} because of cooling. During this period, thermal instability could have contributed to galaxy formation, density fluctuations increasing like $t^{9/5}$ according to equation (117). As the theory is in conflict with the big-bang model we have been discussing (in which $T_D = 20°$ K at 2.5×10^8 years), some means must be found to heat the gas. Doroshkevitch, Zel'dovich, and Novikov (1967) postulate that objects of $\sim 10^5 \mathfrak{M}_\odot$ release energy as they form by gravitational instability acting on isothermal perturbations. However, it is doubtful that they could heat the gas to the required extent. Moreover, the high gas temperatures appear to be in conflict with limits from X-ray astronomy (§ 5.1). The work of Layzer (1963*b*, 1965, 1966) bears upon the possibility of models in which the gas is hotter than the radiation. He defines energy densities associated with density fluctuations:

$$\epsilon_K = \tfrac{1}{2}\langle \rho v^2 \rangle , \qquad \epsilon_G = \tfrac{1}{2}\langle \rho \phi \rangle , \tag{118}$$

where v is peculiar velocity and ϕ the gravitational potential due to density fluctuations. He shows that

$$\epsilon_G = -2\pi G \rho^2 \alpha^2 \int_0^\infty r f(r) dr , \tag{119}$$

where $\alpha^2 = \langle \delta\rho^2 \rangle / \rho^2$ and $f(r)$ is the density autocorrelation function. In the absence of radiation,

$$d(\epsilon R^3) + p dR^3 = 0 , \tag{120}$$

where $\epsilon = \epsilon_K + \epsilon_G$ and $p = p_K + p_G$, with

$$p_K = \tfrac{2}{3}\epsilon_K , \qquad p_G = \tfrac{1}{3}\epsilon_G . \tag{121}$$

According to Layzer, density fluctuations contribute a negative gravitational energy and a negative pressure which can considerably alter the energy balance in cosmology. He argues that often a density fluctuation satisfies the virial theorem,

$$2\epsilon_K + \epsilon_G = 0 , \tag{122}$$

which, together with equation (121), implies that $p = 0$ and $\epsilon R^3 =$ constant. Equation (122) then implies that $\epsilon_K R^3$ and $\epsilon_G R^3$ are individually constant, in contrast to the usual case, where $\epsilon_K R^3$ decreases because of the work done by p_K. The potential energy of the density fluctuations releases energy by increasing α^2, which compensates for the work done by p_K. Since the autocorrelation function can increase no faster than R^2, equation (119) shows that α^2 must increase faster than R to keep $\epsilon_G R^3$ constant. Although it appears not to be necessary to do so, Layzer assumes that $\epsilon_K R^3$, which is proportional to the kinetic energy per gram, is degraded to the molecular level as heat, so its constancy implies that of T_D. From equation (122) and the observed value of ϵ_G due to galaxies and other condensations of matter, Layzer estimates that T_D must lie in the range 10^4 °–10^6 ° K. Kaufman (1965) has calculated the background radiation due to free-free emission by intergalactic gas of this temperature. She obtained agreement with the determination of brightness by Penzias and Wilson (1965) at 7.4 cm with a gas temperature of 4×10^4 ° K. However, since the gas has larger opacity at long wavelengths, saturation is achieved at smaller redshifts, so the predicted brightness temperature increases with the wavelength. The calculated values of 9° and 0.7° K at 21 and 0.9 cm, respectively, disagree with the observed values, 3.2° ± 1° K and 2.6° ± 0.2° K. It may be premature to conclude that Layzer's theory of density fluctuations is therefore incorrect, however, as much of ϵ may be stored at the macroscopic level, as rotation, say. The criticism by Peebles and Dicke (1966) seems unjustified, as no allowance is made for the negative contribution of p_G.

The heating suggested by Layzer could cause a thermal instability to develop, with subsequent galaxy formation, but the phenomenon is of secondary interest because it presupposes large density fluctuations which themselves could lead directly to galaxies by gravitational instability. Fluctuations in the charge distribution (Layzer 1965) might be able to heat the gas without density fluctuations, and so would be of greater interest. Saslaw (1967) searched for instabilities associated with thermonuclear burning and with decoupling. He found an instability associated with the sensitive dependence of the degree of ionization on temperature. However, in the idealized equations he used, this instability does not affect the motion of material. The point warrants further study with more exact equations.

2.4. Galaxy Formation in Other Cosmological Models

Antimatter is usually excluded in conventional models, except at the very early phase when $kT > m_p c^2$, when space was flooded with baryon-antibaryon pairs. The number of pairs is comparable to the number of photons, which is now about 10^9 times the observed number of baryons, so that in this early phase it must be supposed that the net baryon number was everywhere exactly 10^{-9} of the number of baryons if one is to explain the present situation. Harrison (1967*b*, *c*) pointed out that it would be more natural to suppose that, averaged over the entire Universe, the net baryon number is zero, preserving overall matter-antimatter symmetry. To account for the observed presence of baryons, Harrison postulated that there are slight local fluctuations in the ratio of baryons to antibaryons, of the order of 10^{-9} of the total number of baryons. As kT falls below $m_p c^2$, all but the excess particles annihilate, leaving isolated pockets of matter and antimatter which would, upon emergence into the matter-dominated era, behave like the Universe we now see. The present density of the Universe is then a direct result of the amplitude of the fluctuation in the net baryon number. Upon emergence from decoupling, the distribution of gas would be highly inhomogeneous and therefore subject to gravitational instability as explained in § 2.1. If the scale of the fluctuations were that of galaxies, the condensations formed would be of galactic order. Since $\delta\rho/\rho \simeq 1$, the galaxies would form soon after decoupling, all at about the same time, as suggested by the discussion of § 2.1. This theory has a number of appealing features, including the relationship between properties of galaxies and other observable quantities like the mean density of the Universe. In addition, it may be subject to direct test if matter-antimatter annihilation is observed in encounters of galaxies with other galaxies or with intergalactic gas.

Lemaître (1949) considered galaxy formation in a static Einstein model with $\Lambda > 0$, and suggested that the long cosmic lifetime would allow time for gravitational instability to develop. Recently quasar spectra have been interpreted to support a model in which

Λ is only 10^{-5} greater than the critical value for an Einstein universe (Kardashev 1967). Harrison (1967*a*) has, however, shown that such a model has little advantage over Friedmann models, because density perturbations grow exponentially at the rate $(\Lambda - k^2 c_s^2)^{1/2}$, while the model itself is unstable with the rate $\Lambda^{1/2}$, so that, as we have found previously, the time scales are comparable.

Steady-state cosmologists have rejected evolutionary models, and with them the picture of galaxy formation discussed above. Bonnor (1957) showed that the cosmic repulsion characteristic of steady-state theories makes gravitational instability particularly difficult. For instability to occur at all, the parameter $K = 4\pi G\rho/3H^2$ must exceed 5, while Hoyle's (1948) version of the steady-state theory assigns to it the value 0.5. Roxburgh and Saffman (1965) have shown that this requirement may be relaxed (to $K > 2$) if the particle creation characteristic of steady-state models does not occur at rest in comoving coordinates. Although Hoyle and Narlikar (1966*a*) have abandoned steady-state theory in its original form, we consider below some of the ideas on galaxy formation developed by the steady-state cosmologists, because of possible application in a broader context.

Hoyle (1958) and Gold and Hoyle (1959) proposed a mechanism related to thermal instability. They supposed that intergalactic gas is heated by newly created particles to 10^9 ° K. As the density is 10^{-5} cm^{-3}, the resulting pressure (1.4×10^{-12} dyn cm^{-2}) is large, and can compress cooler material into galaxies. This model contradicted the early X-ray observations (Gould and Burbidge 1963; Field and Henry 1964), according to which the temperature of the intergalactic gas is less than 4×10^6 ° K. Burbidge, Burbidge, and Hoyle (1963) suggested that the intergalactic gas is heated only locally by shock waves originating in galactic explosions, new generations of galaxies then originating as a secondary effect. This idea encounters the difficulty that the cooling behind intergalactic shock waves is probably too slow for thermal instability to occur. Sciama (1964) reviewed the problem of galaxy formation in steady-state models, and adopted $T = 10^5$ ° K, which is compatible with more recent X-ray observations (Bowyer, Field, and Mack 1968). By taking a fairly high intergalactic density, 5×10^{-5} cm^{-3}, he found that the cooling time could be reduced to 10^{10} years, as hydrogen emits rather effectively at 10^5 ° K. He postulated that galactic condensations form by thermal instability, and showed that the scale of the unstable fluctuations is bounded from below by the requirement that thermal conduction plays no role, and from above by the requirement that pressure equilibrium can be maintained during the cooling process. These considerations yield $L = 10^{24}$ cm and a mass of $10^{11}\,\mathfrak{M}_\odot$. Hunter (1966) carried out a detailed analysis of thermal instability in steady-state models, including the effects of cosmic repulsion. Thermal instability was found to be effective only if the gas temperature is relatively low ($\sim 10^5$ ° K), in agreement with Sciama. However, he finds that pressure equilibrium cannot be maintained if $\mathfrak{M} > 10^8\,\mathfrak{M}_\odot$, contrary to Sciama's estimate, and it therefore is difficult to form massive galaxies. Even for the small masses, Hunter finds that contraction ceases when T falls to 10^4 ° K, owing to the sharp drop in radiative efficiency of hydrogen below that temperature. Hence the result is only a moderately dense cloud, rather than a galaxy. Perhaps molecules could assist the radiative efficiency at low temperatures and permit contraction to continue.

Another scheme in steady-state models is the reproductive cycle invented by Sciama

(1955). In this mechanism, parent galaxies engender daughter galaxies by gravitational collapse in the high-density wake left by the parent as it moves through the intergalactic gas. Bonnor (1957) pointed out that cosmic repulsion might vitiate this process, and Harwit (1961*a*, *b*) showed that even if the parent galaxy is moving very slowly (the most favorable case), K would have to exceed 5. A later comment by Sciama (1964) appears not to refute Harwit quantitatively, so this mechanism appears to suffer from the same difficulty that ordinary gravitational instability does in steady-state models.

Hoyle and Narlikar (1966*c*) have proposed still another method for forming galaxies. In companion papers (1966*a*, *b*) they investigate the properties of a creation field, C, which they postulate modifies Einstein's field equations. If the (unknown) coupling constant of matter to this field is chosen to be a certain critical value, the presence of massive objects lowers the threshold for particle creation, so that the object can double its mass in a relatively short time. The C-field associated with creation of new particles causes a local expansion which simulates a cosmological expansion; escape to infinity is prevented by the massive object at the center. An instability occurs when a momentary drop in the creation rate reduces C and raises the threshold, cutting off further creation. As the instability is calculated to occur at a mass of $10^{13}\,\mathfrak{M}_\odot$, an interpretation of the observed upper limit to the mass of galaxies is suggested. Hoyle and Narlikar calculate the mass distribution of the galaxies formed in this way, and obtain a projected brightness distribution of $r^{-5/3}$, which they claim fits the observations better than Hubble's r^{-2} law. The elliptical galaxies formed in this way should have massive ($10^{8}\,\mathfrak{M}_\odot$) cores confined within a radius of 30 pc, which the authors associate with the observed nuclear activity of many galaxies. The galaxies do not rotate, having formed by expansion rather than by contraction. King and Minkowski (1966) show that the latter two predictions are contradicted by observation, because at least some massive ellipticals have no core of the predicted size and are rotating at several hundred kilometers per second.

Alfvén (1965) has discussed galaxy formation in the context of the Klein-Alfvén cosmology, according to which the observed expansion of the Universe results from the bounce of an originally collapsing cloud of matter and antimatter. The bounce is caused by radiation pressure generated by annihilations when the cloud reaches high density (10^{-2} particle cm^{-3}). Galaxies form by gravitational instability, and contain matter of both signs, separated by thin layers in which annihilation proceeds. Although the model would explain why large-scale density perturbations exist (the postulate of uniformity having been discarded), more analysis of the model itself, including the effects of general relativity on the dynamics of the gas cloud, must be made before it becomes possible to discuss galaxy formation in detail.

2.5. Critique of Gravitational Instability

We have shown that small density fluctuations can grow until galaxies separate out from the background. However, many authors have pointed out that the statistical fluctuations ($\sim 10^{-34}$) associated with the number of particles ($\sim 10^{68}$) in a mass of galactic order are far too small to grow to 1 percent by decoupling. Peebles (1965) asserts that the assumption that the initial fluctuations are statistical only "appears over-idealized in view of the wide range of possibilities open to the real Universe." This remark points up the fact that the calculation of the expected fluctuations on the basis of the equilibrium statistical mechanics of a classical gas is not a reliable way to estimate the fluctuations in a dynamical self-gravitating system. There are two ways in which one may be in error. First, even the equilibrium statistical mechanics of a self-gravitating gas appears to be an unsolved problem. One would suppose that in a system whose dimensions much exceed the Jeans length, gravitationally bound systems would be present even in thermodynamic equilibrium, and would represent the state of maximum entropy, contrary to the simple $N^{-1/2}$ estimate of density fluctuations. Second, it may be incorrect to apply even a modified statistical mechanics to an evolving system like the Universe, as it would be based on a principle of maximum entropy, or missing information, whereas it is not known even how to define the latter concept for an evolving Universe. Peebles's remark points to the need to assess more realistically the range of possibilities that is open, in order to compute the probability of various states, such as the one in which density perturbations are present to a degree.

While we may hope for progress in this area, we should also not neglect the possible effects of nongravitational forces. Conceivably some of them, like the radiation forces discussed by Gamow, could induce instabilities which would, in any case, contribute to the result we see today. While proposals like Hoyle's and Alfvén's are of interest, more work is required before we understand well the origin of inhomogeneities even in conventional cosmology.

3. CONTRACTION OF A PROTOGALAXY

3.1. Initial Conditions

Considerable development must occur after a gas cloud separates out from the background. We may take the epoch of the latter to be $t_{M,G}$ on the cosmic time scale ($t = 0$ on the galactic time scale). At this moment the cloud is at rest, and is about to begin contracting to galactic dimensions. We consider the contraction, with its associated phenomenon of flattening, in this section. The evolution of the cloud depends on its properties at $t = 0$, including size, shape, angular momentum, temperature, magnetic field, turbulence, and chemical composition (as the last affects the radiative efficiency). Possibly external influences, such as proximity to other clouds, connections with an external magnetic field, and the background radiation field will affect the outcome. If the cloud is initially spherical, cold, uniform, and free of turbulence and magnetic fields, the analysis of § 2.1 predicts collapse into a singularity; the same is true in relativity. While Hoyle and Fowler (1963) have suggested that such dense objects actually exist, they appear not to be related to normal galaxies. Hence galaxies must originate with different initial conditions from the stated ones. The range of initial conditions which has so far been treated is narrow, and includes uniform spheroids with uniform rotation around the minor axis. The temperature has a special spatial dependence and is proportional to a power of the density, the power being adjusted in accordance with the radiation efficiency. Certain magnetic configurations have been considered. We return to the question of the actual initial values of various parameters in § 3.3 after formulating a method to calculate the collapse.

3.2. Calculation of the Collapse

The only completely correct method is of course to solve the partial differential equations of magnetohydrodynamics and radiative transfer for the postulated initial conditions. Instead, we discuss here the similarity method applied by Lynden-Bell (1962, 1964) to uniform spheroids in uniform rotation. We can generalize to include the effects of gas pressure by considering a temperature variation of the form

$$T(r, t) = T_*(t)\left(1 - \frac{r^2}{a^2} - \frac{z^2}{c^2}\right), \tag{123}$$

where r and z are cylindrical coordinates, and a and c ($= \alpha a$) are semiaxes of the spheroid, which depend on t. (Since the law [123] assigns different temperatures to different elements of gas, even though all have the same density, it is not compatible with adiabatic motion of gas from the epoch of decoupling.) The isotherms are all similar to the isotherm $T = 0$, the surface of the spheroid,

$$r^2/a^2 + z^2/c^2 = 1\ . \tag{124}$$

Lynden-Bell showed that the uniform, uniformly rotating spheroid collapses homologously, the density and angular velocity remaining uniform. The same is true if the temperature follows equation (123), as can be seen from the Lagrangian equations of motion:

$$\frac{d^2r}{dt^2} = r\left(\frac{d\varphi}{dt}\right)^2 - 2Ar - \frac{1}{\rho}\frac{\partial p}{\partial r}\ , \tag{125}$$

$$\frac{d}{dt}\left(r^2\frac{d\varphi}{dt}\right) = 0\ , \tag{126}$$

and

$$\frac{d^2z}{dt^2} = -2Cz - \frac{1}{\rho}\frac{\partial p}{\partial z}\ . \tag{127}$$

It is assumed that no torques are present. The gravitational terms A and C can be written from potential theory as functions of a and α ($\equiv c/a$) for a uniform spheroid:

$$\left\{\begin{matrix} A \\ C \end{matrix}\right\} = \frac{3G\mathfrak{M}}{2a^3}\left\{\begin{matrix} F(\alpha) \\ \alpha^{-2}G(\alpha) \end{matrix}\right\}\ , \tag{128}$$

where

$$F(\alpha) = \tfrac{1}{2}(1 - \alpha^2)^{-3/2}\cos^{-1}\alpha - \tfrac{1}{2}\alpha(1 - \alpha^2)^{-1}\ ,$$

$$G(\alpha) = \alpha(1 - \alpha^2)^{-1} - \alpha^2(1 - \alpha^2)^{-3/2}\cos^{-1}\alpha\ . \tag{129}$$

Equations (125)–(127) admit a similarity solution of the form

$$R(t) = r/r_0\ , \qquad Z(t) = z/z_0\ , \tag{130}$$

where r_0 and z_0 are initial values. From equation (126) we see that

$$d\varphi/dt \equiv \Omega = \Omega_0/R^2 \tag{131}$$

is independent of position. If $T_*(t)$ obeys a polytropic law,

$$T_*(t) = T_0(\rho/\rho_0)^{\gamma-1}\ , \tag{132}$$

one sees from the form of equations (123), (125), and (127) that all terms are proportional to the coordinates, so that a similarity solution does exist provided that

$$R\ddot{R} = \xi R^{-2} - \alpha_0 R^{-1} F(\alpha) + \eta (R^{-2} Z^{-1})^{\gamma-1} , \tag{133}$$

and

$$Z\ddot{Z} = -\alpha_0^{-1} R^{-1} G(\alpha) + \alpha_0^{-2} \eta (R^{-2} Z^{-1})^{\gamma-1} . \tag{134}$$

Dots represent differentiation with respect to

$$\tau = (4\pi G \rho_0)^{1/2} t ; \tag{135}$$

$$\xi = \frac{\Omega_0^2}{4\pi G \rho_0} \tag{136}$$

and

$$\eta = \frac{\Re T_0}{2\pi G \mu \rho_0 a_0^2} \tag{137}$$

are dimensionless rotation and pressure parameters. Equations (133) and (134) are a fourth-order system in R and Z (since $\alpha = \alpha_0 Z/R$), and are to be solved with the initial conditions

$$\dot{R}(0) = 0 , \quad \dot{Z}(0) = 0 , \quad R(0) = 1 , \quad Z(0) = 1 , \tag{138}$$

since the initial spheroid is at rest. The solutions are functions of ξ, η, α_0, and γ. Equations (128) and (129) are valid if the cloud remains a uniform spheroid. Since

$$-\frac{1}{\rho}\frac{d\rho}{dt} = \frac{\partial v_z}{\partial z} + \frac{1}{r}\frac{\partial}{\partial r}(r v_r) = \frac{\dot{Z}}{Z} + 2\frac{\dot{R}}{R} \tag{139}$$

depends only on t, ρ remains uniform if it is initially. Since $r/a = r_0/a_0$ and $z/c = z_0/c_0$ by equation (130), the spheroidal shape is preserved by equation (124).

Solutions appear in the literature for the special cases listed in Table 4.

TABLE 4

COLLAPSE SOLUTIONS IN THE LITERATURE

Case	ξ	η	α_0
I.....	0	>0	1
II.....	0	0	<1
III.....	>0	0	1
IV.....	>0	>0	1

Case I represents a spherical cloud which has internal pressure, but no rotation. Since $F(1) = G(1) = \frac{1}{3}$, equations (133) and (134) are initially identical. Hence $R = Z$ throughout, and

$$R\ddot{R} = -\tfrac{1}{3}R^{-1} + \eta R^{-3(\gamma-1)} , \tag{140}$$

which has the energy integral

$$\tfrac{1}{2}\dot{R}^2 - \tfrac{1}{3}(R^{-1} - 1) + \frac{\eta}{3(\gamma - 1)}[R^{-3(\gamma-1)} - 1] = 0 ; \tag{141}$$

the last term is replaced by $-\eta \ln R$ if $\gamma = 1$. The behavior of the second (gravitational) term is the same as that of the third (pressure) term if $\gamma = 4/3$. Hoyle (1953) argued that in practice $\gamma \approx 1$, because radiative cooling keeps the temperature from rising above 10^4 ° K. In this case, the gravitational term dominates, and collapse proceeds to very high densities. Even if the gas were to become opaque, so that radiation is trapped, the main contributor for masses of galactic order would be radiation pressure, so that the effective γ would probably not much exceed 4/3. This is probably not high enough to halt the collapse if general relativistic effects are included. Hence the only way one avoids making a single dense object in case I is for fragmentation to occur (Hoyle 1953). This process, which involves small-scale motions not included in the present treatment, is considered in § 4.1.

Case II is a cold nonrotating cloud which is initially flattened. Lin, Mestel, and Shu (1965) showed that such a cloud becomes increasingly flat, a thin disk being the final result. This means that initial deviations from sphericity are amplified, casting doubt upon the applicability of case I, since it is doubtful that the initial cloud would be exactly spherical.

Case III is a cold rotating cloud which is initially spherical, and was treated by Lynden-Bell (1964). If $\xi = 0$, the cloud collapses to a point as in case I with $\eta = 0$. With a small amount of rotation ($\xi = 1/400$), the collapse initially follows the solution for $\xi = 0$, but centrifugal force finally halts the contraction in R but not in Z. For larger ξ, the disk forms rapidly. Lin, Mestel, and Shu (1965) suggested that the contraction in Z would be halted if η were finite.

Case IV is like case III, but η is finite. Strittmatter (1966) derived equations identical to equations (133) and (134) using the tensor virial equations. In the case $\xi = \frac{1}{6}$, $\eta = \frac{1}{6}$, he found that the collapse was halted by pressure at $Z = 0.014$ (at which time $R = 0.72$), consistent with the conjecture of Lin *et al.* We shall see later, however, that η is probably much smaller, so that the collapse is finally halted only at very small values of Z.

3.3. Collapse of Protogalaxies

To apply the results of the previous section, we need estimates of η, ξ, and α_0. In § 2.1 we found that $t_M \geq 3.5 \times 10^7$ years. Taking $t_M = 5 \times 10^8$ years as an example, we find that the density at maximum extent, ρ_M (called ρ_0 in § 3.2) is 2×10^{-26} g cm^{-3}, so that for $\mathfrak{M} = 10^{11}\, \mathfrak{M}_\odot$, $a_0 = 4.5 \times 10^4$ pc. Adiabatic expansion from the time of decoupling would give $\langle T \rangle = 0.5°$ K. Since the central temperature is 2.5 times the mean value (eq. [123]), $T_0 = 1.3°$ K and $\eta = 10^{-6}$. To estimate ξ, one must know the origin of rotation. Lifshitz (1946) showed that primordial angular velocities tend to decrease with time, so that countering the associated centrifugal forces is more difficult at early phases. It seems more likely that rotation originated later. Von Weizsäcker (1951) attributed both the formation and the rotation of galaxies to intergalactic turbulence, basing his analysis on the observation that the Reynolds number of such large-scale flow is very large. In his model, protogalaxies are the result of turbulent eddies in which turbulent velocities happen to cancel the cosmic expansion, so that a bound system is formed. This picture accounts for the rough equality between the rotational and translational velocities of galaxies. However, no energy source for the turbulent motion is known. One might consider the energy of expansion in this regard, but the Lifshitz analysis shows that at least small-amplitude rotational motions decrease rather than amplify, while the release of gravitational energy leads to compression rather than to rotation.

Hoyle (1951) suggested a mechanism based on the nonlinear terms which occur when gravitational instability has increased the density perturbation to $\delta\rho \simeq \rho$. Thus, when $t = t_M$ (cosmic time scale), the density within clouds (ρ_0) is 5.6 times the mean density, and tidal interaction between neighboring protogalaxies is significant. Hoyle estimates the effects of these tidal interactions as follows. A cloud at the origin will in general have different moments of inertia I around the three axes (as it is not completely spherical), and hence is subject to tidal torques exerted by neighboring clouds at distances d_1, d_2, . . . , etc. If the clouds have mass $\mathfrak{M}$, the x-component of angular acceleration due to cloud j is

$$\dot{\Omega}_x = \frac{3G\mathfrak{M}m_j n_j}{d_j^3}\left(\frac{I_z - I_y}{I_x}\right), \tag{142}$$

where l, m, and n are direction cosines. Because of the factor $1/d_j^3$, only the nearest clouds exert a significant torque. One can then show that

$$\dot{\Omega}_{\max} \simeq 4\pi G\rho A\ , \tag{143}$$

where

$$A = \frac{\bar{\rho}f}{\rho}\left(\frac{\Delta I mn}{I}\right)_{\max} \tag{144}$$

and

$$f = \frac{\mathfrak{M}}{(4/3)\pi\bar{\rho}d^3}. \tag{145}$$

Here $\bar{\rho}$ is the mean density, ρ is the internal density of the central cloud, d is the mean distance to the nearest neighbors, and the maximum is with respect to the directions of neighboring clouds. Even if all matter is in clouds, f is of the order of, but less than, unity. If the central cloud begins to contract at $t = 0$, it will do so on a time scale approximately equal to t_M. Because ρ increases from ρ_0, A decreases from its initial value A_0, and most of the angular velocity is acquired in a time $\frac{1}{3}t_M$, say. Hence $\Omega \simeq \frac{1}{3}t_M\dot{\Omega}_{0,\max} = \pi^2 A_0/8t_M$. If one ignores the action of torques during the later phases of the collapse, it proceeds in the manner described in § 3.2, and this value of Ω may be used to find that

$$\xi = \frac{\Omega_0^2}{4\pi G\rho_0} = \frac{\pi^2 A_0^2}{8}, \tag{146}$$

which is considerably less than unity because A_0 is rather small, according to equation (144). According to equation (133), centrifugal forces halt the contraction when $R = 4\xi/\pi = \pi A_0^2/2$. We show in § 5.3 that the Milky Way may have undergone a homologous contraction with $R \simeq 0.17$, which corresponds to $\xi = 0.13$ and $A_0 = 0.33$. This seems large, but not impossible; smaller values would lead to denser systems, which might be identified with elliptical galaxies. Thus, local variations in the tidal interactions would result in a range of values of ξ, and hence a variety of galactic forms.

If we now consider the collapse of a cloud with $\xi = 0.13$ and $\eta = 10^{-6}$, we see that extremely small values of Z will be reached before gas pressure halts the collapse in the z-direction, even if $\gamma = 5/3$, the maximum value for ordinary gas pressure. If T stops increasing at 10,000° K (Hoyle 1953), the behavior would switch to that of a $\gamma = 1$ gas, and fantastic compression would be required to halt the z-motion. It therefore is unlikely that gas pressure alone is responsible for the finite thickness of galactic disks. As we shall see in § 4.3, gas pressure (including turbulence) may account for the steady-state situation, but here we would require not only that the gravitational forces be balanced, but also that the large velocities associated with collapse be reversed, a much more difficult requirement. Possibly a magnetic field parallel to the disk, as currently observed in the Milky Way, could effectively halt the collapse. Strittmatter has treated this problem and finds that the effective γ exceeds 2. This solution would require a primordial field of unknown origin and a level of ionization during collapse which is sufficient to couple the field to the gas.

3.4. Instability during Collapse

Lin, Mestel, and Shu (1965) showed that spheroids tend to become flatter during collapse even in the absence of rotation, suggesting that the shape of the cloud is sensitive to initial conditions. Lynden-Bell (1964) studied the distortions from sphericity in planes normal to the rotation axis in rotating clouds. The perturbation ψ satisfies the equation

$$\frac{d^2\psi}{d\varphi^2} - \Gamma\psi = 0\,, \tag{147}$$

where $d\varphi = \Omega dt$ and

$$\Gamma(\alpha) = \frac{RH(\alpha)}{\xi\alpha} - 1\,,$$

$$H(\alpha) = \frac{3}{4(1-\alpha^2)^2}\{(1-\alpha^2)^{-1/2}\cos^{-1}\alpha - \alpha[1 + \tfrac{2}{3}(1-\alpha^2)]\}\,. \tag{148}$$

Since $\xi < \frac{1}{3}$ for collapse to occur at all and since $H(\alpha)$ is 2/5 at $\alpha = 1$, $\Gamma(1) > \frac{1}{5}$, and the collapse is unstable initially at least. Although H remains of order unity as $\alpha \to 0$, $R \to \pi\xi/4$, and it is not clear in general whether Γ remains positive. In the specific cases considered by Lynden-Bell it does, and instability persists throughout the collapse. Lynden-Bell concludes that an originally spheroidal cloud may be distorted into a bar-shaped object, possibly related to barred spiral galaxies. Equation (147) differs from equation (33) (which is for density perturbations), in that the stabilizing term (unity in eq. [148]) is due to centrifugal force rather than to pressure. It is scale-independent and therefore constant during the collapse. Instabilities of shape therefore probably occur on all scales, not only the largest one studied by Lynden-Bell.

Hunter (1962, 1964) studied density perturbations in a collapsing spherical cloud. Naturally his analysis is closely related to Bonnor's for the expanding case, the collapse being described by equations (17) and (18) with $\frac{1}{2}\pi \leq \alpha \leq \pi$ and $k = 1$. It is assumed that pressure does not significantly affect the motion of the cloud as a whole ($\eta \ll 1$), but the effects on small scales of motion are included. The results are contained in an equation equivalent to equation (38), so that equations (47) and (48) and table 1 may be employed (with $\alpha \to \pi - \alpha$). If the collapse is adiabatic ($\gamma = 5/3$), stability increases with time; while if it is isothermal ($\gamma = 1$), it decreases. One may also use the concept of Jeans mass, $\mathfrak{M}_J$, instability occurring if $\mathfrak{M} > \mathfrak{M}_J$. It can be written in terms of the instantaneous density and temperature of the cloud as

$$\mathfrak{M}_J = \frac{\pi}{6\rho^{1/2}}\left(\frac{\pi\gamma RT}{G\mu}\right)^{3/2}, \tag{149}$$

so that instability is governed by the temperature-density history of the cloud. If $\gamma = 5/3$, it is seen that $\mathfrak{M}_J$ increases like $\rho^{1/2}$; whereas if $\gamma = 1$, it decreases like $\rho^{-1/2}$. Consider a cloud of $10^{11}\,\mathfrak{M}_\odot$ emerging from decoupling. Depending on the value of $\delta\rho/\rho$ at decoupling, it reaches maximum extent at $t_M \geq 3.5 \times 10^7$ years. We may consider $t_M = 3.5 \times 10^7$ and 5×10^8 years as examples, in which case $\rho_M = 3.6 \times 10^{-24}$ and 2×10^{-26} g cm^{-3}, respectively. Presumably the expansion from decoupling is adiabatic with $\gamma = 5/3$ since the gas and radiation are decoupled. Hence the corresponding gas temperatures when $t = 0$ are 11° and 0.3° K, respectively. This is shown in figure 2

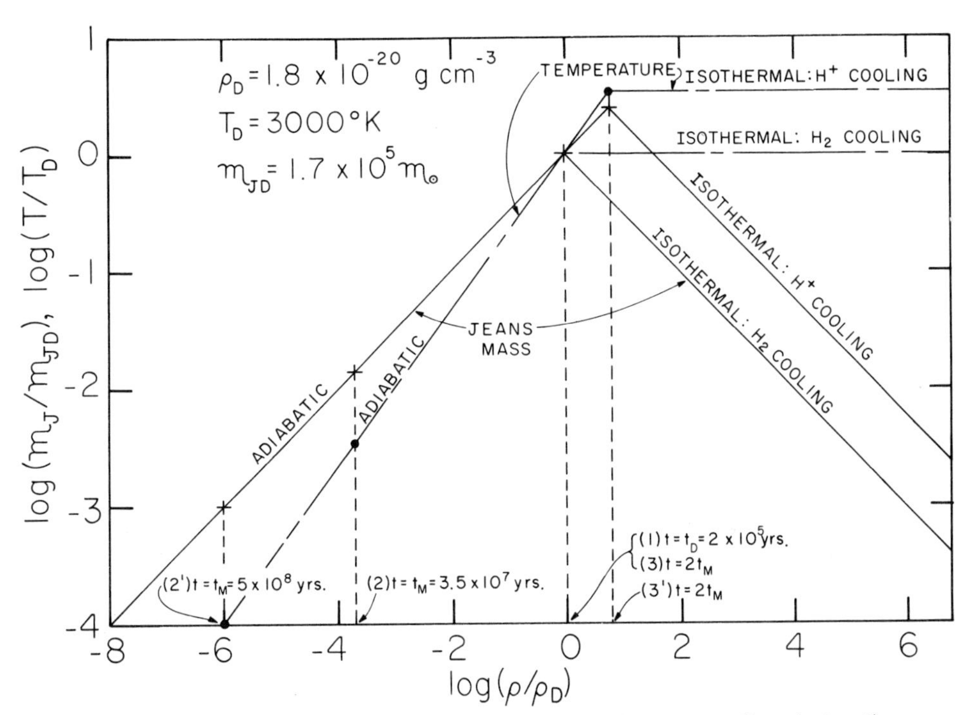

FIG. 2.—Thermal history and Jeans mass after decoupling. Decoupling (point 1) occurs when $T = T_D = 3000°$ K and $\rho = \rho_D = 1.8 \times 10^{-20}$ g cm^{-3}. The galactic cloud expands adiabatically either to point 2, when $T = T_M = 11°$ K and $\rho = \rho_M = 3.6 \times 10^{-24}$ g cm^{-3}, after 3.5×10^7 years (the minimum possible value), or to point 2′, when $T_M = 0.3°$ K and $\rho_M = 1.8 \times 10^{-26}$ g cm^{-3}, after 5×10^8 years. Thereafter it contracts adiabatically toward higher density either to point 3, when $T = 3000°$ K and $\rho = 1.8 \times 10^{-20}$ g cm^{-3}, or to point 3′, when $T = 10{,}000°$ K and $\rho = 4.5 \times 10^{-20}$ g cm^{-3}. At this time it switches to isothermal behavior governed by cooling due to H_2 or to H^+, respectively. The value of the Jeans mass (eq. [149]), which depends upon the temperature and density, is also plotted.

(points 2 and 2′). The value of $\mathfrak{M}_J$ at decoupling is $1.7 \times 10^5\,\mathfrak{M}_\odot$ from equation (149), and it decreases to $2400\,\mathfrak{M}_\odot$ and $170\,\mathfrak{M}_\odot$ in the two cases. The value of $\mathfrak{M}_J$ increases again when the collapse starts, and will continue to do so as long as the collapse is adiabatic. The latter will be true as long as there is little dissipation of the collapse energy and as long as the radiation emitted is negligible. Shock waves would dissipate the collapse energy, but one can show from equation (38) that the relative amplitude of acoustic waves ($\sigma \gg 1$) is constant as the collapse proceeds, so there is no tendency beyond the nonlinear terms to amplify small-amplitude disturbances of short wavelength to the point they begin to dissipate. Radiation is probably negligible at first, when the temperature is low. As Hoyle (1953) pointed out, the emission by atomic hydrogen will increase sharply at about 10,000° K owing to partial collisional ionization. Saslaw and Zipoy (1967) show that H_2 molecules can form via $H^+ + H \rightarrow H_2^+$, $H_2^+ + H \rightarrow H_2 + H^+$, from protons remaining after decoupling. They estimate that H_2/H attains 10^{-5}, and that this is sufficient to radiate effectively as soon as T reaches 3000° K. One concludes that the temperature is unlikely to climb much above 3000° K, or at most 10,000° K, so we have drawn these isotherms in figure 2. The cloud returns to point 1 (= point 3) or to point 3′ along an adiabat, where it switches to isothermal behavior. Then $\mathfrak{M}_J$ decreases again like $\rho^{-1/2}$.

The spectrum of small-scale density perturbations present in the cloud depends upon the excitation of the isothermal mode before decoupling, the adiabatic mode being heavily damped at the scales of interest. The Jeans hypothesis requires such perturbations to be present if we are to understand the present structure of galaxies. The Jeans mass at decoupling, $1.7 \times 10^5\,\mathfrak{M}_\odot$, is critical in that greater masses are unstable during the entire subsequent history of the cloud. It is suggestive that globular clusters have masses of this order (Arp 1965). If the spectrum of perturbations is flat in this mass range, the amplification of the larger masses after decoupling might account for a lack of systems below $10^5\,\mathfrak{M}_\odot$. Peebles (1965) argues that actually the initial spectrum would decrease with mass, so that a peak would form at $10^5\,\mathfrak{M}_\odot$ as a result of the amplification. In his view, galaxies may form by the agglomeration of objects of this size. It has been shown more recently that masses above $10^{11}\,\mathfrak{M}_\odot$ are particularly favored because of the lack of damping of the adiabatic mode. Perhaps both isothermal and adiabatic perturbations are initially present with spectra which decrease with mass. Then the same line of reasoning which would lead to a peak in the isothermal spectrum at $10^5\,\mathfrak{M}_\odot$ would lead to a peak in the adiabatic one at $10^{11}\,\mathfrak{M}_\odot$. The latter would account for the galaxies and the former for globular clusters within galaxies. This general picture has been discussed by Zel'dovich (1967).

4. EARLY EVOLUTION

4.1. Fragmentation

Hoyle (1953) proposed that stars and clusters form by a process of fragmentation, according to which successively smaller masses become unstable in an isothermal collapse, leading to a separation into successively smaller units. He discussed particularly a configuration at 10,000° K in pressure equilibrium, which for a galaxy requires $\rho = 10^{-27}$ g cm^{-3}. If the cloud starts to contract, it continues to do so because the temperature is kept low by radiative processes. Fragmentation proceeds until densities of

10^{-9} g cm^{-3} are approached, at which time H^- opacity becomes important, radiation is trapped, and the process reverts to adiabatic. The process therefore ceases with the formation of opaque bodies (stars). Although Hoyle visualized fragmentation as occurring in discrete stages, it may also be viewed as a continuous process. The density perturbations of small scale oscillate up to the time that the Jeans criterion is satisfied, and thereafter they are gravitationally unstable and separate into discrete objects. Since larger scales become unstable earlier, the smaller scales become unstable not with respect to the behavior of the cloud as a whole, but with respect to a subunit of larger mass which has already begun the process of separation. As isothermal pressure is smaller than gravity, the gravitational energy released by the collapse cannot be stored in the form of internal energy or radiation, but must go into kinetic energy of mass motion. The energy radiated in an isothermal collapse is 4×10^{13} ergs g^{-1} for a galaxy of $10^{11}\,\mathfrak{M}_\odot$ fragmenting into bodies of $1\,\mathfrak{M}_\odot$, while the observed binding energy of such a galaxy is 3.4×10^{14} ergs g^{-1} (Fish 1964).

The problem arises of how the energy released is in fact dissipated to form a strongly bound system. Fish considered that shock waves were the most probable of several available mechanisms. Field (1964) considered compression of magnetic fields and associated relativistic particles ($\gamma = 4/3$). Another possibility is suggested by the discussion of § 2.1, according to which a flat perturbation spectrum of 1 percent amplitude at decoupling would have led to the formation of most galaxies at about $2t_M = 7 \times 10^7$ years, independent of mass. On this hypothesis, the binding energy of galaxies would have been established before decoupling, no further dissipation being required. This may be in keeping with the picture of fragmentation, according to which dissipation is prevented by the separation into noninteracting bodies. In the limiting case of no dissipation, the final equilibrium radius of the cloud is $\frac{1}{2}a_0$.

There are difficulties in the fragmentation hypothesis which we discuss now. In order that collisions among fragments, with resulting reversion to the uniform state, not occur, the individual fragments must contract to a high density (small cross-section), and the galaxy as a whole must be prevented from contracting to a small radius. As a spherical galaxy with $\xi = 0$ contracts from a_0 to a minimum radius $a_m = a_0 R$, individual fragments within the galaxy contract from a_{f0} to $a_{fm} = a_{f0} R_f$. The probability that a given fragment collides with one of the other n fragments is $p = n(a_{fm}/a_m)^2 = n^{1/3}(R_f/R)^2$, where we have used the fact that $a_{f0}/a_0 = n^{-1/3}$. The requirement that $p < 1$ is thus equivalent to

$$R_f/R < n^{-1/6} \,. \tag{150}$$

For example, if the mass of a fragment is $10^5\,\mathfrak{M}_\odot$, $n = 10^6$ and $R_f/R < 10^{-1}$, or $\rho_f/\rho > 10^3$. This seems reasonable, at least at first sight. If the fragment originates in a density perturbation ϵ at $t = 0$, it collapses to high density before the cloud as a whole does (eq. [57]). One can show that the cloud density ρ at the time the density ratio ρ_f/ρ reaches a prescribed value is

$$\frac{\rho}{\rho_M} = \left\{\frac{4}{\pi\epsilon}\left[1 - \left(\frac{\rho}{\rho_f}\right)^{1/6}\right]\right\}^2 . \tag{151}$$

For example, if $\epsilon = 10^{-2}$, the requirement that $\rho_f/\rho > 10^3$ yields $\rho/\rho_M > 7500$ or $R < 0.05$, which is reasonable. Thus, if $t_M = 3.5 \times 10^7$ years, the cloud would contract from $a_0 = 7800$ pc to $a_m = 400$ pc, and then reexpand to 3900 pc.

Layzer (1963*a*) pointed out that nonlinear interactions among fragments, which are neglected in the above calculations, may severely limit the collapse of individual fragments and hence make it impossible for collisions to be avoided. Gravitational interactions with neighboring fragments impart both linear motion (equivalent to orbital angular momentum) and angular motion (equivalent to spin angular momentum). While the former may actually be helpful in limiting the degree of overall collapse, the latter limits the collapse of fragments beyond a certain point. His argument is as follows. Two fragments of mass $m = n^{-1}\mathfrak{M}$ separated by a distance l experience a mutual acceleration Gml^{-2}, which in a characteristic time of collapse $(G\rho)^{-1/2}$ results in a velocity $v = G^{1/2}ml^{-2}\rho^{-1/2}$. Since $\rho = ml^{-3}$, $v = (Gm/l)^{1/2}$, which increases, by conservation of angular momentum as the collapse proceeds, to $v = (Gm/l)^{1/2}R^{-1}$. This equals the systematic velocity of collapse, $(G\mathfrak{M}/a_0)^{1/2}R^{-1/2}$ when

$$R = ma_0/\mathfrak{M}l = n^{-2/3}\,, \tag{152}$$

at which point the collapse is halted and the system begins to reexpand because of the strong centrifugal forces. With $n = 10^6$, equation (152) gives $R = 10^{-4}$, which seems small.

In the meantime, mutual gravitational torques impart spin to the fragments as already described in § 3.3. By the equations of § 3.2, spin prevents collapse of the fragments beyond $R_f = 4\pi^{-1}\xi_f$, and since by equation (146) and the fact that $A_0 \lesssim 1$, $\xi \lesssim 1$; hence $R_f \lesssim 1$. Hence if these two processes (linear and angular accelerations among fragments) are the only ones inhibiting the collapse of the galaxy and of the fragments, one would estimate that

$$R_f/R < n^{2/3}\,, \tag{153}$$

or 10^4 if $n = 10^6$. Clearly for any reasonable value of n there is a contradiction between equation (153) and the requirement (150) which is to be met if collisions are to be avoided. Layzer concludes that fragmentation therefore never is really completed as envisaged earlier, and that instead, there is a continual buildup and decay of density perturbations which constitutes a kind of turbulence driven by gravitational forces. Clearly it is of greatest importance to discover whether Hoyle's or Layzer's analysis is correct. As these analyses deal with nonlinear equations in an approximate way, it is desirable to check them by more detailed calculations. Hunter (1964) investigated higher-order terms in the perturbation analysis, and concluded that they tend to accelerate collapse; however, it does not appear that the tidal terms which form the basis of Layzer's critique were included. Arny (1966) studied an idealized model of fragmentation which included the tidal distortion by a condensed central body, although not that by neighboring fragments. He concluded that the tidal forces retard but do not ultimately prevent fragmentation. Michie (1965, 1966) argued that partial cancellation of opposing torques may reduce the final spin below Layzer's estimate. As both spin and orbital angular momenta are stochastic variables in Layzer's model, a Monte Carlo calculation which takes account of tidal torques exerted by randomly placed fragments would be useful.

Another answer to Layzer's critique may follow from Lynden-Bell's (1967) arguments concerning energy exchange in collapsing systems. He points out that if the distribution of momenta of point masses deviates drastically from the equilibrium value (as would be true in a collapsing, fragmenting system), the system tends to relax on a dynamical time scale, energy exchange occurring through collective motions involving many particles. Applied to the present problem, this may indicate that Layzer underestimated the amount of orbital angular momentum acquired by individual fragments, as he based his calculation on random binary interactions. If collective modes are present, the actual accelerations will be far greater, because many particles are acting coherently. The result of this would be to increase the value of R from the value given in equation (152). From equation (150) and the estimate that $R_f \lesssim 1$, it appears that R cannot be much less than unity, and it remains to be seen whether in fact the arguments of Lynden-Bell would lead to this conclusion. It is interesting to note that very probably Layzer's argument concerning the tidally induced spin momenta is still correct, as collective modes will not exert significant torques.

4.2. Formation of Elliptical Systems

The previous section shows that if certain difficulties are overcome, elliptical galaxies may be formed by fragmentation of a collapsing cloud; if angular momentum is negligible, the result may be an E0 galaxy. Belzer, Gamow, and Keller (1951) showed that the elliptical galaxy NGC 3379 is in dynamical equilibrium, with a Maxwellian velocity distribution at the center. This galaxy appears to be typical, as Miller and Prendergast (1962) have shown that it follows the same distribution of brightness with radius that is common for other systems (de Vaucouleurs 1959). One may tentatively conclude that the elliptical galaxies are dynamically relaxed systems. This property could be due to similar conditions at the time of formation or to a subsequent relaxation process. It is believed that binary encounters are too slow to bring about relaxation in the available time, and this is in accord with the fact that there usually appears to be little segregation of various components (e.g., stars and globular clusters) according to mass. On the other hand, phase mixing and collective interactions at the time of formation may be able to bring about partial relaxation, as suggested by Lynden-Bell (1967). Campbell (1962, 1963, 1966) studied the collapse of a spherical galaxy numerically, using 100 shells with small random internal motions and starting from various arbitrary initial configurations. It was found that the system approached a steady state after several free-fall times, having bounced at a radius fixed by the amount of random motion. The density distribution in the final state depended on the initial density distribution and on the amount of random motion, the best fit to observed galaxies being with a $\gamma = 1$ or $4/3$ polytrope initially, and with an initial value $T/|\Omega| \simeq 0.06$. The radial collapse factor was between 4 and 7, compared with our previous estimates, between 2 and 19. According to Campbell, the observed brightness distribution requires that the stellar orbits in the outer parts be largely radial. Such orbits can be obtained in his model if the initial random velocities are much smaller than the radial velocities acquired during the collapse. The smallness of the random velocities implies a large collapse factor. The restriction to initial polytropes with γ between 1 and 4/3 seems not to be consistent with our initial conditions. Neither does Campbell's initial central temperature, 5×10^6 ° K; we would predict a gas temperature of only 10° K on the basis of adiabatic expansion since decoupling. Hence Campbell's work, to the extent that it implies a unique initial state, is in conflict with the picture developed previously on the basis of evolutionary cosmology. It should be noted that the dynamical relaxation described by Lynden-Bell (1967) cannot occur with full freedom in Campbell's machine computations, as the latter are constrained to radial motion only. Hodge and Michie (1969) have applied such a relaxation mechanism to obtain agreement with the brightness distribution of the dwarf ellipticals in the Local Group.

4.3. Formation of Disk Systems

In § 3.3 we saw that if both rotation and pressure are finite, the final state may be a disk of gas in solid-body rotation. Here we consider the early evolution of such a system. If fragmentation occurs before appreciable flattening has occurred, it is difficult for the large z velocities resulting from collapse to dissipate. If, on the other hand, fragmentation for some reason occurs rather late, gas and/or magnetic pressure can in principle decelerate the infalling gas so that an equilibrium disk may be formed. The question of whether the distinction between spirals and ellipticals can be explained along these lines is considered further in § 4.4. The fact that spirals contain much more gas than do ellipticals is qualitatively consistent with such a picture. Mestel (1963) argued that the very fact that the galaxy is rotating tends to inhibit fragmentation. Even if the mass of a fragment exceeds the Jeans mass, its contraction is inhibited by the spin angular momentum with which it is endowed by its participation in the rotation of the galaxy. (This effect is in addition to the spin imparted by interactions among fragments.) The angular velocity of the fragment Ω_f is initially about equal to that of the galaxy, Ω. Hence the ratio of centrifugal force to gravity, $3\Omega_f^2/4\pi G\rho_f$, is about equal to that of the galaxy as a whole, since initially $\rho_f \simeq \rho$. The fragment begins to flatten when R_f is about equal to this quantity; and since the same is true of the galaxy, the fragment cannot contract much more than the galaxy as a whole, with the result that fragmentation is not permanent. This may explain how disk systems can retain their gas.

Of course, star formation is still possible in disk systems. Mestel points out that once flattening is complete, the rotational parameter becomes about equal to $c/a \simeq 10^{-2}$, so that some contraction of fragments is possible. Furthermore, as Mestel (1965*b*) and Spitzer (1967) have shown, a magnetic field can transport angular momentum away from contracting objects. Unless the field is very strong, this process is likely to take longer than free-fall collapse, and so fragmentation is no longer the proper word to describe the process. Rather, star formation proceeds long after the disk is formed, and hence would be in the regime of contemporary interstellar gas dynamics as described by Spitzer (1967) in this series. In summary, it seems reasonable to suppose that disk systems are initially gaseous for a considerable period, as a consequence of the inhibition of fragmentation by rotation, and that star formation proceeds slowly over a much longer period of time.

Similar considerations apply to the radial collapse of the galaxy. In the final equilibrium state, $R = 4\xi/\pi$. Lynden-Bell (1964) found that if $\xi = 4/15$, $Z \to 0$ when $R = 0.75$, compared to the equilibrium value, 0.34. Hence considerable radial contraction occurs after the disk forms. Hunter (1963) showed that the radial collapse energy causes the radius to overshoot its equilibrium value and to oscillate around it at a frequency about equal to the rotation frequency (see below). Again, the energy of this oscillation can be dissipated only if the matter is gaseous. If fragmentation occurs before the oscillation starts, the resulting fragments will have considerable kinetic energy in both the r- and the z-directions, which after phase mixing will appear as random motion both parallel and perpendicular to the galactic plane. Mestel (1965*a*) considers this to be a possible interpretation of elliptical galaxies having moderate ellipticity.

To discuss the early evolution of a galactic disk, we consider an approximate equilibrium state; some information about this can be gleaned from observations of the Milky Way. In treating the distribution of gas in this system, we can take $g_z = -4\pi G\rho_s z$, since stars (density ρ_s) contribute most of the mass and are distributed more widely in z than the gas. If we represent the pressure (turbulent, thermal, or magnetic) by ρc^2, with c a constant, the condition of hydrostatic equilibrium then yields $\rho = \rho_* \exp(-z^2/2\sigma^2)$, where $\rho_* = \rho(z = 0)$ and $\sigma = c/(4\pi G\rho_s)^{1/2}$. According to Schmidt (1957), this law is in fair accord with the observations if $\sigma \simeq 100$ pc. Since $\rho_s \simeq 10^{-23}$ g cm^{-3}, this implies that $c \simeq 8$ km s^{-1}. The thermal velocity of hydrogen atoms at 100° K is only 1 km s^{-1}, so thermal pressure (at least in ordinary clouds) cannot be responsible for the support of the disk. On the other hand, the turbulent velocity parallel to the galactic plane is close to 8 km s^{-1} (Kerr and Westerhout 1965), so that if the turbulence is isotropic, turbulent pressure can provide a significant fraction of the required support. The role of a magnetic field and cosmic rays trapped in it must also be taken into account. The value of the magnetic field at $z = 0$ required to yield the entire support is $B_* = (8\pi\rho_* c)^{1/2} = 6 \times 10^{-6}$ gauss, a value which is at least roughly consistent with various observations (Woltjer 1965). At the moment, however, the observations are too imprecise to permit one to estimate what fraction of the support is due to turbulence and what fraction is due to magnetic pressure. It seems likely that both contribute significantly.

Initially the disk of the newly formed galaxy is gaseous, so that the gravitational field is due to the gas itself. Spitzer (1942) showed that an infinite gaseous disk obeys

$$\rho = \rho_* \operatorname{sech}^2 (z/h) , \tag{154}$$

where

$$h = \frac{c^2}{2\pi GM} , \tag{155}$$

with M the mass/area above the plane:

$$M = \int_0^\infty \rho dz = \rho_* h . \tag{156}$$

Defining m by

$$m = \int_0^z \rho dz \tag{157}$$

one can show that

$$m/M = (1 - \rho/\rho_*)^{1/2} = \tanh (z/h) . \tag{158}$$

From equations (155) and (156) it follows that

$$\rho_* = 2\pi GM^2/c^2 . \tag{159}$$

Hence for a given value of M (determined by collapse to a disk), an increase in c results in an increase in h and a decrease in ρ_*. This elementary fact has implications for the possible turbulent support of the disk. It is readily shown that the turbulence in the Milky Way cannot be primordial, but must be regenerated by external energy sources such as the expansion of H II regions and/or supernova shells (Spitzer 1967). As the latter depend upon the presence of young stars, the rate of input into turbulence depends upon the rate of star formation and hence upon the properties of the gas, such as ρ_*, c, and B. If the turbulent energy input were lower than average in some region, c would drop and ρ_* would rise. This might well increase the rate of star formation and increase the turbulent input again. Hence a negative feedback would obtain, and the balance would be stable. On the other hand, increasing ρ_* might increase B and decrease the rate of star formation and hence the turbulent input. Feedback would then be positive, and the balance would be unstable. The persistence of gas in the Milky Way during about 10^3 times the turbulent dissipation time suggests that the balance is actually stable, and that the feedback is negative.

Field and Saslaw (1965) and Field and Hutchins (1968) have studied a model satisfying these requirements. They derive the relation

$$\frac{d\rho}{at} = -K\rho^2\,, \qquad K = \frac{ac\sigma_c}{\mathfrak{M}_c}\,, \tag{160}$$

for the rate of decrease of gas density due to star formation, where $\mathfrak{M}_c$ and σ_c are the mass and cross-section of interstellar clouds, c is the turbulent velocity, and a is an eigenvalue of the cloud-collision problem, which they calculate to be $\sim 10^{-3}$. Equation (160) reflects the assumption made in the theory, that star formation occurs in massive clouds which are built up by binary cloud collisions; it was proposed earlier on empirical grounds by Schmidt (1959). If the cloud properties are independent of ρ, it indicates that star formation is proportional to $\rho^2 c$, which by equation (159) is proportional to $\rho^{3/2}$. This is in the sense required to stabilize turbulent support of the disk. Hence a possible model for well-developed disk galaxies is based upon turbulent support of the gaseous disk together with a stable balance between turbulence and star formation.

We turn now to the question of the distribution of mass and angular momentum in a disk galaxy. So far, our discussion includes only the special case of solid-body rotation (§ 3.2). While normal spirals do not now rotate as solid bodies, Mestel (1965*a*) has suggested that they may have done so at some time in the past. Crampin and Hoyle (1964) tested observationally the hypothesis that the distribution of angular momentum with mass in normal spirals is the same as in a uniform spheroid in solid-body rotation. To discuss their results, we rewrite the criterion of centrifugal equilibrium, $R = 4\xi/\pi$, using equation (136), in the form

$$\Omega^2 = \frac{3\pi G\mathfrak{M}}{4a^3}\,. \tag{161}$$

The local mass/area is given by

$$M(r) = \int_0^{z(r)} \rho(z, r)dz = \rho c(1 - r^2/a^2)^{1/2}\,, \tag{162}$$

which, because $\mathfrak{M} = 4\pi a^2 c\rho/3$, may be written,

$$M(r) = \frac{3\mathfrak{M}}{4\pi a^2}(1 - r^2/a^2)^{1/2}\,. \tag{163}$$

(The pressure corresponding to $c \simeq 10^{-2}a$ has a negligible effect on the radial equilibrium.) If L denotes the quantity Ωr^2 (angular momentum per unit mass), one finds from equation (162) the amount of mass $d\mathfrak{M}$ associated with the increment dL of angular momentum:

$$\frac{d\mathfrak{M}}{dL} = \frac{3\mathfrak{M}}{2L(a)}\left[1 - \frac{L}{L(a)}\right]^{1/2}, \tag{164}$$

where $L(a) = \Omega a^2$ is the value of L at the edge of the disk. If there are no torques, both $d\mathfrak{M}$ and dL are conserved, even if the galaxy arranges itself into a new equilibrium such that $\Omega(r) \neq$ constant. Hence expression (164) is conserved and can be compared with observation. Crampin and Hoyle (1964) applied the methods of chapter 3 to determine from observation of several galaxies the values of $d\mathfrak{M}$ and dL associated with an interval dr, and found that equation (164) is well obeyed, at least in the outer parts. They suggest an explanation of why the inner parts are observed to be discrepant. They conclude that indeed galaxies may have evolved from clouds in uniform rotation, conserving angular momentum in the process. This evolution is considered further below.

The equations of § 3.2 are inconsistent with equation (154); they describe a disk of constant density and varying temperature, while the latter describes one with constant temperature and varying density. Moreover, the latter applies strictly only to disks of infinite radius, as no allowance is made for the gravitational field parallel to the disk which exists in finite disks. Goldreich and Lynden-Bell (1965*a*) analyzed the gravitational instability of an infinite disk in solid-body rotation, which locally obeys equation (154). Their results, which have profound implications for the formation of spiral arms, are contained in an equation for the wavenumber parallel to the disk $k(= n/h)$, of marginal gravitational instability:

$$\frac{1 + n + \frac{1}{2}n^2\Psi'(n)}{4n(1 - n^2)} = \frac{\pi G\rho_*}{4\Omega^2}, \tag{165}$$

where Ψ is Hurwitz's generalization of the Riemann zeta-function, $\Psi(n) = \zeta(\frac{1}{2}n, 2)$. The dimensionless ratio on the right-hand side has an interesting physical significance. Although the equation strictly applies only to disks with $a \to \infty$, it may be applied approximately to disks of finite radius. Since $\rho_* = M/h$ and $M = 3\mathfrak{M}/4\pi a^2$ for uniform disks with $a \to \infty$ (eq. [163]), we may set $\rho_* = 3\mathfrak{M}/4\pi a^2 h$ in that limit. Similarly, we may use equation (161) for Ω^2, and write

$$Q \equiv \frac{\pi G\rho_*}{4\Omega^2} = \frac{a}{4\pi h}. \tag{166}$$

Thus, for a given Ω, Q increases with ρ_*, and hence with $1/h$; that is, it increases with the degree of flattening. Goldreich and Lynden-Bell were able to show that equation (165) has no roots if $Q < 1.09$, that is, if $h/a > 0.073$. Under these conditions, the pressure is relatively high, so that wavelengths short enough that they are not stabilized by rotational forces are nevertheless stabilized by pressure, as in the Jeans analysis. If $Q = 1.09$, instability sets in at $n = 0.47$, corresponding to $\lambda = 13.5h$. Although their analysis is not strictly applicable for finite a, equation (166) indicates that this corresponds to $\lambda \simeq a$. When $Q > 1.09$, there is a range of unstable wavelengths, the long-wavelength cutoff being due to rotational effects and the short-wavelength cutoff being due to pressure effects. In the limit of large Q, $\lambda_{\max} \simeq a$ and $\lambda_{\min} \simeq 2\pi h$. The latter result agrees with an earlier analysis by Ledoux (1951) of the special case $\Omega = 0$. Hunter

(1963) considered a finite disk with vanishing pressure, so that the disk is infinitely thin, permitting one to obtain analytic results. He found that all axisymmetric disturbances except one are unstable. The one exception is the lowest possible mode, corresponding to the homologous radial distortion of the whole disk. This result coincides with what might be expected from Goldreich and Lynden-Bell. The latter authors (1965*b*) also considered the case $\Omega \neq$ constant, and found that the same analysis applies locally if Ω is replaced by half the epicyclic frequency, $\Omega(1 + rd\Omega/2\Omega dr)^{1/2}$. Bel and Schatzman (1958) studied infinite cylinders with arbitrary rotation laws, with similar results. Safranov (1960) studied the special case of a Keplerian disk and found Q (crit) = 1.4. Toomre (1964) found similar results for stellar disks using the collisionless Boltzmann equation.

Since the observed flattening of galaxies often corresponds to $h/a < 0.073$, instability is expected in practice. Goldreich and Lynden-Bell suggest that such instability will lead to clumpiness in the density distribution, which can be sheared into spiral features by differential rotation. This suggestion can be extended as follows. Equation (160) states that the rate of star formation increases in regions of higher density which form through gravitational instability. After a delay, star formation generates turbulence, which increases c and distends the gas until ρ_* falls below the critical value for gravitational instability (eq. [165]). The disk then becomes dynamically stable, and the rate of star formation falls until the turbulence from the previous burst of star formation has been dissipated. When that happens, ρ_* increases again, gravitational instability sets in, and the cycle is repeated. The oscillatory cyclic behavior of the system is readily traced to the fact that the relation between star formation and turbulence is stable. Possibly spiral arms, which are a periodic phenomenon, are related to the cycle described.

Let us return to the analysis by Hunter (1963). He found that the largest-scale axisymmetric mode which is still unstable corresponds to a single node in the velocity, with the inner parts moving inward and the outer parts outward. Mestel (1963) discovered a nonlinear transformation of a disk which shares the single-node property. If the gas at r moves to $r' = \pi r^2/2a$ (so that $a' = \pi a/2$), then $r'/r = \pi r/2a$, so that for $r < 2a/\pi$ material moves inward, while for $r > 2a/\pi$ it moves outward. If angular momentum is conserved, $\Omega' r'^2 = \Omega r^2$, so that the rotation law becomes

$$\Omega' = \frac{2a\Omega}{\pi r'} = \left(\frac{3\mathfrak{M}G}{\pi a}\right)^{1/2} \frac{1}{r'}, \tag{167}$$

which implies that $v_\theta' = \Omega' r' = \text{constant} \equiv V$. The new distribution of mass, found by setting $M'r'dr' = Mrdr$, is

$$M' = \frac{3\mathfrak{M}}{4\pi^2 a r'}\left(1 - \frac{2r'}{\pi a}\right)^{1/2}. \tag{168}$$

On the other hand, equation (167) implies a definite variation of the centrifugal force and hence of the gravitational field required to balance it. This in turn implies a unique mass distribution, which Mestel shows is

$$M^* = \frac{3\mathfrak{M}}{4\pi^2 a r'}\left[1 - \frac{2}{\pi}\sin^{-1}\left(\frac{2r'}{\pi a}\right)\right]. \tag{169}$$

Equations (168) and (169) agree at both $r' = 0$ and $r' = a'$, the maximum relative deviation being 11 percent. Mestel's transformation therefore takes a disk in centrifugal equilibrium into a new one also in equilibrium. Mestel suggests that this transformation in fact occurs early in the life of some galaxies as a result of the nonlinear development of Hunter's low-order unstable mode. One observes systems which resemble both models with Ω = constant (barred spirals) and those with V = constant (spirals like M31). Unfortunately, the growth rate for the low-order mode is the lowest for any mode. To some extent this may be compensated by the fact that as the galaxy contracts in z, the first modes to appear as h/a decreases below 0.073 are those of low order. It should be noted that the V = constant model has a strong central concentration; in fact, M is singular at $r = 0$ in the simplified calculations. As the latter do not include the finite thickness of the disk, they are expected to break down as r approaches h. A change in behavior at such a distance is actually observed in the Milky Way (fig. 3 of Kerr and Westerhout 1965).

The above discussion is based on neglect of viscous torques, an assumption which is supported by the observational results of Crampin and Hoyle (1964). One can also argue theoretically that torques are negligible (Mestel 1963). The viscous time scale is $t_v = \Omega/\eta\nabla^2\Omega \simeq r^2/c\lambda$, where η, the turbulent viscosity, is the product of the turbulent velocity c and scale λ. The collapse time scale is $t_c = (G\rho)^{-1/2}$. Since by equations (155) and (159) $h = c/(2\pi G\rho_*)^{1/2}$, the ratio of the time scales is

$$t_v/t_c \sim (r/a)^2(a/\lambda)(a/h) \gg (r/a)^2(a/h)\,, \qquad (170)$$

since turbulence must have $\lambda \ll a$. Since $a/h > 1$ (thin disk), $t_v \gg t_c$ and viscosity has negligible effect on the collapse time scale. This derivation indicates two circumstances in which viscosity may not be negligible, however. Viscous forces may be important at points $r \ll a$ and over times much longer than t_c. Models including viscosity have been investigated by various workers following von Weizsäker's (1951) paper on the effects of turbulence in galaxy formation. Lüst (1952) found that in models with V = constant, turbulent viscosity transports angular momentum outward, reducing Ω in the inner regions so that matter there flows inward because of the reduction of centrifugal force. Von Hoerner (1952, 1953, 1955) found similar results with mass distributions closely approximating observed galaxies. Trefftz (1952) found similar effects even when the gravitational effect of the changing mass distribution (ignored previously) was taken into account.

We stated above that it is not known whether turbulence or a magnetic field is the main support of gas in galaxies. If the magnetic field is strong, it too may transport angular momentum (Woltjer 1965). However, a field large enough to have significant effects, 5×10^{-5} gauss, appears to be much larger than the 6×10^{-6} gauss which would be consistent with the observed thickness of the disk. We conclude that although turbulence and/or a magnetic field can transport angular momentum slowly, they probably cannot keep pace with the collapse. Our equations for treating the collapse, based on the absence of torques, are therefore acceptable as a first approximation. According to them, the Mestel-Hunter transformation can drastically alter the mass distribution and rotation law early in the life of the galaxy, with results agreeing with observation.

4.4. Differentiation between Ellipticals and Disks

The clear division of galaxies into ellipticals and disks calls for interpretation. Ellipticals are characterized by lack of gas and by $\alpha > 0.3$, while disk systems ordinarily possess gas and have $\alpha \simeq 0.01$. (Galaxies of type S0 are not considered, as they may be spirals which have been swept free of gas, according to Spitzer and Baade 1951.) It seems probable that the properties of the two types were established at the time of formation. The parameters which may be significant include mass ($\mathfrak{M}$), thermal energy (η), angular momentum (ξ), magnetic energy, and the epoch of formation (t_M). Mass cannot be a crucial parameter as objects of both types but equal masses are found. We saw in § 3.4 that the value of η is determined by the values of $\mathfrak{M}$ and t_M. The value of t_M also may affect the value of ξ by affecting the degree of interaction with neighboring galaxies. The effects of magnetic energy are discussed by Mestel (1965*b*) and Piddington (1967). Of these possibilities, the most interesting would seem to be ξ, which according to Hoyle's theory (§ 3.3) is a random variable whose value depends upon such things as the chance orientation of neighboring galaxies. Hence a range of values of ξ is expected, which could correspond to the variety of degrees of flattening observed. The mechanism by which this occurs could be that in galaxies of low ξ, fragmentation is unhampered by rotational forces and is rapidly completed, while if ξ is large, fragmentation is hampered and much of the material remains gaseous. Since dissipation is enhanced in the latter case, the high z-velocities acquired during collapse can be dissipated in systems of high ξ, with the result that a thin disk is formed.

This picture has several problems, however. First, even nearly spherical galaxies are observed to be rotating (King and Minkowski 1966), so that it is not clear that as a group the ellipticals have less angular momentum than do the disk systems. Second, the theory as it stands suggests that any finite value of ξ, no matter how small, will inhibit fragmentation. According to Layzer, induced spin inhibits fragmentation even if $\xi = 0$. Finally, it is not clear why states of intermediate flattening ($\alpha \simeq 0.1$) and gas content are not populated, as would be expected if ξ is a continuous variable.

5. OBSERVATIONAL DATA

5.1. Intergalactic Matter

According to some theories (Sciama 1964), galaxy formation can occur even at present if the intergalactic medium is sufficiently dense ($\sim 10^{-29}$ g cm^{-3}) and of the right temperature ($\sim 10^5$ ° K). Some information concerning density and temperature has been obtained by X-ray, ultraviolet, and radio observations. Lyman-α (Gunn and Peterson 1965) and Lyman-band (Field, Solomon, and Wampler 1966) absorption has been searched for in the spectra of distant quasars with negative results, placing stringent upper limits on the neutral component of intergalactic gas at $z = 2$. From these limits Weymann (1967*a*) showed that if the density is as high as required by Sciama, $T > 10^5$ ° K. At such a high temperature the gas is ionized, which explains the lack of Lyman-α and Lyman-band absorption, as well as the observed lack of 21-cm absorption (Field 1962; Davies and Jennison 1964; Penzias and Scott 1968), 21-cm emission (Goldstein 1963; Penzias and Wilson 1968), and X-ray absorption (Bowyer, Field, and Mack 1968). The latter authors also find that $T < 3 \times 10^5$ ° K by requiring that bremsstrahlung by the gas does not exceed the observed flux at 44–60 Å. We conclude that there may be a dense ionized gas with $T = 2 \times 10^5$ °K present, as required by Sciama, although another interpretation is that the gas is of lower density than required. A positive result obtained from 21-cm observations of Fornax A by Koehler and Robinson (1966) and by Koehler (1966) corresponds to $n_{\rm H} = 1.5 \times 10^{-7} T_S$, where T_S is the spin temperature (H is taken to be 100 km s^{-1} Mpc^{-1}). This result is contradicted by the upper limit obtained by Penzias and Scott (1968) from observations of Cygnus A, $n_{\rm H} < 3 \times 10^{-8} T_S$, using the same value of H. Therefore, it should not be accepted until it is confirmed by further observations of Fornax A.

Sciama (1966) has suggested that the ionized component might be detected through its Thomson scattering of the radiation from distant objects. If $q_0 = \frac{1}{2}$, $\tau = 1$ at $z = 7.5$, and there should be a drop in radio source count at that redshift. Magnetic fields would cause Faraday rotation, and the observations indicate that $B < 10^{-8}$ gauss over a scale of 10^{24} cm (Sciama 1964), while $B < 10^{-9}$ gauss over a scale of 10^{26} cm (Parijskij 1967). The former value corresponds to a field of 10^{-4} gauss in a galaxy which contracts homologously from intergalactic gas. This field strength is insufficient to prevent collapse, but it certainly would affect the evolution of the galaxy, so refinement of the upper limit is of interest.

5.2. Observation of Young Galaxies

Direct observation of galaxies in an early evolutionary phase would be very significant. According to equation (61), large galaxies cannot form until $2t_M = 7 \times 10^7$ years in the evolutionary picture. This epoch corresponds to redshifts of 60, 20, and 10 for $q_0 = 0.02$, 0.5, and 2, respectively, so that detecting galaxies soon after their formation may be very difficult in evolving models. In steady-state models the situation is rather different, the ages of galaxies even at small redshifts being distributed like e^{-3Ht}, so that 10 percent of the galaxies would be less than 10^9 years old. Several papers deal with the ages of nearby galaxies. Burbidge and Burbidge (1959*a*, *b*) studied the two binary systems VV 117 and VV 123, each of which contains an elliptical and a spiral, and concluded that the spiral may have formed recently in the wake of the elliptical in accordance with Sciama's (1955) theory. Later Sandage (1963) determined the colors of the four galaxies and found that those of the ellipticals were normal, while those of the nuclei of the spirals ($B - V = 0.68$ and 0.35) indicated average stellar ages of 3×10^9 and 5×10^8 years, apparently supporting the contention of Burbidge and Burbidge. However, as Sandage pointed out, these estimates are only lower limits on the ages of the spirals if star formation has occurred recently within their nuclei.

The narrow range of colors observed for ellipticals suggests a common age; the wide variation for spirals might reflect a spread in age or, alternatively, a common age with variations in the rate of recent star formation. Holmberg (1964) found that average mass density, $\mathfrak{M}/L$, and color are all correlated with the morphological type, as is gas content (chapter 9). He interprets these correlations in terms of Schmidt's (1959) model of star formation. Both the morphological type and the mean density are established at the time of formation. Galaxies of high density evolve into stars more quickly and therefore attain higher values of $\mathfrak{M}/L$, redder colors, and lower gas contents. This interpretation thus permits all galaxies to have a common age and explains the observed differences in terms of differing rates of star formation. In fact, not much variation in the ages of galaxies would be permitted in this picture; otherwise a young, dense system would have the same characteristics as an older, less dense system, contrary to observation. Burbidge, Burbidge, and Hoyle (1963) do not subscribe to this interpretation for all galaxies, and point out that some systems are in configurations which imply rapid dynamical evolution and relatively small age. Filamentary structure, large relative velocities, etc., are taken to imply ages as low as 3×10^7 years.

Steady-state theory predicts a wide distribution of galaxy ages, but there is no reason why the same might not be true in an evolving model also. Hence Holmberg's interpretation of the data is an argument against the steady state, but observations like those of Burbidge *et al.*, which indicate that perhaps some galaxies are young, are not necessarily contrary to evolving models. Indeed, it is plausible that there is a distribution of density fluctuation amplitudes at decoupling and hence a distribution of times of formation (table 2).

Shklovskii (1960) pointed out that initially the rate of star formation in a young galaxy would be very large, owing to the large amount of available interstellar gas. He estimated that supernovae alone would make a massive galaxy brighter than magnitude -23, as is sometimes observed for radio galaxies. Field (1964) showed that if the angular momentum of a galaxy is so low that star formation is extremely rapid, the luminosity from stars can exceed magnitude -25, while Sturrock (1966) showed that gravitational energy can be converted efficiently to intense nonthermal emission during the collapse of a young galaxy. Such processes could conceivably account for some of the unusual galaxies observed even at relatively modest redshifts. For example, Longair (1966) has interpreted radio source counts in terms of a population of powerful sources which peaks sharply at $z = 4$ and which decreases in abundance after that epoch. This behavior is in qualitative agreement with the hypothesis that such objects are young and that their numbers reflect the distribution of epochs of formation, which in turn reflect the amplitude distribution of density perturbations.

The latter hypothesis suggests that one search for bright galaxies at rather large redshifts. Partridge and Peebles (1967*a*, *b*) have investigated the luminosity-time history of massive spiral galaxies in some detail. If one takes $t_M = 1.4 \times 10^8$ years, then the bright phase occurs at $z = 10$–30; it lasts about 3×10^7 years. During this phase the magnitude is roughly -27, with most of the energy due to stars of 30,000° K, so that the redshift energy curve peaks between 1 and 3 μ. Individual galaxies may be marginally detectable in spite of the night-sky emission, while the integrated background may be detected in the 5–15 μ range where confusion with other sources of radiation is less serious. The integrated background would have the advantage of being independent of intergalactic electron scattering, which might completely diffuse the images of individual galaxies. Weymann (1967*b*) found an even larger integrated background in a model which included a wider range of galaxy types, and concluded that a recent upper limit on the visual background of $\nu I_\nu < 3 \times 10^{-5}$ ergs cm^{-2} s^{-1} sterad^{-1} (Roach and Smith 1967) implies that galaxy formation must have occurred at $z > 11$. Future measurements of faint sources may therefore give additional information about galaxy formation.

5.3. History of the Milky Way Galaxy

The analysis of stellar populations and galactic nucleosynthesis may permit a reconstruction of the early history of the Galaxy. It is generally thought that the oldest stars have the smallest metal content and the smallest concentration to the disk (Blaauw 1965). This may be understood as follows, according to Oort (1964). The protogalaxy, originally an inhomogeneity in the expanding Universe, contracts toward higher density, its rotation around an axis which is fixed by the total angular momentum increasing as it does so. Stars form continually up to the present, those which formed earliest retaining the unflattened distribution of the gas at the time they formed, and the ones which formed later retaining the flattened distribution of gas which resulted from the dissipation of z-motion. Nucleosynthesis in young stars continually enriches the gas, so that later generations of stars have more heavy elements. This simple picture cannot be accepted until various points have been investigated in more detail. For example, the lack of unique correlation between age and metal content (Pagel 1966) may be explicable in terms of large spatial variations in the chemical composition of the interstellar medium at any one time. Particularly striking examples of this phenomenon, which may not be easily explained, have been found by Spinrad (1966). Schmidt (1959) and others have attempted a quantitative reconstruction of the evolution of the Galaxy from gas into stars. An interesting variation on such models would be the inclusion of supernova explosions and subsequent diffusion of the elements produced therein by means of a statistical model. It seems evident that nucleosynthesis must have been extraordinarily rapid during the first 10^9 years (Greenstein 1966) in order to account for the rapid thousand-fold increase in heavy elements implied by the observations of old clusters. According to Partridge and Peebles (1967*a*), one would expect some 7×10^9 supernovae during the first 10^8 years—about 2×10^4 times the present rate—which may be sufficiently rapid.

Eggen, Lynden-Bell, and Sandage (1962) discuss evidence that not only did the Galaxy contract from a diffuse state, but it did so by virtually free-fall collapse. They consider the kinematical properties and ultraviolet excess of 221 dwarf stars, and show that small metal content, and hence presumably large age, are correlated with a high value of orbital eccentricity and inclination and a low value of angular momentum. The circular, low-inclination, high-angular-momentum orbits of young stars are readily explained by their birth from interstellar gas near the Sun, which has the same characteristics. The highly eccentric and inclined orbits of low angular momentum observed for old stars are explained as a result of a presumed large inward radial motion of gas at the time they were formed. The radial velocity v_r must be of the same order as the tangential velocity v_θ, because otherwise the orbital eccentricity would be small at the time of formation, and since they show that eccentricity is an adiabatic invariant for slowly varying gravitational potentials (which would exist if $v_r \ll v_\theta$), the present eccentricity would be small also, contrary to observation. The requirement that $v_r \simeq v_\theta$ amounts to stating that the Galaxy collapsed dynamically in a few $\times\, 10^8$ years. Similar arguments apply to the z-motion, so that the correlation of orbital inclination with age is interpreted as the result of collapse in the z-direction at the same epoch.

Another possibility, which is dismissed by Eggen, Lynden-Bell, and Sandage, is that the stars all formed from gas which was distributed much as it is at present. Because the angular momentum of each star is conserved, it would have been formed from gas having the same angular momentum. For example, stars with $\delta(U - B) = 0.2$ having $L = 1.2 \times 10^3$ kpc km s^{-1} must have originated at $r = 5$ kpc. To explain the presence of such stars in the solar neighborhood, one must postulate a large peculiar velocity for the gas, about 50 km s^{-1} in the present example. This model has been considered in some detail by van Wijk (1956). Recent evidence on the turbulence of interstellar gas makes such large velocities unlikely, but one must keep in mind the observed high radial velocity of the 4-kpc arm (Oort 1962) as well as the possibility that turbulent velocities were much larger in the past. We conclude that the interpretation of Eggen *et al.* may not be unique.

If one accepts the model of Eggen, Lynden-Bell, and Sandage, the picture of the formation of the Galaxy strongly resembles that of § 3.2, and one can therefore apply the results of §§ 3.2 and 4.3 to estimate the initial conditions. Since $\mathfrak{M} = 1.8 \times 10^{11} \mathfrak{M}_{\odot}$ (Schmidt 1965), the initial radius a_0 would be 55 kpc for an assumed value $t_M = 5 \times 10^8$ years. If $a' = 15$ kpc is the present radius of the disk and if the disk underwent a Mestel-Hunter transformation, then a, the radius of the collapsed disk in solid-body rotation, would have been 9.6 kpc. Material now at $r' = 5$ kpc would have been at $r = 5.5$ kpc ($0.58a$) before the transformation; and since the collapse prior to the transformation is homologous, $r_0 = 0.58a_0 = 32$ kpc. Hence material now at 5 kpc would have undergone a homologous contraction of 0.17 and a nonhomologous contraction of 0.9, for an overall factor 0.16, in rough agreement with the observational estimate of 0.1 by Eggen *et al.* based on the orbits of stars having the same angular momentum as matter now at 5 kpc. We conclude that $t_M = 5 \times 10^8$ years (or somewhat larger) is in fair agreement with observation; this corresponds to a redshift of 2.5 at the epoch $2t_M$ (if $q_0 = 0.5$). This agreement suggests that the hypothesis that $2t_M = 7 \times 10^7$ years for all galaxies (§ 4.1) is not correct.

5.4. Multiple Systems of Galaxies

Chapters 13, 14, 15, and 16 of this volume present evidence that galaxies often are found in groups containing from two members to several thousand members. The observations may be summarized as follows: (i) Systems of small multiplicity, such as binaries, are extremely common. As Holmberg (1962) finds that the fraction of bright galaxies not in the Virgo cluster which are in systems of multiplicity n is $\simeq 2^{-n}$ for $n \leq 7$, only half of all galaxies are single. (ii) Clusters containing up to a few thousand members are readily observed and catalogued (Abell 1958; Zwicky, Herzog, and Wild 1961; Zwicky and Herzog 1963). It is consistent with observation to suppose that all galaxies may belong to clusters (Scott 1962; Neyman 1962). (iii) Superclusters containing about 10 rich clusters within a diameter of 40 Mpc and having masses of 10^{16}–$10^{17} \mathfrak{M}_{\odot}$ are thought to exist by some workers (Abell 1961; de Vaucouleurs 1961), but not by Zwicky. (iv) There appears to be little clustering on a larger scale.

These data raise the question whether the theories described above for the formation of isolated galaxies are relevant. We consider some aspects of this question below. First, consider systems of small multiplicity. Ambartsumian (1961) argues that the probability of capture is so small that such systems must have existed as such since the time of their formation. This being the case, one may postulate that multiple systems have formed as the result of fragmentation of large clouds, which at their maximum extension must have been larger than the diameter of the final system, some 100–200 kpc (Holmberg 1962). There seems to be no reason why this might not have occurred in an evolutionary model; it would imply a rather late time of formation, however. A somewhat more difficult problem concerns the sign of the energy of systems containing only a few members.

Ambartsumian (1958) pointed out that the mere existence of such a system may not imply negative energy, as energy exchange among the members might lead to evaporation in a short time even if the energy were negative. Indeed, Worrall (1967) has observed ejection of particles after a few dynamical time scales in computer experiments on systems of five members. Hence the energy of such systems must be determined by observation. Burbidge and Burbidge (1961) discuss 16 systems with these results: $E < 0$, 3; $E > 0$, 6; uncertain, 3; and no conclusion, 4. For example, the system of five spirals associated with NGC 55 has $E > 0$ unless $\mathfrak{M}/L > 500$, about 100 times the normal value. It appears that the only way to rescue the Jeans hypothesis, which implies $E < 0$, is for disruptive forces to add energy long after the galaxies form. Vorontsov-Velyaminov (1961) subscribes to the existence of such forces, while Ambartsumian (1958) postulates that multiple galaxies originate in the division of galactic nuclei, a hypothesis for which he finds support in the observation of violent events in such nuclei. Such ideas would lead to the conclusion that the observed galaxies in systems of positive energy should be relatively young. We have already seen (§ 5.2) that the usual observational tests for the youth of such systems may be inconclusive because of the incidence of later phases of star formation. Arp (1968) draws attention to chains of galaxies near powerful radio sources, which he interprets as ejecta from such sources. In many of his cases there is little doubt that the supposedly ejected galaxies have normal colors and forms indicative of old age, raising the necessity for unusual physical hypotheses if the ejection idea is accepted.

In the case of rich clusters of galaxies there is a well-known discrepancy between the masses determined for individual galaxies and those determined by the virial theorem. The Coma cluster may be bound by the mass within galaxies with a reasonable value of $\mathfrak{M}/L$ (Neyman, Page, and Scott 1961), but the Virgo cluster is almost certainly not. It is of interest that Koehler and Robinson (1966) claim to have found atomic hydrogen in the latter cluster. Although the amount is insufficient to bind the cluster, other forms of gas (including ionized gas) could contribute enough mass and not have been detected as yet. Unlike the situation with small groups, there is no alternative to negative energy for rich clusters of galaxies within the framework of the Jeans hypothesis, as it is inconceivable that thousands of galaxies could be given positive energy by some explosive process. Hence observations of the total mass of clusters are crucial to the evaluation of the Jeans picture of galaxy formation.

If one ignores the mass discrepancy, two possibilities emerge for the formation of clusters. If the amplitude of density perturbations on the scale of clusters is larger than that on the scale of galaxies at decoupling, one can apply the theory described above for the formation of isolated galaxies to the formation of clusters from intergalactic gas. In this case, individual galaxies would appear later as fragments in the contracting protoclusters. On the other hand, if the amplitude of large-scale fluctuations were originally smaller (as seems likely on general grounds), formation of individual galaxies may have been completed before gravitational instability on the large scale became manifest. Van Albada (1960, 1961) has considered the clustering of galaxies which have already attained high internal density, using the collisionless Boltzmann equation in both the linear and nonlinear regime. He utilized the moment method and set moments greater than the second (corresponding to heat flow and higher-order effects) equal to zero.

The linear analysis yields an equation like equation (33), which resulted from Bonnor's hydrodynamic treatment, with an effective $\gamma = 3$ and with the zero-order pressure being equal to the density times the velocity variance of galaxies. One can therefore calculate an instantaneous Jeans length separating stable from unstable wavelengths as in § 2.1. If the rms velocity of galaxies is 50 km s^{-1}, perturbations on the scale of clusters are now unstable provided that $\rho > 3 \times 10^{-30}$ g cm^{-3}.

Nonlinear computations were carried out for initially unstable perturbations in a $k = 0$ model. Starting at $t = 7 \times 10^7$ years with a density perturbation of 7 percent (corresponding to $h \simeq 10^{-3}$ at decoupling), van Albada finds that the expansion begins to lag that of the Universe, so that at a time corresponding to about 10^9 years in the past, a nucleus with density several thousand times that of the background has built up. The centrifugal repulsion due to the large contraction factor and angular momentum of individual galaxies keeps the nucleus from attaining infinite density. At the present time, the nucleus would have a radius of 0.23 Mpc and a density of 8×10^{-26} g cm^{-3}, in fair agreement with observations of the Coma cluster. In this model, the cluster would still be contracting, a result tentatively confirmed observationally by Neyman and Scott (1961) by an analysis of the velocity residuals for galaxies in the cluster. Such systematic velocities can affect the value of the mass determined by naïve application of the virial theorem, by as much as a factor of 2 or 3 either way, a result which could be partially responsible for the mass discrepancy. Gilbert (1966) has obtained results quite similar to van Albada's.

The above theory of clustering is highly suggestive, and seems to be roughly in accord with what one might expect if large-scale perturbations were of small amplitude ($\sim 0.1\%$) at decoupling. However, in order to have sufficient orbital angular momentum so that the final cluster would agree with observation, van Albada was forced to assume a random velocity of 50 km s^{-1} now, corresponding to 1100 km s^{-1} at the initial epoch, 7×10^7 years. It is not at all clear how such high random velocities would have arisen so early in the history of the Universe. One possibility is that initially the random velocities were much lower, but that they arose later during the formation of the cluster by the action of gravitational instability on scales smaller than that of the cluster itself. This picture is similar to that suggested by Lynden-Bell (§ 4.1) for the origin of orbital angular momentum of the fragments in collapsing galaxies.

Finally, we consider the origin of superclusters. As we have already pointed out (§ 2.2), masses exceeding $10^{16} \mathfrak{M}_\odot$ are unusual in that gravitational instability for them proceeds uninterrupted from the earliest times. This fact, coupled with a postulated spectrum of initial density perturbations which decreases toward larger wavelengths, might lead to objects having masses near $10^{16} \mathfrak{M}_\odot$, which could be associated with the superclusters. Hence the structure of the Universe today on scales exceeding $10^{11} \mathfrak{M}_\odot$ could be interpreted as the result of gravitational instability acting on a definite spectrum of perturbations. Within this framework, the cosmological principle, which asserts that the Universe is uniform on the largest scales, can be rephrased in terms of an initial perturbation spectrum which approaches zero at very large wavelengths.

Much of this article was written with the support of the National Science Foundation under grant GP-4975.

REFERENCES

Abell, G. O. 1958, *Ap. J. Suppl.*, **3**, 211.
———. 1961, *A.J.*, **66**, 607.
Adams, J. B., Mjolsness, R., and Wheeler, J. A. 1958, in *La Structure et l'Evolution de l'Univers: Onzième Conseil de Physique, Solvay* (Brussels: R. Stoops), p. 113.
Adler, R., Bazin, M., and Schiffer, M. 1965, *Introduction to General Relativity* (New York: McGraw-Hill Book Co.).
Albada, G. B. van. 1960, *B.A.N.*, **15**, 165.
———. 1961, *A.J.*, **66**, 590.
Alfvén, H. 1965, *Rev. Mod. Phys.*, **37**, 652.
Ambartsumian, V. A. 1958, in *La Structure et l'Evolution de l'Univers: Onzième Conseil de Physique, Solvay* (Brussels: R. Stoops), p. 24.
———. 1961, *A.J.*, **66**, 536.
Arny, T. T. 1966, *Ap. J.*, **145**, 572.
Arp, H. C. 1965, in *Galactic Structure*, ed. A. Blaauw and M. Schmidt (Chicago: University of Chicago Press), chap. 19, p. 401.
———. 1968, private communication.
Bel, N., and Schatzman, E. 1958, *Rev. Mod. Phys.*, **30**, 1015.
Belzer, J., Gamow, G., and Keller, G. 1951, *Ap. J.*, **113**, 166.
Blaauw, A. 1965, in *Galactic Structure*, ed. A. Blaauw and M. Schmidt (Chicago: University of Chicago Press), chap. 20, p. 435.
Bonnor, W. B. 1957, *M.N.R.A.S.*, **117**, 104.
Bowyer, C. S., Field, G. B., and Mack, J. E. 1968, *Nature*, **217**, 32.
Burbidge, E. M., and Burbidge, G. R. 1959*a*, *Ap. J.*, **130**, 12.
———. 1959*b*, *ibid.*, p. 23.
———. 1961, *A.J.*, **66**, 541.
Burbidge, E. M., Burbidge, G. R., and Hoyle, F. 1963, *Ap. J.*, **138**, 873.
Callan, C., Dicke, R. H., and Peebles, P. J. E. 1965, *Am. J. Phys.*, **33**, 105.
Campbell, P. M. 1962, *Proc. Nat. Acad. Sci.*, **48**, 1993.
———. 1963, unpublished Ph.D. thesis, Department of Physics, University of Colorado.
———. 1966, *Proc. Nat. Acad. Sci.*, **55**, 1.
Chernin, A. D. 1965, *Astr. Zh.*, **42**, 1124 (English transl. in *Soviet Astr.—AJ*, 1966, **9**, 871).
Crampin, D. J., and Hoyle, F. 1964, *Ap. J.*, **140**, 99.
Davies, R. D., and Jennison, R. C. 1964, *M.N.R.A.S.*, **128**, 123.
Dicke, R. H., Peebles, P. J. E., Roll, P. G., and Wilkinson, D. T. 1965, *Ap. J.*, **142**, 414.
Doroshkevitch, A. G., Zel'dovich, Ya. B., and Novikov, I. D. 1967, *Astr. Zh.*, **44**, 295.
Eggen, O. J., Lynden-Bell, D., and Sandage, A. R. 1962, *Ap. J.*, **136**, 748.
Field, G. B. 1962, *Ap. J.*, **135**, 684.
———. 1964, *ibid.*, **140**, 1434.
———. 1965, *ibid.*, **142**, 531.
Field, G. B., and Henry, R. C. 1964, *Ap. J.*, **140**, 1002.
Field, G. B., and Hutchins, J. 1968, *Ap. J.*, **153**, 737.
Field, G. B., and Saslaw, W. C. 1965, *Ap. J.*, **142**, 568.
Field, G. B., and Shepley, L. C. 1968, *Ap. and Space Sci.*, **1**, 309.
Field, G. B., Solomon, P. M., and Wampler, E. J. 1966, *Ap. J.*, **145**, 351.
Fish, R. A. 1964, *Ap. J.*, **139**, 284.
Gamow, G. 1948, *Phys. Rev.*, **74**, 505.
———. 1949, *Rev. Mod. Phys.*, **21**, 367.
———. 1956, *Vistas in Astronomy*, **2**, 1726.
Gilbert, I. H. 1966, *Ap. J.*, **144**, 233.
Gold, T., and Hoyle, F. 1959, in *Paris Symposium on Radio Astronomy* (*IAU Symp. No. 9*), ed. R. N. Bracewell (Stanford, California: Stanford University Press), paper 104, p. 583.
Goldreich, P., and Lynden-Bell, D. 1965*a*, *M.N.R.A.S.*, **130**, 97.
———. 1965*b*, *ibid.*, p. 125.
Goldstein, S. J., Jr. 1963, *Ap. J.*, **138**, 978.
Gould, R. J., and Burbidge, G. R. 1963, *Ap. J.*, **138**, 969.
Greenstein, J. L. 1966, in *Abundance Determinations in Stellar Spectra* (*IAU Symp. No. 26*), ed. H. Hubernet (New York: Academic Press), p. 354.
Gunn, J. E., and Peterson, B. A. 1965, *Ap. J.*, **142**, 1633.
Harrison, E. R. 1967*a*, *Rev. Mod. Phys.*, **39**, 862.
———. 1967*b*, *Phys. Rev. Letters*, **18**, 1011.
———. 1967*c*, private communication.
Harwit, M. 1961*a*, *M.N.R.A.S.*, **122**, 47.
———. 1961*b*, *ibid.*, **123**, 257.
Hodge, P. W., and Michie, R. W. 1969, *A.J.*, **74**, 587.

Hoerner, S. von. 1952, *Zs. f. Ap.*, **31**, 165.
———. 1953, *ibid.*, **32**, 51.
———. 1955, in *Gas Dynamics of Cosmic Clouds* (*IAU Symp. No. 2*), ed. J. M. Burgers and H. C. van de Hulst (Amsterdam: North-Holland), p. 172.
Holmberg, E. 1962, in *Problems of Extragalactic Research* (*IAU Symp. No. 15*), ed. G. C. McVittie (New York: Macmillan Co.), p. 187.
———. 1964, *Ark. Astr.*, **3**, 387.
Hoyle, F. 1948, *M.N.R.A.S.*, **108**, 372.
———. 1951, in *Problems of Cosmical Aerodynamics* (Dayton, Ohio: Central Air Documents Office), p. 195.
———. 1953, *Ap. J.*, **118**, 513.
———. 1958, in *La Structure et l'Evolution de l'Univers: Onzième Conseil de Physique, Solvay* (Brussels: R. Stoops), p. 53.
Hoyle, F., and Fowler, W. A. 1963, *Nature*, **197**, 533.
Hoyle, F., and Narlikar, J. V. 1966*a*, *Proc. Roy. Soc. London*, **A290**, 143.
———. 1966*b*, *ibid.*, p. 162.
———. 1966*c*, *ibid.*, p. 177.
Hoyle, F., and Wickramasinghe, N. C. 1967, *Nature*, **214**, 969.
Hunter, C. 1962, *Ap. J.*, **136**, 594.
———. 1963, *M.N.R.A.S.*, **126**, 299.
———. 1964, *Ap. J.*, **139**, 570.
Hunter, J. 1966, *M.N.R.A.S.*, **133**, 181.
Irvine, W. M. 1965, *Ann. Phys.*, **32**, 322.
Jeans, Sir J. H. 1928, *Astronomy and Cosmogony* (Cambridge: Cambridge University Press).
Kardashev, N. 1967, *Ap. J.* (*Letters*), **150**, L135.
Kato, S., Nariai, H., and Tomita, K. 1967, *Proc. Astr. Soc. Japan*, **19**, 130.
Kaufman, M. 1965, *Nature*, **207**, 736.
Kerr, F. J., and Westerhout, G. 1965, in *Galactic Structure*, ed. A. Blaauw and M. Schmidt (Chicago: University of Chicago Press), chap. 9, p. 167.
King, I. R., and Minkowski, R. 1966, *Ap. J.*, **143**, 1002.
Koehler, J. A. 1966, *Ap. J.*, **146**, 504.
Koehler, J. A., and Robinson, B. J. 1966, *Ap. J.*, **146**, 488.
Layzer, D. 1963*a*, *Ap. J.*, **137**, 351.
———. 1963*b*, *ibid.*, **138**, 174.
———. 1964, *Ann. Rev. Astr. and Ap.*, **2**, 341.
———. 1965, *Proceedings Amer. Math. Soc. Summer Seminar on Relativity Theory and Astrophysics*, Ithaca, N. Y., 1965 July–August.
———. 1966, *Nature*, **211**, 576.
———. 1968, *Ap. Letters*, **1**, 99.
Ledoux, P. 1951, *Ann. d'Ap.*, **14**, 438.
Lemaître, G. 1949, *Rev. Mod. Phys.*, **21**, 357.
Lifshitz, E. 1946, *J. Phys. USSR*, **10**, 116.
Lin, C. C., Mestel, L., and Shu, F. H. 1965, *Ap. J.*, **142**, 1431.
Longair, M. S. 1966, *M.N.R.A.S.*, **133**, 421.
Lüst, R. 1952, *Zs. f. Naturforschung*, **7A**, 87.
Lynden-Bell, D. 1962, *Proc. Cambridge Phil. Soc.*, **58**, 709.
———. 1964, *Ap. J.*, **139**, 1195.
———. 1967, *M.N.R.A.S.*, **136**, 101.
McCrea, W. H. 1955, *A.J.*, **60**, 271.
Mestel, L. 1963, *M.N.R.A.S.*, **126**, 553.
———. 1965*a*, *Quart. J.R.A.S.*, **6**, 161.
———. 1965*b*, *ibid.*, p. 265.
Michie, R. W. 1965, private communication.
———. 1966, *A.J.*, **71**, 171.
Miller, R. H., and Prendergast, K. H. 1962, *Ap. J.*, **136**, 713.
Misner, C. W., and Sharp, D. H. 1965, *Phys. Letters*, **15**, 279.
Nariai, H., Tomita, K., and Kato, S. 1967, *Progr. Theoret. Phys.*, **37**, 60.
Narlikar, J. V., and Wickramasinghe, N. C. 1967, *Nature*, **216**, 43.
Neyman, J. 1962, in *Problems of Extragalactic Research* (*IAU Symp. No. 15*), ed. G. C. McVittie (New York: Macmillan Co.), p. 294.
Neyman, J., Page, T., and Scott, E. L. 1961, *A.J.*, **66**, 633.
Neyman, J., and Scott, E. L. 1961, *A.J.*, **66**, 581.
Oort, J. H. 1958, in *La Structure et l'Evolution de l'Univers: Onzième Conseil de Physique, Solvay* (Brussels: R. Stoops), p. 163.
———. 1962, in *The Distribution and Motion of Interstellar Matter in Galaxies*, ed. L. Woltjer (New York: W. A. Benjamin, Inc.), p. 14.
———. 1964, *Trans. IAU*, **12A**, 789.

Ozernoy, L. M. 1964, in *Proceedings Symposium on Variable Stars and Stellar Evolution, Moscow*, 1964 February.
Pagel, B. E. J. 1966, in *Abundance Determinations in Stellar Spectra* (*IAU Symp. No. 26*), ed. H. Hubernet (New York: Academic Press), chap. 43.
Parijskij, Yu. N. 1967, *Astr. Zh.*, **44**, 971.
Parker, E. N. 1953, *Ap. J.*, **117**, 431.
Partridge, R. B., and Peebles, P. J. E. 1967*a*, *Ap. J.*, **148**, 377.
———. 1967*b*, *ibid.*, **147**, 868.
Peebles, P. J. E. 1965, *Ap. J.*, **142**, 1317.
———. 1967*a*, *ibid.*, **147**, 859.
———. 1967*b*, in *Proceedings Fourth Conference on Relativistic Astrophysics, New York*, 1967 January.
Peebles, P. J. E., and Dicke, R. H. 1966, *Nature*, **211**, 574.
Penzias, A. A., and Scott, E. H., III. 1968, private communication.
Penzias, A. A., and Wilson, R. W. 1965, *Ap. J.*, **142**, 419.
———. 1968, private communication.
Piddington, J. H. 1967, *M.N.R.A.S.*, **136**, 165.
Roach, F. E., and Smith, L. L. 1967, *Geophys. J.R.A.S.*, **15**, 227.
Roxburgh, I. W., and Saffman, P. G. 1965, *M.N.R.A.S.*, **129**, 181.
Sachs, R. K., and Wolfe, A. M. 1967, *Ap. J.*, **147**, 73.
Safranov, V. 1960, *Dok. Akad. Nauk USSR*, **130**, 53; *Ann. d'Ap.*, **23**, 979.
Sandage, A. 1963, *Ap. J.*, **138**, 863.
Saslaw, W. C. 1967, *M.N.R.A.S.*, **136**, 39.
Saslaw, W. C., and Zipoy, D. 1967, *Nature*, **216**, 976.
Savedoff, M., and Vila, S. 1962, *Ap. J.*, **136**, 609.
Schmidt, M. 1957, *B.A.N.*, **13**, 15.
———. 1959, *Ap. J.*, **129**, 243.
———. 1965, in *Galactic Structure*, ed. A. Blaauw and M. Schmidt (Chicago: University of Chicago Press), chap. 22, p. 513.
Sciama, D. W. 1955, *M.N.R.A.S.*, **115**, 3.
———. 1964, *Quart. J.R.A.S.*, **5**, 196.
———. 1966, in *High-Energy Astrophysics* (*Varenna Lectures*) (New York: Academic Press), p. 418.
Scott, E. L. 1962, in *Problems of Extragalactic Research* (*IAU Symp. No. 15*), ed. G. C. McVittie (New York: Macmillan Co.), p. 269.
Shakeshaft, J. R., and Webster, A. S. 1968, *Nature*, **217**, 339.
Shklovskii, I. S. 1960, *Soviet Astr.—AJ*, **4**, 885.
Silk, J. I. 1967, *Nature*, **215**, 1155.
Spinrad, H. 1966, *Pub. A.S.P.*, **78**, 367.
Spitzer, L., Jr. 1941, *Ap. J.*, **94**, 232.
———. 1942, *ibid.*, **95**, 329.
———. 1967, in *Nebulae and Interstellar Matter*, ed. L. H. Aller (Chicago: University of Chicago Press), chap. 1.
Spitzer, L., Jr., and Baade, W. 1951, *Ap. J.*, **113**, 413.
Strittmatter, P. A. 1966, *M.N.R.A.S.*, **132**, 359.
Sturrock, P. A. 1966, *Nature*, **211**, 697.
Toomre, A. 1964, *Ap. J.*, **139**, 1217.
Trefftz, E. 1952, *Zs. f. Naturforschung*, **7A**, 99.
Vaucouleurs, G. de. 1959, *Handbuch der Physik*, Vol. **53** (Berlin: Springer-Verlag), p. 320.
———. 1961, *A.J.*, **66**, 629.
Vorontsov-Velyaminov, B. 1961, *A.J.*, **66**, 551.
Weizsäcker, C. F. von. 1951, *Ap. J.*, **114**, 165.
Weymann, R. 1966*a*, private communication.
———. 1966*b*, *Ap. J.*, **145**, 560.
———. 1967*a*, *ibid.*, **147**, 887.
———. 1967*b*, preprint, Steward Observatory, University of Arizona.
Wijk, U. van. 1956, *A.J.*, **61**, 277.
Woltjer, L. 1965, in *Galactic Structure*, ed. A. Blaauw and M. Schmidt (Chicago: University of Chicago Press), chap. 23.
Worrall, G. 1967, *M.N.R.A.S.*, **135**, 83.
Zel'dovich, Ya. B. 1966, *Usp. Fiz. Nauk*, **89**, 647 (English transl. in *Soviet Phys.—Usp.*, 1967, **9**, 602).
———. 1967, in *Highlights of Astronomy*, ed. L. Perek, *Proceedings 13th General Assembly IAU, Prague, Czechoslovakia*, 1967 August.
Zwicky, F., and Herzog, P. 1963, *Catalogue of Galaxies and Clusters of Galaxies*, Vol. **2** (Pasadena, California: California Institute of Technology).
Zwicky, F., Herzog, E., and Wild, P. 1961, *Catalogue of Galaxies and Clusters of Galaxies*, Vol. **1** (Pasadena, California: California Institute of Technology).

CHAPTER 11

Stellar Dynamics and the Structure of Galaxies

K. C. FREEMAN
Mount Stromlo and Siding Spring Observatories, Research School of Physical Sciences, The Australian National University

1. INTRODUCTION

THE BASIC questions of galactic dynamics are: (i) What is the present dynamical state of galaxies? And then (ii) How was this state attained? This means that galactic dynamics is deeply involved in the problem of galaxy formation, which is one of the fundamental problems faced by astronomers today. The dynamical approach attempts to reconstruct from present data the history of galaxies, back to the time when they began to collapse from the intergalactic medium. This reverse approach to the problem of galaxy formation complements the direct cosmological approach.

Galaxies are systems containing stars and gas. Mostly the proportion of gas is less than 20 percent, so the dynamical theory of galaxies is largely concerned with the dynamics of their stellar component. Section 2 contains some basic results in the theory of self-gravitating stellar systems. These results will be useful for §§ 3 and 4, where we describe some dynamically important properties of galaxies. The remainder of this chapter is a discussion of several topics which we believe are important for the dynamical theory of galaxies. We do not include hydromagnetic topics, or problems of galactic nuclei, which are discussed at length by others: see, for example, Woltjer (1965) and Burbidge (1970).

2. FOUNDATIONS OF STELLAR DYNAMICS

2.1. THE RELAXATION TIME

Consider a system of stars like a galaxy. Each star moves in the potential field of all the other stars; let this field be Φ. Then the equation of motion of a star is

$$\mathfrak{M}\ddot{r} = -\mathfrak{M}\nabla\Phi \,, \tag{1}$$

where

$$\nabla^2\Phi = 4\pi G\rho \tag{2}$$

Received May 1972.

is Poisson's equation. $\mathfrak{M}$ is the star's mass, and ρ is the total mass density at r; it is really the sum of a large number (10^7–10^{12}) of δ-functions. As the star orbits, it feels the potential of distant stars, which is smooth, and the fluctuations in Φ due to nearby stars. We ask whether these fluctuations have, on the average, any significant effect on a stellar orbit. This is a classical problem in stellar dynamics. Its solution is usually given by evaluating the *relaxation time* T, which is the time required for the fluctuations to affect significantly the orbit of a star. Ostriker and Davidsen (1968) show that

$$T = c^3/[8\pi G^2\mathfrak{M}^2 n \ln (c^3 T/2G\mathfrak{M})] \,, \tag{3}$$

where G is the gravitational constant, n the number density of stars, and c a typical stellar velocity.

Table 1 gives representative values of $\mathfrak{M}$, n, c, and T for three examples; for the E galaxy, n is a typical mean value within the optical diameter. T is much longer than the probable age of the Galaxy and nearby galaxies ($\sim 2 \times 10^{10}$ years). This means that we can neglect the effect of encounters with nearby stars, and regard galaxies as *collisionless stellar systems;* Φ is then the potential of a *smoothed-out mass distribution.*

TABLE 1

THE RELAXATION TIME

Object	$\mathfrak{M}(\mathfrak{M}_\odot)$	$n(\text{pc}^{-3})$	$c(\text{km s}^{-1})$	T (years)
Solar neighborhood	1.0	0.1	20	5×10^{12}
Center of M31	0.4	10^4	200	5×10^{11}
Giant E galaxy	0.15	10	400	10^{16}

2.2. The Collisionless Boltzmann Equation

We can represent a stellar system as (i) a system of n bodies ($n \sim 10^7$–10^{12} for galaxies) or (ii) a material continuum in phase space. Phase space is the direct sum of configuration ($\boldsymbol{r}$) space and velocity ($\boldsymbol{c}$) space; a point in phase space has coordinates $(\boldsymbol{r}, \boldsymbol{c}) = (x, y, z, c_x, c_y, c_z)$. Representation (i) would give far more detail than we want, even if the equations were manageable. Representation (ii) is more convenient: a *mass* density $f(\boldsymbol{r}, \boldsymbol{c}, t)$ of stars at $(\boldsymbol{r}, \boldsymbol{c}, t)$ is used to describe the system, so that $f\, d^3r\, d^3c$ is the mass of stars in the interval $d^3r\, d^3c$ at time t. This density f is usually called the *distribution function* and will satisfy a continuity equation in phase space analogous to the usual fluid continuity equation in configuration space.

To derive this continuity equation, it is convenient to use conjugate variables $(\boldsymbol{p}, \boldsymbol{q})$ instead of $(\boldsymbol{r}, \boldsymbol{c})$. A stellar orbit in $(\boldsymbol{p}, \boldsymbol{q})$-space is then defined by Hamilton's equations

$$\dot{\boldsymbol{q}} = \partial H/\partial \boldsymbol{p} \quad \text{and} \quad \dot{\boldsymbol{p}} = -\partial H/\partial \boldsymbol{q} \,; \tag{4}$$

in a steady-state, conservative system, the Hamiltonian $H(\boldsymbol{p}, \boldsymbol{q})$ is the total energy of the star. Let D/Dt be the convective derivative $\partial/\partial t + \dot{\boldsymbol{q}}\cdot\partial/\partial\boldsymbol{q} + \dot{\boldsymbol{p}}\cdot\partial/\partial\boldsymbol{p}$, so DF/Dt gives the total rate of change of some function $F(\boldsymbol{p}, \boldsymbol{q}, t)$ associated with the star, as the star proceeds along its orbit. Now consider a volume element $\Delta^3p\Delta^3q$ of $(\boldsymbol{p}, \boldsymbol{q})$ phase space: this volume element is defined at time t by a mass of stars $\Delta\mathfrak{M} = f\Delta^3p\Delta^3q$, where f is the distribution function. Note again that f is a mass density, not a number density. Because we are dealing with a collisionless stellar system, the orbits of these stars are defined by the smoothed-out potential field of the whole system, so the form of the Hamiltonian for each star is the same (this would not be true if binary encounters were not negligible). *Liouville's theorem*, which states that $\Delta^3 p \Delta^3 q$ remains constant as the stars defining it proceed along their orbits, is then valid. Since the *same* stars define $\Delta^3 p \Delta^3 q$ at all times, $D(\Delta\mathfrak{M})/Dt = 0$; because $D(\Delta^3 p \Delta^3 q)/Dt = 0$, it follows that

$$Df/Dt = 0 \, . \tag{5}$$

This is the fundamental continuity equation for f in phase space. It is sometimes called the *collisionless Boltzmann equation.* For example, if $\boldsymbol{r}$ and $\boldsymbol{c}$ are the position and velocity of a star, and $\boldsymbol{F}$ the acceleration of a star at $(\boldsymbol{r}, \boldsymbol{c})$, then equation (5) becomes

$$\frac{\partial f}{\partial t} + \boldsymbol{c} \cdot \frac{\partial f}{\partial \boldsymbol{r}} + \boldsymbol{F} \cdot \frac{\partial f}{\partial \boldsymbol{c}} = 0 \, . \tag{6}$$

Its form in conjugate coordinates is useful for transforming the equation into different coordinate systems.

The dynamical theory of collisionless stellar systems is based on equation (5) and Poisson's equation.

2.3. The Hydrodynamical Equations

The successive moments of equation (6) over velocity give a sequence of transfer equations for average quantities like the mean density and the mean velocity at $\boldsymbol{r}$. The mean density at $\boldsymbol{r}$ is

$$\rho = \int f d^3 c \, , \tag{7}$$

where the integral is taken over all velocity space. The mean velocity $\langle \boldsymbol{c} \rangle$ at $\boldsymbol{r}$ is defined by

$$\rho \langle \boldsymbol{c} \rangle = \int \boldsymbol{c} f d^3 c \, ; \tag{8}$$

and the velocity dispersion tensor $\boldsymbol{\sigma}$ at $\boldsymbol{r}$ by

$$\rho \sigma_{ij} = \int (c_i - \langle c_i \rangle)(c_j - \langle c_j \rangle) f d^3 c \, . \tag{9}$$

Assume that the acceleration depends only on $\boldsymbol{r}$ and that f is zero for sufficiently large $\boldsymbol{c}$. The zeroth moment of equation (6) is then

$$\int \frac{Df}{Dt} d^3 c = \frac{\partial \rho}{\partial t} + \frac{\partial}{\partial \boldsymbol{r}} \cdot (\rho \langle \boldsymbol{c} \rangle) = 0 \, . \tag{10}$$

This is the usual fluid continuity equation. The first moment is

$$\int \boldsymbol{c} \frac{Df}{Dt} d^3 c = 0 \, , \tag{11}$$

which, using equation (10), becomes

$$\frac{d\langle \boldsymbol{c} \rangle}{dt} = \boldsymbol{F} - \frac{1}{\rho} \frac{\partial}{\partial \boldsymbol{r}} \cdot (\rho \boldsymbol{\sigma}) \, , \tag{12}$$

where d/dt is the convective derivative $\partial/\partial t + \langle \boldsymbol{c} \rangle \cdot \partial/\partial \boldsymbol{r}$. Compare this with the equation of motion for a fluid

$$\frac{d\boldsymbol{v}}{dt} = \boldsymbol{F} - \frac{1}{\rho} \frac{\partial P}{\partial \boldsymbol{r}} + \text{viscous terms} \, . \tag{13}$$

We could go to higher moments, but the nth moment introduces a new tensor of rank $(n+1)$, so the sequence of equations would remain unclosed. Equations (10) and (12) are called the equations of *stellar hydrodynamics:* they are originally due to Jeans, and are very useful for understanding stellar motions in the Galaxy.

We make two remarks.

1. In a conventional fluid, it is possible to make some physical assumptions which allow the closure of the system of moment equations. The fluid pressure tensor $\boldsymbol{P}$, which corresponds to $\rho\boldsymbol{\sigma}$ in equation (12), can be written $-P_{ik} = -P\delta_{ik} + P'_{ik}$, where $-P\delta_{ik}$ is isotropic and P satisfies an equation of state, and P'_{ik} is velocity dependent: its velocity dependence follows from Hooke's law for the fluid. This leads to

$$-\frac{\partial}{\partial \boldsymbol{r}}\cdot \boldsymbol{P} = -\frac{\partial P}{\partial \boldsymbol{r}} + \text{viscous terms} . \tag{14}$$

In the fluid, the isotropy of P comes from the high frequency of particle collisions, which leads also to the equation of state. The continuity and moment equations are then closed (together with an energy equation). In the stellar fluid, collisions are rare, so $\boldsymbol{\sigma}$ is in general not isotropic (see § 2.5).

2. For a time-independent fluid flow, the "particle" paths and the streamlines coincide. In the stellar fluid, stellar *orbits* and the streamlines of $\langle \boldsymbol{c} \rangle$ do *not* in general coincide.

2.4. Jeans's Theorem

A function $F(p, q, t)$ that remains constant along a stellar orbit is called an *integral* of the motion: $DF/Dt = 0$. For example, in a steady-state conservative system $[\Phi = \Phi(r)]$, the total energy of a star is an integral. Integrals represent hypersurfaces in (p, q, t)-space: in six-dimensional phase space there are obviously only six independent integrals $I_i(p, q, t) = \text{constant}$ $(i = 1, \ldots, 6)$ associated with an orbit.

The collisionless Boltzmann equation $Df/Dt = 0$ means that f is itself an integral of the motion; i.e., it is constant along an orbit. An orbit is a one-parameter path

$$p = p(p_0, q_0; t) , \qquad q = q(p_0, q_0; t) , \tag{15}$$

where t is the parameter and there are six initial values p_0, q_0. This orbit can also be specified by the intersection of six independent integral surfaces

$$I_i(p, q, t) = \text{constant} \qquad (i = 1, \ldots, 6) ; \tag{16}$$

here the six constants take the place of p_0, q_0. Because f is constant along each orbit, it follows that $f = f(I_i)$; i.e., the distribution function depends only on the integrals of the motion. This result is called *Jeans's theorem.*

We now distinguish between two kinds of integrals. Consider the example of the two-dimensional harmonic oscillator (Woltjer 1967):

$$x = A \sin \alpha(t - t_\alpha) , \qquad y = B \sin \beta(t - t_\beta) . \tag{17}$$

If α/β is rational, this orbit is periodic; i.e., if we eliminate the time from equations (17), then the orbit is a one-parameter path in $(x, y, \dot{x}, \dot{y})$-space. If α/β is not rational, then the orbit fills the region $|x| \le A$, $|y| \le B$. There are at least two time-independent integrals:

$$I_\alpha = \dot{x}^2 + \alpha^2 x^2 , \qquad I_\beta = \dot{y}^2 + \beta^2 y^2 ; \tag{18}$$

these obviously define three-dimensional hypersurfaces in the four-dimensional phase space. There can be only one more time-independent integral; we get it by eliminating time from equations (17), and

$$I_3 = t_\beta - t_\alpha = \alpha^{-1}\sin^{-1}(x/A) - \beta^{-1}\sin^{-1}(y/B)\,. \tag{19}$$

Now take a particular value of I_3 and of y so

$$x = A\sin\left\{\alpha I_3 + \frac{\alpha}{\beta}\,\mathrm{Sin}^{-1}\left(\frac{y}{B}\right) + \frac{\alpha}{\beta}\,n\pi\right\} \tag{20}$$

where Sin^{-1} denotes the principal value, and n is an integer. If α/β is rational, x takes only a finite number of values, so I_3 is a proper three-dimensional hypersurface and the orbit is confined to a one-parameter region: we know it is periodic. If α/β is irrational, then all values of $x(n)$ are different and $x(n)$ can come arbitrarily close to any value permitted by I_α. In this case I_3 is infinitely multiple-valued and effectively provides no constraint on the orbit, which fills the region $|x| \leq A$, $|y| \leq B$ permitted by I_α and I_β. This kind of integral is called *nonisolating; I_α* and I_β (and I_3 for rational α/β), which are not infinitely multiple-valued, are *isolating*.

By definition, f is a mass density at (p, q, t). We therefore expect it to be one-valued. This means that f can depend *only on the isolating integrals* of the motion (see Lynden-Bell 1961). The number of isolating integrals obviously depends on the form of the potential. For time-independent systems, the only isolating integral in general is the Hamiltonian itself (which need not be the total energy).

So far, the only constraint on f is that $f = f(I_i)$: it then satisfies $Df/Dt = 0$. We will see later that further constraints are imposed by the density distribution and the mean internal motions and velocity dispersion within the system represented by f, and by the requirement that the system be stable.

2.5. Some Simple Consequences of Jeans's Theorem

In a steady-state collisionless system, $Df/Dt = 0$ and Poisson's equation

$$4\pi G\rho(\boldsymbol{r}) = 4\pi G\int f d^3c = \nabla^2\Phi(\boldsymbol{r}) \tag{21}$$

holds. In *spherically symmetric* systems with $\Phi = \Phi(\boldsymbol{r}) = \Phi(r)$, there are four isolating integrals:

$$E = \tfrac{1}{2}c^2 + \Phi \quad \text{and} \quad \boldsymbol{h} = \boldsymbol{r}\times\boldsymbol{c}\,. \tag{22}$$

For example, the isothermal sphere has $f = A\exp(-E/\sigma^2)$, so $\sigma_{ik} = \sigma^2\delta_{ik}$. Poisson's equation is then $\nabla^2\Phi = \text{constant} \times \exp(\Phi/\sigma^2)$, which can be integrated numerically. This sphere has infinite mass for finite $\rho(0)$: to model a finite system with this distribution function, it is necessary to exclude stars with velocities greater than some cutoff value (see § 6.2).

In the *axisymmetric* case with $\Phi(\boldsymbol{r}) = \Phi(R, z)$, where (R, φ, z) are cylindrical polar coordinates, there are in general only two isolating integrals:

$$E = \tfrac{1}{2}c^2 + \Phi \quad \text{and} \quad h = Rc_\varphi\,. \tag{23}$$

A third integral $I_3(z, c_z)$ will exist if, say, $\Phi(R,z) = \Phi_1(R) + \Phi_2(z)$. However, if E and h are the only isolating integrals, then

$$f = f(E,h) = f(\tfrac{1}{2}c_R{}^2 + \tfrac{1}{2}c_z{}^2 + \tfrac{1}{2}c_\varphi{}^2 + \Phi,\ Rc_\varphi)\,. \tag{24}$$

We see immediately that the only nonzero component of $\langle \mathbf{c} \rangle$ can be $\langle c_\varphi \rangle$, because f is symmetrical in c_R, c_z; i.e., rotation is the only allowed mean motion in an axisymmetric time-independent system. Further, $\sigma_{RR} = \sigma_{zz}$. The second of these predictions is incorrect for stars near the Sun: $\sigma_{RR} \sim 4\sigma_{zz}$. This suggests that E and h are not the only isolating integrals for most orbits passing through the solar neighborhood (see § 5).

2.6. Adiabatic Invariants

In steady-state stellar dynamics there are time-independent integrals associated with time-independent Hamiltonians (see previous section). Now consider a periodic system with Hamiltonian $H[p, q; \lambda(t)]$, where $\lambda(t)$ is a slowly varying (compared with the periods) function of time. Functions which remain approximately constant along an orbit, as the Hamiltonian changes slowly, are called *adiabatic invariants.* It turns out that the action variables

$$J_k = \oint p_k dq_k \tag{25}$$

are adiabatic invariants (see Landau and Lifshitz 1960); the integral is taken over one period. For example, the harmonic oscillator with frequency $\omega = \omega(t)$, such that $\dot{\omega}/\omega^2 \ll 1$, has $J = 2\pi H/\omega$ approximately constant, although the Hamiltonian H itself is not constant.

Adiabatic invariants have already been used successfully in stellar dynamics (see §§ 7 and 10). They promise to be useful for discussing slowly evolving systems: most galaxies probably fit into this family because the dynamical evolution time scales ($\sim 10^{10}$ years) are long compared with the orbital time scales of stars within them ($\sim 10^8$ years).

2.7. The Virial Theorem

For a steady-state system of stars, the *virial theorem* states that $2T + \Omega = 0$, where T is the total internal kinetic energy and Ω is the gravitational potential energy: see Chandrasekhar (1960) for its derivation. If the velocity dispersion and the length scale are known for a stellar system, then this theorem can be used directly to estimate its mass.

3. SOME BASIC PROPERTIES OF THE GALAXY

Before there is any hope of formulating an adequate theory of the dynamics and evolution of galaxies, it is essential to know what their various constituents are, how they are distributed in space, and how they move. In our Galaxy, it is difficult to determine the large-scale structure, mainly because of interstellar absorption; but the kinematics of stars in the region of the Sun can be observed in great detail. In other galaxies, the large-scale structure is much more easily determined, but their great distances mean that many details are unobservable. In this way the Galaxy and external galaxies provide complementary information about their structure and kinematics.

In this section we discuss some of the basic information on the kinematics and structure of our Galaxy. There is more information on this subject in the important article by Oort in Volume **5** of this series.

3.1. The Rotation of the Galaxy

The Galaxy appears to be a spiral system. Most spiral galaxies rotate. This rotation is easy to measure in external systems but more difficult to measure for the Galaxy because of interstellar absorption for optical objects and uncertain distances for H I and H II regions. Galaxies do not generally rotate as rigid bodies, so we need to consider the kinematics of differential rotation.

Figure 1 shows the geometry of the situation. We assume that the Galaxy is axisymmetric and consider only the motion in the galactic plane. Let (R, φ) be polar coordinates with the origin at the galactic center. Let V_R, V_φ be the velocity of the solar neighborhood (i.e., the systematic mean motion of the younger stars around the Sun, with the

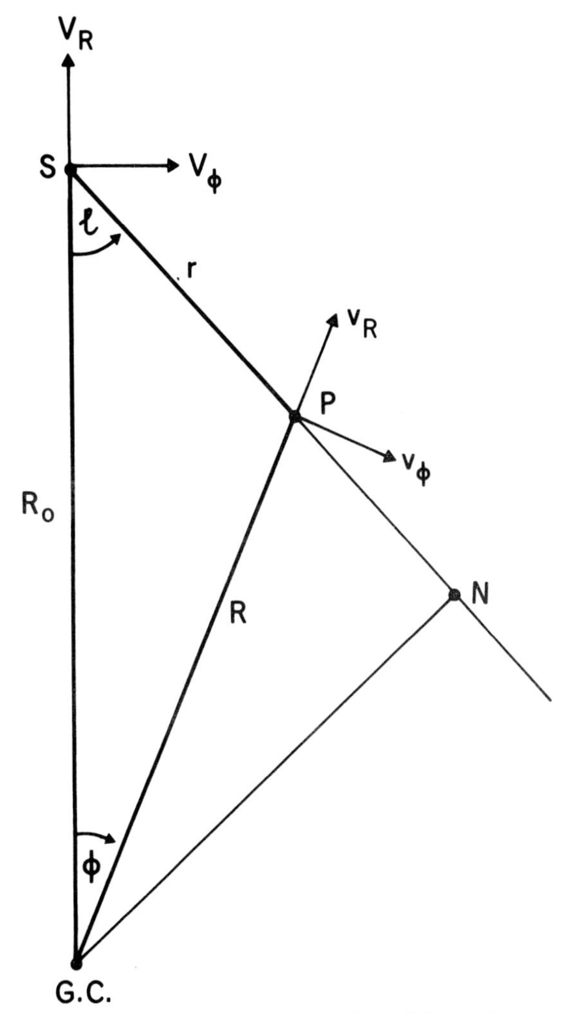

Fig. 1.—The geometry for differential rotation.

peculiar motion of the Sun taken out: see § 3.2), and v_R, v_φ the velocity of a nearby point $P(R, \varphi)$. Write $\Omega = v_\varphi/R$.

The radial velocity, referred to the solar neighborhood, of matter at $P(R, \varphi)$ is

$$V_{\rm rad} = v_\varphi \sin(l + \varphi) - v_R \cos(l + \varphi) - V_\varphi \sin l + V_R \cos l\,, \tag{26}$$

where l is galactic longitude. Now $R_0 \cos l = r + R\cos(l + \varphi)$ and $R_0 \sin l = R \sin(l + \varphi)$, so

$$V_{\rm rad} = R_0 \sin l \left(\frac{v_\varphi}{R} - \frac{V_\varphi}{R_0}\right) - v_R \cos(l + \varphi) + V_R \cos l\,. \tag{27}$$

This result holds for all (R, φ). To the first order in r/R_0, equation (27) becomes

$$V_{\rm rad} = -\tfrac{1}{2}R_0 r \Omega'(R_0) \sin 2l + \tfrac{1}{2}R_0 r \left(\frac{V_R}{R_0}\right)' \cos 2l + \left[\frac{V_R}{R_0} + \frac{R_0}{2}\left(\frac{V_R}{R_0}\right)'\right] r\,, \tag{28}$$

where primes denote differentiation with respect to R, and $(V_R/R_0)'$ is $(v_\varphi/R)'$ evaluated at R_0. It appears from observation that the cos $(2l)$ term and the l-independent term (K-term) are approximately zero, so $V_R \approx 0$ and rotation is the only mean motion near the Sun. Define Oort's constants of galactic rotation by

$$A = \frac{1}{2}\left(\frac{v_\varphi}{R} - v_\varphi'\right)_{R_0} = -\tfrac{1}{2}R_0 \Omega'(R_0)\,, \tag{29}$$

$$B = -\frac{1}{2}\left(\frac{v_\varphi}{R} + v_\varphi'\right)_{R_0} = -\left[\frac{1}{2R}(R v_\varphi)'\right]_{R_0}\,. \tag{30}$$

Then from equation (28)

$$V_{\rm rad} = A r \sin 2l \tag{31}$$

to first order in r/R_0. Higher-order corrections can be calculated. Similarly the proper motion is

$$\mu = A \cos 2l + B\,. \tag{32}$$

The present adopted values for A and B are $A = 15$ km s^{-1} kpc^{-1} and $B = -10$ km s^{-1} kpc^{-1}. The proper motion given by equation (32) is small ($\lesssim 0\farcs005$ year^{-1}), and this makes B difficult to determine: see § 3.3.

The rotation curve $v_\varphi(R)$ is an important function related to the large-scale mass distribution in the Galaxy. The radial velocities of young stars and H I regions (which emit the 21-cm line) are not difficult to measure. It is possible to estimate the distances of optical objects, but very distant optical objects are obscured by the interstellar matter. On the other hand, H I regions can be detected at distances greater than 20 kpc, but there is no direct way of measuring their distances. To derive the rotation curve from 21-cm observations, we consider only circular motions, so

$$V_{\rm rad} = R_0 \sin l[\Omega(R) - \Omega(R_0)] \tag{33}$$

from equation (27); again, this holds exactly for all R. Because $\Omega(R)$ is generally monotone decreasing, $V_{\rm rad}$ should have its maximum absolute value for a given l at N, which is the closest point to the galactic center (see fig. 1). The 21-cm profile for this l should then show a sharp cutoff corresponding to this maximum value of $V_{\rm rad}$, and this is observed (see Kerr and Westerhout 1965). At N, where $R = R_0 |\sin l|$ is *known*, equation (33) becomes

$$V_{\rm rad} = V_{\rm max} = [v_\varphi(R) - v_\varphi(R_0)|\sin l|]\frac{\sin l}{|\sin l|} \tag{34a}$$

or to first order in r/R_0

$$V_{\rm max} = 2AR_0 \sin l(1 - |\sin l|)\ . \tag{34b}$$

From equation (34a) the rotation curve $v_\varphi(R)$ is determined: $V_{\rm max}$ is the observed quantity and comes from the cutoff velocity of the 21-cm profile, and R is given by $R = R_0 \sin l$ at N (provided that there is neutral hydrogen at N). The velocity $v_\varphi(R_0) = V_\varphi$ is not observable directly, so the resulting rotation curve depends on the calibration of $v_\varphi(R_0)$; this is shown in figure 2. In turn, $v_\varphi(R_0)$ depends on the adopted values of A, B and R_0 (see § 3.3).

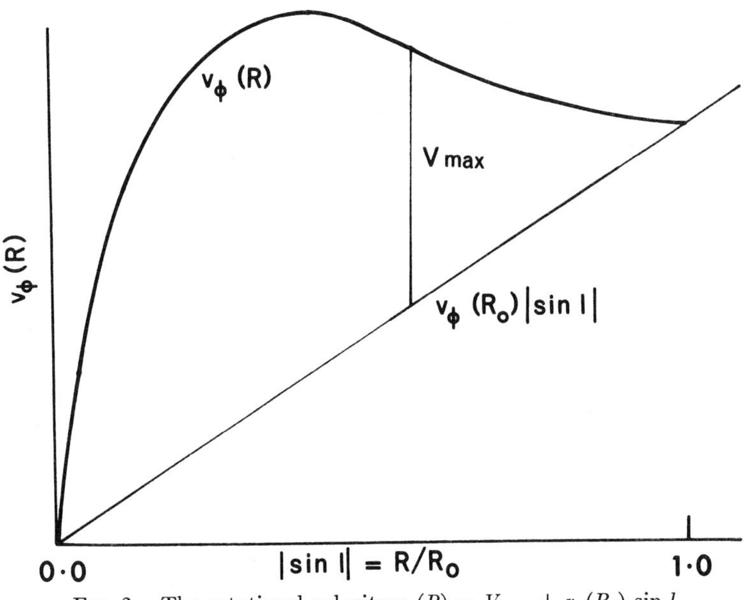

FIG. 2.—The rotational velocity $v_\varphi(R) = V_{\rm max} + v_\varphi(R_0) \sin l$.

Figure 3 (from Kerr 1964) shows the Galactic rotation curve derived in this way, for 4 kpc $< R <$ 10 kpc. There is a systematic difference for the northern ($l < 90°$) and the southern ($l > 270°$) observations, which may reflect some asymmetry in the Galaxy; a similar difference exists in M31 (Roberts 1966). The dips in the galactic rotation curve may be due to local motions associated with the spiral arms (Shane and Bieger-Smith 1966) or to the absence of neutral hydrogen at the tangential points N for the corresponding values of l. For $R > R_0$ there are no observed points because there is no tangential point N for $|l| > 90°$. For $R <$ 4 kpc there are significant noncircular motions; inside $R \approx$ 700 pc, there is a rapidly rotating nuclear disk, with a rotation velocity of about 200 km s^{-1}.

3.2. Stellar Motions near the Sun

The radial velocity, proper motion, and distance of a star give its velocity relative to the Sun. The velocity dispersion of homogeneous sets of stars, and their mean velocities relative to the Sun, yield some interesting information about the dynamics of the galaxy and about the local mass density and its gradient.

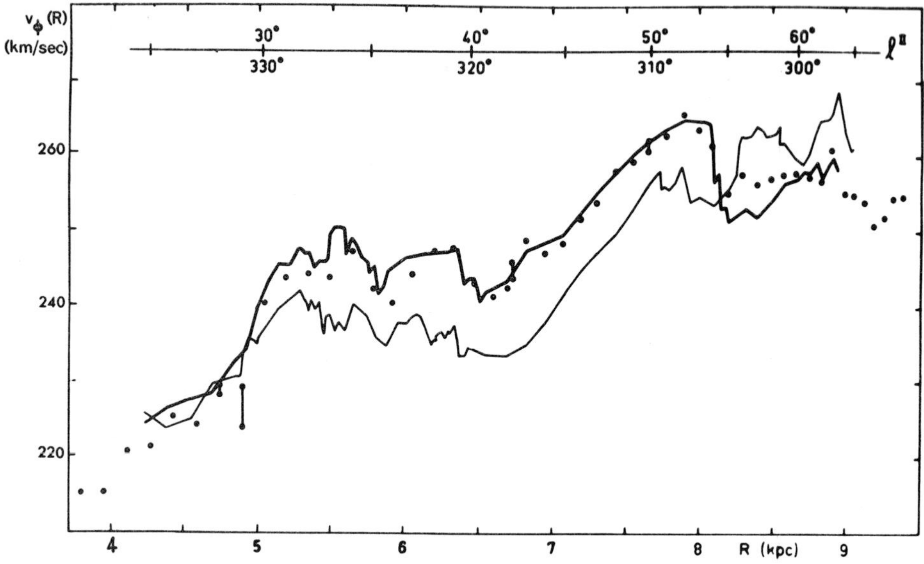

Fig. 3.—The Galactic rotation curve (from Kerr 1964).

We take a cylindrical polar coordinate system (R, φ, z) with $z = 0$ defined by the galactic plane and φ increasing in the direction of galactic rotation. Assume the Galaxy is axisymmetric and time-independent. Define the *circular* velocity Θ at the Sun by

$$\Theta^2 = (R\partial\Phi/\partial R)_{R_0}\,, \tag{35}$$

where $\Phi(R)$ is the gravitational potential. Now take a coordinate system with (i) its origin now at the Sun, and (ii) its origin describing a circular orbit around the galactic center with the circular velocity Θ; this coordinate system defines the *local standard of rest* (LSR). The value of Θ remains uncertain; however, it is possible to measure the velocity of the Sun, and the mean velocity of homogeneous groups of stars, relative to the LSR.

Let (U, V, W) be the velocity components of a star relative to the Sun; they are positive in the direction of the anticenter, of galactic rotation, and of the north galactic pole, respectively. Figure 4 shows examples of the (U, V) velocity distributions for two homogeneous classes of stars. Note how the distribution for the younger F stars is dominated by a few clumps or *moving groups*. The mean velocity of the older F stars, which form a reasonably smooth distribution, is $(\langle U\rangle, \langle V\rangle, \langle W\rangle) = (14, -15, -8)$ km s^{-1}, and their velocity dispersion has components $(\sqrt{\sigma_{RR}}, \sqrt{\sigma_{\varphi\varphi}}, \sqrt{\sigma_{zz}}) = (24.5, 15,$

12) km s^{-1}. Delhaye (1965) lists extensive data for mean velocities and velocity dispersions for many classes of stars: the following results are evident.

i) The ratio $\sigma_{RR}:\sigma_{\varphi\varphi}:\sigma_{zz}$ is about 4:1.5:1 in the mean, with a large scatter about these values. However, $\sigma_{RR} > \sigma_{zz}$ for almost every sample; this suggests that most stellar orbits have a third integral apart from E and h (see §§ 2.5 and 5).

ii) The mean values of $\langle U \rangle$ and $\langle W \rangle$ are about 9 km s^{-1} and -7 km s^{-1}, with no significant deviations from these means. However, $\langle V \rangle$ decreases with increasing velocity dispersion, as shown in figure 5. The reason for this is clear from equation (12): as the velocity dispersion increases, the mean azimuthal velocity needed to maintain radial equilibrium for a class of stars decreases. By extrapolating the data of figure 5 to zero velocity dispersion, we find the value of $\langle V \rangle \approx -12$ km s^{-1} appropriate to a class of stars with $|\boldsymbol{\delta}| = 0$. This class of stars would have its mean azimuthal velocity equal to the circular velocity: the velocity of the Sun relative to the LSR is then $(-9, +12, +7)$ km s^{-1}.

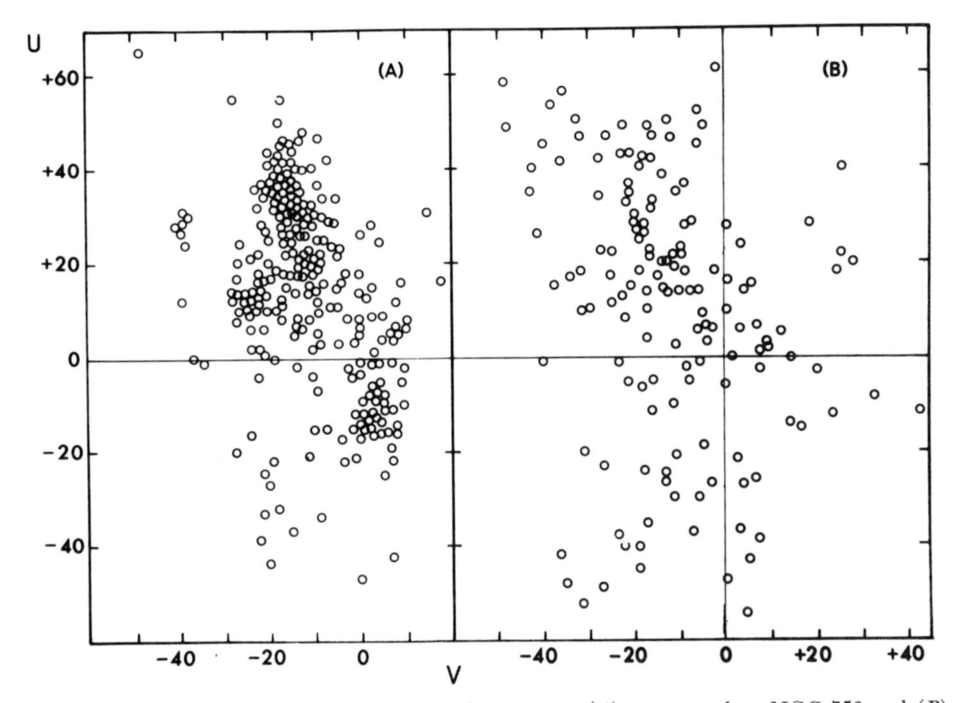

FIG. 4.—Distribution of (U, V) velocities for F stars, (A) younger than NGC 752 and (B) older than NGC 752 (from Eggen 1969).

iii) From equation (24) we would expect $\sigma_{R\varphi} = 0$ in an axisymmetric time-independent galaxy, so the principal axes of the (U, V) velocity distributions should lie along the radial and azimuthal directions. This is evidently not so for the younger stars shown in figure 4: for the young F stars, the angle between the major principal axis and the radial direction (the *vertex deviation*) is about 20°. This is probably the result of the clumping of points representing the moving groups; the velocity distribution for these young stars is not well mixed, i.e., not time-independent.

iv) The velocity dispersion seems to be a function of stellar age. For example, approximate values of $\sqrt{\sigma_{RR}}$ are 10 km s^{-1} for OB stars, 20 km s^{-1} for the A dwarfs, and 30 km s^{-1} for the F dwarfs; it increases to about 200 km s^{-1} for the halo subdwarfs, which are probably among the oldest objects in the Galaxy. We will discuss this further in § 9.5.

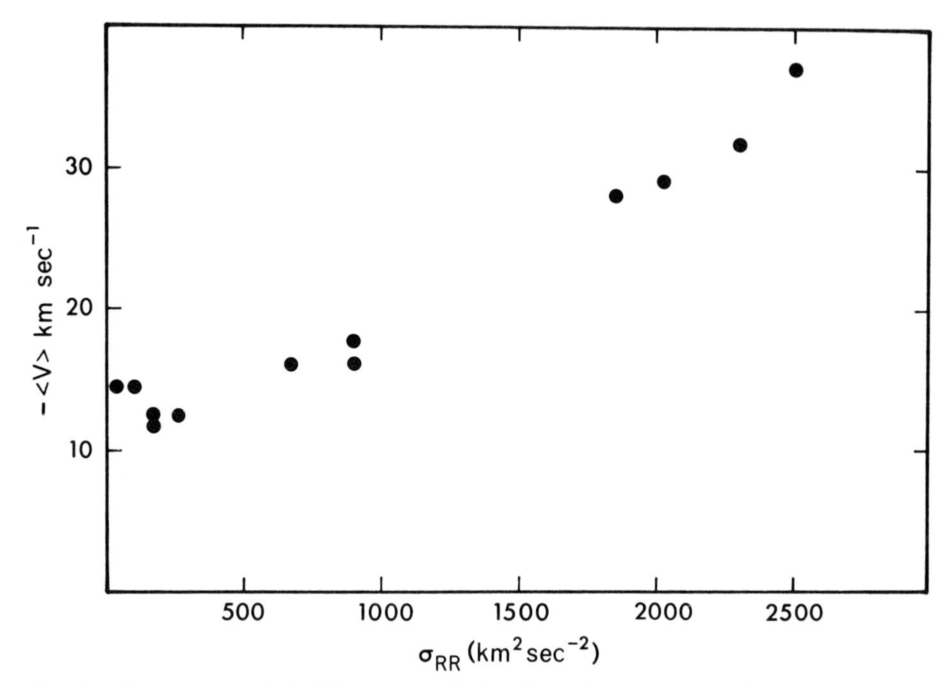

FIG. 5.—The asymmetrical drift against velocity dispersion for various classes of stars (from Delhaye 1965).

To understand result (ii) and to derive some useful relationships, write the hydrodynamical equation (12) in cylindrical polar coordinates. Its R- and z-components are

$$\frac{1}{R}\frac{\partial}{\partial R}(\rho R \sigma_{RR}) + \frac{\rho}{R}(\Theta^2 - \langle c_\varphi \rangle^2 - \sigma_{\varphi\varphi}) + \frac{\partial}{\partial z}(\rho \sigma_{Rz}) = 0\,, \tag{36}$$

$$\frac{\partial}{\partial z}(\rho \sigma_{zz}) + \frac{1}{R}\frac{\partial}{\partial R}(R \rho \sigma_{Rz}) = -\,\rho \frac{\partial \Phi}{\partial z}\,. \tag{37}$$

Here we replaced $\partial\Phi/\partial R$ by Θ^2/R, $\langle c_R \rangle$ and $\langle c_z \rangle$ are taken as zero (see result (ii) above), and $\langle c_\varphi \rangle$ is referred to a nonrotating frame. If the distribution function $f = f(E, h, I_3)$ as in § 2.5, then $\sigma_{Rz} = 0$.

3.2.1. *The asymmetric drift.*—As a first approximation, neglect σ_{Rz}; equation (36) becomes, for small $\Theta - \langle c_\varphi \rangle$,

$$\Theta - \langle c_\varphi \rangle \approx - \frac{\sigma_{RR}}{2\Theta}\left(\frac{\partial \log \rho\sigma_{RR}}{\partial \log R} + 1 - \frac{\sigma_{\varphi\varphi}}{\sigma_{RR}}\right), \tag{38}$$

where Θ is a property of the largescale potential field Φ, while ρ, $\boldsymbol{\sigma}$, and $\langle c_\varphi \rangle$ depend on the distribution function f for the sample of stars that we are studying. The ratio $\sigma_{\varphi\varphi}/\sigma_{RR}$ appears to be independent of the class of star. If we also assume that $\partial \log (\rho\sigma_{RR})/ \partial \log R$ is also independent of class (i.e., that the length scale for variation of $\rho\sigma_{RR}$ is independent of class), then equation (38) predicts $\Theta - \langle c_\varphi \rangle \propto \sigma_{RR}$, as observed. The quantity $\Theta - \langle c_\varphi \rangle$ is called *Stromberg's asymmetrical drift.* Its correlation with velocity dispersion was a major stepping stone to our present concept of galactic structure: see § 3.5.

3.2.2. *The total mass density.*—Now consider equation (37); if $\sigma_{Rz} = 0$, then

$$- \rho \frac{\partial \Phi}{\partial z} = \frac{\partial}{\partial z} (\rho\sigma_{zz}) , \tag{39}$$

so

$$- \frac{\partial^2 \Phi}{\partial z^2} = \frac{\partial}{\partial z}\left[\frac{1}{\rho}\frac{\partial}{\partial z}(\rho\sigma_{zz})\right] . \tag{40}$$

Here σ_{zz} and ρ refer to some homogeneous population of stars, and Φ is the *total* gravitational potential. Poisson's equation is

$$\frac{\partial^2 \Phi}{\partial R^2} + \frac{1}{R}\frac{\partial \Phi}{\partial R} + \frac{\partial^2 \Phi}{\partial z^2} = 4\pi G \rho_{\text{total}} , \tag{41}$$

where ρ_{total} is the *total* matter density. Equation (39) gives the force law $-\partial\Phi/\partial z$ perpendicular to the galactic plane; equations (40) and (41) give the total density near the Sun. The first two terms of equation (41) can be evaluated at $z = 0$ from Oort's constants, and are about 30 times smaller than the third term, which is given by equation (40). This means that the velocity dispersion and density distribution in z for some homogeneous well-mixed class of stars (e.g., A0 stars or K giants) leads to an estimate of the total mass density near the Sun and the force law $-\partial\Phi/\partial z$. For details, see Woolley (1965) and Oort (1965). Present estimates for the total mass density near the Sun lie between 0.10 and 0.15 $\mathfrak{M}_\odot$ pc^{-3}; the density of known stars and interstellar matter near the Sun is about 0.086 $\mathfrak{M}_\odot$ pc^{-3}, and any remainder is unidentified.

3.2.3. *The density gradient.*—The radial density gradient $\partial\rho/\partial R$ measures the concentration of a class of stars to the galactic center: this helps to define the population of the class (see § 3.5). It is also very important for building mass models of the Galaxy; see § 3.3. The gradient $\partial\rho/\partial R$ is difficult to measure directly, but it can be estimated from the kinematical properties of the class of stars, using equation (36); see Oort (1965). We assume that σ_{RR} and σ_{zz} are constant. The σ_{Rz} term is evaluated by assuming that, for points away from the galactic plane, the major principal axis of the velocity distribution points toward the galactic center. It follows that $\sigma_{Rz} \approx zR^{-1} (\sigma_{RR} - \sigma_{zz})$, so equation (36) becomes

$$\frac{\partial \log \rho}{\partial \log R} = - \frac{\Theta^2 - \langle c_\varphi \rangle^2}{\sigma_{RR}} - \left(1 - \frac{\sigma_{\varphi\varphi}}{\sigma_{RR}}\right) - \left(1 - \frac{\sigma_{zz}}{\sigma_{RR}}\right) . \tag{42}$$

For stars with small asymmetric drifts (young Population I), the logarithmic density gradient is small, in the range 0 to -1. For extreme halo objects, the density gradient is more nearly -3. The values of the gradient calculated from equation (42) are in reasonable agreement with the observed values for the few classes of stars which have directly observed density gradient data. See Blaauw (1965) for details.

3.2.4. *The ratio* $\sigma_{\varphi\varphi}/\sigma_{RR}$.—The ratio $\sigma_{\varphi\varphi}/\sigma_{RR}$ is related to Oort's differential rotation constants, and provides a useful constraint on their determination. The collisionless Boltzmann equation is

$$c_R \frac{\partial f}{\partial R} + c_z \frac{\partial f}{\partial z} + \frac{1}{R}(c_\varphi^2 - \Theta^2)\frac{\partial f}{\partial c_R} - \frac{c_R c_\varphi}{R}\frac{\partial f}{\partial c_\varphi} - \frac{\partial \Phi}{\partial z}\frac{\partial f}{\partial c_z} = 0 . \tag{43}$$

Multiply equation (43) by $c_R(c_\varphi - \langle c_\varphi \rangle)$ and integrate over velocity space. Assuming $\sigma_{Rz} = 0$ and $c_\varphi - \langle c_\varphi \rangle \ll \langle c_\varphi \rangle$ (which is generally true for objects with small asymmetric drift), this becomes

$$\sigma_{\varphi\varphi}/\sigma_{RR} \approx \frac{1}{2\langle c_\varphi \rangle}\frac{\partial(R\langle c_\varphi \rangle)}{\partial R} \approx \frac{-B}{A - B}, \tag{44}$$

where $A = \frac{1}{2}(\langle c_\varphi \rangle/R - \langle c_\varphi \rangle')$, $B = -\frac{1}{2}(\langle c_\varphi \rangle/R + \langle c_\varphi \rangle')$ are defined at the Sun for $\langle c_\varphi \rangle = \Theta$. Data for well-mixed classes of stars with small asymmetric drift indicate that $-B/A$ probably lies in the range 0.5–0.73.

3.3. Mass Models of the Galaxy

Mass models of the Galaxy are important for (i) calculating the total mass of the Galaxy, (ii) computing stellar orbits, and (iii) estimating the circular velocity for $R > R_0$ where direct measurement of the rotation curve for neutral hydrogen is not possible; this circular velocity is needed to estimate the distances of radio emitting regions from their radial velocities. These models are constructed to reproduce the observed rotation curve for $R < R_0$, assuming that the system is time-independent and axisymmetric. As pointed out in § 3.1, the values of the velocity and radius for the rotation curve are not absolute until values of A, B, and R_0 are known. Finding these values is a major problem in model construction.

Schmidt's (1965) model is the most widely used at present. His procedure for choosing A, B, and R_0 uses many of the results described in the previous two sections, and is worth summarizing here to show their usefulness.

The *constant* A comes directly from radial velocities of young objects (eq. [31]). It obviously depends on the distance scale used for these objects. The adopted value is 15 km s^{-1} kpc^{-1}.

The *constant* B evaluated from proper motions via equation (32) is -7 km s^{-1} kpc^{-1}, but this is uncertain. The ratio $\sigma_{\varphi\varphi}/\sigma_{RR}$ from equation (44) suggests that $-B/A$ is in the range 0.5–0.73. Trial flat-disk models, with the estimated density gradient near the Sun, suggest $B = -10$ km s^{-1} kpc^{-1}.

The *constant* R_0 is derived directly from the distribution of objects like the RR Lyrae stars and globular clusters, which are concentrated to the galactic center. Again, R_0 depends on the distance scale for these objects; its adopted value is 10 kpc. The product AR_0 can be estimated from equation (34b), at longitudes where 21-cm observations show that there is neutral hydrogen present at the tangent point. AR_0 is probably in the range 135–150 km s^{-1}. The set of values A = 15 km s^{-1} kpc^{-1}, $B = -10$ km s^{-1}

kpc^{-1}, and $R_0 = 10$ kpc is then remarkably consistent, considering the diverse methods used in their estimation.

Schmidt's model to represent this data consists of three components: a central mass point ($0.07 \times 10^{11}\mathfrak{M}_\odot$), a rather flat inhomogeneous spheroid interior to the Sun ($0.82 \times 10^{11}\mathfrak{M}_\odot$), and an exterior shell with $\rho \propto R^{-4}$ ($0.93 \times 10^{11}\mathfrak{M}_\odot$). Its total mass is then $1.8 \times 10^{11}\mathfrak{M}_\odot$. The eccentricity of the system is adjusted to reproduce the total mass density at the Sun: this is taken to be $0.148\mathfrak{M}_\odot$ pc^{-3}. Note that about half the total mass lies in the exterior shell. The mass of this shell, and therefore the force field in the model for $R > R_0$, depend on the choice of its density law. Because there is little direct indication of this density law, it is chosen to match the density gradient at the Sun, which comes mainly from equation (42).

Another model, due to Innanen (1966*a*), includes recent measures of the halo star densities at intermediate galactic latitudes. In this model, which also has $\Theta = 250$ km s^{-1}, $\rho(R_0) = 0.15\mathfrak{M}_\odot$ pc^{-3}, and represents the rotation data for $R < R_0$, the total mass is $1.3 \times 10^{11}\mathfrak{M}_\odot$; 0.3 of this total mass comes from the halo. This model and Schmidt's model obviously differ considerably in their details and total mass; it will be difficult to devise a really reliable model until much more is known about the density variation in the outer parts of flattened galaxies. Surface photometry of edge-on lenticular galaxies, for which the internal absorption is small, will be useful for this problem, because the behavior of the *surface* density (projected density per unit area of the galactic plane) for these systems is already known (see § 4).

3.4. Epicyclic Orbits

Consider the motion of a star in an almost circular orbit which lies almost in the galactic plane. The components of its motion in z and in the plane are then independent, to the first order in the deviations from the plane circular orbit. Let R_0 be the radius of the circular orbit with the same angular momentun per unit mass $h = Rc_\varphi$ as the star, and write $y = R_0(\varphi - ht/R_0^2)$; this is then the coordinate in the direction of rotation, relative to an origin moving with the circular velocity corresponding to h. Then, to first order in $(R - R_0)/R_0$, the stellar orbit relative to this origin is

$$R - R_0 = C \sin(\kappa t + \epsilon), \quad y = \beta C \cos(\kappa t + \epsilon), \tag{45}$$

where

$$\kappa^2 = \frac{3h^2}{R_0^4} + \left(\frac{\partial^2\Phi}{\partial R^2}\right)_{R_0} = -4B(A - B) \tag{46}$$

and $\beta = 2h/(\kappa R_0^2) = [(B - A)/B]^{1/2}$. C and ϵ are arbitrary constants. The complete orbit is then the vector sum of the *direct* circular orbit associated with h and the *retrograde* elliptic motion of equation (45). This is called an *elliptic epicycle*, and κ is the *epicyclic frequency.*

For the adopted values of A and B near the Sun, the axial ratio $\beta \approx 1.6$ and the epicycle period $2\pi/\kappa \approx 1.9 \times 10^8$ years. Compare this with the rotation period for the circular orbit $2\pi/(A - B) \approx 2.5 \times 10^8$ years and the period for oscillation perpendicular to the galactic plane, which is about 0.7×10^8 years (Oort 1965). From equation (45), a maximum peculiar velocity of 25 km s^{-1} in the R-direction corresponds to $C \approx 0.8$ kpc.

It is possible to show from equation (45) that the ratio of the average peculiar velocity of a star in the φ-direction (i.e., the average difference between its φ-velocity and the circular velocity at $R - R_0$) to its average peculiar R-velocity is $\sqrt{[-B/(A - B)]} = 1/\beta$, i.e., the *inverse* of the axial ratio for the epicycle itself. This result was derived in § 3.2.4 from the hydrodynamical equations. Several other consequences of the epicyclic theory, including the Lindblad dispersion orbits, are described by Oort (1965) and Woltjer (1967). Woolley (1965) gives an interesting kinematical analysis of the moving stellar groups (see Eggen 1965), based on epicyclic orbits; some moving groups can be seen in figures 4*a* and 4*b*.

3.5. Populations in the Galaxy

The correlation of the asymmetric drift with velocity dispersion (§ 3.2.1) led Lindblad to his concept of the Galaxy as a superposition of subsystems with varying mean rotational velocities $\langle c_\varphi \rangle$ and velocity dispersion, and therefore with varying degrees of flattening. For references, see Lindblad (1959). For example, the young stars form a very flat subsystem with small velocity dispersion (about 10 km s^{-1}) and negligible asymmetric drift, while the most metal deficient RR Lyrae stars, which are old objects, have an extended spheroidal distribution, a large velocity dispersion ($\sim$100 km s^{-1}), and $\Theta - \langle c_\varphi \rangle \simeq 200$ km s^{-1}; this subsystem hardly partakes of the galactic rotation.

It is now known that the asymmetric drift is related not only to the velocity dispersion but to the metal abundance, and metal abundance in turn is usually believed to be a function of age: see figure 6. The relation is not one-to-one, but is in the sense that, with few exceptions, *only* low metal abundance objects show *extreme* values of the related kinematical parameters $\Theta - \langle c_\varphi \rangle$, $|\delta|$, $|W|$ (z-velocity relative to the Sun), and $|z_{\max}|$ (maximum orbital height above the galactic plane). Objects with low metal abundance, with small values of $\Theta - \langle c_\varphi \rangle$, etc., do exist (see, for example, Sandage 1969). We will discuss these points further in §§ 7 and 11.

The relationship between the chemical and kinematical properties of classes of stars leads to the present concept of stellar populations. This is discussed in detail by Blaauw (1965). Populations are a useful way of describing the main galactic subsystems. In summary, *Population I* includes young ($<5 \times 10^8$ years) metal-rich objects with $\sqrt{\sigma_{zz}} < 10$ km s^{-1} and the interstellar gas and dust; these are associated mainly with the spiral arms. The *old disk population* objects have ages in the range 5×10^8 years to more than 5×10^9 years, metal abundances mostly close to the solar value, and $\sqrt{\sigma_{zz}} \sim 20$ km s^{-1}. This subsystem extends to about 800 pc from the galactic plane, and probably contains about 70 percent of the total galactic mass. *Population II* contains the most extended subsystems: its members are found up to about 10 kpc from the plane, with metal abundances down to about 1 percent of the solar value and asymmetric drifts as large as 200 km s^{-1}. They are probably the oldest objects in the Galaxy. The populations were originally defined by their color-magnitude diagrams: the stars of the solar neighborhood defined the Population I and the globular clusters the Population II.

The correlation between chemical and kinematical properties leads to the following working picture for the formation of the Galaxy. The oldest Population II objects (subdwarfs, globular clusters, metal-poor RR Lyrae stars) formed when the Galaxy was extended and far from equilibrium, before much metal enrichment occurred. The remainder of the Galaxy continued contraction to a disklike configuration, approximately in centrifugal equilibrium, and dissipated the necessary energy before forming stars; rapid metal enrichment took place. This is the disk population. The residual gas settled to a very flat disk in which the star formation is still going on; this is the Population I. We will refer to this working picture several times later in the chapter.

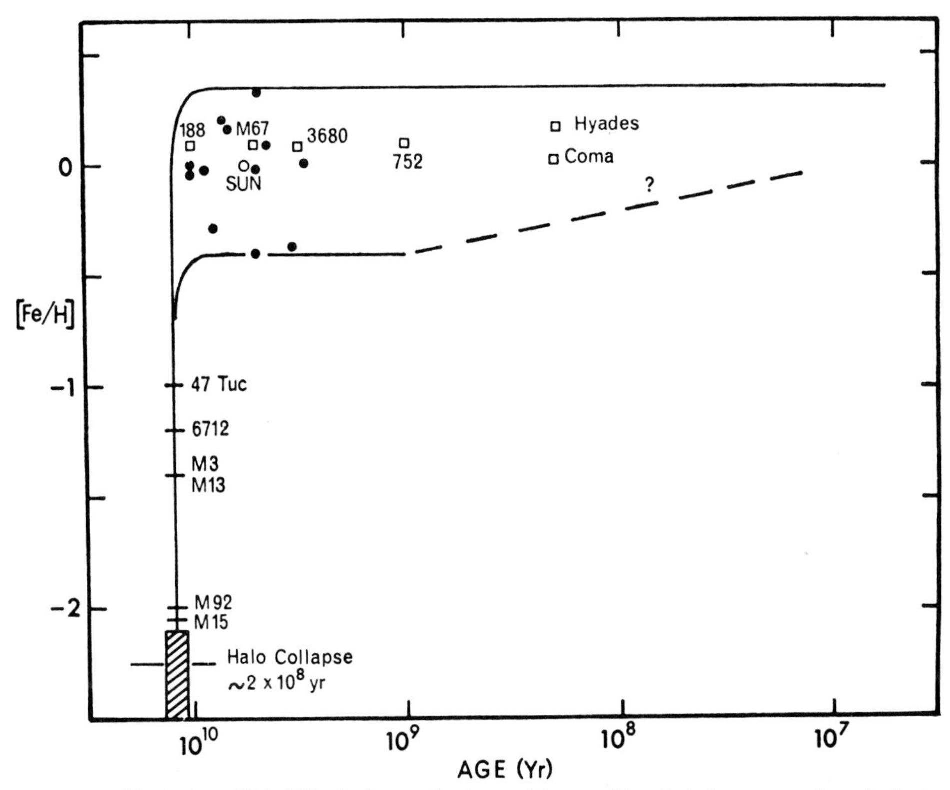

FIG. 6.—Variation of [Fe/H] relative to the Sun, with age. Closed circles are nearby subgiants. (From Eggen and Sandage 1969.)

4. SOME GENERAL PROPERTIES OF GALAXIES

For the purpose of this chapter, the main results coming from the discussion in § 3 of the kinematics and distribution of matter in the Galaxy are the concept of stellar populations and the working picture for the formation of the Galaxy. Now we consider the external galaxies, and describe briefly some of the large-scale properties which seem to be of particular importance for understanding the formation and dynamics of these systems. More detailed accounts of some of their properties are given by de Vaucouleurs (1959*a*) and in chapters 1–5 and 9, in this volume. At the end of this section, there is a summary of some of the main problems in galactic dynamics.

4.1. Morphological Classification of Galaxies

The revised Hubble system of classification described by de Vaucouleurs (1959*a*) seems best suited for the purposes of this chapter; this system recognizes explicitly several morphological properties which appear to be dynamically significant. It includes four *classes* of galaxies: the ellipticals (E), lenticulars (SO), spirals (S), and irregulars (I).

The ellipticals show no structural details except for strong condensation to a nucleus and a smooth outward luminosity decrease to an indefinite edge. Their apparent axial ratios b/a lie between 1 and about 0.3, and they are classified En, where $n = 10(1 - b/a)$. Systems flatter than E7 show a distinct disk (e.g., NGC 3115 [Sandage 1961]). If we assume that ellipticals are oblate spheroids and that they are randomly oriented, then it is possible to compute the intrinsic distribution of axial ratio from the apparent distribution. This has been done, most recently, by Sandage, Freeman, and Stokes (1970). The apparent and intrinsic distributions for 168 ellipticals are shown in figure 7; the intrinsic axial ratios lie anywhere between 1 and 0.3.

The lenticulars show a nucleus in a central lens surrounded by a faint diffuse envelope.

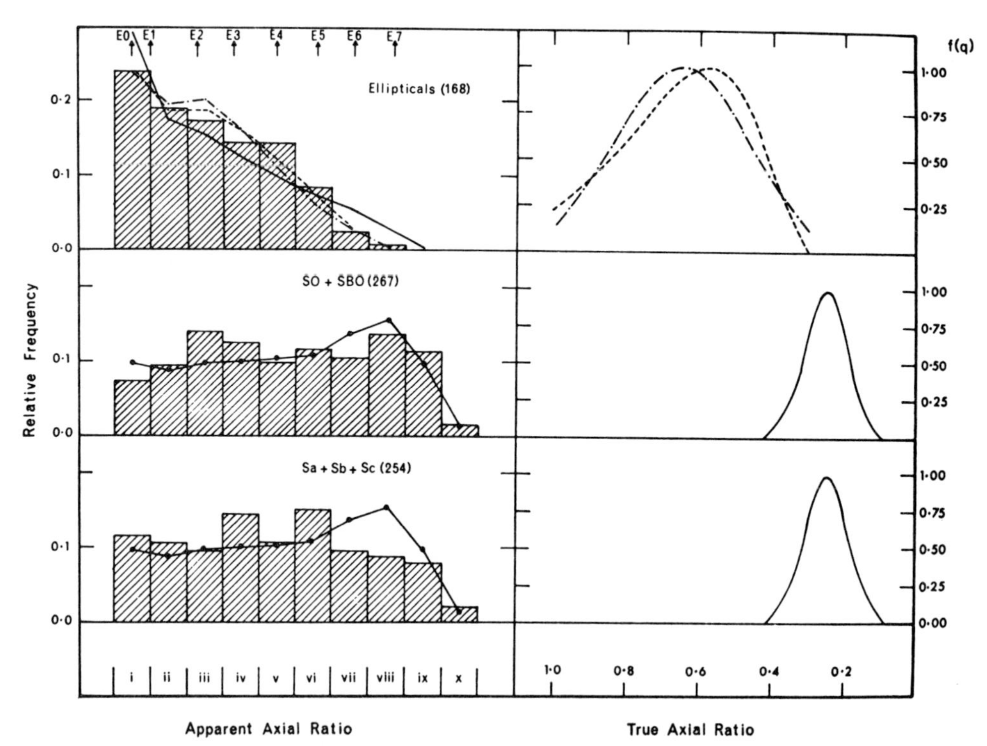

Fig. 7.—*Left*, histograms of the distribution of apparent axial ratios for E, S0, and spiral galaxies. The class intervals i–x are defined approximately by log q = 1.00 (−0.09) 0.10 (see Sandage *et al.* 1970), where q is the axial ratio. The curves are predicted distributions for various assumed intrinsic distributions. *Right*, assumed intrinsic distributions corresponding to the curves on the left. The case of uniform flattening for E galaxies is a straight line truncated at $q = 0.3$ and is not shown.

They have a smooth structure like the ellipticals, but their luminosity distribution is like that for the spirals, although they have no spiral arms. The lenticulars and spirals have similar distributions of apparent and intrinsic axial ratio (fig. 7). Both classes are intrinsically flat disklike systems (their mean axial ratio is about 0.25, with only small dispersion), so we consider them together.

Among the spirals and lenticulars, the classification system recognizes (i) two *families*, designated ordinary (SA, SA0) and barred (SB, SB0), with intermediate objects (SAB, SAB0); and (ii) two *varieties*, the ringed (r) and S-shaped (s) types, with transition type (rs). Along each of the spiral sequences SA(r), etc., there are four main *stages* *a*, *b*, *c*, *d*, from "early" to "late." They are defined by the openness of the spiral pattern, the degree of resolution of the arms into bright stars, and the apparent ratio of the disk

TABLE 2

FREQUENCIES OF MAIN CLASSES

	E	S0	S	Im	I0	Pec
f (percent)......	13.0	21.5	61.1	2.55	0.85	0.9

TABLE 3

FREQUENCIES OF SUBTYPES AMONG 994 SPIRAL GALAXIES

	0/a	a	ab	b	bc	c	cd	d	dm	m	?	Total	f
SA.............	17	25	25	57	57	82	30	9	3	4	2	311	31.3
SAB...........	13	15	23	45	50	71	35	11	3	7	1	274	27.6
SB.............	26	43	33	83	27	55	27	28	9	30	10	366	36.8
S..............	4	1	.....	6	1	13	1	10	.....	.....	7	43	4.3

to the central bulge. All of these increase from *a* to *d*, although there do exist galaxies with early type classification from the first two criteria but with small central bulges (Sandage 1961; Freeman 1970*a*). There are three stages of lenticulars, $S0^-$, S0, and $S0^+$, defined mainly by an increasing amount of dark matter. The transition from S0 to Sa is denoted S0/a.

Two kinds of irregulars are recognized: these are the Magellanic irregulars (Im), like the Small Magellanic Cloud, and the more rare I0 irregulars, like M82. The transition from the spiral sequence to the Im class is denoted Sm.

About 1 percent of bright galaxies cannot be classified within this revised Hubble system, and they are denoted as peculiar.

Table 2 gives the relative frequencies of the main classes. Table 3 gives the frequencies of subtypes among 994 spirals. Both these tables are taken from de Vaucouleurs' (1963*a*) classification of 1500 bright galaxies.

We make two remarks. (i) All transitions—between classes, families, varieties, and stages—are smooth. In particular, there is no clear division between barred and non-barred systems. (ii) Barred spirals are not unusual: the frequency ratio SA:SAB:SB is

31:28:37 (de Vaucouleurs 1963*a*). If one plans to discuss the dynamics of disk galaxies, then it must be recognized that axisymmetry is the exception, not the rule.

4.2. Absolute Luminosities and Colors of Galaxies

All lenticulars and spirals earlier than Sc with known absolute magnitudes M_B have $-16 > M_B > -22$. The later-type spirals and the Magellanic irregulars include systems fainter than $M_B = -16$. For example, the Sm and Im galaxies of the Local Group (see chap. 14) have M_B from -18.1 down to at least -12.3.

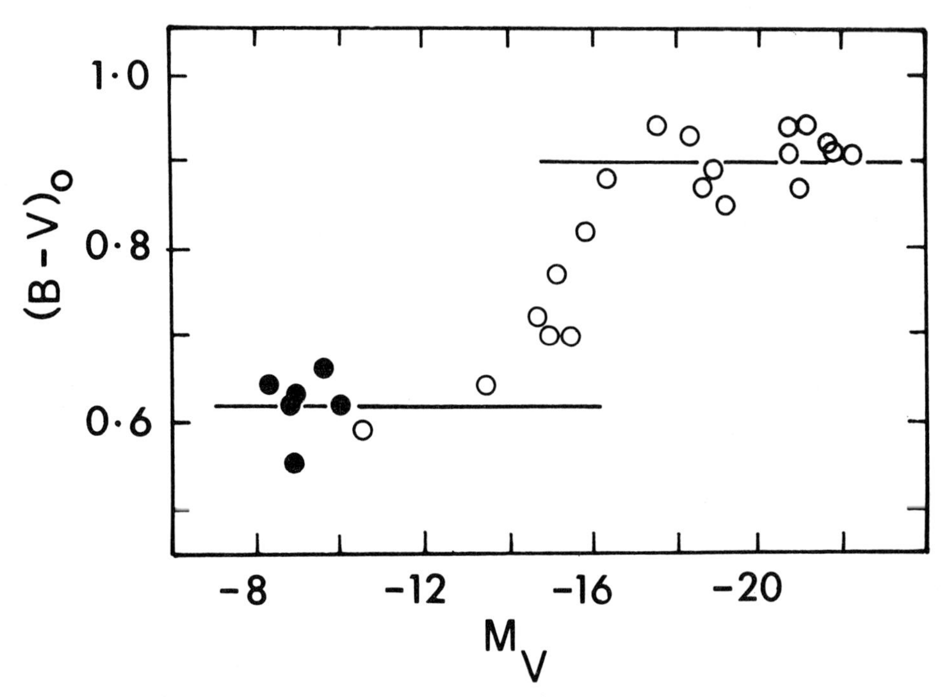

Fig. 8.—Intrinsic color index of elliptical galaxies (*open circles*) and globular clusters (*filled circles*) against their absolute magnitude (from Baum 1959).

The brightest ellipticals have $M_B \approx -22$. The brightest elliptical in a cluster has a mean $M_B \approx -21$ (the value depends on the adopted Hubble constant: our absolute magnitudes here are based on those of chap. 14), with a standard deviation of only 0.3 mag (see, for example, Peach 1969). Ellipticals cover a range of at least 13 magnitudes, from -22 for the brightest cluster giants, through -16 for the dwarf companion M32 of M31, -13 for the low-surface-brightness dwarf in Fornax, and down to -9 for the faintest known extreme dwarf elliptical in the Local Group; this system is no brighter than the bright globular clusters of the Galaxy.

The integrated colors of ellipticals are of particular interest. Figure 8, from Baum (1959), shows the intrinsic color indices of some E galaxies against their absolute magnitude. The giant systems ($M_V < -16$) have $B - V \approx 0.9$. For the dwarfs, $B - V$ decreases rapidly with increasing absolute magnitude, and at $M_V = -14$ the colors are globular-cluster-like. De Vaucouleurs (1961*a*) has a similar result, in $U - B$ and $B - V$, for E and S0 galaxies in the Virgo cluster. This suggests that the population of ellipticals depends on their absolute magnitude: see also § 4.5.

4.3. Luminosity and Color Distributions in Galaxies

4.3.1. *The ellipticals.*—The brighter ellipticals, with $M_B \lesssim -15$, have a characteristic radial surface-brightness distribution $I(R)$ which follows closely the empirical law

$$\log \frac{I(R)}{I(R_e)} = -3.33 \left[\left(\frac{R}{R_e}\right)^{1/4} - 1\right] \tag{47}$$

except in the innermost and outermost parts (see de Vaucouleurs 1959*a*). Here R_e is the radius inside which half the total light is emitted. This means that the brighter ellipticals have all relaxed to a similar dynamical state. For the dwarf ellipticals, the surface brightness drops off more rapidly in the outer parts; their luminosity profiles are like those for the globular clusters (see § 6 and fig. 22).

The radial $U - B$, $B - V$ color distribution is fairly uniform in the brighter ellipticals; neither color changes by more than about 0.1 mag in the mean from the inner to the outer regions of these systems (see de Vaucouleurs 1961*a*). This means that the same types of stars probably dominate the optical luminosity in the inner and outer parts of ellipticals. An exception may be the very extended system M87 (de Vaucouleurs 1969), where the outer corona is rather bluer than the inner parts and may be globular-cluster-like in its stellar content.

Fish (1964) has established an important relationship between the total mass $\mathfrak{M}$ and the potential energy Ω for the brighter ellipticals. Because so few masses are known for ellipticals, he assumed that the ratio of surface brightness to surface density is constant, both within a single system and from galaxy to galaxy. The total mass is then

$$\mathfrak{M} = \kappa L \, , \tag{48}$$

and the potential energy of a galaxy is

$$\Omega \propto \kappa^2 L^2 / R_e \, , \tag{49}$$

where L is the total luminosity and R_e again the effective radius. Fish finds that

$$\Omega \propto \mathfrak{M}^{3/2} \, ; \tag{50}$$

the data are shown in figure 9. It follows from equation (50) that $L \propto R_e^2$, so the mean surface brightness is approximately constant for the brighter ellipticals. Fish's law comes up again in § 11.3.1.

4.3.2. *The spirals and lenticulars.*—We know that the spirals and lenticulars are intrinsically flat galaxies. They have a similar luminosity distribution which is characteristic of the whole family of disklike systems.

Figure 10 shows the radial surface brightness distribution for three disklike galaxies: NGC 4459 [SA(r)0], M83 [SAB(s)c], and M33 [SA(s)cd]. De Vaucouleurs (1959*a*) has pointed out that these $I(R)$ distributions have two main components: an inner *spheroidal* component which follows reasonably closely the $R^{1/4}$ law

$$\log I(R) \propto -R^{1/4} \tag{51}$$

and an outer *exponential* component (disk) with

$$I(R) = I_0 e^{-\alpha R} \, , \tag{52}$$

which contributes a large part of the total light and angular momentum. For example, in M31, which has a fairly prominent spheroidal component, more than 75 percent of the total blue light (de Vaucouleurs 1958) and probably more than 95 percent of the total angular momentum (Takase 1967) come from this exponential disk. The spheroidal component may be prominent as in NGC 4459, or very weak as in M33; see figure 10.

The two-component nature of $I(R)$ for disk systems distinguishes them from the

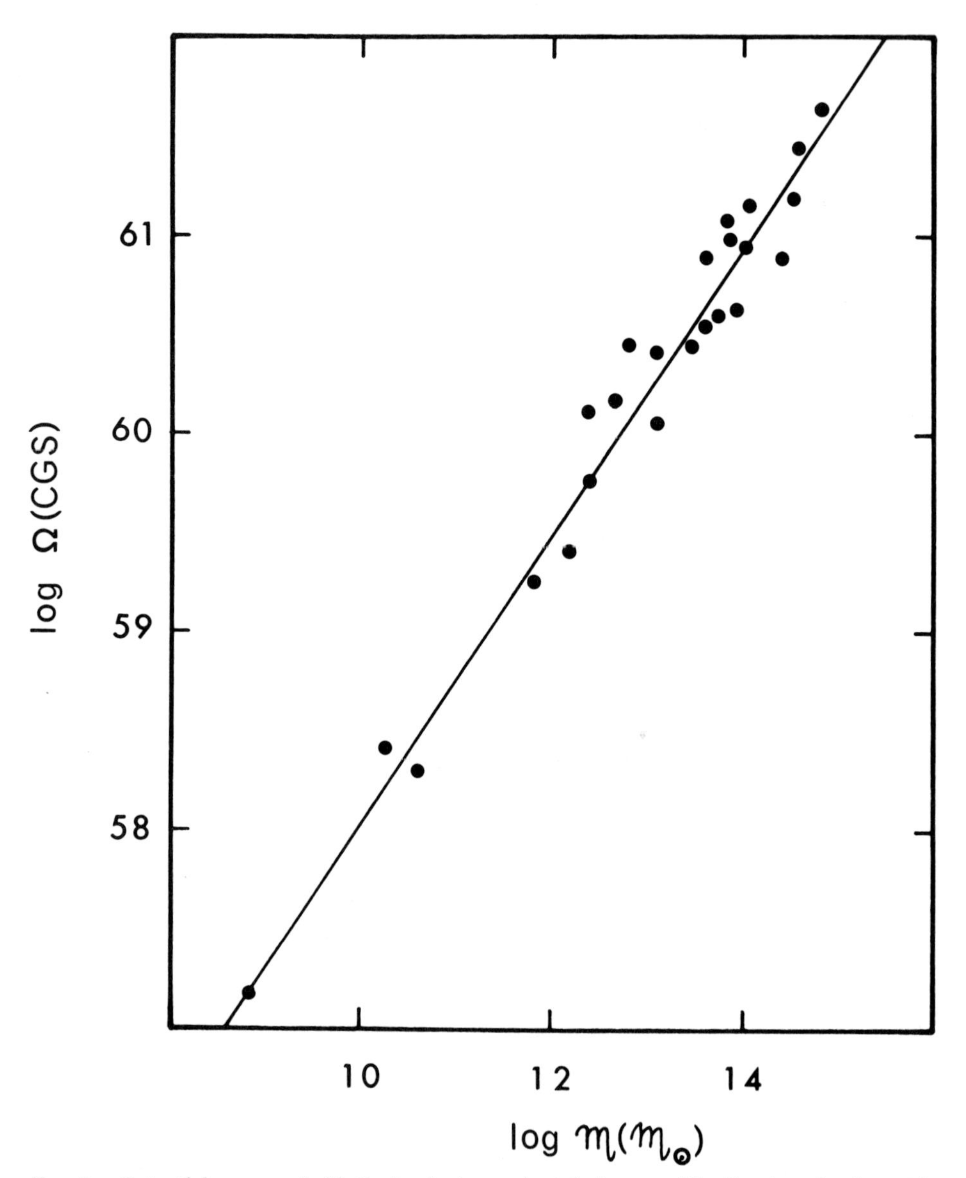

FIG. 9.—Potential energy of elliptical galaxies against their mass. The line has the slope 3/2.

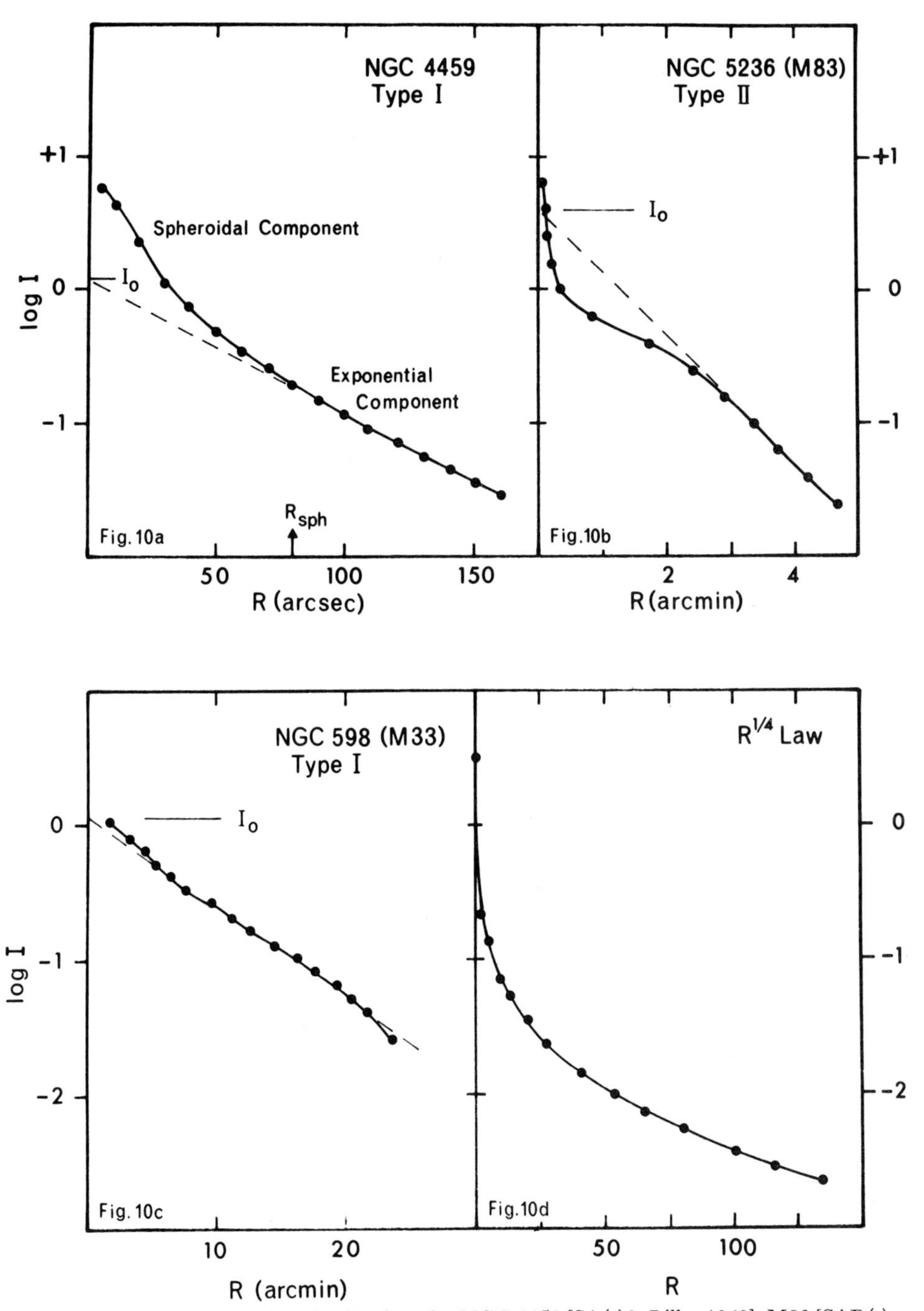

FIG. 10.—Radial luminosity distributions for NGC 4459 [SA(r)0, Liller 1960], M83 [SAB(s)c, Sersic 1969] and M33 [SA(s)cd, de Vaucouleurs 1959*b*]. The $R^{1/4}$ law is also shown. The radial scale in fig. 10d is arbitrary, as is the intensity scale in all four figures.

normal ellipticals. The $R^{1/4}$ luminosity law is also shown in figure 10: it also appears approximately exponential for large R, so the $I(R)$ law for the outer parts of ellipticals is *qualitatively* similar to that shown in figure 10 for NGC 4459. However, the important difference between the $I(R)$ distributions for ellipticals and disk galaxies is that, because the ellipticals are really one-component systems conforming to the $R^{1/4}$ law, the apparently exponential and nonexponential parts of their $I(R)$ profiles are *not* independent. For example, the fraction of the total light emitted within the nonexponential part of the $I(R)$ distribution is approximately the same for all ellipticals; this is certainly not true for disk galaxies.

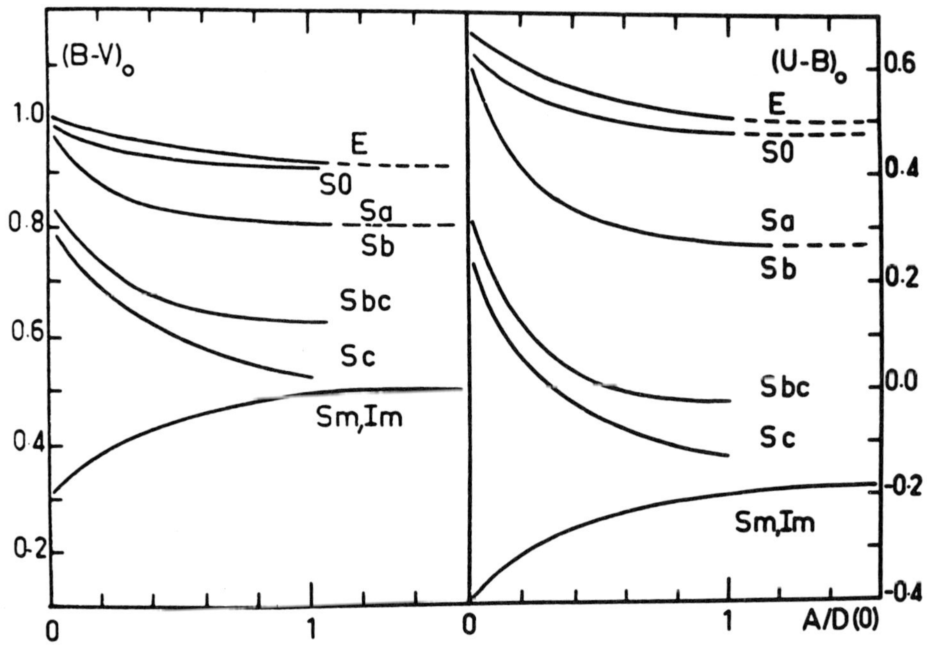

FIG. 11.—Mean normal color-aperture relations for galaxies (from de Vaucouleurs 1961*a*). Here A is the aperture diameter and $D(0)$ the "face-on" diameter.

Almost every disk galaxy with measured $I(R)$ shows an exponential disk (see Freeman 1970*a*); this includes the Magellanic irregulars. This means that, in all these systems, the outer parts which contain most of the angular momentum have settled to a similar dynamical state. We discuss this further in § 9.3.

For the S0 to Sc systems, the colors are usually redder in the inner regions corresponding to the spheroidal component than they are in the disk. The disk colors appear to be fairly uniform within the disk of any one system (de Vaucouleurs 1958, 1959*b*), but vary with type from $(B - V, U - B) \approx (0.9, 0.5)$ at S0 to $(0.5, -0.1)$ at Sc (de Vaucouleurs 1961*a*). Presumably these colors reflect the increasing amount of young Population I activity in the later types. For the Magellanic systems, both colors *increase* by about 0.2 mag from the inner to the outer parts (see fig. 11).

It is worth discussing briefly the luminosity distributions for the barred galaxies. On most photographic reproductions the bar looks dominant in size and luminosity, but this impression is not usually correct. Surface photometry shows very clearly that the bar is a relatively small feature immersed in an approximately axisymmetric disk. Table 4 gives the type, the ratio L_B/L_T of the bar luminosity to the total luminosity, the ratio $a_B/a_{27.0}$ of the bar length to the disk diameter at a surface brightness of 27.0 mag arcsec^{-2}, and the source of the photometry, for three galaxies. Note that the asymmetry associated with the bar is limited to the inner parts of the systems; most of the light comes from the approximately axisymmetric disk. This disk has the exponential luminosity distribution of equation (52) and does not appear to differ in any way from the disk of the SA systems. To summarize this paragraph: the main difference between the A and B families of spirals is the departure from axisymmetry in the *inner* parts of the B galaxies.

What is the nature of the bar; i.e., to what population does it belong? In some earlier-type systems like NGC 6744, the bar follows the $R^{1/4}$ distribution, which appears to be characteristic of the spheroidal component in SA galaxies (de Vaucouleurs 1963*c*). The

TABLE 4

DATA FOR THREE BARRED SPIRALS*

NGC	Type	L_B/L_T	$a_B/a_{27.0}$	Source
1313......	SB(s)d	0.12	0.15	de Vaucouleurs 1963*b*
6744......	SAB(r)bc	0.10	0.10	de Vaucouleurs 1963*c*
1433......	SB(r)a	0.15	0.30	de Vaucouleurs (unpublished)

* From Freeman (1970b).

bar appears to be part of the spheroidal component in the SB0 systems and in spirals earlier than about SBc; its colors and spectra are consistent with this picture (de Vaucouleurs 1961*a*, Code 1967). For later-type galaxies, the bar nature is not clear: in NGC 55, a nearly edge-on SBm system (de Vaucouleurs 1961*b*), the bar is no more extended in the z-direction than is the disk. At least for the earlier-type systems, we identify the bar tentatively with the spheroidal population.

Galaxies earlier than about SBc have the bar central in the disk. At about SBd, a *second* asymmetry appears: the bar becomes displaced from the disk center (see de Vaucouleurs and Freeman 1970*a*). This displacement seems to be important for the dynamical theory of the Magellanic barred spirals (§ 10).

Finally we return to the two-component luminosity profile for the disk galaxies. Freeman (1970*a*) has collected blue-light luminosity profiles for 36 spiral and S0 galaxies. He finds the following:

i) For 28 of the 36 systems, the exponential disks have approximately the same intensity scale I_0 (defined by eq. [52] and fig. 10): 21.65 B-mag arcsec^{-2}, with a standard deviation of only 0.30 mag arcsec^{-2}, despite a range of nearly 5 mag in absolute magnitude. This result is shown in figure 12. Six of the seven high-surface-brightness disks belong to galaxies whose luminosity profiles dip below the projected exponential component; the type example is M83, shown in figure 10. It is interesting that the uniformity of mean surface brightness appears to hold for the ellipticals (§ 4.3.1) and also for the disk galaxies: the reasons are probably not the same because the dynamical situations are so different (see also §§ 4.4 and 11.3.2).

ii) S0 to Sbc systems have any value of the disk length scale α^{-1} between 1 and 5 kpc, while later-type systems have predominantly low values of α^{-1} ($\lesssim$ 2 kpc) (see fig. 13). This means that the late-type spirals and irregulars are smaller in the mean than the early-type systems, and because of observation (i) they are also fainter in the mean.

iii) The relative brightness and size (fig. 14) of the spheroidal and disk components are only weakly correlated with morphological type.

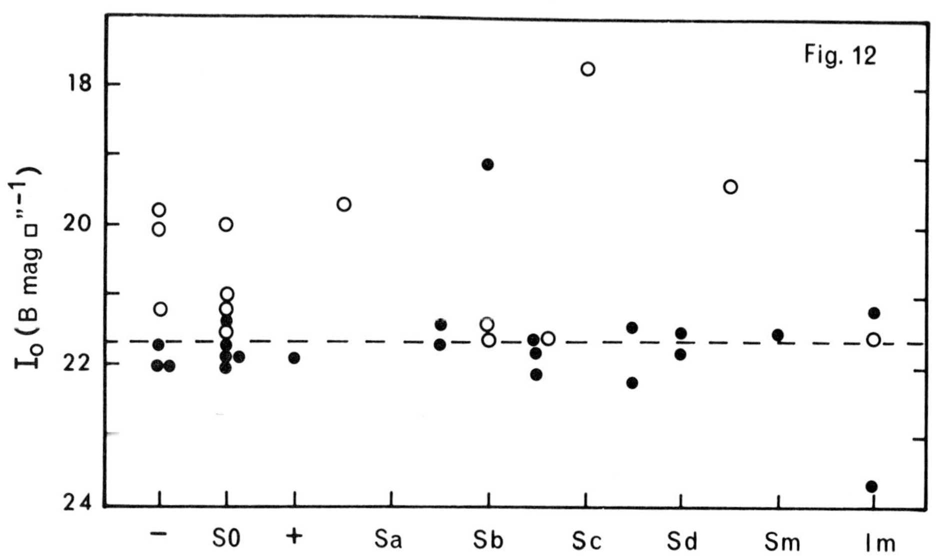

FIG. 12.—Intrinsic distance-independent blue light luminosity scale I_0 for the exponential disks of 36 galaxies against their morphological type. The broken line at $I_0 = 21.65$ is the mean for 28 galaxies. *Filled circles*, galaxies with type I luminosity profiles; *open circles*, type II profiles, as defined in fig. 10. I_0 is also defined in fig. 10.

4.4. Internal Motions and Masses

4.4.1. *The normal ellipticals.*—Several observers (Osterbrock 1960; Walker 1962; King and Minkowski 1966) have detected rotation in the inner few hundred parsecs of some systems, with amplitudes up to 50 km s^{-1} per 100 pc, and also rapid rotation in the nucleus of M32. There is no direct observational information about rotation in the outer parts of ellipticals. However, their intrinsic axial ratios ($\langle b/a \rangle \approx 0.6$) may suggest that they are not in rapid rotation. This is supported by King's (§ 6.2) quasi-isothermal nonrotating models with isotropic velocity dispersion; these models provide an excellent fit to the observed $I(R)$ profiles. The conclusion is that rotation does not play a major part in the support of ellipticals, which is then essentially a balance between self-gravity and stellar pressure. See, however, Prendergast and Tomer (1970) for a discussion of rotating ellipticals.

If we assume that the mass/light ratio is independent of radius, then it is possible to derive the mass $\mathfrak{M}$ of ellipticals from the observed velocity dispersion $\langle v^2 \rangle$ in the central regions:

$$\mathfrak{M} \propto R_e \langle v^2 \rangle \, , \tag{53}$$

where the constant of proportionality depends on the adopted dynamical model for these systems. R_e is readily measurable photometrically, but $\langle v^2 \rangle$ is more difficult to determine. Obviously the relevant $\langle v^2 \rangle$ is that for the approximately isothermal body of the galaxy; however, it is usually possible to measure $\langle v^2 \rangle$ only for the bright nucleus. It should be checked carefully that these two values for $\langle v^2 \rangle$ are the same, because it may be that in some cases the nucleus is a dynamically independent object. From the mass and integrated luminosity, the mass/luminosity ratio follows (table 5). This ratio

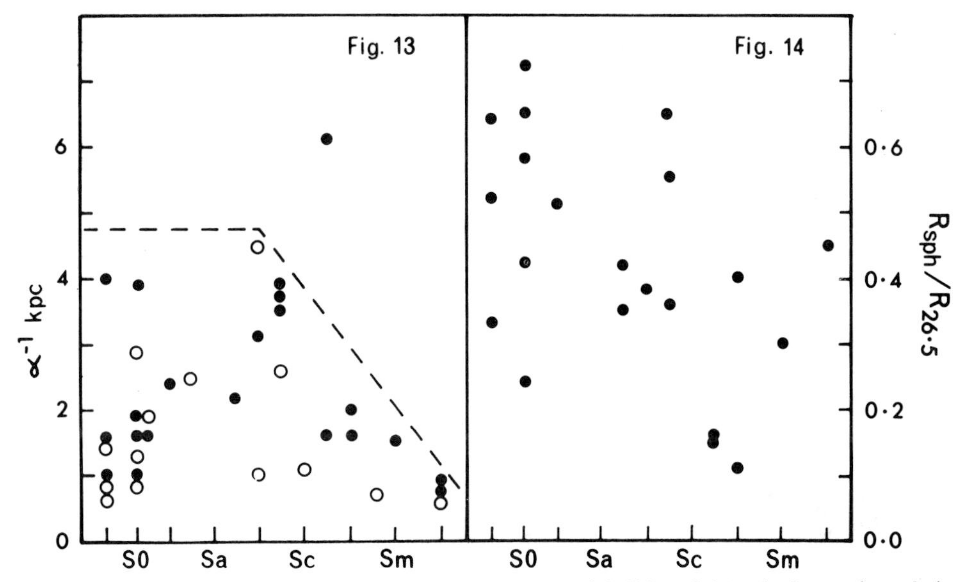

FIG. 13.—The length scale α^{-1} (kpc) for the exponential disks of 36 galaxies against their morphological type. Coding as in fig. 12. *Broken line*, the apparent upper envelope.

FIG. 14.—The ratio $R_{\rm sph}/R_{26.5}$ of the apparent radius of the spheroidal component, as defined in fig. 10*a*, to the disk radius at $B = 26.5$ mag arcsec^{-2} (face-on) for galaxies with type I luminosity profiles.

TABLE 5

MASS/LUMINOSITY DATA FOR ELLIPTICALS*

NGC	Type	$\mathfrak{M}(\mathfrak{M}_\odot)$	$\mathfrak{M}/L$
221	E2	2.9×10^9	11
3379	E1	9.6×10^{10}	11
4406	E3	9.6×10^{11}	39
4472	E2	1.1×10^{12}	19
4486	E1	2.6×10^{12}	59

* Data from Fish 1964.

is an important constraint on the stellar content of these systems. The observed range in mass/luminosity is large: from 11 for NGC 221 to 59 for NGC 4486 (both have been measured by independent observers, and the agreement is good). This is interesting, because the $U - B$, $B - V$ colors for these two systems differ by only 0.02 mag (de Vaucouleurs 1961*a*). There is not yet any direct information about the masses for the dwarf ellipticals, but from their luminosities (chap. 14) their masses are probably in the range 10^7 to $10^9\ \mathfrak{M}_\odot$.

4.4.2. *The spirals.*—These are discussed thoroughly in chapters 3 and 9; we will just summarize the main points here.

The internal motions for spirals are measured almost always for the gaseous component, from the emission line velocities for ionized (optical) or neutral (21-cm) gas. The dominant component of the motion is rotation; however, in many spirals there is clear evidence for noncircular motions. For example:

i) In the Galaxy the 4-kpc arm has a velocity of -53 km s^{-1}, seen in 21-cm absorption against the Sgr A source (see Kerr and Westerhout 1965).

ii) The large-scale 21-cm velocity field in M31 is clearly not consistent with pure rotation (see chap. 9). Many systems including M31 show a difference between the rotational velocities measured along the major axis on the two sides of the nucleus.

iii) Barred spirals frequently show noncircular motions either in the nuclear regions or directed along the bar. The SBm systems appear to have extensive regions of noncircular velocities. These systems are discussed in § 10.

iv) For some spirals, including our own and M31, the "plane" of the disk is bent. Along a line through the galactic center and normal to the Sun-center line, the H I layer in the Galaxy is flat (within about 20 pc) for $R < 10$ kpc; for $R > 10$ kpc, the H I layer flares out and bends up on one side of the nucleus and down on the other. At $R \approx 15$ kpc the layer deviates from the plane by about 1 kpc. (See Kerr and Westerhout 1965 for a description of the H I distribution in the Galaxy.) We can expect z-motions of the gas associated with this bending (see § 9.4).

If we assume that all gas motions are rotational, then it is possible to derive a mean rotation curve $v_\varphi(R)$. Rotation curves are now available for many spirals, and in almost all cases the rotation is highly nonuniform. A typical rotation curve, for NGC 5055, is shown in figure 15*a*. However, in a few spirals the rotation does appear to be nearly uniform throughout the region of spiral structure; an example, for NGC 157, is shown in figure 15*b*. If we further assume that the observed *rotational* velocity field $v_\varphi(R)$ is the *circular* velocity field $\Theta(R)$ defined in equation (35), then it is possible to derive the mass of a galaxy from fitting the observed $v_\varphi(R)$ to the theoretical rotation curve for some adopted model of the mass distribution (see chap. 3 for details). This mass has a fairly large uncertainty because (i) $v_\varphi(R)$ often comes from a velocity field which has an obvious noncircular component; (ii) the angle i between the normal to the galactic plane and the line of sight, and the distance Δ to the galaxy enter into the derived mass as a factor $\Delta/\sin^2 i$, and both are uncertain; and (iii) the rotation curve is unknown beyond some radius R_x, so the derived mass depends on the behavior of the adopted theoretical model for $R > R_x$. For example, the mass of the Galaxy derived from Schmidt's model is $1.8 \times 10^{11}\ \mathfrak{M}_\odot$, while that from Innanen's model is $1.3 \times 10^{11}\ \mathfrak{M}_\odot$.

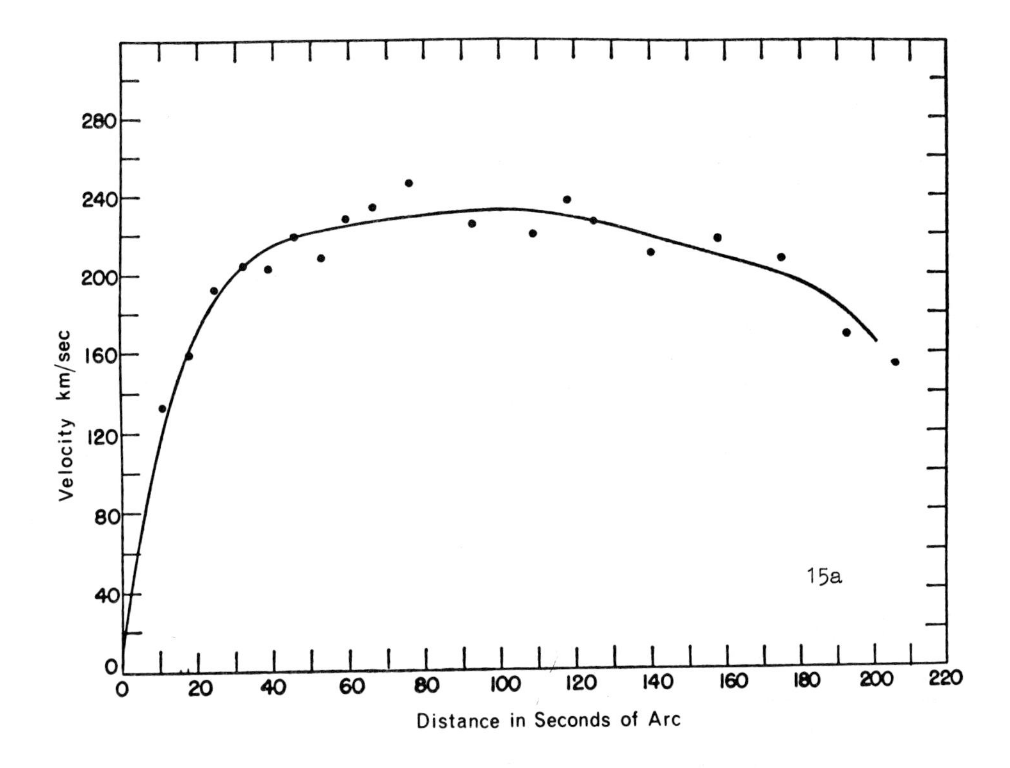

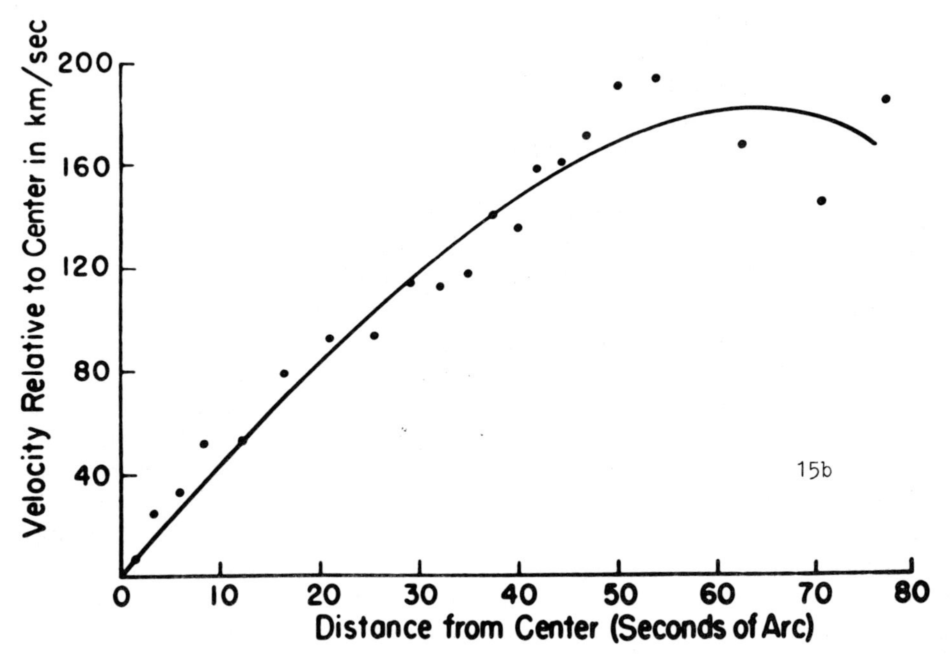

FIG. 15.—(*a*) The rotation curve for NGC 5055 (from Burbidge *et al.* 1960*a*). (*b*) The rotation curve for NGC 157 (from Burbidge *et al.* 1961).

Both models represent the same rotation data for $R < R_0$ and have the same total density at the Sun (see § 3.3).

An interesting feature of the observed rotation curves, pointed out by Crampin and Hoyle (1964), is that the angular-momentum distribution in the outer parts of spiral systems is very similar to the distribution for a uniformly rotating uniform density spheroid. This feature comes up again in § 9.3.

M. S. Roberts (1969) has collected mass and luminosity data for 91 spiral and irregular galaxies. He finds the important result that the mean (total mass)/(total blue light luminosity) ratio at a given morphological type is almost independent of type for these systems, although the scatter at a given type is fairly large. The ratio has a mean value of 7.5 in solar units (cf. table 5).

4.4.3. *The frequency distribution of masses.*—The frequency distribution of galactic masses is not known. Even in the Local Group, the total number of dwarf galaxies is quite uncertain. Only five of the 27 known members are brighter than $M_B = -16$ (chap. 14), so most have masses between about $10^6\ \mathfrak{M}_\odot$ and $5 \times 10^9\ \mathfrak{M}_\odot$.

4.5. The Content of Galaxies

The integrated spectral type in blue light of normal ellipticals and the inner parts of lenticulars and early-type spirals is about K0 III. However, the high $\mathfrak{M}/L$ values for these systems mean that they must contain a large proportion of late-type dwarfs or other low-luminosity objects. This is confirmed by Spinrad's analysis of the absorption-line spectrum of the nucleus of M31 (Spinrad 1966; Spinrad and Taylor 1971). He finds that about 25 percent of the visual light and almost all the mass comes from lower-main-sequence stars. The luminosity function from $M_v = +5$ to $M_v = +16$ rises much more rapidly than the van Rhijn function, and gives an $\mathfrak{M}/L$ value of about 40. Note that in a galaxy lifetime the main sequence burns down only to about K0 V, so the enrichment by late-type dwarfs is probably an initial luminosity function effect.

The Draco and Sculptor dwarf ellipticals have color-magnitude diagrams like globular clusters; also, the luminosity function for the Draco dwarf is very similar to that for the globular cluster M3 (Baade and Swope 1961; Hodge 1965). (The absolute magnitude for Sculptor is -11.2, and that for Draco is unknown.) From the (surface brightness)/(star count) ratio for the outer regions of NGC 205 ($M_B = -15.8$), Baum (1959) argues that this system has a predominantly globular-cluster-like content, i.e., metal deficient. On the other hand, M32 ($M_B = -15.6$) is more like the old population of M31.

There is little quantitative information on the content of late-type spirals and irregulars (see, however, de Vaucouleurs 1959*c*). The surface brightness distributions for these systems show that there is an exponential disk component underlying the spiral-arm population and contributing most of the light. This disk component is visible in the lenticulars without the confusion of the spiral structure: presumably we can identify this component with the old disk population of the Galaxy. However, there is really very little known about the basic stellar content of the disks of any of the disklike galaxies, except for the Galaxy, where the disk appears to be a metal-rich component (see Eggen and Sandage 1969).

It seems that, despite their young appearance, the late-type spirals and Magellanic irregulars are probably old. For example, (i) Walker's (1964) photographs of M33

[SA(s)cd] show a smooth flattened distribution of Population II red giants underlying the distribution of red supergiants associated with the spiral structure, (ii) both M33 and the Magellanic Clouds contain some apparently normal globular clusters (Hiltner 1960; Gascoigne 1966).

The gas content of galaxies has been studied at length: see chapter 9. M. S. Roberts (1969) shows that the mean ratio of (H I mass)/(total mass) rises smoothly from about 0.01 at Sa to about 0.20 at the Magellanic irregular stage. He also shows that the mean surface density in neutral hydrogen rises smoothly with type, in the same sense. Sandage, Freeman, and Stokes (1970) argue that the mean gas *density* is probably the fundamental independent observable which determines the morphological type.

Galaxies of all types show [O II] λ3727 emission. Table 6 gives the frequency of occurrence of this emission line in galaxies of different types. Note that at least some ellipticals and lenticulars contain ionized interstellar gas (although usually in the nuclear regions only).

There is more detailed information on the content of galaxies in chapter 2 in this volume.

TABLE 6

FREQUENCY OF OCCURRENCE OF λ3727 IN GALAXIES*

	E	S0	Sa	Sb	Sc	Irr
(percent)	16	40	56	71	73	94

* From Roberts 1963.

4.6. SOME PROBLEMS

Now we list some general questions which need to be answered before we can understand the formation and structure and evolution of galaxies.

4.6.1. *Problems associated with the different kinds of galaxies.*—What are the physical quantities that determine whether a protogalaxy will collapse to an elliptical or a disklike configuration; and if it is disklike, into which of the families (SA or SB) and varieties (r or s) it will fall?

How did this collapse take place (if it did in fact collapse)?

What are the fundamental variables that change along the spiral sequence?

Is the spiral sequence an evolutionary progression, or are the different spiral types essentially defined at their formation?

Why has star formation ceased in the lenticular systems?

What is the angular momentum content of the elliptical galaxies and the bulge components of the disk galaxies?

Why are the stellar contents of the dwarf and the giant ellipticals so different, and why is the luminosity function for the giant ellipticals apparently so dwarf-enriched when compared with the van Rhijn function?

What is the stellar content of the disks in the disklike galaxies, and at what stage in their chemical history did these disks form?

4.6.2. *Problems associated with the mass and angular momentum distributions.*—What processes define the observed surface brightness distributions for the ellipticals and for the disklike systems, and the angular momentum distribution associated with the rotation curves for the spirals?

What is the origin of the angular momentum for the disk galaxies?

Why are there no ellipticals flatter than E7?

What is the significance of Fish's law for the ellipticals (§ 4.3.1) and of the uniformity of the intensity scale I_0 for the exponential disks of the disklike galaxies (§ 4.3.2)?

4.6.3. *The problem of spiral structure.*—What is the origin of the spiral structure in the spiral galaxies, and how can it persist, if it does, for the lifetime of a galaxy ($\sim 10^{10}$ years), despite the generally observed differential rotation which would wind up any structure partaking of it in a time of order 10^8 years?

Is the spiral structure mainly a gas and young-star phenomenon, or is the background disk significantly involved?

What is the role of the dark matter in spirals, especially in the SB(s) systems which show spectacular straight dust lanes along the leading edges of their bars?

4.6.4. *Some other problems.*—What is the reason for the bending of the plane in many spirals?

What is the origin of the noncircular motions observed in many spirals and Magellanic irregulars?

Are there third integrals associated with most stellar orbits in the Galaxy; and if so, what is their nature?

Why does the velocity dispersion for disk stars near the Sun increase with stellar age?

What is the origin of the large random motions observed for halo objects in the Galaxy?

In the rest of this chapter, we discuss some attempts to answer some of these questions.

5. THE THIRD INTEGRAL

It follows from Jeans's theorem (§ 2.5) that the velocity dispersion components σ_{RR}, σ_{zz} are equal in an axisymmetric time independent galaxy if there are only the two isolating integrals $E = \frac{1}{2}c^2 + \Phi$ and $h = Rc_\varphi$. For stars now near the Sun, $\sqrt{\sigma_{RR}} \approx 2\sqrt{\sigma_{zz}}$. This could result from the fact that the Galaxy, as seen by these stars, is not axisymmetric or time independent. While departures from axisymmetry may contribute to the anisotropy of $\boldsymbol{\sigma}$ (§ 9.5), this could not be the only cause, because $\sqrt{\sigma_{RR}} \approx 2\sqrt{\sigma_{zz}}$ holds also for relatively high-velocity stars such as the subdwarfs which spend most of their time relatively far from the galactic plane and so presumably see an approximately axisymmetric, time-independent potential. It follows that most stellar orbits in the Galaxy probably have a *third integral.*

Unlike the integrals E and h, the existence of this third integral is not obvious from the form of the Hamiltonian. However, numerical experiments show that most stellar orbits do behave as if they have three time-independent isolating integrals (Ollongren 1962). Further, while most or all lower-energy orbits appear to have a third integral, relatively few of the higher-energy orbits do so; at a given E and h, the existence of a third integral for these orbits depends on the initial values of the coordinates and velocities (Hénon and Heiles 1964). The nature of this kind of integral is a deep problem in the theory of dynamical systems, and a proper discussion is quite outside the scope of this chapter. We will just give a qualitative discussion to illustrate some of the main points. Contopoulos (1967*a*) and Ollongren (1965) review the subject from a relatively astronomical viewpoint, and the general dynamical theory is described by Arnold and Avez (1967).

We want to visualize the third integral. Consider the orbit of a star in an axisymmetric time-independent potential $\Phi\ (R,\ z)$. E and h are the only obvious isolating integrals.

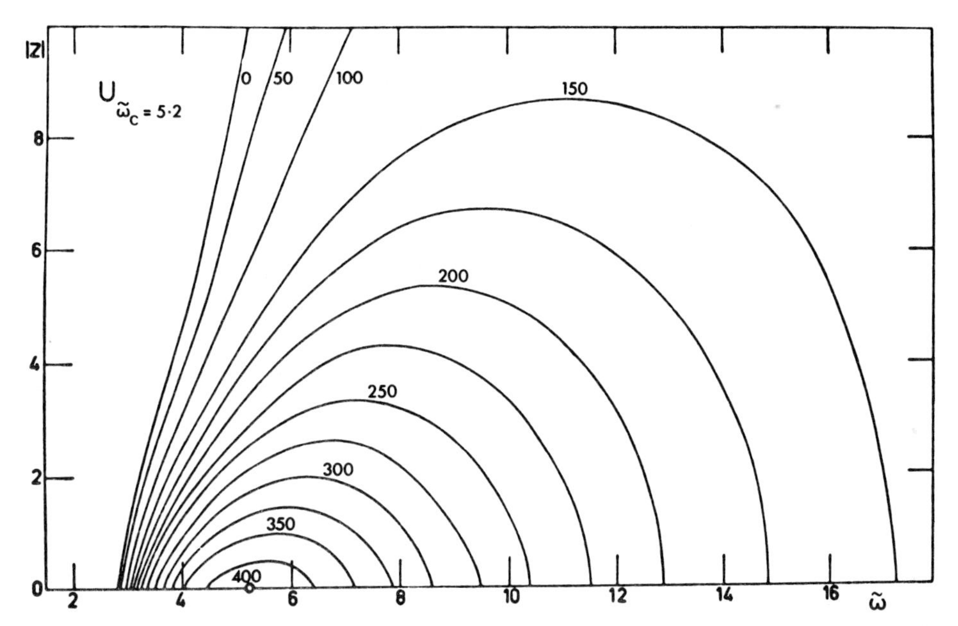

FIG. 16.—Equipotential curves in the $(R,\ z)$ plane for the 1956 Schmidt model. The value of h corresponds to a circular orbit with $R = 5.2$ kpc (from Oort 1965).

Because $h = Rc_\varphi$ is constant, we need study only the motion in the $(R,\ z)$ plane, i.e., in two degrees of freedom. The energy E becomes

$$\begin{aligned} E &= \tfrac{1}{2}c_R{}^2 + \tfrac{1}{2}c_z{}^2 + \tfrac{1}{2}c_\varphi{}^2 + \Phi(R,\ z) \\ &= \tfrac{1}{2}c_R{}^2 + \tfrac{1}{2}c_z{}^2 + \Phi'(R,\ z)\ , \end{aligned} \tag{54}$$

where

$$\Phi'(R,\ z) = \Phi(R,\ z) + h^2/2R^2 \tag{55}$$

is the potential for motion in the $(R,\ z)$-plane. Some equipotentials of Φ' in the Schmidt model and for a given value of h are shown in figure 16; note that circular orbits correspond to the minimum energy for a given angular momentum. For motion in the $(R,\ z)$-plane, E is the one obvious isolating integral.

How many more isolating integrals can there be? We are considering a four-dimensional phase space $(R,\ z,\ \dot{R},\ \dot{z})$; the integral E constrains an orbit to a three-dimensional hypersurface in this space. One more isolating integral would restrict the orbit to a two-parameter surface. A further integral would mean that the orbit is one-parameter, i.e., it is periodic. However, we know that for two degrees of freedom all orbits are periodic only in very special potentials. It follows that there can be only one more isolating integral, except for periodic orbits: if it exists, it is the third integral.

Now consider a star with $E = E_0$, $h = h_0$. If there are no more isolating integrals, this orbit should pass arbitrarily close to any point inside the appropriate equipotential $\Phi'(R, z) = E_0$. (This equipotential is called the zero velocity curve, ZVC, in the $[R, z]$-plane.) If there is a third time-independent integral, then the orbit is constrained to some subregion within the ZVC. For example, the two-dimensional harmonic oscillator $R = A \sin \alpha t$, $z = B \sin \beta t$, has its energy

$$E = \tfrac{1}{2}\dot{R}^2 + \tfrac{1}{2}\dot{z}^2 + \tfrac{1}{2}\alpha^2 R^2 + \tfrac{1}{2}\beta^2 z^2 \tag{56}$$

as an isolating integral, which confines the orbit within the ellipse

$$E = \tfrac{1}{2}\alpha^2 R^2 + \tfrac{1}{2}\beta^2 z^2 \tag{57}$$

(see fig. 17). However there is another independent isolating integral

$$E_\alpha = \tfrac{1}{2}\dot{R}^2 + \tfrac{1}{2}\alpha^2 R^2 \tag{58}$$

and this further confines the orbit within the shaded box region shown in fig. 17. If α/β is irrational, the orbit covers the shaded box (box orbit). If α/β is rational, then all orbits are periodic. This situation holds generally for separable potentials: if the ratio of the two relevant frequencies is irrational, then all orbits are box-type orbits, where the boxes are defined in the appropriate curvilinear coordinate system.

In models of the Galaxy, with nonseparable potentials, lower-energy orbits show a third integral for given values of E and h: these orbits include box orbits, periodic orbits, and quasi-periodic orbits (tube orbits) which stay always close to a stable periodic orbit. Examples are shown in figure 18. The periodic orbits come from initial conditions which result in a rational ratio for the two frequencies.

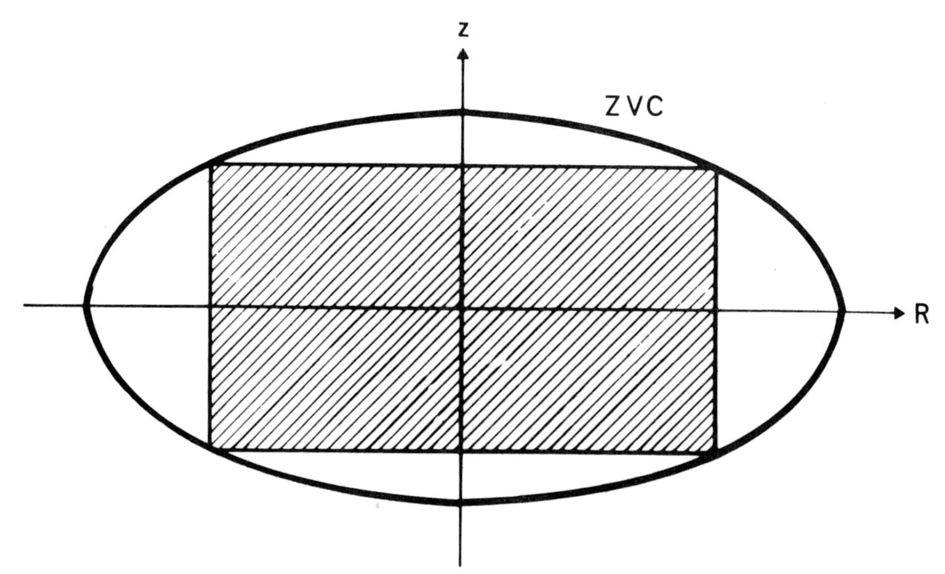

FIG. 17.—A box orbit and the zero velocity curve for the 2-D harmonic oscillator.

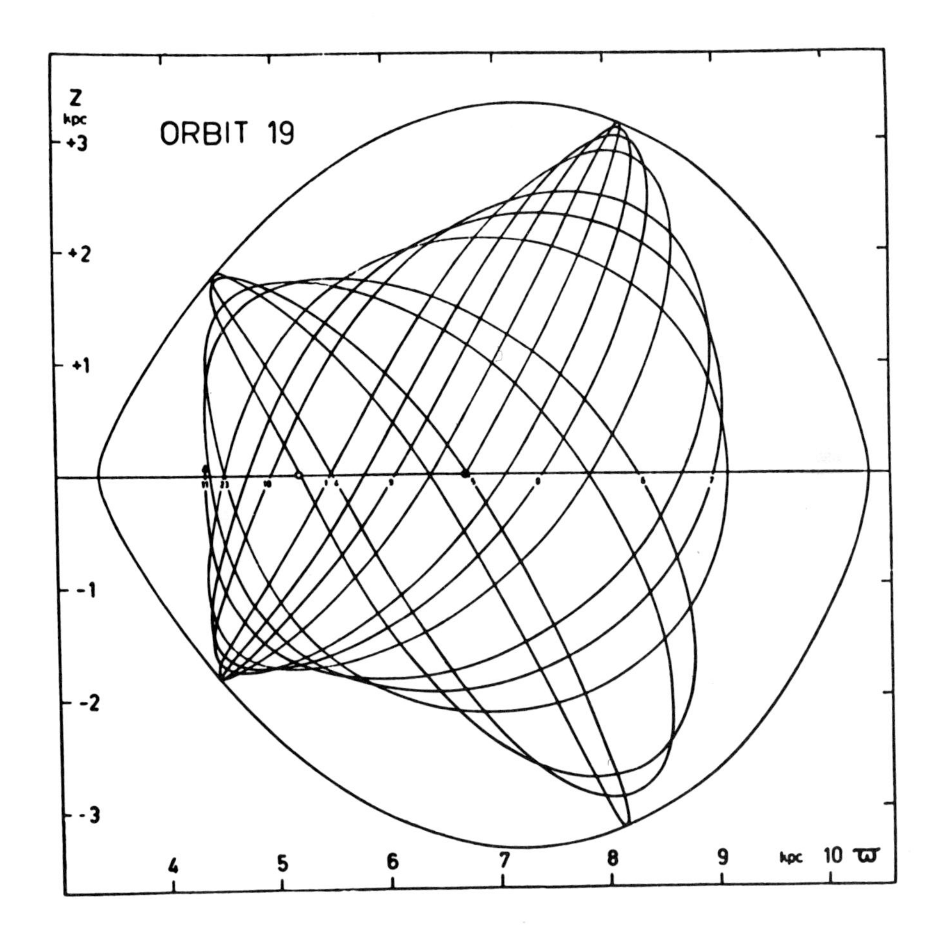

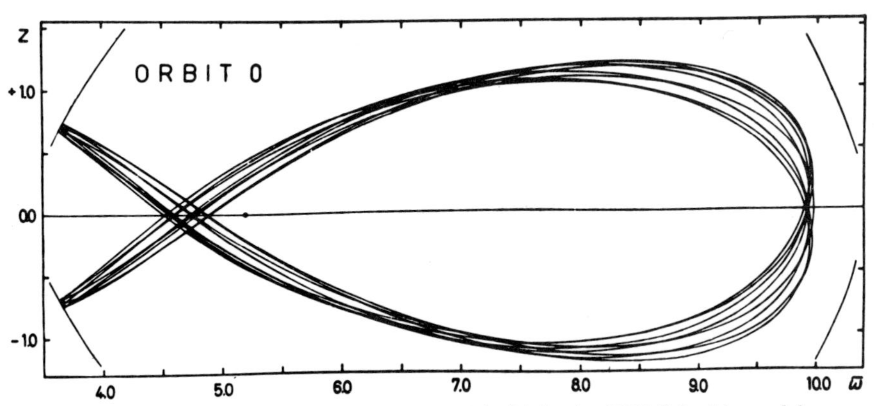

FIG. 18.—A box orbit (19) and a tube orbit (0) in the 1956 Schmidt model.

There is a convenient method of visualizing the integrals associated with a collection of orbits with two degrees of freedom and with the same values of E and h. This is the *surface of section* method. Because $E=E(R, z, \dot{R}, \dot{z})$ is an integral for an orbit, only three of the four coordinates are independent, say $(R, z, \dot{R})$, so we can consider the orbit's trajectory in $(R, z, \dot{R})$-space. The orbit can go anywhere inside the volume defined by $E = \frac{1}{2}\dot{R}^2 + \Phi'(R, z)$ in this space. Now take a section through this volume: say $z = 0$. Then the orbit's successive crossings of the $(z = 0)$-plane generates a set of *points* in the $(z = 0)$-plane, inside the region $E = \frac{1}{2}\dot{R}^2 + \Phi'(R, 0)$ in the $(R, \dot{R})$-plane; if there is no other integral, then this set of points fills the region. Now let there be one more integral. Then the orbit lies on a surface $I_3(R, z, \dot{R}) = \text{constant}$ in $(R, z, \dot{R})$-space, so that in the $(z = 0)$-plane it generates a *curve* $I_3(R, 0, \dot{R}) = \text{constant}$. Figure 19*b* shows a surface of section for some orbits in a simple potential. All orbits have the same energy; all but one appear to have an isolating integral corresponding to the third integral. Note how

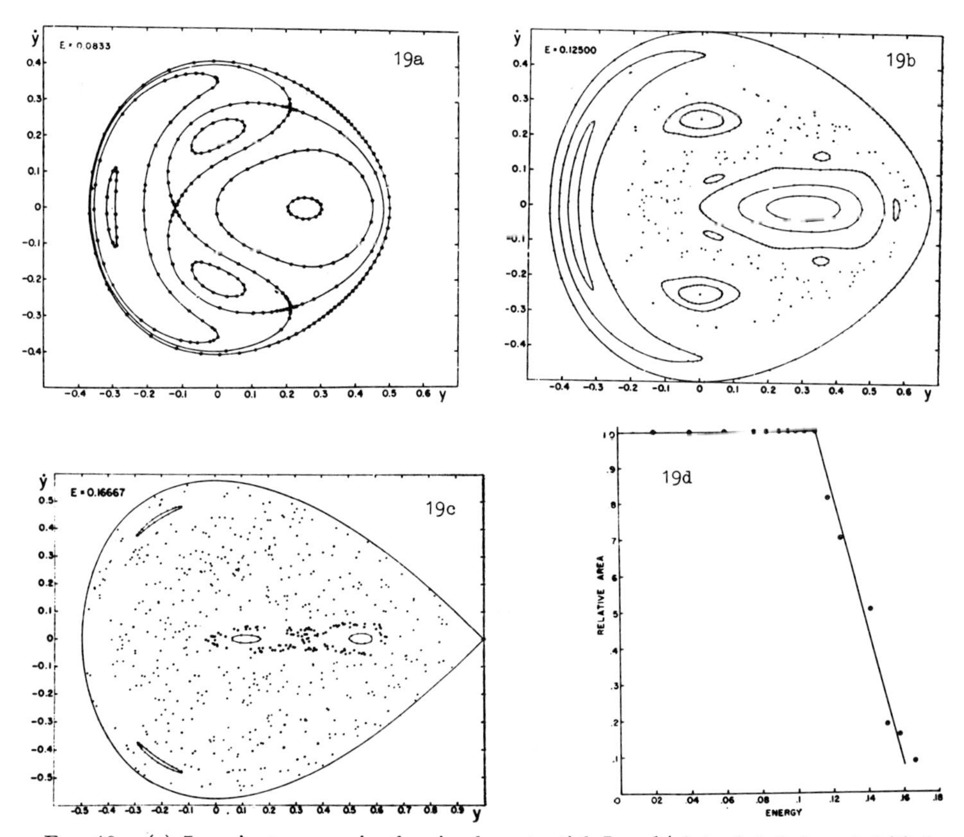

FIG. 19.—(*a*) Invariant curves in the simple potential $\Phi = \frac{1}{2}(x^2 + y^2 + 2x^2y - 2y^3/3)$ for $E = 0.08333$. The points satisfy $x = 0$, $\dot{x} > 0$. (*b*) Invariant curves for $E = 0.12500$. All the isolated points belong to one trajectory. (*c*) Invariant curves for $E = 0.16667$. All the isolated points belong to one trajectory. (*d*) Relative area covered by the invariant curves as a function of energy (from Hénon and Heiles 1964).

the periodic orbits generate a finite number of isolated points in the surface of section. These are called invariant points, because they are invariant under the mapping of the surface of section onto itself that is generated by the orbit. Similarly the closed curves are called invariant curves. Invariant points corresponding to stable periodic orbits have closed invariant curves about them. These closed curves represent quasi-periodic orbits which remain always close to the stable periodic orbit; for a quasi-periodic orbit, the surface I_3 is a torus in the $(R, z, \dot{R})$-space, within which the periodic orbit is embedded.

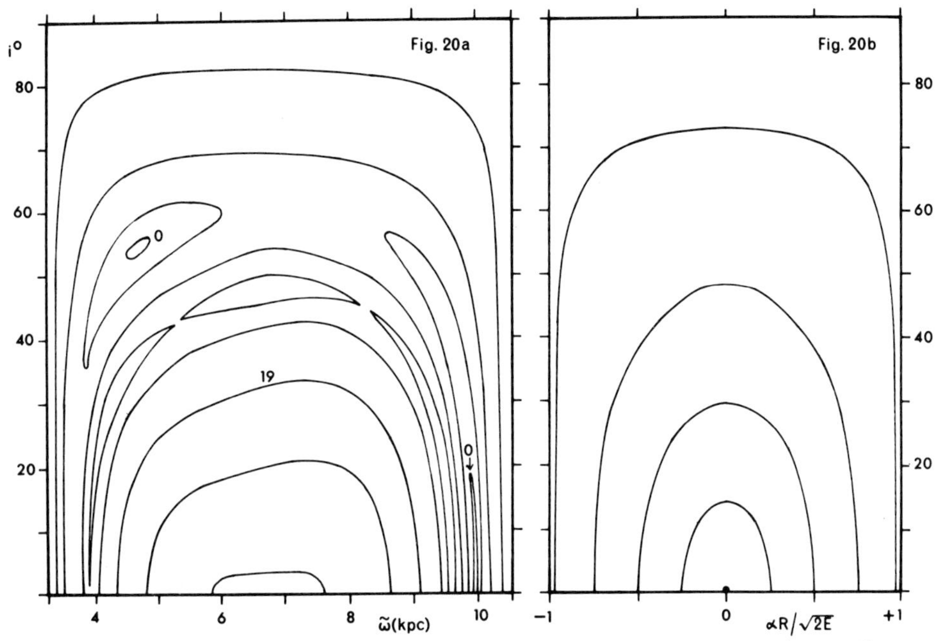

FIG. 20.—(*a*) Inclination diagram for orbits in the 1956 Schmidt potential (from Ollongren 1966). Curves labeled 19 and 0 correspond to orbits shown in fig. 18. (*b*) Inclination diagram for the 2-D harmonic oscillator.

Ollongren (1962, 1966) has published an extensive numerical investigation of orbits in the Schmidt (1956) model Galaxy. He chose a single pair of values for (E, h), which correspond to fairly high velocity orbits ($|z| < 3.5$ kpc, $4 \lesssim R \lesssim 10$ kpc, with $R_0 = 8.2$ kpc), and computed orbits for a large range of initial conditions. All appeared to have a third integral: they included mainly box orbits and tube orbits for many rational ratios of the two frequencies (see fig. 18 for examples). The relationship of the box and tube orbits is seen well in figure 20*a;* this is a surface of section diagram adapted from Ollongren (1966) and shows $i = \tan^{-1}|\dot{R}/\dot{z}|$ at $z = 0$, against R. The outer rectangular box defines the accessible region: it is itself the invariant curve associated with a box orbit which is always arbitrarily close to $z = 0$. Most of the curves shown represent box orbits. The closed invariant curves represent tube orbits (the orbits shown in fig. 18 are identified in fig. 20*a*). For comparison, figure 20*b* shows invariant curves for some orbits in the simple harmonic-oscillator field of equation (56) and figure 17. Again the rectangular box represents the orbit arbitrarily close to $z = 0$. The similarity of figures 20*a* and 20*b* is obvious, except that there are no periodic orbits for the harmonic oscillator apart from the trivial ones $A = 0$ or $B = 0$. The invariant point at the origin in figure 20*b* represents the periodic orbit $A = 0$.

Following this work, Hénon and Heiles (1964) investigated the behavior of the third integral in a simple analytic potential which simulates the essential properties of Φ'. They showed that the relative area of the surface of section covered by invariant curves decreases rapidly with increasing energy above some critical energy (see fig. 19*d*). Hayli's (1965) results for globular-cluster orbits in the Galaxy are consistent with this. The mechanism through which the invariant curves break down has been studied at length: see Contopoulos (1967*b*). It appears to be associated with the complex hierarchy of "chains of islands" shown in figure 19*b*. Note also how intimately the third integral is associated with the existence of stable periodic orbits.

Apart from its intrinsic interest, the third integral is important in galactic dynamics for the following reasons. (i) Most stars in the Galaxy have a third integral, so it is possible for the velocity-dispersion components in R and z to be unequal. (ii) Its existence should be recognized when we attempt to construct self-consistent models of disk galaxies, i.e., models with

$$\rho(R, z) = \int f(E, h, I_3) d^3c \, . \tag{59}$$

The functional form of $f(E, h, I_3)$ determines the local value of σ_{RR}/σ_{zz}. For example, Contopoulos (1967*a*) models the potential field near the Sun with the model

$$\Phi' = \tfrac{1}{2}(Ax^2 + Bz^2) - \epsilon x z^2 - \epsilon' x^3/3 \, , \tag{60}$$

where $x = R - R_0$ and A, B, ϵ, and ϵ' are constants. This leads to a third integral

$$I_3 = \tfrac{1}{2}(Ax^2 + \dot{x}^2) + \frac{\epsilon}{4B - A}\left[(A - 2B)xz^2 - 2x\dot{z}^2 + 2\dot{x}z\dot{z}\right] + \epsilon' x^3/3$$
$$+ \text{ higher-order terms in } \epsilon, \epsilon' \, . \tag{61}$$

Barbanis (1962) takes the distribution function

$$f = A \exp(-2kE - 2lI_3) \tag{62}$$

with I_3 truncated after the first-order terms, and shows that $(k, l) = (19.7, 14.1)$ kpc^{-2} (10^7 years)^2 represents satisfactorily the observed velocity distribution near the Sun for stars with velocities less than 80 km s^{-1}.

6. THE SELF-CONSISTENCY PROBLEM: MODELS FOR GLOBULAR CLUSTERS AND ELLIPTICAL GALAXIES

The problem is to construct self-consistent, self-gravitating stellar systems, i.e., to find solutions to the coupled equations

$$\frac{Df}{Dt} = 0 \quad \text{and} \quad \nabla^2\Phi = 4\pi G\rho(r) = 4\pi G \int f d^3c \, .$$

There are several reasons for doing this: (i) The luminosity distributions observed in elliptical galaxies can be reproduced by making realistic models. This gives important information about the distribution function f which in turn leads to a picture for the formation of these galaxies (see § 6.2). (ii) For systems which are too difficult to model realistically (e.g., barred galaxies), unrealistic models can guide our intuition about the behavior of these systems (see § 10). (iii) Time-dependent models are important for understanding the processes which go on during the collapse phase (see § 8). These models are usually numerical experiments on N-body collisionless systems.

The time-independent self-consistency problem has been approached in two ways. The first (Jeans's problem) assumes the potential $\Phi(r)$ and derives f. The second (inverse Jeans problem) is to assume the functional dependence of f on the integrals of the motion, and then deduce $\Phi(r)$ and hence $\rho(r)$. Perek (1962) reviews the extensive literature on this problem. In this section, we give simple examples of the two approaches to the problem and then describe some important models for globular clusters and elliptical galaxies.

6.1. Two Simple Examples

6.1.1. *The direct problem.*—We wish to construct an infinite slab of stars, with thickness $2D$ and constant density ρ (fig. 21). The variable part of the gravitational potential

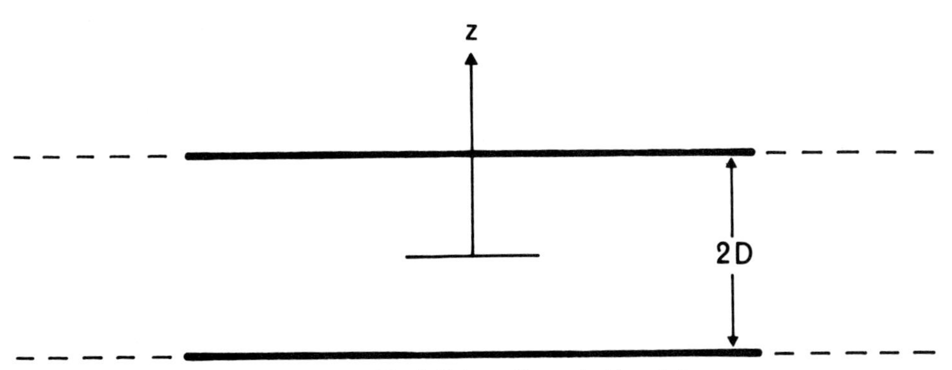

Fig. 21.—The infinite self-gravitating slab.

is $\Phi = 2\pi G\rho z^2$. There is one nontrivial integral, $E = \frac{1}{2}c_z^2 + 2\pi G\rho z^2$; c_x and c_y are also integrals, and we integrate over them immediately. If $f = f(E)$, then $Df/Dt = 0$ and we need only find the function f such that

$$\rho = \int f(E) dc_z . \tag{63}$$

At a point z, $dc_z = dE/[2(E - \Phi)]^{1/2}$. The range of integration over E is $\Phi(z) < E < \Phi(D)$: no star at z has E outside this range. Then equation (63) becomes

$$\rho = \int_{\Phi(z)}^{\Phi(D)} f(E)\{2[E - \Phi(z)]\}^{-1/2} dE . \tag{64}$$

This is an Abel integral equation. The solution is

$$f(E) = \frac{2^{1/2}\rho}{\pi}[2\pi G\rho D^2 - E]^{-1/2} . \tag{65}$$

This distribution function represents the required self-gravitating slab.

6.1.2. *The inverse problem.*—The isothermal sphere has f = const. $\exp(-\beta E)$, where $E = \frac{1}{2}c^2 + \Phi(r)$. $\Phi(r)$ and $\rho(r)$ follow from Poisson's equation

$$\rho(r) = (4\pi G)^{-1}\nabla^2\Phi(r) = \rho(0)e^{-\beta\Phi(r)}, \tag{66}$$

where we take $\Phi(0) = 0$. This equation is integrated numerically, and its solution is given by Emden (1907).

6.2. Models of Globular Clusters and Elliptical Galaxies

The extensive literature on this subject has been reviewed by Michie (1964). We will concentrate here on some models devised by King (1966); these seem to be important for understanding the dwarf and the giant ellipticals, although the models were originally constructed to represent the globular clusters.

In globular clusters, stellar encounters are not negligible. The relaxation time (10^8–10^9 years) is shorter than the lifetime (10^{10} years); however, it is long compared with the orbital time (10^6 years). As a result, the cluster stays well mixed as its distribution tends toward a Maxwellian distribution. However, the Maxwellian distribution is unacceptable because (i) it corresponds to the isothermal sphere and so has infinite radius and mass, and (ii) it has high-velocity stars which are not bound to the cluster. What is needed is a distribution function that is produced by stellar encounters but drops to zero at a finite limiting velocity.

In the presence of encounters, the Boltzmann equation becomes

$$Df/Dt = (\partial f/\partial t)_{\text{enc}}, \tag{67}$$

where the right-hand side represents the rate of change of f due to encounters. In the Fokker-Planck approximation, the right-hand side is replaced by terms which represent a diffusion-like process of f in velocity space (see, for example, Michie 1963). For the lower-energy stars, which are confined to the inner regions, f is approximately Maxwellian; for the higher-energy stars, equation (67) can be written

$$\frac{Df}{Dt} = \frac{1}{c}\frac{\partial}{\partial E}\left[K\left(\frac{\partial f}{\partial E} + \beta f\right)\right], \tag{68}$$

where E is the energy of the star (Lynden-Bell 1967*a*). Michie's (1963) solution to this equation is

$$\partial f/\partial E + \beta f = \text{constant}, \tag{69}$$

which leads to a steady state for the higher-energy stars. It follows that

$$f(E) = A\exp(-\beta E) - A\exp(-\beta E_{\text{esc}}), \tag{70}$$

where A, β, and E_{esc} are parameters: $f = 0$ for $E > E_{\text{esc}}$.

King has used this distribution function to construct a one-parameter sequence of stationary self-consistent models for spherical globular clusters. The one parameter is βE_{esc}. All stars have the same mass in these models. For a central density ρ_0, there is a natural length scale

$$r_c = [9/(4\pi G\rho_0\beta)]^{1/2}. \tag{71}$$

E_{esc} comes in through the *tidal* radius r_t of the cluster in the gravitational field of the Galaxy. The cluster density ρ is zero at the tidal radius. The models are projected on the sky, and are presented as log (Σ/Σ_0) against log (r/r_c), where Σ is the surface *density*. The free parameter corresponding to $\beta E_{\rm esc}$ is log (r_t/r_c): log $(r_t/r_c) = \infty$ is the isothermal sphere.

These curves are an excellent fit to the radial star-count distribution for globular clusters and, if the surface *brightness* measures are scaled to compensate for the ellipticity of the isophotes, they are also an excellent fit to the surface brightness distributions for the elliptical galaxies. Figure 22 shows some of these curves. De Vaucouleurs's $R^{1/4}$ law is similar to the log $(r_t/r_c) = 2.20$ curve (which fits the E galaxy NGC 3379 very closely: see King 1966) but falls off less rapidly in the outer parts.

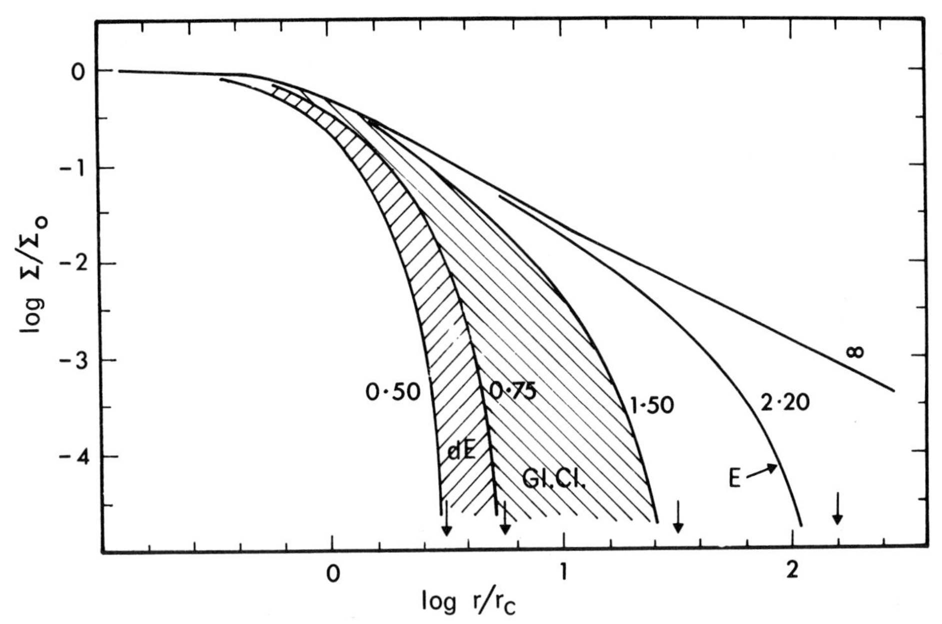

FIG. 22.—Some King curves. The numbers show the value of log (r_t/r_c) for each curve. The arrows indicate the tidal radii r_t. Regions of log (r_t/r_c) corresponding to the dE galaxies and the globular clusters are hatched.

Because the normal ellipticals follow the $R^{1/4}$ law fairly closely, they have log $(r_t/r_c) \approx 2.2$. Star-count distributions for the dwarf ellipticals in the Local Group (Hodge and Michie 1969) are well represented by King's models with $0.5 \lesssim$ log $(r_t/r_c) \lesssim 0.75$ [except for NGC 185 with log $(r_t/r_c) \approx 1.5$]. Most globular clusters fall in the range $0.75 \lesssim$ log $(r_t/r_c) \lesssim 1.50$ (King 1966).

The important feature of these models is the form of the deviation of $f(E)$ from the Maxwellian distribution: this deviation follows from the Fokker-Planck equation. There are of course many other ways to construct a distribution that is approximately Maxwellian for low E but has no stars above a given E. For example, Woolley and Dickens (1961) take

$$f = Ae^{-\beta E} \quad (E < E_{\rm esc})$$
$$= 0 \quad (E > E_{\rm esc}) ; \tag{72}$$

however, King shows that this and some other possible truncations lead to models which fit the data significantly less well than do his models. Another possibility is that the distribution function is anisotropic. For example, Hodge and Michie (1969) compare star counts in some dwarf elliptical systems with models that have predominantly radial orbits in their outer parts; this anisotropy causes the density to fall more rapidly with r than for the isotropic models. Although Hodge and Michie claim that their models show dwarf ellipticals to have anisotropic distribution functions, the star-count data can in fact be fitted well by King's isotropic models with a slightly smaller tidal radius r_t than they used. King (1966) also points out that for the globular clusters it is not at present possible to discriminate clearly between tidal dropoff and anisotropy dropoff of the star counts in the outer parts.

The main conclusion of this section is that King's models fit the dwarf and normal ellipticals well, so these systems probably have the Maxwellian-like distribution function of equation (70). Note, however, that in King's models all stars have the same mass. Lynden-Bell (1967*a*) points out that an approximately Maxwellian distribution (equipartition of *energy*) implies mass segregation in an E galaxy, with the lighter stars at the outside, and this would lead to a greater variation of color with radius than is observed (see § 4.3). It follows that we need a distribution function like that of equation (70) but with the same spatial distribution for stars of all masses. It is the distribution function of equation (70) but with E = energy per unit mass. This represents equipartition of *energy per unit mass*, or temperature proportional to the mass.

Equipartition of energy per unit mass would not be achieved through stellar encounters, even if the relaxation time through encounters for elliptical galaxies were short enough (it is not: see table 1). We will discuss the relaxation of elliptical galaxies in § 8.

7. THE COLLAPSE OF THE GALAXY

In this section we discuss a classic paper by Eggen, Lynden-Bell and Sandage (ELS) (1962): their main result is that the Galaxy collapsed rapidly from its protogalaxy on a time scale of the same order as the free-fall time ($\sim 5 \times 10^8$ years).

Consider first a plane eccentric orbit in the Galaxy. The radial (R) and azimuthal (φ) periods are generally incommensurable, so the orbit is not generally closed, and an eccentricity cannot be defined as usual. However, e', defined by

$$1 - (e')^2 = [1 + \oint p_R dR / \oint p_\varphi d\varphi]^{-2} , \tag{73}$$

where the integrals are action variables defined in § 2.6, is a measure of the eccentricity of the orbit, and is an adiabatic invariant: e' remains approximately constant if the potential field of the Galaxy changes slowly.

ELS discuss the space motions and ultraviolet excesses $\delta(U - B)$ for 221 well-observed nearby dwarf stars. For their argument, they assume that $\delta\,(U - B)$ is a measure of the age of a star. They adopt a simple Galactic potential

$$\Phi = -G\mathfrak{M}/[b + (R^2 + b^2)^{1/2}] , \tag{74}$$

where $\mathfrak{M}$ is the galactic mass and b a length scale. They compute the apo- and perigalactic distances R_1 and R_2 for each star, and define an eccentricity

$$e = (R_1 - R_2)/(R_1 + R_2) , \tag{75}$$

which is fairly close to e' given in equation (73). It turns out that the orbital eccentricity e, the W-component of the space motion, and the angular momentum per unit mass h are correlated with $\delta\ (U - B)$ (i.e., with age), in the sense shown in figure 23. The younger stars [$\delta\ (U - B) < 0.15$] have $e \lesssim 0.5$, $|W| \lesssim 50$ km s^{-1}, and $|h| \gtrsim 1500$ kpc km s^{-1}, while the older stars have predominantly large eccentricities and low angular momenta, and *any* value of $|W| < 400$ km s^{-1}.

Figure 23*b* suggests that the younger objects were formed near the galactic plane while the older ones were formed at almost any height above the plane. This suggests in turn a collapse of the Galaxy to a disk, during or after the formation of the oldest stars.

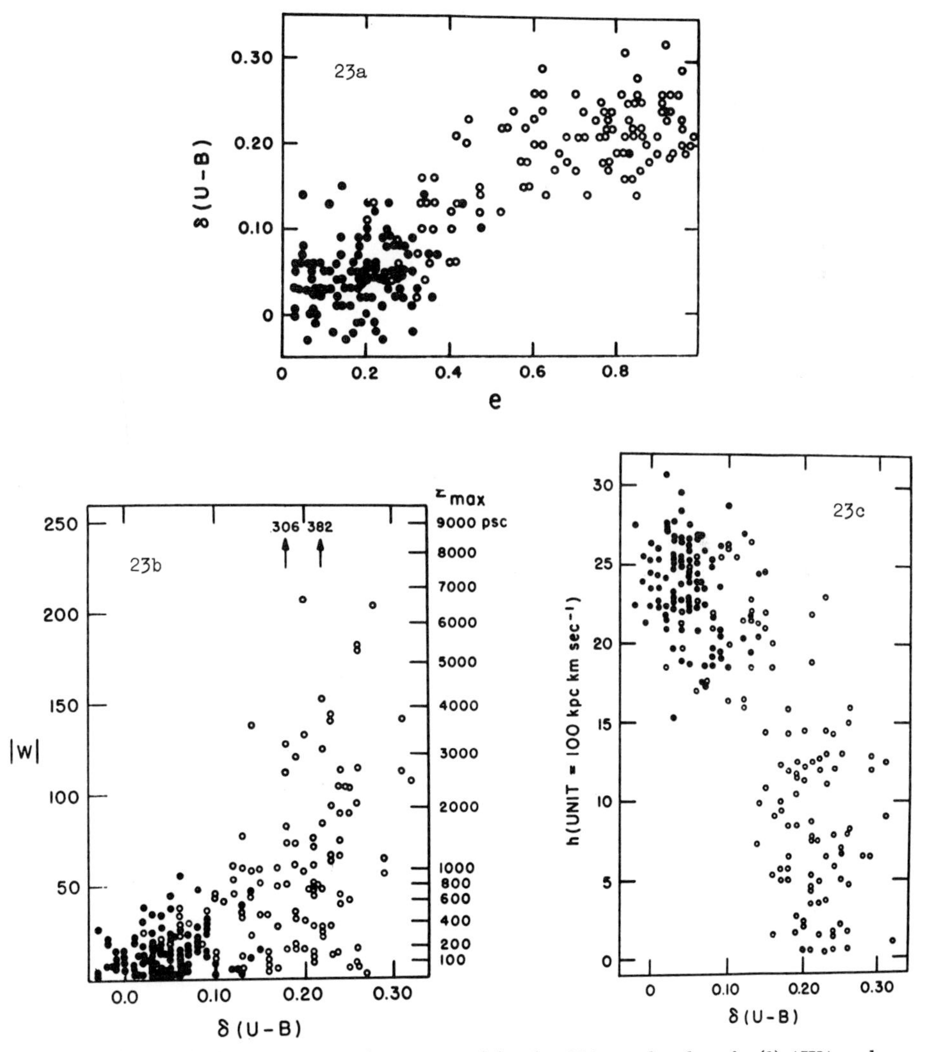

FIG. 23.—(*a*) Ultraviolet excess against eccentricity for 221 nearby dwarfs. (*b*) $|W|$ and $z_{\max}$ against ultraviolet excess. (*c*) Angular momentum against ultraviolet excess (from Eggen *et al.* 1962).

The time scale of this collapse also follows from figure 23. Assume the Galaxy was always axisymmetric, so h is an integral even if Φ changes. If Φ changes slowly, then e remains approximately constant; if Φ changes rapidly compared with the orbital period, then e also changes; for a rapidly *contracting* galaxy, e will mostly increase. Figure 23a shows that the oldest stars have almost rectilinear orbits. It seems unlikely that they formed from matter moving in centrifugal equilibrium, because figure 23c shows that these stars have low angular momenta; their *circular* orbits would have $R \lesssim 5$ kpc. To place them in their present elliptical orbits would require a large amount of added kinetic energy, for which there is no obvious source (see also chap. 10, § 5.3). It is most likely that the Galaxy was in a state of contraction when these stars formed: if this contraction was slow compared with the rotation period (which is of the same order as the orbital period), then the eccentricities of the oldest stars would remain small, because they are adiabatic invariants. It follows that the collapse was rapid.

There are other possibilities for the origin of the high-eccentricity orbits of the oldest stars, but they appear less likely. We discuss them in § 11.2.3.

The scale of the collapse, in z, can be estimated from figure 23b. It is at least a factor 25, from $|z|_{max} \approx 10$ kpc for the oldest stars to $|z|_{max} \approx 400$ pc for the younger ones. ELS attempt to estimate also the scale of the collapse in R. They suggest that after the oldest stars formed, the matter which remained in gaseous form fell in toward the galactic center, conserving its angular momentum and losing its kinetic energy through gas-cloud–gas-cloud encounters, and finally settled into circular orbits corresponding to its original value of h. The mean of h, for the stars with great apogalactic distances R_1, corresponds to a circular orbit with radius R_0, which then leads to the radial collapse scale R_1/R_0. However, as a result of selection effects, the mean value of h for known stars with large apogalactic distances (>50 kpc) is negative (Dixon and Freeman, unpublished), so the radial collapse scale is indeterminate. (Using the mean value of $|h|$, ELS derive a collapse factor ~ 10.) Note that known stars with $\delta(U - B) \geq 0.15$ are almost equally divided between those with direct and those with retrograde orbits; i.e., their mean angular momentum is near zero. The mean angular momentum for *all* stars with $\delta(U - B) \geq 0.15$ in the solar neighborhood is unknown: catalogs of these stars are biased toward low-h objects, because these are more easily found from proper motion surveys.

ELS point out that in figure 23c there are no stars with large $\delta(U - B)$ and large $h \approx h_{\odot} = 2500$ kpc km s^{-1}. These stars should exist in their picture because gas and young stars in circular orbits do exist at the Sun's distance from the galactic center. At its precollapse distance, this matter could have formed metal-weak stars with high angular momentum and with highly eccentric orbits. These stars would be difficult to find because (i) they are near perigalacticum, and are moving through the solar neighborhood at near their maximum velocity, (ii) their large perigalactic distances mean that they were formed in regions of low density, so there are probably few of them, and (iii) their high angular momenta mean that they do not have a large velocity relative to the Sun, and this weighs against their discovery in proper-motion surveys. However, Sandage (1969) finds seven stars with $h \gtrsim h_{\odot}$ and $\delta(U - B) \gtrsim 0.16$ in a sample of about 100 new subdwarf candidates; the Bottlinger diagram for this sample, shown in figure 24, illustrates also the main points of figures 23*a* and 23*c*. The existence of these old stars with high h is important for the discussion in § 11.

To summarize this section: ELS show that stars in the solar neighborhood with large values of e and $|W|$ and low values of h are old, although old stars with low e and $|W|$ and high h do exist. They infer that the Galaxy collapsed rapidly to a disk, by a factor of at least 25 in $\dot{z}$. This important conclusion has had a significant influence on the development of galactic dynamics: see also §§ 8 and 11.

8. COLLECTIVE EFFECTS AND COLLISIONLESS RELAXATION IN STELLAR SYSTEMS

We recall the two basic questions of stellar dynamics: What is the present dynamical state of galaxies? And how was this state attained? This section is relevant to the second

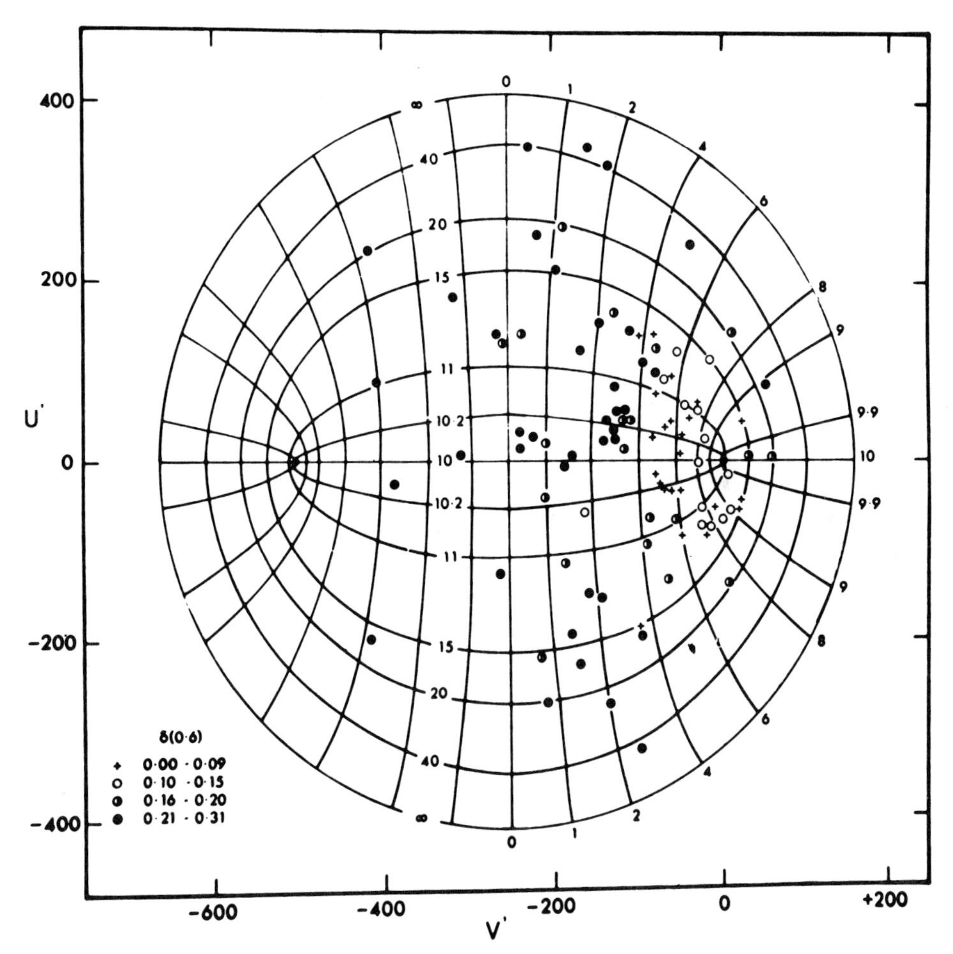

FIG. 24.—Bottlinger diagram for a sample of new subdwarfs (from Sandage 1969). Lines of constant apogalactica and perigalactica are marked in kiloparsecs. Velocities are relative to the local standard of rest.

question. The similarity of the luminosity distribution in elliptical galaxies suggests the existence of some dynamical process that drives these systems toward some sort of natural equilibrium. We know that binary star-star encounters are inadequate to do this. There is, however, a class of interactions which come about through the forces exerted on an individual star by bulk irregularities in the system. A well-known example of these *collective* interactions is the two-stream instability in a collisionless plasma; the counterstreaming beams of electrons are unstable for rapid enough counterstreaming, even though the electrons do not interact through encounters. Collective effects are probably important in galaxies during the collapse phase, and provide the interaction required for the relaxation of the elliptical galaxies. To demonstrate some of these effects, we first consider the behavior of small disturbances in a uniform-density medium of stars and gas. Much of this discussion follows Lynden-Bell (1967*b*) (LB), where more details can be found. Many of these results are due to him.

8.1. The Stability of Homogeneous Media

8.1.1. *The infinite gravitating fluid.*—We perturb an infinite inviscid fluid with uniform density ρ_0 and initially at rest. Although this is not an equilibrium solution (Poisson's equation is not satisfied), it does give useful results to compare with corresponding results for stellar systems. The behavior of the fluid is determined by the momentum equation,

$$\frac{\partial \boldsymbol{u}}{\partial t} + (\boldsymbol{u}\cdot\nabla)\boldsymbol{u} = -\frac{1}{\rho}\nabla p - \nabla\Phi \tag{76}$$

the continuity equation,

$$\partial\rho/\partial t + \nabla\cdot(\rho\boldsymbol{u}) = 0\ ; \tag{77}$$

and Poisson's equation,

$$\nabla^2\Phi = 4\pi G\rho\ , \tag{78}$$

where $\boldsymbol{u}$, ρ, p, and Φ are the velocity, density, and pressure of the fluid, and the gravitational potential, respectively. Now write $\Phi = \Phi_0 + \Phi_1$, etc., where Φ_0 is the unperturbed value of Φ and Φ_1 is a small perturbation, let $\eta = \rho_1/\rho_0$ and neglect second-order terms. Then equations (76)–(78) become

$$\frac{\partial \boldsymbol{u}_1}{\partial t} = -c_s^2\nabla\eta - \nabla\Phi_1\ , \tag{79}$$

$$\frac{\partial \eta}{\partial t} + \nabla\cdot\boldsymbol{u}_1 = 0\ , \tag{80}$$

$$\nabla^2\Phi_1 = 4\pi G\rho_0\eta\ , \tag{81}$$

where $c_s^2 = dp/d\rho$ is the sound speed. Now look for plane-wave solutions $\eta \sim \exp(i\omega t - ikx)$, etc. Nontrivial solutions are possible only if

$$\omega^2 = k^2c_s^2 - 4\pi G\rho_0\ . \tag{82}$$

The fluid is stable ($\omega^2 > 0$) to disturbances with $k^2 > 4\pi G\rho_0/c_s^2 = k_J^2$; this just means that for $k > k_J$ the self-gravity of the disturbance is insufficient to overcome the stabilizing pressure, so the disturbance cannot grow. For $k^2 < k_J^2$, the fluid is unstable ($\omega^2 < 0$) and the disturbances grow exponentially. The quantity k_J corresponds to the

Jeans length, $L_J = c_s/\sqrt{(4\pi G\rho_0)}$. Note that the stable waves are undamped acoustic waves, modified by self-gravity; there are no damped or growing oscillatory modes.

8.1.2. *Collisionless stellar systems.*—Perturbations of a homogeneous collisionless stellar system follow the collisionless Boltzmann equation and Poisson's equation. Derivation of the stability criterion is more complicated than for the gravitating fluid: see LB. A Maxwellian distribution of stars

$$f(c) = \rho_0(2\pi\sigma^2)^{-1/2} \exp(-c^2/2\sigma^2), \tag{83}$$

where ρ_0 is the unperturbed density and σ^2 the velocity dispersion, is unstable to waves of wavenumber k for

$$k^2 < 4\pi G\rho_0/\sigma^2 \tag{84}$$

just as for the gravitating fluid. For the superposition of two Maxwellians, the condition becomes

$$k^2 < 4\pi G\left(\frac{\rho_1}{\sigma_1^2} + \frac{\rho_2}{\sigma_2^2}\right), \tag{85}$$

so a low-density component with small velocity dispersion can be important for determining the stability of the medium.

8.1.3. *Landau damping.*—In the gravitating fluid, the stable waves with $k > k_J$ are undamped acoustic waves (§ 8.1.1). This is not the case for the collisionless stellar system; here the stable waves damp by giving their energy to individual stars. The process is called Landau damping. A stable wave with frequency ω and wavenumber k propagates through the system with phase velocity ω/k; it loses energy to stars moving slightly slower than this velocity and gains energy from the slightly faster ones. Because the distribution functions for stellar systems are usually monotone decreasing functions of velocity, there are mostly more stars with $c < \omega/k$ than with $c > \omega/k$, so the wave loses its collective energy to the individual stars, and is therefore damped. The rate of energy loss from the wave is proportional to $(\partial f/\partial c)_{c=\omega/k}$, where f is the distribution function for the stars; see Stix (1962) for a simple discussion of Landau damping. However, there do exist situations where a wave *gains* energy at the expense of the individual stars. For example, the counterstreaming of two otherwise stable distributions of stars can be unstable through this inverse Landau damping: see LB.

Note that (i) the interaction between individual stars and the wave is really collective: the wave corresponds to density fluctuations which gain or lose energy at the expense of the rest of the stars. (ii) When fluctuations in collisionless systems are considered, these velocity space processes are always present in addition to the usual configuration-space processes mentioned already.

8.1.4. *Gas and star systems.*—For a system containing gas and stars, the behavior of perturbations is determined by the collisionless Boltzmann equation for the stars, the momentum and continuity equations and an equation of state for the gas, and Poisson's equation for the total density and gravitational potential.

If the mean velocity of the stars relative to the gas is greater than the sound speed for the gas, then the system is unstable to disturbances of all wavelengths (see LB). This comes about through an inverse Landau damping process: acoustic waves in the gas gain energy from the individual stars.

8.2. Phase Damping

The collective processes described in § 8.1 depend on the gravitational attraction of the disturbance: e.g., Landau damping works through the gravitational interaction of the wave with the individual stars. There is another important process which damps disturbances in a stellar system but does not depend on the self-gravity of the disturbance itself.

Consider first small perturbations of an equilibrium stellar system: f_0, Φ_0 are the equilibrium distribution function and gravitational potential, and $f_0 + f_1$, $\Phi_0 + \Phi_1$ are the perturbed values. To the first order in f_1, Φ_1, the Boltzmann equation and Poisson's equation become

$$\frac{\partial f_1}{\partial t} + \boldsymbol{c}\cdot\frac{\partial f_1}{\partial \boldsymbol{r}} - \frac{\partial \Phi_0}{\partial \boldsymbol{r}}\cdot\frac{\partial f_1}{\partial \boldsymbol{c}} - \frac{\partial \Phi_1}{\partial \boldsymbol{r}}\cdot\frac{\partial f_0}{\partial \boldsymbol{c}} = 0\,, \tag{86}$$

$$\nabla^2\Phi_1 = 4\pi G \int f_1 d^3c\,. \tag{87}$$

Let λ, Λ be the length scales of the perturbation and of the equilibrium system, and c, C be the velocity scales. From Poisson's equation, $\Phi_1 \sim f_1 c^3 \lambda^2$ and $\Phi_0 \sim f_0 C^3 \Lambda^2$, so the ratio of the third to the fourth term of equation (86) is $(C/c)^4 \Lambda/\lambda$. Then if λ/Λ is small enough, and $C \sim c$, the fourth (self-gravity) term of equation (86) can be neglected. (The length scale λ must of course be much less than the Jeans length.) This means that the perturbation propagates along the unperturbed orbits and is spread out because the periods of the unperturbed orbits are in general not the same.

A simple example illustrates this process, and shows why it is called phase mixing (or phase damping). For a one-dimensional system, equation (86) becomes

$$\frac{\partial f_1}{\partial t} + c\frac{\partial f_1}{\partial x} - \frac{\partial \Phi_0}{\partial x}\frac{\partial f_1}{\partial c} = 0 \tag{88}$$

if we neglect the self-gravity of the disturbance. From Jeans's theorem (§ 2.4), f_1 is a function of the integrals of the motion. Let $\Phi_0(x)$ be a potential well; then the stellar orbits are periodic and there are two integrals. One is the energy, $E = \frac{1}{2}c^2 + \Phi_0$. Let $\tau(E)$ be the period of an orbit, and let x_0 be the initial value of x for the orbit. Then the *phase* of the orbit is

$$\varphi(E, x) = \varphi(E, x_0) + \frac{2\pi t}{\tau}\,, \tag{89}$$

so $\varphi(E, x) - 2\pi t/\tau$ is also an integral of the motion. Then the distribution function is $f_1 = f_1(E, \varphi - 2\pi t/\tau)$.

Figure 25 shows the evolution of a distribution function f_1 in a potential well. The closed curves represent the periodic stellar orbits, and are curves of constant energy E. Initially the stars are confined to an interval ΔE in energy and $\Delta\varphi$ in phase. Because the orbit periods $\tau(E)$ depend on the energy, the stars are spread over all phases after a long enough time, although they remain within the energy interval ΔE (hence the name phase mixing); i.e., f_1 appears to lose its dependence on φ but maintains its dependence on the isolating integral E. Note that the area of (x, c)-space covered by the stars of f_1 remains constant when looked at with high enough resolution, even though the area appears to increase when looked at with low resolution.

Lynden-Bell (1962) discusses the approach to equilibrium of a disturbance as it phase-mixes. He shows that $f_1(E, \varphi, t)$ converges to the average $\bar{f}_1$ of $f_1(E, \varphi, 0)$ taken over all phases, in the following sense. Let $Q(x, c, t)$ be some smoothly varying function of (x, c). Its mean value at any time is

$$\bar{Q}(t) = \int f_1 Q dx dc \; ; \tag{90}$$

f_1 converges to $\bar{f}_1$ as $t \to \infty$ in the sense that

$$\int f_1(E, \varphi, t) Q dx dc \to \int \bar{f}_1(E) Q dx dc \; , \tag{91}$$

although f_1 does not converge pointwise to $\bar{f}_1$. That is, phase mixing makes f_1 converge *in the mean* to a function of the isolating integrals alone.

Our main object in § 8 is to describe processes which act to smooth out irregularities in collisionless stellar systems. So far we have two distinct processes: (i) phase mixing,

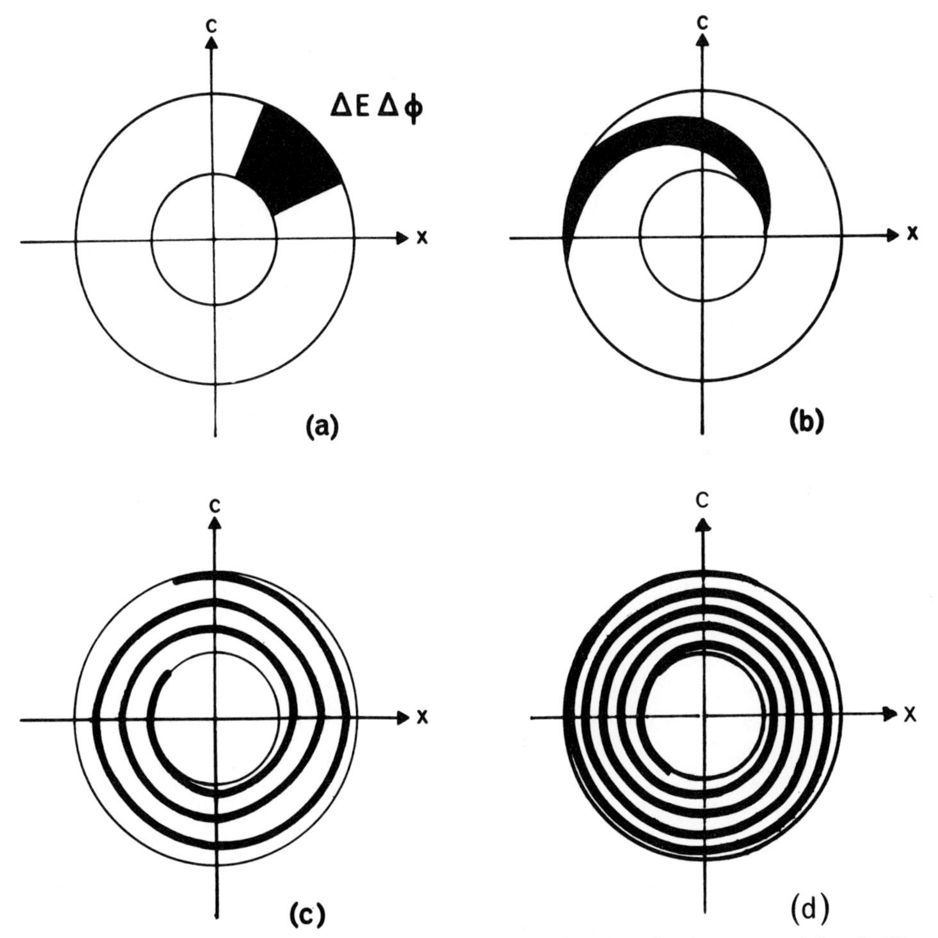

FIG. 25.—Phase damping: the evolution of a distribution function in a potential well. Time increases from (*a*) to (*d*).

which does not depend on self-gravity and which does not change the dependence of f on the isolating integrals of the motion; and (ii) Landau damping, which does both. In a real self-gravitating system, both processes will act together; the density (and gravity) irregularities set up by the phase mixing (see fig. 25*c*) will be damped by the Landau damping. The characteristic time for both these processes to act is of the same order as the orbital time, $\sim(G\rho_0)^{-1/2}$, where ρ_0 is the density of the medium.

Now we describe a third important process called violent relaxation or mean field relaxation. Acting together with phase mixing and Landau damping, it probably determines the present relaxed distribution function for the elliptical galaxies.

8.3. Violent Relaxation

This process is described by Lynden-Bell (1967*a*). It is the relaxation that occurs when the mean large-scale gravitational field changes rapidly with time, as in the collapse phase of our Galaxy (§ 7), so the energies of stars are not constant along their orbits. Let $\Phi(r, t)$ be the time-dependent gravitational potential and $E = \frac{1}{2}c^2 + \Phi$ the energy per unit mass of a star. Then along a star's orbit

$$DE/Dt = \partial\Phi/\partial t\,. \tag{92}$$

Now consider a galaxy so far from equilibrium that the changes in the potential Φ are of the same order as Φ itself. Then for an individual star, E changes on a time scale which is typically the same as the vibration period for the whole galaxy, and this is again of order $(G\rho)^{-1/2}$.

Note (i) how violent this relaxation is, and (ii) that the mass of the individual star does not appear in equation (92), so we would not expect this relaxation process to lead to any segregation of stars by mass.

The birth oscillations of a galaxy which forms far from equilibrium will be severely damped by phase mixing and Landau damping. Lynden-Bell suggests that, although the galaxy would oscillate for only a few periods, this might be long enough for significant relaxation; i.e., in the notation of the previous section, $f(E, \varphi)$ would have its E-dependence entirely changed, as well as its φ-dependence mixed out.

Statistical-mechanics methods can be used to predict the most probable state resulting from the violent relaxation. The basic assumption is that there are so many equivalent elements of phase and such violent variations in Φ that a typical element of phase is equally likely to be found anywhere in phase space, subject to the following restrictions: (*a*) the total energy of the system is conserved; (*b*) the total number of phase elements with a given phase density remains constant; so (*c*) no two elements of phase can overlap in phase space, because otherwise the phase density would be different in the region of overlap.

These restrictions have an overtone of a statistical mechanics with exclusion and distinguishable particles. The usual procedures of statistical mechanics lead to the most probable state

$$\bar{f}(E) = \eta\,\frac{\exp[-\beta(E-\mu)]}{1+\exp[-\beta(E-\mu)]}\,, \tag{93}$$

where η is the initial phase density (assumed constant), and β and μ are constants. Note that (i) equation (93) gives the distribution toward which mean field relaxation leads, rather than that necessarily attained, because the time variations in Φ may die out (i.e., the system finds a stable steady state) before the relaxation process is complete. (ii) In § 8.2 we pointed out that a distribution function $f(E, \varphi, t)$ *converges in the mean* to $\bar{f}(E)$, which is the average of $f(E, \varphi, 0)$ over all phases. If we look at figure 25 with high enough resolution, it is clear that $f(E, \varphi, t)$ itself never reaches an equilibrium; however at lower resolution, $f(E, \varphi, t)$ does appear to converge to $\bar{f}(E)$. The function $\bar{f}(E)$ is called the *coarse-grained* distribution, and equation (93) gives the coarse-grained distribution which is the most probable result of the violent relaxation.

For galactic conditions, there is a simple approximation to equation (93). Assume that the stars formed at t = 0 through the Jeans instability (see § 8.1.1) from the protogalactic gas, and that this medium of stars has phase density η immediately after formation. Then we can easily show that after relaxation $\bar{f} \ll \eta$, so galaxies are in the nondegenerate limit of equation (93) and $\bar{f}$ can be written

$$\bar{f}(E) = A \exp(-\beta E) , \tag{94}$$

where $A = \eta \exp(\beta\mu)$. In other words, the violent relaxation process leads to the isothermal sphere, but with E the energy per unit mass (cf. § 6).

The isothermal sphere is an unacceptable model for a relaxed galaxy because it has infinite mass and radius. It arises from the statistical theory because this theory does not include the finite duration of the galaxy's unsteadiness: relaxing conditions persist only while the galaxy is dynamically unsteady, so high-energy orbits which lie partly outside the relaxing region and have periods longer than the time for which the galaxy is unsteady will not acquire their full quota of stars in a real system. The statistical theory also ignores the tidal influence of other galaxies, which will further depopulate the high-energy orbits. To handle these higher-energy orbits more realistically, Lynden-Bell argues that the high-energy end of the distribution function $\bar{f}$ diffuses in velocity space according to a Fokker-Planck equation in the form

$$\frac{D\bar{f}}{Dt} = \frac{1}{c}\frac{\partial}{\partial E}\left[K\left(\frac{\partial \bar{f}}{\partial E} + \beta\bar{f}\right)\right] , \tag{95}$$

where K is a constant, while at low energies $\bar{f}$ is approximately Maxwellian. A stationary solution of equation (95) is (cf. § 6)

$$\partial\bar{f}/\partial E + \beta\bar{f} = \text{const.}, \quad \text{or} \tag{96}$$

$$\bar{f} = A[\exp(-\beta E) - \exp(-\beta E_{esc})] , \tag{97}$$

where E_{esc} is identified as the escape energy in a tidal field. We expect E_{esc} to be near zero, so equation (97) only modifies the Maxwellian significantly close to the escape energy: however, this modification does result in a model galaxy with finite mass. We have already seen (§ 6.2) how well the model galaxies constructed from equation (97) reproduce the observed surface brightness distribution for the normal elliptical galaxies and the star count distributions for the dwarf ellipticals.

Mean field relaxation cannot so far explain the distribution function for the disk galaxies: these have high angular momentum, and the outer parts are held far from the central regions so have little chance of sharing in any overall relaxation. Formal application of the statistical mechanics gives no difference between the two velocity-dispersion components σ_{RR} and σ_{zz}, so that if relaxation has occurred, it must have been driven by anisotropic processes. See § 11 for more discussion.

We now have the main result of § 8: mean field relaxation, acting together with Landau damping and phase mixing, explains how ellipticals can look approximately Maxwellian despite their very long relaxation times.

8.4. Time-dependent Stellar Systems

Time-dependent self-consistent model systems are useful for investigating the approach to equilibrium of stellar systems which are initially far from equilibrium. It is obviously important that we understand fully the processes that act during this approach to equilibrium, because they probably define many of the large-scale properties of galaxies. Here we summarize briefly some numerical experiments on N-body collisionless systems. In these systems the stars interact only through the smoothed-out time-dependent potential of the system: this potential is calculated from Poisson's equation. Because they are collisionless, systems with large N ($\sim 10^5$) can be managed.

The simplest one-dimensional model has infinite plane sheets for stars. These sheets move along their normal. Note that the gravitational field of an infinite plane sheet is independent of the distance from it, so the interparticle force in this model is of rather longer range than the usual r^{-2} field. As a result, the predictions of this model may differ from those of models with the r^{-2} field. For example, Hohl and Feix (1966) point out that their model with 2000 plane sheets phase damps more slowly than the system of concentric spherical mass shells discussed by Hénon (1966).

Several experiments have been made with these N-body systems to test the statistical mechanics of violent relaxation. They start with systems initially out of equilibrium and compare the coarse-grained distribution function after many free fall times with the predicted $\bar{f}$ of equation (93). The usual result (e.g., Hénon 1968; Hohl and Campbell 1968) is that about 75 to 95 percent of the stars follow equation (93): the percentage depends on the initial conditions. The remainder are usually high energy stars and do not follow equation (93) presumably because these stars have periods long compared with the time in which the system is unsteady.

Miller and Prendergast (1968) have devised a potentially powerful approach to numerical experiments with collisionless systems, in which they allow the particles to move only in discrete steps in phase space. This approach preserves the essential physical features of collisionless systems, and allows $\sim 10^5$ particles to be followed: see Miller, Prendergast and Quirk (1970) for some examples in four-dimensional phase space.

Hohl (1971) has simulated the approach to equilibrium of a disk of stars by numerical experiments on 10^5 particles in four-dimensional phase space: see §§ 9.3 and 9.1.4.

9. THE DISKS OF SPIRAL AND LENTICULAR GALAXIES

We recall from § 4 that the lenticular and spiral galaxies are intrinsically flat systems. Their luminosity profiles show two main components: an inner (bulge) component, which may be prominent or weak, and an outer exponential disk. This disk provides a large part of the total light and most of the total angular momentum. We also recall that these galaxies are rotating rapidly: in particular the asymmetric drift for most stars in the galactic disk is much smaller than the circular velocity (see fig. 5), so the disk is approximately in centrifugal equilibrium. This contrasts with the situation in ellipticals and presumably also in the spheroidal components of the disk galaxies, where the velocity dispersion contributes most of the support against the system's own gravity.

Our understanding of the disklike galaxies is being held back by several fundamental dynamical problems associated with the disks. Almost all disks have the exponential luminosity profile, which means that they are in a similar dynamical state; the reason for this is not yet known. Violent relaxation, which accounts successfully for the dynamical state of the elliptical galaxies, does not seem to work for the disklike systems. The stability of the disks also provides some important problems. These include the origin and maintenance of the spiral structure, and the reason for the bending of the galactic plane as observed in the Galaxy and many external systems. In this section we discuss these problems. First we describe some general results concerning the stability of galactic disks: these will be useful later.

9.1. The Stability of the Disk

9.1.1. *The stabilizing effect of rotation.*—From § 8.1.1 we recall that an infinite homogeneous fluid is unstable to perturbations with wavenumbers $k^2 < 4\pi G\rho/c^2$, where ρ is the density of the fluid and c the sound speed; for these modes the stabilizing pressure cannot support the self-gravity of the disturbance. Now let the fluid rotate uniformly with angular velocity $\mathbf{\Omega}$. Perturbations with wave vectors perpendicular to $\mathbf{\Omega}$ are unstable if

$$k^2c^2 < 4\pi G\rho - 4\Omega^2 \tag{98}$$

(Chandrasekhar 1961). For $\pi G\rho > \Omega^2$, the rotation stabilizes some of the long-wavelength modes which would be unstable in the absence of rotation; however, the longest-wavelength modes remain Jeans-unstable.

Now consider the stability of a uniformly rotating disk of *finite thickness*. Goldreich and Lynden-Bell (1965*a*) analyze this problem in detail and give a simple physical explanation for their results. Let the disk have thickness T. A region of this disk with horizontal scale k^{-1} is stable if its self-gravitational energy is less than the sum of its internal plus rotational energies. The thermal energy per unit mass $\sim c^2$, the rotational energy $\sim\Omega^2k^{-2}$, and from Poisson's equation the gravitational energy $\sim 4\pi G\rho/(k^2 + T^{-2})$, so the perturbation is stable if

$$4\pi G\rho/(k^2 + T^{-2}) < \Omega^2/k^2 + c^2 \,. \tag{99}$$

Figure 26 shows the two sides of this inequality. If Ω is not too large, the two curves intersect twice and there are *two* critical wavenumbers k_1, k_2 such that modes with $k_1 < k < k_2$ are unstable. Short-wavelength modes ($k > k_2$) are pressure-stabilized, as in the infinite medium. The effect of the finite thickness is to limit the growth of the gravitational energy for small k. The long-wavelength modes ($k < k_1$) are then stabilized by the rotation, which was not so for the infinite medium.

In the limit as $T \to \infty$ we recover from equation (99) the condition of equation (98) for the infinite medium, except for numerical coefficients.

It is also possible to introduce an anisotropic pressure into the analysis. The result is that the disk is most unstable to modes propagating in the plane of the disk along the direction corresponding to the smallest pressure component. For the Galaxy, this suggests that tangential modes (i.e., modes propagating in the φ-direction) should be the most unstable. Some numbers from Goldreich and Lynden-Bell (1965*a*) will illustrate this. First we must rearrange equation (99) to be in dimensionless form. The thickness of the disk depends on ρ and c, so we can write $T^2 = Qc^2/(4\pi G\rho)$, where Q is some constant defined by the equation of state for the disk matter. Equation (99) then becomes

$$(1 + k^2T^2)^{-1} < (\Omega^2/4\pi G\rho)/(k^2T^2) + Q^{-1} \tag{100}$$

for stability; i.e., there is a critical value of $4\pi G\rho/\Omega^2$ for which the two curves in figure 26 just touch, with a corresponding critical value for kT. Larger values of $4\pi G\rho/\Omega^2$ allow the existence of unstable modes as shown in figure 26. For the isothermal disk, Goldreich and Lynden-Bell's detailed analysis gives $(4\pi G\rho/\Omega^2)_{\text{crit}} = 11.68$ with $(kT)_{\text{crit}} = 1.4$. For the Galactic disk, they use the anisotropic velocity dispersion observed near the

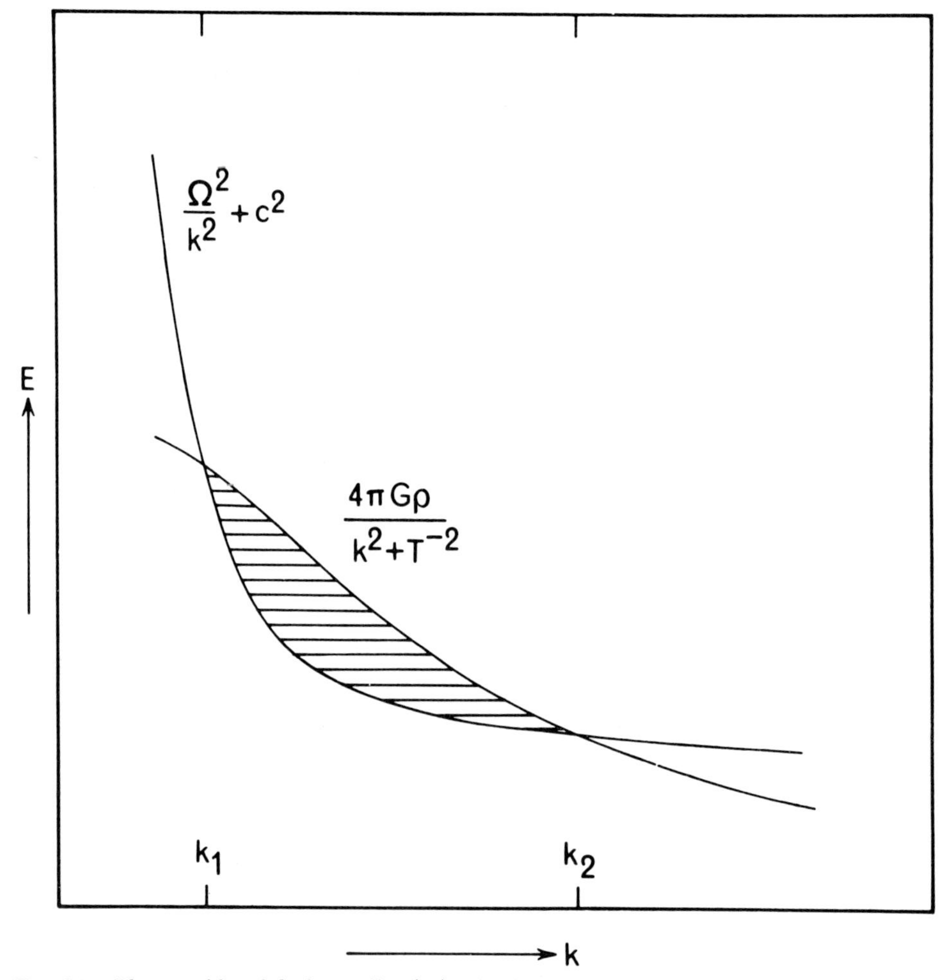

FIG. 26.—The two sides of the inequality (99). The shaded region corresponds to unstable modes.

Sun. They find that for radial modes $(4\pi G\rho/\Omega^2)_{\rm crit} = 28.8$ with $(kT)_{\rm crit} = 0.52$, while for the tangential modes the corresponding values are 21.6 and 0.86; i.e., a galactic disk with $21.6 < 4\pi G\rho/\Omega^2 < 28.8$ would be stable to all radial modes but unstable to tangential modes with wavenumbers between some values k_1, k_2. This may be relevant to the spiral-structure problem, which is discussed in § 9.6.

Most stability investigations of stellar disks consider infinitely thin disks. For the isothermal gas sheet, in the limit as $T \to 0$ with $\rho \to \infty$ and the vertical pressure approaches 0, such that the surface density μ remains finite, the disk is unstable to modes in the plane of the disk with k between some values k_1, k_2 if

$$c < \pi G\mu/2\Omega \,. \tag{101}$$

9.1.2. *Toomre's work on disks of stars.*—For a very thin rotating disk of stars, we can use an order-of-magnitude argument similar to that for equation (99) to show that modes with wavenumber k are stable if

$$2\pi G\mu k^{-1} < \Omega^2 k^{-2} + c^2 \,, \tag{102}$$

where μ is the surface density, c the velocity dispersion, and Ω the angular velocity of the disk. The disk is then unstable to modes with k between some values k_1, k_2 if

$$c < \pi G\mu/\Omega \,; \tag{103}$$

this is the result of equation (101) except for the numerical coefficient.

Toomre (1964) analyzes the stability of a very thin rotating disk with undisturbed distribution function

$$f(c_R, c_\varphi; R) = \frac{\mu}{2\pi(\sigma_{RR}\sigma_{\varphi\varphi})^{1/2}} \exp\left\{-\frac{c_R{}^2}{2\sigma_{RR}} - \frac{[c_\varphi - V(R)]^2}{2\sigma_{\varphi\varphi}}\right\}, \tag{104}$$

where μ is again the surface density, assumed independent of radius R, and σ_{RR}, $\sigma_{\varphi\varphi}$ are the velocity-dispersion components, assumed small compared with the equilibrium rotational speed $V(R)$. We recall from equation (44) that $\sigma_{\varphi\varphi}/\sigma_{RR} = B/(B - A)$. Toomre looks for small-amplitude, short-wavelength *axisymmetric* modes in this equilibrium disk, and shows that the disk is stable to these modes if

$$\sigma_{RR}{}^{1/2} > 3.36 G\mu/(-4\Omega B)^{1/2} \,. \tag{105}$$

This is very close to the result of equation (101) for the isothermal sheet if Ω is uniform. It follows that any disk which has a smooth distribution of stars and is much older than its rotation period must have its velocity dispersion locally greater than this critical value, almost everywhere. For the solar neighborhood, this critical value is about 30 km s^{-1}: this is similar to the dispersion observed for the later-type dwarfs which account for most of the recognized stellar matter in the solar neighborhood (see Blaauw 1965 and Delhaye 1965). The observed velocity dispersion in the solar neighborhood is then consistent with the usual assumption that these stars are distributed fairly uniformly over the galactic disk.

What would happen to a disk that is initially in equilibrium with random motions too small to stabilize it? Toomre argues that it would undergo several generations of instability, each leaving the system with increased velocity dispersion, until the velocity dispersion became high enough to stabilize it in a new smooth equilibrium state. This increase in internal energy requires a readjustment in the large scale distribution of the disk material.

The instability process could also be relevant to the observed inequality of σ_{RR} and σ_{zz} near the Sun. We know already that a third integral would allow this inequality to persist, once it was established. Toomre suggests that σ_{RR} would increase preferentially over σ_{zz} as the instability mechanism drives the initially unstable disk toward its stable equilibrium. Velocity-dispersion problems are discussed again in § 9.5.

Another interesting point about the stability of thin rotating disks of stars is made by Julian (1969). He investigates the possibility that these disks may be overstable, i.e., that there are disturbances with amplitudes growing in proportion to exp (st), where s is complex with positive real part. He finds that there are no such overstabilities whenever the density of stars in phase space is a decreasing function of epicyclic amplitude. This suggests that the velocity distribution must be similar in shape to that of equation (104) if these overstabilities are to be avoided.

Julian and Toomre (1966) consider the response of the thin stellar disk of equation (104) to applied nonaxisymmetric disturbances. Again the theory is local: the unperturbed surface density, epicyclic frequency, and velocity dispersion are all spatially constant. They discuss two particular kinds of disturbance. The first is a spatially sinusoidal density perturbation, impulsively applied at $t = 0$. This perturbation shows a considerable transient growth, the size of which depends on the initial angle between the wavelet's crests or troughs and the circumferential direction. As the differential rotation shears the disturbance from an initially leading one to a trailing one, the amplitude of the disturbance grows rapidly at first, but is then strongly phase damped. This means that, in a *stellar* disk, relatively recent forcing is needed for the amplified disturbance to be substantial: damping occurs within a few epicyclic periods ($\sim 10^8$ years in the Galaxy). By contrast, in a *gas* disk (Goldreich and Lynden-Bell 1965*b*) where there is no phase damping, the amplified response continues to oscillate undamped.

Julian and Toomre then examine the response of the stellar disk to a steadily orbiting point mass. This sets up an extended trailing density response in the disk, even when the velocity dispersion is large enough to avoid any axisymmetric instabilities (see eq. [105]). Again, the time taken to establish this response is about an epicyclic period. Some interesting implications of this work include (i) a stellar bar rotating within a stellar disk would trail density wakes in the disk, (ii) the process of establishing the trailing response should produce a net drag force on a massive particle in epicyclic motion, which would tend to make the orbits of gas clumps in a galaxy more circular, (iii) within the approximations of the theory, a stellar disk is incapable of long maintaining self-consistent disturbances (e.g., spiral structure) except for those which are very nearly axisymmetric.

9.1.3. *Hunter's work on cold disks.*—A particularly simple example of a self-gravitating disk has the surface density distribution

$$\mu(R) = \frac{3\mathfrak{M}}{2\pi R_0^2}\left(1 - \frac{R^2}{R_0^2}\right)^{1/2} \tag{106}$$

and rotates uniformly in centrifugal equilibrium with the angular velocity

$$\Omega = (3\pi G\mathfrak{M}/4R_0^3)^{1/2}\,, \tag{107}$$

where $\mathfrak{M}$ and R_0 are the disk's mass and radius; all particles are moving in circular orbits, so the disk is pressureless or cold. Hunter (1963) has investigated the oscillations and stability of this disk. Ignoring motions perpendicular to the plane, he finds an infinite number of unstable axisymmetric and nonaxisymmetric modes, as we would expect for a cold disk from § 9.1.1. The axisymmetric modes tend to break the disk up into rings. In addition, there are some nonaxisymmetric waves of constant amplitude propagating around the disk, and there is only one stable axisymmetric mode which corresponds to the disk expanding and contracting as a whole. This analysis is interesting because it considers the oscillations of a *complete* self-gravitating disk: the other investigations described in this section are all studies of the *local* stability of more complicated systems.

Hunter (1965) generalizes this study to disks of greater central concentration. These disks have a circular velocity $V(R)$ given by

$$V^2(R) = \text{constant}\,[1 - (1 - R^2/R_0^2)^n]\,, \tag{108}$$

so $n = 1$ corresponds to the uniformly rotating disk. While there is only one stable axisymmetric mode for $n = 1$, for $n = 10$ a second mode becomes stable: this mode corresponds to an expansion and contraction with one node. Increasing central concentration can therefore cause large-scale modes of oscillation to be stabilized.

9.1.4. *Ostriker's and Peebles's Work.*—There is some evidence from theory that the velocity dispersion given by Toomre's criterion (eq. [105]) is not in fact large enough to stabilize the disk. Hohl's (1971) N-body study began with 10^5 particles distributed in a disk with enough velocity dispersion to suppress local axisymmetric instabilities according to Toomre's rule. Hohl found that the disk was unstable to a large-scale barlike mode, and that by the time the system had settled down the velocity dispersion had increased by a factor between 2 and 6. Ostriker and Peebles (1973) considered this further, asking how we can account for the apparent stability of the Galaxy. They collected various examples of rotating disklike systems, including the Maclaurin spheroids, a self-consistent disk constructed by Kalnajs (1972), and their own three-dimensional N-body integrations. They pointed out that each of these examples was stable only for the ratio $T_{\text{mean}}/|\Omega| \lesssim 0.14$: here Ω is the potential energy and T_{mean} is the kinetic energy associated with ordered motion, so from the virial theorem the ratio must lie between 0 and 0.5. This means that these systems are stable only if a large part of their kinetic energy comes from random motions. There are then two ways to stabilize the galactic disk. (1) The disk is really hotter than we believe, with random velocities of the same order as the rotational velocity. (2) The mass of the spheroidal bulge interior to the disk is comparable to the disk mass (this increases $|\Omega|$). However, we do observe galaxies in which the bulge component is apparently weak and the velocity dispersion apparently small (compared with the rotational velocity): their value of $T_{\text{mean}}/|\Omega|$ would then be significantly larger than 0.14. This is obviously an important problem which will hold back our understanding of the dynamics of the disk until it is sorted out.

9.2. Mestel's Hypothesis

Now we turn to the problem of the large-scale mass distribution in the disk galaxies: the origin of the observed mass distributions in the disks is not yet understood, although the corresponding problem for the ellipticals appears to be solved (see § 8.3). We recall from § 4.4.2 and figures 15*a* and 15*b*, that the rotation of spiral galaxies is usually highly nonuniform, although in a few examples the rotation is nearly uniform. Mestel (1963) idealized these two kinds of rotation curve by $V(R)$ = constant and $\Omega(R)$ = constant; he asked whether both could result from the collapse of a primeval gaseous sphere of roughly uniform density ρ and uniform angular velocity Ω, with the hypothesis that each element conserves its angular momentum during the collapse. Support for this simple collapse picture came later from Crampin and Hoyle (1964), who showed that the mass–angular-momentum distribution for eight spiral galaxies was very similar to that for the uniform-density uniformly rotating sphere; see also § 9.3.

In Mestel's picture, the primeval sphere collapses spherically until the increasing centrifugal force halts the collapse normal to the rotation axis. This occurs at a time t_i when $\Omega_i^2 = 4\pi G\rho_i/3$. Subsequent flattening into a disk, without any motion normal to the rotation axis, leads to the surface density law

$$\mu(R) = 2\rho_i R_i(1 - R^2/R_i^2)^{1/2} . \tag{109}$$

The gravitational field in the plane is $-\partial\Phi/\partial R = -\pi^2 G\rho_i R$, which is greater than the centrifugal field $\Omega_i^2 R$, so the disk must now contract. Let this contraction be homologous with each element still conserving its angular momentum, i.e., $R \rightarrow kR$, $R_i \rightarrow kR_i = R_0$, $\Omega_i \rightarrow \Omega_i/k^2 = \Omega_0$, $2\rho_i R_i \rightarrow \mu_0$: this disk will reach centrifugal equilibrium with $\Omega(R)$ = constant for $k = 4/(3\pi)$ (cf. eqs. [106] and [107]). This is the model for one kind of rotation curve.

Now suppose this disk undergoes a highly nonhomologous transformation $R \rightarrow R' = cR^2/R_0$ where c is another constant, with each element again conserving its angular momentum. The rotation law then becomes $V(R') = \Omega'(R')R' = \Omega_0 R_0/c$ = constant, which is the model for the other kind of rotation curve. The transformed surface density is

$$\mu'(R')R' = \frac{\mu_0 R_0}{2c}\left(1 - \frac{R'}{cR_0}\right)^{1/2} , \tag{110}$$

and this is in centrifugal equilibrium, at least not too far from the center, for $c = \pi/2$; so this disk is also derivable *approximately*, under detailed conservation of angular momentum, from the uniform primeval sphere. Mestel then shows that the $V(R)$ = constant disk is derivable *exactly* from a primeval sphere with a slight negative density gradient. In summary, the collapse of a uniformly rotating sphere with roughly uniform density and with every particle conserving its angular momentum can lead to both the V = constant disk and the Ω = constant disk in centrifugal equilibrium. It is not yet known why some galaxies become approximately uniform rotators and others choose the more centrally concentrated configuration with approximately constant $V(R)$. This problem comes up again in the next section.

9.3. The Exponential Disk

We recall from § 4.3.2 that almost all disk galaxies follow the radial surface brightness distribution $I(R) = I_0 \exp(-\alpha R)$ in the disk. Freeman (1970*a*) studied the dependence of the parameters I_0 and α on morphological type: in a sample of 36 galaxies, 28 have approximately the same value of I_0 (fig. 12) despite the wide range of absolute magnitude and morphological type included in this sample. We need to understand this property, and particularly the exponential nature of the disks. The discussion below comes from Freeman (1970*a*).

The color distribution within the disks is approximately uniform, so it seems reasonable to assume that the exponential luminosity distribution is associated with an exponential surface density distribution $\mu(R) = \mu_0 \exp(-\alpha R)$. If we now assume that this disk is in centrifugal equilibrium, it is possible to calculate the circular velocity law $V(R)$: it turns out to be

$$V^2/R = \pi G \mu_0 \alpha R (I_0 K_0 - I_1 K_1) \,, \tag{111}$$

where I and K are modified Bessel functions evaluated at $\frac{1}{2}\alpha R$ (fig. 27). The total

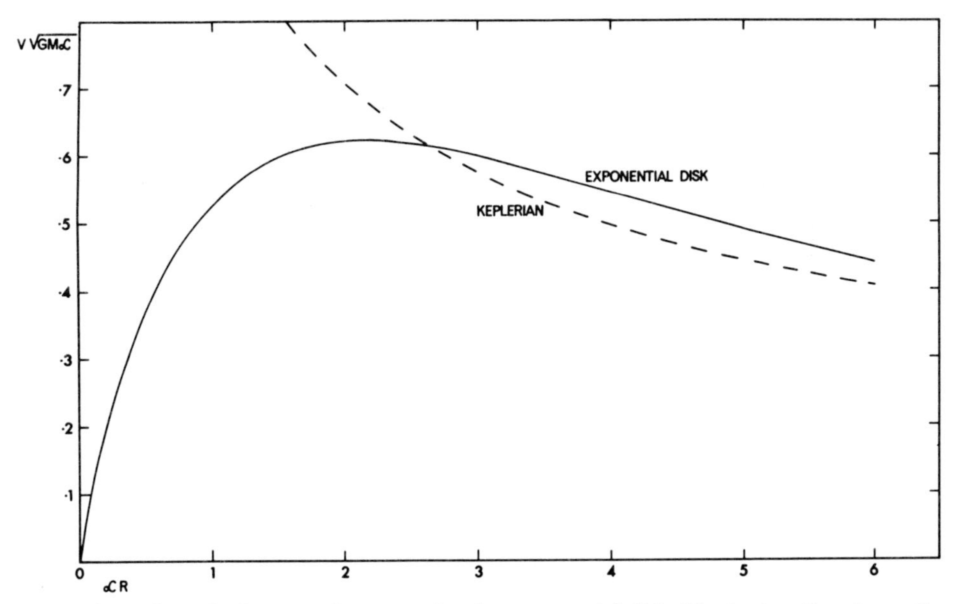

Fig. 27.—Dimensionless rotation curve for the exponential disk. The broken line shows the corresponding Keplerian curve.

mass for the exponential disk is

$$\mathfrak{M} = 2\pi\mu_0/\alpha^2 \,; \tag{112}$$

and the total angular momentum is

$$\mathfrak{H} = 1.109(G\mathfrak{M}^3/\alpha)^{1/2} \,. \tag{113}$$

The approximate constancy of the surface brightness scale I_0 for most disk galaxies has an interesting consequence. Because the mass/(blue light luminosity) ratio is only weakly dependent on morphological type (M. S. Roberts 1969), and because the disk provides a large fraction of the integrated blue light, we can infer that the surface density scale μ_0 is also approximately constant from galaxy to galaxy, although the masses of the galaxies in the sample have a range of about a factor 100 or even more. It then follows from equations (112) and (113) that

$$\mathfrak{H} \propto \mathfrak{M}^{7/4} \tag{114}$$

for the disks. Because most of the mass and angular momentum of disk galaxies probably lies in the disk, we should also expect the *total* mass and angular momentum for most of these systems to follow this law. This result would hold on dimensional grounds for any surface density distribution $\mu = \mu(R; \mu_0, L)$, where L is a length scale, if μ_0 is constant from system to system. Conversely, the total mass and angular momentum are probably endowed properties of a protogalaxy which are conserved as it collapses to form a galaxy; the apparent constancy of μ_0 for most disk galaxies would be explained if we knew why (i) galactic disks have only one characteristic length scale, and (ii) the distribution of mass and angular momentum in the protoclouds followed the 7/4 law.

The 7/4 law could in principle be checked observationally by fitting model rotation curves (e.g., Brandt and Belton 1962) to the observed rotation curves for spiral galaxies, and so deriving total masses and angular momenta. Unfortunately the appropriate equations are ill-conditioned, and a meaningful derivation of the $\mathfrak{H}(\mathfrak{M})$ law is not now possible.

Now we consider the internal distribution of angular momentum within the exponential disk. Define the mass–angular-momentum distribution $\mathfrak{M}(h)$ as the total mass with angular momentum per unit mass less than h. If Mestel's hypothesis is correct, that the angular momentum of each mass element is conserved during the collapse of the primeval cloud, then $\mathfrak{M}(h)$ is invariant during the collapse. This means that the disk contains some important dynamical information about the primeval cloud. We can calculate $\mathfrak{M}(h)$ for the exponential disk from the rotation curve, and it is shown as the full curve in figure 28.

Following Crampin and Hoyle (1964), we can compare $\mathfrak{M}(h)$ for the exponential disk of length scale α^{-1} and mass $\mathfrak{M}$ with the $\mathfrak{M}(h)$ for the disk D with surface density

$$\begin{aligned} \mu(R) &= \mu_0(1 - R^2/a^2)^{1/2}, \qquad R < a\,, \\ &= 0 \qquad\qquad\qquad\qquad\quad R > a\,; \end{aligned} \tag{115}$$

uniform angular velocity Ω, and the same total mass $\mathfrak{M}$. This *disk* D has exactly the same $\mathfrak{M}(h)$ as a uniformly rotating *sphere* with the same $\mathfrak{M}$, Ω, and radius a. It turns out that if

$$\Omega a^2 \simeq 2.80(G\mathfrak{M}/\alpha)^{1/2}\,, \tag{116}$$

the disk D and the exponential disk have almost identical $\mathfrak{M}(h)$ distributions; see figure 28. Only for the outer parts of the disk ($\alpha R > 6$), which contains less than 0.02 of the total mass and 0.05 of the total angular momentum, do the two $\mathfrak{M}(h)$ distributions differ significantly. From equations (107) and (116), the disk D is also in centrifugal equilibrium if $a = 3.33\alpha^{-1}$ and $\Omega = 0.253(G\mathfrak{M}\alpha^3)^{1/2}$; i.e., each exponential disk in centrifugal equilibrium has a corresponding disk D, also in centrifugal equilibrium, with the same mass, almost the same angular momentum, and almost the same $\mathfrak{M}(h)$ distribution (see fig. 29). This is important for at least two reasons. (i) It reinforces Crampin and Hoyle's similarity of $\mathfrak{M}(h)$ for spirals and the uniform sphere. (ii) It means that $\mathfrak{M}(h)$ does not itself uniquely determine the exponential nature of the disk.

Because almost all disk galaxies have exponential disks, the $\mathfrak{M}(h)$ distribution for all these disks is similar to that shown in figure 28. The similarity of this $\mathfrak{M}(h)$ to that for the sphere obviously does not mean that all protodisks were necessarily uniform

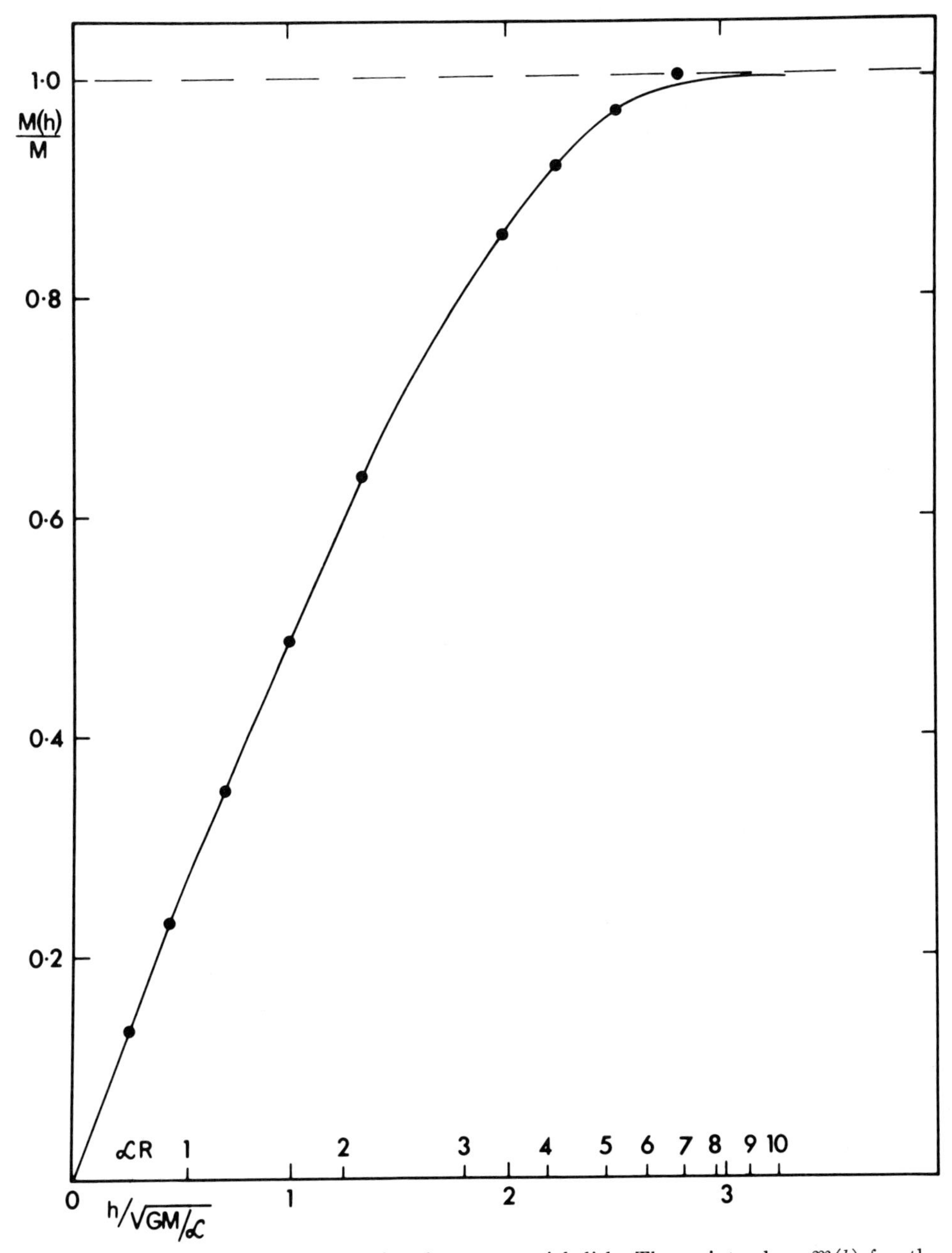

Fig. 28.—The full curve is $\mathfrak{M}(h)$ for the exponential disk. The points show $\mathfrak{M}(h)$ for the disk D of equation (115).

spheres. However, if $\mathfrak{M}(h)$ is invariant during the collapse phase, then almost all protoclouds destined to be S0 or spiral galaxies have acquired a similar $\mathfrak{M}(h)$ (in dimensionless variables). On the other hand, if $\mathfrak{M}(h)$ is not invariant, then there must be a very efficient mechanism which establishes the characteristic $\mathfrak{M}(h)$ for these systems as they form. In either case, the origin of this $\mathfrak{M}(h)$ distribution sets a dynamical problem that must be solved before we can properly understand the present state of disk galaxies.

We know that $\mathfrak{M}(h)$ does not alone define the exponential disk: a disk D in centrifugal equilibrium is also permitted. However, given the invariance of $\mathfrak{M}(h)$, it seems unlikely that the disk D should be the end product of the collapse: Hunter (§ 9.1.3) showed that it is unstable. Nevertheless, there may be other centrifugal equilibrium states, all with nearly the same $\mathfrak{M}(h)$, between the exponential disk and the disk D, and some of these may be stable. In particular, it is important to find out whether the exponential disk itself is the most stable of these states.

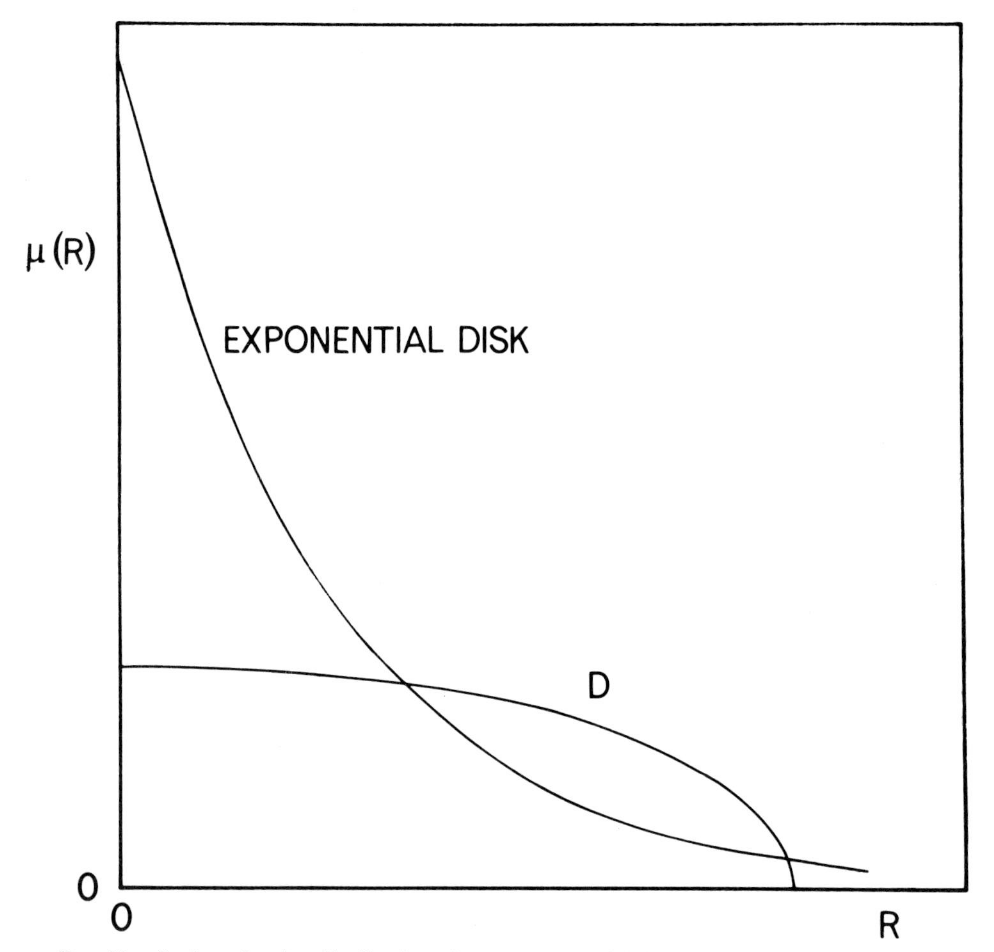

FIG. 29.—Surface density distributions for an exponential disk and a disk D with the same mass, almost the same total angular momentum and almost the same $\mathfrak{M}(h)$ distribution.

Some interesting N-body experiments by Hohl (1971) are relevant to this problem. Hohl starts with 10^5 representative stars, confined to move in the galactic disk, initially with the same surface density distribution as the disk D, with the minimum velocity dispersion according to Toomre's criterion of equation (105), and rotating in equilibrium. He finds that the final mass distribution for this disk gives a high-density central core and a disk population of stars that is closely exponential. This distribution takes a few rotation periods to establish.

9.4. The Bending of the Galactic Plane

In § 4.4.2 we noted that the H I layer in the Galaxy is flat for $R < 10$ kpc; for $R >$ 10 kpc the layer bends up on one side of the nucleus and down on the other, deviating from the plane by about 1 kpc at $R \approx 15$ kpc. There is evidence for similar bending in M31 (Roberts 1966), and it can be seen clearly in the light distribution for the edge-on S0 NGC 4762 (Sandage 1961).

For the Galaxy in particular, at least four explanations of this bending have been offered:

i) It is a tidal distortion due to the Magellanic Clouds at approximately their present distance (Avner and King 1967).

ii) A contracting galaxy may be flattened along an axis at an angle to the angular-momentum vector, because rotation is probably unimportant dynamically in the early stages of the contraction. The disk would then precess freely, and this causes a systematic bending of the galactic disk. Lynden-Bell (1965) shows that the angle need be only 1° to account for the observed bending.

iii) The past orbit of the Magellanic Clouds may have carried them much closer to the Galaxy than they are now. Their tidal effect would have been much larger then than now, and the observed warp might be a transient remnant of a tidal wave set up at that time (Hunter and Toomre 1969).

iv) It is due to the flow of intergalactic gas past the Galaxy and its halo (Kahn and Woltjer 1959).

Hunter and Toomre (1969) have carried out a fully self-consistent linear analysis of the vertical oscillations of some model galactic disks—here "vertical" means perpendicular to the galactic plane. The model disks are infinitely thin and axisymmetric, are in centrifugal equilibrium, and have a finite outer radius R_0. They study the behavior of small perturbations with longitude dependence $\sim\exp(im\varphi)$. The main results are:

1) They show that an arbitrary disk cannot be unstable to axisymmetric ($m = 0$) or to $m = 1$ warping modes. This means that vertical instabilities of the disk are unlikely to cause the observed bending.

2) They discuss the eigenmodes for different model disks and find the important result that the spectrum of possible eigenfrequencies for a disk ceases to be discrete and becomes continuous if the surface density $\mu(R) \rightarrow 0$ sufficiently fast as $R \rightarrow R_0$. An approximate argument suggests that all eigenmodes are discrete only if $\int_0^{R_0}[\mu(R)]^{-1}dR$ converges. The integral does not converge for most realistic model disks.

3) The importance of this result becomes clear when they consider the way a disk responds to an impulsive perturbation. Systems with very discrete modes tend to oscillate indefinitely without radical change of appearance. On the other hand, systems

with a continuum of modes are by nature dispersive; i.e., their response to an impulse evolves with time as a result of adjacent modes drifting further and further from synchronism through different rates of oscillation (cf. § 8.2 on phase damping).

4) Finally Hunter and Toomre, armed with these results, offer a critique of the various suggested bending mechanisms. (i) Steady tidal distortion by the Magellanic Clouds seems to be inadequate: for an LMC mass of 10^{10} $\mathfrak{M}_\odot$ and a Galactic mass and outer radius of 1.2×10^{11} $\mathfrak{M}_\odot$ and 16 kpc, the maximum distortion is only about 100 pc at the edge, compared with the observed distortion of about 1 kpc. (ii) Free modes of oscillation also seem to be unlikely causes of the bending. Because the eigenfrequency spectrum is probably continuous, Hunter and Toomre believe that the principal bending distortions of the galactic disk would damp away within about 10^9 years. (iii) They favor the explanation that the galactic plane remains warped today mainly as a relic of a tide raised during a close passage of at least the LMC. They point out that the last pericenter along any conceivable LMC orbit must have occurred between 4×10^8 and 6×10^8 years ago, which is less than the estimated 10^9-year damping time for the

TABLE 7

AGES AND VELOCITY DISPERSIONS FOR EVOLVED DISK STARS

Region	$\sqrt{\sigma_{RR}}$	$\sqrt{\sigma_{\varphi\varphi}}$	$\sqrt{\sigma_{zz}}$	Age	No.
I	10	12	8	$5\text{–}50 \times 10^7$	280
II	30	9	8	$1\text{–}6 \times 10^9$	37
III	44	25	18	$6\text{–}10 \times 10^9$	66
IV	60	33	36	$>10 \times 10^9$	64
s.d., $\delta > 0.15$	230	84	90	$>10 \times 10^9$	128

NOTE.—Ages in years, $\sqrt{\sigma}$ in km s^{-1}.

relatively slowly evolving $m = 1$ responses. However, a perigalactic distance of about 20 kpc and an LMC mass of at least 2×10^{10} $\mathfrak{M}_\odot$ are needed to make this picture numerically tenable. (iv) It is difficult to prove or disprove the intergalactic gas flow picture by considering disk dynamics alone. However (our comment), the S0 galaxy NGC 4762 has a clearly bent disk which presumably contains very little gas (see § 4.5). For that galaxy at least, the intergalactic gas flow explanation seems unlikely.

The Galaxy, M31, and NGC 4762 all have bent disks and apparently rather close companions. A search for galaxies with bent disks and no nearby companions would provide a useful test for the general applicability of Hunter and Toomre's bending picture: NGC 1448 (see the Palomar *Sky Survey*) may be one example.

9.5. PROBLEMS OF VELOCITY DISPERSION IN THE DISK

We know from § 3.2 that the velocity dispersion of stars in the disk appears to increase with age, even for those stars whose ages are less than a few billion years. Table 7, from Eggen (1970), shows this clearly. The data for this table come from nearby evolved F and G stars with trigonometrical parallaxes that place them at least 0.5 mag above the main sequence (fig. 30). Eggen separates them into four homogeneous age groups: I (bluer than NGC 752), II (between NGC 752 and M67), III (between M67 and NGC 188) and IV (redder than NGC 188). The ages adopted for M67 and NGC 188 are those of Sandage and Eggen (1969). According to Oort (1965), a Gaussian distribution of stars with $\sigma_{zz}^{1/2} = 36$ km s^{-1} has an average distance from the galactic plane of about 500 pc, so groups I–IV are certainly disk stars. It seems unlikely that the mean dynamical conditions in the galactic gas disk have changed significantly in the last few billion years; this means that the older stars probably increased their velocity dispersion *after* formation, rather than that they were formed with larger velocities. We need to understand how this occurs.

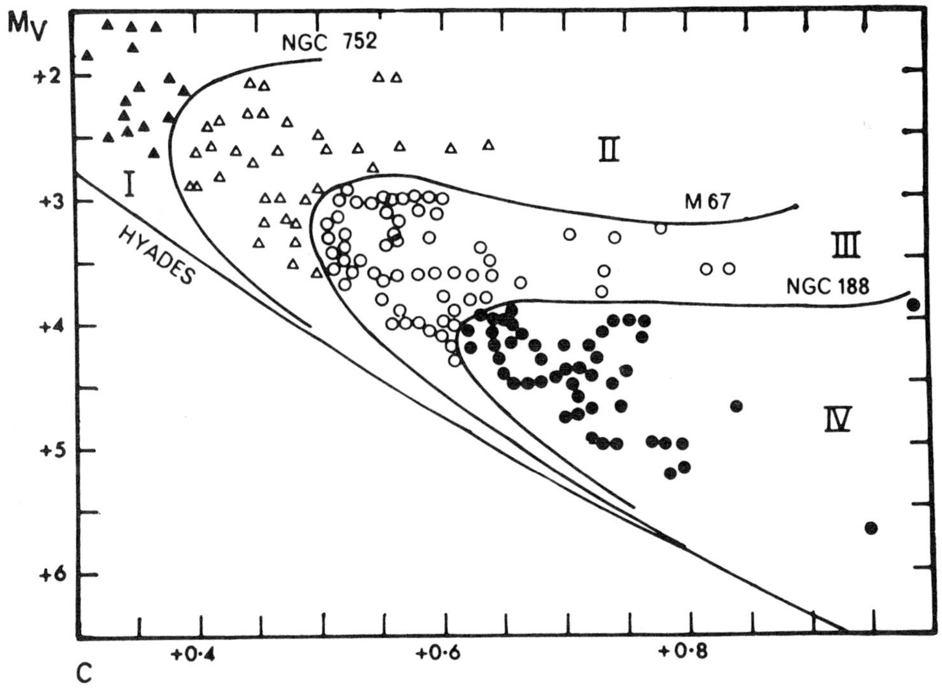

FIG. 30.—The separation of nearby evolved F and G stars into age groups, as described in Section 8.5 (from Eggen 1970).

The problem seems quite distinct from that of the much larger velocity dispersion for the halo stars: this probably originates from the collapse of the Galaxy, as discussed in § 7. Table 7 also includes for comparison data for a sample of 128 subdwarfs with $\delta(U - B) > 0.15$, which are probably all halo stars.

The most obvious source for a secular increase in velocity dispersion comes from the dynamical interaction of individual stars with large gas clouds or with the large-scale spiral structure itself. Spitzer and Schwarzschild (1953) suggested that interaction with large cloud aggregates ($>10^5\ \mathfrak{M}_\odot$) could have increased the velocity dispersion sufficiently. Julian (1967) examined the effect of large-scale gas inhomogeneities on the velocity dispersion, including the forces from the accompanying standing waves in the stellar disk itself (see § 9.1.2): he found that they could produce significant increases in the stellar velocity dispersion during the galactic lifetime.

Barbanis and Woltjer (1967) studied the possibility that the large-scale spiral structure is responsible for the secular increase in velocity dispersion. They used a spiral potential Φ_s superposed on a realistic axisymmetric disk potential, where

$$\Phi_s = \epsilon R^{1/2}(16 - R)e^{-\eta t} \cos\left(2\varphi + 20 \ln \frac{R + 4}{8}\right) (\mathrm{kpc})^2(10^7\,\mathrm{yr})^{-2}\,; \qquad (117)$$

here R is in kiloparsecs and t is in units of 10^7 years. The constant η is inversely proportional to the persistence time of the spiral structure. It turns out that this potential with $\epsilon = 0.001$ gives a reasonable representation of the surface density perturbations due to the spiral arms and observed near the Sun, and also of the bumps observed in the H I rotation curve (see fig. 3). They examined the average properties of low-velocity orbits in this potential, and concluded that the velocity dispersion of the older disk stars can be understood this way if (i) a few billion years ago the spiral arms were about three times as massive as they are now, or (ii) a spiral pattern much like the present one has dissolved and reformed about 10 times. They also concluded that (iii) no spiral pattern very much stronger than the present one could have existed in the Galaxy for any length of time, and (iv) a spiral pattern like the present one could not have dissolved and reformed much more frequently than once every half-billion years; otherwise the velocity dispersions would be larger than observed.

The stars of table 7 again have $\sigma_{RR} > \sigma_{zz}$. We have already discussed in § 5 the implications that this has for the existence of a third integral. Mechanisms for producing the secular increase of velocity dispersion for the disk stars over the last few billion years must also maintain this inequality during the increase.

9.6. The Problem of Spiral Structure

We know from table 2 that about two-thirds of the bright galaxies show some sort of spiral structure in the disk. This is clearly a dynamical manifestation; and if we could understand it, our knowledge of the dynamics and evolution of the disklike galaxies would certainly be greatly enhanced. The basic problem is the origin of this structure and its maintenance against the background differential rotation field which is usually (but not always) found in ordinary spirals.

9.6.1. *Relevant observational data.*

1) Spiral galaxies are probably about 50 rotation periods old, while the characteristic time for the differential rotation to wind up a material structure rotating with the disk is only a few rotation periods. However, there are spiral galaxies in nearly uniform rotation (see fig. 15), so spiral structure is probably not *uniquely* associated with the usual differential nature of the rotation.

2) The spiral arms, as defined by concentrations of gas, young stars, and the associated dust, usually outline a trailing spiral pattern extending over much of the disk: the dust lanes mostly appear on the inside edge of the bright optical arms.

3) While the H I layer in the Galaxy is thin ($\approx$220 pc between half-density points) and fairly flat within $R = 10$ kpc, there is H I, associated by its velocity with the Perseus arm and the outer arm, at heights exceeding 2 kpc from the plane (Oort 1970). At the

distance of the Perseus arm, a star in the galactic plane would need a z-velocity component of about 75 km s^{-1} to reach this height; for gas filaments, this velocity would be rather higher because of deceleration by the interstellar medium.

4) Noncircular gas motions are associated with the spiral structure in the Galaxy. These include (i) the 4-kpc expanding arm, (ii) the streaming motions, of order 8 km s^{-1}, associated with the Sagittarius and Scutum arms (Burton and Shane 1970), (iii) the rolling motions observed in the Perseus and 4-kpc arms (see Westerhout 1970). For example, in part of the Perseus arm, the peak velocity changes from -60 km s^{-1} at $b = 3°$ to -40 km s^{-1} at $b = -3°$.

5) We stress again that all transitions in the morphological classification, between classes (S0, S, I), varieties (r, s), stages (a, b, . . . , m), and particularly between the barred and nonbarred families, are smooth. This strongly suggests that one mechanism produces spiral structure in both SA and SB systems.

6) Sandage, Freeman, and Stokes (1970) studied how the morphological stage (a, . . . , m) depends on large-scale parameters such as the disk surface brightness scale, and its length scale (see § 9.3), the size ratio of the spheroidal component to the disk, the mean *total* mass density, and the mean H I density. Only the H I density varies in a strongly systematic way with morphological stage; since the stage is defined observationally mainly by the openness of the spiral pattern and the degree of resolution of the arms into luminous stars, this suggests that the openness of the arms predicted by any theory should depend on the mean H I density. Note, however, that the Im systems, with the highest gas content, show at most a rudimentary spiral structure.

7) The neutral hydrogen distribution often extends well beyond the bright optical spiral structure: M33 is a good example (Gordon 1971).

8) Finally we note the lack of direct observational information about the degree of involvement of the old disk itself in the spiral structure.

It is obvious from this list that the problem of spiral structure is an excellent example of a dynamical problem for which the Galaxy and external systems provide vital complementary information.

Because of the problem of differential rotation, there appear to be only two possible independent kinds of spiral-structure processes. (*a*) The spiral structure is wound up by the differential rotation but re-forms. (*b*) The spiral structure rotates approximately uniformly while the gas and stars within it are in differential rotation. In this picture, the structure is a density wave propagating uniformly through the differentially rotating disk.

So far the spiral-structure theories that have been properly worked out apply mainly to axisymmetric disks, i.e., to SA spirals: the SB spirals are discussed in § 10. First we describe a type (*a*) theory.

9.6.2. *Goldreich and Lynden-Bell's theory.*—Goldreich and Lynden-Bell (1965*b*) investigate the local instabilities of a differentially rotating gas sheet. They start with a disk having local values of the Oort constants A and B and mean density ρ such that all axisymmetric modes are just stable (see § 9.1.1). Tangential modes may still be unstable. A perturbation, whose density contours initially point radially out from the galactic center, is sheared by differential rotation and may grow in density by up to a factor of 100 in a rotation period to form a trailing spiral arm. Provided the perturbations remain within the linear regime, the growth then gives way to rapid oscillation as the trailing disturbance is stretched out by the differential rotation. The modes that grow most have wavelengths of about $4\pi T$ when they point radially; here T is the thickness of the galactic disk.

In their picture, star formation begins when the perturbation density has grown sufficiently: then the spiral arms turn on as a visible feature. The structure is wrapped up by the differential rotation, the gas disk becomes thicker because of the turbulence associated with star formation, and this reduces the gas density below the level $\pi G\rho \gtrsim 4B(B - A)$ required for instability. When the brightest young stars have died, the turbulence in the gas disk dissipates rapidly, the disk becomes thinner, and the density rises until the growth of sheared perturbations can begin again. Thus generation after generation of spiral arms form, wind up, and disperse. It is interesting that $\pi G\rho/4B(B - A) \approx 1.2$ in the solar neighborhood.

The theory is local. There is no preference for two spiral arms or whole spiral arms. The stellar background will be affected by the oscillating gravitational field, which should have a considerable relaxing effect on the stellar distribution (cf. § 9.5). This relaxing effect appears to be excessively large (Lynden-Bell 1966). In this paper, Lynden-Bell introduces a weak magnetic field which has been sheared until it is parallel to the flow lines. The field destabilizes some long-wavelength modes with wavelength $\sim c^2/G\mu$ (usual symbols) ≈ 5 kpc in the Galaxy, even when all other modes, including the sheared modes, are stable. The modified spiral theory starts with all nonmagnetic modes stable, so $\pi G\rho < 4B(B - A)$. The long-wavelength magnetic modes are able to grow; but as they are wrapped up by the shear, the wavelength is reduced and the modes are pressure stabilized. The excessive relaxation is clearly reduced.

The local nature of the theory does not yet allow its detailed application to a large-scale spiral pattern. The theory also depends on the differential nature of the rotation: see § 9.6.1, paragraph 1.

9.6.3 (i). *The density-wave theory: basic theory.*—The basic concept of spiral structure as a spiral density wave is due to Lindblad (1941 and later papers). The modern form of the theory was developed by Lin and associates: an account of their theory is given by Lin, Yuan, and Shu (1969, LYS). They look for a self-sustained spiral density wave involving the gas and the stars. This wave has a gravitational field with spiral equipotentials. LYS start with this field and carry out the analysis according to the plan shown in figure 31. Using their symbols, the imposed spiral gravitational field has the small potential $\mathfrak{V}_1 = A(\varpi) \exp \{i[\omega t - m\theta + \Phi(\varpi)]\}$: here (ϖ, θ) are polar coordinates in the plane of the disk, t is the time, Φ is a slowly varying monotone function multiplied by a large parameter, ω is a constant (real for neutral waves), m is an integer, and $A(\varpi)$ varies slowly. This represents a spiral pattern with m arms, the pattern rotating with angular velocity $\Omega_p = \text{Re}(\omega)/m$; its radial wavenumber is $|k(\varpi)| = |\Phi'(\varpi)|$. They calculate the *linear* density response of the stars and the gas to this potential, with the further *asymptotic* approximation $[k(\varpi)\varpi]^{-1}$ is small; this corresponds to a small pitch angle ($<15°$) for the spiral field.

First calculate the stellar response. Let κ be the epicyclic frequency and $\nu = (\omega - m\Omega)/\kappa$ and $x = k^2\langle c_\varpi{}^2\rangle/\kappa^2$, where $\langle c_\varpi{}^2\rangle$ is the radial velocity dispersion component. Let

σ_{*0} be the unperturbed stellar surface density. Then the stellar response in this approximation to the imposed potential $\mathfrak{B}_1$ is

$$\frac{\sigma_*}{\sigma_{*0}} = -\frac{k^2 A}{\kappa^2(1-\nu^2)}\,\mathfrak{F}_\nu(x)\exp\{i[\omega t - m\theta + \Phi(\varpi)]\}\,, \tag{118}$$

where $\mathfrak{F}_\nu(x)$ is a reduction factor such that $\mathfrak{F}_\nu(0) = 1$. This reduction factor represents the reduction in the surface density response due to the nonzero velocity dispersion of the stellar disk. Similarly, the response in the gaseous disk is

$$\frac{\sigma_g}{\sigma_{g0}} = -\frac{k^2 A}{\kappa^2(1-\nu^2)}\,\mathfrak{F}_\nu^{(g)}(x_g)\exp\{i[\omega t - m\theta + \Phi(\varpi)]\}\,, \tag{119}$$

where σ_{g0} is the unperturbed gas density and $x_g = k^2a^2/\kappa^2$: here a is the equivalent acoustic velocity for the gas.

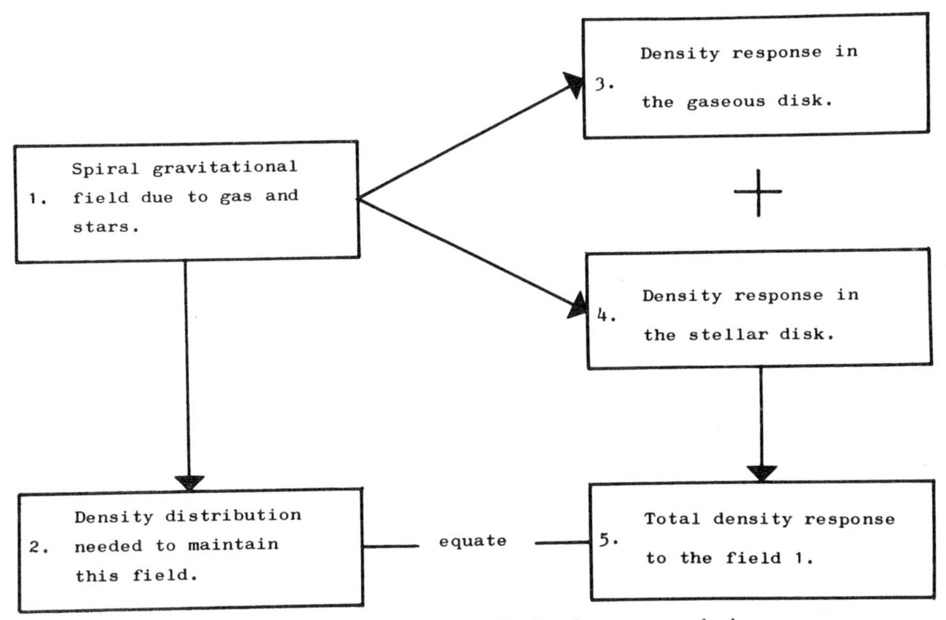

FIG. 31.—Block diagram for the density wave analysis.

Now calculate the surface density associated with the imposed potential through Poisson's equation. It is

$$\sigma_p = -\frac{|k|A}{2\pi G}\exp\{i[\omega t - m\theta + \Phi(\varpi)]\}\,. \tag{120}$$

We can also calculate the velocity disturbance to the gas at some ϖ. It is $\sigma_g/\sigma_{g0}:u_\varpi/\varpi\Omega:u_\varphi/\varpi\Omega = -k\varpi:m(\Omega_p/\Omega - 1):i\kappa^2/2\Omega^2$.

A self-consistent wave requires $\sigma_p = \sigma_* + \sigma_g$. This provides the dispersion relation

$$\frac{\kappa^2}{2\pi G|k|}(1-\nu^2) = \sigma_{*0}\mathfrak{F}_\nu + \sigma_{g0}\mathfrak{F}_\nu^{(g)}\,. \tag{121}$$

For a given galactic model, this connects the wavenumber $|k|$ and the frequency ν. We must have $(1 - \nu^2) > 0$: the cases $\nu = \pm 1$ correspond to the Lindblad resonances $\Omega_p = \Omega \pm \kappa/m$, where the stars see the oscillating field at the same frequency as the epicyclic frequency. The waves defined by this dispersion relation have the following properties. (*a*) They extend over the range in ϖ defined by $\Omega(\varpi) - \kappa(\varpi)/m < \Omega_p < \Omega(\varpi) + \kappa(\varpi)/m$. Note that there are no eigenmodes in this theory: Ω_p can take any value consistent with $|\nu| < 1$. Figure 32 shows $\Omega(\varpi)$, $\kappa(\varpi)$, and $\Omega \pm \kappa/2$ from Schmidt's (1965) model. For example, for $\Omega_p = 11$ km s^{-1} kpc^{-1}, the pattern with $m = 2$ could

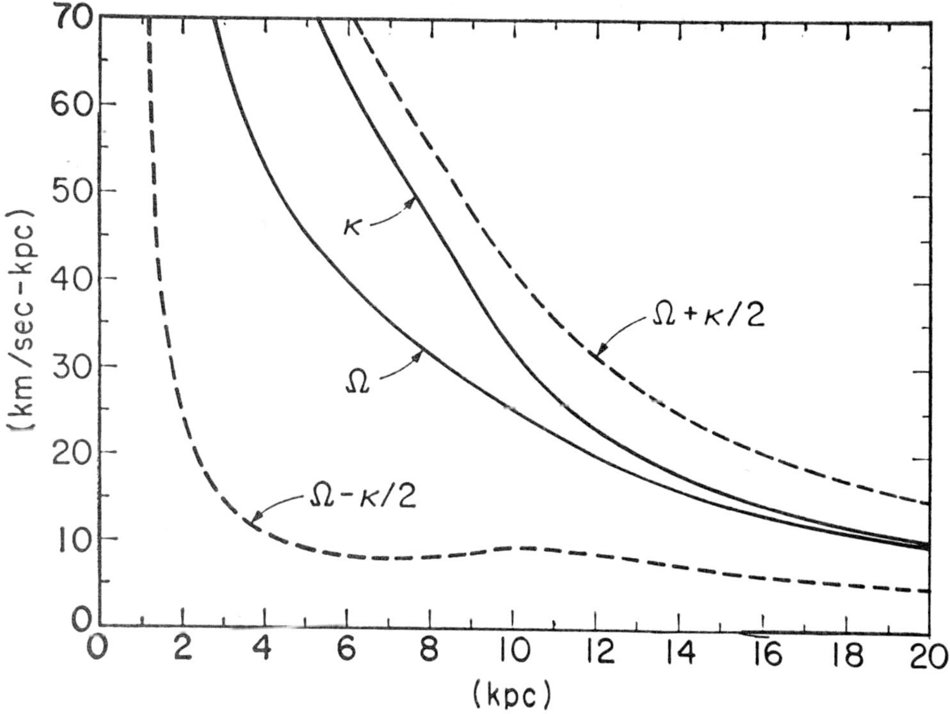

FIG. 32.—Rotation curve, etc., for the Galaxy from the Schmidt model (from Lin *et al.* 1969).

extend from 4 kpc to beyond 20 kpc. For $m > 2$, its extent is very limited; this favors the two-armed spirals (at least in the inner parts of galaxies) because most rotation curves are qualitatively similar. (*b*) The dispersion relation includes the velocity dispersion parameters x and x_g. For a purely stellar disk, the minimum value of x for stability is, from equation (105), $x = 0.2857$. (*c*) To this approximation, there is no indication favoring trailing waves ($k < 0$) or leading waves ($k > 0$). (*d*) The direct observational predictions of the theory include, first, the shape of the arms. This follows from the dispersion relation once $\Omega(\varpi)$, Ω_p, $\langle c_\varpi^2 \rangle$, a, σ_{*0}, and σ_{g0} are chosen, because $|k| = |\Phi'(\varpi)|$: integration gives $\Phi(\varpi)$ and hence the shape of the arms. Conversely the shape can be used to derive Ω_p if the other parameters are known. The direct observational predictions also include the systematic gas motions associated with the arms, once A is known.

Now we describe briefly some different approaches to the density-wave theory by other workers. Each provides various insights into the theory. Contopoulos (1971) considers whether there is any preference for leading or trailing waves near the important inner Lindblad resonance (ILR). He calculates the linear response to an imposed trailing or leading spiral field near the ILR, and finds that it is always trailing if the imposed wave is only slowly growing.

Berry and Vandervoort (1970) look for tightly wound steady spiral waves in a pressureless thin self-gravitating disk. Their analysis is an extension of the Lin-Shu theory in that the amplitude of their waves is finite rather than infinitesimal. They are able to find two kinds of solutions. (i) The pattern has $\Omega_p < \Omega$ and extends to the corotation radius from the origin, and its amplitude vanishes at both these limits. They can find this solution only if the angular velocity Ω is approximately uniform. (ii) For disks with appreciable differential rotation, a solution is found only if the subsystem of gas and low-velocity stars supporting the wave makes only a small contribution to the total galactic mass. They also find that the pattern becomes more tightly wound as the mass fraction in this subsystem decreases (cf. paragraph 6 of § 9.6.1). They identify these type II waves with the density waves of the linear theory. Again, trailing and leading waves are allowed without distinction.

While the Lin-Shu and Berry-Vandervoort analyses are both asymptotic solutions valid for tightly wound arms, Kalnajs (1970) finds numerically a two-armed density wave without this approximation. His equilibrium model is a flat disk derived from the rotation curve for M31, and has sufficiently high random velocities from 4 kpc outward to stabilize it against local axisymmetric collapse. He looks for small-amplitude perturbations by solving Poisson's and the linearized Boltzmann equations. The solutions have the form Re $A(r) \exp [i(m\theta + \omega t)]$, and the modes $[A(r),\omega]$ are given by the eigenfunctions and eigenvalues of an integral equation. This contrasts with the linear asymptotic solutions where the frequency must be determined observationally from several observational constraints (see LYS). Kalnajs derives numerically the largest discrete $m = 2$ mode, largest in the sense that the gravitational interactions associated with it are strongest as measured by the shift of the pattern speed Ω_p from the value $\Omega - \kappa/2$ that it would have if the gravitational effects of the mode were neglected (see eq. [121]). The resulting wave is a trailing barlike distortion of the central regions of the system, growing in amplitude with a growth rate $\sim 5 \times 10^8$ years and driving a fairly tightly wound trailing spiral response in the gas. This mode has a pattern speed $\Omega_p \approx 32$ km s^{-1} kpc^{-1}, which is much higher than the observationally determined value $\Omega_p \approx 13$ km s^{-1} kpc^{-1} in the Galaxy from the asymptotic theory (LYS). The reason for this difference is not clear; the asymptotic theory should be valid in the Galaxy because the pitch angle of the arms appears to be fairly small, and Ω_p deduced from observation via the asymptotic theory is fairly tightly determined in the range 11–14 km s^{-1} kpc^{-1}. It may just mean that the observed mode is not unique.

Another nonasymptotic analysis is made by Fujimoto (1968*a*). He imposes a small trailing spiral surface density perturbation $\sigma_1(r) \cos 2[\theta - g(r)]$ on a disk in centrifugal equilibrium with the rotation curve $V(r) = r(a^2 + r^2)^{-3/4}$ in dimensionless variables: here the unit of length is the radius of the disk with spiral structure and $g(r)$ is monotone increasing. Neglecting gas pressure and linearizing the equations of motion and continuity, he calculates numerically the amplitude of the surface density response $\chi(r)$ in the gas component. For a stationary self-consistent density wave, $\chi(r)$ and $\sigma_1(r)$ must be equal, and so must their phases. Fujimoto compares these two functions graphically. For example, for $a = 0.5$, he is able to match $\chi(r)$ and $\sigma_1(r)$ very well over $0.1 < r < 0.9$ with $\Omega_p \approx 0.7$, independent of the number of windings of the spiral structure. Note that (i) for these parameters there are no Lindblad resonances in the region $r < 1$; and (ii) for $a = 0.5$, the angular velocity at the edge of the disk $\Omega(r = 1) = 0.85$ in the dimensionless units, so $\Omega_p < \Omega$ everywhere within the spiral structure region (cf. the LYS value of Ω_p for the Galaxy). No self-consistent solutions are found for $a < 0.3$. Solutions for the more open spiral perturbations require a larger ratio of gas mass to stellar mass, as observed. Within its approximations this study illustrates particularly clearly the basic dynamics of the stationary density-wave concept away from regions of Lindblad resonance.

In an N-body investigation, Miller, Prendergast, and Quirk (1970) observe spiral density waves in a plane system of 1.25×10^5 gravitating particles. These waves persist for about three revolutions of the spiral pattern, i.e., $\sim 10^9$ years. It is worth noting that the pattern speed for this experimentally observed spiral pattern is only just greater than $\Omega - \kappa/2$ throughout the system (see Quirk 1971): it corresponds to the low Lin-Shu value. Note too that (i) their system has no Lindblad resonances, and (ii) their system and the Galaxy are not comparable in detail: although their spiral pattern lasts only about 10^9 years, the velocity dispersion in their system is relatively higher, which makes it more difficult for gravitational modes to persist.

9.6.3 (ii). *Observational studies of the density-wave theory.*—So far, these studies have served partly to check whether the basic concept of a density wave is correct, and partly to derive parameters such as Ω_p and the amplitude of the wave from the Lin-Shu asymptotic theory.

Yuan (1969*a*) discusses the H I kinematics in the Galaxy. He ascribes the dips in the H I rotation curve of figure 3 to systematic motions of order 8 km s^{-1} associated with the density waves. These dips locate the Sagittarius and Norma-Scutum arms in the Galaxy, and this leads to a typical pitch angle of 6°–7° for the spiral arms in this region. Then, using the Schmidt model and assuming that the velocity dispersion has the value for marginal stability as given by equation (105), he estimates from the dispersion relation of equation (121) that Ω_p is in the range 11–14 km s^{-1} kpc^{-1}.

Yuan (1969*b*) computes the birthplaces in the Galaxy of 25 stars with accurate space motions and well-determined ages in the range $(1–2) \times 10^8$ years. He finds that these stars are born in the gaseous spiral arms, as expected, only if the density-wave concept is adopted and the effect of the propagating spiral field is included in the orbit computation. Again, the required pattern speed is about 13 km s^{-1} kpc^{-1}, and the strength of the spiral field is about 5 percent of the mean field.

Another test of the basic density wave concept is by Dixon (1971). He compares the distribution of red and blue stars in the range $M_V = -4 \pm 0.5$ across a spiral arm in M33. The blue stars are younger than 2×10^7 years while the red ones are between 2×10^7 and 5×10^7 years old. If the spiral arm is a density wave, then the arms as defined by red and blue stars should be relatively displaced, and this is observed. This is a very direct and valuable test of the basic concept. It is limited unfortunately to Local Group objects at present, and this rather restricts its application.

9.6.3 (iii). *Origin of the spiral structure.*—Although there is now some evidence that the density-wave concept is correct, the *origin* of the spiral structure is not yet understood.

Toomre (1969) exposes a serious difficulty in his study of the group velocity of spiral waves in galactic disks. He asks how some postulated spiral wave pattern would evolve with time, and shows that frequency and wavenumber information, and probably also the wave energy, associated with a packet of tightly wound spiral density waves propagate radially with the group velocity $\partial\omega/\partial k$. In the Galaxy, this velocity is about -10 km s^{-1}, which corresponds to a time scale of about 10^9 years; i.e., in this time the waves wrap more tightly and drift in radius toward the corresponding Lindblad resonance (provided that this is in the disk). The wavelength decreases indefinitely, and the wave packet decays through phase mixing. This means that long-lived spiral patterns require replenishing. The problem of maintaining the density wave against the dispersive effect associated with the group velocity may be as serious as the original problem of maintaining the spiral structure against the differential rotation. Toomre suggests three possible processes for maintaining the density wave: (*a*) local instabilities of the disk; (*b*) excitation by tidal forces from outside, e.g., from a companion; (*c*) that spiral waves are the consequence of some more basic density asymmetry like a central barlike wave. He favors possibility (*c*).

Simkin (1970) also studies the possibility (*c*). She takes the simple model of a uniformly rotating infinite uniform isothermal gas disk disturbed by a point mass in circular orbit about the center with a different angular velocity. This produces two outward traveling spiral responses in the gas.

On the other hand, Quirk (1971) also investigates possibility (*c*) in an N-body study of a self-gravitating disk. It was not clear whether the persistence of the spiral density waves observed in an earlier investigation (Miller, Prendergast, and Quirk 1970) was due to the self-gravity of the wave or to a central asymmetric barlike structure. Quirk shows that the bar is not necessary, and that the persistence is due to the self-gravity of the spiral arms themselves. He also attempted to drive a spiral pattern with a central rotating bar, but was unsuccessful.

Another problem comes from the "antispiral" theorems. Lynden-Bell and Ostriker (1967) investigated the normal modes for the steady circular flow of a nondissipative gas. Their theorem states that there exists a complete set of normal modes such that no stable one has a spiral structure. Stable spiral modes exist only if stable modes of the same symmetry are degenerate; they then occur in conjugate pairs, one leading and one trailing. Shu (1970) shows that a similar theorem holds for a collisionless stellar disk in the linear theory, for all neutral modes for which resonances do not occur. He infers that the existence of spiral density waves demands one or more of the following conditions: (*a*) there exist strong stellar resonances; (*b*) finite-amplitude effects are important; (*c*) overstabilities are present; (*d*) the disk structure is externally driven. Shu argues that condition (*d*) may be the significant one; it is interesting to see that Kalnajs's (1970) wave fulfils condition (*c*), while Berry and Vandervoort's (1970) analysis corresponds to condition (*b*).

Lin (1970) suggests that the density waves are initiated by gravitational clumping near the corotation radius, where $\Omega(R) = \Omega_p$. Shu, Stachnik, and Yost (1971) test this possibility for three galaxies by choosing Ω_p equal to the observed angular velocity at the outermost edge of the H II region distribution; they then compare the shape of the predicted density wave with the observed spiral structure. Observational and procedural uncertainties, however, make this test rather inconclusive.

Marochnik and Suchkov (1969) adopt a model galaxy with two subsystems, one differentially rotating and representing Population I, the other nonrotating and representing the halo. They put forward the interesting suggestion that inverse Landau damping of disturbances in this system would lead to finite-amplitude spiral waves. They find no preference for leading or trailing waves, and find that $\Omega_p < \Omega(r)$ is a necessary condition for this process. It is worth pointing out here that, although the Galaxy certainly has a substantial stellar halo, there do exist spiral galaxies like M33 which have no significant visible spheroidal component. On the other hand, the old disk stars contain most of the disk mass, and there is certainly an asymmetric drift of the old disk stars relative to the gas component in *any* spiral system: the relevance of this asymmetric drift to spiral structure does not seem to have been considered yet, and it may be important here.

In summary, while there is now good evidence that the density-wave concept is relevant to spiral structure, the problem of the origin of these waves and their maintenance against the natural dispersion remains wide open.

9.6.3 (iv). *Shock formation in spiral galaxies.*—The narrow dust lanes and H II distributions frequently associated with spiral arms suggest the possibility of shocks in the gas flow associated with the spiral arms. W. W. Roberts (1969) takes as the basic kinematical state a gas flow with circular streamlines and in centrifugal equilibrium. He then imposes a tightly wound spiral gravitational potential and calculates the response of the gas by looking for particular solutions satisfying two conditions: (1) The flow passes through two periodically located shocks which lie along equipotentials of the imposed spiral field. (2) The gas flows along a narrow, closed, nearly concentric stream tube about the galactic center and repeats itself every half-revolution of the gas flow around the disk. The parameters for the flow are the inclination i of the spiral arm to the circumferential direction, the pattern speed Ω_p, the relative amplitude F of the imposed spiral field to the axisymmetric field, the radius ϖ of the stream tube, and the mean turbulent velocity $a(\varpi)$ of the gas. When these parameters are specified and the conditions (1) and (2) met, the shock location relative to the spiral arm is determined. For $\tan i = 1/7$, $\Omega_p = 12.5$ km s^{-1} kpc^{-1}, $F = 5$ percent, $\varpi = 10$ kpc, and $a = 10$ km s^{-1}, he finds that the shocks lie just to the inner edge of the spiral arms when the arms are trailing; the ratio of maximum gas density to mean gas density is about 5. With the Schmidt model, the outer boundary for a two-armed shock pattern is 12–13 kpc for $F = 5$ percent. For $F \lesssim 1$ percent, there is no possible shock solution except near 3 kpc, where free shock modes are possible in the absence of an imposed field. Leading shock patterns are also possible. In this case the shock occurs at the outer edge of the spiral arm, so it may be possible observationally to distinguish leading spiral galaxies from trailing spirals.

Support for this concept comes from a high-resolution radio continuum study of M51 by Mathewson, van der Kruit, and Brouw (1972). They show clearly how the arms defined by the continuum emission lie along the dust lanes on the *inside* of the optical arms. Since the continuum emission is strongly dependent on the gas compression, they identify the radio arms with the shocks of the theory. On the other hand, Bok (1970) points out that the strongest optical absorption associated with the Carina arm of our Galaxy occurs at the *outer* edge of the arm.

10. BARRED SPIRAL GALAXIES

We know from Section 4.3.2 and table 4 that the barred spirals have a relatively small bar immersed in an approximately axisymmetric disk, so the asymmetry associated with the bar is limited to the inner parts of these systems. In this section we describe some theoretical topics associated with this asymmetry; a more detailed discussion of of barred galaxies is given by de Vaucouleurs and Freeman (1970*b*), denoted dVF.

10.1. The Formation of the Bar

Given the absence of a detailed formation theory for disklike galaxies, the particular problem of the formation of a bar in some systems is probably not fundamental. For example, Lin, Mestel, and Shu (1965) have integrated the equations of motion for the cold collapse of nonrotating uniform ellipsoids: they show how any small initial prolateness is rapidly amplified during the collapse. This should be relevant to the early stages of the collapse of a protogalaxy, when self-gravity probably dominates the pressure and centrifugal forces. Lynden-Bell (1964) has shown how the collapse of a cold rotating uniform oblate spheroid is unstable to small perturbations which tend to elongate the spheroid along a diameter. These two analyses suggest that the formation of a barlike structure is an entirely natural gravitational phenomenon, and that it is unnecessary to invoke magnetic or explosive processes to produce these bars (see also Fujimoto 1968*b*).

10.2. Some Properties of Bar-like Stellar Systems

Very little is known about the dynamical properties of rotating barlike stellar systems. To study these properties, realistic self-consistent models of asymmetric systems would be useful. These models should be uniformly rotating because, unless they are short-lived features, the bars they represent should be uniformly rotating; this is borne out by observation (e.g., Burbidge, Burbidge, and Prendergast 1960*b*). So far Freeman's (1966*b*) models are the only ones available: these are a two-parameter family of uniform-density prolate ellipsoids, rotating end over end. Although they are not realistic, they do have an unexpected property which seems to be present in real barred systems.

Consider a general uniformly rotating bar, in a steady state referred to the rotating frame. There is only one isolating integral of the motion in this system, the Jacobi integral

$$J = \tfrac{1}{2}c^2 + \Phi_1 . \tag{122}$$

Here Φ_1 is the potential for the centrifugal plus gravitational force field, so the distribution function $f = f(J)$ depends only on J. The average velocity $\langle \mathbf{c} \rangle$ of stars, referred to the rotating frame, is then identically zero because

$$\int \mathbf{c} f(J) d^3c \equiv 0 \tag{123}$$

by the symmetry of J in $\mathbf{c}$. However, in Freeman's uniform ellipsoids, J is not the only isolating integral, so $\langle \mathbf{c} \rangle$ is not identically zero. It turns out that the mean stellar motion shows a strong circulation *counter* to the sense of rotation, just as in the fluid Riemann ellipsoids (see Chandrasekhar 1969). This circulation is so strong that near the minor axes of the bar the tangential velocity of the stars in the bar (referred to a nonrotating frame) is in the opposite sense to the rotation of the figure of the bar (see fig. 33). In NGC 4027, an ideally oriented late-type barred spiral, this remarkable predicted effect has been observed (de Vaucouleurs, de Vaucouleurs, and Freeman 1968). The effect was evident in seven stellar absorption lines on three independent spectrograms, so it is probably real. If this circulation is a general property of the bars in SB galaxies, then most stars in the bar have two isolating integrals of the motion: the second integral is probably like the third integral in the Galaxy.

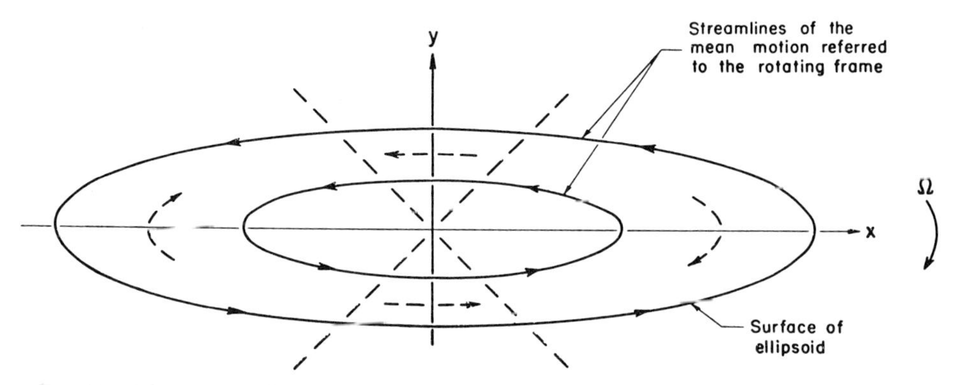

FIG. 33.—The mean stellar motion in a stellar ellipsoid. The broken arrow indicates the sense of the mean circulation referred to an inertial frame. (From de Vaucouleurs and Freeman [dVF] 1970*b*.)

10.3. FORMATION OF SPIRAL STRUCTURE

We have already stressed the need for a unified spiral structure theory, relevant to SA and SB systems, because of the smoothness of the transition SA → SAB → SB. The theories described in § 9.6 are directed at understanding the spiral structure of axisymmetric systems: we know from § 9.6.3 (iii) that the extension of the density-wave picture to the SB systems may not be straightforward. In the same spirit, much of the work on the structures of SB galaxies depends on the properties of test-particle orbits in the gravitational field of a rotating bar and therefore on the essential asymmetry of the bar itself; it is probably of little use for understanding the structure of the SA systems. However, the particle-orbit theory for the spiral structure and rings in barred galaxies and for some of the features of the highly asymmetric SBm systems (§ 10.5) seems to have some success and probably points to some relevant dynamical process. It seems worth describing this work briefly.

Following the discovery of large-scale gas motion of 50–100 km s^{-1} along the bars of some barred spirals (see dVF), several authors have described the possible formation of spiral arms in SB galaxies from the outflowing matter. There are some interesting results: (*a*) Model the bar by a rapidly rotating ellipsoid of typical mass and dimensions. The gravitational torque of this bar can support a steady outflow along the bar, against the

Coriolis force, of order 100 km s^{-1}, as observed. (*b*) Follow numerically these outflowing particles after they leave the bar: they define quite realistic spiral arms as long as the outflow continues (see dVF for details).

The rate of mass loss from the bar ends is estimated at about 1 $\mathfrak{M}_{\odot}$ $year^{-1}$ (de Vaucouleurs and de Vaucouleurs 1963). The outflowing matter slowly carries mass and angular momentum away from the bar, and this causes the bar to evolve dynamically, on a time scale which turns out to be of the same order as the galaxy lifetime ($\sim 10^{10}$ years). It proved possible to follow the evolution of the self-consistent ellipsoidal systems mentioned in § 10.2, as they lost mass and angular momentum, because the evolution time scale is much longer than the orbital time for a star; the adiabatic invariants (see § 2.6) associated with the isolating integrals of the motion are then conserved, and it could be shown (Freeman 1966*b*) that an ellipsoid stayed within the family of uniform ellipsoids as it evolved. The result is that a typical ellipsoid becomes stubbier and more dense, and rotates more rapidly, which suggests that the evolution of individual barred systems would proceed in the sense SBc to SBa. This does not necessarily mean that systems which are now SBa are older than those which are now SBc, because the evolution time scale may itself be a function of the total mass and angular momentum.

Antonov (1961) and Goldreich and Lynden-Bell (1965*b*) have described another possible scheme for the formation of spiral arms in SB systems: A uniformly rotating bar with a realistic density distribution has X-type neutral points of the centrifugal plus gravitational forces near the bar ends. Material is lost from these points, forming trailing arms and removing angular momentum from the bar. This again makes the bar contract, and both its angular velocity and density increase. If the angular velocity increases fast enough, then the neutral points move inward through the bar and the mass loss continues. However, there is evidence (Freeman 1966*a*) that the angular velocity does not increase rapidly enough: the bar appears to become progressively more bound, and the neutral points move outward, rather than inward, from the bar ends.

Julian and Toomre (1966) suggest an interesting mechanism for spiral-structure formation for SB systems (see also § 9.1.2). They consider the response of a disk of stars to a steadily orbiting mass point, and find a local trailing density disturbance in the disk around this mass point. They suggest that similar trailing wakes at the bar ends of SB galaxies could produce the spiral structure. This raises again a fundamental observational question: To what extent is the disk population of a spiral galaxy involved in the spiral structure?

10.4. Formation of Ring Structure

The SB(r) systems have a bright ring encircling the bar (see *The Hubble Atlas* [Sandage 1961] for examples). The composition of these rings is not yet known. Their presence in systems at all stages of the classification from SBO/a to SBcd strongly suggests that they belong to the old disk population. SB(r) systems are not particularly rare: in de Vaucouleurs's (1963*a*) sample of 366 SB spirals, 77 are classified SB(r) and 95 SB(rs). We would like to know the physical parameters which determine whether a barred spiral has the (r) or the (s) form.

It turns out that the simple theory of particle orbits in the gravitational field appears to give a good account of the ring structures. Danby (1965) first investigated this possibility. He showed how the rings could be explained by orbits trapped in an annulus about the bar. However, for his model (a Jacobi ellipsoid) these orbits were all of improbably high energy. The existence around the bar of regions in which particles are preferentially trapped will be fairly sensitive to the dimensionless numbers of the potential field, such as $\Omega^2/4\pi G\rho$, where Ω is the bar angular velocity and ρ the mean density of the bar. This is the drawback to using Jacobi ellipsoids, for example, to study orbit properties. They form a one-parameter sequence: its stable members have $0.0710 < \Omega^2/4\pi G\rho < 0.0935$, and this is rather a small range. The corresponding potential field may not be a fair representation for that of an SB system, and then inferences about particle trapping could be unreliable. In particular, the values of $\Omega^2/4\pi G\rho$ suggested by observation seem to be somewhat smaller than for the Jacobi ellipsoids (dVF).

Freeman and Harrington (unpublished: see dVF for details) studied orbit properties for relatively slowly rotating barred systems. For a prolate spheroidal bar with an axial ratio of 5, mass $\mathfrak{M}$, semilength L, and angular velocity Ω, there is one dimensionless number, defined for convenience as $Q = 3G\mathfrak{M}/(\Omega^2 L^3)$ (the breakup angular velocity corresponds to $Q \approx 1$). For a fairly slowly rotating bar ($Q \approx 15$) there exists a family of approximately circular periodic orbits in the plane of rotation, encircling the bar near the bar ends. These orbits lead the bar slightly, are highly stable, and cover a fairly wide annulus (width $\sim L/4$) around the bar. It can be shown numerically, by surface-of-section studies (see § 5), that it is easy to trap enough matter near these stable periodic orbits to produce the apparent increase in luminosity associated with the observed rings. However, as Q is decreased (angular velocity increased), the annulus covered by the stable circular periodic orbits becomes less wide and the matter trapping efficacy decreases. For $Q \lesssim 8$ the periodic orbits no longer exist and the trapping of matter to form the ring structure is no longer possible; so, in this picture, the SB(r) systems are slower rotators, in a dimensionless sense, than the SB(s) systems. There is no real observational evidence yet either for or against this inference. There is, however, evidence that the barred spirals do rotate more slowly than the ordinary SA spirals (Mayall and Lindblad 1970).

10.5. The Magellanic Barred Spirals

We pointed out in § 4.3.2 that galaxies earlier than about SBc have the bar located centrally in the disk. In systems later than SBd, the bar becomes displaced from the disk center; for the Magellanic [SB(s)m] spirals in particular, this asymmetry is a basic and characteristic property of the mass distribution. It shows up very clearly in the surface distribution of supergiant stars, neutral hydrogen and total luminosity (fig. 34*a*), and is undoubtedly associated with the asymmetry, with respect to the center of the bar, of the rotation curves observed in Magellanic systems (e.g., LMC [fig. 34*b*]).

Figure 35*a* shows schematically the main structural features of SB(m) systems. The separation of the bar and disk centers, C_B and C_D, is of order 1 kpc, as is the distance between C_D and the center of symmetry of the rotation curve, C_R. The spiral structure includes a major arm A_1 which dominates the spiral pattern, and one or more of the minor arms A_2, A_3, A_4, usually in that order of prominence (see reproductions of SBm

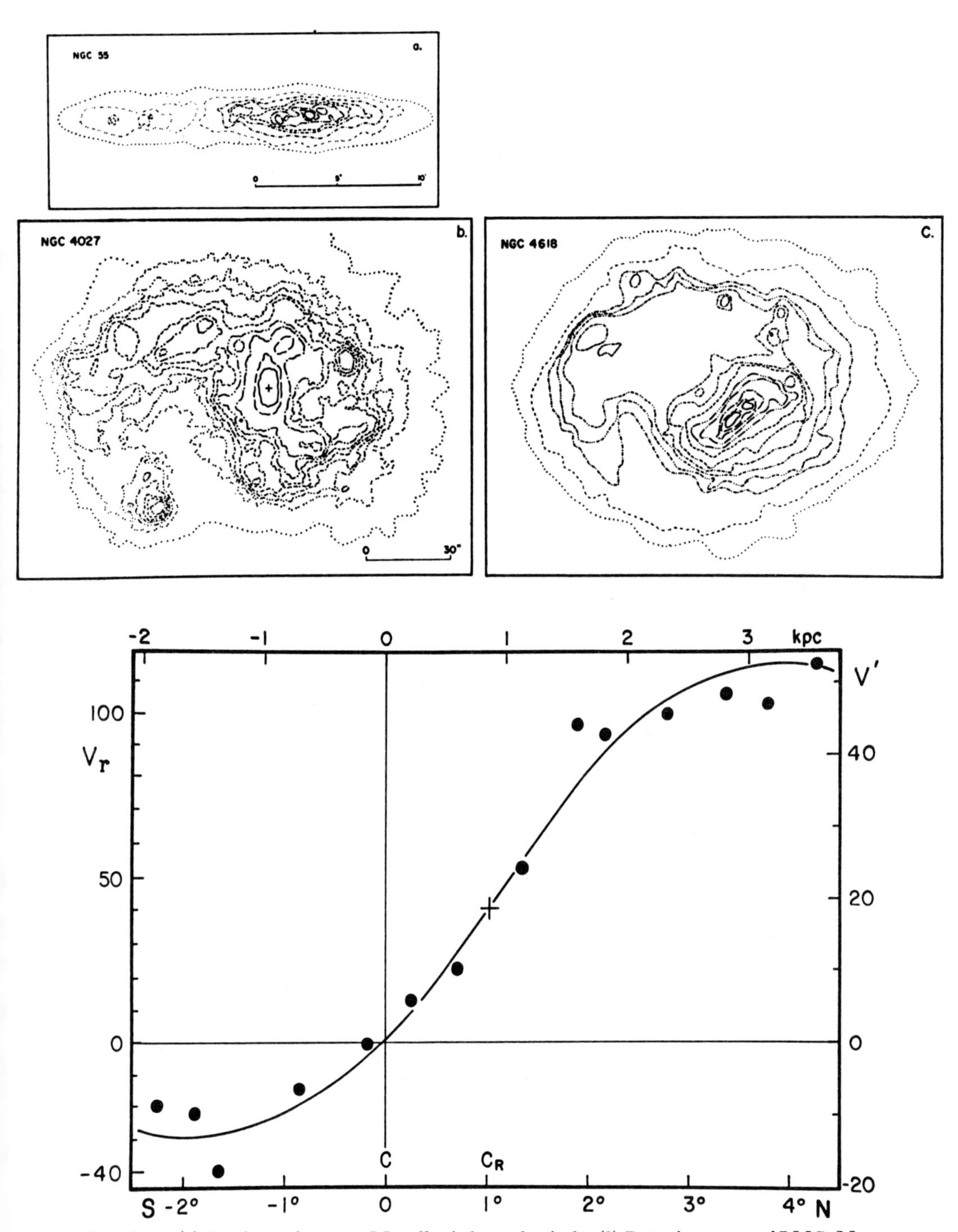

FIG. 34.—(*a*) Isophotes for some Magellanic barred spirals. (*b*) Rotation curve of LMC. Mean optical velocities of H II regions are plotted against distance to optical center *C* of bar. V' is the observed velocity and V the rotational velocity. The center of symmetry defines the rotation center C_R. (From dVF.)

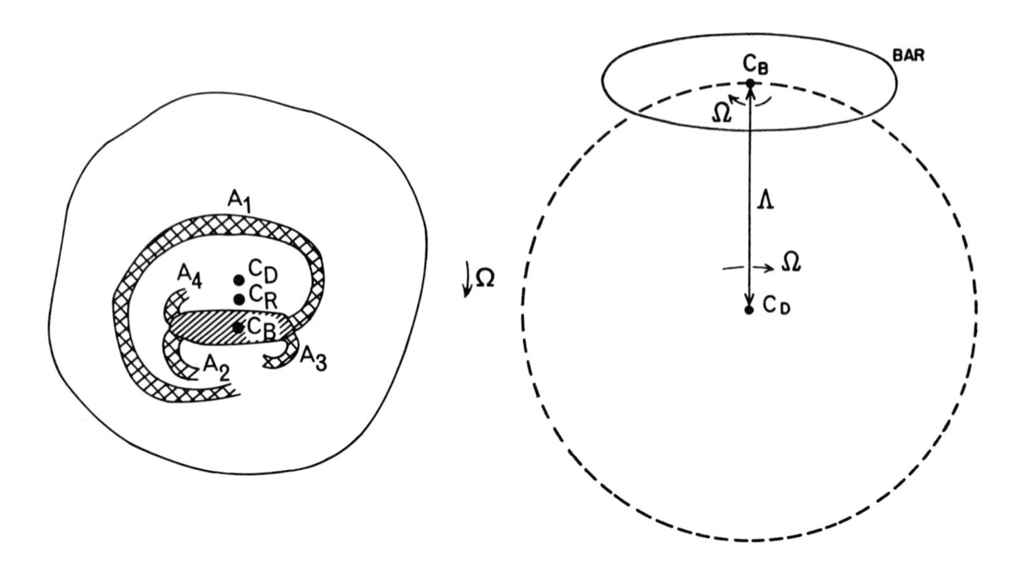

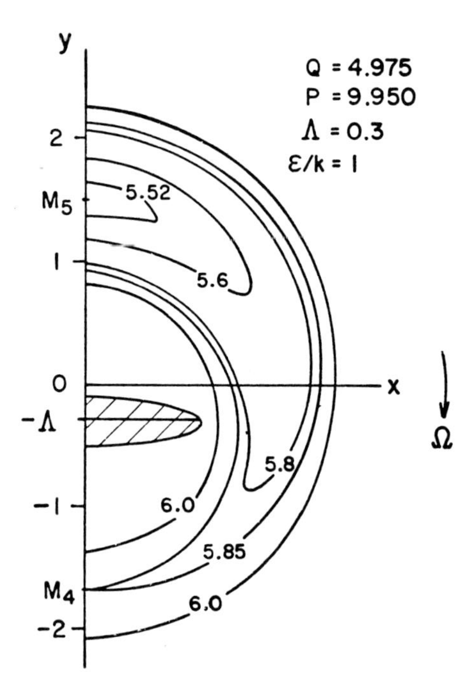

FIG. 35.—(*a*) Schematic structure of the SB(m) systems. (*b*) The model for the SB(m) systems. (*c*) Equipotentials of the combined gravitational + centrifugal force field in the SB(m) model. (From dVF.)

galaxies in dVF). In a sample of 366 barred spirals (de Vaucouleurs 1963*a*), there are 30 SB(m) systems which share this basic asymmetrical structure.

The internal motions in SB(m) spirals are interesting: (i) There is clear evidence for gas motions of order 50–100 km s^{-1} along the bar in several systems. (ii) The H I velocity fields in the LMC and SMC are very complicated and are certainly not consistent with a pure rotation field. (iii) The centers of approximate symmetry of the rotation curves derived for the LMC, the SMC, NGC 4631, NGC 4027, and NGC 55 are in every case displaced from the center of the bar by 0.5–1.5 kpc, in the direction of the major trailing arm (see figs. 34*b* and 36*a*). (iv) A detailed analysis of the velocity field in NGC 4027 provides the mean rotation curve (assuming circular motion) shown in figure 36*a*. On the major arm side of the bar, the rotational velocity appears to decrease from 180 km s^{-1} at $R = 1.2$ kpc to 70 km s^{-1} at $R = 2.4$ kpc and increases again to 180 km s^{-1} at $R = 3.6$ kpc ($R = 0$ is the bar center). There is no corresponding feature in the velocity curve on the other side of the bar (see dVF for references).

A dynamical theory of the Magellanic barred spirals would be particularly worthwhile. The LMC and SMC are only about 60 kpc from the Sun, so it is possible to make very detailed photometric and kinematic studies of these two systems to guide the theory. First we need to choose a model for the gravitational potential in these systems. Because of the observed noncircular motions, the rotation curves give an unreliable measure of the force field. Surface photometry of the LMC, NGC 55, NGC 4027, NGC 4618, etc. (see dVF for references), again suggests that most of the mass lies in the disk and bar populations (cf. § 4.3.2), so we assume that the potential field is determined entirely by these two components. A simple model, consistent with the observed separation of the centers C_D and C_B of the disk and bar, is shown in figure 35*b*. C_D and C_B are separated by a distance Λ, and we assume that the bar rotates (synchronously) about C_D with angular velocity Ω, in such a way that the potential seen by an observer rotating with the bar is time-independent.

This is an unusual model for the potential field of a galaxy. It cannot be justified dynamically at this stage, but is used because it is simple and includes the relative displacement of the bar and disk populations which is the fundamental property of the Magellanic barred spirals. The model is described in detail by dVF: they take for the flat disk the potential (Eggen, Lynden-Bell and Sandage 1962)

$$\Phi_d(R) = \frac{G\mathfrak{M}_d}{k + \sqrt{(k^2 + R^2)}}, \tag{124}$$

where $\mathfrak{M}_d$ is the disk mass and k is a length scale, and the bar is a heterogeneous prolate spheroid of axial ratio $a/c = 5$ and mass $\mathfrak{M}_b$. In a real equilibrium stellar system which this model is intended to represent, the relation between Ω, $\mathfrak{M}_b$, $\mathfrak{M}_d$, Λ, k, and $\epsilon = \sqrt{(a^2 - c^2)}$ will probably not be simple; so the significant dimensionless numbers for the model,

$$Q = \frac{105G\mathfrak{M}_b}{32\Omega^2\epsilon^3}, \quad P = \frac{G\mathfrak{M}_d}{\Omega^2\epsilon^2 k}, \tag{125}$$

and k/ϵ, Λ/ϵ, are chosen from observation. For example, in the LMC, $\Lambda = 0.5$ kpc, $\epsilon = 1.5$ kpc, and $k = 1.3$ kpc. About 15 percent of the total light comes from the bar,

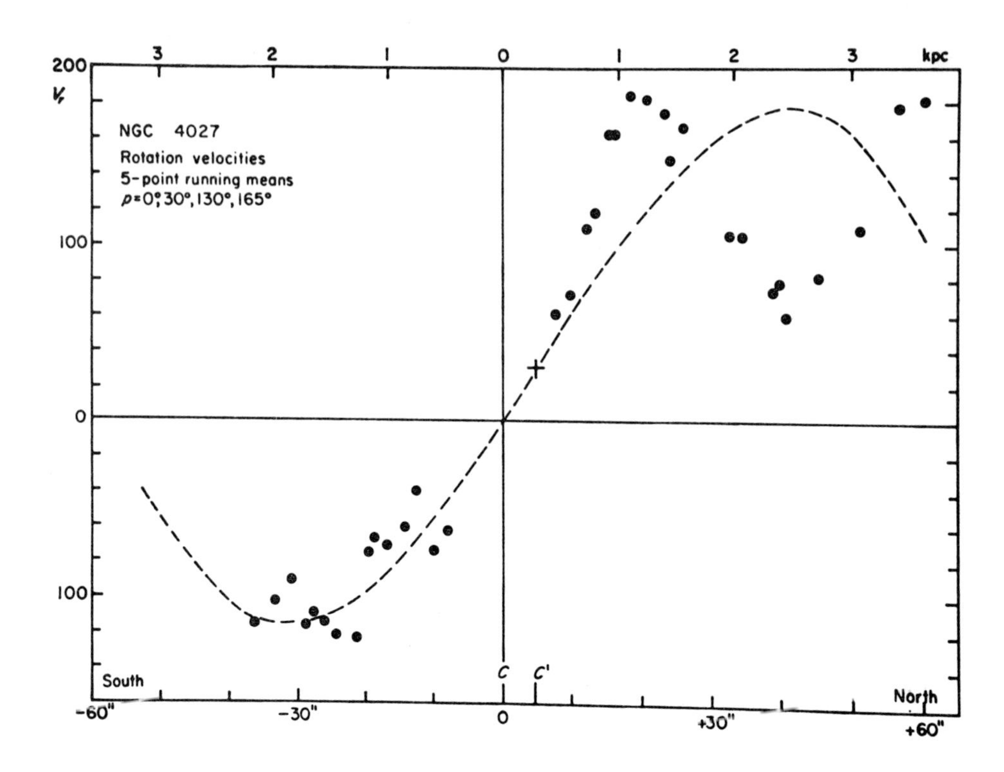

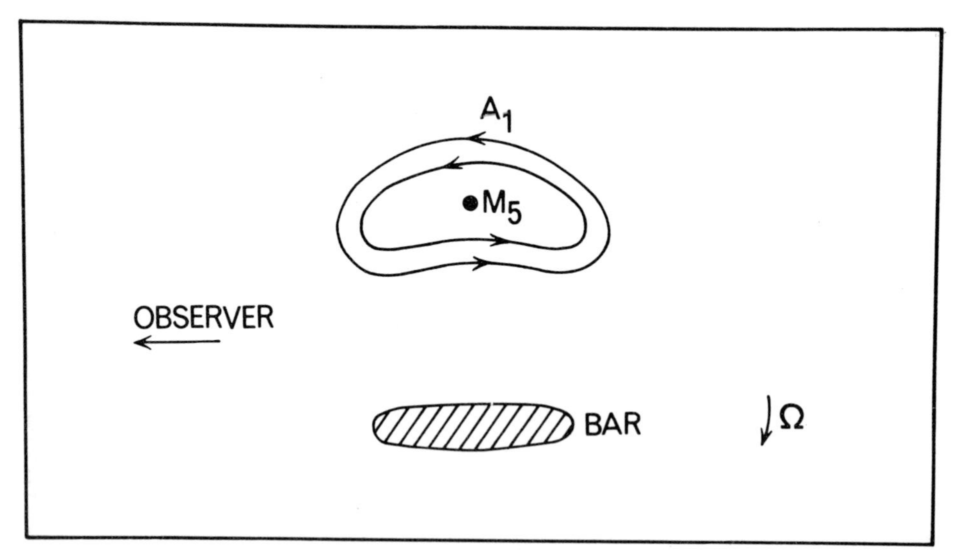

FIG. 36.—(*a*) The rotation curve for NGC 4027. *Broken line*, best-fit cubic to the mean points, showing displacement of center of symmetry C' from the bar center C. (*b*) Trajectories in the plane of rotation of matter trapped about the stable neutral point M_5 in the SB(m) model. (From dVF.)

so $P/Q \approx 2$, if the mass/luminosity ratio is the same in the disk as in the bar. The angular velocity Ω is taken as 50 km s^{-1} kpc^{-1}. The total mass is probably about $1.5 \times 10^{10}\,\mathfrak{M}_\odot$, so $Q \approx 4$. Similar values are found for NGC 4027. The final model adopted by dVF has $(Q, P, k/\epsilon, \Lambda/\epsilon) = (4.97, 9.95, 1, 0.3)$. We see immediately that the bar is rotating relatively slowly; cf. § 10.4. (The breakup angular velocity corresponds to $Q \approx 1$.)

Figure 35c shows the equipotentials of the combined gravitational plus centrifugal field. There are only three neutral points (instead of the five that exist for $\Lambda = 0$): M5 is stable, and M4 is unstable. Freeman and Harrington (unpublished: see dVF for details) studied the properties of stellar orbits in the vicinity of M5 in an attempt to understand the rotation curve of NGC 4027 (fig. 36*a*). They show that matter can be trapped within a limited region about M5, which is located at about 2 kpc north of the bar center of NGC 4027, on the major arm side. This matter circulates in a retrograde sense around M5 (see fig. 36*b*). Now consider the rotation curve that would be observed for this system by an observer located at a great distance to the left-hand side of figure 36*b*; this location is appropriate to NGC 4027. For points in the region enclosed by the orbit A_1, the matter circulating around M5 would cause an increase over the "local circular velocity" in the apparent rotational velocity for $R < 2$ kpc and a corresponding decrease for $R > 2$ kpc, with approximately the observed amplitude. Because the natural point M4 is unstable, there is no corresponding velocity disturbance on the south side of the bar.

The dynamical theory of the Magellanic barred spirals has not progressed far; the general problems of the asymmetrical disk and bar configuration and the spiral structure remain untouched.

11. FORMATION AND EVOLUTION OF GALAXIES

Our present working picture for the formation of the Galaxy comes from the concept of populations and the observed correlation of kinematic and chemical properties of stars. It is usually assumed that the most metal-weak stars, which have the most eccentric orbits, are the oldest. This leads (§ 7) to the concept that the Galaxy collapsed on a time scale $\sim 2 \times 10^8$ years from an extended protosystem. The oldest Population II objects formed early in this collapse, before much metal enrichment had occurred. The remainder of the protosystem continued its contraction. Rapid metal enrichment took place, the less metal-poor Population II stars formed, and the remaining enriched gas, having dissipated much of its energy, settled to a disk in centrifugal equilibrium before forming stars. The metal abundance in this disk has about the solar value, and star formation in the disk is still going on now.

We will first discuss this working picture, and then discuss some further chemical and dynamical problems associated with the formation and evolution of galaxies. Again it will be obvious how the Galaxy and external galaxies provide complementary information for these problems. It should be pointed out now that not much is known about many of these questions, so our discussion will often be limited to stating the problem and some speculation.

11.1. The Metal-Enrichment Picture

We know that the halo of the Galaxy has a spatial gradient of metal abundance. (i) Sandage (1969) shows that the mean W velocity for subdwarfs increases with ultraviolet excess. For example, subdwarfs in the two intervals $\delta(U - B) = 0.16$–0.20 and 0.21–0.31 have $\langle |W| \rangle = 49$ and 93 km s^{-1}, respectively; these correspond to maximum heights above the galactic plane of about 800 pc and 2 kpc. Sandage infers that at greater heights (5–10 kpc) above the plane, only very metal-weak objects would be found, while subdwarfs now close to the plane have a wide range of abundance. (ii) The most metal-deficient RR Lyrae stars, with $\Delta S \geq 5$, form a much more extended subsystem than those with $\Delta S < 5$ (see Blaauw 1965). (iii) Kinman (1959*a*) shows clearly how the subsystem defined by the weak-lined globular clusters (spectral types F2–F6) has a radius about three times greater than that for the strong-lined clusters (G0–G5). This chemical gradient suggests directly that the chemical mixing in the halo was not very efficient.

In the working picture it is assumed that the oldest halo stars are those with the lowest metal content, so that objects which now populate the outer parts of the halo were the first to form. As the protogalaxy became enriched, it formed the more metal-rich objects which now populate the inner parts of the halo. However, there is now evidence (Schmidt 1959; Sanduleak 1969) that the rate R of star formation is related to the gas density approximately by $R \sim \rho_{\mathrm{gas}}^n$, where $n \approx 2$. In a collapsing protosystem, we would expect the gas density to increase toward the center, so star formation should begin at the center, and the metal enrichment should be most rapid there because $n > 1$. We could then expect that some of the relatively *metal-rich* objects in the central regions of the Galaxy are in fact *older* than the more widely distributed *metal-poor* objects. Note that the age differences involved are probably only of the same order (2×10^8 years) as the dynamical time (Sandage 1970): nevertheless, details of the order of events within this short time are vital for understanding the formation and enrichment processes, and it is worth questioning whether the metal-weak stars in the outer parts of the system really formed first. It may be that their low metal abundance results from their location at formation, and their metal abundance or $\delta(U - B)$ is not a unique measure of their age within this short time interval. Note too that while we talk glibly of metal enrichment, the astrophysical details of this enrichment are by no means clearly understood (see, however, Arnett 1971).

We know that the halo shows a substantial gradient in metal abundance among a sample of very old objects whose ages probably differ by only a few times 10^8 years. Contrast this with the disk of the Galaxy. The disk has objects of all ages; however, even the oldest nearby disk stars have nearly the solar abundance in the mean (Eggen and Sandage 1969). Note that these oldest disk objects may in fact be as old as the halo. For example, present estimates for globular-cluster ages are $(10$–$12) \times 10^9$ years (Sandage 1970); the old disk cluster NGC 188 is $(8$–$10) \times 10^9$ years old (Sandage and Eggen 1969), and from figure 30 we see that there are evolved disk stars significantly older than NGC 188. Further, there is no evidence, as far as we know, for a spatial gradient of metal content in the disk of the *Galaxy;* however, Searle (1971) shows that the nitrogen/hydrogen ratio may change by about a factor 10 across the disks of some (external) Sc spirals.

In the working picture, the gaseous disk matter derives much of its metals during the rapid enrichment phase of the halo. This seems unlikely to us, for two reasons. (i) It would need active mixing between the halo and the predisk gas, but from the chemical gradient in the halo there seems to have been little mixing in the halo itself. In particular, most halo stars near the Sun have a much lower metal abundance than the nearby disk stars, so *local* mixing between halo and disk would not have contributed much to the metal content of the disk. On the other hand, the high angular momentum and low random velocities of the nearby disk objects make it unlikely that they were ever in the less metal-poor inner regions of the halo. (ii) There exist objects like M33 and the LMC with apparently normal abundances but with surface photometry showing that they have negligible spheroidal bulge or halo components (de Vaucouleurs 1959*b*, 1960). This would suggest that the disk is capable of processing its own metals during *its own* first burst of star formation.

Because of the relative flatness of the disk, we know that star formation in the disk did not take place before the protodisk had flattened to a shape similar to its present shape, i.e., $\lesssim 1$ kpc half-thickness. If the disk did produce its own metals, we should then see some early generation metal-poor stars with disk kinematics. A few such stars are known (Sandage 1969), but a careful analysis would be needed to decide whether there are enough of them. Note that while almost all halo subdwarfs are found from their proper motions, relatively few metal-weak disk objects would be discovered in this way.

If this concept of the disk doing its own enrichment seems plausible, then much more information, chemical and kinematic, is needed about the oldest disk stars to decide whether or not it is correct. Some useful data for this problem comes from Hearnshaw (1972), who shows that there is some correlation of metal deficiency with age, orbital eccentricity, and maximum height above the galactic plane for a sample of old evolved disk stars whose ages (3 to 11×10^9 years) can be estimated fairly accurately. These objects all have $z_{\max} < 800$ pc and orbital eccentricity $e < 0.5$, so they do appear to be genuine disk stars. Information about the relative metal abundances of the bulge and disk components in external systems is also essential for this problem; for example, in M31 the halo may be more like the old Population I in the Galaxy than a globular-cluster-like population (see chapter 2, and van den Bergh 1969).

Galaxies more massive than about $10^{10}\ \mathfrak{M}_\odot$ seem mostly to have normal abundances. M32 ($\sim 3 \times 10^9\ \mathfrak{M}_\odot$) appears to be slightly deficient, and NGC 205, at about the same mass, is somewhat more metal-poor. The Sculptor- and Fornax-type dwarf ellipticals, with masses below about $10^8\ \mathfrak{M}_\odot$, are very metal-deficient (see chapter 2). The reason for the apparent decrease of metal abundance with galactic mass is not really understood.

It is usually believed that metal enrichment takes place through successive generations of star formation within the first few times 10^8 years of the galaxy's life. Then presumably the low-mass galaxies complete their star formation in relatively few generations; or alternatively, their stellar luminosity function at the time of star formation included relatively few of the massive stars which are believed to be the site of nucleosynthesis (see, e.g., Arnett 1971). Possibly the lower escape velocities in the lower-mass systems are relevant here: they will be less able to retain the high-energy metal-rich ejecta from supernovae through which the enrichment is believed to take place. (This suggestion is due to Sandage.)

One exception to this ordering is the S0 system NGC 5102. Its absolute magnitude $M_B \approx -18.6$, so its mass is probably about $5 \times 10^{10}\ \mathfrak{M}_\odot$: however, its unpublished integrated *UBVRI* colors and its spectrum are like those for the globular cluster M5.

11.2. Some Kinematic Problems

The high velocities of halo objects near the Sun, and the extended nature of this population, have led to many different pictures for the formation of the halo. Suggested sources for the halo's relatively high energy include (i) that it collapsed from a very extended protocloud (Eggen, Lynden-Bell and Sandage 1962), (ii) that the gaseous protocloud was highly turbulent (Schwarzschild 1964), (iii) that the halo stars were shot out from the nucleus of the Galaxy in a giant explosion (Unsöld 1969). There is now a large amount of information on the kinematics of the halo (see Oort 1965), but the interpretation is not yet clear enough to define its formation picture uniquely. We have four comments, mainly on the kinematics of the halo objects.

11.2.1. *The angular momentum/mass ratio for the halo.*—One essential piece of information for this problem is the ratio $\mathfrak{H}/\mathfrak{M}$ of total angular momentum to total mass for classes of halo objects. This quantity is difficult to estimate from the kinematics of nearby subdwarfs: although their space motions are relatively well known, these stars are found mainly by their large proper motions, and this selection effect biases the known subdwarf population strongly toward the objects with the lowest angular momentum. Further, these stars are intrinsically faint, so their spatial distribution is not known. At present, estimates from the RR Lyrae stars and the globular clusters probably give a more reliable estimate for $\mathfrak{H}/\mathfrak{M}$ for the halo: these objects are recognizable at large distances from the Sun, and it is possible to estimate the asymmetric drift for their subsystems from radial velocities alone. Kinman (1959*b*) analyzed the radial-velocity data for the globular clusters. He shows that their asymmetric drift is almost the same as that for the field RR Lyrae stars with periods in the range $0^d.5$–$0^d.65$: this group is the most common in globular clusters, and this suggests a common origin for these RR Lyrae stars in both clusters and the field. From their asymmetric drift (167 ± 30 km s^{-1}) he estimates $\mathfrak{H}/\mathfrak{M}$ for the globular-cluster subsystem and finds that it is almost equal to $\mathfrak{H}/\mathfrak{M}$ for the whole Galaxy, as calculated from the first Schmidt model. On the other hand, the most metal-poor RR Lyrae stars have an asymmetric drift of 220 ± 23 km s^{-1} (see Oort 1965) and it seems unlikely that their $\mathfrak{H}/\mathfrak{M}$ value would be so high. This important problem needs still more attention; however, it is not obvious how these estimates of $\mathfrak{H}/\mathfrak{M}$ could be improved significantly.

11.2.2. *Disk globular clusters.*—The apparent absence of a disklike subsystem of globular clusters in the Galaxy is of interest here. Kinman (1959*a*) shows that even the most flattened sample of globular clusters, those with spectral types between G0 and G5, has an axial ratio of about 2, which is not disklike. Presumably dynamical conditions have not been right for globular-cluster formation in the disk. In contrast, the LMC contains many massive young clusters; these clusters are concentrated to the disk of the LMC, their masses are of order 10^4–$10^5\ \mathfrak{M}_\odot$, and they are dynamically like the globular clusters in the Milky Way, despite their ages of order 10^7 years (Freeman and

Gascoigne, unpublished). Most of the young LMC clusters have radial velocities close to those for the neutral hydrogen near them, so they are probably in nearly circular orbits. However, two are known to have large peculiar velocities ($\sim$80 km s^{-1}), despite their short ages—i.e., they appear to be runaway clusters. In this context we should note also the presence of solar-composition A stars high (up to 12 kpc) in the halo (Rodgers 1971); the origin of these is not yet entirely understood.

11.2.3. *The origin of the orbital eccentricities.*—From Eggen, Lynden-Bell, and Sandage (1962), it seems clear that there is a correlation of the metal abundance with kinematical properties of halo stars in the solar neighborhood, in the sense that stars with the lowest metal abundances have the most eccentric orbits. They argue that the high-eccentricity orbits are associated with a rapid collapse of the protogalaxy (§ 7). Do these high eccentricities simply reflect the infalling orbits, during the collapse phase, of the gas clouds from which these stars formed, or are they the result of mean field relaxation (§ 8.3)? This question is important because, if the first alternative is correct, then the high-eccentricity stars contain more detailed information about the protogalaxy than if the second alternative is the right one; for example, can we reliably infer the size of the protogalaxy, at the time that star formation began, from the maximum observed apogalactic distances of the nearby subdwarfs? It is not obvious how this question can be resolved. However, in favor of the first alternative is that nearly all stars with high eccentricities and large apogalactica have very low metal abundances; because of the observed chemical gradient in the halo, we would probably expect a range of abundances for these kinematically extreme stars if mean field relaxation were responsible for their high energies.

There is the possibility that the halo objects were shot out of the nucleus of the Galaxy. Any subsystem originating in this way must have a very low value of $\mathfrak{H}/\mathfrak{M}$ at the time of its origin. In the absence of any large-scale transfer of angular momentum, it would still have a very low $\mathfrak{H}/\mathfrak{M}$ now. There is no evidence that $\mathfrak{H}/\mathfrak{M}$ for the halo is so low. Kinman (1959*b*) finds a fairly high value for the globular clusters. For the known subdwarfs, the apparent angular momentum is low, but this is severely affected by the selection effects mentioned earlier, and the real value of $\mathfrak{H}/\mathfrak{M}$ for the subdwarf system is surely significantly higher than its apparent value. Further, inspection of the Bottlinger diagram for subdwarfs (e.g., Sandage 1969) shows that few have $|h|$ small enough to penetrate within 1 kpc of the galactic center. Note too that there are subdwarfs in *retrograde* orbits with $|h|$ as large as the Sun's value, and they contribute to the apparently low mean value of $\mathfrak{H}/\mathfrak{M}$ for the known subdwarfs (see Eggen 1970, fig. 17). These large negative values of h can be understood readily as originating from quite weak transverse random motions early in the collapse—for example, a 25 km s^{-1} transverse motion at 100 kpc from the galactic center has the same angular momentum as the Sun's orbit.

Finally we consider the cosmogonic significance of the observed velocity dispersions for various kinds of objects in the Galaxy. For example, is the fact that the disk contains subsystems of different flattening and velocity dispersion significant for understanding the formation of the Galaxy? It seems natural to define disk stars as those with $z_{\max} <$ 800 pc: Eggen (1970) shows how the z-distribution of stars with $\delta(U - B) < 0.15$ (i.e., stars of normal abundance or only mildly metal-deficient) drops abruptly at $z \approx 800$ pc.

For disk objects, we know that the asymmetric drift for subsystems of nearby stars is proportional to their velocity dispersion (§ 3.2.1), and that the velocity dispersion increases with age up to a maximum value for $\sqrt{\sigma_{RR}}$ of about 60 km s^{-1} for the oldest disk stars (table 7). We note that (i) a relatively large asymmetric drift for a sample of disk stars means only that the angular momentum per unit mass for the sample *near the Sun* is relatively low: it does not necessarily mean that $\mathfrak{H}/\mathfrak{M}$ for the whole subsystem is low, because the subsystem may be relatively extended; (ii) the correlation between velocity dispersion and age for the disk stars probably results from dynamical interaction of these stars with the spiral structure (§ 9.5). In contrast, to understand the much larger velocity dispersions for the halo objects (e.g., $\sqrt{\sigma_{RR}} \approx 230$ km s^{-1} for the extreme subdwarfs; see table 7) we need to invoke the processes discussed above. For the purpose of our formation picture, it is then probably meaningful to distinguish only two independent populations, defined by the dynamical process which gave each its random energy: the disk (including stars of all ages) and the halo. In other words, while the galactic subsystems with varying degrees of flattening (as recognized by Lindblad) are obviously dynamically significant, it may be that only two *classes* of these subsystems (i.e., the disk and the halo) are cosmogonically significant.

11.2.4. *The anisotropy of the velocity dispersion.*—For the extreme subdwarfs in table 7, the velocity dispersion has components $(\sqrt{\sigma_{RR}}, \sqrt{\sigma_{\varphi\varphi}}, \sqrt{\sigma_{zz}}) = (230, 84, 90)$ km s^{-1}. While the absolute values of these components are probably larger than the real value because of the selection effects already mentioned, it seems unlikely that their ratios should be far wrong. The velocity dispersion is then highly anisotropic. Note that an isotropic dispersion would not be expected if angular momentum is conserved during the collapse phase. For a nearly spherical distribution of stars, φ and z lose their distinction and together define a tangential sphere: this is reflected in the near-equality of $\sigma_{\varphi\varphi}$ and σ_{zz}. A star's φ- and z-velocity components are then determined mainly from its initial values, by conservation of angular momentum. On the other hand, the R-velocity component is determined by its potential energy loss during the collapse plus mean field relaxation; there is then no reason to expect isotropy of $\boldsymbol{\sigma}$.

If we believe that elliptical galaxies and the halo of the Galaxy both collapsed rapidly during their formation (§ 8.3), then the anistropy of $\boldsymbol{\sigma}$ for the halo subdwarfs could lead us to expect that the ellipticals are similarly anisotropic. It is then interesting to see how well King's isotropic models fit the luminosity profiles of elliptical galaxies (§ 6.2).

11.3. Problems of the Large-Scale Mass Distribution

11.3.1. *The ellipticals.*—We know (§§ 4.3.1, 6.2, 8.3) that most normal ellipticals appear to follow fairly closely the $R^{1/4}$ luminosity law, which in turn is close to the profile for a King model with distribution function $f(E) = A[\exp(-\beta E) - \exp(-\beta E_{esc})]$ and $\log(r_t/r_c) \approx 2.20$; the ratio r_t/r_c is a measure of βE_{esc}. This *form* of the distribution function can be explained as the end product of mean field relaxation: we need also to explain why most normal ellipticals adopt the particular value of βE_{esc} that corresponds to the $R^{1/4}$ law. Is this value of βE_{esc} imposed by the mean tidal field in which these galaxies move (as it is, presumably, for the globular clusters in the Galaxy), or does it follow from the processes which determine how much energy is radiated away by a protogalaxy before star formation occurs (see the discussion of Fish's law below)?

Although most normal ellipticals (i.e., $M_B \lesssim -18$) seem to follow the $R^{1/4}$ law, it is now clear that the cluster giants ($M_B \approx -21.5$) and the dwarf ellipticals ($M_B \gtrsim -15$) do not. De Vaucouleurs (1969, 1970) and Arp and Bertola (1971) showed that several giant ellipticals are much larger than expected. For example, M87 ($M_B = -21.3$) has a diameter of at least 1°, with a possible maximum diameter of 1°.5–2°.0 or 0.2 Mpc. Its faint outer regions are described as a separate corona by its discoverers; however, its luminosity profile, shown in figure 37, is well represented over the entire luminosity range by a King model with $\log(r_t/r_c) \approx 2.50$. The tidal radius for this model is 73′, which is consistent with the diameter detected so far. NGC 4889 in the Coma cluster

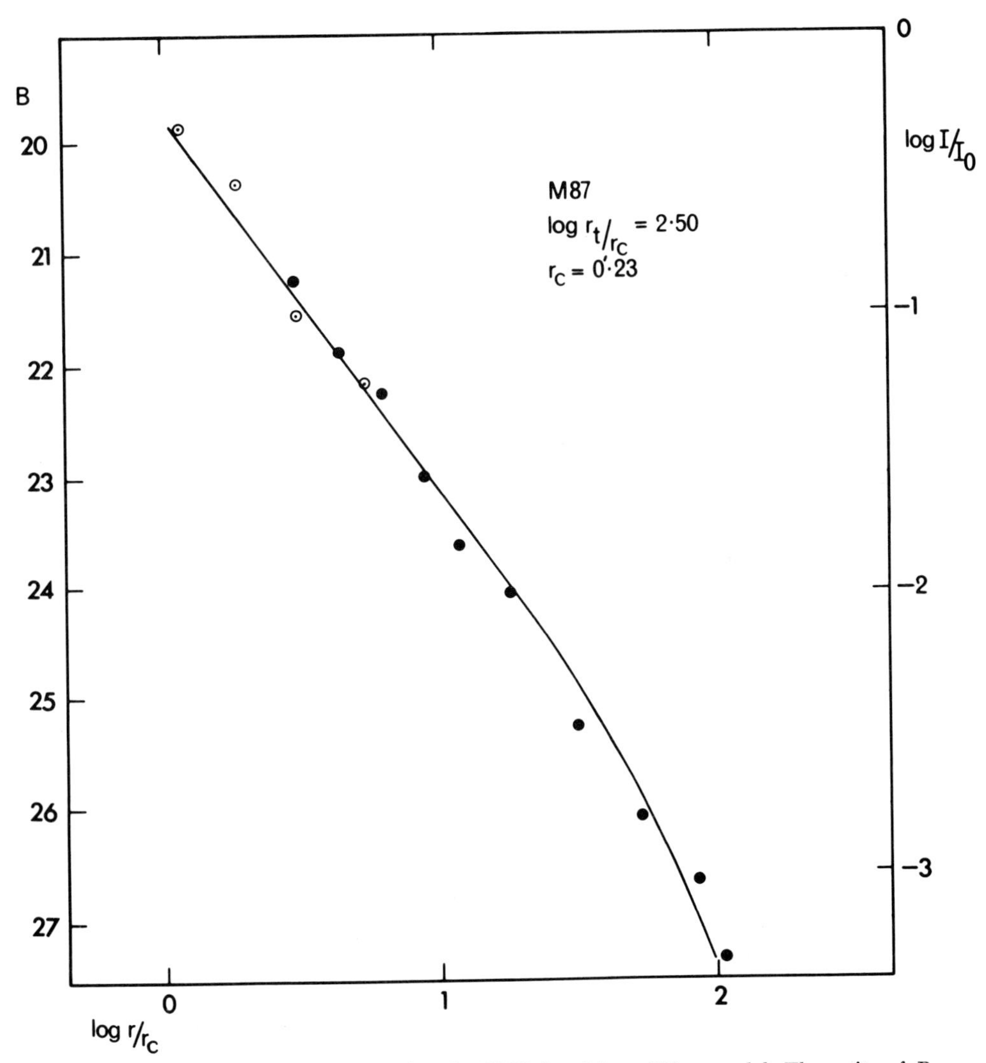

Fig. 37.—The luminosity distribution for M87 fitted by a King model. The units of B are mag arcsec^{-2}.

has a similar profile (de Vaucouleurs 1970). On the other hand, NGC 185 ($M_B = -14.7$) has $\log(r_t/r_c) \approx 1.5$ (Hodge 1963), and the Sculptor-type dwarf ellipticals ($M_B \gtrsim -13$) appear to have $\log(r_t/r_c)$ in the range 0.5–0.75 (derived from de Vaucouleurs and Ables [1968] and Hodge's work; see Hodge 1966 for references). This progression of $\log(r_t/r_c)$ with M_B is summarized in table 8; it is important to establish the reason for this progression.

Next we consider Fish's law (§ 4.3): assuming that the total mass/luminosity ratio is approximately constant for ellipticals, Fish (1964) showed that the potential energy $\Omega \sim \mathfrak{M}^{3/2}$ over about three decades in the total mass $\mathfrak{M}$. The importance of an $\Omega(\mathfrak{M})$ law is that the potential energy now observed in ellipticals is approximately twice the radiation emitted by the protogalaxy during its collapse: this follows directly from the virial theorem. As a result, the form of the $\Omega(\mathfrak{M})$ law characterizes the radiation processes that took place during the collapse, before star formation occurred. Fish argues that $\Omega \sim \mathfrak{M}^{3/2}$ would result if the collapse was stopped by the protogalaxy becoming opaque, by electron scattering, to its own radiation; the protocloud then breaks up

TABLE 8

LOG (r_t/r_c) AND ABSOLUTE MAGNITUDE FOR SOME E GALAXIES

NGC	$-M_B$	log (r_t/r_c)
4486	21.5	2.50
3379	19.5	2.20
185	14.7	1.50
Fornax	12.9	0.50

into stars, and the total energy of the system (and hence its potential energy in equilibrium) is then determined. This is a simple and attractive picture for the collapse and star-formation phases; a detailed study would be well worthwhile, to see if it can be put on a firm theoretical base.

The angular-momentum content of elliptical galaxies is not well known, although it may well be the major physical variable distinguishing them from the disk galaxies. Large-scale rotation in these systems is difficult to measure directly, although the scarcity of intrinsically spherical ellipticals (fig. 7) does suggest (but does not necessarily imply) that most of them are rotating. It may be possible to estimate the $\mathfrak{H}/\mathfrak{M}$ ratio by constructing model systems to reproduce the observed luminosity profiles and the observed change of isophote eccentricity with radius (see Prendergast and Tomer 1970); because there are so many variables, it is not obvious that this approach will be successful.

11.3.2. *The disk galaxies.*—Now we consider the origin of the mass distribution in the spheroidal component or halo of the disk galaxies. Within the framework of the rapid-collapse and mean-field-relaxation picture for the formation of the ellipticals, we might expect the spheroidal components to have mass distributions somewhat like the ellipticals, but modified by the presence of the disk and by rotation; for example, the box-shaped spheroidal component of the S0 galaxy NGC 128 (Sandage 1961) may result from a differential rotation field in this component. So far, there is little relevant theoretical or observational information; however, de Vaucouleurs (1958) shows that the spheroidal component of M31 does follow the $R^{1/4}$ luminosity law over at least the inner 20′ or 4 kpc. The galactic halo contributes so much to our present fragmentary knowledge about galaxy formation that it seems vital to get some detailed understanding of the present dynamical state of spheroidal components in general. This will require a systematic program of surface photometry and model construction, in the same spirit as that already done for the ellipticals; such a program is under way at Mount Stromlo.

The exponential disk of the disk galaxies was discussed in § 9.3. It was pointed out that the mass–angular-momentum distribution $\mathfrak{M}(h)$ does not automatically define its exponential nature; rather, the exponential disk is probably the most stable of all equilibrium disk configurations with that particular $\mathfrak{M}(h)$. The origin of the observed $\mathfrak{M}(h)$ distribution for the disks of these systems is the fundamental problem in understanding their mass distribution. It is not known yet whether this characteristic $\mathfrak{M}(h)$ is an endowed property of all protodisk galaxies, or whether there is some very efficient process which establishes this $\mathfrak{M}(h)$ distribution as the systems form.

Given the observed $\mathfrak{M}(h)$ distribution, it is worth asking at what stage of the galaxy's formation the exponential density distribution was established in the disk. We know that the exponential disk is not just an old disk property; it seems to involve the complete stellar content. For example, in M33 (de Vaucouleurs 1959*b*), which is an almost pure exponential disk, the ultraviolet luminosity distribution $U(R)$ is exponential with the same gradient as the blue and visual light distributions. Since the young Population I presumably contributes significantly to $U(R)$, this population is also exponentially distributed. The gas associated with the young Population I is probably the remains of gas left over after the old disk population formed (see § 11.4 and Sandage, Freeman, and Stokes 1970). This suggests that the exponential disk was established during the settling phase of the disk, *before* significant star formation in the disk began. On the other hand, Toomre (1964) points out that a differentially rotating disk of stars is unstable to axisymmetric disturbances unless the radial component σ_{RR} of the velocity dispersion exceeds some critical value (§ 9.1.2). He suggests that the action of these instabilities may be to increase σ_{RR} until the disk is stabilized. These instabilities could be important as a process capable of establishing the exponential nature of the disk *after* most of the disk stars had formed.

A striking feature of the exponential disks is the apparent constancy of the surface density scale μ_0 for most disk systems (§ 9.3). It is interesting to compare this with Fish's result for the ellipticals (§ 11.3.1): his law $\Omega \sim \mathfrak{M}^{3/2}$ implies that the mass $\mathfrak{M} \sim R^2$, where R is a length scale, so the mean surface density is again approximately constant from elliptical to elliptical. Although these two results are superficially similar, their physical reasons are probably quite different. For the ellipticals, Fish shows that his law would follow if the collapse of protoellipticals were halted by their opacity becoming sufficiently high; i.e., it is established *late* in the collapse phase. On the other hand, for the disk systems, we know that the constancy of μ_0 follows from the relationship $\mathfrak{H} \sim \mathfrak{M}^{7/4}$ between the total mass and angular momentum of the disk; because $\mathfrak{H}$ and $\mathfrak{M}$ are presumably conserved quantities, it seems likely that *this* law is established very *early* in the history of the protogalaxies.

Finally we ask: What physical parameters of a protogalaxy determine the subsequent mass ratio of the spheroidal component to the disk component in the resulting disk galaxy? This question will be easier to answer when more is known about the present dynamical state of the spheroidal component. For example, if this component really has a low $\mathfrak{H}/\mathfrak{M}$, then its relative mass could just follow from the relative amount of low-angular-momentum material in the protogalaxy (see Sandage, Freeman, and Stokes 1970). However, if Kinman's high value of $\mathfrak{H}/\mathfrak{M}$ for the globular-cluster system turns out to be typical of that for the whole spheroidal component in our Galaxy, then the dynamical distinction between predisk- and prebulge-component matter is not at all clear, and this makes the question more difficult.

11.4. Morphological-Type Problems

The two main problems associated with the morphological types of galaxies are: (i) What parameters determine whether a protogalaxy will end up as an elliptical or a disk galaxy? (ii) What parameters define the type (0, a, . . . , m) of the disk system?

11.4.1. *Elliptical or disk.*—The obvious large-scale difference between these two classes is that the disk is approximately in centrifugal equilibrium (i.e., its energy of random motion is low), while the equilibrium in the ellipticals is mainly between self-gravity and the pressure gradient $\nabla \cdot (\rho \boldsymbol{\sigma})$ of equation (12). Given the total mass $\mathfrak{M}$ and angular momentum $\mathfrak{H}$ (the energy of an extended protocloud will be near zero), can we predict from these parameters whether the system will be elliptical or disk? This will depend on how we believe the protocloud's collapse was stopped. While it seems likely that the disk collapse is halted by the increasing centrifugal force, it is not at all clear what process leads to the fragmentation of an elliptical or the spheroidal component of a disk galaxy. One possible process, suggested by Fish (§ 11.3.1), depends on the system becoming opaque to its own radiation as it contracts. A simple order-of-magnitude calculation indicates whether rotation or opacity would dominate in stopping the collapse of a rotating protocloud.

Consider a uniform spherical rigidly rotating protocloud of mass $\mathfrak{M}$, angular momentum $\mathfrak{H}$, and, at some instant, angular velocity Ω and radius L. Let it stay uniform, spherical, and rigidly rotating as it collapses; and, following Fish, assume that the hydrogen is fully ionized and the other constituents contribute a negligible number of free electrons: then free electron scattering is the dominant source of opacity, and the optical depth associated with the length L is

$$\tau = \frac{3}{4\pi} \frac{\sigma_e}{m_p} \frac{\mathfrak{M}}{L^2} , \tag{126}$$

where σ_e is the free-electron scattering coefficient and m_p the proton mass. The ratio B of the gravitational to centrifugal force in the plane of rotation is

$$B = G\mathfrak{M}/L^3\Omega^2 = 4G\mathfrak{M}^3 L/25\mathfrak{H}^2 , \tag{127}$$

so

$$B^2\tau = \frac{3\sigma_e}{4\pi m_p} \left(\frac{4}{25}\right)^2 \frac{G^2\mathfrak{M}^7}{\mathfrak{H}^4} = \Theta , \tag{128}$$

where Θ is a constant for a particular protocloud. Opacity will stop the collapse when τ reaches some value: assume this value is $\tau = 1$. When $B = 1$, the centrifugal force will stop it. Then $\Theta > 1$ means that opacity halts the collapse before the centrifugal force is large enough, and this would produce an elliptical galaxy in Fish's picture: $\Theta < 1$ means that rotation halts the collapse, and the result is a disk system.

We can check whether this picture is reasonable by estimating Θ for real galaxies.

(i) *Disk galaxies.*—From § 9.3 we know that $G^2\mathfrak{M}^7/\mathfrak{H}^4 = 2\pi\mu_0/1.51$ (eqs. [111] and [112]) is approximately constant for most exponential disk systems, because the surface density scale μ_0 is nearly constant from galaxy to galaxy: its value is approximately $5 \times 10^8\ \mathfrak{M}_\odot\ \mathrm{kpc}^{-2}$. Then $\Theta \approx 11 \times 10^{-4}$ for the disks, and this is certainly less than 1.

(ii) *Ellipticals.*—We assume that the nonspherical ellipticals are rotating and model them roughly by the Maclaurin spheroid of the same eccentricity e (Chandrasekhar 1969). The most common ellipticals (intrinsically), from figure 7, are E4, with $e = 0.80$. For this Maclaurin spheroid, $G^2\mathfrak{M}^7/\mathfrak{H}^4 = 192\mathfrak{M}/a^2$, where a is the semimajor axis. From Fish (1964, table 1), a typical value of $\mathfrak{M}/a^2$ for ellipticals is $7 \times 10^{10}\ \mathfrak{M}_\odot\ \mathrm{kpc}^{-2}$, where we tentatively identify the length a with the observed effective radius. With these numbers, $\Theta = 7.25$ for E4 galaxies; opacity would have stopped their collapse, as in the picture.

The flattest ellipticals observed are E7. Flatter systems, like NGC 3115 (Sandage 1961), show a distinct disk; this suggests that rotation and opacity both contribute significantly to halting their collapse, so we would expect $\Theta \approx 1$ for these objects. The calculated value for an E7 model is $\Theta = 1.24$. The assumptions in the model make this agreement fortuitous, but it does suggest that (i) a protocloud's $\mathfrak{M}$ and $\mathfrak{H}$ determine the nature of the resulting galaxy, and (ii) a detailed study of this picture is worth doing. It has obvious extensions, for a protocloud more centrally condensed than our homogeneous model, to the formation of the disk plus spheroidal component subsystems in real disk galaxies.

Another problem in studying the differences between E and disk systems comes from the observed ratio of total mass to total blue-light luminosity, $\mathfrak{M}/L$. M. S. Roberts (1969) shows that typical values of $\mathfrak{M}/L$ for disk systems at various stages are in the range 6 to 9 solar units. On the other hand, for the ellipticals in table 5, $\mathfrak{M}/L$ varies from 11 to 59: this large range is interesting because there is no obvious correlation of $\mathfrak{M}/L$ with total mass, and these systems all have similar colors. However, as pointed out in § 4.4.1, their masses come from the velocity dispersion in the nucleus, and it is important to check in each case that this dispersion is typical of that in the main body of the system.

11.4.2. *The types in the disk family.*—The first problem is to establish the basic observable quantities that determine the morphological type (0, a, . . . , d) for a disk galaxy. As pointed out in §4.1, these types are defined by (i) increasing openness of the spiral pattern, (ii) increasing degree of resolution of the arms into luminous stars (or more generally by increasing Population I activity, as measured by the mass fraction of gas and dust, the number and size of the spiral arm H II regions, and the *UBV* colors), and (iii) increasing apparent ratio of the exponential disk to the central bulge. Of these three criteria, it is now clear that the bulge/disk correlation with type is weak (e.g., Freeman 1970*a*). Further, the criteria (i) and (ii) are probably not independent because the openness of the spiral pattern depends on the relative gas density, at least within the density-wave theory [see § 9.6.3 (i)]; the type is then defined essentially by the Population I content.

This point is made by Sandage, Freeman, and Stokes (1970). They list two pairs of disk galaxies with similar photometric properties but of widely different morphological types (table 9): this shows how the type does not depend strongly on the system's large-scale mass distribution, and therefore on its total mass and angular momentum. However, there is a clear correlation (fig. 38) of type with the mean H I density, ρ_{HI}, so they argue that the probable distinction between the systems in table 9 is their value of ρ_{HI}. The problem is then to account for this ρ_{HI} difference in systems that are otherwise so alike. If the old disk stars formed at about the same time τ_D, then there are two basic possibilities. (*a*) Immediately after τ_D, each pair of systems had similar values of ρ_{HI} and had similar appearances as type Sd or later; the depletion of the interstellar matter then proceeded at different rates in the two systems, despite their apparent structural similarity. The present morphological type of a spiral system is then set up by the evolutionary rate of the individual system. (*b*) After the old disk stars formed at τ_D, there was relatively little gas left, for example, in NGC 4503 (S0), and rather

TABLE 9

TWO PAIRS OF GALAXIES WITH SIMILAR PHOTOMETRIC PARAMETERS*

Parameter	NGC 4503	NGC 1313	NGC 1332	NGC 5005
Class	SBO$^-$	SB(s)d	S(s)O$^-$	SAB(rs)bc
$R_{sph}/R_{26.5}$	0.33	0.40	0.64	0.65
α^{-1} (kpc)	1.5	1.6	4.0	3.5
I_0 (mag arcsec^{-2})	21.7	21.5	22.0	21.8
M_B	−18.4	−18.7	−19.9	−20.1

NOTE.—$R_{sph}/R_{26.5}$ is defined in the legend to fig. 14.
* From Sandage *et al.* (1970). I_0 and α are defined in § 4.3.2.

more (>10 percent by mass) in NGC 1313 (Sd); and subsequent stellar evolution via new star formation has been so slow as not to interchange the initial gas ratios. The present morphological type is then set up at the time of formation of the old disk stars.

While no one understands yet how spiral systems evolve, it does seem very likely that the evolutionary rate—i.e., the rate of depletion of the extreme Population I component —depends on some large-scale parameter of the galaxy. It is then difficult to see how two systems as similar in their background properties as NGC 4503 and 1313, or NGC 1332 and 5005, could have evolved at very different rates if their H I densities also were similar after τ_D, as in possibility (*a*). Sandage *et al.* therefore prefer possibility (*b*), that the morphological type of spiral systems was determined by the relative amounts of gas left over after the stars of the old disk population formed. Since this time, if galaxies are coeval, all spiral systems will have evolved to earlier types by depletion of their interstellar matter, but the extent of this evolution is severely limited by the existence today of Sd and Sm systems.

If this argument is correct, then it means only that the present forms of galaxies along the Hubble sequence are controlled more by the initial conditions of formation than by subsequent evolution along the sequence. It leaves untouched two fundamental questions: (i) Why is the amount of gas left over after the formation of the old disk stars so different in systems of different type which are otherwise apparently fairly similar in their background properties? (ii) We know from § 4.3.2 that most systems of type later than about Scd have small length scales ($\alpha^{-1} \approx 2$ kpc) and therefore low mass and angular momentum. (Earlier-type systems have any value of α^{-1} between about 1 and 5 kpc.) It is these late-type systems that have the largest fraction of gas left over after the epoch of formation of the old disk stars: Why are their length scales predominantly small?

Finally we note the prevalence of ellipticals and lenticulars (rather than spirals) in the compact clusters of galaxies (see Baade and Spitzer 1951). This is probably relevant here.

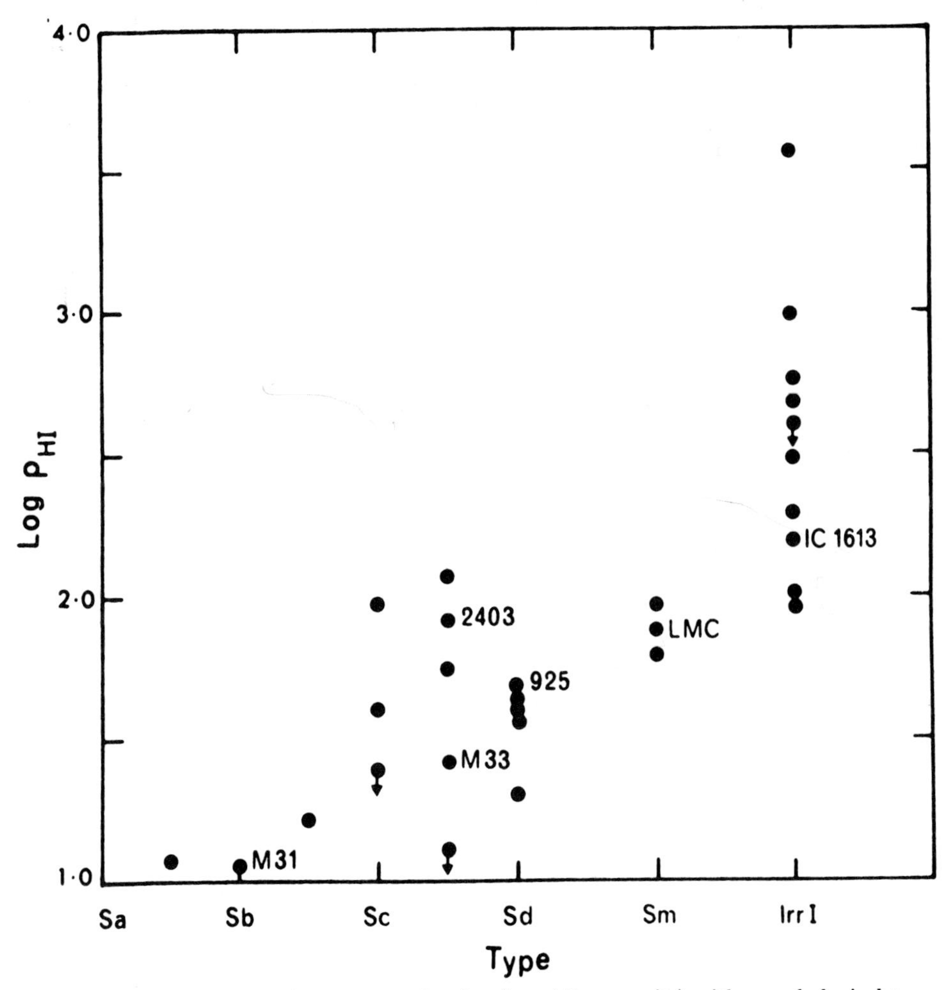

FIG. 38.—Variation of hydrogen mass density (in arbitrary units) with morphological type. Galaxy diameters used to calculate the density are optical values. Arrows denote galaxies whose optical diameters are smaller than H I diameters, causing the true H I density to be smaller than the plotted value. (From Sandage *et al.* 1970.)

11.5. The Origin of Angular Momentum

While little is known directly about the angular momentum of elliptical galaxies, we do know that at least 80 percent of the bright galaxies belong to the disk family (de Vaucouleurs 1963*a*), and these appear to rotate rapidly. The origin of their angular momentum $\mathfrak{H}$ is a major problem that must be considered in any picture of galaxy formation. Possible sources for $\mathfrak{H}$ should also explain why, for most disk galaxies, (i) $\mathfrak{H}$ is associated with a simple state of approximately planar circular motion; (ii) $\mathfrak{H} \sim \mathfrak{M}^{7/4}$; (iii) $\mathfrak{M}(h)$ has the particular form associated with the exponential disk (§ 9.3). The angular momentum is usually attributed either to primordial turbulence in the Universe (e.g., von Weizsäcker 1951) or to the tidal interaction of the protocloud with its environment (e.g., Hoyle 1951). It is not yet clear which of these is correct.

Hunter (1970) points out that the tidal-interaction picture could lead to a far greater variety of disklike states other than the simplest state of pure planar circular motion. Many of the more general disks are rotating, at least in part, about an axis in their plane and have nonzero circulation about an axis in their plane.

Peebles (1969, 1971) investigates by two methods the acquisition of angular momentum by a galaxy in the gravitational-instability picture for galaxy formation. In this picture, large irregularities like galaxies grow under the influence of gravity from small imperfections in the early Universe. He considers the transfer of angular momentum to a developing protogalaxy by the gravitational interaction of the system's quadrupole moment with the tidal field of nearby protosystems. The first method (Peebles 1969) is a perturbation analysis for the phase before the galaxies separated out as distinct systems, plus a quadrupole interaction calculation after separation; he calculates a final value of $\mathfrak{H} \sim 2.4 \times 10^{73}$ g cm^2 s^{-1} for Galactic parameters, compared with 1.5×10^{74} g cm^2 s^{-1} for Innanen's (1966*b*) model Galaxy. The second method (Peebles 1971) is an N-body calculation: when the results are scaled to Galactic parameters, it gives again $\mathfrak{H} \sim 3 \times 10^{73}$ g cm^2 s^{-1}. It is not clear whether the discrepancy of a factor 5 between this value and the "observed" value is significant, considering the uncertainties in the calculations and the model.

In contrast, Oort (1970*b*) estimates the transfer of angular momentum to a protosystem from nearby individual protogalaxies and from groups of galaxies. He finds the calculated angular momentum, within his approximations, to be at least a factor 100 smaller than the observed value, and infers that $\mathfrak{H}$ must come from large-scale turbulence at the time when the protogalaxies became individual units. Oort develops a detailed picture of galaxy formation based on this inference. He deduces that protogalaxies formed from the turbulent cells when the scale of the Universe was about 1/30 of its present scale (compared with 1/1000 of its present scale in the gravitational-instability picture), because only then could they contain the observed amount of angular momentum.

REFERENCES

Antonov, V. 1961, *Vest. Leningrad Gos. Univ.*, Ser. 6, No. 13.
Arnett, W. D. 1971, *Ap. J.*, **166**, 153.
Arnold, V., and Avez, A. 1967, *Problèmes Ergodique de la Mechanique Classique* (Paris: Gauthier-Villars).
Arp, H., and Bertola, F. 1971, *Ap. J.*, **163**, 195.
Avner, E. S., and King, I. R. 1967, *A.J.*, **72**, 650.
Baade, W., and Spitzer, L. 1951, *Ap. J.*, **113**, 413.

Baade, W., and Swope, H. 1961, *A.J.*, **66**, 300.
Barbanis, B. 1962, *Zs. f. Ap.*, **56**, 56.
Barbanis, B., and Woltjer, L. 1967, *Ap. J.*, **150**, 461.
Baum, W. 1959, *Pub. A.S.P.*, **71**, 106.
Bergh, S. van den. 1969, *Ap. J. Suppl.*, No. 171.
Berry, C. L., and Vandervoort, P. O. 1970, *IAU Symposium No. 38*, p. 336.
Blaauw, A., 1965, in *Galactic Structure*, ed. A. Blaauw and M. Schmidt (Chicago: University of Chicago Press), chap. 20.
Bok, B. J. 1970, *IAU Symposium No. 38*, p. 246.
Brandt, J. C., and Belton, M.J.S. 1962, *Ap. J.*, **136**, 352.
Burbidge, E. M., Burbidge, G. R., and Prendergast, K. H. 1960*a*, *Ap. J.*, **131**, 282.
———. 1960*b*, *ibid.*, **132**, 654.
———. 1961, *ibid.*, **134**, 874.
Burbidge, G. R. 1970, *Ann. Rev. Astr. and Ap.*, **8**, 369.
Burton, W. B., and Shane, W. W. 1970, *IAU Symposium No. 38*, p. 397.
Chandrasekhar, S. 1960, *Principles of Stellar Dynamics* (New York: Dover).
———. 1961, *Hydrodynamic and Hydromagnetic Stability* (London: Oxford University Press).
———. 1969, *Ellipsoidal Figures of Equilibrium* (New Haven: Yale University Press).
Code, A. D. 1967, *A.J.*, **72**, 789.
Contopoulos, G. 1967*a*, in *Relativity Theory and Astrophysics*, ed. J. Ehlers (Providence: Am. Math. Soc.), vol. **2**, p. 98.
———. 1967*b*, *Bull. astr.* (Paris), **2**, 223.
———. 1971, *Ap. J.*, **163**, 181.
Crampin, D. J., and Hoyle, F. 1964, *Ap. J.*, **140**, 99.
Danby, J. M. A. 1965, *A.J.*, **70**, 501.
Delhaye, J. 1965, in *Galactic Structure*, ed. A. Blaauw and M. Schmidt (Chicago: University of Chicago Press), chap. 4.
Dennis, T. R. 1966, *Ap. J.*, **146**, 581.
Dixon, M. E. 1971, *Ap. J.*, **164**, 411.
Eggen, O. J. 1965, in *Galactic Structure*, ed. A. Blaauw and M. Schmidt (Chicago: University of Chicago Press), chap. 6.
———. 1969, *Ap. J.*, **155**, 701.
———. 1970, in *Vistas in Astronomy*, ed. A. Beer (Oxford: Pergamon Press), Vol. **12**, p. 367.
Eggen, O. J., Lynden-Bell, D., and Sandage, A. 1962, *Ap. J.*, **136**, 748 (ELS).
Eggen, O. J., and Sandage, A. 1969, *Ap. J.*, **158**, 669.
Emden, R. 1907, *Gaskugeln* (Leipzig and Berlin: Teubner).
Fish, R. 1964, *Ap. J.*, **139**, 284.
Freeman, K. C. 1966*a*, *M.N.R.A.S.*, **133**, 47.
———. 1966*b*, *ibid.*, **134**, 1.
———. 1970*a*, *Ap. J.*, **160**, 811.
———. 1970*b*, *IAU Symposium No. 38*, p. 351.
Fujimoto, M. 1968*a*, *Ap. J.*, **152**, 391.
———. 1968*b*, *ibid.*, p. 523.
Gascoigne, S. C. B. 1966, *M.N.R.A.S.*, **134**, 59.
Goldreich, P., and Lynden-Bell, D. 1965*a*, *M.N.R.A.S.*, **130**, 97.
———. 1965*b*, *ibid.*, p. 125.
Gordon, K. J. 1971, *Ap. J.*, **169**, 235.
Hayli, A. 1965, *Ann. d'Ap.*, **28**, 49.
Hearnshaw, J. B. 1972, unpublished thesis, Australian National University.
Hénon, M. 1966, *14e Colloque International d'Astrophysique* (*Liege*), 243.
———. 1968, *Bull. Astr.* (Paris), **3**, 241.
Hénon, M., and Heiles, C. 1964, *A.J.*, **69**, 73.
Hiltner, W. A. 1960, *Ap. J.*, **131**, 163.
Hodge, P. 1963, *A.J.*, **68**, 69.
———. 1965, *Ap. J.*, **142**, 1390.
———. 1966, *ibid.*, **144**, 869.
———. 1967, *ibid.*, **148**, 719.
Hodge, P., and Michie, R. W. 1969, *A.J.*, **74**, 587.
Hohl, F. 1971, *Ap. J.*, **168**, 343.
Hohl, F., and Campbell, J. W. 1968, *A.J.*, **73**, 611.
Hohl, F., and Feix, M. R. 1966, *Ap. J.*, **147**, 1164.
Hoyle, F. 1951, *Problems of Cosmical Aerodynamics* (Dayton: Central Air Documents Office), p. 195.
Hunter, C. 1963, *M.N.R.A.S.*, **126**, 299.
———. 1965, *ibid.*, **129**, 321.
———. 1970, *Ap. J.*, **162**, 445.
Hunter, C., and Toomre, A. 1969, *Ap. J.*, **155**, 747.
Innanen, K. 1966*a*, *Zs. f. Ap.*, **64**, 158.

Innanen, K. 1966*b*, *Ap. J.*, **143**, 150.
Julian, W. H. 1967, *Ap. J.*, **148**, 175.
———. 1969, *ibid.*, **155**, 117.
Julian, W. H., and Toomre, A. 1966, *Ap. J.*, **146**, 810.
Kahn, F. D., and Woltjer, L. 1959, *Ap. J.*, **130**, 705.
Kalnajs, A. J. 1970, *IAU Symposium No. 38*, p. 318.
Kerr, F. J. 1964, *IAU Symposium No. 20*, p. 81.
Kerr, F. J., and Westerhout, G. 1965, in *Galactic Structure*, ed. A. Blaauw and M. Schmidt (Chicago: University of Chicago Press), chap. 9.
King, I. R. 1966, *A.J.*, **71**, 64.
King, I. R., Hedemann, E., Hodge, S. M., and White, R. E. 1968, *A.J.*, **73**, 456.
King, I. R., and Minkowski, R. 1966, *Ap. J.*, **143**, 1002.
Kinman, T. D. 1959*a*, *M.N.R.A.S.*, **119**, 538.
———. 1959*b*, *ibid.*, p. 559.
Landau, L. D., and Lifshitz, E. M. 1960, *Mechanics* (Oxford: Pergamon Press).
Liller, M. H. 1960, *Ap. J.*, **132**, 306.
Lin, C. C. 1970, *IAU Symposium No. 38*, p. 373.
Lin, C. C., Mestel, L., and Shu, F. 1965, *Ap. J.*, **142**, 1431.
Lin, C. C., Yuan, C., and Shu, F. 1969, *Ap. J.*, **155**, 721. (LYS)
Lindblad, B. 1941, *Stockholm Obs. Ann.*, Vol. **13**, No. 10.
———. 1959, *Handbuch der Physik*, ed. S. Flügge (Berlin: Springer Verlag), **53**, 21.
Lynden-Bell, D. 1961, *M.N.R.A.S.*, **123**, 1.
———. 1962, *ibid.*, **124**, 279.
———. 1964, *Ap. J.*, **139**, 1195.
———. 1965, *M.N.R.A.S.*, **129**, 299.
———. 1966, *Observatory*, **86**, 57.
———. 1967*a*, *M.N.R.A.S.*, **136**, 101.
———. 1967*b*, in *Relativity Theory and Astrophysics*, ed. J. Ehlers (Providence: Am. Math. Soc.), Vol. **2**, p. 131.
Lynden-Bell, D., and Ostriker, J. P. 1967, *M.N.R.A.S.*, **136**, 293.
Marochnik, L. S., and Suchkov, A. A. 1969, *Ap. and Space Sci.*, **4**, 317.
Mathewson, D. S., Kruit, P. C. van der, and Brouw, W. N. 1972, *Astr. and Ap.*, **17**, 468.
Mayall, N. U., and Lindblad, P. O. 1970, *Astr. and Ap.*, **8**, 364.
Mestel, L. 1963. *M.N.R.A.S.*, **126**, 553.
Michie, R. W. 1963, *M.N.R.A.S.*, **125**, 127.
———. 1964, *Ann. Rev. Astr. and Ap.*, **2**, 49.
Miller, R. H., and Prendergast, K. H. 1968, *Ap. J.*, **151**, 699.
Miller, R. H., Prendergast, K. H., and Quirk, W. J. 1970, *Ap. J.*, **161**, 903.
Ollongren, A. 1962, *B.A.N.*, **16**, 241.
———. 1965, *Ann. Rev. Astr. and Ap.*, **3**, 113.
———. 1966, *IAU Symposium No. 25*, p. 98.
Oort, J. H. 1965, in *Galactic Structure*, ed. A. Blaauw and M. Schmidt (Chicago: University of Chicago Press), chap. 21.
———. 1970*a*, *IAU Symposium No. 38*, p. 142.
———. 1970*b*, *Astr. and Ap.*, **7**, 381.
Osterbrock, D. 1960, *Ap. J.*, **132**, 325.
Ostriker, J. P., and Davidsen, A. F. 1968, *Ap. J.*, **151**, 679.
Peach, J. V. 1969, *Nature*, **223**, 1140.
Peebles, P. J. E. 1969, *Ap. J.*, **155**, 393.
———. 1971, *Astr. and Ap.*, **11**, 377.
Perek, L. 1962, *Adv. Astr. and Ap.*, **1**, 165.
Prendergast, K. H., and Tomer, E. 1970, *A.J.*, **75**, 674.
Quirk, W. J. 1971, *Ap. J.*, **167**, 7.
Roberts, M. S. 1963, *Ann. Rev. Astr. and Ap.*, **1**, 149.
———. 1966, *Ap. J.*, **144**, 639.
———. 1969, *A.J.*, **74**, 859.
Roberts, W. W. 1969, *Ap. J.*, **158**, 123.
Rodgers, A. W. 1971, *Ap. J.*, **165**, 581.
Rougoor, G. W. 1964, *B.A.N.*, **17**, 381.
Sandage, A. 1961, *The Hubble Atlas* (Washington: Carnegie Institution of Washington).
———. 1969, *Ap. J.*, **158**, 1115.
———. 1970, *ibid.*, **162**, 841.
Sandage, A., and Eggen, O. J. 1969, *Ap. J.*, **158**, 685.
Sandage, A., Freeman, K. C., and Stokes, N. R. 1970, *Ap. J.*, **160**, 831.
Sanduleak, N. 1969, *A. J.*, **74**, 47.
Schmidt, M. 1956, *B.A.N.*, **13**, 15.
———. 1959, *Ap. J.*, **129**, 243.

———. 1965, in *Galactic Structure*, ed. A. Blaauw and M. Schmidt (Chicago: University of Chicago Press), chap. 22.
Schwarzschild, M. 1964, *Roy. Obs. Bull.*, No. 82, pp. 55–59.
Searle, L. 1971, *Ap. J.*, **168**, 327.
Sersic, J. 1969, *Galaxias Australes* (Cordoba: Universidad Nacional de Cordoba).
Shane, W. W., and Bieger-Smith, G. P. 1966, *B.A.N.*, **18**, 263.
Shu, F. H. 1970, *Ap. J.*, **160**, 89.
Shu, F. H., Stachnik, R. V., and Yost, J. C. 1971, *Ap. J.*, **166**, 465.
Simkin, S. M. 1970, *Ap. J.*, **159**, 463.
Spinrad, H. 1966, *Pub. A.S.P.*, **78**, 367.
Spinrad, H., and Taylor, B. J. 1971, *Ap. J. Suppl.*, No. 193.
Spitzer, L., and Schwarzschild, M. 1953, *Ap. J.*, **118**, 106.
Stix, T. H. 1962, *Plasma Waves* (New York: McGraw-Hill).
Takase, B. 1967, *Publ. Astr. Soc. Japan*, **19**, 427.
Toomre, A. 1964, *Ap. J.*, **139**, 1217.
———. 1969, *ibid.*, **158**, 899.
Unsöld, A. 1969, *Science*, **163**, 1015.
Vaucouleurs, G. de. 1958, *Ap. J.*, **128**, 465.
———. 1959*a*, in *Handbuch der Physik*, ed. S. Flügge (Berlin: Springer-Verlag), **53**, 311.
———. 1959*b*, *Ap. J.*, **130**, 728.
———. 1959*c*, *Pub. A.S.P.*, **71**, 83.
———. 1960, *Ap. J.*, **131**, 574.
———. 1961*a*, *Ap. J. Suppl.*, No. 48.
———. 1961*b*, *Ap. J.*, **133**, 404.
———. 1963*a*, *Ap. J. Suppl.*, No. 74.
———. 1963*b*, *Ap. J.*, **137**, 720.
———. 1963*c*, *ibid.*, **138**, 934.
———. 1969, *Ap. Letters*, **4**, 17.
———. 1970, *ibid.*, **5**, 219.
Vaucouleurs, G. de, and Ables, H. 1968, *Ap. J.*, **151**, 105.
Vaucouleurs, G. de, and Freeman, K. C. 1970*a*, *IAU Symposium No. 38*, p. 356.
———. 1970*b*, *Vistas in Astronomy*, ed. A. Beer, Vol. **14**, 163 (dVF).
Vaucouleurs, G. de, and Vaucouleurs, A. de. 1963, *Ap. J.*, **137**, 363.
Vaucoulcurs, G. de, Vaucouleurs, A. de, and Freeman, K. C. 1968, *M.N.R.A.S.*, **139**, 425.
Walker, M. F. 1962, *Ap. J.*, **136**, 695.
———. 1964, *A.J.*, **69**, 744.
Weizsäcker, C. F. von. 1951, *Ap. J.*, **114**, 165.
Westerhout, G. 1970, *IAU Symposium No. 38*, p. 122.
Woltjer, L. 1965, in *Galactic Structure*, ed. A. Blaauw and M. Schmidt (Chicago: University of Chicago Press), chap. 23.
———. 1967, *Relativity Theory and Astrophysics*, ed. J. Ehlers, (Providence: Am. Math. Soc.), Vol. **2**, p. 1.
Woolley, R. 1965, in *Galactic Structure*, ed. A. Blaauw and M. Schmidt (Chicago: University of Chicago Press), chap. 5.
Woolley, R., and Dickens, R. 1961, *Roy. Obs. Bull.*, No. 42.
Yuan, C. 1969*a*, *Ap. J.*, **158**, 871.
———. 1969*b*, *ibid.*, p. 889.

CHAPTER 12

The Extragalactic Distance Scale

SIDNEY VAN DEN BERGH
David Dunlap Observatory, University of Toronto,
Richmond Hill, Ontario, Canada

1. INTRODUCTION

PERHAPS no event since the Copernican revolution has so profoundly affected human thought as has the discovery of the true nature of extragalactic nebulae. Now, for the first time, the Universe itself has become the subject of observation and scientific study. The first bold exploration of the realm of the nebulae has been described by Hubble (1936*a*). Of all the remarkable discoveries made during the pioneering era of extragalactic research, the discovery of the expansion of the Universe was probably the most fundamental.

Observations by Hubble and Humason (1931) showed that the mean apparent magnitude $\langle m_{pg} \rangle$ of cluster galaxies was related to the mean radial velocity $\langle V \rangle$ of cluster members by the equation

$$\log \langle V \rangle = 0.2 \langle m_{pg} \rangle + 0.5 \, , \quad (1)$$

in which V is measured in km s^{-1}. This equation is equivalent to the linear velocity-distance relation

$$\langle V \rangle = HD \, , \quad (2)$$

in which H is the Hubble constant and D the distance of a cluster in Mpc. (H is expressed in km s^{-1} Mpc^{-1} and has the dimension $[t^{-1}]$.) From equations (1), (2), and the definition of absolute magnitudes one obtains

$$\log H = 0.2 \langle M_{pg} \rangle + 5.5 \, , \quad (3)$$

in which $\langle M_{pg} \rangle$ is the mean absolute magnitude of cluster galaxies. For the seven brightest members of the Local Group, Hubble and Humason (1931) obtained $\langle M_{pg} \rangle = -13.5$. Substitution of this value into equation (3) yields $H = 630$ km s^{-1} Mpc^{-1}. Using more recent values of the distance moduli of the seven brightest Local Group members (Sandage 1958) and integrated photographic magnitudes by Holmberg (1958) and de Vaucouleurs (1960), one obtains $\langle M_{pg} \rangle = -17.0$; substituting this value into

Received September 1965; revised September 1969.

equation (3) yields $H = 125$ km s^{-1} Mpc^{-1}. The difference between these two values is entirely due to an increase in the adopted distance scale within the Local Group. It should perhaps be emphasized that the determination of the apparent magnitudes of the brightest stars in resolved nebulae of known redshift (Hubble and Humason 1931) did not yield an independent calibration of the distance scale. This is so because the adopted absolute magnitudes of the brightest stars were derived from the assumed distance moduli of Local Group members.

It is clear from the discussion presented above that the distances to the galaxies of the Local Group are essential stepping stones in the calibration of the extragalactic distance scale. This will continue to be the case until it becomes possible to make direct comparisons between objects in the Galaxy and similar objects in very distant galaxies with substantial redshifts.

All determinations of the extragalactic distance scale are ultimately based on the assumption that recognizable types of distant objects are similar to nearby objects of the same type. Due to possible differences in age, evolutionary history, and abundance of the elements, this assumption may not be correct for any particular type of distance indicator. The basic philosophy adopted in this review is that systematic errors of this kind will be minimized if the largest possible number of methods of distance determination is used.

2. PERIOD-LUMINOSITY RELATION OF CLASSICAL CEPHEIDS

2.1. Assumption Underlying Use of the Period-Luminosity Relation

Ever since the discovery of the period-luminosity relation of classical Cepheids by Leavitt (1908, 1912) it has been clear that Cepheids provide a powerful tool for the investigation of astronomical distances. It is, however, a tool that should be employed cautiously. Use of the period-luminosity relation for the calibration of extragalactic distances involves the tacit assumption that classical Cepheids in different galaxies have identical properties. A mounting body of evidence suggests that this assumption is probably not correct. For example, Arp and Kraft (1961) have shown that the dependence of pulsation amplitude on period in the Small Magellanic Cloud differs from that which is observed in the Galaxy. Additional evidence is provided by the differences between the period-frequency functions of Cepheids in the Galaxy, the Large Cloud, and the Small Cloud (Shapley and McKibben 1942*a*, *b*). Within M31 and the Galaxy the properties of Cepheids seem to depend on distance from the center (Baade and Swope 1965; van den Bergh 1958; Bahner, Hiltner, and Kraft 1962). Observations by Gascoigne and Kron (1965) and by Gascoigne (1969) appear to show that the Cepheids in the Magellanic Clouds are intrinsically bluer than are those in the Galaxy.

Probably the differences between Cepheids in different stellar systems are due to small differences in chemical composition. Such differences might affect both the structure and the pulsation characteristics of individual Cepheids. Changing the helium or metal abundance will move the position of the Cepheid instability strip in the $M_{\rm bol}$ versus $\log T_e$ plane (Christy 1966). An additional complication is that the depth to which Cepheids penetrate the instability strip depends on composition (Hofmeister 1967*b*). Furthermore, the $B - V$ color of the instability strip might be increased by stronger blanketing resulting from higher metal abundance or higher microturbulence (Bell and Rodgers 1969).

In subsequent sections a universal period-luminosity-color relation will be used to determine the distances of extragalactic systems. The main justification for this procedure is the profoundness of our ignorance!

2.2. The Period-Luminosity-Color Relation

In the color-magnitude diagram, Cepheid variables occupy an instability strip of finite width (Sandage 1958; Kraft 1961). The absolute magnitude of a Cepheid is therefore a function of both its period *and* its intrinsic color. Let it be assumed that, within the instability strip, lines of constant period have a slope

$$d\langle M_V\rangle/d[\langle B\rangle - \langle V\rangle]_0 = \alpha\,, \tag{4}$$

in which $\langle M_V\rangle$ is the *intensity* mean absolute magnitude and $[\langle B\rangle - \langle V\rangle]_0$ is the *intensity* mean of the intrinsic color. Then the quantity

$$M_W \equiv \langle M_V\rangle - \alpha[\langle B\rangle - \langle V\rangle]_0 \tag{5}$$

will be a function of period only. In an unreddened extragalactic system, in which all Cepheids may be considered to be located at the same distance, the quantity

$$W_0 \equiv \langle V\rangle - \alpha[\langle B\rangle - \langle V\rangle]_0 \tag{6}$$

will be a function of period only. From semiempirical considerations Sandage and Tammann (1969) obtain $\alpha = 2.67$.

According to Whitford (1958) a value $A_V/E_{B-V} = 3.0 \pm 0.2$ is appropriate throughout most of the Galaxy. (A higher ratio of total to selective absorption possibly prevails in H II regions.) In the Large Magellanic Cloud, Gascoigne (1969) obtains $A_V/E_{B-V} = 2.8 \pm 0.25$; and in the Andromeda Nebula, Kron and Mayall (1960) and van den Bergh (1968*a*) get $A_V/E_{B-V} \simeq 2.5$. It follows from these observations that reddening lines and lines of constant period have almost the same slope in the color-magnitude diagram. For small and intermediate reddening values it may therefore be assumed that lines of constant period and reddening lines are parallel. Under this assumption,

$$W_0 = W \equiv \langle V\rangle - 2.67[\langle B\rangle - \langle V\rangle] \tag{7}$$

is a function of period only, regardless of the reddening suffered by individual Cepheids in an extragalactic system.

Figure 1 shows a plot of W versus log P for those Cepheids in the Magellanic Clouds for which photoelectric observations (Gascoigne 1969) are now available. The plot shows a remarkably small scatter about the relations

$$W_{\rm LMC} = 16.39(\pm 0.03) - 3.67 \log P \tag{8}$$

and

$$W_{\rm SMC} = 17.00(\pm 0.03) - 3.67 \log P\,. \tag{9}$$

The standard deviations of individual Cepheids from the adopted regression lines are 0.15 and 0.16 mag for the Large Cloud and the Small Cloud, respectively. This dispersion is comparable to that which would be expected from the combined effects of observational errors and the depth of each Cloud along the line of sight. In deriving equation (9)

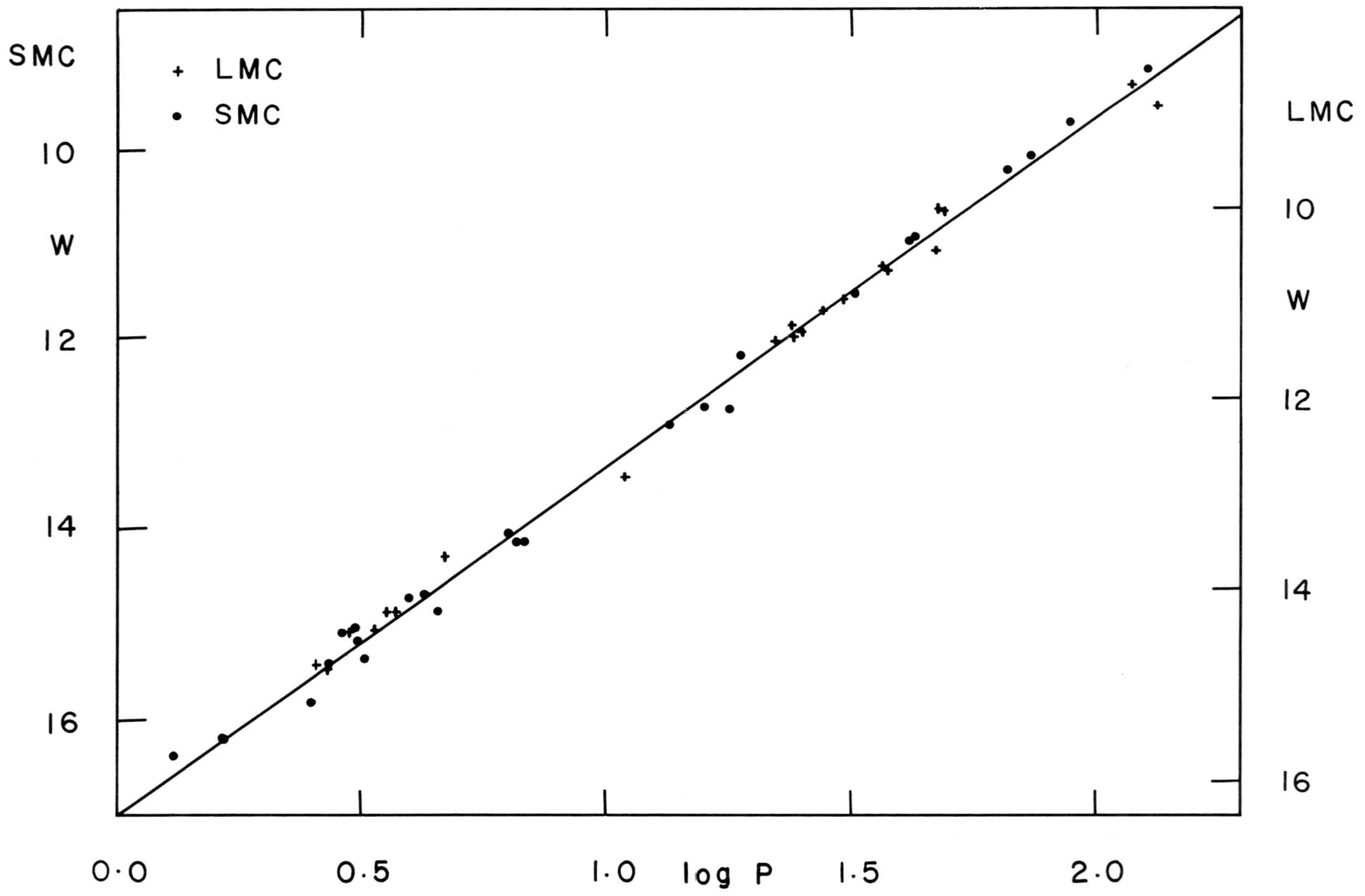

Fig. 1.—Plot of the reddening-free parameter W versus log P for Cepheids in the SMC (*scale left*) and LMC (*scale right*). Two short-period variables with sinusoidal light curves are not plotted. The standard deviation of individual Cepheids about the regression lines given by equations (8) and (9) is only 0.15 mag.

the two short-period, small-amplitude variables HV 1779 and HV 1897 were ignored. These two objects belong to a separate subspecies which is discussed by Gaposchkin (1961) and by Gaposchkin and Gaposchkin (1966).

Figure 2 shows a plot of the W versus log P relation for Cepheids in Variable Star Field IV in M31 (Baade and Swope 1963). The regression line

$$W_{\mathrm{M31}} = 21.87(\pm 0.08) - 3.67 \log P \tag{10}$$

is plotted in the figure. Long-period Cepheids are seen to fall below the line defined by equation (10) whereas short-period Cepheids fall above it. This effect might be accounted for by assuming that small systematic errors exist in the photoelectrically calibrated photographic observations of M31 Cepheids. Such systematic differences, which are probably due to photographic background effects, are also found when Gascoigne's (1969) photoelectric Cepheid observations in the Clouds are compared with photoelectrically calibrated photographic observations by other authors.

A plot of W versus log P for the Cepheids in NGC 6822, which have been observed by Kayser (1967), is shown in figure 3. Kayser's photoelectrically calibrated photographic observations exhibit a standard deviation of 0.66 mag about the relation

$$W_{6822} = 21.75(\pm 0.18) - 3.67 \log P\,. \tag{11}$$

Because of their large dispersion the observations of Cepheids in NGC 6822 do not provide a sensitive test of the hypothesis that period-luminosity-color relations are similar in different galaxies.

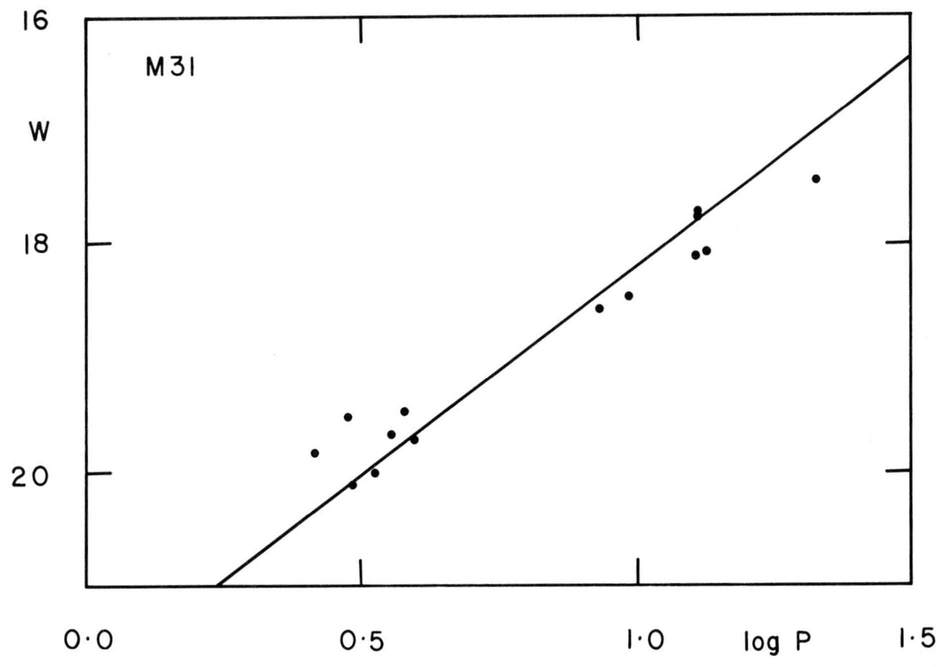

FIG. 2.—Plot of W versus log P for the Cepheids in M31 Variable Star Field IV. The adopted regression line is given by equation (10).

2.3. Galactic Calibration of the Period-Luminosity-Color Relation

Galactic Cepheids that are located in star clusters, OB associations, and R associations can be used to calibrate the period-luminosity-color relation. Table 1 lists data on the most secure calibration standards. Reddening values for individual Cepheids were taken from the work of Fernie (1967*a*). The dependence of the relation between E_{B-V} and A_V on spectral type (Fernie 1963) was taken into account in determining the absolute magnitudes of individual Cepheids. The galactic data, which are plotted in figure 4, are adequately represented by the relation

$$M_W = -2.46(\pm 0.07) - 3.67 \log P . \tag{12}$$

It should perhaps be emphasized that the accuracy of the galactic calibration is too low to provide an independent check on the slope of the M_W versus log P relation. The

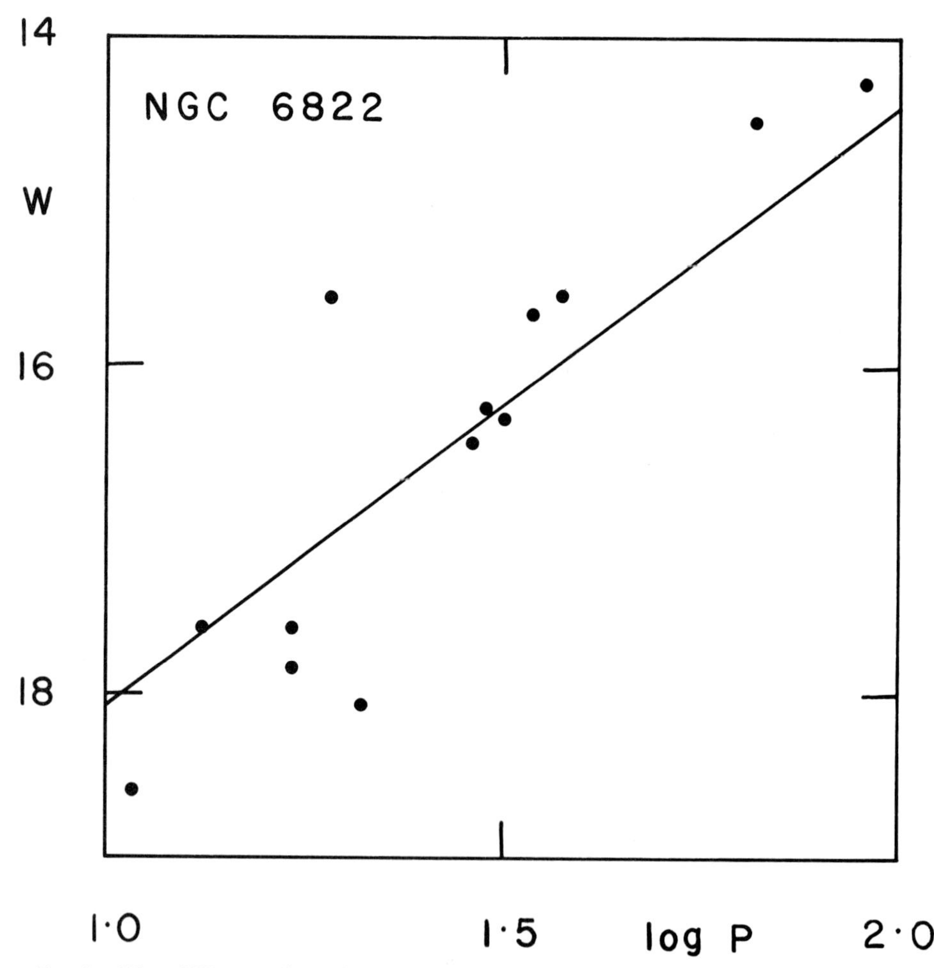

Fig. 3.—Plot of W versus log P for Kayser's Cepheid observations in NGC 6822. The adopted regression line is given in equation (11).

strongest statement that can be made at the present time is that currently available data do not exclude the possibility that the W versus log P relations have the same slope in the Magellanic Clouds, NGC 6822, M31, and the Galaxy.

The assumption that the zero point of the M_W versus log P relation is the same in different galaxies yields the absolute distance moduli $W - M_W = (m - M)_0$ listed in table 2. The formal mean errors quoted in the table assume that the observations are free of any systematic errors and that Cepheids in all galaxies obey identical period-luminosity-color relations.

TABLE 1

CEPHEIDS USED FOR GALACTIC CALIBRATION OF THE PERIOD-LUMINOSITY-COLOR RELATION

Name	log P	$\langle M_V \rangle$*	$\langle B \rangle_0 - \langle V \rangle_0$	Method†	References‡
SU Cas.	0.290	−2.5±0.2	0.39	R	4, 5
EV Sct.	0.490	−3.1±0.7	0.58	C	3, 5
α UMi.	0.599	−3.2±0.3	0.51	B	1, 2
CEb Cas. . . .	0.651	−3.2±0.5	0.60	C	3, 5
CF Cas.	0.688	−3.1±0.5	0.69	C	3, 5
CEa Cas. . . .	0.711	−3.3±0.5	0.68	C	3, 5
U Sgr.	0.829	−3.8±0.1	0.62	C	3, 5
DL Cas.	0.903	−4.1±0.2	0.69	C	3, 5
S Nor.	0.989	−3.9±0.2	0.78	C	3, 5
RS Pup.	1.617	−6.5±0.3	0.80	A	2, 5, 6

* Throughout this chapter angular brackets denote an intensity mean over the light cycle.
† R, in association of reflection nebulae; C, in open cluster; A, in association of OB stars; B, binary.
‡ (1) Fernie 1966; (2) Fernie 1967*b*; (3) Fernie and Marlborough 1965; (4) Racine 1968*a*; (5) Sandage and Tammann 1968; (6) Westerlund 1963.

TABLE 2

DISTANCE MODULI OBTAINED FROM CEPHEIDS

Galaxy	$(m-M)_0$	Galaxy	$(m-M)_0$
LMC.	18.85±0.08*	M31.	24.33±0.11
SMC.	19.46±0.08	M33.	24.55±0.15:
NGC 6822.	24.21±0.19	IC 1613.	24.73±0.3?

* Quoted uncertainties are *internal* mean errors.

2.4. REDDENING VALUES AND APPARENT DISTANCE MODULI FOR MEMBERS OF THE LOCAL GROUP

2.4.1. *The Large Magellanic Cloud.*—From observations of a star cluster in the direction of the LMC, which is located 185 pc below the galactic plane, Sanduleak and Davis Philip (1968) obtain a reddening $E_{B-V} = 0.06 \pm 0.02$. A value $\langle E_{B-V} \rangle = 0.07 \pm 0.04$ is obtained by Feast, Thackeray, and Wesselink (1960) from a study of foreground field stars. Gascoigne (1969) finds $\langle E_{B-V} \rangle \simeq 0.08$ for the Cepheids in the LMC. Van den Bergh and Hagen (1968) obtain $\langle E_{B-V} \rangle = 0.20 \pm 0.05$ for young star clusters, which are presumably still associated with the gas and dust clouds from which they were formed, and $\langle E_{B-V} \rangle = 0.06 \pm 0.01$ for old star clusters. Adopting a foreground redden-

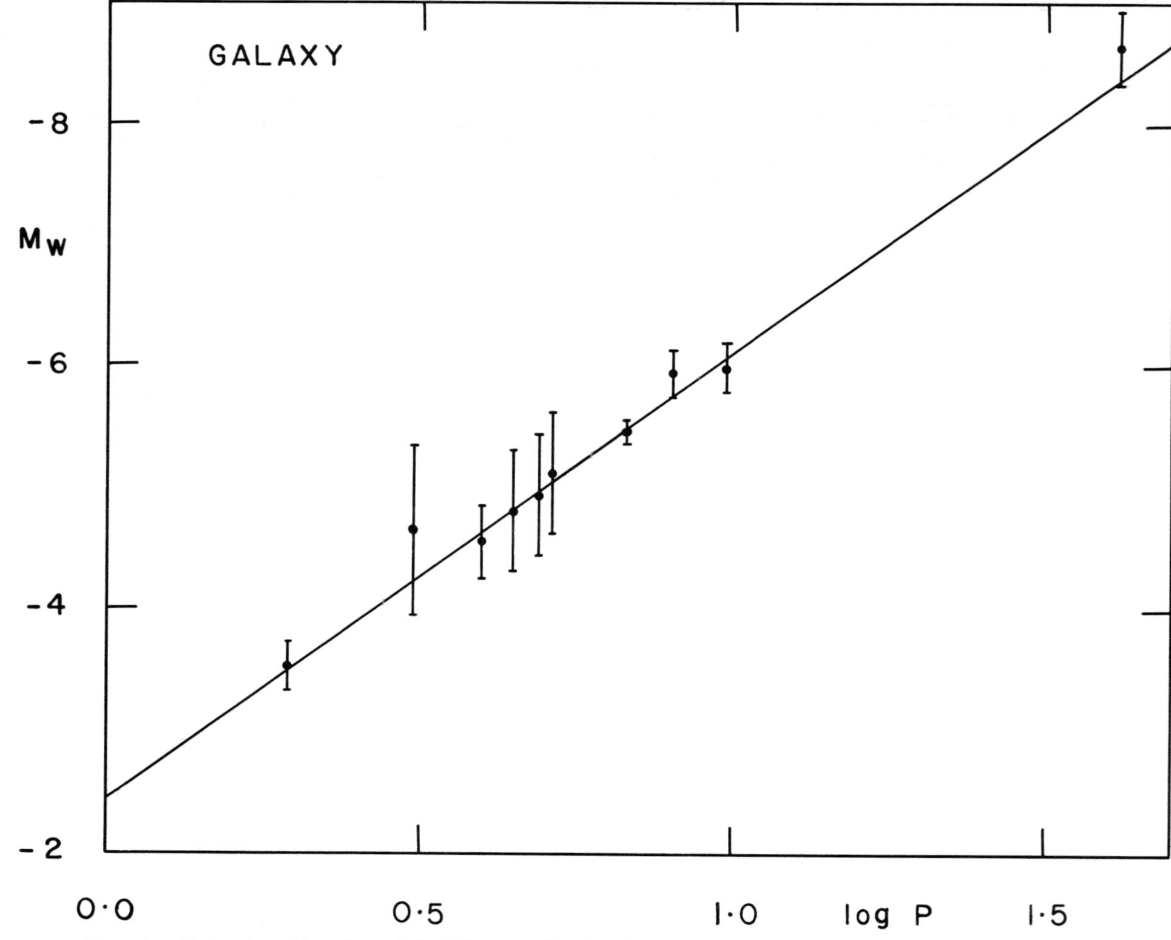

FIG. 4.—Galactic calibration of the W versus log P relation using Cepheids in clusters and associations and a Cepheid that is a member of a binary system.

ing $E_{B-V} = 0.06 \pm 0.02$ and $(m - M)_0 = 18.85 \pm 0.08$ yields $(m - M)_V = 19.03 \pm 0.10$ for the LMC.

2.4.2. *The Small Magellanic Cloud.*—The reddening of the SMC is discussed by Gascoigne (1969), who suggests that $E_{B-V} = 0.02$ represents the best compromise between currently available data. Adopting $E_{B-V} = 0.02 \pm 0.02$ and $(m - M)_0 = 19.46$ yields $(m - M)_V = 19.52 \pm 0.10$.

The true distance moduli given in table 2 yield distances of 59 and 78 kpc for the LMC and SMC, respectively. This difference of 19 kpc in the radial distances of the Clouds may be compared with an angular separation of 22° which corresponds to 26 kpc. These values are compatible with radio observations which indicate that the Magellanic Clouds are embedded in a common hydrogen envelope.

2.4.3. *The Andromeda Nebula.*—From multicolor photometry of distant galactic foreground stars McClure and Racine (1969) obtain a reddening[1] $E_{B-V} = 0.11 \pm 0.02$ m.e. in the direction of the Andromeda Nebula. This value is in good agreement with $E_{B-V} = 0.09 \pm 0.02$ obtained by van den Bergh (1969) from a comparison of the colors of galactic and M31 globular clusters having the same line strengths in their integrated spectra.

Polarization measurements of the nuclei of NGC 205, M31, and M32 by Elvius and Hall (1964) give a mean polarization $\Delta m_p = 0.018 \pm 0.003$ mag. Since the polarization vectors of the members of the M31 group are almost parallel to those of the foreground stars ν And and 32 And (Behr 1959), it appears reasonable to assume that this polarization is produced in the Galaxy. The galactic longitude of M31 suggests that $\Delta m_p/A_V \simeq 0.045 \pm 0.010$ (estimated m.e.), so that $E_{B-V} = 0.13 \pm 0.03$.

It is concluded that the available evidence suggests a galactic foreground reddening of $E_{B-V} = 0.11 \pm 0.02$ in the direction of M31. This value is smaller than the value $E_{B-V} = 0.16 \pm 0.03$ which Baade and Swope (1963) found from the stars in M31 Variable Star Field IV. The reason for this difference is, presumably, that some of the reddening in Field IV is produced by dust clouds within the Andromeda Nebula itself.

Adopting $E_{B-V} = 0.11 \pm 0.02$ and $A_V/E_{B-V} = 3.0$ yields apparent distance moduli $(m - M)_V = 24.66 \pm 0.13$ and $(m - M)_B = 24.77 \pm 0.14$ for objects that do not suffer absorption within M31 itself.

2.4.4. *M33.*—No modern data are available on the Cepheid variables in M33. According to Hubble (1936*a*) the Cepheids in M33 are 0.1 mag brighter than are those in M31, which would make $(m - M)_B = 24.67 \pm 0.14$. From measurements of foreground field stars in the direction of M33 McClure and Racine (1969) obtain a galactic foreground reddening $E_{B-V} = 0.03 \pm 0.02$. Schmidt-Kaler (1967) obtains a mean reddening $E_{B-V} = 0.08 \pm 0.03$ for the star clusters in M33. This value includes some reddening within the Triangulum Galaxy itself. The value $E_{B-V} = 0.03 \pm 0.02$ will therefore be adopted, whence $(m - M)_V = 24.64 \pm 0.14$ and $(m - M)_0 = 24.55 \pm 0.15$.

2.4.5. *NGC 6822.*—From Kayser's (1967) photographic observations of 12 foreground field stars, which are located more than 300 pc below the galactic plane, a value $E_{B-V} = 0.19 \pm 0.03$ is obtained. The formal mean error of this result does not take into account possible systematic errors in the photographic photometry.

[1] Reddening values used in this section have been converted to those which would be observed for a B0 star (Fernie 1963).

Combining the quoted reddening value with the true distance modulus $(m - M)_0 = 24.21 \pm 0.19$, which is obtained from Kayser's Cepheid observations, yields an apparent visual distance modulus $(m - M)_V = 24.78 \pm 0.21$.

2.4.6. *IC 1613.*—From a provisional comparison of the Cepheids in IC 1613 with those in the Small Magellanic Cloud, Sandage (1962) finds that the photographic apparent distance modulus of IC 1613 is 5.31 mag larger than that of the Small Cloud. Adopting $(m - M)_B = 19.54$ for the SMC then yields $(m - M)_B = 24.85$ for IC 1613. The uncertainty of this value is not known. From *UBV* observations of foreground field stars Sandage (1962) obtains a reddening $E_{B-V} = 0.03$ so that $(m - M)_V = 24.82$ and $(m - M)_0 = 24.73$.

3. NOVAE

3.1. Galactic Calibration

Schmidt-Kaler (1957) has used the relation between the magnitudes of novae at maximum light and their rate of decline to determine the distance moduli of M31 and the Magellanic Clouds. Table 3 (Schmidt-Kaler 1965*b*) gives revised data on the absolute

TABLE 3

Data on Galactic Novae (Schmidt-Kaler 1965b)

Nova	Year	log t_2	M_{pg} (max)*	Method†
CP Pup	1942	0.45:	−9.3+0.3	1, 2, 3.
V603 Aql	1918	0.60	−8.7±0.2	1, 2.
Her	1960	0.71	−8.7±0.2	2, 3.
CP Lac	1936	0.85	−8.6±0.2	1, 2.
GK Per	1901	0.98	−8.3±0.3	1, 2.
V476 Cyg	1920	1.04	−8.8±0.2	1, 2.
DK Lac	1960	1.16	−8.0±0.4	2, 3.
T Sco	1860	1.18	−8.6±0.2	4.
EU Sct	1949	1.32	−8.5±0.4	2, 3.
DN Gem	1912	1.40	−7.0±0.4	2.
BT Mon	1939	1.45:	−7.6±0.8	3.
Her	1963	1.50	−6.6±0.4	2.
DQ Her	1934	1.90	−6.5±0.2	1, 2.
T Aur	1891	2.00	−6.1±0.3	1.
RR Pic	1925	2.00	−7.1±0.4	1.

* Quoted uncertainty is the internal mean error of the determination

† Method 1 = expansion π; method 2 = Ca (and Na) π; method 3 = galactic rotation π (A = 15 km s^{-1} kpc^{-1}); method 4 = globular cluster π.

magnitudes and rates of decline of galactic novae. In the table M_{pg}(max) is the absolute magnitude at maximum light and t_2 is the time (in days) which a nova required to decline by 2 magnitudes from maximum light. The data in table 3 are plotted in figure 5.

3.2. The Andromeda Nebula

Table 4, which is based on observations by Arp (1956) and Rosino (1964), gives data on well-observed novae in M31. In some cases the numerical results quoted by these authors have been slightly revised by Schmidt-Kaler (1965*b*). The data given in table 4 are plotted in figure 6. The figure shows that the scatter about the adopted m_{pg}(max) versus log t_2 relation is remarkably small. Only one nova (Rosino No. 36) appears to deviate significantly from the adopted mean relation. Fitting the galactic novae to the mean m_{pg}(max) versus log t_2 relation obtained in M31 yields an apparent distance modulus $(m - M)_{pg} = 24.65 \pm 0.12$. The quoted mean error does not include possible systematic errors in the luminosity calibration of galactic novae. *If* the mean reddening of the novae in M31 is similar to that in Variable Star Field IV, for which a value $E_{B-V} = 0.16$ was previously adopted, one obtains $(m - M)_V = 24.49$ and hence $(m - M)_0 = 24.01$.

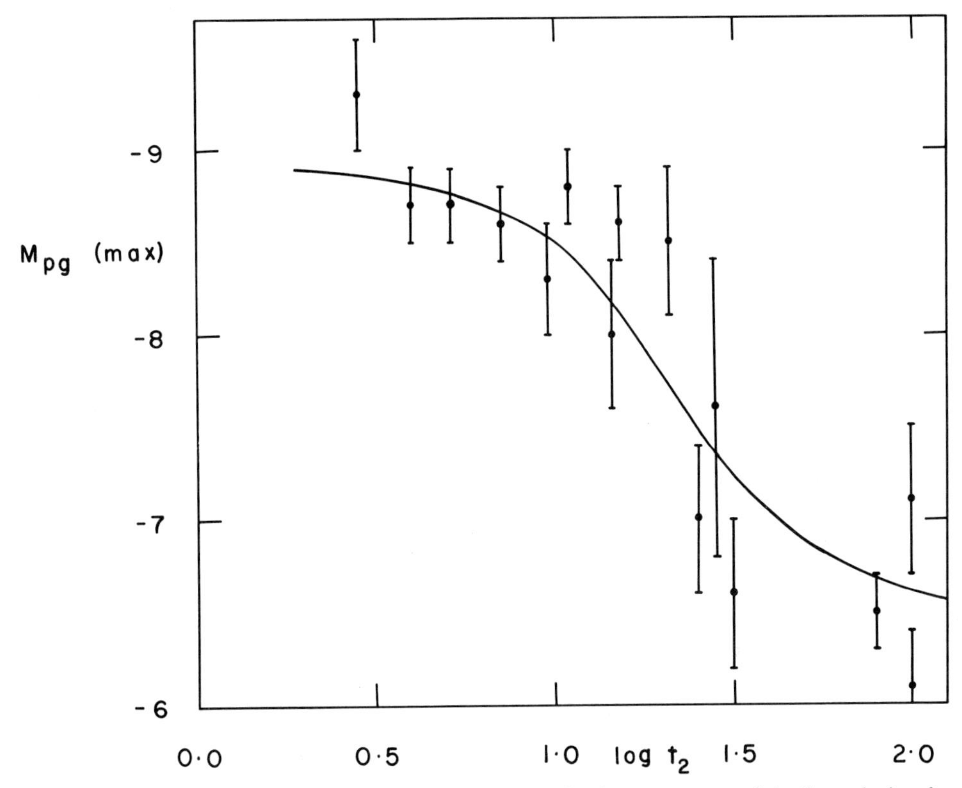

FIG. 5.—Galactic calibration of the maximum magnitude versus rate of decline relation for novae. The smooth curve was taken from the M31 observations, which are plotted in figure 6.

3.3. The Magellanic Clouds

Data on novae observed in the Magellanic Clouds (Henize, Hoffleit, and McKibben Nail 1954; Schmidt-Kaler 1957) are given in table 5. Many of the quoted data are quite uncertain because of the fragmentary nature of the nova light curves. Giving half-weight to uncertain observations the Small Cloud data yields $(m - M)_{pg} = 19.44 \pm 0.15$. Excluding two entries in table 5 which are marked (?) and giving half-weight to uncertain observations yields $(m - M)_{pg} = 19.47 \pm 0.41$ for the LMC. Novae are not concentrated in dusty regions containing young stars and H II regions. Values $E_{B-V} = 0.08$ and $E_{B-V} = 0.02$ may therefore be appropriate for the novae in the LMC and SMC, respectively. If these values are adopted, the apparent visual distance moduli become $(m - M)_V = 19.39$ for the Large Cloud and $(m - M)_V = 19.42$ for the Small

TABLE 4

DATA ON NOVAE IN M31 (Schmidt-Kaler 1965b)

No.	$\log t_2$	m_{pg}(max)	No.	$\log t_2$	m_{pg}(max)
		First-Class Data			
R33*	0.70	16.2	A15	1.20	16.4
R28	0.73	15.8	R37	1.30	16.9
A7	0.90	16.0	R15	1.30	16.3
A6	0.92	16.0	R18	1.33	17.1
A8	0.93	16.0	R20	1.35	16.9
R16	0.96	16.3	R21	1.35	16.7
A12	1.05	16.1	A13	1.42	17.0
A5	1.07	15.9	A21	1.43	17.4
R30	1.07	16.2	A19	1.45	17.6
R6	1.14	16.5	A22	1.48	17.6
R4	1.15	16.5	A20	1.52	17.2
R38	1.15	16.9	A24	1.53	17.8
R9	1.20	17.0	R36	1.88	16.9
		Second-Class Data			
A2	0.29	15.7	R42	1.48	16.9
A1	0.30	15.8	A25	1.51	17.6
A10	0.84	15.9	R12	1.60	17.6
A4	1.00	16.3	A23	1.63	17.4
R43	1.04	16.5	R19	1.80	17.9
A16	1.11	16.7	A28	1.92	17.9
R46	1.22	16.7	A29	1.97	18.0
R34	1.24	17.0	A30	2.05	18.1
A17	1.46	17.2			

* A = Arp 1956; R = Rosino 1964.

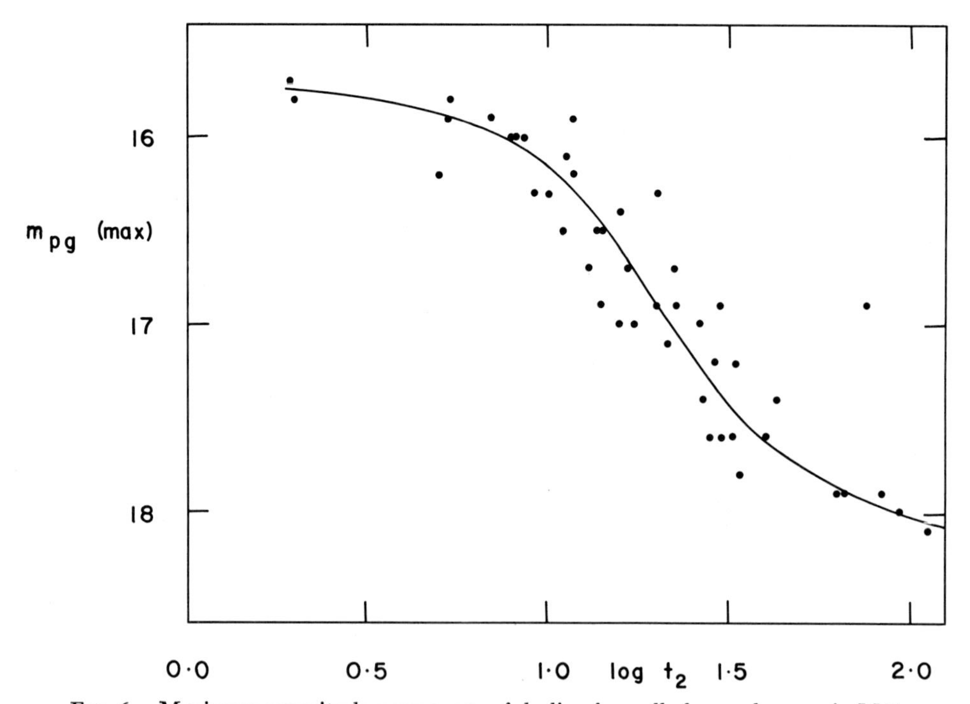

FIG. 6.—Maximum magnitude versus rate of decline for well-observed novae in M31.

Cloud. The corresponding true distance moduli are $(m - M)_0 = 19.15$ for the LMC and $(m - M)_0 = 19.36$ for the SMC. These (very uncertain!) values are seen to be in reasonable agreement with those that are obtained from the period-luminosity-color relation of classical Cepheids.

TABLE 5

NOVAE IN THE MAGELLANIC CLOUDS

Nova	Year	log t_2	m_{pg}(max)	M_{pg}	$(m-M)_{pg}$
			Small Magellanic Cloud		
N Tuc.......	1952	fast	11.0	−8.6:	19.6:
N Tuc.......	1897	1.1	11.0:	−8.3	19.3:
VZ Tuc......	1927	1.1	11.4:	−8.3	19.7:
N Tuc.......	1951	1.3	11.5	−7.8	19.3
			Large Magellanic Cloud		
N Men......	1951	very fast	11.95	−8.85:	20.8:
N Dor.......	1948	0.8	13.0?	−8.7	21.7?
N Dor.......	1937	0.9	10.65	−8.65	19.3
N Dor.......	1936	1.1	10.5:	−8.3	18.8:
N Hyi*......	1935	1.1	11.0	−8.3	19.3
RY Dor.....	1926	1.6	12.0?	−7.0	19.0?

* Possible supernova in NGC 1511.

4. RR LYRAE VARIABLES

4.1. GALACTIC CALIBRATION

The values $\langle M_V \rangle = +0.5$ and $\langle M_B \rangle = +0.8$ will be adopted for the absolute magnitudes of RR Lyrae variables. These values represent a compromise between the values obtained by Sandage (1964) for globular-cluster RR Lyrae variables and by van Herk (1965) for field RR Lyrae stars. The uncertainty of the adopted absolute magnitudes is probably about 0.2 mag.

4.2. DISTANCES OF THE MAGELLANIC CLOUDS

According to Tifft (1963), $\langle V \rangle = 19.6$ for the field RR Lyrae stars in the Small Cloud, from which $(m - M)_V = 19.1$. For the variables in the globular cluster NGC 121 in the Small Cloud, Tifft finds $\langle V \rangle = 19.45$. The difference between the magnitudes of the field and cluster variables might be due to slightly different distances or to background enhancement in the crowded cluster field.

For the RR Lyrae variables in the clusters NGC 1466 and NGC 2257 in the Large Cloud, Gascoigne (1966) finds $\langle V \rangle = 19.1$ and $\langle V \rangle = 19.2$, respectively, from which a mean apparent modulus $(m - M)_V = 18.65$ is obtained. It should, of course, be emphasized that these distance moduli for the Clouds are quite uncertain, because they are based on photometry close to the limit attainable with 74-inch telescopes. It is, however, of interest to note that Cepheids, novae, and RR Lyrae stars all indicate that the Small Cloud is more distant than is the Large Cloud.

5. W VIRGINIS STARS

Available data (Fernie 1964; Demers 1966; Dickens and Carey 1967; Kwee 1968) show that the period-luminosity relation of type II Cepheids exhibits considerable dispersion. This intrinsic dispersion suggests that the period-luminosity relation of W Virginis stars is probably not unique. In the present discussion Kwee's (1968) relation between the light curves and the luminosities of type II Cepheids will be adopted. It should, however, be emphasized that the evidence for a dependence of the luminosity of a W Virginis variable on the shape of its light curve is still very weak.

5.1. The Small Magellanic Cloud

Gaposchkin and Gaposchkin (1966) report observations of three W Virginis stars in the Small Cloud. Two of these variables fall within the period range over which Kwee (1968) gives period-luminosity relations for galactic W Virginis stars. For the "flat-topped" variable HV 12901, Kwee's period-luminosity relation gives $\langle M_B \rangle = -1.67$. Combining this value with $\langle m_{pg} \rangle = 17.36$ yields $(m - M)_{pg} = 19.03$ and hence $(m - M)_V = 19.01$. For the "crested" variable HV 1828, Kwee's period-luminosity relation gives $\langle M_B \rangle = -2.27$. In conjunction with $\langle m_{pg} \rangle = 17.28$ this yields $(m - M)_{pg} = 19.55$, from which $(m - M)_V = 19.53$. Slightly larger distance moduli would have been obtained if small (but uncertain) corrections had been applied to Arp's (1960) photoelectric standards, which were used by Gaposchkin and Gaposchkin (1966).

Tifft (1963) has observed a type II variable star of period $1^d.43$ with $\langle V \rangle = 18.05$ near the SMC globular cluster NGC 121. Comparison with M13 No. 1 ($\langle V \rangle = 13.69$, $P = 1^d.46$) and M15 No. 1 ($\langle V \rangle = 14.89$, $P = 1^d.44$) yields $\langle M_V \rangle = -0.4$ if $\langle M_V \rangle = +0.5$ for the RR Lyrae stars in M13 and M15. With $\langle M_V \rangle = -0.4$ and $\langle V \rangle = 18.05$, a value $(m - M)_V = 18.45$ is obtained for the SMC. This distance modulus is significantly smaller than are the distance moduli obtained from Cepheids, novae, RR Lyrae stars, and long-period variables of Population II. This suggests that the short-period type II variables in the SMC may be brighter than their Galactic counterparts. In this respect they might resemble the short-period type II variables in dwarf spheroidal galaxies. According to van Agt (1967), variables with $P = 1^d.4$ are 0.95 mag brighter in dwarf spheroidal galaxies than are those in globular clusters in the Galaxy. Adopting $\langle M_V \rangle = -1.35$ for the variable in NGC 121 yields a distance modulus $(m - M)_V = 19.40$. Clearly this distance modulus, which is based on observations of a single star, is quite uncertain.

The mean of the three distance moduli obtained above is $(m - M)_V = 19.31 \pm 0.15$.

5.2. The Andromeda Nebula

Baade and Swope (1963, 1965) and Gaposchkin (1962) have observed W Virginis stars in M31. Only variables in Field IV, in which absorption and background enhancement effects are relatively small, will be considered in detail. The mean of the two period-luminosity relations given by Kwee (1968) yields $(m - M)_B = 24.90 \pm 0.10$ for the seven long-period type II variables in Field IV. Adopting the value $E_{B-V} = 0.16$, which is obtained from the Cepheids in Field IV (Baade and Swope 1963), then gives $(m - M)_V = 24.74 \pm 0.10$.

6. RED GIANTS OF POPULATION II

The use of globular-cluster red giants as distance indicators has the advantage that such stars are about 3 magnitudes brighter than RR Lyrae stars. The observations of these stars are therefore much less susceptible to photometric errors than are the observations of the RR Lyrae variables. A disadvantage of this method of distance determination is, however, that the absolute magnitudes of globular-cluster giants are known to be sensitive to metal abundance (Arp 1955; Sandage and Wallerstein 1960).

The Magellanic Clouds contain many old star clusters, some of which are undoubtedly real globulars. Others may be open clusters of the NGC 2158, NGC 7789 variety (Gascoigne 1966). Data on the clusters, which are known to be globular because they contain RR Lyrae stars (Thackeray 1963), are given in table 6. [In the table, V^* denotes the apparent visual magnitude of the giant branch at $(B - V)_0 = 1.40$.] With $\langle M_V \rangle = +0.5$ for RR Lyrae stars, one has $M_V{}^* = -1.6$ in metal-rich globular clusters and $M_V{}^* = -2.6$ in metal-poor clusters (Sandage and Wallerstein 1960). Adopting $M_V{}^* = -2.1 \pm 0.5$ for the clusters in the Magellanic Clouds yields $(m - M)_V = 19.05$ for the Small Cloud and $(m - M)_V = 18.37$ for the Large Cloud.

TABLE 6

CLUSTERS CONTAINING RR LYRAE VARIABLES

Location	NGC	V*	Reference
SMC..........	121	16.95	Tifft 1963; Gascoigne 1966
LMC..........	1466	16.35	Gascoigne 1966
LMC..........	1978	16.4:	Hodge 1960
LMC..........	2257	16.05	Gascoigne 1966

* Magnitude at $(B - V)_0 = 1.40$.

According to van den Bergh and Racine (1967) the brightest stars in NGC 205, which is a companion to M31, have an apparent magnitude $V = 21.8 \pm 0.3$. Comparison of image-tube spectra of NGC 205 with spectra of globular clusters in the Galaxy (van den Bergh 1969) suggests that the dominant stellar population of NGC 205 is quite metal-poor. Accordingly, it will be assumed that $M_V = -2.6$ (Sandage and Wallerstein 1960) for the brightest stars in NGC 205. Adopting this value then yields $(m - M)_V = 24.4 \pm 0.3$ for NGC 205, which is in satisfactory agreement with other determinations of the distance modulus of the Andromeda Galaxy.

7. GLOBULAR CLUSTERS

The brightest and faintest known globular clusters in the Galaxy differ in luminosity by more than two orders of magnitude. Presumably these luminosity differences are due to differing physical conditions in the regions in which individual globular clusters were formed. Somewhat different *average* physical conditions might also be expected to prevail in different protogalaxies. This suggests that one might not be surprised to find differences by factors of, say, 2 or 3 between the mean luminosities of globular clusters in different galaxies. These considerations suggest that considerable caution should be exercised when globular clusters are used to estimate extragalactic distances.

The red clusters in the Magellanic Clouds probably constitute a mixture of true globular clusters and old open clusters (Gascoigne 1966). It therefore appears imprudent to use them to estimate the distance of the Clouds. No such difficulties exist for the outlying clusters associated with the Andromeda Nebula. Magnitudes are available (Hiltner 1960; Kron and Mayall 1960; Racine 1965; van den Bergh 1969) for all 34 known clusters which lie outside the region of M31, in which star formation is currently taking place. For these clusters $\langle B \rangle = 17.1 \pm 0.2$. Hogg (1963) gives distance moduli and apparent magnitudes for 34 globular clusters in the Galaxy with $|b^{\mathrm{II}}| > 16.^{\circ}0$, yielding $\langle M_B \rangle = -6.5 \pm 0.2$ if $\langle M_V \rangle = +0.5$ for the cluster-type variables. Comparison with photometry by Kron and Mayall shows that the integrated apparent magnitudes of high-latitude clusters adopted by Hogg are too faint by 0.3 mag in the mean, so that $\langle M_B \rangle = -6.8 \pm 0.2$. Comparison of this value with $\langle B \rangle = 17.1 \pm 0.2$ for the globular clusters in M31 then yields $(m - M)_B = 23.9 \pm 0.3$. Both the data on the high-latitude clusters and those for the outlying clusters in M31 are affected by selection effects, which discriminate against faint clusters. A comparison of the bright ends of the integral luminosity functions of the sample of halo globular clusters in M31 and the Galaxy, which should be almost independent of selection effects, also yields $(m - M)_B = 23.9$, with an estimated uncertainty of 0.1 mag. Assuming $E_{B-V} = 0.11$, this corresponds to $(m - M)_V = 23.79$. This value is significantly smaller than the values obtained by other methods. The implication of this result is that the globular clusters in M31 are intrinsically brighter than those in the Galaxy. This conclusion is confirmed by observations of the brightest globulars (which are least affected by selection effects) in the Andromeda Nebula and our Galaxy. Assuming $(m - M)_V = 24.6$, van den Bergh (1969) finds that M31 contains 18 bright clusters with $M_V \leq -10.0$ compared with only two such clusters in the Galaxy.

8. SPECTRAL LUMINOSITY DETERMINATIONS

Classifications on the MK system are available for a number of supergiants in the Magellanic Clouds. The data for ordinary supergiants of luminosity class Ia are subject to selection effects, which tend to favor observation of stars with above-average luminosity. For early-type stars of luminosity class Ia-0 the data by Feast, Thackeray, and Wesselink (1960) should, however, be reasonably complete. For seven such stars a true distance modulus $(m - M)_0 = 18.3 \pm 0.3$ and a mean absorption $A_V = 0.4$ mag are obtained. The quoted mean error of the distance modulus for the Large Cloud does not include the uncertainty in the galactic calibration (Schmidt-Kaler 1965*a*) of stars of luminosity class Ia-0. According to Schmidt-Kaler (1965*b*), the galactic calibration for Ia-0 stars may be slightly too faint, which would make $(m - M)_0$ larger than the value adopted above.

Buscombe and Kennedy (1962) have published absolute magnitudes, based on Johnson and Iriarte's (1958) calibration of the equivalent width of $H\gamma$, for eight early-type supergiants in the Small Cloud. Giving half-weight to uncertain observations yields $(m - M)_V = 18.7$ and $(m - M)_0 = 18.4 \pm 0.3$.

Hutchings (1966) has obtained photographic $H\gamma$ observations of 149 stars in the Magellanic Clouds. From these observations he obtains $(m - M)_V = 18.2 \pm 0.4$ for the LMC and $(m - M)_V = 18.3 \pm 0.4$ for the SMC. This unexpectedly small distance modulus of the SMC is confirmed by Westerlund, Danziger, and Graham (1963). From photoelectric Hβ photometry of seven supergiants in the wing of the Small Cloud these authors find $(m - M)_V = 18.47 \pm 0.24$.

The unweighted mean of the Hβ and Hγ observations of the SMC yields $(m - M)_V = 18.49$ compared to $(m - M)_V = 18.2$ for the LMC. It is not yet clear why these distance moduli should be so much smaller than those which are obtained by other methods.

9. SUMMARY OF DATA ON THE LOCAL GROUP

The individual determinations of the distance moduli of Local Group members, which have been obtained in §§ 2–8, are summarized in tables 7 and 8. It will be seen from these tables that different methods of distance determination yield distance moduli that differ by as much as 1 magnitude. Such differences may arise from systematic and random errors in the observations or from the uncertainty of the galactic calibrations. Additional dispersion could also be introduced by differences in age, evolutionary history, and chemical composition.

TABLE 7

APPARENT VISUAL DISTANCE MODULI OF LOCAL GROUP GALAXIES*

Method	Wt	LMC	SMC	M31	M33	NGC 6822	IC 1613
Cepheids........	3	19.0 ±0.10	19.52±0.10	24.66±0.13	24.64±0.14:	24.78±0.2	24.82
Novae..........	2	19.39±0.4	19.42±0.15	24.49±0.12	....	....	....
RR Lyrae stars..	2	18.65±0.2	19.10±0.2	....	....	....	....
W Virginis stars.	1	....	19.31±0.15	24.74±0.10	....	....	....
Population II giants........	1	18.37±0.5	19.05±0.5	24.40±0.3	....	....	....
Globular clusters	1	....	....	23.9 ±0.1	....	....	....
MK classification	1	18.70±0.3	....	....	....	....	....
Hβ + Hγ.......	1	18.20±0.4	18.49±0.3	....	....	....	....
$\langle m-M\rangle_V$.......	....	18.84±0.17	19.24±0.14	24.50±0.13	24.64	24.78	24.82
V.............	....	0.1	2.3	3.5	5.7	8.5	9.6
M_V...........	....	−18.7	−16.9	−21.0	−18.9	−16.3	−15.2
A_V (foreground)..	....	0.18	0.06	0.33	0.09	0.57	0.09

* Errors quoted for each method are *internal* mean errors. The mean error of $\langle m - M\rangle_V$ is derived from the agreement between different methods.

TABLE 8

TRUE DISTANCE MODULI OF LOCAL GROUP MEMBERS

Method	Wt	LMC	SMC	M31	M33	NGC 6822	IC 1613
Cepheids..........	3	18.85	19.46	24.33	24.55	24.21	24.73
Novae...........	2	19.15	19.36	24.01	....	....	....
RR Lyrae stars.....	2	18.47	19.04	....	....	....	....
W Virginis stars.....	1	....	19.25	24.26	....	....	....
Population II giants.	1	18.19	18.99	24.07	....	....	....
Globular clusters....	1	....	....	23.46	....	....	....
MK classifications..	1	18.30	....	....	....	....	....
Hβ+Hγ..........	1	17.90	18.43	....	....	....	....
$\langle m-M\rangle_0$...........	...	18.62±0.18	19.18±0.14	24.10±0.14	24.55	24.21	24.73
Distance (kpc)......	...	53	69	660	810:	700:	880:

The final distance moduli which are adopted for M31 and the Magellanic Clouds depend rather critically on the relative weights that one may wish to assign to different methods of distance determination. More or less arbitrarily a weight of three units has been assigned to moduli derived from Cepheids, a weight of two units has been given to moduli obtained from novae and RR Lyrae stars, and unit weight has been assigned to distance moduli determined from W Virginis stars, Population II giants, globular clusters, MK classifications, and Hβ or Hγ photometry. It should perhaps be emphasized that the resulting distance moduli of Local Group members depend rather critically on the assumption that the Cepheids in all galaxies obey a universal period-luminosity-color relation.

Cepheids, novae, RR Lyrae stars, Population II giants, and Hγ photometry all indicate that the SMC is more distant than the LMC. The adopted difference in the distances of the two Clouds is still quite compatible with radio observations which show that the Clouds are embedded in a common hydrogen envelope. The very small distances that are obtained from Hβ and Hγ photometry of the brightest stars in the Clouds remain unexplained. The distance modulus of M31 obtained from the integrated magnitudes of globular clusters is incompatible with the distance moduli obtained by other methods. This probably indicates that the integrated absolute magnitudes of globular clusters do not have the same frequency distribution in different galaxies.

10. DISTANCES BEYOND THE LOCAL GROUP

Ancient mariners who sailed the distant seas of Earth often took along three chronometers on their voyages. In this way they could always tell if *one* of their clocks did not keep good time. By the same token the astronomer of today is well advised to use as many distance indicators as possible to chart the vast reaches of space beyond the Local Group. Such caution seems particularly important because some of the required measurements are among the most difficult encountered in modern astronomy.

Nine methods that can be used to measure intercluster distances will be discussed in the following sections. Each of these methods is itself uncertain. The degree of internal agreement between the distances obtained by these different methods does, however, provide some indication of the degree of confidence which may be placed in the resulting distance scale.

11. THIRD BRIGHTEST CLUSTER GALAXY

The Palomar *Sky Survey* has been thoroughly searched for small, relatively nearby clusters, which contain a galaxian population similar to that observed in the Local Group. The four best examples of such clusters are listed in table 9. A rough estimate of the value of the Hubble constant may be obtained by comparing M33, which is the third brightest member of the Local Group, with the third brightest galaxy in other small clusters. Such a comparison is made in table 9. In the table $\langle V \rangle$ denotes the mean velocity of recession of the cluster, corrected for galactic rotation. The values of $B_3(0)$, the integrated magnitude of the third brightest cluster galaxy, were taken from de Vaucouleurs and de Vaucouleurs (1964) and corrected for galactic absorption by means of the equation

$$B(0)_{\mathrm{cor}} = B(0) - 0.24 \operatorname{cosec} |b| \, . \tag{13}$$

From the data in table 9 it is found that

$$\langle M_{B3}(0)\rangle = 5 \log (H/100) - 18.7 \pm 0.2 , \quad (14)$$

with a standard deviation of 0.5 mag.

Adopting $(m - M)_B = 24.7 \pm 0.5$ and $B(0) = 6.5$ for M33, one obtains $M_{B3}(0) = -18.2 \pm 0.5$. Substitution of this value into equation (14) yields $5 \log (H/100) = +0.5 \pm 0.7$. The corresponding value of the Hubble constant is $H = 126$ (+48, −35) km s^{-1} Mpc^{-1}.

It should perhaps be emphasized that this value of the Hubble constant is based on the *assumption* that all small clusters of galaxies have similar luminosity functions. Furthermore, this determination involves a subjective choice of clusters, which are believed to be similar to the Local Group.

TABLE 9

DATA ON THIRD BRIGHTEST GALAXY IN SMALL CLUSTERS

Cluster	$B_3(0)_{cor}$	$\langle V\rangle$ (km s^{-1})	$M_{B3}(0)-5 \log (H/100)$
NGC 3227........	12.0	1200±40	−18.4
NGC 3368........	10.5	690±40	−18.7
NGC 3627........	9.6	600±60	−19.3
NGC 4274........	11.4	870±70	−18.3
Mean..........	....	....	−18.7±0.2

12. SURFACE BRIGHTNESS OF GALAXIES

Holmberg (1958) has shown that the surface brightness of spiral galaxies correlates with their intrinsic luminosity. For 36 spirals in the Virgo cluster he finds that the photographic magnitude and surface brightness, both corrected to pole-on orientation and galactic latitude 90°, are related by

$$S_0' = 0.181\, m_0' + 13.19 , \quad (15)$$

in which S_0' is expressed in magnitudes per square minute of arc. The standard deviation of m_0' obtained from equation (15) and S_0' is 1.3 mag. Holmberg finds that $S_0' = 14.80$ for M31 and $S_0' = 15.38$ for M33. These values were obtained by assuming a cosecant absorption relation of the form

$$A_{pg}(b) - A_{pg}(90°) = 0.25\, (\operatorname{cosec} |b| - 1) . \quad (16)$$

This relation overestimates the *difference* between the absorption in the direction of M33 and the Virgo cluster. The values $A_{pg}(\text{M31}) = 0.44$, $A_{pg}(\text{M33}) = 0.12$, and $A_{pg}(\text{Virgo}) = 0.00$ will be assumed. [The value $A_{pg}(\text{Virgo}) = 0.00$, which is based on the observation that the Coma cluster (Crawford and Barnes 1969) is unreddened, is still quite uncertain.] Using the adopted absorption values yields $S_0' = 14.80$ for M31 and $S_0' = 15.50$ for M33. From equation (15) it is then found that, at the distance of the Virgo cluster, M31 and M33 would have corrected apparent photographic magnitudes $m_0' = 8.90 \pm 1.3$ and $m_0' = 12.76 \pm 1.3$, respectively. Comparison with the actual values $m_{pg_0}' = 3.33$ for M31 and $m_{pg_0}' = 5.90$ for M33 then shows that the true distance modulus of the Virgo cluster is 5.57 ± 1.3 mag greater than that of M31 and 6.86 ± 1.3 mag greater than that of M33. Using the Local Group distances quoted in table 8 then yields $(m - M)_0 = 30.5 \pm 0.9$. The rather large mean error of this result is due to (1) the large intrinsic dispersion of galaxies about the relation given in equation (15) and (2) the fact that only two Local Group galaxies are available for calibration.

To derive the Hubble constant from this value of the distance modulus, it is necessary to know the mean radial velocity of the Virgo cluster. According to van den Bergh (1960*c*) the Virgo cluster has a mean redshift $\langle V \rangle = 1175 \pm 64$ km s^{-1}. This is in good agreement with the value $\langle V \rangle = 1136$ km s^{-1} obtained by Humason, Mayall, and Sandage (1956). [De Vaucouleurs's (1961) suggestion that the Virgo cluster consists of two superposed clouds of galaxies does not seem to be supported by Kowal's (1968, 1969) work on the supernovae in the Virgo cluster.] Substituting $\langle V \rangle = 1175 \pm 64$ km s^{-1} and $(m - M)_0 = 30.5 \pm 0.9$ into

$$(m - M)_0 = 15 + 5 \log \langle V \rangle - 5 \log (H/100) \tag{17}$$

yields $\log (H/100) = -0.15 \pm 0.9$ which corresponds to $H = 93$ $(+48, -31)$ km s^{-1} Mpc^{-1}.

13. LUMINOSITY CLASSIFICATION OF GALAXIES

Van den Bergh (1960*b*) has developed a system of luminosity classification of galaxies that is based on the degree to which spiral structure is developed. For giant and supergiant galaxies, which can be observed to large distances, the classification system may be calibrated directly in terms of the Hubble constant. Using the integrated $B(0)$ magnitudes given by de Vaucouleurs and de Vaucouleurs (1964) and assuming the cosecant absorption law of equation (13), the calibration given in table 10 is obtained.

TABLE 10

CALIBRATION OF LUMINOSITY CLASSES

Luminosity Class	$M_B(0) - 5 \log (H/100)$	Standard Deviation (mag)	No. of Galaxies
I–II.........	-19.7 ± 0.1	0.4	13
II–III.......	-18.8 ± 0.2	0.7	10

For M31, which is of luminosity class I–II, $B(0) = 4.6$ and $(m - M)_B = 24.61 \pm 0.13$ so that $M_B(0) = -20.01 \pm 0.13$. For M33, which is of luminosity class II–III, $B(0) = 6.5$ and $(m - M)_B = 24.67 \pm 0.3$(?), so that $M_B(0) = -18.27 \pm 0.3$. From these data and the calibration given in table 10 it is found that $5 \log (H/100) = -0.3 \pm 0.4$ for M31 and $5 \log (H/100) = +0.5 \pm 0.8$ for M33. The weighted mean value $5 \log (H/100)$ is -0.05 ± 0.4, which corresponds to $H = 98$ $(+19, -17)$ km s^{-1} Mpc^{-1}. The main uncertainty of this result is again due to the fact that only two Local Group galaxies are available for calibration. Unfortunately the dwarf irregular galaxies in the Local Group cannot be used because apparent magnitudes and radial velocities are not yet available for a sufficient number of distant dwarf galaxies.

14. DIAMETERS OF H II REGIONS

Sérsic (1960) has made measurements of the diameters of H II regions on plates obtained with the 100-inch and 200-inch telescopes. These measurements were corrected for the size of the seeing disk on each plate. Unfortunately only blue plates, on which the contrast between nebula and background is often quite small, were available for Sérsic's investigation.

If d'' denotes the mean of the true angular diameters (measured in seconds of arc) of the three largest H II regions in a galaxy, then

$$Q = (m - M)_0 + 5 \log d'' \tag{18}$$

will be independent of distance. Inspection of Sérsic's data shows that Q is a function of both the integrated color and the luminosity of a galaxy. This result is to be expected

TABLE 11

DIAMETERS OF H II REGIONS

Location	$(m-M)_0$	$5 \log d''$	Q
M31	24.10	7.4	31.5
M33	24.55	8.4	33.0
LMC	18.62	13.4	32.0
SMC	19.18	13.2	32.4
NGC 6822	24.21	8.3	32.5
IC 1613	24.73	7.6	32.3

because the brightest and the bluest galaxies contain the strongest population of OB stars. Using true distance moduli given by van den Bergh (1960*b*) for 13 distant galaxies, which Sérsic classifies as Sc$^+$, dSc, and Ir, it is found that

$$Q + 5 \log (H/100) = 32.22 \pm 0.27 \,. \tag{19}$$

The standard deviation from the mean is 0.98 mag, which corresponds to a 57 percent dispersion in the linear diameter of the mean of the three largest H II regions. By what appears to be an arbitrary subdivision of intrinsically faint galaxies into types Ir and dSc, Sérsic obtains a standard deviation of only 0.16 mag within each subgroup. This value corresponds to an 8 percent dispersion in the mean diameter of the three largest H II regions in a galaxy. This result is hard to credit if one recalls (Sérsic 1959) that the largest and third largest H II regions in the Large Magellanic Cloud have diameters that differ by more than a factor of 2. In any case Sérsic's division of faint galaxies into classes dSc and Ir is not supported by the classifications given by de Vaucouleurs (1964) and by van den Bergh (1960*b*).

Sérsic's (1960) measurements of the mean diameters of the three largest H II regions in the galaxies of the Local Group are given in table 11. From the data in this table $\langle Q \rangle = 32.44 \pm 0.16$ for the five Sc$^+$ and Ir galaxies in the Local Group. Substitution of this value into equation (19) yields $5 \log (H/100) = -0.2 \pm 0.3$. The quoted mean error is probably somewhat optimistic because it does not take into account possible systematic errors resulting from the fact that the sample of 13 distant galaxies are mainly late-type spirals whereas four out of five Local Group members used to determine $\langle Q \rangle$ are irregulars. A value 5 log $(H/100) = -0.2 \pm 0.4$, which corresponds to $H = 91$ $(+19, -15)$ km s^{-1} Mpc^{-1}, will be adopted.

A check on the results obtained above is provided by Sérsic's measurements of the diameters of H II regions in five late-type galaxies in the M81 cluster. Combining these measurements with the distance modulus $(m - M)_0 = 27.55 \pm 0.13$ given by Tammann and Sandage (1968) yields $\langle Q \rangle = 32.69 \pm 0.26$, which is in satisfactory agreement with the value $\langle Q \rangle = 32.44 \pm 0.16$ obtained for late-type members of the Local Group.

15. MASS-TO-LIGHT RATIOS

Recently Roberts (1969) has made the remarkable discovery that the mass-to-light ratios of spiral and irregular galaxies do *not* depend on their Hubble type. Roberts's data show no clear-cut correlation between the luminosity and the mass-to-light ratio of galaxies. These results suggest that it might be possible to derive the Hubble constant from a comparison of the mass-to-light ratios of nearby and distant galaxies.

The mass of a galaxy that is determined from its rotation curve is proportional to its assumed distance D. The absolute luminosity of a galaxy is proportional to D^2 and hence $(\mathfrak{M}/L) \propto D^{-1} \propto H$. Assuming the distances given in table 8, Roberts's data yield $\langle \log (\mathfrak{M}/L) \rangle = 0.63 \pm 0.11$ for the spiral and irregular galaxies in the Local Group. Substituting this value into

$$\langle \log (\mathfrak{M}/L) \rangle - \log (H/100) = 0.61 \pm 0.05 \,, \tag{20}$$

which is obtained from Roberts's data for 60 galaxies with D(Roberts) > 4.0 Mpc, yields 5 log $(H/100) = 0.1 \pm 0.6$ which corresponds to $H = 105$ $(+33, -26)$ km s^{-1} Mpc^{-1}.

16. BRIGHTEST NONVARIABLE STARS IN GALAXIES

It appears extremely hazardous to use the brightest nonvariable stars of Population I as distance indicators. Available evidence suggests that the magnitude of the brightest stars depends on both nebular type and total Population I content. In distant galaxies the identification of the brightest stars is rendered uncertain because bright star clusters cannot be distinguished from individual stars of high luminosity. Additional difficulties are introduced by galactic foreground stars that are scattered among the truly extragalactic stars of high luminosity. Since the bright end of the stellar luminosity function is thinly populated, the statistical separation of truly extragalactic stars of high luminosity from foreground field stars is uncertain.

Using statistical methods, Hubble (1936*b*) found that $m_{pg} = 16.0$ for the brightest stars in M31. After excluding star clusters, Seyfert and Nassau (1945) find $m_{pg} = 16.5$ for the brightest stars in the Andromeda Nebula. According to Hubble (1936*b*) the brightest stars in the Sb galaxy NGC 4192 in the Virgo cluster have $m_{pg} = 20.2$. This value probably refers to the brightest stars and not to the brightest H II regions because in M31, which is also of type Sb, the brightest stars are more luminous in B than are the brightest H II regions (Racine 1965). If Hubble's magnitude scales are correct, then the data given above suggest that the Virgo cluster has a distance modulus which is 3.7 magnitudes greater than that of the Andromeda Nebula. According to Baade and Swope

(1965) Hubble's magnitude scale in M31 is too bright by 2.3 magnitudes at $m_{pg} = 20.2$. *If* a similar correction applies to Hubble's scale in NGC 4192, the distance modulus of the Virgo cluster would be 6.0 magnitudes larger than that of M31, i.e., $(m - M)_{pg} = 30.6$.

According to Hubble (1936*b*) the brightest stars in M33 have $m_{pg} = 15.6$, which may be compared with $m_{pg} = 20.8$ for the brightest stars in the Sc galaxy M100 in the Virgo cluster (Sandage 1958). Assuming the brightest stars in the Sc giant galaxy M33 to be similar to those in the Sc supergiant M100, one finds that the distance modulus of the Virgo cluster is 5.2 magnitudes greater than that of M33, i.e., $(m - M)_{pg} = 29.9$. A larger distance modulus would, of course, be obtained for the Virgo cluster if the most luminous stars in M100 are brighter than the most luminous stars in M33.

Perhaps the best *guess* that can be made from the bright-star data is that $(m - M)_{pg} = 30.25 \pm 1.0$ for the Virgo cluster. Assuming the Virgo cluster to be unreddened (cf. § 12) and adopting a mean cluster velocity $\langle V \rangle = 1175 \pm 64$ km s^{-1} Mpc^{-1} then yields $5 \log (H/100) = +0.1 \pm 1.0$. The corresponding value of the Hubble constant is $H = 105$ ($+51$, -41). The uncertainty of this value could be significantly reduced by new photoelectrically calibrated measurements of the brightest stars in the Virgo cluster.

Tammann and Sandage (1968) have recently suggested that the brightest *red* stars of Population I might possibly prove to be useful as distance indicators. It should, however, be kept in mind that the absolute magnitudes of red supergiants are possibly sensitive to chemical composition (Hofmeister 1967*a*; van den Bergh 1968*b*; Schlesinger 1969).

17. GLOBULAR CLUSTERS

Because of their high luminosities, globular clusters can be seen out to very large distances. This makes globulars potentially valuable as calibrators of the extra-galactic distance scale (Sandage 1968*a*). Unfortunately the usefulness of globular clusters as distance indicators is impaired by their very large luminosity dispersion (the brightest and the faintest known globulars in the Galaxy differ in luminosity by a factor of about 100). Furthermore, available data, which are still very incomplete, may indicate (cf. § 7) that the mean luminosity of globulars differs somewhat from galaxy to galaxy.

Extra-galactic distances can also be estimated by assuming that the absolute magnitude of the *brightest* globular cluster is the same in different galaxies. It should be emphasized that this assumption is as yet unproven.

The brightest known globular cluster in the Galaxy is ω Centauri (Gascoigne and Burr 1956) for which $M_B = -9.7$ if $\langle M_B \rangle_{RR} = +0.8$. The brightest cluster in the Andromeda Nebula is MII, which is located in the halo of M31. According to van den Bergh (1969), $B = 14.56$ for this cluster. Assuming $(m - M)_B = 24.61$ for M31, this yields $M_B = -10.05$ for the brightest globular cluster in the Andromeda Nebula. (The clusters 39, 42, 93, and 327 in M31 have luminosities that are comparable to MII; the exact absolute magnitudes of these reddened clusters depend on their assumed intrinsic colors.) The data for M31 and the Galaxy yield a mean absolute magnitude $M_B = -9.88$, with an uncertainty of, say, 0.3 mag for the brightest globular clusters in supergiant spirals. This value may be compared with $B = 21.2 \pm 0.2$ (Racine 1968*b*) for the brightest globular clusters in the supergiant E galaxy M87 in the Virgo cluster. If it is *assumed* that the brightest globulars in M87, M31, and the Galaxy have the same luminosity, then $(m - M)_{pg} = (m - M)_0 = 31.08 \pm 0.36$ is obtained for the Virgo cluster. Combining this value with the mean observed cluster velocity $\langle V \rangle = 1175 \pm 64$ km s^{-1} yields $5 \log (H/100) = -0.7 \pm 0.4$, which corresponds to $H = 72$ (+15, −12) km s^{-1} Mpc^{-1}. This value should perhaps be regarded as an upper limit to the Hubble constant derived from globular clusters. This is so because the total population of globular clusters in M87 is about an order of magnitude greater than it is in M31 and the Galaxy. *Other things being equal,* it might therefore be expected that the brightest clusters in M87 would be more luminous than those in M31 and the Galaxy.

18. SUPERNOVAE

Potentially supernovae are the most valuable source of information on the extragalactic distance scale. This is so because very distant supernovae can, in principle, be compared directly with supernovae in our own Galaxy. This method therefore avoids the necessity of having to use the galaxies of the Local Group as intermediate distance standards.

TABLE 12

COLOR OBSERVATIONS OF TYCHO'S SUPERNOVA

Date	Color*	Adopted $B-V$	$(B-V)_0$†	A_V
1572, Nov.....	Like Venus ($B-V=0.83$) and Jupiter ($B-V=0.82$)	0.82	0.3±0.1	1.64
1573, Jan.....	Similar to Mars ($B-V=1.36$) and Aldebaran ($B-V=1.52$)	1.44	1.0±0.1	1.48
1573, Apr.....	Similar to Saturn ($B-V=1.04$)	1.04	0.5±0.1	1.75

* Planetary colors from Harris (1961). † From Mihalas (1963).

Unfortunately, only fragmentary data are available on most of the supernovae that are known to have occurred in the Galaxy during the last 1000 years. Neither the maximum apparent magnitude V(max) nor the distance of the supernova of 1006 in Lupus are accurately known. The Crab supernova of 1054, for which both the distance and the apparent magnitude are reasonably well known, seems to have been an unusual object (Minkowski 1966) for which no extragalactic counterpart is yet known. In the case of Kepler's supernova of 1604, V(max) and the reddening are known but the distance is not. Finally the supernova of 1667, which produced the radio source Cassiopeia A, was not seen by seventeenth century observers.

Tycho's supernova of 1572 appears to be the only object for which sufficient data are available to permit a rough estimate of the extragalactic distance scale. From a definitive study of the original records Baade (1945) finds that Tycho's supernova reached $V = -4.0 \pm 0.3$ at maximum light. The reddening and absorption suffered by this supernova may be estimated from a comparison of the observed colors (see table 12) with the colors of Type I supernovae plotted in figure 5 of Mihalas (1963). The resulting values of A_V, which take into account the dependence (Fernie 1963) of A_V/E_{B-V} on $(B - V)_0$, are given in the last column of table 12. The results are surprisingly consistent, yielding $\langle A_V \rangle = 1.62$ with a formal mean error of about 0.1 mag. Adopting this value for the absorption in the direction of Tycho's supernova yields $V_0(\max) = -5.62 \pm 0.32$. From a comparison of 21-cm observations with rotation models of the Galaxy (Minkowski 1965; Menon and Williams 1966) the distance of the supernova is found to be greater than 3.0 kpc and probably less than 3.5 kpc, i.e., $(m - M)_0 = 12.56 \pm 0.16$. Combining this value with $V_0(\max) = -5.62 \pm 0.32$ yields $M_V(\max) = -18.18 \pm 0.36$. It should be emphasized that this value is based on a very insecure distance estimate for Tycho's supernova!

From a study of extragalactic supernovae of Type I Kowal (1968, 1969) obtains

$$\langle M_{pg}(\max) \rangle = -18.65 + 5 \log (H/100) . \quad (21)$$

Individual supernovae of Type I are found to have a dispersion of 0.61 mag about the adopted mean. Assuming $M_V(\max) \simeq M_{pg}(\max)$ then yields $5 \log (H/100) = +0.45 \pm 0.7$ which corresponds to $H = 123$ (+47, −34) km s^{-1} Mpc^{-1}. Needless to say, this estimate, which is based on ancient observations of a single supernova, is highly uncertain.

19. GALAXY DIAMETERS

It may be shown that the diameters and the luminosities of galaxies are rather closely correlated. Heidmann (1969) has shown that this correlation can be used to estimate the value of the Hubble constant. Using the observations of galaxies in the Virgo cluster by Holmberg (1958), van den Bergh (1961) finds that

$$\log \phi = 2.477(\pm 0.008) - 0.155(\pm 0.006) m_{pv} , \quad (22)$$

in which ϕ is the galaxy diameter in minutes of arc. Nearby galaxies of the same linear diameter, that are located at different distances, obey the relation

$$\log \phi = \text{const.} - 0.200\, m_{pv} . \quad (23)$$

Unfortunately the slopes of the $\log \phi$ versus m_{pv} relations given by equations (22) and (23) are not very different. As a result the cosmic dispersion of 0.4 mag about the relation given by equation (22) translates into a mean error of 1.8 mag in the distance modulus. The adopted diameters of M31, M33, the LMC, and the SMC are listed in table 13. NGC 6822 has not been included in the table because extrapolation of its diameter to zero reddening is both large and uncertain. The dwarf galaxy IC 1613 was excluded because the validity of equation (22) has not been established for such very faint

TABLE 13

ADOPTED GALAXY DIAMETERS AND RESULTING VIRGO CLUSTER MODULI

Galaxy	ϕ	$(m-M)_0$
M31	210′*	32.18
M33	85*	32.72
LMC	850†	29.06
SMC	300†	28.52

* Holmberg's (1958) diameters extrapolated to zero reddening.
† Extrapolated to $B = 26.5$ mag (arc sec)$^{-2}$ from Elsässer (1959).

galaxies. From the data in table 13 and from equations (22) and (23) a value $\langle m - M\rangle_0 = 30.62 \pm 0.90$ is obtained for the Virgo cluster. With $\langle V\rangle = 1175 \pm 64$ km s^{-1} for the Virgo cluster this yields $5 \log (H/100) = -0.27 \pm 0.9$, corresponding to $H = 88$ $(+46, -30)$ km s^{-1} Mpc^{-1}.

20. SUMMARY OF DATA ON THE HUBBLE CONSTANT

The results on the value of the Hubble constant that have been obtained in §§ 11–19 are summarized in table 14. Values of the Hubble constant derived by means of entirely different methods are seen to be in surprisingly good agreement. An unweighted mean of the nine values listed in table 14 yields $\langle 5 \log (H/100)\rangle = -0.02$ compared with $\langle 5 \log (H/100)\rangle = -0.14 \pm 0.19$ for a weighted mean, in which the weights assigned to each method are proportional to the inverse squares of the mean errors listed in the

TABLE 14

SUMMARY OF DETERMINATIONS OF THE HUBBLE CONSTANT

No.	Method	5 log (H/100)
1	Third brightest member of small clusters	+0.5 ±0.7
2	Surface brightness of galaxies	−0.15±0.9
3	Luminosity classifications of galaxies	−0.05±0.4
4	Diameters of H II regions	−0.2 ±0.4
5	Mass-to-light ratios	+0.1 ±0.6
6	Brightest non-variable stars in galaxies	+0.1 ±1.0
7	Brightest globular clusters in galaxies	−0.7 ±0.4
8	Extragalactic supernovae and Tycho's supernova	+0.45±0.7
9	Galaxy diameters	−0.25±0.9
	Adopted mean*	−0.1 ±0.3

* Corresponds to $H = 95$ $(+15, -12)$ km s^{-1} Mpc^{-1}.

table. The dispersion of individual determinations of 5 log (H/100) about the adopted mean is smaller than would be expected from the formal mean errors assigned to each method; i.e., the results obtained by different methods agree fortuitously too well.

A value $\langle 5 \log (H/100)\rangle = -0.1 \pm 0.3$, which corresponds to $H = 95$ $(+15, -12)$ will be adopted. The quoted mean error takes into account the fact that the nine methods listed in table 14 are not entirely independent. This is so because most of the methods are sensitive to changes in the zero point of the Cepheid period-luminosity relation, the absolute magnitude of the RR Lyrae stars, and the distance to the Virgo cluster. For example, making all Cepheids fainter by 0.20 mag would change H from 95 to 100 km^{-1} Mpc^{-1}. Making the RR Lyrae variables brighter by 0.20 mag would only change H from 95 to 94 km s^{-1} Mpc^{-1}. Finally, a change of 10 percent in the adopted distance to the Virgo cluster would change the Hubble constant by approximately 5 percent.

21. REGIONAL VARIATIONS OF THE HUBBLE CONSTANT

In previous sections it has been assumed, for the sake of simplicity, that the expansion of the Universe is both uniform and isotropic. In fact individual galaxies, clusters of galaxies, and perhaps even larger spatial groupings, might be expected to exhibit random motions that are superposed on the general expansion of the Universe. Available optical evidence (Humason, Mayall, and Sandage 1956) suggests that such random motions are of the order of 200–300 km s^{-1}. The velocity of the Galaxy relative to the absolute reference frame provided by the 3° K microwave radiation (Partridge 1969) is also of this order.

De Vaucouleurs (1964) has used radial velocities and luminosity classifications of galaxies to investigate possible regional variations in the expansion of the Universe. After correcting for galactic absorption and luminosity effects he finds that galaxies with reduced magnitude $m_{pg} = 11.5$ have $\langle \log V \rangle = 3.054 \pm 0.015$ in the direction of the center of the "Local Supercluster" (Virgo cluster), whereas $\langle \log V \rangle = 3.108 \pm 0.022$ for galaxies with reduced magnitude $m_{pg} = 11.5$ in the anticenter direction. From a statistical point of view the difference between these two values is only marginally significant. A comparison of the type made by de Vaucouleurs involves a number of assumptions. One of these is that the velocity dispersion of field galaxies is everywhere the same. With a finite velocity dispersion, $\log \langle V \rangle$ will be larger than $\langle \log V \rangle$, for galaxies located at the same distance. Because the space density of nearby galaxies is greater in the direction of the Virgo cluster than it is in the opposite direction (van den Bergh 1960*d*), the velocity dispersion might be highest in the direction of the "Local Supercluster." If this is indeed the case, the value of H derived by de Vaucouleurs in the direction of the Virgo cluster will be too low compared with the value of H obtained in the supergalactic anticenter direction. A more recent investigation by de Vaucouleurs and Peters (1968) suffers from the fact that their method of analysis cannot distinguish between velocity anisotropy and large-scale density inhomogeneities.

In § 20 it was shown that determinations of the Hubble constant are quite sensitive to the assumption that the Virgo cluster does not have a large random radial velocity. (A random radial velocity component of 100 km s^{-1} for the Virgo cluster would change the mean value of the Hubble constant derived in § 20 by about 5 percent.) From a comparison of the luminosity classifications of galaxies in the Virgo cluster with luminosity classifications of field galaxies van den Bergh (1960*a*) finds $(m - M)_0 = 30.46 \pm 0.08 - 5 \log (H/100)$ for the Virgo cluster. This value yields a predicted radial velocity of 1236 ± 46 km s^{-1} for the Virgo cluster, in excellent agreement with the observed value $\langle V \rangle = 1175 \pm 64$ km s^{-1}. It follows that the radial velocity of the Virgo cluster does not differ systematically from the mean Hubble flow which is defined by the northern Shapley-Ames galaxies that are used to calibrate the David Dunlap Observatory luminosity classification system.

In his Halley lecture Sandage (1968*b*) shows that the absolute magnitudes of the brightest members of rich clusters exhibit a dispersion of only 0.3 mag about the relation

$$\langle M_V \rangle = -21.78 + 5 \log [H(\infty)/100] , \tag{24}$$

in which $H(\infty)$ is the Hubble constant beyond the Local Supercluster. This value may be compared with the absolute magnitude of the brightest member of the Virgo cluster to obtain the ratio $H(0)/H(\infty)$, in which $H(0)$ is the local value of the Hubble constant. According to Holmberg (1958), NGC 4472, which is the brightest galaxy in the Virgo cluster, has $m_{pv} = 8.49$. Substituting this value, $V = 1175 \pm 64$ km s^{-1}, and M_V from equation (24) into

$$5 \log [H(0)/100] = 15 - (m_{pv} - M_V) + 5 \log V \tag{25}$$

yields $5 \log [H(0)/H(\infty)] = 0.08 \pm 0.32$ and hence $H(0)/H(\infty) = 1.04 \pm 0.15$. This result indicates that, within the accuracy of currently available data, there is no systematic difference between the Hubble flow inside and outside the Local Supercluster. This result is confirmed, albeit with lower accuracy, by intercomparison of the supernovae in the Virgo cluster (Kowal 1968, 1969) and those in the Coma cluster. A discrepant result is, however, obtained by Abell and Eastmond (1968) who compare the galaxian luminosity function of the Virgo cluster with those of the Coma and Corona Borealis clusters.

22. FUTURE OBSERVATIONS

Currently available telescopes and observational techniques permit a significant improvement in the accuracy of the data from which the Hubble constant is derived. Such improvements can be achieved by (*a*) more accurate determinations of distances within the Local Group, (*b*) extending techniques used in the Local Group to such nearby clusters as the M81 group and the South Polar Group, (*c*) new observations of distant galaxies and clusters, and (*d*) a detailed study of possible anisotropy effects in relatively nearby regions of space.

22.1. Desirable Observations within the Local Group

The distance estimate for M33 which is given in tables 7 and 8 is based on the statement by Hubble (1926) that the Cepheids in the Triangulum Galaxy are 0.1 mag brighter than those in the Andromeda Nebula. After almost half a century new measurements are clearly desirable! Observations of W Virginis stars and Population II giants are also feasible with currently available techniques. Photoelectric observations of the Cepheids in IC 1613 would provide a valuable check on the assumed validity of a universal period-luminosity-color relation. Direct observation of the RR Lyrae variables in M31 now appears possible by applying superposition techniques (Racine 1967) to baked IIIaJ plates obtained at the prime focus of the 200-inch telescope. For example, RR Lyrae stars of period 0.500 days could be detected by blinking composites of all plates obtained at the same time on different nights of one observing run. Use of radio observations of supernova remnants in the Magellanic Clouds (Shklovskii 1960; Poveda and Woltjer 1968; van den Bergh 1968*b*) still appears premature.

22.2. Observations of Nearby Clusters

Distance determinations made by using (*a*) the third brightest members of small clusters, (*b*) the surface brightness of galaxies, (*c*) the luminosity classification of galaxies, (*d*) the diameters of H II regions, and (*e*) the brightest nonvariable stars are primarily limited by the small number of calibration standards available within the Local Group. Such distance determinations would become significantly more accurate if observations of Cepheids and novae, which can be calibrated in the Galaxy, were used to establish distances to such nearby clusters as the M81 group and the South Polar Group. Such observations would help to determine if the brightest red supergiants of Population I are suitable distance indicators. Available data (Hubble and Sandage 1953; Tammann and Sandage 1968) are not yet sufficient to establish if the brightest blue variables of Population I occupy a relatively narrow magnitude range or if they populate a continuum between the brightest blue nonvariable stars and supernovae of Type II.

22.3. Observations of Distant Galaxies

a. Very few magnitudes and redshifts are available for distant dwarf galaxies (van den Bergh 1966). Comparison of such observations with data on dwarf galaxies in the Local Group, the M81 group, and the South Polar Group would greatly strengthen the extragalactic distance scale determinations based on luminosity classifications.

b. The accuracy of distance determinations made by using diameters of H II regions could be improved significantly by obtaining new direct photographs in red light with the 200-inch telescope.

c. Photoelectrically calibrated photographic observations should be obtained of the brightest spiral galaxies in the Virgo cluster. Such observations could provide greatly improved information on the apparent magnitudes of the brightest stars in galaxies of the Virgo cluster.

d. With baked IIIaJ emulsion the 200-inch telescope should be able to photograph large numbers of ordinary novae in the galaxies of the Virgo cluster.

e. New observations and statistical studies should be undertaken to investigate the possible dependence of the luminosity of the brightest cluster galaxy on the total number of cluster galaxies and on the relative frequency of different Hubble types among the cluster members.

f. Additional data should be obtained to see if the luminosity functions of rich clusters of galaxies (Abell and Eastmond 1968) are sufficiently stable to permit estimates of relative cluster distances.

g. The dependence of the integrated colors of elliptical galaxies on their absolute magnitude (Baum 1959; McClure and van den Bergh 1968) might be used to study possible anisotropy effects in the local expansion of the Universe.

The compilation of this review would not have been possible without the assistance of the following colleagues, who graciously made unpublished observational material available to me: G. O. Abell, W. A. Baum, R. J. Dickens, J. D. Fernie, S. C. B. Gascoigne, G. van Herk, P. W. Hodge, C. T. Kowal, R. D. McClure, R. Racine, M. S. Roberts, A. R. Sandage, T. Schmidt-Kaler, J. L. Sérsic, G. de Vaucouleurs, and A. J. Wesselink.

REFERENCES

Abell, G. O., and Eastmond, S. 1968, *A.J.*, **73**, S161.
Agt, S.L.T.J. van. 1967, *B.A.N.*, **19**, 275.
Arp, H. C. 1955, *A.J.*, **60**, 317.
———. 1956, *ibid.*, **61**, 15.
———. 1960, *ibid.*, **65**, 404.
Arp, H. C., and Kraft, R. P. 1961, *Ap. J.*, **133**, 420.
Baade, W. 1945, *Ap. J.*, **102**, 309.
Baade, W., and Swope, H. H. 1963, *A.J.*, **68**, 435.
———. 1965, *ibid.*, **70**, 212.
Bahner, K., Hiltner, W. A., and Kraft, R. P. 1962, *Ap. J. Suppl.*, **6**, 319.
Baum, W. A. 1959, *Pub. A.S.P.*, **71**, 106.
Behr, A. 1959, *Veröff. Univ.-Sternw. Göttingen*, no. 126.
Bell, R. A., and Rodgers, A. W. 1969, *M.N.R.A.S.*, **142**, 161.
Bergh, S. van den. 1958, *A.J.*, **63**, 492.
———. 1960*a*, *Ap. J.*, **131**, 558.
———. 1960*b*, *Pub. David Dunlap Obs.*, **2**, 159.
———. 1960*c*, *M.N.R.A.S.*, **121**, 387.
———. 1960*d*, *Pub. A.S.P.*, **72**, 312.

Bergh, S. van den. 1961, *Observatory*, **81**, 30.
———. 1966, *A.J.*, **71**, 922.
———. 1968*a*, *Observatory*, **88**, 168.
———. 1968*b*, *The Galaxies of the Local Group* (*Comm. David Dunlap Obs.*, No. 195).
———. 1969, *Ap. J. Suppl.*, **19**, 145.
Bergh, S. van den, and Hagen, G. L. 1968, *A.J.*, **73**, 569.
Bergh, S. van den, and Racine, R. 1967, *A.J.*, **72**, 69.
Buscombe, W., and Kennedy, P. M. 1962, *J.R.A.S. Canada*, **56**, 113.
Christy, R. F. 1966, *Ann. Rev. Astr. and Ap.*, **4**, 353.
Crawford, D. L., and Barnes, J. V. 1969, *A.J.*, **74**, 407.
Demers, S. 1966, Ph.D. thesis, University of Toronto.
Dickens, R. J., and Carey, J. V. 1967, *Roy. Obs. Bull.*, No. 129.
Elsässer, H. 1959, *Zs. f. Ap.*, **47**, 1.
Elvius, A., and Hall, J. S. 1964, *Lowell Obs. Bull.*, **6**, 123.
Feast, M. W., Thackeray, A. D., and Wesselink, A. J. 1960, *M.N.R.A.S.*, **121**, 337.
Fernie, J. D. 1963, *A.J.*, **68**, 780.
———. 1964, *ibid.*, **69**, 258.
———. 1966, *ibid.*, **71**, 732.
———. 1967*a*, *ibid.*, **72**, 422.
———. 1967*b*, *ibid.*, **72**, 1327.
Fernie, J. D., and Marlborough, J. M. 1965, *Pub. A.S.P.*, **77**, 218.
Gaposchkin, C. P. 1961, *Vistas in Astronomy*, **4**, 184.
Gaposchkin, S. 1962, *A.J.*, **67**, 334.
Gaposchkin, C. P., and Gaposchkin, S. 1966, *Smithsonian Contr.*, **9**, 1.
Gascoigne, S. C. B. 1966, *M.N.R.A.S.*, **134**, 59.
———. 1969, *ibid.*, **146**, 1.
Gascoigne, S. C. B., and Burr, E. J. 1956, *M.N.R.A.S.*, **116**, 570.
Gascoigne, S. C. B., and Kron, G. E. 1965, *M.N.R.A.S.*, **130**, 333.
Harris, D. E. 1961, in *Planets and Satellites*, ed. G. P. Kuiper and B. M. Middlehurst (Chicago: University of Chicago Press), p. 272.
Heidmann, J. 1969, *C. R. Acad. Sci.*, **268B**, 1782.
Henize, K. G., Hoffleit, D., and McKibben Nail, V. 1954, *Proc. Nat. Acad. Sci.*, **40**, 365 (= Harvard Reprint no. 387).
Herk, G. van. 1965, *B.A.N.*, **18**, 71.
Hiltner, W. A. 1960, *Ap. J.*, **131**, 163.
Hodge, P. W. 1960, *Ap. J.*, **132**, 346.
Hofmeister, E. 1967*a*, *Zs. f. Ap.*, **65**, 164.
———. 1967*b*, *ibid.*, **65**, 194.
Hogg, H. B. S. 1963, *Pub. David Dunlap Obs.*, **2**, 337.
Holmberg, E. 1958, *Lund Obs. Medd.*, Ser. 2, No. 136.
Hubble, E. P. 1926, *Ap. J.*, **63**, 236.
———. 1936*a*, *The Realm of the Nebulae* (New Haven: Yale University Press).
———. 1936*b*, *Ap. J.*, **84**, 158.
Hubble, E. P., and Humason, M. L. 1931, *Ap. J.*, **74**, 43.
Hubble, E. P., and Sandage, A. R. 1953, *Ap. J.*, **118**, 353.
Humason, M. L., Mayall, N. U., and Sandage, A. R. 1956, *A.J.*, **61**, 97.
Hutchings, J. B. 1966, *M.N.R.A.S.*, **132**, 433.
Johnson, H. L., and Iriarte, B. 1958, *Lowell Obs. Bull.*, **4**, 47.
Kayser, S. E. 1967, *A.J.*, **72**, 134.
Kowal, C. T. 1968, *A.J.*, **73**, 1021.
———. 1969, *Pub. A.S.P.*, **81**, 608.
Kraft, R. P. 1961, *Ap. J.*, **133**, 39.
Kron, G. E., and Mayall, N. U. 1960, *A.J.*, **65**, 581.
Kwee, K. K. 1968, *B.A.N.*, **19**, 374.
Leavitt, H. S. 1908, *Harvard Ann.*, **60**, 87.
———. 1912, *Harvard Obs. Circ.*, No. 173.
McClure, R. D., and Bergh, S. van den. 1968, *A.J.*, **73**, 1008.
McClure, R. D., and Racine, R. 1969, *A.J.*, **74**, 1000.
Menon, T. K., and Williams, D. R. W. 1966, *A.J.*, **71**, 392.
Mihalas, D. 1963, *Pub. A.S.P.*, **75**, 256.
Minkowski, R. 1965, private communication.
———. 1966, *A.J.*, **71**, 371.
Partridge, R. B. 1969, *Am. Scientist*, **57**, No. 1, 37.
Poveda, A., and Woltjer, L. 1968, *A.J.*, **73**, 65.
Racine, R. 1965, unpublished M.A. thesis, University of Toronto.
———. 1967, *A.J.*, **72**, 65.
———. 1968*a*, *ibid.*, **73**, 588.

———. 1968*b*, *J.R.A.S. Canada*, **62**, 367.
Roberts, M. S. 1969, *A.J.*, **74**, 859.
Rosino, L. 1964, *Ann. d'Ap.*, **27**, 498.
Sandage, A. R. 1958, *Ap. J.*, **127**, 513.
———. 1962, *Problems of Extra-Galactic Research*, ed. G. C. McVittie (New York: Macmillan), p. 359.
———. 1964, *Observatory*, **84**, 245.
———. 1968*a*, *Ap. J.* (*Letters*), **152**, L149.
———. 1968*b*, *Observatory*, **88**, 91.
Sandage, A. R., and Tammann, G. A. 1968, *Ap. J.*, **151**, 531.
———. 1969, *ibid.*, **157**, 683.
Sandage, A. R., and Wallerstein, G. 1960, *Ap. J.*, **131**, 598.
Sanduleak, N., and Davis Philip, A. G. 1968, *A.J.*, **73**, 566.
Schlesinger, B. M. 1969, *Ap. J.*, **157**, 533.
Schmidt-Kaler, T. 1957, *Zs. f. Ap.*, **41**, 182.
———. 1965*a*, *Astronomie und Astrophysik*, ed. H. H. Voigt (Berlin: Springer), p. 284.
———. 1965*b*, private communication.
———. 1967, *A.J.*, **72**, 526.
Sérsic, J. L. 1959, *Observatory*, **79**, 54.
———. 1960, *Zs. f. Ap.*, **50**, 168.
Seyfert, C. K., and Nassau, J. J. 1945, *Ap. J.*, **102**, 377.
Shapley, H., and McKibben, V. 1942*a*, *Proc. Nat. Acad. Sci.*, **28**, 200 (= Harvard Reprint no. 241).
———. 1942*b*, *Harvard Coll. Obs. Bull.*, no. 916.
Shklovskii, I. S. 1960, *Soviet Astr.—AJ*, **4**, 355.
Tammann, G. A., and Sandage, A. R. 1968, *Ap. J.*, **151**, 825.
Thackeray, A. D. 1963, *Adv. Astr. and Ap.*, **2**, 263.
Tifft, W. G. 1963, *M.N.R.A.S.*, **125**, 199.
Vaucouleurs, G. de. 1960, *Ap. J.*, **131**, 574.
———. 1961, *Ap. J. Suppl.*, **6**, 213.
———. 1964, *A.J.*, **69**, 737.
Vaucouleurs, G. de, and Peters, W. L. 1968, *Nature*, **220**, 868.
Vaucouleurs, G. de, and Vaucouleurs, A. de. 1964, *Reference Catalogue of Bright Galaxies*, (Austin: University of Texas Press).
Westerlund, B. E. 1963, *M.N.R.A.S.*, **127**, 71.
Westerlund, B. E., Danziger, I. J., and Graham, J. 1963, *Observatory*, **83**, 74.
Whitford, A. E. 1958, *A.J.*, **63**, 201.

CHAPTER 13

Binary Galaxies

THORNTON PAGE
Naval Research Laboratory

1. INTRODUCTION

THE fact that galaxies often appear in pairs, groups, and clusters, as discussed in the next three chapters, is undoubtedly related to their origin and evolution (chapters 10 and 11). It is of practical importance in determining masses, diameters, and the luminosity function. The number of galaxies so associated ranges up to 500 or 1000 in large clusters, but the binary pair is simplest for several reasons: dynamical, geometrical, and observational. As in the case of binary stars, pairs of galaxies were first recognized observationally by angular separations that are smaller than the average. From positions given in the NGC and by Shapley and Ames (1934), and from examination of photographs, K. Lundmark collected several hundred cases of close pairs, some of them associated with a third galaxy, or with larger groups.

2. DEFINITION OF OPTICAL AND PHYSICAL PAIRS

In 1937, E. Holmberg published a catalog of 827 double and multiple galaxies found on a wide variety of photographs covering over 52 percent of the sky to about sixteenth magnitude. Unfortunately, many of the plates were of poor quality; a recent check by Zonn (1962) shows that many of the smaller galaxies listed by Holmberg appear to be star images on the Palomar *Sky Atlas* prints. Of course, any two galaxies can be said to form a "pair"; Holmberg selected those for which the projected angular separation $S \leq 2(A_1 + A_2)$, where A_1 and A_2 are the maximum angular dimensions of the two galaxies, and the factor 2 was chosen somewhat arbitrarily in order to limit the number of chance lineups in which one galaxy is far behind the other. With this selection criterion, Holmberg found 695 pairs (and 132 multiple groups) among some 55,000 single galaxies brighter than about sixteenth magnitude.

The number of chance lineups can be estimated from a formula due to G. Polya (1919) for the probability that, if n points are distributed at random on a sphere, none of them will fall within the angle S from an $(n + 1)$th point:

$$p(n, S) = \cos^{2n}(S/2) \simeq e^{-nS^2/4} . \quad (1)$$

Received August 1972.

The expected number of chance separations between S and $S + dS$ is $(n^2S/2)e^{-nS^2/4}dS$, in which each pair is counted twice. Using $N_1 = n\pi/4(180)^2$ to represent the number of galaxies per square degree, one finds the number of chance pairs per square degree with separations less than or equal to S (in degrees) is approximately

$$N_2(N_1, S) = \pi N_1^2 S^2/2 , \qquad (2)$$

from which the number of chance pairs with $S \leq 6' = 0^\circ.1$ expected among 20,000 single galaxies in 15,000 square degrees outside the zone of avoidance surveyed by Holmberg is 42, or 6 percent of the 695 pairs in his catalog. Presumably the remaining 653 (94 percent) are physical pairs, with smaller than average space separations. (Unlike binary stars, orbital motions cannot be used to confirm the physical pairs of galaxies.)

Assuming that the line of centers between galaxies in a pair is randomly oriented relative to the line of sight, Holmberg (1954) determined the distribution of space separations, R, to be

$$\begin{aligned} f(R) &= k[1 - (R/R_m)^3] \text{ for } R \leq R_m \\ &= 0 \text{ for } R \geq R_m , \end{aligned} \qquad (3)$$

where $R_m = (2.3/h) \times 10^5$ pc and $h =$ the Hubble constant in km s^{-1} per 10^4 pc (approximately 1.0). The distribution of double-star separations is also well fitted by equation (3) with $R_m = 2000$ a.u. This similarity, despite the vast difference in scale, probably indicates a common factor that affects the formation or breakup of a pair of masses in stable orbits about the center of mass. The most obvious factor is perturbations caused by field galaxies that have high probability of disrupting a pair whose separation is greater than R_m.

Since the value of R_m in equation (3) corresponds to maximum angular separations about 7.5 times as large as the angular diameter of a galaxy, counts of pairs with wide separations were made on Palomar *Atlas* prints by Page, Dahn, and Morrison (1961). It was found that, for $S \geq 3(A_1 + A_2)$, there were long chains of pairs, each galaxy having two or more others within the selected maximum separation. For this practical reason, in the counts made by Dahn and Morrison and by Zonn and Cook, binary galaxies were defined as pairs with separations

$$S \leq 3(A_1 + A_2) . \qquad (4)$$

Such counts exclude a few physical pairs of large space separation near R_m, but the form of equation (3) limits these few wide pairs to about 2 percent of the physical pairs involved. A larger proportion are missed at very small S because the two images merge when one galaxy is behind the other.

The "chaining" described above shows how pairs of galaxies cannot be considered separately from small groups in projection on the celestial sphere. It is often the case that two galaxies in a group of three or more will have much smaller separation than the others. This led to the definition of an "unambiguous pair": two images with separation S and no other image closer than S to either one. In a study of three areas (one in the Coma cluster, one in the Virgo cluster, and one between clusters) totaling 270 square degrees, Page, Dahn, and Morrison (1961) found 322 pairs satisfying equation (4), 254 being unambiguous pairs, among 8500 galaxies brighter than 18.5 mag. Application of equation (2) showed that 80 percent of these are chance lineups or "optical pairs," but that the density of the remaining 20 percent ("physical pairs") also depends on N_1^2. These counts included pairs with a magnitude difference as large as $m_2 - m_1 = 9$ mag, and with a ratio of angular diameters as large as $A_1/A_2 = 60$. It might seem reasonable to select as physical pairs those with small magnitude differences and nearly equal dimensions or of similar morphological type. Such selection has probably been unconsciously applied in many published lists of binary galaxies, but it is not justified by independent evidence and is negated by the most obvious example of a well-studied physical triple in Andromeda:

	NGC 205	NGC 221	NGC 224	
m.......	9.7	9.4	4.6	($m_1 - m_3 = 5.1$)
A.......	10ʹ.0	3ʹ.6	159ʹ	($A_3/A_2 = 44.$)
Type....	SB0	E2	Sb	

In summary, binary galaxies may best be defined by angular separation, S, and equation (4). Statistical estimates can be made for the number of optical pairs based on the projected density of galaxies in any one region of several square degrees. Physical pairs probably have space separations less than about 230 kpc. The proportion of optical pairs is about 5 percent among binary galaxies brighter than sixteenth magnitude, but increases to 80 percent at limiting magnitude 18.5.

3. OBSERVATIONAL DATA

With allowance for the zone of avoidance and the missing southern polar cap, there should be about 20,000 pairs of galaxies brighter than 18.5 mag on the Palomar *Atlas* prints, including about 4000 physical pairs. For each of these it is possible to measure or estimate the following quantities that are of value in statistical studies of binary galaxies (subscripts 1 and 2 refer to the brighter and the fainter galaxy in each pair):

α_1, δ_1, α_2, δ_2: equatorial celestial coordinates;
A_1, A_2: major angular diameters or largest dimensions;
B_1, B_2: minor angular diameters;
θ_1, θ_2: position angles of major diameters;
θ_{12}: position angle of the line of centers;
S_{12}: angular separation;
T_1, T_2: morphological types;
m_1, m_2: magnitudes (usually photographic, or B-magnitudes);
C_1, C_2: colors.

From other observations one gets:

Spectral types;
V_1, V_2: radial velocities;
F_1, F_2: radio fluxes at various frequencies;
X_1, X_2: X-ray fluxes at various wavelengths.

Of these, the first eight have been catalogued by Holmberg (1937), who included most of the pairs among the single galaxies listed in the Shapley-Ames catalog (1934). Other catalogs of single galaxies also contain pairs: 94 in the radial-velocity list of Humason, Mayall, and Sandage (1956), an estimated 260 in the de Vaucouleurs' catalog (1964), and 48 in the reclassification of 935 northern galaxies by van den Bergh (1960). Vorontsov-Velyaminov's *Atlas of Interacting Galaxies* (1959) lists 355 peculiar and connected pairs found in the Palomar *Atlas;* Zonn and Cook (1963) have measured a sample of 427 pairs from the center square degree of each Palomar *Atlas* print; Arp's *Atlas of Peculiar Galaxies* (1966) lists 64 pairs (Nos. 269 to 332); and Wirtanen (1963) has listed 45 peculiar close pairs noticed during his counts of galaxies on Lick plates.

These data can be applied to the following questions, each considered in more detail below: (*a*) Do binary galaxies differ systematically from single galaxies in form, luminosity, color, or dimensions? (*b*) What is the correlation between morphological types, between luminosities, and between dimensions of the two galaxies in a physical pair? (*c*) Is there any preference for parallel axes of the galaxies in a pair? (*d*) What are the masses of binary galaxies, and are the masses correlated with luminosities and morphological types?

4. TYPES AND MAGNITUDES OF GALAXIES IN PAIRS

The list of 427 pairs brighter than about 17 mag found by Zonn and Cook (1963) on Palomar *Atlas* prints was analyzed by Page (1965) without correction for optical pairs. There is some evidence that the morphological types, all estimated by Zonn following Hubble's schema, tend to err toward elliptical (E) and S0 classification when the image size is less than 1′. Table 1 shows the numbers of cases, $N_2(T_1, T_2)$, where the type, T, can be E (E0–E7), or S0 (including SB0), or Sa (early spirals), or Sb, Sc (late-type spirals), or SB (barred spirals of early or late type), or Ir (irregular type). The entries labeled (S) are the sums of Sa and Sb, Sc.

In table 2 the proportions of these types among the 854 galaxies in the sample are compared with those among individual galaxies and pair members in van den Bergh's (1960) list. It is clear that the galaxies in Zonn's sample of pairs show an excess of E–S0 and Sa types, possibly due to errors in typing very small images. However, the last two columns of table 2 show that the larger images of pair members on van den Bergh's list do *not* differ significantly from the other nonpair galaxies typed by him in a uniform manner.

If pairs were drawn at random from the 854 galaxies in Zonn's sample, one would expect the number of pairs of type T_1, T_2 to be proportional to the product of the numbers of single galaxies, $N(T_1)$ and $N(T_2)$. Actually, there is some tendency for like types in a pair, indicated by the "pairing factor," $Q(T_1, T_2)$ listed in table 3 and calculated from

$$N_2(T_1, T_2)/N_2 = Q(T_1, T_2)N(T_1)N(T_2)/N^2 , \qquad (5)$$

where N_2 is the total number of pairs (427) and N is the total number of galaxies (854). The values of $Q(T_1, T_2)$ along the diagonal in table 3 are significantly larger than 1.0. However, the accuracy is poor because of small N_2 (in some cases), the dilution of optical pairs, and possible errors in types.

Similar statistical defects limit the significance of average magnitude differences or luminosity ratios in binary galaxies. Magnitudes have not yet been estimated for the

TABLE 1

Types of Galaxies in Pairs

A. $N_2(T_1, T_2)$ = Number of Pairs of Types T_1, T_2 in a Sample of 427 Pairs

Type of Larger, T_1	Type of Smaller Galaxy in the Pair, T_2							Total Pairs
	E	S0, SB0	Sa	Sb, Sc	(S)	SB	Ir	
E	13	14	6	6	(12)	0	1	40
S0, SB0	6	74	11	10	(21)	1	4	106
Sa	10	24	46	28	...	1	3	112
Sb, Sc	10	31	24	44	...	2	12	123
(S)	(20)	(55)	...	...	(142)	(3)	(15)	(235)
SB	0	7	2	2	(4)	2	2	15
Ir	0	14	3	10	(13)	0	4	31
Total pairs	39	164	92	100	(192)	6	26	427
Number of galaxies in total sample by type	79 E	270 S0, SB0	204 Sa	223 Sb, Sc	(427) (S)	21 SB	57 Ir	854 all types

B. Proportions of Mixed Pairs in Which One Type Is Larger

Type of Larger	Type That Is More Often Smaller					
	E	S0, SB0	Sa	Sb, Sc	SB	Ir
E	...	0.7	...	...	...	(1.0)
S0, SB0	...	...	...	...	...	...
Sa	0.63	0.69	...	0.54	...	(0.5)
Sb, Sc	0.63	0.76	...	...	(0.5)	0.55
SB	...	0.9	(0.7)	(0.5)	...	(1.0)
Ir	...	0.78	(0.5)	...	...	...

Notes.—(S) = Sa, Sb, or Sc. Ir is mainly Ir I. Counts, N_2, and types, T_1 and T_2, by W. Zonn from Palomar *Sky Survey* prints. Parentheses indicate poorly determined values.

TABLE 2

Frequency of Morphological Types

Hubble Type	DDO Lum. Class	854 Galaxies in 427 Pairs Typed by Zonn	154 Members of Holmberg Pairs on van den Bergh List	935 Galaxies Typed by van den Bergh (1960)
Ir	...	57 (6.7)	2 (1.3)	20 (2.1)
Sc	I, II	...	18 (11.7)	112 (12.0)
Sc	III	...	13 (8.5)	64 (6.9)
Sc	IV, V	...	0 (0)	30 (3.2)
Sc	(all)	...	47 (30.5)	255 (27.3)
Sb	I, II	...	16 (10.4)	117 (12.5)
Sb	III, IV	...	10 (6.5)	74 (7.9)
Sb	(all)	...	40 (26.0)	257 (27.5)
Sb, Sc	(all)	244 (28.6)	87 (56.5)	512 (54.8)
Sa	(all)	204 (23.9)	7 (4.5)	72 (7.7)
E, S0	(all)	349 (40.8)	41 (26.6)	214 (22.9)
Other	(all)	0 (0)	17 (11.0)	117 (12.5)
Total	(all)	854 (100.0)	154 (100.0)	935 (100.0)

Note.—Numbers in parentheses are percentages.

Zonn-Cook (1963) list, and the limiting magnitudes for the other lists mentioned above are not well defined. It is obvious that, as the limiting magnitude is increased, more pairs of large magnitude difference, $m_2 - m_1$, will be found, and a larger proportion of these will be optical pairs of no significance. Only if single galaxies as well as pairs are counted to several different magnitude limits, in the manner of Page, Dahn, and Morrison (1961), can the true distribution of $m_2 - m_1$ be determined for dynamical pairs. For the 695 pairs in Holmberg's (1937) catalog (which is probably complete to 14.4 mag for the area covered, but which contains many objects of 15th magnitude and a few as faint as 16.2 mag), the average magnitude differences are

0.81 mag for 500 pairs of E or S0, range 0.0–4.2 mag;
1.19 mag for 100 mixed pairs of galaxies, range 0.0–2.8 mag;
0.93 mag for 65 pairs of spirals, range 0.0–1.7 mag;
0.93 mag for all 695 pairs, range 0.0–4.2 mag.

As already noted, there is some uncertainty in Holmberg's morphological types, which have been converted here from the Wolf system to Hubble types E and S.

TABLE 3

PAIRING FACTOR, $Q(T_1, T_2)$ IN 427 PAIRS

TYPE OF LARGER, T_1	TYPE OF SMALLER GALAXY IN THE PAIR, T_2							TOTAL
	E	S0, SB0	Sa	Sb, Sc	(S)	SB	Ir	
E	3.56	1.12	0.63	0.58	0.61	0.0	0.4	1.01
S0, SB0	0.48	1.73	0.34	0.28	0.31	0.3	0.44	0.79
Sa	1.06	0.75	1.88	1.05	....	0.4	0.44	1.10
Sb, Sc	0.97	0.88	0.90	1.51	....	0.7	1.61	1.10
(S)	1.04	0.82	....	....	1.33	0.6	1.05	1.10
SB	0.0	2.11	0.8	0.7	0.76	7.4	2.9	1.43
Ir	0.0	1.56	0.44	1.34	0.91	0.0	2.11	1.09
Total	0.99	1.21	0.90	0.90	0.90	0.57	0.91	1.00

Using the absolute photographic magnitudes of pair members in van den Bergh's (1960) list, we find a range from −14.6 mag (for NGC 221) to −21.0 mag (for NGC 4472), a median of about −18.6 mag, and 92 pair members with absolute magnitudes between −17.5 and −19.5 mag. The fainter member of a pair often does not appear on van den Bergh's list, so this median value is biased toward higher luminosity; but since the list was not intended to include pairs, this sample of binary galaxies is comparable to the rest of van den Bergh's list. For these field galaxies the median absolute magnitude is −18.7 mag and the distribution is similar, as shown in figure 1. This result, together with the similar distribution of morphological types, indicates that *the galaxies in pairs form a group statistically similar to the field galaxies*. It is to be expected that the correlation of color index with morphological types (Holmberg 1964) applies to binary as well as to single galaxies.

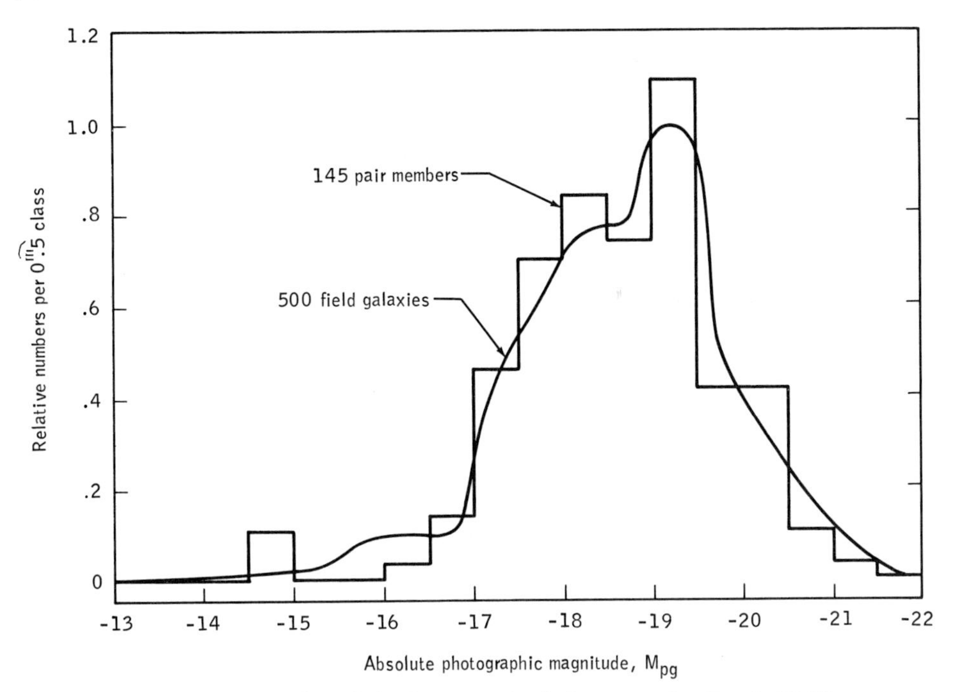

FIG. 1.—Distribution of absolute magnitude from van den Bergh (1960).

5. DIMENSIONS OF GALAXIES IN PAIRS

The data in table 1 indicate that in a majority of mixed pairs S0 and SB0 types are the smallest, E types are somewhat larger, and Sa and barred types the largest. Two factors are involved: (1) the mean absolute dimensions and deviations characteristic of each morphological type; and (2) a possible tendency toward smaller-than-average or larger-than-average galaxies in the formation of a pair.

Holmberg (1946) showed that the measurement of angular dimensions is subject to serious systematic errors, but the de Vaucouleurs (1964) catalog establishes a consistent set of angular dimensions. These have been used by Page (1965) together with radial velocities as distance indicators to calculate the maximum dimensions of 128 randomly selected single galaxies, 98 galaxies in groups, and 51 galaxies in close physical pairs. The results in table 4 show that deviations from the mean values are large, and that differences between single galaxies, group galaxies, and binary galaxies are statistically insignificant.

6. ORIENTATION OF AXES IN PAIRS OF GALAXIES

As shown in chapter 1, all but the Ir type of galaxies have rough symmetry about an axis (the axis of rotation), and can be represented roughly by ellipsoids of "true axis ratio" $e(T)$ with approximate values

$$e(\text{Sc, Sb}) = 0.1\ ; \qquad e(\text{Sa, S0}) = 0.2\ ; \qquad e(\text{E}) = 0.3 \text{ to } 0.4\ .$$

Using the measured angular diameters, A_1, A_2, B_1, and B_2, the position angles, θ_1, θ_2, and θ_{12} defined in § 3, the appropriate values $e(T_1)$ and $e(T_2)$, and the direction of rotation, it is possible to determine the inclinations i_1 and i_2 (angles between our line of sight and the two axes), and the angle ϕ between the axes. From a number of edge-on spiral-type binaries ($i_1 \simeq i_2 \simeq 90°$) it is clear that ϕ is not always close to zero; see, for instance, NGC 2814–20, 3786–88, and 4206–16 in the Holmberg (1937) catalog. However, the distribution of ϕ in a number of binaries, or its correlation with types and separations, probably will be of importance in any theory of the origin and evolution of binary

TABLE 4

MEAN DIMENSIONS OF GALAXIES BY TYPE

Type	128 Single Field Galaxies		98 Galaxies in Groups		51 Galaxies in Isolated Tight Pairs	
	n	hA	n	hA	n	hA
E0 to E7	29	10.8±4.	41	10.1± 4.	21	8.5±4.
S0, SB0	20	11.3±4.	27	15.6± 4.	9	7.0±3.
E, S0, SB0	49	11.0±4.	68	12.3± 4.	30	8.1±4. (6.7±3. in 5 *mixed* pairs)
Sa	12	17.6±7.	8	12.1± 3.	2	12.6±5.
SBa	8	14.5±7.	1	17.8	0	
Sa, SBa	20	16.4±7.	9	12.7± 3.	2	12.6±5.
Sb	13	18.9±5.	4	25.7±12.	5	15.9±7.
SBb	6	24.6±7.	7	27.0± 4.	1	7.1
Sb, SBb	19	20.7±6.	11	26.6± 7.	6	14.5±7.
Sc	18	16.6±4.	4	22.0± 7.	10	16.0±6.
SBc	18	16.4±5.	0		2	13.0±4.
Sc, SBc	36	16.5±5.	4	22.0± 7.	12	15.5±6.
Sa, Sb, Sc	43	17.6±5.	16	18.0± 7.	17	14.8±6.
SBa, SBb, SBc	32	17.5±6.	8	25.9± 4.	3	11.1±4.
S, SB	75	17.6±6.	24	20.6± 6.	20	14.3±6. (13.2±4. in 5 *mixed* pairs) (18.4±7. in 5 others)
Ir	4	5.2±3.	6	12.1± 7.	1	12.3
All types	128	14.6±5.	98	14.4± 5.	51	10.8±5.

NOTE.—h = Hubble constant in km s^{-1} per 10^4 pc (approximately 1.0). Errors listed are mean deviations.
A = diameter in kiloparsecs, averaged for n galaxies.

galaxies. Page (1964, 1965) applied the following steps in an analysis of measurements by Zonn and Cook (1963) not yet completed:

a. The distribution of θ_1 and θ_2 is found to be uniform between 0° and 180°, indicating that the galaxies in pairs are randomly oriented to the line of sight. Thus the distribution of i should be proportional to cos i.

b. From values of A and B corrected for Holmberg's (1946) systematic errors, and from the rough values of $e(T)$, values of i can be determined in 100 or more binaries. The data analyzed so far show a pronounced selection of faint binaries with i near 90°. If this selection effect can be eliminated, the data can be used to modify $e(T)$ and to eliminate observational bias in ϕ.

c. The value of ϕ can be computed from i_1, i_2, θ_1, θ_2, and θ_{12}. In cases where the direction of rotation cannot be determined from spiral forms or spectra, there is a fourfold ambiguity in ϕ, but the distribution of ϕ can be tested statistically for axes near parallel or for random relative orientations.

Unfortunately, the effects of dilution by optical pairs and observational selection have prevented any firm conclusions.

7. THE DYNAMICS OF BINARY GALAXIES

It is generally assumed that binary galaxies are in stable orbits under the influence of gravitational forces only. With the further assumptions that the orbits are nearly circular and randomly oriented to the line of sight, Holmberg (1937) and Page (1960, 1962) were able to estimate the mean masses of galaxies in pairs, as mentioned in chapter 3 and summarized in table 5. The agreement between the average masses of binaries and the averages for single galaxies, shown in table 6, confirms the dynamical model and the statistical similarity between binaries and single galaxies in the field. That is, most or all of the 52 pairs studied by Page are physical pairs. The high mass/luminosity ratio determined for E and S0 types has a bearing on the evolution of galaxies discussed in chapters 10 and 11.

TABLE 5

AVERAGE MASS AND M/L, DOUBLE GALAXIES

No. of Systems n	No. of Galaxies, ΣN_i (by type)				Mean Mass $h\langle M\rangle/10^{10}$ ($M_\odot$)	Mean M/L $\langle M/hL\rangle$ (solar units)	Notes
	Irr	S	S0	E			
52	2	52	17	43	31.2±10.6	38.0±19.9	All systems
33	1	29	10	26	26.0±13.9	31.2±26.0	Pure pairs only
41	1	44	13	33	28.7± 9.0	43.8±15.2	High-weight obs. only
16	2	32	0	0	4.0± 4.2	3.2± 4.2	S and Irr only
10	1	19	0	0	1.6± 2.3	1.4± 1.8	Pure pairs only
13	1	27	0	0	1.5± 1.7	1.3± 1.5	High-weight obs. only
18	0	0	11	26	66.2±29.	98. ±68.	E and S0 only
13	0	0	8	18	63.6±38.	92. ±92.	Pure pairs only
13	0	0	8	19	59.4±15.	90. ±37.	High-weight obs. only
18	0	20	6	17	31.4±17.	46. ±23.	Mixed systems
10	0	10	2	8	27.7±23.	41. ±34.	Pure pairs only
15	0	17	5	14	31.4±18.	46. ±26.	High-weight obs. only
					$h\langle M_E\rangle/10^{10}$, assuming $\langle M_E\rangle=30\langle M_S\rangle$		
15	0	17	5	14	60.7±36.		Mixed only
13	0	0	8	19	59.4±15.		E and S0 only
28	0	17	13	33	60.0±19.		E, S0, and mixed
13	1	27	0	0	43.4±53.		S and Irr only
41	1	44	13	33	59.6±16.		All high-weight obs.

NOTES:
Each system includes N_i galaxies in two groups treated as mass-points.
For "pure pairs," $N_i = 2$, and no other galaxy is nearby.
"High-weight observations" include only those systems for which observed relative velocities have weight greater than 0.5.
h is the Hubble constant in units of 10^{-4} km s^{-1} pc^{-1} ($h \simeq 1$).
$\langle M\rangle$ is the mean mass of one galaxy, in $M_\odot$.
L is the total photographic luminosity of a galaxy, in $L_\odot$.
$\langle M_E\rangle$ is the mean mass of E and S0 galaxies.
$\langle M_S\rangle$ is the mean mass of S, SB, and Irr galaxies.
Each value of $h\langle M\rangle$ and $\langle M/hL\rangle$ results from a least-squares solution from which rms errors of the mean were also determined.
The values of $\langle M/hL\rangle$ for S and Irr galaxies were incorrectly listed in the first publication (Page 1962).

TABLE 6

MASSES OF GALAXIES
(Grouped according to morphological type)

Galaxy (NGC or Name)	m_{pg}	Type	$V/100$ (km s^{-1})	Method	Reference*	$hM/10^{10}$ ($M_{\odot}$)	M/hL (solar units)
55.	7.9	Ir–Sc	1.0	H II	deV 1961	3.	2.
I1613.	10.0	Ir I	− 1.3	21-cm	Volders 1961	0.03	4.
3034 (M82).	9.6	Ir II	3.2	Ls	Mayall 1960	1.	9.
				Ls	BBR 1964	1.	
3432.	11.9	Ir–Sc	5.9	. . . .	Bertola 1966	> 0.5	7.
3556.	10.9	Ir–Sc	7.6	Ls	BBP 1960	1.	1.
6822.	9.7	Ir I	0.7	21-cm	Volders 1961	0.15	0.1
LMC.	0.5	Ir–S	0.2	21-cm, H II	deV 1963	1.1	5.
VV 254.	. . . . . .	Ir	45.9	Ls	BB 1963	9.8	. . .
Mean, 8 Ir.	. . . . . .	. . . .	. . . .	. . . .	. . . .	**1.**	**4.**
157.	11.2	Sc	18.4	Ls	BBP 1961*d*	4.4	1.5
253.	6.9	Sc	1.0	Ls	BBP 1962*b*	20.	2.
598 (M33).	6.2	Sc	−0.1	21-cm, H II	†	1.	3.
					Brandt 1965	3.4	10.
613.	11.0	SBc	15.2	Ls	BBRP 1964	~ 1.5	
1084.	11.1	Sc	14.3	Ls	BBP 1963*a*	0.8	1.
1365.	10.3	SBc	15.1	Ls	BBP 1960	2.5	1.5
1792.	11.3	Sc	. . . .	Ls	RBBP 1964	1.4	1.3
2146.	11.3	Sc	9.9	Ls	BBP 1959*b*	1.3	1.5
2903.	9.5	Sc	5.1	Ls	BBP 1960	4.0	2.1
3646.	11.8	Sc	42.0	Ls	BBP 1961*c*	20.	4.
4490.	10.3	Sc	6.2	. . . .	Bertola 1966	> 0.15	0.4
4535.	10.9	Sc	18.5	. . . .	Davies 1963	2.1	2.7
4605.	11.1	Sc–Ir	2.8	. . . .	Bertola 1966	> 0.07	1.6
4631.	9.7	Sc	6.5	Ls	deV 1963	2.4	1.8
5194 (M51).	8.6	Sc	5.5	Ls	BB 1964	4.3	11.
5248.	11.0	Sc	11.4	Ls	BBP 1962*a*	4.	1.5
5457 (M101). . . .	8.5	Sc	4.2	21 cm	Volders 1959	~ 1.0	~ 17.
6181.	12.7	Sc	23.1	Ls	BBP 1965*b*	3.6	. . .
6503.	10.7	Sc(dwf)	3.5	Ls	BBCRP 1964*a*	0.13	0.8
7320.	13.	Sc	10.7	Ls	BB 1961	~ 4.4	~ 8.
. . . (GB 1).	. . . . . .	SB(pec)	. . . .	. . . .	BBS 1967	> 4.	
Mean, 21 Sc, SBc.	. . . . . .	. . . .	. . . .	. . . .	. . . .	**3.**	**3.**
Mean, 29 Ir, Sc, SBc.	. . . . . .	. . . .	. . . .	. . . .	. . . .	**3.**	**3.**
16 Double systems.	10–13	2 Ir, 30 S	6–76	Circ. orbits	Table 5	**4.0**	**3.2**
224 (M31).	4.3	Sb	− 0.7	21-cm, H II	‡	34.	8.4
972.	12.3	Sb	16.8	Ls	BBP 1965*c*	0.9	1.6
1068 (M77).	10.	Sb(em)	12.0	Ls(em)	BBP 1959*a*	2.0	2.7
1097.	10.4	SBb	12.1	Ls(em)	BB 1960	0.6	0.5
1808.	11.1	Sb	8.2	Ls	BB 1968	2.7	2.
1961.	12.2	Sb	40.3	. . . .	Davies 1963	4.6	3.
3031 (M81).	8.1	Sb	0.9	H II	Münch 1959	12.	6.

* References are abbreviated as follows: BB 1959 = Burbidge and Burbidge 1959; BB 1960 = Burbidge and Burbidge 1960; BB 1961 = Burbidge and Burbidge 1961; BB 1963 = Burbidge and Burbidge 1963; BB 1964 = Burbidge and Burbidge 1964; BB 1968 = Burbidge and Burbidge 1968; BBCRP 1964*a* = Burbidge *et al.* 1964*a;* BBCRP 1964*b* = Burbidge *et al.* 1964*b;* BBF 1961 = Burbidge, Burbidge, and Fish 1961*a, b, c;* BBP 1959*a* = Burbidge *et al.* 1959*a;* BBP 1959*b* = Burbidge *et al.* 1959*b;* BBP 1960 = Burbidge *et al.* 1960; BBP 1961*a* = Burbidge *et al.* 1961*a;* BBP 1961*b* = Burbidge *et al.* 1961*b;* BBP 1961*c* = Burbidge *et al.* 1961*c;* BBP 1961*d* = Burbidge *et al.* 1961*d;* BBP 1962*a* = Burbidge *et al.* 1962*a;* BBP 1962*b* = Burbidge *et al.* 1962*b;* BBP 1962*c* = Burbidge *et al.* 1962*c;* BBP 1963*a* = Burbidge *et al.* 1963*a;* BBP 1963*b* = Burbidge *et al.* 1963*b;* BBP 1963*c* = Burbidge *et al.* 1963*c;* BBP 1965*a* = Burbidge *et al.* 1965*a;* BBP 1965*b* = Burbidge *et al.* 1965*b;* BBR 1964 = Burbidge *et al.* 1964; BBRP 1964 = Burbidge *et al.* 1964; BBS 1967 = Burbidge *et al.* 1967; Davies 1963 = Davies *et al.* 1963; D-A 1960 = Duflot-Augard 1960; RCBBP 1965 = Rubin *et al.* 1965; deV 1961 = deVaucouleurs 1961; deV 1963 = deVaucouleurs and deVaucouleurs 1963.

† References for NGC 598 (M33): Wyse and Mayall 1942; Volders 1959; Dieter 1962.

‡ References for NGC 224 (M31): Wyse and Mayall 1942; Schmidt 1957; Münch 1954.

TABLE 6—*Continued*

Galaxy (NGC or Name)	m_{pg}	Type	$V/100$ (km s^{-1})	Method	Reference*	$hM/10^{10}$ ($M_\odot$)	M/hL (solar units)
3504	11.6	SBb	14.7	Ls(em)	BBP 1960	0.8	1.
3521	9.6	Sb	6.4	Ls(em)	BBCRP 1964*b*	8.	5.
4258	8.9	Sb	5.3	Ls	BBP 1963*c*	10.	2.4
4826	9.6	Sb	3.5	Ls	Rubin 1965	~ 8.	...
5005	10.5	Sb	10.8	Ls(em)	BBP 1961*a*	10.	2.5
5055 (M63)	9.3	Sb	6.0	Ls(em)	BBP 1960	4.5	2.
5383	12.4	Sb	23.7	Ls	BBP 1962*c*	4.	7.
7331	10.6	Sb	10.7	Ls	RCBBP 1965	~ 8.	~ 3.
7479	11.6	Sb	26.6	Ls	BBP 1960	> 0.8	0.5
Milky Way	......	Sb	0.	21-cm, stat.	Schmidt 1965	18.	5.6
Mean, 17 Sb, SBb	......	......	......	......	......	**6.**	**4.**
681	13.0	Sa	18.0	Ls	BBP 1965*a*	1.5	4.8
2782	12.5	Sa	25.1	Ls	D-A 1960	11.	7.5
3623 (M65)	10.2	Sa	6.4	Ls	BBP 1961*b*	20.	7.2
7469	12.7	Sa(em)	50.2	Ls	BBP 1963*b*	0.8	0.5
Mean, 4 Sa	......	......	......	......	......	**6.**	**4.**
Mean, 21 Sb, SBb, Sa	......	......	......	......	......	**6.**	**4.**
Mean, 50 Ir, S, SB	......	......	......	......	......	**4.**	**3.**
221 (M32)	9.7	E2	0.2	......	Fish 1964	0.3	11.
				......	BBF 1961	0.36	27.
3115	10.1	E7	4.2	......	BBF 1961	15.	46.
				......	Mayall 1960	35.	36.
3379	10.5	E1	7.5	......	Fish 1964	13.	20.
4111	11.6	E–S0	8.4	......	Poveda 1961	4.	14.
4278	11.2	E	6.2	......	Poveda 1961	5.	14.
4406 (M86)	10.3	E3	− 3.7	......	Fish 1964	96.	39.
4472 (M49)	10.	E1	8.6	......	Fish 1964	110.	19.
4486 (M87)	9.6	E0	11.9	......	Fish 1964	260.	60.
5128	8.	E(pec)	4.0	Ls	BB 1959	15.	13.
Mean, 9 E	......	......	......	......	......	**60.**	**30.**
28 Double systems	9–14.	33 E, 13 S0	7–48	Circ. orbits	Table 5	**60.**	**90.**
10 Groups of galaxies	......	4S, 4E each	3–91	Virial theorem	Table 7	**250.**	**280.**
9 Clusters of galaxies	......	100 each	7–108	Virial theorem	Table 7	**130.**	**600.**

For close pairs, spectra of both galaxies can be obtained in one exposure with the slit in position angle θ_{12}. Then V_1, V_2, and $V_1 - V_2 = \Delta V$ can be measured on one spectrogram. The mean $V = (V_1 + V_2)/2$ in km s^{-1} is used in Hubble's law to determine the distance of the pair in parsecs:

$$D = (V/h) \times 10^4 . \quad (6)$$

A detailed study of measurement errors in V showed a standard deviation ± 90 km s^{-1}, with weights W of single observations ranging from 0.05 to 20. There is a further dispersion in equation (6), and values of h ranging from over 1.5 down to 0.5 have been used in the literature.

The difference between the two radial velocities in a pair, ΔV, is assumed to be the projection of a circular orbital velocity v. The observed angular separation S (in minutes of arc) is the projection of the line of centers (of length R) divided by the distance, D, and by the number of minutes in a radian. By Kepler's harmonic law of two-body gravitational orbits, Rv^2 is proportional to the sum of the masses; hence

$$M_1 + M_2 = 675\, SV(\Delta V)^2/h \cos^3\phi \cos^2\psi \geq 675\, SV(\Delta V)^2/h\,, \quad (7)$$

where ϕ is the projection angle of R, having an unknown value between 0 and $\frac{1}{2}\pi$, v is assumed to be perpendicular to R (circular orbit), and ψ is another angle involved in the projection of v, having some value between 0 and 2π. When $\phi = \frac{1}{2}\pi$, one galaxy is behind the other, and the pair would not be recognized as a double. At the other extreme, very wide pairs (large S) were not selected for observation. The distribution of R is given by equation (3).

On the assumptions that ϕ, ψ, M, and R are independent of each other in a sample of many double galaxies, and that the errors in ΔV are normally distributed (so that the mean square of measured ΔV must be reduced by the variance, σ^2/W),

$$(\Delta V)^2 - \sigma^2/W = 5.92 \times 10^{-8} h\langle M\rangle(0.19 + 10^4/SV)\,, \quad (8)$$

where σ is the standard deviation and W the weight of measurements of ΔV, $\langle M\rangle$ is the mean mass of a single galaxy in all the pairs, and the relative errors in S and V are negligible compared with $\sigma/\Delta V$. Equation (8) is a regression between observed $(\Delta V)^2$ and observed $(0.19 + 10^4/SV)$, and a least-squares solution for $h\langle M\rangle$ was made from the observations of 33 pairs of galaxies, yielding a value $h\langle M\rangle = 2.6 \times 10^{11} \pm 1.4 \times 10^{11}\, M_\odot$ ($1\, M_\odot = 2 \times 10^{33}$ g). In another 19 cases, observations referred to groups of N galaxies approximating a pair. The simplest of these ($N = 3$) consisted of a close pair of galaxies with a satellite. These were included with the factor $\frac{1}{2}N$ on the right of equation (8) yielding $h\langle M\rangle = 3.1 \times 10^{11} \pm 1.1 \times 10^{11}$, as shown in table 5.

Least-squares solutions of equation (8) were also made for subsets of the data, as shown in table 5, from which it is clear that the mean mass of an elliptical, $\langle M_{\rm E}\rangle = 30\langle M_{\rm S}\rangle$, where $\langle M_{\rm S}\rangle$ is the mean mass of spirals in these pairs and groups. The mixed systems confirm this fact, and it is also indicated in the individual mass determinations of table 6, although these vary widely in the case of elliptical (E) galaxies.

The total luminosity of a large group of stars was at first expected to be proportional to the total mass, even though any one star may be 10,000 times more luminous or 1000 times less luminous than the Sun. However, all theories of stellar evolution show that massive stars of very high luminosity are short-lived, so that an old population of stars should have lower luminosity for a given total mass. The ratio M/L in solar units is as small as 10^{-3} for young giant stars and as large as 1000 for long-lived dwarf stars. The luminosity of a galaxy is defined in these solar units as

$$L = D^2\, 10^{0.104-0.4m} = (V/h)^2\, 10^{8.104-0.4m}\,, \quad (9)$$

where m is the measured apparent photographic magnitude of the galaxy. Introducing the sum of N luminosities into equation (8), we get another regression involving the same left-hand side, the desired mean $\langle M/hL \rangle$, and the observables V/S, V^2, and the sum $\Sigma\, 10^{8.104-0.4m}$ on the right. Least-squares solutions for $\langle M/hL \rangle$ yield the values given in table 5 and show that the mean $\langle M/L \rangle$ for massive E galaxies is 30–60 times the value for spirals (S), somewhat more than would be expected if the E galaxies consist simply of older stars. This may indicate an admixture of nonluminous matter in E galaxies, although optical evidence of obscuring dust clouds, and radio evidence of nonluminous hydrogen, are limited to spirals. It is possible that other forms of matter are involved, such as collapsed masses or stars of very low temperature.

The validity of these results has been discussed (Page 1962), and it is shown that the assumption of circular orbits and the tidal effects neglected in equation (7) are not likely to have affected the results significantly. If M is positively correlated with R, so that more massive pairs are systematically of wider separation than less massive ones (a possible result of the mechanics of galaxy formation or of later perturbations by intruders), then the values of $h\langle M \rangle$ in table 5 are *underestimated*. If the observed pairs are all embedded in an intergalactic medium of uniform density ρ, the mass involved in equations (7) and (8) would be $2M + 4\pi\rho R^3/3$, and this dependence on R or SV again results in an underestimate. Motions in clusters of galaxies and cosmological models fitted to the Hubble law of redshifts imply values of ρ as high as 10^{-28} g cm^{-3}. The resulting increase in $\langle M \rangle$ is approximately $5 \times 10^{35}\ \rho/h^3$ or about $10^7\ M_\odot$, which is insignificant (only one part in 10^3 or 10^4).

Although the selection in S has been accounted for, other effects of selection might influence the means in table 5. Selection of the higher-luminosity pairs is to be expected, although small-diameter galaxies and those of low surface brightness are apt to be overlooked on photographs; high surface brightness is selected for velocity measurements. Because the more luminous galaxies in a class are expected to be the more massive ones, the estimated average masses, $\langle M_E \rangle$ and $\langle M_S \rangle$, are undoubtedly biased toward higher values. However, the large ratio $\langle M_E \rangle / \langle M_S \rangle$ cannot be explained as a result of this selection, and for three reasons: (1) the E galaxies included in the set of pairs (Page 1962) are somewhat fainter than the S galaxies included; (2) in the mixed pairs, E galaxies are as often brighter than S galaxies as they are fainter; and (3) the results for mixed pairs confirm $\langle M_E \rangle / \langle M_S \rangle = 30$. Note, also, that for the pairs selected, the mean luminosity $\langle L_E \rangle \simeq 0.67 \langle L_S \rangle$ if the spread is not extreme.

8. FORMATION AND EVOLUTION OF BINARY GALAXIES

Ambartsumian (1957) questioned the stability of groups and clusters of galaxies, and assumed that some unspecified explosion in the past left each of them with positive total energy. At a special conference on the stability of systems of galaxies (Neyman, Page, and Scott 1961), the stability of binary galaxies, demonstrated by consistent mass determinations, was considered an exception. Small groups consisting of one or more close pairs such as those used by Page (1962) are probably also stable. Ambartsumian's dynamical arguments for instability apply specifically to "trapezoidal groups" of four or more masses, all separated by comparable distances. A strong argument against stability of such groups is that application of the virial theorem to their observed radial velocities yields a total mass far larger than the expected sum of individual masses of the group members (see Chapters 3, 14, and 17).

The large difference in mass between elliptical galaxies and spirals shown in tables 5 and 6 raises questions about the initial conditions probably responsible for forming different types of galaxies, the binaries, groups, and clusters, which might also lead to large differences in mass and in M/L. The concept developed by Holmberg (1964) is that higher initial density in a protogalaxy gas cloud leads to more rapid star formation and an early decrease in L, or increase in M/L. Preliminary calculations by Page (1964) show that M/L is also sensitive to the *size* of stars formed, which is probably set by the turbulence

TABLE 7

MASSES OF GROUPS AND CLUSTERS OF GALAXIES

Group or Cluster	Angular Diameter	Total m_{pg}	$\langle V\rangle/100$ (km s^{-1})	hR (Mpc)	N_b	$N_b h\langle M\rangle/10^{10}$ ($M_\odot$)	$h\langle M\rangle/10^{10}$ ($M_\odot$)	$\langle M/hL\rangle$
VV 115 (Seyfert)	1′.9	....	44.	0.01	5	24.	5.	....
VV 116	....	12.7	64.	....	2E, 3S	100.	20.	....
VV 150	1′.2	....	73	0.02	S	....	....	....
VV 166 (NGC 67–72)	....	....	67.9	....	3E, 3S	....	....	350
VV 288 (Stephan)	3′.7	11.8	67.	0.04	5(E, S)	500.	100.	100.
NGC 55	500′.0	8.7	5.5	0.4	6S	600.	100.	500.
NGC 383 (Pisces)	....	....	....	....	25E	12500.	500.	260.
NGC 3031–77 (M81)	....	6.	2.	....	>4S	....	120.	200.
NGC 6027 (Serpens)	1′.6	14.	45.	0.01	3E, 3S	....	....	....
NGC 7619 (Pegasus)	120′.0	11.	40.	0.3	5E	2500.	500.	300.
Local Group	....	....	....	....	2I, 2S, 2E	....	400	....
Sculptor	950′.0	....	3.	0.37	6	1700	280	....
NGC 6166	2′.5	13.0	9.1	0.03	5E	1400.	280.	175.
Abell 2199	12′.0	....	90.	0.15	>19	....	....	....
(Mean group)	....	....	....	**0.15**	**8**	**2000**	**250.**	**280**
CVn cluster	19°	6.6	6.8	1.1	30S	4500.	150.	400.
Fornax cluster	5°.7	....	15.	0.75	30	4700.	157.	....
Pegasus cluster	2°.0	....	39.	0.67	50	4200.	84.	....
UMa cluster	10°	....	20.	1.8	50	2800.	56.	....
Hercules cluster	1°.4	....	108.	1.3	50S, 30E	5600.	70.	....
Virgo E	11°.5	6.3	11.	1.1	100E	24000.	240.	600
Virgo S	11°.0	....	19.	1.8	100S	<45000.	<450.	....
Coma cluster	9°.0	9.4	67.	5.2	500E	75000.	150.	900
NGC 541	....	....	....	....	500S	5000.	10.	....
Mean cluster	....	....	....	**1.7**	**100**	**13000.**	**130.**	**600**

of the protogalaxy gas cloud. Studies of the orientations of galaxies in pairs are expected to show that physical pairs have high total angular momentum; that is, the formation of binaries depends on the initial angular momentum of the protogalaxy gas clouds. The "pairing factors" in table 3 may therefore reflect the likelihood of a single gas cloud having high enough angular momentum to form a binary galaxy, and turbulence distributed uniformly enough so that the two galaxies formed are of similar morphological type.

The larger gas clouds that formed groups and clusters may have had higher initial density and a wider spectrum of turbulence. In a general way, this is borne out by the larger values of M/L shown in table 7, where the total masses are calculated from the virial theorem on the assumption that these groups and clusters are stable. N_b represents the number of bright galaxies in each group or cluster, so that the calculated average mass $\langle M \rangle$ of a galaxy is slightly overestimated. Note also that if h is as small as 0.5, then all masses are doubled, relative to the Local Group, and $\langle M/L \rangle$ only half of the listed $\langle M/hL \rangle$. However, this applies to all the double-galaxy results in table 5, and to most of the single-galaxy data in table 6, and leaves the trend of increasing $\langle M/L \rangle$ from singles to doubles to groups to clusters of galaxies roughly as shown in figure 2.

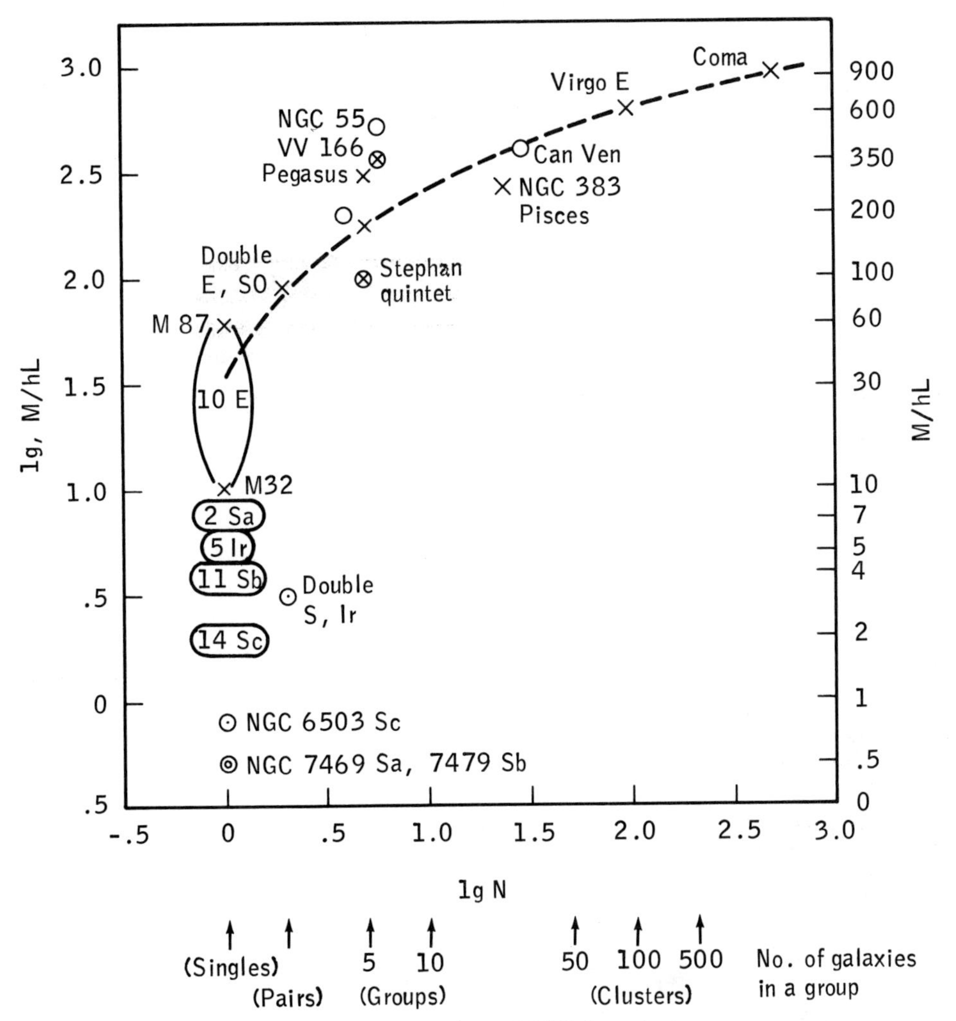

FIG. 2.—Average values of M/L for galaxies.

REFERENCES

Ambartsumian, V. 1957, *Pub. Armenian Nat. Acad. Sci.* Ser. 11, (in Russian).
Arp, H. C. 1966, *Atlas of Peculiar Galaxies* (Pasadena: California Institute of Technology).
Bergh, S. van den. 1960, *Pub. David Dunlap Obs.*, **2**, 159.
Bertola, F. 1966, *Asiago Pub.*, No. 186.
Brandt, J. 1965, *M.N.R.A.S.*, **129**, 309.
Burbidge, E. M., and Burbidge, G. R. 1959, *Ap. J.*, **129**, 271.

Burbidge, E. M., and Burbidge, G. R. 1960, *ibid.*, **132**, 30.
———. 1961, *ibid.*, **134**, 244.
———. 1963, *ibid.*, **138**, 1306.
———. 1964, *ibid.*, **140**, 1445.
———. 1968, *ibid.*, **151**, 99.
Burbidge, E. M., Burbidge, G. R., Crampin, D. J., Rubin, V. C., and Prendergast, K. H. 1964*a*, *Ap. J.*, **139**, 539.
———. 1964*b*, *ibid.*, p. 1058.
Burbidge, E. M., Burbidge, G. R., and Fish, R. A. 1961*a*, *Ap. J.*, **133**, 393.
———. 1961*b*, *ibid.*, p. 1092.
———. 1961*c*, *ibid.*, **134**, 251.
Burbidge, E. M., Burbidge, G. R., and Prendergast, K. H. 1959*a*, *Ap. J.*, **130**, 26.
———. 1959*b*, *ibid.*, p. 739.
———. 1960, *ibid.*, **132**, 640.
———. 1961*a*, *ibid.*, **133**, 814.
———. 1961*b*, *ibid.*, **134**, 232.
———. 1961*c*, *ibid.*, p. 237.
———. 1961*d*, *ibid.*, p. 874.
———. 1962*a*, *ibid.*, **136**, 128.
———. 1962*b*, *ibid.*, p. 339.
———. 1962*c*, *ibid.*, p. 708.
———. 1963*a*, *ibid.*, **137**, 376.
———. 1963*b*, *ibid.*, p. 1022.
———. 1963*c*, *ibid.*, **138**, 375.
———. 1965*a*, *ibid.*, **142**, 154.
———. 1965*b*, *ibid.*, p. 641.
———. 1965*c*, *ibid.*, p. 649.
Burbidge, E. M., Burbidge, G. R., and Rubin, V. C. 1964, *Ap. J.*, **140**, 942.
Burbidge, E. M., Burbidge, G. R., Rubin, V. C., and Prendergast, K. H. 1964, *Ap. J.*, **140**, 85.
Burbidge, E. M., Burbidge, G. R., and Shelton, J. W. 1967, *Ap. J.*, **150**, 783.
Davies, R. D., Gottesman, S., Reddish, V., and Vershuur, G. 1963, *Observatory*, **83**, 245.
Dieter, N. H. 1962, *A.J.*, **67**, 217.
Duflot-Augard, R. 1960, *Pub. Obs. Haute-Provence*, No. 14, **5**, 1.
Fish, R. A. 1964, *Ap. J.*, **139**, 284.
Holmberg, E. 1937, *Lunds Ann.*, No. 6.
———. 1946, *Medd. Lunds Astr. Obs.*, Ser. 2, No. 117.
———. 1954, *ibid.*, Ser. 1, No. 188.
———. 1964, *Ark. Astr.*, **3**, 387.
Humason, M., Mayall, N., and Sandage, A. 1956, *A.J.*, **61**, 97.
Mayall, N. U. 1960, *Ann. d'Ap.*, **23**, 344.
Münch, G. 1959, *Pub. A.S.P.*, **71**, 101.
Neyman, J., Page, T., and Scott, E. 1961, *A.J.*, **66**, 533.
Page, T. 1960, *Ap. J.*, **132**, 910.
———. 1961, in *Proc. 4th Berkeley Symp. Math Stat.*, **3**, 277.
———. 1962, *Ap. J.*, **136**, 685.
———. 1964, *Science*, **146**, 804.
———. 1965, *Smithsonian Ap. Obs. Spec. Rept.*, No. 195.
Page, T., Dahn, C., and Morrison, F. 1961, *A.J.*, **66**, 614.
Polya, G. 1919, *Astr. Nach.*, **208**, 175.
Poveda, A. 1961, *Ap. J.*, **134**, 910.
Rubin, V. 1965, *Ap. J.*, **141**, 885.
Rubin, V. C., Burbidge, E. M., Burbidge, G. R., and Prendergast, K. H. 1964, *Ap. J.*, **140**, 80.
Rubin, V., Crampin, D. J., Burbidge, E. M., Burbidge, G. R., and Prendergast, K. H. 1965, *Ap. J.*, **141**, 759.
Schmidt, M. 1956, *B.A.N.*, **13**, 15.
———. 1957, *ibid.*, **14**, 17.
Shapley, H., and Ames, A. 1934, *Harvard Ann.*, Vol. **88**, No. 4.
Vaucouleurs, G. de. 1961, *Ap. J.*, **133**, 405.
Vaucouleurs, G. de, and Vaucouleurs, A. de. 1963, *Ap. J.*, **136**, 363.
———. 1964, *Reference Catalogue of Bright Galaxies* (Austin: University of Texas Press).
Volders, L. 1959, *B.A.N.*, **14**, 323.
Volders, L., and Högbom, J. A. 1961, *B.A.N.*, **15**, 307.
Vorontsov-Velyaminov, B. 1959, *Atlas of Interacting Galaxies* (Moscow).
Wirtanen, C. A. 1963, mimeographed list of close pairs (Lick Observatory).
Wyse, A., and Mayall, N. U. 1942, *Ap. J.*, **95**, 24.
Zonn, W. 1962, *A.J.*, **68**, 82.
Zonn, W., and Cook, N. 1963, *Pub. A.S.P.*, **75**, 184.

CHAPTER 14

Nearby Groups of Galaxies

G. DE VAUCOULEURS
McDonald Observatory, The University of Texas

1. DEFINITION OF A GROUP

BINARY galaxies discussed in chapter 13 are a special case of multiplicity and clustering among galaxies. Large-scale clustering is discussed in chapter 15. The present chapter discusses multiple systems: groups, clusters, and clouds of intermediate size ($D \simeq 1$–3 Mpc) and population ($n \simeq 10$–100). Such systems can be identified and best studied individually in our immediate neighborhood ($\Delta < 20$ Mpc). The nearer galaxies used in the calibration of the distance scale and in assessing the detailed properties of galaxies are generally members of nearby groups.

By "group" is meant a number $n > 2$ of *distinct* galaxies that are bound gravitationally for periods of the order of 10^9 years or greater (star cluster analogy) or perhaps are genetically related as products of a common explosive formative event during the past 10^9 years or so even though they may not be gravitationally bound at present (stellar association analogy).

These concepts, however, are not suitable as operational definitions. In practice a "group" will be a "small" number of galaxies, say $n \simeq 10$, brighter than $M = -16$ (perhaps of a specified class only, either elliptical and lenticular, or spiral and irregular) occupying a volume of space V in which the average density $\rho_n = n/V$, say $\rho_n \simeq 10$ per Mpc^3, is at least one order of magnitude greater than the average density $\langle \rho_1 \rangle$ in a surrounding volume of space of diameter one order of magnitude greater than that of the group. In other words, the significant factor is the *contrast* with the surrounding field, not the absolute densities or numbers. Note that number density is considered, not mass density (which would exclude the galactic equivalent of stellar associations).

Space density, however, is not an observable, and the basic observational criterion of clustering is surface density in the apparent distribution. Since dwarf galaxies (dE, dIm) are inconspicuous or invisible in all but the nearest groups, a group is first identified by its brightest members. Further criteria of group membership include a small velocity range and a general similarity of morphological types, apparent magnitudes, and diameters.

Received September 1965.

2. CENSUS OF NEARBY GROUPS

In this chapter "nearby" is defined by the condition $m - M < 31.2$ (corresponding to $\Delta < 17$ Mpc and $\langle V \rangle < 3000$ km s^{-1}). This condition excludes the nearest large clusters of the Coma type and implies for the brightest member galaxies an apparent magnitude $m_G < 13$ corresponding to the nominal limit of the Harvard census of 1250 galaxies (Shapley and Ames 1932) and to the estimated 50 percent completeness level of the *Reference Catalogue of Bright Galaxies* including twice as many galaxies (G. and A. de Vaucouleurs 1964).

In order to make the census of nearby groups as complete as possible, the following procedure was adopted:

1. A list was prepared of all galaxy groupings previously described in the literature as "groups" or "clusters" and which have at least one member in the *Reference Catalogue.*
2. A finding list of all possible pairs, multiple galaxies, or groups of bright galaxies was extracted from the *Reference Catalogue* by inspection of listings by coordinates and of distribution maps.
3. All recognized large clusters (Coma, Perseus, etc.) were rejected from further discussion.
4. A tabulation of galaxy types, magnitudes, diameters, and velocities was prepared for all other groupings which appeared likely to fall in the range defined above.
5. The relative distance moduli of the nearest groups ($m - M < 30$) were derived from all available secondary distance criteria (luminosity class, brightest stars, H II regions), using the distance moduli of the Local Group members derived from primary criteria (Cepheids, novae, RR Lyrae stars) for absolute calibration, on the distance scale defined by Sandage (1961) and by van den Bergh in chapter 12 of this volume.
6. Using the nearest groups as standards, tertiary distance indicators (luminosity and diameter) were calibrated as a function of galaxy type and rank and used to compute photometric and geometric moduli of all other identified groups and clusters.

This procedure appears to be successful for all groups including S and I systems, but the scarcity of nearby groups of E and L systems prevents a direct calibration of these classes by secondary criteria; the indirect calibration by means of groups including both S, I and E, L systems is somewhat uncertain because of the high degree of segregation (cf. § 7.6).

3. DISTANCE MODULI

Distance moduli μ were derived from primary and secondary indicators for a dozen nearby groups (Sculptor, M81, M51–M101, CVn I, CVn II, M66, M96, NGC 3190, NGC 1023, etc.) including a total of 68 galaxies ($M_G < -16$) with known magnitudes and diameters in the standard systems. From this sample the following relations were derived for the absolute magnitude $M_G(n)$ and linear diameter $D(n)$ of the nth brightest or largest galaxy in a group.

3.1. Magnitudes

$\mu = B(0) - M_G(n)$, where $B(0)$ is the B magnitude in the standard system of the *Reference Catalogue*, and $M_G(n) = M_0 + 0.5n$ $(n \leq 5)$, with

$M_0 = -20.0$ for types E, L, and S0–S7 (i.e., S0/a to Sd),
$= -19.5$ for types S8, S9 (Sdm, Sm)
$= -19.0$ for type I9 (= Im).

The luminosity classes of spirals (van den Bergh 1960*a*) were also used where available with the following calibration (after van den Bergh, but adjusted to fit the latest data on the Local Group):

Class.......	I	I–II	II	II–III	III	III–IV	IV	IV–V	V
$\mathfrak{L}$..........	1	2	3	4	5	6	7	8	9
$-M_G$.......	19.95	19.45	18.95	18.45	18.00	17.55	16.85	15.65	14.00

Here again M_G is on the $B(0)$ system.

N.B. The total (asymptotic) magnitude is about 0.5 mag less than $B(0)$ depending on galaxy type (0.67 at type E, 0.33 at Im).

3.2. Diameters

$\mu = 27.68 + 5[\log D(n) - \log D(0)]$, where $D(0)$ is the "face-on" diameter (in minutes of arc) in the standard system of the *Reference Catalogue* and $\log D(n) = \gamma_0 - 0.10n$ (D in kpc, $n \leq 5$), with

$\gamma_0 =$	1.10	1.15	1.20	1.25	1.30	1.35	1.30	1.20

for

$\mathfrak{T} =$	E	L	S0,1	S2,3	S4,5	S6,7	S8,9	I9.

3.3. Sampling Correction

The groups used in the calibration are all small nearby groups with the following mean luminosity function:

$M_G <$	-19	-18	-17	-16
$\bar{N}(M_G)$	1.3	3.2	4.5	6.4.

In larger groups, a size of sample correction is required since the larger the sample (population of group) the greater the probability of occurence of outstanding objects. The correction was determined through a comparison of the distance moduli derived for the Virgo cluster taken as a whole or divided into smaller groups and subgroups. If N_{18} is the number of galaxies having $M_G < -18$ in a group, the correction to the modulus derived as explained in §§ 3.1 and 3.2 above is

$$\Delta\mu = 2.0 (\log N_{18} - 0.5) .$$

3.4. Absorption Correction

Finally, photometric apparent moduli must be corrected for galactic absorption: $\mu_0 = \mu - A_B$; the correction was computed through the following equation derived from a new analysis of the Mount Wilson (Hubble's) counts (de Vaucouleurs and Malik 1969):

$$A_B = A_1\left[1 + \frac{P}{B_1} \cos l + \frac{Q}{B_1} \cos 3l\right] \operatorname{cosec} |b - b_0(l)| , \quad (1)$$

where

$$b_0(l) = -0\overset{\circ}{.}25 + 1\overset{\circ}{.}7 \sin l + 1\overset{\circ}{.}0 \cos 3l \tag{2}$$

and

	B_1	P	Q
N.G.H.	−0.188	−0.024	−0.004
S.G.H.	−0.204	−0.010	+0.032 .

The value used here for the optical half-thickness of the Galaxy, $A_1 = 0.2$ mag, applies only to a sample selected by apparent brightness; it agrees well with color-excess data (Holmberg 1958; de Vaucouleurs 1961*a*) down to the lowest galactic latitudes.

4. LOCAL GROUP

The Local Group of galaxies is often arbitrarily defined by a radius $\Delta \simeq 1.0$ Mpc from the Galaxy or, possibly, from the center of gravity of the Galaxy and M31, the two giant members. The Local Group includes two average spirals M33 and the Large Magellanic Cloud, over a half-dozen dwarf Magellanic irregulars (Small Cloud, N6822, IC 1613, IC 10, A2359, A0956, A0957, A1009), and at least a dozen dwarf ellipticals (N147, N185, N205, N221, A0058, A0237, A1003, A1006, A1111, A1127, A1719, and other poorly known globular-like systems). Detailed information on the Local Group members is given in tables 1a and 1b.

Additional S, I members hidden by galactic absorption might be discovered by their 21-cm emission. An apparent obscuration patch in Microscopium, suggested as a possible intergalactic dark cloud within the Local Group (Hoffmeister 1962), has not been confirmed.

The distance moduli in table 1b based on primary and secondary criteria (de Vaucouleurs 1955; van den Bergh 1960*c;* Sandage 1961; and chap. 12 by van den Bergh) define the distance scale used in the present chapter.

Figure 1 shows the apparent distribution of Local Group members in supergalactic coordinates. There is only a slight concentration toward the supergalactic equator, as might be expected for very close objects.

Figure 2 is a map of the Local Group projected onto the supergalactic plane; note the strong "subclustering" tendency around two dominant multiplets (Galaxy–LMC–SMC triplet + dE satellites; M31–M33 pair + dE satellites). The overall diameter of the group is about 2.0 × 1.5 Mpc in the SG plane and 1.0 Mpc at right angles to it (0.5 Mpc if A0956, A1009 are not members).

The Local Group is a typical loose group without central condensation. Except for the satellite dE systems it includes only spirals of type Sb and later and Magellanic irregulars. The distribution of *total* absolute magnitudes (assuming $M_B = -18.8$ for the Galaxy, after Gyllenberg 1937) is as follows:

$M_T(B) \leq$		−20		−19		−18		−17		−16		−15		−14
n	1		0		3		1		1		2		4	
N		1		1		4		5		6		8		12

for a total of 12 members brighter than $M_T(B) = -14$; another dozen or more with less precisely known magnitudes are in the range −14 to −10. For further discussion of dwarfs in the Local Group see § 8.

TABLE 1a

LOCAL GROUP

Object (1)	l^{II} b^{II} (2)	SGL SGB (3)	Type $\mathfrak{L}$ (4)	$B(0)$ V_0 (5)	log D log R (6)	log $D(0)$ (′) (7)	$D(0)$ (kpc) (8)
The Galaxy	...	...	Sbc?	...	...	...	(12)
	...	...		...	...		
N147, D3	119.8	344.0	E5p	11.48	0.78	0.70	1.0
	−14.3	15.3	...	...	0.22		
N185	120.8	344.0	E+3p	10.92	0.73	0.70	1.0
	−14.5	14.3	...	−10	0.07		
N205	120.7	337.2	E+5p	9.71	1.00	0.88	1.5
	−21.1	13.1	...	−6	0.29		
N221	121.2	336.5	E2	9.39	0.56	0.51	0.6
	−22.0	12.5	...	+17	0.12		
N224, M31	121.2	336.9	SA(s)b	4.61	2.20	2.00	20.0
	−21.6	12.6	2	−68	0.49		
N598, M33	133.6	329.2	SA(s)cd	6.47	1.79	1.70	10.5
	−31.3	−0.1	4	−11	0.21		
N6822, D209	25.4	229.8	IB(s)m	9.49	1.22	1.16	2.1
	−18.4	57.1	8	+73	0.13		
I10	119.0	355.1	SB(s)m	12.5:	0.60	0.54	1.3
	−3.3	17.9	7	−92	0.15		
I1613, D8	129.9	299.9	Im	11.02	1.05	1.02	2.0
	−60.6	−1.8	...	−129	0.07		
SMC, A0051	302.8	224.9	IB(s)m	3.1:	2.4:	2.3:	2.9
	−44.3	−14.8	...	−13	0.30		
Scl, A0058	287.8	264.7	dE	...	...	...	...
	−83.2	−9.6	...	...	...		
For, A0237	237.3	266.0	dE	...	...	...	...
	−65.7	−30.2	...	−70	...		
LMC, A0524	280.5	216.5	SB(s)m	1.2:	2.7:	2.65:	6.5
	−32.9	−34.1	...	+16	0.07		
Leo A, A0956, D69	196.9	70.6	Im	...	...	...	...
	52.4	−25.8	...	...	...		
Sex B, A0957, D70	233.2	96.1	Im	...	...	...	...
	43.8	−39.6	...	...	...		
Sex C, A1003	240.1	103.2	dE	...	...	...	...
	41.9	−40.4	...	...	...		
Leo I, A1006, D74	226.0	89.6	E4	...	...	...	...
	49.1	−34.6	...	...	...		
Sex A, A1009, D75	246.2	109.8	IBm	...	0.73	0.70	1.5
	39.9	−40.6	...	+118	0.09		
Leo II, A1111, D93	220.1	87.8	dE	...	...	...	...
	67.2	−16.3	...	...	...		
UMa, A1127	202.3	83.0	dE	...	...	...	...
	71.8	−10.2	...	...	...		
UMi, A1508, D199	105.1	48.4	dE	...	...	...	...
	44.8	27.1	...	...	...		
Ser, A1513	0.9	128.1	dE	...	...	...	...
	45.9	33.7	...	...	...		
Dra, A1719, D208	86.4	44.5	dE	...	...	...	...
	34.7	44.2	...	...	...		
Cap, A2144*	30.5	257.1	dE	12.37	0.68	0.68	...
	−47.7	34.0	...	...	0.00		
Peg, A2304	87.1	303.2	dE	...	...	...	...
	−42.7	29.1	...	...	...		
WLM, A2359, D221	75.7	278.5	Im	...	0.8:	0.7:	1.3
	−73.6	8.1	...	+2	0.3:		

EXPLANATIONS OF COLUMNS.—Col. (1), identification in NGC, IC, BGC, DDO, or Anon. Col. (2), new galactic coordinates l^{II}, b^{II}. Col. (3), supergalactic coordinates SGL, SGB (cf. BGC). Col. (4), revised type and DDO luminosity class coded as in BGC. Col. (5), $B(0)=B$ mag within standard "face-on" diameter $D(0)$ (cf. BGC). V_0 = corrected redshift. Col. (6), log D = log major diameter in standard system (D in min of arc). log R = log D/d = log axis ratio. Col. (7), log $D(0)$ = log "face-on" major diameter (min of arc). Col. (8), $D(0)$ = "face-on" diameter in kpc.

* May be a globular cluster.

TABLE 1b

LOCAL GROUP

Object (1)	μ (2)	A_B (3)	μ_0 (4)	Δ (5)	B_T (6)	$-M_T$ (7)	X (8)	Y (9)	Z (10)
Galaxy...	....	....	....	0.01	....	(18.8)	0	0	0
N147....	25.0:	0.8	24.2:	0.69	10.6	14.4	+0.64	−0.18	+0.18
N185....	25.0:	0.8	24.2:	0.69	10.3	14.7	+0.64	−0.18	+0.17
N205....	24.7	0.5	24.2	0.69	8.9	15.8	+0.62	−0.26	+0.16
N221....	24.7	0.5	24.2	0.69	9.1	15.6	+0.62	−0.27	+0.15
N224....	24.7	0.5	24.2	0.69	4.4	20.3	+0.62	−0.26	+0.15
N598....	24.6	0.3	24.3:	0.72	6.3	18.3	+0.62	−0.37	0.00
N6822...	24.1	0.6	23.5	0.50	9.3	14.8	−0.18	−0.21	+0.42
I10......	29.0	3.5:	25.5:	1.26	11.7	17.3	+1.20	−0.10	+0.39
I1613....	24.3	0.2	24.1	0.66	10.1	14.2	+0.33	−0.57	−0.02
SMC....	18.8	0.3	18.5	0.05	2.8	16.0	−0.03	−0.03	−0.01
Scl......	20.4:	0.2	20.2:	0.11	9.2	11.2	−0.01	−0.11	−0.02
For......	22.0:	0.2	21.8:	0.23	9.1	12.9	−0.01	−0.20	−0.12
LMC....	18.7	0.4	18.3	0.05	0.6	18.1	−0.03	−0.02	−0.03
Leo A....	25.4:	0.2	25.2:	1.10	13.1	12.3	+0.33	+0.93	−0.48
Sex B....	....	0.3	....	....	12.0	....	....	....	....
Sex C....	21.0:	0.3	20.7	0.14	....	....	−0.02	+0.10	−0.09
Leo I....	22.0:	0.2	21.8:	0.23	11.3	10.7	0.00	+0.19	−0.13
Sex A....	25.3:	0.3	25.0:	1.00	11.7	13.6	−0.26	+0.71	−0.63
Leo II....	22.0:	0.2	21.8:	0.23	12.9	9.1	+0.01	+0.22	−0.06
UMa....	20.5:	0.2	20.3:	0.12	....	....	+0.01	+0.12	−0.02
UMi.....	19.6:	0.3	19.3:	0.08	....	....	+0.05	+0.05	+0.04
Ser......	....	0.3	....	....	....	....	....	....	....
Dra......	19.4:	0.4	19.0	0.06	....	....	+0.03	+0.03	+0.04
Cap......	....	0.3	....	....	....	....	....	....	....
Peg......	21.5:	0.3	21.2:	0.17	....	....	+0.08	−0.12	+0.08
WLM....	24.9:	0.2	24.7:	0.87	11.2	13.7	+0.13	−0.85	+0.12

EXPLANATIONS OF COLUMNS.—Col. (1), identification. Col. (2), apparent modulus μ in B system. Col. (3), galactic absorption A_B in B system. Col. (4), corrected (geometric) apparent modulus μ_0. Col. (5), distance Δ in megaparsecs. Col. (6), apparent total (asymptotic) magnitude B_T in B system. Col. (7), absolute magnitude $-M_T$ in B system. Cols. (8), (9), (10), projection of Δ (Mpc) on OX, OY, OZ axes of supergalactic rectangular coordinate system.

The total mass (of galaxies) in the group $\mathfrak{M}_T \simeq 6.5 \times 10^{11}\, \mathfrak{M}_\odot$ is not much larger than the combined masses of M31 and the Galaxy ($5 \times 10^{11}\, \mathfrak{M}_\odot$); M33 and the Magellanic irregulars LMC, SMC, N6822, IC 10, etc., add only 30 percent, and the combined mass of the ellipticals is probably negligible ($<10^{10}\, \mathfrak{M}_\odot$) even if their space density is as high as 100 per Mpc^3. For a volume of the order of 2 Mpc^3, the smoothed mean density is $\bar{\rho}_G \simeq 0.5 \times 10^{-28}$ g cm^{-3}. The total absolute magnitude of the group is $M_T \simeq -20.9$ (B) or -21.6 (V) with some uncertainty due to the indirect estimate for our Galaxy; the average mass-luminosity ratio is $f = \mathfrak{M}_T/\mathfrak{L} \simeq 20$ (B and V). The total mass of neutral hydrogen in galaxies is $\mathfrak{M}_H \simeq 1.0 \times 10^{10}\, \mathfrak{M}_\odot$, and the ratios of hydrogen to total mass and of hydrogen to luminosity are $h = \mathfrak{M}_H/\mathfrak{M}_T \simeq 0.01_5$ and $g \simeq \mathfrak{M}_H/\mathfrak{L}_B \simeq 0.3$.

5. THE NEARER GROUPS WITHIN 10 MEGAPARSECS

The nearer groups are the groups whose distance moduli $\mu_0 \leq 30$ ($\Delta \leq 10$ Mpc) can be derived from secondary distance criteria (H II regions, bright stars, luminosity class). These groups in turn serve to calibrate the tertiary distance criteria (magnitudes,

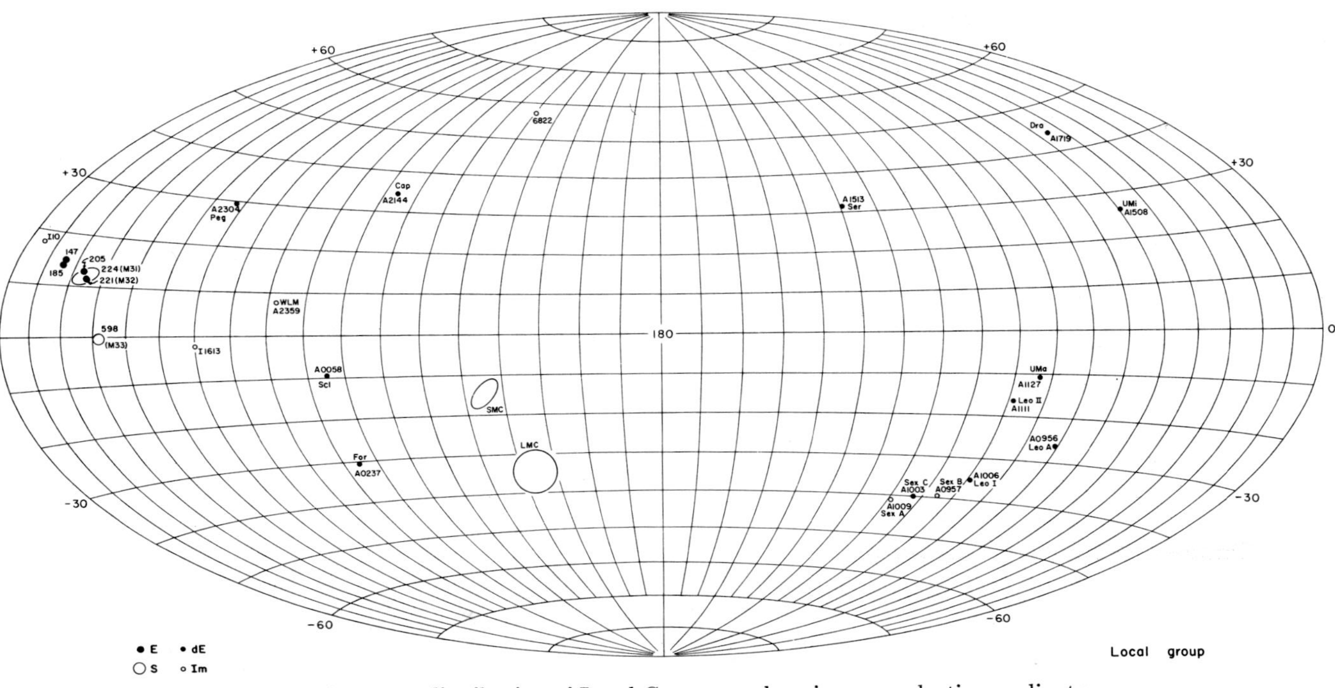

FIG. 1.—Apparent distribution of Local Group members in supergalactic coordinates.

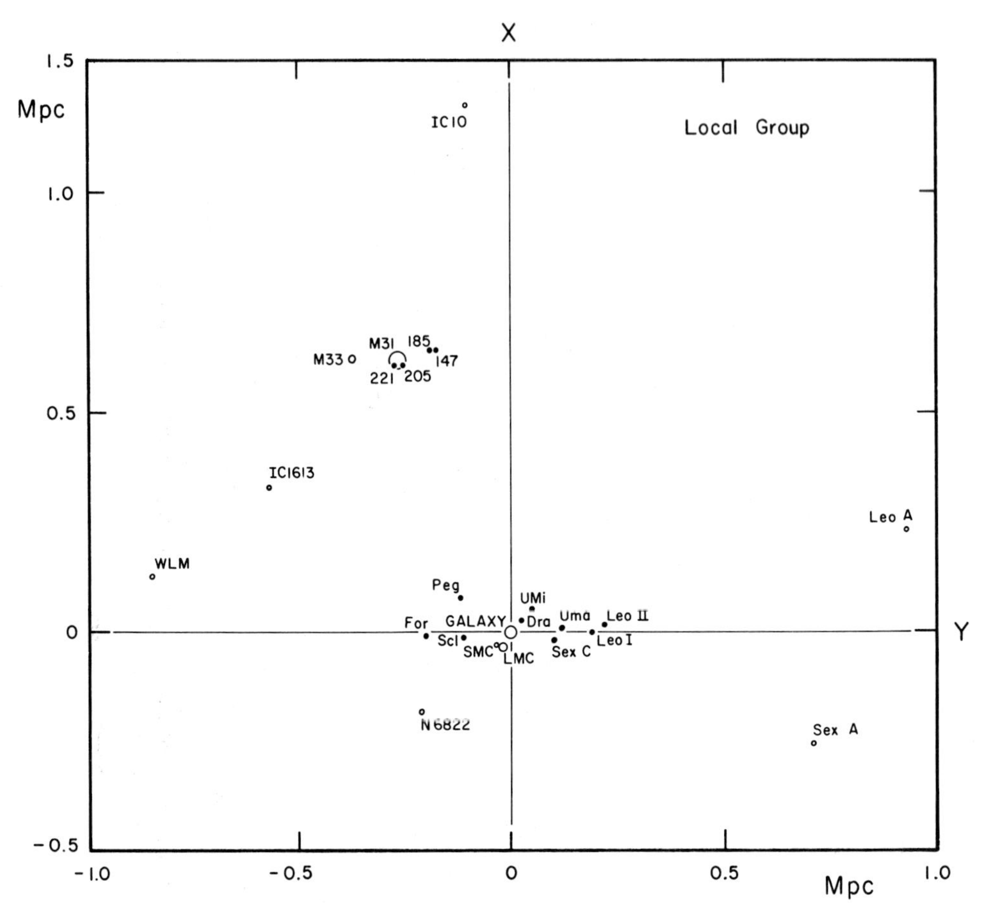

FIG. 2.—Map of Local Group projected onto supergalactic plane.

diameters) which are used to derive the distances of more distant groups or of southern hemisphere groups for which the secondary criteria are still missing. The apparent distribution of the nearer groups in supergalactic coordinates is shown in figure 3. Available data on each group taken mainly from the *Reference Catalogue of Bright Galaxies* (abbreviated BGC) are presented in tabular form (tables 2 and 3). A brief description of individual groups follows.

G1. *Sculptor Group* ($\mu_0 = 26.9$, $\Delta = 2.4$, $V_0 = 220$). This nearest of all nearby groups is a loose association of six or seven late-type spirals Sc to Sm (NGC 45, 55, 247, 253, 300, 7793, and perhaps IC 5332) distributed at the rim of a 20° ring near the south galactic pole (de Vaucouleurs 1956*a*, 1959).

The fainter, smaller member NGC 45 also has the largest velocity, and it may be on the far side of the group. If the group had a depth along the line of sight equal to its apparent diameter, the individual distance moduli could have a range of up to $\Delta\mu = 0.7$ mag from, say, 26.7 to 27.4. However, the nearly circular outline suggests a flat structure

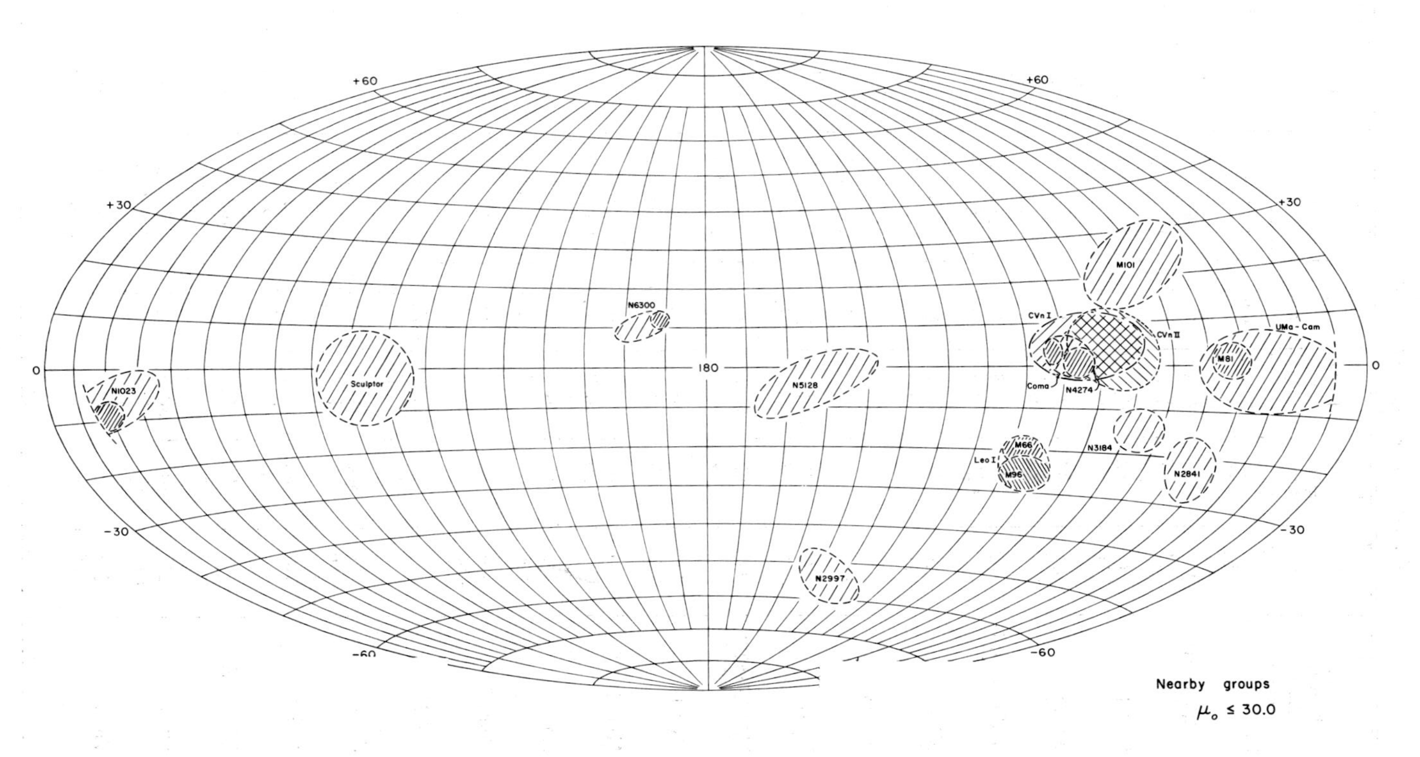

FIG. 3.—Apparent distribution of nearer group ($\mu_0 \leq 30.0$) in supergalactic coordinates.

TABLE 2

ELEMENTS OF 54 NEARBY GROUPS

No.	Group, Cluster, Cloud	l^{II} b^{II} (°)	SGL SGB (°)	μ A_B (mag)	μ_0 Δ (Mpc)	$D\times d$ (°) (Mpc)	V_0 n A.D.	X Y Z
G1......	Scl	5 −80	265 −3	27.1 0.2	26.9 2.4	25×20 1.0×0.8	220 6 102	+ 0.1 − 2.4 − 0.2
G2......	M81 (N3031)	142 +41	42 +1	27.3 0.3	27.0 2.5	40×20 1.8×0.9	160 9 66	+ 1.8 + 1.7 + 0.0
G3......	CVn I	162 +80	82 +5	28.1 0.2	27.9 3.8	28×14 1.9×0.9	342 9 69	+ 0.5 + 3.8 + 0.2
G4......	N5128	310 +20	155 −5	28.5 0.5	28.0 4.0	30×... 2.1×...	319 5 55	− 3.8 + 1.3 − 0.4
G5......	M101 (N5457)	102 +60	64 +23	28.5 0.2	28.3 4.6	23×16 1.8×1.3	508 8 85	+ 1.8 + 3.8 + 1.7
G6......	N2841	170 +34	45 −20	29.2 0.3	28.9 6.0	15×7 1.6×0.8	589 4 122	+ 3.7 + 4.4 − 1.6
G7......	N1023	145 −20	341 −8	29.6 0.6	29.0 6.3	20×10 2.2×1.1	625 8 102	+ 5.9 − 1.9 − 1.0
G8......	N2997	250 +19	133 −53	30.0 0.6	29.4 7.6	14×8 1.9×1.1	534 2	− 3.0 + 3.1 − 6.3
G9......	M66 (N3627)	241 +64	97 −19	29.6 0.2	29.4 7.6	7×4 1.0×0.6	592 5 74	− 0.9 + 7.1 − 2.4
G10.....	CVn II	138 +75	73 +3	29.7 0.2	29.5 8.0	22×12 3.0×1.6	747 15 71	+ 2.4 + 7.7 + 0.3
G11.....	M96 (N3368)	231 +58	93 −26	29.8 0.2	29.6 8.3	11×7 1.6×1.0	741 9 80	− 0.7 + 7.4 − 3.6
G12.....	N3184	176 +60	67 −13	30.1 0.2	29.9 9.6	10×5 1.7×0.8	629 4 122	+ 4.0 + 8.4 − 2.7
G13.....	Coma I	198 +86	89 +3	30.1 0.2	29.9 9.6	11×5 1.8×0.8	944 15 170	+ 0.1 + 9.5 + 0.5
G14.....	N6300	329 −15	195 +10	31.0 1.0	30.0 10.0	10×3 1.7×0.6	1270 3 100	− 9.5 − 2.7 + 1.6
G15.....	Cet I	178 −56	300 −27	30.5 0.3	30.2 11.0	12×9 2.2×1.7	1513 7 200	+ 4.7 − 8.6 − 5.0
G16.....	N1566	265 −43	235 −40	30.6 0.3	30.3 11.5	5×5 1.0×1.0	999 5 156	− 4.9 − 7.2 − 7.5
G17.....	UMa I (Z)	165 +75	77 −1	30.5 0.2	30.3 11.5	10×4 1.8×0.7	979 2	+ 2.3 +10.2 − 0.2
G18.....	Vir S	284 +75	103 −3	30.5 0.2	30.3 11.5	12×12 2.3×2.3	1087 21 780	− 2.7 +11.2 − 0.6
G19.....	Vir E	284 +75	103 −3	30.5 0.2	30.3 11.5	12×12 2.5×2.5	1013 38 413	− 2.7 +11.2 − 0.5

TABLE 2—*Continued*

No.	Group, Cluster, Cloud	l^{II} b^{II} (°)	SGL SGB (°)	μ A_B (mag)	μ_0 Δ (Mpc)	$D \times d$ (°) (Mpc)	V_0 n A.D.	X Y Z
G20.....	Vir Y	304 +52	127 −2	30.7 0.3	30.4 12.0	14×7 2.9×1.4	1307 13 322	− 7.2 + 9.6 − 0.6
G21.....	N1433	253 −49	248 −42	30.7 0.2	30.5 12.5	7×5 1.5×1.1	665 2	− 3.6 − 9.0 − 7.9
G22.....	N1672	270 −38	226 −39	30.8 0.3	30.5 12.5	5×? 1.0×?	?	− 6.3 − 7.1 − 8.1
G23.....	N3672	273 +49	121 −22	30.8 0.3	30.5 12.5	8×6 1.8×1.3	1583 2	− 5.9 + 9.9 − 5.2
G24.....	UMa I (Y)	162 +64	66 −5	30.7 0.2	30.5 12.5	9×4 20.×0.8	998 1	+ 4.9 +11.5 − 1.0
G25.....	Vir S′	287 +70	107 −4	30.7 0.2	30.5 12.5	3×2 0.6×0.4	1541 3 277	− 3.6 +11.9 − 0.8
G26.....	Vir X	293 +65	113 −3	30.7 0.2	30.5 12.5	13×7 2.8×1.5	1272 18 365	− 5.0 +11.4 − 0.5
G27.....	Grus	348 −65	250 +8	30.8 0.2	30.6 13.2	8×6 1.9×1.5	1561 8 112	− 4.4 −12.0 + 2.6
G28.....	UMa I (X)	150 +54	56 −2	30.8 0.2	30.6 13.2	7×5 1.6×1.1	1090 1	+ 7.5 +10.8 − 0.5
G29.....	Vir III	352 +55	122 +24	30.9 0.3	30.6 13.2	15×11 3.5×2.5	1729 11 149	− 6.1 +10.4 + 5.3
G30.....	N5866	90 +52	62 +32	31.0 0.3	30.7 13.8	6×3 1.4×0.7	920 3 130	+ 5.8 +10.3 + 7.2
G31.....	Eri	212 −55	282 −40	30.9 0.2	30.7 13.8	15×12 3.6×2.9	1574 19 160	+ 2.5 −10.8 − 8.2
G32.....	UMa I (S)	143 +69	70 +3	30.9 0.2	30.7 13.8	9×7 2.2×1.7	1016 5 63	+ 4.8 +12.9 + 0.9
G33.....	Cet II	160 −68	292 −15	31.0 0.2	30.8 14.5	9×8 2.3×2.0	1929 6 95	+ 5.1 −13.0 − 4.0
G34.....	UMa I (N)	141 +60	61 +3	31.0 0.2	30.8 14.5	8×5 2.0×1.3	1074 8 95	+ 7.1 +12.6 + 0.5
G35.....	Vir V	298 +57	121 −6	31.0 0.2	30.8 14.5	8×5 2.0×1.2	879 1	− 7.1 +12.6 − 1.5
G36.....	N2207	230 −18	264 −78	31.4 0.5	30.9 15.	6×5 1.5×1.2	1827 3 419	− 0.3 − 2.8 −14.7
G37.....	N5676	88 +60	70 +28	31.1 0.2	30.9 15.	6×2 1.5×0.5	2411 3 49	+ 4.7 +12.4 + 7.0
G38.....	N6876	324 −32	214 +3	31.3 0.4	30.9 15.	6×2 1.5×0.5	?	−12.5 − 8.3 + 0.6
G39.....	N134	340 −84	265 −4	31.2 0.2	31.0 16.	6×5 1.6×1.4	1786 2	− 1.8 −15.9 − 1.0

TABLE 2—*Continued*

No.	Group, Cluster, Cloud	l^{II} b^{II} (°)	SGL SGB (°)	μ A_B (mag)	μ_0 Δ (Mpc)	$D \times d$ (°) (Mpc)	V_0 n A.D.	X Y Z
G40.....	N488	138 −58	302 −6	31.2 0.2	31.0 16.	4×2 1.0×0.5	2336 3 46	+ 8.9 −13.2 − 1.4
G41.....	N2768	154 +40	43 −8	31.3 0.3	31.0 16.	7×4 1.9×1.1	1681 7 160	+11.3 +10.8 − 2.6
G42.....	N2964	194 +50	66 −27	31.2 0.2	31.0 16.	9×4 2.5×1.1	1463 5 89	+ 5.5 +13.1 − 7.4
G43.....	N3396	193 +63	75 −16	31.2 0.2	31.0 16.	1×1 0.3×0.3	1599 4 73	+ 3.9 +14.9 − 4.3
G44.....	N3923	287 +33	142 −20	31.3 0.3	31.0 16.	6×3 1.6×0.8	1635 4 172	−11.7 + 9.0 − 5.7
G45.....	Pavo-Indus	344 −45	230 +14	31.3 0.3	31.0 16.	13×8 3.6×2.2	2224 9 472	− 9.9 −11.7 + 3.8
G46.....	Vir W	283 +69	108 −6	31.2 0.2	31.0 16.	3×2.5 0.8×0.7	2168 5 187	− 5.1 +14.9 − 1.9
G47.....	N3190	213 +55	82 −28	31.3 0.2	31.1 16.5	4×3 1.2×0.9	1198 7 94	+ 2.1 +14.5 − 7.7
G48.....	N3504	204 +64	80 −17	31.3 0.2	31.1 16.5	5×2 1.4×0.5	1437 3 30	+ 2.4 +15.7 − 4.5
G49.....	N3607	232 +67	93 −17	31.3 0.2	31.1 16.5	4×2 1.1×0.5	1057 7 229	− 0.8 +15.4 − 4.6
G50.....	N5846	0 +48	125 +32	31.4 0.3	31.1 16.5	2.5×2.5 0.7×0.7	1806 8 290	− 8.1 +11.4 + 8.7
G51.....	N6643	105 +28	31 +31	31.5 0.4	31.1 16.5	7×3.5 2.0×1.0	1938 5 262	+12.3 + 7.0 + 8.5
G52.....	N6861	355 −33	223 +25	31.4 0.3	31.1 16.5	7×4 2.0×1.2	2909 2	−11.5 − 9.7 + 6.7
G53.....	For I	237 −55	263 −42	31.3 0.2	31.1 16.5	7×7 2.0×2.0	1464 12 260	− 1.5 −12.2 −10.1
G54.....	N3245	202 +58	71 −22	31.4 0.2	31.2 17.5	2×1 0.5×0.2	1256 3 96	+ 3.9 +15.8 − 6.7

EXPLANATION OF COLUMNS.—Cols. (1) and (2), group number and identification. Col. (3), approximate galactic coordinates of center. Col. (4), approximate supergalactic coordinates of center. Col. (5), apparent distance modulus μ and galactic absorption A_B in B system. Col. (6), geometric distance modulus μ_0 and distance Δ in megaparsecs. Col. (7), approximate major and minor diameters $D \times d$ of group in degrees and in megaparsecs. Col. (8), mean corrected velocity V_0, number n of objects in mean, and average deviation A.D. from V_0. Col. (9), components X, Y, Z of distance Δ in Mpc in supergalactic rectangular coordinate system.

TABLE 3—BRIGHTEST MEMBERS OF THE 14 NEARER GROUPS

G1. Sculptor Group

NGC (1)	Type (2)	$\mathfrak{L}$ (3)	$B(0)$ (4)	log $D(0)$ (5)	V_0 (6)	$-M_G$ (7)	log $D(0)$ (8)
253.....	SAB(s)c	....	8.1	1.15	258	19.0	1.04
55.....	SB(s)m sp	....	8.2	1.16	97	18.9	1.05
300.....	SA(s)d	7	9.0	1.24	95	18.1	1.13
247.....	SAB(s)d	7	9.9	1.14	183	17.2	1.03
7793.....	SA(s)dm	6½	10.3	0.85	197	16.8	0.74
45*....	SA(s)dm	8	11.2	0.87	489	15.9	0.76

Possible: I5332 + D1, D6 dwarfs?
* Pair with N24 in background?

G2. M81 (N3031) Group

NGC (1)	Type (2)	$\mathfrak{L}$ (3)	$B(0)$ (4)	log $D(0)$ (5)	V_0 (6)	$-M_G$ (7)	log $D(0)$ (8)
3031.....	SA(s)ab	2	7.88	1.28	88	19.42	1.30
2403.....	SAB(s)cd	5	9.07	1.11	255	18.23	1.03
3034.....	I0 sp	....	9.57	0.74	322	17.73	0.66
4236.....	SB(s)dm	7	10.40	1.04	(186)	16.90	0.96
2976.....	SA(s)m p	....	11.09	0.50	169	16.21	0.42

Probable: 3077, I2574 = D81, A0814 (Ho II) = D50, A0936 (Ho I) = D63.
Possible: 2366 = D42, 2403, I342, N1569, 1560?, I356? (in UMa–Cam cloud).
Other possible (dwarfs): D33, 38, 39, 44, 51, 53, 66, 71, 77, 78, 80, 82, 86, 87, 122, 123, 165.

G3. CVn I Cloud

NGC (1)	Type (2)	$\mathfrak{L}$ (3)	$B(0)$ (4)	log $D(0)$ (5)	V_0 (6)	$-M_G$ (7)	log $D(0)$ (8)
4736.....	(R)SA(r)ab	3:	8.91	0.83	362	19.19	0.91
4258.....	SAB(s)bc	....	9.19	1.11	530	18.91	1.19
4826.....	(R)SA(rs)ab	....	9.60	0.77	352	18.50	0.85
4449.....	IB m	5	10.08	0.59	269	18.02	0.67
4214.....	IAB(s)m	6	10.38	0.79	311	17.72	0.87

Probable: 4136, 4150, 4244, 4395, I4182, A1157 = D105.
Other possible (dwarfs): D99, 125, 126, 129, 133, 141, 143, 154.

G4. N5128 Group

NGC (1)	Type (2)	$\mathfrak{L}$ (3)	$B(0)$ (4)	log $D(0)$ (5)	V_0 (6)	$-M_G$ (7)	log $D(0)$ (8)
5236.....	SAB(s)c	....	8.22	1.00	335	20.16	1.15
5128.....	S0p	....	8.38	1.01	271	20.18	1.19
4945.....	SB(s)cd:	....	9.48	0.90	....	19.30	1.12
5102.....	SA0⁻	....	10.55	0.70	348	17.90	0.85
5068.....	SAB(rs)cd	6	11.05	0.78	410	17.26	0.90
5253.....	I0p	....	11.14	0.38	229	17.26	0.53

Possible dwarf: D174.

G5. M101 (N5457) Group*

NGC (1)	Type (2)	$\mathfrak{L}$ (3)	$B(0)$ (4)	log $D(0)$ (5)	V_0 (6)	$-M_G$ (7)	log $D(0)$ (8)
5457.....	SAB(rs)cd	1	8.58	1.39	415	19.92	1.55
5194.....	SA(s)bc p	1	9.03	0.95	552	19.47	1.11
5055.....	SA(rs)bc	3	9.52	0.92	600	18.98	1.08
5195.....	I0p	....	10.94	0.45	634	17.56	0.61
5585.....	SAB(s)d	7	11.66	0.64	467	16.84	0.80

Other members: 5204, 5474; probable: 5959.
Possible: 4605, 5907, 6503(?), A1353 (Ho IV), A1339 (Ho V).
Possible dwarfs: D167, 168, 169, 172, 175, 176, 177, 178, 181, 182, 183, 191, 193, 194, 205.
* May consist of 2 separate multiplets centered on M51 and M101.

TABLE 3—*Continued*

NGC (1)	Type (2)	$\mathcal{L}$ (3)	$B(0)$ (4)	log $D(0)$ (5)	V_0 (6)	$-M_G$ (7)	log $D(0)$ (8)

G6. N2841 Group

NGC (1)	Type (2)	$\mathcal{L}$ (3)	$B(0)$ (4)	log $D(0)$ (5)	V_0 (6)	$-M_G$ (7)	log $D(0)$ (8)
2841	SA(r)b	1	10.27	0.72	671	18.93	1.02
2681	SAB(rs)0/a	...	11.34	0.41	751	17.86	0.71
2541	SA(s)cd	7	12.14	0.68	634	17.06	0.98
2500	SB(rs)d	7	12.39	0.40	513	16.81	0.70
2552	SA(s)m?	8	(12.54)	0.39	...	16.66	0.69
2537	IB(s)m p	...	12.55	0.14	422	16.65	0.44

Possible: I2233 (Sd sp).
Possible (dwarfs): D40, 41, 43, 46, 48, 49, 52, 55, 59.

G7. N1023 Group

NGC (1)	Type (2)	$\mathcal{L}$ (3)	$B(0)$ (4)	log $D(0)$ (5)	V_0 (6)	$-M_G$ (7)	log $D(0)$ (8)
1023	SB(rs)0$^-$	...	10.65	0.55	729	18.95	0.93
925	SAB(s)d	4	10.96	0.83	718	18.64	1.21
891	SA(s)b:/	...	11.24	0.78	721	18.36	1.16
I239	SAB(rs)cd	...	12.14	0.61	...	17.46	0.99
1058	SA(rs)c	6:	12.26	0.38	583	17.34	0.76
1003	SA(s)cd	...	12.45	0.48	741	17.15	0.96

Possible: N672, I1727, N1156(?).
Probable (dwarfs): D24, 25.
Possible (dwarfs): D11, 17, 19, 22, 26.

G8. N2997 Group

NGC (1)	Type (2)	$\mathcal{L}$ (3)	$B(0)$ (4)	log $D(0)$ (5)	V_0 (6)	$-M_G$ (7)	log $D(0)$ (8)
2997	SAB(rs)c	...	(10.64)	0.80	765	19.36	1.942
2835	SB(rs)c	...	(11.38)	0.71	636	18.62	1.17
2784	SA(s)0^0:	...	11.52	0.37	431	18.48	0.83
2848	SAB(s)c	7:	(12.55)	0.40	...	17.45	0.86
2763	SB(r)cd p	7:	12.91	0.31	...	17.09	0.77

Possible (dwarfs): D56, 57, 60, 61, 62.

G9. M66 (N3627) Group

NGC (1)	Type (2)	$\mathcal{L}$ (3)	$B(0)$ (4)	log $D(0)$ (5)	V_0 (6)	$-M_G$ (7)	log $D(0)$ (8)
3627	SAB(s)b	3:	9.89	0.76	591	19.71	1.44
3628	Sbp	...	10.43	0.87	730	19.17	1.25
3623	SAB(rs)a	3:	10.51	0.70	640	19.09	1.08
3489	SAB(rs)0$^+$	...	11.24	0.29	570	18.36	0.67
3593	SA(s)0/a	5	11.91	0.47	429	17.69	0.85

Probable: 3596, 3666.
Possible: 3485, 3506, 3547; dwarfs: D89, 91, 108.

G10. CVn II Cloud

NGC (1)	Type (2)	$\mathcal{L}$ (3)	$B(0)$ (4)	log $D(0)$ (5)	V_0 (6)	$-M_G$ (7)	log $D(0)$ (8)
4631	SB(s)d	5:	10.04	0.87	646	19.66	1.27
4490	SB(s)d p	5	10.29	0.61	622	19.41	1.01
3675	SA(s)b	3	11.11	0.54	727	18.59	0.94
4656	SB(s)m p	...	11.18	0.76	775	18.52	1.16
4051	SAB(rs)bc	3	11.23	0.60	698	18.47	1.00

Other probable: 3769, 3769 A, 3949, 4088, 4111, 4143, 4242, 4485, 4618, 4625, 4627, 4657.
Possible: 4025, 4288.

TABLE 3—*Continued*

NGC (1)	Type (2)	$\mathfrak{L}$ (3)	$B(0)$ (4)	log $D(0)$ (5)	V_0 (6)	$-M_G$ (7)	log $D(0)$ (8)
			G11. M96 (N3368) Group				
3368.....	SAB(rs)ab	....	10.32	0.67	800	19.48	1.09
3351.....	SB(r)b	3	10.75	0.74	643	19.05	1.16
3379.....	E$^+$1	....	10.83	0.36	746	18.97	0.78
3384.....	SB(s)0$^-$	....	10.84	0.41	636	18.96	0.83
3377.....	E5–6	....	11.75	0.30	593	18.05	0.72

Other members: 3239, 3377A, 3412, 3447, 3447A.
Probable: 3299, 3300, 3306, 3346, 3357, 3419A.
Possible: 3433, 3444, 3466, 3506.
Possible dwarfs: D79, 88, 89, 90.

NGC (1)	Type (2)	$\mathfrak{L}$ (3)	$B(0)$ (4)	log $D(0)$ (5)	V_0 (6)	$-M_G$ (7)	log $D(0)$ (8)
			G12. N3184 Group				
3184.....	SAB(rs)cd	3	10.59	0.80	418	19.51	1.29
3198.....	SB(rs)c	3	11.09	0.80	670	19.01	1.29
3432.....	SB(s)m sp	....	11.94	0.50	594	18.16	0.99
3319.....	SB(rs)cd	3	11.95	0.69	832	18.15	1.18
			G13. Coma I Cloud				
4725.....	SAB(r)abp	1	10.21	0.89	1109	19.89	1.37
4559.....	SAB(rs)cd	4	10.56	0.82	852	19.54	1.30
4565.....	SA(s)b?	1:	10.61	0.84	1171	19.49	1.32
4414.....	SA(rs)c:	3:	11.21	0.41	720	18.89	0.89
4494.....	E1–2	....	11.31	0.28	1305	18.79	0.77

Other members: 4203, 4245, 4251, 4274, 4278, 4283, 4314, 4448, 4670, A1244.
Possible: 4062, 4146, 4286, 4359, 4375, I3330.
Possible (dwarfs): D101, 117, 131, 133, 143, 154.

NGC (1)	Type (2)	$\mathfrak{L}$ (3)	$B(0)$ (4)	log $D(0)$ (5)	V_0 (6)	$-M_G$ (7)	log $D(0)$ (8)
			G14. N6300 Group				
6300.....	SB(r)b	....	11.54	0.49	1120	19.26	1.11
6221.....	SB(s)c	....	11.75	0.42	1281	19.40	1.11
6215.....	SA(s)c	....	12.06	0.24	1410	19.09	0.83
I4662 A..	S	....	....	0.14	....	....	0.72
6215 A...	S:	....	....	0.00	....	....	0.69

Possible: I4710, I4713, I4714.

EXPLANATION OF COLUMNS.—Col. (1), NGC or other identification. Col. (2), revised type. Col. (3), DDO luminosity class. Col. (4), standard "face-on" magnitude $B(0)$. Col. (5), standard "face-on" diameter $D(0)$ (in arc min). Col. (6), corrected redshift V_0 (km s^{-1}). Col. (7), absolute magnitude corresponding to $B(0)$. Col. (8), log linear diameter corresponding to $D(0)$ (in kiloparsecs).

seen face-on. The group has not merely no central concentration, but actually an empty region in its center. This peculiar structure is apparent in several other loose groups of spirals. It suggests the possibility that such groups are the galactic equivalent of old expanding stellar associations (de Vaucouleurs 1959).

The maximum dimensions of the group, $25° \times 20° = 1.0 \times 0.8$ Mpc (or $20° \times 20°$ if IC 5332 is excluded), are of the same order as the Local Group limited to its brighter members. The velocity range is about 400 km s^{-1} (+95 to +489), or 160 (+95 to +258) if N45 is excluded. Detailed photometry, continuum, and H I radio data, optical and radio rotation curves, and mass estimates are available for several of the brighter members (de Vaucouleurs 1961*b;* de Vaucouleurs and Page 1962; Epstein 1964; Robinson and van Damme 1964; and chap. 3 of this volume). An H I cloud sharing the velocity of NGC 300 has been reported by Shobbrook and Robinson (1967) in a position several degrees from the galaxy where no optical luminosity can be seen.

G2. M81 *Group* ($\mu_0 = 27.0$, $\Delta = 2.5$, $V_0 = 160$). This well-known group centered on the M81–M82 pair in Ursa Major includes several late-type spirals and dwarf irregular satellites: NGC 2976, NGC 3077, IC 2574 (= DDO 81), A0936 (Ho I = DDO 63), and A0814 (Ho II = DDO 50) (Holmberg 1950). Several large nearby spirals including NGC 2366, 2403, and possibly 4236 have comparably low velocities and may be outlying members of the group. NGC 2403, NGC 2366, A0814, and two or three other dwarfs (DDO 44, 51, 53) may in fact form a subgroup in a larger cloud extending into Camelopardalis and including also the low-latitude, obscured group formed by NGC 1560, NGC 1569, IC 342, IC 356, Maffei 1 and 2,[1] and possibly several dwarfs (DDO 33, 38, 39). This cloud may be interrupted by the galactic absorption belt. The overall dimensions of the cloud $40° \times 20° = 1.8 \times 0.9$ Mpc are typical of these formations (compare UMa I, Virgo II, Leo II, etc.); the restricted M81 group is only $13° \times 7° = 0.6 \times 0.3$ Mpc, or somewhat smaller than the Local Group.

Because of the large range of galactic latitudes covered by the UMa-Cam Cloud the apparent modulus varies from 27.3 ($A = 0.3$ mag) for M81 to 28.3 ($A = 1.2$ mag) for IC 342. The velocity range is about 350 km s^{-1} (−26 to +322).

The M81 group includes two examples of the rare I0 (or Irr II) galaxy type, NGC 3034 and NGC 3077. Both display well-known optical and/or radio peculiarities (see BGC for references). Radio continuum and H I emission, optical and radio rotation curves, and mass estimates are available for several of the brighter members (Epstein 1964, Heeschen and Wade 1964). (See chap. 3 of this volume by E. and G. Burbidge for detailed references on rotation curves.)

G3. *Canes Venatici I Cloud* ($\mu_0 = 27.9$, $\Delta = 3.8$, $V_0 = 342$). A loose cloud of low-velocity objects may be isolated in the foreground of several more distant, overlapping groups and clouds in the UMa–CVn–Coma area. This whole region is described as the CVn cluster or M94 group by van den Bergh (1960*d*), who points out that many dwarfs are concentrated in this area (van den Bergh 1959); Sersic (1960) describes a subset as the UMa I group and another as the UMa II group (both of which, however, are all within the boundaries of CVn). A detailed comparison of luminosities, diameters, velocities, H II regions, and brightest stars supports the latter interpretation. After much searching and with some hesitation in borderline cases, the following objects were iso-

[1] For more recent discussions of this group, see Bottinelli *et al.* (1971, 1972).

lated as members of the foreground CVn I cloud (or restricted M94 group): NGC 4136, 4150, 4214, 4244, 4258, 4395, 4449, 4736 (M94), IC 4182, A 1157(= DDO 105), and possibly NGC 4826 (M64). Other possible dwarf members include DDO 99, 125, 126, 129, 133, 141, 143, and 156. All members are spirals of type Sb or later and Magellanic irregulars. The overall dimensions of the cloud are 28° × 14° = 1.9 × 0.9 Mpc (including N4826), and the velocity range is about 300 km s^{-1} (236–530). Radio continuum and H I emission has been detected in several members, including NGC 4214, 4244, 4258, 4449, 4736 (Heeschen and Wade 1964; Epstein 1964), and optical rotation curves and mass estimates are available for NGC 4258, 4736, etc. (see chap. 3).

G4. NGC 5128 *Group* (μ_0 = 28.0, Δ = 4.0, V_0 = 319). Several large southern galaxies having low velocities (range 271–410) may form a loose group or chain centered on NGC 5128, and including NGC 4945, 5102, 5236, 5253, and possibly 5068 (de Vaucouleurs 1956*a*, Sersic 1960*b*). Two (NGC 5102 and perhaps 5128 = Cen A) are lenticulars, three are late-type spirals Sc–Scd, and one (N5253) is peculiar, possibly an I0 irregular (the BGC classification Imp is incorrect). The overall length of the chain is 30° = 2.1 Mpc or 20° = 1.4 Mpc (excluding N5068 which has the highest velocity). The velocity range is, however, only 140 km s^{-1} (271–410). Even if this chain does not form a physical (bound) group, it is useful to obtain some estimate of the distance of NGC 5128. Because of the large range of galactic latitudes covered the apparent modulus varies from 28.2 (A = 0.3) for N5068 to 28.8 (A = 0.8) for NGC 4945. In addition to the extensive literature on NGC 5128 (see BGC for references), some optical and radio emission data are available for NGC 4945 and 5236 (de Vaucouleurs 1964; Epstein 1964).

G5. M101 *Group* (μ_0 = 28.3, Δ = 4.6, V_0 = 508). This is a traditional group, but its membership is somewhat uncertain. It is included by van den Bergh (1960*d*) in his extended M94 group, but Holmberg (1950) and Sersic (1960*a*) make it a separate group. An analysis of luminosities, diameters, H II regions, and velocities, as well as the distribution on the sphere, supports the latter view. The major members, then, are M101 (N5457) and its satellites NGC 5204, 5474, 5585, and the wide pair formed by M51 (N5194 and companion N5195) and M63 (N5055); a probable member is N5949, and possibly some outlying systems including N4605, N5907, N6503, A1353 (Ho IV), and A1339 (Ho V). Many dwarfs are concentrated in this region including DDO 175, 185, 186, 191, 193, and 194 around M101, and DDO 167, 168, 169, 172, 176, 177, 178, 181, 182 and 183 around M51–M63. Except for NGC 5195 (type I0), all bright members of the group are spirals of type later than Sb and dwarf Magellanic irregulars. The velocity range is 240 km s^{-1} (395–634). The overall dimensions of the group are 23° × 16° = 1.8 × 1.3 Mpc, and again there is no central condensation. In fact, the group consists mainly of two subgroups centered on M101 and M51, each about 10° × 7° = 0.8 × 0.6 Mpc and are separated by some 10° = 0.8 Mpc. The separation between the M31–M33 and Galaxy–LMC subgroups in the Local Group is about 0.7 Mpc. Detailed photometric, spectroscopic, and radio data are available for M51, M63, and M101 (Holmberg 1950; Epstein 1964).

G6. NGC 2841 *Group* (μ_0 = 28.9, Δ = 6.0, V_0 = 589). Several large spirals near the border of Ursa Major and Lynx have low velocities (range 422–751) and may form a loose group including N2841 and 2681 as the brightest members; the loose triplet of late-type spirals N2500, 2541, 2552; and the close pair formed by N2537, a peculiar

Magellanic irregular, and IC 2233, an edge-on Sd (the L classification in BGC is an error). Several dwarf systems in the area, including DDO 40, 41, 43, 46, 48, 49, 52, 55, and 59, are possible members.

The overall dimensions 15° × 7° = 1.6 × 0.8 Mpc are normal. If the adopted modulus is correct, only one member (N2841) is brighter than −18, but N2681 is at M_G = −17.9. Little detailed information is available for these galaxies, and there are no positive radio data.

G7. NGC 1023 *Group* (μ_0 = 29.0, Δ = 6.3, V_0 = 625). Several large spirals of type Sc and later including NGC 925, 1003, 1058, and IC 239 are clustered around the bright lenticular system N1023 in low galactic latitudes at the border of Perseus and Andromeda. The edge-on Sb system N891 and the close pair of late-type spirals N672, I1727 (V_0 = 496 and 518) are probable or possible outlying members. N1156 (V_0 = 497), a Magellanic irregular some 20° away, is another possible outlying member. The velocity range is 245 km s^{-1} (496–741). The group is at the edge of the galactic absorption belt (which may conceal some members in lower latitudes), and the apparent modulus varies from 29.3 for N672 (A = 0.33) to 29.6 for N1003 (A = 0.64). Several dwarf irregulars are probable (DDO 24, 25 shown in *The Hubble Atlas of Galaxies*, plate 39) or possible (DDO 11, 17, 19, 22, 26) members. The overall dimensions of the group 20° × 10° = 2.2 × 1.1 kpc (including N672) are rather large, but the diameter of the core around N1023 is only 8° = 0.9 Mpc.

There is little detailed optical information and no positive radio data for these galaxies (except for 21 cm line and continuum emission for N891), although two of the brightest supernovae appeared in N1003 and N1058.

G8. NGC 2997 *Group* (μ_0 = 29.4, Δ = 7.6, V_0 = 534). A loose group, consisting of one lenticular N2784 and several large late-type spirals including N2763, 2835, 2848, 2997, may be isolated in low galactic latitudes at the border of Hydra and Antlia in the foreground of the distant Hydra cloud (μ_0 = 31.5, V_0 = 2074). Redshifts are known for only two objects (N2784, 2835) but the large diameters and luminosity classes of the others confirm the existence of a nearby group. The overall dimensions are 14° × 8° = 1.9 × 1.1 Mpc. Several dwarf systems including DDO 56, 57, 60, 61, 62 are in this area. The large Magellanic irregular N3109 (V_0 = 130), which lies about 5° north following N2997, is probably an isolated foreground object. There is very little optical or radio information on members of this group which is one of the only two nearby groups at supergalactic latitudes greater than 50°.

G9. M66 *Group* (μ_0 = 29.4, Δ = 7.6, V_0 = 592). This is the well-known compact triplet of spirals including M65 (N3623), M66 (N3627), and N3628 together with several outlying systems including probably N3593, 3596, and 3666, and possibly N3485, 3489, 3506, and 3547. The velocity range is 300 km s^{-1} (429–730), and the overall dimensions are 7° × 4° = 1.0 × 0.6 Mpc. Only a few dwarfs including D89, 91, and 108 are in this area, but N3628 has an extremely faint companion or appendage discovered by Zwicky.

The M66 group partly overlaps in both projection and velocity range with the richer and larger M96 group (G11) with which it is often combined to form an enlarged Leo group (e.g., as in Humason, Mayall, and Sandage 1956) which will be denoted here as the Leo I cloud. Apart from optical rotation curves for M65 and M66 (cf. chap 3), little detailed information is available for the others outside that referenced in BGC.

G10. *Canes Venatici II Cloud* ($\mu_0 = 29.5$, $\Delta = 8.0$, $V_0 = 747$). This is part of the complex region described as the Canes Venatici (M94) cluster by van den Bergh (1960*d*) and as the UMa II group by Sersic (1960*a*). A detailed analysis of magnitudes, diameters, H II regions, luminosity classes, and velocities over the whole region indicates that a dozen bright objects (N3769, 3769 A, 3949, 4051, 4088, 4111, 4143, 4242, 4485, 4490, 4625, 4618), mainly late-type spirals and irregulars (N4111 and 4143 are lenticulars), form an elliptical core area measuring $15° \times 8° = 2.1 \times 1.0$ Mpc. The overall dimensions of the cloud are increased to $22° \times 12° = 3.0 \times 1.6$ Mpc if several probable or possible outlying members such as 3675, 4627, 4631, 4656–4657, 4800, and the dwarf systems N4025, 4288 are included. The assignment of outlying objects to this cloud rather than to other overlapping or adjacent clouds or groups (CVn I, Coma I, M101) is to some extent a matter of interpretation. This is especially true of the many dwarf galaxies noted by van den Bergh (1959) in this general area (see fig. 6); it is not possible without further detailed study to determine which cloud they belong to. Several of the brighter objects in the CVn II cloud have been the subject of fairly detailed optical and radio investigations (for references see BGC and chap. 3).

G11. M96 *Group* ($\mu_0 = 29.6$, $\Delta = 8.3$, $V_0 = 741$). This is the major condensation in the Leo I Cloud; it is centered on a dense core ($3° \times 1.^\circ5 = 0.4 \times 0.2$ Mpc) including N3351 (M95), 3368 (M96), 3377, 3377 A, 3379, 3384, and 3412. Other probable and possible members listed in table 3 cover an area $11° \times 7° = 1.6 \times 1.0$ Mpc. Several dwarfs including DDO 79, 88, 89, 90 are other possible members. The morphological types show greater variety than nearer groups, ranging from giant ellipticals (N3377, 3379) and lenticular (N3384) to late-type spirals and Magellanic irregulars. As the nearest of the well-mixed groups, the M96 group is one of the best fields for the calibration of distance indicators applicable to E and L systems (N3377 and 3379 are rich in globular clusters). The velocity range is 310 km s^{-1} (593–904), rather small for a dense group.

Extensive photometric and spectroscopic studies have been made of several of the bright members, especially N3379, for which there is also a mass estimate from velocity dispersion (see references in BGC). There are as yet no positive radio data, except marginal continuum emission from M95 and M96.

G12. NGC 3184 *Group* ($\mu_0 = 29.9$, $\Delta = 9.6$, $V_0 = 629$). A few late-type spirals with fairly large diameters and consistent velocities (range: 418–832 km s^{-1}) stand out in Leo Minor in the foreground of more distant clouds. The four listed in table 3 cover an area $10° \times 5° = 1.7 \times 0.8$ Mpc. Two more (N3344 and 3510) 10° south of the group have velocities in the same range and might be included as possible members. There is a remarkable paucity of DDO dwarfs in this area, and the reality of the group is questionable.

There is also very little detailed optical or radio information on the galaxies in this group.

G13. *Coma I Cloud* ($\mu_0 = 29.9$, $\Delta = 9.6$, $V_0 = 944$). This is an elliptical area $11° \times 5° = 1.8 \times 0.8$ Mpc enclosing two main condensations, the denser N4274 group, and a loose grouping around NGC 4565. The N4274 group (van den Bergh 1960*a;* Sersic 1960*a*) includes N4245, 4251, 4274, 4278, 4283, 4314, 4414, 4448, and possibly N4062, N4146, N4203, N4359, I3330, and other insufficiently documented objects; the group has a 3° = 0.5 Mpc core surrounded by scattered objects over a 6° = 1.0 Mpc region. Morphological types are well mixed, including a pair of giant ellipticals (N4278, 4283). The grouping around N4565 includes also N4494, N4559, N4725, and A1244 (the low-velocity anonymous spiral in the outskirts of the great Coma I cluster) and probably some other less well-documented objects. The dimensions of this grouping are 11° × 5° = 1.8 × 0.8 Mpc. Velocities in the N4274 group tend to be lower (range: 622–1078; mean: 829) than in the N4565 area (range: 876–1305; mean: 1134), but there is considerable overlap. A number of DDO dwarfs listed in table 3 are other possible members of the Coma I cloud.

Except for some photometry of N4494 and 4565, there is surprisingly little detailed optical or positive radio information on members of this cloud in spite of the favorable location near the north galactic pole (see references in BGC).

G14. NGC 6300 *Group* ($\mu_0 = 30.0$, $\Delta = 10.0$, $V_0 = +1270$). Three large obscured spirals at low galactic latitudes in Ara—N6300 and the pair N6215–6221—may be the brighter members of a loose group partly hidden by the absorption belt; the pair of late-type spirals I4710, I4713, and perhaps I4662 A and I4714 are other possible members. The large, low-velocity Magellanic irregular I4662 ($V_0 = +237$) is clearly in the foreground. The length of the chain is 10° = 1.7 Mpc. Because of the range of galactic latitudes the apparent modulus varies from $\mu = 31.1$ for N6215–6221 ($A = 1.1$ mag) to 30.8 for N6300 ($A = 0.8$) and 30.6 for I4710–4713 ($A = 0.5$).

The large absorption correction and poorness of the group make the distance modulus quite uncertain.

6. NEARBY GROUPS BEYOND 10 MEGAPARSECS

Table 2 lists 40 other nearby groups whose distance moduli are between 30.2 and 31.2. The apparent distribution is shown in figure 4. The survey is believed to be substantially complete to $\mu_0 = 31.0$ ($\Delta = 16$ Mpc) (see § 7.1).

Table 4 lists the five brightest members. Some further remarks follow:

G15. *Cetus I cloud.*—Includes N1052 and N1068 groups.

G16, 21, 22. *N1566, N1433, N1672 groups.*—Parts of Dorado cloud complex. The mean velocity of the N1433 group depending on two velocities only is uncertain. No velocity is available as yet for the southern group N1672 ,which is partly covered by the Large Magellanic Cloud.

G17, 24, 28, 32, 34. *UMa I (Z, Y, X, S, N) groups and clouds.*—Parts of UMa I cloud complex. UMa I (N) and UMa I (S) are merely the northern and southern halves of the same cloud at $\mu_0 = 30.7$–30.8 and north of the supergalactic plane, while UMa I (X) and UMa I (Y) are apparently distinct subclouds of a nearer cloud at $\mu_0 = 30.5$–30.6 and south of the same plane. UMa I (Z), the southernmost grouping in the greatUMa cloud complex, is actually closer to the CVn II cloud in both direction and distance ($\mu_0 = 30.3$).

G18, 19, 25. *Virgo S, Virgo E, and Virgo S′.*—Parts of Virgo I cluster. For detailed analyses of the Virgo cluster and discussions of possible foreground and background objects see Reaves (1956), de Vaucouleurs (1961*c*), Holmberg (1961).[2]

[2] For a more recent discussion see G. and A. de Vaucouleurs (1973).

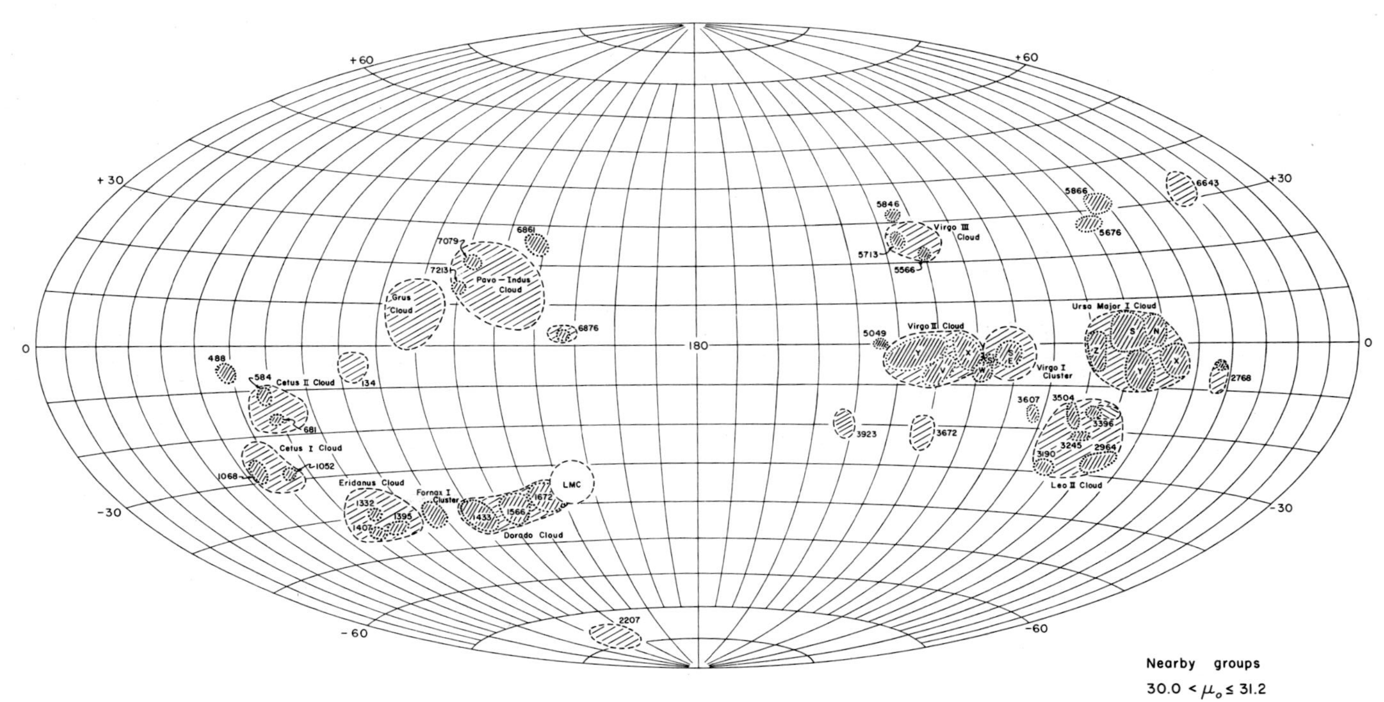

FIG. 4.—Apparent distribution of nearby groups ($30.2 \leq \mu_0 \leq 31.2$) in supergalactic coordinates.

TABLE 4—BRIGHTEST MEMBERS OF 40 NEARBY GROUPS*

NGC	Type	$B(0)$	V_0
G15. Cetus I			
N1068	Sb	9.81	1094
N 936	S0$^+$	11.28	1371
N1084	Sc	11.38	1448
N1087	Sc	11.74	1833
N1055	Sb	11.77	....
G16. N1566			
N1566	Sbc	10.09	1173
N1553	S0^0	10.57	1033
N1549	E0	11.05	942
N1617	Sa	11.40	....
N1574	S0$^-$	11.62	667
G17. Ursa Major I (Z)			
N4151	Sab	11.48	989
N3941	S0^0	11.66	969
N4145	Sd	11.68	....
N4369	Sa	12.84	....
N3813	Sb	12.88	....
G18. Virgo I (S)			
N4321	Sbc	10.26	1552
N4501	Sb	10.49	2056
N4254	Sc	10.52	2397
N4569	Sab	10.58	.........*
N4579	Sb	10.72	1680
G19. Virgo I (E)			
N4472	E2	9.84	855
N4649	E2	10.30	1200
N4486	E0–1p	10.30	1187
N4382	S0$^+$p	10.43	712
N4374	E+1	10.82	878
G20. Virgo Y			
N4697	E6	10.58	1176
N4699	Sb	10.60	1369
N4731	Scd	11.63	1305
N4856	S0/a	11.68	1095
N4939	Sbc	11.70	2862*
G21. N1433			
N1433	Sa	10.90	787
N1512	S0$^+$	11.58	542
N1448	Scd	11.67	....
N1493	Scd	12.10	....
N1411	S0	12.16	....
G22. N1672			
N1559	Scd	11.06	....
N1672	Sb	11.29	....
N1688	Sc	12.66	....
N1796	Sb	13.18	....
N1703	Sb	....	....
G23. N3672			
N3672	Sc	11.80	1845
N3892	S0$^+$	12.70	....
N3637	S0^0	13.01	....
N3865	....	13.07	....
N3818	E5	13.23	1320
G24. Ursa Major I (Y)			
N3776	Sc	11.09	998
N3583	Sb	12.20	....
N3614	Sc	12.42	....
N3415	?	12.98	....
N3478	?	13.04	....
G25. Virgo I (S′)			
N4535	Sc	10.90	1854
N4380	Sb:	12.37	....
N4469	S0/a	12.41	....
N4519	Sd	12.47	1125
N4424	Sa:	12.57	....
G26. Virgo X			
N4303	Sbc	10.28	1559
N4636	E$^+$0–1	11.01	778
N4536	Sbc	11.21	1814
N4517	Scd:	11.43	1095
N4643	S0/a	11.64	1325
G27. Grus			
N7552	Sab	11.44	1639
N7424	Scd	11.46	....
N7410	S0$^+$	11.61	1637
I5267	S0/a	11.65	1695
N7582	Sab	11.84	1428
G28. Ursa Major I (X)			
N3310	Sbc	11.20	1090
N3448	I0	12.42	1476*
N3549	Sc:	12.76	....
N3445	Sm	13.16	2069*
N3458	S0	13.26	....

* See footnote at end of Table 4.

TABLE 4—*Continued*

NGC	Type	$B(0)$	V_0
	G29. Virgo III		
N5566	Sab	11.62	1581
N5746	Sb	11.81	1826
N5713	Sbcp	11.99	1930
N5701	S0/a	12.16	....
N5584	Scd	12.16	....
	G30. N5866		
N5866	S0$^+$	11.19	972
N5907	Sc:	11.40	725
N5879	Sbc:	12.32	1064
N5905	Sb	12.41	....
N5908	Sb	12.96	....
	G31. Eridanus		
N1232	Sc	10.73	1734
N1398	Sab	10.73	1395
N1187	Sc	11.21	1491
N1300	Sbc	11.34	1565
N1407	E+0	11.43	1707
	G32. Ursa Major I (S)		
N3938	Sc	11.02	919
N3893	Sc:	11.10	1065
N4096	Sc	11.31	....
N4157	Sb:	11.80	....
N4217	Sb	11.81	....
	G33. Cetus II		
N720	E5	11.47	1813
N584	E4	11.71	1885
N779	Sb	12.20	1453*
N596	E0	12.31	2097
N615	Sb	12.51	1991
	G34. Ursa Major I (N)		
N3992	Sbc	10.80	1147
N3953	Sbc	11.11	1041
N3631	Sc	11.27	1162
N3898	Sab	11.60	1135
N3718	Sap	11.72	1128
	G35. Virgo V		
N4546	S0$^-$	11.62	879
N4691	S0/a	11.81	994*
N4487	Scd	11.82	....
N4593	Sb	11.87	2561*
N4504	Scd	12.01	....
	G36. N2207		
N2207	Sbc	11.54	2455
N2217	S0$^+$	11.69	1334
N2223	Sb	12.18	....
N2280	Scd	12.24	....
N2139	Scd	12.39	1691
	G37. N5676		
N5676	Sbc	11.87	2395
N5660	Sc	12.44	....
N5633	Sb	13.07	2484
N5689	S0/a	13.17	2355
I1029	S:	....	....
	G38. N6876		
N6943	Scd	12.32	....
I5052	Sd	12.47	....
N6876	E3	12.79	....
N6808	Sa	13.60	....
	G39. N134		
N134	Sb	11.22	1665
N289	Sbc	11.92	1907
N150	Sb	12.34	....
N148	S0?	12.95	....
N254	S0?	12.97	....
	G40. N488		
N488	Sb	11.41	2284
N474	S0^0	12.51	2405
N520	P	12.75	2320
N521	Sbc	12.75	....
N470	Sb	12.75	....
	G41. N2768		
N2768	E$^+$6	11.48	1495
N2805	Sd	11.95	2023
N2742	Sc:	12.41	1380*
N2880	S0$^-$	12.97	1614
N2654	Sa	13.05	1448
	G42. N2964		
N2859	S0$^+$	11.96	1649
N2964	Sbc	12.37	1284
N3003	Sbc	12.52	1429
N3032	S0^0	12.86	1500
N3067	Sab	12.91	1455

* See footnote at end of Table 4.

TABLE 4—*Continued*

NGC	Type	$B(0)$	V_0
	G43. N3396		
N3430.....	Sc	12.39	1709
N3395.....	Scdp	12.46	1622
N3396.....	Imp	12.90	1611
N3427.....	Sb:	....	....
N3413.....	S0 sp	....	....
	G44. N3923		
N3923.....	E4–5	11.40	1551
N4105.....	E3	12.10	1665
N3904.....	E^{+}2–3	12.43	1374
I764.......	Sc:	12.44	....
N4106.....	$S0^{+}$	12.49	1948
	G45. Pavo-Indus		
N7213.....	Sa	11.57	1751
N7205.....	Sbc	11.70	1404
N7049.....	$S0^{0}$	12.04	2153
N7083.....	Sbc	12.14	....
N7144.....	E0	12.15	2085
	G46. Virgo W		
N4261.....	E2–3	11.84	2093
N4281.....	$S0^{+}$	12.41	2492
N4273.....	Sc	12.51	2192
N4260.....	Sa	12.70	1827
N4235.....	Sa	12.86	....
	G47. N3190		
N3227.....	Sa	11.75	1005
N3190.....	Sap	12.20	1255
N3162.....	Sbc	12.30	1361
N3193.....	E2	12.37	1273
N3226.....	E^{+}2p	12.77	1232
	G48. N3504		
N3504.....	Sab	11.80	1473
N3414.....	S0	12.23	1391
N3512.....	Sc	13.12	1449
N3418.....	$S0^{+}$	....	....
N3380.....	S	....	....

NGC	Type	$B(0)$	V_0
	G49. N3607		
N3607.....	$S0^{0}$	11.42	840
N3626.....	$S0^{+}$	12.11	1361
N3686.....	Scd	12.24	930
N3608.....	E2	12.31	1117
N3684.....	Sbc	12.63	1329
	G50. N5846		
N5846.....	E^{+}0–1	11.76	1784
N5813.....	E1–2	12.09	1891
N5838.....	$S0^{-}$	12.14	1441
N5850.....	Sb	12.25	2385
N5806.....	Sb	12.70	1309
	G51. N6643		
N6643.....	Sc	11.97	1790
N6217.....	Sbc	12.17	1616
N6340.....	S0/a	12.21	2351
N6412.....	Sc	12.62	1751
N6654.....	S0/a	12.80	2180
	G52. N6861		
N6868.....	E2	12.31	2734
N6861.....	$S0^{-}$	12.43	....
A2021......	S0:	12.49	....
N6902.....	S0/a	12.67	....
N6893.....	$S0^{0}$	12.85	....
	G53. Fornax I		
N1399.....	E0	11.15	1311
N1380.....	S0	11.30	1712
N1404.....	E1	11.34	1828
N1326.....	$S0^{+}$	11.75	....
N1350.....	Sbc	11.80	1657
	G54. N3245		
N3245.....	$S0^{0}$	12.04	1198
N3254.....	Sbc	12.41	1170
N3277.....	Sab	12.60	1399
N3274.....	....	13.12	....
N3245 A...	Sb	....	....

* Additional velocities (1973 December): G.33 N779 foreground? G.28 N3445 background? N3448 member. G.35 N4593 background? G.20 N4939 background? G.18 N4569, old $V_0 = +893$ in error, new $V_0 = -382$.

G20, 26, 35. *Virgo Y, Virgo X, Virgo V.*—Parts of Virgo II cloud complex.

G46. *Virgo W group.*—Includes the W′, Wa, Wb subgroups (see de Vaucouleurs 1961*c*) in the background of the Virgo I cluster.

G27. *Grus cloud* (de Vaucouleurs 1956*a;* Shobbrook 1966).—Possible foreground object: I5332 (in Sculptor group?).

G29. *Virgo III cloud.*—Includes N5566 and N5713 groups.

G31. *Eridanus cloud.*—Includes N1209 and N1332 groups.

G33. *Cetus II cloud.*—Includes N584 and N681 groups.

G36. *N2207 group.*—This little-known southern group in low galactic latitudes has the highest supergalactic latitude of all the nearby groups.

G39. *N134 group.*—This group is close to the south galactic pole and in the vacant center of the much nearer Sculptor group (G1).

G42, 43, 47, 48, 54. *N2964, N3396, N3190, N3245.*—Parts of Leo II cloud complex.

G45. *Pavo-Indus cloud.*—Includes N7079 and N7213 groups. Foreground: N7090 (V_0 = +730), I5152 (V_0 = +31), I5201. Background: N6970 (V_0 = +5440).

G50. *N5846 group.*—See de Vaucouleurs (1960*a*).

G51. *N6643 group.*—Foreground N6503 (V_0 = +279); background N6621 (V_0 = +6490).

G53. *Fornax I cluster.*—N1316 (V_0 = +1715), N1365 (V_0 = +1571), possibly in foreground? (see de Vaucouleurs 1956*a;* Hodge 1959, 1960).

7. STATISTICAL PROPERTIES OF NEARBY GROUPS

7.1. Completeness of Survey

The frequency distribution of distance moduli of the 54 groups in table 2 plus the Local Group is given in table 5.

A plot of the cumulative frequencies $N = \Sigma n$ (fig. 5) is consistent with the relation $N \propto \Delta^3$, i.e., $\log N = a + 0.6\,\mu_0$, expected for a statistically uniform space density of cluster centers in the range $30 < \mu_0 < 31$. There is an apparent excess of nearby groups for $\mu_0 < 30$ due in part to a genuine higher density near the Local Group and in part to the fact that a finer division of groups is possible in our immediate neighborhood; at larger distances an increasing proportion of groups have been included in larger clouds. To compensate for this effect let us visualize (figs. 8 and 9) the region of the Local Group observed from a great distance, say, from the Virgo cluster; then the M81, Sculptor, and Local Groups might appear merely as condensations in the same cloud (our Local Cloud) and the N5128 chain might be unnoticed. The circles and dashed line in figure 5 corre-

TABLE 5

Frequency of Distance Moduli for 55 Nearby Groups*

μ_0	<26	26.0–26.9	27.0–27.9	28.0–28.9	29.0–29.4	29.5–29.9	30.0–30.4	30.5–30.9	31.0–31.2
n	1	1	2	3	3	4	7	18	16
$N = \Sigma n$	1	2	4	7	10	14	21	39	55

* Including Local Group.

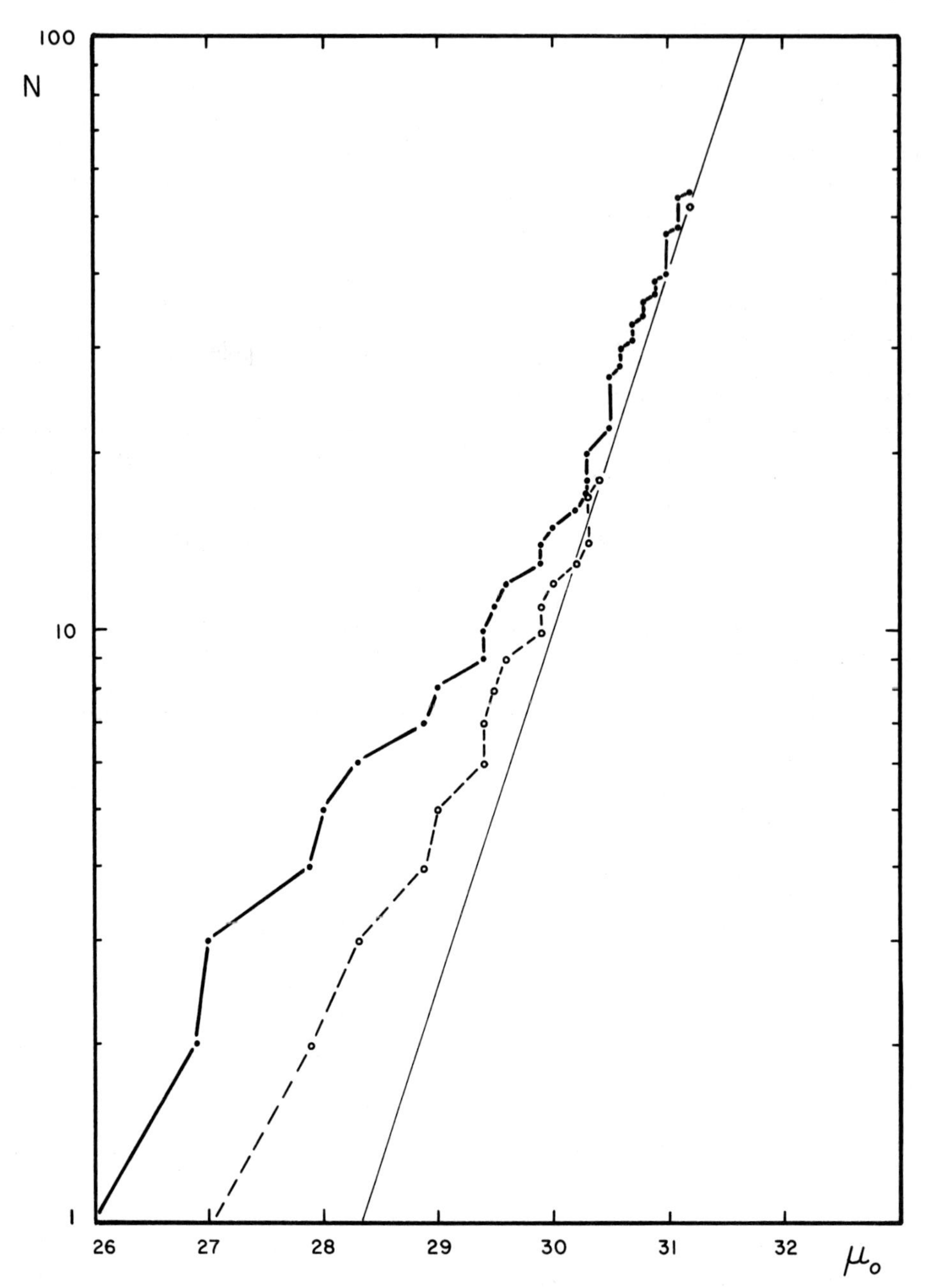

FIG. 5.—Cumulative frequency distribution of distance moduli of nearby groups (including Local Group).

spond to the corrected frequency function allowing for this effect. The presence of a Local Cloud is still in evidence, but the approximation to the Δ^3 law is quite good for $\mu_0 >$ 29 and apparently right up to the limit of the survey at $\mu_0 \simeq 31$. This is because in preparing this chapter all identifiable groups with $\mu_0 < 32$ (and many beyond) were considered and only those with $\mu_0 \leq 31.0$ (31.2 to allow for accidental errors) are discussed here. Hence the survey should be substantially complete out to $\mu_0 = 31.0$ ($\Delta = 16$ Mpc) except, of course, for the obscured galactic belt (about one-fifth of the sky area).

7.2. Space Density of "Cluster" Centers

Allowing for galactic obscuration, some 50 to 60 cluster (or group or cloud) centers should lie within $\Delta = 16.0$ Mpc, or in a spherical volume of 17×10^3 Mpc^3. The average volume of space per cluster or "cluster cell," then, is 300 Mpc^3, and the average distance between adjacent "cluster centers" is on the order of 7 Mpc.

For comparison, the average diameter of the nearby groups, clouds, and clusters is about 2 Mpc; most condensations are well separated by regions of lower density.

7.3. Frequency Function of Diameters

Table 6 gives the distribution of major diameters of groups and clouds or clusters.

TABLE 6

Frequency of Major Diameters of 55 Nearby Groups*

D(Mpc)	0.0	0.5	1.0	1.5	2.0	2.5	3.0	3.5	4.0	Mean
38 groups*...........	2	8	8	17	3	...	...	...		1.5
17 clouds and clusters	...	...	...	6	5	3	1	2		2.5

* Including Local Group.

Median and mean diameters are: groups, 1.5/1.5 Mpc (range 0.3–2.5); clouds and clusters, 2.3/2.5 Mpc (range 1.8–3.6). There is a suggestion of bimodal distributions: small groups (0.3–1.2 Mpc), large groups (1.4–2.5 Mpc); small clouds or clusters (1.8–2.5 Mpc), large clouds (2.8–3.6 Mpc).

7.4. Luminosity Function

The mode of formation of the *Reference Catalogue* prevents the derivation of a meaningful luminosity function for groups other than the Local Group, (§ 4) except for the very brightest galaxies. Table 7 gives the numbers N_{18} and N_{17} of systems with $M_G <$ -18 and $M_G < -17$ in the $B(0)$ system (remember that M_T is about 0.5 mag brighter on the average) for 38 groups (including the Local Group) and 17 clouds or clusters. The counts appear to be complete to -17 for $\mu_0 < 30.0$ only, and to -18 for $\mu_0 < 31.0$. The average population is $\bar{N}_{18} = 4.3$ for groups and $\bar{N}_{18} \simeq 15$ for clouds and clusters. On the average a cloud or cluster (as defined here) comprises three or four groups. This is consistent with the ratio of average diameters $(2.5/1.5)^3$ and the statistical relation between volume and average density in nearby space (de Vaucouleurs 1960*a*, 1961*d*).

7.5. Statistical Masses

The loose structure of most nearby groups does not inspire confidence in their stability or in the validity of total masses derived from velocity dispersion through the virial

TABLE 7

LUMINOSITY FUNCTION AND POPULATION-TYPE INDICES OF 55 NEARBY GROUPS

Group	N_{18}	N_{17}*	$\bar{t}$	Range	S_t	n
0 Local	2–3	3–4	+6.4	(+3, +9)	2.1	5
1 Scl	3	4	+7.3	(+5, +9)	1.0	6
2 M81	2	3	+4.6	(−1, +9)	3.7	5
3 CVn I	4	6	+5.6	(+2, +10)	3.5	5
4 5128	3	6	+3.3	(−4, +10)	4.7	6
5 M101	3	4	+4.0	(−1, +7)	2.0	5
6 2841	1	3	+5.8	(0, +10)	2.9	6
7 1023	2	4	+3.8	(−4, +7)	3.0	6
8 2997	3	5	+3.6	(−3, +6)	2.6	5
9 M66	3	4	+1.0	(−2, +3)	1.6	5
10 CVn II	5	11	+6.0	(+3, +9)	2.0	5
11 M96	2	5	−2.0	(−6, +3)	3.6	5
12 3184	4	4	+6.3	(+5, +9)	1.4	4
13 Coma I	10	12	+2.0	(−6, +5)	3.2	5
14 6300	3	(3)	(+4.3)	(+3, +5)	(0.9)	3
15 Cet I	10	(16)	+2.8	(−2, +5)	1.9	5
16 1566	9	(13)	−1.6	(−6, +4)	3.3	5
17 UMa I (Z)	3	(5)	+2.0	(−3, +7)	2.4	5
18 Virgo S	27	(45)	+3.4	(+2, +5)	0.9	5
19 Virgo E	32	(51)	−5.0	(−6, −2)	1.2	5
20 Virgo Y	16	(22)	+1.4	(−6, +6)	3.5	5
21 1433	8	(8)	+1.6	(−3, +6)	3.5	5
22 1672	3	(4)	+4.0	(+3, +6)	1.2	5
23 3672	2	(6)	−1.5	(−6, +5)	3.2	4
24 UMa I (Y)	3	(6)	(+4.3)	(+3, +5)	(0.9)	3
25 Virgo S′	5	(10)	+3.2	(0, +7)	2.2	5
26 Virgo X	24	(37)	+1.8	(−5, +6)	3.4	5
27 Grus	17	(24)	+1.6	(−2, +6)	2.1	5
28 UMa I (X)	3	(6)	+2.8	(−3, +9)	3.8	5
29 Virgo III	15	(23)	+3.0	(+2, +6)	1.6	5
30 5866	5	(5)	+2.6	(−2, +5)	1.8	5
31 Eri	19	(24)	+2.2	(−5, +5)	3.0	5
32 UMa I (S)	15	(20)	+4.2	(+3, +5)	1.0	5
33 Cet II	6	(8)	−2.4	(−6, +3)	4.3	5
34 UMa I (N)	15	(18)	+3.2	(+1, +5)	1.4	5
35 Virgo *V*	9	(12)	+2.2	(−4, +6)	3.4	5
36 2207	6	(7)	+3.4	(−2, +6)	2.3	5
37 5676	3	(4)	+3.0	(0, +5)	1.2	4
38 6876	3	(4)	+2.0	(−6, +7)	4.5	4
39 134	(5)	(5)	+0.8	(−3, +4)	3.0	5
40 488	5	(6)	+1.8	(−3, +4)	2.3	4
41 2768	5	(6)	+0.8	(−5, +7)	4.2	5
42 2964	4	(8)	+1.0	(−3, +4)	2.8	5
43 3396	3	(3)	+4.2	(−3, +10)	3.4	5
44 3923	8	(8)	−2.8	(−6, +5)	3.4	5
45 Pavo-Indus	24	(24)	0.0	(−6, +4)	3.6	5
46 Virgo *W*	6	(14)	−0.2	(−6, +5)	3.1	5
47 3190	6	(8)	−1.0	(−6, +4)	3.6	5
48 3504	3	(3)	+0.5	(−3, +5)	3.0	4
49 3607	6	(10)	−0.2	(−6, +6)	4.2	5
50 5846	8	(10)	−1.8	(−6, +3)	3.8	5
51 6643	(5)	(5)	+2.8	(0, +5)	2.2	5
52 6861	9	(12)	−3.2	(−6, 0)	1.4	5
53 For I	15	(15)	−2.6	(−6, +4)	2.9	5
54 3245	4	(4)	+1.5	(−3, +4)	2.2	4

* Parentheses indicate incomplete data.

theorem (Local Group: Humason and Wahlquist 1955, Kahn and Woltjer 1959, Godfredsen 1961; Sculptor group: de Vaucouleurs 1959; M81 group: Holmberg 1950, Ambartsumian 1958, Limber 1961; CVn cluster: van den Bergh 1960*d;* Virgo Cluster: Oort 1958, van den Bergh 1960*d*, Holmberg 1961, de Vaucouleurs 1961*c;* N5846 group: de Vaucouleurs 1960*a;* see also Neyman, Page, and Scott 1961).[3]

Although the evidence is perhaps not yet completely conclusive, the overall impression gained from extensive discussions of this topic is that while large, centrally condensed clusters of the Coma type are probably sufficiently relaxed and stable over periods of time long enough to justify an application of the virial theorem, the same cannot be said of the majority of nearby groups and clouds with the possible exceptions of the E components of the Virgo I and Fornax I clusters. Hence masses derived from velocity dispersion are probably meaningless.

TABLE 8

AVERAGE STATISTICAL MASSES OF NEARBY GROUPS AND CLOUDS

Average	Nearer Groups	All Groups	Clouds	Clusters*
Number of groups	10	27	13	3
Velocity dispersion σ_v (km s^{-1})	100†	200	250	650:
Radius R (Mpc)	0.4	0.4	0.6	0.6:
Total mass $\mathfrak{M}_T (10^{11}\ \mathfrak{M}_\odot)$	40	160	480	3000:
Mass per galaxy $\mathfrak{M}_1 (10^{11}\ \mathfrak{M}_\odot)$§	2	5	3	12:
Density ρ (10^{-27} g cm^{-3})	2	5	3	24:

* Vir I (E), Vir I (S), For I.
† Approx. corrected for observational errors.
§ Assuming $N_T \simeq 10 N_{18}$ in first 5 magnitudes.

There is, therefore, little point in applying the virial theorem to each of the 55 nearby groups and clouds. It should be sufficient to list the average masses and densities that would result from a conventional application of the standard method to the mean of all nearby groups and clouds.

The calculations use the crude but sufficient approximation $\mathfrak{M}_T \simeq 5R\sigma_v^2/G$, where $R \simeq \bar{D}/4$, if $\bar{D}$ = mean major diameter of groups, and $\sigma_v \simeq 2\langle \text{A.D.} \rangle$, if $\langle \text{A.D.} \rangle$ = mean of average deviations from V_0 in table 2. The results, shown in table 8, display the familiar discrepancy between average galaxy masses from rotational studies ($\sim 10^{10}\ \mathfrak{M}_\odot$) and from the virial theorem ($\sim 10^{11}$–$10^{12}\ \mathfrak{M}_\odot$). It is more pronounced for the three clusters than for groups or clouds. Rejection of possible foreground or background objects in Virgo reduces the discrepancy only slightly (de Vaucouleurs 1961*c;* Holmberg 1961).

7.6. POPULATION TYPES

There is a remarkable segregation of galaxy types among nearby groups and clouds; this phenomenon, which was first noted in the Grus cloud (de Vaucouleurs 1956*a*), the UMa I cloud (Morgan 1958), and the "M94 group" (van den Bergh 1960*d*), is probably related to differences in age and/or physical conditions (gas density, temperature, composition) at the time of formation of the groups.

[3] For more recent discussions see Rood *et al.* (1970) and Geller and Peebles (1973).

A quantitative index of population type can be obtained by means of a numerical scale attached to the classification stage as follows:

$\mathfrak{T}$	E	E^+	L^-	L	L^+	I0	S0/a	Sa	Sab	Sb	....	Sdm	Sm	Im
t	−6	−5	−4	−3	−2	−1	0	1	2	3	....	8	9	10.

In principle accurate color indices could be used, but in addition to the fact that color indices are not yet available for all galaxies in nearby groups, the color scale is too compressed in the early types ($t < 0$) and it is sensitive to absorption and emission. Table 7 gives the mean type index t, range and average deviation S_t (a "purity" index) of the four to six brightest galaxies in nearby groups.

Nearly all the nearer groups are "S" type, i.e., have a large majority of spirals and Magellanic irregulars with an occasional lenticular or I0 system and practically no giant ellipticals. Among all the nearby groups only the "E" core of the Virgo cluster and the Fornax I cluster have a dominant population of ellipticals.

8. NEARBY DWARF GALAXIES

8.1. Definition

A dwarf galaxy may be defined in terms of absolute magnitude and/or linear size. There is, however, a continuous transition between giant, average, and dwarf galaxies. The separation must rest on some arbitrary demarcation lines. As far as we know, lenticulars and spirals in the Sa–Sc range are all brighter than $M_T \simeq -16$; dwarf systems fainter than -16 occur only among ellipticals and late-type spirals Sd–Sm or Magellanic irregulars Im, but of course not all E or Im systems are dwarfs. For instance, by this definition the Large Magellanic Cloud ($M_T = -18.1$) is certainly not a dwarf and the Small Cloud ($M_T = -16.0$) is a borderline case, but M32 ($M_T = -15.6$) and NGC 205 ($M_T = -15.8$) qualify as dwarfs. Similar systems in the Virgo cluster are N4486B, the bright compact dE0 companion of M87, and IC 3475, the prototype of the low-luminosity diffuse ellipticals described by Reaves (1956), which are both at $M_T = -16.0$. The low-luminosity dwarf ellipticals of the Local Group exemplified by the Sculptor and Fornax systems are much fainter, at $M_T = -11.2$ and -12.9, respectively, and the extreme dwarf globular-like systems discovered by Wilson (1955) and Zwicky (1957) are fainter still—for example, Leo II is at $M_T = -9.1$.

The lower end of the scale is indefinite, and for all we know (or rather do not know) "pygmy" systems of even smaller populations down to isolated star-cluster size might exist and remain undetected throughout intergalactic space as Zwicky has often argued (1957).

8.2. Dwarfs in the Local Group

The provisional and certainly incomplete luminosity function of the Local Group (§ 4) has only six average or giant members ($M_T < -16$) and a score of dwarfs fainter than -16, of which six (or seven including the SMC) are Magellanic irregulars in the range -16 to -12 and 15 are spheroidal systems in the range -16 to -9 or fainter. This is a minimum because extreme dwarf ellipticals of the Sculptor-Fornax type are observable only if $\mu \leq 22$ ($\mu_0 < 21.5$ or $\Delta \leq 0.2$ Mpc) since their discovery on the Palomar *Sky Survey* plates requires $m_* \leq 20.5$ (if $M_* \simeq -1.5$); the fact that a dozen are known within this range suggests that their space density is high (50 to 100 per Mpc[3]), unless they are satellites of our Galaxy, perhaps related to globular clusters (Wilson 1955) rather than independent galaxies. Star counts in several of the largest dE systems (Hodge 1960, 1961, 1963, 1964), which indicate tidally limited radii, favor the second alternative. There is only a marginal possibility of detecting such systems at the distance of the Andromeda group even with the largest reflectors ($\mu \leq 25$, $m_* \leq 23.5$).[4] *A fortiori* such systems are beyond the reach of the largest telescopes even in the nearest groups ($\mu > 27$, $m_* > 25.5$). For all practical purposes we may be missing the most common type of galaxy in the Universe, much as we fail to detect all except a few of the nearest dwarf stars of $M > +10$.

To a lesser extent the same remark applies to the dwarf Magellanic irregulars of the IC 10–IC 1613 type of which at least six are known in the Local Group. The presence of blue supergiants and H II regions, however, makes them more easily detectable and well beyond the Local Group ($\mu < 30$ if $m_* < 21$ to 22 and $M_* \simeq -8$ to -9).

Small ellipticals of the M32–NGC 205 type are also observable beyond the Local Group, but their small diameters make them difficult to distinguish from star images at the distance of the Virgo cluster. For instance, M32 with a standard linear diameter $D(0) = 0.6$ kpc would have an apparent diameter of $0'.2$ only at the adopted distance $\Delta = 12.6$ Mpc of the Virgo cluster and would therefore look very much like NGC 4486 B which is undistinguishable from star images on survey plates.

8.3. Survey of Nearby Dwarfs

The only systematic search for nearby dwarf galaxies was made by van den Bergh (1959) on the 48-inch *Sky Survey* prints. Criteria for identification depend mainly on low surface brightness and density gradient in an image of diameter 1′ or larger. Compact dE systems are therefore excluded. Most objects must be dIm and low-density dE systems except for a few resolved Local Group "pygmies" and some occasional misidentifications. The apparent distribution in supergalactic coordinates of the DDO dwarfs is shown in figure 6; the large gap is the unobserved southern sky ($\delta < -23°$). The symbols correspond to van den Bergh's classification:

DIr.—Dwarf Magellanic irregulars of the N6822, IC 1613, WLM type.

DSp.—Dwarf spirals which from van den Bergh's description are clearly late-type barred spirals of the SBd–SBm types or Magellanic irregulars of the SMC–LMC type.

DEl.—Dwarf ellipticals of the N205 type of which only a few of the nearer and larger ones can be identified on the *Sky Survey* prints.

DSph.—Dwarf spheroidal galaxies of the IC 3475 type in the Virgo cluster (van den Bergh notes that resolved systems of this type in the Local Group such as Draco are difficult to distinguish from clusters of distant galaxies on the *Sky Survey* prints).

Local Group members and objects larger than $1'.0$ are identified by open circles in figure 6.

[4] Three such systems were discovered by van den Bergh (1972*a*) on IIIa-J plates taken with the 48-inch Palomar Schmidt telescope and were resolved into stars ($m_* \leq 24$) with the 200-inch reflector (van den Bergh 1972*b*, 1973); the total magnitude of And I is $B \simeq 14.9$ with $B - V \simeq 0.75$, $U - B \simeq 0.25$ (unpublished McDonald data).

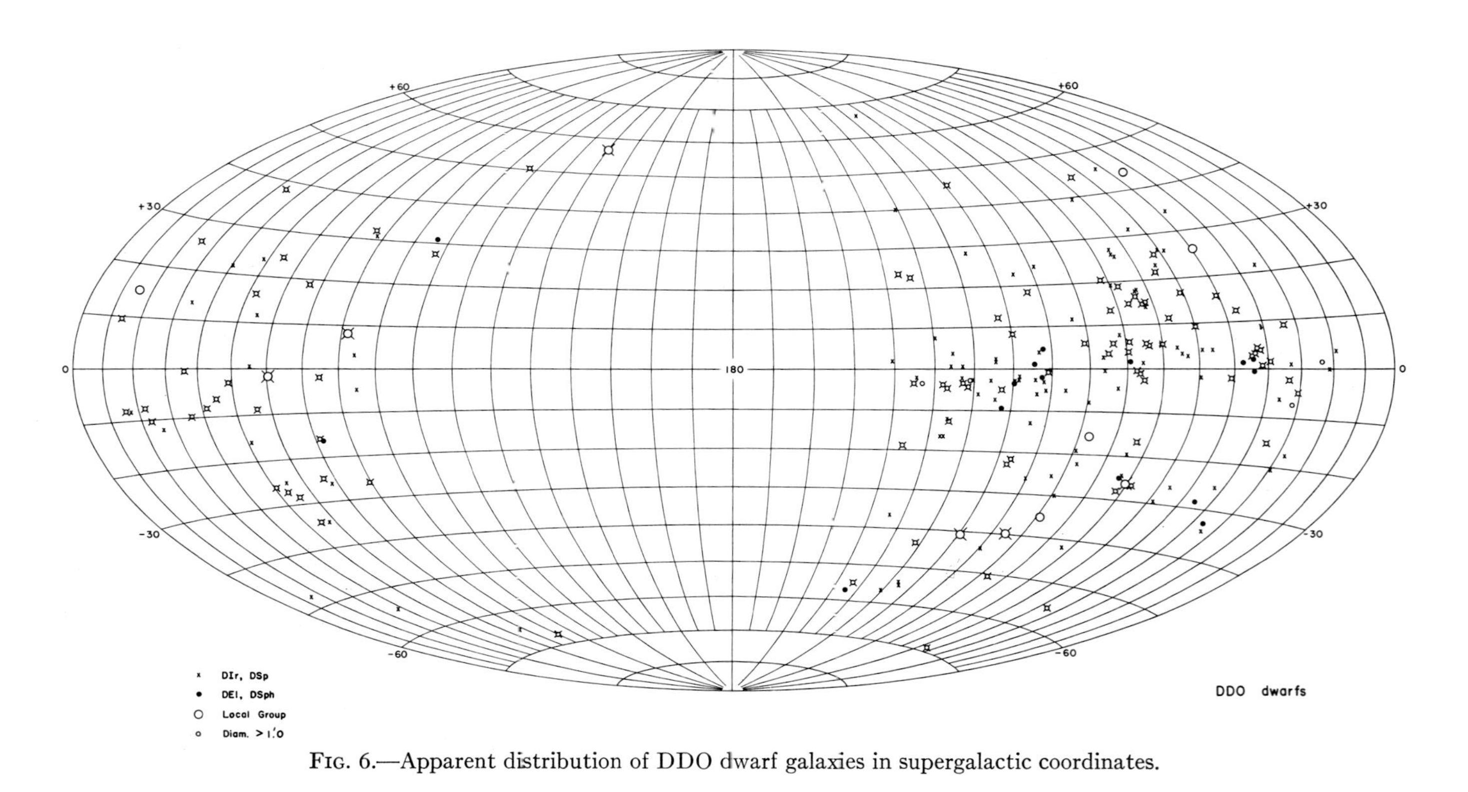

Fig. 6.—Apparent distribution of DDO dwarf galaxies in supergalactic coordinates.

The nearby dwarfs are clearly concentrated to the supergalactic equator, especially in the northern (galactic) hemisphere, and several condensations corresponding to the nearby groups can be recognized (compare with figs. 3 and 4), in particular the M81 and M101 groups, the Virgo I cluster, and the CVn I, CVn II, Vir II clouds. Probable or possible associations of DDO dwarfs with some individual groups are noted in table 3. In most cases, however, definite proof of association will require more detailed studies.

8.4. Completeness of Survey

The frequency function of apparent diameters for the DDO dwarfs is as follows:

ϕ		$16'$		$8'$		$4'$		$3'$		$2'$		$1'.5$		$1'.0$
n	1		6		8		12		29		46		120	
$N = \Sigma n$		1		7		15		27		56		102		222.

A plot of log N versus log ϕ approximates closely the $N \propto \phi^{-3}$ distribution expected for a statistically uniform space density in the range $11' < \phi < 20'$, but falls off rapidly for $\phi < 10'$. Either incompleteness begins at $\phi = 10'$ or there is a local excess of dwarfs associated with the Local Group. The seven largest DDO dwarfs ($\phi \geq 8'.0$) are listed in table 9: six (D8, 74, 199, 208, 209, 221) are in the Local Group and one (D81) is in the M81 group.

TABLE 9

DDO Dwarfs > 8′

DDO	ϕ	Δ(Mpc)	ϕ(kpc)
8=I1613............	$15'.0$	0.66	3.1
74=Leo I............	$8'.5$	0.23	0.6
81=I2574............	$11'.5$	2.5	8.5
199=UMi............	$20'.0$	0.08	0.5
208=Draco............	$8'.0$	0.06	0.14
209=N6822............	$12'.5$	0.50	1.8
221=WLM............	$11'.0$	0.87	2.8

If the smallest DDO dwarfs ($\phi = 1'.0$) have the same mean linear diameter $\bar{\phi} = 2.5$ kpc as the seven above, the limiting range of the survey is $\Delta \simeq 8.5$ Mpc on the average; half the total ($\phi \simeq 1'.5$) is within an average range of 5.5 Mpc. However, the range of linear diameters is large (8.5–0.14 kpc or a 60–1 ratio in the first seven), and the range of the survey in depth is necessarily indefinite.

For a survey of dwarf galaxies in the Virgo cluster, see Reaves (1956).

9. ISOLATED NEARBY GALAXIES

Thirty years ago galaxies were generally regarded as primarily distributed in a so-called general field, i.e., more or less isolated in space, with only a small minority in occasional groups or clusters. More recently there has been some speculation that perhaps the opposite is true and that all galaxies are clustered (even if some stochastic models include "clusters" having $n = 1$ member!). Holmberg (1940) has counted apparent companions of a selected sample of G.C. objects ($10 < m < 13$) in the Reinmuth survey; after statistical correction for optical companions and incompleteness he derived the following relative frequencies:

Multiplicity m	1	2	3	4	5	6	7
Frequency (%) f	47	24	15	7	4	(2)	(1)
$f = 2^{-m}$	50	25	12.5	6.25	3.12	1.56	0.78 .

Thus 47 percent of all galaxies of $10 < m < 13$ appear to be single, 24 percent are members of pairs, 15 percent of triplets, etc. It is perhaps more than a passing curiosity that the observed frequencies are well approximated by $f(m) = 2^{-m}$ which suggests that for all $m \geq 8$, $\Sigma f = 1.3$ percent. Statistics based on apparent magnitudes refer to an indefinite volume of space, and frequencies depend very much on the operational definition of a group. Holmberg's counts refer mainly to dense groups and multiplets and include only a small fraction of the loose groups considered in this chapter.

Ideally counts should refer to a specified volume of space, but this is not practical and even then the problem of dwarf galaxies (how far down the luminosity function should one place the cutoff?) will complicate matters.

The present data on nearby groups may nevertheless help to answer the simpler question: Are there isolated galaxies? Figure 7 shows the distribution in supergalactic coordinates of all galaxies in the *Reference Catalogue* which are either brighter than $B(0) = 10.0$ (or corrected Shapley-Ames magnitude m_c), or larger than $D(0) = 10'.0$ or have corrected radial velocities $V_0 < +200$ km s^{-1} (adding IC 10 and IC 342 with allowance for absorption). Out of the 60 galaxies in this objectively selected sample, only eight have not been associated with one of the 55 nearby groups, viz., NGC 404, 1313, 2903, 3109, 3521, 6744, 6946, and IC 5152. In addition there is a possibility that a few galaxies, such as NGC 1316, 4594, 4826, are not really members of the groups (For I, Vir Y, CVn I) to which they have been tentatively assigned. Furthermore, the reality of the NGC 5128 chain as a physical unit may be questionable; but then it is difficult to know where to stop in this "dismemberment" of loose groups, and the logical outcome of an overconservative attitude would be to exclude from consideration all but a few rich clusters and dense groups (the Local Group itself would not hold too well under this critical approach). By the definitions set up in § 1 we must conclude that not more than 8 to 14 of the 60, i.e., 13 to 23 percent, of the "outstanding" nearby galaxies are isolated in space. This is only one-quarter to one-half of Holmberg's estimate.

On the other hand, several of the eight supposedly isolated galaxies might upon further investigation turn out to be members of some of the nearer groups; in particular NGC 404, NGC 3109, and IC 5152 should be examined for possible membership in the Local Group. Other (more remote) possibilities are N1569, IC 342, and perhaps some heavily obscured systems as yet unrecognized. For example, IC 10, although long suspected, was only recently established as a Local Group member (Roberts 1962; de Vaucouleurs and Ables 1965). If this were the case, the frequency of isolated galaxies might be reduced to 10 percent or less, again depending somewhat on how strictly or loosely a group is defined. Nevertheless it seems difficult to reduce the frequency to zero; to the writer's knowledge NGC 1313 and NGC 6744 in the southern sky, and probably NGC 2903 and NGC 6946 in the northern sky, are truly isolated galaxies not associated with any nearby group, although both are in the larger Local Supercluster discussed in § 10.

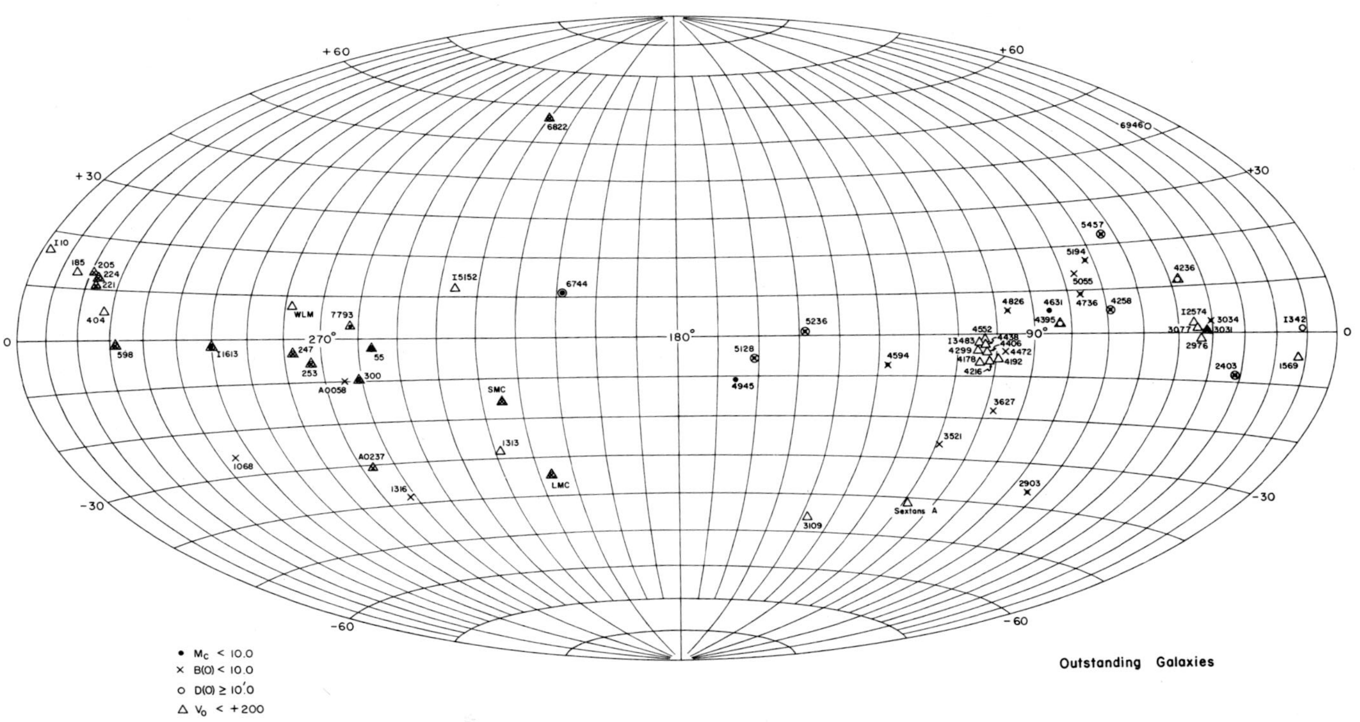

FIG. 7.—Apparent distribution of outstanding galaxies in supergalactic coordinates.

10. APPARENT AND SPACE DISTRIBUTION OF NEARBY GROUPS: LOCAL SUPERCLUSTER

The apparent distribution of nearby groups (figs. 3 and 4) strongly reflects the concentration of bright galaxies toward the plane of the Local Supercluster (Holmberg 1937; de Vaucouleurs 1953, 1956*b*, 1958, 1960*b*; Carpenter 1961; Abell 1961).

Table 10 gives counts of various systems ("outstanding" galaxies, DDO dwarfs, and

TABLE 10

SUPERGALACTIC DISTRIBUTION OF NEARBY GROUPS AND GALAXIES

Systems	Supergalactic Latitude B			
	N(0°, ± 10°)	N(0°, ± 20°)	N(0°, ± 30°)	N_T(0°, ± 90°)
Local Group (fig. 1):				
N	0 (0%)	2 (20%)	4 (40%)	10
S	4 (25%)	11 (69%)	12 (75%)	16
N+S	4 (15%)	13 (50%)	16 (62%)	26
"Outstanding" galaxies (fig. 7):				
N	25 (69%)	30 (83%)	32 (89%)	36
S	10 (42%)	18 (75%)	20 (83%)	24
N+S	35 (58%)	48 (80%)	52 (87%)	60
Dwarf galaxies (fig. 6):				
N	85 (49%)	114 (66%)	146 (85%)	172
S	16 (32%)	26 (52%)	38 (76%)	50
N+S	101 (45%)	150 (63%)	184 (83%)	222
Nearer groups (fig. 3):				
N	6 (75%)	7 (88%)	8 (100%)	8
S	5 (55%)	6 (66%)	8 (89%)	9
N+S	11 (59%)	13 (77%)	16 (94%)	17*
Nearby groups (fig. 4):				
N	5 (36%)	6 (43%)	11 (78%)	14
S	12 (38%)	18 (56%)	24 (75%)	32
N+S	17 (37%)	24 (52%)	35 (76%)	46†
All nearby groups:				
N	11 (50%)	13 (59%)	19 (86%)	22
S	17 (41%)	24 (59%)	32 (78%)	41
N+S	28 (45%)	37 (59%)	51 (81%)	62
Random‡	21%	41%	59%	100%

* Counting UMa–Cam Cloud as three groups, Coma I as two groups.

† Counting Cet I, Cet II, Vir III, Pavo-Indus as two groups each, Eri I as three groups.

‡ Including allowance for effect of galactic obscuration.

nearby groups) as a function of supergalactic latitude separately for the center sector (northern galactic hemisphere) and anticenter sector (southern galactic hemisphere). Comparison of the observed relative frequencies within 10°, 20°, or 30° from the supergalactic equator with the values computed for a random (uniform) distribution (with allowance for galactic obscuration from the Hubble counts of faint galaxies) brings out the strong concentration of all systems to the supergalactic plane. The only possible exception is the Local Group—which is not surprising considering its small volume and our location in it. For all other systems some 75–100 percent of the total population is within 30° of the supergalactic equator (means: 88% in N.G.H., 80% in S.G.H.), as compared with 59 percent for a uniform distribution. The average ratio O/C = (observed)/(computed) varies with supergalactic latitude as follows:

	(0, ±10°)	(0, ±20°)	(0, ±30°)
Galaxies: O/C......	2.3	1.75	1.45
Groups: O/C.......	2.15	1.45	1.35

The flattened local supersystem is also clearly in evidence when the space distribution of the 55 nearby groups and clouds is mapped as in figures 8 and 9. The supergalactic rectangular coordinate system is defined as follows: OX = line of nodes of supergalactic and galactic planes, $X > 0$ in direction $L = 0°$ defined by $l^{\rm I} = 105°$, $b^{\rm I} = 0°$ (de Vaucouleurs 1958, 1960*b*), OY = direction of $L = 90°$ (in Coma), OZ = direction of supergalactic north pole at $l^{\rm I} = 15°$, $b^{\rm I} = +5°$. Figure 8 is a projection in the (Y, Z)-plane; and since the center of the supercluster is apparently in the direction of the Virgo cluster ($L = 104°$), this projection approximates a meridional cross-section. Note the accumulation of groups within 5 Mpc from the supergalactic plane for $Y > -5$ Mpc. The dashed line marks the radius ($\Delta = 16$ Mpc) of the survey. The third coordinate X is shown in each group. The shaded 20° fan along the Z-axis marks the approximate limits of the galactic zone of avoidance. Note the wide gap (not due to obscuration) between the Sculptor group and the more distant groups in the southern hemisphere ($Y < 0$).

Figure 9 is a projection in the supergalactic (X, Y)-plane of the groups and clouds for which $|Z| < 3$ Mpc (*full circles*) and $3 < |Z| < 5$ Mpc (*dashed circles*). The value of Z is shown for each group. This is a first approximation map of that part of the "Local Supergalaxy" which falls within the 16 Mpc radius of the survey. Galaxy counts (Reiz 1941; de Vaucouleurs 1956*b*, 1960*b*; Carpenter 1961) indicate that the Local condensation extends to about twice this distance beyond the Virgo cluster or to $Y \simeq +30$. The Local Supercluster, then, may encompass all groups and clouds within a radius of 15–20 Mpc from a center in the general vicinity of the Virgo cluster (there is no necessity for the center of mass to be within any particular cluster) and within 5–10 Mpc from the supergalactic plane. This includes the majority of the northern nearby groups and clouds but probably excludes most of the southern clouds beyond 7 or 8 Mpc.

Note the strong indications of subclustering within the supersystem; as already noted, there is evidence for a "Local Cloud" (or cloud complex) including the Local Group, Sculptor, M81, M101, N2841, N1023, N5128 groups, and CVn I; another cloud complex could include CVn II, Coma I, and UMa I (Z); other examples are UMa I (N + S, X, Y), Virgo II (X, Y, V), and Leo II.

The Local Cloud includes all the nearer groups within $\Delta \simeq 7$ Mpc and with very few exceptions comprises only spirals and Magellanic irregulars among its 40 or 50 brighter members; the average type index is $\bar{t} = +5.1$ ($\bar{s}_t = 2.9$). For comparison the average type index is $\bar{t} \simeq +3$ for all groups in the range $7 < \Delta < 15$ Mpc and $\bar{t} \simeq 0$ for $\Delta > 15$ Mpc.

The research incorporated in this chapter was supported in part by the National Science Foundation and the U.S. Navy Office of Naval Research. The collaboration of Mrs. A. de Vaucouleurs and the assistance of Mrs. J. Weiss and Mr. H. Corwin greatly expedited the project.

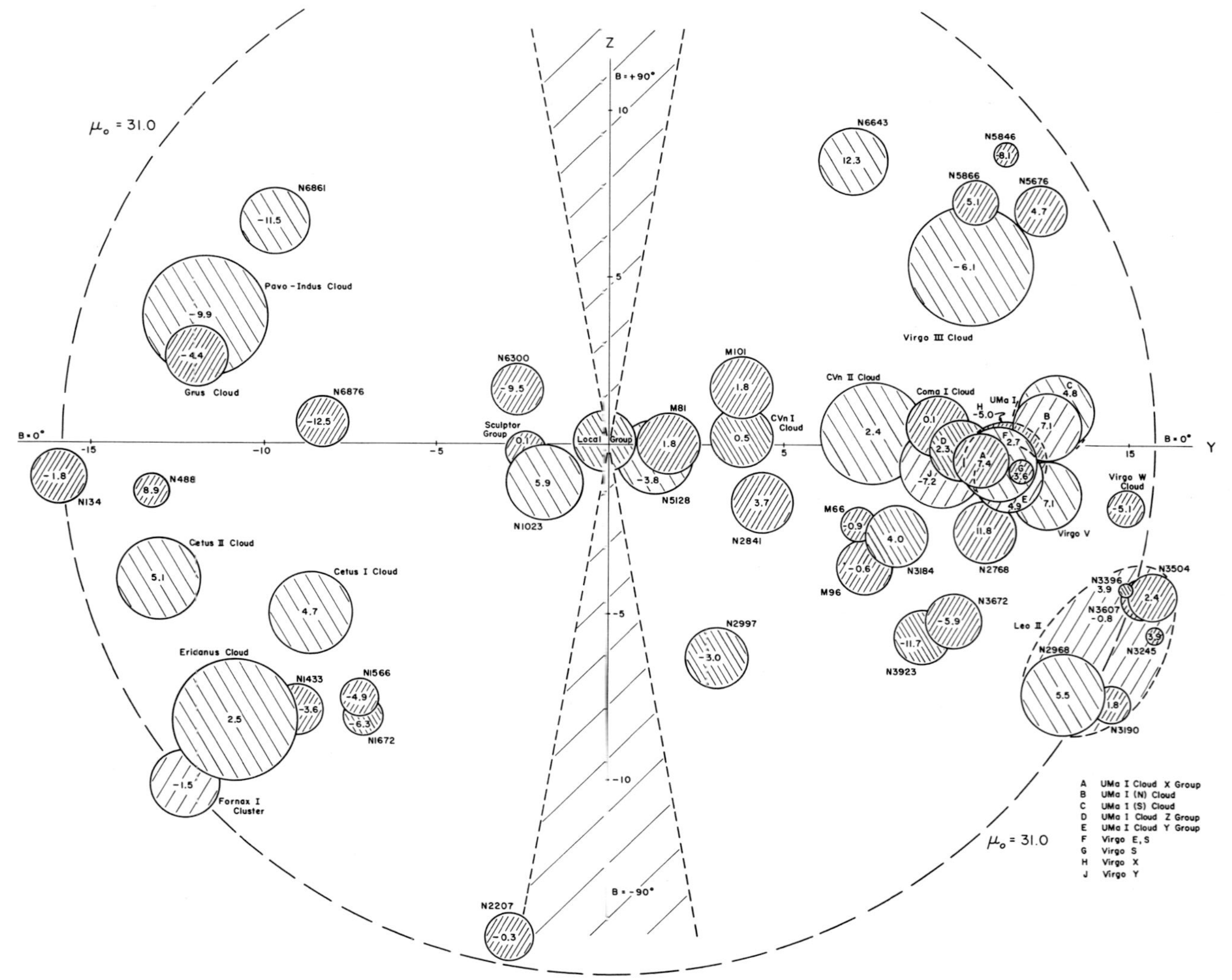

FIG. 8.—Space distribution of nearby groups projected onto the supergalactic (Z, Y) plane.

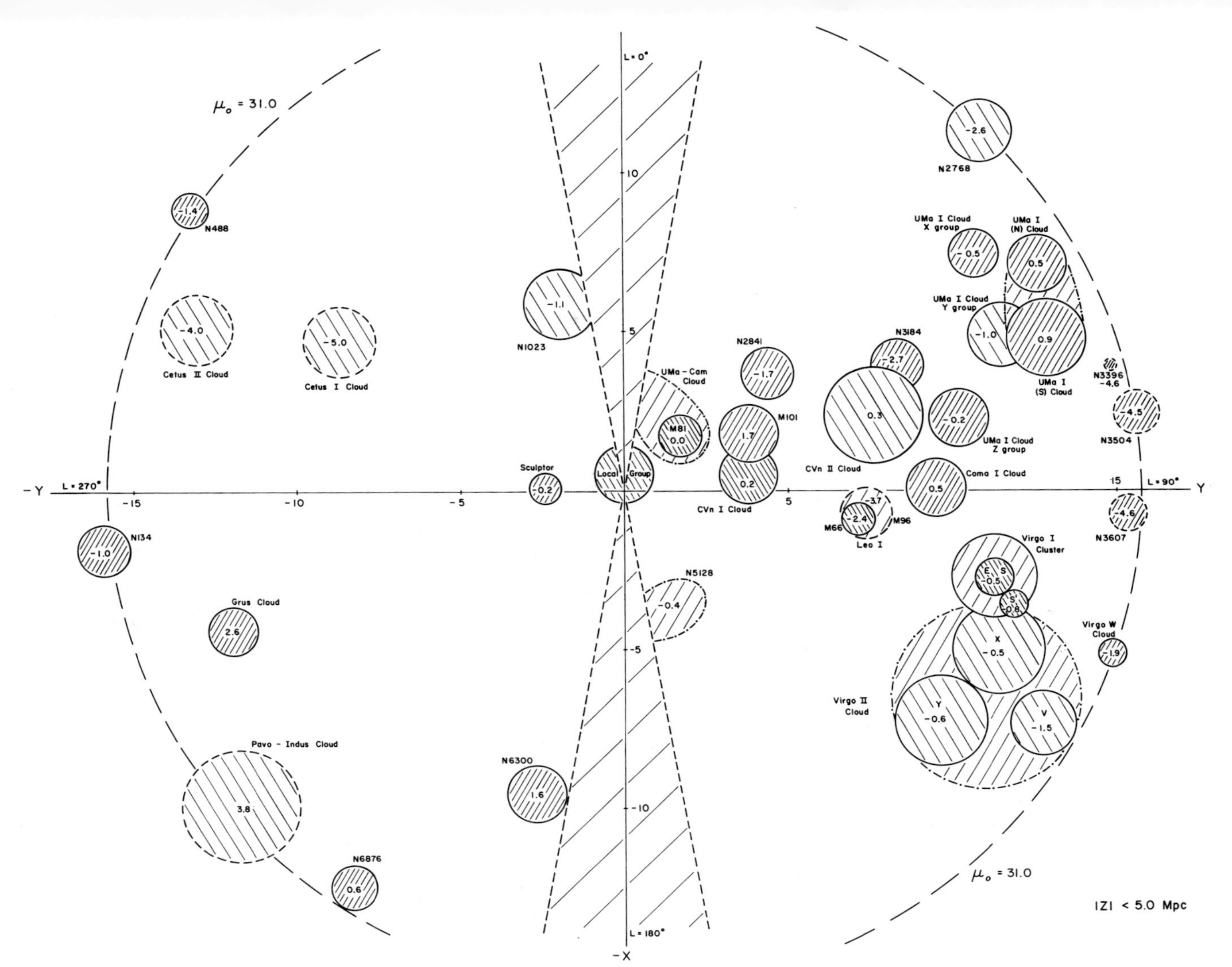

FIG. 9.—Space distribution of nearby groups with $|Z| < 5$ Mpc projected onto the supergalactic (X, Y) plane.

REFERENCES

Abell, G. O. 1961, *A.J.*, **66**, 607.
Ambartsumian, V. A. 1958, in *La Structure et l'Evolution de l'Univers*, Solvay Conference Report (Brussels: R. Stoops), p. 241.
Bergh, S. van den. 1959, *Pub. D. Dunlap Obs.*, Ser. 2, No. 5.
———. 1960*a*, *ibid.*, No. 6.
———. 1960*b*, *Zs. f. Ap.*, **49**, 198.
———. 1960*c*, *J.R.A.S. Canada*, **54**, No. 2, 49.
———. 1960*d*, *Ap. J.*, **131**, 558.
———. 1972*a*, *Ap. J.* (Letters), **171**, L31.
———. 1972*b*, *ibid.*, **178**, L99.
———. 1973, *Bull. A.A.S.*, **5**, No. 4, 448.
Bottinelli, L., Chamaraux, P., Gerard, E., Gougenheim, L., Heidmann, J., Kazes, I., and Lauque, R. 1971, *Astr. and Ap.*, **12**, 264.
Bottinelli, L., Gougenheim, L., and Heidmann, J. 1972, *Astr. and Ap.*, **18**, 121.
Carpenter, R. L. 1961, *Pub. A.S.P.*, **73**, 324.
Epstein, E. 1964, *A.J.*, **69**, 490.
Godfredsen, E. A. 1961, *Ap. J.*, **134**, 257.
Gyllenberg, W. 1937, *Medd. Lunds Astr. Obs.*, Ser. 2, No. 87.
Heeschen, D. S., and Wade, C. M. 1964, *A.J.*, **69**, 277.
Hodge, P. W. 1959, *Pub. A.S.P.*, **71**, 28.
———. 1960, *ibid.*, **72**, 188.
———. 1961, *A.J.*, **66**, 83, 249, 384.
———. 1963, *ibid.*, **68**, 470.
———. 1964, *ibid.*, **69**, 438.
Hoffmeister, C. 1962, *Zs. f. Ap.*, **55**, 46.
Holmberg, E. 1937, *Lund Ann.*, 6.
———. 1940, *Ap. J.*, **92**, 200.
———. 1950, *Medd. Lunds Astr. Obs.*, Ser. 2, No. 128.
———. 1958, *ibid.*, Ser. 2, No. 136.
———. 1961, *A.J.*, **66**, 620.
———. 1964, *Ark f. Astr.*, **3**, 387 = Uppsala Medd., No. 148.
Humason, M. L., Mayall, N. U., and Sandage, A. R. 1956, *A.J.*, **61**, 97.
Humason, M. L., and Wahlquist, H. D. 1955, *A.J.*, **60**, 254.
Kahn, F. D., and Woltjer, L. 1959, *Ap. J.*, **130**, 705.
Limber, D. N. 1961, *A.J.*, **66**, 572.
Morgan, W. W. 1958, in *La Structure et l'Evolution de l'Univers*, Solvay Conference Report (Brussels: R. Stoops), p. 297.
Neyman, J., Page, T., and Scott, E. L. 1961, *A.J.*, **66**, 633.
Oort, J. H. 1958, in *La Structure et l'Evolution de l'Univers*, Solvay Conference Report (Brussels: R. Stoops), p. 163.
Reaves, G. 1956, *A.J.*, **61**, 69.
Reiz, A. 1941, *Lund Ann.*, 9.
Roberts, M. S. 1962, *A.J.*, **67**, 431.
Robinson, B. J., and van Damme, K. J. 1964, in *The Galaxy and the Magellanic Clouds*, IAU-URSI Symp. No. 20, ed. F. J. Kerr and A. W. Rodgers (Canberra: Australian Academy of Science), p. 276.
Sandage, A. R. 1961, in *Problems of Extragalactic Research*, IAU Symp. No. 15, ed. G. C. McVittie (New York: Macmillan), p. 359.
Sersic, J. L. 1960a, *Zs. f. Ap.*, **50**, 168.
———. 1960*b*, *ibid.*, **51**, 64.
———. 1964, *ibid.*, **58**, 259.
Shapley, H., and Ames, A. 1932, *Harvard Ann.*, **88**, No. 2.
Shobbrook, R. R. 1966, *M.N.R.A.S.*, **131**, 365.
Shobbrook, R. R., and Robinson, B. J. 1967, *Aust. J. Phys.*, **20**, 131.
Vaucouleurs, G. de. 1953, *A.J.*, **58**, 30.
———. 1955, *Pub. A.S.P.*, **67**, 350.
———. 1956*a*, *Mem. Mt. Stromlo Obs.*, Ser. 3, No. 13.
———. 1956*b*, in *Vistas in Astronomy*, ed. A Beer (London and New York: Pergamon Press), Vol. 2, p. 1584.
———. 1958, *A.J.*, **63**, 253.
———. 1959, *Ap. J.*, **130**, 718.
———. 1960*a*, *ibid.*, **131**, 585.
———. 1960*b*, *Soviet Astr.—AJ*, **3**, 897.
———. 1961*a*, *Ap. J.*, *Suppl.*, No. 48, **5**, 233.
———. 1961*b*, *Ap. J.*, **133**, 405.
———. 1961*c*, *Ap. J.*, *Suppl.*, No. 56, **6**, 213.

———. 1961*d*, *A.J.*, **66**, 629.
———. 1964, *Ap. J.*, **139**, 899.
Vaucouleurs, G. de, and Ables, H. 1965, *Pub. A.S.P.*, **77**, 272.
Vaucouleurs, G. de, and Malik, G. M. 1969, *M.N.R.A.S.*, **142**, 387.
Vaucouleurs, G. de, and Page, J. 1962, *Ap. J.*, **136**, 107.
Vaucouleurs, G. de, and Vaucouleurs, A de. 1964, *Reference Catalogue of Bright Galaxies* (Austin: University of Texas Press) (BGC).
Vaucouleurs, G. de, and Vaucouleurs, A. de. 1973, *Astr. and Ap.*, **28**, 109.
Wilson, A. G. 1955, *Pub. A.S.P.*, **67**, 17.
Zwicky, F. 1957. *Morphological Astronomy* (Berlin: Springer-Verlag), p. 222.

ADDENDUM TO CHAPTER 14 (JANUARY 1973)

The text and material for this chapter were prepared during the summer of 1965 and based on data in the *Reference Catalogue of Bright Galaxies* (BGC). (A supplementary survey of more distant groups was prepared by Corwin [1967], giving a total of 109 groups with estimated distances and mean velocities.) This material has been used for studies of the motion of the Sun with respect to galaxies (Stewart and Sciama 1967; de Vaucouleurs and Peters 1968) and of the velocity dispersion in groups (Rood, Rothman, and Turnrose 1970; Chamaraux, Montmerle, and Tadokoro 1972; Geller and Peebles 1973; Turner and Sargent 1974). A disturbing feature of the distance scale derived for the groups in this chapter was the implied nonlinearity of the velocity-distance relation at small distances (de Vaucouleurs 1972).

In the meantime better data have become available: (1) systematic errors in estimates of galaxy diameters have been thoroughly investigated (Heidmann, Heidmann, and de Vaucouleurs 1971); (2) additional redshift determinations have accumulated (G. and A. de Vaucouleurs 1967 and other sources), leading to some revisions of group membership; and, especially, (3) much progress has been made in the derivation of isophotal diameters and integrated magnitudes of galaxies (revision of BGC in preparation).

Recently an application of this new material was made (with the collaboration of H. Corwin) to the derivation of revised distances for 77 groups having the best current data, including the 54 groups discussed in this chapter. The revised distances are listed in table A1. From various statistical tests these new distances appear to have very small errors, on the order of 0.1–0.2 mag in the distance moduli. A satisfactory feature of this new distance scale is that it removes almost completely the nonlinearity of the Hubble diagram for nearby groups, at least in the southern galactic hemisphere; a slight nonlinearity persists in the northern hemisphere, however, where the ideal velocity field is apparently perturbed by the local density excess associated with the Local Supercluster (fig. A1). Furthermore, the average value of the Hubble constant indicated by the groups is still about 100 km s^{-1} Mpc^{-1}, in agreement with independent estimates by van den Bergh (1970 and this volume) and Roberts (1972), but in disagreement with the low values of $\sim$50 preferred by Sandage (1972 and this volume) and Abell (1972). This persistent discrepancy between different methods all relying on the same basic calibration (Local Group and nearest groups) indicates a continuing need for further critical analysis of all approaches to this fundamental problem of the distance scale.

TABLE A1

REVISED DISTANCES OF 54 GROUPS*

Group	V	V_0	μ_c	μ_{0c}	Δ	"H"
G01 Scl	230	194	27.16	26.96	2.5	78.7
G02 M81	25	160	27.68	27.38	3.0	53.4
G03 CVn I	280	342	28.47	28.27	4.5	75.9
G04 N5128	530	319	28.49	27.99	4.0	80.3
G05 M101	360	508	28.91	28.71	5.5	91.8
G06 N2841	550	589	29.83	29.53	8.0	73.2
G07 N1023	472	619	29.07	28.47	4.9	125.2
G08 N2997	820	534	30.47	29.87	9.4	56.6
G09 M66	705	592	29.33	29.13	6.7	88.3
G10 CVn II	697	747	29.52	29.32	7.3	102.4
G11 M96	876	741	29.50	29.30	7.2	102.3
G12 N3184	630	629	29.98	29.78	9.0	69.6
G13 Coma I	954	944	29.75	29.55	8.1	115.9
G14 N6300	1411	1270	31.27	30.27	11.3	112.1
G15 Cet I	1500	1513	30.67	30.37	11.8	127.9
G16 N1566	1220	999	31.22	30.92	15.3	65.5
G17 UMa I (Z)	947	979	30.56	30.36	11.8	83.0
G18 Vir S	1152	1087	30.51	30.31	11.5	94.2
G19 Vir E	1093	1013	30.87	30.67	13.6	74.4
G20 Vir Y	1442	1307	30.80	30.50	12.6	103.9
G21 N1433	845	665	31.16	30.96	15.6	42.7
G22 N1672	....	....	31.22	30.92	15.3	....
G23 N3672	1783	1583	31.60	31.30	18.2	86.9
G24 UMa I (Y)	918	998	31.04	30.84	14.7	67.8
G25 Vir S′	1631	1541	31.16	30.96	15.5	99.1
G26 Vir X	1375	1272	30.98	30.78	14.3	88.7
G27 Grus	1583	1561	31.48	31.28	18.0	86.7
G28 UMa I (X)	1025	1090	30.35	30.15	10.7	101.7
G29 Vir III	1749	1729	30.90	30.60	13.2	131.2
G30 N5866	740	920	30.99	30.69	13.7	67.1
G31 Eri I	1680	1574	30.83	30.63	13.4	117.7
G32 UMa I (S)	966	1016	31.11	30.91	15.2	66.7
G33 Cet II	1880	1929	31.19	30.99	15.8	122.4
G34 UMa I (N)	984	1074	30.73	30.53	12.8	84.1
G35 Vir V	1015	879	31.45	31.25	17.8	49.3
G36 N2207	2050	1827	30.54	30.04	10.2	179.6
G37 N5676	2261	2411	31.44	31.24	17.7	136.1
G38 N6876	....	....	31.10	30.70	13.8	....
G39 N0134	1800	1786	31.86	31.66	21.5	83.1
G40 N0488	2236	2336	30.91	30.71	13.9	168.4
G41 N2768	1590	1681	30.98	30.68	13.7	122.8
G42 N2964	1515	1463	31.36	31.16	17.1	85.7
G43 N3396	1635	1599	31.48	31.28	18.0	88.7
G44 N3923	1840	1635	31.97	31.67	21.6	75.7
G45 Pav-Ind	2275	2224	32.20	31.90	24.0	92.6
G46 Vir W	2280	2168	31.83	31.63	21.2	102.4
G47 N3190	1295	1198	31.42	31.22	17.5	68.5
G48 N3504	1490	1437	31.15	30.95	15.5	92.8
G49 N3607	1150	1057	31.67	31.47	19.7	53.7
G50 N5846	1793	1806	31.27	30.97	15.6	115.8
G51 N6643	1690	1938	31.80	31.40	19.1	101.7
G52 N6861	2940	2909	32.24	31.94	24.4	119.3
G53 For I	1615	1464	31.59	31.39	18.9	77.3
G54 N3245	1315	1256	31.82	31.62	21.1	59.5

* V = mean heliocentric velocity; V_0 = mean galactocentric velocity; μ_c = revised apparent modulus; μ_{0c} = revised modulus corrected for extinction in the Galaxy; Δ = revised distance in megaparsecs; "H" = velocity-distance ratio V_0/Δ in km s^{-1} Mpc^{-1}.

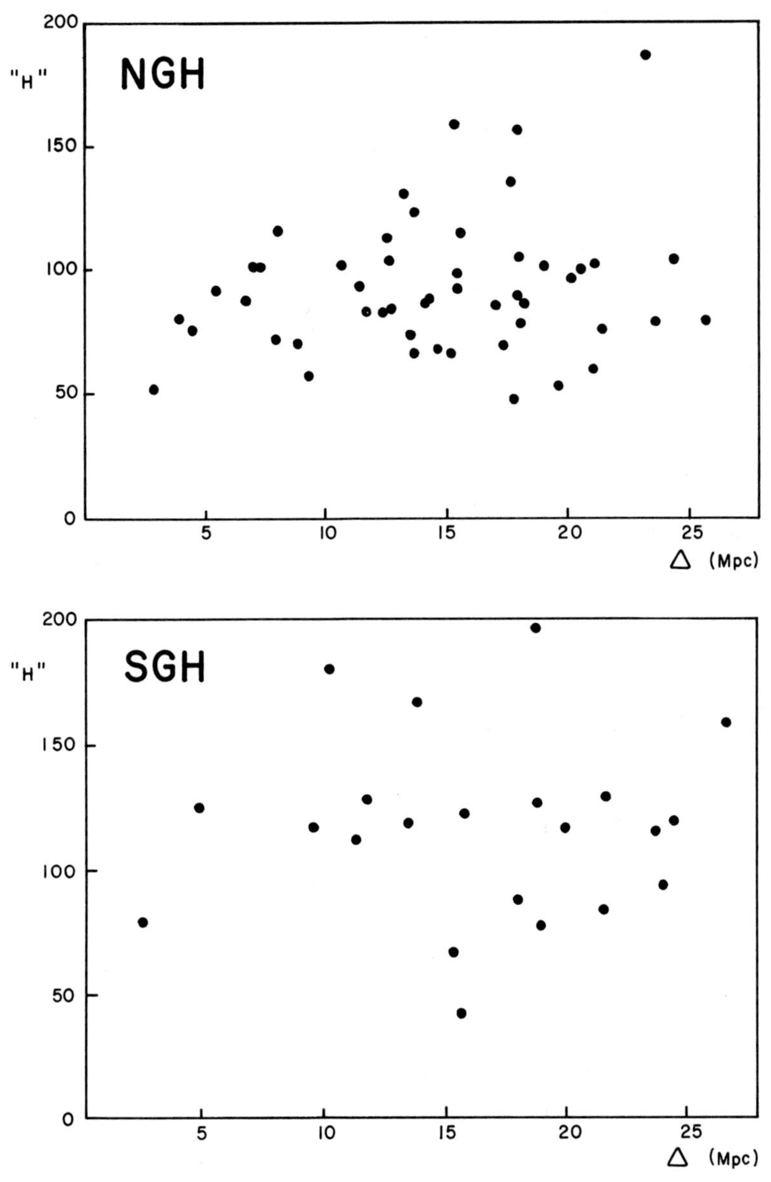

FIG. A1.—Velocity-distance ratio for revised distances of 72 nearby groups in northern and southern galactic hemispheres. Note absence of correlation in SGH (cf. fig. 1 in de Vaucouleurs 1972).

REFERENCES

Abell, G. O. 1972, in *External Galaxies and Quasi-stellar Objects*, IAU Symposium No. 44, ed. D. S. Evans (Dordrecht: Reidel), p. 350.
Bergh, S. van den. 1970, *Nature*, **225**, 503.
Chamaraux, P., Montmerle, T., and Tadokoro, M. 1972, *Ap. and Space Sci.*, **15**, 383.
Corwin, H. G. 1967, unpublished M.A. thesis, University of Kansas.
Geller, M. J., and Peebles, P. J. E. 1973, *Ap. J.*, **184**, 329.
Heidmann, J., Heidmann, N., and Vaucouleurs, G. de. 1971, *Mem. R.A.S.*, **75**, Parts 4–6, 85–104.
Roberts, M. 1972, in *External Galaxies and Quasi-stellar Objects*, IAU Symposium No. 44, ed. D. S. Evans (Dordrecht; Reidel), p. 32.
Rood, H. J., Rothman, V. C. A., and Turnrose, B. E. 1970, *Ap. J.*, **162**, 411.
Sandage, A. 1972, *Ap. J.*, **178**, 1.
Stewart, J. M., and Sciama, D. W. 1967, *Nature*, **216**, 742.
Turner, E. L., and Sargent, W. L. W. 1974, *Ap. J.*, **194**, 587.
Vaucouleurs, G. de. 1972, in *External Galaxies and Quasi-stellar Objects*, IAU Symposium No. 44, ed. D. S. Evans (Dordrecht: Reidel), p. 353.
Vaucouleurs, G. de, and Peters, W. L. 1968, *Nature*, **220**, 868.
Vaucouleurs, G. de, and Vaucouleurs, A. de. 1967, *A.J.*, **72**, 730.

CHAPTER 15

Clusters of Galaxies

G. O. ABELL

Department of Astronomy, University of California, Los Angeles

1. INTRODUCTION

OUR KNOWLEDGE of the large-scale structure of the Universe is limited by our ability to recognize and observe remote objects, and to determine their distances and physical properties. Clusters of galaxies may be the most fundamental condensations of matter in space. In any case, they are the systems which we can most conveniently recognize, and for which estimates of at least relative distances are possible. The study of clusters of galaxies, therefore, in addition to its own intrinsic interest, may well provide the clues to our understanding of the Universe as a whole.

2. NUMBERS AND CATALOGS OF CLUSTERS

Long before the era of extragalactic astronomy it was recognized that the small-scale distribution of "nebulae" in space is not random. Even the small sample of the 35 Messier objects now recognized as galaxies exhibits this nonrandomness; nearly half of these objects are in the vicinity of the Virgo cluster. Plots of the distribution on the sky of "nebulae" listed in John Herschel's *General Catalogue*, or in Dreyer's *New General Catalogue* and *Index Catalogue*, reveal several pronounced concentrations now recognized as nearby clusters of galaxies. Two of these, the Perseus cluster and the Coma cluster, were described in some detail by Wolf (1902, 1906) early in the century. Also, Wirtz (1923, 1924*a*, *b*), who supplemented the distribution of NGC and IC objects with data from the surveys of Fath, Curtis, and himself, called attention to several conspicuous well-defined centers of clustering.

Following Hubble's demonstration of the nature of galaxies in 1924, several systematic surveys of the distribution of galaxies were made. Among these are the extensive surveys by Shapley and his co-workers at Harvard in the two decades following 1930. At least four distinct clusters—the Virgo cluster, the Ursa Major cloud, and two clusters in Fornax—are apparent in the distribution of the bright galaxies in the Shapley-Ames catalog (1932). Shapley himself (1933) catalogued and described 25 individual clusters. The Harvard surveys to faint limiting magnitudes in the southern hemisphere with the Bruce telescope revealed still greater clustering, and Shapley (e.g., 1957) has called

Received January 1966; revised January 1974.

attention to a major unevenness in surface distribution, which apparently cannot be described as local clustering but which suggests "metagalactic structure" or superclustering.

Hubble's (1934) fundamental investigation of the distribution of galaxies over the sky provided further information on the clustering tendency. The survey was conducted on 1283 small selected regions of the sky which were intentionally chosen to avoid most of the well-known clusters; nevertheless, from the scanty evidence available to him, Hubble estimated there to be one great cluster (with hundreds of member galaxies) for every 50 square degrees of the sky. He estimated that if the great clusters averaged 500 members each, they would account for about 1 percent of the total number of observed galaxies. Even after omitting the great clusters from his investigation, however, Hubble found the galaxian distribution to be nonrandom; he interpreted a skewness in the frequency distribution of the numbers of fields with various numbers of counted galaxies as a tendency toward small-scale clustering (see Hubble 1936*a*). According to Hubble, the groups and clusters could not be merely superposed on a random distribution of isolated galaxies; either condensations in the general field produced the clusters, or evaporation of galaxies from clusters populated the general field.

Meanwhile, additional individual clusters were discovered, often by accident. Tombaugh (1937) described three at low galactic latitudes. Zwicky discovered several new clusters with the 18-inch Schmidt telescope on Palomar Mountain, and investigated these and other clusters with that instrument, and later with the 48-inch Schmidt and 200-inch telescopes (for the best summary, see Zwicky 1957). The general prevalence of clustering led him (Zwicky 1938) to propose that all galaxies belong to clusters and that the Universe can be regarded as divided into cluster "cells" with a mean diameter of about 7.5×10^6 pc. (Zwicky's estimate was based on the old distance scale; for the Hubble constant $H = 50$ km s^{-1}, the diameter of a "cluster cell" would be about 7.5×10^7 pc.)

Since World War II, two photographic surveys of the sky have demonstrated conclusively that clusters of galaxies are extremely numerous—far more so than most investigators had believed; these are the Lick 20-inch *Astrographic Survey* and the National Geographic Society–Palomar Observatory *Sky Survey*. From counts of galaxy images on the Lick photographs, Shane and his collaborators (Shane and Wirtanen 1954; Shane 1956*a;* Shane, Wirtanen, and Steinlin 1959; Shane and Wirtanen 1967) have prepared catalogs and charts of the surface density of galaxies (per square degree). They have called attention to many striking clusters and clouds of galaxies, and even to apparent superclusters. The Lick counts have been analyzed statistically under the direction of Scott and Neyman at the Berkeley Statistical Laboratory (Neyman and Scott 1952; Neyman, Scott, and Shane 1953, 1954; Scott, Shane, and Swanson 1954). It was found that the serial correlation between counts in $1° \times 1°$ squares persists to square separations of about 4°, and that it has almost the same value in both galactic polar caps. From the shape of the quasi-correlation function, the statisticians derived parameters that describe the scale and amplitude of the clustering. They then applied these parameters to the manufacture of a "synthetic" field of galaxy images based on a model that assumes that all galaxies are in clusters. Comparison of the synthetic field with fields plotted from actual plates showed a striking similarity between the prediction of the model and the observed distribution of galaxies; if anything, the actual fields displayed a slightly greater clustering tendency than the synthetic one. Neyman, Scott, *et al.* found typical clusters to have populations of the order 10^2 and diameters of a few million parsecs.

Limber (1953, 1954) also investigated the distribution of galaxies from Shane and Wirtanen's counts on the Lick plates and fitted it to a model in which the spatial density of galaxies is a smoothly varying function of position in space. Limber thereby derived an independent estimate of the scale of clustering. Although his procedure was criticized by Neyman and Scott (1955) on the grounds that the discrete nature of galaxies is incompatible with a smoothly varying function describing their distribution among volume cells in space, Limber's results are in qualitative agreement with those of Neyman, Scott, *et al.* A model which takes account of the discrete fluctuations that arise because of the discrete character of the distribution of galaxies as well as those that arise from clustering, and which also applies to a distribution that exhibits clustering but that does not consist entirely of discrete and independent clusters, has been suggested by Layzer (1956).

Tens of thousands of discrete groups and clusters of galaxies are easily identified on the Palomar *Sky Survey* photographs. The writer (Abell 1958) has catalogued 2712 of the very richest of these, and has analyzed the distribution of 1682 clusters that comprise a more or less homogeneous sample chosen from the catalog. Also from the Palomar plates, Zwicky and his associates have prepared a far more extensive *Catalogue of Galaxies and Clusters of Galaxies* (Zwicky, Herzog, Wild, Karpowicz, and Kowal 1961–68). Some statistics on the sizes of the largest of these clusters and on the area of the sky covered by them have been published (Zwicky and Rudnicki 1963, 1966; Zwicky and Berger 1965; Zwicky and Karpowicz 1965, 1966). Finally, Klemola (1969) and Snow (1970) list an additional 78 groups and clusters of southern-hemisphere galaxies discovered on plates taken for the Yale-Columbia proper-motion survey.

Thus it is recognized today that clustering of galaxies is a dominant tendency, and may be fundamental. It is, in fact, not impossible that all or nearly all galaxies belong to clusters, or at least that they were originally formed in clusters. We speak of the general "field" or "background" of noncluster galaxies, but the extent to which such a "field" has physical significance is not known at present. As was shown by the Berkeley statistical investigation of the Lick counts, the impression of a field of noncluster objects can be created by many clusters and groups (in many of which only the one or two brightest members may be visible) seen overlapping in projection. Only those systems that stand out conspicuously against the field (whatever its nature), either because they are unusually rich aggregates or because their members lie in very close proximity to each other in comparison to intergalactic distances, will be recognized as discrete groups or clusters. Obviously small groups will not be identified unless they are relatively nearby. Even rich clusters are increasingly difficult to distinguish from the field at increasing distances. Our knowledge about clusters, therefore, tends to be biased toward the richest systems, and even for them it is never possible to say with certainty which galaxies (except statistically) are cluster members.

The very definition of a cluster or group of galaxies, therefore, is not a trivial matter. Such properties as total sizes, total populations, distributions of member galaxies, and luminosity functions depend very critically on how the clusters are defined and how their members are distinguished from the field. Several investigators (e.g., Zwicky and Abell) have given rather specific operational definitions of clusters for the purposes of their respective studies. Many of the differences in opinion among them concerning the nature of clusters stem from the fact that they have defined clusters in different ways. It is beyond the scope of this review to attempt to provide a definitive definition of a cluster of galaxies. We shall, however, try to be specific about the definitions and assumptions on which the results discussed depend.

3. OBSERVED PROPERTIES OF CLUSTERS

3.1. Types of Clusters

Clusters of galaxies display a wide variety of morphological forms, ranging from rich aggregates of thousands of members to the relatively poor groups, like the Local Group which contains only 17 to 20 known members, or even to double or triple systems if these can be classed as clusters. The smaller groups appear to be by far the most numerous, but at present there exist no reliable data on the actual relative numbers of rich clusters and poor groups. Enough is known, however, to classify the clusters into certain broad categories. For the purposes of the *Catalogue of Galaxies and Clusters of Galaxies*, Zwicky (Zwicky, Herzog, Wild, Karpowicz, and Kowal 1961–1968) classifies clusters as *compact*, *medium compact*, and *open*. He defines a *compact* cluster as one with a single pronounced concentration of galaxies, in which 10 or more galaxies appear (in projection) to be in contact; a *medium compact* cluster is one with a single concentration within which galaxies appear to be separated by several of their diameters, or in which there are several pronounced concentrations of galaxies; an *open* cluster is one without any pronounced peak of population, but which appears as a loose cloud of galaxies superposed on the general field.

A more recent classification of galaxian clusters was suggested by Morgan (1961) on the basis of his study of the 20 nearest clusters in Abell's (1958) catalog. He found that the clusters investigated could be divided into two categories, according to the types of galaxies encountered among their brightest members: (i) those containing appreciable numbers of galaxies of minor central concentration of light (late spiral and irregular galaxies); and (ii) those containing few or none of the latter.

The classifications assigned to clusters by the Morgan and Zwicky systems are not independent, but are strongly correlated. The main features of both systems are preserved by simply classifying clusters as *regular* or *irregular*. Although the demarcation between the two classes is not a sharp one, they do uniquely describe many of the morphological characteristics of most clusters. *Regular* clusters are all rich, having populations of the order of 10^3 or more in the interval of the brightest 6 magnitudes. They show high central concentration and marked spherical symmetry. Their memberships consist entirely, or almost entirely, of galaxies without conspicuous dust—E and S0 galaxies. Examples are the famous clusters in Coma and Corona Borealis (fig. 1) (Abell catalog numbers 1656 and 2065, respectively). Most of Zwicky's compact clusters and Morgan's type ii clusters belong to this class. *Irregular* clusters range from poor groups, like the Local Group, to relatively rich aggregates like the Virgo cluster or the Hercules cluster (fig. 2) (Abell number 2151). They contain little or no spherical symmetry, and

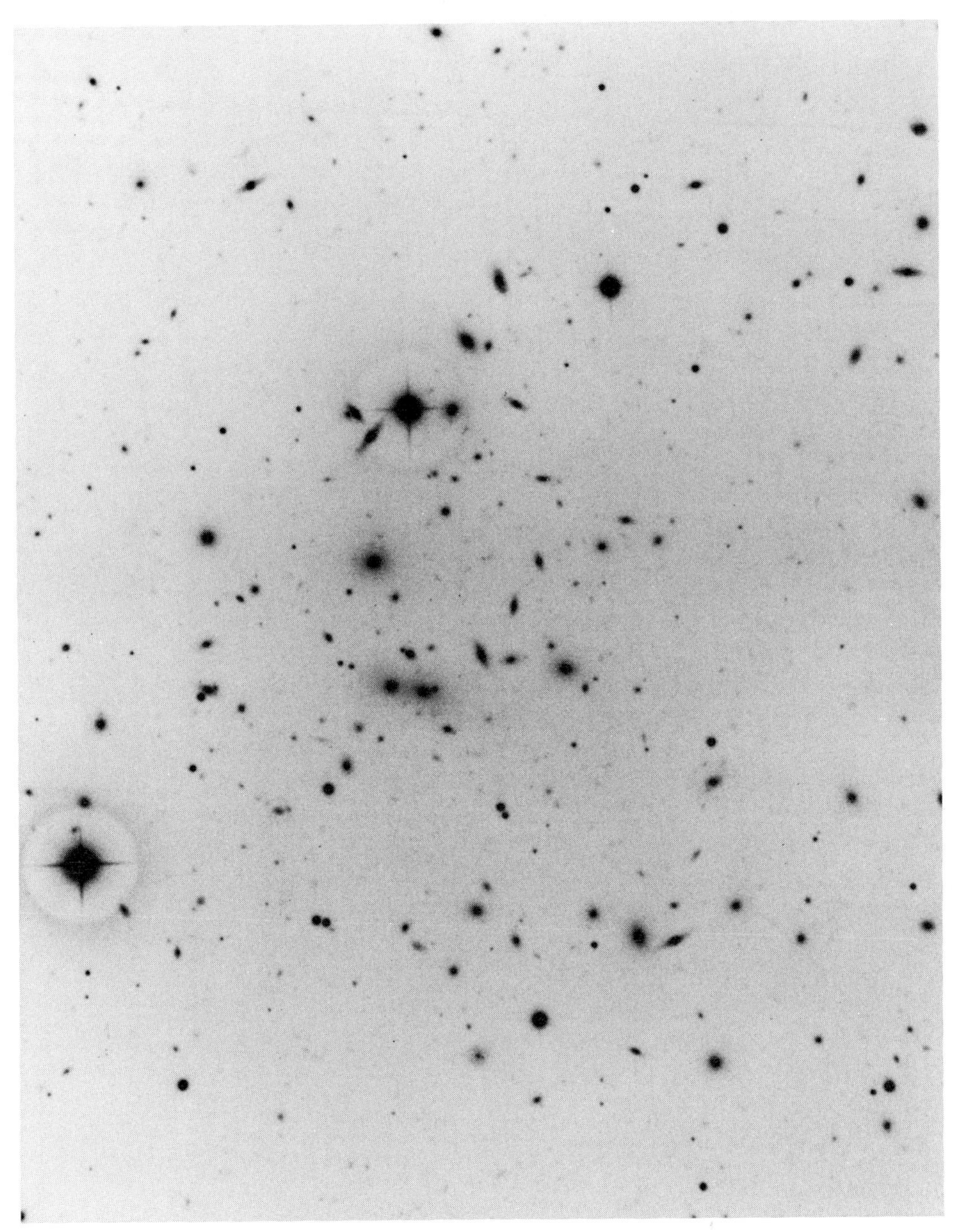

FIG. 1.—The regular cluster, number 2065 (Corona Borealis), photographed with the 200-inch telescope. Scale: 1 mm = 3.9″.

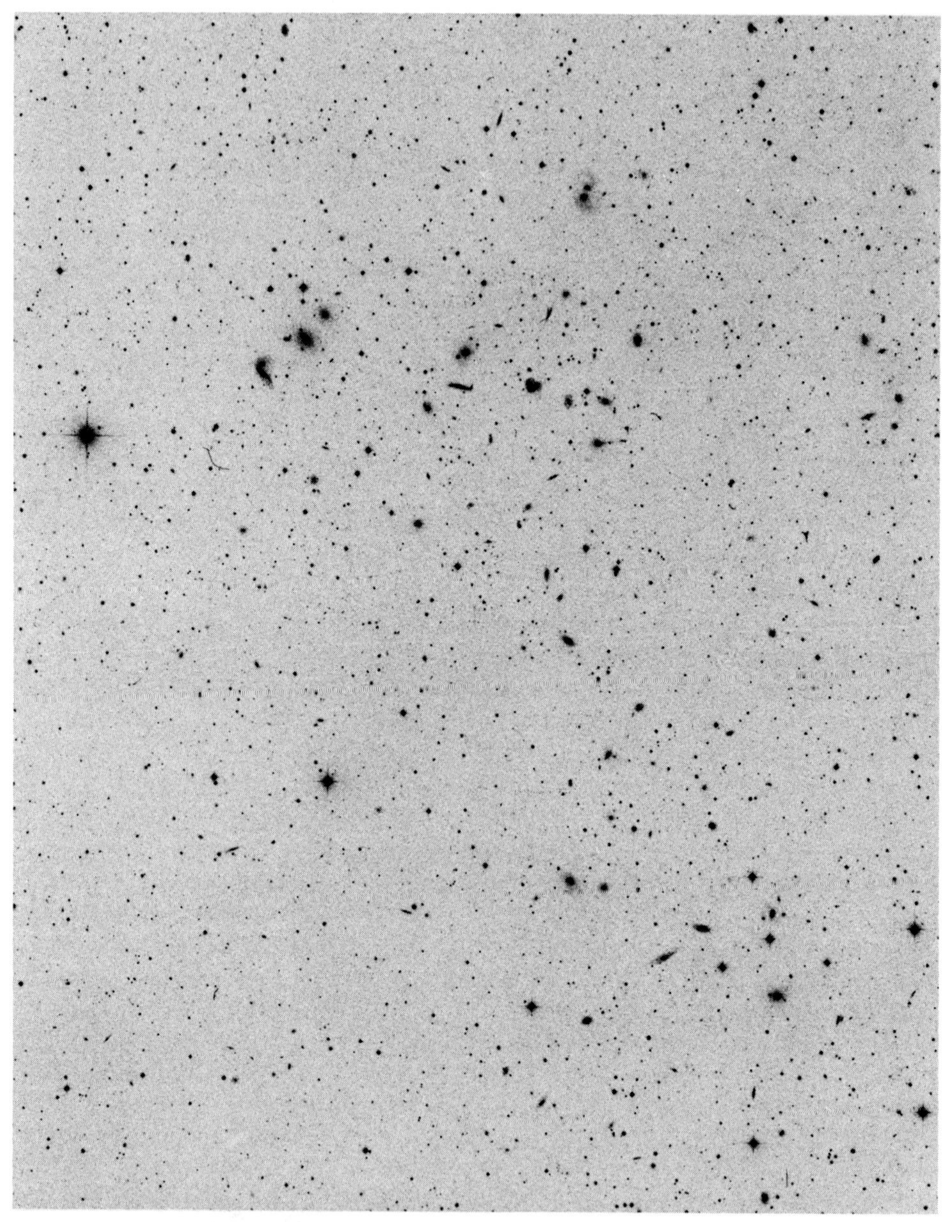

FIG. 2.—The irregular cluster, number 2151 (Hercules), photographed with the 48-inch Schmidt telescope. Scale: 1 mm = 22.3″.

no marked central concentration, although multiple condensations are often present. These clusters usually contain galaxies of all types, including appreciable numbers of late-type spirals and irregulars. To this class belong Zwicky's medium-compact and open clusters, and Morgan's type i clusters. The principal features of regular and irregular clusters are summarized in table 1; explanatory discussion follows.

TABLE 1

TYPICAL CHARACTERISTICS OF REGULAR AND IRREGULAR CLUSTERS

Parameter	Regular Clusters	Irregular Clusters
Symmetry	Marked spherical symmetry	Little or no symmetry
Concentration	High concentration of members toward cluster center	No marked concentration to a unique cluster center; often two or more nuclei of concentration are present
Types of galaxies	All or nearly all galaxies in the first 3 or 4 magnitude intervals are elliptical and/or S0 galaxies.	All types of galaxies are usually present except in the poor groups, which may not contain giant ellipticals. Late-type spirals and/or irregular galaxies present
Number of member galaxies in range of brightest 7 mag	Order of 10^3 or more	Order of 10^1 to 10^3
Diameter (Mpc)	Order of 1–10	Order of 1–10
Presence of subclustering	Probably absent or unimportant	Often present. Double and multiple systems of galaxies common
Dispersion of radial velocities of members about mean for cluster	Order of 10^3 km s^{-1}	Order of 10^2–10^3 km s^{-1}
Mass derived from virial theorem (see §4)	Order of $10^{15}\mathfrak{M}_\odot$	Order of 10^{12}–$10^{14}\mathfrak{M}_\odot$
Other characteristics	Cluster often centered about one or two giant elliptical galaxies	...
Examples	Coma cluster (No. 1656); CrB cluster (No. 2065)	Local Group, M81 group, Virgo cluster, Hercules cluster (No. 2151)

3.2. GALAXIAN CONTENT OF CLUSTERS

As already stated, spiral and irregular galaxies appear to be rare or lacking in regular clusters, but are common in the irregular ones. The Local Group, although a poor irregular cluster, is the one for which the most complete census exists. It contains three spiral galaxies, with M_v in the range -19 to -21, four irregular galaxies ($-14 \geq M_v \geq -18$),[1] four intermediate ellipticals ($-14 \geq M_v \geq -17$),[2] and up to nine dwarf ellipticals of the Sculptor type ($-9 \geq M_v \geq -14$), including three suspected companions of M31 discovered by van den Bergh (1972). There are no giant ellipticals that are certain members of the Local Group, but evidence from the rich clusters suggests that the luminosity function of elliptical galaxies increases with increasing absolute magnitude and giant ellipticals are rare; perhaps we should not expect to find one in a sample of only 17 to 20 galaxies. Undiscovered members of the Local Group may well exist,

[1] Galaxies in the intermediate luminosity range ($-14 \geq M_v \geq -18$)—e.g., NGC 147, IC 1613, and NGC 205—are often referred to as dwarfs; but since galaxies of lower luminosity appear to be even more common, here we shall use the term "dwarf" for the latter.

[2] See n. 1.

especially in the direction of the Milky Way. In particular, there are probably galaxies of lower luminosity than the Ursa Minor system (the least luminous known member of the Local Group). Remote globular clusters are known to exist, for example, at distances up to 10^5 pc (Abell 1955); if such systems are distributed uniformly throughout the Local Group, and are not just outlying members of our own Galaxy, they can properly be classed as galaxies themselves, as Zwicky (1957) and others have proposed. There is no certainty that the Local Group does not contain ever larger numbers of objects of ever smaller mass, ranging down to individual stars.

The irregular Virgo cluster displays a wider range of galaxian types. The majority of the brightest galaxies are spirals; 205 Shapley-Ames galaxies lie in the region of the Virgo cluster: 68 percent are spirals, and only 19 percent are ellipticals, the remainder being irregular or unclassified. The incidence of ellipticals in the Virgo cluster, however, is still significantly higher than among the brightest galaxies in the field. Moreover, the fraction of cluster galaxies that are elliptical increases at fainter magnitudes; of the 272 brightest galaxies in a $6\overset{\circ}{.}6 \times 6\overset{\circ}{.}6$ region centered at (1950) $\alpha = 12^h25^m$, $\delta = +13°$, 53 percent have been classified by the writer as spirals and irregulars, and 47 percent ellipticals and S0's. According to Shapley (1950), just under half of the Virgo galaxies are spirals in the magnitude range 12–14, and only about a quarter in the range 14–16. Reaves (1956, 1967) has searched for faint galaxies in the Virgo cluster on plates taken with the Lick 20-inch astrograph. Of about 1000 objects that Reaves considers possible or probable cluster members, he describes 76; only 11 of these are definitely not of the elliptical type, and cluster membership of some of the 11 is in doubt. The majority of Reaves's objects appear to be similar to IC 3475. They are probably elliptical galaxies of intermediate to low luminosity, ranging from objects like NGC 221 to the brightest Sculptor-type systems in the Local Group; however, Reaves (1962) emphasizes that this interpretation must be viewed with caution until accurate colors are available. Faint dwarf ellipticals (like the Draco and Ursa Minor systems) might be expected to exist in still greater numbers in the Virgo cluster, but they would not have been detected in Reaves's survey. Elliptical galaxies slightly brighter than Sculptor-type dwarfs have also been observed in some other nearby clusters, in particular the M81 group and the Fornax cluster (Hodge 1959, 1960; Hodge, Pyper, and Webb 1965). The scanty evidence available, therefore, suggests that dwarf ellipticals are very common in nearby irregular clusters, and could well be the most numerous kind of galaxy in all clusters.

Whereas spirals are common among the brighter galaxies in irregular clusters, and are actually in the majority among the bright Virgo galaxies, the very brightest cluster members tend to be giant ellipticals—unless the cluster population is small. The four brightest members of the Virgo cluster, according to Holmberg (1958), are the giant ellipticals NGC 4472 ($m_{pv} = 8.5$), NGC 4486 ($m_{pv} = 8.7$), NGC 4649 ($m_{pv} = 9.0$), and NGC 4406 ($m_{pv} = 9.2$). If the distance modulus of the cluster is 31.1 (Sandage 1968), these galaxies have absolute visual magnitudes in the range -21.9 to -22.6. Similarly, Morgan (1961) found that the brightest members in each of the 20 nearest clusters in the Abell catalog tend to have stellar populations of the "evolved" type—yellow giant stars.

Frequently one or two particularly luminous giant elliptical galaxies will be found near the center of a regular cluster; sometimes these "supergiant" ellipticals will exceed the brightness of the next most luminous cluster members by more than a magnitude. The center of the Coma cluster, for example, lies roughly midway between the giant ellipticals NGC 4874 and NGC 4889 [= NGC 4884]. More than 30 clusters containing supergiant galaxies as their brightest members (Morgan cD galaxies) have been described by Matthews, Morgan, and Schmidt (1964) and by Morgan and Lesh (1965). Bautz and Morgan (1970; Bautz 1972) classify clusters I to III according to their brightest elliptical galaxy. A type I cluster contains an extraordinarily large luminous cD galaxy that dominates the cluster, as does NGC 6166 in Abell cluster 2199. The Virgo and Corona Borealis (A2065) clusters, on the other hand, represent type III, in that each has no member that stands out noticeably against the giant galaxy background.

As stated above, the rich regular clusters, in contrast to the irregular ones, appear to be nearly or completely devoid of spiral and irregular galaxies. There are actually several spirals in the field of the Coma cluster, the nearest of the regular clusters. Whether or not they are cluster members, however, deserves discussion. The writer was able to find 47 galaxies that could definitely be classed as spirals on 48-inch Schmidt photographs covering a 70-square-degree region in the vicinity of and including the cluster. The radial velocities of some of these spirals are known, and most lie within 2000 km s^{-1} of the mean cluster velocity of 6866 km s^{-1}. Rood, Page, Kintner, and King (1972) treat these objects as cluster members. Subsequently Rood (1974) has analyzed the radial distribution of 30 spirals in the Coma field and concludes that many are probably members of the cluster, and that spirals and irregulars make up 15 percent of the Coma galaxies in the interval of the brightest 2.7 mag. Rood similarly investigated spirals near cluster A2199, and finds that they are probably field galaxies. The detailed distribution of spirals in and around other more remote regular clusters has not yet been investigated. In § 5.2 we review the strong evidence for superclustering. Quite possibly the spirals in and near a regular cluster (e.g., Coma) share membership in a larger cloud of galaxies but are not properly part of the main condensation of the cluster. The hypothesis cannot be ruled out that at least the inner regions of regular clusters are completely lacking in spirals.

Spitzer and Baade (1951) suggested that the absence of spiral galaxies in rich clusters may be a result of collisions between the cluster galaxies, and the consequent removal of interstellar matter from them. In the Coma cluster, for example, a typical galaxy moves with a speed of the order of 10^3 km s^{-1} with respect to the cluster center; in 5×10^9 years such a galaxy would traverse a distance of 5×10^6 pc. At the time of the Spitzer-Baade analysis it appeared that this was several times the diameter of the cluster, and that most galaxies in passing back and forth through the cluster would have suffered at least one collision since the cluster was formed. With modern estimates of the extragalactic distance scale, however, and the correspondingly larger cluster diameters, it is not certain that the Spitzer-Baade mechanism can have effectively removed interstellar matter from most or all galaxies in a typical rich cluster unless its age is much greater than that usually assumed for the Universe. It is possible, therefore, that spiral galaxies either were never formed in the regular clusters, or have disappeared through other evolutionary processes.

Rich clusters of galaxies also tend to contain strong radio sources. Mills (1960) compared the positions of 1159 sources found in the Sydney survey with those of the 877 rich clusters in the Abell catalog that appear in the same survey region. He found 55 coincidences, whereas only 16 coincidences would be expected by chance. Van den Bergh (1961*a*) similarly compared positions of 282 sources in the 3C catalog with galactic latitudes greater than 25° with positions of clusters in the Abell catalog and found 27 coincidences, whereas nine would have been expected by chance. Moreover, van den Bergh finds that the radio magnitudes of the sources are correlated with the distances of the clusters, further strengthening the significance of the association of sources and clusters. More recently, Pilkington (1964) has rediscussed the coincidences between positions of clusters and of sources in the Cambridge and Sydney catalogs, and has added a search for such coincidences among the sources in the partially complete 4C survey. He finds that the radio positions tend to lie near the projected centers of clusters, and he gives a table of 41 sources that lie (in projection) within about 1 Mpc of cluster centers. He estimates that about 8 of these are chance associations. Rogstad, Rougoor, and Whiteoak (1965) have searched for 21-cm continuum radiation from 39 nearby clusters with the Caltech radio interferometer at Owens Valley, and have detected significant flux from 25 of them. They estimate that nearly 60 percent of the clusters listed in the three nearest distance categories in the Abell catalog contain detectable sources. The small angular sizes of the sources observed suggest that the radiation comes from individual galaxies, and not from sources spread over large intracluster spaces.

According to Minkowski (1963) and Matthews, Morgan, and Schmidt (1964), more than a third of the individual galaxies identified with radio sources are in rich clusters in the Abell catalog. On the other hand, Minkowski estimates that only about 10 percent of all galaxies are in clusters as rich as those catalogued by Abell. Van den Bergh (1961*b*) independently arrived at a similar estimate. Both van den Bergh (1961*a*) and Pilkington find that the number of coincidences between source and cluster positions are not correlated with cluster richness in the manner expected if collisions between galaxies were responsible for the radio emission. On the other hand, Rogstad and Ekers (1969) find that E and S0 galaxies in the field are as likely to be strong radio sources as are those in clusters; their data suggest that the propensity of clusters to be radio sources may simply reflect the tendency for E and S0 galaxies to be in clusters.

Clusters also tend to be X-ray sources. Kellogg, Murray, Giacconi, Tananbaum, and Gursky (1973) give data for 20 clusters, of which 16 are in the Abell catalog. For six clusters with both velocity dispersion and X-ray luminosity known, they find the data consistent with $L_x \propto (\Delta V)^{3.9\pm0.8}$. The X-ray luminosities show a wide spread in clusters of all richnesses; some clusters of low richness are strong X-ray sources, which suggests that some unidentified high-latitude sources may be clusters too poor to have been catalogued.

3.3. The Luminosity Function and Colors of Cluster Galaxies

Hubble (1936*b*) investigated the distribution of absolute magnitudes of 134 spiral and 11 irregular galaxies (most of them not in conspicuous clusters) and represented their luminosity function with a Gaussian curve having a mean absolute photographic magnitude of −14.19 and a standard deviation of 0.85 mag (the mean should be several magnitudes brighter to correspond to the modern distance scale). Hubble's luminosity function applies at best to the type of galaxies represented in his sample (late spirals and irregulars were chosen because in them he could identify what appeared to be individual stars, and hence could estimate distances), and there is no justification for assuming that it holds for elliptical galaxies, which are the major constituents of rich clusters, or, therefore, for cluster galaxies in general. In particular, the great prevalence of dwarf ellipticals was unknown to Hubble at the time of his pioneering investigation. We should, in fact, expect the luminosity functions to vary even among clusters of different morphological type and richness. Considerable confusion has resulted from the failure of some investigators to recognize the inapplicability of Hubble's luminosity function to clusters of galaxies.

Zwicky (1942*a*) may have been the first to summarize the observations favoring the existence of large numbers of galaxies of low luminosity. He also advanced quasi–statistical-mechanical arguments that the galaxian luminosity function must increase with increasing magnitude. His argument is based on the assumption of a stationary universe in statistical equilibrium, and on the application of the Boltzmann principle to determine the relative numbers of galaxies in clusters and in the field, as well as the relative numbers of stars inside and outside galaxies. The analysis leads to the result that most stars (and other matter) should be in the intergalactic space; moreover, he believed that collisions should result in the fragmentation of some galaxies into smaller parts. Zwicky concludes (italics are his): "*. . . individual stars, multiple stars, open and compact star clusters and stellar systems of increasing population will be found in numbers presumably decreasing in frequency as the stellar content of the systems in question increases.*"

Few modern investigators would accept the assumption on which Zwicky's conclusions are based; however, his supposition that the galaxian luminosity function rises at faint absolute magnitudes is almost certainly correct—at least for galaxies in rich clusters to the magnitude limits observed. Later, from a rather ingenious analysis of the mean numbers of galaxies within clusters of various angular diameters, Zwicky (1957) derived the following expression for the integrated luminosity function of cluster galaxies:

$$N(\Delta m) = k(10^{\Delta m/5} - 1) , \tag{1}$$

where $N(\Delta m)$ is the number of galaxies in the range Δm between magnitude m and the magnitude of the brightest cluster galaxy. Unfortunately, as Zwicky defines his cluster angular diameters, they are not related to distance in the manner he assumes in the derivation of equation (1), as has been shown by the writer (Abell 1962) and by Scott (1962). Nevertheless, Zwicky's formula may be qualitatively correct, at least at the fainter magnitudes. He states (Zwicky 1957) that equation (1) is consistent with counts of galaxies in several clusters to two or more different magnitude limits (on plates with different exposure times or taken with different telescopes), and also with the distribution of magnitudes obtained with schraffier photometry among the brightest galaxies in several clusters (e.g., Zwicky and Humason 1964*a*, *b*).

The published data have been reexamined to determine the galaxian luminosity function by Kiang (1961) and by van den Bergh (1961*b*). Kiang, from magnitudes of galaxies both in clusters and in the field, adopts Zwicky's form of the differential luminosity function $\phi(M)$ (derivative of eq. [1]) for faint magnitudes, but finds that the function rises more sharply for bright magnitudes; he adopts a cubic law for $\phi(M)$ over the interval of the brightest 2.5 mag. Van den Bergh analyzes the absolute magnitudes of 240 bright field galaxies, of which 48 are ellipticals and 192 are spirals and irregulars. Although the magnitude range considered by van den Bergh is small (4.5 and 6.5 mag for ellipticals and spirals, respectively), his luminosity functions (found separately for ellipticals and for spirals and irregulars) are both in qualitative agreement with Zwicky's formula at faint magnitudes; however, they rise more rapidly at their bright ends.

Derived luminosity functions are, unfortunately, sensitive to photometric procedures, the difficulty of which are often not appreciated. The surface brightness of an elliptical galaxy drops off very slowly at large distances from the center of its projected image; Hubble's (1930) representation of the brightness is, in fact, an inverse-square function of the radial distance. Hubble's interpolation formula, if extended to infinite radius, leads to an infinite luminosity for a galaxy; thus at some distance the galaxian surface brightness must begin to drop more rapidly. Some investigators (Dennison 1954; Liller 1960; de Vaucouleurs 1948) have found finite luminosities (or "total" magnitudes) for elliptical galaxies, but their measures of surface brightness deviate from Hubble's law only at very large radial distances, where the measurements are extremely difficult. For a few elliptical galaxies, convergence of the total luminosity has not been ascertained observationally; photoelectric measures of M87 by Baum (1955), for example, show no evidence of deviation from Hubble's inverse-square law even to very great distances from the center. Even in spiral galaxies, where the spiral arms seem more sharply defined, the contribution to the total luminosity of the Population II coronal components, which may extend far beyond the spiral arms, is not accurately known.

"Total" magnitudes of galaxies are thus not easy to define unambiguously, and investigators should exercise considerable caution in interpreting published magnitudes. Consistent results have been obtained by Sandage (Humason, Mayall, and Sandage 1956), who defines the magnitude of a galaxy according to the total light contained within a given standard isophote, and by Holmberg (1950, 1958), who integrates the light over a galaxian image from several microphotometer tracings across it. Sandage's magnitudes, however, do not correspond to the same fraction of light in galaxies of similar size but different surface brightness (Abell 1962), and with Holmberg's procedure the contribution to the total light of a galaxy from very faint outer extensions that are below the photographic threshold remains unknown.

On the other hand, if the surface brightnesses of galaxies can be represented by a single model, "total" magnitudes for them can be operationally defined. It appears possible to find a satisfactory representation of the distribution of surface brightness in bright elliptical galaxies. Hubble's formula for the brightness I at a distance r from the center along the major axis of the projected image of an elliptical galaxy,

$$\frac{I}{I_0} = \frac{1}{(1 + r/a)^2} \tag{2}$$

(where I_0 and a are parameters for a particular galaxy), describes the surface brightness over most of its observable image very well. De Vaucouleurs's (1948) formula is

$$\log (I/I_e) = -3.33[(r/r_e)^{1/4} - 1]\,, \tag{3}$$

where r_e is the distance from the center along the major axis to the isophote within which one-half of the total light is emitted, and I_e is the surface brightness at that isophote. It agrees well with the Hubble law for small r, and has the advantage of leading to finite total magnitudes that are in satisfactory agreement with those of Sandage and Holmberg.

The writer has found that a convenient interpolation formula for the distribution of light in an elliptical galaxy is a modification of the Hubble law:

$$\frac{I}{I_0} = \frac{1}{(1 + r/a)^2} \quad \text{for } r/a \leq 21.4$$
$$\frac{I}{I_0} = \frac{22.4}{(1 + r/a)^3} \quad \text{for } r/a > 21.4 \,. \tag{4}$$

Equation (4) agrees with de Vaucouleurs's formula to the precision of present-day photometry, and leads to "total" magnitudes that are (statistically) close to those of Sandage and Holmberg. The writer and Mihalas (Abell and Mihalas 1966) have developed a technique of determining the parameters I_0 and a, and hence the "total" magnitude, of an elliptical galaxy from measures of the brightness of its extrafocal image on each of two or more calibrated photographs taken different amounts out of focus. The total magnitude obtained is actually the magnitude of a fictitious galaxy with surface brightness given by equation (4), and that has extrafocal images of the same brightness as the measured galaxy. The precise form of the surface brightness law is not important except to establish the zero point of the magnitude scale, for in practice it is rare that contributions to an extrafocal image come from parts of a galaxy for which r/a exceeds 21.4. The magnitudes obtained may not always be accurate for individual galaxies, for all elliptical galaxies may not be built on the same model; but they are statistically self-consistent, and are free from systematic effects that depend on the distance of a galaxy measured; in particular, they are suited to the study of luminosity functions of elliptical cluster galaxies, and to the comparison of luminosity functions of different clusters.

To date, the writer has applied the procedure to determine the luminosity functions of six clusters, of which four are regular clusters of known redshift. The luminosity functions of these clusters, which consist predominantly of elliptical galaxies, are all similar. If the logarithmic integrated luminosity functions, $\log N(\Delta m)$, are plotted, they can all be fitted together very satisfactorily with horizontal and vertical shifts that depend only on the relative cluster distances and richnesses. Such a combined integrated luminosity function is shown for four rich clusters in figure 3; they are cluster numbers 1656 (Coma), 2199 (surrounding NGC 6166), 151, and 2065 (Corona Borealis). The coordinates of figure 3 have arbitrary zero points. The smooth curve labeled "Abell" is adopted as the mean function, $\log N(\Delta m)$, for the four clusters. The dashed curve labeled "Zwicky" corresponds to Zwicky's integrated luminosity function, defined by equation (1).

Curves of $\log N(\Delta m)$ compiled from published data of several other observers are plotted in the same manner in figure 4. The points for cluster 1377 (Ursa Major cluster No. 1) are from very old photometry by Baade (1928). The "Zwicky" data for cluster 1656 and the Virgo cluster are from the first two volumes of the *Catalogue of Galaxies and Clusters of Galaxies*. Only the central part of the Virgo cluster is used, and the *Catalogue* data covers only the interval of the brightest 2.7 mag in cluster 1656. The "Holmberg" data for the Virgo cluster are from Holmberg (1958). For comparison, there is also

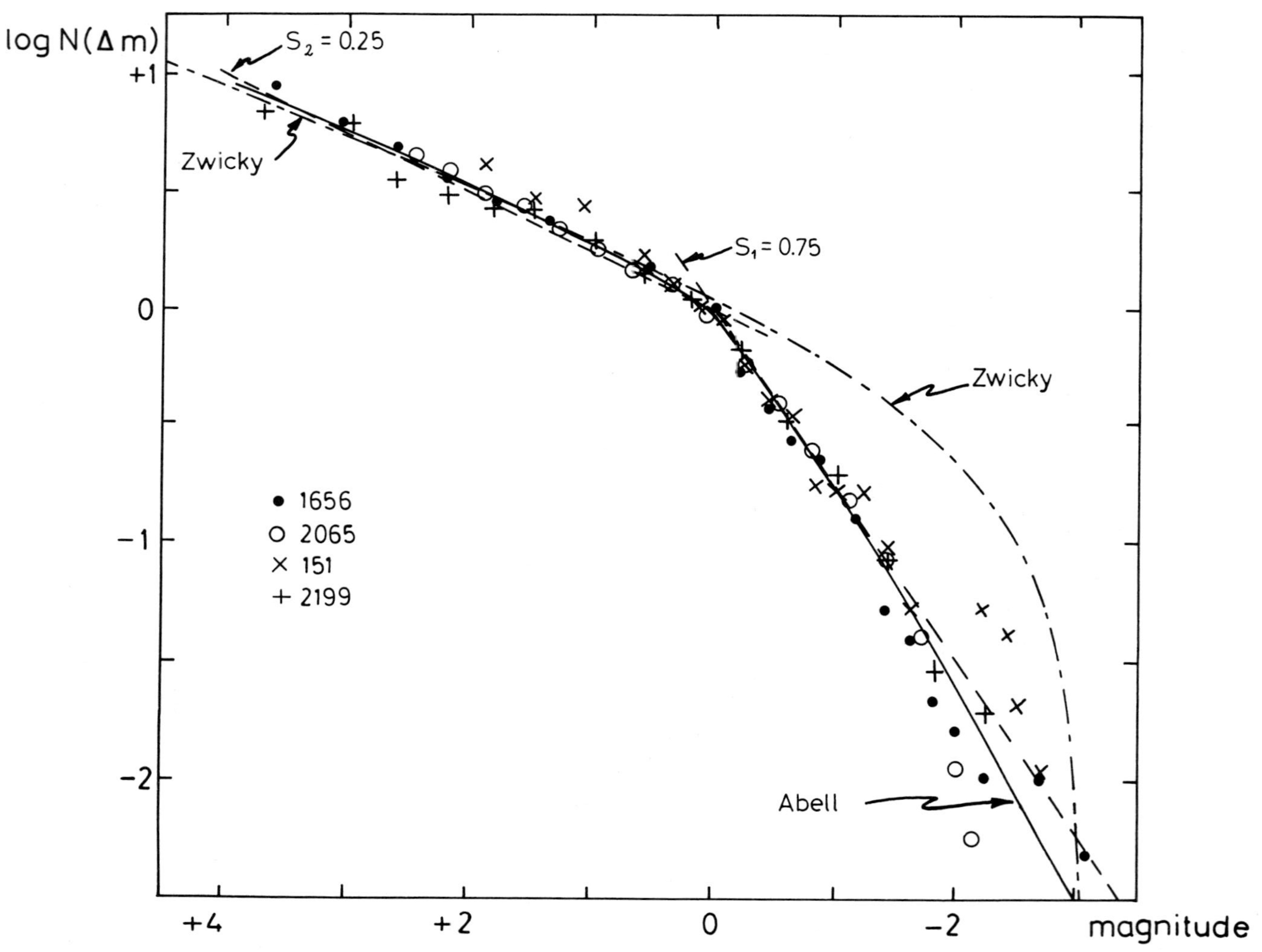

FIG. 3.—The logarithmic integrated luminosity functions of four rich clusters. The symbols and curves are explained in the text. The zero points of the ordinates and abscissae are arbitrary.

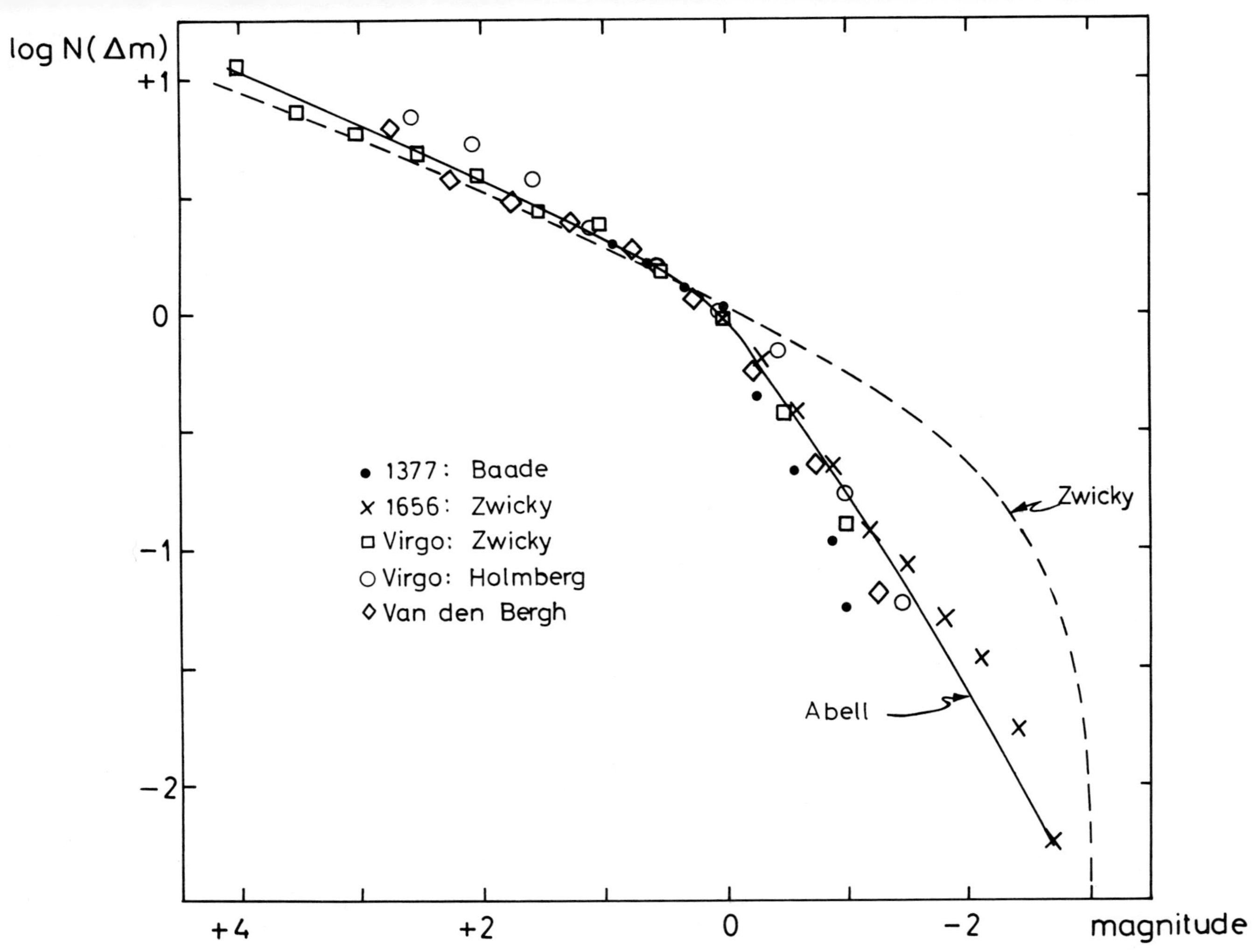

FIG. 4.—Cluster logarithmic integrated luminosity functions adapted from the published data of several observers. See text for explanation. The scale of the abscissae and ordinates is as in fig. 3, and the zero points are arbitrary.

shown the logarithmic integrated luminosity function for 48 elliptical field galaxies, taken from van den Bergh (1961*b*). It must be emphasized that the photometric procedures differed among the various observers; large zero-point differences between the different magnitude systems are expected, and scale errors may exist as well. Zero-point and scale errors are especially likely in the early photographic photometry of Baade. It should also be noted that the Virgo data include many spirals, and the apparently excellent agreement between the Virgo luminosity function and that of the other clusters may be fortuitous. The "Abell" and "Zwicky" curves in figure 4 are the same ones as in figure 3.

The differential luminosity function $\phi(M)$ corresponding to the "Abell" curve in figures 3 and 4 is shown in figure 5. The ordinates are arbitrary; the abscissae are M_{pv}, determined by fitting van den Bergh's luminosity function to the Abell cluster luminosity functions (van den Bergh's absolute magnitudes have been adjusted to correspond to a Hubble constant of 75 km s^{-1} Mpc^{-1}). It appears that $\phi(M)$ for elliptical galaxies can not be represented by a single exponential function. The function log $N(\Delta m)$ shows a very definite change of slope after the interval of the brightest 2 to 3 mag. This abrupt change of slope corresponds to the maximum near the bright end of $\phi(M)$. There is some

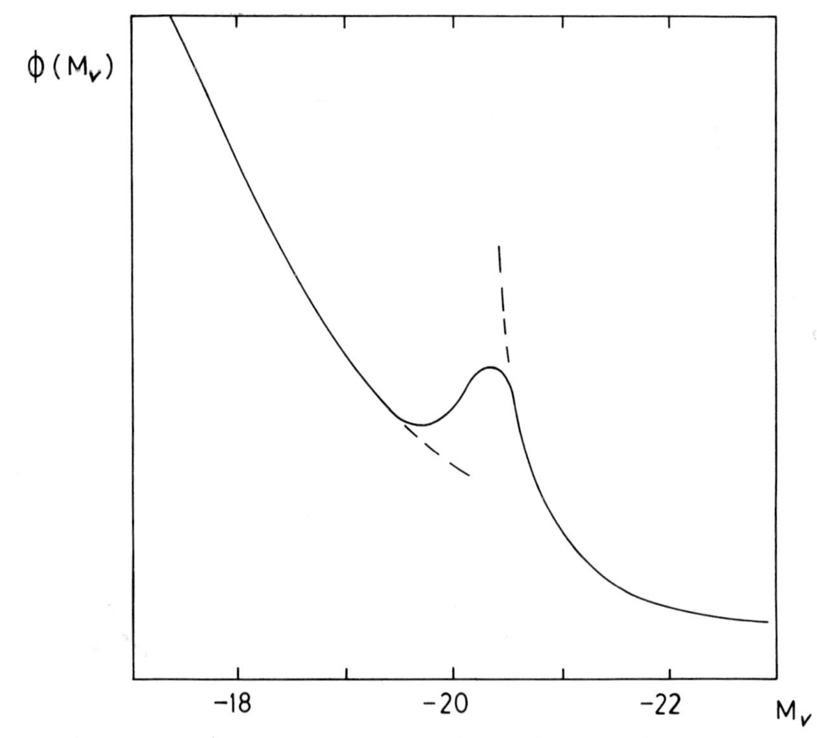

FIG. 5.—The differential cluster luminosity function, derived from figs. 3 and 4. The scale of the ordinates is arbitrary.

evidence that at least in the Coma cluster this peak is contributed mainly by galaxies near the central core of the cluster (Rood 1969; Rood and Abell 1973).

The composite luminosity function for the clusters in figure 3 can be represented rather closely by two intersecting straight lines (*dashed lines* in fig. 3), which represent simple exponential relations over two different magnitude intervals. These lines are merely interpolation formulae, and the approximately exponential character of the integrated luminosity functions should not be taken too literally. For example, a simple exponential function can hardly be expected to match the very brightest galaxies in different clusters of different Bautz-Morgan types (some of which have supergiant cD galaxies and some of which do not). Moreover, if the two straight lines were to apply rigorously the differential luminosity function would have to be discontinuous (*dashed lines* in fig. 5), which is, of course, physically unrealistic. Nevertheless, the magnitude at which the two lines intersect in figure 3, which we define as m^*, is an operationally defined point on the luminosity function. If all galaxies in all clusters are selected from the same population with a parent luminosity function like that in figure 3, m^* should be a useful "standard candle" for determination of relative distances of clusters.

The luminosity functions of all clusters investigated so far show the same characteristic shape (i.e., change of slope as in fig. 3). The best fit of the luminosity function of a cluster to the one in figure 3 determines m^* for that cluster, even if the photometric data are not very complete and if its luminosity function cannot be fitted unambiguously by the two exponential relations. For those clusters of known redshift and for which m^* has been so determined, the plot of *log z* versus m^* has a scatter of about 0.1 mag. (s.d.) (Abell 1962; Bautz and Abell 1973), significantly less than in the corresponding Hubble diagram in which the first brightest cluster galaxies are used as standard candles.

The luminosity function data used to construct figures 3 and 4 are summarized in table 2. The cluster numbers are those in the Abell (1958) catalog. The slopes, s_1 and s_2, of the straight lines used to approximate $\log N(\Delta m)$ are defined by:

$$\begin{aligned} \log N(\Delta m) &= K_1 + s_1 m \quad (\text{for } m \leq m^*) \\ \log N(\Delta m) &= K_2 + s_2 m \quad (\text{for } m > m^*) \,. \end{aligned} \tag{5}$$

Log $N(\Delta m^*)$ is the ordinate of the intersect of the two lines. Note that zero-point and scale differences may exist between the magnitude systems of the different observers, and that m^* is given in photovisual magnitudes for the clusters observed by the writer, and in photographic magnitudes for those observed by others. The slope s_2 is sensitive (particularly for distant clusters) to the corrections applied to the cluster photometry to take account of foreground and background field galaxies. The writer has not yet analyzed the field in a final way, and the values of s_2 given for the clusters investigated by him should be considered provisional. For the purpose of discussion he adopts for the mean luminosity function the values, $s_1 = 0.75$, and $s_2 = 0.25$. (In the limit of faint magnitudes, Zwicky's curve has a slope, $s_2 = 0.20$—in reasonable agreement with the value adopted here.)

Since figures 3–5 and table 2 were prepared for this review, considerable new data on cluster luminosity functions have become available. Gudehus (1971, 1973) has determined luminosity functions for clusters A754 and A1367, and has confirmed that of

Abell for A2065. Abell and Mottmann (in preparation; see Bautz and Abell 1973) have obtained the luminosity function for cluster A2670. Eastmond (Abell and Eastmond 1968; Abell 1972) has new photometry for the elliptical galaxies in the Virgo cluster. Noonan (1971) presents counts in clusters A1656 and A2065 to different magnitude limits that are consistent with the luminosity functions for those clusters shown in figure 3. Rood and Baum (1967, 1968), while not discussing the luminosity function per se, do give very useful photometric and other data for 315 Coma cluster galaxies. Huchra and Sargent (1973) have derived a luminosity function for field galaxies from the data

TABLE 2

DATA PERTAINING TO THE LUMINOSITY FUNCTIONS OF SEVERAL CLUSTERS

Cluster	Mean Radial Velocity (km s^{-1})	Reference for Velocity	s_1	s_2	m^*	log $N(\Delta m^*)$	Observer, or Reference for Photometry
Virgo.....	1136	Humason *et al.* (1956)	0.72	0.20	$m_{pg}=11.9$	1.15	Zwicky, Herzog, and Wild (1961)
Virgo.....	....	....	0.78	0.33	$m_{pg}=10.9$	1.25	Holmberg (1958)
1656.....	6866	Lovasich *et al.* (1961)	0.75	....	$m_{pg}\geq 15.6$	$\geq$2.25	Zwicky and Herzog (1963)
1656.....	....	....	0.78	0.25	$m_{pv}=14.7$	2.25	Abell
2199.....	8736	Minkowski (1961)	0.75	0.26	$m_{pv}=15.4$	2.00	Abell
1377.....	15269	Humason *et al.* (1956)	1.20	0.35	$m_{pg}=16.9$	1.25	Baade (1928)
151......	15781	Humason *et al.* (1956)	0.73	0.33	$m_{pv}=16.4$	1.95	Abell
2065......	21651	Humason *et al.* (1956)	0.72	0.27	$m_{pv}=17.2$	2.23	Abell
Field ellipticals	....	....	0.95	0.27	$M_{pg}=-19.5$†	‡	van den Bergh (1961*b*)

NOTE.—According to de Vaucouleurs (1961*b*), the Virgo cluster is at least two clusters seen in projection. One component, consisting mostly of elliptical galaxies, has a mean radial velocity of 950 km s^{-1}; the other, consisting mostly of spirals, has a mean radial velocity of 1450 km s^{-1}.

† For $H = 75$ km s^{-1} Mpc^{-1}.

‡ For $H = 75$ km s^{-1}Mpc^{-1}, log $N(\Delta M_{pg}{}^*) = -3.79$ galaxies Mpc^{-3}.

in the *Reference Catalogue of Bright Galaxies* (de Vaucouleurs and de Vaucouleurs 1964) that is also in excellent agreement with the cluster luminosity function in figure 3. Because the more recent data are all consistent with those presented here, the figures and table have not been updated.

The cluster luminosity function has not yet been determined observationally at magnitudes fainter than about $m^* + 4$. Galaxies at this observational limit have absolute magnitudes near $M_v \approx -16.5$; these are still relatively luminous objects, brighter, in fact, than all but four Local Group galaxies. Reaves (1966) searched for dwarf galaxies in the central core of the Coma cluster and found 32 probable dwarfs in the range $-14 > M_v > -16$. If the total content of dwarf galaxies in nearby groups and in the Virgo cluster is typical, we would expect the luminosity function to continue to increase at fainter magnitudes, but the precise form of $\phi(M)$—for example, whether it increases monotonically—is not known. In particular, there is no justification for extrapolating the interpolation formula (eq. [5]) for the integrated luminosity function beyond the observed magnitude range. Such an extrapolation of equation (5) would predict that the number of Sculptor-type systems (with M_v in the range -9 to -14) in the Coma cluster is of the order 10^5. Nevertheless, even this large number of dwarfs would contribute only slightly to the total luminosity (and presumably also to the total mass) of the cluster; extrapolation of equation (5) to infinite magnitude leads to a finite luminosity for the cluster of 1.2×10^{13} times the solar luminosity, and only 13 percent of this light would be contributed by galaxies fainter than $m_{pv} = 18.3$, the magnitude limit for which the luminosity function is observed.

Colors of galaxies, at least of their inner portions, can be measured far more easily than total magnitudes, and fairly accurate data are available. Published results of photoelectric photometry by de Vaucouleurs (1961*a*) and photographic photometry by Holmberg (1958) include galaxies both in the field and in the Virgo cluster. Colors of the brighter cluster galaxies appear to be the same as those of galaxies of similar morphological type in the field. For giant ellipticals, $B - V$ color indices are about 0.95 to 1.00; $B - V$ decreases through the sequence of spirals from about 0.8 for Sa's to about 0.5 for Sc's. Both de Vaucouleurs (1961*a*) and Baum (1959), however, report that colors of elliptical galaxies in the Virgo cluster that are successively fainter than $m_{pv} \approx 11.5$ are successively bluer than those of giant ellipticals, and at $m_{pv} \approx 14$, $B - V$ is from 0.6 to 0.8. Sandage (1972) has obtained three-color observations of six dwarf ellipticals in the Virgo cluster, and of 25 elliptical galaxies in the Coma cluster, and confirms the correlation between $B - V$ and magnitude. He also shows that there is an even stronger dependence of $U - B$ on V, in the sense of fainter galaxies being bluer. On the other hand, both de Vaucouleurs's and Holmberg's data include color measures of several intrinsically fainter Local Group ellipticals of intermediate luminosity, which appear to have colors that are about the same as those of giant ellipticals. Holmberg even includes two dwarf ellipticals (the Leo systems in the Local Group), and they also have approximately "normal" colors.

3.4. Populations of Clusters

It is not possible to state with certainty whether particular objects in the field of a cluster of galaxies are actually members of that cluster. Even radial velocities do not provide an unambiguous answer because of the velocity dispersion that always exists within a cluster. Only statistically can we speak of cluster populations, and then we must define: (1) the magnitude range counted; (2) the way the cluster is presumed to be bounded; and (3) the way corrections are applied for the field of foreground and background objects (of course, the meaning of the "field" itself is open to question).

Populations are given by Zwicky and his collaborators in their *Catalogue of Galaxies and Clusters of Galaxies*. The population listed for a cluster is the number of galaxies visible on the red Palomar *Sky Survey* plate that are contained within the isopleth (drawn by eye estimate) at which the surface density of galaxies is twice that in the surrounding field; the cluster counts are corrected for the mean field count itself. It is clear that the populations are not independent of distance, since a greater magnitude range is counted in nearby clusters than in distant ones. Also, since an isopleth contains not all of its cluster, but only that part within which the surface density is twice that of the surrounding field, the entire cluster population is not counted, even in the magnitude range considered. Moreover, as has been shown by Abell (1962) and Scott (1962), the isopleth contains a smaller fraction of the projected image of a distant cluster than of a nearby one. Finally, Zwicky himself has called attention to the fact that his cluster populations may be affected by interstellar and possible intergalactic absorption. Having noted these various selection effects, we summarize the population listed in the first two volumes of the *Catalogue* in table 3. The data are segregated by Zwicky's morphological cluster classes (compact, medium compact, and open), and also according to his estimate of the relative cluster distances. The radial velocities corresponding to the distances

TABLE 3

NUMBERS OF CLUSTERS OF VARIOUS POPULATIONS IN THE FIRST TWO VOLUMES OF THE *Catalogue of Galaxies and Clusters of Galaxies*

Population	<100	100–199	200–299	300–399	400–499	500–599	≥600
Near:							
Compact	0	2	2	2	0	2	6
Medium compact	6	27	27	9	8	3	14
Open	9	41	27	19	7	1	14
MD:							
Compact	1	9	10	3	2	2	1
Medium compact	25	120	48	26	14	7	10
Open	21	109	33	8	7	3	1
D:							
Compact	19	36	9	3	1	1	3
Medium compact	86	198	60	11	5	3	1
Open	50	164	24	5	1	0	0
VD:							
Compact	139	143	30	2	1	0	0
Medium compact	267	190	36	1	4	0	0
Open	95	97	12	0	0	0	0
ED:							
Compact	465	132	11	0	1	0	0
Medium compact	307	159	10	1	0	0	0
Open	36	18	0	0	0	0	0

separating the five distance classes—near, medium distant (MD), distant (D), very distant (VD), and extremely distant (ED)—are quoted as 15,000, 30,000, 45,000, and 60,000 km s^{-1}, respectively.

Another body of data comes from the writer's cluster catalog (Abell 1958). This study concerns only the very richest clusters, but those are chosen in such a way as to comprise a relatively homogeneous sample, and the "populations" are defined in a manner to be as independent as possible of distance. The population of a cluster in the Abell catalog is the number of galaxies brighter than $m_3 + 2$, where m_3 is the photo-red magnitude of the third brightest cluster member. All counts were made on the red *Sky Survey* plates, and include those galaxies in the prescribed magnitude range that are contained within a circle centered on the image of the cluster, minus the number in the same magnitude range contained within a circle of similar size in the nearby field. The circles used were large compared to the main concentrations of galaxies within the cluster. A circle radius, in millimeters, was $4.6 \times 10^5/cz$, where cz, the velocity of recession in km s^{-1}, was estimated by comparing the tenth brightest cluster member with the tenth brightest members of clusters of measured redshift. The counts were thus extended to approximately the same linear distance in space from each cluster center, irrespective of the distance of the cluster. The cluster survey included all clusters north of $\delta = -27°$ that are at galactic latitudes greater than about 30°, and that have redshifts in the range 0.02–0.2. The frequency distribution of clusters of various populations is summarized in table 4.

Just (1959) has called attention to a slight positive correlation between distance and richness of the clusters in the Abell catalog. He believes the effect to be statistically significant, and cites it as evidence for secular evolution in the Universe (ruling out the steady-state cosmology). Actually the apparent dependence of cluster richness on distance is not large, and is very sensitive to slight errors of observations that may also depend on distance. The possible role of such errors has been discussed quantitatively by Paál (1964), who shows that the Just effect can be explained without evolution if Abell failed to include in the homogeneous sample only about 10 percent of the most distant clusters. Paál suggests that such a loss of remote clusters could result from the fact that the diameters of the counting circles used by Abell are inversely proportional to the cluster redshifts, rather than reflecting the dependence of angular diameter on distance that is predicted by any particular cosmological theory. It is the judgment of the writer that further independent observations are required to establish whether the Just effect is real.

TABLE 4

NUMBERS OF CLUSTERS OF VARIOUS POPULATIONS IN THE ABELL CATALOG

Population	Number of Clusters
50– 79	1224
80–129	383
130–199	68
200–299	6
≥300	1

The Abell and Zwicky data on cluster richnesses are not easy to compare because of the different ways in which cluster "populations" are defined. Both tables 3 and 4, however, show clearly that the numbers of clusters increase rapidly with decreasing population. It is unfortunate that data do not exist on the relative numbers of rich clusters and small groups; probably the latter are far more frequent. Multiplicity is common even among what appear to be "field" galaxies. According to Holmberg (1962*a*), only 47 percent of nearby systems have multiplicity 1 (that is, are single galaxies), and 53 percent are either double or multiple systems. It should be noted, however, that all of the systems investigated by Holmberg are within what may be a local supercluster (see § 5); moreover, multiplicity exists even within irregular clusters.

3.5. Sizes and Structures of Clusters

Sizes of clusters are difficult to determine because (*a*) the boundary between a cluster and the field is indefinite, and (*b*) usually only the brighter cluster members are observed whereas if the cluster has approached a state of statistical equilibrium the fainter unseen members should be more broadly distributed in space. On the other hand, as discussed in the next section, there is not yet conclusive evidence that even a single cluster is in a state of statistical equilibrium; if we assume that the bright and faint members of a cluster do, in fact, occupy the same volume of space, and if we can distinguish cluster from field galaxies, then we can obtain an estimate of the cluster size. There is some justification for this assumption, at least for a few irregular clusters. Dwarf elliptical galaxies in both the Virgo cluster (Reaves 1956, 1964) and the Fornax cluster (Hodge, Pyper, and Webb 1965) occupy approximately the same area in the sky as the bright cluster members (although they do not necessarily have the same spatial distribution). The angular diameter of the Virgo cluster (Sandage 1958) is about 7°; the corresponding linear diameter is about 3×10^6 pc (for $H = 50$ km s^{-1} Mpc^{-1}). The irregular Hercules cluster (cluster 2151) has dimensions of about 3.8×10^6 by 2.5×10^6 pc (Burbidge and Burbidge 1959*b*). The Local Group is about 1×10^6 pc in diameter. The sizes of the rich regular clusters are more controversial (see below), but are probably of about the same order.

The sizes of the largest clusters in the *Catalogue of Galaxies and Clusters of Galaxies* are discussed by Zwicky and his collaborators (Zwicky and Rudnicki 1963; Zwicky and Berger 1965; Zwicky and Karpowicz 1965, 1966). Although there is considerable scatter in the data, they report that the largest clusters of all types (compact, medium compact, and open) have about the same size, which is of the order 10^7 pc. This is actually a lower limit for it is the diameter corresponding to Zwicky's isopleth where the surface density of galaxies is twice that of the field, and not to the entire clusters.

The irregular clusters do not display symmetry or regular structure. Multiple condensations of galaxies, however, are often present. A statistical study of the surface distribution of galaxies in seven rich clusters was made to estimate the extent of subclustering within them (Abell, Neyman, and Scott 1964). Counts of galaxies in rings centered on galaxies within the clusters are compared to counts in rings centered on arbitrary grid points. Higher mean counts in the rings surrounding galaxies than in those around the grid points (after correction for any central concentration of the cluster as a whole) indicates subclustering, and analysis of the counts gives information on the extent of the subclustering. Of the clusters investigated to date, all that are irregular show evidence of subclustering; the two most regular clusters (1656 and 2065) do not. The subgroups appear to have characteristic radii of the order 10^5 pc. There are, of course, two subgroups in the Local Group—centered on the Galaxy and on M31. It is also well known that double and multiple galaxies are common in certain other irregular clusters; Holmberg (1937) and van den Bergh (1960) have called attention to probable binary systems in the Virgo cluster; and the Burbidges (1959*b*), to such systems in the Hercules cluster.

The rich regular clusters, which show high central concentration and spherical symmetry, and little or no subclustering, are amenable to representation by theoretical models of galaxy distribution. Data on the surface distributions of galaxies in a fair number of clusters (both regular and irregular) are now published, although the completeness, quality, and format of the data vary considerably from cluster to cluster. Fortunately some clusters have been studied independently by several investigators;

comparison of their results can be a valuable aid in assessing the reliability of the available material. Some references to the distribution of galaxies in clusters (listed by number in the Abell catalog) are:

31...................... N. Bahcall (1972)
234..................... Zwicky (1956)
426 (Perseus)........... Zwicky (1957); Rudnicki (1963)
732..................... Zwicky (1956)
801..................... Zwicky (1956)
1060 (Hydra I).......... Zwicky (1957); Kwast (1966)
1132.................... N. Bahcall (1971)
1185.................... Rudnicki and Baranowska (1966*b*)
1213.................... Rudnicki and Baranowska (1966*b*)
1367.................... Rudnicki and Baranowska (1966*a*)
1643.................... Zwicky (1956)
1656 (Coma)............. Zwicky (1937, 1942*b*, 1957, 1959); Omer, Page, and Wilson (1965); Noonan (1971); Rood *et al* (1972); N. Bahcall (1973*b*)
1677.................... Zwicky (1956)
2065 (Corona Borealis).. Zwicky (1956); Noonan (1971)
2199.................... Clark (1968); Rood and Sastry (1972); N. Bahcall (1973*c*)
Cancer.................. Zwicky (1957)
Pegasus................. Zwicky (1957)

Counts in six clusters by the writer are as yet unpublished, except for a summary of the counts in cluster 1656 by Noonan (1961*a*, *b*). The counts of galaxies on the Lick astrographic plates by Shane and his collaborators reveal many conspicuous clusters, and isopleths for several of these have been published separately by Shane (1956*b*).

Several types of analytical representation of the galaxy distribution in rich clusters have been attempted. Zwicky has fit the projected distribution of galaxies in several clusters to a bounded isothermal distribution projected on a plane. An early solution for the density distribution in the isothermal polytrope was by Emden (1907), who tabulates the density as a function of the dimensionless variable ξ, representing the radial distance from the center of the configuration. Zwicky relates the linear distance r from the center of a cluster of galaxies to Emden's ξ by $r = \alpha\xi$, and calls the scale factor α the ***structural index*** of the cluster. The values he reports for α (1957), translated to the distance scale used in this review ($H = 50$ km s^{-1} Mpc^{-1}), are 24.2, 25.0, 32.3, and 13.8 (in units of 10^{22} cm) for clusters 1656, 1060, 426, and the Cancer cluster, respectively. Zwicky cites the rather small spread in the values for α among different clusters (actually, the spread is about a factor of 2) as evidence that all regular clusters are similar in size and structure. The writer's counts in cluster 1656 have also been compared with the isothermal polytrope, both by himself (unpublished) and by Noonan (1961*a*, *b*). The isothermal representation is a satisfactory one, but the writer finds $\alpha = 48 \times 10^{22}$ cm—almost twice Zwicky's value. More recently N. Bahcall (1972, 1973*a*, *c*) has shown that the galaxy distribution in several clusters can be matched satisfactorily to the isothermal polytrope, and she suggests further that the structural index (or an equivalent scaling factor) can be used, statistically, as a distance indicator for clusters.

In distant clusters of small angular size, Zwicky (1956) explains that α is difficult to determine with precision. For such remote systems, he introduces another parameter, which he calls the *distribution index*, DI, defined by

$$\mathrm{DI} = \frac{N(a_0)}{100\,N(a_0/10) - N(a_0)}, \tag{6}$$

where $N(a_0)$ and $N(a_0/10)$ are the total numbers of cluster galaxies within angular distances a_0 and $a_0/10$ of the cluster center, respectively, after subtraction of n_a galaxies for each square degree. The quantity n_a is the mean number of galaxies per square degree along the circle of radius a_0, and thus represents an upper limit to the density of field galaxies. Zwicky choses a_0 to be inversely proportional to the redshift of the cluster, so

TABLE 5

MINIMUM POPULATIONS AND DISTRIBUTION INDICES IN SEVEN REGULAR CLUSTERS OF GALAXIES*

Cluster (Abell Number)	cz (km s^{-1})	$N(a_0)$	DI
234....	51,700	461	0.115
732....	61,000	451	0.110
801....	57,600	391	0.129
1643....	59,300	394	0.115
1656....	6,850	467	0.084
1677....	54,900	332	0.107
2065....	22,000	551	0.103

* Adapted from Zwicky 1956.

that it presumably corresponds to the same linear distance from the center in a nearby as in a remote cluster; for a cluster of redshift 51,700 km s^{-1}, a_0 has the value 10'.0. Zwicky attempts to adjust the exposure times of the cluster photographs so that the counts are extended through a similar magnitude range in all of them. If there were no central concentration of galaxies in a cluster, DI would be infinite. Actual values of DI for seven clusters are given in table 5, adapted from Zwicky (1956). The observed velocity of recession, cz, for each cluster is given, and also $N(a_0)$, which Zwicky points out is a lower limit to the total cluster population. Counts in clusters 1656 and 2065 were made with the 48-inch Schmidt telescope, and Zwicky states that it is difficult to adjust them to those made in the five more remote clusters on 200-inch telescope plates. Nevertheless, the small spread in the values of $N(a_0)$ and DI suggest a remarkable similarity among the regular clusters investigated.

Other investigators have experimented with different mathematical representations of the density distribution in clusters of galaxies. De Vaucouleurs (1960) and Shane and Wirtanen (1954) find that the surface density of galaxies in the Coma cluster (1656) follows closely a law of the form

$$\log N(r) = ar^{1/4} + b, \tag{7}$$

where $N(r)$ is the projected density at distance r from the center and a and b are constants. Omer, Page, and Wilson (1965) have averaged counts of galaxies in parallel strips across the same cluster by Omer, Page, and Wilson and by Shane and have represented the counts by a series of Hermite polynomials of even order, from which they derived a formula for the spatial density of galaxies in the cluster. Scott (1962) represents Zwicky's counts in the five most distant clusters listed in table 5 by Maxwellian distributions for which the ratio of σ to the distance of each cluster is 4×10^{-4}; for all five clusters, the linear value of σ is about 4.5×10^5 pc. All of these mathematical distributions seem to fit actual clusters about as well as the isothermal one. Each also involves a fitting constant or scale factor, such as α, DI, or σ; and if, as seems possible, the rich regular clusters are similar to one another, the relevant scale factor might prove valuable as a criterion for cluster distances—at least statistically.

An observation important to the study of the dynamics of a cluster is to determine whether there is a segregation of its bright and faint galaxies. If it has reached a state of equipartition of energy, the fainter—presumably less massive—galaxies should fill a larger volume of space than the brighter ones. Zwicky (1942*b*) presents evidence that such a segregation exists among the brighter Virgo cluster galaxies; but there is also a segregation according to galaxy type, the spirals showing a larger radial distribution than the ellipticals. Both Holmberg (1962*b*) and de Vaucouleurs (1961*b*) have shown that the mean radial velocities are different for the spirals and ellipticals in the cluster. De Vaucouleurs argues a strong case that the Virgo cluster is really (at least) two clusters seen in projection—one being a compact cluster of elliptical and S0 galaxies, and the other a loose open cluster consisting mostly of spirals and irregulars. If de Vaucouleurs's interpretation is correct, then differences in the radial distributions of Virgo cluster galaxies of different types and/or magnitude cannot be regarded as evidence of segregation.

Rudnicki (1963), who has counted galaxies in cluster 426, believes that in that cluster the faint galaxies show a wider surface distribution than the bright ones. Unfortunately, one side of cluster 426 is partially hidden by interstellar absorption, and Rudnicki may have adopted too low an estimate for the density of galaxies that are in the field; a moderate and entirely possible increase in the field estimate completely removes the apparent segregation of bright and faint galaxies in that cluster.

Hodge, Pyper, and Webb (1965) have compared the distribution of the 50 faint galaxies they discovered in the Fornax cluster to that of the brighter cluster members. They find that cluster membership is too difficult to ascertain farther than 4° from the center, so they limit their comparison to the inner regions. There, the faint galaxies appear to show less central concentration, suggesting that a real segregation exists. Unfortunately, the number of bright galaxies is small (about 20), and small sample fluctuations or subclustering may influence the picture.

The fact that the peak at the bright end of the luminosity function of the Coma cluster seems to be at least partially due to brighter galaxies at the cluster center (Rood 1969; Rood and Abell 1973) indicates some dependence of the galaxy distribution in the cluster on galaxian mass. N. Bahcall (1972, 1973*c*) finds some possible segregation by mass of galaxies in cluster A31, but no evidence for the same in cluster 2199. Kwast (1966) finds the bright and intermediate galaxies in A1060 to have the same distribution, but the faint galaxies in the cluster to be more broadly distributed. Rudnicki and Baranowska (1966*a*, *b*) find some evidence for mass segregation in A1185 and A1213, but not in A1367.

The best studied cluster, and the best example for discussion, is the Coma cluster (1656). Here again, different investigators disagree on the size of the cluster and the distribution of bright and faint galaxies. Zwicky (1957) holds that the cluster is at least 6° in radius, that it contains about 10^4 members brighter than $m_{pg} = 19$, and that there is marked segregation of its brighter and fainter members. Noonan (1961*a*, *b*), however, has reanalyzed Zwicky's own published counts and finds the cluster to be only 100′ in radius, and to have identical distributions of bright and faint galaxies. Omer, Page, and Wilson (1965), from the analysis of their own and of Shane's counts in the Coma cluster, arrive at the same conclusion as Noonan. These different conclusions result from different interpretations of the surface density of field galaxies. The mean field density is quite constant from 2° to 6° from the cluster center, but is slightly lower in some directions at a distance just beyond 6°. Zwicky interprets the lower average density 6° to 7° from the cluster as the true field and attributes the excess of faint galaxies at distances less than 6° to cluster membership. The other investigators interpret the higher density from 2° to 6° as the correct field, in which case there is not an excess of faint galaxies with a larger radial distribution than the bright ones; the total cluster population to $m_{pg} \sim 19$ is then only about 10^3.

The writer (Abell 1963) finds the cluster to be slightly elongated in a northeast-southwest direction, and to be probably contained within an oval region of dimensions 5°4 × 4°3. The oval shape of the cluster is also apparent in the equal-density contours of Shane and Wirtanen (1954), and in unpublished data of Reaves (Omer, Page, and Wilson 1965). To the magnitude limit $m_{pv} = 18.3$, he has found that in a 70 square-degree area outside this oval region the number of galaxies brighter than magnitude m is given by $\log N(m_{pv}) = \text{constant} + 0.6\, m_{pv}$, which is expected if these are field galaxies distributed at random through space. At most, only a few percent of these galaxies can be faint outlying cluster members. Thus, the cluster cannot contribute appreciably to the field beyond the 5°4 × 4°3 region. The writer adopts 77 galaxies per square degree (with $m_{pv} \leq 18.3$) as a lower limit to this field density; the correct value may be a little higher, because the average is taken over an area which includes the low-density region several degrees away from the cluster center. If the counts within the oval cluster region are corrected for this field density, the luminosity function obtained for the cluster is the one given in table 2. The radial spatial distributions of galaxies brighter than $m_{pv} = 16.0$ and in the range $m_{pv} = 16.0$–18.3, derived from averaged counts in north-south and east-west strips across the cluster, are shown in figure 6. The slight segregation of the brighter and fainter galaxies that appears to be exhibited in figure 6 may result from the adoption of slightly too low a field density; this, in other words, is an upper limit to the segregation of bright and faint galaxies. The writer's results are not incompatible with those of Noonan and of Omer, Page, and Wilson, who find no segregation at all, and a total cluster radius of 100′, or 4×10^6 pc. In addition, they are compatible with results of N. Bahcall (1973*b*), who finds no evidence for segregation of bright and faint galaxies except possibly for a slightly (20 percent) higher concentration of the brightest cluster galaxies near the core.

Rood *et al.* (1972) also concur that there is no marked segregation of bright and faint galaxies in the Coma cluster. However, they find the cluster to have a radius of 200′. The writer thinks it possible that the larger size Rood *et al.* find may result from their treating as members some outlying galaxies that may be part of a superstructure of galaxies that do not really belong to the main Coma cluster concentration. However, this question bears on how a cluster is defined in the first place, and there exists no unambiguous agreed-on operational definition. As remarked earlier, answers to such questions as the memberships, sizes, and distributions of galaxies of different masses within clusters depend on how clusters are defined and how their members are distinguished from the field. Differences between how these matters are handled may well account for the diversification of results described in this section.

3.6. Velocity Dispersions in Clusters

Radial-velocity observations have been made in a number of clusters, especially to obtain data for cosmological tests. Usually, however, the mean radial velocity of a cluster

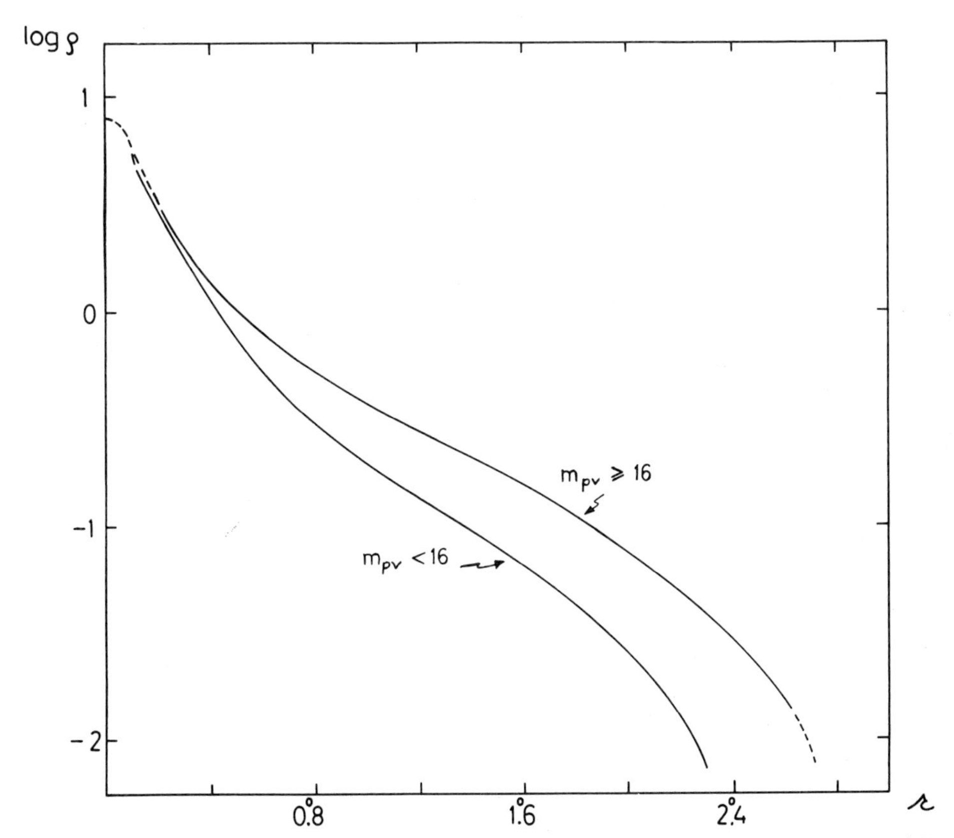

Fig. 6.—The radial density distribution of bright and faint galaxies in the Coma cluster (No. 1656) derived from Abell's counts. The zero points of the ordinates are arbitrary and have been shifted to agree for bright and faint galaxies at the cluster center.

is estimated from redshift measures of only two or three individual members—too few to provide much information about the internal kinematics of the cluster. For a few clusters and groups, however, there are enough data to allow a more or less meaningful estimate of the dispersion of velocities to be made. Data for several such clusters are given in table 6. Successive columns list the name of the cluster or the Abell catalog number (or both), the mean radial velocity (corrected for galactic rotation), the square root of the dispersion in radial velocities, the number of galaxies whose measured redshifts were used in calculating these quantities, and the authority for the observations.

TABLE 6

VELOCITY DATA FOR SEVERAL CLUSTERS

Cluster	$\langle V_r \rangle$ (km s^{-1})	$\langle v_r^2 \rangle^{1/2}$ (km s^{-1})	n	References
Leo group (around NGC 3627)..	787	260	18	1, 2
Virgo cluster..........	1136	643	73	1
Elliptical component.	950	550	33	3
Spiral component....	1450	750	19	3
Fornax cluster ($\alpha \approx 3\frac{1}{2}^h$; $\delta \approx -36°$)...	1452	287	12	1, 2
Pegasus I cluster (around NGC 7619)..	3836	260	6	1
Group around NGC 383	5274	504	9	1
Cluster 194 (around NGC 541)...	5321	406	41	4
Cluster 426 (Perseus)...........	5437	713	7	1, 2
Cluster 1656 (Coma).............	6866	932	46	5
Cluster 2199 (around NGC 6166)..	8736	541	15	6
Cluster 2151 (Hercules)..........	10775	631	15	7
Cluster 1377..........	15269	358	4	1
Cluster 2065 (Corona Borealis)....	21651	1210	8	1

REFERENCES.—(1) Humason, Mayall, and Sandage 1956; (2) Mayall and de Vaucouleurs 1962; (3) de Vaucouleurs 1961*b*; (4) Zwicky and Humason 1964*a* (they do not state whether or not their radial velocities are corrected for galactic rotation); (5) Lovasich *et al.* 1961; (6) Minkowski 1961; (7) Burbidge and Burbidge 1959*b*.

Three points should be noted: (1) The square roots of the dispersion given are in *radial* velocities (i.e., as seen from the galactic center). If the velocity field throughout a cluster is isotropic, the true value should be $\sqrt{3}$ times the value given. If, on the other hand, the motions of member galaxies are largely radial in a cluster, the true square root of the dispersion should be less than $\sqrt{3}$ times the tabulated value because most of the galaxies measured are near the projected center of a cluster where a majority would be expected to be moving nearly in the line of sight. (2) Redshifts can be measured only for the brightest galaxies in most clusters. If a degree of statistical equilibrium exists in a cluster, the fainter galaxies should have a larger dispersion in velocity; if not, the tabulated values may be near the true ones, except for projection effects. (3) Typical individual measures of redshifts of galaxies carry probable errors of from 50 to 200 km s^{-1}. When only a few galaxies in a cluster are measured, the velocity dispersion can be considerably in error for this observational reason alone.

For a more current and complete compilation, see Noonan (1973).

4. DYNAMICS OF CLUSTERS

4.1. General Considerations

The incomplete, and even conflicting, observational data do not permit us to draw a very complete picture of the dynamics of even a single cluster. At best, from observational data alone, we can make only rough estimates of clusters' probable dynamical structures.

The time of relaxation in the central regions of a typical regular cluster, like the Coma cluster, is almost certainly greater than 10^9 years, and probably lies within a small factor of 10^{10} years (Oort 1958; van Albada 1960). It is doubtful, therefore, that in the presumed age of the Universe ($\sim 10^{10}$ years) gravitational encounters between galaxies can have set up anything approaching a state of statistical equilibrium in a regular cluster. If, however, such a state *could* have been reached, we would expect: (1) an equipartition of energy with a consequent segregation of velocities of galaxies according to their masses; (2) a radial segregation of galaxies within the cluster according to mass, with the less massive, faster-moving galaxies occupying a larger spatial distribution; and (3) a mass distribution within at least the central regions of the cluster resembling that of the isothermal polytrope. Observational verification of all of these effects has been reported by various investigators, but these observations are inconclusive.

In cluster 194, Zwicky and Humason (1964*a*) report radial velocities for 41 galaxies that they presume to be cluster members. The mean velocity is 5321 km s^{-1}, with an rms dispersion of 406 km s^{-1}. The brightest 21 galaxies, however, have a mean velocity of 5254 km s^{-1}, with a dispersion of 360 km s^{-1}, while the corresponding figures for the faintest 21 galaxies are 5392 and 439 km s^{-1}. Zwicky and Humason interpret the slightly higher dispersion in velocity of the fainter members to be indicative of a degree of equipartition of energy. The brightest six galaxies do have velocities very near the mean for the cluster, but for the remaining galaxies the effect vanishes; if anything, the dispersion in velocity *decreases* with increasing magnitude, as is seen at once on a plot of Zwicky and Humason's data. Indeed, as those authors themselves point out, there is a larger difference between the velocity dispersion of galaxies in the southwest and northeast halves of the cluster, than between the bright and faint members. In short, the case for a velocity-dispersion dependence on magnitude in cluster 194 is not convincing. Negative results for such segregation have been found for five other groups and clusters (Lovasich, Mayall, Neyman, and Scott 1961; Neyman and Scott 1961; Rood 1965). In their analysis of the Coma cluster, Rood *et al.* (1972) find no evidence for a luminosity dependence of velocity dispersion, with the possible exception that the very brightest galaxies in the central core may have a smaller rms dispersion; they regard these results, however, as statistically inconclusive.

Zwicky (1957) reports a very pronounced radial segregation of bright and faint members in the Coma cluster; but as was pointed out in the last section, this result depends on what assumption is made about the field. The writer's own investigation indicates that a slight segregation possibly exists (fig. 6), but not necessarily so. Rudnicki's (1963) finding of a segregation of bright and faint galaxies in cluster 426 may result from irregular interstellar absorption. The case for the Virgo cluster is unclear because it may be more than one cluster. The slight segregation found by Hodge, Pyper, and Webb (1965) in the Fornax cluster appears to be real, but Fornax is a poor cluster, and the segregation is at best small. Other investigations have also suggested slight radial segregation of bright and faint galaxies in some clusters (N. Bahcall 1972; Kwast 1966; Rudnicki and Baranowska 1966*b;* Rood 1969; Rood and Abell 1973; Noonan 1971), but the evidence is sometimes contradictory, and in no case is there segregation present in the amount required by complete statistical equilibrium.

As we have already seen, both Zwicky and N. Bahcall have shown that observations of the density distributions in several regular clusters are compatible with isothermal distributions, but their data can also be represented about as well by various other distributions that have nothing to do with statistical equilibrium.

In summary, there is no convincing evidence that equipartition of energy exists in any cluster, although there is some evidence for partial radial segregation by mass in a few clusters. Nor would we expect equipartition of energy, because of the long times of relaxation in clusters. Ages of clusters are probably at most only a few relaxation times; some dynamical evolution has probably taken place, but it is unlikely that a final state has been reached.

4.2. Masses of Clusters

If Newton's laws are valid over intracluster distances, then the virial theorem,

$$\tfrac{1}{2}\ddot{I} = 2T + \Omega\,, \tag{8}$$

should apply. If, further, a cluster is in a steady state, the polar moment of inertia should be constant, so $\ddot{I} = 0$, in which case

$$2T + \Omega = 0\,. \tag{9}$$

Masses computed for clusters of galaxies are usually based on equation (9). The requirement that $\ddot{I}$ strictly vanish does not have to be rigorously fulfilled; if only the cluster is on the verge of stability, so that the total energy is zero, we should have

$$E = T + \Omega = 0\,, \tag{10}$$

and the mass derived from equation 10 is only a factor of 2 smaller than that obtained from equation (9).

The kinetic energy T can be approximated by $\frac{1}{2}\mathfrak{M}_{\rm cl}\langle V^2\rangle$, where $\mathfrak{M}_{\rm cl}$ is the total cluster mass and $\langle V^2\rangle$ is the velocity dispersion weighted by mass. Radial velocities are usually observed for the brighter and presumably the relatively massive cluster members; thus little error results from estimating $\langle V^2\rangle$ from the observed dispersion in radial velocities in a cluster, $\langle V_r^2\rangle$. The principal uncertainty is that of projection effects. If the velocity field is isotropic at all points in a cluster, then $\langle V^2\rangle = \langle 3V_r^2\rangle$; if most galaxies move radially through the cluster, then $\langle V^2\rangle$ may only slightly exceed $\langle V_r^2\rangle$.

The potential energy, Ω, is

$$-G\sum_{i\neq j}\frac{m_i m_j}{r_{ij}}\,,$$

where the summation must be carried out over all pairs of galaxies of masses m_i and m_j separated by distance r_{ij}. If the cluster has n members, there are $n(n-1)/2 \approx n^2/2$ such pairs. It is customary to write the potential as $-G\mathfrak{M}_{cl}^2/2R'$, where $1/R'$ is a weighted mean of the $(1/r_{ij})$'s. Attempts have been made to evaluate $1/R'$ directly for a few groups, and even for the Hercules cluster (A2151) by the Burbidges (1959*b*). More often, R' is associated with the apparent radius of a cluster. For rich symmetrical regular clusters, the potential can be expressed more conveniently:

$$-\Omega = G\int_0^R \frac{\mathfrak{M}(r)d\mathfrak{M}(r)}{r}, \tag{11}$$

where $\mathfrak{M}(r)$ is the mass contained within a radial distance r of the cluster center. For a constant mass-to-light ratio, we can write $\mathfrak{M}(r) = fL(r)$, which leads to

$$-\Omega = Gf^2\int_0^R \frac{L(r)dL(r)}{r}. \tag{12}$$

The integral in equation (12) can be evaluated numerically from the density distribution of luminosity in the cluster derived from observations, leading finally to

$$-\Omega = qG\mathfrak{M}_{cl}^2/R, \tag{13}$$

where R is the outer cluster radius and q is a numerical factor that is near unity. For the Coma cluster, for example, Rood (1965), using the writer's observations, finds that $q = 0.92$. The only unknown in equation (9) (or eq. [10]) is thus the total cluster mass. Setting $q = 1$, we find

$$\mathfrak{M}_{cl} = \xi\langle V_r^2\rangle R/G, \tag{14}$$

where the factor ξ allows for the factor of 2 uncertainty in whether equation (9) or equation (10) is more appropriate, and for the projection effects between $\langle V_r^2\rangle$ and $\langle V^2\rangle$; ξ thus lies in the range $\frac{1}{2}$ to 3.

Although the dynamical method of estimating masses of galaxian clusters is well known, the assumptions involved have been reviewed here because the masses found often seem discordant with those estimated from the luminosities of the cluster members. Ratios of mass to light (in solar units) found for individual galaxies (e.g., Burbidge and Burbidge 1959*a*) or for double galaxies (Page 1962) range from 1 to 15 for spirals and from 10 to 70 for ellipticals and S0's. On the other hand, to reconcile the masses estimated for a number of groups and clusters of galaxies from dynamical and luminosity methods, mass-to-light ratios for some of those systems would have to lie in the range from 100 to 1000 (e.g., Limber 1962). Years ago, Zwicky (1933), and Smith (1936) called attention to the unexpectedly high mass obtained for the Virgo cluster from its internal kinematics. The Burbidges (1959*b*) have called attention to the same situation in the Hercules cluster (A 2151). Other discussions of the problem of the high dynamical masses found for groups and clusters include those of de Vaucouleurs (1960), Burbidge and Burbidge (1961), van den Bergh (1961*c*), Limber (1961), and Rood (1965). The problem, then, is to understand why mass-to-light ratios found for clusters and groups of galaxies are often higher—by even a factor of 10 or more—than those found for individual or double galaxies. Among the possibilities proposed are the following:

1. Ambartsumian (1958) has suggested that clusters (at least some of the smaller

groups) have positive energy and are expanding. On the other hand, to reduce the dynamical mass of a cluster even by just a factor of 4 from that given by the virial theorem, we would need to have $T = -2\Omega$; in this case, the observed speeds of galaxies are, in the mean, only $\sqrt{2}$ times greater than the speeds they will have when the cluster has expanded to infinity. In a period of at most a few times 10^9 years, all clusters would dissipate enough to have lost their identity, and it would be difficult to understand why most galaxies are still in clusters.

2. Perhaps the high masses of some clusters are due to a few extremely massive galaxies. There does appear to be a tendency for the mass-to-light ratio to be an increasing function of mass for elliptical galaxies (Rood 1965). However, it is very hard to invent convincing stellar populations that can give the extremely high mass-to-light ratios that would be required. Moreover, many of the smaller clusters and groups for which the mass-to-light ratio appears to be high have only spirals among their massive members.

3. The derived mass-to-light ratios are proportional to the assumed value of the Hubble constant. A greatly expanded distance scale would therefore reduce them. But it would also reduce the corresponding ratios found for most individual galaxies and for double galaxies, so the discrepancy would remain.

4. There is a probability that metagalactic structure of larger size than clusters exists i.e., second-order clusters (see next section). If so, and if gravitational forces exist between clusters in such a system, then two or more clusters seen overlapping in projection may occasionally be mistaken for an isolated system, and their relative motions can lead to a spuriously high observed velocity dispersion. In some cases, such an effect may partially be responsible for the mass-to-light discrepancy. The Hercules cluster, in particular, is in a region rich in groups and clusters of about the same distance and may not be a single dynamic unit. De Vaucouleurs's suggestion that the Virgo cluster is at least two clusters has already been mentioned, as has the possibility of inappropriately assuming cluster membership for outlying galaxies in the Coma cluster. In fact, Gott, Wrixon, and Wannier (1973) have shown that some systems that have been identified as groups may actually be optical alignments of field galaxies that are not gravitationally bound at all, and thus that high mass-to-light ratios reported for these systems may be entirely spurious.

5. Subclustering, or incidence of duplicity within a cluster, results in lowering the average separation of galaxies, thereby increasing the potential energy for a given mass. Van den Bergh (1960) has pointed out that binary galaxies in the Virgo cluster can possibly affect significantly the derived cluster mass. Studies of subclustering in several clusters, however (Abell, Neyman, and Scott 1964), show that the phenomenon can not be important generally.

6. A popular explanation is that clusters contain a large amount of invisible matter which adds to their masses but not to their visual luminosities. Zwicky (1957) believes that intergalactic obscuring matter in clusters hides more remote clusters, although the existence of such absorption is not universally accepted. For the Coma cluster, he estimates an obscuration of 0.6 mag (Zwicky 1959). The dust required to produce this absorption, however, corresponds to a density of only 10^{-30} g cm^{-3}, if the dust has the same obscuring properties as interstellar grains in the Galaxy (de Vaucouleurs 1960);

this would be a negligible contribution to the cluster mass. Evidence for luminous haze in the centers of clusters is inconclusive and controversial; if *visible* luminous matter does exist, it can in no case contribute substantially to the mass of a cluster. Field (1959) has shown that neutral hydrogen cannot be important enough in intergalactic space to affect cluster masses measurably. Observations of the 21-cm line in absorption reported by Robinson, van Damme, and Koehler (1963) suggest the presence of neutral hydrogen in the Virgo cluster, but the total mass indicated is more than two orders of magnitude less than that of the cluster itself. Recent 21-cm observations of three groups by Gott, Wrixon, and Wannier (1973) indicate that neutral hydrogen falls short by at least a factor of 25 of having enough mass to gravitationally bind those systems. Nor is hot plasma likely to be able to contribute appreciably to the potential energies of clusters; an analysis by Davidson, Bowyer, and Welch (1973) of the radio, soft X-ray, and far-ultraviolet observations of the Coma cluster, for example, seems to rule out the possibility that ionized gas contributes enough mass to bind that cluster. Still, intracluster material cannot yet be ruled out in all cases.

TABLE 7

MASS-LIGHT RATIOS FOR SEVERAL RICH CLUSTERS

Cluster	Mass, $\mathfrak{M}$ $\xi=2.1$ ($\mathfrak{M}_\odot$)	Visual Luminosity, L (solar units)	$\mathfrak{M}/L$
Virgo	2.4×10^{14}	1.3×10^{12}	181
194	2.4×10^{14}	3.8×10^{12}	64
1377	1.9×10^{14}	2.7×10^{12}	70
1656 (Coma)	1.7×10^{15}	1.2×10^{13}	144
2065 (Corona Borealis)	2.9×10^{15}	1.2×10^{13}	231
2199 (around NGC 6166)	4.3×10^{14}	6.1×10^{12}	71

Actually, the discrepancy may not be serious for rich clusters. Luminosities, masses, and mass-to-light ratios for six relatively rich to rich clusters are computed from the data in tables 2 and 6 with equation (14), and are listed in table 7. The factor ξ has been set equal to 2.1, following Rood's (1970) analysis of the Coma cluster velocities. The uncertainty in ξ can affect a derived cluster mass by at most a factor of about 2. If, for example, the clusters have zero energy, so that equation (10) rather than (9) applies, the actual mass-to-light ratios would be half those given in table 7. Total radii of 1.2 and 4.0 Mpc are used for the Virgo and Coma clusters, respectively. The Corona Borealis cluster was also assumed to have a radius of 4.0 Mpc, and 3.0 Mpc was arbitrarily chosen for the other clusters, for which the writer has not yet determined radii observationally. A Hubble constant of 50 km s^{-1} Mpc^{-1} is adopted throughout. The visual luminosity is uncertain for cluster 1377, for which only the old photometry of Baade is available, but it is probably correct to within a factor of 2; the velocity dispersion for this cluster, however, is calculated from only four radial velocities, so the mass is very uncertain. For cluster 194, the luminosity was calculated from magnitudes published by Zwicky and Humason (1964*a;* Abell 1964). Zwicky and Humason (1964*b*) later reported that some (they do not say which) of the galaxies whose magnitudes they published were subsequently discovered not to be cluster members, but it is doubtful that elimination of those objects can affect the luminosity given in table 7 by as much as a factor of 2. The $\mathfrak{M}/L$ ratio of 144 found for the Coma cluster can be compared to the value of about 165 found by Rood *et al.* (1972) (adjusted to the Hubble constant used here).

The clusters listed in table 7 are the only rich ones for which the writer has been able to find enough data to compute mass-to-light ratios. The ratios listed are all of the order 10^2 or less. For elliptical galaxies in binary systems, Page (1962) finds $\mathfrak{M}/L \approx 50$ (for $H = 50$ km s^{-1} Mpc^{-1}). There may still be a discrepancy of a factor up to 3 because of the uncertainty of ξ, but this reexamination of the data shows that a *serious* mass discrepancy does not *necessarily* exist in rich clusters.

4.3. Formation and Evolution of Clusters

It is highly improbable that clusters of galaxies have been built up by chance encounters of galaxies in the general field. Close encounters of three or more bodies of comparable mass can lead to the formation of stable pairs. However, the time required to build a cluster is very great; even if the Universe were old enough ($> 10^{12}$ years), most galaxies would still be in the general field rather than in clusters, as the observations suggest. Moreover, if the galaxies in a three-body encounter do not have comparable mass, the energy exchange is too small to produce captures, and observed clusters containing galaxies whose masses range through several orders of magnitude could never be built up by such a process (Ambartsumian 1961). We are nearly forced to conclude, therefore, either that clusters are systems whose member galaxies became gravitationally bound at more or less the same time, or that the clusters represent condensations from pregalaxian material and that subcondensations within them became galaxies.

Many investigators have attempted to determine conditions under which galaxies or clusters can condense from gas or plasma. As yet, there is no successful complete theory for the formation of galaxies and clusters from primordial material. On the other hand, a theory of van Albada (1960, 1961, 1962) gives a rather detailed account of how clusters may form and evolve, given certain initial conditions. His theory is of sufficient interest to warrant a brief review here.

Van Albada assumes that galaxies already existed in a nearly homogeneous expanding universe before the clusters formed. He then considers the conditions under which a region of the expanding universe can become gravitationally unstable. For the sake of mathematical amenability, he considers only spherically symmetrical condensations. For the case of zero cosmological constant, he finds that density fluctuations of only about 2 to 7 percent can lead to gravitational instabilities, which result in the formation of a cluster containing matter originally spread over a region of radius

$$r \approx \sigma/(G\rho)^{1/2}\,, \tag{15}$$

where ρ is the smoothed-out density of the universe at the time the instability commences, and σ is the square root of the velocity dispersion. For the latter, van Albada adopts 50 km s^{-1}. If the radius of the universe is R, ρ varies as R^{-3} and σ as R^{-1}; thus r varies as $R^{-1/2}$. The instabilities, therefore, occur over regions that are relatively smaller as the universe expands. At an early history of the expansion, when R is small, one would expect condensations which would lead to large-scale inhomogeneities, contrary to the cosmological principle.[3] Instabilities must therefore be inhibited somehow at very early epochs, until ρ has decreased to a value of 10^{-24} or 10^{-25} gm cm^{-3}; van Albada suggests that radiation pressure might provide this inhibiting force.

With numerical calculations, van Albada follows the growth of an instability and its subsequent evolution. Since the galaxies already exist when the condensation begins, there is no dissipation of kinetic energy by cooling. A central nucleus develops and steadily increases in density and velocity dispersion. Matter streams into the nucleus from a surrounding corona, which in turn attracts matter from the entire unstable region. As the nucleus contracts, the corona changes only slightly. Meanwhile, the expansion of the universe makes the cluster appear relatively more and more prominent against the surrounding field. The density distribution in the inner regions of the cluster is compatible with those observed—for example, with an isothermal polytrope. At no time, however, is the cluster in statistical equilibrium. The condensations develop most slowly from the initial density perturbations of the smallest amplitude. It is possible therefore, that an early instability over a large relative region containing a large mass could begin with a small-amplitude density excess over the mean, and just now be developing into a great cluster. A model of such a system with an age of several billion years has dimensions and structure roughly resembling the Coma cluster.

A point to be remembered is that in the van Albada picture the galaxies exist before clusters form; it is large-scale instabilities in a "fluid" of point-mass galaxies that produce the clusters. Moreover, at early stages the contraction of a cluster is only "contraction" relative to the expanding universe. At later stages the nucleus, as it increases in mass, may contract in an absolute sense, but the increasing velocity dispersion eventually inhibits further contraction and the nucleus expands again. Because the outer layers of the cluster are always expanding absolutely, they cannot stop the expansion of the nucleus, and van Albada expects the cluster ultimately to dissipate. None of his detailed models, however (at the time of writing), have been carried to such an advanced state; in fact, none of them may even correspond to development as advanced as that of actual clusters.

A different approach to the study of the dynamical evolution of clusters is that of Aarseth (1963, 1966, 1969), who begins with a cluster of n members in an arbitrary configuration and with arbitrary initial velocities, and simply solves numerically the "n-body problem." He has considered clusters of from 25 to 100 members, interacting under purely gravitational forces. To take account of the finite sizes of galaxies, Aarseth chooses for the potential function of a galaxy,

$$\Phi = -\frac{G\mathfrak{M}}{(r^2 + \xi^2)^{1/2}}, \tag{16}$$

where ξ is its effective radius.

One of Aarseth's most interesting models, and one which may provide some approximation to an actual cluster of galaxies, has 100 members, distributed initially with a

[3] The correctness of the cosmological principle has never been observationally verified more than very roughly.

spherically symmetrical density configuration of the form

$$\rho(r) = \rho(0)[1 - r/R] . \tag{17}$$

The 100 objects (representing galaxies) have four different masses: 40 have a relative mass of 0.25; 30 of 0.75, 20 of 1.50, and 10 of 3.75. The complete mass range is thus over a ratio of 15–1. All members are started with random velocities, with the restriction that no object has an initial velocity that will carry it to a greater radial distance from the cluster center than R. In particular, there is no initial correlation between mass and velocity. Aarseth's integration follows the cluster for a period which would correspond in a real galaxian cluster to about 10^{10} years. During that time, he finds that a nucleus develops which is surrounded by an extended halo. A few of the halo members escape. At the end of the integration period there is a slight radial segregation of members by mass, but the least massive objects have a radial distribution less than 40 percent greater than the most massive ones, and the extreme range in mean kinetic energy, originally over a factor of 15, is reduced by only about half; in other words, the cluster is far from a state of statistical equilibrium. Application of the virial theorem in a naïve way to find the mass of the cluster, however, would lead to a mass error of less than 20 percent. Aarseth finds that about 50 collisions would have occurred in the cluster over the 10^{10} years. Only a few stable binary systems are formed, but those have long lifetimes. A similar model cluster with 50 members showed qualitatively similar evolution.

Aarseth also followed the evolution of two irregular model clusters of 50 members each. One was initially V-shaped with its members at rest, and the other had its members spread evenly over an elliptical region, and with small random velocities. In each case a stable "subsystem" formed as a cluster nucleus that retained stability over the entire integration time. Even for these clusters, the mass found from application of the virial theorem would be correct to within 50 percent.

5. THE DISTRIBUTION OF CLUSTERS

5.1. The Evidence for the Local Supercluster

A local irregularity in the distribution of nearby groups and clusters of galaxies has been suspected for several decades. From his analysis of the distribution of double and multiple galaxies, Holmberg (1937) inferred the existence of a metagalactic cloud with a diameter, according to the distance scale used in this review, between 90 and 150 Mpc, and with a center lying in the general direction of the north galactic pole at a distance of about 18 Mpc. Holmberg's conclusion was confirmed qualitatively by Reiz's (1941) study of the distribution in direction and magnitude of 4000 galaxies in the northern galactic polar cap. The idea of a local supercluster of galaxies was revived by de Vaucouleurs (1953, 1956, 1958), who described it in considerable detail. According to de Vaucouleurs, the supercluster has a diameter (with $H = 50$ km s^{-1} Mpc^{-1}) of about 75 Mpc, and it contains, in addition to the Local Group, the Virgo cluster, the Ursa Major cloud, and numerous smaller groups and clusters. He finds the system to be flattened, so that the bright galaxies are seen in the sky highly concentrated toward a great circle (the "supergalactic equator") with its pole at $l^{\rm II} = 47°$, $b^{\rm II} = 5°$. He believes the center of the system to lie within or near the Virgo cluster. The flattening suggests rotation;

from his analysis of the radial velocities of bright galaxies, de Vaucouleurs presents interesting evidence for differential rotation, and derives about 500 km s^{-1} for the rotational velocity of the Galaxy about the center of the system. From this rotational velocity, he derives a total mass for the Local Supercluster of the order of $10^{15}\,\mathfrak{M}_{\odot}$. On the other hand, the present or "instantaneous" period of revolution of the Galaxy is about 2×10^{11} years; and even though de Vaucouleurs believes the system to be expanding slowly, it can hardly have completed even one rotation unless it was formed at a very early epoch in the expansion of the Universe, when its mean density was orders of magnitude higher than at present. Quite possibly, therefore, the apparent flattening of the supercluster may have nothing to do with its presumed rotation.

The dynamical properties of the Local Supercluster may not be well established, but further evidence for its reality as a geometrical entity is provided by an independent investigation by Carpenter. Carpenter (1961; Abell 1961) studied the distribution in magnitude and direction of galaxies brighter than $m_{pg} = 16$ in a large region of the north galactic hemisphere from the Palomar *Sky Survey* prints. At magnitudes brighter than 13.5, he finds a highly significant concentration of galaxies along a 90° sector of an 18° strip along de Vaucouleurs's "supergalactic equator." In the next interval of 1 mag, the number of galaxies drops off very rapidly compared with expectations for a uniform galaxy distribution in depth. For $m_{pg} > 14.5$, however, the logarithm of the number of galaxies brighter than m_{pg} increases as $0.6m_{pg}$, which would be expected if most of these galaxies are remote ones, beyond the limits of the supercluster. Carpenter finds a similar result for galaxies in adjacent 18° strips saddling the strip along the supergalactic equator, except that the total number of bright galaxies in these zones is less than in the central strip.

5.2. Other Evidence of Second-Order Clusters

The tendency of clouds, clusters, and groups of galaxies to form assemblages of higher order than single clusters was noted long ago by Shapley (1933, 1957). The phenomenon of superclustering was demonstrated dramatically, however, by the analysis of the counts by Shane and his associates of galaxies brighter than $m_{pg} \approx 18$ on photographs taken for the Lick *Astrographic Survey*. Shane and Wirtanen (1954) describe six clouds of larger dimension than normal clusters, each containing multiple condensations. Three of the Shane-Wirtanen clouds (Nos. 4, 5, and 6) correspond to apparent groups of two or more clusters, and two (Nos. 2 and 3) to single clusters in the writer's catalog of rich clusters; the remaining clusters in the Shane-Wirtanen clouds apparently are not rich enough for inclusion. More recently Shane (1956*b*) has called attention to additional rich aggregations of galaxies, some of which contain several centers of condensation, suggesting multiple clusters. Some of these systems are described by Dr. Shane elsewhere in this volume. Typical dimensions of these clouds (for $H = 50$ km s^{-1} Mpc^{-1}) lie between 15 and 60 Mpc.

Even the very rich clusters in the writer's catalog (Abell 1958) show a strong tendency for second-order clustering. Of the 2712 clusters catalogued, 1682 were selected as comprising a homogeneous statistical sample. Clusters in the sample all have populations (defined in § 3.4) of at least 50, redshifts in the range $d\lambda/\lambda = 0.02$–0.20, and lie at great enough galactic latitudes that interstellar absorption does not prevent their identification (usually at latitudes greater than about 30°). The surface distribution of these clusters is shown in figure 7. The clusters are classified according to distance, the mean redshifts of clusters in distance groups 1 through 6 being, respectively, 0.027, 0.038, 0.067, 0.090, 0.140, and 0.180.

Superficial examination of figure 7 shows an obvious clustering tendency of the clusters themselves. To test the significance of possible superclustering, the part of the sky covered by the homogeneous sample was divided into grid cells of various sizes, and for each sized cell the frequency distribution, $f(t)$, of cells containing t clusters each was determined. A χ^2 test was used to estimate the probability that $f(t)$ would be obtained in a random sampling from a population whose frequency distribution is the binomial distribution, $B(t)$. The $f(t)$ and comparison with $B(t)$ was determined separately for clusters in distance groups 5 and 6, and for clusters in distance groups 1 through 4 combined. It was found that $f(t)$ approaches $B(t)$ for very small cell sizes, for then every cell contains either one cluster or none. With increasing cell size, $N(t)$ departs more and more from $B(t)$; the probability $P(\chi^2)$, of $N(t)$ being a random sampling from a population with frequency distribution $B(t)$ for the most distant clusters (for which the sample is largest), is as low as 10^{-30} to 10^{-40}. For larger cell areas $P(\chi^2)$ increases again, mainly because the sample size diminishes (fewer large area cells fit into the sky than small area ones) and the deviation of $N(t)$ from $B(t)$ is less significant. $P(\chi^2)$ should also eventually increase with cell size if the cells become large compared to any anisotropics in the cluster distribution—that is, if superclustering is "smoothed out." The writer originally interpreted an observed inverse correlation between the cell diameters for which $P(\chi^2)$ is a minimum with the cluster distance class as an indication that the second-order clustering occurs on the same scale at all distances surveyed. This interpretation is not strictly justified because of the smaller significance of the results for large cell sizes. However, at cell sizes smaller than those for which $P(\chi^2)$ is at a minimum, the descent of $P(\chi^2)$ with cell size is steepest for the most distant clusters and least steep for the nearest, as one would expect for superclusters of a common scale displaying smaller angular sizes at greater distances.

The evidence that second-order clusters may have similar linear sizes at different distances argues against their being illusions produced by interstellar or intergalactic obscuration. Simple inspection of figure 7 would also seem to rule out absorption as the cause of the clumpy cluster distribution; if apparent clumps of relatively near clusters are merely parts of a uniform or random distribution of clusters seen through holes in absorbing material, then apparent clumps of more remote clusters should be seen in the same directions, but certainly not between them, as is the case.

About 50 apparent groupings of clusters—probably second-order clusters—can be identified in figure 7. The writer has described 17 of these groupings (Abell 1961). The mean number of clusters (in the homogeneous sample) among the 17 second-order groups is 10.6 ± 6.0 (s.d.). This number, of course, refers only to the very rich clusters in the Abell catalog; the total number of clusters and groups of all kinds in a typical second-order cluster might be greater by an order of magnitude or more. The mean linear diameter of the 17 groups of clusters is 78.0 ± 23.8 (s.d.) Mpc. The list of 17 systems includes two of the Shane clouds—the Corona cloud and the Serpens-Virgo cloud—described elsewhere in this volume.

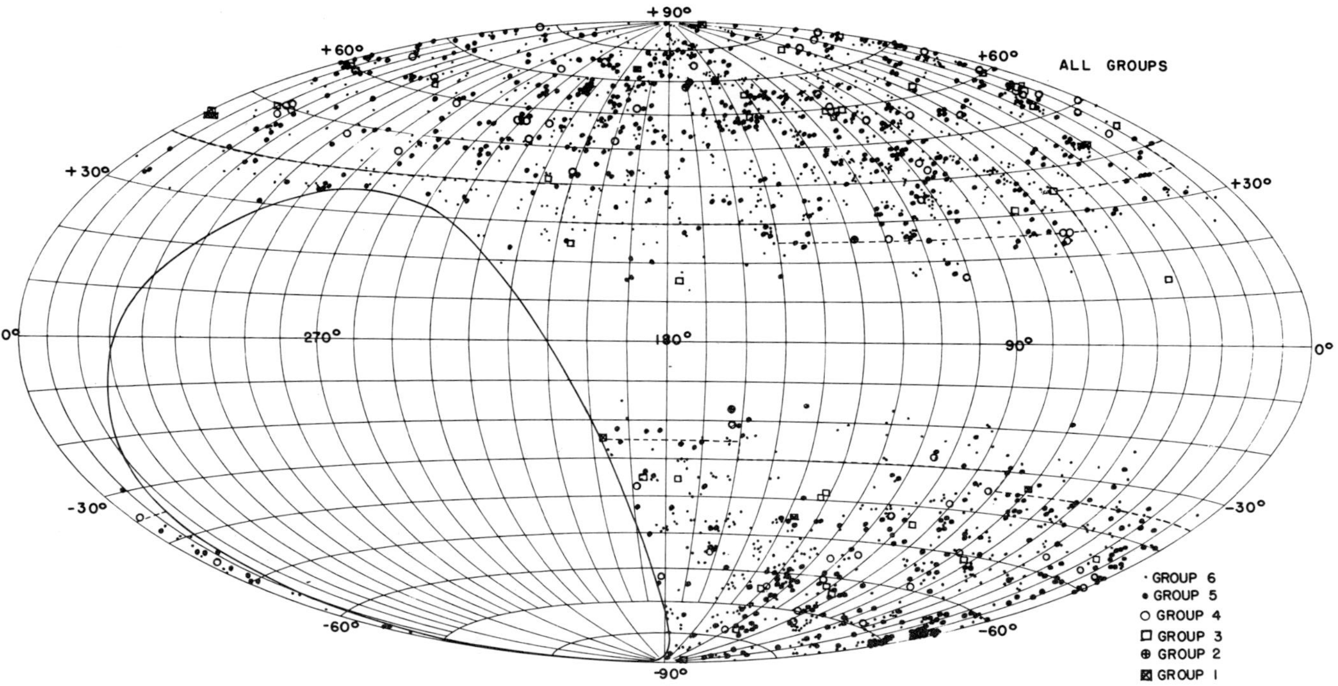

FIG. 7.—The distribution in galactic coordinates (old system) of the clusters in Abell's catalog. Those clusters closer to the galactic equator than the dotted line are not part of the homogeneous statistical sample. The large empty oval region is the part of the sky not covered by the Palomar *Sky Survey*. The plot is on an Aitoff equal-area projection.

The distribution of clusters in Volumes 1, 2, 3, and 5 of the *Catalogue of Galaxies and Clusters of Galaxies* (Zwicky *et al.* 1961–1968), the first four volumes of the catalog to be published, have been analyzed in the same way as were those in the Abell catalog with nearly identical results by Abell and Seligman (1965, 1967).

Numerous other investigators have attempted analyses of the distribution of rich clusters of galaxies in the published catalogs. Among them, Kiang and Saslaw (1969) computed serial correlations of Abell clusters in 50-Mpc cubic cells to determine the three-dimensional cluster distribution, and find correlations over a scale of at least 100 Mpc and possibly to 200 Mpc. Bogart and Wagoner (1973) performed nearest-neighbor tests on the Abell clusters, and found that the distribution of nearest-neighbor distances from half of the clusters (sources) to the other half (objects) has a significantly smaller mean than does the corresponding distribution when a set of random points is used for sources, indicating that the clusters are significantly clustered. Bogart and Wagoner estimated the scale of the clustering by rotating the "object" half of the clusters in galactic longitude until the distribution of nearest-neighbor distances approached the random one. The angular scale found for distance group 5 clusters is slightly greater than for the more distant group 6 clusters, suggesting a physical association of clusters; the corresponding linear scale is $\sim$200 Mpc.

Statistical analyses of three catalogs of extragalactic objects have been carried out recently by the Princeton group (Peebles 1973; Hauser and Peebles 1973; Peebles and Hauser 1974; Peebles 1974). The sources are the Abell (1958) catalog, the galaxies catalogued by Zwicky and his associates (Zwicky *et al.* 1961–68), and the galaxies catalogued from the Lick Astrographic plates (Shane and Wirtanen 1967). Peebles and Hauser have investigated the correlation between objects in the individual catalogs and the cross correlations of objects in different catalogs. They find that the clusters and individual galaxies seem to correlate in direction, both separately and with each other, over angular distances of up to 6°. The linear size of these homogeneities is of the order 100 Mpc.

Peebles has also developed a powerful statistical method for detecting variations over the surface distribution of galaxies or clusters by means of a two-dimensional power spectrum. It was first applied (Yu and Peebles 1969) to test the hypothesis of complete second-order clustering of the Abell catalog clusters. Yu and Peebles found that if second-order clusters contain an average of 10 rich clusters each, then only about 10 percent or less of the Abell clusters can be members of such superclusters, and that in a model of complete superclustering, on the average there could be at most about 2 clusters per supercluster. It should be noted that in these calculations those clusters of distance class 5 in the southern galactic hemisphere, where inspection of figure 7 suggests second-order clustering to be most pronounced, were omitted because that part of the Abell catalog seemed to Yu and Peebles to be atypical.

Peebles (1973) developed the power-spectrum approach further, and he and Hauser reanalyzed the Abell catalog (Hauser and Peebles 1973). They report "clear and direct evidence of superclusters with small angular scale" and that the structure corresponds to an average of 2 to 3 clusters per supercluster.

The early χ^2 tests described above are subject to misinterpretation because of the possibility of a general absorption gradient and other systematic effects, and the results of these tests alone should thus be viewed with caution. However, as we have seen, the same results are obtained with more sophisticated tests, made possible with modern computing equipment, especially the powerful power-spectrum analysis. These studies of the catalogs of observed galaxies and clusters of galaxies show very strong—perhaps overwhelming—evidence for inhomogeneities in the large-scale distribution of matter in space with a scale (for $H = 50$ km s^{-1} Mpc^{-1}) of the order of 10^8 pc.

If Newton's laws are valid over dimensions of second-order clusters, and if the latter do not partake of the general expansion of the Universe, we can use the virial theorem to estimate the velocity dispersion within such a system. We denote the mean separation of its members by R', and have

$$\langle V^2 \rangle^{1/2} = 4.6 \times 10^{-2} (\mathfrak{M}/R')^{1/2} \text{ km s}^{-1} , \tag{18}$$

where $\mathfrak{M}$ and R' are in solar masses and parsecs, respectively. If the mass of a typical rich cluster is $5 \times 10^{14} \mathfrak{M}_\odot$, the entire mass of a typical supercluster probably lies in the range 10^{15} to $10^{17} \mathfrak{M}_\odot$. Adopting 20 Mpc for R', we find that $\langle V^2 \rangle^{1/2}$ should lie in the range 300–3000 km s^{-1}. If the velocity field is isotropic, the observed rms dispersion in radial velocity should actually lie under 2000 km s^{-1}.

Radial velocities are known for six clusters that are suspected of forming a second-order cluster covering an elongated region centered near $\alpha = 16^h14^m$, $\delta = +29°$ (Abell 1961). The total range of these six velocities is about 3000 km s^{-1}. There are not enough data to determine a meaningful velocity dispersion for the system, but at least the observations are not incompatible with the assumption that gravitational interactions occur between its members.

If second-order clusters are expanding, or if they do not have negative total energy, the observed dispersion in radial velocities could be higher than the value derived above. Suppose, for example, that gravitational forces within such a system are negligible, and that it expands at the general universal rate. Then the corresponding spread of radial velocities across a second-order cluster of diameter D Mpc should be $\Delta V \sim H \times D = 50 \times 75 = 3750$ km s^{-1}. Since our estimate of the value of D is proportional to H^{-1}, the derived value of ΔV is independent of the value assumed for H.

5.3. The Large-Scale Distribution of Clusters and the Mean Density of Matter in the Universe

Hubble's (1934) study of the distribution of galaxies revealed by 60- and 100-inch telescope photographs showed that to the accuracy of the observations the galaxies, except for a clustering tendency, appeared to be distributed uniformly in depth in all directions. The distribution of the rich clusters has similarly been investigated by the writer (Abell 1958). The homogeneous sample selected from the Abell catalog yields the frequency distribution $N(z)$, of clusters with (estimated) redshift, z. There is, as we have seen, a small-scale clustering tendency of the clusters themselves, and $N(z)$ cannot be determined (from these data) accurately enough to choose between different cosmological models. There is, however, no evidence for departure from uniformity over regions of space that are large compared with the scale of the second-order clustering. Neither the Hubble nor the Abell data are accurate or complete enough to ensure that the cosmological principle is precisely realized throughout the observable region of space, but both are at least compatible with the assumption of large-scale homogeneity.

An important cosmological datum is the mean density of matter in space. The available velocity-dispersion data permit us to make meaningful estimates of the possible range of the density of the Universe—estimates which include any unseen matter within the clusters that can produce dynamical effects.

The mean density of a rich cluster gives us an upper limit. If the Corona Borealis cluster (number 2065) has a mass of $3 \times 10^{15} \mathfrak{M}_\odot$ (probably an upper limit), it has a mean density of the order of 10^{-27} g cm^{-3}. Because cluster 2065 is an unusually rich and compact one, typical rich clusters probably have mean densities an order of magnitude lower. A more realistic upper limit is the mean density of a typical second-order cluster. For a mass of $10^{16} \mathfrak{M}_\odot$ and a radius of 40 Mpc, such a system would have a mean density of the order 10^{-30} g cm^{-3}.

A lower limit to the density of the Universe is found by assuming that all of its mass is contained in clusters as rich as those in the writer's catalog. After correcting for the fact that the statistical sample does not cover the entire sky, we estimate that about 4000 such clusters probably exist within a distance of 1.2×10^9 pc (corresponding to the distance of distance class 6 clusters). If $10^{14} \mathfrak{M}_\odot$ is a lower limit to the mass of a rich cluster, there are at least 8×10^{50} grams of matter within a distance of 1.2×10^9 pc, corresponding to a mean density of about 4×10^{-33}. This estimate is probably too low for the mean density of the Universe by at least an order of magnitude.

Thus, we estimate that the mean density of that matter in the Universe whose gravitational influence produces observable kinematical effects lies in the range of 10^{-32} to 10^{-30} g cm^{-3}. The best guess, to order of magnitude, is 10^{-31} g cm^{-3}. A Hubble constant, $H = 50$ km s^{-1} Mpc^{-1}, is assumed for these estimates; of course, the density estimate is proportional to H^2. The corresponding value of the deceleration parameter, q_0 (which is independent of H), is 10^{-2}.

Note added 1974 January 18.—The manuscript for this chapter was submitted early in 1966. Because of unavoidable delays in the publication of this volume, the chapter has become outdated in many respects. When the author received the manuscript with copy editing for final printing, he attempted to incorporate some recent references. Time did not permit a complete revision, however, and much of the chapter still has the flavor of an 8-year-old review. In particular, §§ 3.6 and 4.3 should be read with cognizance that much recent work is not reflected therein.

REFERENCES

Aarseth, S. J. 1963, *M.N.R.A.S.*, **126**, 223.
———. 1966, *ibid.*, **132**, 35.
———. 1969, *ibid.*, **144**, 537.
Abell, G. O. 1955, *Pub. Astr. Soc. Pacific*, **67**, 258.
———. 1958, *Ap. J. Suppl.*, **3**, 211.
———. 1961, *A.J.*, **66**, 607.
———. 1962, in *Problems of Extra-galactic Research*, ed. G. C. McVittie (New York: Macmillan), pp. 213–238.
———. 1963, *A.J.*, **68**, 271.
———. 1964, *Ap. J.*, **140**, 1624.
———. 1972, in *External Galaxies and Quasi-stellar Objects*, ed. D. S. Evans (Dordrecht: Reidel), p. 341.
Abell, G. O., and Eastmond, S. 1968, *A.J.*, **73**, S161.
Abell, G. O., and Mihalas, D. M. 1966, *A.J.*, **71**, 635.

Abell, G. O., Neyman, J., and Scott, E. L. 1964, *A.J.*, **69**, 529.
Abell, G. O., and Seligman, C. E. 1965, *A.J.*, **70**, 317.
———. 1967, *ibid.*, **72**, 288.
Albada, G. B. van. 1960, *Bull. Astr. Inst. Netherlands*, **15**, 165.
———. 1961, *A.J.*, **66**, 590.
———. 1962, in *Problems of Extra-galactic Research*, ed. G. C. McVittie (New York: Macmillan), pp. 411–428.
Ambartsumian, V. A. 1958, in *La Structure et l'Evolution de l'Univers*, Solvay Conference (Brussels: Stoops), p. 241.
———. 1961, *A.J.*, **66**, 536.
Baade, W. 1928, *Astr. Nach.*, **233**, No. 5573, 66.
Baade, W., and Spitzer, L. 1951, *Ap. J.*, **113**, 413.
Bahcall, N. A. 1971, *A.J.*, **76**, 995.
———. 1972, *ibid.*, **77**, 550.
———. 1973*a*, *Ap. J.*, **180**, 699.
———. 1973*b*, *ibid.*, **183**, 783.
———. 1973*c*, *ibid.*, **186**, 1179.
Baum, W. A. 1955, *Pub. Astr. Soc. Pacific*, **67**, 328.
———. 1959, *ibid.*, **71**, 106.
Bautz, L. P. 1972, *A.J.*, **77**, 1.
Bautz, L. P., and Abell, G. O. 1973, *Ap. J.*, **184**, 709.
Bautz, L. P., and Morgan, W. W. 1970, *Ap. J.* (*Letters*), **162**, L149.
Bergh, S. van den. 1960, *M.N.R.A.S.*, **121**, 387.
———. 1961*a*, *Ap. J.*, **134**, 970.
———. 1961*b*, *Zs. f. Ap.*, **53**, 219.
———. 1961*c*, *A.J.*, **66**, 566.
———. 1972, *Ap. J.* (*Letters*), **171**, L31.
Bogart, R. S., and Wagoner, R. V. 1973, *Ap. J.*, **181**, 609.
Burbidge, E. M., and Burbidge, G. R. 1961, *A.J.*, **66**, 541.
Burbidge, G. R., and Burbidge, E. M. 1959*a*, *Ap. J.*, **130**, 15.
———. 1959*b*, *ibid.*, **130**, 629.
Carpenter, R. L. 1961, *Pub. Astr. Soc. Pacific*, **73**, 324.
Clark, E. E. 1968, *A.J.*, **73**, 1011.
Davidsen, A., Bowyer, S., and Welch, W. 1973, *Ap. J.* (*Letters*), **186**, L119.
Dennison, E. 1954, Ph.D. thesis, University of Michigan.
Emden, R. 1907, *Gaskugeln* (Leipzig: B. G. Teubner).
Field, G. 1959, *Ap. J.*, **129**, 525.
Gott, J. R., Wrixon, G. T., and Wannier, P. 1973, *Ap. J.*, **186**, 777.
Gudehus, D. H. 1971, Ph.D. dissertation, University of California, Los Angeles.
———. 1973, *A.J.*, **78**, 583.
Hauser, M. G., and Peebles, P. J. E. 1973, *Ap. J.*, **185**, 757.
Hodge, P. W. 1959, *Pub. Astr. Soc. Pacific*, **71**, 28.
———. 1960, *ibid.*, **72**, 188.
Hodge, P. W., Pyper, D. M., and Webb, C. J. 1965, *A.J.*, **70**, 559.
Holmberg, E. 1937, *Ann. Obs. Lund*, No. 6.
———. 1950, *Medd. Lunds*, Ser. 2, Vol. **13**, No. 128.
———. 1958, *ibid.*, Ser. 2, Vol. **14**, No. 136.
———. 1962*a*, in *Problems of Extra-galactic Research*, ed. G. C. McVittie (New York: Macmillan), pp. 187–193.
———. 1962*b*, *ibid.*, pp. 401–410.
Hubble, E. 1930, *Ap. J.*, **71**, 231.
———. 1934, *ibid.*, **79**, 8.
———. 1936*a*, *The Realm of the Nebulae* (London: Oxford University Press), pp. 72–82.
———. 1936*b*, *Ap. J.*, **84**, 158.
Huchra, J., and Sargent, W. L. W. 1973, *Ap. J.*, **186**, 433.
Humason, M. L., Mayall, N. U., and Sandage, A. R. 1956, *A.J.*, **61**, 97.
Just, K. 1959, *Ap. J.*, **129**, 268.
Kellogg, E., Murray, S., Giacconi, R., Tananbaum, H., and Gursky, H. 1973, *Ap. J.* (*Letters*), **185**, L13.
Kiang, T. 1961, *M.N.R.A.S.*, **122**, 263.
Kiang, T., and Saslaw, W. C. 1969, *M.N.R.A.S.*, **143**, 129.
Klemola, A. R. 1969, *A.J.*, **74**, 804.
Kwast, T. 1966, *Acta Astr.*, **16**, 45.
Layzer, D. 1956, *A.J.*, **61**, 383.
Liller, M. H. 1960, *Ap. J.*, **132**, 306.
Limber, D. N. 1953, *Ap. J.*, **117**, 134.
———. 1954, *ibid.*, **119**, 655.
———. 1961, *A.J.*, **66**, 572.

Limber, D. N. 1962, in *Problems of Extra-galactic Research*, ed. G. C. McVittie (New York: Macmillan), pp. 239–257.
Lovasich, J. L., Mayall, N. U., Neyman, J., and Scott, E. L. 1961, *Proceedings of the Fourth Berkeley Symposium on Mathematical Statistics and Probability*, **3**, 187.
Matthews, T. A., Morgan, W. W., and Schmidt, M. 1964, *Ap. J.*, **140**, 35.
Mayall, N. U., and Vaucouleurs, A. de. 1962, *A.J.*, **67**, 363.
Mills, B. Y. 1960, *Australian J. Phys.*, **13**, 550.
Minkowski, R. 1961, *A.J.*, **66**, 558.
———. 1963, *Proc. Nat. Acad. Sci. U.S.*, **49**, 779.
Morgan, W. W. 1961, *Proc. Nat. Acad. Sci. U.S.*, **47**, 905.
Morgan, W. W., and Lesh, J. R. 1965, *Ap. J.*, **142**, 1364.
Neyman, J., and Scott, E. L. 1952, *Ap. J.*, **116**, 144.
———. 1955, *A.J.*, **60**, 33.
———. 1961, *ibid.*, **66**, 581.
Neyman, J., Scott, E. L., and Shane, C. D. 1953, *Ap. J.*, **117**, 92.
———. 1954, *Ap. J. Suppl.*, **1**, 269.
Noonan, T. W. 1961*a*, *Pub. Astr. Soc. Pacific*, **73**, 212.
———. 1961*b*, Ph.D. thesis, California Institute of Technology.
———. 1971, *A.J.*, **76**, 182.
———. 1973, *ibid.*, **78**, 26.
Omer, G. C., Page, T. L., and Wilson, A. G. 1965, *A.J.*, **70**, 440.
Oort, J. H. 1958, in *La Structure et l'Evolution de l'Univers* (Brussels: Stoops), p. 163.
Paál, G. 1964, *Mitt. d. Sternwarte d. Ungarischen Akad. d. Wissenschaften*, No. 54.
Page, T. 1962, *Ap. J.*, **136**, 685.
Peebles, P. J. E. 1973, *Ap. J.*, **185**, 413.
———. 1974, *Ap. J. Suppl.*, **28**, 37.
Peebles, P. J. E., and Hauser, M. G. 1974, *Ap. J. Suppl.*, **28**, 19.
Pilkington, J. D. H. 1964, *M.N.R.A.S.*, **128**, 103.
Reaves, G. 1956, *A.J.*, **61**, 69.
———. 1962, *Pub. Astr. Soc. Pacific*, **74**, 392.
———. 1964, *A.J.*, **69**, 556.
———. 1966, *Pub. Astr. Soc. Pacific*, **78**, 407.
———. 1967, in *Modern Astrophysics*, ed. M. Hack (Paris: Gauthier-Villars), p. 337.
Reiz, A. 1941, *Ann. Obs. Lund*, **9**, 65.
Robinson, B. J., Damme, K. van, and Koehler, J. A. 1963, *Nature*, **199**, 1176.
Rogstad, D. H., and Ekers, R. D. 1969, *Ap. J.*, **157**, 481.
Rogstad, D. H., Rougoor, G. W., and Whiteoak, J. B. 1965, *Ap. J.*, **142**, 1665.
Rood, H. J. 1965, Ph.D. thesis, University of Michigan.
———. 1969, *Ap. J.*, **158**, 657.
———. 1970, *ibid.*, **162**, 333.
———. 1974, *Pub. Astr. Soc. Pacific*, **86**, 99.
Rood, H. J., and Abell, G. O. 1973, *Ap. Letters*, **13**, 69.
Rood, H. J., and Baum, W. A. 1967, *A.J.*, **72**, 398.
———. 1968, *ibid.*, **73**, 442.
Rood, H. J., Page, T. L., Kintner, E. C., and King, I. R. 1972, *Ap. J.*, **175**, 627.
Rood, H. J., and Sastry, G. N. 1972, *A.J.*, **77**, 451.
Rudnicki, K. 1963, *Acta Astr.*, **13**, 230.
Rudnicki, K., and Baranowska, M. 1966*a*, *Acta Astr.*, **16**, 55.
———. 1966*b*, *ibid.*, **16**, 65.
Sandage, A. R. 1958, *Ap. J.*, **127**, 513.
———. 1968, *Ap. J.* (*Letters*), **152**, L149.
———. 1972, *Ap. J.*, **176**, 21.
Scott, E. L. 1962, in *Problems of Extra-galactic Research*, ed. G. C. McVittie (New York: Macmillan), pp. 269–293.
Scott, E. L., Shane, C. D., and Swanson, M. D. 1954, *Ap. J.*, **119**, 91.
Shane, C. D. 1956*a*, *A.J.*, **61**, 292.
———. 1956*b*, in *Vistas in Astronomy*, Vol. 2, ed. A. Beer (London and New York: Pergamon Press), p. 1574.
Shane, C. D., and Wirtanen, C. A. 1954, *A.J.*, **59**, 285.
———. 1967, *Pub. Lick Obs.*, **22**, 1.
Shane, C. D., Wirtanen, C. A., and Steinlin, U. 1959, *A.J.*, **64**, 197.
Shapley, H., 1933, *Proc. Nat. Acad. Sci. U.S.*, **19**, 591.
———. 1950, *Pub. Univ. Michigan Obs.*, **10**, 79.
———. 1957, *The Inner Metagalaxy* (New Haven: Yale University Press), chap. 9.
Shapley, H., and Ames, A. 1932, *Ann. Harvard College Obs.*, **88**, 43.
Smith, S. 1936, *A.J.*, **83**, 23.
Snow, T. P. 1970, *A.J.*, **75**, 237.

Spitzer, L., and Baade, W. 1951, *Ap. J.*, **113**, 413.
Tombaugh, C. W. 1937, *Pub. Astr. Soc. Pacific*, **49**, 259.
Vaucouleurs, G. de. 1948, *Ann. d'Ap.*, **11**, 247.
———. 1953, *A.J.*, **58**, 30.
———. 1956, in *Vistas in Astronomy*, Vol. **2**, ed. A. Beer (London and New York: Pergamon Press), pp. 1584–1606.
———. 1958, *A.J.*, **63**, 253.
———. 1960, *Ap. J.*, **131**, 585.
———. 1961*a*, *Ap. J. Suppl.*, **5**, 233.
———. 1961*b*, *ibid.*, **6**, 213.
Vaucouleurs, G. de, and Vaucouleurs, A. de. 1964, *Reference Catalogue of Bright Galaxies* (Austin: University of Texas Press).
Wirtz, C. 1923, *Medd. Lunds Astr. Obs.*, Ser. 2, No. 29.
———. 1924*a*, *Astr. Nach.*, **222**, No. 5306, 22.
———. 1924*b*, *ibid.*, **223**, No. 5336, 123.
Wolf, M. 1902, *Heidelberg Pub.*, **1**, 125.
———. 1906, *Astr. Nach.*, **170**, No. 4069, 211.
Yu, J. T., and Peebles, P. J. E. 1969, *Ap. J.*, **158**, 103.
Zwicky, F. 1933, *Helv. Phys. Acta*, **6**, 110.
———. 1937, *Ap. J.*, **86**, 217.
———. 1938, *Pub. Astr. Soc. Pacific*, **50**, 218.
———. 1942*a*, *Phys. Rev.*, **61**, 489.
———. 1942*b*, *Ap. J.*, **95**, 555.
———. 1956, *Pub. Astr. Soc. Pacific*, **68**, 331.
———. 1957, *Morphological Astronomy* (Berlin: Springer-Verlag), chaps. 2–5.
———. 1959, *Handbuch der Physik*, ed. S. Flügge (Berlin: Springer-Verlag), **53**, 390.
Zwicky, F., and Berger, J. 1965, *Ap. J.*, **141**, 34.
Zwicky, F., Herzog, E., Wild, P., Karpowicz, M., and Kowal, C. T. 1961–1968, *Catalogue of Galaxies and Clusters of Galaxies* (in 6 volumes) (Pasadena: California Institute of Technology).
Zwicky, F., and Humason, M. L. 1964*a*, *Ap. J.*, **139**, 269.
———. 1964*b*, *ibid.*, **139**, 1393.
Zwicky, F., and Karpowicz, M. 1965, *Ap. J.*, **142**, 625.
———. 1966, *Ap. J.*, **146**, 43.
Zwicky, F., and Rudnicki, K. 1963, *Ap. J.*, **137**, 707.
———. 1966, *Zs. f. Ap.*, **64**, 246.

CHAPTER 16

Distribution of Galaxies

C. D. SHANE
Lick Observatory

1. HISTORICAL NOTE

THE FIRST serious studies of the distribution of nebulae were made by William Herschel in the course of his program to discover and catalog these objects. He noted their tendency to be most numerous near the galactic poles (Herschel 1785) and called attention to a "stratum" of high nebular density extending through the Coma and Virgo regions approximately at right angles to the Milky Way (Herschel 1784). This latter feature is suggestive of the arrangement of bright galaxies indicated on the charts accompanying the Shapley-Ames (1932) catalog. Herschel (1811) also noted an elongated region in which the nebulae were sparsely distributed. This region extended from near β Capricorni through Aquila and Hercules to near γ Draconis. He did not very specifically associate scarcity of nebulae with the Milky Way, perhaps because of his inability to distinguish between gaseous nebulae and those we now recognize as external galaxies. He did note, however, that nebulae tend to lie in regions relatively devoid of stars (Herschel 1784).

The extension of William Herschel's catalog to the southern hemisphere by John Herschel stimulated a number of studies of the distribution of nebulae. Charts based on John Herschel's General Catalog and published by R. A. Proctor (1869) demonstrated their very striking avoidance of the Milky Way. The more extensive *New General Catalogue* by Dreyer was used by S. Waters (1894) to develop an even more specific picture of nebular distribution.

Kant in 1755 had speculated that the nebulae were remote, independent stellar systems, or "island universes," as they were subsequently called. The bearing of the distribution studies on this concept is of considerable interest. The relation of the nebulae to our Galaxy was discussed by William Herschel (1785, 1802). For a time he supported the theory that the nebulae are independent galaxies, but he later became very doubtful of the correctness of this hypothesis (Herschel 1811). During much of the nineteenth and into the early years of the twentieth century, the majority of scientific opinion was opposed to the "island universe" theory (Curtis 1933). It was generally argued that since the nebulae avoided our Milky Way, this peculiar spatial arrangement indicated

Received October 1970.

that they must be in some way associated with our own Galaxy. The argument lost much of its force when Sanford (1917) marshaled evidence to show that the apparent avoidance of the Milky Way could be the consequence of absorbing material distributed near our galactic plane. In the same year Curtis (1917, 1918) pointed out that a number of spiral nebulae seen edgewise exhibit strong obscuration extending along their fundamental planes. If our Galaxy is a fairly normal example of a spiral nebula, we should expect to find just the type of nebular distribution with respect to the Milky Way that is observed. Thus the observed distribution that for many years had been urged against the "island universe" theory, now proved most effective in supporting it.

2. OBSERVATIONS

Distribution studies of galaxies are based either on catalogs or on counts of galaxy images on photographs.

The most reliable data for relatively small intergalactic distances are derived from catalogs of galaxies. The catalogs should be complete, or nearly so, to some specific magnitude, and their magnitude scales should be as free as possible from systematic errors. The Shapley-Ames catalog (Shapley and Ames 1932), especially as greatly extended and calibrated by the de Vaucouleurs (de Vaucouleurs and de Vaucouleurs 1964) fulfills the requirements admirably; but its rather bright limit of completeness, given by de Vaucouleurs (1956) as 98 percent for $m = 11.5$ and 50 percent for $m = 12.5$, restricts its application to our immediate neighborhood where the density of galaxies is known to be above average. The Zwicky catalog (Zwicky and collaborators 1961–1968), which aims at completeness to magnitude 15.5, has a much deeper penetration. Its magnitude scale has been calibrated by Kron and Shane. For Volumes 2 to 6, the systematic errors in the range 13.0–15.7 mag appear not to exceed 0.2 mag on the system used by Kron and Shane which adopts as standard the light within a circle of 150″ radius from a twelfth magnitude galaxy. The corrections are somewhat larger for Volume 1. The Zwicky catalog is a prime source of information for the surface distribution of galaxies as well as for their distribution in depth.

Because of the large numbers of faint galaxies, it is feasible to make complete catalogs only to the brighter magnitudes. The value of such catalogs in distribution studies is therefore restricted to the nearby galaxies. In order to extend the studies of distribution to larger volumes of space, it is necessary to resort to counts of galaxies, either continuous over the sky or in sample areas.

The Harvard program of galaxy counts under the direction of Shapley (1961) was based on photographs taken with the Metcalf 16-inch refractor at the Agassiz station near Cambridge and with the Bruce 24-inch refractor located near Bloemfontein, South Africa. More than half the sky was surveyed on the plates. The method of examining the plates consisted of marking the images identified as belonging to galaxies and in measuring their magnitudes by comparison with stars recorded on the plates. The counts are listed by square degrees and by magnitude intervals. Although fainter galaxies were measured, it is thought that the survey is reasonably complete to mag 17.6. In the interpretation of the counts, it was recognized that the definition toward the edges of the 30-square-degree fields is inferior, so that for certain types of discussion only the central 9 square degrees of each plate were used. The Harvard survey is of special value as it is the only one that covers the southern part of the sky.

The Mount Wilson survey was made by Hubble (1934), who used the 100-inch and 60-inch telescopes of the Mount Wilson Observatory. In order to cover a greater range in depth it was supplemented by a related investigation by Mayall (1934) at the Lick Observatory. Mayall made use of earlier plates taken with the 36-inch Crossley reflector as well as plates he took especially for the purpose.

Hubble adopted as standard an exposure of 1 hour with the 100-inch telescope. However, about half of the 765 survey fields were taken with the 60-inch, and a number of exposures with both telescopes were less than an hour. The survey fields were spaced 5° apart in galactic latitude and 10° in longitude for the lower latitudes, with larger longitude spacings toward the galactic poles. Additional plates taken for special purposes brought the total number of fields to 1283. The plates used were Eastman 40 emulsion. Each plate was examined carefully, and the identifiable galaxies were marked. Through the application of empirically determined corrections, the counts were reduced to the number of galaxies per square degree that could be identified at the center of a 100-inch plate of excellent quality exposed for 1 hour through the Mount Wilson zenith air mass. These values were tabulated and used in the discussion.

Hubble made a careful determination of the limiting magnitude of the counts by schraffierkassette comparisons with stars in Selected Area 57 and with the north polar sequence. Some extrapolation to the faintest magnitudes was necessary. The limiting magnitude corresponding to the tabulated counts was thus determined to be 20.0. Later photoelectric observations by Stebbins, Whitford, and Johnson (1950) showed that the published magnitudes in Selected Area 57 were too bright by the order of a half-magnitude. Hubble's adopted limiting magnitude was therefore correspondingly too bright.

Hubble (1936*a*) subsequently investigated the average number of galaxies per square degree in certain selected fields to various limiting magnitudes based on exposures of different lengths with the 60-inch and 100-inch telescopes. His table of results includes Mayall's measures with the Crossley reflector. It covers a range from 18.47 to 21.03 in apparent magnitude and from 1.89 to 3.16 in $\langle \log N \rangle$, where N is the number of galaxies per square degree. The fainter of these magnitude determinations is recognized to be uncertain.

The Lick Observatory survey by Shane and Wirtanen (1967) is based on 1246 plates taken with the 20-inch Carnegie astrographic telescope of the Lick Observatory. It covers the entire sky north of declination $-23°$. The total number of images counted was 1,250,000, corresponding to 800,000 galaxies. 103aO plates were used, giving a magnitude system quite close to the international P system. The results are published in the form of tables giving the original counts for each square degree together with the necessary reduction constants and are here summarized in the form of six charts (figs. 1–6). The isopleths indicate the numbers of galaxies per square degree after smoothing by fours. The limit of identification was fairly conservative as attested by photographs of a number of fields taken with the 120-inch telescope. For more than seventy 10′ squares tested, there was only one case in which the faintest image counted did not correspond to a galaxy. Approximately 70 faint galaxies at the limit of identification were measured photoelectrically with the 120-inch telescope by Kron and Shane. Each galaxy was the faintest one counted in its 10′ square. The "limiting photographic magnitude" of the counts as derived from these measures is 18.8.

3. GALACTIC EXTINCTION

The most striking feature of the apparent distribution of galaxies is their avoidance of the Milky Way. Hubble's (1934) survey enabled him to outline this zone of avoidance with the exception of the southern portion lying between $l^{II} = 240°$ and $340°$. The zone varies in width from 38° to 12° with its greatest width near the longitude of the galactic center. The central line of the zone departs from a great circle in a somewhat wavy fashion. Within the zone a few galaxies were noted by Hubble, suggesting an irregular structure of the obscuration with occasional partially transparent regions.

The Lick Observatory survey reveals a considerable number of real or suspected galaxies at many points deep within the zone. It was decided to define the zone by the isopleth of five galaxies per square degree. In general the zone thus defined agrees quite well with Hubble's delineation (fig. 7). The Lick Observatory counts were supplemented for the southern portion of the sky by Shapley's (1961) counts. These isopleths were drawn by estimation from his figure 85. They fit closely at the four points of junction with those drawn from the Lick Observatory counts. The zone thus outlined is shown superposed on the Lund Milky Way picture in figure 8.

Although the central line of the zone does not follow a great circle, it is of interest to fit the nearest great circle to it. The solution gives the pole at $l^{II} = 285\overset{\circ}{.}9$, $b^{II} = 88\overset{\circ}{.}0$. Thus the equator of obscuration is inclined 2° to the accepted galactic equator. A rediscussion of Hubble's counts by de Vaucouleurs and Malik (1969) gave $l^{II} = 270°$, $b^{II} = 88\overset{\circ}{.}0$ for the absorption pole.

Steinlin (1962) plotted the Lick Observatory counts in galactic coordinates between $b^{I} = \pm 25°$. The plot was made before the definitive corrections had been determined, but it portrays the zone of avoidance in close agreement with the final Lick Observatory results as shown in figure 8.

If the obscuring matter were arranged in uniform plane layers extending to infinity and parallel to the galactic plane, the effect of extinction on a uniform real distribution of galaxies should be given by

$$\log N = B - \alpha \operatorname{cosec} b\,, \tag{1}$$

where B is the log of the average number of galaxies per square degree that would be observed if there were no galactic obscuration. With equal amounts of obscuration north and south of the observer, α and B would have the same values for both hemispheres. This physical model does not closely approximate the true situation, but the equation serves as a convenient empirical formula that, when applied over large areas of the sky, fits the observations well. The optical thickness of the galaxy from the observer in the direction of the pole equals $\alpha dm/d \log N_m$ mag. Table 1[1] gives $d \log N_m/dm$ as a function of m on the international photographic scale. Its value ranges from 0.60 for the nearer galaxies to 0.38 for those at a distance corresponding to that of the average galaxy of apparent photographic magnitude 21.

[1] See § 5 below for further explanation of Table 1.

Hubble (1934) applied equation (1) to his counts and found $\alpha = 0.15$ with no significant difference between the northern and southern galactic hemispheres. He used $d \log N_m/dm = 0.6$ and thus obtained 0.25 for the optical half-thickness of our galaxy. In his discussion he omitted certain low-latitude fields he believed to be strongly affected by local obscuration.

Oort (1938) used some of Hubble's data and a value of $d \log N_m/dm = 0.55$ based on an approximate correction for redshift. His value for the optical half-thickness was 0.31 mag.

Shane (unpublished) rediscussed Hubble's observations using all of his counts for $|b| > 7°$ except for the 20-minute exposures with the 60-inch. The calculations gave $\alpha = 0.192$. Hubble's assigned limiting magnitude was 20.0, but a reasonable correction to his scale (see § 2 above) gives approximately 20.5. When reduced to the pole, this

TABLE 1

DEPENDENCE OF NUMBERS OF GALAXIES ON LIMITING MAGNITUDE

m (1)	m_c (2)	Δm (3)	z (4)	$\log(N_m/N_{12})$ (5)	$d \log N_m/dm$ (6)
12.0	11.96	0.04	0.005	0.000	0.59
13.0	12.94	0.06	0.009	0.588	0.58
14.0	13.91	0.09	0.014	1.169	0.58
15.0	14.86	0.14	0.022	1.740	0.57
16.0	15.79	0.21	0.034	2.296	0.55
17.0	16.68	0.32	0.051	2.833	0.52
18.0	17.53	0.47	0.075	3.343	0.49
19.0	18.33	0.67	0.109	3.821	0.46
20.0	19.07	0.93	0.153	4.263	0.42
21.0	19.74	1.26	0.208	4.667	0.38

NOTE.—m = observed apparent magnitude on the international photographic scale, reduced to the pole; $m_c = m - \Delta m$, the apparent magnitude corrected for the redshift; $z = d\lambda/\lambda_0$; N_m = the number of galaxies per square degree to magnitude m.

average magnitude limit for an optical half-thickness of 0.5 mag becomes 0.3 mag brighter or 20.2. The redshift, and thus $d \log N_m/dm$, corresponds to this magnitude. From table 1 $d \log N_m/dm = 0.41$. The optical half-thickness therefore is $0.192/0.41 = 0.47$ mag. In their rediscussion of Hubble's counts de Vaucouleurs and Malik (1969) assigned a limiting magnitude of 19.4, derived $\alpha = 0.205$, and with $d \log N_m/dm = 0.44$ they found the optical half-thickness to be 0.47 mag.

The Lick Observatory counts provide material for a more detailed study of the galactic extinction. A general solution based on all this material for $|b| \geq 10°$ gave $B = 1.95$, $\alpha = 0.242$. The limiting magnitude, 18.8, requires a subtraction of 0.3 mag to correct to the pole. With this reduced value, $d \log N_m/dm$ from table 1 is 0.47. The optical half-thickness from the Lick Observatory counts is therefore $0.242/0.47 = 0.51$ mag.[2]

The observed region of the sky was then divided into three areas for which separate solutions were made with the results given in table 2.

[2] NOTE.—There is evidence that the Lick counts do not extend to as faint limiting magnitudes in low latitudes as in high. Thus the value 0.51 mag is probably too large. Counts from the Zwicky catalog to magnitudes 15.5 give 0.40 for the extinction. Further study has been undertaken.

The values of B in table 2 agree as closely as might be expected from the observed irregularities in apparent galaxian distribution. The lack of a significant difference between the extinctions in the north and south galactic latitudes confirms the earlier conclusions, based on galaxy counts, that the Sun is located near the central plane of obscuration.

Holmberg (1958) made an independent determination of galactic extinction based on measures of the surface brightness of 119 spiral galaxies over a large range of galactic latitudes. His value of the optical half-thickness was 0.22 mag, but with the omission of six low-latitude spirals it became 0.26. The large difference between Holmberg's value for the optical half-thickness and the values based on counts probably reflects a very spotty arrangement of the absorbing material. Only galaxies in relatively transparent areas are observed for surface brightness, and thus regions with strong absorption are ignored. This explanation suggests that galaxies observed in low latitudes should be reddened by amounts related to the extinction coefficient based on surface brightness

TABLE 2

DEPENDENCE OF LOG N ON COSEC b

AREA	B	α	EXTINCTION COEFF.
South latitudes	1.93	0.24	0.51
Symmetrically placed north latitudes	1.94	0.22	0.46
Remaining north latitudes	1.98	0.26	0.55
Means	1.95	0.24	0.51

rather than on the extinction derived from counts. The extinction obtained by the surface-brightness method is perhaps more appropriate for use in correcting galaxian magnitudes. However, extinction derived from counts should be used in the analysis of counts.

4. CLOUDS OF GALAXIES

It is well known that galaxies are not distributed in space with statistical uniformity unless possibly on a very large scale. Double and multiple galaxies, groups of a few tens, and clusters with hundreds of members are commonplace. In fact, the isolated galaxy could be exceptional. Application of the Neyman and Scott theory (Neyman, Scott, and Shane 1953) to the Lick Observatory counts yields results consistent with a universe in which all galaxies are clustered. There may be no important background of field galaxies.

Simple clusters are not the largest aggregations of galaxies. There exist complex associations on an even larger scale comprising several centers of condensation. The term "superclusters" has been applied to them, but they may more appropriately be designated "clouds." This section deals with these clouds of galaxies.

Shapley first indicated the occurrence of these large associations. An early examination of the Lick Observatory galaxy counts showed that such associations are a common property of galaxian distribution (Shane and Wirtanen 1954, 1967). Abell (1961) has lent further weight to the concept through his studies of the distribution of clusters. Zwicky (1959), however, is of the opinion that superclustering does not exist and that the apparent groupings suggesting it are clouds "which contain various local concentrations in the numbers of galaxies." Perhaps the difference in viewpoint is one of definition.

A considerable number of these clouds are clearly shown on the charts, figures 1–6. Three apparent clouds in figure 3 have been studied in detail by plotting the Lick Observatory counts by 10′ squares and smoothing by fours. The chart of the Hercules cloud is given in figure 9, of the Corona cloud in figure 10, and of the Serpens-Virgo cloud in figure 11. The further discussion indicates that in the latter two groups, each apparent cloud is actually composed of two separate clouds near the same line of sight but distinctly separated in distance. The initial selection of these superposed clouds was probably due to their more than average prominence.

The Hercules cloud ($\alpha = 16^h02^m$; $\delta = +17°.0$) (fig. 9) consists of the main Hercules cluster and four other prominent clusters from 1° to 2° south of it. The main Hercules cluster and two of the others are listed in Abell's (1958) catalog of rich clusters. The tenth brightest galaxy in each of these three clusters has photo-red magnitude 13.8, as measured by Abell on 103aE plates taken with the 48-inch Schmidt through the red Plexiglas filter. The two remaining clusters, as shown on plates with the 20-inch Carnegie astrograph, reveal no apparent difference in the magnitude of the brighter galaxies from those in the three clusters listed by Abell. The evidence is thus strong that the cloud represents a definite association of clusters in space, not merely in direction.

If the limits of the cloud are defined by twice the general background count of galaxies in the vicinity, its dimensions are 2°.4 in α by 4°.0 in δ. From 15 galaxies in the Hercules cluster measured by Humason (Humason, Mayall, and Sandage 1956) and by Burbidge and Burbidge (1959) the redshift averages 10,776 km s^{-1}, and with the Hubble constant H equal to 100 km s^{-1} Mpc^{-1}, the distance is 107.8 Mpc. The linear dimensions of the cloud perpendicular to the line of sight are thus 4.5 by 7.6 Mpc.

If we use Holmberg's value[3] of 0.26 mag for the galactic extinction coefficient in the blue, the apparent magnitude of the tenth brightest galaxy in each cluster reduced to outside the galaxy is 13.4. At a distance of 107.8 Mpc the absolute photo-red magnitude becomes $M = -21.8$. With zero extinction $M = -21.4$. The best value probably lies between these extremes.

The Corona cloud ($\alpha = 15^h23^m$; $\delta = +29°.8$) (fig. 10) contains 13 clusters of which twelve are listed by Abell (1958). Of these 12, six have their brightest galaxy in the photo-red magnitude range 15.4–15.8, with an average of 15.65 mag, and six in the photo-red magnitude range 16.6–17.0, with an average of 16.8 mag. From inspection, the unlisted cluster appears to belong to the brighter group. Thus the complex of clusters seems to consist of two clouds separated in distance by a factor of 1.7. In view of various possible sources of discordance, the range of 0.4 mag within each cloud is not inconsistent with a spatial association of the component clusters.

For the Corona cluster an average of eight corrected measures in the Humason, Mayall, and Sandage (1956) catalog gives a redshift of 21,651 km s^{-1}, corresponding to a distance of 217 Mpc. The angular dimensions of the nearer cloud, taken to twice the background density, are 3°.7 in α and 5°.2 in δ. The corresponding linear dimensions

[3] Holmberg's value of 0.26 mag is used instead of the Lick 0.51 mag since the former is based on observations of conspicuous galaxies. If the extinction is spotty, these in general would lie in regions of low extinction—as is probably the case with the conspicuous clusters and clouds.

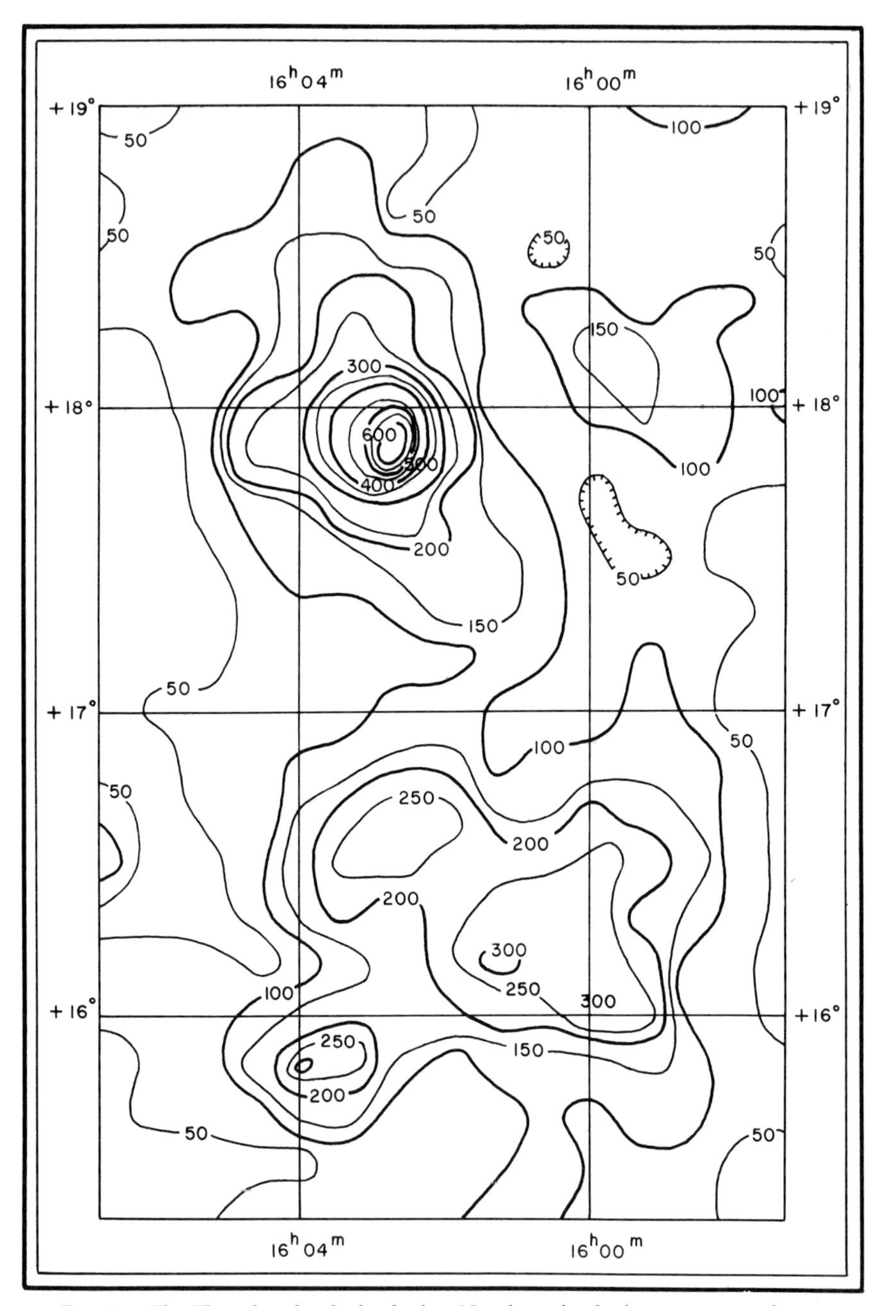

FIG. 9.—The Hercules cloud of galaxies. Number of galaxies per square degree.

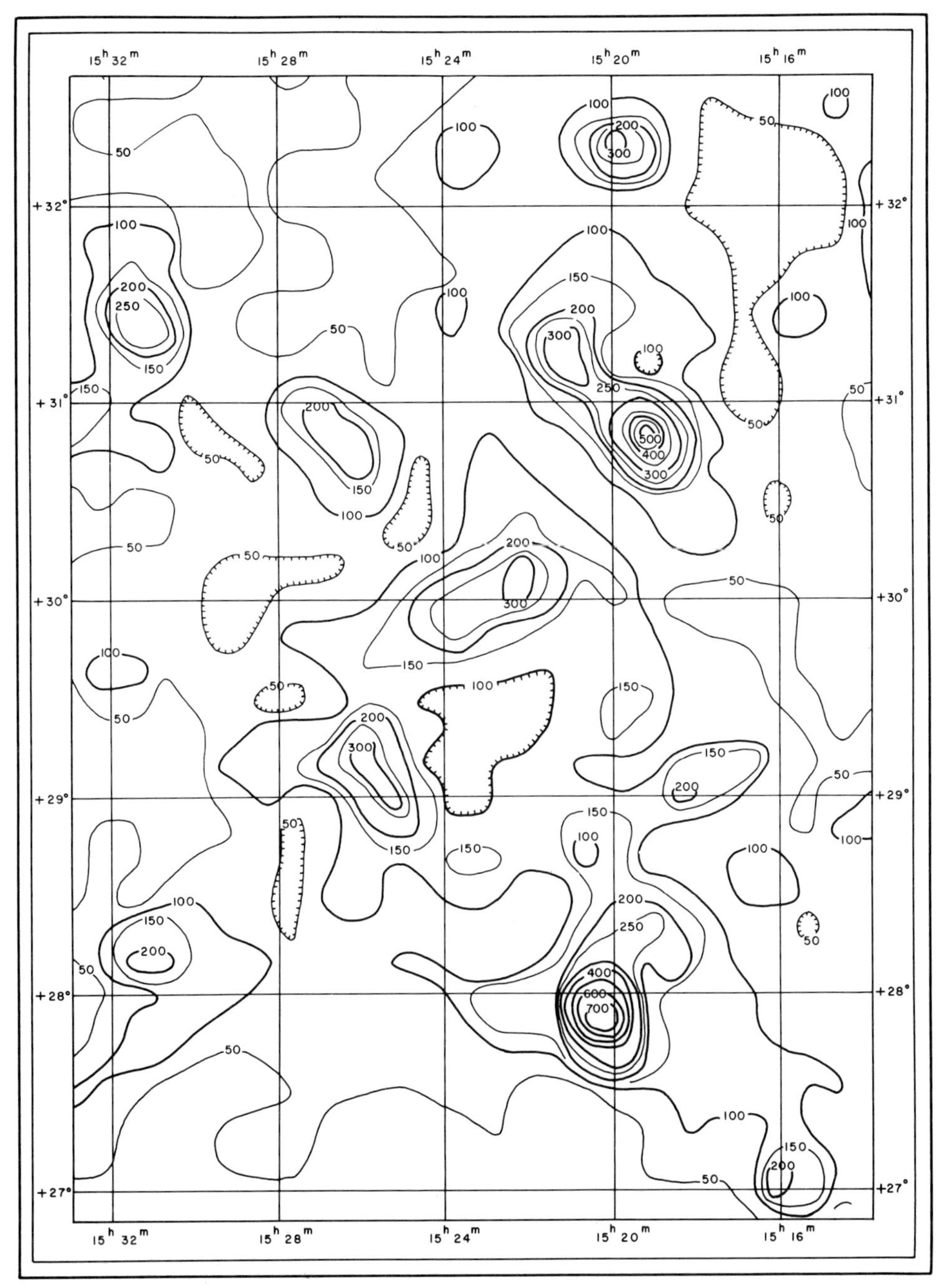

FIG. 10.—The Corona cloud of galaxies. Number of galaxies per square degree.

are 14 by 20 Mpc. The more distant cloud has angular dimensions 2°.6 by 4°.3 in α and δ, respectively, with linear dimensions 17 by 28 Mpc.

The centers of the two clouds, determined by averaging the positions of their clusters, are displaced $1^m.9$ in α and $32'$ in δ in the sense that the more remote one is northwest of the nearer cloud. For the nearer cloud we take the average photo-red magnitude of the tenth brightest galaxy in each cluster to be 15.6. With a distance for the Corona cluster of 217 Mpc and assuming the Holmberg galactic extinction coefficient, we find for the absolute photo-red magnitude $M = -21.4$. If we ignore the galactic extinction, $M = -21.1$.

The Serpens-Virgo cloud ($\alpha = 15^h12^m$; $\delta = +5°.5$) (fig. 11) contains 10 clusters listed by Abell and six not listed by him. Of the 10 listed clusters, seven have their tenth brightest galaxy in the photo-red magnitude range 15.7–16.0 with a mean of 15.9. Three

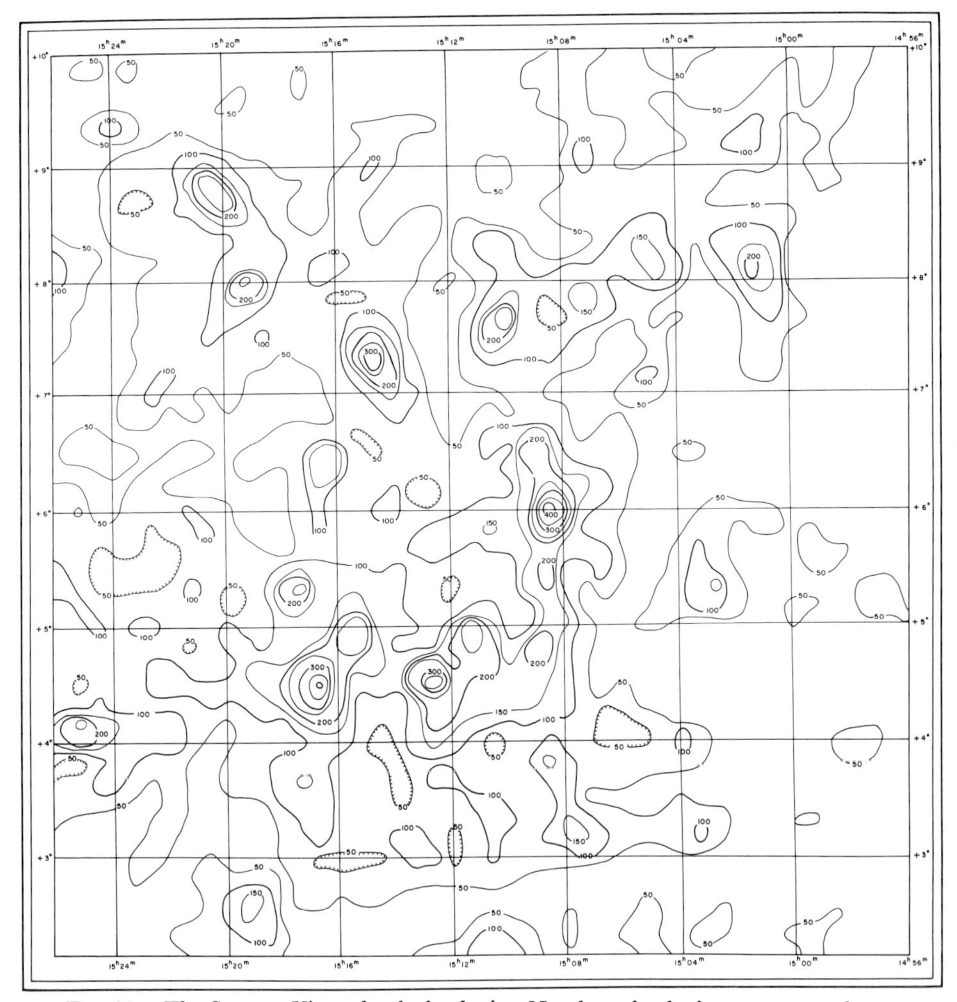

FIG. 11.—The Serpens-Virgo cloud of galaxies. Number of galaxies per square degree.

of the unlisted clusters lie near the same magnitude range. The average of the positions of the clusters in this group is $\alpha = 15^h11^m.2$; $\delta = +6°17'$. Humason (Humason, Mayall, and Sandage 1956) gives a corrected redshift of 28,333 km s^{-1} in one of the main clusters, corresponding to a distance of 283 Mpc. The dimensions to twice the background counts are 6°.5 in α by 5°.6 in δ or 32 by 28 Mpc. The photo-red absolute magnitude of the tenth brightest galaxy in the average cluster is -21.7 if the blue correction for extinction is used, or -21.4 with no correction.

A second complex of clusters apparently lies near the same line of sight but at only seven-tenths the distance of the first group. In two of the Abell clusters the tenth brightest galaxies have magnitudes of 15.0 and 15.1, while two unlisted clusters apparently belong in the same range. The center of this cloud is at $\alpha = 15^h16^m.3$; $\delta = +7°14'$, and its dimensions are 2°.7 by 4°.5 or 9.2 by 15.5 Mpc. Mayall (Humason, Mayall, and Sandage 1956) measured a redshift of about 10,600 km s^{-1} for two neighboring galaxies in a cluster of this group. The redshifts are too small by a factor two to fit with the magnitude data. One of the galaxies is far brighter than any other galaxy in the cluster, and the two probably constitute a foreground pair. One cluster in the area is too faint to belong to either group while another is sparsely populated and of ambiguous magnitude classification.

Abell (1961) has identified many additional clouds from the listings of clusters in his catalog (Abell 1958). He studied 17 of these aggregations in some detail and found an average diameter of 39 Mpc based on $H = 100$ km s^{-1} Mpc^{-1}.

The local supercluster comprises an extensive aggregation of galaxies within which our own Galaxy is eccentrically located. From the time of Herschel it has been recognized that the brighter galaxies in the northern galactic hemisphere tend to concentrate in a zone nearly at right angles to the Milky Way. This northern concentration is very evident on the charts accompanying the Shapley-Ames (1932) catalog, but in the southern hemisphere it is much less well marked. The distribution has been analyzed by de Vaucouleurs (1958) who interprets it in terms of a large flattened supercluster with our Galaxy at a considerable distance from the center. A study of the radial velocities of the brighter galaxies by Rubin (1951), which was further extended by de Vaucouleurs (1958), showed that the measures are consistent with a slow rotation of the system in its plane combined with an expansion. According to de Vaucouleurs our Galaxy is 10–12 Mpc from the center of the system. The overall diameter of the supercluster is about 30 Mpc.

Additional evidence of large clouds of galaxies is derived from general galaxy counts. Rubin (1954) applied the theory of fluctuations developed by Chandrasekhar, Münch, and Limber (Limber 1953) to the Harvard counts. She found that the microscale of fluctuations, which is a measure of the size of aggregations, equals the mean distance to a twelfth-magnitude galaxy. This distance corresponds to $z = 0.006$, which with $H = 100$ gives a linear diameter of 18 Mpc for an average aggregation.

A similar study by Limber (1954) based on the Lick Observatory counts yielded an average diameter of 8 Mpc for $H = 100$.

The discussion by Neyman, Scott, and Shane (1953) of the Lick Observatory counts was designed to test the hypothesis that all galaxies are distributed in clusters. It was assumed that within a cluster the spatial distribution is normal in three dimensions with a standard deviation σ. Their analysis of the observations of Shane and Wirtanen (1954) yielded an average value for σ of 0.37 Mpc. The standard deviation must be multiplied by a factor of 3 or greater to give a realistic value of the diameter. One should also use the improved value of -18.0 for the mean absolute photographic magnitude of a galaxy and of $+18.3$ for the limiting photographic magnitude of the counts corrected for redshift. With these data, the diameter of a typical cloud is about 8 Mpc.

There is some evidence of irregularities in density on an even grander scale. Shapley (1938) noted that the Harvard counts to magnitude 18.2 show a strong transverse density gradient extending across the south polar cap for more than 40°. This gradient suggests a structural feature of perhaps 200 Mpc in size.

Shapley has also pointed out (Shapley 1961) that to magnitude 17.1 there is an excess of approximately 50 percent in the number of galaxies in the northern galactic hemisphere over the number in the southern hemisphere. Yet to magnitude 17.6 the numbers are practically equal, as is also the case with the Lick counts to magnitude 18.8. Again an irregularity in distribution on a scale of nearly 200 Mpc is suggested. Some further evidence to support this conjecture is adduced in the next section.

The evidence relating to large clouds of galaxies may be summarized as follows:

1. Studies of five clouds based on the Lick Observatory counts and on Abell's cluster material show that they have a complicated structure consisting in each case of from four to 10 clusters. In two instances a pair of the five clouds lie practically in the same direction but at very different distances. The linear dimensions of these clouds range from 6 to 30 Mpc.

2. Our Galaxy apparently is located eccentrically in a great cloud about 30 Mpc in diameter.

3. The Harvard and the Lick counts suggest a general tendency toward the occurrence of clouds averaging 8 Mpc or more in diameter.

4. There is some suggestion of irregularities in distribution on an even larger scale, possibly 200 Mpc in extent.

5. DISTRIBUTION IN DEPTH

If galaxies are distributed with statistical uniformity both in luminosity and position in Euclidian space, then

$$\log N_m = 0.6m_c + \text{const.}\,, \tag{2}$$

where m_c is the apparent magnitude calculated from the luminosity and the distance in accordance with the inverse square law. The quantity m_c is related to the observed value, m, through corrections for possible intergalactic obscuration and redshift. There is no convincing evidence that intergalactic obscuration plays a significant role in counts of galaxies. For the fainter magnitudes, however, the redshift is of great importance. By virtue of the so-called energy and number effects, it reduces the apparent bolometric luminosity in the ratio $1/(1+z)^2$, where $z = d\lambda/\lambda_0$ (Hubble 1936*a;* Robertson 1955). In addition, the decreased frequency of each quantum shifts the spectral energy curve toward longer wavelengths, thus producing a change in apparent magnitude by an amount K. The value of K is given for three morphological types of galaxies by Humason Mayall, and Sandage (1956), based on energy distributions in galaxies measured by

Pettit (1954) for E-type galaxies and by Stebbins and Whitford (1948) for E, Sb, and Sc galaxies. The results, based on the Stebbins and Whitford observations, when averaged according to the observed distribution among morphological types (Hubble 1936*b*), may be represented by the following empirical formula for photographic magnitudes over the range $z = 0.00$ to $z = 0.25$:

$$K = 4.1z\ . \tag{3}$$

The total correction to the observed apparent photographic magnitudes due to redshift is $-\Delta m$ where

$$\Delta m = m - m_c = 5 \log (1 + z) + 4.1z\ . \tag{4}$$

From Humason, Mayall, and Sandage (1956) we also have

$$m_c = 5.000 \log cz - 4.235\ . \tag{5}$$

By means of equations (2), (4), and (5) it is possible to calculate N_m and $d \log N_m/dm$. The results are given in table 1.

Studies of distribution in depth require counts of galaxies over extensive areas of the sky taken to several magnitude limits. For the fainter magnitudes the Harvard, Mount Wilson, and Lick surveys are available. The Harvard survey, though not complete, extends over the southern sky and in general lists numbers of galaxies counted in tenth magnitude intervals to 17.6 mag. In the Mount Wilson counts by Hubble, 1-hour exposures of excellent quality were assigned magnitude limits of 20.0 and 19.4 for the 100-inch and the 60-inch telescopes, respectively.

Hubble implicitly assumed that for his counts there existed a magnitude such that all brighter and no fainter galaxies were counted. He called this the "limiting magnitude." He realized, of course, that the existence of such a definite limit only approximated the real situation.

In discussing the Lick counts Kron and Shane (Kron and Shane 1974) have adopted a more realistic definition. Let the number of galaxies in a given area of the sky and in magnitude range m to $m + dm$ be $g(m)dm$. Let the probability of identifying a galaxy of magnitude m in the course of counting images on a photograph be $\Phi(m)$. $\Phi(m)$ decreases from unity to zero with increasing m over a restricted range of m and is at least qualitatively represented by

$$\Phi(m) = (2\pi)^{-1/2} \int_{(m-m_0)/\sigma}^{\infty} \exp\left(-\tfrac{1}{2}t^2\right) dt\ , \tag{6}$$

where m_0 and σ are constants.

The number of identified galaxies in any given area is

$$\int_{-\infty}^{+\infty} g(m)\Phi(m)dm\ .$$

If we could count all galaxies to a specified limit called m_l, the number would be

$$\int_{-\infty}^{m_l} g(m)dm\ .$$

We define the "limiting magnitude" of the actual counts, m_l, by means of the equation

$$\int_{-\infty}^{m_l} g(m)dm = \int_{-\infty}^{+\infty} g(m)\Phi(m)dm\ . \tag{7}$$

Photoelectric measures by Kron and Shane made with the Lick 120-inch telescope on the faintest galaxies counted in the Lick survey gave $\sigma = 0.52$ mag and $m_0 = 18.67$ mag for fields of average richness. Equation (7) then yields $m_l = 18.79$. It is interesting to note that a reduction of the Kron and Shane measures using "limiting magnitude" in the Hubble sense, i.e., $\sigma = 0$, gave $m_l = 19.11$.

It has been found by Kron and Shane (1974) that the error dispersion of the Zwicky magnitudes is 0.26. This dispersion indicates a correction of $+ 0.04$ to the catalogued values to derive m_l on the basis defined above. In addition, it is necessary to add 0.05 to the tabulated values to allow for the tabular interval of 0.1. Finally the systematic corrections to the Zwicky magnitudes must be applied. These corrections vary with magnitude and differ substantially between Volume 1 and the other five volumes. Mean values are used here.

In table 3 the numbers of galaxies in the Zwicky catalog to different limiting magnitudes are compared with the Lick counts in the same portion of the sky reduced to Zwicky's magnitudes. The reduction was made by means of table 1. Columns (1) and

TABLE 3

COMPARISON BETWEEN ZWICKY AND LICK COUNTS

Zwicky Magnitude	Catalog Correction (mag)	m_l	No. per Square Degree		Ratio
			Zwicky	Lick	
15.7	+0.05	15.84	1.47	0.96	1.53
15.5	+0.04	15.63	1.00	0.73	1.37
14.9	−0.04	14.95	0.354	0.307	1.15
13.9	−0.18	13.81	0.093	0.068	1.37
12.9	−0.18	12.81	0.029	0.018	1.61

(2) contain the Zwicky magnitudes and their corrections, and column (3) the corresponding values of m_l. Column (4) gives the average number of galaxies per square degree to each m_l, and column (5) the corresponding numbers derived from the Lick counts. Column (6) lists the ratios of the numbers in the two preceding columns.

If the galaxies were uniformly distributed in depth, these ratios should be unity except for statistical fluctuations. If complete accuracy in the limiting magnitudes is assumed, we conclude that the average density of galaxies in our neighborhood is considerably greater than out to the limit of the Lick counts. This conclusion is in accord with de Vaucouleurs's concept of the local supercluster. The value 1.53 for Zwicky magnitude 15.7 as distinguished from unity could be accounted for by an error in the Lick limiting mag of 0.4 or of 0.33 in the Zwicky calibration. Both of these values seem unreasonably large. One puzzling feature of table 3 is the pronounced minimum of the ratio at mag 14.9. It could arise from a combination of magnitude errors and density fluctuations. It thus appears that the average density of the galaxies to the limit of the Zwicky catalog is substantially greater than to the threefold greater distance of the Lick survey.

We may now summarize what we know of the general features in the distribution of galaxies. Clustering seems to be a general, if not a universal, property among the galaxies. We find larger aggregations comprising numbers of clusters that extend over linear distances up to 30 Mpc. There is suggestive evidence of still larger assemblages of galaxies on a scale of 100 Mpc or more. This evidence consists of Shapley's transverse density gradient, the north-south disymmetry in counts to a magnitude limit of 17, and the relation between the Zwicky and the Shane and Wirtanen counts. One must recognize the possibility suggested by de Vaucouleurs (1970) that among galaxies the irregularities in distribution constitute a hierarchy of such spread that we cannot approximate the distribution as uniform on any observable scale, however large.

The courtesy of Professor C. Schalen in authorizing the use of the Lund picture of the Galaxy is gratefully acknowledged.

REFERENCES

Abell, G. O. 1958, *Ap. J. Suppl.*, **3**, 211.
———. 1961, *A.J.*, **66**, 607.
Burbidge, E. M., and Burbidge, G. R. 1959, *Ap. J.*, **130**, 629.
Curtis, H. D. 1917, *Pub. A.S.P.*, **29**, 145.
———. 1918, *Lick Obs. Pub.*, **13**, 45.
———. 1933, *Handbuch der Physik* (Berlin: Springer), Vol. **5**, Part 2, 835.
Herschel, W. 1784, *Collected Works* (London: R.S. and R.A.S., 1912), **1**, 157.
———. 1785, *ibid.*, **1**, 223.
———. 1802, *ibid.*, **2**, 199.
———. 1811, *ibid.*, **2**, 459.
Holmberg, E. 1958, *Medd. Lunds Astr. Obs.*, Ser. 2, No. 136.
Hubble, E. P. 1934, *Ap. J.*, **79**, 8.
———. 1936*a*, *ibid.*, **84**, 517.
———. 1936*b*, *The Realm of the Nebulae* (New Haven: Yale University Press), p. 55.
Humason, M. L., Mayall, N. U., and Sandage, A. R. 1956, *A.J.*, **61**, 97.
Kron, G. E., and Shane, C. D. 1974, *Ap. and Space Sci.*, **30**, 127.
Limber, D. N. 1953, *Ap. J.*, **117**, 134.
———. 1954, *ibid.*, **119**, 655.
Mayall, N. U. 1934, *Lick Obs. Bull.*, **16**, 177 (No. 458).
Neyman, J., Scott, E. L., and Shane, C. D. 1953, *Ap. J.*, **117**, 92.
Oort, J. H. 1938, *B.A.N.*, **8**, 233.
Pettit, E. 1954, *Ap. J.*, **120**, 413.
Proctor, R. A. 1869, *M.N.R.A.S.*, **29**, 337.
Robertson, H. P. 1955, *Pub. A.S.P.*, **67**, 82.
Rubin, V. C. 1951, *A.J.*, **56**, 47.
———. 1954, *Proc. Nat. Acad. Sci.*, **40**, 541.
Sanford, R. F. 1917, *Lick Obs. Bull.*, **9**, 80 (No. 297).
Shane, C. D., and Wirtanen, C. A. 1954, *A.J.*, **59**, 285.
———. 1967, *Pub. Lick Obs.*, Vol. **22**, Part I.
Shapley, H. 1938, *Proc. Nat. Acad. Sci.*, **24**, 527.
———. 1957, *The Inner Metagalaxy* (New Haven: Yale University Press), p. 132.
———. 1961, *Galaxies* (Cambridge: Harvard University Press), p. 159.
Shapley, H., and Ames, A. 1932, *Harvard Ann.*, **88**, No. 2.
Stebbins, J., and Whitford, A. E. 1948, *Ap. J.*, **108**, 413.
Stebbins, J., Whitford, A. E., and Johnson, H. L. 1950, *Ap. J.*, **112**, 469.
Steinlin, U. 1962, *A.J.*, **67**, 370.
Vaucouleurs, G. de. 1956, *A.J.*, **61**, 430.
———. 1958, *ibid.*, **63**, 253.
———. 1970, *Science*, **167**, 1203.
Vaucouleurs, G. de, and Malik, G. M. 1969, *M.N.R.A.S.*, **142**, 387.
Vaucouleurs, G. de, and Vaucouleurs, A. de. 1964, *Reference Catalogue of Bright Galaxies* (Austin: University of Texas Press).
Waters, S. 1894, *M.N.R.A.S.*, **54**, 526.
Zwicky, F. 1959, *Handbuch der Physik* (Berlin: Springer), **53**, 410.
Zwicky, F., Herzog, E., Wild, P., Karpowicz, M., and Kowal, C. T. 1961–1968, *Catalogue of Galaxies and Clusters of Galaxies*, Vols. **1–6** (Pasadena: California Institute of Technology).

CHAPTER 17

Galaxy Clustering
Its Description and Its Interpretation

DAVID LAYZER
Harvard College Observatory, Cambridge, Massachusetts

1. INTRODUCTION

THE distribution of galaxies in space and velocity is characterized by two fundamental empirical laws established by Hubble in the late 1920's and early 1930's through a series of investigations with the 60-inch and 100-inch telescopes at Mount Wilson: the principle of statistical homogeneity and isotropy, and the approximately linear relation between mean redshift and apparent magnitude. These statistical regularities are superposed on and partially hidden by local structure of great diversity and complexity. Galaxy clusters represent only the most conspicuous of the local departures from statistical uniformity. Early studies often drew a qualitative distinction between clusters and the general field, but Hubble's galaxy counts clearly showed that the spatial distribution of galaxies in the general field is strongly clustered; see especially the discussion by Bok (1934). Holmberg (1975) has estimated that at least 70 percent of the galaxies in a representative volume belong to groups dominated by a giant spiral. Clustering thus appears to be a fundamental property of the spatial distribution of galaxies.

The clustering tendency has a hierarchic character. Aggregates of galaxies often consist of smaller aggregates and are members of larger aggregates. The Local Group, for example, contains two major subgroups, one dominated by M31, the other by the Milky Way. Such subgroups are also found in the Virgo cluster (Holmberg 1975), which also exhibits structure on a larger scale. Finally, according to de Vaucouleurs (1953), the Local Group and the Virgo cluster both belong to a supercluster. Abell (1958) has shown that superclustering is a general phenomenon.

Does the clustering hierarchy extend indefinitely to larger and larger aggregates, as in Charlier's (1908) model of the Universe, or does it eventually terminate? Hubble concluded that there is a finite upper limit to the scale of clustering. Zwicky (1938) reached the same conclusion, and estimated the maximum scale of clustering to be 7.5×10^7 pc (reduced to the extragalactic distance scale defined by $H = 50\,\mathrm{km\,s^{-1}\,Mpc^{-1}}$).

Received August 1969; revised February 1974.

Abell's (1958) study of the spatial distribution of compact galaxy clusters established that for these systems the maximum scale of clustering is about 9×10^7 pc. This implies that the visible Universe contains a large number of "fair samples" (Hubble 1936). By comparing the statistical properties of different fair samples one can in principle check the assumption of statistical homogeneity and isotropy. Although this has not been done in detail, many different kinds of observations support the assumption and none contradicts it. The failure to detect any significant difference between the large-scale statistical properties of the Universe in different directions provides powerful support both for the assumption of statistical homogeneity and isotropy and for the existence of an effective upper bound to the scale of clustering.

A common difficulty in all observational studies of galaxy clustering is that of discriminating between galaxies occupying a given volume of space and background galaxies in the same part of the sky. This tends to limit the study of individual aggregates to the brighter members of relatively rich or compact systems. With a single important exception it has also frustrated attempts to obtain reliable statistical information about clustering through studies of individual aggregates. The exception is Holmberg's (1975) study of groups dominated by giant spirals. Holmberg was able to estimate the frequency of occurrence of these groups in a given volume of space, to describe their statistical properties, and to derive, for the first time, reliable information about the luminosity function of galaxies down to absolute magnitude $M = -11$.

Unfortunately, there is no immediate prospect of overcoming the practical difficulties that prevent the application of similar methods to larger aggregates. To obtain information about clustering at larger scales one must therefore resort to statistical analyses of galaxy counts (or counts of galaxy systems). This is the subject of § 2.

The present discussion of galaxy counts emphasizes an aspect of the subject that has received comparatively little attention in the past. As is true of any data analysis, the procedures of observation, selection, reduction, and analysis all impose limitations on the kind and quantity of information that can be obtained. It is a truism to say that one cannot extract more information than the data contain; but it is not always easy to know what is there or to devise an effective means of getting it out. So the questions I have concentrated on are: What sort of information is potentially present in galaxy counts? What observational procedures are required to preserve it? How can it be extracted?

Section 3 deals with the dynamics of clustering. The spatial distribution of galaxies and their distribution in velocity space are two sides of the same coin. Einstein's theory of gravitation provides the necessary dynamical link. Because the gravitational fields involved are weak and their linear scales are small (compared with the diameter of the visible Universe), a linearized form of Einstein's theory is accurate enough for all practical purposes. This approximation is more general than Newton's, to which it reduces when it is applied to effectively isolated systems. The derivation given in § 3 exploits the connection with Newton's theory and avoids explicit use of Einstein's field equations and equations of motion. There follows a discussion of the energy principle for cosmic distributions. This yields a new physical interpretation of the clustering spectrum function that plays a central role in the considerations of § 2. Finally, we describe an extension of the virial theorem to cosmic distributions. This provides a testable relation between quantities that can be derived from analyses of galaxy counts and from measurements of redshifts.

Section 4 deals with the origin of clustering. A brief sketch is given of a unified theory of the origin of self-gravitating systems, from planets and stars to superclusters of galaxies. The theory makes certain predictions about the form of the clustering spectrum. For example, it predicts an upper bound for the linear scale of clustering. In addition, it offers a framework for future theoretical and observational studies of a more specialized nature.

2. THE ANALYSIS OF GALAXY COUNTS

2.1. The Poisson Distribution

Consider a distribution of points in two- or three-dimensional space. We may identify the points with the centers of mass of galaxies or systems of galaxies or with the centroids of galaxy images on a photographic plate. Let $N(\Omega)$ denote the number of points in a region Ω. N is a random variable. Each of its possible values, $k = 0, 1, 2, \ldots,$ is associated with a probability $\mathcal{P}\{N(\Omega) = k\} = p_k$.

Instead of specifying the set $\{p_k\}$ we may specify the probability-generating function

$$G_{N(\Omega)}(t) = \sum_{k=0}^{\infty} \mathcal{P}\{N(\Omega) = k\} t^k . \tag{2.1.1}$$

The properties of probability-generating functions needed for the present discussion are summarized in the Appendix.

The Poisson distribution may be defined by the following postulates:

i) The occupation numbers $N(\Omega)$, $N(\Omega')$ associated with distinct non-overlapping regions Ω, Ω' are independent random variables. Thus the probability that a given region contains a given number of points does not depend on the distribution of points outside the region.

ii) For sufficiently small values of $V(\Omega)$, the volume of the region,

$$\mathcal{P}\{N(\Omega) = 1\} = \rho V + o(V) , \quad \mathcal{P}\{N(V) > 1\} = o(V) , \tag{2.1.2}$$

where $o(V)/V \to 0$ as $V \to 0$.

iii) $\rho =$ constant.

The first two postulates imply that all statistical properties of the distribution are defined by a probability density, ρ. The third postulate states that this density is uniform.

To derive the generating function for $N(\Omega)$ we divide the region into n subregions each of volume V/n. Since

$$N(\Omega) = \sum_{i=1}^{n} N(\Omega_i) \tag{2.1.3}$$

and the random variables $N(\Omega_i)$ are statistically independent,

$$G_{N(\Omega)} = \prod_{i=1}^{n} G_{N(\Omega_i)} = [1 - \rho V/n + t\rho V/n + o(V/n)]^n . \tag{2.1.4}$$

This holds for all sufficiently large values of n. Letting $n \to \infty$, we get

$$G_{N(\Omega)}(t) = \exp\{(t - 1)\rho V\} . \tag{2.1.5}$$

Notice that neither the dimension of the space nor the shape of the region figures in this formula.

On expanding equation (2.1.5) and setting the coefficient of t^k equal to p_k, we get the usual formula for a Poisson distribution. The Poisson distribution can accordingly be described as a statistically uniform distribution.

2.2. Statistically Homogeneous and Isotropic Distributions

We obtain an obvious generalization of the Poisson distribution by dropping the assumption $\rho = \text{const.}$ in favor of the more general assumption that $\rho(x)$ is a statistically homogeneous and isotropic random function of position. Instead of equation (2.1.5) we then obtain, by the same argument,

$$G_{N(\Omega)}(t) = \exp\{(t-1)P(\Omega)\} \tag{2.2.1}$$

where

$$P(\Omega) \equiv \int_\Omega \rho(x)dV . \tag{2.2.2}$$

The first two moments of N are given by

$$\langle N(\Omega)\rangle = G_N{}'(1) = P(\Omega) , \tag{2.2.3a}$$

$$\langle N^2(\Omega)\rangle = G_N{}''(1) + G_N{}'(1) = P^2(\Omega) + P(\Omega) . \tag{2.2.3b}$$

These moments are themselves random variables. Averaging equations (2.2.3) with respect to the probability distribution associated with the random function $\rho(x)$ gives

$$\langle\langle N(\Omega)\rangle\rangle = \langle P(\Omega)\rangle = \langle\rho\rangle V , \tag{2.2.4a}$$

$$\langle\langle N^2(\Omega)\rangle\rangle = \langle P^2(\Omega)\rangle + \langle\rho\rangle V . \tag{2.2.4b}$$

The double averages in equations (2.2.4) have a simple physical interpretation. Because the scale of clustering is finite, the distribution is ergodic. That is, expectation values can be approximated to any desired degree of accuracy by space averages. This is true of both kinds of expectation values. The inner brackets correspond to a space average over regions with a fixed value of $P(\Omega)$; the outer brackets, to an average in which regions with different values of $P(\Omega)$ contribute in proportion to the associated probabilities. The double expectation value can therefore be approximated by an average over regions of fixed but arbitrary size and shape selected at random from a sufficiently large sample. Denoting this kind of space average by $\mathcal{E}\{\quad\}$, we have

$$\mathcal{E}\{N(\Omega)\} = \langle\rho\rangle V , \quad \mathcal{E}\{N^2(\Omega)\} = \langle P^2(\Omega)\rangle + \langle\rho\rangle V . \tag{2.2.5}$$

The fluctuating components of N, ρ, and P are defined by

$$\tilde{N} = N - \mathcal{E}\{N\} , \quad \tilde{\rho} = \rho - \langle\rho\rangle , \quad \tilde{P} = P - \langle\rho\rangle V . \tag{2.2.6}$$

The variance of N is given by

$$\text{Var}\{N\} = \mathcal{E}\{N^2\} - [\mathcal{E}\{N\}]^2 = \langle\tilde{P}^2\rangle + \langle P\rangle = \text{Var}\{P\} + \mathcal{E}\{N\} . \tag{2.2.7}$$

For a Poisson distribution, $\tilde{\rho} \equiv 0$ and $\text{Var}\{N\} = \mathcal{E}\{N\}$. The additional term on the right-hand side of equation (2.2.7), which is positive unless $\tilde{\rho} \equiv 0$, represents the effect of clustering. As $V \to 0$, $\text{Var}\{N\} \to \mathcal{E}\{N\}$.

To calculate the covariance of $N(\Omega)$ and $N(\Omega')$, where Ω and Ω' are distinct regions, we note that for fixed values of $P(\Omega)$, $P(\Omega')$ the occupation numbers are independent random variables. It follows from equation (A8) in the Appendix that

$$\langle N(\Omega)N(\Omega')\rangle = P(\Omega)P(\Omega') , \tag{2.2.8}$$

whence

$$\begin{aligned}\text{Cov}\,\{N(\Omega), N(\Omega')\} &= \mathcal{E}\{\tilde{N}(\Omega)\tilde{N}(\Omega')\} = \langle\tilde{P}(\Omega)\tilde{P}(\Omega')\rangle \\ &= \text{Cov}\,\{P(\Omega), P(\Omega')\} \qquad (\Omega \neq \Omega') .\end{aligned} \tag{2.2.9}$$

If we consider regions Ω of fixed shape and size, we may regard $N(\Omega)$ and $P(\Omega)$ as random functions of the position of the centroid of the region Ω. Denoting these functions by $N_\Omega(x)$, $P_\Omega(x)$, we may combine formulae (2.2.7) and (2.2.9):

$$\mathcal{E}\{\tilde{N}_\Omega(x)\tilde{N}_\Omega(x')\} = \langle\tilde{P}_\Omega(x)\tilde{P}_\Omega(x')\rangle + \delta(x, x')\mathcal{E}\{N_\Omega\} , \tag{2.2.10}$$

where $\delta(x,x') = 1$ or 0 according as $x = x'$ or $x \neq x'$, it being understood that in the second case the regions $\Omega(x)$, $\Omega(x')$ do not overlap. (The autocovariance of N or P for overlapping regions can obviously be expressed in terms of those for identical and for nonoverlapping regions.)

By virtue of the assumed statistical homogeneity and isotropy of the underlying probability density, the autocovariances of $N_\Omega(x)$ and $P_\Omega(x)$ depend on a single variable, the separation $|x - x'|$. The autocorrelation function of $P(x)$ is defined by

$$\langle\tilde{P}_\Omega(x)\tilde{P}_\Omega(x')\rangle = \langle\tilde{P}_\Omega{}^2\rangle f_{P_\Omega}(|x - x'|) . \tag{2.2.11}$$

To second order, the statistical properties of the random functions $N_\Omega(x)$, $P_\Omega(x)$ are specified by the constants $\langle P\rangle$, $\langle P^2\rangle$, and the function $f_P(r)$.

Relations connecting higher-order moments and correlations of $N_\Omega(x)$ and $P_\Omega(x)$ can readily be worked out, but are not needed in the following discussion.

Finally we note that the complete set of moment equations (2.2.5) can be summarized compactly in the formula

$$G_N(t) = \exp\,\{G_P(t) - 1\} , \tag{2.2.12}$$

where N, P refer to the same region Ω. In this formula the probabilities associated with the random variable N are unconditional probabilities, instead of conditional probabilities defined for a fixed value of P as in equation (2.2.1); and the corresponding average is the one we have denoted by double angular brackets or by $\mathcal{E}\{\quad\}$.

Similarly, statistical properties involving pairs of nonoverlapping regions Ω, Ω' are related by the formula

$$G_{N,N'}(t, t') = \exp\,\{G_{P,P'}(t, t') - 1\} , \tag{2.2.13}$$

where $G_{N,N'}$ denotes the joint-probability-generating function for N and N'.

2.3. The Counted Number

In the preceding discussion $N(\Omega)$ denotes the occupation number of a region Ω. This is not in general the same as the counted number, which we shall denote by $\mathfrak{N}(\Omega)$. For example, one might decide to count galaxies in a certain range of absolute or apparent magnitude. Again, the probability of counting a galaxy in a given magnitude range is normally less than unity for practical reasons.

Let $p(x)$ denote the probability of counting a point at the position x. We assume that $p(x)$ is a smoothly varying function of its argument. Instead of equation (2.1.4) we then have

$$G_{\mathfrak{N}(\Omega)} = [1 - p(x)\rho(x)V/n + tp(x)\rho(x)V/n + o(V/N)]^n , \qquad (2.3.1)$$

provided that the region Ω is sufficiently small. For an arbitrary region Ω we again obtain formula (2.2.1), but with $P(\Omega)$ given by

$$P(\Omega) = \int_\Omega p(x)\rho(x)dV . \qquad (2.3.2)$$

With this modified definition of $P(\Omega)$, the formulae of § 2.2 apply to $\mathfrak{N}$ as well as N—except for (2.2.11): the random function $P_\Omega(x)$ with P given by equation (2.3.2) need not be statistically homogeneous and isotropic.

2.4. A Special Case

Neyman and Scott (1959 and references cited therein) have based extensive analyses and discussions of galaxy counts on a mathematical model belonging to the class of statistically homogeneous and isotropic distributions treated in § 2.2. In its simplest form this model envisages a Poisson distribution of cluster centers, each cluster having a fixed spherically symmetric distribution of probability density and a random number of members.

It is easy to derive a formula for $G_{\mathfrak{N}}$. Let λdV denote the number of cluster centers in a small volume element dV centered on the point x, ν the number of cluster members, and $u(x, \Omega)$ the number of times (0 or 1) a given member of the cluster whose center lies in dV contributes to the counted number. Then

$$\mathfrak{N}(\Omega) = \int \lambda dV \cdot \nu \cdot u(x, \Omega) . \qquad (2.4.1)$$

The random variables λdV, ν, and u are mutually independent. From the rules for calculating generating functions of sums and products of independent random variables (see the Appendix), we get from equation (2.4.1)

$$G_{\mathfrak{N}} = \prod_{dV} G_{\lambda dV}\{G_\nu[G_u(t)]\} . \qquad (2.4.2)$$

The generating function of $u(x, \Omega)$ is

$$G_{u(x,\Omega)} = 1 - p(x, \Omega) + p(x, \Omega)t , \qquad (2.4.3)$$

where $p(x, \Omega)$, the probability that a member of the cluster whose center lies at x will contribute to $\mathfrak{N}(\Omega)$, is given explicitly by

$$p(x, \Omega) = \int_\Omega q(|x - x'|)p(x')d^3x' , \qquad (2.4.4)$$

where $q(|x - x'|)$ is the spherically symmetric probability density in a cluster and $p(x')$ is the probability that a point at position x' within the region Ω contributes to $\mathfrak{N}(\Omega)$.

Since the cluster centers have a Poisson distribution,

$$G_{\lambda dV} = e^{(t-1)\lambda dV} . \qquad (2.4.5)$$

From equations (2.4.2), (2.4.3), and (2.4.5) we get

$$G_{\mathfrak{N}} = \exp \lambda \int dV\{G_\nu[1 + p(t - 1)] - 1\} . \qquad (2.4.6)$$

If Ω and Ω' are nonoverlapping regions, a similar argument gives for the joint-probability-generating function of $\mathfrak{N}$ and $\mathfrak{N}' = \mathfrak{N}(\Omega')$,

$$G_{\mathfrak{N},\mathfrak{N}'} = \exp \lambda \int dV\{G_\nu[1 + p(t-1) + p'(t'-1)] - 1\} . \tag{2.4.7}$$

By comparing equation (2.4.6) with (2.2.12) and equation (2.4.7) with (2.2.13) one can obtain explicit formulae for $G_{\mathfrak{N}}$ and $G_{\mathfrak{N},\mathfrak{N}'}$; and formulae for $G_{\mathfrak{N},\mathfrak{N}',\mathfrak{N}''}$, etc., can be found by a similar procedure.

The Neyman-Scott model is of mathematical interest as a simple example of the class of distributions described in § 2.2. However, it is of limited practical value. From analyses of galaxy counts one can hope to determine, at most, the first- and second-order statistical properties of the probability density $\rho(x)$. (In fact, as the following discussion will show, practical considerations impose rather severe limits on what can be learned about these properties from galaxy counts.) This information is obviously insufficient to determine the arbitrary parameters and functions needed to specify the model—even if the luminosity function is assumed to be known. Thus the model is underdetermined.

But it is also overdetermined. The most conspicuous characteristic of galaxy clustering is its wide range of linear scales, extending from perhaps 10^5 to 10^8 pc. A model with only a single scale of clustering (or even two scales) obviously cannot adequately represent this essential feature of the observed distribution of galaxies.

Finally, because the model is so artificial, the quantities that define it have little physical significance.

By contrast, the variance and the autocorrelation function of $\rho(x)$ not only have direct physical significance but are also the statistical quantities most directly related to the information that can be extracted from galaxy counts. The dominant role of second-order statistical quantities is of course a feature common to a great many physical problems.

2.5. The True and the Projected Distributions; the Function $p(x)$

Let $\lambda(\theta, \phi)$ denote the probability per unit area on a sphere of unit radius, centered on the observer, that images meeting given criteria will be registered by an ideal photographic plate, let $p(r, \theta, \phi)$ denote the probability that a galaxy belonging to some specified class and having the indicated coordinates will produce an image, and let $\rho(r, \theta, \phi)$ denote the probability density characterizing the spatial distribution of galaxies belonging to this class. By equation (2.3.2)

$$\Lambda(\omega) = \int_\omega \lambda(\theta, \phi) d\omega = \int_0^\infty \int_\omega p(r, \theta, \phi)\rho(r, \theta, \phi) r^2 d\omega dr , \tag{2.5.1}$$

where $d\omega = \sin \theta d\theta d\phi$. Hence

$$\lambda(\theta, \phi) = \int_0^\infty p(r, \theta, \phi)\rho(r, \theta, \phi) r^2 dr . \tag{2.5.2}$$

The analysis of galaxy counts involves two logically distinct steps (which have usually not been separated in practice), indicated by the following scheme:

$$\{\Lambda_\omega(\theta, \phi)\} \rightarrow \lambda(\theta, \phi) \rightarrow \rho(x) . \tag{2.5.3}$$

Statistical information about the random functions Λ for various choices of ω contains information about the random function λ, and this information in turn contains information about the probability density ρ. The operations leading from Λ to λ and from λ to ρ are formally similar: both may be described as smoothing. The composite smoothing operation is described by equation (2.5.1). But the practical considerations governing how much information is lost are entirely different for the two steps, and it is for this reason that the distinction is worth making.

The effects of smoothing are discussed in § 2.6. In the remainder of this section we discuss the function $p(x)$.

Let us first consider some special cases.

i) $\rho = \bar{\rho} =$ const., $p(x) = p(r)$. In this case the spatial distribution is statistically uniform and the absorption is the same in all directions. Formula (2.5.2) gives

$$\lambda(\theta, \phi) = \rho \int_0^\infty p(r) r^2 dr = \text{const} . \tag{2.5.4}$$

That is, the projected distribution is a Poisson distribution whatever the form of the function $p(r)$—a well-known result.

ii) $\rho = \bar{\rho} =$ const., *variable galactic absorption.* Suppose, for the sake of illustration, that all images brighter than a certain apparent magnitude, and only such images, are counted; and that absolute magnitude does not vary systematically with distance. Then the function $p(x)$ is determined by the luminosity function, and it follows from the form of the well-known relation between apparent magnitude, absolute magnitude, and distance, that $p(x)$ has the form

$$p(x) = p[k(\theta, \phi) r] , \qquad k(\theta, \phi) \geq 1 , \tag{2.5.5}$$

where $p(r)$ is the probability density corresponding to zero absorption. That is, absorption merely increases the effective distance of an object. Inserting equation (2.5.5) into (2.5.2) gives

$$\lambda(\theta, \phi) = \bar{\rho}[k(\theta, \phi)]^{-3} \int_0^\infty p(r) r^2 dr . \tag{2.5.6}$$

iii) $\rho \neq$ const., $p(x) = p(r)$. In this case the absorption is uniform and the spatial distribution is nonuniform. We then have

$$\lambda(\theta, \phi) = \int_0^\infty \rho(r, \theta, \phi) p(r) r^2 dr . \tag{2.5.7}$$

From equations (2.5.6) and (2.5.7) it is clear that, for a given function $p(r)$, a given projected distribution $\lambda(\theta, \phi)$ can result from absorption without clustering, from clustering without absorption, or from appropriate combinations of clustering and absorption. The only secure way to disentangle the effects of clustering and absorption is to examine distributions corresponding to different functions $p(r)$, e.g., distributions to different limiting magnitudes. If variations in the projected probability density are due mainly to variable galactic absorption, changing the form of $p(r)$ merely multiplies the function $\lambda(\theta, \phi)$ by a constant factor. If they are due mainly to clustering, both the detailed structure of the function and its statistical properties are changed.

Similar but somewhat subtler considerations apply to variable intergalactic absorption.

If the physical scale of galaxy clustering has a finite upper limit, it should be possible to map out the variations of galactic absorption by studying the projected distribution of sufficiently distant galaxies. Such a project, at least on a limited scale, would seem to be a necessary preliminary to reliable observational studies of clustering.

The precise form of the function $p(r)$ depends on the selection criteria, and will not be discussed here. One point deserves brief comment, however. We have implicitly assumed that $p(r, \theta, \phi)$ and $\rho(r, \theta, \phi)$ refer to a fixed time, whereas in fact the integrand in equation (2.5.2) should be evaluated at the appropriate retarded time. In practical applications the retarded time is given with sufficient accuracy by the first-order relation

$$t(r) = t - r/c\,, \tag{2.5.8}$$

where t denotes the epoch of observation. This simple correction is valid so long as the linear relation between mean redshift and apparent bolometric magnitude obtains. If one were to use galaxy counts to fainter limits than this criterion permits, one would need to take account of higher-order relativistic effects (which depend on the mean spatial curvature) as well as evolutionary effects.

2.6. Harmonic Analysis; Smoothing

In the previous section we referred to the loss of information that results from smoothing. To formulate this notion in a more precise way we represent $\tilde{\rho}(x)$ by a Fourier-Stieltjes integral:

$$\tilde{\rho}(\boldsymbol{x}) = \int \exp\,(i\boldsymbol{k}\cdot\boldsymbol{x})d^3Z_\rho(\boldsymbol{k}) \tag{2.6.1a}$$

$$\equiv \int \exp\,(i\boldsymbol{k}\cdot\boldsymbol{x})\tilde{\rho}_{\boldsymbol{k}}d^3k\,, \tag{2.6.1b}$$

where, because $\tilde{\rho}$ is real,

$$d^3Z_\rho(\boldsymbol{k})^* = d^3Z_\rho(-\boldsymbol{k})\,, \quad \tilde{\rho}_{\boldsymbol{k}}{}^* = \tilde{\rho}_{-\boldsymbol{k}}\,. \tag{2.6.2a, b}$$

These formulae require a word of explanation. Functions like $\tilde{\rho}(\boldsymbol{x})$ do not have Fourier transforms in the ordinary sense because they are not absolutely integrable. On the other hand, they can be represented by Stieltjes integrals, as in equation (2.6.1a), where $Z_\rho(\boldsymbol{k})$ is a nondifferentiable function. Alternatively, we can use the conventional Fourier representation (2.6.1b), where the transform $\tilde{\rho}_{\boldsymbol{k}}$ is a "generalized function." Functional analysis provides a rigorous foundation for the calculus of generalized functions.

The most familiar example of a generalized function is the Dirac delta function, which occurs, for example, in the generalized Fourier representation of cos k_0x:

$$Z_\rho(\boldsymbol{k}) = \tfrac{1}{2}[E(\boldsymbol{k} - \boldsymbol{k}_0) + E(\boldsymbol{k} + \boldsymbol{k}_0)]\,, \tag{2.6.3a}$$

$$\rho_{\boldsymbol{k}} = \tfrac{1}{2}[\delta(\boldsymbol{k} - \boldsymbol{k}_0) + \delta(\boldsymbol{k} + \boldsymbol{k}_0)] \tag{2.6.3b}$$

where the function E and the generalized function δ are defined by

$$E(\boldsymbol{k}) = E(k_x)E(k_y)E(k_z)\,; \quad E(t) = \begin{cases} 0 & (t < 0) \\ 1 & (t \geq 0)\,, \end{cases} \tag{2.6.4}$$

$$\delta(\boldsymbol{k}) = \delta(k_x)\delta(k_y)\delta(k_z)\,; \quad f(0) = \int_{-\infty}^{\infty} \delta(t)f(t)dt\,, \tag{2.6.5}$$

where f is an arbitrary function. The equations

$$\delta(t) = \frac{dE(t)}{dt}, \tag{2.6.6}$$

$$\tilde{\rho}_k = \lim \frac{d^3 Z_\rho(k)}{dk_x dk_y dk_z}, \tag{2.6.7}$$

are nonsense in the framework of ordinary analysis, but are given meaning by the rules governing generalized functions.

The autocovariance of the random function $\tilde{\rho}(x)$ is given by

$$\langle \tilde{\rho}(x)\tilde{\rho}(x')\rangle = \int\int \exp\,[i(k\cdot x + k'\cdot x')]\langle \tilde{\rho}_k \tilde{\rho}_{k'}\rangle d^3k d^3k' . \tag{2.6.8}$$

We shall assume that the autocorrelation function is absolutely integrable; this is a convenient mathematical formulation of the assumption that there is a finite upper limit to the scale of clustering. It follows that the autocorrelation function has a Fourier transform in the ordinary sense. In addition, the postulate of statistical homogeneity and isotropy requires the autocorrelation function to depend on x and x' only through the scalar $|x - x'|$. We must therefore have

$$\langle \tilde{\rho}_k \tilde{\rho}_{k'}\rangle \propto \delta(k + k')F(k) . \tag{2.6.9}$$

It is convenient to set the constant of proportionality equal to the variance of $\rho(x)$:

$$\langle \tilde{\rho}_k \tilde{\rho}_{k'}\rangle = \langle \tilde{\rho}^2\rangle \delta(k + k')F_\rho(k) . \tag{2.6.10}$$

Inserting this formula with equation (2.6.8) gives

$$\langle \tilde{\rho}(x)\tilde{\rho}(x')\rangle = \langle \tilde{\rho}^2\rangle \int \exp\,(ik\cdot x)F_\rho(k)d^3k . \tag{2.6.11}$$

Thus $F_\rho(k)$ is the (three-dimensional) Fourier transform of the autocorrelation function $f_\rho(r)$:

$$f_\rho(r) = \int \exp\,(ik\cdot x)F_\rho(k)d^3k . \tag{2.6.12}$$

From equation (2.6.11) or (2.6.12) it follows that

$$\int F_\rho(k)d^3k = 1 . \tag{2.6.13}$$

From equations (2.6.2b) and (2.6.10) it follows that the function $F(k)$ is nonnegative.

Equations (2.6.10) and (2.6.11) are the fundamental mathematical results on which the following discussion of smoothing is based. For a rigorous justification of these formulae we refer the reader to the mathematical literature.

The angular part of the integration in equations (2.6.11) and (2.6.12) can be carried out explicitly:

$$rf_\rho(r) = 4\pi \int_0^\infty F_\rho(k) \sin kr\, dk \tag{2.6.14}$$

which shows that, apart from a constant factor, $kF_\rho(k)$ is the Fourier sine transform of $rf_\rho(r)$.

The corresponding formula in two dimensions is easily shown to be

$$f_\rho(r) = 2\pi \int_0^\infty F_\rho(k)J_0(kr)kdk , \tag{2.6.15}$$

where J_0 is the Bessel function of order 0. The inverse formula is

$$F_\rho(k) = \frac{1}{2\pi} \int_0^\infty f_\rho(r) J_0(kr) r dr \,. \tag{2.6.16}$$

We define the smoothed probability density $\tilde{\rho}_w(x)$ by the formula

$$\tilde{\rho}_w(x) = \int \tilde{\rho}(x + y) w(y) d^3y \,, \tag{2.6.17}$$

where $w(y)$ denotes an arbitrary weighting function, so defined that its centroid lies at $y = 0$;

$$\int y w(y) d^3y = 0 \,. \tag{2.6.18}$$

The Fourier transforms of the smoothed and unsmoothed densities are related in a very simple way:

$$(\tilde{\rho_w})_k = \Phi(k; w) \tilde{\rho}_k \tag{2.6.19}$$

where the smoothing function $\Phi(k; w)$ is given by

$$\Phi(k; w) = \int \exp(ik \cdot y) w(y) d^3y \,. \tag{2.6.20}$$

Thus the effect of smoothing is to multiply the Fourier transform of the probability density by the smoothing function.

The autocovariance of the density is affected in an equally simple way. From equations (2.6.10) and (2.6.17) we obtain

$$\langle \tilde{\rho}_w(x) \tilde{\rho}_w(x') \rangle = \langle \tilde{\rho}^2 \rangle \int \exp[ik \cdot (x - x')] |\Phi(k; w)|^2 F_\rho(k) d^3k \,. \tag{2.6.21}$$

Thus the Fourier transform of the autocovariance of the smoothed density differs from that of the unsmoothed density by the factor $|\Phi(k; w)|^2$.

On comparing equation (2.5.17) with (2.5.1), we see that the random function $\Lambda_\omega(x)$, which defines the statistical properties of galaxy counts in cells ω, is a smoothed probability density. The corresponding smoothing function is given by

$$\Phi(k; \omega) = \int_0^\infty \int_\omega \exp[ik \cdot (x - x_0)] p(r) r^2 dr d\omega \tag{2.6.22}$$

with

$$x_0 = \int_0^\infty \int_\omega x p(r) r^2 dr d\omega \Big/ \int_0^\infty \int_\omega p(r) r^2 dr d\omega \,, \tag{2.6.23}$$

where we have replaced $p(x)$ by $p(r)$, i.e., we have assumed that any effects of galactic absorption have been removed.

The smoothing function defined by equations (2.6.22) and (2.6.23) depends in general on the coordinate r of the centroid. This contradicts the assumption underlying formula (2.6.21), that the smoothing functions associated with the points x, x' are identical. In the application of this formula, however, we shall always be concerned with pairs of points that have a common radial coordinate. For such pairs the assumption is valid.

In general, the function Φ defined by equations (2.6.22) and (2.6.23) is an extremely complicated function of k. One could of course evaluate it numerically and then determine the desired spectrum function $\langle \tilde{\rho}^2 \rangle F_\rho(k)$ from the inverse of equation (2.6.21):

$$\langle\widetilde{\rho^2}\rangle|\Phi(\boldsymbol{k};w)|^2F_\rho(k) = (2\pi)^{-3}\int \exp(-i\boldsymbol{k}\cdot\boldsymbol{r})\langle\widetilde{\rho}_w(\boldsymbol{x})\widetilde{\rho}_w(\boldsymbol{x}+\boldsymbol{r})\rangle d^3r \quad (2.6.24)$$

or

$$\langle\widetilde{\rho^2}\rangle F_\rho(k) = |\Phi(\boldsymbol{k};\omega)|^{-2}\langle\widetilde{\Lambda}_\omega{}^2\rangle F_{\Lambda_\omega}(k) . \quad (2.6.25)$$

This procedure would be difficult and time-consuming, however, and would require us to treat smoothing in angle and smoothing in distance, which depend upon distinct observational factors, as mutually dependent and inseparable components of a single smoothing process.

The smoothing factor simplifies considerably if the width of the shell that contributes substantially to the integral in equation (2.6.22) is much less than its radius. If this condition is satisfied, we can set $r = r_0 + (z - z_0)$ in equation (2.6.22). The function Φ can then be written as the product of a radial factor and a tangential factor:

$$\Phi(\boldsymbol{k};\omega) = \Phi_1(k_z;p)\Phi_2(\boldsymbol{\kappa},\omega) , \quad (2.6.26)$$

where

$$\Phi_1(k_z;p) = \int_{-\infty}^{\infty} \exp(ik_z z)p(r_0+z)(1+2z/r_0)dz \quad (2.6.27)$$

and

$$\Phi_2(\boldsymbol{\kappa},\omega) = \int_\omega \exp(i\boldsymbol{\kappa}\cdot\boldsymbol{R})r_0{}^2d\omega = \int_\omega \exp(i\boldsymbol{\kappa}\cdot\boldsymbol{R})d^2R . \quad (2.6.28)$$

The two-dimensional vector $\boldsymbol{R}$ is the projection of the vector $\boldsymbol{x}$ onto a plane at right angles to the line of sight, passing through the point $\boldsymbol{x}_0$.

The shape of the region ω is at our disposal. The best choice from a practical as well as a theoretical standpoint is clearly a circle. It avoids the unnecessary introduction of spurious preferred directions—the unavoidable complications that result from having to treat the radial and tangential directions differently are troublesome enough—and makes possible a uniform procedure for counting galaxies in different parts of the sky. The smoothing function for a circle is easily found to be

$$\Phi_2(\boldsymbol{\kappa},\omega) \equiv \Phi_2(\kappa R) = 2\pi R^2J_1(\kappa R)/\kappa R , \quad (2.6.29)$$

where J_1 is the Bessel function of order 1 and R is the radius of the corresponding cylindrical region centered on the point $\boldsymbol{x}_0$.

Returning now to equation (2.6.21), we have

$$\langle\widetilde{\Lambda}_\omega(\hat{\boldsymbol{x}}_0)\widetilde{\Lambda}_\omega(\hat{\boldsymbol{x}}_0{}')\rangle = \langle\widetilde{\rho^2}\rangle\int_0^\infty\int_{-\infty}^{\infty}\int_0^{2\pi} \exp(i\boldsymbol{\kappa}\cdot\boldsymbol{R})|\Phi_1(k_z;p)|^2|\Phi_2(\kappa R)|^2F_\rho(k)\kappa d\kappa dk_z d\phi$$

$$= 2\pi\langle\widetilde{\rho^2}\rangle\int_0^\infty\int_{-\infty}^{\infty} J_0(\kappa R)|\Phi_1(k_z;p)|^2|\Phi_2(\kappa R)|^2F_\rho[(k_z^2+\kappa^2)^{1/2}]\kappa d\kappa dk_z , \quad (2.6.30)$$

where $\hat{\boldsymbol{x}}_0$ denotes the unit vector $\boldsymbol{x}_0/|\boldsymbol{x}_0|$ and $\hat{\boldsymbol{x}}_0{}' = \boldsymbol{x}_0 + \boldsymbol{R}$. Consider now the (two-dimensional) Fourier transform of the left-hand term in equation (2.6.30):

$$\langle\widetilde{\Lambda}_\omega(\hat{\boldsymbol{x}}_0)\widetilde{\Lambda}_\omega(\hat{\boldsymbol{x}}_0{}')\rangle = \langle\widetilde{\lambda^2}\rangle\int \exp(i\boldsymbol{\kappa}\cdot\boldsymbol{R})|\Phi_2(\kappa R)|^2F_\lambda(\kappa)d^2\kappa$$

$$= 2\pi\langle\widetilde{\lambda^2}\rangle\int_0^\infty J_0(\kappa R)|\Phi_2(\kappa R)|^2F_\lambda(\kappa)\kappa d\kappa . \quad (2.6.31)$$

Comparing the last two equations, we see that

$$\langle\widetilde{\rho^2}\rangle \int_{-\infty}^{\infty} |\Phi_1(k_z; p)|^2 F_\rho[(k_z^2 + \kappa^2)^{1/2}] dk_z = \langle\widetilde{\lambda^2}\rangle F_\lambda(\kappa) . \tag{2.6.32}$$

Thus the spectrum function of the two-dimensional random field λ is related to the spectrum function of the three-dimensional random field ρ, modified by smoothing in the radial direction, through an integral equation.

Through a change of variables equation (2.6.32) can be transformed into an equation of Abel's type which may be solved numerically by standard methods. However, it is more instructive—and probably sufficiently accurate for all practical purposes—to obtain an approximate analytic solution as follows.

We approximate the function $|\Phi_1(k_z; p)|^2$ by a Gaussian:

$$|\Phi_1(k_z; p)|^2 = Z^2 \exp(-\sigma_z^2 k_z^2) , \tag{2.6.33}$$

where

$$Z = \int_0^{\infty} p(r)(r/r_0)^2 dr \simeq \int_{-\infty}^{\infty} p(r_0 + z)(1 + 2z/r_0) dz . \tag{2.6.34}$$

By assigning an appropriate value to the parameter σ_z one should be able to represent the function $|\Phi_1(k_z; p)|^2$ tolerably well in the interval between its first positive and negative zeros. Outside this range the precise form of the smoothing function is of no consequence (see § 2.7), and all that needs to be required of an approximating function is that its amplitude be sufficiently small and rapidly decreasing—requirements that are amply satisfied by the Gaussian.

As a further justification of the approximation (2.6.33), it may be noted that the probability $p(r)$ which figures in the definition (2.6.27) of Φ_1 depends in practice on observational considerations, some of which are only of a semiquantitative character. So there are practical as well as theoretical reasons for using a simple representation of the smoothing function, though of course the actual form of the approximation (2.6.33) is mathematically motivated.

Abel's integral equation and its solution are:

$$f(v) = \int_0^{v} \phi(u)(v - u)^{-p} du , \tag{2.6.35}$$

$$\phi(v) = \frac{1}{\pi} \sin(p\pi) \int_0^{v} f'(u)(v - u)^{p-1} du , \tag{2.6.36}$$

where $0 < p < 1$ and $f(0) = 0$. The substitutions

$$f(v) = \kappa Z^{-1} \exp(-\sigma_z^2 \kappa^2) \langle\widetilde{\lambda^2}\rangle F_\lambda(\kappa) , \quad v = \kappa^{-2} ; \tag{2.6.37a}$$

$$\phi(u) = k^3 \exp(-\sigma_z^2 k^2) \langle\widetilde{\rho^2}\rangle F_\rho(k) , \quad u = k^{-2} ; \tag{2.6.37b}$$

transform equation (2.6.32) into Abel's equation with $p = \frac{1}{2}$. Hence

$$\exp(-\sigma_z^2 k^2) k^2 \langle\widetilde{\rho^2}\rangle F_\rho(k) = -(\pi Z)^{-1} \langle\widetilde{\lambda^2}\rangle \int_{\kappa=k}^{\infty} \kappa (r^2 - k^2)^{-1/2} d[\kappa F_\lambda(\kappa) \exp(-\sigma_z^2 \kappa^2)] . \tag{2.6.38}$$

This completes the formal solution of the problem we set out to solve.

One step in the preceding derivation requires further comment. In passing from equations (2.6.30) and (2.6.31) to equation (2.6.32) we canceled the common smoothing factor $|\Phi_2|^2$. Similarly, by multiplying equation (2.6.38) by the reciprocal of the smoothing factor $|\Phi_1|^2$ we obtain a formula for the unsmoothed spectrum function $\langle\tilde{\rho}^2\rangle F_\rho(k)$. But this formula and equation (2.6.32) are meaningful only within a finite range of wavenumbers, defined by the main peak of the smoothing factor concerned. If the smoothed spectrum function could be determined precisely outside this range, it would be found to have zeros at the same wavenumbers as the smoothing factor, so that dividing the smoothed spectrum function by the smoothing factor would yield a singularity-free quotient, the unsmoothed spectrum function. In practice, however, the two sets of zeros will not coincide precisely and the quotient will have spurious singularities. In effect, smoothing is an irreversible operation for wavenumbers exceeding a certain critical value which depends on the quality of the data as well as on the nature of the smoothing and which is best determined by trial and error.

To recapitulate: One may derive an estimate of the three-dimensional spectrum function $\langle\tilde{\rho}^2\rangle F_\rho(k)$ in a limited range of wavenumbers through the following sequence of steps. (i) By appropriate averaging procedures (discussed in greater detail in § 2.8), estimate the autocovariance of the function $\mathfrak{N}_\omega(x)$, the counted number of galaxies in a two-dimensional cell ω centered on x. (ii) Calculate the autocovariance of the random function $\tilde{\Lambda}_\omega(x)$ from the formula

$$\langle\tilde{\Lambda}_\omega(x)\tilde{\Lambda}_\omega(x')\rangle = \mathcal{E}\{\mathfrak{N}_\omega(\hat{x})\mathfrak{N}_\omega(\hat{x}')\} - \delta(\hat{x}, \hat{x}')\mathcal{E}\{\mathfrak{N}_\omega\} . \tag{2.6.39}$$

(iii) Calculate the corresponding spectrum function

$$\langle\tilde{\Lambda}_\omega{}^2\rangle F_{\Lambda_\omega}(\kappa) = \frac{1}{2\pi}\int_0^\infty \langle\tilde{\Lambda}_\omega(x)\tilde{\Lambda}_\omega(x + R)\rangle J_0(\kappa R)RdR . \tag{2.6.40}$$

(iv) Correct for smoothing in angle to obtain the spectrum function

$$\langle\tilde{\lambda}^2\rangle F_\lambda(\kappa) = |\Phi_2(\kappa R)|^{-2}\langle\tilde{\Lambda}_\omega\rangle F_{\Lambda_\omega}(\kappa) \tag{2.6.41}$$

in a range of κ defined by the main peak of the smoothing factor:

$$\Phi_2(\kappa R) = \pi R^2 \frac{2J_1(\kappa R)}{\kappa R} . \tag{2.6.42}$$

(v) Calculate the spectrum function $\langle\tilde{\rho}^2\rangle F_\rho(k)$ from equation (2.6.38), again in a finite wavenumber range defined by the smoothing factor.

2.7. The Clustering Amplitude $\alpha_\lambda(\omega)$

To measure the two-dimensional covariance function, one needs to have counts of uniform quality, corrected for the effects of variable galactic absorption and covering a region of the sky whose dimensions are substantially greater than the maximum angular scale of clustering for the objects under consideration. If these requirements cannot be met, it may still be possible to get information about the two-dimensional spectrum functions from a less demanding kind of measurement.

To second order, the statistical properties of the random function $\lambda(x)$ are defined by the two constants $\langle\lambda\rangle$, α_λ, and the autocorrelation function $f_\lambda(r)$. Here α_λ is the *clustering amplitude*, defined by

$$\alpha_\lambda^2 = \langle\tilde{\lambda}^2\rangle/\langle\lambda\rangle^2 . \tag{2.7.1}$$

In the same way we define clustering amplitudes for the random functions ρ and Λ_ω. From equation (2.6.31) we obtain

$$\alpha_\Lambda^2(\omega) = \mathrm{Var}\,\{\Lambda_\omega\}/\langle\Lambda_\omega\rangle^2 = 2\pi\alpha_\lambda^2 \int_0^\infty \Psi(\kappa R)F_\lambda(\kappa)\kappa d\kappa , \tag{2.7.2}$$

where

$$\Psi(u) = [2J_1(u)/u]^2 . \tag{2.7.3}$$

Let us consider the variation of $\alpha_\Lambda(\omega)$ with $\omega(\propto \pi R^2)$. The limiting forms of equation (2.7.2) for small and large R are:

$$\alpha_\Lambda^2(\omega)/\alpha_\lambda^2 \sim \begin{cases} 1 & (\omega \to 0) \\ 4\pi R^{-2}F_\lambda(0) & (\omega \to \infty) , \end{cases} \tag{2.7.4}$$

where we have made use of the formula

$$\int_0^\infty \Psi(u)udu = 2 . \tag{2.7.5}$$

The behavior of α_Λ at intermediate values of ω depends on the form of the function $F_\lambda(\kappa)$. (*a*) Suppose that $F_\lambda(\kappa)$ is a monotonically decreasing function of κ. Because $F_\lambda > 0$, it follows that the quantity

$$\omega\alpha_\Lambda^2(\omega) = 2\pi^2\alpha_\lambda^2 \int_0^\infty F_\lambda(u/R)\Psi(u)udu \tag{2.7.6}$$

increases monotonically with increasing R toward the limiting value $4\pi^2F_\lambda(0)$. (*b*) If $F_\lambda(\kappa)$ has a single peak, then as R increases $\omega\alpha_\Lambda^2$ will at first increase and then decrease toward the same limiting value as in case (*a*), achieving its maximum when the peak of $F_\lambda(u/R)$ coincides approximately with the peak of the function $u\Psi(u)$.

From equation (2.6.32) we see that the behavior of the two-dimensional spectrum function $F_\lambda(\kappa)$ depends on both the three-dimensional spectrum function and the smoothing factor $|\Phi_1(k_z;p)|^2$. If $F_\rho(k)$ is a monotonically decreasing function of k, then $F_\lambda(\kappa)$ is also a monotonically decreasing function of κ. But if $F_\rho(k)$ is a peaked function, $F_\lambda(\kappa)$ may be either peaked or monotonically decreasing, depending on the form of the smoothing factor. It will be peaked only if the smoothing width (the width of the smoothing factor at half-maximum, say) is sufficiently small.

The quantity $\alpha_\Lambda(\omega)$ is related to the so-called index of clumpiness, defined by

$$K(\omega) = \mathrm{Var}\,\{\mathfrak{N}_\omega\}/\mathfrak{N}_\omega . \tag{2.7.7}$$

From equation (2.6.39) we have

$$K(\omega) = 1 + \langle\lambda\rangle\omega\alpha_\Lambda^2(\omega) . \tag{2.7.8}$$

We may obtain an approximate solution of equation (2.7.2), regarded as an integral equation for F_λ, by introducing the following approximation for the smoothing factor:

$$\Psi(u) \simeq \exp\,(-\tfrac{1}{4}u^2) . \tag{2.7.9}$$

Because the approximate function and its first three derivatives have the correct values at $u = 0$, the approximation is accurate for small values of u, which are the most important. In addition, the approximate function satisfies the integral condition (2.7.5).

Equation (2.7.2) now takes the form

$$\alpha_{\Lambda}^2(\omega) = \pi\alpha_{\lambda}^2 \int_0^{\infty} \exp(-tR^2/4)F_{\lambda}(t^{1/2})dt \, . \tag{2.7.10}$$

The inversion of this equation can be carried out by standard techniques for calculating inverse Laplace transforms. The remarks about smoothing made at the end of § 2.6 apply here also.

The present method of obtaining information about the spectrum function $F_{\rho}(\kappa)$ is less direct, and hence probably less accurate, than the method described earlier. However, as was mentioned earlier, its observational requirements are less stringent than those of the more direct method.

2.8. Estimating Space Averages

Let us now take a closer look at the first step in the sequence enumerated at the end of § 2.6: the estimation of space averages indicated by $\mathcal{E}\{\ \}$. These averages refer to an idealized infinite sample. In practice one must of course use finite samples. In considering the resulting errors we shall assume that the sample, though finite, is random and that it is drawn from a region whose dimensions are substantially larger than the maximum scale of clustering. Both these requirements can be satisfied in practice.

Let Q denote a measurable quantity such as $\mathfrak{N}_{\omega}$ or $\mathfrak{N}_{\omega}\,\mathfrak{N}_{\omega}'$ and let Q_i denote the measured value of Q at the ith measurement. By virtue of our assumptions, we may regard successive measurements of Q as measurements of independent random variables with identical probability distributions. The n-point average

$$\langle Q\rangle_n = n^{-1}\sum_{i=1}^{n} Q_i \tag{2.8.1}$$

is also a random variable. The de Moivre-Laplace limit theorem tells us that, for sufficiently large values of n, the possible values of the n-point average have a Gaussian probability distribution about $\mathcal{E}\{Q\} \equiv \bar{Q}$ with variance

$$\begin{aligned} \text{Var}\,\{\langle Q\rangle_n\} &= \langle(\langle Q\rangle_n - \bar{Q})^2\rangle = \left\langle\left[\frac{1}{n}\sum_{i=1}^{n}(Q_i - \bar{Q})\right]^2\right\rangle \\ &= n^{-1}\,\text{Var}\,\{Q\}\, . \end{aligned} \tag{2.8.2}$$

The square root of this quantity is a measure of the amount by which the n-point average may be expected to depart from the theoretical expectation value.

Applying this formula to $\mathfrak{N}_{\omega}$ and $\mathfrak{N}_{\omega}\,\mathfrak{N}_{\omega}'$, we get

$$\text{Var}\,\{\langle\mathfrak{N}_{\omega}\rangle_n\} = n^{-1}\,\text{Var}\,\{\mathfrak{N}_{\omega}\} = n^{-1}\langle\tilde{\mathfrak{N}}_{\omega}^2\rangle\, , \tag{2.8.3}$$

$$\text{Var}\,\{\langle\mathfrak{N}_{\omega}\mathfrak{N}_{\omega}'\rangle\} = n^{-1}\{\langle\tilde{\mathfrak{N}}_{\omega}^2\tilde{\mathfrak{N}}_{\omega}'^2\rangle - \delta(x, x')\langle\tilde{\mathfrak{N}}_{\omega}^2\rangle^2\}\, , \tag{2.8.4}$$

where $\mathcal{E}\{\ \}$ has been replaced by angular brackets. To obtain a rough estimate of the errors resulting from the use of n-point averages, it is sufficient to evaluate the right-hand side of equation (2.8.4) under the assumption that the fourth-order average properties of the field $\mathfrak{N}_\omega(x)$ are related to the second-order average properties as for a Gaussian joint-probability distribution. This gives

$$\langle \tilde{\mathfrak{N}}_\omega{}^2 \tilde{\mathfrak{N}}_\omega{}'^2 \rangle = n^{-1}\{\langle \tilde{\mathfrak{N}}_\omega{}^2\rangle + 2\langle \tilde{\mathfrak{N}}_\omega \tilde{\mathfrak{N}}_\omega{}'\rangle^2\}$$
$$= n^{-1}([\mathrm{Var}\ \{\mathfrak{N}_\omega\}]^2 + 2[\mathrm{Cov}\ \{\mathfrak{N}_\omega, \mathfrak{N}_\omega{}'\}]^2)\ . \tag{2.8.5}$$

From equation (2.8.3) we see that the errors associated with estimates of the mean occupation number depend in the expected way on n, and also, in the rather complicated way indicated by the discussion of the preceding section, on the area ω. The fractional error to be expected in an n-point estimate of $\langle \mathfrak{N}_\omega \rangle$ is of order

$$\delta\langle \mathfrak{N}_\omega \rangle / \langle \mathfrak{N}_\omega \rangle \simeq \alpha_\Lambda(\omega) n^{-1/2}\ , \tag{2.8.6}$$

where α_Λ is the dimensionless clustering amplitude defined by equation (2.7.2). Since the clustering amplitude may be much larger than unity, the number n of measurements required to produce an acceptable fractional error can be quite large.

For the fractional error of the estimated variance of $\mathfrak{N}_\omega$ we obtain from equations (2.8.4) and (2.8.5)

$$\delta\langle \tilde{\mathfrak{N}}_\omega{}^2 \rangle / \langle \tilde{\mathfrak{N}}_\omega{}^2 \rangle \simeq (n/2)^{-1/2}\ . \tag{2.8.7}$$

From equations (2.8.6) and (2.8.7) we see that when $\alpha_\Lambda > 4$, the variance of the counted number can be estimated more accurately than the mean counted number. However, this conclusion depends on the assumption (2.8.5), which may substantially underestimate the fourth-order average in question.

The estimated variance of $\Lambda_\omega(x)$, which is of more interest than that of $\mathfrak{N}_\omega(x)$, is given by

$$\langle \tilde{\Lambda}_\omega{}^2 \rangle_n = \langle \tilde{\mathfrak{N}}_\omega{}^2 \rangle_n - \langle \mathfrak{N}_\omega \rangle_n\ . \tag{2.8.8}$$

The variance of this estimate is the sum of the variances of the quantities on the right-hand side:

$$\mathrm{Var}\ \{\langle \tilde{\Lambda}_\omega{}^2 \rangle_n\} = n^{-1}\{\langle \tilde{\mathfrak{N}}_\omega{}^2 \rangle + 2\langle \tilde{\mathfrak{N}}_\omega{}^2 \rangle^2\}\ . \tag{2.8.9}$$

For sufficiently large values of ω, the second term on the right-hand side dominates and the fractional error is given by equation (2.8.7). As ω decreases in magnitude, the relative importance of the two terms shifts in favor of the first term, which ultimately becomes dominant. Thus for a given value of n there is a lower limit to the cell size that can be employed in the counts if the fractional error in the estimated value of $\langle \tilde{\Lambda}_\omega{}^2 \rangle$ is to be kept reasonably small. Notice that the two contributions are comparable when $\langle \tilde{\mathfrak{N}}_\omega{}^2 \rangle \simeq 1$.

Finally, the variance of the estimated autocovariances (for distinct cells) is

$$\mathrm{Var}\ \{\langle \tilde{\Lambda}_\omega \Lambda_\omega{}' \rangle_n\} = \mathrm{Var}\ \{\langle \tilde{\mathfrak{N}}_\omega \tilde{\mathfrak{N}}_\omega{}' \rangle\} = n^{-1}([\mathrm{Var}\ \{\mathfrak{N}_\omega\}]^2 + 2[\mathrm{Cov}\ \{\mathfrak{N}_\omega \mathfrak{N}_\omega{}'\}]^2)\ ; \tag{2.8.10}$$

and the fractional error of the estimate is

$$\frac{\delta\langle \tilde{\Lambda}_\omega \tilde{\Lambda}_\omega \rangle}{\langle \tilde{\Lambda}_\omega \tilde{\Lambda}_\omega \rangle} \geq n^{-1/2} \frac{\mathrm{Var}\ \{\mathfrak{N}_\omega\}}{\langle \tilde{\Lambda}_\omega \tilde{\Lambda}_\omega \rangle} = \frac{n^{-1/2}}{f_{\mathfrak{N}}(|x - x'|)}$$
$$= n^{-1/2}(1 + \langle \Lambda_\omega \rangle / \langle \tilde{\Lambda}^2 \rangle)/f_\Lambda(|x - x'|)\ . \tag{2.8.11}$$

This formula brings out what is probably the most important limitation imposed by the finite sample size: *The autocovariance of $\Lambda_\omega(x)$ can be accurately measured only at separations for which the autocorrelation function is substantially larger than $n^{-1/2}$; and the maximum admissible separation diminishes with decreasing cell size.* Now, the tail of the autocorrelation function contains information primarily about the low-wavenumber end of its Fourier transform, the two-dimensional spectrum function. So decreasing the cell size effectively removes information about the low-wavenumber end of the spectrum, while making it possible to gain more information about the high-wavenumber end. These considerations indicate that more than a single cell size should be employed in the analysis of counts to extract all the information they contain.

The effective truncation of the measured autocorrelation function at a value of r where $f \simeq n^{-1/2}$ may be regarded as a kind of smoothing. Thus if the truncated autocorrelation function is given by

$$\begin{aligned} f^\dagger(r) &= f(r)\,, \quad r \leq R\,, \\ &= 0\,, \qquad r > R\,, \end{aligned} \tag{2.8.12}$$

its Fourier-Bessel transform is given by

$$F^\dagger(k) = 2\pi \int (\kappa - \kappa')G(\kappa')d^2\kappa'\,, \tag{2.8.13}$$

where

$$G(\kappa) = \frac{1}{2\pi} \int_0^R J_0(\kappa R)RdR = \frac{\kappa R J_1(\kappa R)}{2\pi\kappa^2}\,. \tag{2.8.14}$$

The preceding considerations may be useful in the planning of measurements and as an indication of what kind of errors to expect. To estimate the actual errors in a given set of measurements, one should not, of course, rely on such *a priori* considerations. Rather, one should follow the standard procedure of dividing the data into a number of independent samples and estimating their statistical properties separately. Eventually, by combining all the samples, one obtains the best possible set of estimates together with estimates of the errors.

2.9. Abell's Analysis of Superclustering

To study the spatial distribution of galaxy clusters, Abell (1958) selected a homogeneous statistical sample of 1682 clusters from his catalog of 2712 very rich galaxy clusters recorded by the Palomar *Sky Survey*. The restriction to very rich clusters enabled him to apply essentially uniform selection criteria, independent of distance.

From a special study of clusters with measured redshifts, Abell derived a statistical relation between the redshift and the apparent magnitude of the tenth brightest cluster member. (Sandage [1967, 1968] has shown that a tighter relation obtains between the redshift and the apparent magnitude of the *brightest* cluster member.) On the basis of this relation he assigned approximate redshifts to all the clusters in the sample, which could then be split up into six distance groups corresponding to mean redshifts $z = 0.0027$, 0.038, 0.067, 0.090, 0.140, 0.180. These groups were studied separately.

The study revealed no significant differences between the distributions in the two galactic hemispheres and no systematic variations with distance, thus confirming the statistical isotropy of the sample and the absence of systematic errors in the assignment of distance classes.

For each distance class, Abell studied the variation of the index $K(\omega)$ defined by equation (2.7.7). He found that K had a well-defined maximum corresponding to an angular cell diameter that varied inversely as the mean distance. This result shows that the clustering of the apparent distribution is not caused by galactic obscuration. Abell was also able to rule out intergalactic absorption as a major cause of the observed clustering by showing that gaps in the distribution of clusters in a given distance class were often filled by concentrations of clusters belonging to a more distant class.

From the angular separations corresponding to the peak values of K in the various distance classes, Abell estimated the scale of second-order clustering to be 9×10^7 pc (reduced to the value $H = 50$ km s^{-1} Mpc^{-1}). On a substantially greater scale, the spatial distribution of galaxies appeared to be statistically uniform. We shall discuss the theoretical significance of these results in § 4.

Zwicky and Rudnicki (1966) and Karpowicz (1967) have studied the apparent distribution of galaxy clusters in Zwicky's very extensive catalog of these objects. These studies do not provide clear evidence of second-order clustering. But neither do they support the authors' contention that second-order clustering does not occur. In the light of the discussion of §§ 2.7 and 2.8, inspection of the data they present shows that the data themselves are consistent with the conclusions reached by Abell from his analysis of a more limited statistical sample of higher quality.

2.10. The Lick Galaxy Counts

Shane and Wirtanen (1954) have published counts of galaxies recorded on plates of the Lick *Astrographic Survey*. In considering this material one should bear in mind that it was not primarily intended for analyses of clustering.

The counts have a nominal limiting photographic magnitude of 18.4. They were made in $10' \times 10'$ squares and then combined to give the occupation numbers of $1° \times 1°$ squares; only these numbers have been published. Each plate covers an area $6° \times 6°$. Adjacent plates overlap in strips 1° wide, which makes it possible to estimate some of the systematic and random errors.

The actual limiting magnitude of the counts varies in a random way from plate to plate by as much as several tenths of a magnitude. In addition there is a large systematic variation, amounting to about 30 percent, between the east and west edges of each plate. Although this effect was discovered early in the survey program, its origin could not be determined, so the effect could not be eliminated.

To compensate in some measure for systematic instrumental effects, Shane and Wirtanen worked out a set of correction factors. Unfortunately, such corrections obviously cannot undo the structural distortion of the projected distribution. For example, on a plate whose limiting magnitude is 0.2 mag fainter than the nominal value, the scale of the two-dimensional spectrum function is contracted by about 10 percent. Dividing the counted numbers by 1.3 will reduce their expectation values to approximately the correct value, but will not rescale the spectrum function. The same is true of corrections for the east-west effect. Moreover, because the effective distances of the galaxies vary from plate to plate, as well as across a given plate, the measured correlations between occupation numbers are smaller than they would be in the absence of these variations. In effect, the systematic instrumental errors introduce a complicated kind of smoothing in distance and thus destroy a significant portion of the potentially available information.

Another important source of error in the Lick counts is the unavoidable use of subjective and nonreproducible criteria for the recognition of galaxy images. Different observers, or the same observer on different occasions, do not count the same images. The resulting errors probably have a systematic character.

After all the corrections that can be applied have been applied, one can compare the occupation numbers assigned to cells that lie in the strips common to adjacent plates. Such a comparison shows that the rms fluctuations are comparable to the square root of the mean occupation number! The most obvious—though not necessarily the correct—interpretation of this result is that the galaxy images contributing most to the counts have only a small probability of being counted. If this interpretation is correct, one would clearly need to know the value of this probability in order to extract quantitative information from an analysis of the counts.

It is illuminating to study the variation of mean occupation numbers with galactic latitude outside the zone of avoidance ($b > 30°$). If the spatial distribution is statistically homogeneous and the effective thickness of the absorbing layer is uniform in the vicinity of the Sun, the mean occupation numbers should vary according to the cosecant law:

$$\log_{10} \bar{\mathfrak{N}}_\omega = A - 0.6\Delta m \operatorname{cosec} b\,. \tag{2.10.1}$$

Hubble found that this law was accurately obeyed in the vicinity of the north galactic pole, and obtained for Δm the value 0.25 ± 0.05, a result subsequently confirmed by Holmberg. Least-squares analyses of the Lick counts yield the following values of Δm in Areas I–IV (each area covers a range of 6 hours in right ascension between declinations $-20°$ and $-23°$): $\Delta m = 0.62$ (Area I), 0.30 (Area II), 0.88 (Area III), and 0.52 (Area IV). In each of the plots of mean occupation number against the cosecant of galactic latitude the plotted points scatter widely about the line of regression. These results signal the presence of additional systematic errors, and make it even more doubtful that information of a quantitative character can be extracted from these data.

Because the Lick counts were not designed for clustering analyses, they embody a number of other features that it would be desirable to avoid, if possible, in future counts. (1) The counts do not refer to a homogeneous population defined by selection criteria that are independent of distance. (2) The distance discrimination is inadequate for quantitative analysis. In future work apparent magnitude should be supplemented by other criteria that will permit finer distance discrimination. (3) The use of a square counting grid with one set of lines parallel to the celestial equator detracts from the usefulness of the counts. For geometrical reasons, only the counts in a narrow equatorial zone can profitably be analyzed, and within this zone the anisotropy introduced by smoothing over square cells needlessly complicates and weakens the analysis. Instead of square cells aligned with the plane of the Earth's equator, one could perhaps use circular—or even octagonal—cells uniformly distributed over relatively absorption-free regions of the celestial sphere. (4) The Lick counts refer to a single distance class. For such counts it is impossible to distinguish with confidence between the effects of clustering and absorption.

Extensive analyses of the Lick counts have been attempted by Neyman and Scott (1959) and by Limber (1957 and references cited therein). These proceed from measurements of the autocovariance of the counted numbers. To gain an idea of the accuracy with which this function can be determined from the published counts, Kaftan-Kassim and the writer (unpublished) divided the material for Areas I–IV into strips 5° wide bounded by circles of constant galactic latitude, discarding material referring to latitudes less than 30°. In each strip the autocovariance of the counted numbers (corrected according to the prescription given by Shane and Wirtanen 1954) was calculated at intervals of 1° from the formula

$$C_{\mathfrak{N}_\omega}(\theta) = \langle \mathfrak{N}_\omega^2 \rangle f_{\mathfrak{N}_\omega}(\theta) = \frac{1}{n(\theta)} \sum_{|\theta - \theta_{ij}| \leq 1/2} (\mathfrak{N}_i - \overline{\mathfrak{N}}_i)(\mathfrak{N}_j - \overline{\mathfrak{N}}_j) \, . \qquad (2.10.2)$$

Here $\overline{\mathfrak{N}}_i$ and $\overline{\mathfrak{N}}_j$ denote respectively the averages of the occupation numbers $\mathfrak{N}_i$ and $\mathfrak{N}_j$ that actually figure in the sum, which runs over all pairs of cells whose central separation is within 0°5 of θ. The arithmetic was done by a digital computer.

The autocovariance of $\Lambda_\omega(\theta)$ was calculated from the formula

$$C_{\Lambda_\omega}(\theta) = C_{\mathfrak{N}_\omega}(\theta) - 2\delta_{\theta,0}\overline{\mathfrak{N}}_\omega \, . \qquad (2.10.3)$$

The extra term $-\delta_{\theta,0}\overline{\mathfrak{N}}_\omega$ on the right-hand side is an empirical correction for the residual counting errors which, as has already been mentioned, approximate the fluctuations in a Poisson process. The autocovariance calculated in this way turned out to vary reasonably smoothly near $\theta = 0$.

The calculated autocovariance and autocorrelation functions varied widely in shape and scale. The scale θ_0 may be defined by

$$C_{\Lambda_\omega}(\theta_0) = e^{-1} C_{\Lambda_\omega}(0) \, . \qquad (2.10.4)$$

At the galactic poles, in each of the four areas, the computed values of θ_0 scattered more or less uniformly over the range $1° \lesssim \theta_0 \lesssim 2°$. At middle galactic latitudes, for which there were considerably fewer data points, most of the measured values of θ_0 fell between 2° and 4°, though a few smaller values were also measured.

The tendency of θ_0 to increase with decreasing galactic latitude, and hence decreasing mean distance of the galaxies being counted, indicates that the scale of the autocorrelation function is not wholly determined by the smoothing factor, as one might be inclined to suspect from the near coincidence between the measured scale at high galactic latitudes and the dimension of the cells.

There is also some evidence bearing on the form of the spectrum function at larger wavenumbers. Neyman and Scott (1959) have presented two-dimensional autocorrelation functions based on the unpublished Lick counts in $10' \times 10'$ squares. The scale of these autocorrelation functions is approximately equal to the grid dimension, 10′. This suggests that the two-dimensional spectrum function is tolerably flat over the wavenumber range corresponding to the linear range 10′–1°, though of course the measurements could also be explained in other ways.

In view of the uncertainties in the data, inferences of a more quantitative nature cannot be drawn from them with a reasonable degree of confidence. Once better data are available, the methods outlined in preceding sections can be used to extract quantitative information about the clustering spectrum.

2.11. Other Analyses

Peebles and Hauser (Peebles, 1973, 1974; Peebles and Hauser 1974; Hauser and Peebles 1973) have recently undertaken a comprehensive analysis of the Abell, Shane-Wirtanen, and Zwicky catalogs.

Peebles's (1973) model for the spatial distribution of galaxies is similar to that of § 2.2, but its statistical properties are not completely defined. Thus Peebles defines a covariance function $\xi(r)$ in two distinct ways:

$$\delta P = n[1 + \xi(r)]\delta V \,, \tag{2.11.1}$$

where δP is the probability that an object is found within volume element δV at distance r from a randomly chosen object, n being the mean number density of objects; and

$$\langle C(r) - \delta(r)\rangle = n[1 + \xi(r)] \,, \tag{2.11.2}$$

where $C(r)$ is the spatial covariance function of the density function $\sigma(x) = \Sigma_i \delta(x - x_i)$ for a collection of points x_i, and the angular brackets indicate an ensemble average. However, the statistical properties of the ensemble are not specified, and the equivalence of the two definitions of $\xi(r)$ is not demonstrated.

It is interesting to calculate the probability δP defined above for the statistical model defined in § 2.2. For small nonoverlapping regions Ω, Ω' of volume V, V' respectively, we have

$$\begin{aligned} \mathcal{P}\{N(V) = 1, N(V') = 1\} &= \left\langle \int_V \rho dV \cdot \int_{V'} \rho dV \right\rangle \\ &= \bar{\rho}^2 V V' + \langle \tilde{\rho}(x) \tilde{\rho}(x') \rangle V V' \\ &= \bar{\rho}^2 V V' (1 + \mathrm{Cov}\,\{\rho\}) \,. \end{aligned}$$

Now $\delta P = \mathcal{P}\{N(V') = 1 | N(V) = 1\} = \mathcal{P}\{N = 1, \; N' = 1\}/\mathcal{P}\{N = 1\} = \bar{\rho} V(1 + \mathrm{Cov}\,\{\rho\})$. Thus

$$n = \bar{\rho} \,, \quad \xi(r) = \mathrm{Cov}\,\{\rho\} \,; \tag{2.11.3}$$

Peebles's covariance function is the covariance of the probability density ρ introduced in § 2.2. Thus the considerations of § 2.2 provide a theoretical basis for Peeble's model.

Yu and Peebles (1969) and Peebles (1973) analyze distributions on the celestial sphere in terms of spherical harmonics $Y_l^m(\theta, \phi)$ rather than in terms of trigonometric functions in the tangent plane. When angular separations of a large fraction of a radian are involved, this procedure is clearly called for. For angular separations $\lesssim 5°$, however, the use of spherical harmonics does not add significantly to the accuracy of the analysis and is mathematically unwieldy. As we have already seen, there are important practical reasons (e.g., the difficulty of allowing adequately for Galactic absorption) for restricting analyses of galaxy counts to small angular separations.

The qualitative and semiquantitative inferences drawn by Peebles and Hauser from their analyses of the abovementioned data are identical with the conclusions of §§ 2.9 and 2.10. In particular, Hauser and Peebles (1973) confirm the conclusion, earlier questioned by Yu and Peebles (1969), that superclustering is a real phenomenon.

3. DYNAMICS OF CLUSTERING

3.1. Field Equations and Equations of Motion for a Cosmic Distribution in the Newtonian Approximation

Newton's theory applies to effectively isolated dynamical systems in which the gravitational fields are sufficiently weak and slowly varying. The last two conditions are satisfied by intergalactic gravitational fields. (A weak gravitational potential is much smaller than c^2, a slowly varying field is one in which $l/\tau \ll c$, where l and τ are length and time scales characterizing the field variations.) On the other hand, not all galaxies can be considered to belong to effectively isolated systems. We therefore require a theory that is somewhat more general than Newton's. The required generalization is easy if the scale of local irregularities in the cosmic distribution of matter is much smaller than the radius of the (theoretically) visible Universe. As we have seen, this condition is satisfied at the present time; the maximum scale of local irregularities is less than 10^8 pc, while the radius of the visible Universe ($\simeq ct$ in a universe of zero mean spatial curvature) is about 10^{10} pc.

The following argument uses certain principles of general relativity but avoids explicit use of its mathematical formalism. Consider a sphere whose radius is much greater than the scale of local irregularities but much smaller than the radius of the visible Universe. The uniform component of the mass distribution outside this sphere produces no gravitational field inside the sphere, by symmetry. The fluctuating component of the external mass distribution gives rise to a field which, as may be verified afterward, is very small near the center of the sphere. So the gravitational field near the center of the sphere is nearly the same as it would be if the sphere were embedded in empty space. By virtue of our assumption about the radius of the sphere, we may use Newton's theory to evaluate the field and to obtain the equations of motion.

We write the matter density in the form

$$\rho(x, t) = \bar{\rho}(t) + \tilde{\rho}(x, t) , \tag{3.1.1}$$

where $\bar{\rho}(t)$ is the mean cosmic density and $\tilde{\rho}(x, t)$ is a statistically homogeneous and isotropic random function. In the application to the observed distribution of galaxies, ρ may also be interpreted as a probability density, as in § 2.

With the fluctuating density ρ we associate the Newtonian potential

$$\phi_R(x) = -G\int_{V(r)} \frac{\tilde{\rho}(x')d^3x'}{|x - x'|}, \tag{3.1.2}$$

where R denotes the radius of the sphere under consideration. We now sharpen our assumption about the scale of density fluctuations by requiring the random function $\phi_R(x)$ to converge uniformly to a function $\phi(x)$ as $R \to \infty$. The function $\phi(x)$ is given by

$$\phi(x) = -G\int \frac{\tilde{\rho}(x')d^3x'}{|x - x'|}, \tag{3.1.3}$$

where the integration extends over all space. This equation is equivalent to Poisson's equation

$$\Delta\phi = 4\pi G\tilde{\rho} \tag{3.1.4}$$

plus the auxiliary condition $\langle\phi\rangle = 0$.

The gravitational field may be split up into a part resulting from the uniform component of the mass distribution and a part resulting from the fluctuating component. Thus we obtain

$$\boldsymbol{F} = -\nabla\phi - (4\pi/3)G\bar{\rho}\boldsymbol{x}\,. \tag{3.1.5}$$

The Newtonian equations of motion for a test particle are

$$d\boldsymbol{V}/dt = \boldsymbol{F}\,, \tag{3.1.6}$$

where $\boldsymbol{V}$ denotes the velocity of the particle with respect to the center of the sphere.

This velocity may be split up into a mean velocity and a peculiar velocity,

$$\boldsymbol{V} = \bar{\boldsymbol{V}} + \boldsymbol{v}\,. \tag{3.1.7}$$

The mean velocity $\bar{\boldsymbol{V}}$ is the local velocity of the "substratum." By virtue of the uniformity and isotropy of the substratum, the mean velocity field must have the form

$$\bar{\boldsymbol{V}} = H(t)\boldsymbol{x}\,. \tag{3.1.8}$$

To determine the function $H(t)$, we apply equation (3.1.6) to a particle locally at rest in a uniform distribution of matter. This gives

$$\frac{d\bar{\boldsymbol{V}}}{dt} = \dot{H}\boldsymbol{x} + H\frac{d\boldsymbol{x}}{dt} = (\dot{H} + H^2)\boldsymbol{x} = -\frac{4\pi}{3}G\bar{\rho}\boldsymbol{x}\,. \tag{3.1.9}$$

The equation of continuity gives a second relation connecting H and $\bar{\rho}$:

$$d\bar{\rho}/dt = -3H\bar{\rho}\,, \tag{3.1.10}$$

which may be integrated in the form

$$\bar{\rho}S^3 = \text{const.}\,, \qquad H \equiv \dot{S}(t)/S(t)\,. \tag{3.1.11}$$

From equation (3.1.8) we see that $S(t)$ may be interpreted as the distance between two arbitrarily chosen elements of the substratum. Thus we obtain from equation (3.1.9)

$$\ddot{S} = -(4\pi/3)G\bar{\rho}S\,, \tag{3.1.12}$$

which could also have been written down at once as the equation of motion for a test particle of coordinate S.

The corresponding energy integral is

$$\tfrac{1}{2}\dot{S}^2 - \frac{4\pi}{3}G\bar{\rho}S^2 = -\tfrac{1}{2}kc^2\,, \tag{3.1.13}$$

where k is a constant of integration which, in view of the fact that S is determined only up to a constant multiplier, may be set equal to 0, +1, or −1. For all values of k, the solution to equation (3.1.13) has a singularity at some finite value of t, which may be taken to be the epoch $t = 0$; the function $S(t)$ is then completely determined. For $k = 0$,

$$S \propto t^{2/3}\,, \qquad H = \tfrac{2}{3}t^{-1}\,. \tag{3.1.14}$$

The solutions for $k = \pm 1$ are given in parametric form by

$$k(S/S_0) = 1 - \cos(k^{1/2}u)\,,$$

$$k(ct/S_0) = u[1 - \sin(k^{1/2}u)/k^{1/2}u]\,. \tag{3.1.15}$$

In principle the value of k can be found from observations of sufficiently distant objects (provided their intrinsic properties are known), but so far this has not proved possible in practice. For the purposes of this discussion the precise form of the function $S(t)$ does not matter.

Returning to the equations of motion (3.1.6), we can now eliminate the mean velocity and the mean density to obtain the following equation for the peculiar velocity:

$$dv/dt + Hv = -\nabla\phi\,. \qquad (3.1.16)$$

Equations (3.1.4) and (3.1.16) involve only fluctuating quantities and make no reference to any fixed point or direction. Together with equations (3.1.11) and (3.1.12) which govern the mean density, they include Newton's theory as a special case.

The simplest example of a system to which Newton's theory does not apply is a uniform medium ($\tilde{\rho} \equiv \phi \equiv 0$), for which the equations of motion (3.1.16) have the solution[1]

$$v(t) \propto S^{-1}(t)\,. \qquad (3.1.17)$$

A more rigorous derivation of the preceding results can be obtained by the approach described in § 4. The present derivation, however, may afford greater insight into the connection between Newton's theory in the restricted sense and the Newtonian approximation to Einstein's theory as applied to a cosmic distribution of matter.

3.2. The Energy Equation and the Clustering Spectrum

If we calculate the total kinetic or gravitational energy of a spherical region of radius R, we find that it is proportional to R^5. Thus it is impossible to define a kinetic or gravitational energy per unit mass. On the other hand, it is possible to define mean kinetic and gravitational energies per unit mass for the fluctuating velocity and mass distributions, and these quantities satisfy an energy-balance equation.

The derivation is analogous to the conventional one. Multiplying the equations of motion (3.1.15) by v and using the relation

$$dx/dt = v + Hx\,, \qquad (3.2.1)$$

which follows from equations (3.1.6) and (3.1.7), we obtain

$$\frac{d}{dt}\left(\tfrac{1}{2}v^2\right) + Hv^2 = -\nabla\phi\cdot\frac{dx}{dt} + Hx\cdot\nabla\phi\,. \qquad (3.2.2)$$

Next we multiply this equation by the quantity $dm = \rho dV$, which commutes with the operator d/dt, and integrate. To avoid an infinite result we integrate over a finite region, divide by the mass of the region, and then let the size of the region increase without limit. With the help of equations (3.1.2) and (3.1.10) we obtain finally

$$\frac{d}{dt}\left\{S^3\left[\tfrac{1}{2}\langle\rho v^2\rangle + \tfrac{1}{2}\langle\rho\phi\rangle\right]\right\} + HS^3\left[\langle\rho v^2\rangle + \tfrac{1}{2}\langle\rho\phi\rangle\right] = 0\,, \qquad (3.2.3)$$

where the angular brackets indicate space averages. Since we are assuming, as in § 2, that the distribution is statistically homogeneous and isotropic and that the local irregu-

[1] A generalization of this result to particles moving at relativistic speeds shows that the *momentum* of a free particle varies like S^{-1}. Applied to photons, this rule states that the frequency varies like S^{-1}. So if a photon is emitted at time t_1 and absorbed at time t_0, its frequencies at these two times are related by $\nu(t_1)/\nu(t_0) = S(t_0)/S(t_1)$.

larities have a finite linear scale, the space averages are equivalent to averages with respect to the probability distributions defining the random functions ρ, $\boldsymbol{v}$.

The kinetic, gravitational, and total energy densities are defined respectively by

$$e^{\text{kin}} = \tfrac{1}{2}\langle \rho v^2 \rangle\,, \qquad e^{\text{grav}} = \tfrac{1}{2}\langle \rho\phi \rangle\,, \qquad e = e^{\text{kin}} + e^{\text{grav}}\,. \tag{3.2.4}$$

By analogy with the kinetic theory of gases we define a corresponding set of pressures:

$$p^{\text{kin}} = \tfrac{2}{3}e^{\text{kin}}\,, \qquad p^{\text{grav}} = \tfrac{1}{3}e^{\text{grav}}\,, \qquad p = p^{\text{kin}} + p^{\text{grav}}\,. \tag{3.2.5}$$

We can then write the energy-balance equation (3.2.3) in the illuminating form

$$\frac{d}{dt}(eS^3) + p\frac{d}{dt}(S^3) = 0\,. \tag{3.2.6}$$

Thus if we interpret e and p as the internal energy density and pressure of the cosmic medium, equation (3.2.6) asserts that the cosmic expansion is adiabatic; the entropy per unit mass stays constant.

The kinetic energy density is clearly nonnegative. Moreover, the gravitational energy is nonpositive:

$$e^{\text{grav}} = \tfrac{1}{2}\langle \rho\phi \rangle = \tfrac{1}{2}\langle \tilde{\rho}\phi \rangle = \frac{1}{8\pi G}\langle \phi\Delta\phi \rangle = -\frac{1}{8\pi G}\langle |\nabla\phi|^2 \rangle\,, \tag{3.2.7}$$

where we have made use of equation (3.1.3) and of Green's theorem.

As in § 2, we may represent the fluctuating density field by a generalized Fourier integral:

$$\tilde{\rho}(x) = \int \exp(i\boldsymbol{k}\cdot\boldsymbol{x})\tilde{\rho}_{\boldsymbol{k}} d^3k\,. \tag{3.2.8}$$

It follows from equation (3.1.4) that the generalized Fourier coefficient $\phi_{\boldsymbol{k}}$ is given by

$$\phi_{\boldsymbol{k}} = -4\pi G k^{-2}\tilde{\rho}_{\boldsymbol{k}}\,. \tag{3.2.9}$$

With the help of equation (2.6.10) we now obtain

$$e^{\text{grav}} = \tfrac{1}{2}\langle \tilde{\rho}\phi \rangle = -8\pi^2 G\langle \tilde{\rho}^2 \rangle \int_0^\infty F_\rho(k)dk\,. \tag{3.2.10}$$

This formula shows the spectrum function in a new light. The quantity of gravitational energy stored in the wavenumber range dk is proportional to $F_\rho(k)$.

If we require the gravitational energy density to be finite, the integral in equation (3.2.10) must converge. This integral has the dimensions (length)^2 and defines an average clustering scale:

$$l^2 \equiv 4\pi \int_0^\infty F_\rho(k)dk\,, \qquad e^{\text{grav}} = -2\pi G\langle \tilde{\rho}^2 \rangle l^2\,. \tag{3.2.11}$$

3.3. The Virial Theorem for Cosmic Distributions

If the inertial radius of a self-gravitating system is constant in time (or, more generally, stays within certain bounds that vary sufficiently slowly), the mean kinetic and gravitational energies of the system are related. It is of interest to inquire under what circumstances a similar relation may be expected to obtain between the kinetic and gravitational energy densities of a cosmic distribution. In principle both quantities are observable.

The conventional derivation of the virial theorem proceeds from Lagrange's identity, which relates the second derivative of a system's moment of inertia to its kinetic energy and its virial. The assumption that the moment of inertia has bounded variation entails a relation between the kinetic energy and the virial (which for a self-gravitating system reduces to the potential energy). A similar derivation clearly does not apply to a cosmic distribution of particles that are not permanently bound in well-defined systems. Here a typical particle performs a kind of random walk, its rms displacement from a given point of reference increasing indefinitely with time.

An alternative way of deriving the virial theorem for a closed system (Layzer 1963) proceeds from the equations of motion for a test particle in the integrated form

$$V(t) = \int_{t_0}^{t} F(t')dt' + V(t_0) . \tag{3.3.1}$$

Averaging this equation over the particles belonging to the system, we obtain

$$\langle V^2(t)\rangle_m = \int_{t_0}^{t} \langle V(t)\cdot F(t')\rangle_m dt' + \langle V(t)\cdot V(t_0)\rangle_m , \tag{3.3.2}$$

where $\langle \ldots \rangle_m$ denotes a mass-weighted average. We now assume (*a*) that the system is statistically stationary, so that if u and v are any dynamical quantities,

$$\langle u(t)v(t')\rangle = \langle u(t+\tau)v(t'+\tau)\rangle \tag{3.3.3}$$

for all values of τ; and (*b*) that $\langle u(t)v(t+\tau)\rangle \to 0$ as $\tau \to \infty$. It follows that the second term on the right-hand side of equation (3.3.2) $\to 0$ as $t \to \infty$. And if we use equation (3.3.3) to rewrite the first term, setting $\tau = t - t'$, we obtain

$$\int_{t_0}^{t} \langle V(t)\cdot F(t')\rangle_m dt' = \int_{t_0}^{t} \langle V(2t-t')\cdot F(t)\rangle_m dt'$$

$$= -\langle x(t)\cdot F(t)\rangle_m + \langle x(2t-t_0)\cdot F(t)\rangle_m \to -\langle x(t)\cdot F(t)\rangle_m , \tag{3.3.4}$$

whence

$$\langle V^2\rangle_m + \langle x\cdot F\rangle_m \to 0 \quad \text{as} \quad t \to \infty . \tag{3.3.5}$$

This is one version of the classical virial theorem.

The preceding derivation can easily be adapted to a cosmic distribution of gravitating particles. In place of equation (3.3.1) we have

$$v(t) = S^{-1}(t)\int_{t_0}^{t} S(t')F(t')dt' + (S_0/S)v(t_0) , \tag{3.3.6}$$

whence

$$\langle v^2(t)\rangle_m = S^{-1}(t)\int_{t_0}^{t} \langle v(t)\cdot F(t')\rangle S(t')dt' + (S_0/S)\langle v(t)\cdot v(t_0)\rangle . \tag{3.3.7}$$

Because the cosmic distribution is expanding, it would be physically unrealistic to impose the requirement of stationarity. It turns out, however, that a weaker requirement suffices. We write

$$F(t,\tau) \equiv \langle v(t)\cdot F(t-\tau)\rangle_m S(t-\tau)/S(t)$$

$$= F(t^*+\tau,\tau) + (t-t^*-\tau)\left[\frac{\partial F}{\partial t}\right]_{t=t^*+\tau} + \ldots , \tag{3.3.8}$$

choosing t^* so as to make the integral of the second term in the expansion vanish when equation (3.3.8) is inserted into equation (3.3.7).

If the distribution were stationary, only the first term in the expansion (3.3.8) would be nonzero. We now assume that the expansion converges so rapidly that only the first two terms need to be kept. We assume further, as before, that all correlations of the type $\langle u(t)v(t_0)\rangle$ tend to zero as t increases. Then a routine calculation analogous to (3.3.4) gives

$$\langle v^2(t)\rangle_m + \langle x(t^*)\cdot F(t^*)\rangle_m = 0\,, \tag{3.3.9}$$

where the displacement $x(t^*)$ is reckoned for each particle from an arbitrary initial position.

It follows from formula (3.1.3) that the virial is given by

$$\langle x\cdot F\rangle_m = \tfrac{1}{2}\langle\phi\rangle_m\,. \tag{3.3.10}$$

Noting that

$$\langle \rho u\rangle = \bar{\rho}\langle u\rangle_m \tag{3.3.11}$$

and writing

$$\epsilon = e/\dot{\rho}\,, \tag{3.3.12}$$

we can rewrite equation (3.3.9) in the form

$$2\epsilon^{\mathrm{kin}}(t) + \epsilon^{\mathrm{grav}}(t^*) = 0\,. \tag{3.3.13}$$

This equation relates the specific kinetic energy at a given instant to the specific gravitational energy at a somewhat earlier instant. The interval $\eta = t - t^*$ represents the local time of relaxation—the time required for the fluctuating gravitational fields to set their stamp on the distribution of peculiar velocities.

The assumption that the expansion (3.3.8) converges rapidly requires that $\eta \ll t$. We may therefore replace $\epsilon^{\mathrm{grav}}(t^*)$ in equation (3.3.13) by the first two terms in its Taylor expansion:

$$2\epsilon^{\mathrm{kin}}(t) + \epsilon^{\mathrm{grav}}(t) - \eta\,\frac{d\epsilon^{\mathrm{grav}}(t)}{dt} = 0\,. \tag{3.3.14}$$

The energy-balance equation (3.2.6) can be written in the form

$$\frac{d}{dt}\,(S^2\epsilon^{\mathrm{kin}}) + S\,\frac{d}{dt}\,(S\epsilon^{\mathrm{grav}}) = 0\,. \tag{3.3.15}$$

By eliminating ϵ^{kin} between the last two equations we obtain the following second-order differential equation for ϵ^{grav}:

$$\tfrac{1}{2}\eta\,\frac{d}{dt}\,S^2\,\frac{d\epsilon^{\mathrm{grav}}}{dt} - \frac{1}{2}\frac{d}{dt}\,(S^2\epsilon^{\mathrm{grav}}) + S\,\frac{d}{dt}\,(S\epsilon^{\mathrm{grav}}) = 0\,. \tag{3.3.16}$$

The first integral of this equation is

$$\frac{d\epsilon^{\mathrm{grav}}}{dt} \propto S^{-2}e^{-t/\eta}\,. \tag{3.3.17}$$

It follows from equations (3.3.15) and (3.2.5) that

$$p \propto S^{-2}e^{-t/\eta}\,. \tag{3.3.18}$$

Thus the cosmic pressure decays exponentially in a time short compared with the expansion time (since, by assumption, $\eta \ll$ t). Equation (3.3.13) can therefore be replaced by

$$2\epsilon^{\text{kin}}(t) + \epsilon^{\text{grav}}(t) = 0 \qquad (t \gg t_0)\,, \tag{3.3.19}$$

which is the final form of the cosmological virial theorem.

When the virial theorem is satisfied, the energy-balance equation (3.2.6) shows that the specific energy is constant. It then follows from equation (3.3.19) that

$$\epsilon = \tfrac{1}{2}\epsilon^{\text{grav}} = -\epsilon^{\text{kin}} = \text{const.} < 0\,. \tag{3.3.20}$$

So the assumption $\eta \ll t$ can be satisfied only if the specific energy is negative.

With the help of equation (3.2.10) we can write the virial theorem in the form

$$\langle v^2 \rangle_m = 2\pi G\bar{\rho}\alpha_\rho{}^2 \int_0^\infty F_\rho(k)dk\,. \tag{3.3.21}$$

The left-hand term of this relation can in principle be evaluated from redshift measurements; the right-hand term, from galaxy counts. Unfortunately, the relevant observational evidence is not yet of sufficiently high quality to permit a meaningful test.

The preceding derivation depends on the assumption that the relaxation time η is much less than the epoch t. Self-consistency demands that this condition be satisfied when the relation (3.3.21) actually holds. We then have

$$\eta \simeq l/v \simeq \alpha^{-1}(2\pi G\bar{\rho})^{-1/2}\,. \tag{3.3.22}$$

For a universe with zero mean spatial curvature,

$$6\pi G\bar{\rho}t^2 = 1\,. \tag{3.3.23}$$

Hence

$$\eta/t \simeq \alpha^{-1}\,. \tag{3.3.24}$$

Thus the condition $\alpha \gg 1$ is necessary for the validity of the virial theorem.

4. ORIGIN OF CLUSTERING

4.1. The Phenomenon of Clustering

What must a theory of galaxy clustering explain? The data analyses reviewed in §§ 2.9 and 2.10 support the following broad generalizations. (*a*) The clustering of galaxies is hierarchical, with at least three well-defined levels: multiple galaxies, galaxy clusters, and superclusters. (*b*) There is a maximum scale of clustering. Between the linear scales characteristic of clusters and of superclusters, the amplitude of the clustering spectrum (the Fourier transform of the density autocorrelation function) diminishes markedly, and at still larger scales it is too small to be detected by existing analyses. (*c*) The measured peculiar velocities of galaxies are consistent with the assumption that clustering is a dynamical phenomenon, i.e., that peculiar velocities are related to number-density fluctuations as predicted by the cosmological virial theorem (§ 3.3).

Galaxies themselves may be regarded as occupying an intermediate level in a hierarchy of self-gravitating systems that extends from subplanetary masses to galaxy superclusters. The former are just massive enough to qualify as self-gravitating systems; the latter are just dense enough to stand out from the cosmic background. While it is true that clustering is not uniformly pronounced over this entire range, we know from both theory and observation that clusters belonging to larger self-gravitating systems are continually being disrupted by tidal forces and mutual encounters. In our own

Galaxy, for example, globular clusters are depleted by successive passages through the Galactic plane, open clusters are torn apart by tidal forces, and the nonoccurrence of double stars in dense clusters can plausibly be attributed to their disruption by encounters. These considerations suggest the following working hypothesis:

The hierarchy of gravitationally bound systems has resulted from a continuing interaction between two opposing processes: hierarchic construction, which progressively generates new kinds of self-gravitating systems; and tidal disruption, which tends to break down existing structure in gravitationally bound systems.

This hypothesis raises two distinct theoretical problems. The first is to elucidate the process of hierarchic construction and to predict the initial clustering spectrum. The second is to explain why systems belonging to different levels of hierarchy differ systematically in their ability to survive disruptive interactions. Why, for example, have protostars and protogalaxies proved so much more durable than systems of intermediate mass? And why, among the latter, have globular clusters fared better than both less massive and more massive systems? The first question motivates §§ 4.2–4.13, the second is touched upon in § 4.4.

4.2. The Cosmological Principle

Any discussion of origins must begin with a statement of the auxiliary (initial, boundary, or symmetry) conditions that serve to define the Universe as a physical system. The simplest and commonest assumption about the large-scale structure of the Universe is the symmetry postulate called by Einstein the cosmological principle, and by Hubble the principle of uniformity. In its strongest form this assumption states that no statistical feature of the Universe serves to define a preferred position or direction in space: the Universe is statistically homogeneous and isotropic. The cosmological principle is consistent with all current observational knowledge. It has received especially strong support from recent studies of the cosmic microwave background, which place an upper limit of about 10^{-3} on the magnitude of any possible anisotropy in the angular intensity distribution over a wide range of angular scales.

Does the cosmological principle express a basic symmetry of the Universe, or is it merely a convenient simplifying assumption appropriate to our present limited state of knowledge? So long as one is concerned with the description and interpretation of events that are not too remote in space and time, the distinction between these two views is philosophical rather than practical. It assumes practical importance, however, when one studies the early Universe or seeks to understand features of the present-day Universe that were shaped by processes occurring near the beginning of the cosmic expansion.

Consider, for example, the currently popular hypothesis of primordial hydrodynamic or hydromagnetic turbulence. A uniform, isotropic, expanding fluid is stable against the onset of turbulence. If one takes the view that the cosmological principle is an exact symmetry property of the Universe, one must assume that any large-scale, statistically homogeneous and isotropic velocity fluctuations were built into the initial state. Such an assumption seems highly contrived. On the other hand, if one regards the cosmological principle as merely a convenient approximation, one can show that currently undetectable departures from statistical homogeneity and isotropy would have been much stronger in the early Universe. In these circumstances primordial turbulence seems a considerably more attractive hypothesis.

There are *a priori* arguments in favor of both views of the cosmological principle. The argument supporting its approximate character runs as follows. The Universe is a macroscopic physical system. Experience teaches us that macroscopic physical systems are never precisely symmetrical. Symmetry is a property of physical laws, not of the initial and boundary conditions that define physical systems. To this argument one might reply that the Universe and the auxiliary conditions needed to define it are unlike other macroscopic systems and auxiliary conditions. They are just as unique as the laws of physics themselves. If invariance under spatial rotations and translations is a fundamental symmetry of physical laws, why should it not be a fundamental symmetry of the Universe in which these laws apply?

Although *a priori* arguments of this kind may be suggestive, they can hardly be decisive. Like any other scientific hypothesis, the cosmological principle must ultimately be judged by the success of predictions based on it. In this article I adopt the view that the principle, in the strong form quoted above, is an exact symmetry property of the Universe. Alternative hypotheses and their observable consequences have recently been reviewed by Novikov and Zel'dovich (1973).

Einstein's theory of gravitation predicts that a universe satisfying the cosmological principle must expand from (or contract towards) a singular state of infinite density in the finite past (or finite future).[2] Observation tells us that the Universe is in fact expanding, not contracting. Its structure and evolution are therefore determined by the conditions that prevail at very high densities.

Two broad alternatives now present themselves. One may assume that macroscopic density, velocity, magnetic field, or other fluctuations were present at the earliest epochs at which current physical theories apply, or one may assume that the initial state was macroscopically uniform. Let us consider the second alternative. If the early Universe is assumed to be macroscopically uniform, all that remains to be specified are its temperature, its curvature, and the lepton-baryon ratio. By far the most important of these initial data is the temperature. The value of the lepton-baryon ratio has important but limited observational consequences, while the effects of spatial curvature are entirely negligible during the earliest stages of the cosmic expansion.

In the "standard cosmological model" the temperature at some conveniently selected early epoch is fixed by the assumption that the cosmic microwave background, whose present temperature $T_0 \simeq 2.7°$ K, is the redshifted remnant of a primordial radiation field (the primeval fireball). The standard model has been rather thoroughly investigated; see Weinberg (1972), Peebles (1971), and Harrison (1973). So far no process has been found that gives rise to significant macroscopic density fluctuations under the conditions postulated by the model. The high gas temperature, as well as interactions between matter and radiation fields, stabilize the cosmic medium against the growth of density fluctuations. Unless some essential physical process has been overlooked, it would appear that if one interprets the cosmic microwave background as the remnant of an initial radiation field, one is forced to postulate that the initial state was highly nonuniform.

Although initial nonuniformity appears to be a *necessary* feature of the standard

[2] Strictly speaking, the theory does not predict that the singularity is actually attained, because the theory is not valid for values $t \lesssim (Gh/c^5)^{1/2} \simeq 10^{-43}$ s, when quantal fluctuations of the metric are important.

model, this assumption has not yet been shown to be *sufficient* to account for the existence of gravitationally bound systems. In view of this unresolved difficulty with the standard model, it may be premature to abandon the hypothesis of a uniform initial state. But then two questions need to be answered: (i) What physical processes could have given rise to suitable density fluctuations (i.e., density fluctuations capable of evolving into gravitationally bound systems) in an initially uniform, but not necessarily hot, universe? (ii) What physical processes could have given rise to the cosmic microwave background in such a universe?

The origin of cosmic nonuniformity and the origin of the thermal radiation background are thus closely related problems. The assumption of a primordial radiation field apparently forces the assumption of primordial density fluctuations; the assumption of primordial uniformity forces the assumption that the cosmic microwave background is nonprimordial. The symmetry suggested by this formulation is slightly deceptive, however. The first hypothesis requires a set of initial conditions that incorporate a substantial quantity of information concerning the present state of the Universe. The second hypothesis has minimal information content. It is considerably more vulnerable to observational disproof than the first hypothesis, but its potential explanatory power is correspondingly greater.

4.3. Initial Growth of Small-Amplitude Density Fluctuations

The idea that a uniform, unbounded medium is unstable against the growth of self-gravitating density fluctuations goes back to Newton (1692). It was formulated mathematically by Jeans (1902), who concluded that sinusoidal density fluctuations with wavenumbers greater than a certain critical value propagate as modified sound waves, while those with wavenumbers smaller than this value grow or decay exponentially. Jeans's discussion is technically incorrect, because it proceeds from what we now know to be an incorrect assumption about the undisturbed medium, namely, that it is at rest. Lifshitz (1946) gave a correct relativistic treatment of the problem, taking the undisturbed medium to be a Friedmann universe. He found that Jeans's criterion does divide the wavenumbers of oscillatory modes from those of monotonically growing and decaying modes, but the growth and decay rates turned out to be algebraic rather than exponential. Using this result, he argued that gravitationally bound systems cannot have arisen from thermal fluctuations in a macroscopically uniform medium. The argument (slightly modified) runs as follows:

For the sake of simplicity we suppose that the unperturbed Friedmann universe has zero spatial curvature and zero pressure. (The effects of spatial curvature are negligible at sufficiently early epochs, and thermal motions retard or prevent the growth of density fluctuations, so that the assumption of zero pressure will lead us to overestimate the growth rates of fluctuations.) The density of the unperturbed medium is $\rho \propto t^{-2}$. Lifshitz showed that, for the growing mode, the fractional density fluctuation $\delta\rho/\rho \equiv s \propto t^{2/3}$. Hence

$$s_0/s_1 = (\rho_1/\rho_0)^{1/3}\,, \tag{4.3.1}$$

where the suffixes 0, 1 refer respectively to the final and initial epochs t_0, t_1. A fluctuation is gravitationally bound at time t_0 if $s_0 \gtrsim 1$. If we assume that the initial fluctuations are thermal, we may set $s_1 = N^{-1/2}$, where N is the average number of particles per fluctuation, provided (as shown below) that the temperature of the gas exceeds the degeneracy temperature of the electrons. The condition $s_0 \gtrsim 1$ can then be written in the form

$$N \lesssim (\rho_1/\rho_0)^{2/3} \equiv N_{\rm max} . \tag{4.3.2}$$

Lifshitz remarked, in effect, that setting $\rho_1 \simeq 10^{15}$ g cm^{-3} (the density of nuclear matter) and $\rho_0 \simeq 10^{-30}$ g cm^{-3} (the present value of the mean cosmic density) gives $N_{\rm max} \simeq 10^{30}$. Thus the largest mass that could have condensed from thermal fluctuations at nuclear density is of order 10^6 g—far too small to be of astronomical interest.

The preceding argument is incomplete in several respects. Most obviously, it leaves open the possibility that thermal fluctuations at supernuclear densities could have given rise to significant density fluctuations. In addition, it neglects the effects of finite temperature. One cannot simply assume that the temperature is low enough to permit the growth of density fluctuations corresponding to any given value of N, because the temperature of the medium must exceed the degeneracy temperature of the electrons. For if the electron gas is strongly degenerate, then fluctuations in electron number-density are negligible, and fluctuations in the ion density will excite longitudinal plasma oscillations. Because the ion plasma frequency $\omega \simeq (4\pi ne^2/m_p)^{1/2}$ always greatly exceeds the gravitational growth rate $\omega_g \simeq (4\pi Gnm_p)^{1/2}$ (the ratio of the two frequencies is the square root of Eddington's number $e^2/Gm_p{}^2 \simeq 10^{36}$), all modes that can be excited by fluctuations of the ion density are oscillatory.

The condition that a nonrelativistic electron gas be nondegenerate is

$$v_e > hn^{1/3}m_e{}^{-1} , \tag{4.3.3}$$

where v_e is the electron thermal velocity, h is Planck's constant, n is the electron or ion density, and m_e is the electron mass. Jeans's criterion is

$$v_p{}^2 < GM/L = Gm_pN^{2/3}n^{1/3} , \tag{4.3.4}$$

where the suffix p refers to protons. It follows from the two preceding equations and the relation $m_ev_e{}^2 = m_pv_p{}^2$, valid in thermal equilibrium, that

$$N^{2/3} > (e^2/Gm_p{}^2)(h^2/m_ee^2)n^{1/3} . \tag{4.3.5}$$

Now, inequality (4.3.2) may be written in the form

$$N^{2/3} < n_0{}^{-4/9}n^{4/9} . \tag{4.3.6}$$

At nuclear density the right-hand side of inequality (4.3.5) exceeds the right-hand side of inequality (4.3.6) by six orders of magnitude. Hence both inequalities cannot be simultaneously satisfied at subnuclear cosmic densities—a considerably stronger result than that given by Lifshitz.

At supernuclear densities the gas is relativistic, and for the purpose of the present calculation we may assume that it consists entirely of neutrons. In place of inequality (4.3.3) we now have the nondegeneracy condition

$$p > hn^{1/3} , \tag{4.3.7}$$

where p denotes the thermal momentum of a neutron and n denotes the number density of neutrons. Combining this inequality equality with the Jeans condition gives

$$N^{2/3} > (e^2/Gm^2)(hc/e^2) \simeq 10^{39} . \tag{4.3.8}$$

Combining inequalities (4.3.6) and (4.3.8), we get

$$n > 10^{88} n_0 > 10^{82} \quad \text{cm}^{-3} . \tag{4.3.9}$$

The corresponding restriction on the initial epoch is $t_1 < 10^{-27}$ s. These estimates could be in error by many orders of magnitude, because they presuppose a greatly oversimplified description of the cosmic medium at supernuclear densities. Even so, it seems impossible to avoid the conclusion that only thermal fluctuations at strongly supernuclear densities are of potential cosmogonic significance.

But now a fresh difficulty arises. On physical grounds, one might expect that the diameter L of a gravitationally coherent density fluctuation could not exceed the diameter of the visible Universe $\simeq 2ct$. (This expectation will be justified in § 4.5.) Now, the mass of the observable Universe (and hence of the largest gravitationally coherent fluctuation) varies like $\rho(ct)^3 \propto t$. At nuclear density the mass of the visible Universe is of order 1 $M_\odot$. As one goes backward in time, the mass of the visible Universe decreases, while the upper mass limit for a gravitationally bound fluctuation implied by inequality (4.3.2) increases. The two limits coincide when $M \simeq 10^{-12} M_\odot$, which accordingly represents an absolute upper limit for the mass of a gravitationally coherent fluctuation. This limit, however, does not take thermal motions into account. If we allow for thermal motions, we must use inequality (4.3.9), which gives an upper mass limit of $10^{-22} M_\odot$.

We conclude that gravitationally bound systems are unlikely to have originated from thermal fluctuations in a uniform cosmic medium.

4.4. The Weak-Field Approximation

The Newtonian approximation (§ 3.1) depends on two assumptions: that the maximum scale of local irregularities is small compared with the radius of the visible Universe; and that the fluctuating gravitational field is weak. Although both conditions are satisfied now (§ 2.9), the first condition was presumably not satisfied in the past, since the mass of the visible Universe is given by (see eq. [4.12.8])

$$M^{\text{vis}} = 2^{3/2}\pi^3(ct)^3\bar{\rho} \simeq 2 \times 10^{39} t \quad \text{grams} , \tag{4.4.1}$$

which shows that the mass of the visible Universe was comparable to the mass of a typical galaxy at $t \simeq 10^5$ s. If we wish to discuss the origin of galaxies and galaxy clusters, we must therefore give up one of the two simplifying assumptions underlying the Newtonian approximation.

Fortunately, we may keep the second, more important, assumption, that local gravitation fields are weak enough to justify a linearized description. The cosmological virial theorem (§ 3.3) shows that local gravitational fields and peculiar velocities do not tend to increase systematically as t decreases, and the approximate theory described below allows one to verify *a posteriori* that this generalization remains valid for density fluctuations whose scale is comparable to or greater than the diameter of the observable Universe.

The standard procedure for deriving the weak-field approximation is to linearize the nonlinear field equations and equations of motion of Einstein's theory. This was first done by Lifshitz (1946). The linearization procedure allows considerable freedom in the choice of coordinates. Lifshitz's coordinates have certain mathematical advantages, but do not reduce to Newtonian coordinates in the Newtonian limit. This creates difficulties in the physical interpretation of the results, especially as they concern fluctuations with scales comparable to or greater than the radius of the visible Universe, for which relativistic effects are important. The use of asymptotically Newtonian coordinates (Layzer 1971) simplifies the physical interpretation. In these coordinates, for a universe with zero mean spatial curvature ($k = 0$), the field equations take the form (Layzer 1971)

$$(\Box - 3H\partial_t)\psi - (9/2)H^2\psi + H^2\phi = 4\pi G\tilde{\rho}\,, \tag{4.4.2a}$$

$$(\Box - 3H\partial_t)\phi - (9/2)H^2\psi - 5H^2\phi = 4\pi G\tilde{\rho}\,, \tag{4.4.2b}$$

$$(\Box - 3H\partial_t)\boldsymbol{A} - 5H^2\boldsymbol{A} = 4\pi G\rho\boldsymbol{v}\,. \tag{4.4.2c}$$

The derivatives with respect to space coordinates are defined by

$$\frac{\partial}{\partial x_\alpha} = S^{-1}\frac{\partial}{\partial \bar{x}_\alpha}\,, \tag{4.4.3}$$

where the $\bar{x}_\alpha$ are comoving coordinates and ∂_t denotes a time derivative with the comoving space coordinates held fixed; $\Box$ is the d'Alembertian; and we have set $c = 1$.

When the scale of local irregularities is much smaller than ct, the field equations (4.4.2) reduce to their Newtonian forms:

$$\Delta\begin{cases}\phi\\ \psi\\ \boldsymbol{A}\end{cases} = 4\pi G\begin{cases}\tilde{\rho}\\ \tilde{\rho}\\ \rho\boldsymbol{v}\end{cases}. \tag{4.4.4}$$

In the Newtonian approximation the two scalar potentials ϕ and ψ coalesce. In both approximations the vector potential $\boldsymbol{A}$ is determined by the peculiar-velocity field and is not coupled to the scalar potentials.

The weak-field potentials satisfy the following auxiliary conditions:

$$\nabla\cdot\boldsymbol{A} + \tfrac{1}{4}(3\dot{\psi} + \dot{\phi}) + H\phi = 0\,, \tag{4.4.5a}$$

$$\nabla(\psi - \phi) - 4(\dot{\boldsymbol{A}} + 2H\boldsymbol{A}) = 0\,, \tag{4.4.5b}$$

which in the Newtonian limit reduce to

$$\phi = \psi\,, \quad \nabla\cdot\boldsymbol{A} + \dot{\phi} = 0\,. \tag{4.4.6}$$

The vector potential $\boldsymbol{A}$ does not appear in Newton's theory because the Newtonian equations of motion involve only the potential ϕ $(=\psi)$. In the weak-field approximation, however, the equations of motion do involve $\boldsymbol{A}$:

$$(d/dt + H)\boldsymbol{v} = -\nabla p/\rho - \nabla\psi + 4H\boldsymbol{A}\,. \tag{4.4.7}$$

These are the fluid equations of motion in Lagrangian form, with the viscous terms omitted. In discussing the dynamics of clustering we also neglected pressure effects, but in the present context these are important.

The energy-balance equation in the weak-field approximation is found to be

$$\frac{d}{dt}(eS^3) + p\frac{dS^3}{dt} = \frac{S^3}{4\pi G}\langle\dddot{\psi}(\psi + 7H\ddot{\psi} + 12H^2\dot{\psi})\rangle\,, \tag{4.4.8}$$

where

$$e = e^{\text{gas}} + e^{\text{kin}} + e^{\text{grav}}\,, \qquad p = p^{\text{gas}} + p^{\text{kin}} + p^{\text{grav}} \tag{4.4.9}$$

with e^{kin} given by the same formula as before (eq. [3.2.4]) and

$$e^{\text{grav}} = 3p^{\text{grav}} = -\frac{1}{8\pi G}\langle|\nabla\psi|^2\rangle\,. \tag{4.4.10}$$

The quantities e^{gas} and p^{gas} are the ordinary molecular contributions to the energy density and the pressure.

The non-Newtonian terms on the right-hand side of equation (4.4.8) are not important in practice. They are comparable to other terms in the equation only for fluctuations whose wavelength is comparable to or greater than ct. But we shall find that such fluctuations do not contribute significantly to the energy in any case.

4.5. Initial Growth of Density Fluctuations in the Weak-Field Approximation

In equation (4.4.2) the quantity $\tilde{\rho}$ represents the (peculiar) mass density, because it satisfies the equation of continuity, whatever the linear scale of the fluctuations. The linearized equation of continuity and the linearized equations of motion are

$$\partial_t s + \nabla\cdot \boldsymbol{v} = 0\,, \tag{4.5.1}$$

$$\partial_t \boldsymbol{v} + H\boldsymbol{v} = -\nabla p/\rho - \nabla\psi + 4H\boldsymbol{A}\,, \tag{4.5.2}$$

where $s \equiv \tilde{\rho}/\bar{\rho}$. Keeping in mind that the space and time derivatives denoted by ∇ and ∂_t do not commute (see eq. [4.4.3]) and using the relation $a^2 = \delta p/\delta\rho$, where a denotes the speed of sound, we can eliminate $\boldsymbol{v}$ between equations (4.5.1) and (4.5.2) to obtain

$$a^2\Delta s - \ddot{s} - 2H\dot{s} - 4H\nabla\cdot\boldsymbol{A} + \Delta\psi = 0\,. \tag{4.5.3}$$

Using the field equations and auxiliary conditions, one can rewrite this equation in the form

$$\ddot{s} + 2H\dot{s} + (-a^2\nabla^2 - \tfrac{3}{2}H^2)s = 3H\dot{\psi} + \tfrac{9}{2}H^2\psi + 3H^2\phi\,. \tag{4.5.4}$$

In the Newtonian approximation the right-hand side of this equation vanishes. By expressing s as a spatial Fourier integral, one can easily solve the resulting linear, homogeneous equation.

In the zero-pressure limit ($a = 0$), the general solution of the Newtonian approximation to equation (4.5.4) is

$$s = \int s_{\boldsymbol{k}} \exp{(i\boldsymbol{k}\cdot\boldsymbol{x})}d^3k\,, \tag{4.5.5}$$

$$s_{\boldsymbol{k}} = A_{\boldsymbol{k}}(ckt)^2 + B_{\boldsymbol{k}}(ckt)^{-3}\,, \tag{4.5.6}$$

where the wavenumber k varies with time according to

$$k = \bar{k}/S(t)\,; \quad \bar{k} = \text{const.}\,; \quad S \simeq t^{2/3}\,. \tag{4.5.7}$$

In equation (4.5.6) $A_{\boldsymbol{k}}$ and $B_{\boldsymbol{k}}$ are arbitrary integration constants. Thus a given Fourier component is the sum of a growing mode that varies like $t^{2/3}$ and a decaying mode that varies like t^{-1}.

In the weak-field approximation, Burke and Layzer (Burke 1971) obtained the exact solution (again in the zero-pressure limit)

$$s_k = A_k[(ckt)^2 + 10/3] + B_k(ckt)^{-3} . \tag{4.5.8}$$

The decaying mode has the same behavior at all wavenumbers, but the behavior of the growing mode differs qualitatively at small wavenumbers from its behavior in the Newtonian limit. The formula $s \propto t^{2/3}$ is valid only for wavenumbers $k \gg (10/3)^{1/2}(ct)^{-1} \equiv k_c$. For wavenumbers $k \ll k_c$, $s \propto$ const. Thus the mathematical theory confirms the physical expectation that the diameter of the visible Universe represents an upper limit for the wavelength of a gravitationally coherent density fluctuation.

According to equation (4.5.8), a mode whose wavenumber $k \ll k_c$ remains quiescent until the visible Universe has grown large enough to contain it; thereafter it grows at an algebraic rate. But this prediction must not be taken literally. Our analysis takes no account of processes that tend to disrupt density fluctuations. In the absence of detailed arguments to the contrary, there is no reason to suppose that density fluctuations with $k \ll k_c$ could survive long enough to become gravitationally coherent. In any case, the present theory imposes an absolute upper limit $M \simeq 10^{-12}\, M_\odot$ on the mass of gravitationally coherent density fluctuations (§ 4.4).

The effects of finite pressure are qualitatively the same in the weak-field approximation as in the Newtonian approximation. We are primarily interested in gravitationally coherent fluctuations, for which the right-hand side of equation (4.5.4) may be set equal to zero. The corresponding equation for s_k, first given by Bonnor (1956, 1957), is

$$\ddot{s}_k + 2H\dot{s}_k + (a^2k^2 - \tfrac{3}{2}H^2)s_k = 0 . \tag{4.5.9}$$

Bonnor pointed out that a sufficient condition for a permanently growing mode is that the coefficient of s_k be negative; for if an initially growing mode were to attain a finite maximum, the conditions $\dot{s}_k = 0$, $\ddot{s}_k < 0$ would hold there, and this cannot happen if the coefficient of $s_k < 0$. This argument yields the instability criterion

$$k < k_J \equiv (3/2)^{1/2}H/a = (2/3)^{1/2}(at)^{-1} . \tag{4.5.10}$$

This may be compared with the criterion

$$k > k_c \equiv (10/3)^{1/2}(ct)^{-1} , \tag{4.5.11}$$

which follows from equation (4.5.8).

A detailed discussion of the growth of small-amplitude fluctuations in the weak-field approximation, including numerical solutions of the exact differential equations and a discussion of the effects of finite spatial curvature, has been given by Burke (1971).

Although the existence of a low-wavenumber cutoff for gravitationally coherent density fluctuations seems physically reasonable, previous treatments, employing different coordinates, gave no indication of a cutoff. Is the appearance (or nonappearance) of a cutoff simply an artifact of the coordinate system employed? Burke (1971) has made a careful study of this question. He finds that analyses employing different coordinate systems are, as expected, mathematically equivalent, and he derives explicit transformation formulae connecting them. For example, Lifshitz showed that the relative fluctuation $\delta e/e$ of the invariant energy-density e has a growing mode that varies like $t^{2/3}$ for *all* wavenumbers. The same result can easily be shown to follow from the present analysis —as, of course, it must do, since δe and e are invariants. However, δe coincides with $\delta\rho$, the matter-density fluctuation, only in the Newtonian limit. For fluctuations with wavenumbers $k \leq k_c$, δe and $\delta\rho$ behave quite differently, and it is the behavior of $\delta\rho$ that is relevant to our present concerns, since $\delta\rho = \tilde{\rho}$ is the quantity that satisfies the equation of continuity (eq. [4.5.1]). For a fuller discussion of the somewhat delicate mathematical points involved, see Burke (1971).

4.6. Thermal Instability

The preceding considerations make it seem unlikely that thermal fluctuations at subnuclear or supernuclear densities or exotic instabilities at strongly supernuclear densities could have played a significant cosmogonic role. The question therefore arises: Under what physical conditions (if any) can cosmogonically significant density fluctuations arise in an initially uniform cosmic medium? Evidently, the fluctuations must be initiated by some nongravitational instability. Thermal instability—the spontaneous growth of nearly isobaric temperature fluctuations in an initially uniform medium, owing to inequalities in the radiative-loss rate—has often been nominated for this role.

When radiation and thermal conduction are taken into account, the energy-balance equation for the gas component of the cosmic medium takes the form

$$\begin{aligned} T d\sigma/dt &= d(e/\mu)/dt + p d(1/\mu)/dt \\ &= -\mathfrak{L} + \operatorname{div}(K \operatorname{grad} T) , \end{aligned} \tag{4.6.1}$$

where σ, e, and μ denote, respectively, the densities of specific entropy, internal energy, and rest mass. The radiative-loss function $\mathfrak{L} \equiv q - r$ is the specific emission rate q minus the specific absorption rate r. For a perfect gas,

$$e = \tfrac{3}{2} n k_B T , \quad p = n k_B T , \quad \mu = nm ; \tag{4.6.2}$$

and equation (4.6.1) takes the form

$$d(5k_B T/2m)/dt - (k_B T/m) d(\ln p)/dt = -\mathfrak{L} + \operatorname{div}(K \operatorname{grad} T) . \tag{4.6.3}$$

Setting

$$T = \bar{T}(1 + \theta) , \quad \mathfrak{L} = \bar{\mathfrak{L}} + (\partial\mathfrak{L}/\partial T)_p \theta \bar{T} , \tag{4.6.4}$$

where the overbar refers to the uniform, unperturbed state, and bearing in mind that $p = \bar{p} \propto \bar{n}\bar{T} \propto S^{-(3+a)}$, where the mean temperature $\bar{T}$ is assumed to vary like S^{-a}, we obtain from equation (4.6.3)

$$(d/dt + bH)(5k_B\bar{T}/2m) = -\bar{\mathfrak{L}} , \tag{4.6.5}$$

$$(d/dt + bH - \bar{\lambda} + \tilde{\lambda} + K'k^2)\theta_k = 0 , \tag{4.6.6}$$

where

$$b = 2(a + 3)/5 , \quad K' = 2mK/5k_B ,$$

$$\lambda = 2m\mathfrak{L}/5k_B\bar{T} , \quad \tilde{\lambda} = (2m/5k_B)(\partial\mathfrak{L}/\partial T)_p , \tag{4.6.7}$$

and the suffix k indicates a Fourier component corresponding to wavenumber k. A complete theoretical description would need to include, along with equations (4.6.5) and

(4.6.6), a pair of equations for the radiation field; these will not be needed in the present discussion.

Equation (4.6.6) differs from its noncosmological analog (for a static medium radiating into empty space) through the second and third terms on the left-hand side. The second term tends to stabilize temperature fluctuations, since $b > 0$. The third term is stabilizing if the mean free path of emitted photons is short compared with the radius of the visible Universe (see below), and destabilizing in the opposite case. It follows from equation (4.6.6) that a necessary condition for thermal instability is

$$\tilde{\lambda} < \lambda - bH\,, \tag{4.6.8}$$

and that, if this condition is fulfilled, the range of unstable wavenumbers is given by

$$k^2 < k_c{}^2 = (-\tilde{\lambda} + \lambda - bH)/K'\,. \tag{4.6.9}$$

For variations in the radiative-loss rate to be of order H, as required by the instability criterion (4.6.8), the gas must be ionized. It is then opaque to its own radiation, and $\bar{\mathfrak{L}}$ is given by

$$\begin{aligned}\bar{\mathfrak{L}} &= (d/dt + H)(\overline{e^{\mathrm{rad}}}/\bar{\rho}) \\ &= (d/dt + H)[B(\bar{T})/\bar{\rho}] = -4(a-1)H\bar{B}/\bar{\rho}\,.\end{aligned} \tag{4.6.10}$$

The first line of this equation is a generally valid formula expressing the radiative-loss rate of the gas as the rate of change of the specific energy of the radiation field. The opacity of the gas to its own radiation ensures that it is in thermal equilibrium with the matter, so that we may set $e^{\mathrm{rad}} = B(T) = (\pi^2/15)\,(hc)^{-3}(k_{\mathrm{B}}T)^4$.

We must now distinguish two extreme cases, according to the value of the ratio

$$\Gamma \equiv e^{\mathrm{rad}}/e^{\mathrm{gas}} = 2B(T)/3nk_{\mathrm{B}}T \propto n^{-1}T^3\,. \tag{4.6.11}$$

a) $\Gamma \gg 1$.—From equations (4.6.5) and (4.6.10) it follows that

$$a - 1 = (4\Gamma)^{-1}\,, \qquad b = 8/5\,, \qquad \lambda = -3H/5 \qquad (\Gamma \gg 1)\,. \tag{4.6.12}$$

Since a is only slightly greater than unity, Γ, which varies like $S^{3(1-a)}$, decreases very slowly as the Universe expands and the temperature drops. Ultimately the temperature becomes so low that the gas recombines and becomes transparent to its own radiation. Thereafter the radiation temperature decreases like S^{-1} while the gas temperature decreases like S^{-2}.

b) $\Gamma \ll 1$.—Equations (4.6.5) and (4.6.10) now give

$$a = 2\,, \qquad b = 2\,, \qquad \tilde{\lambda} = -12\Gamma H/5 \qquad (\Gamma \ll 1)\,. \tag{4.6.13}$$

In this case Γ decreases like S^{-3} as the Universe expands, and the gas-temperature and radiation-temperature decrease like S^{-2} until the gas recombines.

In the two extreme cases, the instability criterion (4.6.8) takes the form

$$\begin{aligned}\tilde{\lambda} &< -11H/5 \qquad (\Gamma \gg 1) \\ &< -2H \qquad\quad (\Gamma \ll 1)\,.\end{aligned} \tag{4.6.14}$$

We must now distinguish between optically thin and optically thick fluctuations. A fluctuation that is opaque to its own radiation is in local thermodynamic equilibrium with the radiation field. If $\Gamma \gg 1$, the gas temperature cannot vary appreciably because the thermal content of the gas is negligible compared with that of the radiation field. On the other hand, if $\Gamma \ll 1$, then $\tilde{\lambda}$ is of order ΓH and hence cannot satisfy the instability criterion (4.6.14). Thus fluctuations opaque to their own radiation cannot become thermally unstable.

For optically thin fluctuations, $\mathfrak{L}$ is given by

$$\bar{\rho}\mathfrak{L} = \kappa(T, p)[B(T) - B(\bar{T})], \tag{4.6.15}$$

where κ is a mean absorption coefficient; κ^{-1} is the free-flight time of an energy-containing photon. Hence

$$\tilde{\lambda} \propto (\partial\mathfrak{L}/\partial T)_p = \kappa B'/\bar{\rho} = 4\kappa B/\rho T > 0\,. \tag{4.6.16}$$

It follows from the instability criterion (4.6.14), which requires $\tilde{\lambda}$ to be negative, that in this case, too, stability always prevails.

In short, so long as the Universe remains opaque to its own radiation, it is thoroughly stable against the growth of thermal fluctuations, whether they are optically thin or optically thick.

If the Universe were transparent to its own radiation, the term $B(\bar{T})$ on the right-hand side of equation (4.6.15) would be absent and $\tilde{\lambda}$ could be negative for sufficiently large negative values of $\partial\kappa(T, p)/\partial T$. As mentioned above, the condition that the Universe be transparent to its own radiation conflicts with the demand that the cooling rate be great enough to offset the stabilizing effect of the cosmic expansion. However, let us assume for the sake of the argument that the criterion for thermal instability can somehow be satisfied. What then? It has been suggested that, in a wide variety of astronomical contexts, a thermally unstable medium separates into two thermally stable phases in pressure equilibrium, the temperature of one of the two stable phases being lower than the initial temperature of the uniform medium, that of the other, higher (see Field 1965 for a unified discussion). The following considerations suggest that the consequences of thermal instability may be quite different.

The growth of temperature fluctuations in an initially uniform medium creates pressure gradients, which accelerate the fluid. Even if only a small range of wavenumbers is initially excited, the resulting velocity field must be highly disordered. For the instability tends to create cool, contracting, pockets in a hot, expanding matrix. At each interface between a hot and a cool region, there is an acceleration directed from the (expanding) region of lower density into the (contracting) region of higher density. All such interfaces are unstable against Stokes-Rayleigh-Taylor instability at wavenumbers smaller than a certain limit set by viscosity and thermal convection. Thus thermal instability may be expected to generate a turbulent velocity field. The energy of this field will increase at the expense of the gas's thermal energy until the enhanced thermal conductivity reaches such a high value that a further increase would result in suppression of the driving thermal instability. In the equilibrium state, the rate of turbulent dissipation equals the rate at which the turbulence extracts energy from the gas's thermal reservoir via thermal instability. This equilibrium is self-regulating, since changes in the energy of the turbulent field produce changes in the effective thermal conductivity that alter the turbulent-energy supply so as to oppose the original changes.

To summarize, thermal instability probably cannot occur under physical conditions relevant to the present discussion. If it does occur, the preceding considerations suggest that it does not give rise to a separation of the medium into distinct phases but rather to a state of self-regulating turbulence.

4.7. Early History of the Cold Universe

The preceding discussion strongly suggests that cosmogonically significant density fluctuations cannot arise spontaneously in a hot universe. Even if strong density fluctuations are built into the initial state, it is by no means clear that they can survive long enough to become cosmogonically significant. In this and the following sections I discuss the implications of an alternative hypothesis: that the Universe was initially in a state of global thermodynamic equilibrium at zero temperature.

This hypothesis immediately raises the question: How can finite temperatures and complex structure arise in an initially structureless medium at zero temperature? The emergence of order from initial chaos would appear to violate the spirit, if not the letter, of the second law of thermodynamics. In fact, there is no contradiction. Local thermodynamic equilibrium is maintained by microscopic processes, each with a characteristic rate that depends on temperature and density. So long as the rates of all the controlling microscopic processes exceed the expansion rate, significant departures from thermodynamic equilibrium cannot develop; the Universe expands quasi-adiabatically through a sequence of near-equilibrium states.

On the other hand, if the rate of an equilibrium-maintaining reaction falls below the expansion rate, the aspect of equilibrium controlled by that reaction can no longer adjust itself to the instantaneous global state of the medium, and substantial departures from local thermodynamic equilibrium may develop. In particular, the medium may be heated by particle decays and exothermic nuclear reactions; and nonequilibrium particle concentrations may be frozen in when the characteristic rates of the reactions that control them become comparable to the expansion rate. Both kinds of departures are important in the present discussion.

4.7.1. *Thermal History of the Early Cold Universe.*—At strongly supernuclear densities the cold universe consists, presumably, of massive hyperons. When the cosmic density becomes low enough, unstable particles will begin to decay. A particle decay becomes energetically possible when the decay energy can supply the Fermi energy ($\propto n^{1/3}$ for relativistic particles) of the daughter fermions. Part of the decay energy will be available to heat the medium if either (*a*) none of the decay products is a fermion, or (*b*) the decay rate $\lesssim H$. The second condition is satisfied only by neutrons, which, however, are consumed in nuclear reactions long before they have a chance to decay; see § 4.7.2. Among the unstable particles that satisfy the first condition are neutral pions and neutral K-mesons.

Kaufman (1970) has made a detailed study of particle decays in the early stages of a cold universe. She concluded that the resulting temperature increase might be comparable to or greater than the nucleon degeneracy temperature but was probably much smaller than the electron degeneracy temperature. Since the temperature of an adiabatically expanding gas varies with density in exactly the same way as the degeneracy temperature, the ratio is independent of density. Thus the electron gas remains strongly degenerate.

4.7.2. *Helium Production in the Cold Universe.*—At subnuclear densities, the equilibrium concentrations of neutrons, protons, electrons, and neutrinos are maintained by the reaction $n + \nu \rightleftharpoons p + e^-$. The rate of this reaction becomes comparable to the expansion rate at a cosmic density of order $\rho \simeq 10^{13}$ g cm^{-3}, corresponding to the epoch $t \simeq 10^{-4}$ s. The proton-neutron ratio prevailing when the reaction rate becomes comparable to the expansion rate is determined by the concentration of neutrinos, assumed to be degenerate. The higher the degenerate-neutrino concentration, the greater the rate of production of protons and the higher the proton-neutron ratio. Once the reaction has ceased, the surviving neutrons quickly combine with protons to form deuterons, most of which are quickly consumed in reactions that yield α-particles as their final product. The resulting helium-hydrogen ratio, y, depends sensitively on the value of the lepton-baryon ratio $L = (n_e + n_\nu)/(n_p + n_n)$; as L varies from 1.2 to 1.35, y varies from 0.25 to 0 (Kaufman 1970). Thus the primordial abundance of helium inferred from astronomical observation (reviewed by Danziger 1970) is considerably less critical for the present theory than for the "standard model," which predicts a primordial helium abundance of 28 percent by mass.

Nuclear reactions also heat the cosmic medium. However, they occur too early to lift the electron degeneracy, provided that the electron temperature $T < 0.1\ T_{\rm deg}$ beforehand (Kaufman 1970).

4.7.3. *The Metallic Phase.*—When the cosmic medium emerges from the stage of particle decays and nuclear reactions ($\rho \simeq 10^7$ g cm^{-3}, $t \simeq 10^{-1}$ s), it may be described as a liquid metal. The nucleons are distributed without long-range correlation in a uniform, strongly degenerate sea of electrons. The liquid-metallic phase endures during a certain period of adiabatic expansion, and terminates in one of two ways. Either the nucleons and electrons recombine, or else the liquid metal freezes (i.e., undergoes a phase transition to the solid metallic state).

A freezing phase transition occurs if the temperature falls below the melting temperature of the solid while the medium is still dense enough to be pressure-ionized. A theoretical discussion by Hively (1971) showed that this condition *may* be satisfied if the electron gas is strongly enough degenerate. Because experimental information about metallic hydrogen is lacking (even the existence of a stable metallic phase has not yet been definitely established), one must rely on theoretical calculations. Unfortunately, some of the approximations underlying existing theories are at best marginally valid under the conditions relevant to the present discussion. Consequently, no secure inference concerning the occurrence or nonoccurrence of a freezing phase transition can now be drawn.

In the absence of an *ab initio* theory of melting, the semiempirical Lindemann criterion is often used to estimate the melting temperature. According to this criterion, which has proven highly reliable in practice, a lattice melts when the mean square vibration amplitude of the nucleons exceeds a fixed, empirically determined, fraction of the square of the lattice spacing. (For the alkali metals this fraction $\simeq 1/16$.) For nuclei with charge Z and atomic number A, the Lindemann criterion yields the following formula for the melting temperature T_m (Hively 1971, p. 32):

$$T_m = 1.9 \times 10^3 Z^2 A^{-1/3} \rho^{1/3}\,{}^\circ K\,. \tag{4.7.1}$$

Now, the electron gas becomes nonrelativistic at cosmic densities $\rho \lesssim 10^6$ g cm^{-3}. Thereafter its temperature decreases according to the nonrelativistic adiabatic law $T \propto S^{-2} \propto \rho^{2/3}$. Initially $T \gg T_m$, but equation (4.7.1) shows that adiabatic cooling will ultimately cause T to fall below T_m. Will the medium then be pressure-ionized?

Pressure-ionization persists so long as the mean spacing of nuclei $\lesssim a_0$ $(= \hbar^2/me^2)$, i.e., for densities $\rho \gtrsim 1$ g cm^{-3}. The Lindemann criterion (4.7.1.) predicts that freezing occurs before recombination if $T < 0.064 T_{\text{deg}}$. As mentioned above, Kaufman's discussion of heating by particle decays and nuclear reactions indicates that this condition is satisfied. Thus the available theoretical evidence, though inconclusive, is compatible with the hypothesis that a universe initially at zero temperature undergoes a freezing phase transition. According to calculations by Hively (1971), freezing occurs at cosmic density $\rho \simeq 10^2$ g cm^{-3} $(t \simeq 10^2$ s).

The solid-metallic phase persists at least until the density has dropped to $\rho \simeq 1$ g cm^{-3}. At this point, according to calculations by Hively (1971, p. 53), the insulating molecular solid becomes the thermodynamically favored phase. Hively has argued, however, that recombination cannot actually take place until a substantially lower density, $\rho \simeq 0.17$ g cm^{-3}, has been reached. Before this occurs, the metallic phase becomes unstable in another way.

The equation of state for metallic hydrogen at zero temperature may be approximated by the formula (Hively 1971, p. 47)

$$E = 2.21 R_e^{-2} - 2.716 R_e^{-1} - (0.115 - 0.031 \ln R_e) + 0.07 R_e^{-3/2}\,. \tag{4.7.2}$$

This formula represents the ground-state energy per electron, measured in rydbergs (1 Ry $= e^2/2a_0$), of a lattice of protons embedded in a uniform, degenerate electron gas. The interelectronic distance R_e is defined by $(4\pi/3)R_e^3 n = 1$, and is measured in units of the Bohr radius a_0. The four terms on the right-hand side of formula (4.7.2) represent, respectively, the Fermi energy for free electrons; the electrostatic energy of the lattice; the electron correlation energy (as given by the Pines-Nozieres interpolation formula); and the zero-point energy of the protons.

From this formula, which is approximately valid in the range $1 \lesssim R_e \lesssim 2$, one can calculate the pressure $p = dE/d(n^{-1})$. The pressure vanishes when $\rho \simeq 0.62$ g cm^{-3} and thereafter assumes negative values. Under laboratory conditions, fracture would begin to occur at this point; but because the cosmic expansion is uniform and isotropic, shear stresses are absent and the onset of fracture may be delayed. However, when the density has dropped to a value $\rho \simeq 0.31$ g cm^{-3}, its (isothermal) compressibility $[\propto dp/d(n^{-1})]$ becomes negative. The medium is now thermodynamically unstable and must shatter into fragments.

Although Hively's calculations support the picture just sketched, it would be premature to exclude alternative histories. For example, suppose that, contrary to theoretical expectations, the metal-insulator transition occurs as soon as the molecular solid becomes the thermodynamically favored phase. The preceding discussion of fragmentation remains qualitatively valid. The pressure now vanishes at density $\rho = 0.089$ g cm^{-3}. A short time later the isothermal compressibility vanishes, the uniform state becomes thermodynamically unstable, and the medium shatters into stable solid fragments—a history first suggested by Zel'dovich (1963).

Again, it is conceivable that, instead of solidifying, the cosmic medium remains in the liquid metallic phase. The liquid medium would then shatter into liquid fragments, like an oil film on the surface of an expanding balloon.

Although these histories differ in important respects, they all predict the formation of large-amplitude, large-scale density fluctuations at epochs $\simeq 1\text{–}3 \times 10^3$ s.

4.8. Thermal Fluctuations Revisited

Zel'dovich pointed out that a "gas" composed of macroscopic solid-hydrogen fragments would have much larger thermal fluctuations than an ordinary gas. A fluctuation of galactic mass ($\simeq 10^{11}\, M_\odot$) can begin to grow at an epoch $t \simeq 10^5$ s, when it would just fill the visible universe. Between that epoch and the epoch of galaxy formation ($\lesssim 10^{17}$ s), its amplitude could increase by a factor $(10^{17}/10^5)^{2/3} = 10^8$. If the initial fluctuation is to be of thermal origin, it must contain fewer than 10^{16} "particles" (so that $N^{-1/2} > 10^{-8}$); hence the mass of a "particle" must exceed $10^{-5}\, M_\odot$.

Now, the masses of fragments into which the solid (or liquid) medium shatters is limited by the cohesion energy of the solid (or liquid) phase. The internal kinetic energy of a newly formed, expanding fragment is proportional to the fifth power of its radius, while its cohesion energy increases only as the cube of the radius. If a fragment is too large, cohesion cannot check its expansion and it will break up into smaller fragments. The maximum mass $M = (4\pi/3)\rho R^3$ of a fragment is determined by the condition that the specific binding energy ϵ_0 equal the expansion energy per unit mass of a sphere of radius R:

$$E_0 = \tfrac{3}{5}R^2H^2 = (8\pi G/5)\rho R^2 . \tag{4.8.1}$$

The binding energy and zero-pressure density of the molecular solid are given by

$$E_0 = 3 \times 10^{-4}(e^2/2a_0m_p) , \qquad \rho = 0.089 \text{ g cm}^{-3} , \tag{4.8.2}$$

and the corresponding value of M is

$$M = 5 \times 10^{25} \text{ g} = 2.5 \times 10^{-8} M_\odot . \tag{4.8.3}$$

This mass is too small by almost three orders of magnitude.

For the metallic solid phase, the specific binding energy and zero-pressure density, as calculated by Hively, are

$$E_0 = 0.03(e^2/2a_0m_p) , \qquad \rho = 0.65 \text{ g cm}^{-3} . \tag{4.8.4}$$

The corresponding value of M is

$$M = 2 \times 10^{28} \text{ g} = 10^{-5} M_\odot . \tag{4.8.5}$$

This result is encouraging, but there is a difficulty.

The preceding argument assumes that $\delta\rho/\rho$ is of order $N^{-1/2}$ for a thermal fluctuation of mass NM. Yet immediately before the onset of fragmentation, thermal density fluctuations were smaller by a factor $(M/m_p)^{1/2} \simeq 10^{26}$. Now, the time t required for thermal motions of order v to produce a fluctuation of order $(nL^3)^{1/2}$ in a region of diameter L is given by

$$(nL^3)^{1/2} = (nL^2vt)^{1/2}, \tag{4.8.6}$$

or $t = L/v$. For fluctuations of galactic mass, $L = cH^{-1}$ initially. Hence $t = (c/v)H^{-1}$; the time required to produce thermal fluctuations of the assumed magnitude greatly exceeds the expansion time. In fact, the initial amplitude of fluctuations of galactic mass is not $N^{-1/2}$ but $(v/c)^{1/2}N^{-1/2}$. Since $v \simeq 10^5$ cm s^{-1}, thermal fluctuations in the fragment-gas are nearly three orders of magnitude too small to explain the condensation of protogalaxies, even if the metallic-hydrogen fragments are as massive as we have supposed.

4.9. Gravitoturbulence

The immediately preceding discussion of thermal fluctuations rests on the implicit assumption that the cosmic medium after fragmentation behaves like an ordinary gas far from its critical point. The following considerations suggest that the cosmic medium, at this stage of its evolution, may behave quite differently from an ordinary gas.

Unlike an ordinary gas or an ionized plasma, a gravitating gas has no characteristic interaction distance. In the statistically uniform state, the characteristic scale of local gravitational fields is comparable to the interparticle distance; but if macroscopic density fluctuations are present, local gravitational fields are also present at corresponding scales, in accordance with Poisson's equation $\phi_k = 4\pi G k^{-2}\tilde{\rho}_k$. The range of interaction scales extends from the interparticle distance to the diameter of the visible Universe. If the internal energy of a uniform gravitating gas is strongly positive, its behavior may not differ significantly from that of an ordinary gas. Macroscopic density fluctuations need not be present, and if they are present at a given instant, they need not persist. On the other hand, if $e < 0$, the formula

$$e = e^{\text{kin}} + e^{\text{grav}} = \tfrac{1}{2}\langle\rho v^2\rangle - 2\pi G\langle\widetilde{\rho^2}\rangle l^2 \tag{4.9.1}$$

shows that both $\langle\tilde{\rho}^2\rangle$ and l^2 must be finite. Thus the condition $e < 0$ is both necessary and sufficient for persistent clustering. Astronomical observations strongly suggest that $e < 0$ now. Yet even under the present assumption that the initial state had zero temperature, the internal energy density must have been strongly positive during the earliest stages of the cosmic expansion, owing to zero-point motions. How did the transition to a state of negative internal energy take place?

As the Universe expands, e decreases; and just before fragmentation it becomes negative by virtue of the electrostatic cohesive forces. After fragmentation, however, cohesion contributes to the internal energy of the fragments but not to the internal energy of the fragment-gas as given by equation (4.9.1). At the "moment" of fragmentation, the fragments have zero peculiar velocity and uniform spatial distribution; hence $e = 0$.

As the fragments separate, a fluctuating gravitational potential ϕ of magnitude $\sim GM/L$ and scale L comes into being, and the fragments acquire peculiar velocities of order $v \simeq (GM/L)^{1/2}$, where M is the mass of a typical fragment and L the interfragment separation.

Now, the energy-balance equation

$$d\epsilon/dt + 3Hp/\bar{\rho} = d(\epsilon^{\text{kin}} + \epsilon^{\text{grav}})/dt + H(2\epsilon^{\text{kin}} + \epsilon^{\text{grav}}) = 0 \tag{4.9.2}$$

shows that the specific energy $\epsilon = e/\bar{\rho}$ diminishes while $p > 0$. But the magnitude of ϵ^{grav} cannot fall below GM/L, and the adiabatic cooling rate for ϵ^{kin} is twice the adiabatic cooling rate for ϵ^{grav}. Thus after a time of order H^{-1} from the onset of fragmentation, ϵ^{kin} and ϵ^{grav} are given by

$$\epsilon^{grav} \simeq -2\epsilon^{kin} \simeq 2\epsilon = -GM/L . \qquad (4.9.3)$$

The internal energy of the "gas" is now strongly negative, and it is energetically possible for persistent large-scale density fluctuations to form. The following considerations suggest that this is in fact what happens next.

The thermodynamic inequality

$$(\partial p/\partial \rho)_T > 0 \qquad (4.9.4)$$

is a necessary condition for the stability of a uniform medium in thermodynamic equilibrium. If it should fail, the medium would become unstable against the spontaneous growth of local density fluctuations at all linear scales. In an ordinary vapor the isothermal compressibility vanishes at the critical point; the resulting instability gives rise to the phenomenon of opalescence.

We may calculate the isothermal compressibility of a gravitating gas from the formula (see eq. [3.2.5]):

$$p = \tfrac{1}{3}\langle \rho v^2 \rangle - \frac{2\pi}{3} G\langle \widetilde{\rho^2} \rangle l^2 . \qquad (4.9.5)$$

By analogy with the kinetic theory of gases, we equate the temperature to a constant multiple of the mean kinetic energy per unit mass:

$$T \propto \bar{\rho}^{-1}\langle \rho v^2 \rangle . \qquad (4.9.6)$$

Now consider a "local" compression whose linear scale greatly exceeds the fluctuation scale l. Under such a compression

$$\delta(l^3\bar{\rho}) = \delta(\widetilde{\rho}/\bar{\rho}) = 0 . \qquad (4.9.7)$$

Hence

$$\bar{\rho}(\partial p/\partial \bar{\rho})_T = \tfrac{1}{3}\langle \rho v^2 \rangle - \tfrac{4}{3} \times \frac{2\pi}{3} G\langle \widetilde{\rho^2} \rangle l^2 . \qquad (4.9.8)$$

With the help of equation (4.9.5) and (4.9.1), we can rewrite this equation:

$$\bar{\rho}(\partial p/\partial \bar{\rho})_T = p + \tfrac{1}{9}e^{grav} . \qquad (4.9.9)$$

Since $e^{grav} < 0$, it follows from the preceding formula that the pressure must exceed a finite positive value,

$$p_c = -\tfrac{1}{9}e^{grav} , \qquad (4.9.10)$$

if the "gas" is to be stable against the growth of density fluctuations. On the other hand, the cosmological virial theorem (§ 3.3) indicates that p tends to relax to the value zero. Suppose that, as a result of the relaxation process discussed in § 3.3, the pressure falls below p_c, so that the right-hand side of equation (4.9.9) goes negative. The resulting instability will give rise to large pressure gradients and to correspondingly large density fluctuations and motions. If the time scale for the development of the instability is short compared with the expansion time, the resulting changes in e^{kin} and e^{grav} satisfy the condition

$$\delta e = \delta e^{\text{kin}} + \delta e^{\text{grav}} = 0 , \tag{4.9.11}$$

and the corresponding pressure change will be given by

$$\delta p = \tfrac{2}{3}\delta e^{\text{kin}} + \tfrac{1}{3}\delta e^{\text{grav}} = \tfrac{1}{3}\delta e^{\text{kin}} > 0 . \tag{4.9.12}$$

Thus the instability causes the pressure to increase and thus tends to lift the instability.

These considerations suggest that the "gas" will relax into a self-regulating turbulent state, analogous to the state of an atmosphere that is unstable against convection, the pressure remaining slightly below its critical value. The condition $p \simeq p_c$ implies that

$$p \simeq -\tfrac{1}{3}e . \tag{4.9.13}$$

Inserting this formula into the energy-balance equation

$$d\epsilon/dt + 3Hp/\bar{\rho} = 0 , \tag{4.9.14}$$

we obtain

$$\epsilon \propto S \propto t^{2/3} . \tag{4.9.15}$$

This condition is analogous to the constant-flux condition in a convective atmosphere. It controls the total quantity of energy resident in the turbulent field and thus determines the pressure difference $(p_c - p)$ that drives the turbulence.

It is crucial for the preceding discussion that the cosmic medium be capable of transmitting pressure forces. However, a "gas" composed of gravitating particles is essentially collisionless, because its relaxation time is comparable to or greater than the expansion time. Now, the preceding considerations indicate that gravitational interactions and macroscopic motions make the principle contributions to the energy-balance equation. In this respect the cosmic medium resembles an ideal gravitating gas. But this model may be inadequate for discussing the dynamics of the medium. In particular, if a substantial fraction (by mass) of the cosmic medium after fragmentation were in the form of small solid grains or gas molecules, it would be capable of transmitting pressure forces in much the same way as an ordinary gas.

The state of the cosmic medium after fragmentation depends on the initial mass spectrum of the fragments, the competition between accretion and collisional disruption, and the specific binding energy of the metallic-hydrogen fragments. Consider the last point. As calculated from equation (4.7.2), the specific binding energy of metallic hydrogen at zero pressure is less than the specific binding energy of the dispersed atoms. Hively (1971) has reviewed and extended earlier calculations and has concluded that the zero-pressure binding energy of the metallic solid relative to the dispersed atoms may in fact be positive. However, the calculated difference ($0.03\ m_p^{-1}$ Ry $\text{g}^{-1} \simeq 4 \times 10^{11}$ ergs g^{-1}) is small, and the calculation depends on approximations whose validity cannot be assessed with complete confidence. Thus we cannot rule out the possibility that the metallic-hydrogen fragments would begin to disintegrate immediately after separating out. Even if the newly formed fragments are stable, enough matter may be in highly divided form to permit the application of thermodynamic considerations.

A closely related question concerns the time available for the growth of large-scale density fluctuations. The motions responsible for the growth of density fluctuations are driven by pressure gradients. The preceding discussion tacitly assumes that the time required to establish the spectrum of density fluctuations is small compared with the expansion time. Now, first-order perturbation theory predicts that only fluctuations of dimensions $L < (p/\rho)^{1/2}H^{-1}$ can grow appreciably in a period less than the expansion time H^{-1}. Yet we have assumed that fluctuations with dimensions $L \simeq ct = cH^{-1}$ are present. Of course, in the cold universe $(p/\rho)^{1/2} \ll c$.

This difficulty is not, however, peculiar to the cold universe; for consider fluctuations that do satisfy the requirement $L \lesssim aH^{-1}$, where a denotes the speed of sound in the cosmic medium. Such fluctuations cannot satisfy the Jeans criterion for gravitational instability, which may be written in the form $L \gtrsim aH^{-1}$.

The resolution of this dilemma depends on the nonlinear character of gravitoturbulence. Consider the equation of continuity

$$\partial_t s + \nabla \cdot \boldsymbol{v} + \nabla \cdot (s\boldsymbol{v}) = 0 \tag{4.9.16}$$

and its Fourier transform

$$\partial_t s_{\boldsymbol{k}} + \boldsymbol{k} \cdot \boldsymbol{v}_{\boldsymbol{k}} + \boldsymbol{k} \cdot \int s_{\boldsymbol{k}'} \boldsymbol{v}_{\boldsymbol{k}-\boldsymbol{k}'} d^3k' = 0 \,. \tag{4.9.17}$$

If the fluctuations at high wavenumbers have sufficiently large amplitudes, the nonlinear i nteractions represented by the third term on the left-hand side of equation (4.9.17) wil control the growth of the low-wavenumber components of s. If the high-wavenumber components have amplitudes of order unity, the low-wavenumber components can have large growth rates. In fact, the initial amplitude of fluctuations whose scale is comparable to the distance between fragments is of order unity.

4.10. Generalized Pressure and Internal Energy

In the preceding discussion we tried to distinguish—not entirely consistently—between the fluctuations causing the instability and those resulting from it. To avoid having to maintain this artificial distinction, we seek to define pressure and internal energy not merely for scales much greater than the maximum scale of local irregularities, but for all scales. The simplest procedure is to consider Fourier components of the velocity and density fields with wavenumbers greater than k as contributing to the internal energy and pressure at wavenumber k. Thus we write

$$P_k = \int_k^\infty p_k dk \,, \qquad E_k = \int_k^\infty e_k dk \,. \tag{4.10.1}$$

We have already defined the gravitational parts of e_k and p_k. The kinetic parts may be defined in an analogous way in terms of the Fourier transform of the field $\rho^{1/2}\boldsymbol{v}$.

We can now apply the stability criterion (4.9.6) at any wavenumber:

$$(\partial P_k/\partial \rho)_{T_k} > 0 \,, \tag{4.10.2}$$

where the temperature T_k is defined by

$$P_k{}^{\text{kin}} \propto \bar{\rho} T_k \,. \tag{4.10.3}$$

Instead of equation (4.9.10), we now have

$$\begin{aligned} \bar{\rho}(\partial P_k/\partial \rho)_{T_k} &= P_k{}^{\text{kin}} + \tfrac{4}{3} P_k{}^{\text{grav}} \\ &= P_k + \tfrac{1}{3} P_k{}^{\text{grav}} = P_k + \tfrac{1}{9} E_k{}^{\text{grav}} \,. \end{aligned} \tag{4.10.4}$$

If relaxation occurs at wavenumber k, then

$$P_k \to 0 . \tag{4.10.5}$$

From this point onward, the discussion is identical with that of § 4.9, except that E_k occurs in place of e and P_k in place of p.

4.11. The Fluctuation Spectrum and the Binding-Energy Spectrum

The Ornstein-Zernike theory of critical-point fluctuations (Landau and Lifshitz 1958, p. 366), based wholly on thermodynamic considerations, predicts the following universal forms for the density autocorrelation functions and its Fourier transform in a gas of vanishing isothermal compressibility:

$$f_\rho(r) \propto r^{-1} , \quad F_\rho(k) \propto k^{-2} . \tag{4.11.1}$$

These formulae are valid (insofar as the underlying theory is valid) for a finite range of linear scales and wavenumbers. For linear separations smaller than the interparticle distance, the density autocorrelation function may be set equal to unity; for linear separations exceeding the diameter of the container (in the present application, the visible Universe), it may be set equal to zero:

$$\begin{aligned} f_\rho(r) &= 1 && (r < r_{\min}) \\ &= r_{\min}/r && (r_{\min} < r \leq r_{\max}) \\ &= 0 && (r_{\max} < r) , \end{aligned} \tag{4.11.2}$$

where

$$(4\pi/3)r_{\min}^3 n = 1 , \quad r_{\max} = ct . \tag{4.11.3}$$

The corresponding spectrum function is given by

$$F_\rho(k) = (2\pi)^{-2} r_{\min} k^{-2} [(kr_{\min})^{-1} \sin (kr_{\min}) - \cos (kr_{\max})] . \tag{4.11.4}$$

The low-wavenumber limit $F_\rho(0)$ is given by

$$F_\rho(0) = (2\pi)^{-2} r_{\min} r_{\max}^2 . \tag{4.11.5}$$

For wavenumbers $k \gg r_{\max}^{-1}$, the term $\cos (kr_{\max})$ in formula (4.11.4) may be neglected; it arises from the artificial discontinuity at $r = r_{\max}$ in formula (4.11.2). Similarly, the long-period oscillations in $F_\rho(k)$ at values of $k > \pi r_{\min}^{-1}$ are an artifact introduced by the assumed form of the autocorrelation function $f_\rho(r)$ for $r < r_{\min}$. The essential features of $F_\rho(k)$ are adequately represented by the formula

$$F_\rho(k) = (2\pi k_{\max})^{-1}(k^2 + \kappa^2)^{-1} , \tag{4.11.6}$$

where

$$k_{\max} = \pi r_{\min}^{-1} , \quad \kappa^2 = 2(ct)^{-2} . \tag{4.11.7}$$

Finally, the clustering scale l is given by

$$\begin{aligned} l^2 &= \int_0^\infty f_\rho(r) r dr = \int_0^\infty F_\rho(k) dk \\ &= r_{\min} r_{\max} = (4k_{\max}\kappa)^{-1} . \end{aligned} \tag{4.11.8}$$

Of greater interest than the spectrum of density fluctuations is the gravitational-energy spectrum, defined by

$$e^{\text{grav}} = -(8\pi G)^{-1}\langle|\nabla\psi|^2\rangle$$

$$= \int_0^\infty e_k^{\text{grav}} dk \,. \tag{4.11.9}$$

Thus

$$4\pi k^2 \delta(\boldsymbol{k} + \boldsymbol{k}') e_k^{\text{grav}} = (8\pi G)^{-1}(\boldsymbol{k}\cdot\boldsymbol{k}')\langle\psi_{\boldsymbol{k}}\psi_{\boldsymbol{k}'}\rangle \,. \tag{4.11.10}$$

To evaluate the quantity $\langle\psi_{\boldsymbol{k}}\psi_{\boldsymbol{k}'}\rangle$, we note that

$$\psi_{\boldsymbol{k}} \simeq -4\pi G k^{-2}\tilde{\rho}_{\boldsymbol{k}} \qquad (kct \gg 1)$$

$$\simeq -\tfrac{1}{3}s_{\boldsymbol{k}} \qquad (kct \ll 1) \,, \tag{4.11.11}$$

where $s_{\boldsymbol{k}} = \tilde{\rho}_{\boldsymbol{k}}/\bar{\rho}$. (One can establish the second asymptotic formula by studying the behavior of $\psi_{\boldsymbol{k}}$ and $s_{\boldsymbol{k}}$ in the neighborhood of $k = 0$.) An interpolation formula exhibiting the correct asymptotic behavior at both large and small wavenumbers is

$$\psi_{\boldsymbol{k}} = -\frac{4\pi G\tilde{\rho}_{\boldsymbol{k}}}{k^2 + \kappa^2} \,. \tag{4.11.12}$$

Inserting this formula into equation (4.11.10), one obtains, with the help of equation (2.6.10),

$$e_k^{\text{grav}} = -8\pi^2 G\langle\tilde{\rho}^2\rangle F_\rho(k)[1 + (\kappa/k)^2]^{-2} \,. \tag{4.11.13}$$

Using formula (4.11.6) to represent $F_\rho(k)$, we obtain finally

$$e_k^{\text{grav}} = -2\pi G\langle\tilde{\rho}^2\rangle \mathfrak{F}(k;\kappa, k_{\max}) \,, \tag{4.11.14}$$

$$\mathfrak{F}(k;\kappa, k_{\max}) = 2k_{\max}^{-1}k^{-2}[1 + (\kappa/k)^2]^{-3} \,. \tag{4.11.15}$$

Note that a factor 4π has been absorbed in the definition of $\mathfrak{F}$.

Integrating equation (4.11.4), we obtain

$$e_k = -2\pi G\langle\tilde{\rho}^2\rangle l_g^2 \tag{4.11.16}$$

where, by formula (4.11.8),

$$l_g^2 = (3\pi/8)(k_{\max}\kappa)^{-1} = \tfrac{3}{8}l^2 \,. \tag{4.11.17}$$

Formula (4.11.14) represents the gravitational-energy spectrum in the range $k_{\min} \lesssim k \lesssim \kappa$. The predictions that e_k^{grav} varies like k^{-2} for $k \gg (ct)^{-1}$ and peaks at a value $k \simeq (ct)^{-1}$ do not depend on the precise forms of the interpolation formulae (4.11.10) and (4.11.12), which merely enable us to express these features in a mathematically convenient way. The behavior of e_k^{grav} at low wavenumbers, $k \ll (ct)^{-1}$, is not predicted by the present theory; presumably it falls off quite rapidly in this range.

4.12. The Gravitoturbulent Regime

The gravitoturbulent state is characterized by strong interactions among the Fourier components of the density. These preserve the form of the clustering spectrum, given by equations (4.11.6) and (4.11.15), while the clustering amplitude α increases monotonically with time. According to equation (4.9.17),

$$\epsilon^{\text{grav}} = -2\pi G\bar{\rho}\alpha^2 l_g^2 \propto S \,. \tag{4.12.1}$$

But

$$l_g^2 \propto l^2 = r_{\min} r_{\max} \propto n^{-1/3} t \propto S^{5/2} . \quad (4.12.2)$$

Since $\bar{\rho} \propto S^{-3}$, it follows from the two preceding equations that

$$\alpha^2 \propto S^{3/2} \propto t . \quad (4.12.3)$$

The gravitoturbulent regime commences at an epoch t_1 when the mean separation between the initial fragments (of mass M and diameter D) is of order $2D$. Assuming that the binding energy per electron of the metallic-hydrogen fragments is of order 1 rydberg per proton and recalling that the fragments begin to separate at $t \simeq 10^3$ s, $\rho \simeq 0.65$ g cm^{-3}, we may write down the following approximate values for relevant variables at the commencement of the gravitoturbulent regime:

$$t \simeq t_1 \simeq 3 \times 10^3 \text{ s} , \quad r_{\max} = ct_1 , \quad \bar{\rho}_1 \simeq 0.08 \text{ g cm}^{-3} .$$

$$r_{\min} = 2D , \quad \epsilon_1^{\text{grav}} \simeq -GM/2D \simeq \tfrac{1}{2}\epsilon_0 ,$$

$$\epsilon_0 \simeq -e^2/2a_0 m_p = 1.3 \times 10^{13} \text{ ergs g}^{-1} ,$$

$$M \simeq 5 \times 10^{30} \text{ g} , \quad D \simeq 2.5 \times 10^{10} \text{ cm} . \quad (4.12.4)$$

We can now calculate the initial value of α^2. Equating the expressions for ϵ^{grav} given by equations (4.12.1) and (4.12.4), we have

$$2\pi G \bar{\rho} \alpha^2 l_g^2 = (3\pi/4) G \bar{\rho} \alpha^2 l^2 = (3\pi/2) G \bar{\rho} \alpha^2 D r_{\max}$$

$$= GM/2D = (\pi/12) G (8 \bar{\rho}) D^2 ,$$

whence

$$\alpha^2 = 4D/9r_{\max} \simeq 1.2 \times 10^{-4} . \quad (4.12.5)$$

Thus the initial clustering amplitude $\alpha \simeq 10^{-2}$.

The first gravitationally bound systems begin to separate out when α becomes of order unity. With the formation of gravitationally bound systems, the cosmic medium ceases to behave like a gas and relaxes into the essentially collisionless, pressure-free state described by the virial theorem (§ 3.3). This epoch, which we denote by t_2, marks the end of the gravitoturbulent phase. For the purposes of the present semiquantitative discussion, we may assume that the transition from the gravitoturbulent state to the pressure-free state occurs abruptly when $\alpha = 1$. Then t_2 is given by

$$t_2 = \alpha^{-2} t_1 \simeq 2.4 \times 10^7 \quad \text{s} . \quad (4.12.6)$$

The mass of the visible Universe is given by

$$M^{\text{vis}} = (2\pi/\kappa)^3 \rho \simeq 2 \times 10^{39} t \quad \text{grams} , \quad (4.12.7)$$

where κ is defined by equation (4.11.7) and we have used the relation $6\pi G \rho t^2 = 1$. Inserting t_2 from equation (4.12.6) into this formula, we obtain

$$M_2^{\text{vis}} = 4.8 \times 10^{46} \text{ g} = 2.4 \times 10^{13} M_\odot . \quad (4.12.8)$$

4.13. The Clustering Process

The clustering spectrum is frozen in at time t_2. This means that its characteristic parameters vary according to

$$k_{\max} , \kappa \propto S^{-1} , \quad l^2 \propto S^2 \quad (t > t_2 \simeq 6 \times 10^7 \text{ s}) , \quad (4.13.1)$$

and that

$$\epsilon = \text{const.}\,, \qquad \alpha^2 = S/S_2 \qquad (t > t_2)\,. \tag{4.13.2}$$

The formula for α^2 follows from the constancy of ϵ and the condition $\alpha(t_2) = 1$, which defines the epoch t_2. The spectrum function $\mathfrak{F}$ scales according to

$$\mathfrak{F}(k, t)dk = (S/S_2)^2\mathfrak{F}^*(k_2)dk_2\,, \tag{4.13.3}$$

where

$$k_2 \equiv (S/S_2)k\,, \qquad \mathfrak{F}^*(k_2) = \mathfrak{F}(k_2; \kappa(t_2), k_{\max}(t_2))\,. \tag{4.13.4}$$

The reduced, or comoving, wavenumber k_2 is associated with a fixed mass (see eq. [4.13.7]) and a fixed internal energy, given by

$$\begin{aligned} E(k) &= -\tfrac{1}{2} \times 2\pi G\bar{\rho}\alpha^2 \int_k^\infty \mathfrak{F}(k', t)dk' \\ &= -\pi G\bar{\rho}_2 \int_{k_2}^\infty \mathfrak{F}^*(k_2')dk_2'\,. \end{aligned} \tag{4.13.5}$$

The energy of an isolated, gravitationally bound system of mass M and uniform density ρ is given by

$$E_{\text{g.b.}} = -0.3(4\pi/3)^{1/3}GM^{2/3}\rho^{1/3}\,. \tag{4.13.6}$$

The mass associated with wavenumber k_2 is defined by the relation

$$M(k_2) = (2\pi/k_2)^3\bar{\rho}_2\,. \tag{4.13.7}$$

We may also express ρ in terms of $\bar{\rho}_2$ through the relation

$$\rho \simeq (1 + \alpha)(S_2/S)^3\bar{\rho}_2 \simeq \alpha(S_2/S)^3\bar{\rho}_2 \simeq (S_2/S)^{5/2}\bar{\rho}_2\,, \tag{4.13.8}$$

valid for $\alpha \gg 1$; this inequality prevails during the period when gravitationally bound systems are forming.

Inserting formula (4.13.7) and (4.13.8) into (4.13.6), we obtain:

$$E_{\text{g.b}} = -19G\bar{\rho}_2k_2^{-2}(S_2/S)^{5/6}\,, \tag{4.13.9}$$

which shows that the binding energy of a gravitationally bound system of reduced wavenumber k_2 decreases as the Universe expands. Since the actual internal energy associated with reduced wavenumber k_2 remains constant, it will ultimately be comparable to the energy of a gravitationally bound system of the same mass and mean density. Systems of mass $M(k_2)$ will then begin to separate out.

By equations (4.13.9) and (4.13.5), the condition $E_{\text{g.b.}} = E(k_2)$ becomes

$$k_2^2 \int_{k_2}^\infty \mathfrak{F}^*(k_2')dk_2' = 6(S_2/S)^{5/6}\,. \tag{4.13.10}$$

The right-hand side of this equation is a monotonically decreasing function of the time. The left-hand side approaches 0 as $k_2 \to 0$ and as $k_2 \to \infty$, and has a single maximum at a finite value of $k_2 = k_2^{(1)}$, given by

$$k_2^{(1)}\mathfrak{F}^*(k_2^{(1)}) = 2\int_{k_2^{(1)}}^\infty \mathfrak{F}^*(k_2)dk_2\,. \tag{4.13.11}$$

Having solved this equation for $k_2^{(1)}$, we can now use equation (4.13.10) to find the value of S— call it $S^{(1)}$— at which the first gravitationally bound systems separate out:

$$(k_2^{(1)})^3\mathfrak{F}^*(k_2^{(1)}) = 12(S_2/S^{(1)})^{5/6}\,. \tag{4.13.12}$$

After the first generation of gravitationally bound systems has been formed, a second generation, characterized by a lower value of the reduced wavenumber, will begin to separate out. The discussion of this process is exactly analogous to the preceding discussion. In place of equation (4.13.10), we have

$$k_2^2 \int_{k_2}^{k_2^{(1)}} \mathfrak{F}^*(k_2')dk_2' = 6(S_2/S)^{5/6} , \tag{4.13.13}$$

where allowance has been made for the fact that the internal energy of systems that have already separated out is no longer available to bind more massive systems. The wavenumber $k_2^{(2)}$ of the most tightly bound second-generation systems is given by

$$k_2^{(2)}\mathfrak{F}^*(k_2^{(2)}) = 2\int_{k_2^{(2)}}^{k_2^{(1)}} \mathfrak{F}^*(k_2)dk_2 , \tag{4.13.14}$$

and the epoch at which they separate out is given by

$$(k_2^{(2)})^3\mathfrak{F}^*(k_2^{(2)}) = 12(S_2/S^{(2)})^{5/6} . \tag{4.13.15}$$

The three preceding equations generalize immediately to any stage of the clustering process. The general form of equation (4.13.14) enables us to express the internal energy of systems at the nth level of the hierarchy in the form (see eq. [4.13.5])

$$E_n = -\tfrac{1}{2}\pi G\bar{\rho}_2 k_2^{(n)}\mathfrak{F}^*(k_2^{(n)}) . \tag{4.13.16}$$

Using equation (4.13.7) to pass from wavenumber to mass as the independent variable and introducing the explicit form of the spectrum function given by equation (4.11.15), we obtain from equation (4.13.16) the following relation between internal energy and mass:

$$E(M_n) = -\frac{c^2}{12}(M_{\min}/M_{\max})^{1/3}(M_n/M_{\max})^{1/3}[1 + (M_n/M_{\max})^{2/3}]^{-3} , \tag{4.13.17}$$

where

$$M_{\min} = M(k_{\max}) , \quad M_{\max} = M(\kappa) \equiv M_2^{\text{vis}} . \tag{4.13.18}$$

The numerical values of $M_{\min}$ and $M_{\max}$ are given by

$$M_{\min} \simeq 6 \times 10^{32}\ \text{g} , \quad M_{\max} \simeq 5 \times 10^{46}\ \text{g} . \tag{4.13.19}$$

Thus

$$E(M_n) = -1.7 \times 10^{15}(M_n/M_{\max})^{1/3}[1 + (M_n/M_{\max})^{2/3}]^{-3}\ \text{ergs g}^{-1} . \tag{4.13.20}$$

The initial mass spectrum is determined by the solutions to equation (4.13.14). For $k \gg \kappa$, $\mathfrak{F} \propto k^{-2}$ so that

$$k_2^{(n)} = \tfrac{1}{2}k_2^{(n-1)} , \quad M_n = 8M_{n-1} \quad (k \gg \kappa, M \ll M_{\max}) . \tag{4.13.21}$$

Thus the levels of the initial hierarchy are closely spaced in mass. The total internal energy per unit mass is given by

$$\begin{aligned} \epsilon &= -\pi G\bar{\rho}_2 \int_0^\infty \mathfrak{F}^*(k)dk = -(3\pi^2/8)G\bar{\rho}_2(k_{\max}\kappa)^{-1} \\ &= -(3\pi/96)(M_{\min}/M_{\max})^{1/3}c^2 \simeq -2 \times 10^{15}\ \text{ergs g}^{-1} . \end{aligned} \tag{4.13.22}$$

4.14. Effects of Evolution; Comparison with Observation

According to the working hypothesis advanced in § 4.1, the hierarchy of gravitationally bound systems as we observe it today has resulted from the interplay of two opposing processes: hierarchic construction through gravitational clustering, which continually generates new bound systems; and disruption through tidal interactions and encounters. In the preceding pages I have sketched a tentative theory of the clustering process. To confront this theory with observation, we need to consider the effects of tidal disruption on newly formed self-gravitating systems. This is a complex and as yet little explored topic, but a few general remarks may help to relate the preceding theoretical considerations to astronomical observations.

4.14.1. *Simplification of the Hierarchy.*—Every gravitationally bound system (after the first generation) is both a cluster and a cluster member. As a member of a larger cluster, it tends to gain energy (and hence lose binding energy) through inelastic encounters with other members of the same cluster, as well as through other kinds of tidal interactions. As a cluster, it tends to gain binding energy from the disruption of its members. Thus binding energy tends to flow from smaller to larger clusters in the hierarchy.

This flow is unstable, however. The disruption of clusters at a given level of the hierarchy tends to retard the disruption of clusters at the adjacent levels, since it causes a reduction in the tidal forces acting on member clusters and an increase in the binding energies of parent clusters. Conversely, increased stability at a given level of the hierarchy has a destabilizing effect on clusters at the adjacent levels. This suggests that tidal interactions will gradually thin out the hierarchy, leaving fewer and fewer more or less evenly spaced levels populated by increasingly stable clusters. The thinning-out process continues until the lifetimes of the surviving clusters are comparable to the expansion time. Thus at a given epoch t, we may expect to find clusters whose dynamical lifetimes are comparable to and greater than t.

This conclusion refers, of course, only to gravitationally bound systems formed by the clustering process discussed in the preceding pages. To account for young stars and star systems, one may either suppose that they are formed by a distinct process from diffuse interstellar material, or that they have evolved from more ancient protosystems. For a discussion of the second alternative, see Layzer (1964, 1967).

4.14.2. *Stars and Galaxies.*—The ability of a gravitationally bound system to withstand tidal disruption depends on the ratio between its specific binding energy and the tidal gravitational potential to which it is subjected. The stablest clusters are those whose specific binding energies lie close to or beyond the peak of the initial binding-energy spectrum (4.13.17). We may identify the systems of maximum specific binding energy with galaxies and galaxy clusters. Galaxy superclusters are then predicted to be less tightly bound than the most tightly bound galaxies and galaxy clusters.

Among systems less massive than galaxies, those capable of radiating away energy efficiently enjoy preferred status. Since the tidal forces acting on a system in a given gravitational field is proportional to its diameter, a system whose gravitational-contraction time is comparable to or shorter than the tidal-disruption time corresponding to its initial diameter has an excellent chance of surviving tidal disruption. Protostars are such systems. Systems of smaller mass are solid, and hence susceptible to disruption if they belong to dense clusters. Ultimately, such clusters may be expected to coalesce into single stars or planetary systems.

These considerations explain the preferred status of stars and galaxies in the hierarchy of astronomical systems. Among systems of intermediate mass, the most stable may be expected to occur in the neighborhood of the median level of the initial hierarchy. Since mass increases geometrically along the hierarchy (see eq. [4.13.21]), the masses of the most stable systems in the intermediate mass range should be close to the geometric mean of stellar and galactic masses: 10^5–$10^6\ M_\odot$. The next most stable systems should lie in the vicinity of the one-quarter and three-quarter points of the initial hierarchy, corresponding to the mass ranges 10^2–$10^3\ M_\odot$ and 10^8–$10^9\ M_\odot$.

4.14.3. *Initial Diameters of Galaxies and Protostars.*—The binding-energy spectrum (4.13.20) determines the initial diameters of gravitationally bound systems through the approximate relation $E \simeq - GM/2D$. The energy spectrum peaks at

$$M\dagger = 5^{-3/2} M_{\rm max} \simeq 4 \times 10^{45}\ {\rm g} \simeq 2 \times 10^{12}\ M_\odot\ ; \qquad (4.14.1)$$

and the corresponding value of the specific binding energy, given by equation (4.13.20), is

$$E\dagger = -4.5 \times 10^{14}\ {\rm ergs\ g^{-1}}\ . \qquad (4.14.2)$$

Thus the initial diameter is

$$D\dagger = -GM\dagger/2E\dagger \simeq 3 \times 10^{23}\ {\rm cm} = 100\ {\rm kpc}\ . \qquad (4.14.3)$$

Similarly, from equation (4.13.20) we obtain the following values for the initial energy and diameter of systems of $10^6\ M_\odot$ and $1\ M_\odot$:

$$M = 1\ M_\odot\!: \quad E = -6 \times 10^{10}\ {\rm ergs\ g^{-1}}\,, \qquad D \simeq 2 \times 10^{15} {\rm cm} = 150\ {\rm a.u.} \qquad (4.14.4)$$

$$M = 10^6\ M_\odot\!: E = -6 \times 10^{12}\ {\rm ergs\ g^{-1}}\,, \qquad D \simeq 2 \times 10^{19}\ {\rm cm} = 7\ {\rm pc}\ . \qquad (4.14.5)$$

In assessing these predictions, one should bear in mind that the underlying theory is semiquantitative only, and that even on this level it involves important uncertainties. On the other hand, the predictions concerning the binding-energy spectrum and its low-mass and high-mass cutoffs do not involve any adjustable parameters; though semiquantitative, they are *ab initio* predictions.

As their internal structure is disrupted by tidal interactions, the binding energies of protogalaxies increase at the expense of the binding energy of their member clusters. The maximum change in binding energy and diameter is given by the factor

$$2 \int_{k\dagger}^{\infty} \mathfrak{F}^*(k) dk / k\dagger \mathfrak{F}^*(k\dagger) = 3.4 \qquad (k\dagger = 5^{1/2})\ . \qquad (4.14.6)$$

For less massive systems, the maximum contraction factor is of order 2.

4.14.4. *Origin of Galactic Spin.*—When protogalaxies first separated out of the expanding cosmic medium, their mutual tidal interactions would have generated a statistical distribution of spins even if the medium had previously been substantially free from vorticity. However, this attractive explanation of galactic spin, first proposed by Hoyle (1949) and subsequently studied by Field (1968), Peebles (1969), Oort (1969, 1970), and Harrison (1971), seems unable to account quantitatively for the high spin-to-mass ratios characteristic of spiral galaxies. Thus under highly favorable assumptions about the transfer of angular-momentum between interacting protogalaxies, the estimated spin-to-mass ratio for the Galaxy falls short of the observed value by an order of magnitude, while more conservative estimates fall several orders of magnitude short of the mark (Harrison 1971).

Cosmogonic theories that postulate primordial vorticity are obviously capable of predicting higher spin-to-mass ratios, but they encounter other difficulties. To begin with, such theories merely defer the problem of accounting for galactic spin. Moreover, there is reason to question whether primordial vorticity could survive the initial phases of the cosmic expansion. The last point has been discussed by several authors, whose work is reviewed by Harrison (1971).

As discussed above, tidal interactions may be expected to obliterate much of the detailed hierarchic substructure initially imprinted on galaxies by the clustering process. However, one important trace of this initial structure may be expected to have survived: If a small number of subsystems come together to form a larger system, the resultant angular momentum is likely to be large. Thus if the composite system has mass M and diameter D, the resultant angular momentum is likely to be of order $n^{-1/2}MVD$, where $V^2 = GM/D$. The probability distribution of resultant angular momentum is given by the elementary theory of random walks. Since $n \simeq 8$, typical values of galactic spin predicted by these considerations are less than an order of magnitude smaller than MVD, in agreement with estimates based on observation.

These considerations predict a negative correlation between galactic mass and spin-to-mass ratio, because the mass M increases on the average with n. On the other hand, the range of galactic masses probably extends over two or more levels of the initial hierarchy, so that M and n should not, according to the present theory, be *strongly* correlated.

It is commonly assumed that the morphological differences between ellipsoidal and spiral galaxies result chiefly from the higher spin-to-mass ratio of the latter. Although this view seems likely to be correct, we are not justified in drawing the additional inference that the abrupt changes in morphological properties marking the transition from ellipsoidal to spiral galaxies imply a bimodal distribution of initial spin-to-mass ratios. To deduce the distribution of initial spin-to-mass ratios in galaxies we need a more complete theory of galactic evolution than is currently available. All that can be said with some confidence at present is that the distribution seems to be rather broad and that values of order VD are not uncommon. The theory sketched above affords a simple and natural explanation for these features.

4.15. Star Formation and the Microwave Background

The preceding theory suggests that the preferred mass range for protostars corresponds to a small range of gravitational binding forces just sufficient to overwhelm the solid-state cohesive forces of the solid fragments. The preceding theory does not afford a reliable prediction of this mass range.[3] Nor is the present mass spectrum of stars a reli-

[3] The melting temperature of metallic hydrogen at the calculated zero-pressure density $\rho \simeq 0.65$ g cm^{-3}, as predicted by the Lindemann criterion $T_m = 1.9 \times 10^3\ \rho^{1/3}$ ° K, is $T_m \simeq 1.6 \times 10^3$ ° K. The predicted temperature of a protocluster of mass M, given by equation (4.13.20) and the virial theorem, is

$$T \simeq (2m_p/3k) \times 6 \times 10^{10}(M/M_\odot)^{1/3}$$
$$\simeq 4 \times 10^2(M/M_\odot)^{1/3} .$$

Thus $T > T_m$ if $M \gtrsim 60\ M_\odot$; but this calculation is very uncertain.

able indicator of the primordial mass spectrum, since we cannot directly observe first-generation stars whose initial masses substantially exceeded 1 $M_{\odot}$, and there is no reason to suppose that the stars that happen to be visible now are representative of the primordial stellar population.

Suppose that the preferred mass range for stars was 5–10 $M_{\odot}$. Stars in this mass range would have radiated most of their energy between $\sim 3 \times 10^7$ and $\sim 10^8$ years after the beginning of the cosmic expansion. According to current theories of stellar evolution, such stars would have converted a significant fraction of their mass into heavy elements and would have passed through an explosive phase. Layzer and Hively (1973) have elaborated the hypothesis that the light emitted by this early generation of massive stars was subsequently thermalized by dust grains formed from heavy elements synthesized in the same stars, and that this thermalized radiation is the cosmic microwave background. The observed energy density, thermal spectrum, and high degree of isotropy of the microwave background impose severe constraints on any nonprimordial theory of its origin. These requirements can apparently be met, however, without straining probabilities. The energy produced by the stars is adequate, and it is produced at a time when the Universe could have been sufficiently opaque to long-wavelength radiation to account for the observed degree of thermalization and isotropy under reasonable assumptions about grain size and composition. For the detailed argument, see the paper cited above.

The hypothesis that the preferred mass range for protostars was 5–10 $M_{\odot}$ has other observable consequences. It predicts, for example, that a large fraction of the mass of the Universe resides in burnt-out supernova cores. This population of essentially nonluminous collapsed objects could perhaps resolve the dilemma of the "missing mass" and related dynamical anomalies. Thus the theory of gravitational clustering sketched in the preceding pages offers the possibility of constructing a unified explanation for diverse aspects of cosmic structure.

APPENDIX

We summarize here some properties of probability-generating functions needed in § 2.

Definition. Probability-generating functions are useful for positive-valued random variables. If the random variable u has a discrete set of possible values $a_k > 0$, we define

$$G_u(t) = \sum_{k=0}^{\infty} p_k t^{a_k}, \qquad p_k = P\{u = a_k\} . \tag{A.1}$$

If the possible values are continuous, we define

$$G_u(t) = \int_0^{\infty} p(x) t^x dx . \tag{A.2}$$

If the possible values are both positive and negative, it is more convenient to use the characteristic function.

Moments.

$$\langle u^n \rangle = \left[\left(t \frac{d}{dt} \right)^n G_u(t) \right]_{t=1} . \tag{A.3}$$

This formula follows at once from the definition of $G_u(t)$.

The random variables u, v are independent if, for all pairs of possible values,

$$P\{u = a, v = b\} = P\{u = a\}P\{v = b\} . \tag{A.4}$$

It follows from this definition and the definition of G that

$$G_{u+v}(t) = G_u(t)G_v(t) . \tag{A.5}$$

If c is a constant

$$G_{cu}(t) = G_u(t^c) . \tag{A.6}$$

The product of two independent random variables has the generating function

$$G_{uv}(t) = G_u[G_v(t)] = G_v[G_u(t)] . \tag{A.7}$$

This formula and the preceding one follow directly from the definition of G.

If u, v are dependent (or independent) random variables, their joint generating function is defined by

$$G_{u,v}(t, t') = \sum_{a,b} P\{u = a, v = b\} t^a t'^b \tag{A.8}$$

and

$$\langle u^m v^n \rangle = \left[\left(t\frac{d}{dt}\right)^m \left(t'\frac{d}{dt'}\right)^n G_{u,v}(t, t')\right]_{t=t'=1} . \tag{A.9}$$

REFERENCES

Abell, G. O. 1958, *Ap. J. Suppl.*, **3**, 211.
———. 1965, *Ann Rev. Astr. and Ap.*, **3**, 1.
Bok, B. J. 1934, *Harvard Bull.*, No. 895.
Bonnor, W. B. 1956, *Zs. f. Ap.*, **39**, 143.
———. 1957, *M.N.R.A.S.*, **177**, 104.
Burke, J. R. 1971, Harvard Ph.D. thesis, "On Gravitational Instability in Friedmann Cosmologies."
Charlier, C. V. L. 1908, *Ark. f. Mat., Astr., Fys.*, **4**, 1.
Danziger, I. J. 1970, *Ann. Rev. Astr. and Ap.*, **8**, 167.
Field, G. B. 1965, *Ap. J.*, **142**, 531.
———. 1968, in *Stars and Stellar Systems*, Vol. 9, *Galaxies and the Universe*, ed. A. and M. Sandage and J. Kristian (Chicago: University of Chicago Press, 1975), chap. 10.
Harrison, E. R. 1971, *M.N.R.A.S.*, **154**, 167.
———. 1973, *Ann. Rev. Astr. and Ap.*, **11**, 155.
Hauser, M. G., and Peebles, P. J. E. 1973, *Ap. J.*, **185**, 757.
Hively, R. 1971, Harvard Ph.D. thesis, "Cosmology of a Cold Universe."
Holmberg, E. 1975, in *Stars and Stellar Systems*, Vol. 9, *Galaxies and the Universe*, ed. A and M. Sandage and J. Kristian (Chicago: University of Chicago Press, 1975), chap. 4.
Hoyle, F. 1949, *Problems of Cosmical Aerodynamics* (Dayton: International Union of Theoretical and Applied Mechanics and Int. Astr. Union, Central Air Documents Office), p. 195.
Hubble, E. P. 1936, *The Realm of the Nebulae* (New Haven: Yale University Press).
Jeans, J. H. 1902, *Phil. Trans.*, **199A**, 49.
———. 1961 (1928), *Astronomy and Cosmogony* (3d ed.; New York: Dover Publications), p. 345.
Karpowicz, M. 1967, *Zs. f. Ap.*, **66**, 301.
Kaufman, M. 1968, Harvard Ph.D. thesis, "Thermal History of the Early Stages of a Matter-dominated Universe."
———. 1970, *Ap. J.*, **160**, 459.
Landau, L. D. and Lifshitz, E. M. 1958, *Statistical Physics* (London: Pergamon Press), p. 366.
Layzer, D. 1956, *A.J.*, **61**, 383.
———. 1963, *Ap. J.*, **138**, 174.
———. 1964, *Ann. Rev. Astr. and Ap.*, **2**, 341.
———. 1967, *Mém. Soc. Roy. Sci. Liège*, **15**, 135.
———. 1968, *Ap. Letters*, **1**, 99.
———. 1971, "Cosmogonic Processes," in *Astrophysics and General Relativity*, Vol. 2, edited by M. Chrétien, S. Deser, and J. Goldstein (New York: Gordon and Breach, 1971).
Layzer, D. and Hively, R. 1973, *Ap. J.*, **179**, 361.
Lifshitz, E. M. 1946, *J. Phys. USSR*, **10**, 116.
Limber, D. N. 1957, *Ap. J.*, **125**, 9.

Newton, I. 1692–1693, "Four Letters to Richard Bentley" in *Theories of the Universe*, ed. M. K. Munitz (New York: Free Press), p. 211.
Neyman, J. and Scott, E. L. 1959, *Handbuch der Physik* (Berlin: Springer Verlag), p. 53.
Novikov, I. D. and Zel'dovich, Ya. B. 1973, *Ann. Rev. Astr. and Ap.*, **11**, 387.
Oort, J. H. 1969, *Nature*, **224**, 1158.
———. 1970, *Astr. and Ap.*, **7**, 381.
Peebles, P. J. E. 1969, *Ap. J.*, **155**, 393.
———. 1972, *Physical Cosmology* (Princeton: Princeton University Press).
———. 1973, *Ap. J.*, **185**, 413.
———. 1974, *Ap. J. Suppl.*, **28**, 37.
Peebles. P. J. E. and Hauser, M. G. 1974, *Ap. J. Suppl.*, **28**, 19.
Sandage, A. 1967, *Ap. J.* (*Letters*), **150**, L9, L177.
———. 1968, ibid., **152**, L149.
Shane, C. D., and Wirtanen, C. A. 1954, *A.J.*, **59**, 285.
Vaucouleurs, G. de. 1953, *A.J.*, **58**, 30.
Weinberg, S. 1972, *Gravitation and Cosmology* (New York: John Wiley and Sons).
Yu, J. T. and Peebles, P. J. E. 1969, *Ap. J.*, **158**, 103.
Zel'dovich, Ya. B. 1963, *Soviet Phys. M.J.E.T.P.*, **16**, 1395.
Zwicky, F. 1938, *Pub. A.S.P.*, **50**, 218.
Zwicky, F. and Rudnicki, K. 1966, *Zs. f. Ap.*, **64**, 246.

CHAPTER 18

Radio Astronomy and Cosmology

PETER A. G. SCHEUER
Mullard Radio Astronomy Observatory, University of Cambridge

1. INTRODUCTION

THERE are three distinct ways in which radio observations can be used in attempts to test cosmological models: by determining limits for the mean density of the Universe (1.1); by measurement of any residual radiation which may be left over from an early highly condensed state, in an exploding cosmology (1.2); by studying the development of the radio source population (1.3).

1.1. THE MEAN DENSITY OF THE UNIVERSE

A large class of cosmological models require that the mean density of the Universe should be

$$\rho_0 = \sigma_0 \frac{3H^2}{4\pi G} = 3.76 \times 10^{-29} \sigma_0 \text{ g cm}^{-3} \tag{1}$$

(H = Hubble's constant, taken as 100 km s^{-1} Mpc^{-1} throughout this chapter, G = gravitational constant), where σ_0 is a numerical constant of order 1. The quantity ρ_0 is equivalent to a mean density of 2.26×10^{-5} σ_0 hydrogen atoms cm^{-3}. Such a density is about two orders of magnitude greater than the mean density of matter in galaxies, averaged over all space (Oort 1958). A part of the discrepancy may be accounted for by the intergalactic material in clusters of galaxies which must be postulated if the clusters are gravitationally bound. The contribution of quasi-stellar objects (Sandage 1965) is quite negligible unless QSOs are at least 10^5 times more massive than galaxies. Thus, these cosmological models require that most of the matter should be intergalactic, with a density indicated by equation (1). It is generally assumed that this extragalactic matter is a gas, and that most of it consists of hydrogen; there is no such *a priori* confidence about the temperature and the state of ionization of the gas.

Received January 1973. This chapter, originally written in 1965, was extensively rewritten in 1972, but parts of the original (§§ 1, 2, 4.1, 4.2, 6.3.1) survive substantially unchanged. Also the old value of 100 km s^{-1} Mpc^{-1} for Hubble's constant, and the notation σ_0 for the density parameter instead of the more fashionable $\Omega = 2\sigma_0$, have been retained, rather than risk adding to the number of arithmetical errors.

1.2. Residual Radiation

In the relativistic models which begin with rapid expansion from a state of very high density, some relics of the early stages may remain observable. Gamow (1948) and, independently, Dicke, Peebles, Roll, and Wilkinson (1965) have considered the early stages of the relativistic models; they point out that, if these early stages of the models are to be taken seriously, nearly all the matter would have been converted into helium unless the Universe consisted almost entirely of radiation at the stage when the composition became fossilized. The radiation would have to be a result of the Creation; in these early stages, the matter plays no significant part in the energetics, or even the mass density—it merely acts as a thermometer for the radiation temperature. However, by the time the Universe has expanded by a linear factor of $\sim 10^8$, to its present size, the radiation temperature (corresponding to a helium content, before stellar nucleogenesis, of about 30%) is reduced to about 3° K by the combined effects of dilution and redshift, while the matter density is reduced by dilution only. At wavelengths below 10 cm, this is far greater than the background radiation expected from an extrapolation of the power-law spectrum observed at meter wavelengths. Therefore, great interest was aroused by the simultaneous appearance of the predictions of Dicke *et al.* and the discovery by Penzias and Wilson (1965) that the brightness temperature of the sky is as high as $3.5° \pm 1°$ K at 4080 MHz. A brief indication of the present state of this field is given in § 5.

1.3. The Radio Source Population

The third type of cosmological test by radio observations arises from the fact that individual radio sources can be observed at great distances, and therefore at significantly earlier epochs than the present. The prospect of investigating cosmology through radio sources has been an attractive one ever since the second brightest radio source in the sky was identified with a galaxy with redshift $z = 0.05$; and Schmidt's (1965) measurement of the redshifts of quasi-stellar sources, including 3C 9 with $z = 2.012$, showed that the early promise was not an empty one.

For many years, the only information available for an appreciable number of sources was the flux density, and the only possible cosmological investigation was to count the number $N(S)$ of sources per steradian with flux densities exceeding various limits S. The source counts turned out to be incompatible with a uniform unchanging population of radio sources, whether the metric is taken to be that of a steady-state model or of a Friedmann model. The source counts require that there were more powerful sources at earlier epochs (or at large distances) than there are now in our neighborhood, though they cannot by themselves show how the radio luminosity function evolved. During the last five years a more direct line of attack has become possible, because the redshifts of some 150 quasi-stellar sources have been measured. That no instant solution of the cosmological problem has emerged is due to a variety of reasons: (i) There is no simple interpretation of a "Hubble plot" of radio (or optical) flux density against redshifts. Evidently there is a huge scatter in the luminosities of quasi-stellar sources, so that observational selection is of basic importance, and we cannot study separately the variation of flux density with redshift and the variation of number per cubic gigaparsec with redshift. (Other properties of quasi-stellar objects, in particular their linear sizes, also show a very large scatter.) (ii) As already shown by the source counts for radio sources as a whole, so for quasi-stellar sources the data are incompatible with any unchanging population, which could be used as a standard object whose appearance is modified according to the metric. In fact, the effects of varying the assumed metric are relatively small; what one really studies is the evolution of the luminosity function of quasi-stellar objects. (iii) The analysis is greatly complicated by the importance of both optical and radio flux density limits in the selection of quasi-stellar objects. (iv) It may be that the redshifts of quasi-stellar objects are largely due to causes other than the Hubble expansion, and therefore cannot be used as distance indicators at all.

The radio source counts are discussed in § 6, and the evolution of the luminosity function for quasi-stellar objects in § 7.

2. SUMMARY OF BASIC FORMULAE

The comparison with theory will be confined to homogeneous isotropic world models, described by a Robertson-Walker metric. In order to establish a notation, and for ready reference, the basic formulae are given below.

The Robertson-Walker metric will be used in the form

$$ds^2 = dt^2 - \frac{R^2(t)}{c^2}\left[dr^2 + \left(\frac{\sin Ar}{A}\right)^2(d\theta^2 + \sin^2\theta\, d\phi^2)\right], \tag{2}$$

where ds = interval between events; dt = time interval in a local frame of reference chosen to make the universe appear isotropic; θ, ϕ are spherical polar coordinates on the celestial sphere; r is the coordinate distance from the origin at the present epoch, measured along a geodesic; A is the curvature of space at the present epoch; $R(t)$ is the linear scale of the universe, relative to the scale at the present epoch; and c = speed of light. Light emitted from a distant source with wavelength $\lambda_{\rm em}$, frequency $\nu_{\rm em}$, arrives now with wavelength $\lambda_{\rm rec} = \lambda_{\rm em} + \Delta\lambda = (1 + z)\lambda_{\rm em}$ (definition of z), where

$$(1 + z) = \frac{\nu_{\rm em}}{\nu_{\rm rec}} = \frac{\lambda_{\rm rec}}{\lambda_{\rm em}} = \frac{R(\text{time of reception})}{R(\text{time of emission})} = \frac{1}{R(\text{time of emission})}; \tag{3}$$

and, since $ds = 0$ on a light path, and $d\theta = d\phi = 0$ on a light path to the origin (taken in our Galaxy),

$$r = \int c\,dt/R(t)\,. \tag{4}$$

Equations (3) and (4) together determine r in terms of z once the particular cosmological model is specified by giving $R(t)$. From equation (2) it is easy to deduce the following expressions for observable properties of a celestial object moving with the metric:

$$\text{angular diameter} = \frac{\text{proper diameter}}{(\sin Ar/A)}(1 + z) \tag{5}$$

$$\text{flux density of a radio source} = S(\nu_{\rm rec})$$

$$= \frac{P(\nu_{\rm em})}{(\sin Ar/A)^2(1 + z)} \tag{6}$$

$$= \frac{P(\nu_{\rm rec})}{(\sin Ar/A)^2(1 + z)^{1+\alpha}} \tag{7}$$

if the source has a "straight" spectrum with spectral index α, that is, $S \propto \nu^{-\alpha}$, and $P(\nu)$ is the power emitted by the source per hertz per steradian, at frequency ν.

Number of objects of a given type, per steradian, at "distance" between r and $r + dr$ is

$$(a)\ \rho_n \left(\frac{\sin Ar}{A}\right)^2 dr \tag{8}$$

in cosmologies in which the objects are conserved, and is

$$(b)\ \frac{\rho_n r^2 dr}{(1+z)^3} \tag{9}$$

in steady-state cosmologies (ρ_n = present number of objects per unit volume).

These general formulae are easily specialized to particular models by using the relations:

Steady State:

$$r = \frac{c}{H} z\,, \quad A = 0\left(\text{i.e., replace } \frac{\sin Ar}{A} \text{ by } r\right). \tag{10}$$

Relativistic, pressure negligible, cosmological constant = 0:

$$\dot{R}^2 = \tfrac{8}{3}\pi G\rho R^2 - A^2c^2 = \tfrac{8}{3}\pi G\rho_0 R^{-1} - A^2c^2 = H^2(2\sigma_0 z + 1)\ ; \tag{11}$$

in particular $H^2 = \frac{8}{3}\pi G\,\rho_0 - A^2c^2$, where ρ_0 and H are the present values of mean density and Hubble's constant.

From equations (3), (4), and (11)

$$r = \int (c/R)(dR/\dot{R})$$
$$= (2c/H)(2\sigma_0 - 1)^{-1/2}\{\sin^{-1}(1 - 1/2\sigma_0)^{1/2} - \sin^{-1}[(1 - 1/2\sigma_0)/(1+z)]^{1/2}\}\,, \tag{12}$$

which can be rewritten in the more useful form

$$\frac{\sin Ar}{A} = \frac{c}{H\sigma_0(1+z)}\{z - (\sigma_0^{-1} - 1)[(2\sigma_0 z + 1)^{1/2} - 1]\}\,. \tag{13}$$

In relativistic models with zero pressure and zero cosmological constant the "density parameter" σ_0 is equal to the "deceleration parameter" $q_0 = -R\ddot{R}/\dot{R}^2$.

3. ATTEMPTS TO DETECT INTERGALACTIC ATOMIC HYDROGEN

Three ways have been used to search for atomic hydrogen in intergalactic space. No hydrogen has been found, but significant upper limits have been set to its density.

3.1. Absorption in the 21-Centimeter Line

Radio waves from an extragalactic radio source at redshift z emitted in the frequency range 1420 MHz $< \nu <$ 1420$(1 + z)$ MHz, will somewhere along the line of sight to our Galaxy have frequency 1420 MHz relative to the local standard of rest. If there are n_{H} hydrogen atoms cm^{-3} in that neighborhood, the optical depth will be

$$\tau = 1.7 \times 10^4(n_{\mathrm{H}}/T_s)(100 \text{ km s}^{-1} \text{ Mpc}^{-1}/H)\,. \tag{14}$$

To obtain equation (14), the number of atoms per unit velocity range n_{H}/H has been substituted into the standard formula for τ (e.g., Spitzer 1968, p. 25, eq. [2.56]); T_s is the spin temperature. Thus a trough should appear in the observed spectrum of the source, over the frequency range $1420(1 + z)^{-1}$ MHz $< \nu <$ 1420 MHz, with the optical depth given by equation (14). The step in the spectrum at 1420 MHz will be somewhat polluted by absorption in the Galaxy, and the sharpness of the step at $1420(1 + z)^{-1}$ MHz will be limited by the random motions of the hydrogen near the source. Note that the optical depth does not increase in proportion to the thickness of hydrogen traversed; thus the most sensitive measurements can be made by using the brightest source, Cygnus A ($z = 0.05$).

The first attempt to observe the absorption trough was made by Field (1958). At present the best upper limit to the optical depth is that found by Penzias and Scott (1968):

$$\tau < 5 \times 10^{-4} ,$$

whence

$$n_{\rm H} < 3 \times 10^{-8} T_s (H/100) \text{ atoms cm}^{-3}$$

and the atomic-hydrogen contribution to σ_0 is

$$(\sigma_0)_{\rm H} < 1.3 \times 10^{-3} T_s (100/H) .$$

An indication of hydrogen in the Virgo cluster, found in absorption spectra of Virgo A (M87) by Koehler and Robinson (1966) has not been confirmed by more recent observations (Allen 1968, 1969). Allen also found that the optical depth between the Galaxy and Virgo A was less than 6×10^{-4}.

Unfortunately T_s is very uncertain. At the low densities concerned it is most likely that the spin temperature interacts with the kinetic temperature mainly through the scattering of Lyman-α(Lα) photons rather than direct collisions (Wouthuysen 1952; Field 1959). The resulting estimate of T_s is very insensitive to the kinetic temperature if that is high, but it does depend on the density of Lα photons. Direct observations of the ultraviolet background just longward of Lα are still in an uncertain state; the current theoretical estimate (Field 1972) is that $T_s < 18$°K, corresponding to $(\sigma_0)_{\rm H} < 0.025$ $(100/H)$.

3.2. 21-Centimeter Line Emission from Intergalactic Atomic Hydrogen

Substituting the number of hydrogen atoms per unit velocity range, $n_{\rm H}(z)/H(z)$, into the standard formula for emission from optically thin hydrogen (e.g., Spitzer 1968), we find that the brightness temperature in the neighborhood of the emitting hydrogen is $5.49 \times 10^{-14}\, n_{\rm H}(z)/H(z)$, and therefore the contribution to brightness temperature observed from the Galaxy is

$$\frac{5.49 \times 10^{-14}}{(1 + z)} \frac{n_{\rm H}(z)}{H(z)} = 1.7 \times 10^4 (1 + z)^{-2} (2\sigma_0 z + 1)^{-1/2} n_{\rm H}(z)(100/H) \qquad (15)$$

at frequency $1420(1 + z)^{-1}$ MHz. Thus the background radiation at frequencies just below that of the Galactic hydrogen radiation should have a brightness temperature higher by

$$\Delta T_b = 0.38 (\sigma_0)_{\rm H} (H/100) \qquad (16)$$

than the background just above 1420 MHz. (Absorption of the microwave background is not important, since the optical depth is so small [§ 3.1].)

Goldstein (1963), Davies and Jennison (1964), and Penzias and Wilson (1969) found no detectable step in the background spectrum near 1420 MHz; the best limit is that of Penzias and Wilson who found $\Delta T_b = 0 \pm 0.08°$ K, corresponding to $(\sigma_0)_H = 0$ with an uncertainty of $0.2(100/H)$.

3.3. Absorption in the Lyman-α Line

As in the case of the 21-cm line, so also absorption in the Lα line will cause an absorption trough to appear, in the ultraviolet continuum of an extragalactic source, between observed wavelengths 1216 and $1216(1 + z)$ Å. However, the absorption cross-section for Lα is vastly greater than that for the 21-cm hyperfine transition. Also, all the hydrogen atoms are in the ground state (the decay from even the metastable $2s$ state being fast compared with any conceivable excitation rate), so that the absorption does not depend on an assumed temperature. The optical depth of hydrogen at the epoch corresponding to redshift z is

$$\tau_{L\alpha} = \int n_H(z)\sigma\{\nu + [H(z)r/c]\nu\}\,dr$$
$$= [n_H(z)\lambda/H(z)]\int \sigma(\nu)\,d\nu\ ,$$

since $d\nu/\nu = Hdr/c$. The integrated scattering cross-section $\int \sigma(\nu)d\nu$ is well known from atomic physics (e.g., Allen 1962), so

$$\tau_{L\alpha} = 4.15 \times 10^{10} n_H(z)[100/H(z)]$$
$$= 4.15 \times 10^{10} n_H(z)(1 + z)^{-1}(1 + 2\sigma_0 z)^{-1/2}(100/H) \tag{17}$$

for Friedmann models with zero pressure and zero cosmological constant (cf. eq. (11)). Thus nearly all the radiation emitted shortward of Lα would be scattered out of the line of sight if $n_H(z)$ were greater than 10^{-10} cm^{-3} (Gunn and Peterson 1965; Scheuer 1965). In the spectra of quasi-stellar objects with $z \gtrsim 1.9$, emitted Lα is within the spectral range accessible to ground-based telescopes, and these spectra show no noticeable difference in intensity on the two sides of the Lα emission line. If we take $\tau_{L\alpha} < 0.5$ as representing the observations at $z = 2$, equation (17) shows that

$$(\sigma_0)_H < 10^{-7}(100/H)\ ; \tag{18}$$

and a similar limit applies at all redshifts up to 2.88, the largest known redshift (Oke 1970).

It is difficult to believe that all matter has congealed into galaxies so effectively that much less than 1 part in 10^5 is left over, so it is implausible to regard inequality (18) as a true limit to the total hydrogen density. The alternatives are to suppose either (i) that the redshifts of quasi-stellar objects are not cosmological, so that (18) is not valid at all, or (ii) that intergalactic hydrogen is completely ionized.

Regarding the first possibility, it will clearly be important to search for the Lα cutoff in galaxies, using telescopes above the atmosphere. Owing to the great width of Lα absorption in the Galaxy, observations of fairly distant galaxies are needed. Furthermore stellar radiation is weak in the ultraviolet, and those galaxies that have abnormally strong ultraviolet radiation are also those for which the cosmological interpretation of the redshift is most questionable; thus it may be some time before this problem is resolved.

Similar methods have been used to search for other gases. Field, Solomon, and Wampler (1966) have searched the spectrum of 3C 9 for absorption associated with the Lyman band of molecular hydrogen, and have concluded that its density is less than 4×10^{-9} $(H/100)$ molecules cm^{-3} at $z = 2$. Shklovskii (1964), who was the first in this field, discussed the Mg II doublet at 2800 Å; and Bahcall and Salpeter (1965, 1966) have considered resonance lines of several other elements. If intergalactic space is filled with nonthermal ultraviolet radiation (cf. § 4), heavy elements may be predominantly in highly ionized states, in which case the absence of absorption does not yield good upper limits to the densities of these elements.

4. ATTEMPTS TO DETECT INTERGALACTIC IONIZED HYDROGEN

Intergalactic ionized hydrogen is much harder to detect than is atomic hydrogen. The only significant restriction from observation is the upper limit on the density of hot ($T > 10^7$ °K) hydrogen obtained from observation of the X-ray background; for a recent review, see Field (1972) or Longair (1971). Effects which might be observable in the radio band are considered below.

4.1. Thomson Scattering

Thomson scattering weakens the radiation from a discrete source by a factor $\exp(-\tau_{TS})$, with

$$\tau_{\mathrm{TS}} = \int \sigma_{\mathrm{TS}} n_e R(t) dr \, .$$

Here σ_{TS} is the Thomson scattering cross-section and n_e the electron density at the epoch when the radiation traveled along the line element of proper length $R(t)\, dr$.

In steady-state theories, $n_e = n_{e0}$, the present electron density; using equation (10), one finds

$$\tau_{\mathrm{TS}} = \sigma_{\mathrm{TS}} n_{e0} c H^{-1} \ln(1 + z) = 0.14\, \sigma_0 \ln(1 + z)(H/100) \tag{19}$$

if most matter is in the form of ionized hydrogen. Evidently the effects are small.

In Friedmann models, $n_e = (1 + z)^3 n_{e0}$; using equations (4) and (11) one finds

$$\tau_{\mathrm{TS}} = (\sigma_{\mathrm{TS}} n_{e0} c / H\sigma_0)\{(2\sigma_0 z + 1)^{1/2}[1 + \tfrac{1}{3}z - 1/(3\sigma_0)] - [1 - 1/(3\sigma_0)]\}$$

$$\simeq 0.046(2\sigma_0)^{1/2} z^{3/2}(H/100) \text{ for } z \gg 1 \tag{20}$$

(Bahcall and May 1968). The optical depths for these model universes consisting chiefly of ionized hydrogen are plotted in figure 1.

The detection of Thomson scattering would give an unambiguous measure of electron density, since it is independent of temperature. However, it is also quite independent of frequency, and this lack of recognizable features makes it very difficult to distinguish from, say, evolutionary changes in the mean powers of sources. If large numbers of sources with very large redshifts ($z \gtrsim 6$) were detected in any frequency range, the ratio of isotropic background radiation to radiation from individual sources could set a significant upper limit to n_e.

4.2. Free-free Emission and Absorption

The free-free absorption coefficient depends strongly on the temperature, and so (to a lesser extent) does the emissivity. The ionization process, whatever it is, must supply at least 13.6 eV, and will almost invariably supply a little more, thereby raising the temperature to an appreciable fraction of 10^5 ° K. In the absence of rapid cooling processes, the temperature is unlikely to be much below 10^4 ° K. According to Field (1972) much higher temperatures, up to at least 10^8 ° K, would now be consistent with the observations of the X-ray background.

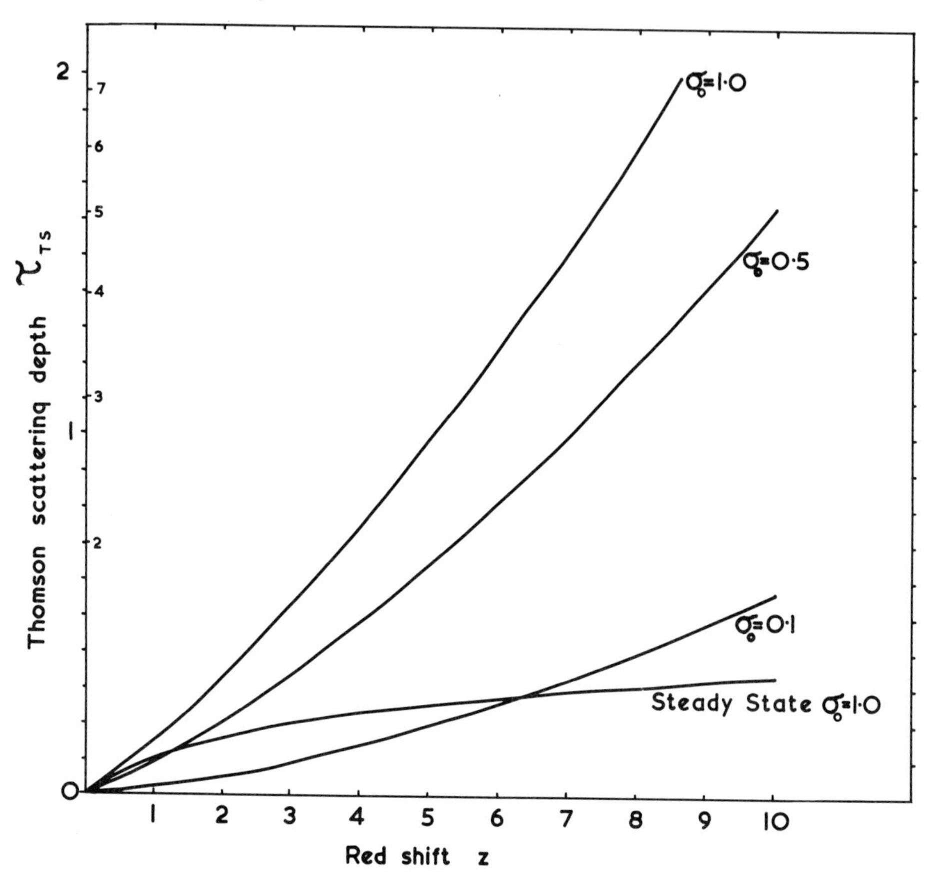

FIG. 1.—Thomson scattering in a fully ionized intergalactic medium.

The absorption coefficient of ionized hydrogen will be taken as

$$0.2T^{-3/2}\nu^{-2}[(1+z)^{-2}n_e^2] \tag{21}$$

at the emitted frequency $\nu(1+z)$. Expression (21) is correct for hydrogen at 10^8 Hz and 10^6 ° K; the slow variation of the Gaunt factor with temperature and frequency is ignored, as is the presence of helium. Then the free-free contributions to optical depth and brightness temperature from an element of light path of proper length $R(t)dr$ are, respectively,

$$d\tau_{\mathrm{FF}} = 0.2T^{-3/2}\nu^{-2}(1+z)^{-2}n_e^2[(1+z)^{-1}dr] \tag{22}$$

and

$$dT_{\mathrm{FF}} = 0.2T^{-1/2}\nu^{-2}(1+z)^{-2}n_e^2[(1+z)^{-1}dr][(1+z)^{-1}\exp(-\tau_{\mathrm{FF}})]\,. \tag{23}$$

The last factor of $(1 + z)^{-1}$ in the expression for dT_{FF} arises, as in equation (6), from the combined effects of the reduced energy of each photon, the reduced rate of arrival of photons due to time dilation, and the compression of the redshifted frequency interval relative to the frequency interval into which the photons were emitted.

Using equations (3), (4), (10), and (11) one can integrate equation (22) to find τ_{FF}, and (23) to find T_{FF} so long as the gas is optically thin ($\tau_{\mathrm{FF}} \ll 1$), with the following results:

Steady-state models:

$$\tau_{\mathrm{FF}} = \tfrac{1}{2}\tau_0[1 - (1 + z)^{-2}]\,, \tag{24}$$

$$T_{\mathrm{FF}} = \tfrac{1}{3}T\tau_0[1 - (1 + z)^{-3}]\,. \tag{25}$$

Relativistic models, $p = \Lambda = 0$, constant temperature:

$$\tau_{\mathrm{FF}} = \tfrac{1}{4}\tau_0\sigma_0^{-3}[(1 + 2\sigma_0 Z)^{1/2}\{\tfrac{1}{5}(1 + 2\sigma_0 Z)^2 + \tfrac{2}{3}(2\sigma_0 - 1)(2\sigma_0 Z + 1) + (2\sigma_0^{-1})^2\}]_0^z$$

$$\simeq 3 \times 10^{-5} T_4^{-3/2} \nu_8^{-2} \sigma_0^{3/2} z^{5/2} (H/100)^3 \text{ if } z \gg 1\,; \tag{26}$$

$$T_{\mathrm{FF}} = \tfrac{1}{3}T\tau_0\sigma_0^{-2}[(2\sigma_0 z + 1)^{1/2}(3\sigma_0 - 1 + \sigma_0 z) - (3\sigma_0 - 1)$$

$$\simeq 0.4 T_4^{-1/2} \nu_8^{-2} (\sigma_0 z)^{3/2} (H/100)^3 \text{ if } z \gg 1\,. \tag{27}$$

In these formulae, τ_0 represents the optical depth of a length c/H of gas at the present density of matter:

$$\tau_0 = 9.7 \times 10^{17} T^{-3/2} \nu^{-2} \sigma_0^2 (H/100)^3\,. \tag{28}$$

In steady-state models both T_{FF} and τ_{FF} are quite negligible for all z if $T > 10^{4\circ}$ K and $\nu > 1$ MHz.

In the relativistic models the optical depth is also much too small to have observable effects at frequencies above about 1 MHz; below 1 MHz any extragalactic absorption is overshadowed by stronger free-free absorption within the Galaxy, as well as by synchrotron self-absorption in many extragalactic sources. It may ultimately be worthwhile to look for systematic differences between the low-frequency spectra of extended extragalactic sources with large redshifts and those with small redshifts in the same region of the sky.

Since $T_{\mathrm{FF}} \propto \nu^{-2}$, the contribution to the flux density per steradian is independent of frequency. The observed diffuse background is lower near 10^9 Hz than at any other frequency up to 10^{11} Hz, and perhaps up to the ultraviolet (cf. fig. 2 of Longair 1971). There, the background is fully accounted for by the sum of the Rayleigh-Jeans spectrum extrapolated from the microwave background and the extrapolation of the nonthermal spectrum from low frequencies. The uncertainty is about 2° K (Howell and Shakeshaft 1966, 1967) at $\nu_8 = 4$ and at $\nu_8 = 6$, so equation (27) requires

$$\sigma_0 z \lesssim 20\ T_4^{1/3}\ (100/H)^2\,, \tag{29}$$

where z refers to the epoch of reheating of the intergalactic medium, and T_4 refers to the temperature immediately after that epoch (since the largest contribution to T_{FF} comes from the earliest epoch).

We have to conclude that, apart from the not very exciting limit (29), free-free emission and absorption in the radio band give no significant information on the density of intergalactic hydrogen.

4.3. Dispersion

The dispersion of time-varying signals, due to the variation with frequency of the group velocity of electromagnetic waves in an ionized gas, is now familiar from observations of pulsars. Haddock and Sciama (1965) suggested that this effect might be detected in extragalactic variable sources. However, the time delay is quite short, and converges for arbitrarily large redshifts to:

$$(1/4H)\,(\nu_p^2/\nu^2) \text{ in steady-state universes} \tag{30}$$

$$H^{-1}(\nu_p^2/\nu^2)(1 - 2\sigma_0)^{-1/2} \ln\,[(1 + (1 - 2\sigma_0)^{1/2})/(2\sigma_0)^{1/2}] \tag{31}$$

in relativistic models with $p = \Lambda = 0$.

Here ν_p represents the present plasma frequency of the intergalactic medium, and ν the observing frequency. In convenient units, the time delay is limited to

$$1500\,\lambda_m^2\,\sigma_0 \text{ seconds} \quad \text{(steady-state models)}\,,$$

$$3000\,\lambda_m^2 \text{ seconds} \quad \text{(relativistic model with } \sigma_0 = 0.5)\,,$$

where λ_m is the wavelength of the observed radiation in meters. Evidently there is little hope of detecting this effect in extragalactic variable sources of the known kinds, whose variations have a time scale of months even at centimeter wavelengths. A pulsar would have to be detected at least at the distance of the Virgo cluster to show effects clearly in excess of dispersion in our Galaxy and its own.

4.4. Faraday Rotation

If there is an intergalactic magnetic field with a component that is ordered on a very large scale, as well as an ionized intergalactic gas, the Faraday rotation of the linear polarization of radio sources should increase with distance. The intergalactic contribution to the rotation measure would be

$$5.5 \times 10^{11} B \cos\theta\;\sigma_0(H/100) f(z) \tag{32}$$

radians m^{-2} in a fully ionized universe, where θ is the angle between the line of sight and the (unknown) direction of the magnetic field B (gauss), and $f(z)$ depends on the cosmological model and the assumed variation of field strength with epoch. For example, in a relativistic model with $\sigma_0 = 0.5$ and an adiabatically expanding field frozen into the ionized gas,

$$f(z) = \tfrac{2}{3}[(1 + z)^{3/2} - 1]\,. \tag{33}$$

Some evidence of a variation in rotation measure with redshift has been produced (Sofue, Fujimoto, and Kawabata 1968; Reinhardt and Thiel 1970), but the effects are not of overwhelming statistical significance and are not easy to disentangle from the effects of Faraday rotation in the Galaxy. At present their reality remains an open question (Mitton and Reinhardt 1972).

4.5. Scattering by Irregularities

Scintillations of radio sources due to electron density irregularities in the ionosphere, in the interplanetary medium, and in the interstellar medium have been discovered, in that order. Why not intergalactic scintillation too? The reason is that no sources of sufficiently small diameter are known, which would take the place of pulsars in the interstellar medium.

In the phase-thin regime (total phase deviation $\lesssim 1$ radian) the source must be at a distance $L \gtrsim a^2/\lambda$ for interference between different scattered beams to take place and hence for amplitude fluctuations to develop. Here a represents the scale size of the irregularities. Therefore, the scale of the diffraction pattern near the observer, which is the same as the scale of the irregularities, is

$$a < (L\lambda)^{1/2} . \tag{34}$$

Since each spot on the source gives rise to such a diffraction pattern, the intensity distribution is smeared out over regions comparable with the source size, if the irregularities are roughly midway between source and observer. Thus, to retain observable intensity fluctuations, we require a source whose linear diameter is less than $(L\lambda)^{1/2}$. Now L is limited to $\sim 10^{28}$ cm by the size of the Universe, so we require a source diameter below $10^{14}\lambda^{1/2}$ cm. This is several orders of magnitude smaller than the self-absorbed cores of quasi-stellar sources, and known sources of smaller size (pulsars, molecular masers) are too faint.

Now consider the phase-thick regime. The condition for interference is now $L\theta_{\text{scatt}} > a$, where θ_{scatt} is the width of the angular spectrum, and the source size must be less than the scale of the diffraction pattern, $\lambda/\theta_{\text{scatt}}$, which is now less than a. Thus

$$\text{source size} < \lambda/\theta_{\text{scatt}} < \lambda L/a$$

and also

$$\text{source size} < a .$$

Therefore

$$\text{source size} < [(\lambda L/a)\, a]^{1/2} = (\lambda L)^{1/2} ,$$

the same condition as before.

On the other hand, it is just possible that the apparent angular diameters of distant sources are broadened by amounts that can be measured with very long baseline interferometers. If the ratio of electron density fluctuation to total baryon density is $\Delta n_e/n$, then the expected broadening for a path length c/H is roughly

$$10^4\, a^{-1/2}\sigma_0\, (\Delta n_e/n)\, \lambda^2\, (H/100) \text{ arc seconds} , \tag{35}$$

where a and λ are both in meters. Current measurements do not yield interesting limits on Δn_e except for $a < 10^{10}$ cm, and the broadening (35) cannot exceed the observed interstellar broadening (Readhead and Hewish 1972) unless $a < 10^{14}$ cm.

4.6. Is There Any Intergalactic Gas?

None of the observations exclude the possibility of a "cosmological density" in the form of black holes, red dwarfs, rocks, or even reprints from the *Astrophysical Journal* (the last being a suggestion due to R. J. Allen); all these might easily have escaped detection. Arp (1965) published an instructive diagram which shows how the notion of a "galaxy" would arise naturally as a selection effect; there may be many other kinds of extragalactic aggregations. Nevertheless, it is interesting to ask whether a "cosmological density" of gas exists, particularly as the hot big-bang models lead us to expect a great deal of hydrogen and helium gas. (For simplicity the helium has been ignored in §§ 3 and 4 of this chapter; it makes no gross difference to numerical estimates.)

The Lα test shows that atomic hydrogen is very rare—so rare that even in a low-density universe the intergalactic medium must be ionized. If the redshifts of quasi-stellar objects are not cosmological, that test fails, and then the only solid upper limit is provided by the 21-cm emission observations (§ 3.2). That limit requires $(\sigma_0)_{\mathrm{H\,I}} < 0.2$ if $H = 100$ km s^{-1} Mpc^{-1}, but is no longer very interesting if $H \leq 50$ km s^{-1} Mpc^{-1}.

A "cosmological density" of ionized hydrogen with $10^4\,° < T < 2 \times 10^8\,°$ K and extending beyond $z = 3$ is consistent with all the observations. It is not free of theoretical difficulties. In 10^{10} years, a fraction of the order of $0.05\, n_{-6} T_4^{-1/2}$ of initially ionized hydrogen recombines. Thus it is not good enough to start a theoretical model with fully ionized hydrogen at some epoch before $z = 3$; there must be an ionizing agent that promptly breaks up each hydrogen atom that recombines. One possibility is collisional ionization in a hot gas. Temperatures above $3 \times 10^5\,°$ K are sufficient in steady-state models (Rees and Sciama 1966); in relativistic models, temperatures of several million degrees K are needed (Weymann 1967; Bergeron 1969) as the density at $z = 2$ is much higher. A few years ago such high temperatures were embarrassingly close to the upper limits set by the observations of the X-ray background, but high temperatures now appear to be permissible (Field 1972). The energy required to heat gas with $\sigma_0 = 0.5$ above $10^6\,°$ K is rather a heavy load for galaxies, but might be available from quasi-stellar objects.

The other possibility is photoionization (Arons and McCray 1969; Rees and Setti 1970; Bergeron and Salpeter 1970; Arons and Wingert 1972). At this point the author must confess that he hastily abandoned this idea on finding that a Strömgren sphere around 3C 9 would have a radius of only a few Mpc. In fact, Lyman continuum radiation from quasi-stellar objects following Schmidt's (1970) evolving luminosity function is not very far below the flux required to ionize hydrogen with $\sigma_0 \simeq 0.5$. The Lyman continuum from quasi-stellar sources would ionize a lower density of hydrogen—equivalent to 2.8×10^{-7} atoms cm^{-3} at $z = 0$ according to Arons and Wingert (1972)—as completely as the Lα test demands. Alternatively, one could postulate a higher ultraviolet energy density, of unknown origin but still probably compatible with observations of the present ultraviolet background radiation, which would sufficiently ionize hydrogen with a "cosmological density." The temperature resulting from photoionization would be in the region of $10^4\,°$ K and the total energy requirements are much lower than those for collisionally ionized gas.

Some astronomers believe that there simply is no intergalactic gas, and we ought to stop worrying about it. Indeed, we have discussed only upper limits. Is there any positive evidence for its existence? According to taste, one might cite: (i) the high-velocity clouds of hydrogen which appear to be falling into the Galaxy (Oort 1970); (ii) X-ray bremsstrahlung from clusters of galaxies (Gursky *et al.* 1972) (but there are rival interpretations of the X-ray emission from clusters [Brecher and Burbidge 1972]); (iii) the diffuse soft X-ray background may (but need not) be interpreted as bremsstrahlung from a hot intergalactic medium (Field 1972); (iv) the shapes of extragalactic double radio sources. In general the components of these sources are bright and compact at the ends farthest from the central object, becoming fainter and more diffuse closer to the central object. It would be most unwise to base arguments on any particular model of radio sources, but whatever the nature of the "engine" concerned may be, one would expect it to shed "smoke" impartially in all directions unless it were moving through some resisting medium.

5. THE MICROWAVE BACKGROUND

5.1. The Observed Spectrum

Diffuse microwave radiation far exceeding the extrapolation of the meter-wave nonthermal background was discovered by Penzias and Wilson (1965), and their paper was

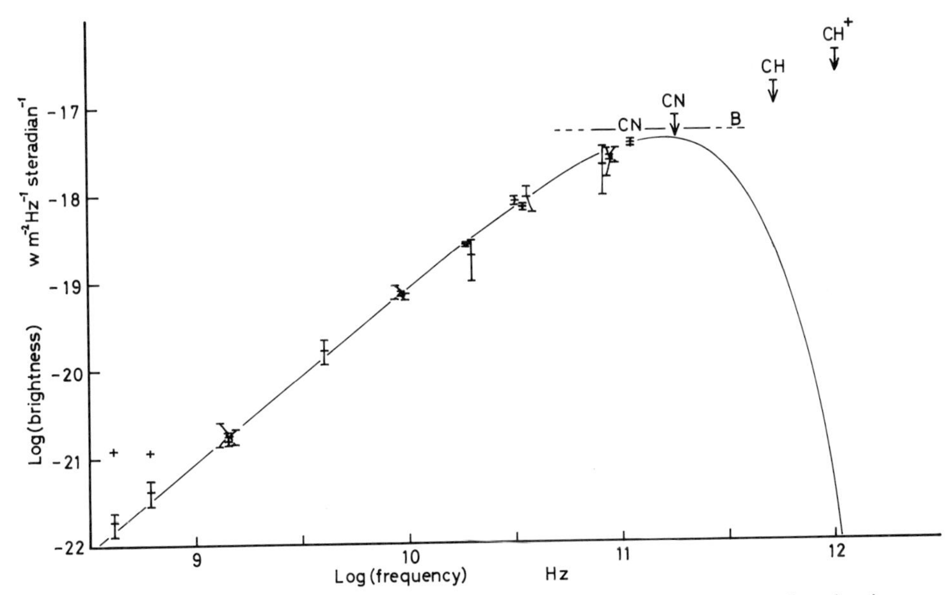

Fig. 2.—The microwave background. The line marked B indicates a mean flux density over the frequency range of the measurements by Blair *et al.* (1971). The points deduced from observations of interstellar CN, CH, and CH^+ are from Thaddeus (1972); references for the other observations may be found there or in Longair (1971). For 408 and 610 MHz, the total emission at the galactic pole is also plotted. The continuous curve represents a 2.75° K blackbody spectrum.

published at the same time as the discussion of residual blackbody radiation from a primeval fireball by Dicke, Peebles, Roll, and Wilkinson (1965). Gamow's (1948) earlier prediction had been forgotten because the detection of the predicted isotropic microwave background had been quite beyond the technical resources of radio astronomy at the time, even though Gamow had overestimated its intensity.

Since 1965, many measurements in the frequency range 0.4 to 90 GHz have established convincingly that the radiation has a Rayleigh-Jeans spectrum up to 40 GHz, with departures toward a 2.8° K blackbody spectrum becoming significant at the highest frequencies (fig. 2).

Many attempts have been made to measure the background at higher frequencies, where the difference between a power law and a blackbody spectrum becomes large. Ground-based observations become impracticable above 100 GHz because of atmospheric emission. The background intensity at 114 GHz has been derived from the ratio of intensities of the components of the CN 3874 Å interstellar line, as observed in absorption in the spectra of 10 stars. The ground state of CN is split, and the populations of the two lowest levels, separated by 114 GHz, are found to correspond to an excitation temperature of 2.78° ± 0.1° K. The next two levels of the CN ground state give an upper limit of 3.38° K at 228 GHz; similar methods applied to interstellar lines of CH and CH^+ give upper limits at still higher frequencies.

Direct measurements from rockets and balloons have given conflicting results. Experiments up to 1971 showed a flux far in excess of 2.8° K blackbody radiation, and also above a smooth curve drawn below the limits obtained from the interstellar lines of CN, CH, and CH^+. More recently, Blair *et al.* (1971) measured an intensity in the range 1.7–12.5 cm^{-1} consistent with blackbody radiation at 3.1° (+ 0.5°, − 2.0°) K, and at the time of writing it seems quite likely that a 2.8° K blackbody spectrum will eventually be confirmed. For a detailed review of the state of measurements in 1972, and in particular a discussion of the use of interstellar molecules as radiation thermometers, see Thaddeus (1972).

Measurements of the variation of intensity over the sky are much more precise than absolute measurements. The radiation is found to be isotropic with a precision approaching (and in some cases better than) $\Delta T/T < 10^{-3}$, on angular scales ranging from 1′.4 × 20′ to 180°. For a list of observations up to 1971, see Longair (1971).

5.2. Interpretation

The Universe is very transparent to microwave radiation. Accordingly, only two explanations of the microwave background suggest themselves in steady-state theory:

(i) Continuous creation of photons, with a spectrum such as to maintain the observed spectrum. I know of no strictly scientific (as distinct from aesthetic) objections to this hypothesis, but it has not been put forward seriously.

(ii) Emission by discrete sources. If these are galactic (e.g., dust grains), the precise isotropy of the background observed at cm wavelengths requires that the Galaxy be optically thick, but the observation of extragalactic sources in this part of the spectrum requires that the Galaxy be transparent. The only way to reconcile these requirements is to postulate that the radiation arises in numerous sharp lines; very artificial hypotheses are then needed to explain a smooth Rayleigh-Jeans spectrum. If the sources are extragalactic (e.g., Wolfe and Burbidge 1969), then the observed isotropy requires that they be much more numerous than normal galaxies if they are uniformly distributed in space (e.g., Hazard and Salpeter 1969).

In models starting from a singularity, the natural interpretation is that of radiation left over from the primeval fireball. The synthesis of elements in such models has been computed in detail, notably by Wagoner, Fowler, and Hoyle (1967). The radiation temperature at given redshift z is fixed at $2.8(1 + z)$ ° K, but the density at given z depends on the assumed value of σ_0; consequently the composition of matter which remains when the temperature and density become too low for further nuclear reactions depends on σ_0. Only hydrogen and helium are produced in large concentrations, though interesting amounts of D and ^{3}He would also be produced in low-density ($\sigma_0 < 0.1$) models. The helium abundance is not sensitive to density, being around one-third by mass for $10^{-2} < \sigma_0 < 10$. If observations of the helium abundance in stars or in gaseous nebulae make it necessary to accept a low value of the primeval helium abundance, it is still possible to reconcile this with the fireball model by (i) decreasing the time-scale of the expansion; one way of doing so is to use metrics that are strongly anisotropic at early epochs (Hawking and Tayler 1966; Thorne 1967; Doroshkevish, Zel'dovich, and Novikov 1967); (ii) postulating a large excess of leptons, in particular a large excess of neutrinos over antineutrinos, thus inhibiting the reaction $p + e^- \rightarrow n + \nu$ (Fowler 1970).[1]

6. COUNTS OF RADIO SOURCES

6.1. Basic Ideas

Consider the simplest case: radio sources all of the same power P watts Hz^{-1} steradian^{-1} distributed through Euclidean space with number density $\rho_n(r)$ sources per unit volume. Then the flux density P/r^2 would be a good indicator of distance r, and the observed relation between S and the number of sources $N(S)$ per steradian with flux density exceeding S would lead immediately to the relation between ρ_n and r. In particular, for the brighter (nearer) sources the density ρ_n should be the density in our immediate neighborhood, and therefore constant; the sources with flux density exceeding S are those in a sphere of radius $(P/S)^{1/2}$, and therefore

$$4\pi N(S) = (4\pi/3)(P/S)^{3/2}\rho_n = \text{constant} \times S^{-3/2} . \tag{36}$$

Thus a plot of log $N(S)$ versus log S is expected to be a straight line of slope -1.5, at any rate for the brightest sources. At large distances the density must decrease (if only to escape Olbers' paradox) so that the (log N, log S) graph should become less steep for small S.

Of course, radio sources do not have one value of P, but a vast range of powers. Let there be ρ_i sources per unit volume of power P_i; then the number of sources per steradian with flux density exceeding S is

$$N(S) = \sum_i N_i(S) = \sum_i \tfrac{1}{3}(P_i/S)^{3/2}\rho_i = \Big(\sum_i \tfrac{1}{3}\rho_i P_i^{3/2}\Big) S^{-3/2} . \tag{37}$$

Numbers of sources add; equation (37) is merely the sum of many equations like (36), one for each type of source. Its form is exactly that of equation (36), only the constant

[1] *Note added 1973 December.*—Observations of the 377-MHz line of atomic deuterium (D. A. Cesarsky, A. T. Moffet, and J. M. Pasachoff, 1973, *Ap. J.* [Letters], **180,** 1), of lines of DCN (K. B. Jefferts, A. A. Penzias, and R. W. Wilson, 1973, *Ap. J.* [*Letters*], **179,** L57; R. W. Wilson, A. A. Penzias, K. B. Jefferts, and P. M. Solomon, 1973, *ibid.*, **179,** L107; P. M. Solomon and N. J. Woolf, 1973, *ibid.*, **180,** L89.) and of HD (L. Spitzer, J. F. Drake, E. B. Jenkins, D. C. Morton, J. B. Rogerson, and D. G. York, 1973, *Ap. J.* [*Letters*], **181,** L116; J. H. Black and A. Dalgarno, 1973, *ibid*, **184,** L101) and of the Lyman lines of atomic deuterium (J. B. Rogerson and D. G. York, 1973, *Ap. J.* [*Letters*], **186,** L95) indicate a cosmic abundance of deuterium of the order of 10^{-5} relative to hydrogen. Rogerson and York's (1973) paper contains a short critical appraisal of these observations and their interpretation. While it is conceivable that deuterium may be produced in supernova explosions (S. A. Colgate, 1973, *Ap. J.* [*Letters*], **181,** L53), these observations favor a low-density Universe with $\sigma_0 < 0.1$.

factor has changed. We still expect a straight ($\log N$, $\log S$) plot of slope -1.5 for bright sources. But we must expect the change in slope for sources of lower flux density to spread out over a larger range in S, for a change in density at a certain distance now corresponds to different S for different powers P_i.

The use of the ($\log N$, $\log S$) relation to investigate the distribution of radio sources in space goes back at least to F. G. Smith (1951; see also Ryle 1950). At that time it was thought that the galactic radio background might be due to numerous discrete sources, whose number density would decrease toward the exterior of the Galaxy. The later observations tell a different story: the observed ($\log N$, $\log S$) relation has a slope steeper than -1.5, indicating an increase in density outward, and the increase in density seems to be much the same in all directions. Such an increase in density is incompatible with a plausible galactic distribution; most of the sources must be extragalactic, a conclusion that was abundantly confirmed by the identification of radio sources with distant galaxies. The isotropic increase of density with distance can then be interpreted naturally as a change in the source density (in any given range of P) with epoch. Furthermore, as there is no statistically significant range of large S for which the slope of the ($\log N$, $\log S$) relation is -1.5, there is a large contribution from sources so powerful and so rare that they have no meaningful "local" density ρ_i—a conclusion which is in good accord with the discovery of many quasi-stellar sources with large redshifts.

The early history of source counts was dominated by disputes over the reliability of source catalogs; fortunately, great advances in technique have now made these irrelevant. The present (1972) status of the observations is reviewed below.

6.2. Observations

Note.—Statements such as "$S_{408} = 7.5$ f.u." will be used to mean that the flux density of a source at 408 MHz is 7.5 flux units $= 7.5 \times 10^{-26}$ watts m^{-2} Hz^{-1}; similarly, P_{1410} will represent the power of a source at 1410 MHz, and powers will always be expressed in watts Hz^{-1} steradian^{-1}. Note also that the adjectives "powerful" (or "strong") and "weak" will be used to describe P, while "bright" and "faint" refer to S.

As radio sources have different spectra, it is essential to use data at one frequency; the most extensive data are still those of Pooley and Ryle (1968) at 408 MHz, and we therefore concentrate on these. For very large flux densities S, the precision of the ($\log N$, $\log S$) relation is limited by the statistical uncertainty in N; therefore, it is essential to use data from the whole or a very large part of the sky. For the smallest flux densities sources are very numerous, and whole-sky surveys are impracticable, so small sample regions are surveyed and isotropy is tested by comparing samples from various parts of the sky.

Pooley and Ryle's data consist of (i) a compilation of various surveys at 408 MHz and 400 MHz covering the whole sky (except for the galactic plane region $|b| < 10°$) between $-30° < \delta < +44°$ for sources with $S_{408} > 10$ f.u., and smaller regions down to 6 f.u. and 4 f.u.; (ii) a survey with a $100'' \times 3°$ fan beam, covering 0.09 steradians for sources with $S_{408} > 1.4$ f.u. and 0.04 steradians for sources with $S_{408} > 0.5$ f.u. (Bailey and Pooley 1968); (iii) the 5C 2 survey (Pooley and Kenderdine 1968) of a circular region approximately 3° in diameter, giving observations complete for $S_{408} > 0.012$ f.u. over 3×10^{-4} steradians and for $S_{408} > 0.06$ f.u. over 3×10^{-3} steradians.

Their composite (log N, log S) relation is shown in Figure 3.

These data could now be supplemented by using two further fan-beam surveys (Wilson 1972) for $S_{408} > 0.6$ f.u., the Bologna catalogs (Braccesi *et al.* 1965; Braccesi, Ceccarelli, Fanti, and Giovannini 1966; Grueff and Vigotti 1968) for sources with $S_{408} > 0.25$ f.u., and the 5C 3 survey (Pooley 1969) for the faintest sources. (The 5C 4

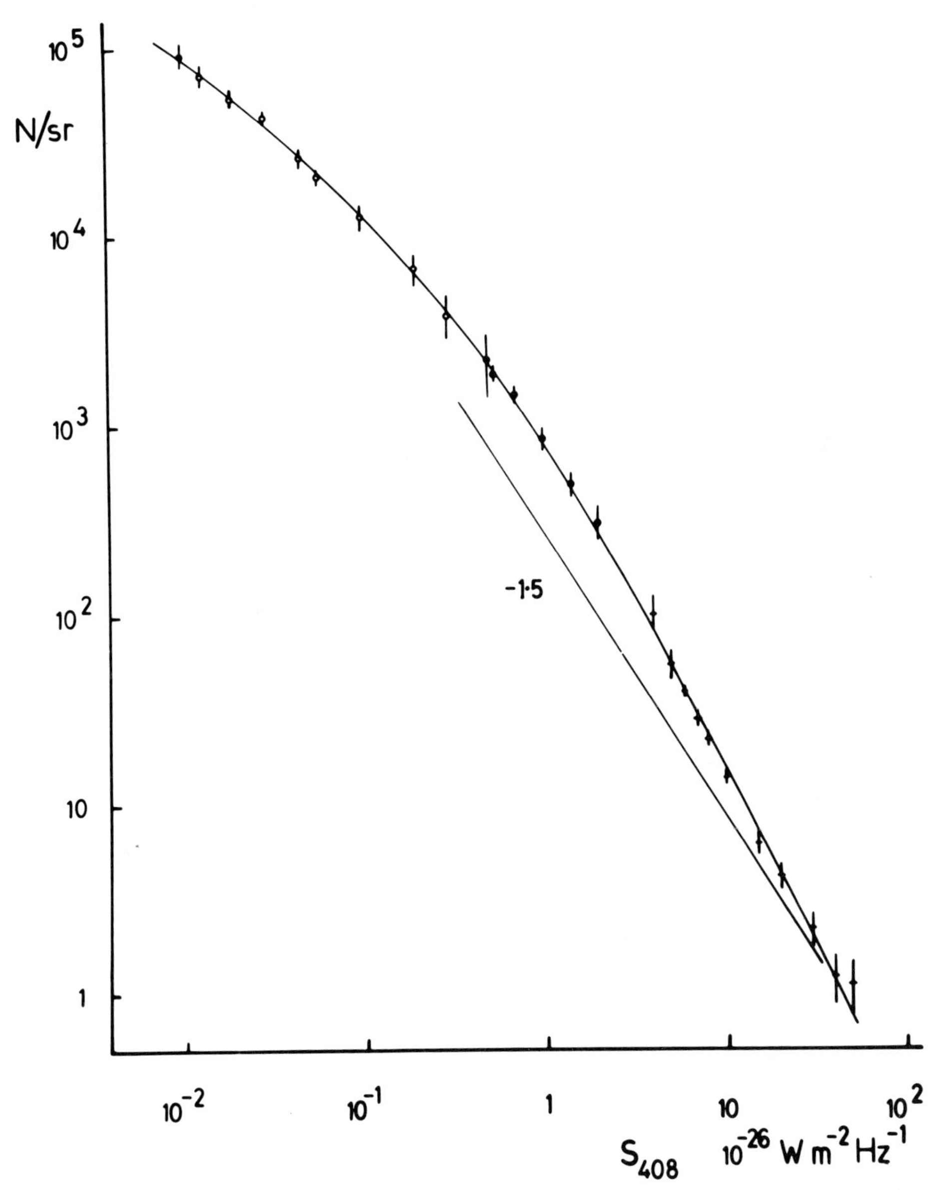

FIG. 3.—Counts of radio sources at 408 MHz (Pooley and Ryle 1968). A straight line of slope −1.5 is also plotted for comparison. Reproduced by kind permission of the authors and the Royal Astronomical Society.

survey [Willson 1970] was deliberately centered on the Coma cluster, and one cannot be certain which sources belong to the cluster.) A systematic incorporation of these data has not been performed. However, the source counts from the 5C surveys are in good mutual agreement, and the principal features of the source counts at higher flux densities have been confirmed by source counts from other surveys (fig. 4).

The difference between the observed source counts and the line of slope -1.5 predicted for a uniform distribution in Euclidean space is shown more clearly by plotting $\log(NS^{3/2})$ against $\log S$, as in figure 5. Near $S_{408} = 1$ f.u. the value of $NS^{3/2}$ rises to about twice its value for large S; for lower flux densities it declines below that value. It is probably fair to say that these features of the $(\log N, \log S)$ relation are generally agreed among observers. There is an apparent exception in the work of Shimmins, Bolton, and Wall (1968) who reported a slope of -1.4 for their source counts at 2700 MHz. Their survey covers only 0.34 steradians of sky, and the statistically best determined part of their curve corresponds to a range of N where figure 3 also has a slope of -1.4 (Pooley 1969). However, some real differences are to be expected between source counts at low and at high frequencies, since above $\sim$2000 MHz very compact sources form a large proportion of the observed sample, while these are self-absorbed at low frequencies.

The reader must refer to the papers cited for details of observational methods, but attention should be directed here to points that require particular care.

6.2.1. *Linearity of the flux-density scale.*—Figure 5 covers a range of 5000:1 in flux density but only 10:1 in $NS^{3/2}$; even small deviations from linearity in the flux-density scale could seriously warp the curve. Fortunately, the linearity of systems employing aperture synthesis is inherently good, since sources over most of the flux-density range produce signals which are below the noise level of any one of the single observations that are digitally recorded. The effective receiver gain is then determined by the mean gradient of the input-output characteristic of the receiver over the range of the system noise, and is independent of the flux density of the source, whatever the law of the detector. Subsequent Fourier synthesis (in which the source contributions add while the noise contributions take a random walk) is a purely digital computation and is strictly linear. These conditions obtained over most of the range covered by the 5C surveys and the fan-beam surveys used by Pooley and Ryle (1968). The linearity of the electronics becomes important when observing calibration sources in order to link the flux-density scale of these observations to the surveys used for bright sources, $S_{408} > 4$ f.u. In Pooley and Ryle's observations the measured upper limit on the nonlinearity of the receiver was 3 percent.

6.2.2. *Effects of confusion and noise.*—For sources near the lower limit of a survey, an additional very faint source within one aerial beamwidth, or a positive noise deflection, will increase the apparent source flux-density appreciably, while an adjacent positive deflection raises the baseline and so diminishes the apparent flux density. There are two systematic effects: (*a*) the estimate of S is biased upward, as the apparent source position is shifted toward positive deflections; (*b*) since there are more faint sources than bright ones, the estimate of $N(S)$ is raised by a scatter in S, even if the estimates of S were unbiased.

In the earlier Cambridge source counts at 178 MHz, these effects were of some importance, and were evaluated in detail using a Monte Carlo method (Pilkington and

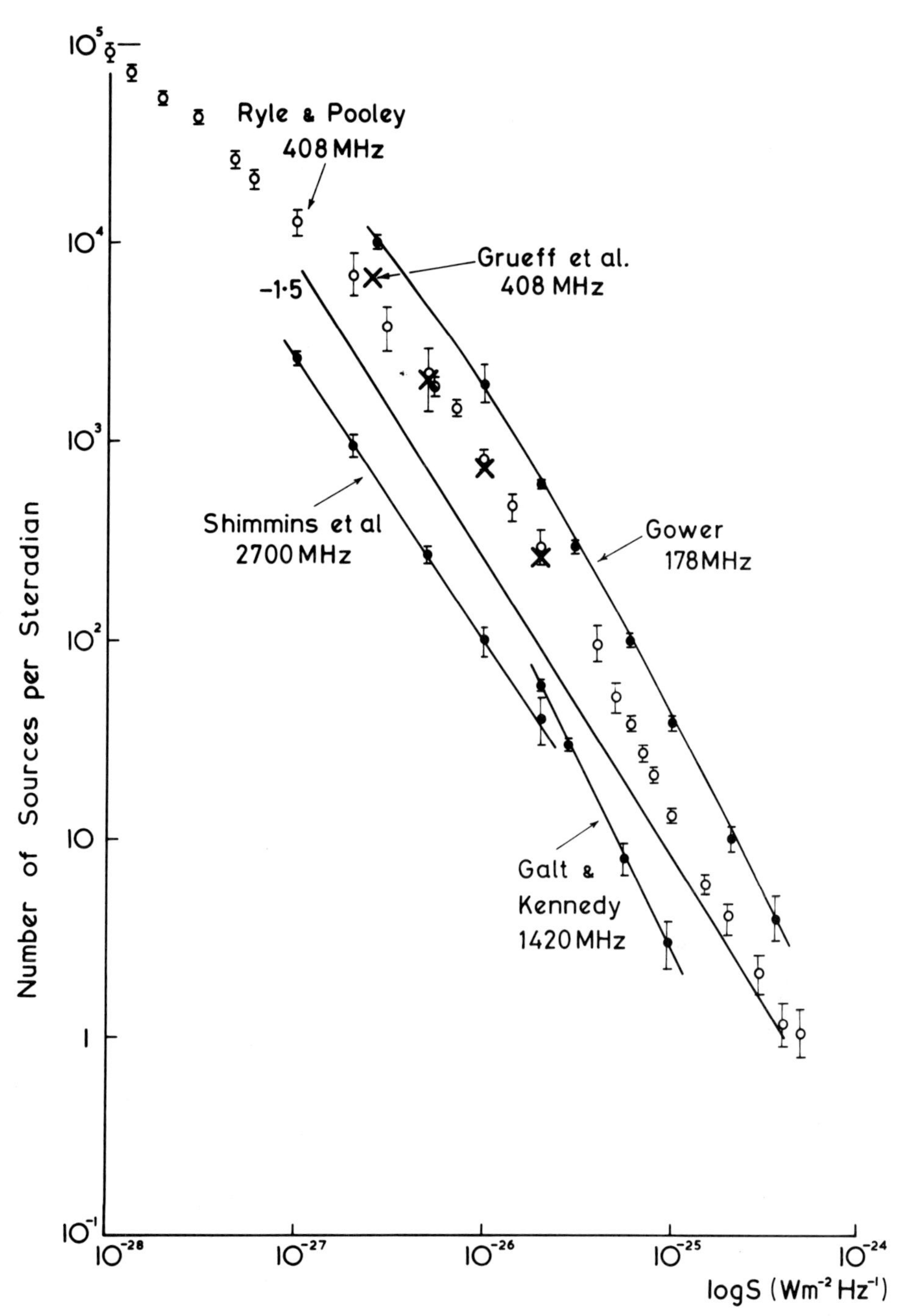

FIG. 4.—Comparison of various source counts (Shakeshaft 1969). The curves for different frequencies are displaced from each other since the flux densities of most sources fall with frequency, but their slopes are similar at a given number of sources per steradian. Reproduced by kind permission of the author.

Scott 1965; Gower 1966). In Pooley and Ryle's work the corrections were smaller (the greatest being 6%, at the limit of the 5C 2 survey), and were estimated from the known beam widths of the surveys and the measured number of sources per steradian at each flux level (involving a slight extrapolation below the limit of 5C 2).

There is one difficulty of a more basic kind: most radio sources are double, and often the components are separated by many times their diameters. When two "point" sources are a few minutes of arc apart, it is not always clear whether they are components of one source or two unrelated sources. Fortunately this question becomes significant only among the faintest sources, where an appreciable number of pairs with small separations occur by chance.

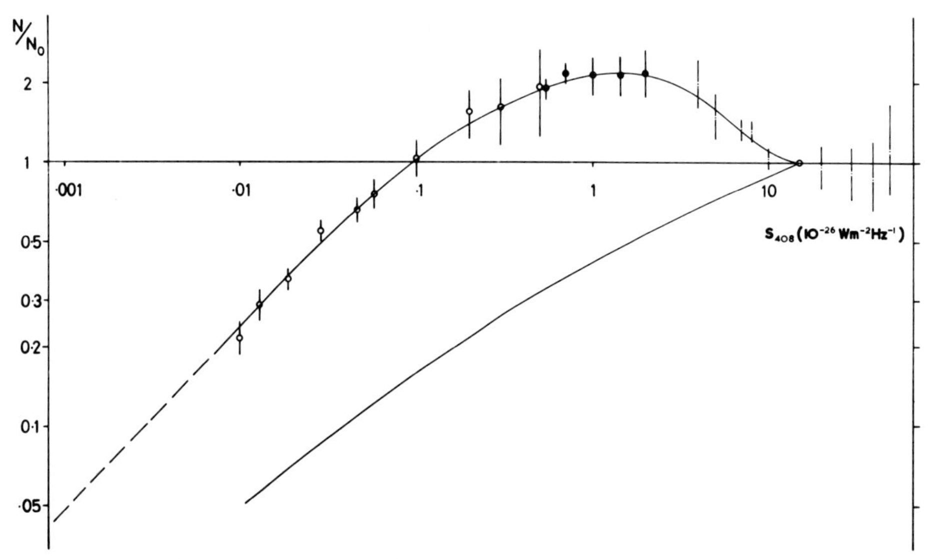

FIG. 5.—Variation of N/N_0 with S (Pooley and Ryle 1968). Here $N_0 = S^{-3/2} \times$ arbitrary constant, the predicted number for a uniform source distribution in Euclidean space. The lower curve represents the variation of N/N_0 for a uniform source distribution with a luminosity function similar to that of figure 6, in a Friedmann model with $\sigma_0 = 0.5$.

6.2.3. *Resolution of sources of large angular size.*—Many of the early surveys were made with interferometers, and it could be argued that some sources were omitted from the catalogs, or recorded with reduced flux density, because they were partially resolved at the interferometer baselines used. More recent detailed knowledge of the angular structure of sources has shown that these effects made little numerical difference to the source counts. The more recent surveys, including those used by Pooley and Ryle (1968), the Bologna surveys at 408 MHz (Braccesi *et al.* 1965; Braccesi *et al.* 1966; Grueff and Vigotti 1968), Galt and Kennedy (1968) (1400 MHz), Shimmins, Bolton, and Wall (1968) (2700 MHz), and the NRAO 5000-MHz survey (Bridle and Davies 1972), all used pencil beams. The integrated flux densities are used in those rare cases where a source is broader than the beam.

6.2.4. *Statistical uncertainties.*—Jauncey (1967; also Crawford, Jauncey and Murdoch 1970) has emphasized that it is not legitimate to fit a straight line to the observed points on a (log N, log S) plot, since the larger values of N include the data for the smaller values of N, so that the observed N are not independent; he has given a procedure for making a proper maximum likelihood estimate of the slope and its uncertainty. In fact, the (log N, log S) relation is a curve, but the same general principles apply. Clearly, a (log N, log S) plot may be misleading over a short range in N, but over a range of 10:1 in N the correlation between the values of N becomes slight.

6.2.5. *Isotropy.*—In some of the nearer clusters of galaxies, several intrinsically weak sources have been found in addition to one strong source (Virgo cluster; Perseus cluster [Ryle and Windram 1968]; Coma cluster [Willson 1970]). Apart from this, no convincing evidence of radio-source clustering has been found.

For the brightest sources ($S_{408} \gtrsim 10$ f.u.) there are fewer than 200 sources in the sky, and the statistical uncertainties are necessarily large as soon as one subdivides the sky into smaller regions.

Throughout the middle range of flux densities the isotropy of the source distribution is fairly well established. The 4C catalog of about 5000 sources at 178 MHz has been examined by Holden (1966), and over a range of angular scales from $0\overset{\circ}{.}5$ to 60° and for flux density limits from 2 to 10 f.u. she found the distribution of sources to be Poissonian within the limits of statistical uncertainty. This work was extended to limiting flux densities $S_{178} = 0.7$ f.u. and $S_{178} = 0.2$ f.u. by Hughes and Longair (1967), who used the statistics of the observed amplitudes in the resolution-limited observations for the 4C survey.

Among the faintest sources, the only valid test so far available is to compare source counts from the 5C surveys with each other ($\sim$ 200 sources in each). Since 5C 3 may contain a few sources associated with M31, and 5C 4 certainly does contain sources in the Coma cluster, the procedure is not ideal; nevertheless, most are background sources, and there are no significant differences when sources identified with Coma cluster galaxies are removed from 5C 4. Recently, large differences have been found between the spectral index distributions of sources in the 5C 1 and 5C 2 fields (Maslowski 1971, 1972; Pauliny-Toth, Kellermann, Davies, Fomalont, and Shaffer 1972); the number of sources involved is not yet very large, and it remains to be seen whether these differences have some prosaic explanation.

Though it does not concern only radio sources, one other indication of anisotropy may be mentioned here. Strittmatter, Faulkner, and Walmsley (1966) found that quasi-stellar objects with redshifts exceeding 1.5 occurred preferentially in two regions of sky near the galactic poles. None of the numerous selection effects that were suggested could satisfactorily account for the effect, but additional data gathered more recently have not confirmed the anisotropy (Wills 1971).

6.3. Interpretation of Source Counts

Replacing Euclidean geometry by any of a very wide range of Robertson-Walker metrics uniformly reduces the predicted value of $NS^{3/2}$ for faint sources, so that the observed source counts must imply some inhomogeneity or evolution of the source population. This inference is quite independent of any knowledge of distances, and, in particular, of the cosmological interpretation of the redshifts of QSOs; it is discussed in § 6.3.1. On the other hand, to specify the form of the evolution we need to be able to turn the S scale in some sense into a distance scale, and efforts in this direction are indicated in § 6.3.2.

6.3.1. The simple 3/2 power law (37) gives $N(S)$ for a uniform distribution of sources in Euclidean space. For a uniform source distribution in a Robertson-Walker metric, equations (7), (8), (9), and (10) give

Steady State:

$$N(S)S^{3/2} = \sum_i \frac{P_i^{3/2}}{r_i^3(1+z_i)^{3/2(1+\alpha_i)}} \int_0^{z_i} \frac{\rho_{ni} r^2 dr}{(1+z)^3}$$

$$= \tfrac{1}{3}\Sigma \rho_{ni} P_i^{3/2} \left[\frac{\ln(1+z_i) + 2(1+z_i)^{-1} - \frac{1}{2}(1+z_i)^{-2} - 3/2}{z_i^3(1+z_i)^{(3/2)(1+\alpha_i)}}\right] \tag{38}$$

and

Source-conserving:

$$N(S)S^{3/2} = \sum_i \rho_{ni} P_i^{3/2} \left[\frac{2Ar_i - \sin 2Ar_i}{4\sin^3 Ar_i}\right](1+z_i)^{-(3/2)(1+\alpha_i)}$$

$$= \tfrac{1}{3}\sum_i \rho_{ni} P_i^{3/2} \frac{f(Ar_i)}{(1+z_i)^{(3/2)(1+\alpha_i)}}, \tag{39}$$

where the last equation defines $f(Ar_i)$. In each case, z_i and r_i are the values appropriate to a source with radio luminosity P_i and spectral index α_i, at the limiting flux density S.

In equation (38) the function of z_i is a monotone decreasing function of z_i for all positive α_i, and, indeed, for all $\alpha_i > -2.5$. Since α is in the neighborhood of $+0.7$ for most sources (e.g., Kellermann, Pauliny-Toth, and Williams 1969), $N(S)S^{3/2}$ is a sum of functions each of which is a monotone decreasing function of z_i. Thus $NS^{3/2}$ must decrease as S decreases according to steady-state theory.

The predictions of evolutionary cosmologies are not so clear-cut. In deriving equation (39), it was assumed that the sources themselves are immutable (or that they are continually replaced by an equal number of similar sources), and such an assumption is quite unwarranted. The purely geometrical effects described by equation (39) are as follows:

i) *Open models.* A is imaginary—$A = iA'$, say—and then, in equation (39),

$$f(Ar_i) = \frac{3}{4}\frac{\sinh 2A'r_i - 2A'r_i}{\sinh^3 A'r_i},$$

which is a monotone decreasing function of r_i. In all open models z increases with r, so that $(1+z_i)^{-(3/2)(1+\alpha_i)}$ is also a decreasing function of r_i (provided that $\alpha_i > -1$); thus in open models the geometrical factors all conspire to make $N(S)S^{3/2}$ decrease for decreasing S.

ii) *Closed models.* The function $f(Ar_i)$ is no longer a decreasing function of r_i. Indeed, $f(Ar_i)$ rises to arbitrarily large values when Ar_i is nearly a multiple of π; this behavior occurs because we are then looking halfway round the universe and observing radiation from the "antipodes" focussed upon us ($Ar_i \simeq \pi$), or looking once round the universe at the "ghosts" of radio sources close to us ($Ar_i \simeq 2\pi$), and so on. Equation (39) is then no longer valid, for it was assumed in its derivation that S diminishes continually as r increases, whereas a glance at equation (6) or (7) shows that S becomes large whenever Ar approaches $n\pi$.

In relativistic models with the cosmological constant $\Lambda = 0$, we are effectively prevented from looking even halfway round the universe; equation (12) shows that, as z approaches infinity, Ar approaches a limit $2 \sin^{-1} (1 - 1/2\sigma_0)^{1/2}$; that is to say, we always see the beginning at a value of Ar less than π. The following argument then shows that, for any positive α, S is a monotone decreasing function of distance, so that equation (39) is valid, and $N(S)S^{3/2}$ is a monotone decreasing function of S.

Equation (12) can be rewritten as

$$(1 + z)^{-1/2} = \cos \tfrac{1}{2}Ar - \frac{H}{Ac} \sin \tfrac{1}{2}Ar \,, \tag{40}$$

and the expression (7) for flux density as

$$AP^{1/2}S^{-1/2} = 2 \sin \tfrac{1}{2}Ar(1 + z)^{1/2+\alpha}(1 + z)^{1/2} \cos \tfrac{1}{2}Ar \,.$$

Since $Ar < \pi$, $\sin \frac{1}{2}Ar$ increases with increasing z. By equation (40),

$$(1 + z)^{1/2} \cos \tfrac{1}{2}Ar = 1 + \frac{H}{Ac} (1 + z)^{1/2} \sin \tfrac{1}{2}Ar$$

and is also an increasing function of z. Therefore S is a decreasing function of z.

Provided all the α_i are positive, equation (39) shows that $N(S)S^{3/2}$ decreases with decreasing S if $f(Ar)(1 + z)^{-3/2}$ is a decreasing function of distance. Using equation (40),

$$f(Ar)(1 + z)^{-3/2} = f(Ar) \cos^3 \tfrac{1}{2}Ar \left(1 - \frac{H}{Ac} \tan \tfrac{1}{2}Ar\right)^3 ,$$

and $f(Ar) \cos^3 \frac{1}{2}Ar$ can be shown to be a monotone decreasing function of Ar in the range 0 to π, either by differential calculus or (less rigorously but more easily) by direct computation.

The following is a summary of the geometrical effects:

Steady-state models.—$NS^{3/2}$ decreases as S decreases, provided only that the spectral indices α of sources exceed -2.5. (The latter happens to be the limiting spectral index for a homogeneous synchrotron self-absorbed source. It is possible to construct inhomogeneous model sources with still more steeply rising spectra, but no such spectrum has been observed.)

Relativistic models.—Geometrical effects alone make $NS^{3/2}$ decrease as S decreases in all open models if $\alpha_i > -1$, and in all closed models with $\Lambda = 0$ if $\alpha_i > 0$.

In some closed models with $\Lambda > 0$, light can propagate many times round the universe, and complications associated with "ghost" sources can arise, which may in principle lead to an increase in $NS^{3/2}$ for small S (Kardashev 1967; McVittie and Stabell 1967; Rowan-Robinson 1968*b*). Longair and Scheuer (1970) confirm that $NS^{3/2}$ can rise above its value for large S by a factor approaching $3\pi/4 = 2.36$ in Lemaître models with a very long "coasting" phase. Such models require most of the observed radio sources to have the cosmological redshift corresponding to the "coasting" phase, and are therefore acceptable only if one adheres rigidly to the terms of reference of this section by disregarding the observed redshifts.

6.3.2. As shown in the previous section, the geometrical properties of a Robertson-Walker metric do not account for the rise in $NS^{3/2}$ with decreasing S; on the contrary, they merely aggravate the problem. We are forced to postulate that (for given P) the density of sources increases initially with z. So much follows from the source counts alone. But if we wish to find the form of this increase, we need to know the luminosity function of the radio sources.

In a catalog of radio sources, covering some large solid angle Ω of sky unobscured by dust, and complete to flux density S_0, we may be sure that we have all radio sources of power P up to distance $(P/S_0)^{1/2}$, i.e., in a volume $\frac{1}{3}\Omega\ (P/S_0)^{3/2}$. Even though we have precise positions for most of the bright sources, not all are identified with optical objects; and to be sure that all such sources within, say, 500 Mpc are identified, we must assume that no extragalactic source is fainter than some limiting absolute magnitude. This is an act of faith; but since all the identified powerful radio sources are QSOs or giant elliptical galaxies, and the weaker extragalactic sources are so much commoner that they need only be sampled in a nearby region, it is one that would be accepted by most astronomers. Not all of the identified radio galaxies have measured redshifts, but reasonable distance estimates may be made from their apparent magnitudes, since their dispersion in absolute visual magnitude is only about 0.5 mag (Peach 1970). The luminosity function is shown in figure 6.

If all sources had the same luminosity, each S would correspond to one distance, so that, given a geometry, the source counts would immediately give the density as a function of epoch.

It might appear at first sight that the evolution of the source population can still be derived uniquely from the source counts, though by a more complicated computation, provided that we have confidence in the luminosity function derived from our local sample of sources with known distances. That is not so, for a number of reasons:

i) The "local luminosity function" is not well defined for the most powerful sources, which are so rare that even the nearest examples have appreciable redshifts. The initial rise in $NS^{3/2}$ is largely due to such sources, for there is no statistically significant part of the observed counts at large S for which $NS^{3/2}$ is constant.

ii) There is no reason to suppose that the luminosity function does not change with epoch; indeed, it cannot remain constant, as Longair (1966) pointed out. If weak sources evolved as rapidly as strong sources must to produce the rise between $S_{408} = 10^{-26}$ and $S_{408} = 10^{-25}$ W m^{-2} Hz^{-1} in figure 5, then the curve could not fall as rapidly as the observed curve at lower flux densities. The contribution of sources with $P_{408} = 10^{24}$ would reach its peak near $S_{408} = 3 \times 10^{-28}$ if the sources with $P_{408} = 3 \times 10^{26}$ W Hz^{-1} steradian^{-1} are responsible for the rise in $NS^{3/2}$ from $S_{408} = 10^{-25}$ W m^{-2} Hz^{-1} downward. Since the total extragalactic background with a spectral index of 0.75 is no more than about 3° K at 408 MHz (Bridle 1967) and the sources with $S_{408} > 10^{-28}$ W m^{-2} Hz^{-1} account for 1.4° K according to the data of figure 5, it is clear that strong evolution of the sources of low luminosity would at the same time have the effect of increasing the background brightness above the observed value.

Thus we have to interpret the source counts in terms of a perfectly general evolving luminosity function $\rho(P, z)$, of which we know only (i) the local values $\rho(P, 0)$ for small P, (ii) something of the values for large P up to $z \simeq 2$, but only for QSOs. Redshifts for galaxies have not been measured beyond $z = 0.46$ (3C 295), and very few are known beyond $z = 0.25$.

There is now an infinite variety of possible models for $\rho(P, z)$ which will all fit the source counts, and many models have indeed been proposed; see Longair (1966);

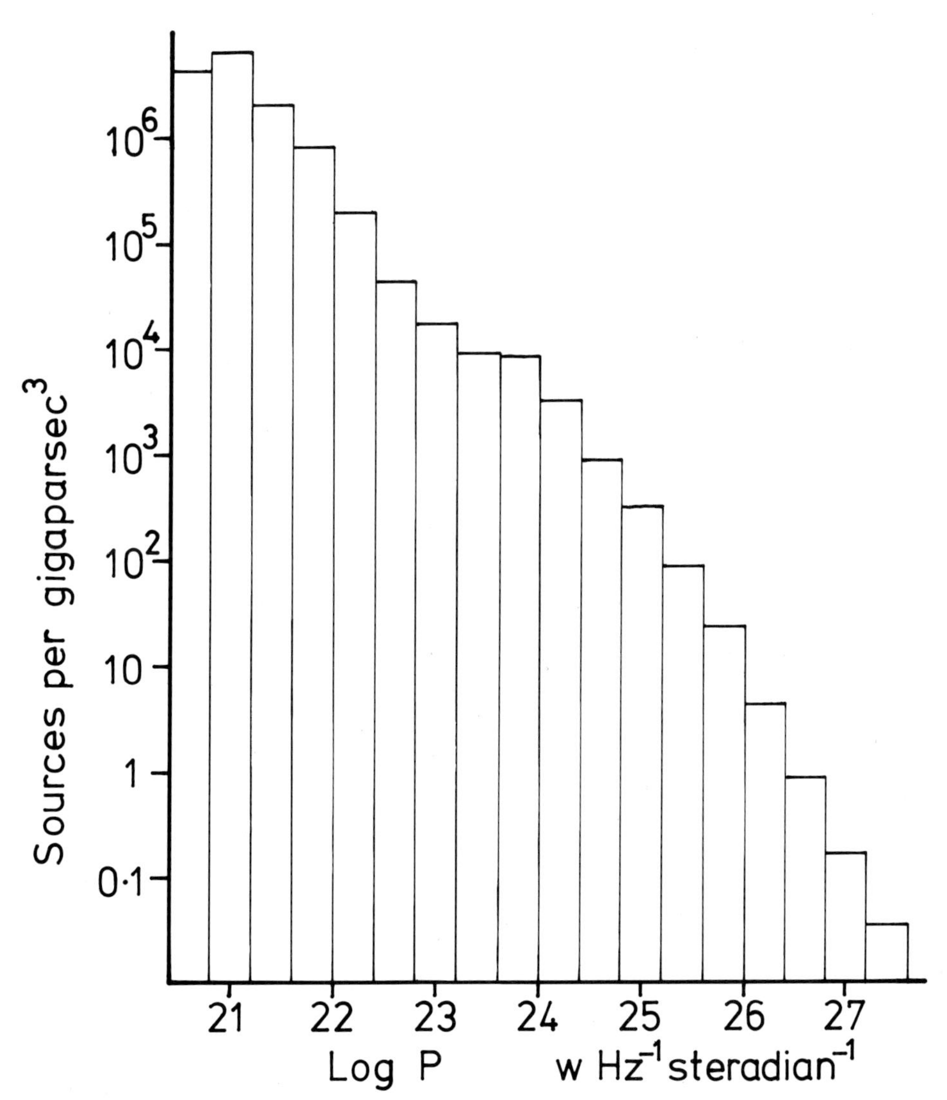

FIG. 6.—The luminosity function of radio sources, according to Longair (1971).

Davidson and Davies (1966); Doroshkevich, Longair, and Zel'dovich (1970); Rowan-Robinson (1970); Ringenberg and McVittie (1970); Davidson (1971), and other papers cited in these. Two special cases of $\rho(P, z)$ are often discussed:

$$\text{i)}\quad \begin{aligned} \rho(P, z) &= \rho(P)\psi(z)\,, && P > P_0\,, \\ &= \rho(P)\,, && P < P_0\,, \end{aligned} \tag{41}$$

called "density evolution"; and

$$\text{ii)}\quad \begin{aligned} \rho(P, z) &= \rho(P/\phi(z))/\phi(z)\,, && P > P_0\phi(z)\,, \\ &= \rho(P)\,, && P < P_0\,, \\ &= 0\,, && P_0 < P < P_0\phi(z)\,, \end{aligned} \tag{42}$$

called "luminosity evolution." Here $\rho(P)$ represents the local luminosity function, $\psi(0) = \phi(0) = 1$, and the need to limit evolution to strong sources, $P > P_0$, has already been discussed. "Luminosity evolution" is what would occur if each strong source with power P at the present epoch t_0 had had power $P\phi(z)$ at the earlier epoch corresponding to redshift z. It is disputed in the literature whether "density" or "luminosity" evolution better represents the observations, and whether such a distinction is even possible. For some forms of luminosity function, e.g. $\rho(P) = KP^{-m}$, the two are strictly equivalent until one reaches the lower limit P_0 of the evolving component of the source population, and the observed local luminosity function is not greatly different from such a power law over a large range of P.

Expressing the evolution in terms of epoch t instead of z, expressions such as

$$\begin{aligned} \psi &= (t/t_0)^{-m}\,, && t > t_1\,, \\ &= 0\,, && t < t_1\,, \end{aligned} \tag{43}$$

$$\begin{aligned} \psi &= (t/t_0)^{-m}\,, && t > t_1\,, \\ &= (t_1/t_0)^{-m}\,, && t < t_1\,, \end{aligned} \tag{44}$$

$$\psi = \exp\,[m(t_0 - t)] \tag{45}$$

(where t_0 is the present epoch, and m and t_1 are constants) and similar expressions for ϕ, have been fitted successfully to the source counts. On reading the literature one may well get the impression that the theories are in violent conflict; in fact, they share the most important features:

i. The source density, for given P, rises rapidly to 1000–10,000 times $\rho(P, 0)$ at $z = 3$–6.

ii. For larger z, $\rho(P, z)$ either declines or rises much less rapidly with z.

iii. Only strong sources evolve.

I mention one further model, because its brazen simplicity makes some features of the model-fitting process clearer. Suppose we take

$$\rho(P, z) = \rho(P, 0) + \rho'(P)\,\delta\,(z - 3), \tag{46}$$

where the second term represents a population of radio sources all at $z = 3.0$ with a luminosity function $\rho'(P)$ chosen to fit precisely the excess source counts. This model, too, has the three features listed above, and it is, by definition of $\rho'(P)$, consistent with the source counts. It is really very similar to the models appearing in the literature, but the latter smooth the rough edges of the delta function, and give the evolving part some sort of continuity with the present-day local luminosity function.

6.4. Are We in a Hole?

The principal conclusion from the source counts is sometimes expressed as: "There is an excess of faint sources"; at other times as: "There is a deficiency of bright sources." The content of the two sentences is the same, but the second version is often used with the implication that the lack of bright sources is a local (though necessarily nonstatistical) fluctuation in the number density of sources around us. Indeed, if one asks how many bright sources would have to be added to the observed sky in order to make the ($\log N$, $\log S$) relation begin with a slope -1.5 at large S and never become steeper, that number of sources is of the order of 100 (various authors' estimates range from about 50 to about 200). Such numbers are too small to affect the isotropy of the source counts in a statistically convincing way, so we cannot argue *directly* from the isotropy of the source counts that the "hole" we live in is isotropic and thus offends against the belief that we are not in a special position in the Universe. Though I do not favor this interpretation, I think it is the most reasonable of the criticisms leveled against the evolutionary interpretation of the source counts, and would draw attention to a good discussion of this point of view by Kellermann (1972).

The argument against the "local hole" interpretation of the source counts proceeds on the following lines:

i) The "hole" cannot be very local ($z \ll 1$), otherwise (*a*) the total flux density from more distant sources would exceed the observed radio background, and (*b*) the source counts would not converge for $S_{408} < 1$ f.u., as they are observed to do. The convergence below 1 f.u. is almost certainly due to cosmological effects, since the number of sources involved is large in that case, and the observed isotropy is statistically meaningful.

ii) Since the "hole" extends to appreciable redshifts, it is incorrect to compare the observed source counts with the law $N \propto S^{-3/2}$ appropriate to a Euclidean metric; one must compare them with a curve such as that in figure 5 appropriate to a Friedmann or a steady-state model. Evidently the discrepancy then becomes much larger, and extends to lower flux densities, if we normalize the computed curves to fit the observed source counts at low S, where the "typical" source population outside our "local hole" predominates. But with the larger number of missing sources the isotropy of the source counts again becomes significant, and we find ourselves once more in the center of a large spherically symmetrical "hole" in the source distribution.

The second part of the above argument becomes complicated because we have no simple, useful, and at the same time rigorous way to limit the range of luminosity functions that may be used in computing the curves for non-evolving source populations. There can be little doubt about the local luminosity function for radio galaxies, but at the 1 f.u. level most sources are not even identified and may consist largely of some other kind of source. As an extreme example, it might be argued that the "hole" is in the distribution of quasi-stellar sources, which are at distances unrelated to their redshifts, and all have the same low radio luminosity, the lowest allowed by the restrictions cited in (i). Short of a very lengthy analysis, it seems best to bypass these complexities with a rather crude argument (iii):

iii) The source counts for a non-evolving population of sources cannot behave like curve 2 in figure 7; they must fit smoothly onto the source counts for small S, like curve 3. Note that, in changing from curve 2 to curve 3, the number of "missing sources" is increased chiefly by the change near the junction of the curves, and depends very little on the behavior of curve 3 at very large flux densities.

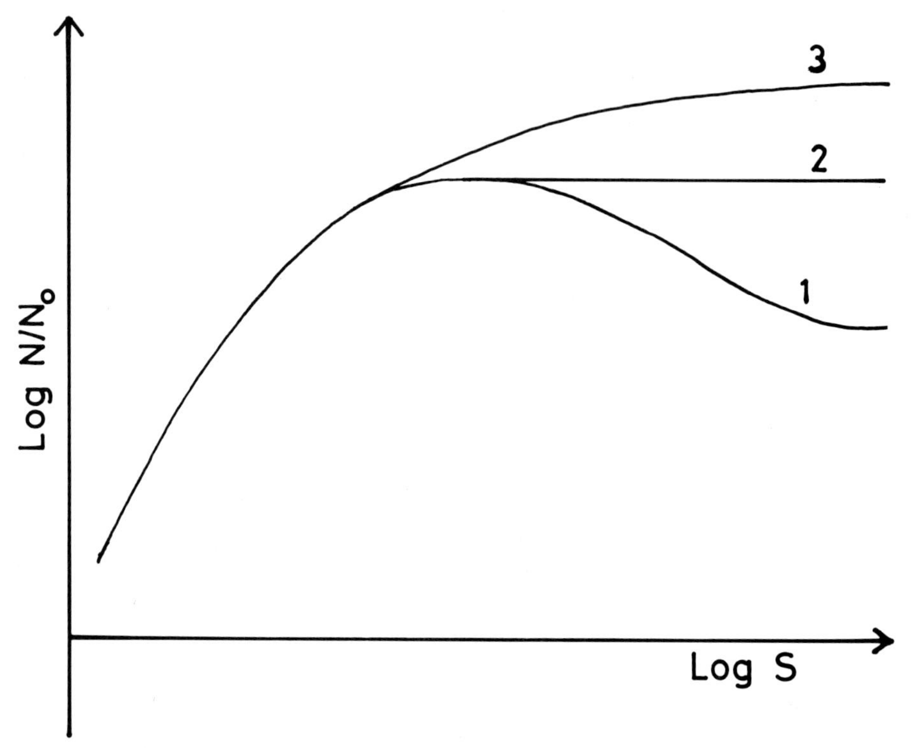

FIG. 7.—The variation of N/N_0 interpreted as a local scarcity of sources.

7. THE EVOLUTION OF THE LUMINOSITY FUNCTION OF QUASI-STELLAR OBJECTS

7.1. THE LUMINOSITY FUNCTION

The emission-line redshifts of over 150 quasi-stellar objects have been measured, and redshifts have also been measured for a comparable number of radio galaxies. Most of the published redshifts of quasi-stellar objects can be found in E. M. Burbidge's (1967) review, supplemented by Schmidt's (1969) review and more recent measurements by Tritton (1971) (4 QSOs); Arp, Burbidge, Mackay, and Strittmatter (1972) (3C 455); Burbidge and Strittmatter (1972*b*) (20 QSOs); Lowrance, Morton, Zucchino, Oke, and Schmidt (1972) (PHL 957); Lynds and Wills (1972) (19 QSOs); and Peterson and Bolton (1972) (9 QSOs and 1 galaxy). Most of the published redshifts of radio galaxies can be found in G. R. Burbidge's (1970) review, supplemented by more recent measurements by Burbidge and Burbidge (1972) (10 galaxies); Burbidge and Strittmatter (1972*a*) (20 galaxies); Tritton (1972) (20 galaxies); and Whiteoak (1972) (27 galaxies).

Given the redshifts of all quasi-stellar objects whose optical and radio flux densities exceed given limits, one could plot a three-dimensional diagram such as that sketched

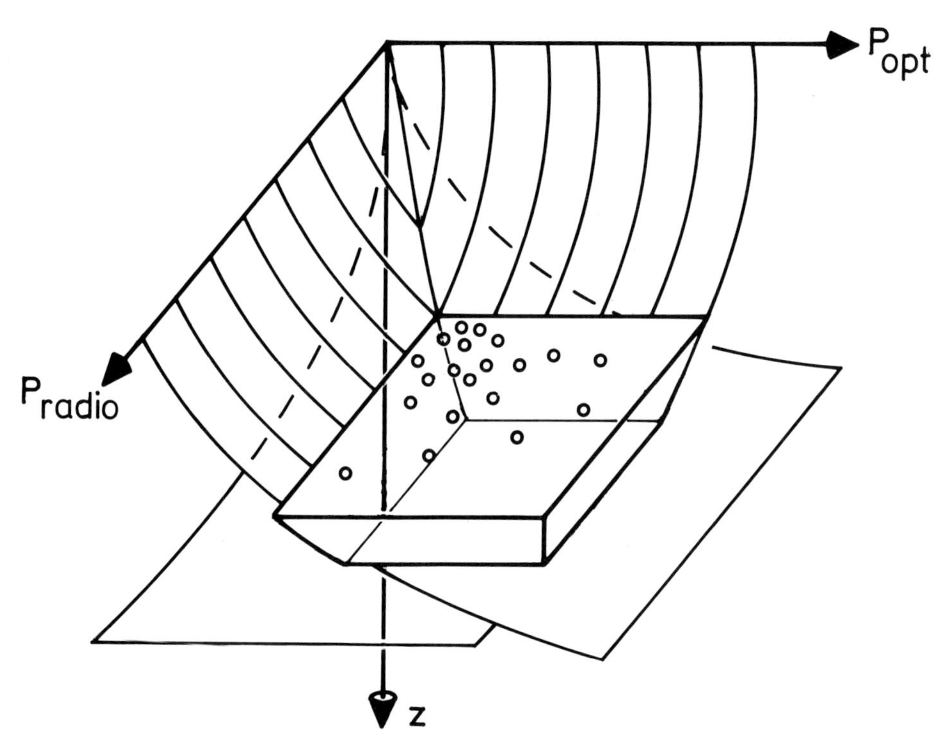

FIG. 8.—Diagram illustrating the joint optical and radio luminosity function. The two intersecting curved sheets represent the optical and radio flux-density limits of the observed sample; the slab shows schematically the luminosity function at one redshift z.

in figure 8, representing $\Phi(P_{\text{opt}}, P_{\text{rad}}, z)$ the joint optical and radio luminosity function, as a function of redshift z. The following complications arise:

i) If Φ is to represent numbers per unit coordinate volume in the ranges dP_{opt}, dP_{rad} of luminosity, then the points in the diagram must be assigned weights inversely proportional to $d(\text{coordinate volume})/dz$.

ii) To convert the observed flux densities and redshifts into powers P, and also to assign weights as described in (i), one must assume a metric. It has been found that the choice of metric does not make a great difference to the results, at any rate within the range of Friedmann models with $0 \leq \sigma_0 \leq 1$.

iii) To find $P(\nu)$ one requires the observed flux density $S[\nu/(1 + z)]$, which Schmidt denotes by f_ν. For $\nu <$ 2GHz most sources have $S \propto \nu^{-\alpha}$, but at higher radio frequencies and in the optical spectrum no such simple rule is generally applicable. Both Schmidt

(1968, 1970, 1972*a*) and Lynds and Wills (1972) use the optical flux density at emitted wavelength 2500 Å, as the redshifted wavelength is then in the region accessible to ground-based observation for most quasi-stellar objects.

iv) The redshifts of quasi-stellar objects are not known over the whole sky down to any useful optical and radio limits. Two complete samples have been obtained:

Schmidt's sample of quasi-stellar sources with $f_{\rm opt} = S(2500(1 + z)\ \text{Å}) > 10^{-30}$ W m^{-2} steradian^{-1} and $S(178\ \text{MHz}) > 10^{-25}$ W m^{-2} steradian^{-1}; there are 34 such objects (Schmidt 1972*a*) in the 5.7 steradians of sky that is above 10° galactic latitude and covered by the 3CR catalog of radio sources (Bennett 1962).

Lynds and Wills's (1972) sample of quasi-stellar sources with $f_{\rm opt} = S(2500(1 + z)\ \text{Å}) > 10^{-30.4}$ W m^{-2} steradian^{-1} and $S(178\ \text{MHz}) > 2 \times 10^{-26}$ W m^{-2} steradian^{-1}. Lynds and Wills found 31 such objects in a selected area of 0.35 steradians.

Note that the radio limit is in each case a limit on S_{178} in the observer's frame, and does not correspond to a well-defined limit on $P_{\rm rad}$ at a given redshift, since the spectra of the sources are not identical. Schmidt (1972*a*) and Lynds and Wills (1972) minimize this uncertainty by using the flux density at emitted frequency 500 MHz for $f_{\rm rad}$, so that only small corrections are required for sources in the middle of the range of redshifts.

(The samples cited above are not quite complete, as there are one or two sources in whose spectra no lines have yet been identified, but they are so few that their absence is unlikely to distort the statistics seriously.)

Sources in complete samples can be plotted in figure 8, or equivalent representations, with weights inversely proportional to the area of sky searched. Most known redshifts are not the result of any exhaustive search over a well-defined area of sky and range of flux densities, and cannot be plotted in figure 8 because we do not know what weight to assign to them.

The next and most serious difficulty is that the straightforward presentation of figure 8 dilutes the available data so much that no immediately comprehensible picture appears. With a very modest division of $P_{\rm opt}$, $P_{\rm rad}$, and z into five classes each, the $(34 + 26)$ objects[2] represent an average of less than half an object per cell. Furthermore, the distribution is very uneven, since most of the objects lie near the surfaces representing the optical and radio flux-density limits. Thus we have some information on $\Phi(P_{\rm opt}, P_{\rm rad}, z)$ for small $P_{\rm opt}$ and $P_{\rm rad}$ in our neighborhood because weak objects are relatively common, but not at large z because they then lie below the flux-density limits. We also have some information on powerful sources for $1 < z < 2$, but know very little about the local distribution for those sources, because there simply is not a statistically significant sample of them near us. Yet there is some significant information in the data, if only we can group them together in meaningful ways. For a start, the data are not compatible with the hypothesis that Φ is independent of z (no evolution), as was shown by the "luminosity-volume test" which is described next.

7.2. The Luminosity-Volume Test

Suppose a quasi-stellar source with redshift z has been picked up in a search of some solid angle Ω of sky, and its $P_{\rm opt}$ and $P_{\rm rad}$ (calculated according to the assumed metric)

[2] Five of Lynds and Wills's sources are also in Schmidt's sample.

are such that it would have been above the optical limit of the search at any redshift up to z_{opt}, and above the radio limit up to z_{rad}. Then an object with the same luminosities would be detectable in the search anywhere inside the coordinate volume V_{max} of the cone of solid angle Ω up to the distance corresponding to $\min(z_{opt}, z_{rad})$. Now, if the sources were distributed uniformly in coordinate volume (no evolution of Φ), then all values of (V/V_{max}) should be equally likely, where V is the volume of the cone extending as far as the distance corresponding to the observed redshift z. Thus the values of (V/V_{max}) of the sources in the sample should be distributed uniformly between 0 and 1; in particular, the mean value $\langle V/V_{max}\rangle$ should be 0.5, with a standard deviation $(12N)^{-1/2}$ for a sample of N sources.

Sciama and Rees (1966) had pointed out that there were fewer quasi-stellar sources with small redshift than one would expect from a uniform distribution in space. The luminosity-volume test, more or less as described above, was used and further elaborated by various authors, including Rowan-Robinson (1968*a*) and Schmidt. Schmidt (1968) using the sample described above (except for one object that was added later), found $\langle V/V_{max}\rangle = 0.7$. This result is virtually independent of the kind of Friedmann model adopted: for $\sigma_0 = 0.0$, $\langle V/V_{max}\rangle = 0.69$, while for $\sigma_0 = 1.0$, $\langle V/V_{max}\rangle = 0.70$. For their sample, Lynds and Wills (1972) found $\langle V/V_{max}\rangle = 0.64$ if $\sigma_0 = 0.0$, and $\langle V/V_{max}\rangle = 0.67$ if $\sigma_0 = 1.0$. For the metric of the steady-state model Schmidt (1968) found $\langle V/V_{max}\rangle = 0.75$, where V and V_{max} now refer to proper volumes, since sources are uniformly distributed in proper volume according to steady-state theory. All these results are significantly above 0.5, and are therefore incompatible with a steady-state model, and also with a non-evolving luminosity function in Friedmann models. Rowan-Robinson (1968*b*) and Petrosian (1969) found certain Lemaître models in which the distribution of V/V_{max} appears to be more uniform, but according to Longair and Scheuer (1970) the distribution is not uniform even in these models if we restrict attention to the region of the model in which the flux density of a source of given power falls with increasing redshift.

One may now test other models of $\Phi(P_{opt}, P_{rad}, z)$ by comparing the distribution of V/V_{max} computed for the model with the observed distribution. This is a valid and useful test. However, it does not use all the available information, essentially because in using the luminosity-volume test one mixes up the data for various ranges of redshift (or, equivalently, for strong and weak sources). In the simple case of a Euclidean metric and only one flux-density limit, the luminosity-volume test would give just the same information as counts of the sources in question to various flux density limits S_0. In that case

$$V = \tfrac{1}{3}\Omega(P/S)^{3/2}\,, \qquad V_{max} = \tfrac{1}{3}\Omega(P/S_0)^{3/2}$$

and therefore

$$V/V_{max} = (S_0/S)^{3/2} \tag{47}$$

for each source, so the distribution of (V/V_{max}) is uniquely determined by the distribution of S, and vice versa. In a Friedmann model the exact equivalence breaks down, since the coordinate volume is not simply proportional to a power of P/S, but something remarkably close to this equivalence seems to persist, even with the added complication of two flux-density limits. Longair and Scheuer (1970) made a numerical experiment, re-assigning the observed redshifts at random among the objects in Schmidt's sample and then performing the luminosity-volume test on these artificial samples, which have the same distribution of S and the same range of redshifts as the original. The values of $\langle V/V_{\max}\rangle$ were almost unchanged. (See Carswell and Weymann 1971 for further discussion. The reader's attention is also drawn to papers in which the views expressed above are condemned: Rees and Schmidt 1971; Lynds and Petrosian 1972.) One way of recovering some of the additional information is to apply a luminosity-volume test separately to sources in various ranges of radio luminosity, or optical luminosity, or both (Schmidt 1968, 1972*a*).

7.3. More General Procedures

More generally, the kind of procedure one has to aim at is the following. Take figure 8, or some equivalent representation of the data. Invent a model of $\Phi(P_{\text{opt}}, P_{\text{rad}}, z)$ containing a reduced number of free parameters or arbitrary functions, so that the rather sparse data become sufficient to define the model. Find the maximum likelihood estimate of the free parameters of the model, and show (by some indication of goodness of fit, or otherwise) that the model is consistent with the data. Lynden-Bell (1971*a*) has shown how to find the best fit for luminosity functions of the form

$$\Phi = \phi_1(P_{\text{rad}}/P_{\text{opt}})\phi_2(P_{\text{opt}})\rho(z) \,. \tag{48}$$

Before one can use such systematic procedures, one has to find a suitable form for Φ. Schmidt (1970) reached the form (48) by considering the great similarity in the redshift distributions for two samples of quasi-stellar objects: one selected by optical methods alone, the other from 3CR radio sources. Schmidt (1970, 1972*a*) found two forms of $\rho(z)$ to fit his data, in a $\sigma_0 = 1$ Friedmann model,

$$\rho(z) = (1 + z)^6 \qquad (z < 2.5) \tag{49}$$

with a much slower increase beyond $z = 2.5$ (actually $\rho(z) = 0$ in the numerical analysis), and

$$\rho(z) = 10^{-5Ht} \tag{50}$$

(t = epoch), the latter giving a slightly better fit. The laws (49) and (50), with suitable scale factors, behave similarly in the range $0.5 < z < 2$ where most of the data lie, but then differ by a factor around 30 at $z = 0$; accordingly the functions ϕ_2 corresponding to the two laws will differ for weak sources, a few of which are observed at small z. Schmidt tests his model Φ against his 3CR sample by a modified luminosity-volume test, using fictitious volumes V' in which each element of coordinate volume is weighted by the density $\rho(z)$; a good fit to the model is then shown by a uniform distribution of $V'/V'_{\max}$ over the interval (0, 1) and a mean, $\langle V'/V'_{\max}\rangle$, close to 0.5. He also shows that his model is tolerably consistent with such information as is available on smaller samples of fainter objects. Lynden-Bell (1971*a*), using Schmidt's 3CR sample, finds a law of evolution close to the exponential law (50). Lynds and Wills (1972), using Schmidt's methods of analysis on their sample of rather fainter objects, found that formula (49) did not fit well, and prefer a law of the form

$$\rho(z) = V^a \,, \tag{51}$$

which fits their data with $a = 1$. The luminosity functions found by all these authors increase rapidly toward low P_{opt}, and this fact may indicate that quasi-stellar objects are the high-luminosity tail of a class of much commoner objects—perhaps N galaxies (Lynden-Bell 1971*b*), particularly since a few objects originally called quasars have already been reclassified as N galaxies.

It is clear that the analysis would become much simpler for a sample selected by purely optical means, so that there is only one flux limit; indeed, much valuable work has been done in this direction (e.g., Sandage and Luyten 1969; Braccesi and Formiggini 1969; Braccesi, Formiggini, and Gandolfi 1970). Unfortunately there seem to be great difficulties in obtaining a truly complete sample of optically selected quasi-stellar objects. So far, the procedure has been to select candidate objects from an optical field by their colors, and take spectra of these candidate objects individually. However, the colors corresponding to a typical quasi-stellar spectrum vary with its redshift, and for some redshifts, notably near $z = 3$, the $U - B$ and $B - V$ colors become very similar to those of some main-sequence stars, and there is likely to be a serious selection effect against objects with these redshifts.

Subject, as always, to the cosmological interpretation of the redshifts, the necessary evolution of the luminosity function seems to be agreed in general terms, though the precision is perhaps typically a factor of 2 either way.

Attempts have also been made to detect evolution in the luminosity function of radio galaxies using the luminosity-volume test (Rowan-Robinson 1970; Schmidt 1972*b*) but have not yielded convincing evidence of evolution; since most of the redshifts concerned are quite small, this result is hardly surprising.

7.4. Other Correlations

Various attempts have been made to find correlations with redshift of (i) the flux-density distribution and (ii) the angular diameter distribution of quasi-stellar sources, chiefly with a view to detecting a connection between redshift and distance, and hence proving (or disproving) the cosmological interpretation of redshifts.

The radio flux densities show no correlation at all with redshift; the optical flux densities do show a correlation with redshift which is statistically significant. Longair and Scheuer (1967) confirmed the latter correlation for the redshifts available at the time, but found that it could be accounted for entirely by the necessary effects of redshift on the energies and arrival rates of photons, without postulating any relation between redshift and distance—for example, the redshifts might be due to various sources residing in gravitational potential wells of various depths. This conclusion is no longer correct for the data assembled by Lynds and Wills (1972); their plot of log (f_{2500}) against log z is fitted appreciably better by the line representing a Friedmann model with $\sigma_0 = 1$ than by a curve representing the effect of redshift alone, though the evidence is perhaps not strong enough to convince a skeptic.

In principle, measurements of angular diameter should provide good discrimination between cosmological models. For an object of given proper diameter, the angular diameter tends to a finite limit for large redshifts in steady-state theory, but increases with redshift for large redshifts in Friedmann models (since the Universe used to be much smaller than it is now; cf. eq. [5]). The maximum angular extent of an extragalactic double radio source is generally quite well defined, since the outer edges of the radio components tend to be sharp. The difficulty (as in tests involving flux density primarily) is that the range of proper diameters is very large, from less than a parsec to hundreds of kiloparsecs. Two attempts have been made to use angular diameters in cosmological tests. In one (Longair and Pooley 1969; Fanaroff and Longair 1972) the sources of large angular diameter among bright sources with known redshifts (radio galaxies as well as quasi-stellar objects) were used to predict how many sources with diameters greater than 70″ arc should be found in surveys of sources two orders of magnitude fainter, if the source population were the same at all redshifts. The number of faint sources of large angular diameter exceeds the predicted numbers, implying that there has been evolution, though it is not nearly as strong as the evolution of the source population as a whole (required to explain the radio source counts) nor as strong as the evolution of quasi-stellar sources. In the other, Miley (1971) has plotted the maximum angular extent of about 50 quasi-stellar objects and a similar number of radio galaxies against redshift: though there is a large scatter in angular size at each redshift, there appears to be a reasonably well defined upper envelope. He finds that (i) there is continuity between the distribution of points representing radio galaxies and quasi-stellar sources, (ii) the upper envelope falls with redshift more rapidly than *either* Friedmann models with $\sigma_0 > 0$ *or* steady-state models predict; the angular diameters appear to fall as z^{-1} up to $z = 2$. This behavior finds a natural explanation in Friedmann models if radio source components interact significantly with the intergalactic medium; the latter has a density proportional to $(1 + z)^3$, and source components would travel less far in a dense medium. Radio-source components might also have shorter lifetimes at large redshifts, for the energy loss of relativistic electrons by Compton scattering of the microwave background was faster at earlier epochs. Qualitatively, at any rate, Miley's result is also consistent with Longair and Pooley's finding that the proportion of large-diameter sources is lower among faint sources than one would expect if sources of all diameters evolved at the same rate.

REFERENCES

Allen, C. W. 1962, *Astrophysical Quantities* (2nd ed., London: Athlone Press).
Allen, R. J. 1968, *Nature*, **220**, 147.
———. 1969, *Astr. and Ap.*, **3**, 316, 382.
Arons, J., and McCray, R. 1969, *Ap. Letters*, **5**, 123.
Arons, J., and Wingert, D. W. 1972, *Ap. J.*, **177**, 1.
Arp, H. 1965, *Ap. J.*, **142**, 402.
Arp, H. C., Burbidge, E. M., Mackay, C. D., and Strittmatter, P. A. 1972, *Ap. J.* (*Letters*), **171**, L41.
Bahcall, J. N., and May, R. M. 1968, *Ap. J.*, **152**, 37.
Bahcall, J. N., and Salpeter, E. E. 1965, *Ap. J.*, **142**, 1677.
———. 1966, *ibid.*, **144**, 847.
Bailey, J. A., and Pooley, G. G. 1968, *M.N.R.A.S.*, **138**, 51.
Bennett, A. S. 1962, *Mem. R.A.S.*, **67**, 163.
Bergeron, J. 1969, *Astr. and Ap.*, **3**, 42.
Bergeron, J., and Salpeter, E. E. 1970, *Ap. Letters*, **7**, 115.
Blair, A. G., Beery, J. G., Edeskuty, F., Hiebert, R. D., Shipley, J. P., and Williamson, K. D., Jr. 1971, *Phys. Rev. Letters*, **27**, 1154.
Braccesi, A., Ceccarelli, M., Fanti, R., Gelato, G., Giovannini, C., Harris, D., Rosatelli, C., Sinigaglia, G., and Volders, L. 1965, *Nuovo Cimento*, **40B**, 267.
Braccesi, A., Ceccarelli, M., Fanti, R., and Giovannini, C. 1966, *Nuovo Cimento*, **41B**, 92.
Braccesi, A., and Formiggini, L. 1969, *Astr. and Ap.*, **3**, 364.
Braccesi, A., Formiggini, L., and Gandolfi, E. 1970, *Astr. and Ap.*, **5**, 264.
Brecher, K., and Burbidge, G. R. 1972, *Nature*, **237**, 440.

Bridle, A. H. 1967, *M.N.R.A.S.*, **136**, 219.
Bridle, A. H., and Davies, M. M. 1972, *IAU Symposium No. 44*, ed. D. Evans (Dordrecht: Reidel), p. 437.
Burbidge, E. M. 1967, *Ann. Rev. Astr. and Ap.*, **5**, 399.
Burbidge, E. M., and Burbidge, G. R. 1972, *Ap. J.*, **172**, 37.
Burbidge, E. M., and Strittmatter, P. A. 1972*a*, *Ap. J.* (*Letters*), **172**, L37.
———. 1972*b*, *ibid.*, **174**, L57.
Burbidge, G. R. 1970, *Ann. Rev. Astr. and Ap.*, **8**, 369.
Carswell, R., and Weymann, M. N. 1971, *M.N.R.A.S.*, **156**, 19P.
Crawford, D. F., Jauncey, D. L., and Murdoch, H. S. 1970, *Ap. J.*, **162**, 405.
Davidson, W. 1971, *M.N.R.A.S.*, **154**, 339.
Davidson, W., and Davies, M. 1966, *M.N.R.A.S.*, **134**, 405.
Davies, R. D., and Jennison, R. C. 1964, *M.N.R.A.S.*, **128**, 123.
Dicke, R. H., Peebles, P. J. E., Roll, P. G., and Wilkinson, D. T. 1965, *Ap. J.*, **142**, 414.
Doroshkevich, A. G., Longair, M. S., and Zel'dovich, Ya. B. 1970, *M.N.R.A.S.*, **147**, 139.
Doroshkevich, A. G., Zel'dovich, Ya. B., and Novikov, I. D. 1967, *Zh. Eksper. Teoret. Fiz.*, **53**, 644 (English transl. *Soviet Phys. —JETP*, **26**, 408).
Fanaroff, B. L., and Longair, M. S. 1972, *M.N.R.A.S.*, **159**, 119.
Field, G. B. 1958, *Proc. I.R.E.*, **46**, 240.
———. 1959, *Ap. J.*, **129**, 536.
———. 1972, *Ann. Rev. Astr. and Ap.*, **10**, 227.
Field, G. B., Solomon, P. M., and Wampler, E. J. 1966, *Ap. J.*, **145**, 351.
Fowler, W. A. 1970, *Comments Ap. and Space Phys.*, **2**, 134.
Galt, J. A., and Kennedy, J. E. D. 1968, *A.J.*, **73**, 135.
Gamow, G. 1948, *Phys. Rev.*, **74**, 505.
Goldstein, S. J., Jr. 1963, *Ap. J.*, **138**, 978.
Gower, J. F. R. 1966, *M.N.R.A.S.*, **133**, 151.
Grueff, G., and Vigotti, M. 1968, *Ap. Letters*, **2**, 113.
Gunn, J. E., and Peterson, B. A. 1965, *Ap. J.*, **142**, 1633.
Gursky, H., Solinger, A., Kellogg, E. M., Murray, S., Tananbaum, H., Giacconi, R., and Cavaliere, A. 1972, *Ap. J.* (*Letters*), **173**, L99.
Haddock, F. T., and Sciama, D. W. 1965, *Phys. Rev.* (*Letters*), **14**, 1007.
Hawking, S. W., and Tayler, R. J. 1966, *Nature*, **209**, 1278.
Hazard, C., and Salpeter, E. E. 1969, *Ap. J.* (*Letters*), **157**, L87.
Holden, D. J. 1966, *M.N.R.A.S.*, **133**, 225.
Howell, T. F., and Shakeshaft, J. R. 1966, *Nature*, **210**, 1318.
———. 1967, *ibid.*, **216**, 753.
Hughes, R. G., and Longair, M. S. 1967, *M.N.R.A.S.*, **135**, 131.
Jauncey, D. L. 1967, *Nature*, **216**, 877.
Kardashev, N. 1967, *Ap. J.* (*Letters*), **150**, L135.
Kellermann, K. I. 1972, *A.J.*, **77**, 531.
Kellermann, K. I., Pauliny-Toth, I. I. K., and Williams, P. J. S. 1969, *Ap. J.*, **157**, 1.
Koehler, J. A., and Robinson, B.J. 1966, *Ap. J.*, **146**, 488.
Longair, M. S. 1966, *M.N.R.A.S.*, **133**, 421.
———. 1971, *Rept. Prog. Phys.*, **34**, 1125.
Longair, M. S., and Pooley, G. G. 1969, *M.N.R.A.S.*, **145**, 121.
Longair, M. S., and Scheuer, P. A. G. 1967, *Nature*, **215**, 919.
———. 1970, *M.N.R.A.S.*, **151**, 45.
Lowrance, J. L., Morton, D. C., Zucchino, P., Oke, J. B., and Schmidt, M. 1972, *Ap. J.*, **171**, 233.
Lynden-Bell, D. 1971*a*, *M.N.R.A.S.*, **155**, 95.
———. 1971*b*, *ibid.*, **155**, 119.
Lynds, R., and Petrosian, V. 1972, *Ap. J.*, **175**, 591.
Lynds, C. R., and Wills, D. 1970, *Nature*, **226**, 532.
———. 1972, *Ap. J.*, **172**, 531.
McVittie, G. C., and Stabell, R. 1967, *Ap. J.* (*Letters*), **150**, L141.
Maslowski, J. 1971, *Astr. and Ap.*, **14**, 215.
———. 1972, *ibid.*, **16**, 197.
Miley, G. K. 1971, *M.N.R.A.S.*, **152**, 477.
Mitton, S. A., and Reinhardt, M. 1972, *Astr. and Ap.*, **20**, 337.
Oke, J. B. 1970, *Ap. J.* (*Letters*), **161**, L17.
Oort, J. H. 1958, *Solvay Conf. on "Structure and Evolution of the Universe"* (Brussels: Stoops), p. 163.
———. 1970, *Astr. and Ap.*, **7**, 381, 405.
Pauliny-Toth, I. I. K., Kellermann, K. I., Davies, M. M., Fomalont, E. B., and Shaffer, D. B. 1972, *A.J.*, **77**, 265.
Peach, J. V. 1970, *Ap. J.*, **159**, 753.
Penzias, A. A., and Scott, E. H., III. 1968, *Ap. J.* (*Letters*), **153**, L7.
Penzias, A. A., and Wilson, R. W. 1965, *Ap. J.*, **142**, 419.
———. 1969, *ibid.*, **156**, 799.

Peterson, B. A., and Bolton, J. G. 1972, *Ap. J.* (*Letters*), **173**, L19.
Petrosian, V. 1969, *Ap. J.*, **155**, 1029.
Pilkington, J. D. H., and Scott, P. F. 1965, *Mem. R.A.S.*, **69**, 183.
Pooley, G. G. 1969, *Nature*, **218**, 153.
Pooley, G. G., and Kenderdine, S. 1968, *M.N.R.A.S.*, **139**, 529.
Pooley, G. G., and Ryle, M. 1968, *M.N.R.A.S.*, **139**, 515.
Readhead, A. C. S., and Hewish, A. 1972, *Nature*, **236**, 440.
Rees, M. J., and Schmidt, M. 1971, *M.N.R.A.S.*, **154**, 1.
Rees, M. J., and Sciama, D. W. 1966, *Ap. J.*, **145**, 6.
Rees, M. J., and Setti, G. 1970, *Astr. and Ap.*, **8**, 410.
Reinhardt, M., and Thiel, M. 1970, *Ap. Letters*, **7**, 101.
Ringenberg, R., and McVittie, G. C. 1970, *M.N.R.A.S.*, **149**, 341.
Rowan-Robinson, M. 1968*a*, *M.N.R.A.S.*, **138**, 445.
———. 1968*b*, *ibid.*, **141**, 445.
———. 1970, *ibid.*, **149**, 365.
Ryle, M. 1950, *Rept. Prog. Phys.*, **13**, 184.
Ryle, M., and Windram, M. D. 1968, *M.N.R.A.S.*, **138**, 1.
Sandage, A. R. 1965, *Ap. J.*, **141**, 1560.
Sandage, A. R., and Luyten, W. J. 1969, *Ap. J.*, **155**, 913.
Scheuer, P. A. G. 1965, *Nature*, **207**, 963.
Schmidt, M. 1965, *Ap. J.*, **141**, 1295.
———. 1968, *ibid.*, **151**, 393.
———. 1969, *Ann. Rev. Astr. and Ap.*, **7**, 527.
———. 1970, *Ap. J.*, **162**, 371.
———. 1972*a*, *ibid.*, **176**, 273.
———. 1972*b*, *ibid.*, p. 289.
Sciama, D. W., and Rees, M. J. 1966, *Nature*, **211**, 1283.
Shakeshaft, J. R. 1969, *Colloquium on Cosmic Ray Studies*, ed. Daniel *et al.* (Bombay: Tata Institute), p. 259.
Shimmins, A. J., Bolton, J. G., and Wall, J. V. 1968, *Nature*, **217**, 818.
Shklovskii, I. S. 1964, *Astr. Tsirk.*, No. 303, p. 3.
Smith, F. G. 1951, Ph.D. thesis, Cambridge University.
Sofue, Y., Fujimoto, M., and Kawabata, K., 1968, *Pub. Astr. Soc. Japan*, **20**, 388.
Spitzer, L. 1968, *Diffuse Matter in Space* (New York: Interscience).
Strittmatter, P. A., Faulkner, J., and Walmsley, M. 1966, *Nature*, **211**, 141.
Thaddeus, P. 1972, *Ann. Rev. Astr. and Ap.*, **10**, 305.
Thorne, K. S. 1967, *Ap. J.*, **148**, 51.
Tritton, K. P. 1971, *M.N.R.A.S.*, **155**, 1P.
———. 1972, *ibid.*, **158**, 277.
Wagoner, R. V., Fowler, W. A., and Hoyle, F. 1967, *Ap. J.*, **148**, 3.
Weymann, R. 1967, *Ap. J.*, **147**, 887.
Whiteoak, J. B. 1972, *Australian J. Phys.*, **25**, 233.
Wills, D. 1971, *Nature Phys. Sci.*, **234**, 168.
Willson, M. A. G. 1970, *M.N.R.A.S.*, **151**, 1.
———. 1972, *ibid.*, **156**, 7.
Wolfe, A. M., and Burbidge, G. R. 1969, *Ap. J.*, **156**, 345.
Wouthuysen, S. A. 1952, *A.J.*, **57**, 31.

CHAPTER 19

The Redshift

ALLAN SANDAGE

Hale Observatories, Carnegie Institution of Washington, California Institute of Technology

1. INTRODUCTION

ONE of the premier facts of science is that galaxies show redshifts in their spectra which increase linearly with distance in the way discussed by Hubble (1929). Necessarily, Hubble's early description of the kinematic field was confined to only a few of the nearest galaxies; the first data available did not extend far beyond the Virgo cluster. But studies by Hubble and Humason at Mount Wilson, by Humason at Palomar, and by Mayall at Lick between 1929 and 1956 showed that the redshift phenomenon is general, and extends to the observable limits of the largest telescopes.

Because the Universe has evolved from an earlier different state, the phenomenon is related to historical events in the distant past. The redshift-distance effect is the natural vehicle with which to study the timing of certain events associated with "creation" that occurred either near the Friedmann (1922) singularity (such as the 3° K blackbody radiation), or near the epoch of galaxy formation (such as may be signaled by a cutoff in quasar redshifts near $z = 4$).

This chapter is concerned in part with a few particular aspects of the measurement of redshifts, and in part with a simplified Friedmann theory of the redshift–apparent-magnitude and diameter relations in a nonstatic universe. An outline of the available observations is made in the final sections where a preliminary comparison is made with predictions from the models.

2. EARLY RESULTS ON THE SPECTRA AND REDSHIFTS OF GALAXIES

2.1. THE FIRST OBSERVATIONS

William Huggins was the first to visually study the spectra of nebulae using a prism spectroscope attached to an 8-inch refractor at his Tulse Hill Observatory, London, in the 1860s. He showed that in a number of nebulae (e.g., the planetaries) there exist bright emission lines rather than absorption lines that would have been present had such objects been resolvable into stars. (For commentary on the importance of the discovery see Huggins and Huggins 1899, *Atlas of Representative Stellar Spectra*, pp. 10–11).

Received August 1972.

Although the discovery of emission-line nebulae settled the controversy that no true gaseous nebulae would remain in lists of telescopic objects if sufficient instrumental power were available—a view that had support from the observations of Lord Rosse with his 6-foot telescope at Parsonstown—Huggins also showed that other nebulae (the galaxies) did not show a bright line spectrum. In the same year (1864) that he first saw the emission lines in planetary nebulae (NGC 6543), Huggins observed the spectrum of M31 visually and believed that it showed absorption lines, but the observation was exceedingly difficult (Huggins and Huggins 1899, pp. 119–121). Although Huggins obtained a photograph of the M31 spectrum in 1888, the absorption lines were so feeble that he reproduced the photograph only much later (1899), and then only to illustrate a historical series of first spectrograms in his *Atlas* (plate II, following p. 165).

Scheiner (1899) published a description of a photographic prism spectrogram of M31 that conclusively showed the existence of absorption features. The first radial velocity of a galaxy (M31) was measured only in 1912 by V. M. Slipher (1914) and published two years later. Slipher's result was a singular achievement because of the slowness of photographic plates, the relatively low speed of spectrographic cameras, and the very faint surface brightness of galaxies.

Slipher obtained four spectrograms of M31 with the Lowell 24-inch refractor in 1912, from which he measured concordant radial velocities, leading to a high-weight mean value. The result, astounding at the time, was $v = -300$ km s^{-1} and was the largest measured velocity for any astronomical object then known.

Slipher continued the work into 1914, where, in the first of his few publications on the subject (Slipher 1915), he presented velocities for 13 galaxies, most of which were positive; the large redshifts were for NGC 1068 and NGC 4594, both with $c\Delta\lambda/\lambda_0 = +1100$ km s^{-1}. By 1922 February, Slipher could give Eddington (1923) velocities for 41 galaxies, ranging from -300 km s^{-1} for M31 to $+1800$ km s^{-1} for NGC 584. An expanded list for Stromberg's (1925) paper on "Radial Velocities of Globular Clusters and Non-Galactic Nebulae" had grown to 45 galaxies, at which point Slipher discontinued the exceedingly difficult observations after proving that the distribution of the velocities of nongalactic nebulae was different from that for galactic objects. This difference showed the presence of a fundamentally new phenomenon.

Except for his 1915 note in *Popular Astronomy*, and for a small piece in the *Proceedings of the American Philosophical Society* in 1917, Slipher did not discuss the data in any detail. But his obviously important results became the starting point for all calculations of the solar motion relative to the nebulae (Truman 1916; Wirtz 1918, 1921, 1922, 1924, 1925; Lundmark 1920, 1924, 1925; Stromberg 1925, and undoubtedly others). A historically interesting review by Hubble is contained in chapter 5 of *Realm of the Nebulae*, written seven years after the discovery of the general expansion effect which Hubble had made by following Wirtz and Lundmark in introducing a Kr term in the classical equation for solar motion. At the time of his discovery, Hubble ascribed the phenomenon to the so-called de Sitter effect which is something quite different from a general expansion given by a nonstatic line element for the Universe as a whole.

The de Sitter effect arose in a much different climate and along a different route from that of the astronomer's fascination with velocities for use in the solar-motion problem. The two streams of thought, one from Slipher and the other that emerged from Einstein's gravitational theory, did not meet inside the observatory until the early 1930s, and then only with considerable confusion.

2.2. De Sitter's Static Solution

Shortly after Einstein's gravitation theory appeared in 1916, particular solutions of the equations were found that gave specific values to the gravitational potentials g_{ij} (and therefore to the metric) as functions of the mass-density and pressure of the Universe as a whole. One of these, by de Sitter, was especially interesting because it appeared to be connected in some manner with the redshift observations of galaxies.

Under the assumption that the large-scale features of the Universe are homogeneous and isotropic, it could be shown that the metric has the form (see, e.g., Robertson 1929; Tolman 1929)

$$ds^2 = R^2 d\sigma^2 - f^2 dt \, , \tag{1}$$

where $d\sigma^2$ is the three-space line element $d\sigma^2 = g_{ij}dx^i dx^j$ which must be taken to have spherical symmetry, and where R and f are functions to be found from the Einstein field equations that connect geometry with dynamics.

Static solutions require that none of the metric coefficients be functions of time. This condition was imposed as an apparently necessary condition in 1916–1917 by Einstein, de Sitter, and others because of the naturalness of the assumption in view of the astronomical data on stars (considered then to constitute the Universe). No large-scale systematic velocity field exists in stellar motions.

Three, and only three, static solutions exist (cf. Tolman 1929). Two contain nothing surprising; but the third, discovered in 1917 by de Sitter, had a particularly interesting place for those concerned with the existence of galaxies and with redshifts before 1931. The three static solutions have line elements

$$ds^2 = \frac{dr^2}{1 - r^2/R^2} + r^2(d\theta^2 + \sin^2\theta d\varphi^2) - dt^2 \tag{2}$$

which was the solution originally discovered by Einstein (1917);

$$ds^2 = dr^2 + r^2(d\theta^2 + \sin^2\theta d\varphi^2) - dt^2 \tag{3}$$

which is that of flat Minkowski space-time; and

$$ds^2 = \frac{dr^2}{1 - r^2/R^2} + r^2(d\theta^2 + \sin^2\theta d\varphi^2) - (1 - r^2/R^2)dt^2 \, . \tag{4}$$

which is that discovered by de Sitter (1917). In forms (2) and (4), R^2 is the radius of curvature of the three-space that is everywhere orthogonal to the time dimension.

Although all three solutions are static in the sense that none of the four metric coefficients g_{ij} (i, j from 1 to 4) depend on time, only (2) and (3) exhibit static properties for test objects placed within their particular manifolds. The third contains the scandalous term $(1 - r^2/R^2)$ as coefficient of dt^2. This means that for a clock at rest, i.e., at a particular place within the de Sitter world (hence $dr = d\theta = d\varphi = 0$), the timelike interval $ds^2 = -(1 - r^2/R^2)dt^2$ depends upon the place r within the manifold. The interval becomes smaller for larger r, meaning that clocks would appear to slow down the greater the distance from the observer. Atomic clocks (i.e., the frequency of radiation) would appear to do the same, and a redshift that increases with distance would occur. This part of the "de Sitter effect" could erroneously be regarded as a motion of recession. A "redshift-distance relation" would follow from equation (4) by noting that the time interval for a clock at distance r from the origin is retarded relative to one at $r = 0$ by

$$dt_1/dt_0 = (1 - r^2/R^2)^{1/2} . \quad (5)$$

This retardation, being the ratio of frequencies ν_0/ν_1, gives the ratio of wavelengths of emitted atomic lines λ_1/λ_0 as

$$\lambda_1/\lambda_0 \equiv 1 + \Delta\lambda/\lambda_0 = (1 - r^2/R^2)^{1/2} , \quad (6)$$

or

$$\Delta\lambda/\lambda_0 = \tfrac{1}{2}r^2/R^2 ,$$

as the predicted relation. This was de Sitter's first result in 1917. Note the quadratic rather than the linear form.

There is, however, a second effect due to the tendency of particles placed into a de Sitter metric to accelerate. The details are not repeated here (see, e.g., Eddington 1923, § 70; de Sitter 1933, § 35, eqs. [79] and [80]); but the result, when the arbitrary assumption was made that there were no initial negative velocities of test particles put into the manifold, was that the actual motion of particles away from the observer just canceled the effect of equation (6), leaving a linear redshift-distance relation (de Sitter 1933, p. 196). But this second effect was not appreciated by all observers in the 1920s, and many looked for a *quadric relation.*

The de Sitter effect does not represent a true expansion of space where the relative distances change between the galaxies. It is rather a mathematical curiosity exhibited by a particular static solution that is now believed to have no physical reality. Nevertheless, before the discovery of true nonstatic solutions by Friedmann (1922) and Lemaître (1927) and before the understanding of these discoveries by the community in the early months of 1930, many astronomers looked for the de Sitter effect in the astronomical data, especially using the redshifts of Slipher. Explicit discussions were given by Stromberg (1925), Lundmark (1925), Wirtz (1925), Silberstein (1924), and undoubtedly others. Indeed, Hubble (1929), in his discovery announcement of the velocity-distance relation, believed at that time that the de Sitter effect had been found. The concept of a general expansion of the Universe had not reached beyond Friedmann and Lemaître, as is evident from Hubble's disclaimer in his discovery paper that the linear relation that emerged then might represent only the most local approximation of a more general dependence of redshift on distance. He was aware of the early de Sitter prediction (eq. [6]) that $z \propto r^2$, and he implied by the final paragraph (Hubble 1929) that his figure 1 could be the tangent at the origin to a relation that was more general than linear.

The truly nonstatic solutions of Friedmann and of Lemaître, where $R = R(t)$, have dominated theoretical thinking since their retrieval from the archive literature of 1922 and 1927 by Eddington (1930). Since that time, the de Sitter effect has not been considered as an explanation of the redshift problem; rather the observational data are taken to describe the expansion of the nonstatic Universe. De Sitter wrote in 1933, "We know now, because of the observed expansion, that the actual universe must correspond to one of the non-static solutions"—or later (de Sitter 1933, p. 191), "the static universes are, so to say, of only academic interest."

Why? The chief reasons for this position today would be the direct astronomical evidence for an evolving universe. Such evidence includes (1) the 3° K blackbody radiation interpreted as the primeval fireball cooled by expansion in the enlarging cavity of the world; (2) the observed high regularity of the physical properties of E galaxies interpreted in terms of equal age for such galaxies, hence galaxy formation appears to have occurred only in a small, finite time interval in the distant past (Sandage 1973); (3) the agreement of the age of our Galaxy (as determined from the globular clusters and subdwarf field stars) with the Friedmann time for the expanding Universe as determined from the value of the Hubble constant ($H_0 \simeq 50$ km s^{-1} Mpc^{-1}) and the deceleration parameter (Sandage and Tammann 1975*b*, Paper VI).

The chief reason for abandoning the de Sitter solution in 1930 was the straightforward nature of the Friedmann-Lemaître evolving solutions with their lack of mathematical paradoxes, and with their physically reasonable $R(t)$ supposition, rather than the physically uninterpretable metric of equation (4).

2.3. Hubble and Humason's Observational Extension of the Redshift Law

Hubble's 1929 announcement of a redshift-distance effect followed an important observational note by Humason (1929) in which his observation of the largest velocity then known was reported. Slipher's largest velocity had been 1800 km s^{-1} for NGC 584, and it was of the utmost importance in testing for a general redshift effect to determine if fainter galaxies had larger redshifts. Recall that Lundmark's (1925) $Kr + Lr^2$ solution required that an upper limit of $\sim$ 3000 km s^{-1} existed.

Humason observed at the Cassegrain focus of the 100-inch reflector and obtained two spectrograms of the brightest E galaxy in a small group centered on NGC 7619. That the de Sitter effect was a motivation for the work is shown by Humason's comment, "About a year ago Mr. Hubble suggested that a selected list of fainter and more distant extra-galactic nebulae, especially those occurring in groups, be observed to determine, if possible, whether the absorption lines in these objects show large displacements toward larger wavelengths as might be expected on de Sitter's theory of curved space-time."

Humason obtained two spectra of NGC 7619 using a two-prism spectrograph with a 3-inch camera, the combination that gave a dispersion of 183 Å mm^{-1} at 4500 Å. The exposure times were 33 hours and 45 hours, each spread over five consecutive nights. The difficulty of these first measurements is emphasized not only by the long exposure times (modern image-tube spectrographs on telescopes of 100-inch aperture can give spectra of NGC 7619 at the same dispersion in 10 minutes), but also by the fact that the H- and K-lines of Ca II were not resolved on the spectrograms due to instrumental problems such as flexure during the long exposure. Normally these lines are easily resolved at this dispersion. Nevertheless, both spectra were measured, and gave the enormous velocity (for that time) of 3779 km s^{-1}, which is in excellent agreement with the modern value of 3757 km s^{-1} (Humason, Mayall, and Sandage 1956; hereafter called HMS). Humason's large value undoubtedly confirmed to both Hubble and himself that galaxies indeed had remarkable kinematic properties, and that something quite fundamental was happening. Preliminary evidence that a linear velocity-distance relation might exist had been given earlier by Robertson (1928).

Following Hubble's 1929 announcement of the redshift effect, Humason began a large observational effort on redshifts of fainter galaxies—mostly in clusters, because even in 1930 Hubble suspected that the brightest galaxies in clusters were among the brightest galaxies anywhere and, therefore, that such objects would be among the most distant systems at any given apparent magnitude.

By 1931 Humason (1931) had observed clusters whose redshifts were as large as 19,700 km s^{-1} for a small group of E galaxies in Leo. From these data, Hubble's 1929 linear distance effect was fully confirmed. In a resulting fundamental and classical paper, Hubble and Humason (1931) not only extended the original velocity-distance relation fifteenfold in distance, but also began the study of galaxy clusters which has become of overriding importance in modern work on the problem. It was in this paper that the properties of clusters of galaxies were first discussed. In particular it was shown that galaxies in the large, regular, populous clusters are virtually all elliptical or S0 systems. This fact, not presently understood, is clearly of great importance in the unsolved problem of galaxy and cluster formation.

By 1936 the reconnaisance redshift program on clusters was finished (Humason 1936). Redshifts had been determined with the Mount Wilson 100-inch reflector for galaxies as distant as those in the Ursa Major No. 1 cluster (cz = 15,520 km s^{-1}), Corona Borealis (21,600 km s^{-1}), Bootes (39,367 km s^{-1}), and Ursa Major No. 2 (40,360 km s^{-1}). The redshifts, combined with estimated apparent magnitudes of the first 10 brightest cluster members gave a redshift–apparent-magnitude relation (hereafter called the Hubble diagram) whose slope was 5 (the requirement for a linear velocity-distance relation) and where the dispersion of the individual points about the mean line was very small (Hubble 1936*b*; Humason 1936).

The remarkable feature of small scatter about the line of slope 5 proved even in 1936 that (1) redshift is a unique and linear function of distance to within very small limits (currently put at $\sigma[\Delta z/z] \lesssim 0.05$ [Sandage 1972*a*, *b*, *c*]), and (2) the absolute magnitude of the bright cluster galaxies has a very small intrinsic dispersion (the modern value for first-ranked cluster members is $\sigma[M_v] \lesssim 0.25$ mag), or a combination of these conclusions with even lower limits if both the ordinate (redshift), and abscissa (magnitude) of the diagram are subject to cosmic scatter separately.

Because of these results in 1931 and 1936, the redshift-distance relation was widely recognized as the most fundamental aspect of the large-scale structure of the Universe. This view has been reinforced by all subsequent work. The redshift relation at once gains us the means to study the past history of the Universe, and, if nothing else, to estimate the time scale of creation.

But it was already clear by the late 1930s (Hubble 1938) that the Hubble diagram could be used to determine the deceleration (q_0) of the expansion. Once the deceleration and the expansion rate are known, all properties of idealized Friedmann models with $\Lambda = 0$ are known, including the mean density, space curvature, volume, dynamical history, oscillation time if $q_0 > \frac{1}{2}$, and age (see, e.g., Sandage 1961*a*, *b*, 1962).

Determination of the deceleration requires that the observations be carried to very large distances such that the look-back times are a significant fraction of the Friedmann time. The corresponding redshifts are very large. To carry out the program, two problems exist: (1) distant clusters must be discovered; (2) their redshifts and magnitudes must be measured.

When the 200-inch telescope went into operation in 1949—20 years after the expansion had been discovered—Humason began again to measure faint cluster redshifts. In a chance discovery by Hubble in 1935 November, a faint cluster in Hydra (0855+0321) had earlier been found at the edge of a plate taken at the Newtonian focus of the Mount Wilson 100-inch, centered on NGC 2716. Humason had attempted to measure its redshift in 1936 at Mount Wilson with a 22-hour exposure but had obtained a spectrum that was not positively readable. Subsequently in 1949, the Hydra cluster was one of the first tried by Humason with the Hale telescope, and he immediately obtained an easily readable redshift of 60,500 km s^{-1}. This was the largest at the time and was still among the largest known in the early 1970s.

New distant clusters were discovered in great abundance as the Palomar Schmidt *Sky Survey* progressed, and by 1956 Humason had measured redshifts of a preliminary sample of these, and had remeasured many of the clusters that he and Hubble had used in their 1936 discussion. The data were given and discussed in HMS in 1956.

It has turned out that the effective limit for redshift determinations with spectrographs that use photographic plates is only slightly larger than the Hydra cluster with $z \equiv \Delta\lambda/\lambda_0 = 0.203$. This is because the heavy night-sky emission bands contaminate the galaxy spectrum so strongly that the signal-to-noise ratio rapidly becomes unfavorable. Because of the finite storage capacity of photographic plates and because of their nonlinearity, it is difficult to recover the signal by standard linear subtraction techniques and long integrations. Consequently, Humason did not carry the redshift determination to clusters beyond Hydra, although he tried long exposures on several distant clusters that had been found by Humason and Sandage in special searches using the 48-inch Schmidt and the 200-inch prime focus (Humason 1957). The problem is well illustrated in Humason's reproduction of spectra of Hydra cluster members published as plate III, 8 of HMS.

In a later development Minkowski (1960) obtained a much larger redshift for the brightest galaxy in a cluster associated with the radio source 3C 295. This E galaxy has $\lambda 3727$ in emission, and two spectra showed a redshift from this feature alone of $z = 0.461$. The absorption lines were degraded beyond detection by heavy night-sky emission noise. Minkowski's redshift is likely to be among the largest that will be determined by spectrographic methods that use photographic plates because photoelectric subtractive spectrographs began to be used in 1971. The integration and subtraction of signals by such linear devices will permit the Hubble relation for galaxies to be extended to redshifts near $z \simeq 1$ in the 1970s. It is in this range that the deceleration parameter will eventually be found by the look-back method. Although results to 1972 have been inconclusive on this point, a great deal of information about the galaxies themselves has been found from the Hubble diagram, as discussed in § 6 (cf. Sandage 1972*c* and later papers of the series).

Even though the measurements of redshifts of very faint galaxies are no longer attempted by the use of photographic techniques, such data for bright galaxies will continue to be obtained with conventional spectrographs for many years. As an aid in making such measurements, we give in the next section lists of expected galaxy lines, representative photographs, and identified comparison lines in He and Ne glow-tube sources.

3. AIDS IN THE PRACTICAL MEASUREMENT OF REDSHIFTS BY PHOTOGRAPHIC METHODS

The surface brightness of galaxies is small (fainter than 15 mag per square arc sec in B-light at the center, and declining as r^{-2}), and the angular diameters are large compared with the width of the spectrograph slit. Except, then, for the very brightest galaxies, low-dispersion spectra are required for routine redshift programs. Dispersions at the plate that range from ~ 150 Å mm^{-1} to ~ 500 Å mm^{-1} are common. Fast cameras of focal ratio f/1 or f/0.5 are also common. Bowen (1962) has described in chapter 2 of Volume **2** various spectrographs and cameras in use for galaxy spectroscopy.

3.1. Lists and Charts of Comparison Spectra

The dispersions are so small that care must be taken in the choice of the comparison spectrum. The number of comparison lines per millimeter on the plate must be small enough to avoid serious blends. It has been common practice at Mount Wilson to use a comparison of helium plus slight hydrogen contamination. The lines are mostly unblended, are well spaced between λ3185 and λ5047, and are in adequate numbers to provide good control of the reductions. The He spectrum is ideal when used in conventional photographic spectrographs with blue plates whose redward cutoff is about λ5100. Lines of He are sparse redward of this cutoff, but are well spaced blueward.

Figure 1 shows a typical comparison spectrum, taken with the Palomar 200-inch prime-focus spectrograph of the N galaxy Ton 256 on a IIaD plate, using an f/1 thick mirror Schmidt camera at an original dispersion of 400 Å mm^{-1}. The spectrum extends to the wavelength cutoff of the aD sensitivity at $\sim$ λ6500, and the strong comparison lines redward of λ5800 are due to Ne, whose characteristics are discussed later in this section. Note the scarcity of lines between the last He line at λ5047.74 and the onset of the major Ne features at λ5852. A line of Ne does exist at λ5400.56, but this and the [O I] night-sky line at λ5577.35 form the only strong features in this wavelength interval. Listed in the first half of table 1 are the adopted wavelengths of the comparison lines, ranging from the He line at λ3185, near the atmospheric cutoff, to the Ne line at λ5400.56. These lines are identified in figure 1. A particularly useful pair to test the resolution of a spectrograph are the Ne lines at λλ5330.78 and 5341.10.

Helium has a few strong lines redward of λ5800. The He lines at λλ5875.6 and 6678.15 would be useful if it were not for their severe blending with the nearby Ne lines at λλ5881.90 and 6678.28 when He and Ne are used together. With these two lines omitted, all of the lines listed in table 1 redward of λ5764 should be free from blending when used with spectrograph and grating combinations capable of resolutions greater than ~ 10 Å.

The identification of the Ne spectrum listed in table 1 is given in figure 2, which is from an image-tube spectrogram of NGC 1511 taken at Mount Stromlo in 1972 at an original dispersion of 215 Å mm^{-1} on baked IIaD plates behind an S20 cathode of a Carnegie image tube.

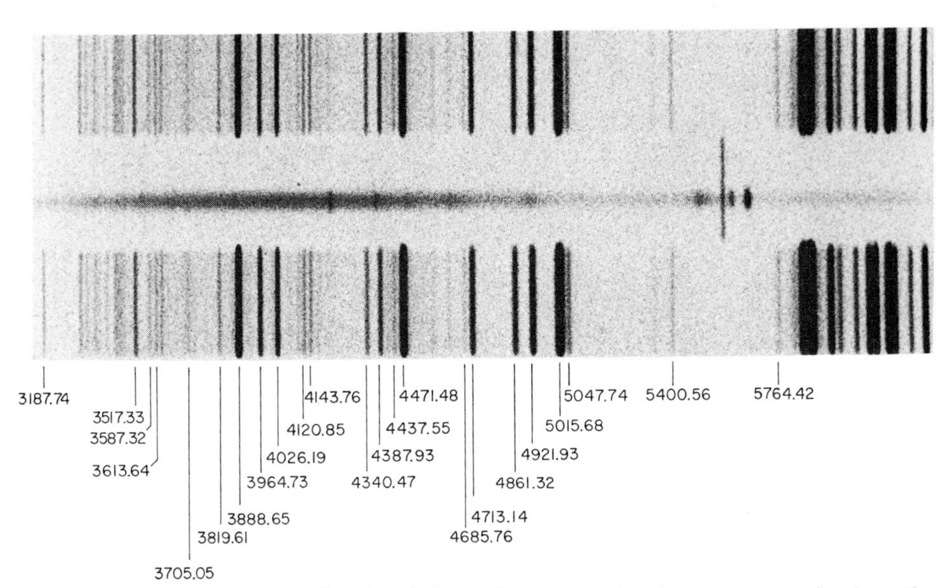

FIG. 1.—Identification of the lines in a helium-plus-neon comparison source ranging from the blue helium line at λ3187 to the neon line at λ5764. The original spectrogram was made with the 200-inch prime-focus spectrograph using an f/1 solid Schmidt camera with IIaD yellow-sensitive emulsion at an original dispersion of 400 Å mm^{-1}. The spectrogram is for the N galaxy Ton 256 which has redshifted emission lines of Hβ, N1, and N2 straddling the night-sky line at λ5577.35.

TABLE 1

COMPARISON-LINE WAVELENGTHS FOR A HELIUM-PLUS-NEON SOURCE WITH STRONG NIGHT-SKY LINES IN THE INTERVAL λλ3187–7032 Å

HELIUM PLUS NEON				NIGHT SKY	
λ	Source	λ	Source	λ	Source
3187.74	He	5341.10	Ne	3650.14	Hg
3517.33	He	5400.56	Ne	4046.96	Hg
3587.32	He	5764.42	Ne	4358.34	Hg
3613.64	He	5852.49	Ne	5460.74	Hg
3705.05	He	5944.83	Ne	5577.35	[O I]
3819.61	He	5975.53	Ne	5769.60	Hg
3888.65	He	6030.00	Ne	5790.66	Hg
3964.73	He	6074.34	Ne	5891.94	Na
4026.19	He	6096.16	Ne	6300.23	[O I]
4120.85	He	6143.06	Ne	6363.88	[O I]
4143.76	He	6266.50	Ne		
4340.47	Ne	6334.43	Ne		
4387.93	He	6382.99	Ne		
4437.55	He	6402.25	Ne		
4471.48	He	6506.53	Ne		
4685.76	He	6532.88	Ne		
4713.14	He	6562.82	H		
4861.32	H	6598.95	Ne		
4921.93	He	6717.04	Ne		
5015.68	He	6929.47	Ne		
5047.74	Ne	7032.41	Ne		
5330.78	Ne				

3.2. Night-Sky Contamination

Extensive airglow and aurora emission lines and bands exist in the night-sky radiation. On long-exposure plates, and on plates taken immediately after sunset, the brighter of these are nearly always recorded in galaxy spectra. In addition to the lines from natural causes, spectrograms taken at observatories near cities began in the 1960s to show lines due primarily to mercury (Hg) from mercury-vapor street lamps. Mount Wilson, Palomar, Lick, Kitt Peak, and Mount Stromlo galaxy spectra always show Hg lines in varying strength depending on the cloud cover reflecting the city lights, the time of night, and the hour angle.

Wavelengths of the principal night-sky emission lines are listed in the last column of table 1 and are identified in figures 3 and 4, which are spectrograms from Palomar and from Mount Stromlo.

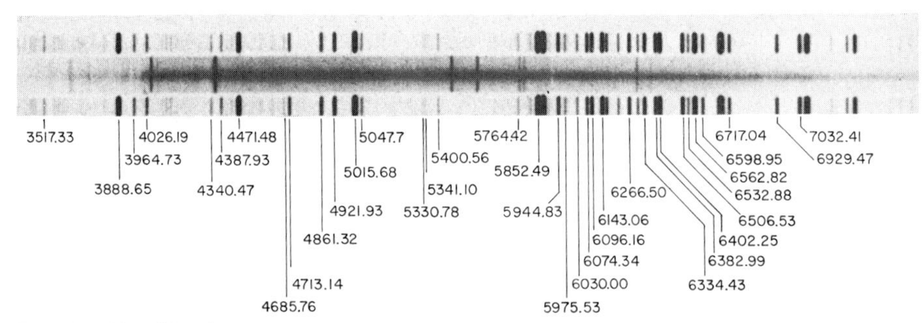

Fig. 2.—Identification of the comparison lines in a helium-plus-neon comparison source. The spectrogram was taken with the Mount Stromlo 74-inch Cassegrain image-tube spectrograph at an original dispersion of 215 Å mm^{-1}. A two-stage Carnegie image tube and a lens transfer imaging system was used. The galaxy spectrum is for NGC 1511. The bright emission lines are generally Hg contamination from Canberra city lights.

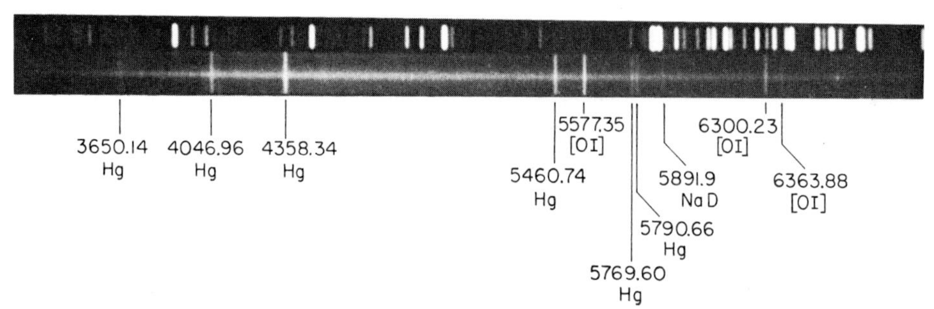

Fig. 3.—Positive print of a Stromlo image-tube spectrogram of IC 1954 with night-sky emission lines marked. Original dispersion was 215 Å mm^{-1}. Most of the contaminating lines are due to Hg from Canberra city lights. The only natural lines that are marked in this reproduction are λ5577 [O I], λ5891 of Na D, λ6300 of [O I], and λ6363 of [O I].

A comprehensive discussion of the natural aurora and airglow spectrum is given by Chamberlain (1961).

3.3. Absorption and Emission Lines in Galaxies

The spectral characteristics of galaxies change along the sequence of classification. The characteristic absorption lines in E, S0, Sa, and the central regions of many Sb systems are similar to the grossest features in G and K stars (see fig. 2 by Humason in HMS). Emission lines are the exception rather than the rule. The most prominent absorption lines are H and K of Ca II, usually of equal strength; occasionally 4226.7 of Ca I; often 4383.56 of Fe I; usually the G-band at 4304.4; occasionally Hβ in absorption; and often a series of faint lines in the green consisting of the Mg I blend at $\langle\lambda\rangle \simeq$ 5175.3 which can sometimes be resolved into two components at 5168.6 and 5183.3, Ca + Fe blend at 5268.9; and two unidentified lines at $\lambda\lambda$5331.5 and 5401.4. The strongest line in the near-red is Na D at $\langle\lambda\rangle \simeq$ 5892.5, which is almost always present. Faint lines at $\lambda\lambda$5856.8, 6025.3 and Hα in absorption at λ6562.8 are also seen occasionally.

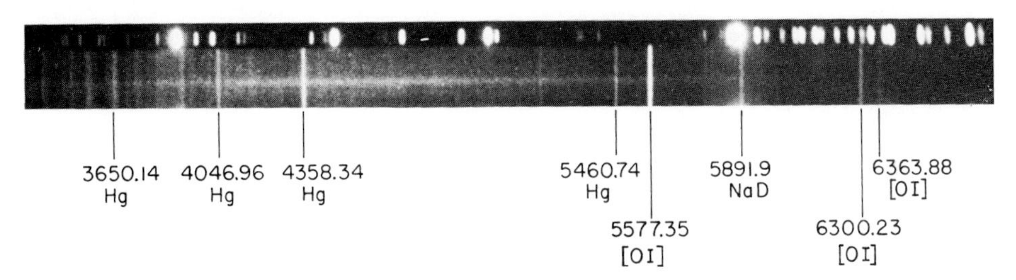

Fig. 4.—Same as fig. 3 above for Palomar. Spectrogram was taken with the 200-inch Cassegrain image-tube spectrograph at an original dispersion of 200 Å mm^{-1} using a helium-plus-neon source. The Hg contamination is from the combined lights of Los Angeles to the northwest and San Diego to the southwest. There was heavy cloud cover over these cities at the time.

In late type Sc-Sd-Sm galaxies an earlier mean composite spectral type is usually present with the Balmer series of hydrogen prominent. The λ3727 line of [O II] is often in emission. Just longward at λ3742 is a deep absorption feature. Moderately strong well-defined hydrogen lines of Hθ at λ3798.6, Hη at λ3835.57, and Hδ at λ4101.7 are sometimes present and, if so, are easily measurable. The H- and K-lines can be present, but usually K(λ3933.7) is weaker than H(3968.5), which is blended with Hϵ(3970).

The usual pattern of emission lines is 3727.3[O II], Hβ(4681.33), 4958.9 of [O III], 5006.84[O III], the triplet complex of [N II] at 6548.1, Hα at 6562.80, [N II] at 6583.60, and the [S II] doublet. Higher-excitation lines of He and [Ne III] can be present in exceptional cases.

In modern spectrographs it is possible to obtain spectra that are in focus from λ3200 to λ7200 at the convenient dispersion of $\sim$ 200 Å mm^{-1}. All of the features mentioned are in this wavelength range, including the night-sky lines listed in table 1. Redshifts are obtained by measuring the plates with a traveling microscope in the direct and reverse direction to avoid systematic bias in setting on the comparison and the galaxy lines. Measurements are recorded to 1 μ. At a dispersion of 200 Å mm^{-1} a setting error of $\pm$ 10 μ corresponds to a velocity error of $\pm$ 120 km s^{-1}. The normal rms error achieved per plate on normal galaxies with an average of six lines is usually $\pm \sim 40$ km s^{-1}, which corresponds to an rms setting accuracy of $\pm$ 3 μ on the plates. This is considerably less than the width of the absorption lines in most galaxies, which shows the stringent necessity to avoid systematic errors in the plate measurements.

It is also important to determine the effective wavelengths of the galaxy lines for each spectrograph and camera because blends of the lines and details of the background spectrum causes $\langle\lambda\rangle$ to differ according to the instrument used. The usual procedure is to obtain spectra of galaxies that have both emission and absorption features. The emission lines can, to first approximation, be assumed to be at their laboratory wavelength because they are normally sharp and unaffected by blends. The effective $\langle\lambda\rangle$ of absorption lines in the same spectrum can be obtained therefrom; and, if enough

TABLE 2

PRELIMINARY ADOPTED WAVELENGTHS FOR ABSORPTION AND EMISSION LINES FOR A CARNEGIE IMAGE-TUBE SPECTROGRAPH OF 8 Å RESOLUTION WITH DISPERSION 200 Å PER MILLIMETER

Absorption Lines						Emission Lines	
λ_0	Line	Visibility	λ_0	Line	Visibility	λ_0	Line (Lab λ)
3472.0...	....	3	5268.98	Ca + Fe	5	3727.3....	[O II]
3798.6...	Hθ	3	5331.5	....	3	3868.74...	[Ne III]
3835.6...	Hη	3	5401.4	....	2	4101.74...	Hδ
3933.7...	K	10	5892.5	Na	7	4340.47...	Hγ
3968.5...	H	10	....		....	4861.33...	Hβ
4101.7...	Hδ	2	PROVISIONAL			4958.91...	[O III]
4226.7...	Ca I	2	5024.6		1	5006.84...	[O III]
4304.4...	G	5	5103.8		1	5875.63...	He I
4383.6...	Fe	3	5591.3		1	6548.10...	[N II]
4863.9...	Hβ	2	5782.8		2	6562.82...	Hα
5175.36..	Mg I	5	5856.8		1	6583.60...	[N II]
5168.6...	Mg I	....	6025.3		1	6717.00...	[S II]
5183.3...	Mg I	....	....		....	6731.30...	[S II]

spectra are available to average out random effects, good $\langle\lambda\rangle$ for all absorption lines can be found by overlap methods. In programs where emission-line objects are rare, the $\langle\lambda\rangle$ of the system must be optimized to give zero mean residuals for all the lines used, with the assumption that this zero is close to the laboratory zero. The procedure was used by Mayall in HMS (1956) and by de Vaucouleurs and de Vaucouleurs (1967), but by only a few others in large routine redshift programs. The adopted $\langle\lambda\rangle$ for the measured lines in the programs of Humason (HMS, 1956, p. 99), Mayall (HMS, p. 133), and the de Vaucouleurs (1967, table 1) are tabulated in the listed references.

In the Stromlo-Palomar redshift program on bright galaxies by Sandage, where image-tube spectra are used at a dispersion of ~ 200 Å mm^{-1}, effective wavelengths of the absorption lines were determined relative to emission lines. The preliminary adopted system is listed in table 2, which can be used as a starting point for other programs. Many lines, known to be real because of their occurrence in at least five spectra, are listed in the first part of table 2; those lines whose existence is still considered somewhat tentative, in the last part. To indicate in a most sketchy way the ease with which various lines can be detected, we list a "visibility number" on a scale of 10. While it is true that the line strengths vary from galaxy to galaxy, this index is useful as a first indication of those lines whose measurement is usually easy, or usually difficult. The strongest lines have an index of 10.

4. REDSHIFTS OF BRIGHT GALAXIES; CONSTANCY OF $\Delta\lambda/\lambda_0$; SYSTEMATIC ERRORS

In 1956 Humason and Mayall (HMS 1956) published redshifts of 620 galaxies observed at Mount Wilson and Palomar (including redshifts of galaxies in 26 clusters) and 300 galaxies observed at Lick. There were 114 galaxies in common between the Mount Wilson–Palomar and the Lick lists. The bright galaxies were selected from the Shapley-Ames catalog (1932) of systems brighter than $m_{pg} \simeq 13.0$. A supplement to the Lick list was published by Mayall and de Vaucouleurs (1962) which contained 92 new redshifts.

A new redshift program was begun in 1960 at McDonald Observatory by the de Vaucouleurs. Redshifts of 113 galaxies, of which 94 were new, were published in 1967. A redshift program on southern galaxies was begun by Sandage at Mount Stromlo, Australia, in 1967 to observe the Shapley-Ames galaxies south of $\delta = 30°$ S which had no redshifts.

Redshift programs by Evans at Pretoria, Shobbrook at Stromlo, Freeman at Stromlo, and others at Pretoria have also produced redshifts of southern galaxies. References to these and other programs on bright galaxies can be found to 1963 in the Introduction to the *Reference Catalogue of Bright Galaxies*. Extensive references to redshift measurements of bright galaxies both in systematic programs and for individual galaxies are given in Tammann (1973), and in Sandage and Tammann (1975*a*) in a discussion of nearby galaxy groups. A complete catalog of redshifts published to 1971 has been prepared by Fredrick and Gutsch (1972).

Beginning in the early 1960s, redshifts of many nearby late-type galaxies (Sb^+, Sc, Sd, Sm) have been measured by radio mapping of the kinematic field in 21-cm radiation. Roberts discusses the technique in chapter 9. Data are listed by Roberts in this chapter, many of which were measured by him. A number of redshifts were measured in the first extensive investigation by Epstein (1964) in a thesis supervised by Roberts. References to other work are given in chapter 9 and in the notes to table 2 of Ford, Rubin, and Roberts (1971).

If redshifts are caused by Doppler motions, the relative shift $\Delta\lambda/\lambda_0$ must be strictly independent of the laboratory wavelength λ_0; i.e., $\Delta\lambda/\lambda_0$ must be constant. Wilson (1949) made an early good test of the requirement using emission lines in NGC 4151 over a wavelength range 3425–6583 Å on a very high-dispersion plate (10.3 Å mm^{-1}) taken at the coudé spectrograph of the Mount Wilson 100-inch. His result was that over this observed interval (about one octave), $\Delta\lambda/\lambda_0$ was constant to within ~ 1 percent. A more stringent test by Minkowski and Wilson (1956) using the galaxy associated with the Cygnus A radio source put a limit on the change of redshift with wavelength at less than 3×10^{-4} per 1000 Å.

The constancy has now been tested over the interval of 19 octaves by comparing optical and radio redshifts of nearby Sb-Sm galaxies by McCutcheon and Davies (1970), and especially by Roberts (1972) over the redshift range from $0 \leq cz \lesssim 6000$ km s^{-1} with the result that $\Delta\lambda/\lambda_0$ is constant over the enormous wavelength range from 4×10^{-5} cm to 21 cm to better than the errors of measurement. The slope coefficient of the regression of V_{optical} on $V_{21\text{ cm}}$ is 1.00 to within $\sigma/2$ of the determination.

But the result is true only if a significant deviation in the correlation over the interval $1200 \lesssim cz \lesssim 2400$ km s^{-1} is neglected. Roberts (1972) has shown that there is a wave of amplitude ~ 100 km s^{-1} in the regression plot in this interval, and he reasonably interpreted this as systematic errors in the optical redshifts listed by HMS in the stated interval. The optical redshifts measured from blue plates are measured too *large* by ~ 100 km s^{-1} near the center of the range (i.e., $cz \simeq 1800$ km s^{-1}), decreasing systematically toward zero near the ends of the range. That the effect is in the optical redshifts is shown by new measurements by Ford and Rubin, using much higher dispersion than in HMS (Ford, Rubin, and Roberts 1971). The cause of the systematic corrections to HMS has not been investigated thoroughly, but it seems likely that it is caused by some effect of blending of galaxian and night-sky lines in the blue region of the spectrum, especially near the precipitous decline of intensity of the $I(\lambda)$ distribution shortward of $\lambda \simeq 4000$. The presence of the error amounts to a systematic deviation of 1.3 Å at λ4000 on the plates, or only 3 μ on plates taken at 400 Å mm^{-1}. This illustrates the extreme care that must be taken to ensure proper effective wavelengths (i.e., a correct version of table 2) for any spectrograph, and in the measurement of the plates taken at such a low dispersion. Random errors of $\pm$ 150 km s^{-1} for optical redshifts are common (see discussions by both Humason and Mayall in HMS; also de Vaucouleurs and de Vaucouleurs 1964) due to problems of measurement alone. Furthermore, as regards the system of velocities, there is little reason to believe at present that the zero point of optical velocities is known systematically to better than perhaps $\pm$ 30 km s^{-1} in any catalog.

5. CORRECTIONS TO MEASURED REDSHIFTS

As with stellar radial velocities, it is always necessary to reduce measured redshifts *to the Sun* by the standard reduction for the Earth's motion.

A second correction to galaxy redshifts is always applied to account for the motion of the Sun relative to the centroid of the Local Group. Justification of this second correction is twofold:

1. A model for galactic rotation is adopted using the standard assumptions of 300 km s^{-1} motion directed toward $l^{\text{II}} = 90°$, $b^{\text{II}} = 0°$. This gives a correction of

$$v_{\text{corr}} = 300 \sin l^{\text{II}} \cos b^{\text{II}} \quad \text{km s}^{-1} , \tag{7}$$

or

$$\Delta z = 0.001 \sin l \cos b , \tag{8}$$

and would, if the scale of 300 km s^{-1} is correct, reduce measured redshifts to that which would be observed from the center of the Galaxy.

2. Actual solutions for solar motion relative to either (*a*) members of the Local Group alone (Humason and Wahlquist 1955) or (*b*) nearby galaxies following Hubble (1939) have been made, for example, by de Vaucouleurs and Peters (1968). The results to date suggest that the motion is dependent on the sample of galaxies used and dependent also on the value of the Hubble constant adopted if galaxies outside the Local

Group are included. The solution by Humason and Wahlquist was surprisingly close to equation (8), which, if valid, would mean that the motion of the center of our Galaxy relative to the Local Group would be negligible.

In any case, application of equations (7) and (8) to observed redshifts has been standard procedure for catalogs extant in 1970. However, one of the particularly interesting programs is to reinvestigate the problem of the local kinematic field by obtaining precise distances to the nearby galaxies using the new methods now available, and then to solve for the solar motion more precisely. The problem is to determine the asymmetry in the nearby redshifts due to the solar motion, and to compare this with the direction and size of the asymmetry in the 3° K radiation (if it can be detected; see Conklin 1969, Henry 1971). Agreement of the two experiments would say something fundamental about Mach's principle concerning the presence of and the reason for a fundamental inertial system (cf. Sciama 1972).

6. ABSOLUTE LUMINOSITY AND INTRINSIC DIAMETER AS FUNCTIONS OF OBSERVABLES IN SPECIFIC MODELS

6.1. Two Approaches

It is often necessary to find intrinsic properties of galaxies or quasars when the redshifts, apparent magnitude, and angular dimensions are known. Because redshift is a unique function of distance (with q_0 as a parameter), the calculation follows directly once the equations are known that account for the effects of space curvature, aberration, deceleration over the look-back time, and the evolution of the properties in the light-travel time as functions of $z \equiv \Delta\lambda/\lambda_0$.

Two approaches have been used.

1. The method of most generality is to assume no model of the universe, but to adopt the reasonable postulate that the four-space manifold $g_{ij}dx^i dx^j$ and the expansion function $R(t)$ are smooth enough that a Taylor expansion in space and time can be made about the present place and epoch—the series reaching to the distance that corresponds to the observed redshift. Observables such as luminosity, angular size, and number of objects per unit volume can be expressed as a series that involves z, R_0, $\dot{R}_0$, $\ddot{R}_0$, and higher derivatives if needed, where R_0 is the present-epoch value of the space dilatation factor (eq. [1]), and $\dot{R}_0$ and $\ddot{R}_0$ are the first and second time derivatives. The ratio $\dot{R}_0/R_0$ is the expansion rate—i.e., the Hubble "constant" H_0, which is of course a function of cosmic time. The deceleration of the expansion is related to $\ddot{R}_0$.

A dimensionless quantity, the deceleration parameter, $q_0 \equiv - \ddot{R}_0 R_0/\dot{R}_0^2 = - \ddot{R}_0/R_0H_0^2$, can be formed which appears prominently in the series expansions for $\theta = f(z, q_0)$, $m = g(z, q_0)$, $N(m) = h(z, q_0)$, $\tau \equiv$ light travel time $= k(z, q_0)$, etc. Heckmann (1942) was among the first to derive such series equations connecting observables. Robertson (1955), McVittie (1956), Davidson (1959), among others, have vigorously developed this approach. An extension and generalization with no commitment to a model is given in a fundamental paper by Kristian and Sachs (1966).

2. A less general approach, but one where the final equations are in closed form, is to adopt a particular dynamical model of the expansion, such as that which follows from integration of the Friedmann-Lemaître differential equations:

$$\frac{\dot{R}^2}{R^2} + \frac{2\ddot{R}}{R} + \frac{8\pi G p}{c^2} = -\frac{kc^2}{R^2} + \Lambda c^2 \tag{9}$$

$$\frac{\dot{R}^2}{R^2} - \frac{8\pi G\rho}{3} = -\frac{kc^2}{R^2} + \frac{\Lambda c^2}{3} \tag{10}$$

(see, e.g., Robertson 1933; Gamow 1952; Bondi 1952; McVittie 1956, for derivations), where p and ρ are the (smeared out) isotropic pressure and density of matter and radiation, and Λ is Einstein's cosmological constant. The equations can be integrated to give the function $R(t)$ at all times. This, then, gives the history of the expansion. The $R(t)$ when combined with an adopted metric and a theory of light propagation gives closed expressions connecting observables that hold for arbitrarily large z values (until eqs. [9] and [10] cease to be valid). The closed equations can be made to yield global properties of the model such as the creation (Friedmann) time, the oscillation time of the Universe as a whole where applicable, the type of intrinsic four-space curvature, the mean density, etc. The Universe (or rather its model) is then fully described.

The disadvantage of the method is that the models are highly idealized, usually starting from the Robertson-Walker homogeneous, isotropic metric. The real world differs to some extent from the models because deviations from homogeneity clearly exist at some level (i.e., the distribution of matter is lumpy on some scale, small as this may be relative to the radius of curvature). Nevertheless, the predictions of the models and the observations agree remarkably well over the range of z-values available for test (cf. Sandage, Tammann, and Hardy 1972). Because of this, and because the formal equations possess many advantages over the series expansions in numerating global properties, we adopt the second method in the remainder of the chapter.

Rigorous derivations from first principles are not given. Rather, we shall set out the final equations with only sketchy derivation which, together with a series of step-by-step recipes for various corrections (K-term including bandwidth, aperture correction, isophotal to metric angular diameter), can be used for the practical calculation of intrinsic properties of objects from their observed luminosities, angles, and redshifts.

It is to be particularly emphasized that the final equations contain observables only. For example, the *construct* of "distance" is suppressed as it must be, because this heuristic variable has different meanings in different contexts, according to a variety of operational definitions. A large number of functions can be invented such as "distance by apparent luminosity," "by angular size," "distance when light left," "distance when light is received," "distance as the light-travel time multiplied by c," "distance by parallax," etc., all of which differ, except in the $z \rightarrow 0$ limit. We adopt the view here that physically meaningful equations must contain only directly observed quantities. "Distance" is not a directly observed quantity and will not be found in any of the final operational equations.

6.2. The Redshift–Apparent-Magnitude Relation

Robertson (1929) and Walker (1936) showed that the most general expression for the line element of a homogeneous and isotropic model is

$$ds^2 = c^2dt^2 - R^2(t)\left[\frac{dr^2}{1-kr^2} + r^2(d\theta + \sin^2\theta d\varphi^2)\right] \equiv c^2dt^2 - R^2du^2\,, \tag{11}$$

where k is the signature of the spatial curvature with values of $+1, 0$, or -1 for elliptic, Euclidean, or hyperbolic space, and where r, θ, and φ are dimensionless, comoving coordinates that are invariant for a given galaxy. Expansion of the three-dimensional manifold Rdu occurs if R is a function of time. When this metric is substituted in the field equations of general relativity, equations (9) and (10) result, the solution of which gives the space dilatation function $R(t)$ for all time.

Let R_0 be the value of R at the present epoch (the time light is received on Earth from any galaxy), and R_1 be its value when light was radiated by the source. The more distant the source from Earth, the larger will be the ratio R_0/R_1. The redshift, $\Delta\lambda/\lambda_0 \equiv z$, is given quite generally by

$$1 + z = R_0/R_1 , \tag{12}$$

which follows from the null geodesic $ds^2 = 0$ for a light track. This equation, derived first by Lemaître (1927, 1931), follows from equation (11) by noting that radial light tracks along the null geodesic require that

$$u = c\int_{t_1}^{t_0} \frac{dt}{R(t)} = \int_0^{r_i} \frac{dr}{(1 - kr^2)^{1/2}} . \tag{13}$$

The function u is *constant* for a given galaxy i, because r_i is an invariant coordinate. Here t_0 is the time of reception and t_1 the time of light emission. To derive equation (12), consider crests of adjacent waves at the source to be separated by Δt_1 seconds, emitted at times t_1 and $t_1 + \Delta t_1$.

These crests are received at the Earth at time t_0 and $t_0 + \Delta t_0$. Hence, by equation (13),

$$\int_{t_1}^{t_0} \frac{dt}{R(t)} = \int_{t_1+\Delta t_1}^{t_0+\Delta t_0} \frac{dt}{R(t)} . \tag{14}$$

Now the right-hand member can be summed as

$$\int_{t_1+\Delta t_1}^{t_0+\Delta t_0} \frac{dt}{R(t)} = \int_{t_1}^{t_0} \frac{dt}{R(t)} + \int_{t_0}^{t_0+\Delta t_0} \frac{dt}{R(t)} - \int_{t_1}^{t_1+\Delta t_1} \frac{dt}{R(t)} . \tag{15}$$

Subtracting equation (15) from equation (14), and noting that

$$\int_{t_i}^{t_i+\Delta t_i} \frac{dt}{R(t)} \equiv \frac{\Delta t_i}{R(t_i)} ,$$

gives

$$\Delta t_0/\Delta t_1 = R_0/R_1 , \tag{16}$$

showing that the time intervals will be unequal if the scale dilatation factors (R) are unequal. In particular, if $R_0 > R_1$, the crests will be received at a lower frequency than they were emitted, because $\Delta t_0 > \Delta t_1$. From this it follows directly that

$$1 + \Delta\lambda/\lambda_0 = R_0/R_1 , \tag{12$'$}$$

which is equation (12), showing that there is a redshift in the light of distant sources when R increases.

To proceed beyond this point requires a theory of light propagation in an expanding manifold of specified geometry. By solving the energy flow problem in an expanding universe whose metric is equation (11), Robertson (1938) showed rigorously that the apparent (bolometric) observed luminosity l of a source of intrinsic luminosity L in such a manifold is given by

$$l = \frac{L}{4\pi R_0^2 r^2 (1+z)^2} . \tag{17}$$

If $R(t)$ is known by the solution of Friedmann's problem (eqs. [9] and [10]), equation (13) can be solved for $r = f(t_0/t_1)$, which, by equation (12) can be given as $g(R, z)$, which can then be used to throw equation (17) into a function of observables only, together with a parameter q_0 that tells in what part of the $R(t)$ curve the present epoch lies. Mattig (1958), by an elegant reduction via this route, showed that

$$R_0 r = \frac{c}{H_0 q_0^2 (1+z)} \{q_0 z + (q_0 - 1)[(1 + 2q_0 z)^{1/2} - 1]\} , \tag{18}$$

for $q_0 > 0$, where $H_0 \equiv \dot{R}_0/R_0$, and $q_0 \equiv -\ddot{R}_0/R_0 H_0^2$, where Λ is put to zero.[1]

Equations (17) and (18) solve the problem of finding L once l and z are measured, if H_0 and q_0 are known. The final equation[2] follows directly as

$$L = 4\pi l c^2 H_0^{-2} q_0^{-2} \{q_0 z + (q_0 - 1)[(1 + 2q_0 z)^{1/2} - 1]\} . \tag{19}$$

It is easy to show that in the special case of $q_0 = 0$ (no matter, hence no deceleration)

$$R_0 r = cH_0^{-1}[z/(1+z)](1 + z/2) . \tag{20}$$

Hence,

$$L = 4\pi l c^2 H_0^{-2} z^2 (1 + z/2)^2 \tag{21}$$

if $q_0 = 0$.

If $q_0 = -1$, as in the steady-state model,

$$R_0 r = czH^{-1} \tag{22}$$

and

$$L = 4\pi l c^2 H_0^2 z^2 . \tag{23}$$

At this point some writers introduce the concept of "luminosity distance D" such that $l = L/4\pi D^2$, and hence $D = R_0 r(1+z)$ by equation (17) which, by equation (18), can be found as a function of z and q_0 alone. But, as with all explicit discussions of distance, the concept is an unnecessary complication because equations (19), (21), and (23) contain all that is necessary to derive the intrinsic luminosity of the object.

Equation (19) can be put into the familiar redshift-magnitude relation using $m \propto -2.5 \log l$, which, upon absorbing $4\pi H_0^{-1}$ into the constant, gives

$$m_{\text{bol}} = 5 \log q_0^{-2} \{q_0 z + (q_0 - 1)[(2q_0 z + 1)^{1/2} - 1]\} + M + C , \tag{24}$$

where C is a constant and M is the mean absolute magnitude. Similarly for $q_0 = 0$,

$$m_{\text{bol}} = 5 \log z(1 + z/2) + M + C , \tag{25}$$

[1] A different derivation of Mattig's equation (18) via the parametric cycloid relations for $R(t)$, where the variation of H_0 with time is accounted for, is given elsewhere (Sandage 1962, eqs. [23] and [31]). The more general form shown in equation (23) in those references must be used when considering the change of l with time due to the increase of R_0 and the decrease of H_0 as the Universe unfolds.

[2] Note that if $q_0 = +1$, equation (19) becomes

$$l = LH_0^2/4\pi c^2 z^2 ,$$

which agrees with intuition because "distance" by Hubble's law in the $z \to 0$ limit is $D = czH_0^{-1}$.

and for $q_0 = -1$,

$$m_{\rm bol} = 5 \log z(1 + z) + M + C \,. \tag{26}$$

6.3. The K-Correction to Measured Intensities

Equations (19), (21), and (23) relate l and L correctly only when both are defined over the same *proper* band width, and are centered at the same *proper* wavelength. However, the actual apparent luminosity that is measured, called here l', differs from the l of these equations if the detector is kept fixed in wavelength and bandwidth regardless of the redshift of the source. For example, such a detector would be a photometer of fixed spectral response. The measured intensity l' differs from the l required in equations (19), (21), and (23) by the K-factor defined in HMS (Appendix B), in Oke and Sandage (1968) in an equivalent but physically clearer way, and measured properly for giant E galaxies by Whitford (chapter 5 here and in 1971), and by Schild and Oke (1971).

The K-term is a purely technical correction. A perfect bolometer above the Earth's atmosphere would require no such correction, nor would a detector whose acceptance bands are moved in λ by $(1 + z)$ for every galaxy and whose bandwidth, at the same time, is widened by a factor $(1 + z)$. For fixed-band detectors and giant E galaxies, Whitford's tabulated corrections have generally been adopted as standard.

The measured flux density l' for *radio* sources must also be corrected for the effect of redshift. For a source whose radio spectrum is characterized by $F(\nu) \propto \nu^{-n}$ it is easy to show that the total K-correction including the bandwidth term for such sources is

$$l = l'(1 + z)^{1-n} \,. \tag{27}$$

The procedure to find L from measurement of l' is then:

1. Correct l' to that which would have been observed had the detector been tuned to the specific proper mean wavelength and bandwidth specified for L. This is the K-correction tabulated by Whitford for E galaxies, or by equation (27) for radio galaxies with spectral index n.

2. For Friedmann models with $\Lambda = 0$, use equation (19), (21), or (23) to find L. For models with $\Lambda \neq 0$, use tables calculated by Refsdal, Stabell, and de Lange (1967).

6.4. The Difference from Hubble's Procedure

The recipe of the last section differs from that used by Hubble (1936*d*; 1953; Hubble and Tolman 1935) because his K-correction contained two additional factors of $1 + z$ which were called the "number" and "energy" effects. In the present formulation we include these already in the *theory* through the $(1 + z)^2$ term in equation (17), and we must not put them again into the K-correction as defined in the last section.

It should also be noted that Hubble (1936*d*) neglected the *bandwidth* term in his K-correction, and this must be considered an oversight. His complete K-term *should* have been either $\Delta m = 7.5 \log (1 + z) + 2.5 \log I(z)/I(0)$ or $\Delta m = 5 \log (1 + z) + 2.5 \log I(z)/I(0)$ depending on "the inclusion or exclusion of the expansion factor," rather than with coefficients of only 5 or 2.5 for the $(1 + z)$ term.

But even with this correction, Hubble's discussion of the redshift-magnitude diagram "with and without recession factors" (Hubble 1938, 1953) cannot be correct within the present formalism because Hubble explicitly assumes that $R_0 r \equiv czH_0^{-1}$ for all z and for all geometries as his equivalent to equation (17). This, of course, is the intuitive choice (see footnote 2), but it has no basis in theory except in the $z \rightarrow 0$ limit for any geometry. Hubble's discussion of the redshift phenomenon "without a recession factor" is then not founded on a theoretical base; the correct form of equation (17) is totally unknown because the physics of a redshift without recession is itself unknown.

6.5. Intrinsic Diameters from Angular Measurements

Because of the effects of space curvature and of aberration due to the recession, the angular diameter of a rigid rod taken to various distances does not generally decrease as z^{-1}. The exact equations follow from the metric of § 6.2 (eq. [11]) by noting that two rays coming from the edges of a galaxy in the plane of the sky at metric distance r satisfy $dt = dr = d\theta = 0$. Hence

$$R^2 du^2 = R^2(t) r^2 d\theta^2 ;$$

or, integrating over the object whose linear metric diameter is $y \equiv Ru$, gives

$$\theta = \frac{y}{R(t_1) r} , \tag{28}$$

where $R(t_1)$ is the spatial dilatation scale length at the time t_1 that light left the galaxy. Using equation (12), equation (28) becomes

$$\theta = y(1 + z)/R_0 r , \tag{29}$$

which, with equations (18), (20), and (22) give

$$\theta(\text{metric}) = \frac{y H_0 q_0^2 (1 + z)^2}{c\{q_0 z + (q_0 - 1)[(1 + 2q_0 z)^{1/2} - 1]\}} \tag{30}$$

for $q_0 > 0$,

$$\theta(\text{metric}) = \frac{y H_0 (1 + z)^2}{cz(1 + z/2)} \tag{31}$$

for $q_0 = 0$, and

$$\theta(\text{metric}) = \frac{y H_0 (1 + z)}{cz} \tag{32}$$

for $q_0 = -1$ for the answer. Note that in no case does the angular diameter decrease as fast as z^{-1}, as would happen in static Euclidean geometry.

Equations (30), (31), and (32) can be used to find the linear separation y of two components of a radio source, once their angular separation is known by measurement. *Any* metric separation can similarly be calculated. However, some dimensions in astronomy are not metric distances, but refer to some isophotal property of an object such as the size to a given surface brightness. Estimated diameters of galaxies on photographic plates are of this type. By a curious theorem discovered by Hubble and Tolman (1935) that surface brightness of similar objects varies as $(1 + z)^{-4}$ rather than being constant with distance, isophotal diameters will differ from the expressions in equations (30), (31), and (32). To prove the theorem, one needs only to divide equation (17) by θ^2 from equation (29) to give

$$\text{S.B.} \propto \frac{l}{\theta^2} = \frac{L}{4\pi y^2 (1 + z)^4} .$$

The model-independent nature of this result is emphasized by Kristian and Sachs (1966).

Isophotal diameters will therefore differ from metric diameters by the ratio of sizes such that the ratio of surface brightness is $(1 + z)^{-4}$. For elliptical galaxies where the radial intensity function is $I(r) \propto r^{-2}$ for large r, the two types of sizes will be related by

$$\theta(\text{metric}) = \theta(\text{isophotal})(1 + z)^2 . \tag{33}$$

A more complete discussion of diameter measurements as a test of cosmological models has been given elsewhere (Robertson 1955; Hoyle 1959; Sandage 1961*a*, 1972*b*) where, in the last reference, a K-like correction to isophotal diameters, caused by the observed wavelength differing from the proper wavelength, is derived.

7. THE OBSERVATIONS AND WORLD MODELS

It is the extremely good agreement of the observations with the predictions of equations (24)–(26) and (30)–(32) that constitutes the modern evidence that (1) either the Universe expands in the way predicted by all nonstatic solutions (i.e., linear in the $z \to 0$ limit, a realm where the effects of $q_0 \neq 0$ can be neglected), or (2) if the Universe does not expand, then the redshift effect, due to whatever unknown mystical cause, imitates the nonstatic solutions precisely. In the face of evidence such as the existence of the 3° K fireball radiation and the agreement of the Hubble time, H_0^{-1}, with the age of our galactic system (cf. Sandage 1970) there seems little question that the Universe does, in fact, expand.

The observations of redshifts and magnitudes of cluster galaxies is shown in figure 5, using observations made at the close of the 1960s. The line is equation (24) with $q_0 = +1$ and the agreement of theory and observation is excellent. The small scatter about the line is evidence that absolute luminosities of first-ranked cluster galaxies have a very small dispersion, and therefore, by carrying the observations of clusters to large redshifts, a good determination of q_0 might be made. A solution from the data in figure 1 (Sandage 1972*b*) gives a highly uncertain value of $q_0 = 1 \pm 1$ (at the 2σ level), where no account is taken of the evolutionary correction. The problem was clearly unresolved at this writing, although many new programs for finding distant clusters and measuring their redshifts, magnitudes, and energy distribution curves were begun in the early 1970s.

Figure 6 shows the diameter-redshift relation (Sandage 1972*a*) for first-ranked cluster galaxies, and the theoretical relation

$$\theta(\text{isophotal}) = yH_0c^{-1}z^{-1}$$

for $q_0 = +1$ that follows by combining equations (30) and (33). Again, the fit is excellent, showing that the kinematic properties of uniformly expanding models (i.e., $cz \propto HD$) fit these local redshift, magnitude, and diameter data very well.

When more extensive data become available to larger redshifts, there is every belief that the parameters of the Universe will be determined, either by model fitting using the relevant $m = f(z, \theta, q_0, \Lambda)$ equations, or by Kristian-Sachs–like solutions for the kinematic field, including shear and rotation. Once accomplished, global properties of particular world models follow, and the history of the unfolding from the Friedmann singularity might then be available. The literature on these possibilities is extensive. Two review papers from the 1960s might usefully be consulted (Sandage 1961*a*, Zel'dovich 1965) concerning some of the model properties.

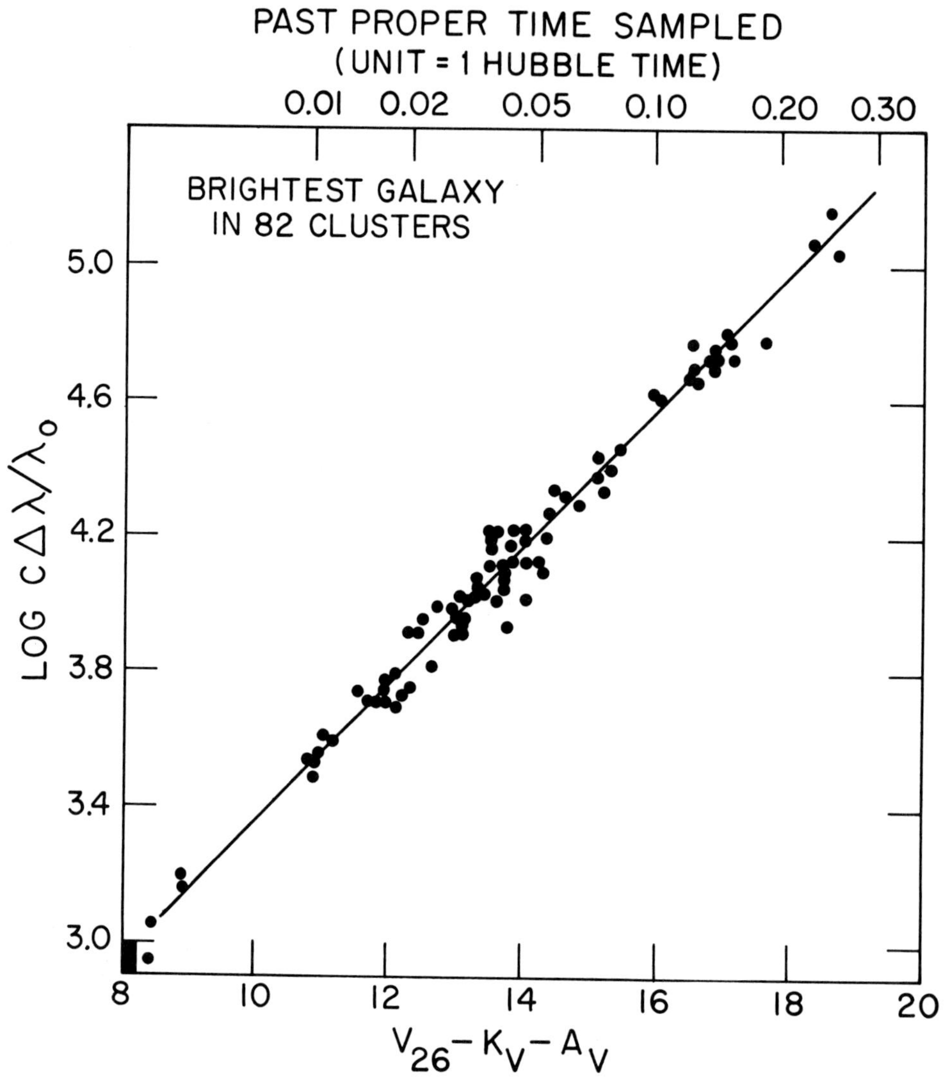

FIG. 5.—The Hubble diagram for 82 first-ranked cluster galaxies. The line is equation (24) of the text with $q_0 = +1$. The box in the lower left corner is the range within which Hubble's first formulation of the velocity-distance relation lay.

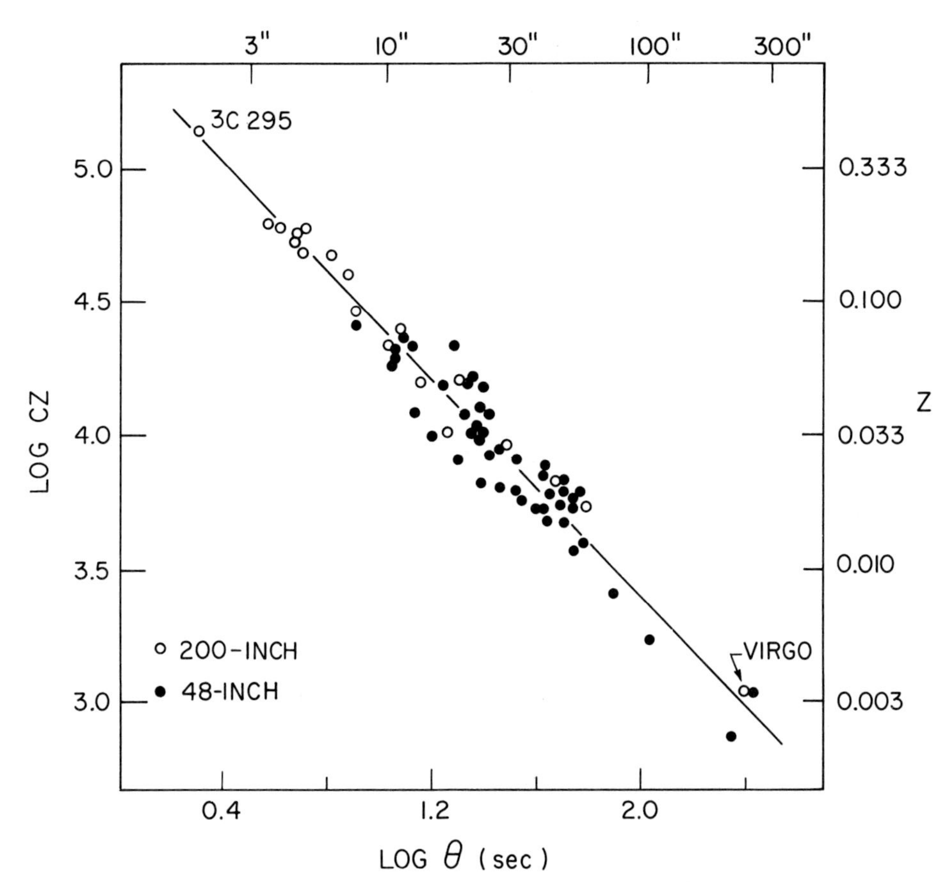

FIG. 6.—The variation of isophotal angular diameter with redshift for first-ranked cluster galaxies. The line is the theoretical relation for $q_0 = +1$, but the data here are not sufficiently accurate to constitute a determination of q_0 in this way yet (Sandage 1972*b*).

Easiest to visualize is the problem of the time scale. The present results ($H_0 = 55 \pm 5$ km s^{-1} Mpc^{-1} or $H_0^{-1} = 17.7 \times 10^9$ years; $q_0 = 1 \pm 1$; $T_{\text{glob cl}} = 12 \times 10^9$ years in 1972) suggest that there is no discrepancy between the age of our Galaxy and the Friedmann time of a big-bang universe. This, plus the evidence that the Universe evolves (equality of age of other galaxies, similarity of H_0^{-1} and the age of chemical elements, and the possible observation of the time horizon for the birth of quasars), is consistent with the view that a universal creation event did, in fact, occur.

REFERENCES

Bondi, H. 1952, *Cosmology* (Cambridge: Cambridge University Press).
Bowen, I. S. 1962, *Astronomical Techniques*, ed. W. A. Hiltner (Chicago: University of Chicago Press), chap. 2.
Chamberlain, J. W. 1961, *Physics of the Aurora and Airglow* (New York: Academic Press).
Conklin, E. K. 1969, *Nature*, **222**, 971.
Davidson, W. 1959, *M.N.R.A.S.*, **119**, 54.

de Sitter, W. 1917, *M.N.R.A.S.*, **78**, 3.
———. 1933, *The Astronomical Aspect of the Theory of Relativity*, University of California Pub. in Mathematics (Berkeley: University of California Press), Vol. **2**, No. 8.
Eddington, A. S. 1923, *Mathematical Theory of Relativity* (Cambridge: Cambridge University Press).
———. 1930, *M.N.R.A.S.*, **90**, 668.
Einstein, A. 1917, *Sitzungsberichte Preussischen Akad. d. Wissenschaften zu Berlin*, p. 142.
Epstein, E. E. 1964, *A.J.*, **69**, 490.
Ford, W. K., Rubin, V. C., and Roberts, M. S. 1971, *A.J.*, **76**, 22.
Fredrick, L., and Gutsch, W. A. 1972, unpublished catalog.
Friedmann, A. 1922, *Zs. f. Phys.*, **10**, 377.
Gamow, G. 1952, *The Creation of the Universe* (New York: Viking Press).
Heckmann, O. 1942, *Theorien der Kosmologie* (Berlin: Springer).
Henry, P. S. 1971, *Nature*, **231**, 516.
Hoyle, F. 1959, *Paris Symposium on Radio Astronomy* (Stanford, Calif.: Stanford University Press), p. 529.
Hubble, E. 1929, *Proc. Nat. Acad. Sci.*, **15**, 168.
———. 1936*a*, *The Realm of the Nebulae* (New Haven: Yale University Press).
———. 1936*b*, *Ap. J.*, **84**, 158.
———. 1936*c*, *ibid.*, **84**, 270.
———. 1936*d*, *ibid.*, **84**, 517.
———. 1938, *Observational Approach to Cosmology* (Oxford: Oxford University Press).
———. 1939, *J. Franklin Inst.*, **228**, 131.
———. 1953, *M.N.R.A.S.*, **113**, 658.
Hubble, E., and Humason, M. L. 1931, *Ap. J.*, **74**, 43.
Hubble, E., and Tolman, R. 1935, *Ap. J.*, **82**, 302.
Huggins, W., and Huggins, Lady. 1899, *Atlas of Representative Stellar Spectra* (London: William Wesley & Son).
Humason, M. L. 1929, *Proc. Nat. Acad. Sci.*, **15**, 167.
———. 1931, *Ap. J.*, **74**, 35.
———. 1936, *ibid.*, **83**, 10.
———. 1957, *Carnegie Institution of Washington Year Book*, **56**, pp. 61–62.
Humason, M. L., Mayall, N. U., and Sandage, A. 1956, *A.J.*, **61**, 97 (HMS).
Humason, M. L., and Wahlquist, A. D. 1955, *A.J.*, **60**, 254.
Kristian, J., and Sachs, R. K. 1966, *Ap. J.*, **143**, 379.
Lemaître, G. 1927, *Ann. Soc. Sci. Bruxelles*, **47**, 49.
———. 1931, *M.N.R.A.S.*, **91**, 483.
Lundmark, K. 1920, *Stock. Acad. Hand.*, Bd. 60, No. 8.
———. 1924, *M.N.R.A.S.*, **84**, 747.
———. 1925, *ibid.*, **85**, 865.
Lynds, C. R. 1972, *Ap. J.* (*Letters*), **168**, L87.
McCutcheon, W. H., and Davies, R. D. 1970, *M.N.R.A.S.*, **150**, 337.
McVittie, G. C. 1956, *General Relativity and Cosmology* (London: Chapman & Hall).
Mattig, W. 1958, *Astr. Nach.*, **284**, 109.
Mayall, N. U., and Vaucouleurs, A. de. 1962, *A.J.*, **67**, 363.
Minkowski, R. 1960, *Ap. J.*, **132**, 908.
Minkowski, R., and Wilson, O. C. 1956, *Ap. J.*, **123**, 373.
Oke, J. B., and Sandage, A. 1968, *Ap. J.*, **154**, 21.
Refsdal, S., Stabell, R., and de Lange, F. G. 1967, *Mem. R.A.S.*, **71**, 143.
Roberts, M. S. 1972, *External Galaxies and Quasi-Stellar Objects*, *IAU Symposium No. 44*, ed. D. S. Evans (Dordrecht: Reidel), p. 12.
Robertson, H. P. 1928, *Phil. Mag.*, **5**, 835.
———. 1929, *Proc. Nat. Acad. Sci.*, **15**, 822.
———. 1933, *Rev. Mod. Phys.*, **5**, 62.
———. 1938, *Zs. f. Ap.*, **15**, 69.
———. 1955, *Pub. A.S.P.*, **67**, 82.
Sandage, A. 1961*a*, *Ap. J.*, **133**, 355.
———. 1961*b*, *ibid.*, **134**, 916.
———. 1962, *ibid.*, **136**, 319.
———. 1970, *ibid.*, **162**, 841.
———. 1972*a*, *Quart. J. R.A.S.*, **13**, 282.
———. 1972*b*, *Ap. J.*, **173**, 485 (Paper I).
———. 1972*c*, *ibid.*, **178**, 1 (Paper II).
———. 1973, *ibid.*, **183**, 711 (Paper V).
Sandage, A., and Tammann, G. A. 1975*a* (Paper V of Hubble constant series), *Ap. J.*, **196**, 313.
———. 1975*b*, *ibid.* (Paper VI of Hubble constant series), **197**, 265.
Sandage, A., Tammann, G. A., and Hardy, E. 1972, *Ap. J.*, **172**, 253.
Scheiner, J. 1899, *Ap. J.*, **9**, 149.

Schild, R., and Oke, J. B. 1971, *Ap. J.*, **169**, 209.
Sciama, D. W. 1972, *Comments Ap. and Space Phys.*, **4**, 35.
Shapley, H., and Ames, A. 1932, *Ann. Harv. Coll. Obs.*, **88**, 43.
Silberstein, L. 1924, *M.N.R.A.S.*, **84**, 363.
Slipher, V. M. 1914, *Lowell Obs. Bull.*, Vol. 1, No. 58.
———. 1915, *Popular Astr.*, **23**, 21.
———. 1917, *Proc. Am. Phil. Soc.*, **56**, 403.
Stromberg, G. 1925, *Ap. J.*, **61**, 353.
Tammann, G. A. 1973, *Astr. and Ap.*, **21**, 355.
Tolman, R. C. 1929, *Proc. Nat. Acad. Sci.*, **15**, 297.
Truman, O. H. 1916, *Popular Astr.*, **24**, 111.
Vaucouleurs, G. de, and Peters, W. L. 1968, *Nature*, **220**, 858.
Vaucouleurs, G. de, and Vaucouleurs, A. de. 1964, *Reference Catalogue of Bright Galaxies* (Austin: University of Texas Press).
———. 1967, *A.J.*, **72**, 730.
Walker, A. G. 1936, *Proc. London Math. Soc.*, **42**, 90.
Whitford, A. E. 1971, *Ap. J.*, **169**, 215.
Wilson, O. C. 1949, *Pub. A.S.P.*, **61**, 132.
Wirtz, C. 1918, *Astr. Nach.*, **206**, 109.
———. 1921, *ibid.*, **215**, 349.
———. 1922, *ibid.*, **216**, 451.
———. 1924, *ibid.*, **222**, 21.
———. 1925, *Scientia*, **38**, 303.
Zel'dovich, Ya. B. 1965, *Adv. Astr. and Ap.*, **3**, 241.

Author Index

Pages of chapters by contributors are shown in **boldface.** Author listings in the reference sections are shown in *italics*. Numbers in parentheses indicate multiple citations.

Aarseth, S. J., 635(3), 636(3), *642*
Abell, G. O., *33*, 125, 134(2), 147(2), *155*(2), 181, 186, 187, 188, 189(2), *195*, *196*, 373, 402(2), *405*, 536, 537(2), *537*, 592, *596*, 597, 600, **601–45**, 603, 604(2), 608, 610, 611, 612, 613, 617(4), 618(9), 620, 621(3), 622, 625, 626, 627, 630, 632, 633, 637(2), 638, 639, 640(2), 641(3), *642*(3), *643*(3), *644*, 654, 655(4), 658, 659(2), 660, *663*, 665, 666, 682(2), 683(4), *722*
Ables, H., 498, *507*, 590, *597*
Adams, J. B., 365, *405*
Adgie, R. L., 184, *195*, *210*(2), 247, 251, *277*, *280*
Adler, R., 361, *405*
Agt, S. L. T. J. van, 522, *537*
Aizenman, M. L., 47, *77*
Aĭzu, K., 185, *195*
Albada, G. B. van, 403, 404(3), *405*, 629, 634(3), 635(3), *643*
Alfvén, H., 306(2), *306*, 380, 381, *405*
Allan, D., 323, *357*
Allen, C. W., 730, *758*
Allen, L. R., 245, 259, *277*, *279*
Allen, R. J., 341, 353, *355*, 729(2), 735, *758*
Aller, H. D., 296, *306*
Aller, L. H., 64, 65, 66(3), 67(6), 69(3), 70(3), 74, *77*(5), *78*(2), *80*, 87(2), 89, 101, *119*, *121*
Altenhoff, W. J., 332, *356*
Ambartsumian, V. A., 117, *119*, 402, 403(2), *405*, 553(2), *555*, 585, *596*, 631, 634, *643*
Ames, A., 4, *34*, 124, 129, 151(2), 152, *156*, 541, 543, *556*, 558, *596*, 601, *644*, 647, 648, 659, *663*, 773, *785*
Anderson, B., 247, 251, *277*, *280*
Angione, R., 255, 272, *281*
Antonov, V., 485, *504*
Appenzeller, I., 294, *306*
Archer, S., 179, *196*
Argyle, E., 317, 341, *355*
Arkipova, V. P., 4, 27, *35*
Arnett, W. D., 492, 493, *504*
Arnold, V., 441, *504*
Arny, T. T., 390, *405*
Arons, J., 274, *280*, 736(3), *758*(2)
Arp, H. C., 4, 5(2), 21, 24(2), *33*, 63, 67, 77, *77*(2), 243(2), 272, *277*, 285, 287, 299(3), *306*, 388, 403, *405*, 497, *504*, 510, 518, 520, 522, 523, *537*(2), 544, *555*, 735, 752, *758*(2)
Arsac, I., 251, *277*
Astier, N., 76, *78*
Avez, A., 441, *504*
Avner, E. S., 471, *504*
Axford, I., 274, 275, *278*, 306, *307*

Baade, W., 4, 5(2), 6(2), 13(3), 22(2), 26, 32(2), *33*(2), *34*, 61, 63, 67, *77*(3), 133, *155*, 159(2), *175*(2), 178(5), *195*, 211, 212, 255, 268, 272, 273, *277*(2), 330, 331, *355*, 398, *407*, 438, 503, *504*, *505*, 510, 513, 517, 522(2), 530, 532, *537*(2), 609(2), 613, 616, 618, 633, *643*(2), *645*
Baars, J. M. W., 326, *355*
Babcock, H. W., 87(2), 88, 98(2), 99, 101(2), 112(2), *119*
Bacon, F., 1, *33*
Bahcall, J. N., 66, *77*, 287(2), 290, 300, 301(7), 302(3), *306*(8), 731(2), *758*(2)
Bahcall, N. A. 66, *77*, 287(2), 623(5), 625, 626, 630(2), *643*
Bahner, K., 510, *537*
Bailey, J. A., 740, *758*
Baines, J., 323, *357*
Baker, J. G., 65, *77*
Baldwin, J. A., 287, 288, 289
Baldwin, J. E., 179(2), *196*, *197*, 311, 332, *355*
Balnaves, K. M., 329, *355*
Baranowska, M., 623(3), 626, 630, *644*
Barbanis, B., 446, 474, *505*(2)
Barbieri, C., *210*
Bardeen, J. M., 274(2), *277*(2)
Bare, C., 251, *277*, 296, *307*
Barnes, J. V., 527, *538*
Barnothy, J., 299, *306*
Barnothy, M. F., 299, *306*
Bartlett, J. F., 184, *196*
Bash, F. N., 243, 251, 254(2), 255, 259, 270, *277*, *279*(2)

Baum, W. A., 38, 58, 59, 61(2), *77*(2), 134(2), *155*, 159(2), 161, 166(2), 169, 171, *175*(2), 428(2), 438, *505*, 537(2), *537*, 612, 618, 619, *643*, *644*

Bautz, L. P., 609(2), 617, 618, *643*(3)

Baylis, W. E., 238, 239, *277*

Bazin, M., 361, *405*

Beale, J. S., 334, 355

Beaujardière, O. de la, 232, *277*

Becklin, E. E., 45(2), *79*, 167, *176*, 293, 295, *307*

Beehler, D., 323, *357*

Beery, J. G., 737, 738, *758*

Behr, A., 517, *537*

Bel, N., 396, *405*

Bell, R. A., 511, *537*

Belton, M. J., 90(2), 91, 105, 107, *119*, 468, *505*

Belzer, J., 391, *405*

Bennett, A. S., 179, 180, 181, *195*(2), 242, *277*, 754, *758*

Berge, G. L., 260, 261(2), 262, 263, *277*, *280*

Berger, J., 603, 622, *645*

Bergeron, J., 736(2), *758*(2)

Bergh, S. van den, 1, 13, 22, 23(2), 24(3), 25(5), 28(3), *33*, 38, 44, 46(3), 59, 61, 63, 75(2), 77, *77*, *78*, 146(4), 153, *155*, 166, *176*, 181, 188, *195*, 493, *505*, **509–39**, 510, 511, 515, 517, 523(2), 524(2), 528(2), 529(2), 531(2), 533, 535(2), 536, 537(3), *537*, *538*(3), 544(3), 545(2), 546(3), 547, *555*, 558, 559(2), 560(2), 572(2), 573, 575(3), 585(3), 587(4), *596*, 597, *600*, 607, 610(4), 611, 612(2), 616(2), 618, 622, 631, 632, *643*

Berry, C. L., 479(2), 481, *505*

Bertola, F., 106(2), 108, 109(2), *119*, 341, *355*, 497, *504*, 550(3), *555*

Bieger-Smith, G. P., 115, *121*, 417, *507*

Bigay, J. H., 125, 126(3), *155*, *156*(2)

Biraud, F., 263, *277*

Blaauw, A., 401, *405*, 422, 424, 463, 492, *505*

Black, J. H., 739

Blair, A. G., 737, 738, *758*

Blake, G. M., 275, *277*

Blum, E. J., 250, *277*

Bodily, L., 323, *357*

Boesgaard, A. M., 71, *77*

Bogart, R. S., 640(2), *643*

Bok, B. J., 483, *505*, 665, *722*

Bolton, J. G., 177(4), 179, 181(3), 185(2), 186, 187(3), 188(2), 190, *195*(5), *196*, 211, 241(2), 243(3), 247, 250, 263, 268(2), *277*(4), *280*(2), 742, 744, 752, *760*(2)

Bologna, J. M., 261, *277*

Bondi, H., 297, *306*, 776, *783*

Bonnor, W. B., 360(2), 365(2), 379, 380, 386, 404, *405*, 701(2), *722*

Bottinelli, L., 319, 327, 333(2), 334(3), 341, *355*(3), 572, *596*(2)

Bottlinger, K. F., 88, 90, *119*

Bowen, I. S., 768, *783*

Bowyer, C. S., 379, 398, *405*

Bowyer, S., 633, *643*

Braccesi, A., 180, 184, 195(2), 242, *277*, 287, 305, *306*, 741(2), 744(2), 757(2), *758*(4)

Bracewell, R. N., 241, 251, *277*

Brandt, J. C., 90(5), 91(2), 101, 105, 106, 107, 112(2), 113, *119*(3), 318, *355*, 468, *505*, 550

Branson, N. J. B. A., 180, *195*

Bray, A. D., 185, *196*, 253, *278*

Brecher, K., 736, *758*

Bridle, A. H., 242, *277*, 744, 748, *759*(2)

Broadfoot, A. L., 43, *77*

Brosche, P., 118, *119*

Broten, N. W., 251, *277*

Brouw, W. N., 482, *506*

Brown, R. Hanbury, 178, 179(2), 180, *195*, 259, *277*

Brundage, R. K., 242, *277*

Brundage, W. D., 330, *355*

Budden, K. G., 253, *277*

Burbidge, E. M., 5, 24, 29, *33*(3), 49, 57, 62, 67, 71(2), 74(2), 76(2), 77, *78*(2), **81–121**, 84, 85, 86, 89, 90(2), 92(2), 98, 100, 103(2), 106(14), 107(11), 108(2), 109(12), 110, 111, 113(9), 114, 115(3), 117, 118, *119*(3), *120*(4), *121*(3), 240, 268(5), *277*(4), 283(2), 286, 288, 289, 299(3), 300(3), 301(2), *307*(4), 318, 321(2), 341(22), *355*(6), *356*(2), 379, 399(2), 400(2), 403, *405*(2), 437(2), 483, *505*, 550(22), 551(13), *555*, *556*(8), 572, 622(2), 628, 631(4), *643*(2), 655, *663*, 752(3), 753(2), *758*, *759*(3)

Burbidge, G. R., 5, 24, 29, *33*(3), 38, 46, 49, 62, 63, 67, 71(2), 74(3), 76(2), 77, *78*(5), *80*, **81–121**, 84, 85, 86, 89, 90(2), 92(2), 98, 100, 103(2), 106(14), 107(11), 108(2), 109(12), 110, 111, 113(9), 114, 115(3), 117, 118, *119*(3), *120*(4), *121*(3), 234, 240, 268(2), 270, 272, 274, 276, *277*(4), 283, 287, 293(2), 298(2), 299(4), 300(3), *307*(4), *308*, 318, 321(2), 341(22), *355*(6), *356*(2), 379(2), 399(2), 400(2), 403, *405*(3), 409, 437(2), 483, *505*(2), 550(22), 551(13), *555*, *556*(8), 572, 622(2), 628, 631(4), *643*(2), 655, *663*, 736, 738, 753(2), *758*, *759*(2), *760*

Burgess, D. D., 291, *307*

Burke, B. F., 330, 332(2), 334, *355*(2), *356*

Burke, J. R., 701(3), 702, *722*

Burke, W. L., 288

Burns, W. R., 318, 332, 334, 341, *355*

Burr, E. J., 531, *538*

Burton, W. B., 475, *505*

Buscombe, W., 524, *538*

Callan, C., 360, *405*
Cameron, A. G. W., 274, *277*, 283, 306, *307*(2)
Campbell, J. W., 460, *505*
Campbell, P. M., 391(2), 392, *405*
Cannon, W., 247, *278*
Capaccioli, M., *210*
Carey, J. V., 522, *538*
Carpenter, R. L., 592, 593, *596*, 637(3), *643*
Carranza, G., 76, *78*, 102, 113, 115, *120*
Carswell, R. F., 289, 756, *759*
Caswell, J. L., 185, 192(2), 193, *195*, 243(2), 268, *277*, *281*
Caughlan, G. R., 74, *78*
Cavaliere, A., 274, *277*, *278*, 305, 736, *759*
Ceccarelli, M., 180, *195*, 242(2), *277*, *278*, 741(2), 744(2), *758*(2)
Cesarsky, D. A., 739
Chamaraux, P., 341(2), *355*(2), 572, *596*, 597, *600*
Chamberlain, J. W., 43, *78*, 771, *783*
Chambers, R. H., 61, *78*
Chandrasekhar, S., 92, *120*, 414, 461, 484, 501, *505*, 659
Charlier, C. V. L., 665, *722*
Chernin, A. D., 370, *405*
Chertoprud, V. E., 306, *307*
Chester, C., 146, *156*, 349, *355*
Chincarini, G., 109(2), 113(2), *120*, *121*
Chisholm, R. M., 251, *277*
Chiu, H., 283
Christiansen, W. N., 241, *278*
Christy, R. F., 510, *538*
Ciurla, T., 295, *307*
Clark, B. G., 184, *197*, 247, 251, 255, *277*, *278*, 296, *307*, 327, 329, *355*
Clark, E. E., 623, *643*
Clark, T. A., 296
Clarke, M. E., 184, *197*
Clarke, R. W., 193, *195*
Clements, E. D., *210*
Code, A. D., 75, 77, *78*, 164, 167, 171, 173, *175*(2), 433, *505*
Cohen, M. H., 245, 247(4), 251, 253(2), 255, 269, 270, *277*, *278*(4), *280*, 296(3), *307*(2)
Cole, D. J., 181(2), *195*, *197*, 249, 260, 261, 262(2), *278*
Colgate, S. A., 274, *278*, 306(2), *307*(2), 739
Colla, G., 242, *278*
Collins, R. A., 243(2), *281*
Conklin, E. K., 775, *783*
Contopoulos, G., 441, 446(2), 479, *505*
Conway, R. G., 243, 245, 255(2), 259, 263, 269, 273, *278*(2), *279*
Cook, N., 542, 544, 546, 548, *556*
Cooke, B. J., 263, *280*
Cooper, B. F. C., 249, 260, 261, 262(2), *278*
Corwin, H., 597(2), *600*
Costero, R., 66(3), 67, 70, *79*
Courtès, G., 67, 76, *78*(2), 102(3), 113, 115, *120*(2)
Crampin, D. J., 106, 107(2), 118(2), 119, *119*, *120*, *121*, 341(2), *355*(2), *356*, 394, 395, 397, *405*, 438, 466, 468(2), *505*, 550, 551(2), *556*(2)
Crawford, D. F., 745, *759*
Crawford, D. L., 527, *538*
Crowther, J. H., 193, *195*, *210*, 251, *280*
Cruvellier, P., 67, *78*
Csiro Radiophysics Div. Staff, 242, 263, *278*
Curtis, H. D., 2, 4(4), *33*, 178, *195*, 601, 647, 648, *663*
Cutler, L., 323, *357*
Czyzak, S. J., 67(2), 69, 70, *77*

Dahn, C. C., 104, *121*, 542(3), 546, *556*
Dalgarno, A., 739
Damme, K. J. van, 341(2), *356*(2), 572, *596*, 633, *644*
Danby, J. M. A., 486, *505*
Danver, C. G., 2, 3(2), *33*
Danziger, I. J., 525, *539*, 706, *722*
Darchy, B. F., 341, 353, *355*
Das Gupta, M. K., 182, *196*, 212, *279*
Davidsen, A. F., 410, *506*, 633, *643*
Davidson, C. (K.), 161, *175*, 290
Davidson, W., 750(2), *759*(2), 775, *783*
Davies, M. M., 744, 750, *759*(3)
Davies, R. D., 260(2), 261, *278*(2), 317, 318, 319, 330, 332, 334, 341, *355*(4), *356*, 398, *405*, 550(2), *556*, 730, 745, *759*, 773, *784*
Davis, M. M., 242, 264, *277*, *280*
Day, G. A., 181(2), *195*, *197*, 242, *280*
De Jong, M. L., 185, *195*
de Lange, F. G., 779, *784*
Delhaye, J., 419, 420, 463, *505*
Demarque, P., 47, *77*
Demers, S., 522, *538*
Demoulin, M. -H., 38, 62, *78*(2), 106, *120*
Dennis, T. R., *505*
Dennison, E., 134(2), *156*, 612, *643*
Dent, W. A., 247(2), *278*(2)
de Sitter, W., 763(5), 764(5), 765(2), *784*
Deutsch, A. J., 43, 44(4), 62, *78*
De Veny, J. B., 288
Dewhirst, D. W., 181, *195*
De Young, D. S., 275(2), *278*(2)
Dicke, R. H., 360, 369, 378, *405*(2), *407*, 726(2), 737, *759*
Dickel, H. R., 67, *78*
Dickel, J. R., 180, *196*, 242(4), *278*(2), *279*, *281*

Dickens, R. J., 449, *507*, 522, 537, *538*
Dieter, N. H., 106(2), 113, *120*, *195*, 333, 341, *355*, 550, *556*
Dixon, M. E., 452, 480, *505*
Dixon, R. S., 242(4), *277*, *278*(3)
Donaldson, W., 247, 251, *277*, *280*
Doroshkevitch, A. G., 377, *405*, 739, 750, *759*(2)
Drake, F. D., 241
Drake, J. F., 739
Dreyer, J. L. E., 2(2), 601, 647
Duchesne, M. 98(3), 112(2), *120*
Duflot-Augard, R., 107, 108, 109(3), *120*, 551, *556*
Dumont, R., 126(2), *156*(2)
Duncan, J. Ch., 5

Eastmond, S., 536, 537, *537*, 618, *642*
Eddington, A., 762, 764(2), *784*
Edeskuty, F., 737, 738, *758*
Edge, D. O., 179(2), *195*, *196*
Eggen, O. J., 32, *33*, 401(2), 402(3), *405*, 419, 424, 425, 438, 450, 451, 452(3), 453, 472(2), 473(2), 489, 492(2), 494, 495(3), *505*(3), *506*
Ehman, J. R., 242(2), *277*, *278*
Eidman, V. I., 230, *278*
Einstein, A., 763(2), *784*
Eissner, W., 69, *78*
Ekers, R. D., 181(2), 184, *195*, *197*(2), 254, 264, 277, *278*(2), 610, *644*
Ellder, J., 296, *307*
Elsässer, H., 533, *538*
Elsmore, B., 178, 179, 180, 183, 184(2), *196*, *197*(4), *210*(2), 257, *280*
Elvius, A., 143, *156*, 273, *278*, 321, *355*, 517, *538*
Emden, R., 448, *505*, 623, *643*
Epstein, E. E., 89(3), 106(9), 108(3), *120*, 319(2), 320, 326, 329, 333, 341, *355*, 572(2), 573(3), *596*, 773, *784*
Epstein, R. I., 215, *278*
Evans, D. S., 13, *33*, 134, *156*, 178, *196*, 268, *278*, 773

Faber, S. M., 166, *175*
Fanaroff, B. L., 758, *759*
Fanti, C., 242, *278*
Fanti, R., 180, *195*, 242(2), *277*, *278*, 741(2), 744(2), *758*(2)
Fath, E. A. 601
Faulkner, D. J., 67(3), 69(2), 70(2), *77*, *78*(2), 287, 297, *307*, *308*, 745, *760*
Feast, M. W., 23(2), *33*(2), 67(2), *78*, 108(2), *120*(2), 341, *355*, 515, 524, *538*
Feix, M. R., 460, *505*
Feldman, P. A., 215, *278*, 306, *308*
Felten, J. E., 237, 272(2), *278*(2)
Fernie, J. D., 514(2), 515(3), 517, 522, 532, 537, *538*(2)
Ficarra, A., 242, *278*
Fichtel, C. E., 67, *78*
Field, G. B., 274, *278*, **359–407**, 369, 372(2), 373, 376, 379(2), 389, 394(2), 398(3), 400, *405*(7), 633, *643*, 719, *722*, 729(3), 731(2), 732, 736(2), *759*(2)
Fish, R. A., 48, 49, *78*(2), 103(2), 104(4), 107(6), *120*(2), 138, 143, 145, *156*, 367(2), 389(2), *405*, 429(3), 435, 440, 498(2), 499(2), 500(2), 501, *505*, 551(7), *556*(2)
Fitch, L. T., 242, *278*
Flather, E., 67, *79*
Fomalont, E. B., 184(2), 189, *196*(4), *210*(2), 242, 251, 254, 258, 259, 264, 269, 270, *277*, *278*, *280*, 745, *759*
Ford, W. K., 62, 71, 74(2), *78*, *79*, 107, 113, 115, *121*, 321, 341(2), *356*, 773, 774, *784*
Formiggini, L., 242, *278*, 757(2), *758*(2)
Fowler, W. A., 70, 74, *78*, *80*, 274, *279*, 293, 297, 305, 306, *307*(3), 381, *406*, 738, 739, *759*, *760*
Fraser, C. W., 255, 272, *281*
Fredrick, L., 773, *784*
Freeman, K. C., 6, 22(2), 25, 26, 29, 32, *33*, *34*, 92(2), 106, 113(2), *120*, *121*, **409–507**, 426(2), 427, 432, 433(3), 439, 452, 466, 467, 475, 483(2), 484(3), 485(3), 486(3), 487, 488, 489(4), 490, 491(3), 494, 499, 500, 501, 502(3), 503, *505*, *506*, *507*(2), 773
Friedmann, A., 761, 764(3), *784*
Frost, A. D., 247, *277*
Fujimoto, M., 479, 480, 483, *505*, 734, *760*

Galt, J. A., 242, 251, *277*, *278*, 744, *759*
Gamow, G., 369(3), 375, 381, 391, *405*(2), 725, 737, *759*, 776, *784*
Gandolfi, E., 242, *278*, 757, *758*
Gaposchkin, C. P., 513(2), 522(2), *538*(2)
Gaposchkin, S., 513, 522(3), *538*(2)
Gardner, F. F., 69, 70(2), *78*, 181, *195*, 236(2), 243(2), 259, 260(10), 261(2), 262(2), *277*, *278*(2), *281*
Gascoigne, S. C. B., 439, 495, *505*, 510(2), 511(2), 513, 515, 517, 521, 523(4), 524, 531, 537, *538*(3)
Gelato, G., 180, *195*, 242, *277*, 741, 744
Geller, M. J., 585, 597, *600*
Gent, H., 184, *195*, *210*(2), 247, 251, *277*, *280*
Georgelin, Y., 76, *78*, 102, 113, 115, *120*
Gérard, E., 572, *596*
Giacconi, R., 610, *643*, 736, *759*
Gilbert, I. H., 404, *405*
Gilbert, J. A., 263, *278*
Ginzburg, V. L., 211, 213(2), 215, 216(2), 220(3), 221, 225, 226, 237, 238, *278*(3)
Giovannini, C., 180, *195*, 242, *277*, 741(2), *758*(2)
Glaze, D., 323, *357*
Godfredsen, E. A., 585, *596*

Gold, T., 274, *278*(2), 306, *307*, 379, *405*
Goldberg, L., 66(2), *78*
Goldreich, P., 274, *278*, 395, 396(2), *405*, 461, 462(2), 464, 475, 485, *505*
Goldstein, R. M., 296
Goldstein, S. J., Jr., 398, *405*, 730, *759*
Golnev, V. Ya., 262, *278*
Gordon, C. P., 333, *355*
Gordon, K. J., 318(2), 332, 334(3), 341(2), *355*(2), 475, *505*
Gott, J. R., 632, 633, *643*
Gottesman, S. T., 317, 318, 319, 330, *355*(2), 550(2), *556*
Gouguenheim, L., 319, 327(2), 333, 341(2), *355*(3), 572, *596*(2)
Gould, R. J., 74, *78*, 379, *405*
Gower, J. F. R., 180, *196*(2), 242, *278*, 744, *759*
Graham, J., 525, *539*
Granz, R., *210*
Greaves, W. M. H., 161, *175*
Greenstein, J. L., 160, 161, *175*, 184, 185(2), *196*(3), 268, *278*, 285, 290, 296, 297, 301, *306*, *307*(2), 401, *405*
Gregory, C. C. L., 4
Grewing, M., 38, *78*
Griffin, R., *210*
Grueff, G., 242(2), *278*(2), 741, 744, *759*
Gudehus, D. H., 617, *643*
Guélin, M., 318, 332, 334(2), *355*
Gulkis, S., 185, *196*, 247, 253, 255, *278*(2)
Gundermann, E. J., 253, 270, *278*
Gunn, J. E., 58, 59, 74, 75(2), *79*, 243, 274(2), 276, *278*(2), *280*, 287, 297, 298, 300, 305, *307*(2), 398, *405*, 730, *759*
Gursky, H., 610, *643*, 736, *759*
Gush, H. P., 251, *277*
Gutsch, W. A., 773, *784*
Gyllenberg, B., 560, *596*

Haddock, F. T., 178, *196*, 247, *278*, 296, *306*, 734, *759*
Hagen, G. L., 515, *538*
Hall, J. S., 517, *538*
Hansson, B., 296, *307*
Hardy, E., 776, *784*
Harlan, E., 295, *307*
Haro, G., 67, *78*
Harrach, R., 323, *357*
Harrington, R. G., 4, *33*, 486, 491
Harris, D. E., 180, 181, *195*, *196*, 242, 243, 253, 270, *277*, *278*(2), 532, *538*, 741, 744, *758*
Harrison, E. R., 365, 378(2), 379, *405*, 695, 719, 720(2), *772*
Harwit, M., 380(2), *405*
Hauser, M. G., 640(5), *643*, *644*, 686(4), *722*, *723*
Hawking, S. W., 739, *759*
Hayes, S., 113, *121*, 167(4), *175*
Hayli, A., 446, *505*
Hazard, C. V., 178, 179(2), 180, 185(5), *195*, *196*(3), 253, *278*, 284, 287(2), 288, *307*, 738, *759*
Haze (Hase), V. T., 67, *79*
Hearnshaw, J. B., 493, *505*
Heckmann, O., 775, *784*
Hedemann, E., *506*
Heeschen, D. S., 75(2), *78*, 179, 180, 184, 192, 193, *196*, *197*, 243, 245, 247, 264, *278*, *280*, 298(2), *307*, 572, 573, *596*
Heidmann, J., 319, 327, 341, *355*, 533, *538*, 572, *596*(2), 597, *600*
Heidmann, N., 319, 327, *355*, 597, *600*
Heiles, C., 440, 444, 446
Henderson, A. P., 67, *78*
Henize, K. G., 20, 67, *78*, 519, *538*
Hénon, M., 440, 444, 446, 460(2), *505*(2)
Henry, P. S., 775, *784*
Henry, R. C., 379, *405*
Herk, G. van, 521, 537, *538*
Herschel, Sir J., 2(4), 601, 647(2)
Herschel, W., 2(3), 647(8), *663*
Herzog, E., 125, *157*(3), 402, *407*, 603, 604, 618(2), 640(2), *645*, 648, *663*
Herzog, P., 402, *407*
Hewish, A., 179, 180, *197*(2), 245, 251, 253(3), *279*(2), *280*, 735, *760*
Hey, J. S. 177(2), *196*(2)
Hiebert, R. D., 737, 738, *758*
Hill, E. R., 180, 181, *196*(2), 242(2), *279*(2)
Hiltner, W. A., 124, 134, *156*, 272, *279*, 294, *306*, 439, *505*, 510, 524, *537*, *538*
Hindman, J. V., 108(2), *120*, 309(2), 316, 318(2), 319, 329(4), 334, 335, 341, *355*(3), *356*(3)
Hitchcock, J. L., 23, *33*, 108, *120*, 153, *156*
Hively, R., 706, 707(6), 708, 711, 721, *722*(2)
Hobbs, R. W., 296, *307*
Hodge, P. W., 5, 6(2), 22, 23, 29, *33*(4), 38, 67(2), 77, *78*, 108(2), *120*, 126, 134, 137, 153(4), *156*(2), 159, *175*, 392, *405*, 438, 449, 450(2), 498(2), *505*(2), 523, 537, *538*, 581, 587, *596*, 608(2), 622, 625, 630, *643*(2)
Hodge, S. M., *506*
Högbom, J. A., 108(2), *121*, 241, *278*, 319(2), 320(2), 321, 341, *357*, *556*
Höglund, B., 113, 115, *120*, *196*, 296, *307*, 341, *356*
Holden, D. J., 242, 243(2), *279*, *281*, 745, *759*
Hoerner, S. von, 137, *156*, 253(2), *279*, 397, *406*
Hoffleit, D., 519, *538*
Hoffmeister, C., 560, *596*
Hofmeister, E., 510, 531, *538*
Hogg, D. E., 184, *197*, 255(2), 259, 269, 273, *279*
Hogg, H. B. S., 524(2), *538*
Hohl, F., 460(3), 465(2), 471(2), *505*(3)

Hollinger, J. P., 296, *307*
Holmberg, E., 1, 3, 8, 22, 23, 24, 28(3), 29, *33*, 38, 89, 95(2), 104(2), 105, 108(2), 112, 113, 119, *120*, **123–57**, 124, 125, 126(2), 128, 129, 131, 132, 133, 137, 138, 139, 143(3), 145(2), 146(3), 147(3), 149(2), 151(4), 153(3), 154, 155, *156*, 159, *175*, 319, 327, 336, 341, 349, *356*, 399, 400, 402(2), *406*, 509, 527(2), 533(2), 535, *538*, 541(4), 542(2), 543, 546(3), 547, 548(2), 549, 554, *556*, 560, 572, 573(2), 576, 585(3), 589, 590(2), 592, *596*, 608, 612(2), 613(3), 618, 619(3), 621(2), 622, 625, 636(2), *643*, 654(2), 655, *663*, 665(2), 666(2), 684, *722*
Houten, C. J. van, 124, 134(3), 137, 138(2), 145, *156*(2)
Howell, T. F., 733, *759*
Hoyle, F., 5, *33*, 70, *80*, 118(2), 119, *120*, 274(2), *279*(2), 293, 297, 298, 299(3), 300, 305, *307*(4), 369, 379(5), 380(2), 381(2), 384(2), 385, 386, 388(2), 389, 390, 394, 395, 397, 398, 400(2), *405*(3), *406*(4), 438, 466, 468(2), 504, *505*, 719, *722*, 738, *760*, 781, *784*
Hubble, E., 1, 4(2), 5(4), 6(7), 8(3), 10, 11, 13, 22(2), 23(4), 24, 26, 27(8), 28, 29, *33*(2), 103, 124, 125, 131, 133(3), 134, 135(2), 136, 137(2), 144, 147, 153(2), *156*, 159, 160, 170, 172, *175*(2), 509(3), 510, 517, 530(2), 531, 536(2), *538*(3), 559, 602(5), 610, 611, 612, 641(2), *643*, 649(5), 650(4), 651, 653(5), 660, 661(3), *663*, 665(2), 666, 684, *722*, 761(3), 762(3), 764(3), 765(2), 766(6), 767(3), 774, 779(4), 780(2), 782, *784*(3)
Huchra, J., 618, *643*
Huggins, Lady, 761, 762, *784*
Huggins, W., 761(2), 762(4), *784*
Hughes, R. G., 745, *759*
Hulst, H. van de, 85, *120*, 323, 343, *356*(2)
Humason, M. L., 5, 13, 26(2), 27, *33*, 38, 39, 77, 78(2), 98, 105, *120*, 143, *156*, 159, 160, 163, 171(2), 175, 183, 191, *196*, 268, *279*, 321, *355*, 509(2), 510, 535, *538*(2), 544, *556*, 574, 585, *596*(2), 611, 612, 618(4), 628(2), 629(3), 633(2), *643*, *645*, 655(2), 659(2), 660, 661, *663*, 761(2), 765(6), 766(5), 767(7), 771, 772, 773, 774(2), 775, *784*(4)
Hunstead, R. W., *210*
Hunter, C., 365, 379(2), 386, 390, 392, 395, 396, *406*(2), 465(2), 470, 471(2), 472(3), 504, *505*(2)
Hutchings, J. B., 524, *538*
Hutchins, J., 394, *405*

Iben, I., *78*
Innanen, K. L., 86, 106, 118, *120*, 423, 436, 504, *505*
Iriarte, B., 172, *176*, 524, *538*
Irvine, W. M., 360, *406*

Janes, K., 288
Jauncey, D. L., 247(2), 251, 255, *277*, *278*, *280*, 287, 296(2), *307*, 745, *759*(2)
Jeans, Sir J. H., 359(5), 364, 388, 403(2), *406*, 412, 696(2), *722*
Jefferts, K. B., 739(2)
Jenkins, J. B., 739
Jennison, R. C., 182(2), *196*(2), 212, *279*, 398, *405*, 730, *759*
Johnson, H. L., 38, 45, 64, *78*, 160, 164(3), 168, 169(5), 170, 172, *175*, *176*(2), 295, *307*, 524, *538*, 649, *663*
Johnson, H. M., 6, *33*, 64, 67(2), *78*, 137, 138, 145, *156*
Jokipii, J., 297, *307*
Julian, W. H., 274, *278*, 464(3), 473, 485, *506*(2)
Just, K., 621, *643*

Kaftan-Kassim, 685
Kahn, F. D., 471, *506*, 585, *596*
Kaidanovskii, N. L., 259, *281*
Kalnajs, A. J., 465, 479(2), 481, *506*
Kant, I., 647
Kardashev, N. S., 232, 233, 244(2), *279*, 379, *406*, 747, *759*
Karpowicz, M., 125, *157*, 603(2), 604, 622, 640(2), *645*(2), 648, *663*, 683, *722*
Katem, B. N., *210*
Kato, S., 370, 376, *406*(2)
Kaufman, M., 369, 377, *406*, 705, 706(2), 707, *722*
Kawabata, K., 734, *760*
Kayser, S. E., 513(2), 514, 517, 518, *538*
Kazès, I., 572, *596*
Keeler, J. E., 4
Keller, G., 391, *405*
Kellermann, K. I., 181, *196*, 222, 239, 241, 243(6), 244(2), 245(7), 246, 247(3), 248, 251, 255, 260, 264, 274, *277*, *278*(2), *279*(4), *280*(4), 288, 295(5), *307*(2), 745, 746, 750, *759*(3)
Kellogg, E., 610, *643*, 736, *759*
Kendall, K. R., 43, *77*
Kenderdine, S., 180(2), 183(2), *196*(2), *197*, *210*, 242(2), 254(2), 255(2), 256, 259, 268, 269, 270, 273, *279*(2), *280*, *281*, 740, *760*
Kennedy, J. E. D., 242, *278*, 744, *759*
Kennedy, P. M., 524, *538*
Kerr, F. J., 86, *120*, 309(2), 318, 334, 335, 343, *355*, *356*(4), 393, 397, *406*, 417(2), 436(2), *506*(2)
Kiang, T., 150, *156*, 611(2), 640, *643*(2)
King, I. R., 48, 49, 77, *78*, 94(2), 100, 104, 107, *120*(2), 136, 137, *156*, 380, 398, *406*, 434(2), 448(2), 449(4), 450(5), 471, 496(2), 497(2), *504*, *506*(3), 609, 623, 627(2), 629, 634, *644*
Kinman, T. D., 112(3), 295(3), *307*(3), 492, 494(2), 495, 500, *506*
Kintner, E. C., 609, 623, 627(2), 629, 634, *644*

Kleinmann, D. E., 295, 299, *307*
Klemola, A. R., 603, *643*
Knight, C. A., 296
Knox-Shaw, H., 4(2), *33*
Koehler, J. A., 353, *356*, 399(2), 403, *406*(2), 633, *644*, 729, *759*
Koehler, S. A., 58, *79*
Kokin, Y. F., 189, *197*
Kollberg, E., 296, *307*
Kowal, C. T., 125, *157*(2), 528, 533, 536, 537, *538*, 603, 604, 640(2), *645*, 648, *663*
Kozlovsky, B., 290, *306*
Kraft, R. P., 510(2), 511, 537(2), *538*
Krasnogorskaja, A., 4, 27, *35*
Kraus, J. D., 180, *196*, 241, 242(4), *277*, *278*(3), *279* 330, 332, *355*, *356*
Kron, G. E., 161(4), 162, *176*, 510, 511, 524(2), *538*(2), 648(2), 650, 661, 662(3), *663*
Kristian, J., **199–210**, 199, *210*(2), 285, 298, 775, 780, *784*
Kruit, P. C. van der, 242(2), 262, 276, *279*, 482, *506*
Kulsrud, R., 274, *280*
Kuzmin, G., 90, *120*
Kwast, T., 623, 625, 630, *643*
Kwee, K. K., 522(6), *538*

Laan, H. van der, 242(2), 243, 245, 247, 262(2), 275, 276(3), *279*(4)
Lallemand, A., 98(3), 112(2), *120*
Lambert, D. L., 66(3), 67, *78*
Lamla, E., 295(2), *307*(2)
Landau, L. D., 414, *506*, 713, *722*
Langebartel, R. D., 23, *34*
Lari, C., 242, *278*
Lasker, B. M., 164, 166, *176*
Lauqué, R., 341(3), 353, *355*(3), 572, *596*
Layzer, D., 360, 361, 369, 377(2), 378(3), 390(7), 391(2), 398, *406*, 603, *643*, **665–723**, 691, 699(2), 701, 718, 721, *722*
Leavitt, H. S., 510, *538*
Ledoux, P., 395, *406*
Legg, M. P. C., 217, 263, *279*
Legg, T. H., 251, *277*
Lemaître, G., 378, *406*, 764(3), 777, *784*
Lenouvel, F., 126, *156*
Lequeux, J., 242, 243, 253, 254, *277*, *279*
Le Roux, E., 213, 225, *279*
Lesh, J. R., 189, *197*, 271, *279*, 609, *644*
Leslie, P. R. R., 180, 184, *196*(2)
Lewis, B. M., 334, 341, *356*
Lifshitz, E. M., 360(2), 365, 369, 371(3), 372, 375(2), 384(2), *406*, 414, *506*, 696(2), 697(2), 698, 701(2), 713, *722*(2)
Liller, M. H., 6, *34*, 134(2), 135(2), 136, 137(2), 138(2), 144(2), *156*, 231, *506*, 612, *643*
Liller, W., 64, 66, *77*
Lilley, A. E., 66, *79*, 323(2), *356*(2)
Limber, D. N., 24, *34*, 585, *596*, 603(3), 631(2), *643*, 659(2), *663*, 685, *722*
Lin, C. C., 333, *356*, 384(3), 386, *406*, 476(2), 478, 479(4), 480(3), 481, 483, *506*(3)
Lindblad, B., 23, *34*, 424(2), 476, 486, 496, *506*(2)
Lingenfelter, R. E., 234, *280*
Little, L. T., 245, 253(2), *279*
Locke, J. L., 251, *277*
Lockhart, I. A., 250(2), 259, *279*(2)
Lodén, K., *80*
Lodén, L. O., *80*
Lohmann, W., 89, 90, *120*
Long, R. J., 243(2), 245(2), *278*, *279*(2)
Longair, M. S., 183, 185(2), 189, 192, 193, *196*, *197*, 242, 276(2), *279*, *280*, 302, *307*, 400, *406*, 731, 733, 737, 738, 745, 747, 748, 749(2), 750, 755(2), 757, 758(3), *759*(6)
Lovasich, J. L., 618, 628, 629, *644*
Low, F. J., 295(2), 298, *307*(2)
Lowrance, J. L., 752, *759*
Lundmark, K., 3, 4(2), 8, *34*, 541, 762(2), 764, 765, *784*
Lunel, M., 126, *156*
Lüscher, E., 238, 239, *277*
Lüst, R., 397, *406*
Luyten, W. J., 195, *197*, 287(2), *308*, 757, *760*
Lynden-Bell, D., 32(4), *33*, *34*, 382, 384, 386(4), 390, 391(2), 392(2), 395, 396(2), 401(2), 402(3), 404, *405*(2), *406*, 413, 448, 450(2), 451, 452(3), 453, 454, 455(3), 457, 458(2), 459, 461, 462(2), 464, 471, 475, 476(2), 481, 483, 485, 489, 494, 495, *505*(2), *506*(2), 756(2), 757
Lynds, C. R., 113, 115, *120*, 268, 273, *279*, 287, 293, 300(2), 301(3), 305, *306*, *307*(3), *308*, 752, 754(5), 755, 756, 757, *784*
Lynds, R., 755, *759*
Lyngå, G., 138, *156*

McAdam, W. B., 179, *196*
MacAlpine, G. M., 290
McClain, E. F., 261, *277*, 323, *356*
McClure, R. D., 38, 44, 46(3), 58(2), 59, 61, 74, 75(4), 77, *78*(2), *79*, 166, *176*, 517(2), 537(2), *538*(2)
McCray, R., 736, *758*
McCrea, W. H., 293, *307*, 360, *406*
McCutcheon, W. H., 341, *356*, 773, *784*
McDonald, F. B., 67, *78*, 587
Macdonald, G. H., 182(3), *196*(2), *210*, 254(2), 255(4), 256, 259(2), 269(2), 270, 273(2), *279*(2)

McGee, R. X., 319, 329(4), 333(2), 335, 341, *355*, *356*(2)
Mack, J. E., 379, 398, *405*
Mackay, C. D., *210*(2), 254, 259, 269, 271, *279*, 286, *307*, 752, *758*
Mackey, M. B., 181, 185, *195*, *196*, 243(2), *277*, 284, *307*
McKee, C. F., 287
McKibben, V. (McKibben Nail), 510, 519, *538*, *539*
McLeish, C. W., 251, *277*
MacLeod, J. M., 180, *196*, 242, *279*
McVittie, G. C., 182, 191(4), 192(3), 193, *197*, 242, *278*, 298, *307*, 747, 750, *759*, *760*, 775, 776, *784*
Madwar, M. R., 4
Malik, G. M., 559, *597*, 650, 653, *663*
Maltby, P., 182, 183, 184, 190, *196*(2), 236, 251, 253, 260, 261(2), 268, 269, 270, *279*(4)
Maran, S. P., 283, *307*
Marandino, G. E., 296(2), *307*
Marlborough, J. M., 515, *538*
Marochnik, L. S., 482, *506*
Martin, E., 161, *175*
Martins, P., 69, *78*
Maslowski, J., 745, *759*
Mathewson, D. S., 179, *196*, 482, *506*
Mathis, J. S., 67, 70(2), *78*
Matthews, T. A., 27, 29, *34*, 178, 181, 183(3), 184(2), 185(5), 189, 190, 193, *196*(7), 236, 249, 268, 269, 270, *279*, 284(3), 285(2), 294, *307*(2), *308*, 353, *356*, 609, 610, *644*
Mattig, W., 778, *784*
May, R. M., 731, *758*
Mayall, N. U., 13, 22, 26(2), *33*, *34*, 38(2), 39, 41, 42(3), 44, 56, 71(2), 77, *78*(2), *79*, 87(4), 88(2), 89(2), 98(3), 101(3), 105(2), 108, 116, *120*(2), *121*(3), 143, 153, *156*, *157*, 159, 163, 168, 169(2), 171(2), *175*, *176*, 183, 191, *196*, 268, *279*, 321(3), *356*(3), 486, *506*, 511, 524(2), 528, 535, *538*(2), 544, 550(3), 551, *556*(3), 574, *596*, 612, 618(5), 628(3), 629, *643*, *644*(2), 649(3), 655(2), 659(2), 660, 661, *663*(2), 761, 765, 772(2), 773(2), 774, *784*(2)
Mayer, C. H., 178, *196*
Meinel, A. B., *78*
Melrose, D. B., 263, *279*
Méndez, M., 66, *78*
Meng, Y. S., 332, *356*
Menon, T. K., 333, *356*, 533, *538*
Menzel, D. H., 65, *77*
Merchant, A., 6, *33*
Merkelijn, J. K., 242, 263, 264, *281*
Merkelin, J., 187, *195*
Mestel, L., 384(3), 386, 392(3), 393, 394, 396, 397(2), 398, *406*(2), 466(2), 468, 483, *506*(2)
Mezger, P. G., 66, 67(2), 69, 70(2), 77, *78*(3), *79*, *196*, 326, 332, *355*, *356*
Michie, R. W., 390, 392, *405*, *406*, 448(3), 449, 450(2), *506*
Mihalas, D. M., 134, *155*, 532(2), *538*, 613, *642*
Miley, G. K., *210*, 242(2), 251, 262, 269, 276, *279*, *280*, 296, 299, 758(2), *759*
Miller, R. H., 47, 71, 77, *77*, *78*, 134, *156*, 391, *406*, 460(2), 480, 481, *506*(2)
Miller, W. C., 287, *308*
Mills, B. Y., 179(4), 180, 181(3), 188, 191(4), *196*(4), 242, 275, *249*(2), 609, *644*
Milne, D. K., 69, 70(2), *78*, 181(2), *196*, *197*
Milton, J. A., 319, 329, 333, 341, *356*
Minkowski, R., 40, 43, 48, 49, 61, 71, 73(2), 74, 77(2), *78*(3), 98, 100(2), 103(2), 112(2), 113, 115, *120*, *121*, **177–97**, 178(6), 181(3), 182(2), 183, 186, 187, 189, 191(2), *195*, *196*(2), **199–210**, 199(2), 211(2), 212, 213, 258, 268(5), 273, *277*, *279*, 380, 398, *406*, 434, *506*, 532, 533, *538*, 610(2), 618, 628, *644*, 767(2), 773, *784*(2)
Misner, C. W., 374, *406*
Mitchell, R. I., 172, *176*
Mitton, S. A., 212, 213, 255(2), 262, *279*(2), 734, *759*
Mjolsness, R., 365, *405*
Moffet, A. T., 182, 183, 184, 190, *196*(2), *210*, **211–81**, 236(2), 247, 249, 251(2), 253, 255(2), 258, 259, 268, 269, 270, 277, *278*, *279*(4), 288, 296, *307*
Moiseev, I. V., 189(2), *197*
Monnet, G., 76, *78*, 102, 113, 115, *120*
Montmerle, T., 597, *600*
Moore, E., 38, 39(2), 56, *78*
Moran, M., 245, *279*
Morgan, W. W., 1, 13(2), 22, 26(3), 27(10), 28(2), 29(3), *34*(4), 38(5), 41, 42(8), 43, 44, 56, 63(2), 74, *79*(4), 116, *121*, 163, 168(2), 169(2), *176*(3), 178, 181(2), 183, 185, 189(3), 193, *196*, *197*, 271(2), 272, *279*(2), 353, *356*, 585, *596*, 604(2), 607, 608, 609(3), 610, *643*, *644*(3)
Morimoto, M., 250, 259, *279*
Morris, D., 184, *196*, 260(3), 261(2), 262, *278*(2), *280*(2)
Morrison, F. F., 104, *121*, 542(3), 546, *556*
Morrison, P., 237, 274, *277*, *278*, *280*, 305
Morton, D. C., 739, 752, *759*
Mottmann, 618
Muller, C. A., 323, *356*
Müller, E. A., 66(2), *78*
Münch, G., 71, *79*, 101(2), 107, 112, 113(3), *121*, 185, *196*, 341, *356*, 550(2), *556*, 659
Murdoch, H. S., 745, *759*
Murray, C. A., *210*, 263, *280*
Murray, S., 610, *643*, 736, *759*

Nariai, H., 370, 376, *406*(2)
Narlikar, J. V., 369, 379, 380(2), *406*(2)
Nassau, J. J., 530, *539*

Ne'eman, Y., 306(2), *307*(2)
Neugebauer, G., 45(2), *79*, 167, *176*, 293, 295, *307*
Neville, A. C., 180(2), 183(3), 187, *196*(2), *197*(2), *210*, 251, 254(2), 255(2), 256, 257, 259, 268, 269, 270, 273, *279*, *280*(2)
Newton, Sir I., 696, *723*
Neyman, J., 117, 118, *121*, 402, 403, 404, *406*(3), 553, *556*, 585, *596*, 602(3), 603(3), 618, 622, 628, 629(2), 632, *643*, *644*(3), 654, 659, *663*, 670, 671, 685(2), *723*
Noerdlinger, P., 297, *307*
Noonan, T. W., 618, 623(4), 626(3), 629, 630, *644*
Novikov, I. D., 306, *307*, 377, *405*, 695, *723*, 739, *759*
Nussbaumer, H., 69, *78*

O'Connell, D. J. U., 29, *34*
O'Connell, R., 38, 42, 61(2), 77, *79*
O'Dell, S. L., 287, 298
Oemler, A., Jr., 287, 298
Oke, J. B., 43, *79*, 166, 167(9), 171(6), 172(3), 173(3), *176*(4), 182, 191, *197*, 284, 285(2), 287, 290, 293(2), 294(2), 295(2), 298(2), *307*(3), *308*, 730, 752, *759*(2), 779(2), *784*, *785*
Oliver, N. J., 43, *78*
Ollongren, A., 440, 441, 445(3), *506*
Olsen, E. T., 184, 185, *197*, 243, *280*, 301, *308*
Omer, G. C., 623, 625(2), 626(3), *644*
Oort, J. H., 84(2), 88, 113, *121*, 124, 134(2), *156*(2), 170, *176*, 192, *197*, 323, 332, *356*(2), 359, 401, 402, *406*, 414, 421(2), 423, 424, 441, 473, 474, 494(2), 504(2), *506*, 585, *596*, 620, *644*, 653, *663*, 719, *723*, 725, 736, *759*
Osborn, W. H., 288
Osmer, P. S., 301, *306*
Oster, L., 224, *280*
Osterbrock, D. E., 27, *34*, 38, 42(3), 43, 63(3), 64, 67, 71(3), 73(5), 74(2), 75, 77, *78*, *79*(3), 168, 176, 290, *307*, 434, *506*
Ostriker, J. P., 274(2), *278*, *280*, 305, 410, 465, 481, *506*(2)
Ozernoy, L. M., 118, 119(2), *121*, 274, *280*, 306, *307*, 374, *407*

Paál, G., 621(2), *644*
Pacholczyk, A. G., 29, *34*, 213, 226(2), 276, *280*(2)
Pacini, F., 271, *277*, *278*, 305, 306, *307*
Padrielli, L., 242, *278*
Page, T. L., 48, 67, *79*, 94(2), 95, 104(4), 115(2), 116(4), 117, 118, *121*(3), 164, *175*, 321, 403, *406*, **541–56**, 542(2), 544, 546, 547, 548, 549(3), 553(4), 554, *556*(3), 572, 585, *596*, 609, 623(2), 625(2), 626(3), 627(2), 629, 631, 634(2), *644*(3)
Pagel, B. E. J., 401, *407*
Palmer, H. P., 247, 251, 259(2), *277*(2), *279*, *280*
Palmer, P., 66, *79*
Paraskevopoulos, J. S., 8, 9, *34*
Pariiskii, Yu. N., 259, *281*, 399, *407*
Parker, E. A., 184(2), 185, *197*(2), *210*
Parker, E. N., 376, *407*
Parker, R. A. R., 290, *307*
Parsons, S. J., 177(2), *196*(2)
Partridge, R. B., 400, 401, *407*, 535, *538*
Pasachoff, J. M., 739
Pastoriza, M., 144, *156*
Pauliny-Toth, I. I. K., 180, 184, *197*, 239, 241, 243(2), 245(3), 246, 247, 248, 260, 264, 274, *279*(2), *280*(4), 288, 295(3), *307*, 745, 746, *759*(2)
Payne, A. D., 242, *280*
Peach, J. V., 428, *506*, 748, *759*
Pease, F. G., 4, *34*
Peebles, P. J. E., 70, *79*, 302(2), *306*, *307*, 360, 366(2), 369(2), 374(2), 375, 378, 380, 381, 388, 400, 401, *405*, *407*, 465, 504(3), *506*, 585, 597, *600*, 640(11), *643*, *644*(2), *645*, 686(10), 695, 719, *722*, *723*(3), 726(2), 737, *759*
Peimbert, M., **37–80**, 63, 65, 66(4), 67(4), 69(6), 70(5), 71(2), 73(3), 74(3), 75, 76, *79*(3), 343, *356*
Penfield, H., 66, *79*
Pengelly, R. M., 65, *79*, *307*
Penston, M. V., 268, *280*
Penzias, A. A., 369, 377, 398(2), 399, *407*(2), 726, 729, 730(2), 737, 739(2), *759*(2)
Perek, L., 83(2), 84, 85(2), *121*, 447, *506*
Perola, G. C., 242(2), 245, 247, 262, 276(2), *279*(2)
Perrine, C. D., 4
Peters, H., 323, *357*
Peters, W. L., 535, *539*, 597, *600*, 774, *785*
Peterson, B. A., 264, 268, *280*, 297, 300, *307*(2), 398, *405*, 730, 752, *759*, 760
Petrosian, V., 755, 756, *759*, *760*
Pettit, E., 28, *34*, 38, 125, 126(3), *156*, 661, *663*
Philip, A. G. D., 515, *539*
Phillips, J. W., 177(2), *196*(2)
Piddington, J. H., 274(2), 275, *280*, 306, *307*, 398, *407*
Pilkington, J. D. H., 180, 184, 185, *197*(2), 242, *280*, 610(2), *644*, 742, *760*
Pinto, G., *210*
Polya, G., 541, *556*
Pooley, G. G., 180(3), 183, *196*, *197*(2), 242(3), *279*, *280*(2), 740(4), 741(2), 742(3), 744(3), 758(2), *758*, *759*, *760*(3)
Pottasch, S. R., 64, 74, *78*, *79*
Pourcelot, A., 76, *78*, 102, 113, 115, *120*
Poveda, A., 93(2), 103, 104, 107(3), *121*, 336, *356*, 536, *538*, 551(2), *556*

Prendergast, K. H., 84, 85, 86, 89, 90(2), 91(2), 92(2), 98, 100, 106(10), 107(8), 109(8), 110, 111, 113(6), 114, 115, 118, *119*, *120*(2), *121*(3), 134, *156*, 318, 341(19), *355*(4), *356*(2), 391, *406*, 434, 437(2), 460(2), 480, 481, 483, 498, *505*, *506*(3), 550(15), 551(11), *556*(5)
Price, R. M., 181, *197*, 236, 249, 260(2), 261, 262(2), *278*
Proctor, R. A., 647, *663*
Prozorov, V. A., 259, *281*
Purcell, G. H., 247, *278*
Pyper, D. M., 608, 622, 625, 630, *643*

Quirk, W. J., 460, 480(2), 481(3), *506*(2)

Racine, R., 515, 517(2), 523, 524, 530, 531, 536, 537, *538*(3)
Radhakrishnan, V., 260, *280*
Raimond, E., 85, *120*, 263, *278*, 343, *356*
Ramaty, R., 234, *280*
Randers, G., 9, *34*
Rapchak, D., 276
Ray, E. C., 274, *278*, 306, *307*
Razin, V., 230, *280*
Read, R. B., 184(2), *197*(2)
Readhead, A. C. S., 735, *760*
Reaves, G., 22, *34*, 576, 586, 589, *596*, 608(5), 618, 622, 626, *644*
Reber, G., 177(2), *197*
Reddish, V. C., 49, 62, *79*, 330, 354, *355*, *356*, 550(2), *556*
Redman, R. O., 124, 125, *156*(2)
Rees, M. J., 263, 274, 276, *280*(3), 305, 736(2), 755, 756, *760*(4)
Refsdal, S., 779, *784*
Reifenstein, E. C., 332, *356*
Reinhardt, M., 734(2), *759*, *760*
Reinmuth, K., 3, *34*, 151, *156*
Reiz, A., 3, *34*, 593, *596*, 636, *644*
Remage, N. H., 318, 341, *355*
Reynolds, J. H., 4, 5(3), 11, 13, 17, 28, *34*
Ribes, J. C., 263, *280*
Richards, R. S., 251, *277*
Ringenberg, R., 750, *760*
Ritchey, G. W., 4
Roach, F. E., 400, *407*
Robbins, R. R., 65, *79*
Roberts, I., 4
Roberts, J. A., 243, 263, *278*, *280*
Roberts, M. S., 58, 63, 67, *79*, 105, 106, 113, 115, *120*, *121*, 146, *156*, 170, *176*, **309–57**, 312, 318(3), 319(4), 320, 327(3), 330, 332(2), 333(2), 334(3), 335(3), 336(4), 341(13), 343, 349, 353, 354, *355*(3), *356*, 417, 438, 439(2), 466, 471, 501, *506*, 530(5), 537, *539*, 590, *596*, 597, *600*, 773(4), 774(3), *784*(2)
Roberts, W. W., 482, *506*
Robertson, D. S., 296
Robertson, H. P., 170, *176*, 660, *663*, 763, 766, 775, 776(2), 777, 781, *784*
Robinson, B. J., 58, *79*, 309, 334(2), 341(2), 353, *356*(4), 399, 403, *406*, 572(2), *596*(2), 633, *644*, 729, *759*
Robinson, I., *278*, 283, *307*
Robinson, L. B., 288, 289
Rodgers, A. W., 495, *506*, 511, *537*
Roeder, R. C., 61, *78*, 182, 191(4), 192(3), 193, *197*
Roffi, G., 242, *278*
Rogers, A. E. E., 296
Rogerson, J. B., 739(3)
Rogstad, D. H., 89, 106(9), 107, 108(2), 110, 111(2), *121*, 189(2), *196*, *197*, 311, 315, 332, 334, 341, *356*(2), 610(2), *644*(2)
Roll, P. G., 369, *405*, 726(2), 737, *759*
Rome, J. M., 179, *196*
Rood, H. J., 107, *121*, 585, 597, *600*, 609(3), 617(2), 618, 623(2), 625(2), 627(2), 629(2), 630(2), 631(2), 632, 633, 634, *644*(5)
Rosatelli, C., 180, *195*, 242, *277*, 741, 744, *758*
Rose, W. K., 261, *277*
Rosino, L., 518, 520, *539*
Rosse, Lord, 4(2), 762
Rothman, V. C. A., 585, 597, *600*
Rougoor, G. W., 89, 106(9), 107, 108(2), 110, 111(2), 113, *121*(2), 189, *197*, 315, 341, *356*, *506*, 610, *644*
Rowan-Robinson, M., 747, 750, 755(2), 757, *760*
Rowson, B., 247, 251(2), *277*, *280*(2)
Roxburgh, I. W., 379, *407*
Rubin, R. H., 66, *79*, 356(3)
Rubin, V. C., 62, 71, 74(2), 76, 77, *78*, 79(2), 106(2), 107(3), 108, 109(2), 113(2), 115(2), *119*, *120*(2), *121*(4), 321(2), 341(7), *355*(3), 550(4), 551(3), *556*(6), 659(2), *663*, 773, 774, *784*
Rudnicki, K., 603, 622, 623(4), 625(2), 626, 630(2), *644*(2), *645*, 683, 723
Rydbeck, O., 296, *307*
Ryle, M., 177, 178, 179(2), 180(5), 183(3), 184, 187, *196*(3), *197*(8), 212, 213, 241, 242(2), 250, 251(7), 252, 257, 268, 270, 276, *279*(2), *280*(6), 285, *307*, *308*, 740(3), 741, 742(2), 744(3), 745, *760*(3)

Sachs, R. K., 373, *407*, 775, 780, *784*
Saffman, P. G., 379, *407*
Safranov, V., 396, *407*
Salpeter, E. E., 253, *280*, 302, *306*, 731, 736, 738, *758*(2), *759*
Sandage, A. R., **1–35**, 4, 5, 6, 8, 11, 13(3), 15, 22(3), 23, 24, 25(2), 28, 29(3), 32(3), *33*(3), *34*(3), 38(2), 39, 43(2), 44, 58(2), 61, 77(2), *78*, *79*(3),

95(2), 105(3), 108(4), 110, 113, 115, *120*(2), *121*(2), 125, 126(3), 138, 144, 146, *156*(2), 159(2), 163, 166, 167(6), 170, 171(7), 172(2), 173(5), 174(4), 175, *175*, *176*(4), 177, 182(2), 183(2), 185(4), 189, 191(5), 195, 196(3), *197*(4), 209, *210*(2), 268(7), 269(2), 273, *279*(2), *280*, 284(2), 285(5), 286, 287(4), 293, 294(2), 298, 305, *306*, *307*, *308*(5), 321, 354, *356*(2), 399(2), 401(2), 402(3), *405*, *407*, 424, 425, 426(3), 427, 438, 439, 450, 451, 452(4), 453(2), 471, 473, 475, 485, 489, 492(6), 493, 494(2), 495(2), 498, 499, 500, 501, 502(3), 503, *505*(2), *506*(3), 509, 511(2), 515, 518(2), 521, 523(3), 528, 530, 531(3), 535, 536(2), 537, *538*(2), *539*(4), 544, *556*, 558, 560, 574, *596*(2), 597, *600*, 608, 612(2), 613(2), 618(4), 619, 622, 628, *643*, *644*, 655(2), 659(2), 660, 661, *663*, 682, *723*, 725, 757, *760*(2), **761–85**, 765(3), 766(2), 767(2), 772, 773(2), 776, 778, 779, 781(5), 783, *784*(5)

Sanduleak, N., 492, *506*, 515, *539*

Sanford, R. F., 648, *663*

Sanitt, N., 287

Saraph, H. E., 69, *78*

Sargent, W. L. W., 287, 294(2), 298, 300, 301(3), *306*(2), *307*(2), *308*, 597, *600*, 618, *643*

Saslaw, W. C., 58, 306, *308*, 378, 388, 394, *405*, *407*(2), 640, *643*

Sastry, Ch. V., 260, *280*

Sastry, G. N., 623, *644*

Savedoff, M., 365, *407*

Sazonov, V. N., 213, 215, 226, *278*

Schalén, C., 663

Schatzmann, E., 396, *405*

Scheer, L. S., *119*, 318, *355*

Scheiner, J., 762, *784*

Scheuer, P. A. G., 179, 181, 185(2), *195*, *196*, *197*(2), 221(2), 224, 247, 252, 253, 275, *280*(2), 295, 300, *308*(2), **725–60**, 730, 747, 755(2), 757, *759*, *760*

Schiffer, M., 361, *405*

Schild, A., 283(2), *307*, *308*

Schild, R., 167(3), 171, 172, 173, *176*(2), 779, *785*

Schlesinger, B. M., 531, *539*

Schmid, W. M., 238, 239, *277*

Schmidt, M., 27, 29, *34*, 85(5), 86, 106, *121*, 178, 181, 183(2), 185(2), 189, 193(2), *196*, *197*, 268(3), 271, 272, 276, *280*, **283–308**, 283, 284(2), 285(4), 286(2), 287(2), 288, 290, 291(2), 293, 294, 296, 297, 299, 300, 301(5), 302(3), 303, 304(3), 305(3), *306*(2), *307*, 308(3), 353, *356*, 393, 394, 399, 401, 402, *407*, 422, 423(2), 436, 441(2), 443, 445, 478, 480, 482, 492, 494, *506*, 550, 551, *556*, 609, 610, *644*, 726, 736, 752(2), 753(2), 754(4), 755(4), 756(7), 757, *759*, *760*(2)

Schmidt-Kaler, T., 517, 518(4), 519, 520, 524(2), 537, *539*

Schott, G. A., 213, *280*

Schraml, J., 67, *78*

Schucking, E. L., 283(2), *307*, *308*

Schwarzschild, M., 61, *77*, 87, 88, *121*, 159, *175*, 473, 494, *507*(2)

Sciama, D. W., 263, *280*, 379(4), 380, 398(2), 399(4), *407*, 597, *600*, 734, 736(2), 755, *759*, *760*(2), 775, *785*

Scott, E. H., 398, 399, *407*, 729, *759*

Scott, E. L., 117, 118, *121*, 402, 403, 404, *406*(2), *407*, 553, *556*, 585, *596*, 602(4), 603(3), 611, 618, 620, 622, 625, 628, 629(2), 632, *643*, *644*(5), 654, 659, *663*, 670, 671, 685(2), *723*

Scott, P. F., 180(2), 184, *196*, *197*(2), 242(2), 253, *278*, *279*, *280*, 744, *760*

Seaquist, E. R., 263, *280*

Seares, S., 26, *34*

Searle, L., 71, 77, *79*, 268, *280*, 294, *308*, 492, *507*

Sears, R. L., 42, 161, 163, *176*

Seaton, M. J., 64, 65, 69, *78*, *79*

Seielstad, G. A., 89, 111, *121*, 260(4), 261(2), 262, 263, *277*, *279*, *280*(2), 315, *356*

Seligmann, C. E., 640, *643*

Sérsic, J. L., 13, 24, *34*, 108, *121*, 144, *156*, 431, *507*, 529(7), 530, 537, *539*, 572, 573(2), 575(2), *596*

Setti, G., 274, 276, *278*, *280*, 288, 291(2), 305, *308*, 736, *760*

Seyfert, C. K., 29, *34*, 530, *539*

Shachbazian, R. K., 189, *197*

Shaffer, D. B., 247(2), 255, 264, *278*(2), *280*(2), 296, 745, *759*

Shakeshaft, J. R., 179(3), 184, 185, *195*, *196*, *197*(3), *210*, 369, *407*, 733, 743, *759*, *760*

Shajn, G. A., 67, *79*

Shane, C. D., 182, *197*, 602(6), 603(3), 623(2), 624, 625, 626(2), 637(4), 640, *644*(5), **647–63**, 648, 649, 649/650, 650, 653, 654(2), 659, 660, 661, 662(3), *663*(3), 683(2), 685, *723*

Shane, W. W., 115, *121*, 417, 475, *505*, *507*

Shapiro, I. I., 296

Shapley, H., 4(2), 8(2), 9, 22, *34*(3), 124, 129, 151(2), 152, *156*, 510, *539*, 541, 543, *556*, 558, *596*, 601(4), 608, 637, *644*(2), 647, 648(2), 650, 652, 654, 659, 660(2), *663*(2), 773, *785*

Sharp, D. H., 374, *406*

Shaviv, G., 66, *77*

Shelton, J. W., 550, *556*

Shepley, L. C., 369, 372(2), 373, *405*

Sheridan, K. V., 250, *279*

Shimmins, A. J., 181(2), 184(2), 185, 187, *195*(2), *196*, *197*(3), 242(3), 263(2), 264(2), *280*(2), *281*, 284, *307*, 742, 744, *760*

Shipley, J. P., 737, 738, *758*

Shipman, H. L., 70, *79*
Shirley, E. G., 125, *156*
Shklovskii, S. I., 177, *197*, 211, 240, 276, *281*, 290, 291, *308*, 400, *407*, 536, *539*, 731, *760*
Shobbrook, R. R., 334, 341, *356*, 572, 581, *596*(2), 773
Sholomitskii, G. B., 189, *197*
Shostak, G. S., 311, 332(2), 334, *356*(2)
Shu, F. H., 384(3), 386, *406*, 476(2), 478, 479(4), 480(3), 481(3), 483, *506*(2), *507*(2)
Silberstein, L., 764, *785*
Silk, J. I., 374(4), *407*
Simkin, S. M., 138, *156*, 481, *507*
Simpson, E., 66, *79*
Sinclair, M. W., 329, *356*
Sinigaglia, G., 180, *195*, 242, *277*, 741, 744, *758*
Sinnerstead, V., *80*
Slee, O. B., 177, 179(3), 180, *195*, *196*(2), 211, 242, 247, 251, *277*(2), *279*, *280*
Slipher, V. M., 762(9), 764, 765, *785*
Slish, V. I., 228(2), *281*
Sloanaker, R. M., 178, *196*, 261, *277*
Smerd, S. F., 225, *281*
Smith, F. G., 177, 178(2), 179, 181, *195*, *197*(3), 740, *760*
Smith, H. J., 294, *308*
Smith, L. F., 268, *281*
Smith, L. L., 172, *176*, 400, *407*
Smith, M. A., 243, *279*
Smith, S., 95, *121*, 631, *644*
Snider, C., 323, *357*
Snow, T. P., 603, *644*
Soboleva, N. S., 262, *278*
Sofue, Y., 734, *760*
Solheim, J. E., 38
Solinger, A., 736, *759*
Solomon, P. M., *277*, 398, *405*, 731, 737(2), *759*
Spinrad, H., **37–80**, 38(3), 43(2), 44(7), 45, 46(3), 47(3), 48, 49, 55(3), 58(2), 59, 62(3), 63, 67(2), 69(4), 70(4), 71, 73(2), 74(2), 75(3), 76, 77, *79*(6), 112, 117, *121*, 167, 170, 174, 175, *176*(3), 343, *356*, 401, *407*, 438(3), *507*(2)
Spitzer, L., 4, 6, *34*, 302, 306, *306*, *308*, 359, 375, 376, 392(2), 393, 394, 398, *407*(2), 473, 503, *504*, *507*, 609(2), *643*, *645*, 728, 729, 739, *760*
Stabell, R., 747, *759*, 779, *784*
Stachnik, R. V., 481, *507*
Stanley, G. J., 177(3), 179, *195*(2), 211, 247, *277*(2)
Stebbins, J., 38, 125, 126(3), *156*, 160(2), 161(6), 162, 163, 171, 172, *176*(2), 649, 661(2), *663*(2)
Steinlin, U., 602, *644*, 650, *663*
Stewart, J. M., 597, *600*
Stewart, P., 243(2), *279*, *281*
Stix, T. H., 455, *507*
Stockton, A. N., 287, 288, 300, 301(2), *307*, *308*
Stokes, N. R., 22, 25, 29, 32, *34*, 268, *281*, 426(2), 439, 475, 499, 500, 502(3), 503, *506*
Strittmatter, P. A., 268, *277*(2), 286(2), 289 293(2), 294, *308*(3), 384, *407*, 745, 752(2), 753, *758*, *759*, *760*
Strom, S. E., 70, *79*
Stromberg, G., 762(2), 764, *785*
Sturrock, P. A., 274, 275, *279*, *281*, 306, *308*, 400, *407*
Suchkov, A. A., 482, *506*
Swanson, M. D., 602, *644*
Swenson, G. W., 180, *196*, 242(2), *278*, *279*
Swihart, T. L., 226, *280*
Swope, H. H., 63, 77, 159, *175*, 438, *505*, 510, 513, 517, 522(2), 530, 537
Syrovatskii, S. I., 213(2), 215, 216, 220(3), 221, 225, 226, 237, 238, *278*(2)

Tadokoro, M., 597, *600*
Takase, B., 430, *507*
Tammann, G. A., 25, *34*, 108, *121*, 511, 515, 530, 531, 536, *538*(2), *539*(2), 765, 773(2), 776, *784*(2), *785*
Tananbaum, H., 610, *643*, 736, *759*
Tauber, G., 306, *307*
Tayler, R. J., 739, *759*
Taylor, B. J., 38(2), 44, 45, 46, 47(3), 49, 55, 58, 59, 62, 74, 75(3), 77, *79*(2), 175, *176*, 438, *507*
Taylor, J. H., 185, *197*
Teller, E., 306
Terrell, J., 297(4), *308*
Terzian, Y., 67, *78*
Thackeray, A. D., 23, *33*, 108, *120*, 515, 523, 524, *538*, *539*
Thaddeus, P., 737, 738, *760*
Thiel, M., 734, *760*
Thomson, J. H., 179, *197*
Thorne, K. S., 739, *760*
Tifft, W. G., 38, 45, *79*, 126, *156*, 164, *176*, 521, 523, *539*
Tinsley, B. M., 49(2), 61, 62, *79*, 174(4), 175(3), *176*(2)
Tolman, R. C., 172, *175*, 763(2), 779, 780, *784*, *785*
Tomasi, P., 242, *278*
Tombaugh, C. W., 602, *645*
Tomer, E., 434, 498, *506*
Tomita, K., 370, 376, *406*(2)
Toomre, A., 24(2), *34*, 63, *79*, 90, *121*, 396, *407*, 463(3), 464(3), 471(2), 472(3), 481(2), 485, 499, *505*, *506*, *507*
Toomre, J., 24(2), *34*, 63, *79*
Torres-Peimbert, S., 47, 74, 77, *79*(2)
Tovmassian, G. M., 189(3), *197*(2)
Trefftz, E., 397, *407*
Tritton, K. P., 268, *281*, 752, 753, *760*

Truman, O. H., 762, *785*
Tsytovitch, V. N., 230(2), *281*
Tuberg, M., 95, *121*
Tucker, R. H., *210*
Turland, B. D., 273(3)
Turner, E. L., 597, *600*
Turner, K. C., 330, 334, 335, *355*, *356*
Turnrose, B. E., 585, 597, *600*
Tuve, M. A., 330, 334, *355*
Twiss, R. Q., 225, *281*

Ulam, S. M., 306, *308*
Ulrich, R. K., 66, *79*
Unsöld, A., 494, *507*
Uscinski, B. J., 253, *277*

Vandenberg, N. R., 296
Vandervoort, P. O., 479(2), 481, *505*
Vanier, J., 323, *357*
Vardya, M. S., 74, *79*
Vaucouleurs, A. de, 12, 13, *34*, 40(2), *79*, 92, 105, 106, 108, 113(3), *121*(2), 126, *156*, 163, 170, *176*, *197*, 268, *281*, 320, *356*, *357*, 484, 485, *507*(2), 526, 528, *539*, 544, 547, 550(2), *556*, 558, 576, *597*(2), 597, *600*, 618, *644*, *645*, 648, *663*, 772(2), 773(2), 774, *784*, *785*
Vaucouleurs, G. de, 1, 2(2), 5(2), 6(3), 8(6), 9, 10(2), 11(3), 12(2), 13(4), 22, 23(8), 25, 26, 27(3), 28(2), 29, *34*(2), 38(3), 39(2), 40(2), 41, 43, 77, *79*(2), 86(2), 92, 101(2), 103, 104, 105, 106, 108(3), 113(3), *120*, *121*(3), 124, 126(2), 133, 134(3), 136, 137(3), 138, 151, *156*(2), 160(2), 161, 163, 164, 166(3), 167, 170, 173(2), *176*(2), *197*, 255, 268, 272, *281*(2), 318, 320, 321(2), 349, *356*, *357*(2), 391, 402, *407*, 425, 426, 427, 428(2), 429(4), 430, 431, 432, 433(7), 436, 438, 449, 483, 484(3), 485(2), 486(2), 487, 488, 489(5), 490, 491(2), 493, 497, 498(2), 499, 504, *507*(5), 509, 526, 528(2), 529, 535(4), 537, *539*(3), 544, 547, 550(3), *556*(2), **557–600**, 558, 559, 560(2), 564, 572(3), 573(2), 576(2), 581(4), 583, 585(5), 590, 592, 593(2), *596*, *597*(5), 597(4), 599, *600*(4), 612(2), 618(2), 619(3), 624, 625(3), 628(2), 631, 632(2), 636(2), 637(3), *645*(2), 648(2), 650, 653, 659(3), 662, 663, *663*(3), 665, *723*, 772(2), 773, 774(2), *785*(2)
Véron, P., 183, 185(2), 187(6), 188, *197*(2), *210*, 241, 263(2), *277*, *281*, 286(2), *308*(2)
Vershuur, G., 550(2), *556*
Vessot, R., 323, *357*
Vigotti, M., 242, *278*, 741, 744, *759*
Vila, S., 365, *407*
Visvanathan, V., 294, *308*
Volders, L., 106(2), 108(2), 113, *121*(2), 180, *195*, 242, *277*, 319(2), 320(2), 321, 341(2), *357*(2), 550(4), *556*(2), 741, 744, *758*
Vorontsov-Velyaminov, B. A., 1, 3, 4(2), 5, 9, 13, 21, 24, 26, 27, 28(2), *34*, *35*, 183, *197*, 403, *407*, 544, *556*

Wade, C. M., 75(2), *80*, 179, 180, 184(2), 192, 193, *196*, *197*(2), *210*(2), 243, 255(2), 259(2), 269(2), 273, *279*(2), *280*, *281*, 572, 573, *596*
Wagoner, R. V., 70, *80*, 242, 274, *277*, *281*, 302, *308*, 640(2), *643*, 738, *760*
Wahlquist, H. D., 585, *596*, 774, 775, *784*
Walden, W. E., 306, *308*
Walker, A. G., 776, *785*
Walker, M. F., 67(2), 69, 70, *77*, 94, 98(4), 109(2), 112(2), 113(5), 115, *120*(2), *121*(3), 434, 438, *507*
Wall, J. V., 242, 263, 264, *281*, 742, 744, *760*
Wallerstein, G., 523(3), *539*
Walmsley, M., 287, *308*, 745, *760*
Wampler, E. J., 284, 287, 288, 289, 290, 294, *308*(2), 398, *405*, 731, *759*
Wannier, P., 632, 633, *643*
Wares, G. W., 67, *80*
Warner, B., 66, *78*
Warren, J. L., 334, 341, *356*
Waters, S., 647, *663*
Webb, C. J., 6, *33*, 608, 622, 625, 630, *643*
Webber, J. C., 242, *278*
Webster, A. S., 369, *407*
Weiler, K. W., 260, 262, 263, *278*, *280*
Weinberg, S., 695, *723*
Weiss, A. A., 225, *281*
Weizsäcker, C. F. von, 384, 397, *407*, 504, *507*
Welch, G. A., 164, *175*
Welch, W., 633, *643*
Weliachew, L., 318, 332, 334(2), 335, *355*, *357*
Wendker, H. J., 242, *281*, 326, *355*
Wesselink, A. J., 23, *33*, 108, *120*, 515, 524, 537, *538*
Westerhout, G., 343, *356*, 393, 397, *406*, 417, 436(2), 475, *506*, *507*
Westerlund, B. E., 268(3), *281*(2), 515, 525, *539*(2)
Westfold, K. C., 213, 216(2), 217, 219, 220, 263, *279*, *281*
Westphal, J. A., 294, *308*
Weymann, R., 29, *34*, 367, 369, 376, 398, 400, *407*, 736, 756, *759*, *760*
Wheeler, J. A., 365, *405*
Whipple, F. L., 38(3), *80*
White, R. E., *506*
White, R. H., 306, *307*
Whiteoak, J. B., 89(2), 106(9), 107, 108(2), 110, 111(3), *121*(2), 189, *197*, 236, 259, 260(7), 261, 262(2), 268, *278*(3), *281*(2), 315(2), 341, *356*(2), 610, *644*, 753, *760*

Whitford, A. E., 38, 39, 43, 44(2), 45, 47, 61(2), 77, *80*, 125(2), 126(3), *156*, *157*, **159–76**, 160(2), 161(3), 163(2), 167(5), 168(2), 169(2), 171(2), 172(3), 173(2), *176*(3), 268, *281*, 511, *539*, 649, 661(2), *663*(2), 779(3), *785*
Whitney, A. R., 296
Wickramasinghe, N. C., 369(2), *406*(2)
Wielen, R., 174, *176*
Wijk, U. van, 402, *407*
Wild, J. P., 225, 250, *281*(2), 323, *357*
Wild, P., 125, *157*, 402, *407*, 603, 604, 618, 640(2), *645*, 648, *663*
Wilkinson, D. T., 369, *405*, 726(2), 737, *759*
Wilkinson, P. N., 273
Williams, D. R. W., 533, *538*
Williams, P. J. S., 221, 228(3), 243(5), 245(3), 246, 247, *279*(2), *280*, *281*(3), 295, *308*, 746, *759*
Williams, R. E., 290
Williamson, K. D., Jr., 737, 738, *758*
Wills, D., 75(3), *80*, 180, 184, 185(5), 192(2), 193, *195*, *196*, *197*(3), 242, 253, 268(2), *277*, *278*, *279*, *281*, 287, 745, 752, 754(5), 755, 756, 757, *759*, *760*
Willson, M. A. G., 242(2), *281*, 741, 742, 745, *760*
Wilson, A. G., 4, 22, *33*, *35*, 586, 587, *597*, 623, 625(2), 626(3), *644*
Wilson, O. C., 773(2), *784*, *785*
Wilson, R. W., 369, 377, 398, *407*, 726, 730(2), 737, 739(2), *759*
Wilson, T. L., 69, 70(2), *78*, 332, *356*
Windram, M. D., 242(2), 268, 270, *280*, *281*, 745, *760*
Wingert, D. W., 736(2), *758*
Wirtanen, C. A., 182, *197*, 295(2), *307*(2), 544, *556*, 602(3), 603, 624, 626, 637, 640, *644*(2), 649, 649/650, 654, 660, *663*, 683(2), 685, *723*
Wirtz, C., 601, *645*, 762(2), 764, *785*
Wisniewski, W. Z., 172, *176*
Wlerick, G., 286, *308*
Woerden, H. van, 85, *120*, 332, 343, *356*
Wolf, M., 1, 2, 5, *35*, 601, *645*
Wolfe, A. M., 46, *80*, 373, *407*, 738, *760*
Woltjer, L., 29, *35*, 240, *281*, 288, 291(2), 297, 298, 306, *307*, *308*(2), 393, 397, *407*, 409, 413, 424, 471, 474, *505*, *506*, *507*, 536, *538*, 585, *596*
Wood, D. B., 38(2), 43, 44, 45, 46(2), 49, *79*, *80*, 166, *176*
Woolf, N. J., 737
Woolley, R., 421, 424, 449, *507*(2)
Worrall, G., 403, *407*
Wouthuysen, S. A., 729, *760*
Wrixon, G. T., 632, 633, *643*
Wyndham, J. D., 183, 184(3), 185(4), 187, *196*(2), *197*(4), 209, *210*, 241, 263(2), *281*, 286, *308*
Wyse, A. B., 87(3), 88, 105, *121*, 153, *157*, 550(2), *556*

Yang, K. S., 180, *196*, 242(4), *278*(2), *279*, *281*
Yen, J. L., 251, *277*
Young, J., 58, 59, 74, 75(2), *79*
York, D. G., 739(3)
Yost, J. C., 481, *507*
Yu, J. T., 640(3), *645*, 686(2), *723*
Yuan, C., 476(2), 478, 479(2), 480(3), *506*, *507*

Zapolsky, H. S., 297(2), *308*
Zakharenkov, V. F., 259, *281*
Zel'dovich, Ya. B., 375(2), 377, 388, *405*, *407*, 695, 708(2), *723*(2), 739, 750, *759*(2), 781, *785*
Zipoy, D., 388, *407*
Zonn, W., 541, 542, 544(5), 545(2), 546, 548, *556*(2)
Zucchino, P., 752, *759*
Zuckerman, B., 66, *79*
Zwicky, F., 4, 24, *35*, 95, *121*, 125, 147(4), 148, 150, 154, *157*(7), 236, 271, 273, *281*, 402(3), *407*(2), 574, 586(2), *597*, 602(4), 603(5), 604(4), 607, 608, 611(8), 612, 617, 618(2), 619, 620(2), 621, 622(4), 623(14), 624(7), 625(2), 626(3), 628, 629(4), 630, 631, 632, 633(2), 640(2), *645*(6), 648, 654, 662, *663*(2), 665, 683, *723*(2)

Subject Index

Absolute magnitude; *see* Luminosity
Absorption
  synchrotron self-absorption, 75, 221, 225–29, 239, 240, 245, 270, 733, 742
  thermal, 221, 224–25
  Tsytovitch effect, 221, 225, 229–32
  see also Intergalactic absorption; Interstellar absorption
Absorption lines, 40, 47, 96–100, 112, 170, 283, 286, 300–302, 321, 438, 484, 761–62, 767, 771–73
Age
  of Galaxy, 61, 410, 765, 783
  of galaxies, 399–400, 765, 783
  of stars, 472–73
  of Universe, 766, 783
Airglow; *see* Nightsky emission
Angular dimensions
  of clusters, 554, 611, 626, 627
  of galaxies, 245, 247–59, 263, 310–11, 319, 332, 541, 543, 553, 557–60, 564, 569–71, 575, 590, 597, 768, 775, 780–81, 783
  of groups, 554, 566–68, 572–76
  of radio sources, 213, 245, 247–59, 263, 269, 610, 727, 735, 757, 780
  see also Diameters
Angular momentum
  of galaxies, 1, 32, 81–82, 115, 118–19, 381, 390–92, 394–95, 397–98, 400–404, 430, 451, 459–60, 467–71, 485, 496, 498, 500, 502–4, 554, 720
    distribution, 438–40, 466, 468–70, 499, 504
    per unit mass, 394, 423, 451, 494–96, 498–500, 504
  of stars, 452, 493, 495
Antimatter, 306, 378, 380
Aperture correction, 776
Aperture of diaphragm, 161, 163, 167, 432
Aperture synthesis, 180, 193, 251, 253, 269, 742
Arecibo, 185
Asiago Observatory, 101
Atlas of Peculiar Galaxies, 5, 77, 544
Axial ratio of galaxies, 83–86, 94, 104, 118, 127–29, 132, 143, 151, 153, 426–27, 548, 561
  see also Flattening; Inclination

Background radiation, 369, 377, 381, 535, 694–96, 720–21, 725–26, 729, 731, 733, 737–39, 758, 761, 765, 781
  isotropy, 369, 721, 775
Bar, 40, 41, 91–92, 386, 433, 464, 481, 483–86, 489–91
Barred spirals, 4–10, 12, 18–19, 23, 27, 40, 81, 91–92, 110, 113–15, 349, 352–53, 386, 397, 427, 432–33, 436, 483–91, 544–48, 550
Blanketing, 167–68, 170, 511
Blueshifts, 297
Blue stellar objects (BSO), 283, 286–87
Boltzmann equation, 410–11, 422, 448, 455–56, 479
Bottlinger diagram, 452–53, 495
Bottlinger's model, 88–90, 319
Brightest stars in galaxies, 29, 510, 530–31, 534, 536–37, 558, 562, 572
Burr Castle, 4

Calibration of magnitudes, 123–25, 167, 171
Cambridge Radio Observatory, 184, 213, 253–54, 310–11, 610, 742
Cannon shell guided by rainbow, 275
Cape Photographic Atlas, 13
Catalogue of Galaxies and Clusters of Galaxies, 125, 402, 603–4, 613, 619–20, 622, 640, 648, 662, 683, 686
cD galaxies, 27, 200, 265–68, 271, 609, 617
Cepheids, 40, 110, 510–18, 525–26, 536, 558
  period-luminosity-color relation, 511–14, 534
    calibration, 514–16
Cerenkov radiation, 230
Chains of galaxies, 542, 573, 576, 590
Chemical abundances, 39, 42–47, 57, 63, 65–67, 69, 74, 76–77, 166, 290, 354, 381, 401, 424–25, 510–11, 523, 531
  gradients, 46, 59, 75–76, 492–93, 495
  see also Helium abundance; Metal enrichment; Metal-poor stars; Metal-rich stars; Super-metal-rich stars
Classification of galaxies, 1–35, 147
  early systems, 2–5
  van den Bergh's system, 24–25, 146, 528–29, 534–37, 544
  Hubble's system, 6–8, 105, 131, 160, 336, 544
  Morgan's system, 26–27, 33, 41, 46
  de Vaucouleurs' system, 8–13, 426–28, 529
  Vorontsov-Velyaminov's system, 27–28
Clouds of galaxies, 557, 570–76, 581, 583, 585, 593, 602, 604, 637
  see also Groups; *for a listing of individual clouds see* Index of Galaxies

Clouds (second-order clustering); *see* Superclusters
Clustering, 665–722
  amplitude, 678–80, 712
  dynamics, 666, 687–93
  of galaxies, 359, 403–4, 557, 662
  origin, 667, 693–721
  of radio sources, 242, 745
  process, 715–17
  scale, 637–38, 640, 660, 663, 666, 671, 674, 680, 690, 693
  spectrum, 689–90, 693, 715
Clusters of galaxies, 25, 27, 46, 58, 81–83, 95–96, 115, 117–18, 125, 146–47, 159, 166, 173, 183, 188–90, 200–207, 242, 271, 273, 275–76, 287, 298, 301, 402–4, 503, 536, 541–42, 553–54, 558, 583, 585, 387, 589, 601–45, 654–55, 658–60, 665, 682–83, 693, 698, 718, 725, 766–67, 781, 783
  Bautz-Morgan type, 609, 617
  brightest galaxy, 189, 200–207, 269, 535, 537, 609, 611, 617, 682, 765–66, 781–83
  catalogues, 601–4
    Abell's, 181, 188–89, 402, 603–4, 610, 620–21, 628, 637–38, 640–42, 655, 659–60, 682, 686
    Zwicky's; see *Catalogue of Galaxies and Clusters of Galaxies*
  cells, 602
  density distribution in clusters, 623–27, 630–31, 635
  diameters; *see* Angular dimension; Clusters, size
  dynamics, 629–39
  evaporation of members, 602
  formation and evolution, 403, 634–36, 766
  galaxian content, 607–10
  luminosity, 619, 633
  mass, 117, 607, 619, 630–34, 641–42
  mass-to-light ratio, 117, 631, 633–34
  number, 601–4
  population, 602–4, 607, 619–21, 624, 626, 637
  richness, 189, 610–11, 613, 621
  segregation of members, 625–26, 629–30, 636
  size, 189, 603, 607, 609, 621–27
  structure, 621–27
  types, 604–7, 620, 622
  velocity dispersion in clusters, 607, 610, 619, 627–30, 633
  see also Clouds; Groups; *for a listing of individual clusters see* Index of Galaxies
Clusters, stellar; *see* Globular clusters; Star clusters
CN index, 46, 58, 75, 166, 168
Cold Universe, 369, 705–8
Collapse; *see* Gravitational collapse
Collapsed massive objects, 46, 274, 721
Collapsed stars, 117, 553
Collisions
  of galaxies, 273–74, 609–11, 636
  of stars, 61, 306, 448, 450, 454
Colors of galaxies, 2, 26, 29, 38, 45, 55, 82, 119, 125–127, 132–33, 137–39, 160, 164–66, 173–75, 284, 286, 293–94, 335–41, 343, 352, 354, 403, 428, 467, 501, 529, 537, 543–44, 546, 608, 610–19
  pg system, 125–26, 129, 130–33, 140–45
  UBV system, 126, 133, 137, 160, 164, 166–67, 170–74, 184, 286–87, 293, 399, 428–29, 432, 436, 494, 501
Color-color diagram, 164, 286–87, 349
Color distribution in galaxies, 133–44, 146, 167, 429–34
Color-magnitude diagram, 42, 59, 63, 153, 424, 438
Compact galaxies, 38, 154, 201, 271, 283, 285–86, 299
Companion galaxies, 147, 329, 334–35, 472, 481, 574
  see also Pairs of galaxies
Comparison spectra, 768–69
Compton effect, 232, 237–40, 272, 298, 373, 376
Continuum radiation; *see* Energy distribution
Contour maps
  of radio sources, 212, 236, 249, 252, 254–58, 273
  of 21-cm line emission, 325, 330–34
Cosecant law; *see* Interstellar absorption; Interstellar reddening
Cosmic microwave background; *see* Background radiation
Cosmic rays, 234, 393
Cosmological constant (Λ), 191, 194, 379, 634, 728, 730, 733–34, 747, 766, 778–79
Cosmological effects, 188, 190–91
Cosmological models, 61, 172–73, 182, 191, 194, 241, 283, 290, 298, 302, 305, 359–61, 366, 374, 377–81, 391, 399–400, 553, 621, 641, 665, 695–96, 725–27, 734–35, 738–39, 749, 775, 781–83
  de Sitter, 763–65
  dust-filled, 370
  Friedmann, 362, 379, 696, 726, 728, 730–31, 733–34, 744, 746–47, 751, 753, 755–58, 761, 764–66, 775, 778–79, 781
  radiation-filled, 369–70, 375
  steady-state, 359, 379–80, 399–400, 621, 726, 728, 731, 733-34, 738, 746–47, 751, 755, 757–58, 778
Cosmological principle, 665–66, 694–96, 763, 776
Counts; *see* Galaxy counts; Quasars, counts; Radio sources, counts; Star counts
Creation field, 380
Cutoff function of radio spectra, 221, 227–28

D galaxies, 27, 181, 189–90, 265–68, 271
Damping of perturbations, 374–75, 388, 472
  Landau, 455–56, 458, 460, 482
  phase, 456–58, 460, 464

Deceleration parameter (q_0), 61, 159, 164, 173–74, 182, 191, 194–95, 368, 642, 728, 765–67, 775, 778, 781–83
Decoupling of radiation and matter, 369, 373–75, 378, 384, 387–89, 391, 400, 403–4
Density
  critical cosmological, 293, 371
  mean universal, 61, 378, 641–42, 725–26, 728, 735–36, 739, 763, 766, 776
  of luminous matter in the Universe, 378, 725
  see also HI, density; HI, surface density; Intergalactic medium, density; Luminosity density (spatial); Mass density; Space density; Surface density; Surface density of galaxies; Surface brightness
Density evolution of radio sources; *see* Luminosity evolution
Density fluctuations, 377–78, 380–81, 400, 403, 455, 634, 659, 662, 693, 695–702, 708–12
  spectrum, 713–14
Density inhomogeneity within the Local Supercluster, 535
Density perturbations, 359, 362–63, 366, 368–69, 371, 375, 379, 388–90, 400, 403–4, 454–58, 461, 464, 471, 474–76, 479–80, 483, 499, 635
Density wave theory, 333, 476–83
Diameters
  metric, 776, 780
  of clusters; *see* Clusters, size
  of galaxies, 123, 142, 146–47, 149, 151–55, 161, 163, 177, 271, 285, 292, 305, 432, 497, 502–3, 533–34, 541, 544, 547–48, 558–60, 569–74, 586, 719, 727, 757–58, 775–81
  of radio sources; *see* Angular dimensions; Linear dimensions
Disk of spiral galaxies, 1, 22, 87, 90, 110, 126, 137, 140, 142, 152, 159, 170, 359–60, 384, 392–98, 401–2, 425–30, 432, 434, 438–39, 446, 451, 453, 460–83, 485–86, 491–93, 496, 498–502
  exponential, 6, 429, 432–33, 438, 440, 460, 466–71, 499, 501, 504
  formation, 392–98, 501
  length scale, 435–36, 468, 475, 502–3
  stability, 392–93, 461–65
  thickness, 152, 393–94, 397, 461–62, 476
  see also Population (disk)
Dispersion measure of radio sources, 734
Distance of the galactic center (R_0), 416–17, 422–23
Distances of galaxies, 105–9, 145–46, 153–54, 160, 166, 171, 182, 190–91, 194, 213, 233, 236, 239, 263, 269, 272, 290–93, 296–98, 302, 309, 318–19, 326, 329, 336–41, 346, 348–50, 352–53, 436, 509, 515–34, 552, 558–60, 562, 564, 566–68, 572–76, 581–82, 587, 589, 597–99, 601, 608, 613, 619–21, 632, 638, 642, 655, 658–60, 682–84, 748, 757, 761, 766, 775–76
Distance indicators, 510–37, 547, 558, 575, 624–25
Distance modulus; *see* Distances
Distance scale, 44, 172, 509–37, 357–58, 560, 597, 609–10, 632, 636
  see also Hubble constant
Distribution
  homogeneous and isotropic, 668–70
  of clusters, 601–2, 636–42
  of galaxies, 591–95, 603, 622, 625, 647–63, 665–86
  Poisson, 667–68, 670, 672
Distribution index of clusters, 624–25
Double galaxies; *see* Pairs of galaxies
Double radio sources; *see* Radio sources, double
Dumbbell galaxies, 200–202, 265–68, 271
Dust, 29, 82, 98, 100, 126, 143, 164, 272, 376, 421, 424, 427, 440, 474–75, 482, 501–2, 553, 604, 609, 632, 721, 738
  in elliptical galaxies, 13
Dwarf galaxies, 572–76, 586–90, 592, 607, 618–19
  distribution, 588–89
  elliptical, 4, 13, 14, 22, 42, 44, 49, 61, 63, 133, 137, 146–47, 153, 159, 392, 428–29, 436, 438–39, 448–50, 459, 497–98, 537, 557, 560, 587, 607–8, 611, 622
  irregular, 528, 537, 557, 572, 574, 587
  spiral, 4, 22, 25, 146, 153, 438, 537, 587
Dynamics of galaxies, 409–504
  early phases, 359–404

Early-type stars, 26, 40–42, 47, 49–53, 58, 61, 63, 70, 73, 169, 349, 354, 394, 401, 416, 440, 474, 476, 480, 524, 529
Earth rotation synthesis, 251
Ejection of mass, 259, 272–73, 275–76, 301, 402
Elliptical galaxies, 4–7, 11, 13, 25, 27, 29–33, 38, 40–63, 72–73, 93–96, 98, 103–5, 107, 115–16, 124–26, 129, 131–37, 144–45, 149–51, 153–55, 160–68, 172–75, 181, 183, 189, 247, 264–69, 271, 299, 336, 353, 360, 367, 380, 385, 391–93, 398–99, 426–35, 438–40, 446–50, 454, 458–59, 466, 496–504, 537, 544–54, 557–59, 562, 576, 586, 604, 607–8, 610–13, 616, 618–19, 625, 631, 634, 661, 720, 765–67, 771, 781
  see also cD galaxies; D galaxies; Dwarf ellipticals; Giant elliptical galaxies
Emission lines, 40, 45, 63–77, 93, 96, 98, 100–102, 113, 144, 181, 183, 263, 271–72, 285, 288–300, 321, 436, 439, 752, 761–62, 767, 771–73
  see also 21-cm line measurements
Emission nebulosities; *see* HII regions
Empty fields, 199–207, 209–10
Energy distribution (spectral) of galaxies, 45, 56, 159–75, 191, 293–96, 298, 300, 660–61
Envelopes of galaxies, 6, 27, 181, 189, 271, 335, 426, 497, 517, 526, 612
Epoch of galaxy formation, 398–400, 402–3, 708, 765, 783
Evolution of galaxies, 29–33, 160, 172–75, 310, 353–54, 359–404, 414, 439, 541, 549

Explosions in galaxies, 115, 118, 144, 160, 297–98, 402, 483, 494–95, 553, 557, 721

Fabry method, 123–25
Fabry-Perot interferometer, 102, 113
Faraday depolarization, 261–62
Faraday depth, 261
Faraday rotation, 236, 260–62, 399, 734
Field galaxies; *see* Isolated galaxies
Flattening (intrinsic) of galaxies 4, 23, 25, 29, 134–36, 144–45, 152–53, 384–86, 392, 395–96, 398, 401, 424, 426–27, 429, 434, 439–40, 460, 493, 496, 501, 547–48
  see Axial ratio; Inclination
flux densities, 179–82, 185–88, 193, 199–207, 213, 228–29, 237, 239, 241, 245, 247, 258, 264–69, 287, 295, 303–4, 326, 348, 726–27, 733, 737, 739–40, 742–48, 751–58
Formation
  of galaxies, 29–33, 354, 359–404, 424–25, 439, 447, 491–504, 553, 603, 765–66
  of stars, 32, 38, 47, 58, 75, 143, 174, 354, 359, 389, 392, 394, 396, 399–402, 425, 476, 491–93, 496, 498–99, 524, 554, 718, 720–21
  see also Clusters, formation
4-kpc arm, 436, 475
Fragmentation, 359, 384, 388–92, 398, 402–4, 500, 611, 708–9, 720
Frequency
  critical, 215, 219, 229–31, 237, 272
  cutoff, 219, 221, 225–41, 245
  cyclotron, 213
  epicyclic, 423, 464, 476, 478
  gyration, 214–15
  plasma, 224
Friedmann cosmology; *see* Cosmological models
Friedmann time, 765, 776, 783

Galactic radio background, 181
Galaxy counts, 182, 542–44, 559, 583, 590, 592, 648–55, 659–61, 665–86, 693
  smoothing, 649–50, 655, 673–78, 684
  see Log N, m relation
Galaxy formation; *see* Formation
Gas
  density, 142, 439, 492
  in galaxies, 29, 32, 37, 58, 63–77, 81–82, 91-92, 96, 107, 116, 118, 126, 174, 275–76, 291–93, 297, 306, 393–94, 397–402, 421, 424, 436, 439–40, 452, 454, 464, 474–76, 480–82, 484, 489, 491, 493, 499, 501–2, 609
  in nuclear regions, 58, 71–76
  intergalactic, 359, 376–80, 398–99, 403
  intracluster, 243, 275, 403
  ionized, 387–88, 398–99, 403, 436, 439
  mass, 67, 77
  primordial, 359–404
  see also HI, HII, H_2
Gaunt factor, 224–25, 732
Giant ellipticals, 47, 58, 100, 153, 159, 161–62, 164, 167–69, 171–73, 269–71, 410, 439, 448, 497, 575–76, 586, 607–8, 619, 748, 779
  see also cD galaxies
Giant spirals, 42, 153, 528, 665–66
Giant stars, 27, 37, 40–42, 44–45, 48, 50–53, 55, 56, 74, 159, 167–70, 174–75, 439, 523, 525–26, 536, 552, 608
Globular clusters, 13, 32, 42–43, 46, 50–53, 63, 74, 110, 116, 136–37, 144, 166, 388, 391, 422, 424, 428–29, 438–39, 446–50, 492–96, 500, 517, 521, 523–26, 531–32, 534, 561, 575, 608, 694, 762, 765, 783
  M3, 59, 61–62, 73
  M5, 61, 494
  NGC 6356, 46
  ω Centauri, 531
Gravitational collapse, 32, 118, 359, 367, 380, 382–93, 397–404, 439, 447, 450–54, 458, 466, 468, 470, 473, 479, 483, 494–500
  time scale for Galaxy, 401, 452, 491–92
Gravitational instability, 359–85, 395–96, 404, 504, 634–35
Gravitational lens, 299
Green Bank, 184, 310, 318, 321
Groups of galaxies, 25, 81, 96, 117, 147, 166, 173, 189, 200–207, 402–3, 504, 541, 547, 553–54, 557–99, 602–4, 618, 621, 628, 631–33, 636–37, 654, 765–66
  diameter, 560, 566–68, 572–76, 583, 585, 587, 589
  distribution, 576–77, 592–95
  evaporation of members, 402
  formation, 585
  rank of member, 558
  space density of groups, 583
  space density in groups, 585
  statistics, 581–86
  third brightest member, 526–27, 534, 536
  type index, 584–86
  velocity dispersion, 566–68
  see also Clouds; Clusters; Multiple galaxies; *for a listing of individual groups see* Index of Galaxies
Gyration time, 240

Halo 32, 59, 61, 110, 420–21, 423, 440, 471, 473, 482, 492–95, 498–99
HI
  clouds, 327, 329, 394, 415–16, 473
  contour maps; *see* Contour maps
  density, 333, 475, 502–3
  distribution, 310, 313–15, 318, 320–21, 329–35, 475
    asymmetry, 334–35
  in galaxies, 2, 40, 58, 76, 96, 110, 309–54, 417, 422, 436, 553, 562, 572–73
  intergalactic, 300, 633, 725, 728–31, 735–36

mass, 310, 323–29, 334–41, 343–49, 353–54
to luminosity ratio, 327–28, 336–43, 349–54, 562
to total mass ratio, 336–43, 346–47, 352–53, 439, 562
surface density, 313, 330, 332–35, 439
HII regions, 1, 2, 26, 29, 42, 63–77, 82, 86, 88, 100–102, 110, 113, 178, 199–207, 224, 316, 320, 329, 332–33, 354, 415, 482, 487, 501, 511, 519, 530, 558, 562, 572–73, 575, 587, 739, 762
diameter, 529–30, 534, 536–37
distribution, 67
electron temperature, 65, 69, 74, 77
filling factor, 64, 67
in disks, 76–77
in nuclei, 63–64, 71–76
in spiral arms, 63–71
interarm, 76–77
M8, 66, 68, 70
M17, 66, 68–70, 72
mass, 64, 67, 71, 73–74, 77
NGC 604, 69, 70
NGC 2070, 69, 70
NGC 5461, 67, 69–71
NGC 5471, 67, 69–71
Orion nebula, 42, 66, 68, 69, 72, 73, 224, 333
30 Doradus, 67
W49, 67
see also Chemical abundances; Intergalactic medium; Ionization
H_2, 46, 333, 387–88, 731
Harvard, 309, 648–49, 652, 659–61
Haute Provence Observatory, 101, 126
Helium abundance, 69, 354, 510, 706, 726, 736, 739
Helwan Observatory, 4
Hertzsprung gap, 40, 61
Hertzsprung-Russell diagram, 33, 40, 42, 47, 48, 57, 61
Hierarchical structure, 663, 665, 693–94, 717–20
High-velocity clouds, 110, 736
Horizontal-branch stars, 46, 62, 73–75
Hubble Atlas, 5, 6, 9–11, 13, 15–16, 22–24, 28, 138, 574
Hubble constant, 25, 61, 95, 105, 108, 110, 116, 132, 146, 150, 154, 173, 190–94, 236, 264, 268, 290, 298, 341, 368, 371, 399, 428, 509, 526–36, 542, 548–49, 551–55, 597, 602, 616, 618, 622–23, 632–34, 636–37, 641–42, 655, 659, 665, 683, 725, 728, 736
regional variations, 534-36, 597, 765, 774–75, 778, 783
Hubble diagram; *see* Redshift–apparent-magnitude diagram
Hubble time, 781, 783

Identification of radio sources, 177–95, 199–213, 263–64, 268, 284–88, 748
Inclination of galaxies, 27, 82, 89, 94, 98, 100, 105, 108, 126–28, 134–36, 139, 143–45, 152–53, 311, 313, 318, 325–28, 335–41, 349, 436, 548
see Flattening (intrinsic)
Inclination of spectral lines, 98–99
Index Catalogue (IC), 2, 601
Index of clumpiness, 679
Infrared, 39, 55–56, 58, 61, 161, 164, 167, 169–70, 184, 239, 285, 287, 295, 298–99, 305
Instability, 375–80, 459, 471, 475, 710–12
criterion, 359, 365, 372–73, 454, 461–63, 465, 476, 701, 703–4, 712
thermal, 702–5
two-stream, 454
see also Gravitational instability
Instability strip, 510–11
Intensity gradient 6
Interacting galaxies, 21, 24, 137, 334, 385, 398, 459, 471–72, 481, 505, 718–20
Interferometer, 177, 179, 182–84, 212–13, 250, 253–55, 269, 273, 286, 310–11, 315, 610, 744
very long baseline (VLB), 247, 251, 269, 275, 296, 735
Intergalactic absorption, 126, 620, 638, 660, 683
Intergalactic medium, 275–76, 300, 471–72, 553, 560
density, 379, 758
HI, 725, 728–31, 735–36
HII, 730–36
reprints of the *Astrophysical Journal*, 735
see also Gas, intergalactic
Interlopers, 283
Interplanetary scintillation, 253, 269, 296, 734–35
Interstellar absorption in the Galaxy, 105, 108, 124, 126–27, 131–32, 139, 143, 152, 161, 181, 268, 327, 336, 414–16, 483, 511, 514, 524–28, 532, 535, 559–60, 562, 566–68, 574, 576, 598, 620, 625, 630, 637–38, 648, 650–55, 658, 672, 675, 678, 683–84, 686, 729
Interstellar absorption in other galaxies, 105, 108, 126–33, 138, 143, 145, 152, 327, 336, 517
Interstellar clouds; *see* HI, clouds
Interstellar matter; *see* Dust; Gas
Interstellar molecules, 737–38
Interstellar reddening, 46, 47, 49, 56, 61, 69, 73, 129–31, 143, 166, 327, 511–12, 514–19, 522, 524, 527, 532–33, 560
Intracluster medium, 276, 632–33
Inverse Compton effect; *see* Compton effect
Invisible matter, 46, 112, 117, 553, 632, 721
Ionization, 76–77, 378, 386
collisional, 73–74, 388, 736
gradients, 71
radiative, 73–74, 736
Ionized hydrogen; *see* HII

Irregular galaxies (type I), 4–5, 8, 20, 22–23, 38, 40, 42, 63, 72, 81, 94, 105, 108, 115–16, 129, 131–32, 146, 149–51, 153–55, 170, 164, 178, 264, 319, 327, 332–54, 426–27, 432, 434, 439–40, 475, 528–30, 541, 545–50, 557–60, 573–76, 586–87, 593, 604, 607–612, 625
Irregular galaxies (type II), 24, 131, 133, 150, 273, 327, 427, 572–73, 586
Isolated galaxies, 147, 589–90, 602–3
Isophote, 123, 125, 139, 151, 182, 333, 433, 498, 597, 612, 776
Isotropy, 682
of radio sources, 241–42, 740, 745, 751
see also Background radiation, isotropy; Cosmological principle; Velocity anisotropy

Jeans length, 359, 384, 404, 455–56, 696–97
Jeans mass, 366, 373, 386–88, 392
Jean's theorem, 412–14, 440, 456
Jets, 178, 183, 255, 272–73, 284
Jodrell Bank, 310, 336

K-correction, 61, 133, 139, 143, 160, 164, 170–73, 182, 191, 268, 660–61, 776, 779, 781
Kitt Peak, 770

L galaxies, 27
Late-type dwarfs, 40–41, 44–46, 48–56, 61–62, 160, 169, 175, 349, 438
Leiden, 309, 321
Lick Observatory, 4, 46, 57, 61, 72, 98, 101, 320, 544, 602–3, 623, 637, 649, 649–50, 650, 652–55, 659–61, 683–86, 761, 770
Linear dimension of radio sources, 213, 228, 237, 241, 243–44
Light travel time, 160, 172–74, 775
Lindblad dispersion orbits, 424
Lindblad resonance, 478–81
Linear dimensions of radio sources, 263, 269–71, 274, 276, 726, 780
Local Cloud of galaxies, 583, 593
Local standard of rest (LSR), 418–19, 453
Local Supercluster, 535, 590, 592–95, 597, 621, 659–60, 662
rotation, 363–64
Log N, m relation, 187, 287, 302, 626, 653, 660
Log N, log S relation, 187–88, 241, 287, 302, 304, 726, 739–41, 744–51
Lowell Observatory, 126, 762
Luminosities of galaxies, 24, 44, 105, 108–9, 116, 127, 132–33, 139, 145–51, 154–55, 159, 164, 166, 173–75, 181, 189–90, 239, 265–69, 271, 276, 298–99, 302, 305, 327–28, 335–41, 348, 353, 400, 428, 430, 435–36, 438, 486, 494, 498, 501–2, 509, 528–30, 534, 537, 541, 544, 546, 549, 552–53, 558–60, 562, 569–73, 608, 611, 616, 618, 631, 633, 660
see Radio luminosity; Synchrotron radiation, luminosity
Luminosity classes, 14, 24, 146, 528, 534–37, 545, 558–60, 562, 569–71, 574–75
Luminosity density (spatial) in galaxies, 132, 144–45
Luminosity distribution in galaxies, 4, 26–27, 91–92, 102–4, 112, 123, 125, 133–45, 151–52, 391–92, 426–27, 429–35, 454, 460–61, 467, 496–99, 612–13
see also Radio brightness distribution
Luminosity evolution
optical, 58–61, 191, 775, 781
of quasars, 727, 736, 752–58
radio, 188, 190, 213, 726, 731, 745, 748–51
Luminosity function
of galaxies, 147–51, 193, 527, 536, 559, 583–84, 586, 604, 607, 610–19, 625–26, 632, 671
of quasars, 194, 727, 752–54, 756–57
of radio sources, 189–95, 726, 744, 748–51, 753–55, 757
of stars, 40, 46, 55, 58–59, 61, 170, 174–75, 336, 438–39, 493, 530, 537
Luminosity-volume test, 302, 304, 754–56
Lunar occultations, 184–85, 251–53, 284, 286

Magellanic Cloud–type galaxies; *see* Irregular galaxies (type I)
Magnetic fields, 213–40, 260–63, 270–72, 274–76, 298, 306, 381, 386, 389, 392–93, 397–99, 476, 483, 695, 734
Magneto-bremsstrahlung; *see* Synchrotron radiation
Magnitudes of galaxies, apparent (*for absolute magnitudes, see* Luminosity), 27, 105–9, 123–55, 164, 181–82, 191, 200–207, 209, 264–69, 320, 509, 527–28, 535, 543–46, 550–52, 554, 557–59, 561–62, 569–71, 575, 578–80, 586, 590, 597, 612–13, 617, 629, 633, 636, 649–50, 653, 655, 658–62, 683, 766–67, 775, 781
Main-sequence stars, 27, 37–38, 40, 42–43, 46, 50–53, 57, 75, 170, 174–75, 284, 438, 757
Mass
of clusters of galaxies, 403–4
determinations, 82–96
distribution in galaxies, 82, 85, 98, 118, 318, 394, 396–97, 410, 416, 436, 439, 446
of galaxies, 62, 77, 81–119, 137, 146, 190, 315, 318–19, 332–48, 353–54, 384–385, 389, 398, 402–3, 414, 421–23, 429–30, 434–38, 467–68, 485, 493–94, 498, 500–503, 541, 544, 549–55, 562, 572–73, 583, 585
Mass density in galaxies, 82–90, 92–93, 97, 100, 112, 118, 132, 136, 319, 399, 410–14, 418, 421–23, 438, 447–49, 475
distribution, 413, 499

gradient, 421
in solar neighborhood, 421
Mass flow in galaxies; *see* Non-circular motions
Mass function of stars, initial, 354
Mass loss
from bar, 485
from stars, 58, 63, 75
Mass-to-light ratio, 41, 46, 48–55, 62, 82, 87–88, 96, 98, 103–10, 112–13, 115–17, 119, 170, 174, 336–44, 353, 399, 403, 467, 491, 498, 501, 530, 534, 549–55, 562, 631-34
Massive objects, 276, 305–6, 380, 384
see also Collapsed massive objects; Spinars
Maximum rotation velocity, 89–90
McDonald Observatory, 101, 104, 161, 587, 773
Metal abundance; *see* Blanketing; Chemical abundances
Metal enrichment, 401, 491–94
Metal-poor stars, 42, 58, 62–63, 167, 401, 491–95, 523
Metal-rich stars, 42, 116, 173, 492
Missing mass; *see* Invisible matter
Mock gravitational force, 375–76, 381
Molecular bands, 39–47, 175
Morphological Catalogue of Galaxies, 4, 26, 27
Morphological type; *see* Classification
Mount Stromlo, 770, 772–73
Mount Wilson Observatory, 4–8, 14–17, 20, 22, 125, 320, 559, 649, 653, 661, 665, 761, 765–68, 773
Moving groups, 418, 420, 424
Multiple galaxies, 83, 360, 402–4, 541, 557, 590, 604, 621–22, 636, 654, 693

N-body systems, 460, 465, 471, 480–81, 504
N galaxies 27-29, 37, 181, 200–205, 207, 265–68, 271–72, 283, 285–86, 298, 757, 768–69
Nançay, 311, 318, 336
Narrow-band photometry, 44–58
National Radio Astronomy Observatory, 184, 251
Neutral hydrogen; *see* HI
Neutron stars, 117
New General Catalogue (NGC), 2, 601, 647
Nightsky emission, 43–44, 100, 124, 166, 321–22, 400, 767–71, 774
Non-circular motions in galaxies, 82, 105, 109, 112–15, 317–18, 417, 436, 440, 475, 489
Nonthermal radiation; *see* Synchrotron radiation
North Polar Sequence (NPS), 125, 649
Novae, 512–21, 525–26, 536–37, 558
Nuclear bulge; *see* Spheroidal component
Nuclear region, 46–60, 71–76, 112, 142–44, 167, 175, 320, 330, 410, 427, 436, 439
Nuclei of galaxies, 28, 37, 63, 116, 144, 178, 181, 240, 271–76, 285, 297–99, 306, 380, 399, 402, 409, 423, 426, 435, 438, 471, 501
double, 271–72
Nucleosynthesis, 354, 369, 400–401, 493, 721, 738

OB associations, 70, 514–15
Olbers' paradox, 739
Oort's constants, 416, 421–23, 475
Open clusters; *see* Star clusters
Optical depth, 221-24, 227–29, 238–40, 245, 247, 270, 290–91, 296, 325-27, 329–30, 333, 349, 353, 369, 373–74, 377, 389, 498–500, 704, 721, 729–33, 738
Owens Valley, 310–11, 318, 610

Pairs of galaxies, 81, 83, 96, 104–5, 112, 115–16, 181, 202–3, 288, 402, 541–55, 621–22, 631–32, 634, 636, 654, 659
angular separation, 542–43, 552
dynamics, 549–53
formation and evolution, 548, 553–55
see also Companion galaxies
Palomar Observatory, 11, 15–17, 20, 101, 103, 125–26, 182, 212, 587, 602, 609, 641, 655, 761, 767–73
Palomar Sky Survey, 23–26, 28, 147, 185–86, 203, 206, 209, 271, 286, 302, 472, 526, 541–45, 586–87, 602–3, 619–20, 637, 639, 682, 767
Peculiar galaxies, 4–5, 42, 61, 63, 154, 178, 181, 183, 264–65, 272, 287, 299, 400, 427, 572–73
Perseverance of radio astronomers, 309
Photographs of galaxies, 2, 4, 13–21, 24, 26, 151, 273, 285, 433, 438, 544, 553, 573, 575–76, 602, 609, 624, 637, 641, 649, 661
Photometry of galaxies, 39, 45, 123–26, 161–73, 191, 199, 433, 489, 493, 499, 502, 537, 612–13, 616, 618, 662
photoelectric, 124–26, 160, 286, 293, 619
photographic, 123-25, 619
see also Narrow-band photometry; Spectrophotometry
Pitch angle, 427, 476, 479–80, 482, 501
Planetary nebulae, 3, 73–74, 76, 290, 761–62
Polarization, 216, 220, 226, 236, 239, 259–63, 272–73, 294, 296, 369, 517
circular, 263
Population, stellar, 491, 496
disk population, 159, 424, 438, 471, 485, 489, 492
Population I, 59, 133, 169, 421, 424–25, 432, 438, 482, 493, 499, 501–2, 530
Population II, 59, 61, 133, 159, 424, 439, 491, 493–94, 612
see Stellar content
Position
of galaxies, 27, 284, 286, 311
of radio sources, 177–90, 199, 209–10, 211, 284, 286

Protogalaxy, 32–33, 366–67, 381–88, 401, 468, 483, 492, 495–96, 498–501, 504, 554, 694, 709, 719
 see also Young galaxies
Pulsars, 274–75, 305, 734–35
Pygmy galaxies, 586–87

q_0; *see* Deceleration parameter
Quarks, 306
Quasars, 28–29, 239–40, 243, 283–306, 725, 735–36, 745–46, 775, 783
 colors, 284, 286, 293–94, 757
 counts, 287, 296, 302, 304, 727
 energy source, 305
 evolution, 302–5
  see also Luminosity evolution
 definition, 285
 distance, 290–93, 296–98, 302
 distribution, 287–88
 identification, 284–88
 lifetime, 298, 305
 luminosity, 298–99, 302, 305, 756
 mass, 291–93, 296–98
 models, 305
 with nebulosities, 284–85, 298
 pairs, 288
 proper motion, 283, 297
 quasistellar objects (QSO, QSG), 195, 283–87, 302, 725
 quasistellar radio sources (QRS, QSS), 182–88, 190, 192–93, 195, 199–207, 210, 241, 247, 263–64, 274, 283, 286, 725–27, 740, 748–58
 spectra, 185, 283–90, 293–302, 378, 398, 730, 754, 757
 structure, 285, 290, 292, 296, 299, 305
 total energy, 305
variability, 283–85, 288–90, 294–99, 305

R_{max}; *see* Turnover radius
Radial velocity; *see* Redshift
Radiation, primordial, 360–77, 380, 387–89, 635, 776
 see also Decoupling; Background radiation
Radio brightness distribution, 183, 247–60, 262, 269
Radio contours; *see* Contour maps of radio sources
Radio galaxies, 27–28, 38, 75, 115, 178–79, 181–83, 187–93, 199–207, 209–10, 211–77, 299, 353, 400, 726, 740, 749, 752, 757–58
Radio heliograph, 250
Radio luminosity, 189–93, 213, 237, 245, 264–71, 276, 299, 302, 305, 739–40, 748–49, 751, 756
 see also Luminosity evolution, radio; Luminosity function of radio sources; Radio brightness distribution; Synchrotron radiation, luminosity
Radio magnitudes, 190–91
Radio sources, 177–95, 211–77, 285, 287, 543, 572–75, 609, 633
 Class I, Class II, 179
 complex, 201–4, 255, 258
 core-halo sources 255, 258-59, 262, 269–71
 and cosmology, 725–58
 counts, 179, 181, 187–88, 190, 241, 399–400, 726–27, 739–52, 758
 double, 182–83, 200, 202, 206, 212, 249, 253–55, 258, 262, 269, 273–76, 299, 736, 744, 758, 780
 energy sources, 273–75
 evolution; *see* Luminosity evolution, radio
 expansion, 240–41, 273–76
 extended, 181, 199, 201, 206, 249, 255, 259, 269–71, 276, 733
 life time, 275–76
 nonthermal, 200–209, 211–77
 optical properties, 271–73
 properties, 241-73
 quasistellar; *see* Quasars
 space density; *see* Space density of radio sources
 spectra, 187–88, 241, 243–47, 270, 276, 284, 288, 295–96, 740, 779
 classification, 245–47
 statistics, 243–45
 strong, 178, 181–82, 192, 195, 211–77
 structure, 182–83, 212–13, 247–59, 261, 269–70, 744
 thermal, 224
 total energy, 264, 270–71
 unidentified, 186–87
 variable, 205, 207, 239, 247–48, 274, 298
 weak, 178, 195, 242, 247, 329
 see also Synchrotron radiation
Radio surveys, 178–81, 241–42
 1C, 178
 2C, 179
 3C, 179–84, 189, 191–92, 610
 3CR, 180–81, 184–88, 193, 195, 199–210, 263, 286–87, 289, 302–5, 754, 756
 4C, 180, 184–85, 242, 286–87, 610, 745
 5C, 180, 242, 740–42, 744–45
 Bologna, 180, 242, 741
 MSH, 180–84, 191–92, 242
 Ohio, 180, 242
 Parkes, 181, 184–87, 242, 263, 265–69, 286, 610
 Vermilion River, 180, 242
Ram pressure, 275–76
Random velocities of galaxies, 104, 404, 535, 693
Razin effect; *see* Absorption, Tsytovitch effect
Redshift of galaxies, 45, 95, 104, 126, 133, 144, 146, 149–50, 160, 164, 170–73, 178, 181–182, 185, 187, 190–91, 193–95, 199–207, 209–12, 233, 236, 243, 263–69, 276, 283–303, 305, 310–11, 317–18, 320–23, 326–27, 329, 334, 378, 399–400, 402, 509–10, 526, 528, 531, 535–36, 543–44, 547, 552–53, 561, 569–76, 578–80,

585, 590, 597–99, 609, 613, 617, 619–21, 624–30, 632, 653, 655, 659–61, 667, 682, 693, 726–35, 740, 745–58, 761–83
cosmological, 290–92, 298–99
errors, 320–21, 773–74
gravitational, 292–93, 296–97, 300
local Doppler, 297–98, 299–300
multiple, 301
Redshift-apparent-magnitude diagram, 24–25, 58, 61, 147, 159, 164, 170, 174, 189, 299, 302, 368, 597, 617, 665, 682, 726, 757, 761, 766, 776–79, 782
Redshift-diameter relation, 726, 757–58, 761, 781
Reference Catalogue of Bright Galaxies, 12–13, 126, 544, 558, 562, 575–76, 583, 590, 597, 618, 648, 773–74
Relaxation of stellar systems, 453–60, 476, 495–96, 629, 692, 710, 712
time, 409–10, 448, 450, 460, 630, 693, 711
violent, 458–61
Resolution
angular, 179, 247, 259, 269, 273, 310, 313, 318, 329–30, 333–34, 744
into stars, 1, 22–23, 26, 427, 475, 501, 611, 761
spectral, 159–60, 166, 173, 293, 768
Ring features in galaxies, 9–10, 12, 19, 23, 28, 330, 333, 427, 485–86
Rotation of galaxies, 23, 81–94, 96–102, 110, 112, 114, 116, 272, 274, 319, 334, 380–85, 392, 394–97, 401, 414–17, 423–24, 434, 436–38, 460–63, 465–66, 471, 474–76, 483–89, 498–504, 526, 533, 548, 719–20
Rotation curve of spirals, 81–93, 96–102, 105, 112, 118, 310–15, 318–21, 418, 422, 436, 440, 467–68, 478–79, 486, 489–91, 530, 572–73
asymmetry, 417
Rotation measure, 260, 262
RR Lyrae stars, 422, 424, 492, 494, 521, 523, 525–26, 531, 534, 536, 558

S0 galaxies, 4, 6–7, 10, 15, 25, 27, 29–30, 44, 58, 72, 77, 81, 94, 96, 98, 126, 129, 131–34, 137–38, 143, 145, 149–51, 154, 164, 166, 181, 183, 264–67, 269, 327, 354, 398, 426–27, 429, 431–34, 438–39, 460–83, 494, 544–51, 557–59, 573–75, 586, 604, 607–8, 610, 625, 631, 766, 771
Sagittarius A, 177, 436
Scanners, 46–47, 58, 61, 166–69, 171, 173, 293
see also Spectrophotometry
Scattering, 127, 164, 224, 237–40, 253, 290–91, 399–400, 731–32, 734–35, 758
Schraffierkassette, 123–25, 611
Schwarzschild radius, 117
Scintillation; *see* Interplanetary scintillation
Sculptor-Fornax type galaxies; *see* Dwarf ellipticals
Second-order clusters; *see* Superclusters
Selected Areas, 124–25, 649
Seyfert galaxies, 28–29, 37, 113, 144, 203, 239–40 247, 265, 267, 270, 272, 283, 285–86, 295, 298–99
Shapley-Ames Catalogue, 6, 25, 124, 129, 146, 151–52, 179, 535, 543, 558, 590, 601, 608, 647–48, 659, 773
Shock front, 333
Shock waves, 379, 388–89, 482–83
Slit spectra; *see* Spectra
Solar motion, 762, 774
Solar neighborhood, 116, 133, 170, 401, 410, 415, 424, 452–53, 463, 476, 495
Space density
of galaxies, 124, 149–50, 557, 562, 587, 589, 597, 603, 648, 775
of quasars, 193–94, 304–5
of radio sources, 190, 193–95, 726, 728, 739–40, 748, 750
Spectra of galaxies, 37–44, 57, 62, 71–72, 96–102, 143–44, 160, 166, 169–70, 175, 178, 183, 263, 271–72, 283–90, 293–302, 494, 548, 551, 573, 575, 761–74
see also Quasars, spectra; Radio sources, spectra
Spectral energy distribution; *see* Energy distribution
Spectral index, 182, 186, 188, 219, 221, 225, 227–28, 232, 234, 237–38, 240, 243–44, 264, 293, 295, 728, 745–48, 753
Spectral lines, 39–77, 161, 175, 765, 771–73
see also Absorption lines; Emission lines
Spectral type of galaxies, 2, 26, 438, 543
Spectrophotometry, 160–66
see also Scanners
Spheroidal component, 1, 22–24, 32, 110, 137–38, 164, 429–30, 433, 460, 465, 475, 482, 493, 498–502
Spinars, 274–76
Spindle galaxies (edge-on spirals), 28
Spiral arms, 1–2, 5–6, 11–12, 17, 22–24, 26, 29, 92, 98, 126, 138, 145, 159, 170, 330–33, 335, 395–96, 424, 438, 474–75, 480, 482, 486
trailing, 474, 476, 478–79, 481–82, 485, 489
Spiral galaxies, 4–6, 11, 14–20, 22–23, 25, 27, 29–33, 38, 40–44, 46, 58, 62–63, 72, 81, 85–86, 92, 94, 96, 98, 100, 105–7, 110, 115–18, 125, 128–33, 137–47, 149–50, 152, 159, 161–64, 167, 169–70, 178, 264, 266–67, 269, 272, 276, 327–28, 332–54, 392, 394, 397, 399–400, 402, 415, 426–27, 429, 431–40, 460–83, 492, 498–504, 527, 529–30, 537, 544–54, 557–60, 572–76, 586, 593, 604, 607–10, 612, 616, 619, 625, 631–32, 654, 661, 719–20, 771, 773
see also Barred spirals; Giant spirals; Supergiant spirals

Spiral structure, 4, 272, 330, 333, 335, 396, 436, 439–40, 461, 463, 473–83
formation, 484–85
origin, 481–82
Star clusters, 2, 40, 42, 514–15, 521, 524, 530, 611, 694
h and χ Persei, 61
M11, 58
M67, 46–47, 59, 62, 116, 169, 174, 472–73
NGC 188, 47, 59, 169, 173–74, 473, 492
NGC 752, 419, 472
NGC 6791, 47
Pleiades, 58
Star counts, 59, 61, 137, 144, 153, 159, 459, 587
Star formation; *see* Formation
Steady-state theory; *see* Cosmological models
Stellar content of galaxies, 26–27, 29–33, 37–63, 159, 168–70, 172–75, 354, 429, 436, 439, 499
radial gradients, 58–59
see Population, stellar
Stellar dynamics, 409–540
Stellar hydrodynamics, 412
Stellar motions, 418–22, 434–38, 763
Stellar orbits, 409–13, 418, 422, 440–46, 452, 456, 458–59, 474, 486, 491, 495
epicyclic, 423–24, 464
Stromberg's asymmetrical drift, 420–21
Strong-lined stars; *see* Metal-rich stars
Structural index of clusters, 623–25
Structure of galaxies, 409–504
Subdwarf stars, 424, 440, 452–53, 473, 492–96, 765
Subgiant spirals, 25, 146
Subgiant stars, 37, 42, 425
Sun, 40, 55, 66, 161, 173, 211, 253, 332, 414–16, 418, 421–24, 438, 440, 446, 452, 463–64, 474, 489, 493–96, 552, 597, 654, 684, 774
Superclusters, 373, 402, 404, 602, 609, 636–41, 654–60, 663, 665, 667, 682–83, 686, 693, 718
Supergalactic coordinates, 560–68, 576–77, 587–90, 592, 593–95, 636
Supergiant elliptical galaxies, 189, 271, 608–9
Supergiant spirals, 24, 146, 528, 531
Supergiant stars, 32, 40, 61, 164, 170, 439, 486, 536, 587
Super–metal-rich (SMR) stars, 46–47, 50–53, 55, 57
Supernovae, 270, 274, 306, 394, 400–401, 494, 521, 528, 532–36, 574, 739
Supernova remnants, 178, 180, 199–201, 240, 245, 275, 536, 721
CasA, 178, 200, 532
Crab Nebula, 178, 200, 211, 245, 532
IC 443, 200
Tycho, 200, 532–34
Surface brightness, 23, 123–33, 137, 139, 145–46, 152, 433, 438, 449, 475, 527–28, 534, 536, 553, 587, 612–13, 654, 762, 768, 781
distribution, 13, 380, 391–92, 438, 440, 449, 459, 466, 587
Surface density in galaxies, 87–88, 90–91, 118, 423, 429, 449, 464, 466–67, 477, 479
distribution, 467, 470–71, 499, 501
Surface density of galaxies, 542–43, 557, 602, 619–20, 622, 624, 626, 649, 649–50, 656–58, 662
Synchrotron radiation, 167, 213–41, 294, 296, 298
energy losses, 232, 237–41, 244–45
equipartition field, 236–37, 270–72
from a single particle, 213–17
from an ensemble of particles, 218–20
lifetime, 217, 237, 272
luminosity, 233–37, 240–41
radiation transfer, 221–24
self-absorption; *see* Absorption
spectrum, 216–41
total energy, 213, 220, 233–37
volume emissivity, 218–228, 238

Tails of radio galaxies, 242–43, 276
Temperature
antenna, 324–26, 353
brightness, 229, 296, 316, 320, 323–26, 330–31, 348–49, 729, 732
color, 160, 169
electron, 290
kinetic, 228
spin, 326, 335, 399, 729
surface brightness, 239
Thermal fluctuations, 708–9
Third integral of stellar orbits, 440–46, 474
Tidal interactions; *see* Interacting galaxies
Turbulence
gravito-, 709–12, 714–15
in galaxies, 393–94, 396, 402, 482, 494
intergalactic, 384, 705
primordial, 694
in protogalaxies, 386, 390, 397, 504, 554
Turnover radius (R_{max}), 89–90, 111, 319
21-cm line measurements, 82, 85–86, 89, 96–97, 105, 110–13, 115–16, 398–99, 416–77, 422, 436, 533, 560, 574, 610, 633, 728–30, 736, 773–74
Types of galaxies; *see* Classification

Ultraviolet, 164, 167–69, 171–72, 239, 285, 398, 633, 729–31, 736
Ultraviolet excess, 285, 401, 450–52, 473, 492, 495
Unidentified sources; *see* Radio sources, unidentified

Velocity anisotropy, 535–36
Velocity dispersion
of galaxies, 634–35, 641–42
of stars, 82, 92–94, 100, 102–4, 107, 112, 418–21, 424, 435, 440, 463–65, 471–80, 495–96, 499, 501

see also Clusters, velocity dispersion; Random velocities of galaxies
Velocity-distance relation, 764–66, 782
Vertex deviation, 419
Very large array (VLA), 251
Very massive objects; *see* Massive objects
Violent events; *see* Explosions
Virial theorem, 81–82, 93–96, 102, 105, 117, 319, 377, 403–4, 414, 465, 498, 553–54, 583, 585, 630–32, 641
cosmological, 666, 690–93, 698, 710, 715, 720
Visibility function, 183, 250–51

W Virginis stars, 522, 525–26, 536
Warp of galaxy plane, 329, 334–35, 436, 440, 461, 471–72
Weak-lined stars; *see* Metal-poor stars
Westerbork, 262, 264, 276, 310–11
White dwarfs, 49–53, 58, 117, 184, 287
Width of spectral lines, 45

X-rays, 377, 379, 398, 543, 610, 633, 731–33, 736

Yerkes Observatory, 4
Young galaxies, 399–400

Z (metal content); *see* Chemical abundances
Zone of avoidance, 542–43, 572, 583, 593, 650–52

Galaxy Index

Individual galaxies (including radio sources and quasars) are listed, as well as clouds, clusters, and groups of galaxies. References to illustrations are in *italics*.

A1244, 571, 575
A2021, 580
Andromeda I dwarf, 587, 607
Anon 0220 + 4058, *21*
B234, 305
B264, 285
BW Tauri; *see* 3C 120
Capricornus system (A2144), 561–62
Centaurus A; *see* NGC 5128
Clouds of galaxies
  Canes Venatici I cloud (G3), 554, 558, 566, 569, 572, 575, 584, 589–90, 593, 598
  Canes Venatici II cloud (G10), 558, 566, 570, 575–76, 584, 589, 593, 598
  Cetus I cloud (G15), 566, 576, 578, 584, 598
  Coma I cloud (G13), 566, 571, 575, 584, 592–93, 598
  Dorado cloud, 576
  Grus cloud (G27), 567, 578, 581, 584–85, 598
  Leo I cloud, 575
  Ursa Major I cloud (G17), 106, 554, 566, 572, 576, 578, 584–85, 593, 598, 601, 636
  Ursa Major–Camelopardalis cloud, 572, 592
  Virgo II cloud (G20, G26, G35), 572, 581, 589, 593
Clusters of galaxies
  A31, 623, 625
  A115, 203
  A151, 613, 618
  A194, 201, 628–29, 633
  A234, 623–24
  A347, 201
  A400, 201
  A426; *see* Perseus I cluster
  A732; *see* Hydra II cluster
  A754, 617
  A801, 623–24
  A1060; *see* Hydra I cluster
  A1132; *see* Ursa Major II cluster
  A1185, 623, 626
  A1213, 623, 626
  A1367, 617, 623, 626
  A1377; *see* Ursa Major I cluster
  A1643, 623–24
  A1656; *see* Coma cluster
  A1677, 623–24
  A2052, 200
  A2065; *see* Corona Borealis cluster
  A2151; *see* Hercules cluster
  A2199, 201, 354, 609, 613, 618, 623, 625, 628, 633
  A2634, 201
  A2670, 618
  Bootes cluster (A1930), 766
  Cancer cluster, 623
  Canes Venatici cluster, 572, 575, 585
  Coma cluster (A1656), 81, 117, 166, 242, 403–4, 497, 527, 536, 542, 554, 558, 576, 585, 601, 604, 607, 609, 613, 617–19, 622–29, 632–35, 742, 745
  Corona Borealis cluster (A2065), 536, 604, *605*, 607, 609, 613, 618, 622–24, 628, 633, 642, 766
  Fornax cluster (G53), 354, 568, 580–81, 585–86, 590, 601, 608, 622, 625, 628, 630
  Hercules cluster (A2151), 117, 554, 604, *606–7*, 622, 628, 631–32, 655
  Hydra I cluster (A1060), 623, 625
  Hydra II cluster (A732), 623–24, 767
  Leo cluster (A1020), 766
  NGC 541 cluster, 554
  Pegasus I cluster, 354, 623, 628
  Perseus I cluster (A426), 166, 188, 200–201, 242, 276, 558, 601, 623, 625, 628, 745
  Ursa Major I cluster (A1377), 613, 618, 628, 633, 766
  Ursa Major II cluster (A1132), 623, 766
  Virgo cluster, 13–14, 22, 44, 108–10, 117, 133, 146, 161, 163, 166–68, 188, 200, 202, 242, 272, 402–3, 429, 527–28, 530–37, 542, 559, 576–77, 581, 585–87, 589, 593, 601, 604, 607–8, 616, 618–19, 622, 625, 628, 630–33, 636, 665, 729, 734, 745, 761
  Virgo E cluster (G19), 554, 566, 576, 585
  Virgo S cluster (G18), 554, 566, 576, 585
Coma A; *see* 3C277.3
Cygnus A; *see* 3C405
DDO 1, 569
DDO 6, 569
DDO 11, 570, 574
DDO 17, 570, 574
DDO 19, 570, 574
DDO 22, 570, 574
DDO 24, 570, 574
DDO 25, 570, 574
DDO 26, 570, 574
DDO 33, 569, 572
DDO 38, 569, 572
DDO 39, 569, 572
DDO 40, 570, 574
DDO 41, 570, 574
DDO 43, 570, 574
DDO 44, 569, 572
DDO 46, 570, 574
DDO 48, 570, 574
DDO 49, 570, 574

DDO 51, 569, 572
DDO 52, 570, 574
DDO 53, 569, 572
DDO 55, 570, 574
DDO 56, 570, 574
DDO 57, 570, 574
DDO 59, 570, 574
DDO 60, 570, 574
DDO 61, 570, 574
DDO 62, 570, 574
DDO 66, 569
DDO 71, 569
DDO 77, 569
DDO 78, 569
DDO 79, 571, 575
DDO 80, 569
DDO 82, 569
DDO 86, 569
DDO 87, 569
DDO 88, 571, 575
DDO 89, 570, 571, 574, 575
DDO 90, 571, 575
DDO 91, 570, 574
DDO 99, 569, 573
DDO 101, 571
DDO 105, 569, 573
DDO 108, 570, 574
DDO 117, 571
DDO 122, 569
DDO 123, 569
DDO 125, 569, 573
DDO 126, 569, 573
DDO 129, 569, 573
DDO 131, 571
DDO 133, 569, 571, 573
DDO 141, 569, 573
DDO 143, 569, 571, 573
DDO 154, 569, 571, 573
DDO 165, 569
DDO 167, 569, 573
DDO 168, 569, 573
DDO 169, 569, 573
DDO 172, 569, 573
DDO 174, 569
DDO 175, 569, 573
DDO 176, 569, 573
DDO 177, 569, 573
DDO 178, 569, 573
DDO 181, 569, 573
DDO 182, 569, 573
DDO 183, 569, 573
DDO 185, 573
DDO 186, 573
DDO 191, 569, 573
DDO 193, 569, 573
DDO 194, 569, 573
DDO 205, 569
Draco system (DDO 208), 22, 137, 153, 159, 438, 560–62, 587, 589, 608
5C 4.51; *see* NGC 4839
5C 4.81; *see* NGC 4869
5C 4.85; *see* NGC 4874
4C 03.01; *see* NGC 193
4C 05.10; *see* NGC 741
4C 07.61; *see* NGC 7503
4C 08.11; *see* NGC 1044
4C 14.37; *see* NGC 3367
4C 15.19, 206
4C 16.21, 206
4C 17.52; *see* NGC 3801
4C 17.66; *see* NGC 6047
4C 25.35; *see* NGC 3689
4C 25.5, 301
4C 34.09; *see* NGC 1167
4C 35.6, 264
4C 36.24; *see* NGC 5141
4C 37.21; *see* NGC 2484
4C 37.43, 287
4C 39.11; *see* NGC 1233
4C 54.17; *see* NGC 2656
4C 55.33.1; *see* NGC 6454
Fornax A; *see* NGC 1316
Fornax system, 137, 428, 493, 498, 560–62
Galaxy, 32, 43, 47, 49, 57, 61, 64–71, 82–83, 85–88, 97, 105–6, 108, 110, 113, 115, 118, 126–27, 137, 159–60, 164, 169, 177, 211, 224–25, 234, 241, 261, 273, 291, 296–98, 309, 325–29, 332–35, 343, 354, 385, 393–94, 397, 400–402, 414–25, 436, 438, 440–46, 448, 450–53, 458, 562–65, 471–76, 479–83, 491–96, 500, 504, 510–11, 515, 518–19, 523–24, 531–36, 551, 560–62, 573, 587, 608, 622, 632, 637, 647–48, 650, *652*, 659–60, 665, 694, 719, 727–30, 733–34, 738, 740, 765, 774–75, 781, 783
GB 1, 550
Groups of galaxies
  G5; *see* M101 Group
  G9; *see* M66 Group
  G11; *see* M96 Group
  G47; *see* NGC 3190 Group
  Leo Group, 574–75, 628
  Local Group, 4, 13, 22, 32, 63, 110, 146–47, 150, 153, 159, 392, 428, 438, 449, 480, 509–10, 515–18, 525–32, 536–37, 554–55, 558–60, 563–64, 572–73, 581–87, 589–90, 592–94, 597, 604, 607, 618–19, 622, 636, 665, 774–75
  M31 Group, 517, 560, 622, 665
  M66 Group (G9), 527, 558, 566, 570, 574, 584, 598
  M81 Group (G2), 22, 108, 147, 150, 530, 537, 554, 556, 558, 566, 569, 572, 581, 584–85, 589, 593, 598, 607–8
  M94 Group, 572–73, 585
  M96 Group (G11), 527, 558, 566, 571, 574–75, 584, 598
  M101 Group (G5), 147, 558, 566, 569, 573, 584, 589, 593, 598
  NGC 1023 Group (G7), 558, 566, 570, 574, 584, 593, 598
  NGC 2841 Group (G6), 566, 570, 573, 584, 593, 598
  NGC 2997 Group (G8), 566, 570, 574, 584, 598
  NGC 3184 Group (G12),566, 571, 575, 584, 598
  NGC 3190 Group (NGC 3227 Group; G47), 527, 558, 568, 580–81, 584, 598
  NGC 4274 Group, 527
  NGC 5128 Group (G4), 566, 569, 573, 584, 593, 598
  NGC 6300 Group (G14), 566, 571, 576, 584, 598
  Pegasus Group, 554
  Pisces Group (NGC 383 Group), 554, 628
  Sculptor Group, 554
   see also South Polar Group
  Serpens Group, 554
  South Polar Group (Sculptor Group; G1), 403, 536–37, 554, 558, 564, 566, 569, 581, 584–85, 593, 598
  Ursa Major Groups I and II (Sérsic), 572, 575
  for a list of additional groups see pp. 566–68, 576–81, 584, *and* 598
Hercules A; *see* 3C348
Ho I, 8, 569, 572
Ho II, 8, *20*, 30–31, 341, 569, 572
Ho IV, 569, 573

Ho V, 569, 573
IC 10, 340, 560–62, 587, 590
IC 239, 570, 574
IC 310, 242
IC 342, 340, 569, 572, 590
IC 356, 569, 572
IC 764, 580
IC 1029, 579
IC 1613, 8, 341, 515, 518, 525, 529, 533, 536, 550, 560–62, 587, 589, 607
IC 1727, 570, 574
IC 1954, 770–71
IC 2233, 570, 574
IC 2574, 8, *20*, 30–31, 341, 569, 572, 589
IC 3330, 571, 576
IC 3475, 586–87, 608
IC 4182, 569, 573
IC 4296, 267
IC 4662, 576
IC 4662A, 571, 576
IC 4710, 571, 576
IC 4713, 571, 576
IC 4714, 571, 576
IC 5052, 579
IC 5152, 581, 590
IC 5201, 581
IC 5267, 578
IC 5332, 564, 569, 572, 581
IC 5532 (3C296), 267
Large Magellanic Cloud (LMC), 8, *20*, 23, 30–31, 37, 39–42, 101, 108, 118, 159, 170, 309–10, 325, 329–30, 333, 335, 341, 432, 439, 471, 486–87, 489, 493–95, 510–12, 515–19, 521–26, 529, 533, 550, 560–62, 573, 576, 586
Leo I system (DDO 74; A 1006), 22, 137, 153, 560–62, 589, 619
Leo II system (DDO 93; A 1111), 22, 137, 153, 155, 560–62, 586, 593, 619
Leo A (DDO 69; A 956), 560–62
Maffei 1 and 2, 572
M31; *see* NGC 224
M32; *see* NGC 221
M33; *see* NGC 598
M49; *see* NGC 4472
M51; *see* NGC 5194
M60; *see* NGC 4649
M64; *see* NGC 4826
M66; *see* NGC 3627
M77; *see* NGC 1068
M81; *see* NGC 3031
M82; *see* NGC 3034
M83; *see* NGC 5236
M84; *see* NGC 4374
M86; *see* NGC 4406
M87; *see* NGC 4486
M94; *see* NGC 4736
M96; *see* NGC 3368
M100; *see* NGC 4321
M101; *see* NGC 5457
M104; *see* NGC 4594
M105; *see* NGC 3379
Milky Way; *see* Galaxy
NGC 24, 569
NGC 45, 27, 337, 564, 569, 572
NGC 55, 101, 108, 337, 403, 433, 489, 550, 564, 569
NGC 128, 498
NGC 134, 579, 581
NGC 147, 22, 32, 560–62, 607
NGC 148, 579
NGC 150, 579
NGC 157, *17*, 24, 30–31, 106, 337, 436–37, 550
NGC 185, 22, 32, 449, 498, 560–62
NGC 193 (4C 03.01), 265
NGC 205, 13, 22, 32, 42, 44, 47, 61, 63, 438, 493, 517, 523, 543, 560–62, 586–87, 607
NGC 221 (M32), 22, 32, 39, 40, 43–44, 46–47, 49, 63, 94, 98, 103–4, 107, 113, 162–63, 167, 171, 173, 428, 434–36, 438, 493, 517, 543, 546, 551, 560–62, 586–87, 608
NGC 224 (M31; 1C 01.01), 26, 32, 39–40, 42–44, 47–61, 67, 71, 74–76, 85, 87–90, 98–99, 101, 105–6, 110, 112–13, 116, 118, 131, 159, 161, 163–64, 167, 175, 178, 310, 316–18, 330–35, 337, 397, 410, 417, 428, 430, 436, 438, 471–72, 479, 499, 510–11, 513, 515, 517–33, 536, 543, 550, 560–62, 573, 607, 745, 762
NGC 247, 106, 337, 564, 569
NGC 253, 106, 113, 115, 264, 269, 337, 550, 564, 569
NGC 254, 579
NGC 289, 579
NGC 300, 106, 111, 334, 337, 564, 569, 572
NGC 309, 9
NGC 382/383 (3C 31), 203, 265, 628
NGC 404, 590
NGC 428, 337
NGC 470, 579
NGC 474, 579
NGC 488, 11, *15*, 27, 30–31, 579
NGC 520, 5, 24, 579
NGC 521, 579
NGC 524, 138, 145
NGC 541, 554, 628
NGC 545/547 (3C40), 201, 265
NGC 584, 579, 581, 762, 765
NGC 596, 579
NGC 598 (M33), 11, *17*, 27, 30–31, 67, 71, 76, 87–89, 101–2, 105–6, 112–13, 115–16, 118, 131, 178, 318, 332–34, 337, 429–31, 438–39, 475, 480–82, 493, 499, 515, 517, 525–29, 531, 533, 536, 550, 560–62, 573
NGC 613, 109, 337, 550
NGC 615, 579
NGC 628, 8, 11, *16*, 27, 30–31, 106, 139, 142, 337
NGC 672, 570, 574
NGC 681, 107, 337, 551, 581
NGC 720, 579
NGC 741 (4C 05.10), 265
NGC 772, 337
NGC 779, 579–80
NGC 891, 152, 570, 574
NGC 925, *18*, 30–31, 106, 113, 115, 337, 570, 574
NGC 936, 578
NGC 972, 106, 337, 550
NGC 1003, 570, 574
NGC 1023, 337, 570
NGC 1042, 140
NGC 1044 (4C 08.11), 265
NGC 1052, 73, 75–76, 247, 269, 576
NGC 1055, 337, 578
NGC 1058, 570, 574
NGC 1068 (M77, 3C71), 11, 28, 109, 113, 115, 203, 265, 270, 337, 353, 550, 576, 578, 762
NGC 1073, 9, *19*, 30–31
NGC 1084, 106, 337, 550, 578
NGC 1087, 578
NGC 1097, 109, 337, 550
NGC 1140, 337
NGC 1156, 337, 570, 574
NGC 1167 (4C 34.09), 265
NGC 1187, 579
NGC 1201, 138
NGC 1209, 581
NGC 1218 (3C78), 203, 265
NGC 1232, 11–12, 140, 142, 579

NGC 1233 (4C 39.11), 265
NGC 1265 (3C 83.7), 201, 242, 265
NGC 1275 (3C84, Perseus A), 113, 115, 188, 200, 242–43, 247–48, 265, 270, 274
NGC 1291, 337
NGC 1300, 8–9, *19*, 23, 30–31, 337, 579
NGC 1302, *15*, 22, 30–31
NGC 1313, 433, 502, 590
NGC 1316 (Fornax A), *236*, 249, 259–60, 262, 265, 272, 399, 581, 590
NGC 1326, 337, 580
NGC 1332, 337, 502, 581
NGC 1350, 580
NGC 1365, 109, 113, 115, 337, 550, 581
NGC 1380, 580
NGC 1398, *19*, 30–31, 579
NGC 1399, 265, 580
NGC 1404, 580
NGC 1407, 579
NGC 1411, 578
NGC 1433, 433, 576, 578
NGC 1448, 472, 578
NGC 1493, 578
NGC 1507, 337
NGC 1511, 521, 768, 770
NGC 1512, 578
NGC 1518, 337
NGC 1532, 337
NGC 1549, 578
NGC 1553, 578
NGC 1559, 578
NGC 1560, 569, 572
NGC 1566, 576, 578
NGC 1569, 337, 569, 572, 590
NGC 1574, 578
NGC 1617, 578
NGC 1637, 337
NGC 1672, 576, 578
NGC 1688, 578
NGC 1703, 578
NGC 1744, 337
NGC 1792, 106, 337, 550
NGC 1796, 578
NGC 1808, 106, 337, 550
NGC 1961, 550
NGC 1964, 338
NGC 2139, 338, 579
NGC 2146, 90–91, 106, 269, 550
NGC 2188, 338
NGC 2207, 579, 581
NGC 2217, 10, 338, 579
NGC 2223, 579
NGC 2280, 338, 579
NGC 2366, 8, *20*, 30–31, 338, 569, 572
NGC 2403, 106, 108, 111, 131, 318, 332, 334, 338, 569, 572
NGC 2444/2445, *21*
NGC 2484 (4C 37.21), 266
NGC 2500, 338, 570, 573
NGC 2523, 9, *19*, 30–31
NGC 2537, 570, 573
NGC 2541, 338, 570, 573
NGC 2552, 570, 573
NGC 2613, 338
NGC 2654, 579
NGC 2656 (4C 54.17), 266
NGC 2663, 266
NGC 2681, 570, 573–74
NGC 2683, 338
NGC 2685, 5, 11
NGC 2716, 767
NGC 2742, 579
NGC 2763, 570, 574
NGC 2768, 579
NGC 2776, 338
NGC 2782, 109, 551
NGC 2784, 570, 574
NGC 2805, 579
NGC 2811, *15*, 22, 30–31
NGC 2814, 548
NGC 2820, 548
NGC 2835, 338, 570, 574
NGC 2841, 11, *15*, 30–31, 163, 338, 570, 573–74
NGC 2848, 570, 574
NGC 2859, 6, 9, 11, 579
NGC 2880, 579
NGC 2903, 106, 338, 550, 590
NGC 2911, 269
NGC 2950, 6
NGC 2964, 579, 581
NGC 2976, 569, 572
NGC 2997, 338, 570, 574
NGC 3003, 579
NGC 3027, 338
NGC 3031 (M81), 42, 47–48, 70–77, 101, 107–8, 131, 163, 335, 338, 550, 569, 572
NGC 3032, 579
NGC 3034 (M82; 3C231), 5, 24, 40, 42, 61–62, 70, 72–73, 77, 108, 113, 115, 131, 204, 264, 269, 271, 273, 320–21, 335, 338, 427, 550, 569, 572
NGC 3067, 579
NGC 3077, 5, 24, 335, 569, 572
NGC 3079 (4C 55.19), 269, 338
NGC 3081, 10
NGC 3109, 111, 338, 574, 590
NGC 3115, 39, 47, 98, 107, 163, 338, 426, 501, 551
NGC 3162, 580
NGC 3169, 338
NGC 3184, 140, 142, 571
NGC 3185, 10, *19*, 23, 30–31
NGC 3190, 580–81
NGC 3193, 580
NGC 3198, 338, 571
NGC 3226, 580
NGC 3227, 62, 107, 113, 115, 338, 527, 580
NGC 3239, 571
NGC 3245, 580–81
NGC 3245A, 580
NGC 3254, 580
NGC 3274, 580
NGC 3277, 580
NGC 3299, 571
NGC 3300, 571
NGC 3306, 571
NGC 3310, 109, 338, 578
NGC 3319, 338, 571
NGC 3344, 140, 142, 338, 575
NGC 3346, 571
NGC 3351, 571, 575
NGC 3357, 571
NGC 3359, 110, 338
NGC 3367 (4C 14.37), 266
NGC 3368 (M96), 11, 163, 571, 575
NGC 3377, 571, 575
NGC 3377A, 571, 575
NGC 3379 (M105), 39, 47, 62, 103, 107, 134, 163, 391, 435, 449, 498, 551, 571, 575
NGC 3380, 580
NGC 3384, 571, 575
NGC 3395, 580
NGC 3396, 580–81
NGC 3412, 571, 575
NGC 3413, 580
NGC 3414, 580
NGC 3415, 578
NGC 3418, 580
NGC 3419A, 571
NGC 3427, 580
NGC 3430, 580
NGC 3432, 106, 338, 550, 571
NGC 3433, 571
NGC 3444, 571
NGC 3445, 578–80
NGC 3447, 571
NGC 3447A, 571
NGC 3448, 578–80
NGC 3458, 578
NGC 3466, 571
NGC 3478, 578
NGC 3485, 570, 574

NGC 3489, 570, 574
NGC 3504, 8, 10, *18*, 30–31, 109, 338, 551, 580
NGC 3506, 570–71, 574
NGC 3510, 575
NGC 3512, 580
NGC 3521, 107, 338, 551, 590
NGC 3547, 570, 574
NGC 3549, 578
NGC 3556, 100, 106, 338, 550
NGC 3583 (5C 2.203), 269, 578
NGC 3593, 570, 574
NGC 3596, 570, 574
NGC 3607, 580
NGC 3608, 580
NGC 3614, 578
NGC 3621, 338
NGC 3623, *17*, 22, 24, 30–31, 107, 338, 551, 570, 574
NGC 3626, 580
NGC 3627, 339, 527, 570, 574
NGC 3628, 339, 570, 574
NGC 3631, 339, 579
NGC 3637, 578, 628
NGC 3646, 106, 339, 550
NGC 3666, 570, 574
NGC 3672, 578
NGC 3675, 570, 575
NGC 3684, 580
NGC 3686, 580
NGC 3689 (4C 25.35), 266
NGC 3718, 5, 339, 579
NGC 3726, 339
NGC 3769, 570, 575
NGC 3769A, 570, 575
NGC 3776, 578
NGC 3786, 548
NGC 3788, 548
NGC 3801 (4C 17.52), 266
NGC 3810, *16*, 30–31
NGC 3813, 578
NGC 3818, 578
NGC 3862, 201
NGC 3865, 578
NGC 3892, 578
NGC 3893, 579
NGC 3898, *15*, 30–31, 579
NGC 3904, 580
NGC 3923, 580
NGC 3938, 139, 142, 339, 579
NGC 3941, 578
NGC 3949, 570, 575
NGC 3953, 131, 579
NGC 3992, 131, 339, 579
NGC 3998, 76
NGC 4025, 570, 575
NGC 4027, 92, 106, 113, 484, 489–91
NGC 4051, 28, 144, 339, 570, 575
NGC 4062, 571, 576
NGC 4088, *18*, 24, 30–31, 570, 575
NGC 4096, 579
NGC 4105, 580
NGC 4106, 580
NGC 4111, 98, 107–8, 138, 145, 551, 570, 575
NGC 4136, 569, 573
NGC 4143, 570, 575
NGC 4144, 339
NGC 4145, 578
NGC 4146, 571, 576
NGC 4150, 569, 573
NGC 4151, 28, 115, 339, 578, 773
NGC 4157, 579
NGC 4192, 530–31
NGC 4203, 571, 576
NGC 4206, 548
NGC 4214, 42, 111, 339, 569, 573
NGC 4216, 548
NGC 4217, 579
NGC 4235, 580
NGC 4236, 106, 108, 339, 569, 572
NGC 4242, 570, 575
NGC 4244, 106, 111, 152, 335, 339, 569, 573
NGC 4245, 571, 576
NGC 4251, 571, 576
NGC 4254, 578
NGC 4258, 76, 107–8, 113, 115, 144, 339, 551, 569, 573
NGC 4260, 580
NGC 4261 (3C270), 200, 242, 266, 580
NGC 4273, 580
NGC 4274, 9, 527, 571, 575–76
NGC 4278, 71–73, 75, 551, 571, 576
NGC 4281, 580
NGC 4283, 571, 576
NGC 4286, 571
NGC 4288, 570, 575
NGC 4293, *17*, 22, 30–31
NGC 4303, 139–40, 142, 339, 578
NGC 4314, 571, 576
NGC 4321 (M100), 9, *14*, 22, 42, 339, 531, 578
NGC 4350, 136
NGC 4359, 571, 576
NGC 4365, 135–36
NGC 4369, 578
NGC 4374 (M84, 3C272.1), 135–36, 163, 171, 202, 242, 266, 578
NGC 4375, 571
NGC 4380, 578
NGC 4382, 578
NGC 4395, *20*, 30–31, 569, 573
NGC 4406 (M86), 46, 107, 135–36, 163, 435, 551, 608
NGC 4414, 571, 576
NGC 4417, 136–38
NGC 4424, 578
NGC 4438, 76
NGC 4448, 571, 576
NGC 4449, 27, 41–42, 67, 69, 70, 339, 569, 573
NGC 4457, 11
NGC 4459, 135–36, 429–33
NGC 4469, 578
NGC 4472 (M49), 39, 42, 47, 58–59, 107, 135–36, 163, 353, 435, 535, 546, 551, 578, 608
NGC 4473, 135–36
NGC 4485, 570, 575
NGC 4486 (M87, 3C274, Virgo A), 107, 113, 115, 135–36, 163, 168, 177–78, 188, 200, 211, 242, 247, 255, 266, 269, 271–73, 429, 435–36, 497–98, 532, 551, 578, 608, 612, 729
NGC 4486B, 107, 586–87
NGC 4487, 579
NGC 4490, 106, 339, 550, 570, 575
NGC 4494, 571, 576
NGC 4501, 578
NGC 4503, 502
NGC 4504, 579
NGC 4517, 578
NGC 4519, 578
NGC 4526, 138
NGC 4535, 139, 142, 550, 578
NGC 4536, 578
NGC 4546, 579
NGC 4552, 135–36
NGC 4559, 339, 571, 576
NGC 4564, 136
NGC 4565, 152, 571, 575
NGC 4567, 11
NGC 4569, *17*, 22, 24, 30–31, 578–80
NGC 4570, 135–36
NGC 4578, 138
NGC 4579, 8–9, *18*, 30–31, 578
NGC 4593, 579–80
NGC 4594 (M104), 43, 47, 163, 590, 762

NGC 4605, 107, 339, 550, 569, 573
NGC 4618, 489, 570, 575
NGC 4621, 100, 135–36
NGC 4625, 570, 575
NGC 4627, 570, 575
NGC 4631, 107, 113, 152, 313, 327, 335, 339, 489, 550, 570, 575
NGC 4636, 578
NGC 4643, 6, *19*, 30–31, 578
NGC 4649 (M60), 44, 135–36, 163, 578, 608
NGC 4651, 183
NGC 4656, 335, 339, 570, 575
NGC 4657, 570, 575
NGC 4670, 571
NGC 4676N, 109
NGC 4676S, 109
NGC 4691, 579
NGC 4696, 266
NGC 4697, 100, 578
NGC 4699, 578
NGC 4725, 571, 576
NGC 4731, 578
NGC 4736 (M94), 11, 109–10, 113, 115, 163, 339, 569, 573, 575
NGC 4750, 11
NGC 4762, 145, 471–72
NGC 4782/4783 (3C278), 266, 271
NGC 4800, 575
NGC 4826 (M64), 109, 163, 551, 569, 573, 590
NGC 4839 (5C 4.51), 266
NGC 4856, 578
NGC 4866, 22
NGC 4869 (5C 4.81), 242, 266
NGC 4874 (5C 4.85), 242, 266, 609
NGC 4889, 497, 609
NGC 4939, 72, 578–80
NGC 4941, 22, 26
NGC 4945, 269, 569, 573
NGC 5005, 107, 339, 502, 551
NGC 5033, 339
NGC 5055, *16*, 22, 30–31, 90, 107, 110–11, 339, 436–37, 551, 569, 573
NGC 5068, 569, 573
NGC 5077, 269
NGC 5101, 11
NGC 5102, 339, 494, 569, 573
NGC 5128 (Centaurus A), 107, 178, 211, *249*, 259–62, 266, 270–72, 353, 551, 569, 573, 581, 590
NGC 5141 (4C 36.24), 266
NGC 5194 (M51), 4, 11, *17*, 24, 27, 30–31, 42, 47, 57, 63, 70–77, 106, 113, 115, 131, 162–63, 170, 269, 334, 339, 482, 550, 569, 573
NGC 5195, 24, 42, 63, 334, 569, 573
NGC 5204, 15, *16*, 22, 24, 27, 30–31, 339, 569, 573
NGC 5236 (M83), 8–9, *18*, 30–31, 106, 334, 339, 429, 431, 433, 569
NGC 5248, *17*, 30–31, 106, 339, 550
NGC 5253, 569, 573
NGC 5273, *15*, 22, 27, 30–31
NGC 5301, 339
NGC 5371, 339
NGC 5383, 109, 113–14, 339, 551
NGC 5457 (M101), 9, *16*, 30–31, 71, 106, 131, 178, 318, 332–34, 340, 550, 569, 573, 575
NGC 5474, 340, 569, 573
NGC 5523, 340
NGC 5532, 204
NGC 5566, 579, 581
NGC 5584, 579
NGC 5585, 340, 569, 573
NGC 5614, *21*
NGC 5633, 579
NGC 5660, 579
NGC 5668, 340
NGC 5676, 579
NGC 5689, 579
NGC 5701, 579
NGC 5713, 340, 579, 581
NGC 5746, 152, 579
NGC 5774, 340
NGC 5806, 580
NGC 5813, 580
NGC 5838, 580
NGC 5846, 580–81, 585
NGC 5850, 580
NGC 5866, 62, 163, 579
NGC 5879, 579
NGC 5899, 340
NGC 5905, 579
NGC 5907, 152, 340, 569, 573, 579
NGC 5908, 579
NGC 5921, 340
NGC 5949, 569, 573
NGC 5962, 340
NGC 6015, 340
NGC 6047 (4C 17.66), 267
NGC 6166 (3C338), 201, 267, 271, 554, 609, 613, 628, 633
NGC 6181, 109, 113, 340, 550
NGC 6207, 340
NGC 6215, 571, 576
NGC 6215A, 571
NGC 6217, 340, 580
NGC 6221, 571, 576
NGC 6300, 571, 576
NGC 6340, 340, 580
NGC 6412, 580
NGC 6454 (4C 55.33.1), 267
NGC 6503, 106, 340, 550, 569, 573, 581
NGC 6621, 581
NGC 6643, *16*, 22, 30–31, 340, 580–81
NGC 6654, 580
NGC 6744, 433, 590
NGC 6808, 579
NGC 6814, *17*, 30–31, 340
NGC 6822, 67, 69–70, 108, 333, 340, 513–17, 525, 529, 533, 550, 560–62, 587, 589
NGC 6835, 340
NGC 6861, 580
NGC 6868, 580
NGC 6876, 579
NGC 6893, 580
NGC 6902, 580
NGC 6943, 579
NGC 6946, 106, 340, 590
NGC 6951, *18*, 30–31
NGC 6970, 581
NGC 7013, 340
NGC 7049, 580
NGC 7079, 581
NGC 7083, 580
NGC 7090, 581
NGC 7137, 340
NGC 7144, 580
NGC 7205, 579
NGC 7213, 579, 581
NGC 7217, 340
NGC 7218, 340
NGC 7236/7237 (3C442), 202, 268
NGC 7314, 340
NGC 7320, 550
NGC 7331, 107, 340, 551
NGC 7332, 138, 145
NGC 7361, 340
NGC 7385, 268
NGC 7410, 578
NGC 7413 (3C455), 268
NGC 7424, 578
NGC 7448, 340
NGC 7469, 109, 551
NGC 7479, 92, 107, 551
NGC 7503 (4C 07.61), 268
NGC 7552, 578

NGC 7582, 578
NGC 7619, 628, 765
NGC 7626, 268
NGC 7640, 107, 335, 340
NGC 7679, 70, 72–73, 77
NGC 7720 (3C465), 201, 255–56, 268
NGC 7741, *19*, 30–31, 340
NGC 7743, *19*, 23, 30–31
NGC 7793, 8, 340, 564, 569
NRAO 512, 288
OH 471, 289
1C 01.01; *see* NGC 224
1C 07.01; *see* 3C48
1C 14.01; *see* 3C295
OQ 172, 289
Pegasus system, 341, 561–62
Perseus A; *see* NGC 1275
PHL 938, 301
PHL 5200, 301
Pictor A, 265
PKS 0237–23, 301
for a list of additional PKS sources see pp. 265–68
Sculptor system, 137, 438, 493, 498, 560–62, 586, 607–8
Serpens system (A1513), 561–62
Sextans A (A1009), 341, 561–62
Sextans B, 560–62
Sextans C (A1003), 560–62
Seyfert's Sextet, 554
Small Magellanic Cloud (SMC) *20*, 23, 30–31, 37, 70, 108, 159, 309–10, 316, 318, 325, 329–30, 333, 335, 341, 427, 432, 439, 471–72, 489, 510–12, 515–19, 521–26, 529, 533, 560–62, 586
Stephan's Quintet, 117, 554
3C9, 203, 285, 300, 726, 731, 736
3C31; *see* NGC 382/383
3C33, 200, 255, *258*, 265, 271, 275–76
3C40; *see* NGC 545/547
3C47, 192, 202, 285, 296, 305
3C48, 178, 185, 200, 284–85, 287, 290–92, 294
3C66, 183, 201, 265
3C71; *see* NGC 1068
3C76.1, 206, 261, 265
3C78; *see* NGC 1218
3C83.7; *see* NGC 1265
3C84; *see* NGC 1275
3C120 (BW Tauri), 247–48, 265, 270, 274, 298
3C129/3C129.1, 202, 206, 262, 276
3C147, 200, 245, 247, 284–85
3C191, 206, 300–301
3C196, 192, 200, 284–85
3C212, 185, 203
3C219, 200, 266, 271
3C231; *see* NGC 3034
3C234, 201, 266, 271
3C270; *see* NGC 4261
3C270.1, 205, 301
3C272.1; *see* NGC 4374
3C273, 185, 200, 284, 287, 290–91, 294–95, 297
3C274; *see* NGC 4486
3C277.3 (Coma A), 205, 266
3C278; *see* NGC 4782/4783
3C279, 294–95
3C280, 192, 202
3C286, 202, 284
3C295 (1C 14.01), 172, 178, 188, 200, *251–52*, 267, 271, 767
3C296; *see* IC 5532
3C338; *see* NGC 6166
3C343, 207, 288
3C343.1, 207, 288
3C345, 206, 288, 294–95
3C348 (Hercules A), 183, 188, 200, 259, 267, 269, 276
3C371, 207, 247, 267, 285
3C405 (Cygnus A), 177–78, 181–82, 188, 200, 211, *212*, 247, 249, 259, 267, 269–71, 276, 323, 399, 726, 729, 773
3C442; *see* NGC 7236/7237
3C446, 294–95
3C452, 200, 255, 257, 268
3C454, 207, 288
3C455; *see* NGC 7413
3C465; *see* NGC 7720
3C275.1, 183, 203
for a list of additional 3C sources see pp. 200–209 *and* 265–68
Ton 256, 285, 768–69
Ton 730, 285
Ton 1530, 301
Ursa Major system (A 1127), 560–62
Ursa Minor system (DDO 199, A1508), 22, 137, 561–62, 589, 608
Virgo A; *see* NGC 4486
VV116, *21*, 106, 554
VV117, 399
VV123, *21*, 399
VV150, 554
VV166, 554
VV254, 108, 550
Wolf-Lundmark-Melotte (WLM; DDO 221, A2359), 560–62, 587, 589
I Zw 114, 341
II Zw 40, 341

Stars and Stellar Systems

Volume IX: Galaxies and the Universe

Edited by Allan Sandage,

Mary Sandage, and Jerome Kristian

This volume reviews the present state of knowledge of the optical and radio properties of galaxies and quasars. Each chapter is contributed by a leading authority in the field. Detailed discussions are given of galaxy types, stellar and gaseous content, brightness, masses, sizes, formation and evolution, distances, and redshifts. Finally, the expansion and structure of the universe as a whole is analyzed in the context of the classical cosmological problem.

ALLAN SANDAGE and JEROME KRISTIAN are astronomers at the Hale Observatories. MARY SANDAGE is a former teacher and research worker in astronomy.